Organism	Gram Stain	Basic Morphology	Disease/ Anatomic Syndrome	Pages
nomadura spp.	+	rod, some filamentous forms	mycetoma	276-277
nomyces israelii	+	filamentous, diptheroid, and coccal	actinomycosis, **head and neck abscesses**	275
illus anthracis	+	rod, encapsulated	anthrax	13, 162, 365, 670
illus cereus	+	rod, encapsulated	**food poisoning**	565
teroides gingivalis	−	small rod	**periodontal disease**	552, 553-554
detella pertussis	−	coccobacillus	pertussis (whooping cough)	267, 363, 367, 368, 372, 379, 382, 535
elia burgdorferi	−	spiral	Lyme disease	365, 668, 670
elia spp.	−	large spiral	relapsing fever	670
cella spp.	−	coccobacillus	brucellosis (Malta fever, undulant fever)	365
mmatobacterium ranulomatis	−	rod, encapsulated	granuloma inguinale	596
pylobacter jejuni	−	rod	campylobacteriosis, gastroenteritis	272, 563
amydia psittaci	NA	coccoid, very tiny	ornithosis, psittacosis	365, 533
amydia trachomatis	NA	coccoid, very tiny	chlamydia, **conjunctivitis,** lympho-, granuloma venereum, **pelvic inflammatory disease,** psittacosis, trachoma, **urethritis**	2, 596, 600, 649
stridium botulinum	+	rod	botulism, **food poisoning**	75, 226, 233, 565, 620
stridium difficile	+	rod	**diarrhea**	559, 564
stridium perfringens	+	rod	gas gangrene, **food poisoning**	565, 638-639
stridium tetani	+	rod	tetanus	372, 619
nebacterium phtheriae	+	rod, club-shaped, pleomorphic, forms palisades	diphtheria	275, 325, 352, 379
ella burnetii	NA	coccobacillus	colds, Q fever	373, 534
erichia coli	−	rod	**diarrheal infections, meningitis, urinary tract infections**	74, 87, 90, 94, 116, 117, 120
cisella tularensis	−	small rod (coccobacillus)	tularemia	270, 365, 664-667
nerella vaginalis	−	small rod	**vaginitis**	599-600
mophilus ducreyi	−	slender rod (coccobacillus)	chancroid	596
mophilus influenzae	−	coccobacillus, some strains form	**meningitis, epiglottitis upper respiratory infections**	10, 355, 375, 526
cobacter pylori		curved rod	**gastritis,** peptic ulcer	271, 563-564
siella spp.	−	rod, encapsulated	**pneumonia, urinary tract infections**	532
onella pneumophila	−	coccoid rod	legionellosis (Legionnaire's disease)	270, 534-535, 694
ospira interrogans	−	spiral	leptospirosis	279, 587-588
ria monocytogenes	+	rod	listeriosis	228, 602
obacterium bovis	A-F	rod	bovine tuberculosis	537
obacterium leprae	A-F	rod	Hansen's disease (leprosy)	78, 87, 640-642
Mycobacterium tuberculosis	A-F	rod, branching forms	tuberculosis	57, 58, 84, 85, 87, 193, 372, 537-538
Mycoplasma hominis			**nongonococcal urethritis, pelvic inflammatory disease**	596, 600
Mycoplasma pneumoniae	NA	too small to be visualized by light microscope	mycoplasmal pneumonia	274, 532-533
Neisseria gonorrhoeae	−	cocci in pairs	blindness, **conjunctivitis,** gonorrhea, neonatal gonorrheal ophthalmia, **pelvic inflammatory disease, urethritis**	93, 267, 353, 375, 376, 589-591, 602
Neisseria meningitidis	−	cocci in pairs; capsules formed in young cells	**meningitis**	50, 267, 376, 379, 612, 615
Nocardia asteroides	+	rod, some filamentous forms	**lung infections**	525
Propionibacterium acnes	+	rod	acne	275, 352
Pseudomonas aeruginosa	−	rod	**contact lens conjunctivitis, folliculitis,** other infections	270, 351, 476, 638
Rickettsia prowazekii	NA	coccobacillus	epidemic typhus	672-674
Rickettsia rickettsii	NA	coccobacillus	Rocky Mountain spotted fever	90, 672
Rickettsia typhi	NA	coccobacillus	murine typhus	674
Rochalimaea henselae	NA	coccobacillus	cat scratch disease	672
Salmonella choleraesuis, Salmonella enteritidis	−	rod	salmonellosis (**food poisoning**)	558, 558
Salmonella typhi	−	rod	typhoid fever	75, 268, 367, 556-558
Shigella spp.	−	rod, generally single	shigellosis (bacillary dysentery)	555-556
Staphylococcus aureus	+	cocci in clusters	abscesses, boils, **endocarditis, food poisoning,** impetigo, osteomyelitis, **pericarditis,** pneumococcal pneumonia, toxic shock syndrome	352, 355, 532, 637
Streptococcus agalactiae (Group B Streptococci)			**neonatal meningitis, pneumonia,** sepsis; postpartum endometritis	602-603
Streptococcus mutans	+	cocci in chains	dental caries	95, 355, 552, 553
Streptococcus pneumoniae	+	lancet-shaped diplococci	pneumococcal pneumonia, **meningitis**	96, 148, 373, 375, 380, 616
Streptococcus pyogenes	+	cocci in chains	childbed fever, **endocarditis,** erysipelas, glomerulonephritis, impetigo, **pharyngitis** (strep throat, scarlet fever)	16, 75, 368, 375, 490, 661, 662, 634-635
Treponema pallidum	−	spiral	syphilis	63, 78, 95, 279, 366, 592-595

ORGANISMS AND DISEASES/ANATOMIC SYNDROMES CAUSED BY THEM

Bacteria (continued)

Organism	Gram Stain	Basic Morphology	Disease/ Anatomic Syndrome	Pages
Ureaplasma urealyticum	NA	very small rod	**nongonococcal urethritis, postpartum fever urethritis**	596
Vibrio cholerae	−	comma-shaped rod	cholera	366, 371, 561-562
Vibrio parahaemolyticus	−	rod	**gastroenteritis**	562
Yersinia enterocolitica	−	rod	**enterocolitis**	562

Organism	Gram Stain	Basic Morphology	Disease/ Anatomic Syndrome	Pages
Yersinia pestis	−	short, thick rod; exhibits bipolar staining	bubonic plague, pneumonic plague, septicemic plague	2-3, 268, 36 664

Key to Gram stain:
- − Gram-negative   A-F acid-fast
- + Gram-positive   NA not applicable

Helminths, Protozoa, and Arthropods

Organism	Type	Disease/ Anatomic Syndrome	Pages
Ancyclostoma duodenale	roundworm	Old World hookworm	569
Ascaris lumbricoides	roundworm	ascariasis	569
Babesia spp.	protozoan	babesiosis	684
Balantidium coli	protozoan	balantidiasis	302, 568
Brugia malayi	roundworm	filariasis	685
Cryptosporidium spp.	protozoan	cryptosporidiosis	224, 568
Echinococcus granulosus	flatworm	dog tapeworm, hydatid disease	306, 570
Entamoeba histolytica	protozoan	amoebic dysentery	566
Enterobius vermicularis	roundworm	pinworm	568-569
Fasciola hepatica	flatworm	sheep liver fluke infection	572
Giardia lamblia	protozoan	giardiasis	9, 365, 383, 566-567
Loa loa	roundworm	loaiasis	651
Necator americanus	roundworm	New World hookworm	569
Onchocerca volvulus	roundworm	onchocerciasis (river blindness)	650-651
Opisthorchis sinensis	flatworm	Chinese liver fluke infection	572-573

Organism	Type	Disease/ Anatomic Syndrome	Pages
Paragonimus westermani	flatworm	lung fluke	307
Pediculus humanus	louse	pediculosis (body lice), head lice	648
Phthirus pubis	louse	crab lice (pubic lice)	648
Plasmodium spp.	protozoan	malaria	301, 373
Sarcoptes scabiei	mite	scabies	308, 648
Schistosoma spp.	flatworm	shistosomiasis	307
Strongyloides stercoralis	roundworm	strongyloidiasis	10, 569-570
Taenia saginata	flatworm	beef tapeworm	306
Taenia spp.	flatworm	tapeworm	365
Toxoplasma gondii	protozoan	toxoplasmosis	365
Trichinella spiralis	roundworm	trichinosis	307, 365, 570
Trichomonas vaginalis	protozoan	trichomoniasis	56, 601-602
Trichuris trichiura	roundworm	whipworm	570
Trypanosoma brucei	protozoan	African trypanosomiasis (sleeping sickness)	8, 422, 625-626
Trypanosoma cruzi	protozoan	Chagas' disease	663
Wuchereria bancrofti	roundworm	filariasis	685

Introduction to
Microbiology

A Case History Approach
THIRD EDITION

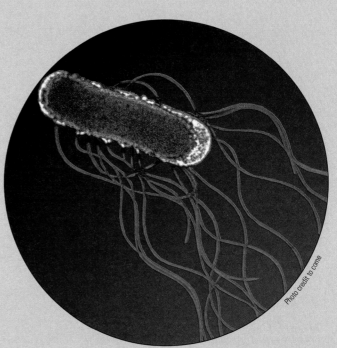

Photo credit to come

John L. Ingraham

University of California, Davis

Catherine A. Ingraham

The Permanente Medical Group, Inc., Rancho Cordova

THOMSON
™
BROOKS/COLE

Australia • Canada • Mexico • Singapore • Spain
United Kingdom • United States

THOMSON

BROOKS/COLE

Editor in Chief: Michelle Julet
Publisher: Peter Marshall
Life Sciences Editor: Nedah Rose
Development Editor: Marie Carigma-Sambilay
Assistant Editor: Christopher Delgado
Editorial Assistant: Jennifer Keever
Technology Project Manager: Travis Metz
Marketing Manager: Ann Caven
Marketing Assistant: Sandra Perin
Advertising Project Manager: Linda Yip
Project Manager, Editorial Production: Teri Hyde

Print/Media Buyer: Kris Waller
Permissions Editor: Sarah Harkrader
Production Service: Graphic World Publishing Services
Text Designer: Andrew Ogus
Photo Researcher: Meyers Photo-Art
Copy Editor: Stavra Ketchmark
Cover Designer: Andrew Ogus
Cover Image: USDA/© Visuals Unlimited, Inc.
Cover Printer: Lehigh Press
Compositor: Graphic World, Inc.
Printer: RR Donnelley/Willard

For more information about our products, contact us at:
Thomson Learning Academic Resource Center
1-800-423-0563
For permission to use material from this text, contact us by:
Phone: 1-800-730-2214
Fax: 1-800-730-2215
Web: http://www.thomsonrights.com

Library of Congress Control Number: 2003102148

Student Edition with InfoTrac College Edition: ISBN 0-534-39465-5
Student Edition without InfoTrac College Edition: ISBN 0-534-39491-4

Instructor's Edition: ISBN 0-534-39469-8

Brooks/Cole—Thomson Learning
511 Forest Lodge Road
Pacific Grove, CA 93950
USA

Asia
Thomson Learning
5 Shenton Way #01-01
UIC Building
Singapore 068808

Australia/New Zealand
Thomson Learning
102 Dodds Street
Southbank, Victoria 3006
Australia

Canada
Nelson
1120 Birchmount Road
Toronto, Ontario M1K 5G4
Canada

Europe/Middle East/Africa
Thomson Learning
High Holborn House
50/51 Bedford Row
London WC1R 4LR
United Kingdom

Latin America
Thomson Learning
Seneca, 53
Colonia Polanco
11560 Mexico D.F.
Mexico

Spain/Portugal
Paraninfo
Calle/Magallanes, 25
28015 Madrid, Spain

We dedicate this book to our family:
Marge, Tom V., Tom I., Luanne, Anna,
Lisa, Dana, Ian, Elena, and John.

BRIEF CONTENTS

CONTENTS

PREFACE

Since its inception, microbiology has been an applied as well as a basic science. These two faces of microbiology are reflected in the principal changes we've incorporated into this third edition: expansion of our use of Case Histories and recognition of the increasing impact of genomics on all biology, perhaps especially microbiology.

Possibly the most important application of microbiology, certainly the most personally relevant, is using the vast knowledge gained from studying microorganisms to treat an individual patient suffering from a particular infectious disease. In previous editions, we emphasized this application in the form of Case Histories, adopting them as a learning tool and our book's trademark. (Just as basic microbiological knowledge is essential for rational treatment of infectious diseases, Case Histories remind, correlate, and stimulate interest in that basic knowledge.) In this third edition we've expanded our use of Case Histories, adding the subtitle—A Case Histories Approach—as emphasis. We've attempted to draw more clearly the link between Case Histories and underlying, basic microbiology by adding a series of "Case Connections" after each chapter-opening Case History.

With respect to basic science, biologists have long known that all a cell is and does is encoded in and directed by the DNA of its genome. Now, with ever-increasing precision, biologists are learning how to read the information encoded in a cell's DNA and interpret its meaning. This new field of "genomics" is rapidly changing all aspects of microbiology, from knowledge of the extent of the diversity of various microorganisms to means of devising rapid ways of identifying microorganisms and rational means of treating infectious diseases. In this third edition we dedicate half of Chapter 7 to the methods of genomics and refer throughout the book to their application to various fields of microbiology.

We'll still visit all aspects of microbiology—basic and applied, including medical, environmental, agricultural, and industrial. One primary emphasis is medical: two entire parts (Parts III and IV) of the book deal with microbial diseases. These follow a consideration of the basic science of microbiology (Part I) and a survey of the various kinds of microorganisms (Part II). Finally, we examine the way humans use microorganisms and the essential roles microorganisms play in Earth's ecology (Part V).

We maintain the approach and attitude of previous editions, including emphasis on the why of metabolism and genetics as well as the what. However, there are changes.

Some of these reflect the scientific progress that has occurred since publication of the second edition, such as the great advances that have been made utilizing new molecular methods to identify mcroorganims and the new effective vaccines that have been introduced. We've altered the presentation of immunology to refelect the major advances in the component now called innate immunity (formerly called nonspecific defenses).

Some Specific Changes Incorporated into the Third Edition

- The powerful observational new tools offered by scanned-proximity probe microscopes, including scanning tunneling microsopes and atomic force microscopes.
- The much used molecular staining method employing green florescent protein.
- Expansion of the coverage of the molecular structure of peptidoglycan.
- Introduction of the concepts of anoxgenic and oxygenic photosynthesis.
- A half chapter on genomics with accompaning figures and tables.
- A section on biofilms.
- Change of the presentation of bacterial taxonomy to reflect the phylogenic approach in the latest *Bergey's Manual.*
- Discussion of "black mold" and "sick buildings."
- A section with accompanying table on emerging diseases.
- Introduction of the concepts of innate and adaptive immunity, reflecting modern usage and expanded knowledge.
- Expansion of the coverage of anthrax and bioterrorism.
- Expansion of the coverage of AIDS, describing the continuing pandemic.

Student-Support Features

The book contains a number of features designed to assist your learning and organize your study.

Learning Goals. Each chapter begins with a succinct list of Learning Goals to help you focus on its overall aims.

Boldface Terms. Certain terms are highlighted in boldface type the first time they occur. We consider them important and suspect they might be new to you. They are defined in the context in which they occur, and those that might not be adequately defined in a dictionary are included in the Glossary at the back of the book.

Boxes. We've included three kinds of boxes—Case Histories, Sharper Focus, and Larger Field.

- Case Histories explore real clinical cases (most from Catherine Ingraham's medical practice). We hope these stories will help you remember the details of the text material more easily. These Case Histories occur throughout the text. Those at the beginning of most chapters play a critical role: they set the stage for much that follows in the chapter. You'll probably want to refer back to these from time to time as you progress through the material that follows.
- Sharper Focus boxes examine a chapter topic in more detail. Just as you can see greater level of detail by focusing a microscope, these boxes (which appear in every chapter) reveal more information about interesting chapter topics.
- Larger Field boxes connect text material to our world. Just as you can expand a microscope's field to see more background around an organism, these boxes expand a chapter topic to its larger context—environmental aspects, history, news events, etc. They emphasize the connections of microbiology to all facets of our lives—beyond the medical aspects explored in the Case Histories.

CD Connection. In most chapters there are one or more "CD Connections" marked by a CD-ROM icon and carrying directions to particular sites on the CD-ROM as well as a brief account of what the site shows. For the most part the designated sites are film loops that quickly convey information that might be difficult to grasp from words alone. For example, one shows how the many flagella on certain bacterial cells coalesce into a single structure that propels the cell through its liquid environment. One look explains it all.

Suggested Readings. Each chapter contains suggested readings. To save space these references to books and scientific journals have been pruned to include only the most helpful and informative sources of information, most of which will lead you to more highly specialized information.

Chapter Summaries. You'll find some summaries within chapters that reiterate in a different way material that you might find difficult to grasp in a single reading. In addition, each chapter ends with a somewhat detailed summary. These are provided to help you recall and organize the material you read in the chapter.

Questions. We've included three kinds of questions at the end of each chapter—review, correlation, and essay.

- Review Questions are just that. They're not designed to make you think, only to recall and to reassure yourself that you've assimilated the material contained in the chapter.
- Correlation Questions are meant to offer a greater challenge. They ask you to collate different pieces of information and compare them or apply them to a new problem. A correct answer to these questions should assure you that you have gained a thorough understanding of the material. Congratulations are in order.
- Essay Questions ask you to deal with larger portions of a subject and add your own opinions.

Clinical Science

Because Part IV considers infectious diseases and how they affect various organ systems, it necessarily employs the terms and approaches used by clinicians who treat infectious diseases. So an introduction to clinical science is needed in order to understand the material presented in Part IV. This is presented at the beginning of Chapter 22.

ADDITIONAL LEARNING AIDS*

Study Guide

This helpful manual contains chapter outlines, key terms, a new feature called "Study Tips," study exercises of various question types, multiple-choice questions, discussion questions, and an answer key (ISBN: 0-534-55225-0).

InfoTrac® College Edition

Available exclusively from Brooks/Cole, this online library offers you unlimited access to more than 700 publications—at any time of the day. With *InfoTrac*, you can search for complete articles from periodicals including *Science, Discover, Annual Review of Microbiology*, and *Bioscience*. The password-protected site is updated daily, and a 4-month subscription is offered free to students who purchase new texts. An online student guide correlates each chapter in this text to InfoTrac articles (ISBN: 0-534-55224-2).

---

*Available to qualified adopters. Please consult your local sales representative for details.

Virtual Molecular Biology Lab CD/Web Site

Explore the world of molecular biology with a fun, accessible, and uniquely powerful CD-ROM. This virtual laboratory environment allows you to make the connection between the lab and the classroom by performing virtual experiments and correlating the experiments to concepts presented in the classroom. The virtual environment allows you to "use" equipment to conduct ten of the most common molecular techniques, giving you a sense of being in a live lab to either augment or replace live lab time. Concept and textbook correlations presented throughout the experiments reinforce the relationship between scientific concepts and how science is conducted. Manipulate live data from GenBank to perform cutting-edge research!

Electronic Companion to Accompany *Beginning Microbiology,* from Cogito Learning Media, Inc.

This interactive CD-ROM, authored by John Ingraham, includes a comprehensive tutorial, quizzes, and exercises about microbiology. This CD-ROM allows you to study the details of microbiology and to visualize the complex processes you learn about in the course. "CD Connections" in the textbook guide you to the CD-ROM for further study.

Microbiology remains, particularly with the advent of genomics, a vital and fast-moving field. Although it presents serious challenges in the form of emerging infectious diseases and the increased resistance of dangerous pathogens to our aramentarium of antibiotics, powerful tools are now available to address these threats. It will be a contest. The words of the most famous microbiologist of all, Louis Pasteur (1822–1895), which seemed appropriate at the time of the first edition, are even more relevant now: *"Messieurs, c'est les microbes qui auront le dernier mot."* ("The microbes have the last word.") We'll have to wait to see what they tell us.

ACKNOWLEDGMENTS

We gratefully acknowledge the indispensable help, encouragement, and guidance of Development Editor Marie Carigma-Sambilay; of Editor Nedah Rose for her involvement, innovation, and decisiveness; and of Senior Project Manager Teri Hyde for her skills of presentation.

John Ingraham
Catherine Ingraham

REVIEWERS

David Alexander
University of Portland

Arden Aspedon
Southwestern Oklahoma State University

Clare Bailey
Florida Community College at Jacksonville

Karen Ballen
Augsburg College

Sookie Bang
South Dakota School of Mines and Technology

Art Barbeau
West Liberty State College

Ted Drouin
Keyano College

David Giron
Wright State University

Anne Heise
Washtenaw Community College

Cristi Hunnes
Rocky Mountain College

Randall Jeter
Texas Tech University

Ted Johnson
St. Olaf College

Steven Kuhl
Indiana University Purdue University Fort Wayne

Dennis Lye
Northern Kentucky University

Joan McCune
Idaho State University

Annette Muckerheide
College of Mount St. Joseph

John Natalini
Quincy University

Kenneth Roth
Eastern Mennonite University

Carl Sillman
Penn State University

Deb Stai
Lake Superior State University

Valerie Watson
West Virginia University

Tit-Yee
University of Memphis

ABOUT THE AUTHORS

Catherine Ingraham Vigran, M.D.
Senior Physician
The Permanente Medical Group, Northern California

Catherine's interest and knowledge of microbiology derive from her training and practice as a board-certified pediatrician dealing principally, as most pediatricians do, with cases of infectious disease. With three children of her own—all girls (see Figure 1.4b)—and those of neighbors, her practice often continues after leaving the clinic. Her conviction that witnessing the experiences of patients suffering from infectious diseases stimulates an inquiring interest in their microbiological basis encouraged her to use them as a learning tool. That's why this book is liberally sprinkled with case histories taken from her practice.

Her commitment to medicine hasn't set limits on her interests. Her premedical training was a double major— Spanish Literature as well as Biological Sciences, graduating with honors in both. Although at the time pursued for its own sake, the study of Spanish literature led to fluency and professional value: about a third of her patients speak only Spanish. Her achievements as a medical student were recognized by election to the Alpha Omega Alpha Honor Medical Society. Music (the piano), mountain going, and, of course, textbook writing fill the time left after medical practice and family—a psychiatrist husband, next-door parents and co-author, as well as the girls.

John Ingraham, Ph.D.
Professor of Microbiology, Emeritus
University of California, Davis

John's career has been academic microbiology at a research university, pursuing genetically based studies on microbial physiology and metabolism, including such topics as the malo-lactic fermentation, fusel oil formation, factors precluding microbial growth at low temperature, biosynthesis of pyrimidine nucleosides, and denitrification. He's published over a hundred research papers as well as five textbooks and was honored by being elected president of the 40,000-member American Society for Microbiology.

John loves teaching, often using the prerogatives of a departmental chairperson to assign himself the beginning class in microbiology.

He continues to consult for the pharmaceutical industry (Genentech) and for a Web site database (EcoCyc) on microbial metabolism, as well as to write on microbiological topics.

ONE

The Science of Microbiology

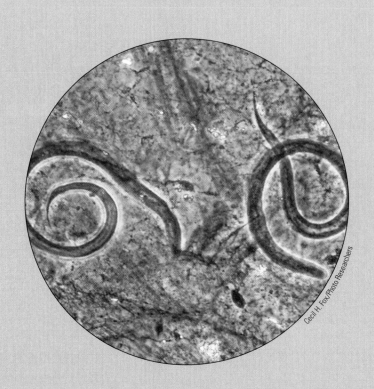

Cecil H. Fox/Photo Researchers

CHAPTER OUTLINE

LEARNING GOALS

To understand:

- *The impact of microorganisms on human affairs*
- *Advances and challenges in applied microbiology*

- *Careers available to trained microbiologists*
- *The scope of microbiology and why it is a separate science*

- *The development of microbiology as a science*
- *Microbiology today and where it is headed in the future*

Microbiological Advances Make a Big Difference

April 1999

J. D., a nineteen-year-old white female, visited an urgent care clinic complaining of abdominal pain and fever. A week earlier she had noticed lower abdominal pain on the right and the left side. Mild at first, the pain worsened gradually, becoming so severe on the night of her visit to the clinic that she could barely walk. Three days earlier, she had developed a fever, lost her appetite, and become too sick to get out of bed. Her concerned parents insisted on taking her to a doctor.

At the clinic, the evaluating physician noted J. D.'s 103°F fever, her pale complexion, and her general appearance of being ill. Because a sexual history was essential to her evaluation, the physician asked her parents if he could speak privately with J. D. Assured that the information would remain confidential, J. D. disclosed that she had been sexually active for a year, with two different partners, and had not used condoms consistently.

J. D.'s physician performed a physical examination. Pressure over the lower part of her abdomen was moderately uncomfortable for her. A pelvic examination revealed a thick yellow discharge coming from the cervical **os** (opening). Moving her cervix was painful, causing her to wince and cry out. Samples of her cervical discharge and of her blood were sent to the microbiology laboratory to see if they contained live microbial cells. Microscopic examination of her blood revealed a markedly elevated number of white blood cells (a count of 23,000).

J. D.'s history, physical examination, and blood cell count strongly suggested pelvic inflammatory disease (PID), a sexually transmitted disease (STD) that spreads upward through the uterus and fallopian tubes to the pelvis and the ovaries. Seriously ill patients such as J. D. need to be hospitalized.

J. D.'s hospitalization was long and complicated. She was given three antibiotics (gentamicin, metronidazole, and a cephalosporin) directly into her veins. Because these medications irritate, the site of insertion of the **line** (the plastic tube delivering them into her vein) had to be changed repeatedly. Her fever and pain continued unabated for 4 days. Concerned that her fever might be caused by an abscess in her fallopian tubes or around her ovaries, a complication that would require surgery, her physician ordered an ultrasound examination of her pelvis. No such abscess was found. Antibiotic therapy along with medications for fever and pain were continued. Finally after 7 days J. D.'s fever declined. After 10 days she was released from the hospital. The cervical culture, taken at the time of her admission, grew a variety of microorganisms, including the sexually transmitted **pathogen** (disease-causing organism) *Chlamydia trachomatis*.

A week later, when J. D. returned to see her doctor, she was feeling

THE UNSEEN WORLD AND OURS

Microbiology is the study of microorganisms (sometimes called **microbes**)—living things that are too small to be seen by the unaided eye.

In spite of their tiny size, they have enormous impact: They maintain our environment in life-sustaining balance, and they are some of our most valuable industrial tools. But unchecked they can destroy our livestock and crops, and they can spoil our food. A few of them can make us sick and even kill us.

Before the development of microbiology as a science, **pathogenic** (disease-causing) microbes often controlled human events. With the epidemics they caused came social and political change, even chaos, as well as untold human suffering. Let's look briefly at a few examples.

Microbes, Disease, and History

The **bubonic plague** (often called plague or the black death) that swept through Europe during the Middle Ages killed about 25 million people—one-third of the population (**Figure 1.1**). The social and political dislocations were immense, and the resulting terror was particularly intense because the cause of the disaster was unknown. Not until 500 years later, in 1890, did microbiologists identify the causative organism (a bacterium called *Yersinia pestis*)

Microbiological Advances Make a Big Difference (continued)

much better. Her fever and abdominal pain were gone. Her appetite was good, and her strength was returning. Her doctor shared his concerns with her: The widespread inflammation of her pelvic organs had probably scarred one or both of her fallopian tubes. This could cause infertility or lead to an **ectopic pregnancy** (a life-threatening condition in which the fetus develops inside a fallopian tube).

June 2002

L. S., a nineteen-year-old white female, visited her doctor for a routine preparticipation sports physical. She was feeling well. While taking her confidential medical history, her doctor learned that L. S. was sexually active. She urged L. S. to consent to be tested for chlamydia and gonorrhea, the two most common STDs. L. S. agreed. Only a urine specimen was necessary, because a rapid, highly sensitive procedure, based on **genomics** (the study of an organism's DNA), for detecting

chlamydia had recently become available. The laboratory report that L. S.'s physician received the next day showed that L. S. had tested positive for chlamydia. She was informed confidentially and asked to return to receive one dose of oral antibiotic, which would cure her infection. L. S. was advised about her relatively low risks of later infertility and ectopic pregnancy.

J. D. and L. S. suffered the same infection, but their outcomes differed dramatically. Because J. D.'s infection was diagnosed only after symptoms appeared, it led to serious complications. Such late diagnosis of chlamydia was the usual situation in 1999. Diagnosis required a pelvic examination and culturing of the causative microorganism. Such time-consuming, invasive procedures seem warranted only for a good reason, usually symptoms. Then the genomics-based test for chlamydia, which was sensitive enough to detect the few organisms present in an infected person's urine, became available, making routine screening for chlamydia feasible. So only 3 years after J. D.'s serious illness, L. S.'s infection could be diagnosed before symptoms appeared and was relatively inconsequential.

Case Connections

- As we'll see in this chapter, the microbiological advance that benefited L. S. so dramatically is merely a recent example of microbiology's long history of contributions to human betterment.
- The new diagnostic test that lead to L. S.'s chlamydia being cured before she developed symptoms is based on genomics, a topic we'll discuss in Chapter 7 (Recombinant DNA Technology and Genomics) and refer to throughout the book.
- What's the significance of J. D.'s elevated white blood cell count? We'll discuss that in Chapter 17 (The Immune System: Adaptive Immunity).

and piece together this story: Infected fleas carry the bacterium. Infection spreads among the rats as fleas bite them. But rats have little resistance to plague and usually die. When rats become scarce, the fleas move to humans and a plague epidemic is under way. Plague is a painful and ugly disease, characterized by swollen lymph glands called **buboes.** Fever soars as the cells of *Y. pestis* proliferate throughout the body. Although rare today, bubonic plague still occurs in parts of the world, including among ground squirrels in the western and southwestern United States (Chapter 27).

The Irish Famine. Potato blight, a disease of plants rather than humans and caused by a fungus rather than a

bacterium, had even greater impact than the plague. Potato blight was responsible for the great Irish famine of the 1800s. Potatoes were the staple of the Irish diet, so when the fungus *Phytophthora infestans* infected potatoes, causing them to rot in the fields, the result was devastating. By 1846 the potato harvest was so meager that starvation and hunger-based disease were widespread. An estimated 1,240,000 people had died, and 1,200,000 more had emigrated to other countries. Potato blight remains an economic threat to potato farmers today.

The Conquest of the Americas. One of the most devastating explosions of infectious disease occurred when Europeans first came to the Americas. When the Spanish

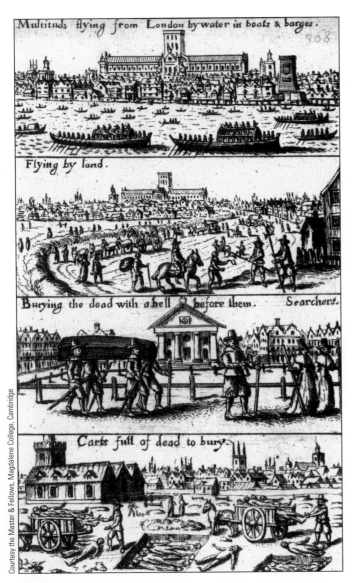

Courtesy the Master & Fellows, Magdalene College, Cambridge

FIGURE 1.1 The bubonic plague that swept through Europe in the Middle Ages was also called the Black Death because internal hemorrhaging caused black patches on the skin.

conquistador Hernando Cortés landed in Mexico in 1519, the Native American population of the central region was 25 to 30 million. But within 50 years, the ravages of diseases brought from Europe shrunk the population to 3 million. Almost 90 percent of the population had died! Europeans, after centuries of exposure, had developed a certain level of tolerance to infectious diseases such as smallpox and measles. But Native Americans, with no previous exposure, were tragically vulnerable.

Once established in Mexico, European diseases spread with lightning speed. When smallpox arrived in Peru, it killed millions, including the reigning Inca and his heir. By 1525, when the Spanish conquistador Francisco Pizarro

arrived, the Incan Empire was in social and political chaos, enabling Pizarro's tiny band to conquer the vast Inca nation without significant military resistance.

Disease and Warfare. Warfare and infectious diseases have always been intimately connected. The poor sanitation, movement of peoples, and malnutrition war brings all foster outbreaks of disease. For example, in 1812 when Napoleon invaded Russia he lost more of his troops to **typhus** (a bacterial disease) than to all other causes combined. And more of his soldiers died from wound-related bacterial infections (such as tetanus and gas gangrene) than from the wound itself.

Microbes and Life Today

Pathogenic microorganisms are still responsible for a vast spectrum of human illnesses and suffering. But the development of microbiology as a science has made astounding progress. Advances in medical microbiology have made it possible to identify the various pathogens that cause infectious diseases and devise ways to control most of them. Scientists have developed vaccines to prevent infections and drugs to treat them once they have become established. Moreover, we have made dramatic progress in preventing infectious disease through public hygiene, such as water treatment, sewage treatment, and better living conditions. But immense challenges still exist. We still cannot completely control microbial disease. We have many effective vaccines, but inoculating every child is not yet possible. We still lack effective vaccines for some of the most devastating infectious diseases, including acquired immunodeficiency syndrome **(AIDS)** and malaria (Chapter 27). We need new and better drugs to replace those that have lost their effectiveness and to treat many deadly diseases, including tuberculosis.

But only a small percentage of microorganisms are pathogens. As the American microbiologist Otto Rahn pointed out, the fraction of microorganisms that cause disease is far less than the fraction of humans who commit first-degree murder. Microbiology has discovered ways to use these other microorganisms to improve the quality of our lives. Let's look at some of the advances that have occurred and the challenges that remain in three areas of applied microbiology: environmental, industrial, and agricultural.

Environmental Microbiology. The study of how microorganisms affect the earth and its atmosphere is called **environmental microbiology** or **microbial ecology.** Our very existence depends on activities of the microbes that maintain our life-supporting environment (Chapter 28). We have learned how to use microorganisms to improve our environment. One of the earliest practical uses was to provide safe and palatable drinking water. Among the more recent uses is the development of ways to break down and eliminate the mounting toxic waste we produce.

Industrial Microbiology. The first human use of microbes was to make and preserve food. People learned how to use microorganisms to make bread, wine, vinegar, cheeses, and olives, as well as many other foods. We also learned to preserve food for lean winters by controlling the growth of microorganisms—by drying, salting, and canning food. Today microorganism-dependent industries are widespread. In addition to processing food, we use microorganisms to make a variety of useful materials, including vitamins, antibiotics, and other pharmaceuticals, such as insulin to treat diabetes (**Table 1.1**).

Agricultural Microbiology. Thanks to research in agricultural microbiology, livestock and crop plants are now largely protected from microbial diseases. Microorganisms that kill insects are used as natural pesticides, and others are used to maintain soil fertility. Current research in this field is directed toward using microorganisms to produce food supplements for animals and humans. Although in the United States we still enjoy agricultural abundance, people in many parts of the world don't. They would benefit from improved agricultural productivity and inexpensive, appetizing sources of food supplements, protein in particular.

Undoubtedly the greatest advances in industrial microbiology—and economic opportunities—will come from applying genetic engineering to medical, environmental, and agricultural problems.

Careers in Microbiology

Among the many reasons to study microbiology is to gain a basic knowledge of the field needed to pursue careers in medical science, ecology, agriculture, or biotechnology. Some students simply want to better understand the world we live in. Others intend to make microbiology their life's work. Careers in microbiology are challenging, rewarding, and varied. The American Society for Microbiology, the national professional microbiologists' organization, has about 40,000 members and is still growing.

Career opportunities in microbiology depend on training, as well as interest. Important contributions come from microbiologists who choose careers in general microbiology, studying microorganisms for their intrinsic interest. General microbiologists do pure research to find new microorganisms, new microbial activities, and new relationships among microorganisms. A major recent achievement in general microbiology was the discovery of archaea, a group of microorganisms we discuss later in this chapter. When general microbiology flourishes, the applied branches of microbiology also flourish.

For information about careers in microbiology, see the Board of Education and Training, at the American Society for Microbiology Web site, http://www.asmusa.org.

TABLE 1.1 Some Industrial Uses of Microorganisms

Product	Contribution of Microorganisms
Cheese	Growth of microorganisms contributes to ripening and flavor. The flavor and appearance of a particular cheese are due in large part to the microorganisms associated with it.
Alcoholic beverages	Yeast is used to convert sugar, grape juice, or malt-treated grain into alcohol. Other microorganisms may also be used; a mold converts starch into sugar to make the Japanese rice wine sake.
Vinegar	Certain bacteria are used to convert alcohol into acetic acid, which gives vinegar its acidic taste.
Citric acid	Certain fungi are used to make citric acid, a common ingredient in soft drinks and other foods.
Vitamins	Microorganisms are used to make vitamins, including C, B_2, and B_{12}.
Antibiotics	With only a few exceptions, microorganisms are used to make antibiotics.
Amino acids	Microorganisms are used to make many amino acids, including monosodium glutamate (MSG), a flavor enhancer.
Human growth hormone, insulin	Human growth hormone, insulin, and other medically useful proteins are made by genetically engineered bacteria.

NOTE: Chapter 29 discusses these processes in detail.

THE SCOPE OF MICROBIOLOGY

Microorganisms are usually divided into six subgroups: bacteria, archaea, algae, fungi, protozoa, and viruses. These subgroups are not closely related. Only a single property links them—their small size. In fact, the diversity of form and function found among the subgroups of microorganisms is as great as the total diversity of all living things. Bacteria, for example, are less like archaea or algae or fungi or protozoa or viruses than a shark is like a giraffe or an orchid is like an eagle.

Why, then, are such unrelated organisms grouped for common study as a subject called microbiology? The answer is a practical one. The techniques for identifying, cultivating, and studying the groups of microorganisms are similar (Chapter 3). Microbiology is a cohesive science because of the methods it uses and approaches it takes to problems, not because of the relatedness of the organisms it studies.

A primary distinction among the subgroups is cell structure. Bacteria and archaea are **prokaryotes** (meaning "before a nucleus"). Their cells lack internal membrane-bound structures. In contrast, algae, fungi, and protozoa, such as plants and animals, are **eukaryotes** (meaning "true nucleus"). Their cells contain a membrane-bound nucleus, as well as numerous other membrane-bound structures called organelles. Viruses are acellular. That is, a virus is not a cell. It is merely a small packet of nucleic acid (the chemical form of genetic information) wrapped in a coat, usually made of protein.

We'll look at each group in detail in Chapters 11, 12, and 13, but here we'll consider some of their general properties (**Table 1.2**). We'll also consider certain helminth (worm) species that are traditionally part of microbiological study.

Bacteria

Bacteria (*sing.*, bacterium) are distinguished by their size and prokaryotic cell structure. Instead of the elaborate internal membrane-bound structures seen inside eukaryotic cells, prokaryotes are filled with a uniform grainy material. Most bacteria are unicellular (single cells) and quite small, even for microorganisms (**Figure 1.2**). A typical bacterial cell has only about one one-thousandth the volume of a typical eukaryotic cell.

Bacteria are highly diverse. Most species have a characteristic cell shape (**Figure 1.3**). They can be spherical, rod-shaped, helical, comma-shaped, star-shaped, or even square. Some bacteria are motile. Others are not. Some obtain energy by processing organic compounds (foods), as animals do. Others utilize light energy through photosynthesis, as plants do. Still others process inorganic materials, such as sulfur or iron, for energy. Some bacteria can grow at temperatures as low as $-20°C$, lower than the freezing point of water; others thrive at temperatures of $110°C$, higher than the boiling point of water. Some grow best under conditions more acidic than dilute hydrochloric acid. Others thrive in environments more alkaline than household ammonia.

Bacteria cause a vast spectrum of diseases—from food poisoning and toxic shock syndrome to syphilis and typhoid fever. But they also make plant and human, as well as other animal, existence possible by keeping the earth's environment and atmosphere in life-sustaining balance (Chapter 28).

Archaea

Archaea (*sing.*, archaeon) were discovered to be a separate group of microorganisms only in the 1970s. Superficially

TABLE 1.2 Subgroups of Microorganisms

Subgroup	Cell Type	Contains Representatives That Are		
		Photosynthetic	Motile	Macroscopic
Bacteria	Prokaryotic	Yes	Yes	No
Archaea	Prokaryotic	Yes[a]	Yes	No
Algae	Eukaryotic	All are	Yes	Yes
Fungi	Eukaryotic	No	No[b]	Yes
Protozoa	Eukaryotic	No	Yes	No
Viruses	Acellular	No	No	No

[a]A few have a primitive mechanism of photosynthesis.
[b]The reproductive cells (spores) of some fungi are motile.

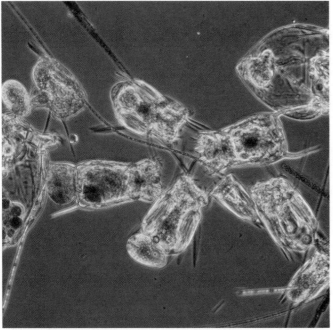

FIGURE 1.2 Microorganisms in pond water. The larger cells are protozoa and unicellular algae. The small specks are bacteria.

they resemble bacteria: They are small prokaryotic cells that usually occur singly. When first discovered they were called **archaebacteria** (ancient bacteria). But they are as distantly related to bacteria as they are to eukaryotes, including humans. To emphasize this difference their name has been changed to archaea. Many archaea live in hostile environments that would be deadly to most living things—places that are extremely hot or acidic or that contain high concentrations of salt. One major group of archaea makes **methane** (natural gas). These microorganisms produce the gas that bubbles up from many quiet ponds. No archaea are human pathogens.

Algae

Algae (*sing.*, alga) are eukaryotic organisms that carry out plantlike photosynthesis. Like all eukaryotes, they have a nucleus and membrane-bound organelles, including chloroplasts (the structures in which photosynthesis takes place). Some algae are unicellular and microscopic. Others, however, consist of many cells and are macroscopic (visible without the aid of a microscope). Kelp, the large brown seaweed that washes up on Pacific Ocean beaches, is an example of a macroscopic alga (**Figure 1.4**). Such multicellular algae may look superficially like higher plants, but they lack characteristic plant organs, including stems, roots, and leaves.

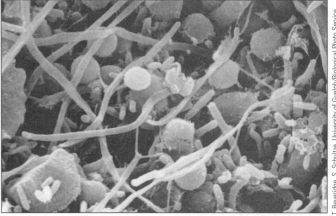

1μm

FIGURE 1.3   A micrograph of various bacteria showing some of their diverse shapes.

Microscopic algae make up the mass of organisms called phytoplankton, which are found near the surface of marine and fresh water. Phytoplankton are at the base of all aquatic food chains. Thus algae are critically important to global ecology; however, they have negligible medical importance.

Fungi

Fungi (*sing.*, fungus) include organisms we call mushrooms, yeasts, and molds. They are eukaryotic, nonphotosynthetic, and either microscopic or macroscopic (**Figure 1.5**). Most fungi are scavengers. They are ecologically important because they decompose dead organisms. A few fungi are pathogenic to animals and humans. Some cause trivial infections, such as ringworm and athlete's foot. Others cause life-threatening infections. An example is *Pneumocystis carinii*, which invades the lungs of immunologically weakened individuals, such as AIDS patients, causing pneumocystis pneumonia. Many fungi are pathogenic to plants. They cause such economically important diseases as corn smut, wheat rust, and potato blight.

Most fungi grow as multibranched tubes that make up a structure called a **mycelium.** Mushrooms are the aboveground structures formed from an extensive underground mycelium. **Yeasts** are unicellular fungi. **Molds** are primitive fungi that infect plants but rarely humans.

Protozoa

Protozoa (*sing.*, protozoon) means "first animals." As the name suggests, they are superficially animal-like. They

FIGURE 1.4   (a) Phytoplankton—the base of the marine food chain—are single-celled algae such as those shown here and cyanobacteria (Chapter 11). They are sometimes so abundant they make water cloudy. (b) Children looking at kelp that has washed up on a beach. Kelp is a multicellular macroscopic brown alga that grows in the Pacific Ocean.

are nonphotosynthetic and usually motile (**Figure 1.6**). Amoebae are the class of protozoa that move by extending tubelike structures called pseudopods. Flagellates and ciliates are the classes that move by long (flagella) or short (cilia) hairlike extensions that wave or beat. All protozoa are microscopic. Protozoa are the height of unicellular complexity. Some have many intracellular organelles that are almost as complex in form and function as some tissues in higher organisms.

Protozoa cause diseases, such as malaria and African sleeping sickness, that kill millions of people every year. Protozoan- and helminth-caused diseases (which we'll dis-

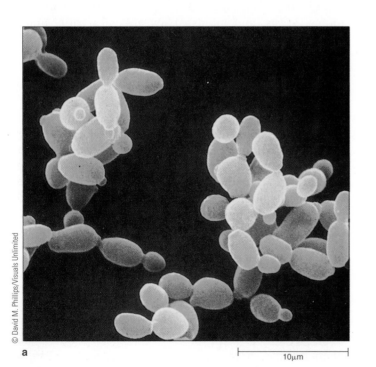

FIGURE 1.5   (a) The microscopic fungus *Epidermophyton floccosum* is one of many associated with athlete's foot. (b) The mushroom *Amanita muscaria,* which causes hallucinations if eaten, is a macroscopic fungus.

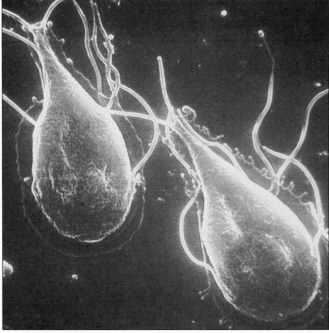

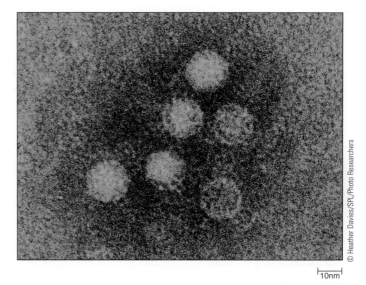

FIGURE 1.7   The common cold is caused by a virus, usually a rhinovirus, like the one pictured here. There is no effective vaccine.

FIGURE 1.6   The protozoan *Giardia lamblia* is a common cause of diarrhea in humans. It was first seen by Van Leeuwenhoek during a bout of intestinal distress.

cuss shortly) are called **parasitic diseases,** although in fact every infectious disease is a case of parasitism (a relationship in which one organism benefits at the expense of another). The study of protozoan- and helminth-caused diseases is called parasitology.

Viruses

Viruses are not cells. They are merely particles of nucleic acid, either ribonucleic acid (RNA) or deoxyribonucleic acid (DNA), packaged in a protein coat and sometimes surrounded by a membrane. Viruses are incapable of reproducing themselves. They can reproduce only inside a host cell. In other words, viruses are obligate intracellular parasites that force their hosts to make more viruses. Viruses infect animals, plants, and other microorganisms (**Figure 1.7**).

Viruses are extremely small, even compared with bacteria. The largest viruses are about one-tenth the size of a typical bacterial cell. The smallest are about one one-thousandth the size. Viruses cannot be seen through an ordinary microscope, but even the smallest can be seen with an electron microscope.

Viruses cause many major diseases of plants, humans, and other animals. Smallpox, yellow fever, and polio are examples of viral diseases that have been particularly deadly in the past. AIDS is an example of a virus-caused disease that is still a worldwide scourge.

As chemically and structurally simple as viruses are, there are even simpler infectious agents called **prions.** They are composed exclusively of protein. How they reproduce remains incompletely understood. Prions cause rare neurological diseases in humans, mad cow disease in cattle, and scrapie in sheep.

Helminths

Helminths are worms, and as such they belong to the animal kingdom. Most are also macroscopic. We study them in microbiology because they cause infectious diseases.

The two types of disease-causing helminths are flatworms and roundworms (**Figure 1.8**). Flatworms include the beef tapeworm, which can grow to lengths of 30 feet in human intestines, and liver trematodes (flukes), which are microscopic. Roundworms include hookworms, parasites that were common in the southern United States until about 50 years ago, and *Trichinella*, which humans acquire from eating contaminated pork.

A BRIEF HISTORY OF MICROBIOLOGY

Microorganisms were discovered more than 300 years ago, but we didn't know much about them until the mid 1800s, when microbiology became an experimental science. Then a period of accelerating progress began that continues today with no end in sight (**Table 1.3**).

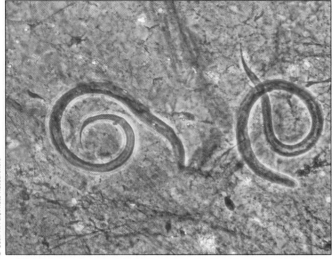

© Cecil H. Fox/Photo Researchers

⊢10μm⊣

FIGURE 1.8 In its microscopic stage, the roundworm *Strongyloides stercoralis* invades human intestines and sometimes migrates to different parts of the body, causing painful rashes.

Van Leeuwenhoek's Animalcules

Microorganisms were discovered about 200 years before Lister treated James Greenlees (Case History: The Case of James Greenlees). Antony van Leeuwenhoek, a Dutch merchant (**Figure 1.9**), made small hand-held microscopes as a hobby. Squinting through the lens at specimens held on a pin, he discovered a world of invisible creatures he called animalcules (small animals). He found them almost everywhere he looked—in water droplets, particles of soil, his teeth scrapings. In 1674 Van Leeuwenhoek communicated his discoveries to the Royal Society of London, sending detailed drawings.

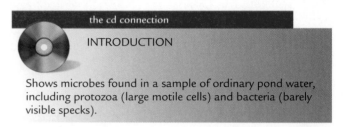

the cd connection

INTRODUCTION

Shows microbes found in a sample of ordinary pond water, including protozoa (large motile cells) and bacteria (barely visible specks).

All his drawings and nine of the estimated 500 microscopes that Van Leeuwenhoek made still exist. The most powerful of these has a magnification of 266×—powerful enough to magnify an average-sized bacterial cell to the size of the period at the end of this sentence. But judging from the details of his sketches, he must have made considerably more powerful microscopes that have been lost.

Van Leeuwenhoek was not the first to make and use microscopes. But he made better ones and used them with greater skill than anyone else. He jealously guarded his

TABLE 1.3 Highlights in the History of Microbiology

1674	Van Leeuwenhoek discovers microorganisms.
1796	Jenner creates a vaccine for smallpox.
1847	Semmelweiss establishes the cause of childbed fever.
1859	Pasteur disproves spontaneous generation of microorganisms.
1865	Lister introduces antiseptic techniques.
1876	Koch proves that specific microorganisms cause specific diseases.
1881	Koch uses agar to obtain a pure culture.
1892	Iwanowski discovers viruses.
1894	Ehrlich articulates the principle of selective toxicity.
1929	Fleming discovers penicillin.
1977	Carl Woese discovers the archaea.
1977	Smallpox is eradicated worldwide.
1982	Stanley Prusiner presents evidence that prions cause neurological diseases in animals and humans.
1983	Luc Montaigner and Robert Gallo announce the discovery of human immunodeficiency virus (HIV), the cause of AIDS.
1995	Craig Ventner, Hamilton Smith, Claire Fraiser, and colleagues at The Institute for Genomic Research (TIGR) announce the first complete sequence of a genome, that of the bacterium *Haemophilus influenzae*.

CASE HISTORY

The Case of James Greenlees

While walking down a street in Glasgow, Scotland, on August 12, 1865, an 11-year-old boy named James Greenlees was struck by a horse-drawn cart. One of its wheels ran over James's leg just below the knee, breaking the bone and pushing its jagged edges through the skin.

Young James's injury was virtually a death sentence. Such skin-piercing broken bones (called **compound fractures**) are usually dirty wounds, and people who suffered contaminated wounds of flesh and bone during the 1800s almost always died from them. Soon the wounds festered and exuded large quantities of pus. Then the tissue around them began to decay and stink. The victim developed fever, and death followed in a matter of days. The only treatment was immediate amputation, replacing the dirty wound with a comparatively clean surgical incision. But even then, nearly half the patients died.

James was taken to the Royal Infirmary, where he came under the care of a young surgeon, Joseph Lister. Lister had a theory about treating open wounds and decided to try it to help James. He splinted the broken bone and dressed the wound in bandages soaked in a solution of phenol. Lister believed that the **phenol** (which we still use as a disinfectant today) would prevent infection. He watched James's wound closely for signs of festering. If it started, Lister planned to amputate.

Four days after the accident, Lister removed the phenol-soaked dressing to examine James's wound. There was no festering, pus, or other evidence of infection. Encouraged, Lister applied new bandages soaked in a more dilute solution of phenol. The wound continued to heal. Six weeks later the broken fragments of bone had become reunited and the wound had healed. James Greenlees was released from the hospital with two sound legs.

Lister called his phenol technique **antisepsis,** meaning "against infection." It had profound practical implications, especially for the developing art of surgery. During the mid-1800s, operations performed in a hospital were almost as likely to lead to infection and death as were contaminated wounds.

What made Lister think of using phenol to treat a contaminated wound? From reading scientific papers, he knew that microorganisms caused decay and **putrefaction** (producing foul-smelling products). He also knew that microorganisms were everywhere—in soil, water, and air. Finally, he knew that phenol killed microorganisms. He concluded that applying phenol to James Greenlees's open wound would kill the microorganisms that had almost certainly entered it, thereby preventing decay and allowing the wound to heal. Lister's reasoning was completely correct.

Joseph Lister.

The Wellcome Centre Medical Photographic Library, London

---

simple (single-lens) microscopes, refusing to sell them or teach others how to make them. Not for 200 years were superior microscopes developed.

Spontaneous Generation

It was immediately obvious to Van Leeuwenhoek that the microbes he observed with his microscope were alive. Some of them moved swiftly, and under certain conditions their numbers increased rapidly. But the source of microorganisms was perplexing to him and others. Microorganisms always appeared suddenly in certain materials (meat juices or plant extracts, for example) that had previously been free of them. It seemed logical that microbes were products of **spontaneous generation** (the formation of living things from inanimate matter). From ancient times, spontaneous generation was thought to be the origin of many organisms (such as rats and flies) that routinely appeared in certain materials.

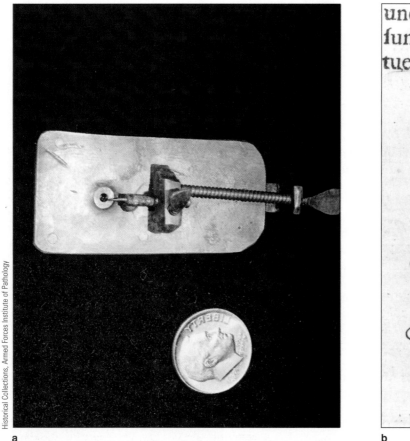

a

b

FIGURE 1.9 (a) A tiny hand-held microscope used by Antony van Leeuwenhoek to see microorganisms. (b) Drawings he published in 1684 of bacteria that he called "animalcules."

Gradually, spontaneous generation was rejected as the origin of visible organisms. Francesco Redi, a physician in Italy, played a major role. In 1665 he did experiments with covered and uncovered jars of meat. He showed that maggots (fly larvae) developed only in meat that flies could reach to lay eggs on. Apparently, spontaneous generation did not occur, at least in the case of flies. Instead, flies and by extension all living things come only from preexisting living things.

Still, many people believed that microorganisms were an exception to this rule. After all, they are very simple and they always appear, in large numbers, soon after a plant or animal dies. Might decomposition form microorganisms instead of microorganisms causing decomposition?

Needham versus Spallanzani.

For 80 years the debate continued. Then the proponents of spontaneous generation seemed to gain ground when in 1745 an English clergyman named John Needham did a well-publicized experiment. Everyone knew boiling killed microorganisms. So he boiled chicken broth, put it in a flask, and sealed it. Microorganisms could develop in it only by spontaneous generation. Indeed,

microorganisms did appear. But an Italian priest and professor named Lazzaro Spallanzani was not convinced. Perhaps microorganisms entered the broth after boiling but before sealing. So Spallanzani put broth in a flask, sealed it, and then boiled it. No microorganisms appeared in the cooled broth. Still the critics were not persuaded. Spallanzani didn't disprove spontaneous generation, they said, he just proved that spontaneous generation required air.

Pasteur's Experiments.

Remarkably, the controversy continued another 100 years and became a significant barrier to the development of microbiology as a science. Finally, in 1859 the French Academy of Science sponsored a competition to prove or disprove the theory of spontaneous generation of microbes. A young French chemist named Louis Pasteur entered. To counter the argument that air was necessary for spontaneous generation, Pasteur used barriers that allowed free passage of air but not microorganisms.

In his most famous experiment, Pasteur boiled meat broth in a flask and then drew out and curved the neck of the flask in a flame (**Figure 1.10**). No microorganisms de-

© Corbis/Bettmann

a

b

FIGURE 1.10   (a) Louis Pasteur in his laboratory. (b) In 1861 he devised this swan-necked flask to prove that microorganisms did not generate spontaneously in sterilized broth exposed to air.

veloped in the flask. But when he tilted the flask so some broth flowed into the curved neck and then tilted it back so the broth was returned to the base of the flask, the broth quickly became cloudy with the growth of microbial cells. Gravity had caused the microbial cells that had entered the flask to settle at the low point of the neck. They never

reached the broth in the base until they were washed into it. Thus Pasteur convinced the scientific world that spontaneous generation of microorganisms does not occur even in the presence of air.

Pasteur was brilliant and lucky. Many early experiments to disprove spontaneous generation of microbes failed because the samples contained **endospores** (highly heat-resistant bacterial structures that are not killed by boiling). Had Pasteur done his experiments with vegetable rather than meat broth, they too certainly would have failed, because plant materials usually carry bacterial endospores. Meat broths rarely do.

Pasteur's simple but elegant experiments grounded microbiology in scientific reality. Microorganisms could now be studied by rational scientific means.

The Germ Theory of Disease

Once spontaneous generation of microbes was disproved, microbiology exploded. It changed from an observational science to an experimental science. The way was opened to study the cause of infectious diseases. Building on Pasteur's work, a German physician, Robert Koch (**Figure 1.11**), proved that microorganisms (**germs,** as they were and are still sometimes called) cause disease. He showed further that specific microorganisms cause specific diseases. Koch also introduced higher scientific standards of rigor to microbiology, as exemplified by those called Koch's postulates.

Koch's Postulates. In 1876 Koch was studying **anthrax,** a disease of cattle and sheep that also affects humans. First, he observed that the same microbial cells were present in all blood samples of infected animals. Second, he cultivated these cells, which today we know to be cells of the bacterium *Bacillus anthracis*, in a pure form outside the infected animal. Third, he injected a healthy animal with the cultured bacterial cells. Fourth, he observed that the animal became infected with anthrax and that its blood contained the same microorganism as did the originally infected animals. These four steps have become known as **Koch's postulates** (although he never presented them as such). If fulfilled, they provide absolute proof that a particular microorganism causes a particular disease. (See the box on Koch's postulates in Chapter 15, One Microbe, One Disease.)

During his work on anthrax, Koch made another critically important contribution to microbiology. He developed a technique to obtain a pure culture of a bacterium (one that contains only a single kind of bacteria) and propagate it. In nature many kinds of bacteria are found growing together (mixed cultures). It is difficult to do valid experiments using mixed cultures because they are so complicated, although

SHARPER FOCUS

FRAU HESSE'S PANTRY

Microbiologists long realized the importance of pure cultures, but obtaining them was another matter. The few successes, based mainly on diluting mixed cultures, were time-consuming and unreliable. Koch conceived a better idea. He reasoned that if he separated a single bacterial cell on a solid surface, it would multiply and form a visible collection of bacterial cells (called a **colony** or a **clone**). These cells, all progeny of the single cell, would be a pure culture. Because multiplication was essential, nutrients had to be supplied. Koch tried everything from a slice of potato to gelatin. The gelatin worked best except for one thing: It melted at 37°C, body temperature, the best temperature for growing pathogens.

A neighbor of Koch's, Frau Hesse, heard of his problem and brought him a jar of agar from her pantry. Agar is a powder made from seaweed that she used to thicken jam. When mixed with nutrients such as meat broth, it became an ideal, nutrient-rich surface that remained solid up to 100°C. Koch now had a simple way to obtain pure cultures, which opened the way for rapid advances in microbiology. Koch's method is still used today.

© Corbis/Bettmann

FIGURE 1.11   Robert Koch, another giant in microbiology, in his laboratory.

such studies are now becoming more common. But early studies on mixed cultures led to great confusion.

Koch's postulates along with his technique for obtaining pure cultures (Greater Focus: Frau Hesse's Pantry) led to spectacular advances in microbiology. Between 1882 and 1900 the microbes that caused almost all the bacterial diseases then prevalent in Europe were isolated, including typhus, dysentery, syphilis, gonorrhea, pneumonia, and—by Koch himself—tuberculosis.

If microorganisms caused infectious diseases, then it should be possible to prevent disease by controlling microorganisms. Medical microbiology advanced rapidly by taking two routes to preventing disease. The first was **immunity**—stimulating the body's own ability to combat infection. The second was **public hygiene**—promoting cleanliness and reducing exposure to disease-causing microorganisms.

Immunity

The phenomenon of immunity was recognized from ancient times: People who suffered from certain diseases never got them again. Apparently, some protective change occurred in the body. In other words, infection produced immunity. Might it be possible to confer immunity without having to suffer disease? Edward Jenner was the first to do so.

Jenner and Smallpox. Edward Jenner, an English physician, knew that dairymaids who had naturally contracted a mild infection called **cowpox** seemed to be protected against smallpox, a horribly disfiguring disease and a major killer. In 1796 Jenner inoculated an 8-year-old boy with fluid from cowpox blisters on the hand of a dairymaid, Sarah Nelms (**Figure 1.12**). The boy contracted cowpox. Then Jenner inoculated him with fluid from a smallpox blister. There was no reaction. Apparently the child had become immune to smallpox by exposure to cowpox. This technique of inducing immunity became known as **vaccination** (from *vacca*, Latin for "cow") or later immunization. The agent that induces such immunity, in this case extract of cowpox blisters, is called a **vaccine.**

Making Vaccines. Jenner's vaccine was naturally occurring. About a hundred years after Jenner, in the early 1880s, Pasteur set about making a vaccine. He did so by culturing disease-causing microorganisms in a way—at near lethal temperature, for example—that they became **attenuated** (weakened). Such attenuated strains lost the ability to cause disease, but they could still confer immunity. First Pasteur developed an attenuated form of the bacterium that caused fowl cholera. Using this strain as a vaccine, he was able to protect chickens from the disease. Then he applied this principle of attenuation to develop vaccines for anthrax and, in 1885, for rabies.

But it was not possible to develop attenuated strains of all pathogens. About this same time, two American microbiologists, Daniel E. Salmon and Theobald Smith, developed a new way to create vaccines. They found that killed microbial cells were effective as vaccines. This discovery led to the development of vaccines against many infectious diseases.

Public Hygiene

Immunization has prevented many diseases and saved many lives, but even more lives have been saved by improved public hygiene. Before the germ theory of disease gained prevalence, pathogen-containing sewage regularly mixed with drinking water. Improving sewage disposal and ensuring a clean public water supply prevented mass outbreaks of many diseases, including cholera and typhoid fever. Similar improvements came in food preservation and eventually inspection. Pasteurization, which kills most pathogens by a brief exposure to heat, is merely one example. (Pasteur originally developed pasteurization to keep wine from spoiling.)

Personal hygiene, especially careful hand washing, also proved effective in preventing disease (Larger Field: Childbed Fever).

MICROBIOLOGY TODAY

The late 1800s became known as the golden age of microbiology. Advances came rapidly and life was dramatically

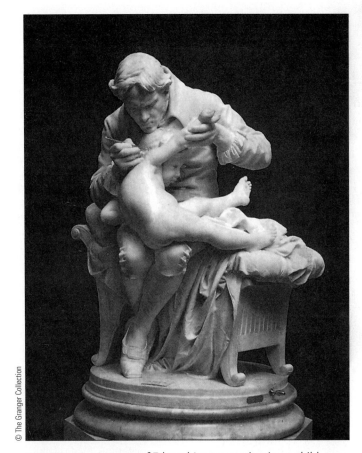

© The Granger Collection

FIGURE 1.12   A statue of Edward Jenner vaccinating a child against smallpox.

improved. But the advances that continued to be made during the twentieth century were no less striking. Let's look briefly at four key areas: chemotherapy, immunology, virology, and genetic engineering.

Chemotherapy

Probably the most significant advance in medical microbiology during the 1900s was the development of **chemotherapy** (treatment of disease with chemicals called drugs). Nineteenth-century microbiologists discovered ways to prevent many infections, but not until the twentieth century did they acquire the ability to treat infections once they had started.

The German physician-chemist Paul Ehrlich is called the father of chemotherapy because he articulated its guiding principle, **selective toxicity.** For a drug to be effective against infection, it must be selectively toxic. That is, it must be deadly or inhibitory to the infecting microorganism but relatively harmless to the affected human. Ehrlich conceived of drugs as being "magic bullets," agents that

LARGER FIELD

CHILDBED FEVER

In the mid 1800s, before knowledge of the germ theory of disease, giving birth in a hospital was risky. Many women who delivered normal, healthy infants never survived to take them home because of an illness known as **childbed fever** or **puerperal sepsis.** At that time, physicians had no idea what caused childbed fever. We now know that it is caused by the bacterium *Streptococcus pyogenes.* Infection begins in the uterus and spreads rapidly through the body. *S. pyogenes* can infect any wound; untreated, the infection causes fever, chills, delirium, and death.

In 1847 Ignaz Semmelweis, a physician in the obstetrics ward of a Viennese hospital, was horrified by the number of women he saw dying. He was also struck by the strange pattern of their deaths. Women in labor were admitted either to the First or Second Clinic of the hospital, depending on when they arrived. Although the two clinics were seemingly identical, almost all the childbed fever deaths occurred in the First Clinic. All Vienna knew about it. Women in advanced labor would wait in the halls, announcing their arrival at a time they hoped would allow them to enter the Second Clinic.

One clear difference between the clinics was staffing. Medical students staffed the First; midwives staffed the Second. But why would that make a difference? The clue Semmelweis needed came when a medical student cut his finger during an autopsy and died of an illness identical to childbed fever. Medical students began their days in the morgue, doing autopsies to learn anatomy. Then they went to the First Clinic. Semmelweis reasoned they must be taking something deadly with them on their hands to mothers in the clinic.

Semmelweis ordered the medical students to wash their hands in a chlorine solution before entering the clinic, a practice not yet routine in hospitals. Physicians often had blood and pus on their hands while treating patients. Semmelweis chose a chlorine solution because it removed the characteristic odor of the morgue, but chlorine also killed the deadly streptococcal bacteria and thus prevented infection. The number of deaths from childbed fever in the First Clinic soon fell to the prevailing low levels in the Second.

Nevertheless, Semmelweis was ridiculed. Three decades later the work of Pasteur and Koch would provide a scientific explanation for Semmelweis's findings and lead to a revolution in controlling infection. But Semmelweis did not live to see it. Ironically, he died of a streptococcal infection in 1865, an outcast from the medical community.

would strike and kill the microbe but not the host. Ehrlich gave the fledgling science of chemotherapy its first success in 1908 by discovering a drug for the treatment of syphilis. He called his drug salvarsan, from the Latin word meaning "to save." For more than 20 years syphilis was the only infectious disease that could be treated by chemotherapy.

Stimulated by Ehrlich's success, research on chemotherapy expanded. Sulfa drugs were the first major class of drugs to come into widespread clinical use. These **synthetic drugs** (organic chemicals manufactured in the laboratory) were discovered in the 1930s when a German chemical company, I. G. Farben, began systematically testing various compounds as possible chemotherapeutic agents.

Antibiotics (chemotherapeutic agents produced by microorganisms) were discovered at about the same time as sulfa drugs, but they proved to be more effective. The first medically useful antibiotic, penicillin (**Figure 1.13**), was discovered by the Scottish microbiologist Alexander Fleming (1881–1955) in 1929. But it did not come into widespread clinical use for another decade because of technical problems in purifying and mass-producing it. In the 1940s, however, during World War II, intensive research made penicillin readily available. The dramatic effectiveness of penicillin gave it the well-deserved title "wonder drug."

Immunology

In the days of Pasteur and Koch, immunology was a branch of microbiology devoted to developing vaccines for preventing infectious diseases. Today immunology is an independent and fast-developing science. In the intervening years we've learned that the immune system is extremely complicated, delicate, and prone to defects. In Chapter 17 we'll discuss the immune system. In Chapter 18 we'll discuss how its defects can cause disease. And in Chapter 19 we'll discuss the clinical applications of immune reactions and how they can be used to diagnose disease.

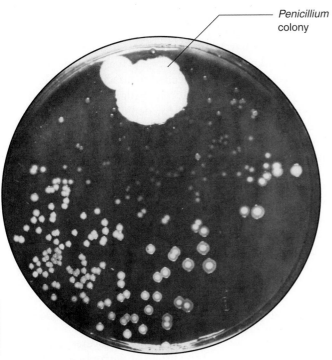

Penicillium colony

© Corbis/Bettmann

FIGURE 1.13 This is the actual Petri dish that led to the discovery of penicillin in 1929. The plate on which the disease-causing bacterium *Staphylococcus aureus* was being cultivated accidentally became contaminated by a fungal spore that developed into a colony of *Penicillium*. By producing penicillin the fungus killed or damaged colonies closest to it.

Virology

Virology, the study of viruses, began in 1892 when the Russian microbiologist Dmitri Iwanowski discovered the tobacco mosaic virus (Chapter 13). Iwanowski was studying a disease of tobacco plants called tobacco mosaic disease. To identify its cause, he forced juice from diseased plants through filters that retained the smallest bacteria. He found the filtered juice still caused disease. Because bacteria were believed to be the smallest microorganisms, Iwanowski first thought his methodology might be flawed. But repeated experimentation convinced him that minute disease-causing agents were passing through the filter. He called these tiny agents "filterable viruses." They could not be seen, even under the most powerful microscopes of that time. Until the electron microscope was developed in the 1930s, we knew viruses existed, but little more.

Basic Biology

Two properties of microorganisms led to particularly exciting developments in the second half of the twentieth century. First, the metabolism and genetic properties of microorganisms are remarkably similar to those of plants and animals, including humans. Thus what we learn about microorganisms is often directly applicable to higher forms of life. In fact, much of our knowledge of the fundamental properties of all living things came first from studies on bacteria. Second, microorganisms, especially bacteria, are especially suitable for experimental investigation: They are easy to culture and they grow rapidly. Under proper conditions certain bacteria can double their numbers every 20 minutes. The number of individual organisms in a single milliliter of a bacterial culture can exceed the number of human beings on Earth. We can thus study enormous numbers of organisms in an extremely short period and undertake experiments that would be impossible using larger, slower-growing organisms.

Genetic Engineering and Genomics

Intensive laboratory studies on microorganisms have led to the development of a remarkable set of techniques, collectively called **genetic engineering** or **recombinant DNA technology.** With this technology researchers can obtain **DNA** (the cell's genetic material) from one organism, manipulate it in the laboratory, and introduce the modified DNA into another cell where it will exert its effect. The DNA can be taken from any organism, and fragments of it from different organisms can be joined together. The organism that is most frequently used as the host for manipulated DNA is the bacterium *Escherichia coli*. For example, industry uses *E. coli* to produce human insulin for the treatment of diabetes and human growth hormone for the treatment of pituitary dwarfism.

The potential of recombinant DNA technology seems almost limitless. We'll examine genetic engineering techniques in Chapter 7 and its uses in Chapter 29.

Recombinant DNA technology has also lead to the ability to decipher and read an organism's genetic material (DNA). We will also consider this topic, called **genomics,** in Chapter 7.

The Future

As an active experimental science, microbiology is little more than a hundred years old. Its pattern of accelerating progress seems likely to continue. And the need is great. Medical microbiology and virology will need to solve pressing problems of resistance of pathogens to antibiotics and evolution of new viruses that cause new diseases. Daily we read about newly discovered strains of bacterial pathogens with increased resistance to known antibiotics. New antibiotics and other chemotherapeutic agents may be the answer, although some researchers emphasize enhancement of the natural immune system. Whatever route researchers follow, they will depend heavily on recombinant DNA technology and genomics.

The other area that will expand rapidly in the future is environmental microbiology. We will rely increasingly on **bioremediation** (use of microorganisms to degrade toxic chemicals) to clean up our environment.

The world of microorganisms is huge and diverse. The vast majority—probably more than 95 percent—of microorganisms haven't yet been identified or named. Some of these unknown microorganisms may benefit us. Many new microbiologists are needed to attack the problems the twenty-first century will surely bring. You might consider the career yourself. It asks for dedication; it demands interest. But whether or not you plan to make microbiology your career, we hope you enjoy this trip that we are about to undertake through the world of microbes.

SUMMARY

Case History: Microbiological Advances Make a Big Difference (p. 2)

1. Recent advances in microbiology based on genomics have made diagnosis of many infectious diseases faster and less invasive, often leading to early cures.

The Unseen World and Ours (pp. 2-5)

2. Microorganisms are organisms that are too small to be seen by the unaided eye.

3. Microbiology is the study of microorganisms.

Microbes, Disease, and History (pp. 2-4)

4. Throughout history, pathogens (disease-causing microorganisms) have had a tremendous negative impact on human affairs. Epidemic diseases have decimated populations and brought social and political chaos along with human suffering.

5. Only a small fraction of microorganisms are pathogenic.

Microbes and Life Today (pp. 4-5)

6. The development of microbiology as a science has allowed us to control harmful microbes and use others for our benefit.

7. Environmental microbiology is the study of how microorganisms affect the earth and its atmosphere.

8. Industrial microbiology deals with microorganism-dependent industries, such as those producing foodstuffs, fermented beverages, and pharmaceuticals.

9. Research in agricultural microbiology has led to healthier livestock and more disease-free crops.

Careers in Microbiology (p. 5)

10. Microbiology offers a diversity of career opportunities, which depend on training, as well as interest.

The Scope of Microbiology (pp. 6-9)

11. Microorganisms are usually divided in to six subgroups: bacteria, archaea, algae, fungi, protozoa, and viruses.

12. Microbiology is a cohesive science because of its methodology and approach to problems, not because of the relatedness of the organisms it studies.

13. Bacteria and archaea are prokaryotes; they lack internal membrane-bound structures.

14. Algae, fungi, and protozoa are eukaryotes; their organelles are membrane bound.

15. Viruses are acellular.

Bacteria (p. 6)

16. Bacteria are extremely small, even for microorganisms.

17. Bacteria vary in shape, motility, and how they get energy. There are species that can withstand freezing, boiling, and extreme acidity or alkalinity.

18. Some bacteria cause disease, but others keep our environment in life-sustaining balance.

Archaea (pp. 6-7)

19. Archaea were discovered as a separate group of microorganisms in the 1970s.

20. At first they were called archaebacteria.

21. Being prokaryotes and small, they resemble bacteria superficially, but they are as distantly related to bacteria as they are to eukaryotes.

22. Many archaea live in extremely hostile environments.

Algae (p. 7)

23. Algae are eukaryotic organisms that carry out photosynthesis.

24. Some algae are unicellular and microscopic. Others consist of many cells and are macroscopic.

25. Algae are not significant medically, but they are critically important to global ecology.

Fungi (p. 7)

26. Fungi include mushrooms, yeasts, and molds. They are eukaryotic and nonphotosynthetic. Some are microscopic; others are macroscopic.

27. A few fungi are pathogenic to humans, and many are pathogenic to plants, causing, for example, corn smut, wheat rust, and potato blight.

Protozoa (pp. 7-9)

28. Protozoa are eukaryotic microorganisms that are superficially animal-like, nonphotosynthetic, and usually motile.

29. Examples of protozoa are amoebae, flagellates, and ciliates.

30. The study of protozoan (and helminth-caused) diseases is called parasitology.

Viruses (p. 9)

31. Viruses are particles of nucleic acid (either RNA or DNA), usually enclosed in a protein coat and sometimes surrounded by a membrane.

32. Viruses are obligate intracellular parasites.

33. Viruses are extremely small, even compared with bacteria.

34. Viruses can infect animals, plants, and microorganisms.

35. Prions are even smaller infectious agents than viruses. They are composed entirely of protein.

Helminths (p. 9)

36. Helminths are macroscopic worms, but some go through microscopic stages in their life cycle; they cause parasitic diseases in plants and animals, including humans.

37. The helminths important to health studies are flatworms and roundworms.

A Brief History of Microbiology (pp. 9-15)

38. Once microbiology became an experimental science in the mid-1800s, a period of accelerating progress began.

Van Leeuwenhoek's Animalcules (pp. 10-11)

39. Antony van Leeuwenhoek, whose hobby was making microscopes, was the first to see microorganisms (about 1674). He called them animalcules.

Spontaneous Generation (pp. 11-13)

40. In 1665 Francesco Redi's experiment with covered and uncovered meat jars proved that maggots do not generate spontaneously.

41. Many scientists were convinced by John Needham's 1745 experiment that spontaneous generation of microorganisms did occur. They said that Lazzaro Spallanzani's experiment with sealed flasks proved only that microorganisms needed air for spontaneous generation.

42. In 1859 Louis Pasteur finally disproved spontaneous generation. Using special swan-necked flasks, he performed experiments that allowed the free passage of air but prevented the entry of microorganisms.

43. Pasteur's experiments set the stage for rapid progress in microbiology.

The Germ Theory of Disease (pp. 13-14)

44. Robert Koch developed the germ theory of disease: Microorganisms (germs) cause infectious diseases, and specific microorganisms cause specific diseases.

45. Koch developed four postulates that, if fulfilled, provide absolute proof that a particular microorganism causes a particular disease.

46. Koch also developed a technique using an agar-solidified nutrient medium for obtaining pure cultures.

Immunity (pp. 14-15)

47. Immunity is stimulating the body's own ability to combat infection. The idea of immunization was based on the observation that people who suffered once from certain diseases did not get them again.

48. Edward Jenner used fluid from cowpox blisters to provide protection against smallpox. Inducing immunity for protection again infectious diseases came to be known as vaccination.

49. Pasteur developed vaccines for anthrax and rabies, using attenuated forms of the disease-causing microorganism. D. E. Salmon and Theobald Smith demonstrated that killed microbial cells were also effective as vaccines.

Public Hygiene (p. 15)

50. Acceptance of the germ theory advanced the idea of public hygiene, promoting cleanliness and reducing exposure to disease, which saved even more lives than immunization.

51. Concern for public hygiene led to clean drinking water, improvements in food preservation (such as pasteurization), and hand washing in hospitals and for personal hygiene.

Microbiology Today (pp. 15-18)

52. Advances in twentieth-century microbiology have been striking in the areas of chemotherapy, immunology, virology, and genetic engineering.

Chemotherapy (pp. 15-16)

53. Chemotherapy is the treatment of disease with chemicals called drugs.

54. Paul Ehrlich articulated the guiding principle of chemotherapy, selective toxicity: To be effective against infection, a drug must kill or inhibit the infecting microorganism without damaging the host.

55. The first major class of drugs to gain widespread clinical use was the sulfa drugs, which are synthetic chemicals.

56. Antibiotics are natural chemotherapeutic agents produced by microorganisms. Penicillin, the first medically useful antibiotic, was discovered in 1929 by Alexander Fleming.

Immunology (p. 16)

57. Immunology studies the immune system, which provides us protection against pathogens.

Virology (p. 17)

58. Virology, the study of viruses, began in 1892 when Dmitri Iwanowski discovered the tobacco mosaic virus; viruses couldn't be seen until the electron microscope was developed.

Basic Biology (p. 17)

59. Microorganisms lend themselves to experimentation because (1) the metabolism and genetics of microorganisms, particularly bacteria, are remarkably similar to those of plants and animals and (2) microorganisms are easy to culture and multiply rapidly, so that enormous numbers can be studied in short periods.

Genetic Engineering and Genomics (p. 17)

60. Genetic engineering (recombinant DNA technology) is a group of techniques for manipulating DNA outside the organism from which it was obtained and introducing it into another cell where it will exert its effect.

61. Genomics is a group of techniques for reading and deciphering an organism's genetic material.

The Future (pp. 17-18)

62. Much microbiological research will be needed to control emerging diseases and successfully treat old ones.

63. Genetic engineering and bioremediation, using microorganisms to clean up toxic chemicals added to the environment, promise to be areas where rapid progress will be made.

REVIEW QUESTIONS

Case History: Microbiological Advances Make a Big Difference

1a. Explain how a new diagnostic procedure could have a major impact on the consequences of pelvic inflammatory disease.

The Unseen World and Ours

1. Define microorganisms. Give some examples of how microorganisms have affected human history.

2. Give some examples of advances in the field of medical microbiology and challenges to be faced.

3. What is environmental microbiology?

4. Name some microorganism-dependent industries. In what directions is industrial microbiology moving today?

5. Give some examples of advances and challenges still to be faced in agricultural microbiology.

The Scope of Microbiology

6. Name the six subgroups of microorganisms.

7. What is the one property all microorganisms have in common? Explain this statement: Microbiology is a cohesive science because of its methodology.

8. Explain the difference between prokaryotes and eukaryotes.

9. What do bacteria and archaea have in common? How are they different?

10. Explain this statement and give examples: Bacteria are extremely diverse.

11. What are algae? What do they look like? Of what importance are they?

12. What are fungi? Name the different types. How are fungi important medically? Ecologically?

13. What are protozoa? Give some examples of motile protozoa and tell how they move. What is parasitology?

14. Explain this statement: Viruses are acellular. Why are viruses called obligate intracellular parasites? How are viruses studied? Why are viruses medically important?

15. What are prions?

16. Why do we study helminths in microbiology?

A Brief History of Microbiology

17. Who was the first person to see microorganisms? When did this occur?

18. Define spontaneous generation. What roles did Redi, Needham, and Spallanzani play in dispelling or propagating the theory?

19. When and by whom was spontaneous generation finally disproved? How did he do it?

20. Explain this statement: Once spontaneous generation was disproved, microbiology changed from an observational science to an experimental science.

21. Discuss Pasteur's contributions to the field of microbiology.

22. What is the germ theory of disease?

23. Discuss Koch's contributions to the field of microbiology.

24. Define these terms: infectious diseases, immunity, immunization, vaccination, vaccine. What is the difference between attenuated and killed vaccines?

25. Compare Jenner's approach to developing a vaccination for smallpox with Lister's development of the phenol method of antisepsis.

26. Define public hygiene. Give some examples of progress made in public hygiene after acceptance of the germ theory of disease.

Microbiology Today

27. Define chemotherapy. What is Ehrlich's principle of selective toxicity? What is the difference between synthetic drugs and antibiotics? Give an example of each.

28. Define immunology, and describe how the field today differs from that of Pasteur and Koch's era.

29. What is virology? How were viruses discovered and by whom?

30. What is genetic engineering? Why is it also called recombinant DNA technology?

31. What role does *Escherichia coli* play in genetic engineering?

32. Name some products of genetic engineering.

33. What is genomics?

34. Explain the two reasons microorganisms lend themselves so well to experimentation.

ESSAY QUESTIONS

1. Pasteur is sometimes called a precarious genius. Why?

2. Many microbiologists feel it will become increasingly more difficult to find effective new antibiotics. Why?

3. Do you think bacteria as small as viruses will ever be discovered? Explain.

SUGGESTED READINGS

Brock, T. D. 1988. *Robert Koch, a life in medicine and bacteriology.* Madison, Wis.: Science Tech.

Desowitz, R. S. 1997. *Who gave Pinta to the Santa Maria?* New York, London: W. W. Norton and Company.

Dubos, R. J. 1988. *Pasteur and modern science.* Edited by Thomas Brock. Madison, Wis.: Science Tech.

McNeill, W. H. 1977. *Plagues and people.* Garden City, N.Y.: Anchor Press.

Roueché, B. 1984. *The medical detectives.* 2 vols. New York: Times Books.

For additional readings, go to InfoTrac College Edition, your online research library at: http://www.infotrac.thomsonlearning.com

TWO

Basic Chemistry

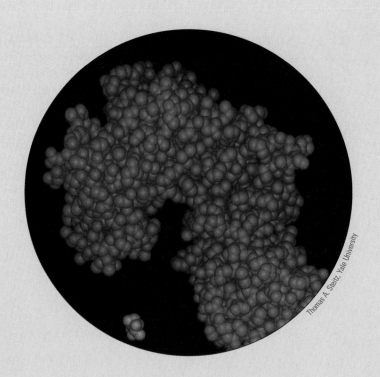

Thomas A. Steitz, Yale University

CHAPTER OUTLINE

LEARNING GOALS

To understand:

- *The basic building blocks of matter—subatomic particles, atoms, elements, and molecules*
- *The ways in which atoms bond together to make molecules and compounds*

- *What a chemical reaction is; what determines whether a reaction occurs; and what determines the rate of a chemical reaction*
- *The unique chemical properties of water*

- *The structure of organic molecules*
- *The structures of the macromolecules—proteins, nucleic acids, carbohydrates, and lipids—from which all cellular microorganisms are built*

THE CHEMICAL SOUP OF LIFE

Living cells are chemical factories that make more chemical factories like themselves. In fact, almost all **organic** (carbon-containing) chemicals, such as those that compose wood, leaves, coal, and petroleum, were made by living cells. Cells also make the organic compounds that become more cells. It's a self-perpetuating cycle. Cells make organic compounds that become more cells that make more organic compounds. But where did the organic compounds that formed the first cells come from? Charles Darwin, best known for his writings on evolution, might have been the first to formulate a plausible explanation. In 1871 he wrote to a friend, "If we could conceive in some warm little pond, with all sorts of ammonia and phosphoric salts, light, heat, electricity, etc., present, that a protein compound was chemically formed ready to undergo still more complex changes. . . ." In other words, Darwin speculated that the organic compounds that make up cells might form spontaneously from simple inorganic compounds under the right conditions. If no living things were present to consume them, they would accumulate. Then they might assemble themselves into a self-reproducing structure—a primitive cell.

About 80 years later, Stanley Miller, an American graduate student working with the Nobel Prize–winning chemist Harold Urey, put Darwin's speculation to an experimental test. He put a chemical mixture of methane, ammonia, and water vapor (the simple gases presumed to be present in Earth's atmosphere before life appeared) in a flask and exposed the mixture to electrical discharges. He found, as he expected, that organic compounds formed in the flask. But quite unexpectedly he found the mixture included a group of amino acids—including glycine, alanine, aspartate, and glutamate. These compounds are found in the proteins of all living things.

Miller's results had a great impact on scientists who study the origin of life. His results supported Darwin's speculations with scientific fact. The organic compounds that compose cells can be made by ordinary chemical reactions from materials that were probably present on Earth before life appeared.

© Chesley Bonestell/Space Art International

This is probably how the Earth looked about 4 billion years ago. Only inorganic compounds were present, but they formed the "soup" of life from which the first cells would appear about 500 million years later.

THE BASIC BUILDING BLOCKS

We need to know a little chemistry to study microbiology. Knowing some chemistry is essential to understanding all biology, but it is particularly important to understanding microbiology. The multiplication of microorganisms and the ways they affect our lives are largely chemical. We rarely see microorganisms because their individual cells are so small. But we are constantly aware of their presence by the chemical changes they cause. We see a tomato rot or a slice of bread turn moldy. We find that the milk we left too long in the refrigerator has turned sour. We suffer from a microbial infection. We are treated with antibiotics and get better. Understanding these processes depends on knowing some basic chemistry.

Chemistry is the science that studies the composition of matter and the changes it undergoes. **Matter** is an all-inclusive term describing everything (other than a complete vacuum) that fills space. In this chapter we'll concentrate on matter found in organisms, particularly microorganisms. The basic chemistry in this chapter is enough to understand the rest of the book.

We'll survey chemistry from the bottom up, starting with atoms, working up to small organic molecules, and

TABLE 2.1   Properties of Subatomic Particles

Subatomic Particle	Unit Charge	Relative Weight	Location in Atom
Proton	+1	1	Nucleus
Neutron	0	1	Nucleus
Electron	−1	1/1837	Cloud surrounding nucleus

finally to the huge macromolecules that make up the major part of all cells. Along the way we'll consider chemical bonds, which hold atoms together to make molecules. We'll examine how they break and form during chemical reactions. We'll pay special attention to water. Its chemistry is essential to life.

Atoms

Atoms are the smallest particles that have the properties of a particular **element** (a substance such as carbon, oxygen, or gold). But atoms aren't the smallest particles in nature. Atoms are composed of three kinds of smaller particles: **protons, neutrons,** and **electrons.** The numbers of protons, neutrons, and electrons in an atom determine what kind of element it is. A gold atom, for example, has 79 protons, 118 neutrons, and 79 electrons. An oxygen atom, on the other hand, has eight protons, eight neutrons, and eight electrons.

Protons, neutrons, and electrons differ in mass, electrical charge, and location within the atom (**Table 2.1**). Protons and neutrons have approximately equal mass. They are enormous compared with an electron. Each of them has 1837 times more mass than an electron. Although identical in mass, protons and neutrons can be distinguished by their electrical charge. A proton carries a single positive (+) charge. A neutron carries no charge. The tiny electron carries a single negative (−) charge. Intact atoms have no net charge. That's because the number of protons and electrons they contain is always equal. (As we'll see later, certain atoms can gain or lose electrons and become **ions.**) As we've seen, gold has 79 of each and oxygen has 8 of each. As a result, both (and all) intact atoms' net electrical charge is zero because the total positive charges of their protons exactly balance the total negative charges of their electrons.

All atoms have the same basic structure. At their center they have a **nucleus** composed of densely packed protons and neutrons (**Figure 2.1a**). Electrons orbit around the nucleus but not randomly. They orbit the nucleus in distinct energy levels or **shells.** These shells fill from the inside out. That is, inner shells are completely filled before electrons enter more distant shells. The innermost shell can hold one or two electrons. For example, hydrogen (the smallest element) has only one electron, which is in the first shell (**Figure 2.1b**). The second shell can contain as many as eight additional electrons. The third shell can contain another eight electrons. Thus a chlorine atom, which has 11 electrons, has two in its first shell, eight in its second, and one in its third. When filled, the first three shells can accommodate up to 18 electrons. Most of the atoms that make up microorganisms and other living things have fewer than 18 electrons and therefore contain all their electrons within the first three shells. The electrons in an atom's outermost shell (the first in the case of hydrogen and the third in the case of chlorine) are called **valence electrons.** They have special significance. We'll return to them when we discuss chemical bonds.

The number of protons (and electrons) an atom contains is called its **atomic number.** Every kind of atom has its own unique atomic number, from 1 to 106. So the atomic number of hydrogen is 1 and sodium is 11 (Figure 2.1b). Oxygen and gold are 8 and 79, respectively.

The mass of atoms is measured in atomic weight units (awu). Because protons and neutrons are each 1 awu, we can come close to determining an atom's **atomic weight** (its mass in awus) just by adding up the number of protons and neutrons it contains. The sum isn't exactly the actual atomic weight. There are other complications. Electrons weigh something and atomic components interact to change mass. But the answer is close. For example, hydrogen has one proton and no neutrons, but its actual atomic weight is 1.00797, not 1. Sodium has 11 protons and 12 neutrons, but its atomic weight is 22.9898, not 30.

In some cases, however, the system of adding weights of protons and neutrons doesn't seem to work at all. The atomic weight of some elements isn't even close to being a whole number. For example, the atomic weight of chlorine is 35.45. That's because chlorine occurs as two different forms, called **isotopes.** Chlorine's atomic number is 17, so both isotopes have 17 protons. But one isotope has 18 neutrons (its atomic weight is about 35) and the other has 20 neutron (its atomic weight is about 37). The mixture of the two that occurs in nature gives chlorine its observed atomic weight of 35.45. Isotopes are relatively common. For example, there are four isotopes of carbon. The most

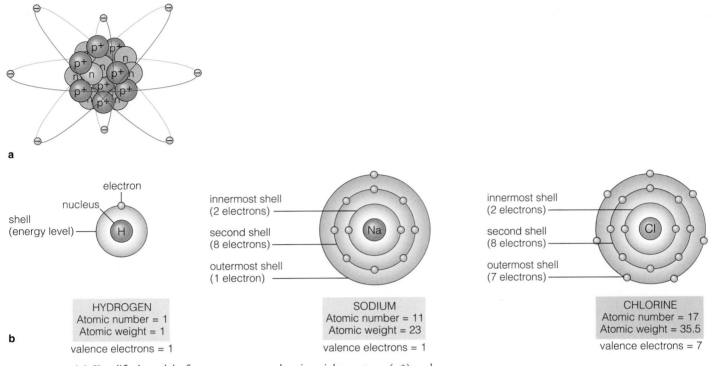

FIGURE 2.1 (a) Simplified model of an oxygen atom showing eight protons (p$^+$) and eight neutrons (n) in nucleus and eight electrons ($^-$) orbiting. (b) Structures of hydrogen, sodium, and chlorine atoms showing the arrangement of their electrons in shells.

abundant of them has 6 protons and 6 neutrons for an atomic weight of 12. But there are carbon atoms with 7, 8, 9, and 10 neutrons. Many isotopes are stable, but some are unstable. An unstable isotope is radioactive. It releases radiation as it **decays** (changes) into a stable element (see the box Sharper Focus: Radioactive Tracers).

Elements

Now we can define an **element** a little more precisely. It is matter composed of atoms all of which have the same atomic number. But as we've seen, all the atoms in an element don't necessarily have the same atomic weight. Ninety-two different elements occur naturally. Fourteen more have been created artificially in the laboratory. Each of these 106 elements has distinctive physical and chemical properties. And each is designated by one- or two-letter abbreviation. For example, H stands for hydrogen, C for carbon, and Fe for iron.

Only about 25 elements are essential to life. Of these, by far the most abundant in living organisms are carbon, hydrogen, nitrogen, oxygen, phosphorus, and sulfur. These six elements account for more than 99 percent of the weight of most living things. The rest is composed of **trace elements.** They are essential to life, but they're needed only in minute amounts. Iron is an example of a trace ele-

ment. It constitutes less than 0.2 percent of the dry weight of most living things, but life cannot exist without it. Some trace elements are listed in **Table 2.2.**

Now we'll look at how atoms are joined together to form larger structures.

Molecules

Two or more atoms joined together are called a **molecule.** The atoms in a molecule may be the same or different. For example, the atoms in a molecule of hydrogen gas are both hydrogen atoms. Water contains two hydrogen atoms and one oxygen atom. Five kinds of atoms are present in DNA: carbon, hydrogen, oxygen, nitrogen, and phosphorus. Molecules vary enormously in size and complexity from something as simple as hydrogen gas to something as complicated as DNA. The hydrogen molecule is the smallest molecule. It's smaller than any atom except the hydrogen atom itself. The DNA molecule that makes up a bacterial chromosome is huge. It's a millimeter long—a thousand times longer than the cell itself. (It has to be highly folded just to fit inside the cell.)

Molecular Weight. The atomic composition of a molecule is usually written as a **molecular formula,** in which the kinds of atoms the molecule contains are listed (by

SHARPER FOCUS

RADIOACTIVE TRACERS

Not until the 1940s did we have the technology to produce relatively pure isotopes. Then researchers learned how to make some and purify others from naturally occurring mixtures. Ordinary hydrogen, for example, with its one neutron could be separated from its heavier but stable two-neutron form (called **deuterium**) and its radioactive three-neutron form could be made (called **tritium**).

Isotopes enable us to label and trace biological structures. Both heavy (stable) and radioactive (unstable) isotopes can be used to label and to trace, allowing researchers to locate them at different places in an organism or a molecule. Isotopes are particularly useful in biology because radioactive isotopes exist for three biologically important elements: ^{3}H for hydrogen, ^{14}C for carbon, and ^{32}P for phosphorus. (Only heavy isotopes exist for oxygen. Nitrogen has a heavy and a radioactive isotope, but the radioactive element is so highly unstable that it is inconvenient.) Even minute amounts of radioactive isotopes can be located by the radiation they emit. An instrument called a scintillation counter can detect even a single radioactive isotope molecule releasing radiation.

Using isotopes to label and trace has led to many important discoveries in microbiology. For example, microbiologists discovered how bacterial chromosomes replicate by labeling them with heavy isotopes. Radioactive tracers have also been used to find where individual atoms move in chemical reactions. The first chemical reaction in photosynthesis was discovered this way. An alga was grown in a radioactive carbon dioxide environment. Then researchers traced the carbon dioxide through the chemical reactions by which it was incorporated into other carbon-containing molecules.

Radioactive tracers play an important role in clinical medicine. Radioisotopes, for example, are routinely used to diagnose thyroid gland abnormalities. The thyroid is the only body structure to take up iodine as a trace element. So the radioisotope for iodine is intravenously injected, and then the thyroid is scanned with a radiation detector. See the illustration for examples of diagnostic scanner images. Radioisotopes are also used to do whole-body scans for cancer. The radioactive isotope ^{67}Ga of the metallic element gallium is used. Isotope ^{67}Ga has the unusual property of being taken up selectively by cancer cells. The isotope is intravenously injected, and then the entire body is scanned with positron emission tomography (PET) to detect radioactivity. The scanner image locates any cancer cells.

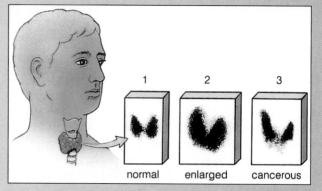

Images of thyroid glands that have taken up radioisotopes of ^{123}iodine.

their abbreviations); the number of each is shown by subscripts. For example, hydrogen gas is written as H_2 and water as H_2O. The size of a molecule is measured by its **molecular weight.** Molecular weight is determined by adding up the atomic weights of the atoms it contains, such as 2 (1 + 1) for H_2 and 18 for H_2O.

(1	×	2)	+	16	= 18
Atomic Weight of Hydrogen		Number of Hydrogen Atoms in the Molecule		Molecular Weight of One Oxygen	

Moles. As we've seen, atomic and molecular weights are indicated in atomic weight units (awu, also called **Daltons**). These are minuscule units. One awu is 1.66×10^{-24} (166 with 23 zeros in front of it). That's not a very practical unit to deal with in the laboratory. Instead, chemists often work with a more useful unit called a **mole.** A mole of a particular kind of molecule is its molecular weight measured in grams. (The equivalent term for atoms instead of molecules is **gram atom.**) A mole of hydrogen weighs 2 grams, and a mole of water weighs 18 grams. In other words, a mole is the weight of a very large number of molecules called **Avogadro's number.** Its value is 6.02×10^{23}. Or to put it the other way around: A mole is

TABLE 2.2 Elements That Occur Most Abundantly in Microorganisms

Element	Symbol	Atomic Number	Atomic Weight	Abundance in *Escherichia coli* (% of dry weight)
Hydrogen	H	1	1.01	8
Carbon	C	6	12.01	50
Nitrogen	N	7	14.01	14
Oxygen	O	8	16.00	20
Sodium	Na	11	22.99	1
Magnesium	Mg	12	24.31	0.5
Phosphorus	P	15	30.97	3
Sulfur	S	16	32.06	1
Chlorine	Cl	17	35.45	0.5
Potassium	K	19	39.10	1
Calcium	Ca	20	40.08	0.5
Iron	Fe	26	55.85	0.2

an amount of a chemical that contains 6.02×10^{23} molecules. This statement is no more mysterious than saying a dozen eggs is the amount that contains 12 eggs.

CHEMICAL BONDS AND REACTIONS

We've seen that molecules are made up of specific numbers of specific atoms. But what holds the atoms together? They are held together by forces called **chemical bonds.** The several kinds of chemical bonds are quite different, but they share one important property. They are low-energy states; therefore they are more stable than the separated atoms. Thus energy is released (largely as heat) when any chemical bond forms, and energy is required to break any chemical bond. In spite of this, biochemists call certain bonds **high-energy bonds** (Chapter 5). They do this because (1) only small amounts of energy are needed to break these bonds and, (2) when they are broken, other bonds form that release even larger amounts of energy.

An atom's valence electrons are critically important in forming chemical bonds. That's because an atom is most stable (in its lowest energy state) when its valence (outermost) shell is completely filled with electrons. And forming bonds is one way to fill the valence shell. This happens in one of three ways: (1) by sharing electrons between atoms, (2) by gaining electrons to fill the valence shell, or (3) by losing all the electrons in the valence shell (**Table 2.3**). De-

pending on how this happens, one of three kinds of bonds is formed: covalent, ionic, or hydrogen (**Table 2.4**). Now we'll discuss each of these three kinds of bonds.

Covalent Bonds

A **covalent bond** is formed when two atoms share pairs of electrons. Each atom contributes one electron to a pair. The structure of water illustrates this principle (**Figure 2.2a**). Hydrogen atoms have one valence electron and oxygen has six. So when two hydrogen atoms share electrons with an oxygen atom, all three atoms can fill their valence shells. Each hydrogen atom gets one shared electron from oxygen, giving it two electrons, which completes its valence shell. The oxygen gets one shared electron from each hydrogen atom, giving it eight, which completes its valence shell. As a result of sharing these electrons, two extremely stable covalent bonds are formed.

Some atoms can share more than a single pair of electrons. For example, oxygen can share two pairs of electrons. Carbon can share up to three pairs. Each shared pair forms a **single covalent bond.** So sharing two pairs forms a **double bond,** and sharing three pairs forms a **triple bond (Figure 2.2b).**

Sometimes the shared electrons in a covalent bond are not equally spaced between the two atoms they join. When this occurs, the part of the molecule with the greater concentration of electrons carries a more negative charge than the other part of the molecule. Such a molecule has a

TABLE 2.3 Electron Shells and Covalent Bonds of Biologically Important Elements

Element	Atomic Number	Number of Electrons in:				Number of Single Covalent Bonds It Can Form
		First Shell	Second Shell	Third Shell	Valence Shell	
Hydrogen	1	1	0	0	1	1
Carbon	6	2	4	0	4	4
Nitrogen	7	2	5	0	5	3 or 5
Oxygen	8	2	6	0	6	2
Phosphorus	15	2	8	5	5	3 or 5
Sulfur	16	2	8	6	6	2

TABLE 2.4 Important Kinds of Chemical Bonds in Living Systems

Bond	Properties	Notes
Covalent	Atoms share a pair of electrons	Strong bonds. Can be single, double, or triple.
Nonpolar	Electrons shared equally	Most covalent bonds are nonpolar.
Polar	Electrons shared unequally	Polar covalent bonds give molecules a positive and negative pole.
Ionic	Attraction between positive and negative charges	Strong but not as strong as covalent. They occur in salts and between charged parts of macromolecules.
Hydrogen	A hydrogen atom is shared between two other atoms	Weak bonds but very important to living things. They give water its unusual properties. They link the two strands of DNA and produce the secondary and quaternary structures of proteins.

positive **pole** (end) and a negative pole. It's called a **polar molecule (Figure 2.3)**. Molecules in which electrons (and hence charge) are evenly distributed are called **nonpolar molecules.** Water is an example of a polar molecule, ethane of a nonpolar molecule.

Thus covalent bonds are formed by sharing pairs of electrons between atoms. They can be single, double, or triple, and they can be polar or nonpolar.

Ionic Bonds

Before considering the formation of ionic bonds, we have to consider the formation of ions. **Ions** are atoms that have completed their outermost (valence) shell of electrons by gaining or losing electrons. Atoms with only a few electrons in their valence shell can loose them to become an ion. The next inner shell then becomes the outermost shell. Because this inner shell is complete, the atom acquires a stable state. Sodium is an example of such an atom. It loses its single valence electron to become a positively charged **sodium ion** (because it now has one more proton than electrons; **Figure 2.4 a, b**). In contrast, atoms with an almost complete valence shell can gain enough electrons to fill their valence shell. Chlorine is an example of such an atom. It can gain one electron to fill its valence shell. The result is a negatively charged **chloride ion** (because it now has one more electron than protons). Positively charged ions (such as sodium ions) are called **cations** because in solution they migrate to the **cathode** (the negative pole) of an electric field. Negatively charged ions (such as chloride ions) are called **anions** because they migrate to the **anode** (positive pole). Unlike chloride and sodium ions, some ions are composed of more than a single kind atom. For example, phosphate ion (PO_4^{3-}) is composed of phosphorus and oxygen atoms. The same is true of sulfate (SO_4^{2-}) and nitrate (NO_3^{-}) ions.

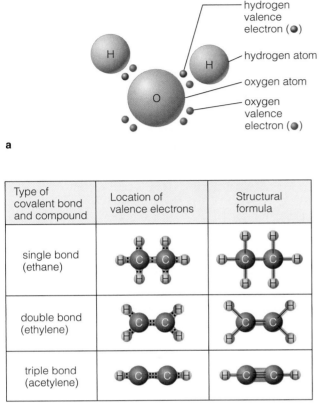

a

Type of covalent bond and compound	Location of valence electrons	Structural formula
single bond (ethane)		
double bond (ethylene)		
triple bond (acetylene)		

b

FIGURE 2.2   Covalent bonds. (a) The two covalent bonds in water. (b) Single, double, and triple covalent bonds.

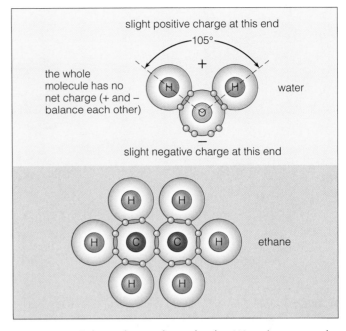

FIGURE 2.3   Polar and nonpolar molecules. Water is an example of a polar molecule (its molecules are bent, forming a 105-degree angle). Ethane is an example of a nonpolar molecule.

Ions play important roles in living systems. For example, magnesium (Mg^{2+}) activates certain enzymes, and calcium (Ca^{2+}) signals environmental changes to eukaryotic cells. Other ions play nonspecific roles. For example, they neutralize charges on other molecules.

Now that we know about ions, we can consider **ionic bonds.** They are simply the electrical attraction between oppositely charged ions. For example, the single positive charge of a sodium ion (Na^+) attracts the single negative charge of chloride (Cl^-), forming sodium chloride (NaCl), table salt (**Figure 2.4c**). Ionic bonds also form between different charged regions of large molecules such as proteins. Ionic bonds are not as strong as covalent bonds.

Hydrogen Bonds

A **hydrogen bond** forms when a hydrogen atom interacts with two different molecules or with parts of the same molecule (**Figure 2.5a**). The hydrogen atom that forms the bond has a tendency to share a pair of electrons with

each of two different atoms. In doing so, it holds them together. As a result, hydrogen bonds form only with atoms that have available pairs of nonbonding electrons. Because such pairs of electrons are present in nitrogen, oxygen, or fluorine atoms, they can form hydrogen bonds. Hydrogen bonds are polar because the hydrogen atom has a slight positive charge and the nitrogen, oxygen, or fluorine atoms have a slight negative charge (**Figure 2.5b**).

Hydrogen bonds are weak. The energy required to break one is only about one-twentieth as much as is required to break a covalent bond. Nevertheless, hydrogen bonds are critically important to living things. They stabilize the structure of many large biochemical molecules, including DNA. They form the basis of DNA's ability to encode genetic information. Hydrogen bonds also help large molecules retain the shape they need to fulfill their biological roles. The hydrogen bonds that form between water molecules give liquid water its unusual properties that make life possible (Figure 2.5b).

Chemical Reactions

A **chemical reaction** occurs when atoms or molecules (called **reactants**) collide and are changed into different combinations of the same atoms (called **products**). In this process, some chemical bonds break and some new ones

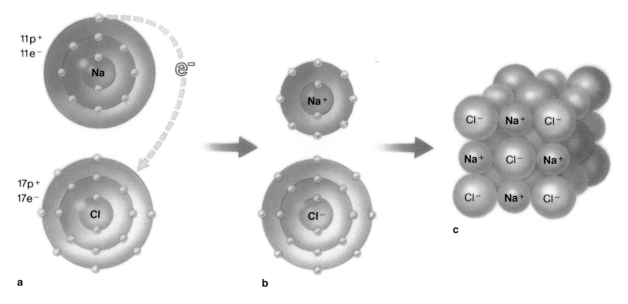

FIGURE 2.4 Ions and ionic bonds. (a) A sodium atom loses an electron and a chlorine atom gains one to become (b) a sodium ion and a chloride ion, respectively. (c) The positive charges of sodium ions attract the negative charges of chloride ions, forming ionic bonds in a salt crystal.

(From *Biology: The Unity and Diversity of Life*, 6th ed., by C. Starr and R. Taggart, Brooks/Cole, 1992. All rights reserved.)

form. As we've already learned, energy is taken up (required or consumed) when bonds are broken and it's released when they form. As a result, chemical reactions occur only if the energy released by forming the new bonds exceeds the amount required to break the old ones. This energy difference is called **free energy** (ΔG). The free energy of a reaction determines whether a reaction *can* occur but not *how fast* it will occur. And, for practical considerations, a very slow reaction is about the same as no reaction at all. Now we'll turn our attention to what determines how fast a reaction occurs.

Reaction Rates. Reactions occur only when certain molecules of a reactant collide—namely, those molecules with sufficient energy (called **activation energy**) to enter an **activated state.** In a sense, attaining sufficient activation energy is a barrier that reactants must overcome before a reaction can take place (**Figure 2.6**). The size of this barrier (the amount of **activation energy** required) and the number of collisions determine how fast a reaction occurs. The reasons are straightforward. Only molecules that collide can react. And molecules become activated randomly mostly by acquiring thermal energy. Molecules are more likely to acquire smaller amounts of energy. Therefore if the activation energy barrier is low, reactant molecules will acquire the necessary activation energy more often and the reaction will occur more rapidly.

Three conditions change rate of reaction: (1) concentration of reactants, (2) temperature, and (3) presence of a catalyst (**Table 2.5**). We'll consider these conditions one at a time.

1. *Concentration of reactants.* Higher concentrations of reactants make collisions more frequent. More frequent collisions speed reactions. Therefore the rate of a reaction increases if the concentration of reactants is increased.

2. *Temperature.* Higher temperature provides more thermal energy to raise reactants to an activated state. More activated reactants speeds reactions. Therefore the rate of a reaction increases if temperature is increased.

3. *Presence of a catalyst.* **Catalysts** are materials that increase the rate of a reaction by decreasing the activation energy barrier. Catalysts bring together the parts of the molecules that must interact for the reaction to occur. Catalysis can be highly effective. Some reactions that occur rapidly in the presence of a catalyst do not occur at perceptible rates in its absence. The distinctive feature of a catalyst is that it is not used up or inactivated by the reaction it promotes. As a result, just a small amount of a catalyst can convert large amounts of reactants into new products. A familiar example of a catalyst is the platinum in the catalytic converters of modern automobiles. It catalyzes a reaction between

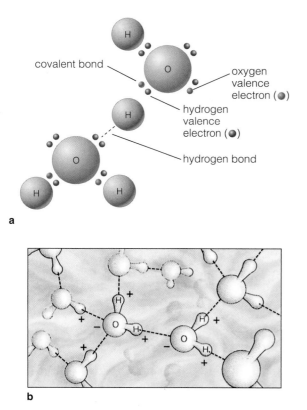

a

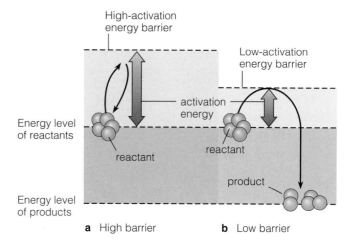

a High barrier   **b** Low barrier

FIGURE 2.6 Activation energy. (a) If the activation energy barrier of a reaction is high, it is unlikely that reactants will acquire sufficient energy to pass over it. (b) If the activation energy barrier of a reaction is low, reactants will readily acquire sufficient energy to pass over it and products will form rapidly.

b

FIGURE 2.5 Hydrogen bonds. (a) A hydrogen bond joining two water molecules. Note how the hydrogen bond results from the tendency of the hydrogen atom to associate with the pair of nonbonding electrons in the lower water molecule (●●), as well as the covalent bond pair it forms with the upper water molecule (● ●). (b) The network of hydrogen bonds (- -) that form in water.

(From Biology: The Unity and Diversity of Life, 6th ed., by C. Starr and R. Taggart, Brooks/Cole, 1992. All rights reserved.)

TABLE 2.5 Factors That Increase the Rate of Chemical Reactions

Factor	Effect
Concentration of reactants	Higher concentrations speed reactions by making collisions more probable
Temperature	Higher temperature speeds reactions by adding thermal energy to achieve an activated state
Presence of catalysts	Presence speeds reactions by lowering activation energy

unburned gasoline and oxygen, thereby decreasing smog-causing emissions. In living things the catalysts that promote chemical reactions are **enzymes.**

Enzymes. Almost all enzymes are proteins. A few are RNA molecules; they are also called **ribozymes.** Most nonenzyme catalysts (such as platinum) can catalyze many different reactions. But enzymes are highly specific. In general, enzymes catalyze only one reaction, but they do it very well. For example, a single molecule of the enzyme acetylcholine esterase (found in nerve tissue) can catalyze a million reactions in a second.

Enzymes catalyze specifically and rapidly because of pockets in their surface called **catalytic sites.** Reactants enter these catalytic sites and become precisely positioned to react with one another (**Figure 2.7**). Because catalytic sites are tailored to specific reactions, nearly every chemical reaction that occurs in a cell requires a different enzyme. Moreover, most cellular reactions have such high activation energies that they

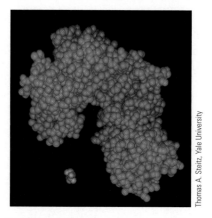

FIGURE 2.7 Catalytic action of an enzyme. In this model the glucose molecule (orange) is about to enter the catalytic site of the enzyme hexokinase (blue).

TABLE 2.6 Some Enzymes and the Reactions They Catalyze

Enzyme	Reaction Catalyzed
Agarase	Hydrolyzes agar
Dehydrogenase	Class of enzymes that remove hydrogen atoms
Alcohol dehydrogenase	Removes hydrogen atoms from alcohol
Lactate dehydrogenase	Removes hydrogen atoms from lactate
Kinase	Class of enzymes that add phosphate groups
Hexokinase	Adds a phosphate group to a hexose (glucose)
Aspartokinase	Adds a phosphate group to aspartate
Polymerase	Class of enzymes that polymerize monomers
DNA polymerase	Polymerizes deoxyribonucleotides to form DNA
RNA polymerase	Polymerizes ribonucleotides to form RNA
Cellulase	Hydrolyzes cellulose
Urease	Hydrolyzes urea

would not occur in the absence of enzymes. So organisms usually require one enzyme for each of the biochemical reactions. The bacterium *Escherichia coli*, for example, requires about 1000 different enzymes to catalyze the approximately 1000 different reactions it needs to grow and reproduce.

Most enzyme names end in "-ase" and suggest the reaction they catalyze. For example, the enzyme that liquefies agar is named agarase (**Table 2.6**).

WATER

Let's pause in our progression from atoms to macromolecules to discuss water, because it plays such an important role the chemistry of life. Water is essential for life. About 70 percent of the weight of microorganisms consists of water molecules. Moreover, most vital chemical reactions occur only in an **aqueous** (watery) environment. Two properties of water make it so important: (1) its capacity to form hydrogen bonds with other water molecules, and (2) it's being a polar molecule. Let's look more closely at these two properties.

Special Properties of Water

Because water molecules form hydrogen bonds with other water molecules, water has properties usually associated with much larger molecules. It has a higher boiling point and a relatively lower freezing point than similarly sized molecules. Thus pure water at sea level exists as a liquid over a wide range of temperatures—from 0° to 100°C (32° to 212°F). Altered pressure and the presence of dissolved substances extend the temperature range of liquid water. Because organisms need liquid water, they live only within this range of temperature.

Hydrogen bonding gives water several other life-sustaining properties:

Hydrogen bonding increases water's **specific heat** (the amount of heat necessary to raise 1 gram of material 1°C). Higher specific heat means water-filled cells can absorb considerable heat before their temperature rises dangerously.

Hydrogen bonding increases water's heat of vaporization (the amount of heat necessary to convert a liquid to a gas). Thus evaporation of water takes up enough heat to cool cells significantly.

Hydrogen bonding makes water cohesive. This cohesive property allows water to be drawn to the tops of trees.

Extensive hydrogen bonding in ice gives it an open structure that makes it less dense than water. As a result, ice floats and is exposed to the sunlight on water surfaces, so it melts in the spring. Were ice denser than water, it would sink to the bottom and accumulate, making Earth largely uninhabitable.

Water as a Solvent

Being a polar molecule makes water an excellent solvent. And biochemical reactions occur in **solution:** individual molecules of a substance (the **solute**) dispersed in a liquid (the **solvent**). Not all molecules dissolve in water. Those that do are called **hydrophilic** (water loving); those that don't are called **hydrophobic** (water fearing).

The way salts dissolve in water illustrates the importance of water's polar properties. If you put sodium chloride (NaCl; table salt), for example, in water, individual Na$^+$ and Cl$^-$ ions **dissociate** (separate) and disperse. Water molecules cluster around each ion in a particular way. The negative poles of water molecules orient toward Na$^+$ ions, while the positive poles of water molecules orient toward Cl$^-$ ions (**Figure 2.8**). In this way, water molecules form **spheres of hydration** around each ion, keeping it in solution.

Colloids. Not all materials that remain stably dispersed in water are dissolved solutes. Some are tiny solid particles, called **colloids**. They range in diameter from about 0.001 to 1 μm (micrometer). These particles remain dispersed because ions attach to them. Then water molecules form spheres of hydration around the ions. This protective layer of ions and spheres of hydration keeps the particles apart. They don't aggregate into chunks. But unlike solutions, which are completely clear, most colloidal dispersions are **turbid** (cloudy). Some colloids have the capacity to form **gels** (open networks of interconnected colloidal particles) when they are concentrated or cooled. Although gels are semisolid and sometimes quite firm, they are mainly water. For example, the firm gels made of agar, which we discussed in Chapter 1, are more than 98 percent water. Cells contain colloidal dispersions and gels. The volume of some parts of microbial cells—for example, the periplasm of bacteria (Chapter 4)—is maintained by gels.

Hydrophobic Interactions. We've seen that hydrophilic compounds dissolve in water because of their polarity. In contrast, hydrophobic compounds, such as oils and gasoline, don't dissolve in water because, being nonpolar, they lack charged regions to interact with water molecules. When nonpolar molecules are put in water, their hydrophobic regions interact, causing the molecules to clump together. For example, when we rinse out a pan used to make candy, the polar sugar dissolves but the nonpolar margarine floats as a blob to the top.

Hydrophobic and hydrophilic interactions help to organize a cell. Cellular membranes are a prime example. They are composed of molecules that are hydrophobic at one end and hydrophilic at the other. In water these molecules spontaneously position themselves so that their hydrophobic ends interact with one another and their hydrophilic ends interact with water. As a result, they aggregate spontaneously to form membranes. We'll discuss membrane formation in greater detail in the section on phospholipids later in this chapter.

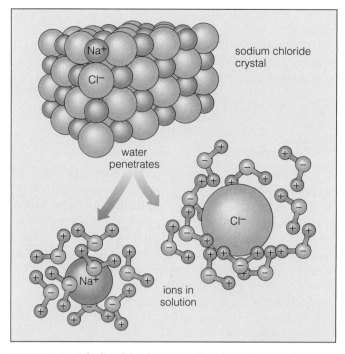

FIGURE 2.8 Salt dissolving in water. Note how the negative poles of water molecules surround the positively charged sodium ion (Na$^+$) and their positive poles surround the negatively charged chloride ion (Cl$^-$).

Hydrogen and Hydroxide Ions. Water has another important quality: It is prone to **ionize** (dissociate into ions).

H_2O	$\longleftrightarrow$	H^+	$+$	OH^-
Water		Hydrogen Ion (Proton)		Hydroxide Ion

Water dissociates into positively charged hydrogen ions (H$^+$, also called **protons**) and negatively charged hydroxide ions (OH$^-$). When this **reversible reaction** (it proceeds in both directions as indicated by the double-headed arrow) reaches equilibrium, only a few ions of each type are present in a large amount of water. Nevertheless, these few ions play crucial roles in the chemistry of life. *Hydrogen and hydroxide ions participate in almost every chemical reaction that occurs in living things.*

Acids, Bases, and Salts. In pure water the concentrations of hydrogen and hydroxyl ions are equal. But many solutes shift this equilibrium by adding to or using up one or the other of these ions. Solutes that add hydrogen ions when they dissociate in water are called **acids**. The dissociation of any acid can be represented by the general equation:

HA	$\longleftrightarrow$	H^+	$+$	A^-
Acid		Hydrogen Ion		Anion

A⁻ represents whatever anion the particular acid contains.

Solutes that add hydroxide ions when they dissociate in water are called **bases.** The dissociation of any base can be represented by the general equation:

BOH	$\longleftrightarrow$	B⁺	+	OH⁻
Base		Cation		Hydrogen Ion

Solutions that contain more H^+ than OH^- ions are said to be **acidic,** and solutions that contain more OH^- than H^+ are said to be **basic** or **alkaline.** Adding a hydrogen ion is equivalent to removing a hydroxide ion and vice versa. So we can broaden our definitions of acids and bases to say that acids are compounds that add hydrogen ions or accept hydroxide ions; bases are compounds that add hydroxide ions or accept hydrogen ions.

Salts are compounds that dissociate in water but do not add hydrogen or hydroxide ions. The general equation for dissociation of a salt is:

BA	$\longleftrightarrow$	B⁺	+	A⁻
Salt		Cation		Anion

In summary, acids, bases, and salts are compounds that dissociate in water. They are distinguished by how they change the concentration of hydrogen and hydroxide ions. Acids increase the concentration of hydrogen ions. Bases increase the concentration of hydroxide ions. Salts add neither.

The pH Scale. We've discussed how acids and bases cause the concentrations of hydrogen and hydroxide ions in water to rise or fall to new values. Now let's assign numbers to these values. Then we can describe quantitatively how acidic or basic a particular solution is. The **pH scale** is used for this purpose. It is based on the concentration of hydrogen ions in a solution. That might sound inadequate because we also want to know about hydroxide ions, but it is not. If we know the concentration of hydrogen ions, we also know the concentration of hydroxide ions because the two are inversely related. As a result, the pH of a solution also indicates its concentration of hydroxide ions, as well as hydrogen ions.

The pH scale runs from 0 to 14. Zero describes highly acidic solutions, whereas 14 describes highly basic solutions (**Figure 2.9**). Because the possible concentrations of hydrogen ions span such a vast range, the pH scale is logarithmic. That is, instead of using numbers themselves to indicate concentration of hydrogen ions, the scale uses their exponents. It is logarithmic scale with base 10. The **value of pH** is the negative exponent of the concentration of H^+ ions. (Concentration is measured in **molarity,** moles per liter, abbreviated as M.) If we know the concentration of H^+ in a solution, we can state its pH. For example, a so-lution that contains 10^{-3} M H^+ has a pH value of 3. Because the pH scale is logarithmic, a change of one pH unit represents a tenfold change in hydrogen ion concentration. Thus vinegar, with a pH of 3, is 10 times more acidic than tomato juice, with a pH of 4.

The key to understanding the pH scale is understanding the relationship between H^+ and OH^- ions. Namely, the product of the concentration of hydrogen and hydroxide ions is always 10^{-14}.

$$(H^+) \times (OH^-) = 10^{-14}$$

Thus in pure water, which has an equal concentration of H^+ and OH^-, the concentrations of both ions are 10^{-7} M and the pH is 7.0. If the concentration of H^+ is 10^{-3} M (in vinegar, for example), then the concentration of OH^- is 10^{-11} M (14-3).

Buffers. Because most enzymes are stable and function best at a particular pH, most cellular reactions function only within a relatively limited range of pH. But the value of the range varies. For example, the interior of human cells must be near pH 7.4 for their enzymes to function well. In contrast, some species of bacteria that live in acid environments, such as leaching gold mines, have an intracellular pH as low as 6.0. And bacteria that live in highly basic environments, such as desert lakes, have an intracellular pH as high as 8.5.

For a cell to survive, its intracellular pH must be maintained near its optimal value. But many factors act continuously to change it. Therefore cells have a variety of mechanisms to stabilize their intracellular pH. One of these is a buffer system.

A **buffer** is a solution of a **weak acid** or a **weak base** or a mixture of them. *Weak* means that the acid or base dissociated only partially. In contrast, strong acids or bases dissociate almost completely. To illustrate this difference, let's look again at the general equation for dissociation of an acid.

HA	$\longleftrightarrow$	H⁺	+	A⁻
Acid		Hydrogen Ion		Anion

In solution, a strong acid exists almost completely as H^+ and A^-, but a weak acid exists mainly as HA. That's how the weak acid (or base) stabilizes or "buffers" the pH of a solution against change. In the case of a weak acid, the anion (A^-) takes up hydrogen ions (become HA) if their concentration rises and the acid (HA) releases them if their concentration falls. Buffers don't maintain a completely constant pH, but they do moderate the consequences of adding H^+ or OH^- ions.

In addition to their natural role in maintaining constant intracellular pH, buffers are important in the laboratory (Chapter 3). They are used to moderate pH changes caused by microbial growth or chemical reactions.

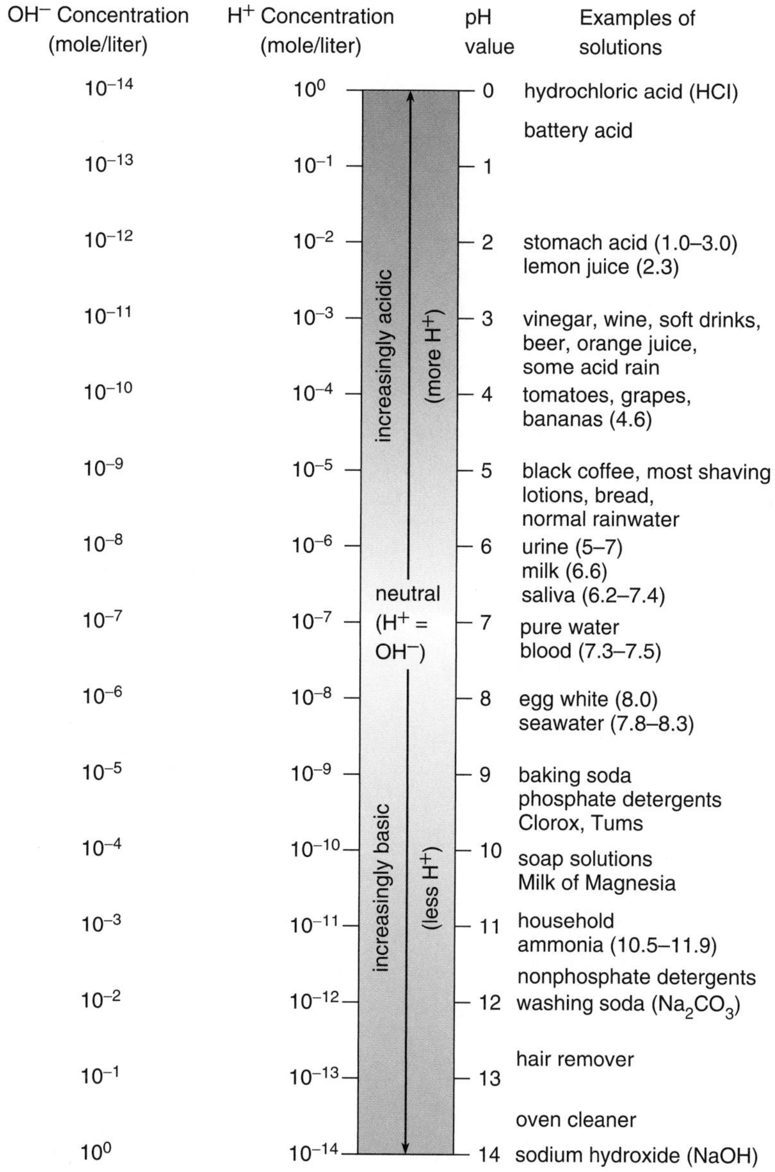

OH⁻ Concentration (mole/liter)	H⁺ Concentration (mole/liter)	pH value	Examples of solutions

The pH scale table:

10^{-14}	10^0	0	hydrochloric acid (HCl)
			battery acid
10^{-13}	10^{-1}	1	
10^{-12}	10^{-2}	2	stomach acid (1.0–3.0) lemon juice (2.3)
10^{-11}	10^{-3}	3	vinegar, wine, soft drinks, beer, orange juice, some acid rain
10^{-10}	10^{-4}	4	tomatoes, grapes, bananas (4.6)
10^{-9}	10^{-5}	5	black coffee, most shaving lotions, bread, normal rainwater
10^{-8}	10^{-6}	6	urine (5–7) milk (6.6) saliva (6.2–7.4)
10^{-7}	10^{-7}	7	pure water blood (7.3–7.5)
10^{-6}	10^{-8}	8	egg white (8.0) seawater (7.8–8.3)
10^{-5}	10^{-9}	9	baking soda phosphate detergents Clorox, Tums
10^{-4}	10^{-10}	10	soap solutions Milk of Magnesia
10^{-3}	10^{-11}	11	household ammonia (10.5–11.9)
10^{-2}	10^{-12}	12	nonphosphate detergents washing soda (Na_2CO_3)
10^{-1}	10^{-13}	13	hair remover
			oven cleaner
10^0	10^{-14}	14	sodium hydroxide (NaOH)

increasingly acidic (more H⁺) — neutral (H⁺ = OH⁻) — increasingly basic (less H⁺)

FIGURE 2.9 The pH scale. (From Biology: The Unity and Diversity of Life, 6th ed., by C. Starr and R. Taggart, Brooks/Cole, 1992. All rights reserved.)

ORGANIC MOLECULES

Now we'll return to our progression from atoms to macromolecules. We left off discussing molecules. Now we'll discuss the major kind of molecules, called organic molecules, that occur in organisms. The term **organic** comes from the word *organism*, or "living thing." Nineteenth-century chemists believed the linkage was absolute—that organic compounds could be made only biologically. Then chemists learned how to make organic molecules in the laboratory. Now the best definition of an **organic molecule** is one that contains carbon atoms. (Carbon dioxide is an exception. It and its simple derivative are usually not considered to be organic.) All other non–carbon-containing molecules are called inorganic. **Biochemical** is a useful term to describe the organic compounds found in organisms.

Carbon Atoms

Why are carbon-containing compounds considered separately in chemistry? It's because a carbon atom can form four covalent bonds with other atoms (Figure 2.2). This gives it great bonding versatility. The number of ways carbon atoms can be linked together and to other atoms is almost infinite. They are the Lego toys of chemistry. Some of the structures they form are quite complex: they may be linear, branched, or in the form of rings (**Figure 2.10**). The flexibility of these structures depends on the nature of the bonds that join their carbon atoms. Carbon atoms joined by single bonds can rotate around each other. In contrast, the relative position of carbon atoms joined by double or triple bonds is fixed.

Some organic compounds, called **hydrocarbons,** consist of only carbon and hydrogen atoms. In these molecules each carbon atom is joined to one or more other carbons; the remainder of its four covalent bonds connect to hydrogen atoms. Hydrocarbons are found mostly in petroleum and coal deposits (the fossilized remains of living things). Most organic compounds in living things, however, contain additional atoms—primarily nitrogen, oxygen, phosphorus, and sulfur.

Functional Groups

To make sense of the huge variety of organic compounds, chemists look for patterns in which atoms are arranged. Some of these are called **functional groups.** The presence of a particular functional group tells us something about the molecule that contains it because every functional group has characteristic properties. For example, a polar

functional group makes a compound more soluble in water. Some functional groups determine if an organic compound is an acid or a base. Moreover, the functional groups in a compound (there can be more than one) determine which other compounds it can react with.

Many different functional groups exist. Two common ones are the carboxyl group (—COOH) and the amino group (—NH₂). Molecules that have a carboxyl group are acids, because they readily ionize to produce hydrogen ions.

$$R-COOH \longleftrightarrow R-COO^- + H^+$$

(R represents the rest of the organic molecule.) Molecules that have an amino group are bases, because they readily accept hydrogen ions.

$$R-NH_2 + H^+ \longleftrightarrow R-NH_3^+$$

Functional groups are the basis for classifying organic molecules into major groups that share important characteristics. Examples of organic compounds and the functional groups they contain are presented in **Figure 2.11.**

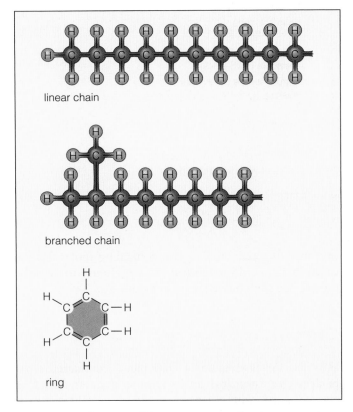

FIGURE 2.10   Some possible arrangements of carbon atoms in organic molecules. This single ring is a benzene ring.

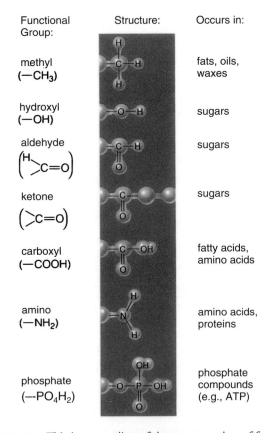

FIGURE 2.11   This is a sampling of the great number of functional groups that occur in biologically important molecules.

MACROMOLECULES

The size of the organic molecules varies enormously. Some small organic molecules contain only a few carbon atoms. Others are enormous, containing thousands or even millions of atoms. These extremely large molecules, called **macromolecules,** play central roles in biology. They determine most of a cell's structures and functions. The four main classes of macromolecules found in cells are proteins, nucleic acids, polysaccharides, and lipids.

Most macromolecules are **polymers** (many parts), built from smaller repeating units called **monomers** (one part), much like a highly complex building might be constructed from one type of bricks. Thus monomers are the building blocks from which macromolecular polymers are made. The process of joining monomers together to form polymers is termed **polymerization.** With the exception of lipids, macromolecules are distinguished by the building blocks from which they are made and the kinds of bonds that join them together (**Table 2.7**). Lipids are not defined by chemical structure. Instead, they are defined by their solubility. Any biochemical molecule that dissolves in a nonpolar solvent is classified as a lipid.

The bonds that join monomers into macromolecules all look as though they might have been formed the same way—by taking a hydrogen ion from one monomer and a hydroxide ion from a neighbor. That's a net loss of one molecule of water. For this reason, before the actual chemical reactions forming macromolecules were known, the process was called **dehydration** (removal of water) **synthesis.** But as you will see in Chapter 5, monomers are not themselves polymerized into macromolecules. Instead, chemical groups are added to the monomers before polymerization, and removed during polymerization.

The reverse of dehydration is **hydrolysis** (meaning "breaking with water"). This reaction actually occurs in cells. By hydrolysis, an organism breaks down macromolecules into monomers to obtain nutrients or building blocks for new cell structures. As the covalent bond is broken, one part of the molecule combines with a hydrogen ion, while the other combines with a hydroxide ion. Thus the outcome of hydrolysis is the splitting of a macromolecule by a water molecule that is also split.

Proteins

Next to water, proteins are the most abundant component of cells. Proteins make up more than half the dry weight of microorganisms and perform many essential biological functions. Most of a cell's proteins are enzymes that catalyze the thousand or more different chemical reactions that occur in a cell. Other proteins play structural roles; they hold parts of the cell together or build cellular organelles. Other proteins carry molecules in and out of the cell across the cell membrane. Still others fulfill highly specialized functions such as cell movement. For example, the corkscrewlike flagella that move many bacteria are composed entirely of protein.

TABLE 2.7 The Major Macromolecules

Macromolecule (Polymer)	Building Block (Monomer)	Bonds That Join Them
Proteins	Amino acids	Peptide
Nucleic acids		Phosphodiester
DNA	Nucleotides (a phosphate, deoxyribose, and a base—adenine, guanine, thymine, or cytosine)	
RNA	Nucleotides (a phosphate, ribose, and a base—adenine, guanine, uracil, or cytosine)	
Polysaccharides	Monosaccharides	Glycosidic
Lipids	Unlike other macromolecules, lipids are not defined by chemical structure Lipids are any organic nonpolar molecule	Some lipids are held together by ester bonds; some are huge aggregates of small molecules held together by hydrophobic interactions

Amino Acids. Twenty different amino acids are the monomers from which proteins are polymerized (**Figure 2.12**). Their quantity and arrangement determine the properties of a protein, such as whether or not it is an enzyme, for example, and if so, which biochemical reaction it catalyzes. All 20 naturally occurring amino acids share the same basic structure. They have one carbon atom that is attached to four different groups: a carboxyl group (—COOH), an amino group (—NH₂), a hydrogen atom (H), and an R group. (*R* stands for the rest of the molecule, whatever it may be.) The R group is different in each of these 20 amino acids. The R group may be polar or nonpolar; acidic, basic, or neutral; ionizable or not. All R groups contain hydrogen; 19 contain carbon; some also contain oxygen and nitrogen. Two contain sulfur.

Any four different atoms or groups of atoms can attach to a carbon atom in two ways. These alternative arrangements produce two different forms of each amino acid, called the L-isomer and the D-isomer. L- and D-isomers are mirror images of each other (**Figure 2.13**). Only the L-isomers of amino acids are found in proteins. The D-isomers are rare in nature, but some occur in **peptidoglycan,** the macromolecule from which bacterial walls are made (Chapter 4); they also occur in a few antibiotics.

Peptide Bonds. The bonds that join amino acids together to form proteins are called **peptide bonds.** They attach the amino group of one amino acid to the carboxyl group of another (**Figure 2.14**). Two amino acids joined together by a peptide bond are called a dipeptide. Three joined together are called a tripeptide, and four or more, a polypeptide. Proteins are long polypeptides. The average protein in the bacterium *Escherichia coli* contains 270 amino acids.

Protein Structure. The specific amino acids and the order in which they are joined are called a protein's **primary structure.** Because most proteins are strings of hundreds of amino acids and each position can be filled by 1 of 20 different amino acids, an almost infinite number of different primary structures is possible.

But proteins do not exist as a simple extended chain of amino acids. Rather, they are folded precisely into complex three-dimensional shapes. The amino acids in the primary structure interact to create a particular protein's secondary, tertiary, and quaternary structures (**Figure 2.15**).

A protein's **secondary structure** is determined by the hydrogen bonds that form at intervals between nonadjacent amino acids. The amino acids of certain proteins twist around a central axis, forming an **alpha helix.** This alpha helix is held together by hydrogen bonds between every fourth amino acid. Other stretches of amino acids assume the conformation of a **beta sheet** (also called a **pleated sheet**). A beta sheet is also held together by hydrogen bonds, but these bonds join adjacent peptide chains. Some sequences of amino acids tend to form alpha helices, whereas others are more likely to form beta sheets. Hemoglobin, the oxygen-carrying protein in blood, is composed of alpha helices. Silk proteins are composed of beta sheets. Some proteins—for example, the hormone insulin—have regions of alpha helices and regions of beta sheets.

A protein's **tertiary structure** is determined by interactions among the R groups of its various amino acids. For example, hydrophobic R groups tend to lie next to one another to minimize their interaction with water. Oppositely charged regions of a protein are held together by ionic bonds. Some proteins (for example, insulin) that contain more than one molecule of the amino acid cysteine are held together by covalent disulfide bonds (—S—S—). (The R group of each cysteine unit ends in a sulfhydryl group, —SH (Figure 2.12), which react with one another

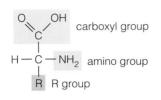

carboxyl group

amino group

R group

a

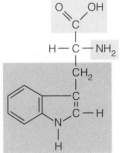

tryptophan (trp)

tyrosine (tyr)

valine (val)

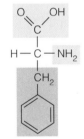

phenylalanine (phe)

methionine (met)

cysteine (cys)

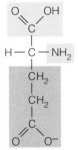

glutamate (glu)

glycine (gly)

b

FIGURE 2.12 Amino acids. (a) General structure of an amino acid. (b) The structures of eight representative amino acids. Some R groups contain carboxyl groups (glutamate); some contain benzene rings (tryptophan, tyrosine, and phenylalanine); some contain sulfur (methionine and cysteine); the R group of one (glycine) is a hydrogen atom.

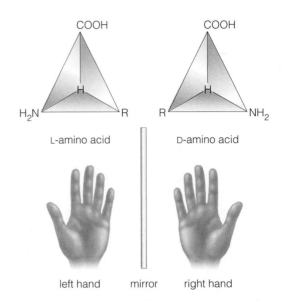

L-amino acid

D-amino acid

left hand   mirror   right hand

FIGURE 2.13 D- and L-isomers. The D- and L-isomers of an amino acid are mirror images of each other, differing as our left and right hands do. They can exist in two mirror-image forms because the four bonds of a carbon atom are equally spaced. Thus they form the shape of a regular tetrahedron, as shown here.

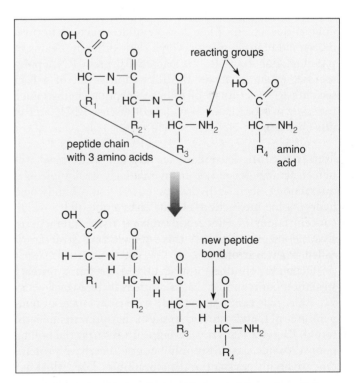

FIGURE 2.14 Peptide bonds. Peptide bonds (shown in red) form between the amino group of one amino acid and the carboxyl group of the next. Note how it appears that one molecule of water is released for each peptide bond formed because the carboxyl group loses an OH and the amino group loses an H.

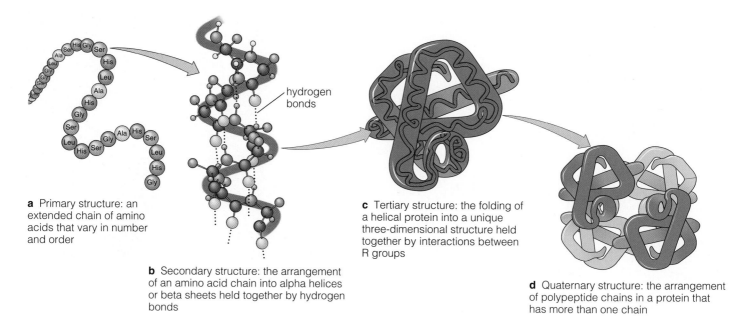

a Primary structure: an extended chain of amino acids that vary in number and order

b Secondary structure: the arrangement of an amino acid chain into alpha helices or beta sheets held together by hydrogen bonds

hydrogen bonds

c Tertiary structure: the folding of a helical protein into a unique three-dimensional structure held together by interactions between R groups

d Quaternary structure: the arrangement of polypeptide chains in a protein that has more than one chain

FIGURE 2.15   Protein structure.

to form disulfide bonds.) The chemical interactions that determine tertiary structure cause protein chains to fold back on themselves, creating a unique shape.

Some complex proteins consist of more than one polypeptide chain. They have a **quaternary structure,** determined by the way in which their different chains fit together. For example, the bacterial protein RNA polymerase, which catalyzes the formation of all of a bacterium's RNA, consists of four polypeptide chains joined together in a particular way. (We'll discuss RNA later in this chapter.)

Denaturation. Secondary, tertiary, and quaternary protein structures depend on many relatively weak molecular interactions, such as hydrogen bonds, ionic bonds, and hydrophobic interactions. Heat, unfavorable pH, or high concentrations of salts readily disrupt these interactions. Destruction of a protein's three-dimensional structure is called **denaturation.**

Denaturation does not change a protein's primary structure, but it almost always destroys its biological activity. Thus cells can function only within the range of temperature, pH, and osmotic strength that prevents denaturation. Bacteria that grow at temperatures near the boiling point of water can do so only because they have proteins that are highly resistant to denaturation by heat. Most microorganisms have proteins that are much less stable and are readily denatured at temperatures as low as 55°C (131°F). Once proteins are denatured they usually cannot be returned to their original state. A familiar example of denatured protein is cooked egg white. Heat irreversibly

changes the normally clear and viscous protein albumin into a solid white mass of denatured protein.

We'll consider denaturation again in Chapter 9, when we discuss **sterilization** (killing all the microorganisms in a particular place). One of the most effective sterilization methods is heating to denature a microorganism's vital proteins.

Nucleic Acids

Nucleic acids are the third most abundant component (after water and protein) of microbial cells. There are two types of nucleic acid molecules, deoxyribonucleic acid (DNA) and ribonucleic acid (RNA). Although DNA makes up only a small fraction a typical microorganism (a few percent of its dry weight), it is of central importance. DNA is the molecule that encodes a cell's genetic information. DNA ensures that progeny resemble parents.

RNA plays a critical role in building proteins. It interprets the information for protein construction encoded in DNA. The organelles on which proteins are manufactured, **ribosomes,** are composed partly of RNA. RNA is much more abundant than DNA, particularly in microorganisms. It can constitute as much as a third of the dry weight of a bacterial cell.

Nucleotides. **Nucleotides** are the building blocks of nucleic acid. Each nucleotide contains a phosphate group and a **nucleoside,** which is composed of a five-carbon sugar and a nitrogen-containing base (**Figure 2.16**). There are

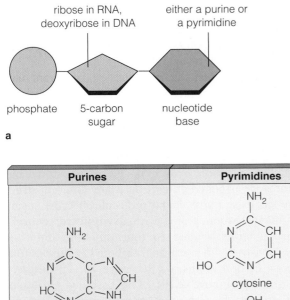

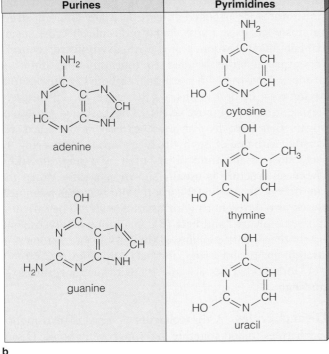

a

Purines	Pyrimidines
adenine	cytosine
guanine	thymine
	uracil

b

FIGURE 2.16 Nucleotides. (a) Components of a nucleotide. (b) The nucleotide bases.

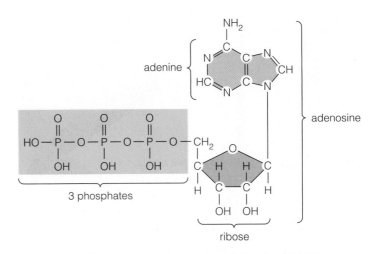

FIGURE 2.17 The structure of adenosine triphosphate (ATP).

oside triphosphate, adenosine triphosphate (ATP), is the major carrier of cellular energy (**Figure 2.17**). It is indispensable in many biochemical reactions, as are other nucleotides, as you will see in Chapter 5.

DNA. DNA is a long covalently bonded polymer of **deoxyribonucleotides.** Alternating units of deoxyribose molecules and phosphate groups form a long strand with bases (cytosine, thymine, adenine, and guanine) sticking out from it. Most molecules of DNA contain two such strands wrapped around each other in the form of a **double helix (Figure 2.18).** The two strands of the DNA molecule are held together by hydrogen bonds that form between pairs of bases. Adenine always bonds to thymine, and guanine always bonds to cytosine. The deoxyribose-phosphate backbone is the same in all molecules of DNA, but the sequence of bases that stick out from it is unique to each molecule. Some molecules of DNA have more adenine-thymine base pairs than guanine-cytosine base pairs, whereas the reverse is true in other molecules.

The sequence of base pairs in a molecule of DNA encodes its genetic information. The discovery of the structure of DNA in 1953 earned the Nobel Prize for biologists James Watson and Francis Crick. Once the structure of DNA was known, it became possible to decipher the genetic code. This achievement led to enormous advances in all biology, including microbiology. In Chapter 6 we'll discuss how genetic information is encoded and preserved in DNA.

RNA. RNA is a polymer of **ribonucleotide** monomers (Figure 2.16). Both DNA and RNA include nucleotides containing adenine, cytosine, and guanine. But RNA has nucleotides containing uracil instead of thymine. RNA differs from DNA in another respect: It is usually single stranded.

two types of nucleotides—**deoxyribonucleotides** (found in DNA) and **ribonucleotides** (found in RNA). In either case the nitrogen-containing base (often simply called a base) may be either a **pyrimidine** (a single-ringed molecule) or a **purine** (a double-ringed molecule). There are three kinds of pyrimidines (**cytosine, uracil,** and **thymine**) and two kinds of purines (**adenine** and **guanine**). Uracil occurs only in ribonucleotides and thymine occurs only in deoxyribonucleotides, but the other bases (cytosine, adenine, and guanine) occur in both kinds of nucleotides. Nucleotides in nucleic acids are joined by **phosphodiester bonds.**

The number of phosphate groups a nucleotide can carry varies from one to three. They are termed nucleoside monophosphates, nucleoside diphosphates, and nucleoside triphosphates, respectively. Some of these fulfill other essential cellular functions. For example, one nucle-

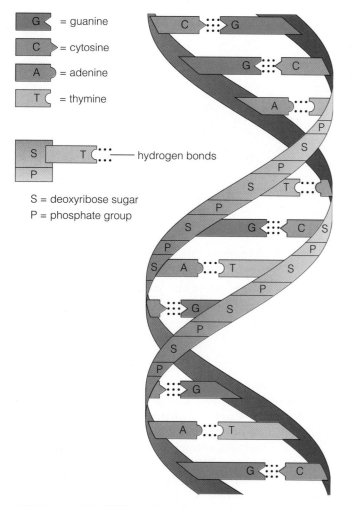

G = guanine
C = cytosine
A = adenine
T = thymine

hydrogen bonds

S = deoxyribose sugar
P = phosphate group

FIGURE 2.18 The DNA double helix. Note that two hydrogen bonds form between G-C pairs and three form between A-T pairs.

Different kinds of RNA perform different functions in building proteins from the genetic blueprint provided by DNA. We'll examine these different types of RNA and the roles they play in Chapter 6.

Polysaccharides

Polysaccharides (meaning "many sugars") are polymers of sugar monomers. Both polysaccharides and sugars are **carbohydrates** (meaning "carbon with water"). Carbohydrates contain the equivalent of one molecule of water for each molecule of carbon (**Table 2.8**). Thus the general formula for a carbohydrate is $(CH_2O)_n$ (*n* is the number of carbon atoms). Before discussing polysaccharides, let's consider the sugars from which they're built.

Monosaccharides. The simplest sugars, **monosaccharides** (meaning "one sugar") are small organic molecules composed of carbon, hydrogen, and oxygen. Approxi-

mately 20 monosaccharides occur in nature. Any molecule with a three- to seven-atom carbon backbone and at least one carbonyl group (=O) and several hydroxyl groups (—OH) is classified as a sugar.

Names of individual sugars and classes of sugars have the suffix -*ose*. For example, a five-carbon sugar is called a **pentose** (meaning "five-ose"). A common example is ribose. Six-carbon sugars are called **hexoses.** Glucose is a hexose. Another six-carbon sugar, fructose, is an **isomer** of glucose (**Figure 2.19**). That is, the two molecules have the same number of carbon, hydrogen, and oxygen atoms, but those atoms are arranged differently. Many different structural isomers exist among sugar molecules because sugars contain many carbon atoms with **asymmetric centers** (one carbon atom bonded to four different atoms or groups). A hexose with a terminal carbonyl group has four asymmetric carbon atoms, so there are 16 (2^4) different hexose isomers (Figure 2.19). The carbonyl group of sugars that have five or more carbon atoms reacts reversibly with one of the hydroxyl groups to form a ring. In fact, most of the time sugars exist as a ring, and their chemical structure is usually shown as a ring. When the ring forms, the carbonyl carbon atom becomes asymmetric. So two different ring forms are possible. They are designated alpha (α) and beta (β). A solution of glucose and most other sugars contains all three forms of monomers—straight chain, alpha ring, and beta ring (**Figure 2.20**).

Monosaccharides are important to microorganisms as nutrients.

Disaccharides. A **disaccharide** consists of two monosaccharides joined together by a glycosidic bond. Disaccharides are plentiful in nature and, like monosaccharides, serve as nutrients. One disaccharide is the common table sugar sucrose. It is composed of a molecule of glucose bonded to a molecule of fructose. Another disaccharide is lactose, or milk sugar. It is composed of one molecule of glucose and one molecule of another monosaccharide, galactose. Many microorganisms that make their home in the human intestine use lactose as a source of energy because this is one of the few natural environments where it is plentiful.

Glucose-Containing Polysaccharides. Polysaccharides, like disaccharides, are built from monosaccharides joined by glycosidic bonds, but polysaccharides contain many monosaccharide units. Some polysaccharides are quite complex. They can be linear or branched chains. They can be built from one or several different monosaccharide monomers. Polysaccharides serve important structural functions in cells. They also serve as storage forms of simple sugars that can be broken down for nutrients as need arises. Glucose alone is built into three widespread polysaccharides—cellulose, starch, and glycogen.

TABLE 2.8 Classes of Carbohydrates

Class	Examples	Composition and Distribution
Monosaccharides	Glucose	A hexose; found in most organisms
	Fructose	A hexose; large amounts occur in fruit
	Ribose	A pentose; constituent of RNA
	Deoxyribose	A pentose; constituent of DNA
Disaccharides	Sucrose	Glucose-fructose; table sugar
	Lactose	Glucose-galactose; milk sugar
	Trehalose	Glucose-glucose; occurs in most organisms; protects membranes from damage caused by desiccation
Polysaccharides	Cellulose	Straight-chain polymer of alpha glucose; occurs mainly in cell walls of plants and algae
	Starch	Straight-chain polymer of beta glucose; storage material in plants and some microorganisms
	Glycogen	Branched-chain polymer of beta glucose; storage material in animals

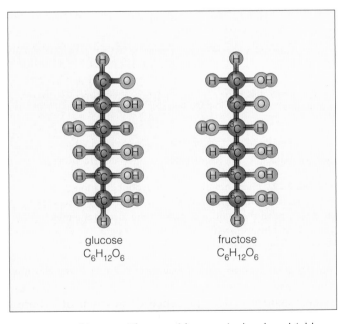

glucose
$C_6H_{12}O_6$

fructose
$C_6H_{12}O_6$

FIGURE 2.19 Hexoses. Glucose with a terminal carbonyl (aldehyde) group (—HC=O) has 4 asymmetric carbon atoms (numbers 2, 3, 4, and 5) and therefore 16 (2^4) isomers. Fructose with a keto group (=C=O) has three asymmetric carbon atoms (numbers 3, 4, and 5) and therefore eight isomers (2^3).

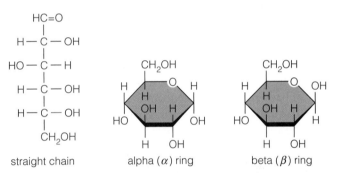

straight chain   alpha (α) ring   beta (β) ring

FIGURE 2.20 The structure of glucose. Three forms of glucose occur in solution—a straight chain and alpha and beta rings. Rings form when the carbonyl group (=O) reacts with a hydroxyl group (—OH).

These polysaccharides are distinguished by having linear or branched chains and glucose rings in either the alpha or beta form (**Figure 2.21**).

Cellulose is the most abundant polysaccharide, and probably the most abundant organic compound in nature, because it is the principal component of plant cell walls. It is also found in most algae and a few bacteria. Cellulose is composed of glucose rings in the beta form that are covalently linked in long linear chains. Cellulose is a strong material that resists breakdown by most microorganisms. Wood is largely cellulose. Cotton fibers are almost pure cellulose.

Starch differs from cellulose only in the form of its glucose monomers. The glucose rings in starch are in the alpha form instead of the beta form. The properties and biological roles of starch and cellulose are very different. Starch is the reserve material that plants and some protozoa make to store excess nutrients for later use.

Glycogen, like starch, is a polymer of glucose rings in the alpha form, but its polymer strands are multiply branched. Glycogen is the animal equivalent of starch. That is, animals, including human beings, store excess nutrients as glycogen, which can provide energy when needed. Algae, fungi, and

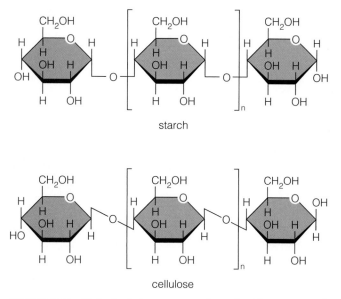

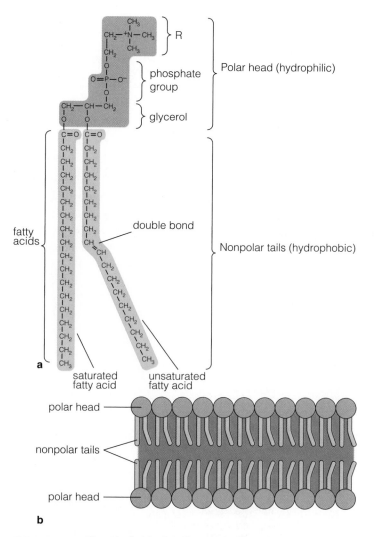

FIGURE 2.21 Biologically important polysaccharides composed of glucose monomers joined by glycosidic bonds. Starch is a long chain of glucose monomers in the alpha form. Cellulose is a long chain of beta glucose monomers.

FIGURE 2.22 Phospholipids. (a) Phospholipid molecules have a polar head and nonpolar tail. Double bonds in unsaturated fatty acid tails bend the molecule, making the membrane they form more fluid. Unsaturated fatty acids are straight. (b) Phospholipid molecules aggregated to form a membrane.

certain bacteria, including *Escherichia coli,* also make glycogen and accumulate it as intracellular storage granules.

Finally, polysaccharides are a major structural component of bacterial cells. They form the protective outermost layer, the **capsule,** of many bacterial cells. We'll talk more about capsules and their medical importance in Chapter 4. The cell walls of some fungi contain chitin, a polymer of a modified glucose molecule, *N*-acetylglucosamine. The chitin in fungi is the same tough material found in the shells of crabs and lobsters.

Lipids

Unlike the other types of macromolecules, lipids are not classified by their building blocks. Some lipids are true polymeric macromolecules (for example, the poly-beta-hydroxyalkanes that some bacteria make to store excess nutrients). But other lipids are simply huge aggregates of small molecules held together by hydrophobic interactions (for example, the lipid components of most cell membranes). Therefore lipids can be defined as being any nonpolar molecule. As such they are soluble in nonpolar solvents, such as gasoline or kerosene, and insoluble in water.

In microbial cells, lipids are most important as components of membranes. They also play other roles, including serving as an intracellular reserve food. Lipids make up about 10 percent of the dry weight of most microorganisms. The presence or absence of subunits called fatty acids categorizes them.

Lipids Containing Fatty Acids. **Fatty acids** are long chains of carbon and hydrogen atoms, with a carboxyl (—COOH) group at the end. Fatty acids can be saturated or unsaturated. **Unsaturated fatty acids** contain double bonds. **Saturated fatty acids** do not.

When three fatty acids are covalently joined to the alcohol glycerol, they form **fats** or **oils,** depending on whether they are solid or liquid at room temperature. Fats and oils are reserve food storage products for most organisms and cellular microorganisms but not for bacte-

BREAD IN THE DESERT

During World War II, American troops in North Africa found a curious-looking powder among captured German supplies. The powder was dried yeast cells, but they were still alive and could make bread dough rise. Researchers in the American food industry had tried for years to produce active dried yeast without success. Under the gentlest drying conditions, the yeast cells died. Even with the German samples in hand, Americans still couldn't duplicate the process. When the war ended, so did the mystery. The yeast had to be grown under conditions that caused cells to accumulate the disaccharide **trehalose.** Then they remained alive and active when dried and could be stored for long periods. Trehalose protects cells during desiccation (drying) by fitting precisely into molecular crevices on the surface of biological membranes, keeping these sensitive structures intact as water is removed. Trehalose plays a similar role in the water bear, a tiny animal that lives in soil and ponds. The water bear contains huge amounts of trehalose, allowing it to survive complete desiccation when its environment dries up.

A USEFUL POLYSACCHARIDE

Cellulose is the tough, light polysaccharide component of the walls of plants and some algae. A few bacteria, including *Acetobacter xylinum*, the bacterium we use to make vinegar, also produce cellulose, but not for cell walls. *A. xylinum* releases its cellulose into its aqueous environment where it floats to the surface, carrying *A. xylinum* cells with it. There they have access to the air they need to grow. The cellulose produced by *A. xylinum* is not like plant cellulose. It is pure and, as such, is nontoxic, strong, and ideal for dressing wounds because it serves as a firm matrix (form-giving base) to promote healing. *A. xylinum* is also a "hi-fi" microbe. Because the cellulose it makes is strong and resilient, it is used in acoustical diaphragms for high-quality stereo speakers.

ria. Fatty acids in fats and oils are joined to glycerol by ester linkages—a bond formed between a carboxyl group on the fatty acid and hydroxyl groups on the glycerol molecule.

Phospholipids are more complex lipids. Like fats and oils, they are composed of a glycerol backbone, but only two fatty acids are attached to it. A phosphate group to which another constituent (R, for "rest of molecule") is attached is linked to the third hydroxyl group on the glycerol molecule (**Figure 2.22**). Phospholipids are the essential components of membranes in plants, animals, and microorganisms. The fatty acid component of the molecule is hydrophobic and constitutes the nonpolar "tail" (insoluble in water). The phosphate group constitutes the small polar "head" (soluble in water). In the watery environment of a cell, phospholipids spontaneously arrange themselves by hydrophobic interactions to form membranes.

Lipids Not Containing Fatty Acids.
Some lipids do not contain fatty acids. The **sterols,** for example, are lipids composed of hydrocarbon rings. Most bacteria and archaea do not have sterols, but sterols are found in the membranes of all eukaryotes, including eukaryotic microorganisms. Sterols are clinically important because they are the targets of one group of antibiotics, the polyenes. The polyenes bind to sterols, damaging them and weakening the microbial membrane. **Cholesterol,** which is associated with heart disease, is a sterol produced only by animals.

SUMMARY

The Basic Building Blocks (pp. 23-27)

1. Matter is composed of small particles called atoms. Atoms are composed of protons, neutrons, and electrons.

2. Protons carry a positive charge, and electrons carry a negative charge. Neutrons have no charge.

Atoms (pp. 24-25)

3. All atoms have the same basic structure. A cloud of electrons orbits a nucleus of densely packed neutrons and protons. Electrons orbit at energy levels, called shells.

4. The atomic number of an atom is the number of protons it contains. The atomic weight of an atom is close to sum of the number of protons and neutrons it contains.

5. Atoms with the same atomic number but different atomic weights are called isotopes.

Elements (p. 25)

6. Matter composed of only one kind of atom is called an element. The most abundant elements in living organisms are carbon, hydrogen, nitrogen, oxygen, phosphorus, and sulfur.

Molecules (pp. 25-27)

7. A molecule consists of two or more atoms, of the same or different elements, joined by chemical bonds. A compound is a molecule that contains more than one type of atom.

8. A molecular formula tells which atoms and how many of each kind form a particular molecule.

9. Avogadro's number (6.02×10^{23}) of any type of molecule is called a mole of that substance. A mole is the molecular weight of the substance in grams.

Chemical Bonds and Reactions (pp. 27-32)

10. Chemical bonds form if the resultant molecule will be at a lower energy state and therefore more stable than the original configuration. Energy is released when a chemical bond forms. Energy is required to break a chemical bond.

11. Valence electrons determine an atom's capacity to form chemical bonds.

12. A covalent bond is formed when atoms share a pair of electrons. Covalent bonds are extremely stable.

13. Covalent bonds form nonpolar molecules (shared electrons are equally spaced) or polar molecules (shared electrons are not equally spaced). Polar molecules, such as water, have a negative and positive pole.

14. Ionic bonds are formed by the attraction between oppositely charged ions or molecules. Ionic bonds are not as strong as covalent bonds.

15. Hydrogen bonds form when hydrogen atoms are shared between two molecules or between different parts of the same molecule. Hydrogen bonds are weak.

16. A chemical reaction occurs when reactants are transformed into different combinations of the same atoms or molecules; these combinations are called products.

17. The rate of a chemical reaction is influenced by the energy state of the reactants, the concentrations of reacting molecules, and temperature.

18. Catalysts increase the rate of a reaction by decreasing the activation energy barrier. A catalyst is not altered by the reaction it brings about.

19. Catalysts in living things are called enzymes.

Water (pp. 32-34)

20. Water is essential for life.

Special Properties of Water (p. 32)

21. Water's capacity to form hydrogen bonds gives it a relatively high boiling point and relatively low freezing point; stabilizes temperature inside cells; makes water cohesive; and makes ice less dense than water. Together these properties make life possible.

Water as a Solvent (pp. 32-34)

22. Water's polarity makes it an excellent solvent. Compounds that dissolve in water are hydrophilic and those that do not are hydrophobic.

23. Colloids are tiny solid particles that create turbid liquids. Some colloidal dispersions when concentrated or cooled form gels.

24. When nonpolar compounds a re put in water, their hydrophobic regions interact; such interactions cause biological membranes to form.

25. Almost every chemical reaction that occurs in living things involves the participation of either hydrogen or hydroxide ions.

26. Water ionizes (dissociates) into an equal number of positively charged hydrogen ions and negatively charged hydroxide ions.

27. Solutes that increase the concentration of hydrogen ions are acids. Solutes that increase the concentration of hydroxide ions are bases.

28. Salts dissociate into anions and cations but not hydrogen or hydroxide ions.

29. The pH scale (logarithmic with base 10) describes how acidic or basic a solution is by assigning it a number from 0 to 14.

30. Buffers are weak acids or weak bases that stabilize the pH of a solution.

Organic Molecules (p. 35-36)

31. Carbon-containing compounds, other than carbon dioxide, are termed organic; organic compounds in living things are termed biochemicals.

32. A carbon atom can form four covalent bonds with other atoms, including carbon.

33. In an organic compound, atoms other than carbon and hydrogen are joined in patterns called functional groups, which give a compound its characteristic properties and determine which other compounds it will react with.

Macromolecules (pp. 37-45)

34. Macromolecules are large molecules formed through polymerization, a process that removes added chemical groups from monomers, the building blocks of macromolecules.

35. In the opposite process, called hydrolysis, macromolecules are broken down into monomers to obtain nutrients or build a new cell structure.

Proteins (pp. 37-40)

36. Proteins function as enzymes, as components of cell structures, in membrane transport, and in cell movement.

37. The building blocks of proteins are amino acids, which are joined by peptide bonds.

38. The primary structure of a protein is the specific order of amino acids it contains. The secondary structure is determined by hydrogen bonds that arrange amino acids into an alpha helix or a beta sheet. The tertiary structure is determined by interactions among R groups. Quaternary structure is the way the different chains fit together.

39. Denaturation is the destruction of a protein's three-dimensional structure by various means, including extremes of temperature, pH, and osmotic strength.

Nucleic Acids (pp. 40-42)

40. The two kinds of nucleic acid molecules are deoxyribonucleic acid (DNA) and ribonucleic acid (RNA). DNA encodes all the cell's genetic information. RNA interprets information encoded in DNA to construct proteins.

41. Nucleic acids are composed of monomers called nucleotides. The five-carbon sugar in DNA is deoxyribose. In RNA it is ribose. DNA also contains phosphate and the bases cytosine, adenine, guanine, and thymine. RNA contains phosphate and the first three bases plus uracil in place of thymine. Nucleotides are joined by phosphodiester bonds.

42. Adenosine triphosphate (ATP) is the major carrier of energy in the cell.

43. The DNA molecule is a double helix, but RNA is mostly single stranded.

Polysaccharides (pp. 42-44)

44. Polysaccharides are polymers composed of monosaccharides. Monosaccharides (also called sugars) are small organic molecules made up of carbon, hydrogen, and oxygen.

45. Two monosaccharides joined by glycosidic bonds form a disaccharide (also called sugars), such as lactose.

46. Polysaccharides can be linear or branched and built from one or several different monosaccharides. Glucose is built into three polysaccharides—cellulose, starch, and glycogen.

Lipids (pp. 44-45)

47. Lipids are nonpolar molecules that are soluble in nonpolar solvents, such as gasoline, and insoluble in water.

48. Lipids are classified according to whether or not they contain fatty acids. Fatty acids in fats and oils are joined to the alcohol glycerol by ester linkages. Lipids that do not contain fatty acids include the sterols.

49. Phospholipids are complex lipids that form membranes.

REVIEW QUESTIONS

The Basic Building Blocks

1. Describe the structure of an atom. How do the mass and electrical charge of the subatomic components differ?

2. What is the special significance of valence electrons?

3. Explain the difference between atomic number and atomic weight.

4. Explain this statement: An element cannot be reduced chemically to a simpler substance.

5. Define these terms: *molecule, macromolecule, compound, organic compound, mole.*

6. What is the molecular formula for water? How does knowing a molecular formula allow you to calculate molecular weight?

7. What is Avogadro's number?

Chemical Bonds and Reactions

8. What are chemical bonds? Explain this statement: Chemical bonding has to do with energy states. What determines an atom's capacity to form chemical bonds?

9. What is a covalent bond? Explain the difference between a polar molecule and a nonpolar molecule. What is the most important polar molecule in living things?

10. Define these terms: *ion, cation, anion.*

11. Name some functions of ions in living systems, including microorganisms.

12. What are ionic bonds? Give an example.

13. What is a hydrogen bond? How are hydrogen bonds important to living things?

14. Which type of bond is most stable? Which type is least stable?

15. Define these terms: *chemical reaction, reactant, product.*

16. Complete this sentence: When chemical bonds are broken, energy is _____, and when chemical bonds are formed, energy is _____.

17. Name some factors that affect chemical reaction rates.

18. Why are enzymes important?

19. What is a ribozyme?

Water

20. What special properties does water have because of its capacity to form hydrogen bonds?

21. Explain how water's polarity makes it an excellent solvent.

22. What are hydrophobic and hydrophilic groups, and why are they vital to living things?

23. What are colloids and gels?

24. Why are hydrogen and hydroxide ions so crucial to the chemistry of life?

25. Distinguish among acids, bases, and salts.

26. What is the pH scale? Why do we say it is logarithmic?

27. What is a buffer, and how do buffers function in living things? How are they used in the laboratory?

Organic Molecules

28. What is biochemistry?

29. Explain the following statement: A carbon atom has great bonding versatility.

30. What are functional groups? Give an example.

Macromolecules

31. Define these terms: *polymers, monomers, polymerization.*

32. How are macromolecules formed? What is the reverse process, and what functions does it serve in a cell?

33. What are the building blocks of proteins? What kinds of bonds hold them together? Give some examples of how proteins function in a cell.

34. What are L-isomers and D-isomers? Why is each important?

35. Explain protein structure. What is denaturation? Define and give an example.

36. Name the two kinds of nucleic acid molecules and explain what role each plays in a cell.

37. What are the basic building blocks of nucleic acids, and what kinds of bonds hold them together?

38. How are DNA and RNA alike chemically? How are DNA and RNA different?

39. What are the functions of these nucleotides: ATP, cAMP, NAD, FAD?

40. Define these terms: *polysaccharide, monosaccharide, carbohydrate.* Give some examples of monosaccharides, and explain why they are important to living things.

41. What kinds of bonds hold polysaccharides together? Explain how each of these polysaccharides is important to plants, animals, and microorganisms: cellulose, starch, glycogen. How are polysaccharides important to bacteria?

42. How are lipids defined? How are lipids that contain fatty acids different from lipids that do not? What are ester linkages?

43. What are phospholipids, and why are they important to living things?

44. How is peptidoglycan different from other macromolecules? What are its basic building blocks? How does it function?

CORRELATION QUESTIONS

1. Are there more kinds of atoms or elements? Explain.

2. What unique properties of hydrogen allow it to form hydrogen bonds?

3. What is the actual concentration and fold difference of hydroxide ion concentrations in solutions of pH 7 and pH 10?

4. Is it possible for the molecular weight of an element to be less than its atomic number? Explain.

5. From what you know about the bonding of carbon atoms and functional groups, write a structural formula for formic acid. Its molecular formula is CH_2O_2.

6. The atomic number of magnesium is 12. Would it form an anion or a cation? What would be the ion's charge?

ESSAY QUESTIONS

1. Discuss what the world would be like if water did not form hydrogen bonds.

2. Discuss the importance of catalysts in biology.

SUGGESTED READINGS

Alberts, B., D. Bray, A. Johnson, J. Lewis, P. Walter, K. Roberts. 1997. *Essential cell biology.* New York: Garland.

Miller, G. T. 1991. *Chemistry: A contemporary approach.* Belmont, Calif.: Wadsworth.

Scientific American. 1985. *Molecules of life.* Special issue. October.

Stryer, L. 1995. *Biochemistry.* 4th ed. San Francisco: W. H. Freeman.

For additional readings, go to InfoTrac College Edition, your online research library at: http://www.infotrac.thomsonlearning.com

THREE

The Methods of Microbiology

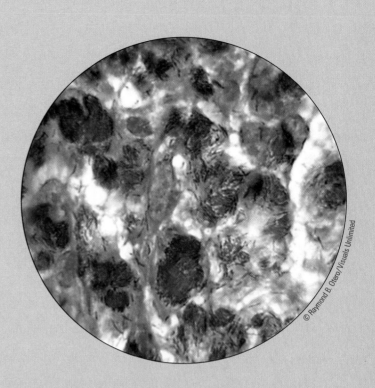

© Raymond B. Otero/Visuals Unlimited

CHAPTER OUTLINE

LEARNING GOALS

To understand:

- *The fundamental properties of light: reflection, transmission, absorption, diffraction, and refraction*

- *The principles of microscopy: how magnification, contrast, and resolution are achieved*

- *The preparations necessary to view microorganisms by microscopy*

- *How the compound light microscope works*

- *How the phase-contrast, darkfield, fluorescence, and Nomarsky modifications of the light microscope increase contrast*

- *How transmission and scanning electron microscopy work*

- *How scanning tunneling and atomic force microscopes work*

- *The advantages and uses of various microscopes*

- *What pure cultures are, why they are important, and how they are obtained*

- *How to cultivate a pure culture: types of media and the laboratory environment*

- *How cultures are preserved*

First, Do a Gram Stain

T. M., a nineteen-year-old male student, left his college dormitory to spend the weekend at home studying for finals. He spent Saturday morning studying effectively, but by early afternoon he began to behave strangely. He paced around the house in his underwear, swearing and muttering. When his parents asked what was wrong, he became agitated and hostile. Finally his father, with the help of a neighbor, forced him into their car and took him to the emergency room of a nearby hospital.

T. M. was completely uncooperative with the emergency room staff. The sudden onset of T. M.'s uncharacteristic behavior strongly suggested to the staff that he had suffered some damage to his central nervous system. The list of possible causes was long. T. M.'s symptoms could have been the first signs of mental illness, or of his having taken drugs or sustaining a head injury. T. M. was restrained and his vital signs were taken. His temperature of 104°F strongly suggested an infection.

Infections of the central nervous system are diagnosed by examining cerebrospinal fluid. To obtain a sample of T. M.'s cerebrospinal fluid by spinal tap, he had to be sedated. While sedated, specimens of his blood and urine were also taken. The emergency room physician did a Gram stain on a sample of T. M.'s cerebrospinal fluid, examined it under a microscope, and within minutes made a definitive diagnosis. He saw that the specimen contained numerous Gram-negative diplococci (spherical bacteria that occur in pairs) and knew that they were almost certainly *Neisseria meningitidis*, because of the way they stained, their shape, their tendency to form pairs of cells, and the fact that *N. meningitidis* is a common cause of bacterial meningitis in young adults. Immediately T. M. was given an intravenous dose of antibiotics. He recovered rapidly and completely.

T. M. might have avoided this illness had he been vaccinated. Most students entering college dormitories are immunized to prevent just such infections, but T. M. had decided not to receive a vaccination.

Neisseria meningitidis infections cause life-threatening disease, and people who do survive may suffer permanent damage, such as the loss of fingers or toes. In T. M.'s case, however, his parents' early observation of his symptoms and prompt action, along with his rapid diagnosis aided by Gram staining his cerebrospinal fluid, allowed early treatment that lead to complete recovery.

Case Connections

■ In this chapter we'll learn about staining, one of the most valuable methods of microbiology. We'll consider various stains including the Gram stain, possibly the most commonly employed and informative stain available.

■ In Chapter 25 we'll examine in greater detail diseases of the central nervous system, including the serious one caused by *Neisseria meningitidis*.

Like all life scientists, microbiologists need to see the organisms they study. Their indispensable tool is the microscope. **Microscopy** (the construction and use of microscopes) began in the seventeenth century with Antony van Leeuwenhoek and his little hand-held microscopes (Chapter 1). Since its humble beginnings, microscopy has undergone many spectacular advances, some of them quite recent, to become the sophisticated science it is today.

Microbiologists also need to do experiments on microorganisms, and to do so they usually have to **culture** (providing suitable conditions for their growth and multiplication) them in the laboratory.

In this chapter we'll consider the different kinds of microscopes—how they work and how to use them. We'll also consider the methods of culturing and preserving microorganisms.

VIEWING MICROORGANISMS

Because our eyes and most microscopes employ light to form images, we'll start with a review of some of its properties.

Properties of Light

Light is part of a continuous spectrum of electromagnetic waves. Radio waves are at the long-wave end of this spectrum (**Figure 3.1**); they can be as long as 2000 meters (more than a mile). **Gamma rays** (produced in the nucleus of radioactive atoms) are at the other end of the spectrum; they are shorter than 0.01 nm (nanometer), smaller than a virus. **Light** (to which the retinas of our eyes are sensitive) is near the middle of the spectrum.

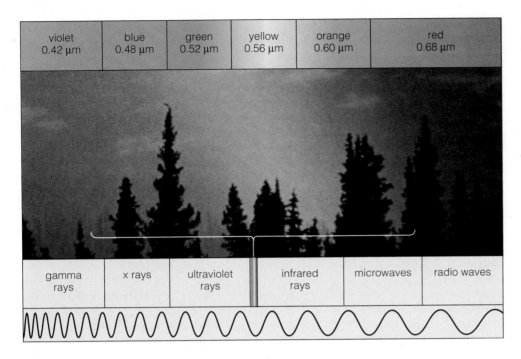

violet 0.42 µm	blue 0.48 µm	green 0.52 µm	yellow 0.56 µm	orange 0.60 µm	red 0.68 µm

gamma rays	x rays	ultraviolet rays	infrared rays	microwaves	radio waves

FIGURE 3.1 Visible light makes up only a tiny part of the electromagnetic spectrum, the part between 400 and 700 nm.

can be **reflected** if it bounces back, **transmitted** if it passes through the object, or **absorbed** if its intensity is diminished because it transferred some or all of its energy to the object (**Figure 3.2**).

Reflected rays bounce off a smooth surface as a ball would. Rays that strike straight on bounce directly back. Those that strike at an angle to the surface bounce off at that same angle.

When a ray of light hits a clear object, such as pure water or clear glass, it is transmitted undiminished. But if the object is not clear (smoked or colored, for example), some light is absorbed. If all wavelengths of light are absorbed equally, the intensity of the light is diminished but the color is unchanged. However, if certain wavelengths are absorbed more than others, the color changes. We see the color of the transmitted or reflected light. For example, a solution of red dye is red because it absorbs the blue component of white light and allows the red to pass through. A leaf is green because it absorbs the blue and red components of sunlight and reflects the green component.

In addition to being reflected, transmitted, or absorbed, the light ray can be bent, either by refraction or diffraction.

Refraction. Lenses in microscopes (or telescopes, cameras, and other optical instruments) form images because they bend light rays by refraction. **Refraction** occurs when a ray of light meets at an angle a substance of a different density. When the light ray enters the denser substance (from air to the glass of a lens, for example), it slows down, but the edge that enters slows first and the opposite edge slows later (**Figure 3.3**). This bends the ray. The amount of bending that occurs depends on the angle and how much the light is slowed down. The greater the angle and the greater the slowing, the greater the refraction (bending) is. Refraction also occurs when light enters a medium of lesser density, but then it binds in the opposite direction. Note in Figure 3.3 that when the light ray passes from air to glass, it bends down, but it bends back up when it reenters the air.

The speed of light in any material is determined by the material's **refractive index** (the ratio of the speed of light traveling in a vacuum to its speed in that particular material). In general, the higher the density, the higher the refractive index is. For example, the refractive index

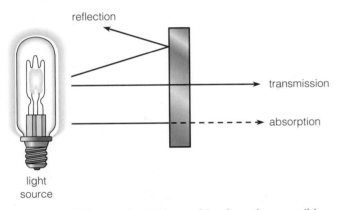

FIGURE 3.2 Light rays that strike an object have three possible fates: reflection, transmission, or absorption.

Wavelength. The properties of light and all electromagnetic radiation resemble those of ocean waves. The distance between two peaks (high points) or two troughs (low points) of a wave is the **wavelength.** The wavelengths of visible light lie between about 400 and 700 nm. Blue light is the shortest wavelength, and red is the longest. Yellow and green are intermediate. White light is a mixture of all wavelengths. Absorption or refraction (which we'll examine later) can reveal its component colors. The **intensity** of a light wave is the height of the wave.

In addition to describing light as waves, we can describe it as a stream of particles called **photons** or as a narrow beam called a **ray.**

Reflection, Transmission, and Absorption. When a ray of light strikes an object, three things can happen: It

SHARPER FOCUS

MEASURING MICROORGANISMS

As in science generally, microbiology uses the **metric system** of measurement. The metric system consists of basic units with prefixes that differ by powers of 10. **Meter** is the basic unit of length, **liter** of volume, and **gram** of mass (weight).

A **decimeter** (**dm**) is one-tenth of a meter, a **centimeter** (**cm**) is one-hundredth of a meter, and a **millimeter** (**mm**) is one one-thousandth of a meter.

The two units most often used for measuring microorganisms are smaller yet. A **micrometer** (**μm**) is one-millionth of a meter (1/1,000,000 or 10^{-6} meter). A

The Metric System (International System of Units)

Quantity	Basic Unit	Symbol	English Equivalent
Length	Meter	m	39 inches—about a yard
Mass	Gram	g	About 1/30 of an ounce
Volume	Liter	L	About 1.06 quarts

Prefix	Symbol	Means Multiply by
	basic unit	10^{0} (1)
Deci	d	10^{-1} (0.1)
Centi	c	10^{-2} (0.01)
Milli	m	10^{-3} (0.001)
Micro	μ	10^{-6} (0.000001)
Nano	n	10^{-9} (0.000000001)
Pico	p	10^{-12} (0.000000000001)

Commonly Used Units

1 meter	=	10^{2} cm	1 gram	=	10^{3} mg	1 liter	=	10^{3} mL
	=	10^{3} mm		=	10^{6} μg		=	10^{6} μL
	=	10^{6} μm		=	10^{9} ng			
	=	10^{9} nm		=	10^{12} pg			
	=	10^{10} Å (Angstroms)						

Note: For example, *Escherichia coli* is about 1 μ or 0.000001 m wide and weighs about 1 pg or 0.000000000001 g.

---

of pure water (at room temperature) is 1.33 (light travels 1.33 times slower in water than in a vacuum). The refractive index of air is 1.0002.

Diffraction. Light rays also bend by diffraction, a process that interferes with the ability of a lens to form a sharp image. Diffraction occurs when rays of light pass through a small opening or pass by the edge of an **opaque** object (one that transmits no light). Diffraction occurs be-

cause of the wave nature of light. Picture ocean waves coming toward a breakwater with an opening through which they can pass. As the waves approach the breakwater, they are parallel to it. But as they pass through the opening, they spread out in semicircles (**Figure 3.4**). For the same reason, light rays (which are perpendicular to the wave front) bend when they go through a small opening or around an obstacle. Because of diffraction, a sharp edge of an opaque object casts a fuzzy shadow.

SHARPER FOCUS

MEASURING MICROORGANISMS (continued)

nanometer (nm) is one-billionth of a meter (1/1,000,000,000 or 10^{-9} meter).

Atomic sizes are measured in **picometers**

(pm), which are one-trillionth of a meter (1/1,000,000,000,000 or 10^{-12} meter), or **angstroms (Å),** which are one ten-billionth

of a meter (1/10,000,000,000 or 10^{-10} meter). The angstrom is not officially part of the

metric system, but chemists and physicists use it often.

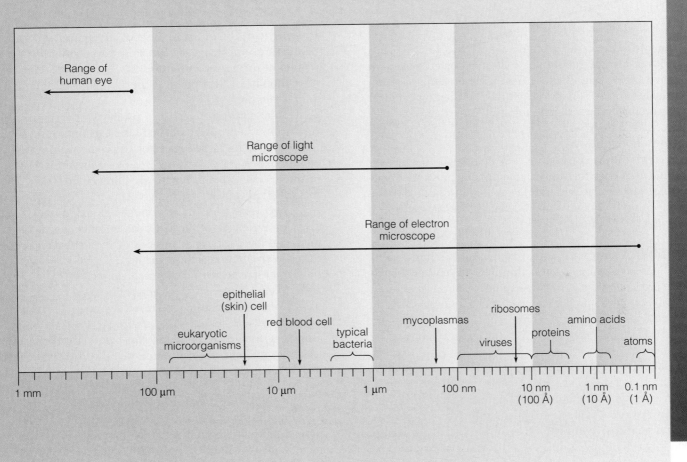

Now let's turn to how light microscopes exploit the properties of light to form magnified images.

Microscopy

We see the world around us because the lens in the front of our eye produces images on the light-sensitive retina at the back of it. But most of us cannot see objects smaller than

about 100 μm in diameter. To see microorganisms, which are 10 to 100 times smaller, we need the help of the lenses in a microscope. There are many types of microscopes, but all of them depend on three factors—magnification, contrast, and resolution—to produce a clear image.

Magnification. We use a microscope to achieve **magnification** (enlargement of the image). Any **convex** lens (thicker in the center than at the edge) can magnify. By

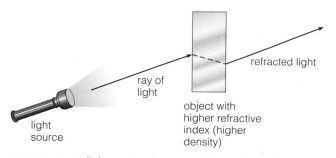

FIGURE 3.3 A light ray is refracted as it leaves air and enters glass (an object with higher refractive index) and again when it reenters air. Note how it bends down when it enters the glass and up when it reenters air.

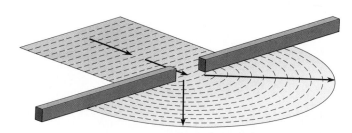

FIGURE 3.4 Diffraction causes light rays to bend as they pass through a small opening or go around an opaque object. The parallel waves of light diffract much as ocean waves do when passing through a breakwater. The light rays (arrows), which are perpendicular to the wave front, bend as they pass through the narrow opening.

refraction, it bends parallel rays of light from an object so that they meet at a single point, the **focal point,** and form an enlarged image of the object behind it (**Figure 3.5**). Magnification can be increased by making the lens more convex or by bringing the object closer to the lens. Even a single highly convex lens is adequate to see microorganisms as small as 1 μm, as Van Leeuwenhoek showed.

Magnification is essential, but it is not enough. The object we're viewing must stand out against its background, and to form a useful image, we must be able to see detail within it. The two main factors necessary to achieve these goals are contrast and resolution.

Contrast. Contrast refers to a difference in light intensity. Such differences within the field of view of microscope are essential for us to see an image of the **specimen** (the object viewed under a microscope). Contrast is usually created by light absorption: Different amounts of light are absorbed as light rays pass through different regions of the specimen.

Obtaining adequate contrast is a special problem when viewing bacteria because, being almost colorless, they absorb very little visible light and therefore generate very little contrast. To obtain useful images of bacteria and most microorganisms, contrast must be increased artificially. One way is to apply **stains,** dyes that color microbial cells or parts of them. But as we'll see later, some kinds of microscopes are able to generate adequate contrast of unstained specimens.

the cd connection

METHODS IN MICROBIOLOGY ➡
Microscopy

Micrographs of a diatom with insufficient, optimal, and excessive contrast.

Resolution. Resolution, the second requirement for obtaining a useful image, is the ability to distinguish detail within an image. For example, a television set with high resolution has a clear picture; one with poor resolution has a fuzzy picture.

Unlike contrast, which is a property of the specimen, resolution is a property of the lens system used to view it. Resolution is measured in terms of **resolving power** (the closest that two points can be to one another and still be distinguished as being separate). Three factors determine the resolving power of a microscope: (1) the size of its **objective lens** (the microscope's first magnifying lens), (2) the wavelength of the light passing through the specimen, and (3) the refractive index of the material between the objective lens and the specimen.

Larger lenses have greater resolving power because they allow a larger cone of light to enter. But how does the wavelength of the light influence resolving power? Recall that light diffracts (bends) as it passes through small openings or around opaque objects. When two points in the specimen lie close together, they produce wave fronts that tend to merge and make the two points appear to be a single point, thereby obscuring detail. But diffraction increases with the wavelength, so resolution can be improved simply by illuminating the specimen with a shorter-wavelength blue light instead of yellow or red light, for example. However, the improvement gained by using blue light is not great. To increase resolution significantly, electromagnetic radiation with much smaller wavelength, such as an electron beam, must be used. That's why electron microscopes, which we'll consider later in this chapter, have such high resolving power and therefore can offer more useful magnification.

The third factor that influences resolving power of a light microscope is the refractive index of the material in the space between the objective lens and the specimen. The greater its refractive index, the greater the resolving power. Most lenses are designed to have air in this space. But some lenses, called **oil immersion lenses,** are designed to have this space filled with oil, called **immersion oil.** Because the

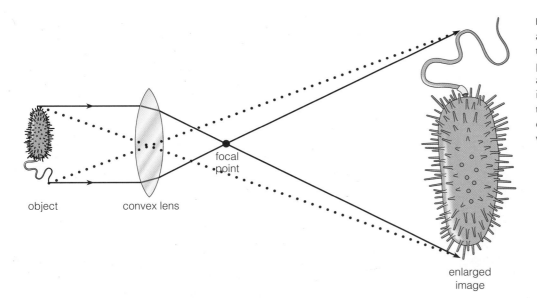

FIGURE 3.5 When parallel rays from an object pass through a convex lens, they converge at the focal point. Rays passing through the center of the lens are unbent. Rays (not shown) passing through intermediate points on the lens are bent to an intermediate extent. An enlarged image forms where all rays converge.

object

convex lens

focal point

enlarged image

refractive index of immersion oil is about 1.5 and that of air about 1.0, an oil immersion lens increases resolving power about 1.5-fold, thus permitting more useful magnification.

Two of the factors that determine resolving power—lens size and use of immersion oil—are properties of the lens. A measurement of these two factors, called **numerical aperture (NA)**, is stamped on the side of a microscope's objective lens. Knowing *NA* and the wavelength of light used to illuminate the specimen, you can calculate resolving power, *d*, the distance between two points that can just be distinguished as being separate.

$$d = \frac{\text{wavelength}}{2\text{NA}}$$

For example, using blue light, with a wavelength of 450 nm, and a high-quality oil immersion lens, with a numerical aperture of 0.65, generates a resolving power of 346 nm or 0.346 μm (450/[2 × 0.65] = 346). That's about one-third the length of an ordinary-size bacterial cell. This means that you can't see many internal details within a bacterial cell even with a good light microscope.

The Compound Light Microscope

Microscopes that use visible light as a source of illumination are called **light microscopes.** Light microscopes, such as Van Leeuwenhoek's, that have a single lens are called **simple microscopes.** Using such a microscope requires skill because the specimen must be held extremely close to the viewer's eye. (Some say that Van Leeuwenhoek's success with his microscopes was due in part to his nearsightedness.) Today's light microscopes are **compound microscopes (Figure 3.6).** They provide greater magnification

and are much easier to use. Compound microscopes have two lenses, an **objective** lens and an **ocular lens,** both of which are **corrected lenses** (several individual lenses bonded together) that compensate for the **aberrations** (defects) inherent in ordinary lenses. Uncorrected aberrations produce images that are surrounded by colored rings (**chromatic aberration**), and not all parts of the field of observation are simultaneously in focus (**spherical aberration**).

The Parts of a Microscope. Most modern compound microscopes have a built-in source of light. First this light passes through a series of lenses called the **condenser,** which concentrates the light onto the specimen, where some of it is absorbed. Then the transmitted light enters the **objective lens,** which forms an image of the specimen within the **body tube** of the microscope. Sometimes a prism is located within the body tube, allowing it to be bent to create a more comfortable viewing angle. The **ocular lens** further magnifies the image within the body tube and projects it to the last lens in the series—the one within the eye. The lens of the eye forms an image on the retina.

The compound microscope just described gives brightfield illumination. That is, the background is more brightly lit than the specimen. A compound microscope can be adapted, however, to view a specimen in other ways by (1) phase-contrast, (2) darkfield, (3) fluorescence, or (4) differential interference contrast (Nomarsky) methods. We'll discuss them later in this chapter.

Total Magnification. Most compound microscopes offer a choice of several objective lenses mounted on a rotating turret; each lens has a different magnifying power. Typically the **low-power lens** magnifies an object 10-fold (10×), the **high-power** (also called **high-dry**) **lens** 40-fold (40×), and the **oil-immersion lens** 100-fold (100×).

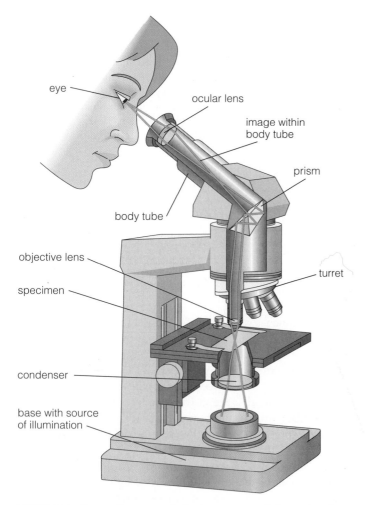

FIGURE 3.6 Parts of a compound microscope and the paths of light rays through it.

to get a sense of its overall appearance. High-power and oil-immersion lenses cover smaller fields but reveal greater detail within the specimen (**Figure 3.7**).

Now we'll consider some of the many ways of preparing specimens for viewing under a microscope.

Wet Mounts

A **wet mount** is probably the simplest way to prepare a specimen for microscopic examination. Just place a drop of liquid containing microorganisms on a microscope slide and place a **coverslip** (a square of thin glass) on top of it. But such **simple wet mounts** dry out quickly. For more extended viewing, a **hanging drop mount** is made by placing a drop of liquid containing microorganisms on a coverslip and suspending it over a depression slide (**Figure 3.8**). Petroleum jelly around the well of the depression slide seals the mount to prevent drying.

Wet mounts are used to observe living microorganisms. For example, wet mounts are routinely used to examine vaginal secretions for *Trichomonas vaginalis*, a highly motile protozoan that causes inflammation of the vagina and urethra. If the specimen contains fish-shaped cells that move jerkily across the field, they are most probably *T. vaginalis*. But many microorganisms lack sufficient contrast to be clearly visible in a wet mount viewed with an ordinary brightfield light microscope. To visualize them, it's necessary to increase contrast either by staining the specimen or using a microscope that generates contrast by different means.

Stains

Stains are dyes that increase contrast by binding selectively to certain cells or to certain parts of them. As a result staining reveals structures that would otherwise be invisible or nearly so. During the nineteenth century, advances in the

Most ocular lenses magnify the image another 10-fold (10×). Total magnification is the product of the power of both lens systems in use (**Table 3.1**). Because its field of view is large, low power is best for scanning a specimen

Microscope	Objective Lens		Ocular Lens		Total Magnification
Light Microscopes					
Low power	10×	×	10×	=	100×
High power	40×	×	10×	=	400×
Oil immersion	100×	×	10×	=	1000×
Electron Microscopes					
Transmission (TEM)					~200,000×
Scanning (SEM)					~10,000×

TABLE 3.1 Total Magnification

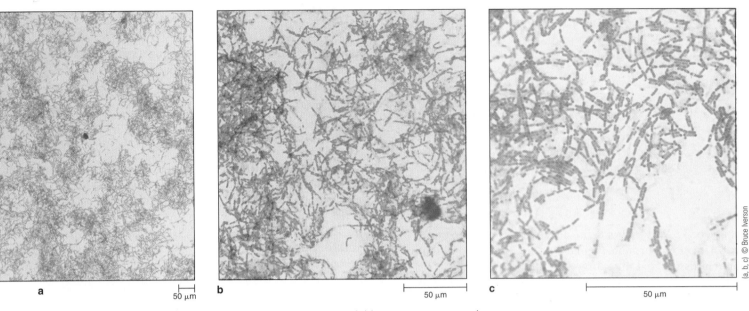

a   50 μm   b   50 μm   c   50 μm

(a, b, c) © Bruce Iverson

FIGURE 3.7   Cells of *Bacillus subtilis* viewed at magnifications available on most compound microscopes. (a) Low power (100×). (b) High power (400×). (c) Oil immersion (1000×).

science of microbiology went hand-in-hand with the development of better ways to stain specimens. For example, Robert Koch was able to discover the cause of tuberculosis largely because he had developed a stain that enabled him to see the causative bacterium, *Mycobacterium tuberculosis*, by itself or within host tissue.

A few stains, called **vital stains** because they stain living cells, can be added directly to a wet mount. But most stains are effective only after microorganisms have been **fixed** (killed and attached to a microscope slide). Microorganisms can be fixed either by heating or treating with chemicals. **Heat fixation** is the usual method: A drop of liquid containing microorganisms is spread in a thin film (a **smear**) on a microscope slide and allowed to air dry; then the slide is passed rapidly through an open flame. Heat from the flame kills the microbial cells by denaturing their protein, which sticks the cells to the slide. **Chemical fixation** does less damage than heat. Osmic acid, formaldehyde, or glutaraldehyde is most often used. One simply adds a drop of the chemical to the liquid containing the microorganisms.

Fixation brings difficulties. It often distorts the cell's appearance, and of course motility can no longer be studied. After fixation, stain is applied to the slide and left there long enough to be absorbed by the specimen. Then excess stain is washed away, usually by flooding the slide with water.

Types of Dyes. Most dyes used for staining are ionic. Those with positive charges are termed **basic dyes** and those with negative charges are acidic dyes (**Table 3.2**). Because the surfaces of most microorganisms carry a slight negative charge, basic dyes are attracted to them and stain

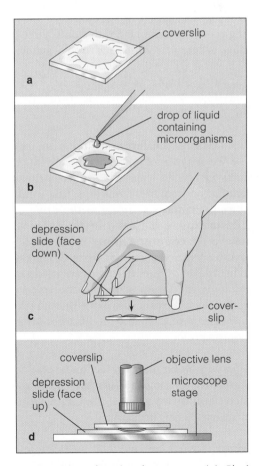

FIGURE 3.8   Preparing a hanging drop mount. (a). Placing petroleum jelly on a coverslip. (b) Adding a suspension of microorganisms. (c) Inverting a depression slide over the coverslip. (d) Viewing the hanging drop mount.

TABLE 3.2 Some Dyes Commonly
Used in Staining

Basic Dyes	Acidic Dyes
Safranin	Eosin
Basic fuchsin	Acid fuchsin
Crystal violet	Congo red
Methylene blue	

them best. Acidic dyes stain negatively charged parts of cells, including proteins; acidic dyes are often used to stain animal tissues that microorganisms have invaded.

Although they are not dyes, compounds called **mordants** are important to some staining procedures. Most mordants intensify staining by increasing a cell's affinity for a dye. Some mordants are used to coat cell appendages, such as flagella, making them thicker and therefore more readily visible when stained.

Staining procedures can be grouped into three classes—simple stains, differential stains, and special stains.

Simple Stains. **Simple stains** use basic dyes to make cells visible. They stain all cells that absorb the dye the same color.

Differential Stains. **Differential stains** are used to distinguish between types of microorganisms. Such procedures usually involve at least three steps—primary staining, destaining, and counterstaining. Often a mordant is also employed. **Primary staining** is the same as simple staining; **destaining** is a treatment that removes stains from certain cells; **counterstaining** is the application of another dye to reveal the cells or parts of cells that have been destained. Two differential stains are used extensively in microbiology, the Gram stain and the acid-fast stain. We'll describe the basic procedures for these strains, keeping in mind that there are many modifications that certain microbiologists prefer.

The Gram Stain. The Gram-stain technique was developed by the Danish bacteriologist Christian Gram in 1884. It divides bacteria into two groups, the Gram-positives and the Gram-negatives (**Figure 3.9**):

Step 1. Primary stain: Gentian violet (made from crystal violet, alcohol, ammonium oxalate, and water) is applied to a fixed specimen of bacteria. Both Gram-positives and Gram-negatives become purple.

Step 2. Mordant: Iodine is applied, which sets the stain but does not change the general appearance of the cell.

Step 3. Decolorization: A decolorizing agent, usually ethanol, is added. Gram-negative bacteria lose their purple color. Gram-positive bacteria retain it.

Step 4. Counterstain: The red dye safranin is added, which turns the decolorized Gram-negatives pink and the Gram-positive bacteria a slightly deeper violet color.

The staining difference between Gram-positives and Gram-negatives reflects an important difference in their outer surface, which we'll discuss in Chapter 4.

The Acid-Fast Stain. The acid-fast stain was developed by Paul Ehrlich in 1882 to identify *Mycobacterium tuberculosis*, the bacterium that causes tuberculosis. This stain colors only mycobacteria and some actinomycetes (Chapter 11). Because of this, these bacteria are called **acid-fast bacteria.** All other bacteria and host tissues become destained, but their presence in a specimen can be detected by counterstaining (**Figure 3.10**). Acid-fast bacteria resist destaining because they have a waxlike material in their cell envelope (Chapter 4). A type of acid-fast stain called the Ziehl-Neelsen stain is performed as follows:

Step 1. Primary stain: Carbolfuchsin (made from basic fuchsin, phenol, ethanol, and water) is applied to the fixed specimen on a microscope slide. All cells stain red.

Step 2. Mordant: The slide is heated to steaming for 5 minutes, which drives the stain into the cells.

Step 3. Decolorization: A decolorizing solution consisting of hydrochloric acid in ethanol is added. This removes the red dye from all cells except acid-fast bacteria.

Step 4. Counterstain: Methylene blue is added as a counterstain. It colors all decolorized bacterial cells and host tissue blue. Acid-fast bacteria remain red.

Special Stains. Microbiologists also have at their disposal an elaborate set of stains that are used for special purposes (**Figure 3.11**). Some special stains reveal specific parts of a microbial cell, such as the wall, the nucleoid, an endospore (**Figure 3.11a**), membranes, flagella, or the capsule (Chapter 4). Combinations of special stains are sometimes used (**Figure 3.12**). Other special stains are used to reveal certain hard-to-stain microorganisms such as spirochetes and rickettsiae (Chapter 11). Here we'll describe two of the many kinds of special stains: **flagella stains** and **negative stains.** They illustrate different principles of staining.

The Leifson Flagella Stain. Flagella are the long, threadlike cellular extensions that allow some bacteria to swim (Chap-

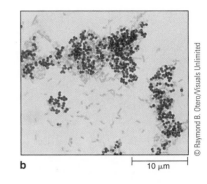

FIGURE 3.9 Gram-staining procedure. (a) Steps of the procedure. (b) A Gram-stained specimen containing both *Staphylococcus aureus* and *Escherichia coli*. The Gram-positive *S. aureus* retains the purple stain of crystal violet. The Gram-negative *E. coli* loses the purple color during decolorizing and is turned pink by the counterstain.

b   10 μm

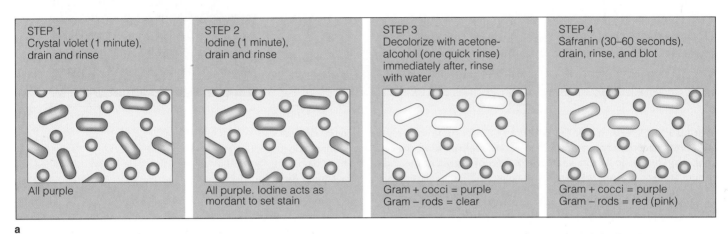

STEP 1
Crystal violet (1 minute), drain and rinse

All purple

STEP 2
Iodine (1 minute), drain and rinse

All purple. Iodine acts as mordant to set stain

STEP 3
Decolorize with acetone-alcohol (one quick rinse) immediately after, rinse with water

Gram + cocci = purple
Gram – rods = clear

STEP 4
Safranin (30–60 seconds), drain, rinse, and blot

Gram + cocci = purple
Gram – rods = red (pink)

a

ter 4). They are too thin to be visible by light microscopy, but flagella staining renders them visible by adding a material that sticks to them, making them thicker. Then the thickened flagella are stained with a dye. All steps in flagella staining are designed to be gentle because flagella break off easily. The Leifson flagella stain involves the following steps:

1. The suspension of bacteria is fixed chemically, with formalin, and spread on a glass slide.

2. It is allowed to air dry without heating.

3. A freshly prepared mixture of tannic acid and rosaniline dye is then added to the slide. The tannic acid thickens the flagella. The rosaniline colors them.

4. Excess stain is washed off by flooding the slide with water.

5. The slide is again allowed to air dry before being examined under the microscope.

Flagella staining reveals the number and arrangement of flagella on bacteria, which is vital information for identifying many species (**Figure 3.11b**). Successful flagella staining is an art that develops only with practice.

Negative Staining. Some bacteria are surrounded by a protective structure called a capsule (Chapter 4). These struc-

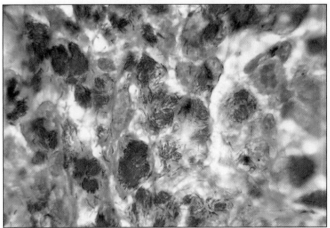

FIGURE 3.10 The acid-fast stain. The masses of *Mycobacterium leprae* cells stain red. The host cells and tissue are counterstained blue.

tures are colorless and therefore invisible in unstained preparations. But their presence can be revealed by negative staining, a process by which the capsule becomes visible because only it remains uncolored. Negative staining involves the following steps:

Step 1. A wet mount of the specimen is prepared.

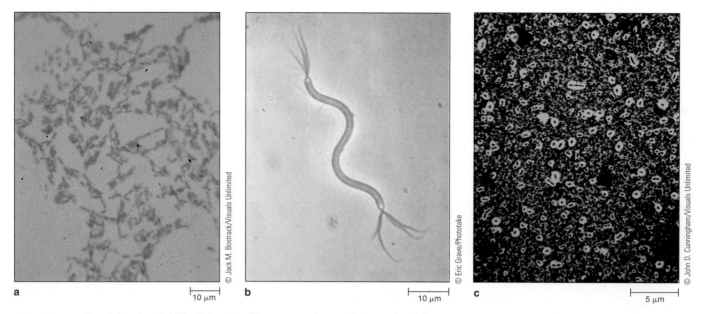

a   10 μm   © Jack M. Bostrack/Visuals Unlimited

b   10 μm   © Eric Grave/Phototake

c   5 μm   © John D. Cunningham/Visuals Unlimited

FIGURE 3.11 Special stains. (a) The Wirtz-Conklin spore stain reveals bacteria with en-dospores. Here the endospores of *Bacillus cereus* appear blue; the cells are counterstained pink. (b) The *Leifson* flagella stain reveals tufts of flagella at the end of the huge bacterium *Spirillum volutans*. (c) Negative staining with India ink reveals capsules around *Klebsiella pneumoniae*. The background is black with India ink. The capsules are clear but slightly pink from the simple stain. The cells with the capsule are a deep pink.

Courtesy of Peter J. Lewis, from Bacterial chromosome segregation, *Microbiology* 147:519-526, 2001

FIGURE 3.12 Triple-stained micrograph of the Gram-positive bacterium, *Bacillus subtilis*. The DNA is stained blue with 4,6-diamino-2-phenylindole. The membrane is stained red with FM4-64. The ribosomes are green as a consequence of one of their proteins being joined to green fluorescent protein (see Sharper Focus: Molecular Stains: A New Use for Jellyfish).

Step 2. India ink is added. The carbon particles in the ink cannot penetrate the capsule, so only the background is blackened. The capsule and cell within are revealed as a clear zone.

Step 3. A simple stain may then be applied to make the cell itself visible (**Figure 3.11c**).

Light Microscopy: Other Ways to Achieve Contrast

Brightfield microscopy of stained specimens is used in most laboratories (**Figure 3.13a**). But other kinds of compound light microscopes can be used for special purposes to view unstained microorganisms. These microscopes depend on various physical principles to generate contrast and produce remarkably detailed images of microorganisms. We'll consider three examples: phase-contrast, darkfield, fluorescent, and Nomarsky microscopy (Figure 3.13).

Phase-Contrast Microscopy. Recall that contrast sufficient to form a visible image in a brightfield microscope can be achieved only if various parts of the field absorb significantly different amounts of light. Phase-contrast microscopes overcome this requirement by using a completely different method of achieving contrast. They can generate

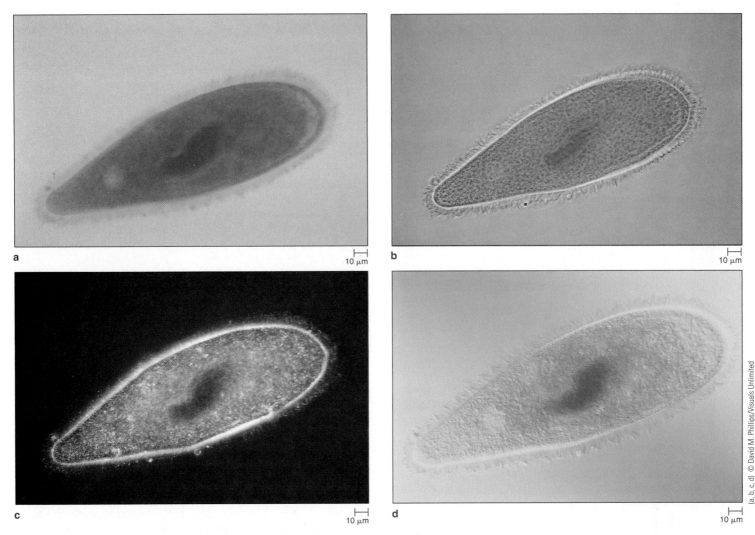

FIGURE 3.13   Images of the protozoan *Paramecium* at the same magnification (1000×) viewed by different microscopy. (a) Brightfield microscopy (a bacterium would be almost invisible). (b) The phase-contrast image shows greater internal detail. (c) Darkfield microscopy shows a bright image in a black background. (d) The Nomarsky image is almost three-dimensional.

contrast from variations in refractive index throughout the field. Physically a phase-contrast microscope differs from a brightfield light microscope only by having an opaque ring in the objective lens and another one in the condenser. The purpose of these rings is to send some light rays through the specimen and others around it.

The phase-contrast microscope is based on the principle that rays of light move at different speeds through materials of different refractive index. So the light rays that pass through the specimen are **out of phase** (peaks of the waves occur at different times) to varying degrees with the rays that go around the specimen. When the out-of-phase rays are brought back together, they generate contrast by **interference,** meaning that peaks of light waves that arrive at a dif-

ferent time cancel each other out (interfere with one another). Those that arrive at or near the same time augment one another. As we have all seen, the same thing happens when ocean waves come to shore from different directions. Augmenting and canceling out produce different light intensities and thus the contrast needed to form a useful image.

The major advantage of phase-contrast microscopy is avoidance of the necessity of fixing and staining. So it can be used to study activities and properties of live cells, including cellular movement and internal cell structures that have not been distorted. A slight disadvantage is the halo of light that surrounds objects in the field of view, an unavoidable consequence of generating contrast by interference (**Figure 3.13b**).

LARGER FIELD

JUST BY LOOKING

It was 1962 and Heinz Stolp, a young German microbiologist, was spending a year in the United States studying the *Erwinia* genus of bacteria. Species of *Erwinia* are major plant pathogens. They cause serious diseases, including fire blight of apples and pears. Stolp was using bacterial viruses to figure out the relationships between various species of *Erwinia*. He knew that one particular virus would attack only closely related strains; strains or species attacked by the same virus must be closely related. To accomplish his goal, he needed many strains of virus.

Stolp went about isolating them in the usual way. He made pour plates and inoculated them with strains of *Erwinia*. Then he added extracts of soil that were suspected to contain viruses he wanted. The *Erwinia* grew as a confluent layer or lawn covering the plate. But where a virus was present, it destroyed *Erwinia* cells in a small region, forming a circular clear zone, called a **plaque.** Stolp obtained his virus strains by picking them from plaques with an inoculation loop.

Every day Stolp prepared many plates and isolated new strains of virus from the previous day's plates. The project went well. One day, while discarding some old plates, he noticed something strange. Some of the plaques had gotten bigger. He knew very well that plaques formed by viruses don't enlarge. Their size is fixed when bacterial growth stops, which usually occurs the day after they are prepared. Why were these plaques getting bigger? Were they caused by something other than a virus?

Stolp took a direct approach. With an inoculating loop, he picked a region from the clear plaque, made a wet mount, and observed it under a phase-contrast microscope. He saw only a few *Erwinia* cells, nothing else. But from time to time an *Erwinia* cell moved abruptly, as though an unseen missile had struck it. Maybe the missile was a predator that destroyed *Erwinia* cells, forming plaques. If so, the predator was too small to be seen with a phase-contrast microscope. So Stolp put a sample from the plaque under a transmission electron microscope. He saw *Bdellovibrio*, as he named them (*bdello*, meaning "leach," and *vibrio*, describing their commalike shape). These bacteria had never been seen before. Now we know the genus to have many species and be widespread. *Bdellovibrio* spp. attack Gram-negative soil bacteria by ramming into them at such speed that the host cells recoil from the impact. Then the *Bdellovibrio* bacterium enters its still-living host, and once inside, kills it. By producing more *Bdellovibrio* cells that consume more bacterial prey, each *Bdellovibrio* cell makes a plaque. These plaques continue to enlarge even after bacterial growth stops because *Bdellovibrio*, unlike bacterial viruses, attack nongrowing cells.

Stolp had discovered a major bacterial group. He did so just by looking.

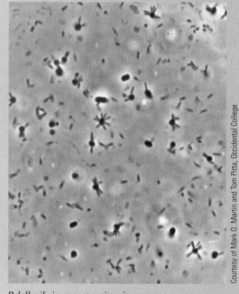

Bdellovibrio penetrating its prey.

Courtesy of Mark O. Martin and Tom Pitta, Occidental College

Nomarsky Microscopy. Nomarsky or **differential interference contrast microscopy** is a variation of phase-contrast microscopy that has come into use during the past 30 years. Like phase-contrast microscopy, Nomarsky microscopy uses differences in refractive index to produce contrast by interference. The two methods differ in how the interfering rays are separated within the microscope. As we've seen, phase-contrast microscopes use a complementary pair of opaque rings: one located below the condenser and the other on the backside of the objective lens. Nomarsky microscopes use a pair of prisms in similar locations. Nomarsky microscopy produces an almost three-dimensional image of finer detail than phase contrast (**Figure 3.13d**).

Darkfield Microscopy. Living unstained cells can also be viewed with a darkfield microscope. It operates on the principle of light **scattering,** the abrupt change of direction of a light ray that occurs when it strikes and bounces off a small object. (The luminous appearance of dust particles in the beam of a flashlight shown through a darkened

SHARPER FOCUS

MOLECULAR STAINS: A NEW USE FOR JELLYFISH

With the advent of recombinant DNA technology (Chapter 7), it has become possible to combine genes and thereby attach a foreign protein to one of a cell's own proteins. If the foreign protein can be easily detected either chemically or visually, it is termed a **reporter** because it reports the presence or location of the protein to which it is attached. One of the most valuable and widely used such reporter proteins is one called **green fluorescent protein** (GFP), which was obtained from the jellyfish *Aequorea victoria*. GFP absorbs light of 395 and 475 nm and reemits (fluoresces) at 475 nm. So if a specimen containing GFP is illuminated with blue light, the location of the GFP becomes bright green.

GFP has become an extremely valuable research tool: Microbes tagged with GFP can be easily detected in a mass of other microbes or within the tissues of a host; tagged molecules can be located within a living cell, and their changing location can be followed as a cell grows or develops.

Francis Wong and John Meeks recently did an experiment that illustrates the value of GFP. They were studying the filamentous cyanobacterium *Nostoc punctiforme* (Chapter 11), which makes special cells called **heterocysts** when it utilizes atmospheric nitrogen as a nutrient. Wong and Meeks knew that a signal protein, HetR, is needed for *N. punctiforme* to be able to form heterocysts. They wondered if HetR might concentrate in these special cells, and they addressed their question with GFP. Using recombinant DNA methods, they combined the HetR and GFP genes, causing the cells that contained the combination gene to make HetR with GFP attached to it. The answer to their question was abundantly clear when they examined these cells under a microscope illuminated with blue light. The heterocysts glowed bright green; the other cells did not. HetR does indeed occur only in heterocysts.

room is a consequence of light scattering.) A darkfield microscope has a special condenser that redirects the light beam coming from the light source so that it goes through the specimen but misses the objective lens. The only light rays that enter the objective lens are those that have been scattered by striking the specimen. The result is a bright image against a dark background (**Figure 3.13c**). Darkfield microscopy is most effective for seeing surface structures. It is commonly used to detect *Treponema pallidum*, the highly motile bacterium that causes syphilis (Chapter 24).

the cd connection

MICROBIAL BEHAVIOR ⟶ *Types of motility/Flagella*

Darkfield microscopy of moving *Escherichia coli* cells revealing how the 6 to 10 peritrichous flagella coalesce to form a single propellerlike structure that drives the cell.

Fluorescence Microscopy. In fluorescence microscopy, contrast is increased by fluorescence. Fluorescent materials absorb light of one wavelength and give off light of a higher wavelength. When fluorescent materials are illuminated by short-wave (invisible) ultraviolet light, they give off visible light, glowing brightly against a dark background (**Figure 3.14**). Unlike phase-contrast and darkfield techniques, fluorescence microscopy depends on a property of the specimen, not of the microscope. Nevertheless, a microscope must be modified with special equipment to illuminate the specimen with short-wave ultraviolet light.

Some microorganisms, including photosynthetic bacteria and algae, are naturally fluorescent. But most microorganisms must be made fluorescent by adding dyes called **fluorochromes.** Fluorescence microscopy can be used to identify specific microorganisms by chemically attaching fluorochromes to antibodies, proteins produced by the immune system that bind to specific microorganisms (Chapter 17). Such fluorescent antibody, or immunofluorescence, tests are important diagnostic uses of fluorescence microscopy. We'll discuss them in detail in Chapter 19.

A modified fluorescence microscope, the **epifluorescence microscope** has dramatically improved fluorescence microscopy. The light source in these microscopes is mounted within the bodytube of the microscope and focused by the objective lens on the specimen. The only light used to form the image is the fluorescent light emitted by the specimen (Figure 3.14).

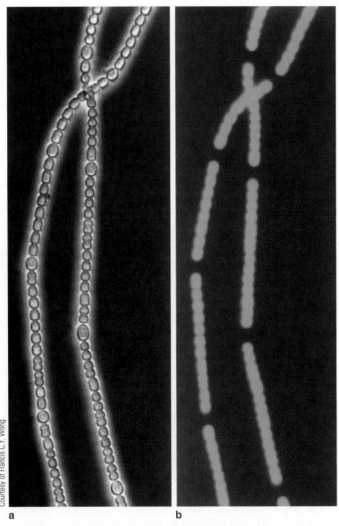

Courtesy of Francis C.Y. Wong

a   b

FIGURE 3.14 Two filaments of a cyanobacterium viewed with phase and epifluorescence microscopy. (a) Phase microscopy (note halos) shows vegetative cells and eight larger, specialized cells called heterocysts. (b) With epifluorescence microscopy the chlorophyll which vegetative cells contain fluoresces red when illuminated with blue light; heterocysts, which lack chlorophyll, do not.

Light Microscopy: Scanning Microscopes.

Scanning light microscopes exploit the intrinsic optical advantages of concentrating on a small field of view: less aberration and less interference from scattered light. These microscopes scan the specimen with a focused beam of light and monitor the many small fields thus formed with a **photodetector** (a device that converts light into an electric signal). The output of the photodetector is processed electronically to form one large field from the many small ones.

Confocal microscopy is a powerful new modification of scanning microscopy. In essence a confocal microscope is two microscopes focused on the same object from oppo-

site sides. One is an illuminating microscope with a small aperture that lights a tiny spot within the specimen. The other, the receiving microscope, is provided with a photodetector connected to a computer. The light spot scans the specimen, and the computer generates an image of one particular slice through the specimen at the depth at which both microscopes are focused. Multiple scans can generate remarkable images: Scans at different depths can generate three-dimensional images; scans with different wavelengths of illuminating light can reveal objects within the specimen that are tagged with different fluorochromes.

Other Light Microscopes

Our survey of light microscopes certainly hasn't been all inclusive. Many other types exist, including stereoscopic microscopes (which form three-dimensional images), polarizing microscope (for examining substances that polarize light), metallographic microscopes (for examining opaque substances), and reflecting microscopes (which can utilize nonvisible infrared or ultraviolet illumination). But none of these are particularly useful to microbiology.

Electron Microscopy

Each of the modified light microscopes we have just discussed has its own special advantages. None, however, has any more resolving power than an ordinary brightfield microscope. As a result, none can be used to form useful images with greater magnification. In contrast, electron microscopes, by utilizing a beam of electrons (which has a very short wavelength), have vastly greater resolving power than light microscopes. Therefore they can be used to produce much higher useful magnification. There are two kinds of electron microscopes (**EM**) in common use, the **transmission electron microscope** (**TEM**) and the **scanning electron microscope** (**SEM**). Each has special advantages.

Transmission Electron Microscopy. The transmission electron microscope (**Figure 3.15**) can resolve points as close together as 1 nm, as compared with about 300 nm for a light microscope. Thus TEM produces a clear, detailed image at magnifications of up to 300,000×, as compared with about 1000× for a light microscope.

A TEM forms an image much as a light microscope does, with some important differences:

- Instead of a beam of light, a beam of electrons is transmitted through the specimen.
- Instead of glass lenses, electromagnetic lenses focus the image.

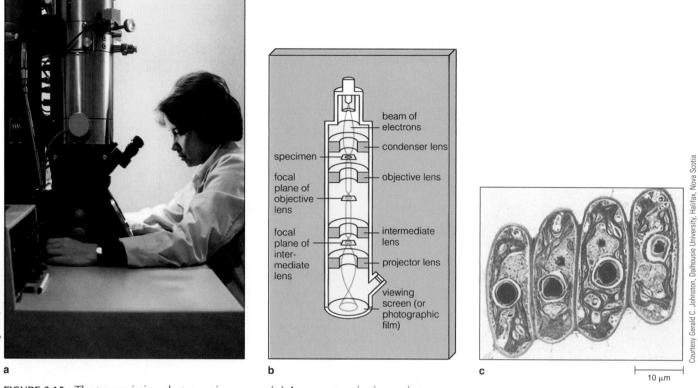

FIGURE 3.15 The transmission electron microscope. (a) An operator viewing an image on the screen. (b) Schematic view of the electron path. (c) A transmission electron micrograph of the green alga *Scenedesmus*. Compare this TEM image with the SEM image of the same specimen in Figure 3.18.

(Part b is from Biology: The Unity and Diversity of Life, 6th ed., by C. Starr and R. Taggart, Brooks/Cole, 1992. All rights reserved.)

- Instead of producing the image on the retina of the eye, it produces one on a fluorescent screen or a photographic plate. (The image is called a **transmission electron micrograph.**)
- Instead of the specimen's being mounted on a glass slide, it is held on a copper grid that allows electrons to pass through.
- Instead of using dyes that absorb light to increase contrast, heavy metals, such as lead, tungsten, and uranium, which absorb electrons, are used.

Using an electron beam presents special technical problems because it has very little penetrating power. As a result, the entire electron path, including the specimen, must be in a vacuum. Also the specimen must be **ultrathin** for the beam to pass through it. The specimen is usually cut with an instrument called a **microtome** into slices no thicker than 0.1 μm—small enough to produce 10 slices from a bacterial cell.

In addition to slicing, specimens for electron microscopy can be prepared by **freeze-fracturing** and **freeze-etching (Figure 3.16)**. These techniques are especially good for examining intracellular membranes because the fracture often runs through membranes, exposing their internal structure. Details of the structure can then be enhanced by freeze-etching: The frozen sample is put in a vacuum and water is removed (by sublimation). This leaves solid material protruding above the surface. Then carbon is sprayed on the still-frozen specimen, forming a replica of the surface. The thin replica, which the electron beam can pass through, is then removed and viewed by TEM.

When objects are viewed by TEM, it is usually necessary to increase their contrast. Sometimes this is done by **shadow-casting.** The specimen is exposed to a shower of a heavy metal, such as gold or platinum, which is deposited at an angle (**Figure 3.17**). Objects in the specimen produce "shadows" in the metal layer, giving the image a three-dimensional appearance that reveals shape and height.

Because specimens must be examined in a vacuum, they must be completely desiccated, which sometimes introduces **artifacts** (structures or details in the image that do not exist in the specimen). For example, some bacteria were once believed to contain complex convoluted membranes called **mesosomes.** These structures are now known to

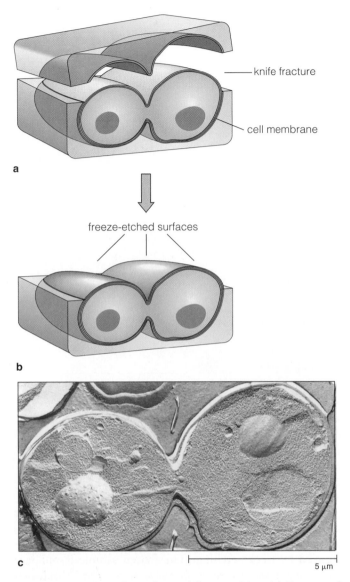

FIGURE 3.16 Freeze-fracturing and freeze-etching. (a) Fracture of frozen sample. (b) Freeze-etching by subliming water from the surface. (c) Transmission electron micrograph of a carbon replica of the etched sample.

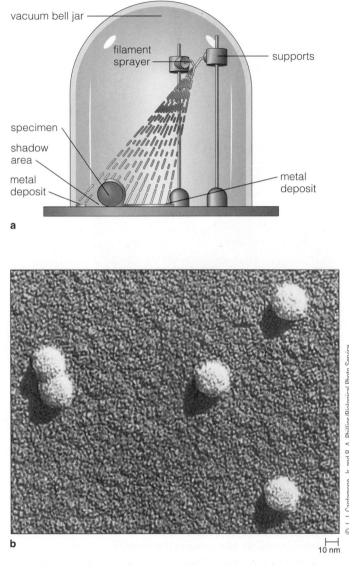

FIGURE 3.17 Shadow-casting. (a) The specimen is showered, at an angle, with a thin layer of heavy metal. (b) A transmission electron micrograph of a shadow-casted poliovirus.

be artifacts of sample preparation. Considerable experience is required to interpret electron micrographs.

Scanning Electron Microscopy.
The scanning electron microscope (SEM) resembles the transmission electron microscope in its use of a beam of electrons, but in most other ways it is different (**Figure 3.18**). Most of the microscopes we have discussed so far pass of a beam of light or electrons through the specimen. In contrast, the SEM only bombards the surface of the specimen with a beam of electrons. These "primary" electrons that strike the specimen eject "secondary" electrons. The number of secondary electrons ejected varies with the composition and shape of different parts of the surface that the primary electrons strike. The secondary electrons are collected and used to generate a signal that is processed electronically. The processed signals produce an image of the surface on a cathode-ray tube (like the picture tube of a television set). So the SEM "sees" only the object's surface. But it has great depth of focus and produces striking images. The SEM can resolve objects as close together as 20 nm. Thus it produces clearly defined detail at magnifications of up to 10,000×.

The scanning electron microscope is used to view surfaces, for example surfaces of intact cells. Specimens are freeze-dried (desiccated in a vacuum while frozen). Then they are coated with a thin layer of heavy metal, such as gold or platinum. Because coating prevents electrons from penetrating the specimen, it sharpens the image (called a **scanning electron micrograph**). Like TEM, SEM views specimens only in a vacuum.

Scanned-Proximity Probe Microscopes: Viewing Atoms and Molecules

In recent years the ultimate of magnification and resolution (at least for the present) seems to have been achieved, namely the ability to resolve and visualize single atoms and molecules. The microscopes that form these images are called **scanned-proximity probe microscopes.** There are about two dozen types of such microscopes. Unlike traditional microscopes, scanned-probe microscopes do not use lenses, so instead of being controlled by the laws of diffraction their limit of resolution is set by the size of the probe they use. The probe is usually sharpened electrically to end at a point composed of a single atom and held very close to the surface of the specimen, at the distance of about the diameter of an atom. The tip scans the surface of the specimen in a regular pattern while measuring some property of the specimen. The measurement is then converted into an electrical signal and used to generate an image on a television-like screen. The challenges faced by the developers of these microscopes were largely mechanical: rigorous elimination of vibration and ultra-precise movement of the probe. The most well known scanned-probe microscopes are the **scanning tunneling microscope** and the **atomic force microscope.** We'll discuss each of them briefly.

Scanning Tunneling Microscopy.
Scanning tunneling microscopy (STM) is used to view surfaces that can conduct electricity; these include metals and semiconducting materials such as silicon used in computers. STM uses a metal probe that carries a slight electrical charge. In regions of the specimen where there are lots of electrons, there is an increased probability of electrons flowing to the tip of the probe instead of to an adjoining atom in the specimen. The flow of electrons to the tip generates the signal that is used to form the image.

Atomic Force Microscopy.
Atomic force microscopy (AFM) is now used commonly to view biologically important molecules. AFM utilizes the attractive and repulsive forces between atoms to form an image of them. A sliver of diamond is used as the probe to scan the specimen. It is held close enough to the surface to detect these forces and be attracted or repulsed by them. The probe is mounted at the end of a highly flexible cantilevered (overhanging)

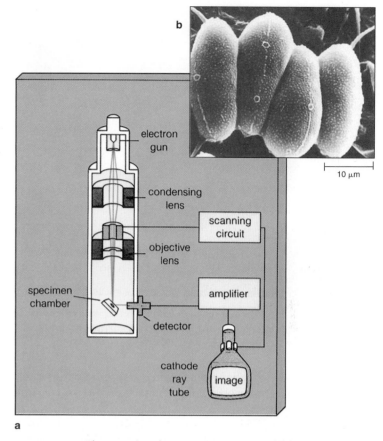

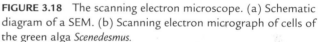

FIGURE 3.18   The scanning electron microscope. (a) Schematic diagram of a SEM. (b) Scanning electron micrograph of cells of the green alga *Scenedesmus*.

beam, which bends in response to the atomic forces. A laser beam detects the beam's flexing, and the signal thus generated is used to form an image of the specimen being scanned.

AFM can achieve a resolution of about 10 pm. That's a hundredth of a nanometer, a hundred thousandths of a micrometer, about 300 thousand times smaller than the limit of resolution of the best light microscope. Moreover, a vacuum is not necessary. The specimen can be in air or even under water. AFM is often used to detect various shapes that DNA molecules form (**Figure 3.19**).

Uses of Microscopy

Each of the various kinds of microscopes has its own particular advantages, disadvantages, and uses (**Table 3.3**). In general, light microscopes are for everyday use. Specimen preparation and operation are relatively rapid and simple. Light microscopy provides valuable information

FIGURE 3.19 Atomic force micrograph circular molecules of DNA linked together in vitro using purified enzymes (helicase and topoisomerase III) and coated with a protein (RecA).

Courtesy Steven Kowalczykowski

about the size, shape, and general appearance of cells. However, resolution is limited and therefore so is useful magnification.

Electron microscopy is principally a research tool. The practical problems of working with electrons rather than visible light make these instruments large and complex compared with light microscopes. But electron microscopes allow microbiologists to view **ultrastructure** (fine details of cells) that are not visible by light microscopy. EM offers sufficient resolution and useful magnification to view viruses and smaller objects such as large molecules.

With scanning tunneling microscopy and atomic force microscopy, individual atoms and molecules can be viewed.

CULTURING MICROORGANISMS

The small size and rapid growth of microorganisms makes them ideal experimental subjects. Billions of organisms (as many as there are people on Earth) can be subjected to study within a single milliliter of culture. It's possible to observe as many generations of some bacteria in an hour as human generations in a century. And the lessons learned from studying microorganisms can often be applied to plants and animals, including humans. To do experiments with microorganisms, it's usually necessary to **culture** (cultivate) them in the laboratory.

Obtaining a Pure Culture

Most experiments are done with a **pure culture** (one consisting of a single type microorganism derived from a single cell). Such pure cultures exist rarely in nature. In their natural environment—for example, in soil, water, or the human body—we find **mixed cultures** (many different microorganisms living together). A pure culture must be obtained artificially.

For the rest of this chapter, we'll discuss methods of obtaining and cultivating pure cultures. These methods were developed for bacteria and fungi, but with slight modifications they can be used for algae and protozoa. Methods of obtaining pure cultures of viruses are discussed in Chapter 13.

Obtaining a pure culture is a two-step process. First, materials are sterilized to eliminate all microorganisms present. Second, one single microbial cell is isolated and cultivated to produce a **clone** (descendants of a single organism).

Sterilization. Eliminating all microorganisms is called **sterilization.** We'll discuss the principles of sterilization in Chapter 9. Here we'll consider only how it relates to cultivating bacteria in the laboratory.

All apparatus and materials used to obtain a pure culture must be sterilized. That includes the **medium** (*pl.*, media, the liquid or solid [gelled] material that supplies nutrients to the culture). It also includes the flasks, test tubes, and dishes that hold the media, as well as the tools, such as pipettes and inoculating needles, used to transfer a culture from one container to another. In the laboratory, microbiologists use heat, filtration, and chemicals to sterilize.

Heat Sterilization. Heat is usually used to sterilize all laboratory materials that are not damaged by high temperatures. How high a temperature and for how long depend on whether **moist** or **dry heat** is used. Moist heat means exposure to steam. Dry heat means heating in an oven or by direct exposure to a flame.

Moist heat is more effective than dry heat. Moist heat at 121°C (250°F) for 20 minutes reliably sterilizes most laboratory materials, but bulky materials and large volumes of liquid must be heated longer in order to heat up the entire mass. For example, it takes only 15 minutes to sterilize 10 ml of liquid in a test tube, but it takes over an hour to sterilize 6 L of liquid in a flask. Because 121°C is higher than the boiling point of water (100°C [212°F] at sea level), liquids must be heated in a pressurized chamber. Steam at 15 pounds per square inch is at a temperature of 121°C. But pockets of air within a pressurized chamber do not reach this temperature. To achieve

TABLE 3.3 The Uses of Microscopes

Type of Microscopy	Principles	Images	Uses in Microbiology
Light microscopy	Glass lenses refract light to form magnified images		
Brightfield	Contrast achieved by absorption	Image slightly darker than background; resolution limited to about 300 nm	Specimens usually must be stained; shows whole organism
Darkfield	Only light scattered by specimen enters microscope, producing high contrast	Brilliantly lit against a dark background	For viewing live cells or flagella that are too thin to be seen by phase contrast; reveals little internal detail
Phase contrast	Contrast achieved by interference; differences in refractive index shift phase of light from specimen	Clear detailed images surrounded by halos	Staining not required, so live cells can be viewed in wet mounts
Nomarsky	Uses the same principle as phase-contrast but splits interfering rays with prisms instead of rings	Image has a three-dimensional appearance	Particularly useful for specimens with a complex background, such as host tissues
Fluorescence	Depends on fluorescence to give off visible light when illuminated with ultraviolet light	Specimen is colored against a dark background	Can be used to view on type of microbe in a complex background using fluorescent probes
Confocal	Uses one microscope to illuminate and another to view a small field as specimen is scanned	With multiple scans, can produce three-dimensional images by computer manipulation	Useful for the results of multiple scans to produce three-dimensional images and multiple fluorescent scans
Electron	Uses a beam of electrons instead of light		
Transmission	Electromagnetic lenses form an image from beams of electrons that pass though specimen	Images reveal precise detail at high magnification	Used to reveal the internal ultrastructure of microbial cells and the morphology of viruses
Scanning	Scans specimen with a beam of electrons, ejecting secondary electrons that generate a signal, which in turn forms an image	Produces high-resolution images with remarkable depth of focus	Useful for viewing microbial surfaces
Scanned-proximity probe	Scans surface of specimen with sharp-pointed probe		
Scanning tunneling	Slightly charged metal probe draws electrons from electron-rich regions of specimen, creating a signal used to form an image	Produces images at the atomic level of conducting and semiconducting surfaces	Useful for examining surfaces of metals and silicone
Atomic force	Deflection of a diamond probe by attractive or repulsive intra-atomic atomic forces flexes its cantilevered support, creating a signal used to form an image	Produces images of molecules	Useful for examining shapes of molecules such as DNA

complete sterilization, such pockets must be eliminated by replacing all the air in the chamber by steam.

A pressurized container designed to sterilize materials with moist heat is called an **autoclave (Figure 3.20)**. An autoclave resembles a larger version of an ordinary home pressure cooker. In fact, some small laboratories use pressure cookers instead of autoclaves. Both are closed metal containers with walls strong enough to contain pressurized steam. Modern autoclaves automatically introduce pressurized steam so that it sweeps all air out of the chamber. They also automatically time the period of sterilization. But still the operator must load the

FIGURE 3.20 Schematic diagram of an autoclave fitted with automatic controls.

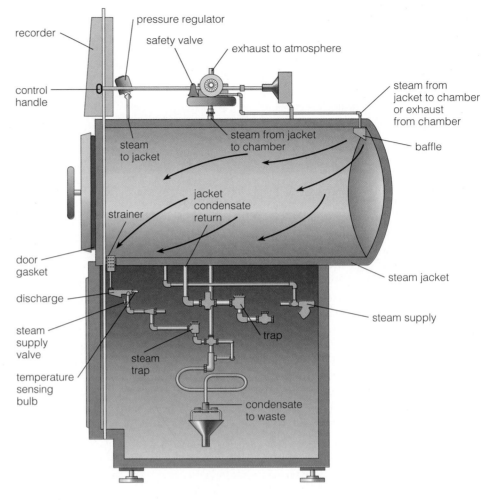

autoclave so that steam comes in contact with all objects inside.

Microbiological media and fabrics, such as towels, are sterilized in an autoclave. So is glassware such as pipettes and empty flasks or test tubes. Alternatively, glass and metal instruments may also be sterilized by dry heat in a hot-air oven. It must be heated to 170°C (338°F) for 90 minutes. Fabrics cannot be sterilized by dry heat because 170°C is hot enough to char them.

An open flame is also used to sterilize. Any time a flask or test tube is opened to add or remove materials, the neck is first passed through an open flame to kill any microorganisms that might have landed there. This procedure minimizes the chances of microbial cells falling into the vessel and contaminating its contents. Empty pipettes are also passed through a flame before use. The wire inoculating loops used to pick up microorganisms from one surface and move them to another are often sterilized by an open flame. They are placed in the flame of a Bunsen burner until the metal glows red. This method of sterilization is quick and convenient, but it can be dangerous when working with disease-causing microorganisms. Sudden heating in a flame can form an aerosol containing live microbial cells that might be inhaled by people in the laboratory. Using a small furnace to heat the loop avoids this hazard.

Filtration. Microbial cells can be removed from liquids or gases by filtration. Filtration is more time-consuming and expensive (a new filter must be used each time) than autoclaving. So only heat-sensitive liquids or solutions are filtered. Filtration does not remove most viruses, but the process is adequate for most routine laboratory purposes. The filters are called **membrane filters.** They are sheets of uniformly porous **nitrocellulose** (a chemically modified form of cellulose also used as smokeless gunpowder). Filters with pores approximately 0.45 μm in diameter are used because they are small enough to remove almost all cellular microorganisms. A liquid is filtered by pouring it onto a membrane filter fitted to a filter flask (**Figure 28.17**). Then a vacuum is applied to pull the liquid

through the filter, leaving microorganisms behind on the filter's surface. Solutions of vitamins, antibiotics, and other heat-sensitive compounds are usually filtered before they are added to a medium.

Chemicals. Most chemicals cannot be used to sterilize culture media because residues remain that are toxic to microorganisms. But one chemical, sodium hypochlorite (household bleach) is commonly added to cultures after experiments are completed. Such treatment is a safety measure used when dealing with dangerous microorganisms. Sodium hypochlorite quickly kills most microorganisms. It's also used to sterilize the surface of certain biological materials, such as seeds and plant tissues. Other chemicals are used to minimize contamination throughout the laboratory. Quaternary ammonium salts are used to disinfect laboratory surfaces such as benches and tables. In large institutions the toxic gas ethylene oxide is used in autoclave-like pressurized chambers to sterilize certain heat-sensitive solid materials, such as clothing and plastic containers.

LARGER FIELD

THE IMPORTANCE OF BEING PURE

All microbiologists know they must have pure cultures to do meaningful experiments. But even skilled microbiologists have been fooled. It happened to Ralph Wolfe, one of our country's most distinguished general microbiologists.

In 1960 Wolfe obtained a culture of the well-studied bacterium *Methanobacillus omelianskii. M. omelianskii* is a strict anaerobe that makes methane (natural gas). Wolfe wanted to research how *M. omelianskii* converts ethanol and carbon dioxide into methane gas. The procedure was clear. He had to lyse (burst) the bacterial cells

and isolate the enzymes that catalyze the conversion in a test tube. It sounded simple enough. But doing it had frustrated all previous researchers and, for about a year, Wolfe as well. Then Wolfe succeeded. One solution of enzymes made methane gas in a test tube! It was a spectacular breakthrough—until a colleague wondered if the culture were pure.

Wolfe set about repurifying the culture. But instead of adding ethanol and carbon dioxide to the culture, he added hydrogen gas in place of ethanol. Colonies appeared on the new medium, but when they

were transferred to a medium with ethanol and carbon dioxide, they did not grow. What had happened? A colleague, M. J. Wolin, realized what was going on. Although *M. omelianskii* formed uniform isolated colonies on a medium with ethanol and carbon dioxide, it was really a mixture of two different bacterial species. One, designated strain S, converted ethanol to hydrogen gas. The other, designated strain M, converted hydrogen gas and carbon dioxide to methane. Neither one alone could grow on ethanol and carbon dioxide. All Wolfe's experi-

ments had been done with a mixed culture. They had to be repeated to determine which enzymes came from strain S and which from strain M.

What lessons are to be learned from Wolfe's story? First, technical problems (such as obtaining a pure culture) that are so basic they are explained in an introductory textbook are very real and they plague even the most advanced research scientists. Second, as in this particular case, even a good microbiologist can be fooled. Pairs of species tend to grow together (to form a consortium) and appear to be a pure culture when they are not.

Isolation. With all the sterile media and equipment we need, now we can go on to isolate a pure culture. The principle is simple: (1) separate a single cell from all others and (2) provide it with the nutrients and environment it needs to grow. Microorganisms usually occur in huge numbers. To isolate a single cell, the population must be diluted (reduced). Dilution is usually done in one of three ways: by the streak plate, pour plate, or spread plate method.

The Streak Plate Method. The easiest and most commonly used method of diluting a microbial population is the **streak plate method (Figure 3.21)**. A population of microorganisms is picked up with a sterile wire inoculation loop. It's diluted by moving the loop back and forth on the surface of an agar-solidified medium in a **petri dish** (petri dishes are also called **plates**). As the loop is streaked back and forth, fewer and fewer microorganisms are deposited on the surface. Then the loop is sterilized in a flame and used to streak the plate again. This time the streaks overlap the first set of streaks so that some microbial cells are dragged onto a fresh, sterile region of the surface. Repeating this process several times suffi-

ciently dilutes the microbial population so that individual cells are deposited separately on the agar surface.

Then the plates are **incubated** (allowed to grow in a suitably warm place) until the individual cells have multiplied sufficiently to form **colonies** (visible masses of cells). Although each colony is probably a clone derived from a single cell, we can't be sure. Perhaps two cells were deposited close enough together on the plate to form a single mixed colony. Therefore to be certain that the culture is pure, a second streak plate is made starting with an isolated colony on the first streak plate. Isolated colonies that develop the second plate are almost surely pure cultures.

The Pour Plate and Spread Plate Methods. As we just saw, in the streak plate method dilutions are made right on the plate. In the pour plate and spread plate methods, dilutions are made before samples are put on the plate. These dilutions must be enormous. For example, it's not unusual to start with a suspension containing a billion (10^9) cells per milliliter. And for these two methods to work best, only about 100 (10^2) cells are put on a plate. That means the suspension has to be diluted about 10 million–fold (10^7). That's

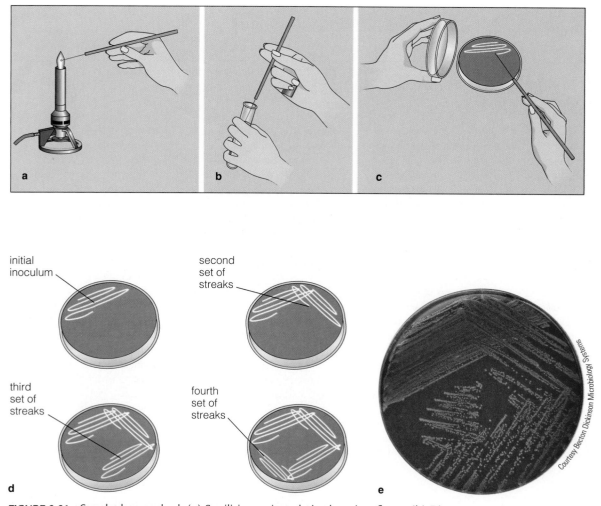

FIGURE 3.21   Streak plate method. (a) Sterilizing an inoculating loop in a flame. (b) Dipping the sterilized loop into a suspension of microbial cells. (c) Streaking the cells on solid medium of a petri dish. (d) Pattern of the first set of streaks and the overlapping second, third, and fourth streaks. (e) Streak plate after incubation. Some isolated colonies are present in the second, third, and fourth set of streaks.

comparable to adding a tablespoon of suspension to a swimming pool. The practical solution is to make **serial dilutions** (multistep dilutions). Usually, serial 10-fold or 100-fold dilutions are made. For 10-fold dilutions (**Figure 3.22**), 1 ml of cells is added to 9 ml of sterile culture medium or **saline** (salt) solution. Then the mixture is shaken thoroughly, and the process is repeated. After two such dilutions, the suspension has been diluted 100-fold ($10 \times 10 = 10^2$). After six, it has been diluted a million-fold (10^6). The dilutions are performed the same way for both the pour plate and spread plate methods. The two methods differ only in the way the diluted suspension is added to the plate.

In the pour plate method, the diluted sample is added to melted agar-containing medium, mixed, and poured into a petri dish. In the spread plate method, the diluted sample is poured onto the surface of an agar-solidified medium in a plate and spread evenly with a sterile glass rod. The liquid and dissolved materials are absorbed into the agar, leaving microbial cells on the surface. In either method the plates are incubated until individual colonies appear. Even pure cultures produce two kinds of colonies on a pour plate. The colonies that develop on the agar surface spread out and become larger than the colonies that are embedded in the agar. As with the streak plate method, there is no assurance that isolated colonies on pour or spread plates are pure cultures. The procedure has to be repeated.

The pour plate and spread plate methods usually produce more isolated colonies than the streak plate method does. Therefore they are preferred when isolating a minority strain from a mixed population.

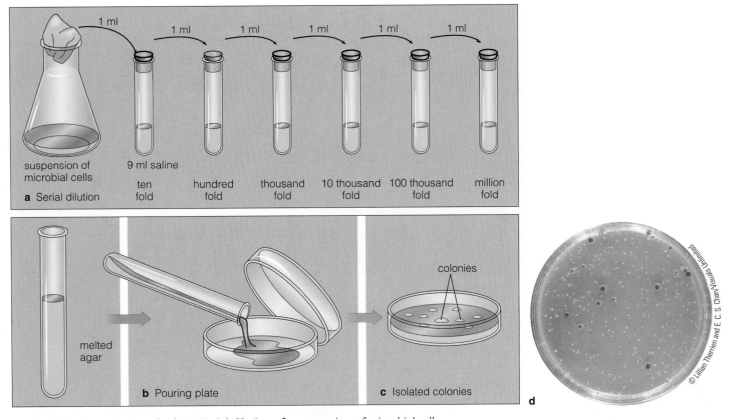

FIGURE 3.22 Pour plate method. (a) Serial dilution of a suspension of microbial cells. (b) The diluted sample is mixed with melted agar and poured into a plate. (c) After incubation, isolated colonies develop. (d) Poured plate after incubation. The small colonies are imbedded in the agar; the large ones are on its surface.

Growing a Pure Culture

If a culture is going to be used, it must be propagated (microbiologists usually say "grown"; so will we.). To grow microorganisms in the laboratory, the nutrients they need must be supplied in the culture medium. The medium can be liquid or solid (most solid media are liquid media that have been gelled by adding agar; see the Sharper Focus: Frau Hesse's Pantry in Chapter 1 to review agar's special properties).

Types of Culture Media. The medium a microbiologist uses depends on the microorganism and why it is being cultivated. Microorganisms have vastly different nutritional requirements. And the reasons for growing them are equally varied. In the following sections we'll discuss some of the kinds of media that microbiologists use.

Defined Media. A **defined medium** is prepared from pure chemicals, so its exact chemical composition is known. A defined medium that has just enough ingredients to support growth is called a **minimal medium.** The number of

ingredients that must be added to a minimal medium varies enormously depending on which microorganism is being grown. *Escherichia coli,* for example, needs relatively simple minimal medium (**Table 3.4**); it must contain only an organic source of carbon (for example, glucose) and a few inorganic salts. In contrast, some organisms, termed **fastidious,** need many organic ingredients. *Leuconostoc citrovorum* is an example of such a fastidious bacterium (**Table 3.5**). Defined media are important for genetic and other precise studies. But there are disadvantages to using them. Preparing a defined medium is time-consuming, and microorganisms grow relatively slowly on defined media. Defined media can be prepared only for microorganisms with known nutritional requirements.

Complex Media. A complex medium is made from extracts of natural materials, such as beef, blood, **casein** (milk protein), yeast, and soybeans. Such extracts contain large numbers of components, so their precise chemical composition is not known. A liquid complex medium is called a **broth.** Casein is a common component of complex media. Usually it is hydrolyzed with enzymes or acid to make it

TABLE 3.4 Ingredients of a Minimal Medium Suitable for Cultivating *Escherichia coli*

Ingredient	Amount	Comments
KH_2PO_4	13.6 g	Source of phosphate and buffers pH changes.
$(NH_4)_2SO_4$	2.0 g	Source of nitrogen and sulfur.
$CaCl_2$	0.01 g	Source of calcium; chloride is not required by bacteria.
$FeSO_4 \cdot 7H_2O$	0.0005 g	Source of iron.
$MgSO_4 \cdot 7H_2O$	0.02 g	Source of magnesium. Because this compound is relatively impure, it also serves as a source of trace elements.
Glucose	1.0 g	Source of carbon.
Distilled water	1000 mL	
Agar	15 g	Added if a solidified medium is desired.

Note: Medium is adjusted to pH 7.4 by adding NaOH.

TABLE 3.5 Ingredients of a Minimal Medium Suitable for Cultivating *Leuconostoc citrovorum*

Ingredient	Amount	Ingredient	Amount
Water	1 L	L-Lysine·HCl	250 mg
Energy Source		DL-Methionine	100 mg
Glucose	25 g	DL-Phenylalanine	100 mg
Nitrogen Source		L-Proline	100 mg
NH_4Cl	3 g	DL-Serine	50 mg
Minerals		DL-Threonine	200 mg
KH_2PO_4	600 mg	DL-Tryptophan	40 mg
K_2HPO_4	600 mg	L-Tyrosine	100 mg
$MgSO_4 \cdot 7H_2O$	200 mg	DL-Valine	250 mg
$FeSO_4 \cdot 7H_2O$	10 mg	Purines and Pyrimidines	
$MnSO_4 \cdot 4H_2O$	20 mg	Adenine sulfate·H_2O	10 mg
NaCl	10 mg	Guanine·HCL·$2H_2O$	10 mg
Organic Acid		Uracil	10 mg
Sodium acetate	20 g	Xanthine·HCl	10 mg
Amino Acids		Vitamins	
DL-Alanine	200 mg	Thiamine·HCl	0.5 mg
L-Arginine	242 mg	Pyridoxine·HCl	1.0 mg
L-Asparagine	400 mg	Pyridoxamine·HCl	0.3 mg
L-Aspartic acid	100 mg	Pyridoxal·HCl	0.3 mg
L-Cysteine	50 mg	Calcium pantothenate	0.5 mg
L-Glutamic acid	300 mg	Riboflavin	0.5 mg
Glycine	100 mg	Nicotinic acid	1.0 mg
L-Histidine-HCl	62 mg	p-Aminobenzoic acid	0.1 mg
DL-Isoleucine	250 mg	Biotin	0.001 mg
DL-Leucine	250 mg	Folic acid	0.01 mg

Source: H. E. Sauberlich and C. A. Baumann, "A Factor Required for the Growth of *Leuconostoc citrovorum*," *Journal of Biological Chemistry* 176 (1948): 166.

LARGER FIELD

AND THEY GREW HAPPILY EVER AFTER

Douglas Nelson, a general microbiologist working today, wanted to grow strains of the bacterium *Beggiatoa*. *Beggiatoa* are interesting for the way they obtain metabolic energy. They oxidize reduced sulfur compounds. Nelson knew these bacteria grow and multiply in the presence of hydrogen sulfide and oxygen, plus a few other nutrients. But he also knew these gases were incompatible. Hydrogen sulfide and oxygen react spontaneously. They exist together for only a short time. But Nelson had an idea. He would use an agar gel. Although an agar gel is quite firm, it is composed almost entirely of water, so dissolved substances diffuse through it readily. Why not let hydrogen sulfide diffuse to the bacteria from one direction and oxygen from another, meeting where *Beggiatoa* could use them? Nelson filled a test tube with agar and provided hydrogen sulfide at the bottom. He left the tube open to air at the top and inoculated it with *Beggiatoa*. His plan worked. Hydrogen sulfide diffused up from the bottom, oxygen from the air diffused down from the top, and *Beggiatoa* grew in a thin line where they met—about halfway down the tube.

more soluble and therefore nutritionally more readily available. Partial hydrolysis breaks proteins into peptides (Chapter 2). Complete hydrolysis breaks them down to amino acids. Partially hydrolyzed casein is called a **peptone.** Commercially available peptones include proteose peptone, tryptone, and tryptose. Completely hydrolyzed casein is called **casein hydrolysate.**

The various ingredients of complex media are commercially available as dried powders. Hundreds of mixtures already formulated into specific complex media are also available. One of these, **nutrient broth,** is probably the most often used complex medium (**Table 3.6**). When solidified with agar, it is called **nutrient agar.** In addition to being easy to prepare, complex media support rapid growth of most microorganisms.

Selective Media. Selective media favor the growth of particular microorganisms. They are used to isolate or detect the favored species in a complex mixture of other microorganisms. For example, a selective medium can be used to isolate *Salmonella typhi* (the bacterium that causes typhoid fever) from feces, which contain huge numbers of different microorganisms. Some selective media contain toxic chemicals, such as sodium azide, potassium tellurite, or crystal violet, that inhibit growth of some microorganisms but not others. For example, SPS agar contains sulfadiazine and polymyxin sulfate. It is used to isolate and identify *Clostridium botulinum* (a bacterium that causes lethal food poisoning) because it suppresses the growth of most *Clostridium* species, but not *C. botulinum.* Other selective

TABLE 3.6 Ingredients of Nutrient Broth, a Complex Medium Suitable for Cultivating Many Species of Bacteria

Ingredient	Amount	Comments
Peptone	5 g	Casein that has been partially hydrolyzed by the enzyme trypsin
Beef extract	3 g	Dried solids of a hot water extract of beef
NaCl	8 g	Added to keep cells from clumping
Distilled water	1000 ml	

Note: When nutrient broth is solidified by adding 15 g of agar per liter, it is called nutrient agar.

media employ an extreme pH value or an unusual carbon source to favor growth of a particular microorganism.

Differential Media. Differential media are used to identify microorganisms by the appearance of their colonies. For example, blood agar (an agar medium containing red blood cells) can be used to identify *Streptococcus pyogenes* (the bacterium that causes strep throat; Chapter 22). On this medium *S. pyogenes* colonies are surrounded by a clear zone because they **lyse** (destroy by bursting) nearby red blood cells (**Figure 3.23**). There are many kinds differential media. All exploit some ability of a particular organism to change the appearance of the medium.

Selective-Differential Media. Some media are both selective and differential. MacConkey agar is an example. It is used to detect strains of *Salmonella* and *Shigella* (intestinal bacteria that cause dysentery; Chapter 23). MacConkey agar is selective because it contains crystal violet and bile salts, which inhibit the growth of many bacteria, but not coliform bacteria and species of *Salmonella* and *Shigella*.

MacConkey agar is differential because it contains neutral red (a pH indicator) and lactose (milk sugar). Coliform bacteria make acid from lactose. The acid turns neutral red (which is colorless at the normal pH of the medium), so colonies of coliform bacteria become brick red. Colonies of *Salmonella* and *Shigella* are easily distinguished because their colonies are uncolored.

Enrichment Culture. Enrichment culture is used to isolate a particular microorganism or type of microorganism from a large, complex natural population. For example, endospore-forming bacteria can be isolated by boiling a sample of soil and culturing the survivors because only endospores survive boiling temperatures. Nitrogen-fixing bacteria can be isolated by culturing a soil inoculum in a nitrogen-free medium. Only nitrogen-fixing bacteria will grow because

they derive their nitrogen from the atmosphere. Environmental microbiologists often use enrichment cultures. For example, they might use one to find a microorganism that can break down a particular toxic chemical. To do so, they inoculate soil into a medium in which the toxic chemical is the only source of carbon. The microorganism that flourishes is the one that might be used to get rid of the toxic chemical (Chapter 29).

Providing a Suitable Environment. To grow a microorganism, we need a proper medium. But that's not enough. We must also provide a suitable environment: Temperature and pH must be maintained within tolerable ranges; oxygen must be provided or excluded.

Temperature. Different microbial species grow best over a particular range of temperature (Chapter 8). In general, most bacteria grow over a range of about 40°C (72°C). They grow most rapidly near the top of the range. To foster rapid growth, then, bacteria are cultivated in as warm an environment as they can tolerate, which usually reflects their natural environment. *Escherichia coli*, the bacterium used most commonly for microbiological experimentation (Chapter 5), exists naturally in the intestine of humans and other mammals. It grows best in the laboratory at 37°C, the temperature of the human body.

Cultures are maintained at a constant temperature in the laboratory in thermostatically controlled incubators or water baths. Incubators are air-filled chambers. They're used to grow cultures on solid or liquid media in any sort of container, including petri dishes, test tubes, or flasks. Liquid media in test tubes or flasks can be held in **water baths** (containers of warm water). Water baths are convenient because they can be kept on the laboratory bench, so cultures can be readily sampled.

pH. Optimal pH of microbial species varies, but any particular species can grow only within a relatively narrow range. Most bacteria grow best at pH values near neutrality—in the range of 6.5 to 7.5. Most fungi, on the other hand, grow better in a pH range between 4.5 and 6.0.

Although the pH of a medium may be ideal for a species when the culture is first inoculated, microbial growth often changes it. This happens because some microorganisms selectively use acidic or basic components from the medium. They can also produce acidic or basic materials as by-products of their growth. To minimize changes in pH value in defined media, buffers (Chapter 2) are usually added. Buffers are added less often to complex media because the natural materials they contain act as weak buffers. The most effective buffers at near-neutral values of pH are phosphate-containing salts and calcium carbonate.

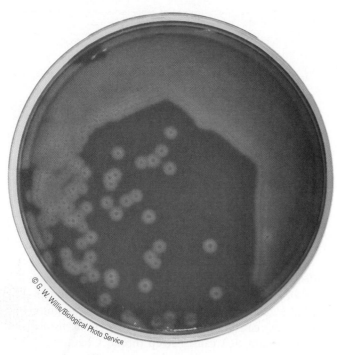

FIGURE 3.23   Differential medium. *Streptococcus pyogenes* is distinguished on this blood agar plate by clear zones of lysed blood cells around each colony.

SUMMARY

Viewing Microorganisms (pp. 50-68)

1. Microscopes allow microbiologists to see the organisms they study.

Properties of Light (pp. 50-53)

2. Light is the part of the spectrum of electromagnetic waves that is visible to the human eye.

3. Wavelength is the distance between two peaks or two troughs. The intensity of a light wave is the height of a wave.

4. If a ray of light hits an object and bounces back, it is reflected. If it passes through the object, it is transmitted. If it transfers some of its energy to the object, it is absorbed.

5. Refraction is bending that occurs when a ray of light enters at an angle an object with a different refractive index.

6. When light rays pass through a small opening or by the edge of an opaque object, they bend. This kind of bending is called diffraction.

Microscopy (pp. 53-55)

7. All microscopes depend on three factors to produce a clear image: magnification (enlargement of an image), contrast (differences in light intensity), and resolution (being able to distinguish two parts of an image that are close to each other).

8. The resolving power of a microscope can be increased by using a larger magnifying lens, by illuminating the specimen with a shorter wavelength, and by using a material (such as immersion oil) with a higher refractive index than air between the objective lens and the specimen.

9. Numerical aperture (NA) is a measurement of an objective lens's resolving power.

The Compound Light Microscope (pp. 55-56)

10. A light microscope uses visible light as a source of illumination. A compound light microscope has two lens systems: an objective lens and an ocular lens. Another series of lenses, called the condenser, directs light through the specimen. The objective lens forms an image of the specimen in the tube of the microscope. The ocular lens magnifies and projects the image to the eye.

11. Compound light microscopes typically are provided with three objective lenses: low power (10×), high power (40×), and oil immersion (100×).

12. An ordinary compound light microscope gives brightfield illumination. The background is more brightly lit than the specimen.

Wet Mounts (p. 56)

13. A simple wet mount is used to observe living microorganisms. A hanging drop mount keeps the microorganisms from drying out, allowing longer viewing.

Stains (pp. 56-60)

14. Stains are dyes used to increase contrast. Most stains are effective only after microorganisms are fixed—killed and attached to a microscope slide. Basic dyes are composed of positively charged ions, and acidic dyes are composed of negatively charged ions.

15. A mordant is a treatment that aids staining.

16. A simple stain uses only one dye. A differential stain involves at least three steps: primary staining, destaining, and counterstaining.

17. The Gram stain, a type of differential stain, distinguishes between Gram-positive and Gram-negative bacteria, reflecting differences in their outer surface.

18. The acid-fast stain—or Ziehl-Neelsen stain—is a differential stain that colors mycobacteria (and actinomycetes) red and all other cells blue.

19. The *Leifson* flagella stain uses stains and mordants to thicken flagella, threadlike appendages used for motility. Negative staining is used to reveal the protective capsule some bacteria have.

Light Microscopy: Other Ways to Achieve Contrast (pp. 60-63)

20. In phase-contrast microscopy, contrast comes from phase shift and interference. Because it can be used with living unstained cells, it can be used to study cellular movement. There is an unavoidable halo of light around the image.

21. Nomarsky microscopy produces contrast by interference, as phase contrast does, but gives a three-dimensional-like image.

22. A darkfield microscope views scattered light. The image is bright against a dark background.

23. Fluorescence microscopy increases contrast through fluorescence and can be used to identify, as well as observe, microorganisms.

Light Microscopy: Scanning Microscopes (p. 64)

24. Confocal microscopy uses two microscopes focused at the same point within the specimen. One microscope illuminates a small spot; the other views it. By scanning the specimen, composite images are formed.

Electron Microscopy (pp. 64-67)

25. The electron microscope provides much greater resolving power and therefore greater useful magnification than a light microscope.

26. The transmission electron microscope (TEM) uses a beam of electrons rather than visible light to form an image. It produces magnifications up to 300,000×.

27. The TEM requires special preparation of specimens, such as ultrathin sectioning or freeze-fracturing/freeze-etching. Sometimes contrast is increased through shadow-casting (coating with a heavy metal).

28. The scanning electron microscope (SEM) bombards the surface of the specimen with a beam of electrons to produce an electronic signal. It produces images with apparent three-dimensional depth.

29. SEM is usually used with intact cells, but freeze-fracturing and freeze-etching allow interior structures to be viewed.

Scanned-Proximity Probe Microscopes: Viewing Atoms and Molecules (p. 67)

30. Scanned-proximity probe microscopes form images with atomic dimensions by scanning the specimen with a sharp probe and using a property of the surface to generate an image-forming signal.

31. Scanning tunneling microscopy draws electron from the specimen to form a signal.

32. Atomic force microscopes generate images by sensing atomic attractions and repulsions.

Uses of Microscopy (pp. 67-68)

33. In general, light microscopes are for everyday use. They give valuable information about the size, shape, and general appearance of cells. But because resolution is limited, so is magnification.

34. Electron microscopes, which allow a view of the ultrastructure of cells, are used in research. Viruses and smaller objects, such as macromolecules, can only be seen with electron microscopes.

35. Scanned-proximity probe microscopes can form images at atomic scales.

Culturing Microorganisms (pp. 67-78)

36. Microorganisms are ideal laboratory subjects because billions can be studied in a single milliliter of culture, because they multiply rapidly, and because what we learn can often be generalized to cell systems, plants, and animals (including humans).

Obtaining a Pure Culture (pp. 68-72)

37. A pure culture consists of a single type of microorganism. Mixed cultures occur in nature.

Sterilization (pp. 68-70)

38. The first step in obtaining a pure culture is sterilization—eliminating all microorganisms from an area. All instruments and materials used in obtaining a pure culture must be sterilized.

39. Moist heat kills faster than dry heat, but the temperature required is higher than the boiling point of water; an autoclave is usually used.

40. Dry heat sterilization requires temperatures of at least 171°F for a minimum of 1 hour. Direct flaming is another form of dry-heat sterilization.

41. Filtration is passing a fluid through a barrier with holes too small for cellular microorganisms. Because viruses pass through filters, filtration does not remove them.

42. Chemical sterilization is used mostly to destroy cultures after an experiment and to minimize contamination in the laboratory.

Isolation (pp. 71-72)

43. Isolation is introducing a single cell of a microorganism into the sterilized medium. A population of cells descended from a single cell constitutes a clone. Clones grown on a solid medium large enough to be visible are called a colony.

44. Before a single cell can be isolated, the cells in a culture must be diluted.

45. In the streak plate method, an inoculation loop is dipped into a mixed culture and streaked across a solid medium in a petri dish. The process is repeated until individual cells are isolated. The cells are incubated until colonies form. Then the entire process is repeated.

46. In the pour plate and spread plate methods, suspensions of microbial cells are diluted before they are put on the plate. In the pour plate method, the diluted sample is added to melted agar, mixed, and poured into a petri dish. In the spread plate method, the diluted sample is poured onto the surface of an agar plate and spread evenly with a glass rod. Plates are incubated until colonies appear. Then the process is repeated.

Growing a Pure Culture (pp. 73-77)

47. A defined medium is prepared from pure chemicals. Fastidious organisms require a very complex chemical medium. Defined media are time-consuming to prepare, and bacteria usually grow on them more slowly than on complex media.

48. Complex media are made from extracts of natural materials such as beef, blood, or casein. A liquid complex medium is called a broth. Complex media are easy to prepare and promote rapid growth.

49. Selective media favor the growth of certain microorganisms. They are used to isolate a particular species from a complex mixture.

50. Differential media are used to identify colonies of a particular type of microorganism.

51. An enrichment culture is used to isolate from a natural population a microorganism with a particular capacity.

Providing a Suitable Environment (pp. 76-77)

52. In general, microorganisms grow more rapidly at warmer temperatures. To minimize pH changes, buffers are usually added to media.

53. Strict aerobes require oxygen. Strict anaerobes cannot grow in the presence of oxygen. Facultative anaerobes use oxygen if it is available but can do without it. Aerotolerant anaerobes cannot use oxygen but it does not adversely affect them. Microaerophiles need low concentrations of oxygen. Providing, restricting, or excluding oxygen from the environment all present technical problems.

Preserving a Pure Culture (pp. 77-78)

54. Cultures that are preserved for study and reference are called stock cultures.

55. A culture can be preserved by desiccation (removing all water). Usually a culture is desiccated by lyophilization, or freeze-drying. Cultures can also be preserved by low-temperature storage. After freezing, the culture is stored below −50°C, either in liquid nitrogen or in an ultra–deep-freeze.

Microorganisms That Cannot Be Cultivated in the Laboratory (p. 78)

56. Many, probably most, microorganisms cannot be cultivated in laboratory media. These include the bacteria that cause syphilis and Hansen's disease (leprosy).

57. The rickettsia and chlamydia bacterial families and all viruses can be cultivated only in living host cells. Intact animals or tissue cultures are used.

REVIEW QUESTIONS

Viewing Microorganisms

1. Why is understanding the properties of light important to the study of microorganisms? What is light?

2. What are the three possible fates of a ray of light that strikes an object head-on?

3. What are reflection, diffraction, refraction, and refractive index?

4. What are magnification, contrast, and resolution?

5. What is the resolving power of a microscope? How can resolving power be increased? What is the numerical aperture (NA) of a microscope lens?

6. Where do the objective lens, ocular lens, condenser, and body tube fit on a light microscope?

7. What is a corrected lens? What is brightfield illumination?

8. How you would prepare a wet mount and a hanging drop mount? What are the advantages and disadvantages of this type of viewing?

9. What are the advantages and disadvantages to staining a sample?

10. What do these terms mean? basic dye, acidic dye, mordant

11. What are the differences between these types of stains?
    a. Simple stain   c. Gram stain
    b. Differential   d. Acid-fast stain
       stain   (Ziehl-Neelsen
                             stain)

12. What is a special stain? What is negative staining used for?

13. What are the underlying principle, type of image produced, advantages, and disadvantages of these forms of microscopy?
    a. Brightfield   d. Fluorescence
       illumination   microscopy
    b. Phase-contrast   e. Nomarsky
       microscopy   microscopy
    c. Darkfield
       microscopy

14. Answer question 13 for the two types of electron microscopy. How are ultrathin sectioning, freeze-fracturing, freeze-etching, and freeze-drying done?

15. How does a scanning tunneling microscope form an image? What sorts of specimens can it examine?

16. How does an atomic force microscope work?

17. What, in general, are the uses of light and electron microscopy?

Culturing Microorganisms

18. What is a pure culture? What are the two basic steps in obtaining a pure culture?

19. Define sterilization. What are the different uses of heat sterilization and chemical sterilization in the laboratory? What are the advantages and disadvantages of moist heat and dry heat? What special precautions must be taken to ensure sterilization in an autoclave?

20. Does filtration sterilize? Explain. When is filtration used in the laboratory?

21. Define these terms: inoculate, clone, colony, incubate, dilute, serial dilutions.

22. Describe how you would carry out these methods to isolate a microorganism for culturing: streak plate method, pour plate method, spread plate method.

23. What are the similarities and differences among these types of media:
    a. Defined   d. Differential
       medium   medium
    b. Complex   e. Selective-
       medium   differential
                         medium
    c. Selective   f. Enrichment
       medium   culture

24. What special considerations must be given to temperature, pH, and oxygen in cultivating a microorganism?

25. Define these types of microorganisms:
    a. Strict   d. Aerotolerant
       aerobes   anaerobes
    b. Strict   e. Microaerophiles
       anaerobes
    c. Facultative
       anaerobes

26. How would you preserve a culture by desiccation and by low-temperature storage?

27. Are there microorganisms that cannot be cultivated in the laboratory? Explain and give examples.

CORRELATION QUESTIONS

1. If microbial cells were suspended in saline instead of water in preparation for viewing with a phase-contrast microscope, would it increase or decrease contrast? Why?

2. Is it more important to have a high numerical aperture for a high-power or a low-power objective lens? Why?

3. We have discussed several ways to obtain a pure culture from a mixed one. There are many other possible ways to do the same thing. Can you think of one? Explain what you would do.

4. You have two microorganisms that grow well in nutrient agar. Does that mean they would both grow well in the same minimal medium? Explain.

5. What's the fundamental difference between a differential medium and a selective medium? Explain.

6. If you wanted to know how many microbial cells were in a sample of soil, would it be better to use microscopy or cultivation? Explain.

ESSAY QUESTIONS

1. If you could have only one kind of microscope, which one would you choose? Discuss the reasons for your decision.

2. Discuss how you would develop a selective medium for a particular microorganism.

SUGGESTED READINGS

Gephardt, P. (ed.). 1994. *Methods of general microbiology.* Washington, D.C.: ASM Press.

Murray, P. R. (ed.). 1999. *Manual of clinical microbiology.* 7th ed. Washington, D.C.: ASM Press.

Morris, V. J., A. P. Gunning, and A. R. Kirby. 1999. *Atomic force microscopy for biologists.* London: Imperial College Press.

Gerhardt, P. (ed.). 1994. *Methods of general microbiology.* Washington, D.C.: ASM Press.

Murray, P. R. (ed.). 1995. *Manual of clinical microbiology.* 6th ed. Washington, D.C.: ASM Press.

Block, S. S. 1991. *Disinfection, sterilization, and preservation.* Philadelphia: Lea and Febiger.

James, J., and Tanke, H. J. 1991. *Biomedical light microscopy.* Boston: Kluwer.

For additional readings, go to InfoTrac College Edition, your online research library at: http://www.infotrac.thomsonlearning.com

FOUR

Prokaryotic and Eukaryotic Cells: Structure and Function

CHAPTER OUTLINE

LEARNING GOALS

To understand:

- *The principal differences between prokaryotic and eukaryotic cells*
- *The structural differences among Gram-positive bacteria, Gram-negative bacteria, and mycoplasmas*

- *The structural and chemical differences between bacteria and archaea*
- *The structure of eukaryotic cells and the functions of their organelles*

- *How molecules cross cell membranes: simple diffusion, osmosis, facilitated diffusion, active transport, group translocation, and engulfment*

An Infectious Examination

C. S. H. was a 25-year-old medical student completing her first clinical clerkship on the internal medicine ward of a large county hospital. She was asked to take a medical history and do a physical examination on a patient who was entering the hospital to be treated for advanced colon cancer. Her patient was a pleasant, cooperative man who appeared ill and older than his stated age of 65 years. He told C. S. H. that he had a long history of alcohol abuse and had developed a worsening cough over the past several months.

While performing the physical examination, C. S. H noticed that her patient coughed deeply and frequently, producing thick sputum. Although he had no fever, C. S. H. suspected that he had a lung infection. She collected a sample of his sputum for microbiological examination. She put some of the sputum sample on a slide and sent the rest to the microbiology laboratory to be cultured. C. S. H. took the slide to a tiny lab on the medical ward, where she fixed it, applied a Ziehl-Neelsen stain (Chapter 3), and examined it under the microscope. When she saw numerous bacilli that were stained bright red (proving they were acid-fast organisms), C. S. H. knew that her patient had active tuberculosis.

Following C. S. H's probable diagnosis, her patient was moved to a room where respiratory isolation precautions were imposed. He was started on a regimen of multiple drugs to treat his disease. A chest x-ray showed abnormalities typical of active tuberculosis. Therefore 6 weeks later C. S. H. was not surprised when the microbiology laboratory reported that *Mycobacterium tuberculosis* had grown in the medium inoculated with the sample of her patient's sputum. Fortunately it was not resistant to any of the antibiotics usually used to treat tuberculosis. Soon, however, C. S. H. became a patient herself.

Like all hospital employees, C. S. H. took annual skin tests for tuberculosis. All the results of her previous tests had been negative. The result of her next skin test, however, was not. An **induration** (firmness) in excess of 15 mm in diameter developed around the site of the skin test site on her forearm. She had no symptoms of illness, and her chest x-ray was normal. This meant that although she had been infected by *M. tuberculosis*, she had not yet developed the disease tuberculosis. But the chances of serious disease developing later in life were significant. To prevent this, her physician recommended that she take the antituberculous drug isoniazid (INH) every day for the next 9 months. After several months of therapy, she noticed tingling in her feet, a symptom of

neuropathy (nerve damage), which can be a side effect of INH treatment. Patients whose diets are poor, as C. S. H.'s probably was during her exhausting time as a medical student, are particularly vulnerable to this complication. C. S. H then took the B vitamin pyridoxine along with her INH. Her symptoms improved. She faithfully finished her 9 months of treatment, finished her medical training, and remains well as a practicing physician 25 years later.

Case Connections

- C. S. H.'s experience relates to the unusual waxy layer of the bacterium *Mycobacterium tuberculosis* discussed later in this chapter.
- The waxy layer traps and holds stain, causing *M. tuberculosis* to be acid fast, as discussed in Chapter 3.
- The waxy layer impedes the entry of nutrients into the cell, which accounts for the slow growth *M. tuberculosis* and therefore C. S. H.'s having to wait 6 weeks for a report from the microbiology laboratory.
- Isoniazid, the drug that cured C. S. H., is uniquely effective against *M. tuberculosis*, as discussed in Chapter 21, because it inhibits synthesis of mycolic acids, which are components of *M. tuberculosis*'s waxy layer (discussed in this chapter).

STRUCTURE AND FUNCTION

To understand the structure of living things, we must think of function, because natural selection acts by changing structure to improve function. In the course of evolution, useful structures persist. Less useful ones improve. Useless ones are eventually lost. This linkage of structure and function operates at every level, including the cell itself.

All organisms except viruses are cellular. (We'll discuss viruses in Chapter 13.) In spite of the enormous diversity

SHARPER FOCUS

A MATCHED TEAM

We can usually figure out how a machine works just by looking at it—but not so with a microbial cell. The primary activities of a microbial cell are chemical, not mechanical. So knowing how it works means finding out what its molecules do and how they are arranged. Success in this endeavor often leads to new insights about the entire cell—and sometimes to important medical advances.

A major advance in understanding microbial structure was by Hiroshi Nikaido, a microbiologist at the University of California, Berkeley. Nikaido discovered how **mycolic acids** (long-chain fatty acids with 70 or more carbon atoms)

are arranged to form a membrane surrounding *Mycobacterium tuberculosis,* the bacterium that causes tuberculosis. Nikaido's discovery explains some of the unusual properties of *M. tuberculosis.*

Only *M. tuberculosis* and closely related bacteria have mycolic acids. These unusual fatty acids have long been assumed to form an amorphous waxy layer around the *M. tuberculosis* cell, giving it some of its unusual properties— slow growth rate, acid-fast staining, resistance to many antibacterial drugs, and unique sensitivity to others. A waxy layer would account for most of these properties by impeding passage of nutrients, dyes, and antibacterial drugs.

Still, some nutrients and antibacterial compounds do enter the cell. How can they pass through the outer waxy layer of mycolic acids?

Using x-ray diffraction (a procedure that reveals molecular structure), Nikaido discovered that the mycolic acid molecules are arranged in two layers with their hydrophobic tails directed toward the space between them. He found that the mycolic acids form a highly ordered membrane, not just a disorganized waxy layer. The membrane surrounds the cell, but proteins are embedded in the layer. They form water-filled pores through which nutrients and certain drugs pass slowly.

Nikaido's discovery explains a good deal about the unusual properties of acid-fast bacteria. It also illustrates an important principle: Cellular structure and function are always related. Like a matched team of horses, one depends on the other.

Hiroshi Nikaido

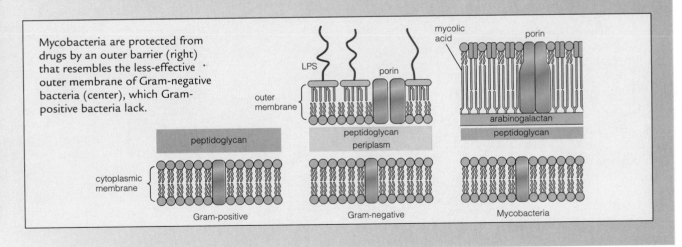

Mycobacteria are protected from drugs by an outer barrier (right) that resembles the less-effective outer membrane of Gram-negative bacteria (center), which Gram-positive bacteria lack.

of living things, in all biology there are only two general types of cells—prokaryotic and eukaryotic. Bacteria and archaea are prokaryotic cells. All other organism—plants, animals, fungi, algae, and protozoa—are made of eukaryotic cells. In this chapter we'll look at the structure of these two types of cells. First we'll consider the simpler structure of prokaryotic cells. Then we'll compare their structure with the more complicated structure of eukaryotic cells. We'll consider the functions of these structures here and look at them in greater detail in the next two chapters.

THE PROKARYOTIC CELL

Viewed under the electron microscope, prokaryotic cells have a grainy but fairly uniform interior (**Figure 4.1**). Eukaryotic cells, on the other hand, are much more complex in appearance (**Figure 4.2**). They contain many internal membranes and membrane-bound structures called **or-** **ganelles** (little organs). Prokaryotic cells do contain a few internal structures called organelles, but they are quite different than the organelles in eukaryotic cells: Prokaryotic organelles are not bounded by lipid membranes. For example, the DNA in a prokaryotic cell occupies an irregularly shaped region called a **nucleoid.** In contrast, the DNA of a eukaryotic cell is located in a well-defined, membrane-bound nucleus. (Recall that *prokaryote* in Greek means "before a nucleus" and *eukaryote* means "true nucleus.")

The most obvious difference between prokaryotic and eukaryotic cells, then, is structural. But there are important chemical differences as well. Virtually all prokaryotic cells (with the exception of some archaea) have a cell wall made of a macromolecule called peptidoglycan (a part polysaccharide and part protein molecule that we'll discuss later in this chapter). No eukaryotic cell contains peptidoglycan.

A third important difference between prokaryotes and eukaryotes is size. Most prokaryotic cells are from slightly less than 1 to several micrometers (μm) across, about the size of organelles within eukaryotic cells. Whole eukaryotic cells are about 10 times larger. Being small may seem trivial, but it is not. The consequences are profound. Size affects the rate at which nutrients can enter and the rate at which they are distributed within the cell. As a result, size affects the rate at which a cell can grow.

Small size permits faster growth rate because a small cell has a large surface area compared with the volume of its contents. Surface area determines how fast nutrients

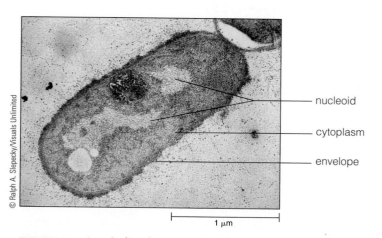

© Ralph A. Slepecky/Visuals Unlimited

— nucleoid

— cytoplasm

— envelope

1 μm

FIGURE 4.1 A typical prokaryotic cell (the bacterium *Bacillus subtilis*). Electron micrograph showing grainy cytoplasm containing an irregular-shaped nucleoid not surrounded by a membrane. The nucleoid appears to consist of two parts because this slice of the cell cut through two lobes of the single nucleoid.

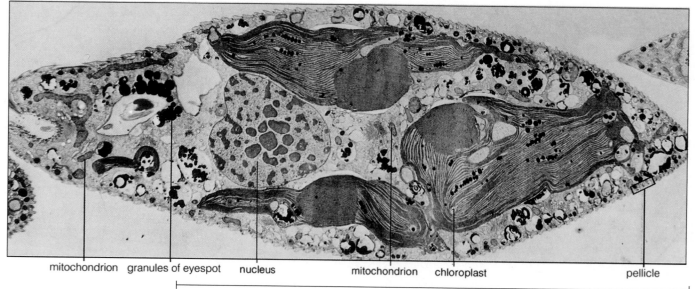

mitochondrion   granules of eyespot   nucleus   mitochondrion   chloroplast   pellicle

10 μm

FIGURE 4.2 A typical eukaryotic cell (the alga *Euglena*). Electron micrograph showing cytoplasm packed with membrane-bound organelles. Compare the well-defined, membrane-bound nucleus with the more diffuse prokaryotic nucleoid and the scale bar, which shows the eukaryotic cell to be more than ten times larger than the typical prokaryotic cell.

TABLE 4.1 Comparison of a Typical Bacterial Cell and a Typical Human Cell

	Bacterial Cell	Human Cell	Comparison
Diameter	1 μm	10 μm	Bacterium is 10 times smaller
Surface area	3.1 μm²	1257 μm²	Bacterium is 405 times smaller
Volume	0.52 μm³	4190 μm³	Bacterium is 8057 times smaller
Surface-to-volume ratio	6	0.3	Bacterium is 20 times greater

can enter the cell from the environment: The larger the surface area, the more rapidly nutrients can enter. Volume determines the cell's need for nutrients: The larger the volume, the more nutrients are needed. Small cells have a higher **surface-to-volume ratio** (relatively more surface area for the same volume) than large cells. The surface-to-volume ratio of a typical bacterial cell is about 20 times greater than that of a typical human cell (**Table 4.1**). Thus prokaryotic cells meet their needs for nutrients quickly and grow rapidly. For example, under ideal conditions, the bacterium *Escherichia coli* can double its size and divide about every 20 minutes—much faster than any eukaryotic cell.

Now we'll discuss the structure of prokaryotic cells, first the bacterial cell and then the archaeon cell.

Structure of the Bacterial Cell

All bacterial cells have an outer covering called the **envelope,** and some species have additional structures, **capsules** and **appendages,** that extend beyond the envelope. The cytoplasm lies within the envelope. It contains the cell's DNA and all its metabolic machinery. We'll examine the structure of a typical bacterial cell starting with the envelope. Then we'll consider the capsule and appendages that are attached to it. And finally we'll examine the cytoplasm and its contents (**Figure 4.3**).

The Envelope. The bacterial envelope can have as many as three layers. Moving from the outside toward the inside of the envelope we encounter an **outer membrane,** a **cell wall,** and a **cytoplasmic membrane** (also called the inner membrane or the plasma membrane). The three major groups of bacteria (Gram-positive, Gram-negative, and mycoplasmas) differ with respect to the number of layers that comprise their envelopes.

The **Gram-negative bacteria** contain all three layers. Because they have two membranes (outer and cytoplasmic), they have an extra cellular compartment: the one that lies between the two membranes. This compartment is

called the **periplasm.** The **Gram-positive bacteria** lack the outer membrane. The **mycoplasmas** lack the outer membrane and the cell wall as well. **Figure 4.4** shows the envelopes of these three groups of bacteria.

The outermost covering of the **acid-fast bacteria,** which includes the human pathogens *Mycobacterium tuberculosis* (Chapter 22) and *M. leprae* (Chapter 26), is a highly unusual waxy layer (see "A Matched Team").

The Outer Membrane. Like all lipid membranes, the outer membrane of Gram-negative bacteria is a bilayer (two attached layers of molecules; **Figure 4.4a**). The inner layer is composed of phospholipid, and the outer layer is composed of mainly **lipopolysaccharide** (**LPS**); the outer layer also contains small amount of phospholipid to which some components of capsule are attached. LPS is an unusual compound found only in the outer membrane of Gram-negative bacteria. As its name implies, lipopolysaccharide is a combination molecule with a lipid moiety on one end and polysaccharide on the other. The lipid

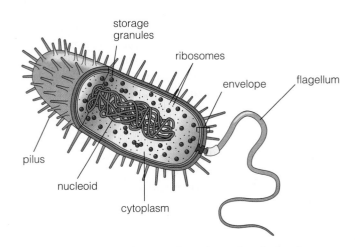

FIGURE 4.3 Structural parts of a prokaryotic cell, showing appendages (pilus and flagellum) exterior to the envelope and the cytoplasm and its contents within. The capsule that surrounds some prokaryotes is not shown.

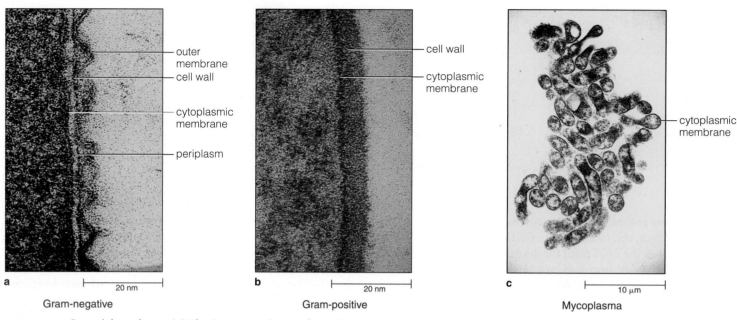

FIGURE 4.4 Bacterial envelopes. (a) The Gram-negative envelope. The wavy appearance of the outer membrane is an artifact. (b) The Gram-positive envelope. (c) Mycoplasma cells. None of these cells has a capsule.

positive

end is **lipid A,** which, like all lipids, is hydrophobic (water-fearing). The polysaccharide tail is hydrophilic (water-loving). Thus lipopolysaccharide has a hydrophilic head and a hydrophobic tail. In this respect LPS is like phospholipid (Chapter 2). It, too, has the capacity to form membranes. But membranes made from LPS have unique properties. They are barriers to polar molecules, as well as nonpolar molecules. As a result, only water and a few gases can cross the lipid part of the outer membrane.

All the other molecules that a Gram-negative cell needs pass through **pores** (small holes) in the outer membrane (**Figure 4.5**). These pores are formed by proteins, called **porins,** that span the membrane. Small molecules (less than about 600 Daltons) diffuse through these pores. Surprisingly, some pores are selective. They let only certain molecules pass through. For example, one porin in *E. coli* allows vitamin B_{12} to enter. Another allows the disaccharide maltose to enter.

The outer membrane is anchored to the rest of the envelope by molecules of **lipoprotein.** These molecules are bonded to the cell wall at one end. The other end carries a lipid that inserts into the inner surface of the outer membrane (Figure 4.5).

The most important function of the outer membrane is protection. And it does a good job. Because they have an outer membrane, Gram-negative bacteria are generally more resistant than other bacteria to toxic substances in the environment, including antibiotics because they can't get through the outer membrane. For example, the antibiotic

rifampin is equally effective against Gram-negative and Gram-positive bacteria if it reaches its target enzyme (RNA polymerase) in the cytoplasm. But because of their outer membrane, intact Gram-negative cells are 1000 times more resistant than Gram-negative cells to rifampin.

The LPS component of the outer membrane is also known as **endotoxin** because it is harmful to humans and other animals. Most of the toxicity of LPS is due to one of its components, lipid A, as we'll discuss in Chapter 15.

Periplasm. The periplasm lies just inside the outer membrane of Gram-negative bacteria. In older microbiological literature, the periplasm was called the **periplasmic space** because electron micrographs showed it as an empty region between the outer membrane and the cytoplasmic membrane (Figure 4.4a), with only the cell's peptidoglycan wall within it. But the periplasm is not an empty space. It's an important organelle of Gram-negative bacteria that's filled with gelatinous material, including two types of proteins. One type is enzymes. Most of them break down certain nutrients so they can pass through the cytoplasmic membrane. The second type is **binding proteins.** They bind to certain nutrients and thereby facilitate their passage across both the outer membrane and the cytoplasmic membrane.

The Cell Wall. All bacteria except mycoplasmas (Chapter 11) have strong, rigid walls made of peptidoglycan. In Gram-negative bacteria, this wall lies within the periplasm. In Gram-positives, it lies just outside the cytoplasmic mem-

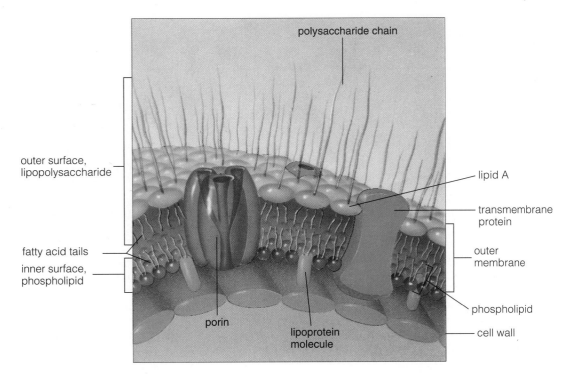

FIGURE 4.5   Outer membrane.

(Art by Carlyn Iverson.)

brane. Now we'll consider both the structure of the bacterial cell wall and its major functions: (1) giving the cell its shape and (2) withstanding turgor pressure. We'll also consider how some bacteria—those that have no walls (mycoplasmas) and those that have lost their walls (L forms)—manage without them.

Structure. Bacterial walls are made of peptidoglycan. Peptidoglycan is a general term, like protein or polysaccharide. It simply refers to a molecule that is part protein (peptido-) and part polysaccharide (-glycan). The specific kind of peptidoglycan that occurs in bacterial walls is **murein,** but it is usually just called peptidoglycan. The polysaccharide (glycan) part of peptidoglycan consists of long chains of alternating units of the two sugars *N*-acetylglucosamine (NAG) and *N*-acetylmuramic acid (NAM). These chains encircle the cell rather like hoops on a barrel. They are held together (cross-linked) by the protein part of the molecule: peptides (short proteins) that connected to NAM units on adjacent glycan strands (**Figure 4.6**). In all bacteria the polysaccharide component consists of alternating units NAG and NAM. The peptide component varies among bacterial species, but the most common type is four amino acids long. Two of the amino acids are L-isomers (the form normally found in proteins). The other two are D-isomers (a form that is rare in nature; Chapter 2). When linked together, the glycan and peptide

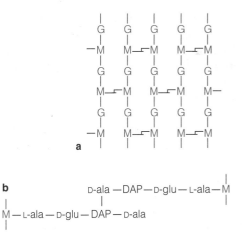

FIGURE 4.6   Peptidoglycan. (a) The meshlike structure of the molecule showing how long glycan strands (red) composed of alternating molecules of NAM (M) and NAG (G) are held together by short peptide bridges (black) that join NAM molecules. (b) Structure of the peptide bridges that occur in *Escherichia coli*. Two four-amino-acid strands, composed of D-alanine (D-ala), diaminopimelic acid (DAP), D-glutamate (D-glu), and L-alanine (L-ala), are linked between the bifunctional DAP of one strand and the L-ala of the other.

chains create a single rigid meshlike molecule, similar to a coat of mail, that encircles the entire cell.

The cell walls of Gram-negative and Gram-positive bacteria are slightly different. A Gram-negative cell wall

SHARPER FOCUS

BE PREPARED

Pore size in the outer membrane is a compromise between letting nutrients in and keeping toxic substances out. Both nutrient and toxin molecules come in all sizes. Prokaryotes have mechanisms for breaking down nutrient molecules so they can enter through smaller pores. However, small molecules diffuse more rapidly through a large pore than through a smaller one and more rapidly if the concentration gradient across the membrane is high.

It is advantageous for a microorganism to have pores as large as possible if it lives in an environment with low concentrations of nutrients. But if there are toxic compounds in the environment, it is better to have smaller pores. Thus bacteria that live in lakes or streams with low nutrient concentrations and few toxic compounds have larger pores than bacteria that live with concentrated nutrients and many toxic substances.

The bacterium *Escherichia coli* encounters both kinds of environment. Normally *E. coli* is a harmless inhabitant of the intestinal tract, where there are concentrated nutrients and toxic compounds. When *E. coli* is shed in feces, however, it usually finds itself in an aqueous environment, poor in nutrients and toxic compounds. But *E. coli* can adapt to either environment because it can make two different porins. One (OmpC) forms small pores, and the other (OmpF) forms larger pores. Warm temperature and moderate osmotic strength (conditions in the intestinal tract) signal it to make OmpC. Cooler temperatures and lower osmotic strength (conditions typical of aqueous environments) signal it to make OmpF. Thus *E. coli* can adapt to either of two quite different environments.

---

consists of a peptidoglycan mesh that is only one layer thick. In contrast, a Gram-positive wall is many layers thick. Gram-positive walls also contain a second component, **teichoic acid**—glycerol or ribitol (a five-carbon polyalcohol) units linked by phosphate groups. Teichoic acids penetrate the multilayered wall, adding to the structural integrity of Gram-positive cell walls.

Cell Shape. The peptidoglycan cell wall confers shape on a bacterial cell wall. A simple experiment shows clearly that this is true. Exposing a rod-shaped bacterial to **lysozyme** (an enzyme that destroys peptidoglycan) converts it into a sphere (the shape of wall-less objects such as soap bubbles). The sphere must be protected by being suspended in a medium with osmotic strength that matches the cell's interior or it will burst.

Bacteria come in many shapes. Some are square, some are star-shaped, and some have irregular shapes. But most are either spherical (called **cocci**, *sing.,* **coccus**), rod-shaped (called **bacilli**, *sing.,* **bacillus**), or spiral-shaped, (called **spirilla**, *sing.,* **spirillum**). There are also intermediate shapes. The most common are a short bacillus (called a **coccobacillus**) and a short, coma-shaped spirillum (called a **vibrio**). (The term "bacillus" is somewhat confusing. If lowercase, it refers to a cell's shape. If capitalized, it refers to the genus *Bacillus*.)

Bacterial cells of some species stick together after division, forming distinctive clusters. Cocci can form three patterns, depending on whether they divide in one, two, or more planes. Bacilli always divide in just one plane; if they don't separate, they form a simple pattern of rods joined end to end. Spiral bacteria usually remain as single microorganisms. Their variability comes from having more or fewer corkscrew turns. **Figure 4.7** shows types and arrangements of bacterial cells.

Turgor Pressure. The lysozyme experiment just described shows that the shape of the peptidoglycan wall determines the shape of a bacterial cell. It also demonstrates the wall's second major function—containing the cell's turgor. **Turgor** (the osmotic pressure within the cell) of a bacterial cell is high because its contents are concentrated. That's why an unprotected lysozyme-treated cell bursts. The turgor that the bacterial wall must withstand can be huge. Some micrococci, for example, have turgor pressures of 350 pounds per square inch—enough to rupture a steel water tank.

Now let's consider the problem of how a bacterial cell that is completely enclosed within a rigid peptidoglycan wall can grow. To get bigger, a bacterial cell has to enlarge its peptidoglycan shell. To do this, bacteria produce enzymes, called **autolysins,** that break cross-linking bonds in peptidoglycan and others, called **transpeptidases,** that reseal the breaks by adding new peptidoglycan monomers. This enlarges the peptidoglycan sac and increases cell size. Of course, the process must be carefully regulated. If autolysins were too active, the cell would burst.

SHARPER FOCUS

NO ONE KNOWS FOR SURE, BUT A LONG TIME

How long can an endospore survive in its state of suspended animation? That's a tough question to answer. Many experiments have shown that more than half of a population of endospores will survive a year's storage at 10°C (50°F). Dried plant samples wrapped and stored without opening for more than 100 years in botanical museums have been found to contain viable endospores.

So most microbiologists are willing to say that endospores can last at least 100 years. Finding viable endospores on older, unwrapped samples is problematic. It is always possible for endospores to have entered during storage or for endospores to have germinated and produced new endospores. This is the problem with the claim that viable endospores have been recovered from Egyptian tombs that are

more than 7000 years old. How do we know that these endospores are first generation? We don't. More convincing is a claim for 2000-year-old endospores. In 1976 the Roman fort Vindolanda, dated A.D. 90–95, had been drained and was being excavated. Viable endospores of *Thermoactinomyces vulgaris* were found in debris that was compacted in layers of clay. *T. vulgaris* is a thermophilic, aerobic

bacterium. It requires warmth and oxygen to grow. But Vindolanda was sealed, cold, and anaerobic (because it was flooded). It seems unlikely that *T. vulgaris* entered later or that vegetative cells formed during the nearly 2000-year-long interim. So what is the answer to the question we started out with? No one knows for sure how long endospores remain viable, but it is a very long time.

The action of transpeptidases is remarkably precise. Some transpeptidases join monomers in a manner that forms a spherical wall and thereby build cocci. Others form cylinders. Both sphere-forming and cylindrical-forming enzymes are needed to build rod-shaped bacteria (They are made up of hemispherical ends connected a straight cylinder). Mutants of the rod-shaped bacterium

E. coli that have lost their cylinder-forming enzyme are cocci. Other transpeptidases build spirilla and vibrios.

Penicillin and related antibiotics inactivate transpeptidases by binding to them. For this reason, transpeptidase enzymes are also called **penicillin-binding proteins** (Chapter 21). Penicillin kills growing bacteria because inactivated transpeptidases cannot close the breaks in the cell

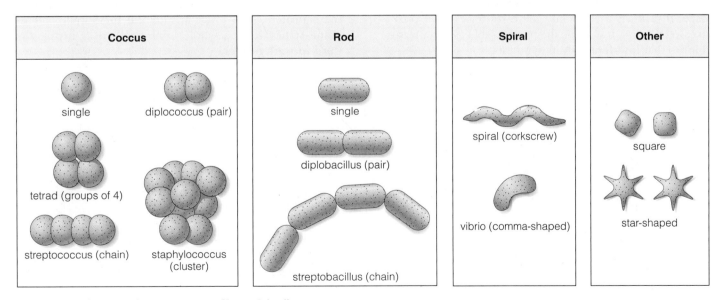

FIGURE 4.7   Shapes and arrangements of bacterial cells.

wall that autolysins make in order to enlarge. The wall becomes too weakened to hold the turgor pressure and the cell **lyses** (bursts).

Mycoplasmas. The mycoplasmas lack a cell wall, but they don't lyse because they don't have significant turgor pressure. They maintain a nearly equal concentration of solutes in their cytoplasm as exists in their external environment by pumping sodium ions out of the cell. If mycoplasmas are deprived of an energy source and are therefore unable to pump out sodium ions, they swell and lyse. The turgor of mycoplasmas doesn't have to be maintained exactly at zero because their cytoplasmic membranes are strengthened by containing **sterols.** Sterols are lipids found in membranes of eukaryotes but not in membranes of most bacteria (Chapter 2).

L Forms. Strains of bacteria sometimes lose the ability to form walls. These wall-less strains, called **L forms,** can grow and multiply if provided osmotic protection. L forms were named for the Lister Institute, where they were first observed more than 60 years ago. Since then, L forms of many genera of bacteria have been observed. Some L forms (called β-stable) can't return to a normal state. Others (called unstable) can. L forms occur naturally in infections, and they can be made artificially by treating normal bacteria with lysozyme or penicillin. Lysozyme removes walls. Penicillin prevents their synthesis.

Bacteria that have lost their walls are called **protoplasts** or **spheroplasts.** Protoplasts lack all traces of the wall, whereas spheroplasts lack most of it. Unlike L forms, protoplasts and spheroplasts do not multiply.

In summary, the peptidoglycan wall of bacteria is their hallmark. It exists nowhere else. It gives bacteria their shape. It protects them from their own huge turgor pressure. In addition, it renders them vulnerable to certain antibiotics that interfere with its synthesis.

The Cytoplasmic Membrane. The membrane that encloses the cytoplasm of any cell is called the cytoplasmic (or plasma) membrane (**Figure 4.8**). The major functions of this membrane are to contain the cytoplasm and to regulate what comes in and what goes out of the cell. The cytoplasmic membranes of bacteria and eukaryotes, called **unit membranes,** are quite similar, but the outer membranes of Gram-negative bacteria and cytoplasmic membranes of archaea are chemically distinct.

The lipid component of unit membranes is composed of phospholipids (Chapter 2). These remarkable molecules form membranes spontaneously. They have a hydrophilic head, made of glycerol and a phosphate group, and a hydrophobic tail, made of two fatty acid chains. In the watery environment they line up in two rows. The hydrophobic tails face one another where the rows touch each other (**Figure 4.8a**). The hydrophilic heads stick out into the water on both sides of the membrane. Under the electron microscope, a cross-section of a unit membrane appears to have three layers: the two rows of phospholipid molecules and space (an artifact of drying between them; **Figure 4.8b**).

Both eukaryotic and prokaryotic phospholipid bilayers are studded with proteins. Most of these are transporter proteins (also called carrier proteins or permeases). They penetrate through the membrane and carry nutrients across it (Chapter 5). Regardless of their function, all proteins that penetrate through the membrane are called **transmembrane proteins.** Other proteins (called **peripheral membrane proteins**) are merely stuck to the surface of the membrane. There are more proteins in bacterial than in eukaryotic cytoplasmic membranes. Typically, proteins make up about half the dry weight of a bacterial cytoplasmic membrane. Because these proteins can move within the membrane, membrane structure as a whole is referred to as fitting the **fluid mosaic model.**

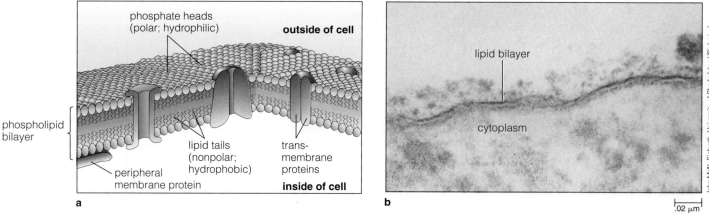

FIGURE 4.8 Cytoplasmic membrane. (a) Schematic representation. (b) Transmission electron micrograph.

Besides transporter proteins, bacterial membranes have proteins for generating metabolic energy. We'll discuss them in Chapter 5.

A cross-sectional view of most bacterial cells shows the cytoplasmic membrane as a bag around the cytoplasm. In some bacteria, however, the membrane folds back on itself, forming intricate structures that penetrate into the cytoplasm (**Figure 4.9**). Bacteria that produce these structures do so to meet their needs for more proteins to generate metabolic energy.

Now we'll consider the structures that lie outside the envelope—appendages and capsules.

Structures Outside the Envelope

Appendages. Bacteria can have two types of appendages: pili (*sing.*, pilus) and flagella (*sing.*, flagellum). A bacterium may have one, both, or neither of these. The presence or absence of appendages depends on the species and the conditions under which it is cultivated. One group of bacteria, the spirochetes, has flagella that are not appendages. They have a modified flagellum, called an axial filament, or endoflagellum, that lies within the periplasm.

Pili. Pili are straight hairlike appendages. Most are short, but some are up to several cell lengths long. Pili are made up of protein molecules called **pilins,** which are arranged helically around a central hollow core. Differences between various pili—width or pitch (angle of the helix)—are due to differences in pilin molecules. Another word for pili is **fimbriae** (*sing.*, fimbria). Once, some microbiologists made a distinction between pili and fimbriae on the bases of length and number per cell. Now the two terms are used interchange-

ably. The principal function of pili is attachment—to surfaces, to bacteria, or to other cells. Certain proteins called **adhesins,** which are interspersed among pilin molecules in pili, actually make the attachment. In some pili, adhesins are located at the tips. In others they are located on the sides.

One type of pilus, called a **sex pilus,** attaches one bacterial cell to another during mating (see Figure 6.20). We'll consider this process in Chapter 6. *E. coli* that are capable of mating bear about six of these sex pili. Other types of pili attach bacteria to plant or animal cells. Some strains of *E. coli* have 100 to 300 of this type of pili. Many disease-causing bacteria have pili that attach to specific host cells. If these pili are lost (through mutation, perhaps), the bacteria can no longer establish an infection (Chapter 15). *Neisseria gonorrhoeae* (the causative agent of gonorrhea) produces several types of pili. Each type adheres to a particular surface of the human body—the urogenital tract, the eye, the rectum, or the throat. Virtually all Gram-negative bacteria have pili. Many Gram-positives do not.

Flagella. Like pili, flagella are thin structures that extend out from the surface of the envelope. But that's as far as the similarity goes. The two appendages differ in both function and structure. The function of flagella is locomotion, not attachment. Flagella are helical (corkscrew-shaped), not straight. Flagella rotate and thereby propel the cell through its watery environment in much the same way as a screw (propeller) drives a ship (**Figure 4.10**).

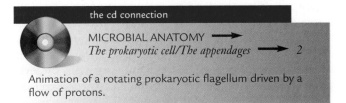

the cd connection

MICROBIAL ANATOMY →
The prokaryotic cell/The appendages → 2

Animation of a rotating prokaryotic flagellum driven by a flow of protons.

Flagella have three distinct parts. The outermost is the helical-shaped filament. It's like the blades of a propeller. It's composed of protein subunits called **flagellin,** which are arranged within the filament much like pilin molecules are arranged within pili. The relatively long filament is attached to a short, thickened, bent region called a **hook,** which acts as a universal joint, allowing the filament to point in different directions. The hook is anchored to an elaborate structure called the **basal body,** which penetrates the cell's envelope and causes the flagellum to rotate. The basal body has a motor that turns the flagellum and a set of bearings, called rings, that allow the flagellum to turn within the stationary envelope.

At their outer edges these rings anchor the flagellum to the various layers of the envelope. At their centers the core of the flagellum, called the **rod,** which is free to rotate, passes through them. Gram-negative bacteria have rings:

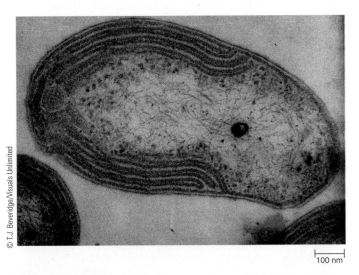

100 nm

FIGURE 4.9 Transmission electron micrograph of the cytoplasmic membrane of *Nitrobacter*. Note how it folds in the cytoplasm, creating a larger area to accommodate proteins that generate metabolic energy.

FIGURE 4.10 The flagellum of a Gram-negative bacterium.

(Art by Carlyn Iverson.)

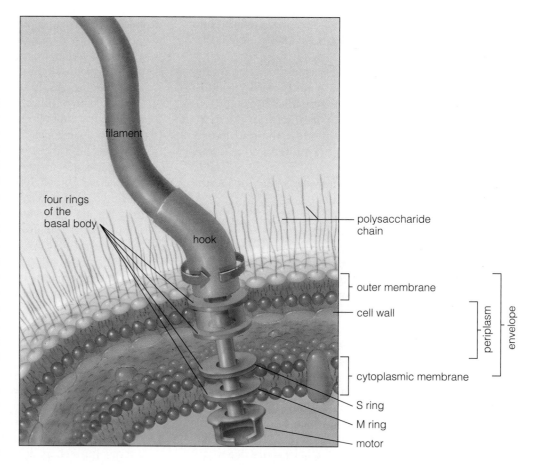

the L ring (for LPS), which is embedded in the outer membrane; the P ring (for peptidoglycan), which is embedded in the wall; and two rings, the S ring (for superficial) and the M ring (for membrane), associated with the cytoplasmic membrane. The S ring sits on it, and the M ring is embedded in it. Gram-positive have only three rings; they lack the L ring because they lack an outer membrane. Both Gram-negative and Gram-positive bacteria have a bell-shaped motor that sticks into the cytoplasm. It rotates the flagellum.

Different types of bacteria have different numbers and arrangements of flagella, called **monotrichous, amphitrichous, lophotrichous,** or **peritrichous** (**Figure 4.11**).

Flagella allow bacteria to seek out favorable environments and avoid harmful ones. All such behavior is called **taxis** (*pl.*, taxes). By **chemotaxis,** bacteria sense certain chemicals and swim toward regions that contain more nutrients and away from regions with toxic materials. By **aerotaxis,** they swim to regions that contain favorable concentrations of dissolved oxygen. By **phototaxis,** photosynthetic bacteria swim to regions of optimal light intensity and quality. A few bacteria are capable of **magnetotaxis,** which allows them to travel along magnetic lines of force. In doing so they enter sediments at the bottoms of marine or fresh waters, where conditions are most favorable for them.

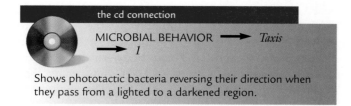

the cd connection

MICROBIAL BEHAVIOR → *Taxis* 1

Shows phototactic bacteria reversing their direction when they pass from a lighted to a darkened region.

The most thoroughly studied tactic response is chemotaxis of *Escherichia coli. E. coli* is peritrichous. It has 6 to 10 flagella spread over its surface. When all these flagella rotate in a counterclockwise direction, they form a single

ropelike structure that turns as a unit and propels the cell in a straight line through the medium—a swimming motion called a **run** (**Figure 4.12**). Periodically all the flagella reverse their rotation and turn clockwise. Then the ropelike structure flies apart and the cell tumbles. When rotation reverses again, the cell again swims in a straight line, but usually in a different direction. Normally runs last a second or two and tumbles last only a fraction of a second. Chemotaxis (and other taxes as well) changes the time of the straight-swimming run. When swimming in a favorable direction (toward a nutrient or away from a toxic compound), the duration of the run is extended up to several minutes. In an unfavorable direction (away from a nutrient or toward a toxic compound), the run lasts even less than a second. By spending more time going in the right direction and less in the wrong direction, the cell eventually gets to an optimal location.

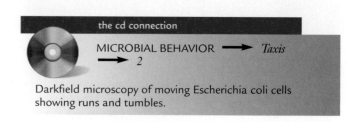

the cd connection

MICROBIAL BEHAVIOR → *Taxis* 2

Darkfield microscopy of moving Escherichia coli cells showing runs and tumbles.

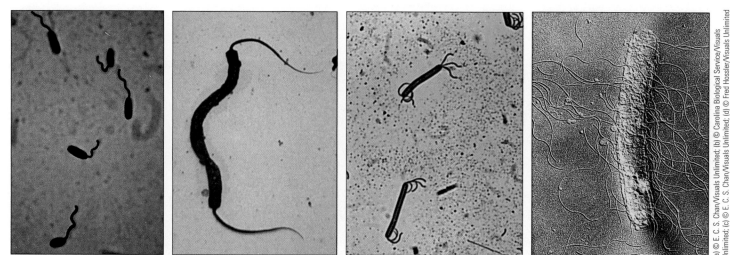

a Monotrichous: Single flagellum at one pole.

b Amphitrichous: Single flagellum at each pole.

c Lophotrichous: Two or more flagella at one or both poles.

d Peritrichous: Flagella all over the surface of the cell.

FIGURE 4.11   Arrangements of flagella on bacteria.

Axial Filaments. The flagella of one group of bacteria, the spirochetes (Chapter 11), are arranged in a unique way. Instead of sticking out from the cell, they are trapped within the periplasm (**Figure 4.13**), where they form two bundles called **axial filaments.** Each is anchored near one end of the cell and wraps around the cell body, extending about halfway up the cell. Together they form a helical bulge in the outer membrane that moves like a corkscrew as the entrapped flagella turn. The moving bulge propels the cell. The spirochete's unique form of movement is well suited to viscous environments such as mud and the surface of mucous membranes. Some spirochetes live in mud; others such as *Treponema pallidum*, which causes syphilis, live on mucous membranes (Chapter 24). Spirochetes swim fastest in liquids that are about as viscous as heavy motor oil. Ordinary flagellated bacteria are almost immobilized in such thick liquids; they swim best in less viscous material, such as blood or water.

Capsule. Most bacteria, both Gram-positive and Gram-negative, secrete slimy or gummy substances that become the outermost layer of the cell. The layer is variously called a **capsule,** a **slime layer,** or a **glycocalyx.** Some capsules are quite thin. Others are thick, up to several times as thick as the rest of the cell. Some bacteria shed their capsule, making the entire culture gel-like. The composition of the capsule varies with the organism that produces it. Most capsules are made of polysaccharides. A few are made of protein (Chapter 2).

The principal function of the capsule is protection. It protects the cell against drying out and phagocytosis (see below). It can also help a cell adhere to a surface where conditions are favorable for growth. For example, the capsule of *Streptococcus mutans*, the bacterium principally responsible for tooth decay, anchors the bacterium to the surface of the tooth. If it weren't for its capsule, *S. mutans* would be dislodged by saliva and we would suffer much less tooth decay.

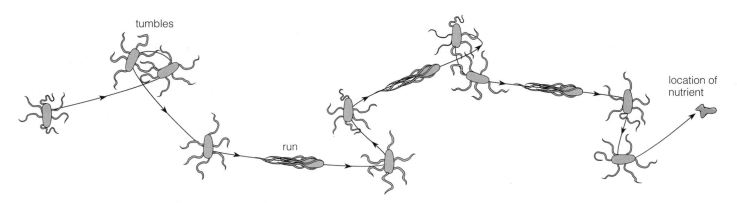

FIGURE 4.12   Chemotaxis. This sketch shows a possible route of a bacterial cell to a nutrient source. Runs toward the nutrient are longer than those away from it.

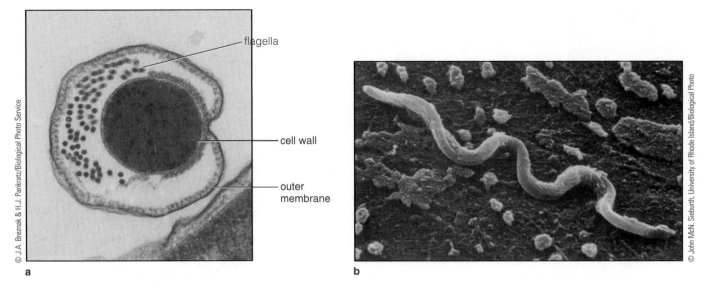

flagella

cell wall

outer
membrane

a

b

FIGURE 4.13 Axial filaments. (a) TEM of a cross-section through a spirochete showing the flagella of the axial filament lying inside the outer membrane. (b) SEM of a spirochete showing the helical bulge on the cell surface formed by the axial filament.

By protecting disease-causing bacteria against **phagocytosis** (engulfment and destruction by host cells; Chapter 16), capsules pay critical roles in infections. A dramatic example is pneumococcal pneumonia (the life-threatening lung infection caused by *Streptococcus pneumoniae*, Chapter 22). *S. pneumoniae*'s ability to cause disease depends entirely on its having a capsule. Strains of *S. pneumoniae* that lack a capsule are harmless (they are quickly consumed by phagocytes). Strains with thin capsules usually cause a mild pneumonia. Strains with the thickest capsules cause fatal pneumonia.

In summary, capsules are highly variable structures with many functions. Different bacteria have capsules for different functions. One function, which we'll return to often in subsequent chapters, is protecting pathogens from phagocytosis so they can cause disease.

Cytoplasm. The cytoplasm is a material that lies inside the cytoplasmic membrane. It is composed primarily of water (90 percent). In electron micrographs of prokaryotic cells, cytoplasm appears uniformly grainy. In spite of its drab appearance, the cytoplasm is highly active. Most of the cell's vital chemical reactions, called metabolism (Chapter 5), take place in the cytoplasm. Some internal structures such as the nucleoid and ribosomes can be distinguished within the cytoplasm. In addition, the cytoplasm contains various structures called inclusion bodies, namely all visible structures in the cytoplasm other than the nuclear region and ribosomes.

Nucleoid. The **nucleoid** or **nuclear region** is an irregular mass of DNA within the cytoplasm. A transmission electron microscope shows that it is relatively well defined (Figure 4.1), even though it is not surrounded by a membrane as the nucleus of a eukaryotic cell is. Bacterial DNA—like all DNA—carries the cell's genetic information. In most bacteria the majority of DNA is arranged in a single circular molecule called the bacterial chromosome. Usually these bacteria also contain smaller circular DNA molecules called plasmids. A few bacteria have linear rather than circular chromosomes. Such bacteria also have linear plasmids.

Ribosomes. The cytoplasm of bacteria is packed with **ribosomes** (small structures that manufacture proteins; Chapter 8). Their small size and great number give bacterial cytoplasm its characteristic grainy appearance. A single bacterial cell may contain as many as 20,000 ribosomes. Whether prokaryotic or eukaryotic, ribosomes are composed of two subunits—a smaller one and a larger one. Both are composed of protein and ribonucleic acid (Chapter 2). Intact prokaryotic ribosomes are smaller than eukaryotic. They are called 70S ribosomes (S for Svedberg units, the way they are measured); eukaryotic ribosomes are called 80S ribosomes. This difference is important in antibiotic actions, which often target ribosomes (Chapter 21). Many antibiotics that bind to 70S ribosomes and interfere with their function are harmless to 80S ribosomes. As a result, such antibiotics stop the growth of disease-causing bacteria without damaging their host.

Inclusion bodies. The most abundant inclusion bodies are the **storage granules** that hold the cell's reserve supplies of nutrients. Their function is similar to that of the globules of fat that some human cells accumulate. Some bacterial species store carbon in granules composed of **glycogen** (a polymer of glucose). Others store carbon in granules composed of lipids called **poly-beta-hydroxyalkanes.** And some don't store any carbon. A few bacterial species store reserves of sulfur or nitrogen, and many bacterial species, as well as algae, fungi, and protozoa, store phosphate as **polyphosphate** (also called volutin). Polyphosphate is used for many cellular functions, including producing adenosine triphosphate (ATP) for energy (Chapter 5). Polyphosphate granules have unusual staining characteristics. They become red when stained with methylene blue dye. For this reason, they are called **metachromatic** (changing color) granules. Many bacterial species have several kinds of storage granules.

Storage granules aren't the only kind of inclusions in bacteria. Some of these other inclusions have highly specialized functions. For example, some photosynthetic aquatic bacteria contain **gas vacuoles.** These are gas-filled regions surrounded by a protein membrane that allow the bacterium to float at the water level that provides the best conditions for photosynthesis. Some photosynthetic bacteria contain **chlorosomes,** structures just inside the cell membrane that house pigments necessary for photosynthesis. An unusual group of bacteria has iron-containing structures called **magnetosomes (Figure 4.14).** These magnetlike structures are needed for magnetotaxis.

Endospores.
Some bacteria produce **endospores,** the most resistant biological structures known. They are **resting structures** (nongrowing) that can survive up to hundreds of years, withstanding extreme heat, dehydration, toxic chemicals, and radiation. They form only when nutrients are exhausted or other conditions become unfavorable for growth. Usually when an endospore-forming species stops growing, it starts forming endospores. Then when favorable conditions return, endospores germinate to produce new **vegetative cells** (cells that grow and reproduce). Endospores are sometimes simply called spores, but they are quite different from the spores fungi produce (Chapter 12).

Endospores can withstand harsh environmental conditions in part because they contain so little water. Most cells consist of more than 90 percent water, but endospores have less than 15 percent water. It's not clear how endospores become so dry, but it is clear that endospores' ability to withstand intense heat is related to their low moisture content They also have high concentrations of calcium and dipicolinic acid (a compound not found in vegetative cells).

When starvation or other unfavorable conditions trigger a bacterial cell to **sporulate** (form an endospore), its cytoplasm divides into two unequal parts (**Figure 4.15**). The larger part then engulfs the smaller one, which is

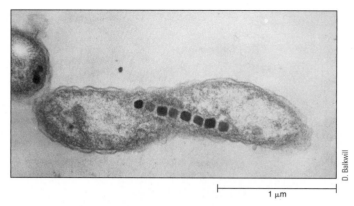

FIGURE 4.14 *Aquaspirillum magnetotacticum,* a magnetotactic bacterium. The square dark spots are magnetosomes.

called the **forespore.** The forespore eventually becomes the endospore. The encircling vegetative cell provides materials for it to build a thick protective wall.

This wall consists of an inner **cortex** and an outer **spore coat.** The cortex is composed largely of peptidoglycan. But it differs from the peptidoglycan in a vegetative cell in that it has fewer cross-linkages. Some of the peptidoglycan backbone also contains a **muramic lactam** (a derivative of muramic acid). The spore coat is composed of a protein called keratin-like, because it resembles the tough protein in outer layers of skin. Inside lies the spore body. It contains all the essential cell parts, including ribosomes and DNA. When sporulation is complete, the surrounding vegetative cell lyses, releasing the endospore.

When conditions again become favorable for growth, endospores **germinate** (grow into new vegetative cells). Most endospores cannot germinate immediately after they are formed, however. They must rest for several days or be **heat-shocked** (heated to 60°C for about 30 minutes). Then, in the proper environment, the permeability of the protective wall changes. Water enters and causes the spore to swell. The spore coat ruptures, and a germ tube grows out. The germ tube is a developing vegetative cell surrounded by a peptidoglycan wall, which contains all necessary cell components. Sporulation and germination are mechanisms of survival that enable these bacteria to withstand long periods of very hash conditions.

Some of the Gram-positive bacteria that produce endospores are pathogenic to humans. These include species that cause anthrax (Chapter 27), as well as species of *Clostridium* that cause botulism, tetanus, and gas gangrene (Chapter 25).

Because endospores can withstand such extremely harsh treatments, all procedures to eliminate bacteria must be designed with endospores in mind. They are the toughest biological structures to kill.

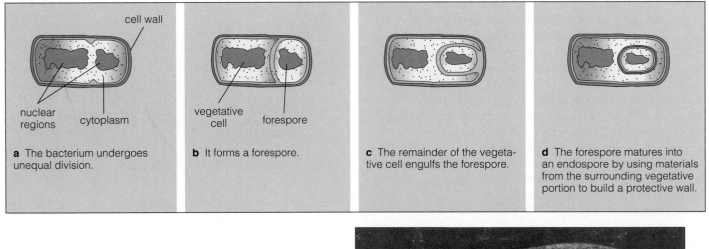

a The bacterium undergoes unequal division.

b It forms a forespore.

c The remainder of the vegetative cell engulfs the forespore.

d The forespore matures into an endospore by using materials from the surrounding vegetative portion to build a protective wall.

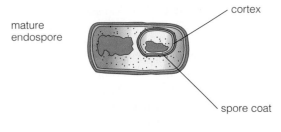

e The wall is composed of an inner cortex and an outer spore coat. The endospore also dehydrates during maturation.

f Transmission electron micrograph of *Clostridium tetani,* the bacterium that causes tetanus. The endospore has enlarged one end of the cell. Later, the cell will lyse, releasing the endospore.

880 nm

© T.J. Beveridge, University of Guelph/Biological Photo Service

FIGURE 4.15 Sporulation.

Structure of the Archaeon Cell

In electron micrographs, archaea and bacteria look quite similar. But there are profound differences in the compounds that form their structures. For example, murein, the peptidoglycan that is the major component of bacterial

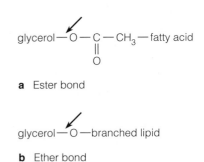

a   Ester bond

b   Ether bond

FIGURE 4.16 Chemical linkages of fatty acids in cytoplasmic membranes. (a) In all organisms except archaea, cytoplasmic membranes contain fatty acids joined to glycerol by ester linkages. (b) In archaea, branched lipid molecules are joined to glycerol by ether linkages.

cell walls, is never found in walls of archaea. Instead, archaea cell walls are composed of either protein or a peptidoglycan called **pseudomurein.** As is the case with bacteria, some archaea do not have walls.

The cytoplasmic membranes of archaea also differ. Cytoplasmic membranes of bacteria (including mycoplasmas) and of all eukaryotes are composed of phospholipids, in which unbranched fatty acids are joined to glycerol by ester linkages (Chapter 2). In archaea, branched hydrocarbon molecules are joined to glycerol by **ether linkages** (**Figure 4.16**). Because ether bonds are stronger than ester bonds, archaea have highly resistant cytoplasmic membranes that help archaea survive in extremely hostile environments—those with high temperature, low pH, and high salt.

THE EUKARYOTIC CELL

All cells except bacterial and archaeal cells are eukaryotic. Thus the cells of algae, fungi, and protozoa are structurally similar to human cells. Unlike prokaryotic cells, which are

TABLE 4.2 Comparison of Prokaryotic and Eukaryotic Cells

Structure	Prokaryote	Eukaryote
Appendages	Pili, flagella that turn	Flagella with whiplike motion
Capsule	On most bacteria; thickness varies greatly	On some algae, protozoa, and fungi
Outer membrane and periplasm	Only on Gram-negative bacteria	Never present on cells; structure present on mitochondria and chloroplasts
Cell wall	All bacteria except mycoplasmas have peptidoglycan (murein) walls; most archaea have walls of peptidoglycan (pseudomurein) or protein	Algae, fungi, protozoa, and plants have walls composed of various materials but never peptidoglycan; no walls on animal cells
Cytoplasmic membrane	All; a phospholipid bilayer without sterols	All; a phospholipid bilayer without sterols
Cytoplasm	Grainy appearance	Contains cytoskeleton; cytoplasmic streaming
DNA	Usually one circular chromosome	Paired chromosomes associated with histones
Organelles	Not unit membrane-bound; include nucleoid, 70S ribosomes, various inclusions	Always unit membrane-bound; include nucleus, endoplasmic reticulum, Golgi apparatus, 80S ribosomes (except 70S in mitochondria and chloroplasts), mitochondria in all aerobes, chloroplasts in all capable of photosynthesis
Cell division	Usually binary fission, rarely budding	Mitosis for asexual division; meiosis to form gametes
Endospores	Present in a few bacterial genera	Never present
Size	Typically one or a few micrometers	Typically 10 μm or more

uniformly grainy under the electron microscope, eukaryotic cells are subdivided into many compartments, each bounded by a unit membrane.

Structure

We'll examine the eukaryotic cell as we did the prokaryotic cell—from the outside in. **Table 4.2** summarizes the principal differences between these cell types.

Appendages. Some eukaryotic cells have flagella on their outer surface. Flagella of eukaryotes serve the same purpose as flagella of prokaryotes—motility—but their structure and movement are completely different. Eukaryotic flagella are 100 to 200 μm long. They are much longer and more complex than prokaryotic flagella, and they do not turn. Instead they move the cell by flexing in various undulating patterns. They are membrane-bound structures about 0.2 μm in diameter. Inside they have nine **microtubule** (proteins capable of movement) doublets arranged in a circle around two central microtubules, an arrangement referred to as 9 + 2. A eukaryotic flagellum bends when the doublets slide along one another. Rhythmic sliding back and forth causes the flagellum to beat at about 10 to 40 strokes per second.

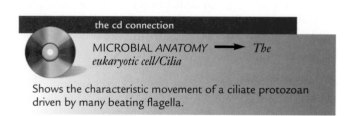

the cd connection

MICROBIAL *ANATOMY* ⟶ *The eukaryotic cell/Cilia*

Shows the characteristic movement of a ciliate protozoan driven by many beating flagella.

Some eukaryotes have shorter appendages called **cilia.** They have the same structure, the same diameter, and the same bending mechanism as eukaryotic flagella. They differ from flagella in length and numbers. Cilia are only 5 to 20 μm long, but there are hundreds of them on ciliated cells such as certain protozoa. Cilia move these cells by beating in waves coordinated by fibers (**kinetodesma**) that interconnect the bases of the cilia. Cilia also aid feeding of protozoa by sweeping particulate matter into their mouthlike **buchal** cavities.

The presence of cilia or flagella distinguishes two phyla of protozoans, the ciliates (Ciliophora) and the flagellates (Sarcomastigophora), which we'll discuss in Chapter 12.

Cell Wall. Although virtually all prokaryotes (except mycoplasmas and some archaea) have cell walls, eukaryotes show much more variation. Most animal cells don't have a cell wall. Human cells never do. Fungi and algae typically do have a cell wall. Protozoa don't have cell walls, but some have a **pellicle** outside their cytoplasmic membrane which functions as a flexible wall.

The chemical structure of eukaryotic cell walls (when they have them) varies. The cell walls of plants and of most algae and some fungi consist of the polysaccharide cellulose (Chapter 2). Some primitive fungi have cell walls composed of **chitin,** a nitrogen-containing polysaccharide. Others contain nonnitrogenous polysaccharides other than cellulose.

Cell walls in eukaryotes function in the same ways they do in prokaryotes. They give shape to a cell, and they resist turgor pressure. Like the wall-less mycoplasmas, unwalled protozoal and animal cells lack turgor pressure.

Cytoplasmic Membrane. The cytoplasmic membrane of eukaryotic cells and the cytoplasmic membrane of bacterial cells are structurally similar (archaeal cells are the exception). Both are unit membranes composed of phospholipids. Both have proteins embedded in them that can move within the lipid matrix. Both regulate what enters and leaves the cell. But there are two important differences between prokaryotic and eukaryotic membranes.

The prokaryotic cytoplasmic membrane contains the proteins involved in generating metabolic energy (Chapter 5). In eukaryotic cells these proteins are located in organelles, called **mitochondria** (found in all aerobic eukaryotes) and **chloroplasts** (found in plants and algae).

Eukaryotic cell membranes contain sterol lipids (Chapter 2). Most prokaryotes do not; only the wall-less mycoplasmas have sterols.

Cytoplasm and Its Contents. Like prokaryotic cytoplasm, eukaryotic cytoplasm is mostly water. Also like prokaryotic cytoplasm, eukaryotic cytoplasm houses the cell's organelles. The eukaryotic cell, however, has an internal structure, called a **cytoskeleton,** that prokaryotes lack. We'll look at the cytoskeleton first and then at the other contents of the cytoplasm: the nucleus, the cytomembrane system, and the mitochondria and chloroplasts.

Cytoskeleton. The cytoskeleton is composed of three types of threadlike protein structures: microtubules, microfibrils, and intermediate filaments. **Microtubules** are the largest and are made of the protein tubulin, as flagella and cilia are. **Microfibrils** are composed largely of the protein actin. Intermediate filaments are composed of a variety of proteins, including keratin, which adds rigidity to the cytoskeleton (**Figure 4.17**). Most of the cytoskeleton is unchanging, but other parts appear and reappear only as needed, such as during cell division.

The cytoskeleton has one main function: to allow the cytoplasm of the larger, more complex eukaryotic cell to move in an orderly way. Microtubules, for example, form during

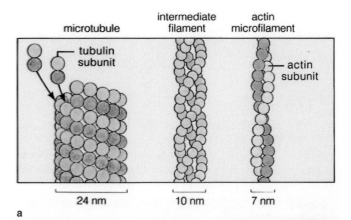

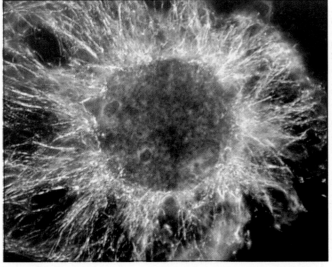

FIGURE 4.17   Cytoskeleton. (a) Eukaryotes have a cytoskeleton composed of three types of protein structures: microtubules, intermediate filaments, and microfilaments. (b) This is a cell of the African blood lily plant. The green structures are microtubules. The purple center is the nucleus.

(From Biology: The Unity and Diversity of Life, 6th ed., by C. Starr & R. Taggart, Brooks/Cole, 1992. All rights reserved.)

cell division, ensuring that each daughter cell receives one complete copy of the genetic blueprint. Microfibrils cause a characteristic cytoplasmic movement called **cytoplasmic streaming,** the means by which nutrients reach all parts of the cell. Streaming also allows one type of protozoan, the amoebae, and one type of fungus, the slime molds, to move. Amoebae and slime molds project blobs of cytoplasm to form structures called **pseudopods** ("false feet") that move these organisms across a surface (see Figure 12.15).

Nucleus. Unlike the prokaryotic nucleoid, the eukaryotic nucleus is enclosed within a double-membrane structure called the **nuclear envelope (Figure 4.18).** This envelope

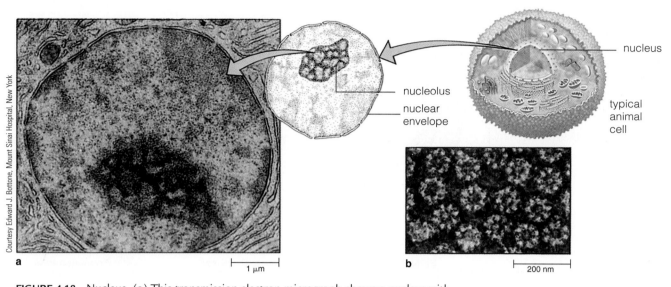

a   1 μm   b   200 nm

FIGURE 4.18   Nucleus. (a) This transmission electron micrograph shows a nucleus with one nucleolus. (b) A close-up of the pores that cover the nuclear envelope.

(Art (L) by D. & V. Henninges; art (R) by Leonard Morgan, both from Biology: The Unity and Diversity of Life, 6th ed., by C. Starr & R. Taggart, Brooks/Cole, 1992. All rights reserved.)

has many pores that allow communication between the nucleus and the cytoplasm. The interior of the nucleus contains **nucleoplasm** (the gelatinous matrix of the nucleus) and several **nucleoli**—dense masses of RNA and protein that manufacture ribosomes.

The nucleus is the largest organelle in a eukaryotic cell. Its principal function is to house almost all the cell's DNA (a small amount of DNA is housed in mitochondria and chloroplasts), which is chemically identical to prokaryotic DNA but structurally different. Also, unlike the DNA of prokaryotic cells, eukaryotic DNA is associated with histones, which are basic proteins that form a DNA-protein structure called a nucleosome. Nucleosomes, in turn, are organized into paired structures called chromosomes, which are visible under the microscope around the time of cell division.

The organization of DNA into chromosome allows the massive amount of DNA in a eukaryotic cell to be separated efficiently during cell division. Unlike prokaryotic cells, which merely duplicate their single molecule of DNA and distribute one copy to each daughter cell, eukaryotic cells undergo two complex types of nuclear division, mitosis and meiosis. Mitosis occurs during cell division for growth. Meiosis is an essential feature of cell division leading to sexual reproduction of eukaryotes.

Mitosis consists of four phases: interphase, prophase, metaphase, and anaphase (**Figure 4.19**). In mitosis the nucleus divides and each daughter cell receives one pair of each type of chromosome that the mother cell contains. During meiosis (**Figure 4.20**), the chromosome pairs split

and each new cell receives only one copy of each chromosome. Thus the products of meiosis, called **gametes,** are haploid. They contain only half as many chromosomes as the original diploid vegetative cell. Later the haploid gametes fuse, producing a **zygote,** a diploid cell that develops into a new organism.

Cytomembrane System and Ribosomes. In addition to the cytoskeleton, a cytomembrane system runs through the eukaryotic cell. The complex membrane system sorts, organizes, and packages the different kinds of molecules that the cell produces. The cytomembrane system connects with the nuclear envelope, but not with the cytoplasmic membrane. The cytomembrane system forms two major structures within cytoplasm—the endoplasmic reticulum and the Golgi apparatus.

The **endoplasmic reticulum (ER)** is a double membrane that folds back on itself, creating a complex pattern of tubes and layered sacs. Some regions of the ER appear smooth under the electron microscope. Others appear rough. Each has a different function. The smooth regions, called the **smooth endoplasmic reticulum (SER),** are the major sites of phospholipid synthesis. Thus the SER produces the components for building and repairing membranes. The rough regions, called the **rough endoplasmic reticulum (RER),** owe their grainy appearance to the presence of ribosomes on their outer surface (**Figure 4.21**), where some of the cell's proteins are made. The RER keeps some of the newly formed proteins and sends others to the **Golgi apparatus,** an organelle made of flattened stacks of membranes (**Figure 4.22**).

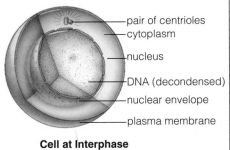

Cell at Interphase

The cell prepares for mitosis by replicating (duplicating) its DNA.

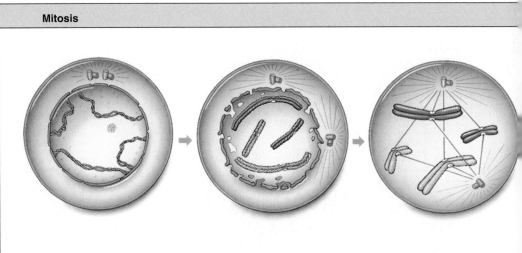

Early Prophase
Duplicated chromosomes start to condense. The nucleus is 2*n*; one set of 2 chromosomes is shaded purple and the other blue.

Late Prophase
Chromosomes condense more, spindle starts to form.

Transition to Metaphase
Condensed chromosomes attach to spindle apparatus.

FIGURE 4.19 Mitosis. Mitosis is division for cell growth. Each new cell receives a pair of each type of chromosome.

(Art by Raychel Ciemma from Biology: Concepts and Applications, 2nd ed., by C. Starr, Brooks/Cole, 1994. All rights reserved.)

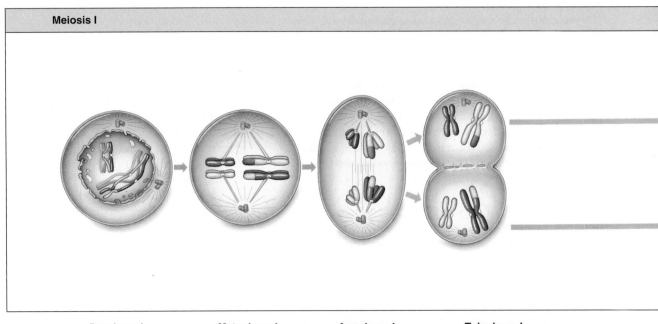

Prophase I
Duplicated chromosomes start to condense, spindle starts to form; duplicate chromosomes pair.

Metaphase I
Pairs of chromosomes align at equator; the two are attached to opposite poles.

Anaphase I
Pairs of chromosomes separate and move to opposite poles.

Telephase I
A haploid number (*n*) of chromosomes is clustered at each pole.

FIGURE 4.20 Meiosis, sexual reproduction in eukaryotic cells. Only reproductive cells undergo meiosis. Chromosomes split and each new cell receives one copy of each chromosome.

(Art by Raychel Ciemma from Biology: Concepts and Applications, 2nd ed., by C. Starr, Brooks/Cole, 1994. All rights reserved.)

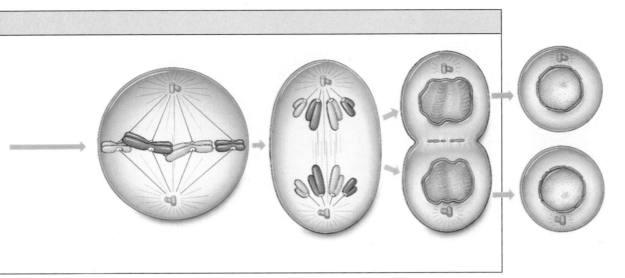

Metaphase
All chromosomes line up at spindle equator.

Anaphase
Chromosomes are separated and moved to opposite poles.

Telophase
Chromosomes decondense; nuclear membranes form and the cell starts to divide.

Interphase
The product of mitosis is two daughter cells, each of which is diploid (*2n*).

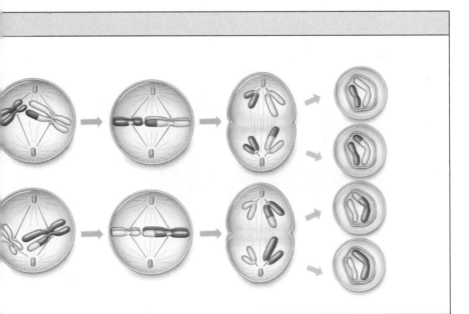

Prophase II
Brief stage; each chromosome still duplicated.

Metaphase II
Chromosomes line up at spindle equator.

Anaphase II
Chromosomes separate, move to opposite poles.

Telophase II
Each daughter nucleus is *haploid* (*n*), with one of each pair of homologous chromosomes that was present in the parent nucleus.

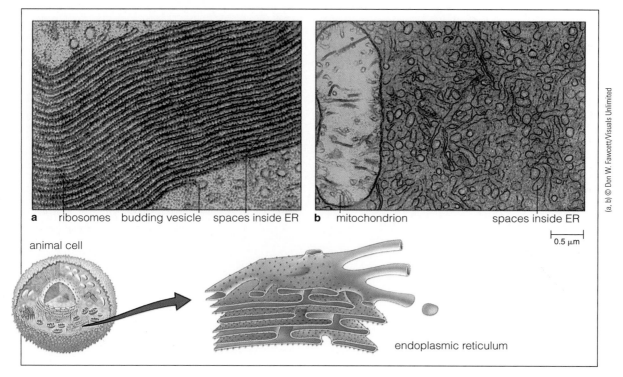

FIGURE 4.21 Endoplasmic reticulum.

(Art (L) by Leonard Morgan; art (R) by Robert Demarest, both from Biology: The Unity and Diversity of Life, 6th ed., by C. Starr & R. Taggart, Brooks/Cole, 1992. All rights reserved.)

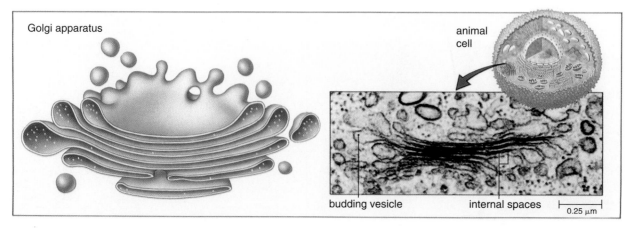

FIGURE 4.22 Golgi apparatus.

(Art (L) by Robert Demarest; art (R) by Leonard Morgan, both from Biology: The Unity and Diversity of Life, 6th ed., by C. Starr & R. Taggart, Brooks/Cole, 1992. All rights reserved.)

Proteins that leave the RER become packaged in membrane-bound vesicles that bud from the ER and then fuse with the Golgi apparatus. Within the Golgi apparatus the molecules are modified and concentrated. Then they are repackaged in new vesicles, which leave the Golgi apparatus on their way to a final destination, either inside or outside the cell. The Golgi apparatus functions as a packaging and distribution center.

One specialized type of vesicle that leaves the Golgi apparatus is called a **lysosome**—a small but powerful packet of enzymes that can destroy many kinds of molecules and microbial cells captured by phagocytes.

Mitochondria and Chloroplasts. In prokaryotic cells, the proteins involved in generating metabolic energy (ATP; Chapter 2) either through respiration or photosynthesis (Chapter 5) are located in cytoplasmic membrane. In eukaryotic cells these energy-generating proteins are located in the intracellular membranes that surround organelles, called mitochondria (respiration) and chloroplasts (photosynthesis). In this respect and many others, mitochondria and chloroplasts resemble Gram-negative bacteria, because that's just what they are—derivatives of Gram-negative bacteria that were captured by an ancient organism when eukaryotic cells first evolved (see Larger Field: Endosymbiosis: Something Old, Something New).

Both are bounded by double membranes: The outer membrane encloses the highly folded system composed of the inner membranes (**Figure 4.23**). The inner membrane of a chloroplast is arranged in stacks called **thylakoids.** Thylakoids contain the pigment chlorophyll, which is involved in photosynthesis. Plants and all but a very few colorless forms of algae have chlorophyll. The inner membrane of mitochondria is deeply folded to form structures called the **cristae,** which are imbedded in the **matrix,** the gelatinous ground material.

FIGURE 4.23 Mitochondria and chloroplasts. (a) Mitochondrion in an animal cell showing its location, its appearance in TEM, and its parts. (b) Chloroplast from a plant cell showing the same features as in part a.

(Art (a) by Leonard Morgan; art (b) by Palay Beaubois, all from Biology: The Unity and Diversity of Life, 6th ed., by C. Starr & R. Taggart, Brooks/Cole, 1992. All rights reserved.)

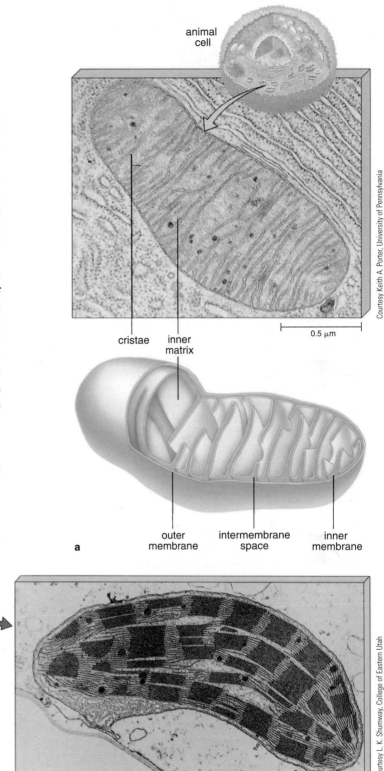

0.5 μm

cristae   inner matrix

outer membrane   intermembrane space   inner membrane

a

typical plant cell

b

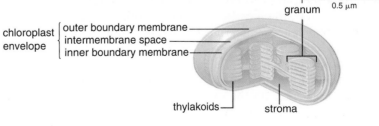

granum

0.5 μm

chloroplast envelope
- outer boundary membrane
- intermembrane space
- inner boundary membrane

thylakoids   stroma

LARGER FIELD

ENDOSYMBIOSIS: SOMETHING OLD, SOMETHING NEW

Mitochondria and chloroplasts resemble prokaryotic cells in many ways. Both these organelles are approximately the same size as a prokaryotic cell. They are the only eukaryotic organelles other than the nucleus that contain DNA. In both, the DNA is arranged like a single circular prokaryotic chromosome. Both organelles contain 70S ribosomes, like those found in prokaryotic cells, rather than the 80S ribosomes found elsewhere in eukaryotic cells. Also the double membranes of mitochondria and chloroplasts resemble the cytoplasmic and outer membranes of Gram-negative bacteria. The external membrane of these organelles even has pores that resemble the

porins in the outer membrane of Gram-negatives. Moreover, mitochondria and chloroplasts multiply the way prokaryotic cells do, by binary fission.

On the basis of the similarities of mitochondria and chloroplasts to prokaryotes, Lynn Margulis proposed in 1981 the **endosymbiotic theory** of cellular evolution (an organism that lives within another is called an endosymbiont). Prokaryotes first appeared about 3.5 billion years ago and eukaryotes only about a billion years ago. According to the theory, at that time a primitive eukaryotic cell engulfed an ancient prokaryotic cell. The two cells continued to evolve, one within the other, in a mutually benefi-

cial symbiotic relationship. The prokaryote benefited by having a place to live and a ready source of nutrients. The eukaryote benefited by having organelles that make ATP by respiration or photosynthesis. Eventually the modern eukaryotic cells, with its membrane-bound mitochondria and chloroplasts, evolved. Recent information on DNA similarities has established the truth of the theory. Indeed, it's now clear which bacteria were engulfed a billion years ago: Mitochondria are the descendants of a Gram-negative bacterium belonging to the alpha proteobacteria group (Chapter 11); chloroplasts are descendants of cyanobacteria (Chapter 11).

How do we know it wasn't another prokaryote that engulfed a smaller prokaryote? We don't, but it seems improbable because only eukaryotes are capable of phagocytosis. In engulfing nutrients, the primitive eukaryote also engulfed prokaryotes.

Mitochondria and chloroplasts have changed enormously during their billion-year stay inside eukaryotic cells. They lost properties that do not benefit their eukaryotic host—their cell wall, many enzymes, and much of their DNA. But they kept capacities needed by their eukaryotic hosts—capturing energy through respiration or photosynthesis.

Are mitochondria and chloroplasts the only rem-

MEMBRANE FUNCTION

As we've seen, unit membranes play critical roles in all types of cells: surrounding the cytoplasm of all cells and the internal organelles of eukaryotic cells. They determine which molecules enter and leave a cell or a membrane-bounded organelle. All unit membranes allow certain molecules to cross freely but block, sometimes completely, the passage of others. In other words, these membranes are semipermeable. Because unit membranes are composed primarily of lipids, which are hydrophobic (polar), hydrophobic molecules can pass directly through the membrane by dissolving in it. Hydrophilic molecules, on the other hand, such as sugars and amino acids, cannot. Neither can molecules that bear a positive or negative charge, such as ions—for ex-

ample, potassium (K^+) and phosphate (PO_4^{3-}) ions. Water is an exception. Although it is a polar molecule, it passes rapidly through water-transmitting pores that form in unit membranes. These pores appear and disappear as the phospholipid molecules constantly move. Biologically important molecules that can cross unit membranes freely include water, carbon dioxide, oxygen, ethanol, and medium-length fatty acids. Most other molecules cannot cross the unit membrane without the help of carrier proteins.

Now we'll look briefly at six ways that molecules pass across unit membranes. Both prokaryotic and eukaryotic cells exhibit four of these ways: simple diffusion, osmosis, facilitated diffusion, and active transport. Prokaryotes alone use group translocation, and only eukaryotes use engulfment.

LARGER FIELD

ENDOSYMBIOSIS: SOMETHING OLD, SOMETHING NEW *(continued)*

nants of prokaryotic endosymbionts in eukaryotes? Decidedly not. Eukaryotes constantly phagocytize prokaryotes, and some persist as endosymbionts when both partners benefit. Aphids and related insects (whiteflies and mealy bugs) have prokaryotic symbionts. Because the relationship is fairly recent—only about 200 million years old—the endosymbionts have not yet lost their Gram-negative envelope. But they can no longer grow on their own outside their host. Some species are closely related to *Escherichia coli.*

Aphids need these endosymbionts. When they are eliminated by antibiotic treatment, the aphid dies. The endosymbionts supply their aphid host with essential nutrients.

Apparently aphids have lost the ability to synthesize about 10 amino acids, which are not present in adequate amounts in an aphid's diet of plant juices. What evolutionary scenario could lead to aphids losing the capacity to synthesize so many amino acids, only to regain it by capturing a prokaryote? Most probably, aphid ancestors lived in a more nutrient-rich environment

(most insects do) and benefited by being spared from synthesizing their own amino acids. But in order to adapt to a diet of

plant juices, aphids suddenly had to reacquire this ability. They did it secondhand, by capturing a prokaryote.

Endosymbiont in an aphid cell.

1 μm

Simple Diffusion

Any molecule that crosses a membrane freely will eventually reach equal concentrations on either side it because of **diffusion,** the random movement by which molecules collide with molecules nearby and eventually distribute themselves uniformly throughout the space available to them (**Figure 4.24**). Any molecule that can cross a membrane will diffuse through it until it reaches equal distribution on both sides. Thus a net flow of molecules in or out of a cell by diffusion occurs only when driven by a **concentration gradient,** a greater concentration of the molecule on one side than the other. No energy is needed to drive such a flow.

The principles of diffusion apply to the passage of water molecules, as well as others across a membrane. However, in the case of water, the process is called **osmosis.**

FIGURE 4.24 Simple diffusion. Dye molecules become evenly distributed throughout a volume of water by simple diffusion.

(Art by Raychel Ciemma from Biology: Concepts and Applications, 2nd ed., by C. Starr, Brooks/Cole, 1994. All rights reserved.)

Osmosis

By osmosis, water crosses a membrane toward the side with the higher concentration of solute molecules and therefore lower concentration of water molecules. That is, the presence of other molecules decreases the concentration of water. And water molecules move in a direction that will tend to equalize their concentration on both sides of the membrane. The total concentration of solute molecules in a membrane-bound space, such as a cell, is referred to as its **osmotic strength.** As water flows into a closed space, it causes an increase in internal pressure, called **osmotic pressure.** In a cell it is called **turgor pressure** or just turgor.

A cell's external environment affects osmotic movement and the integrity of the cell (**Figure 4.25**). A hypertonic environment has a higher concentration of solutes than does the cell's interior. It therefore draws water out of the cell by osmosis, causing the membrane-bounded volume to decrease. This collapse is called **plasmolysis.** In contrast, a hypotonic environment has a lower concentration of solutes than the cell's interior. So water enters the cell's interior, which increases the cell's turgor pressure. If the turgor pressure formed is greater than the cytoplasmic membrane and the cell wall (if the cell has one) can withstand, the cell **lyses** (bursts). An isotonic environment is the happy medium. It has the same concentration of solutes as the cell's interior, so it exerts no osmotic effect on the cell.

Facilitated Diffusion

Substances necessary to sustain life that cannot diffuse across unit membranes must somehow still enter. Such passage is mediated by carrier proteins embedded in the membrane. Carrier proteins that mediate facilitated diffusion bind a compound on one side of the membrane and release it on the other (**Figure 4.26a**). However, facilitated diffusion transports molecules only from regions of higher concentration to regions of lower concentration. In this respect it resembles diffusion.

Carrier proteins that mediate facilitated diffusion are highly specific. They usually transport only one kind of molecule. Not many compounds enter prokaryotes by facilitated diffusion. But many compounds, including most sugars, enter most eukaryotic microorganisms by facilitated diffusion.

Active Transport

Some carrier proteins transport molecules from regions of lower concentration to regions of higher concentration. That's the opposite of what occurs with diffusion.

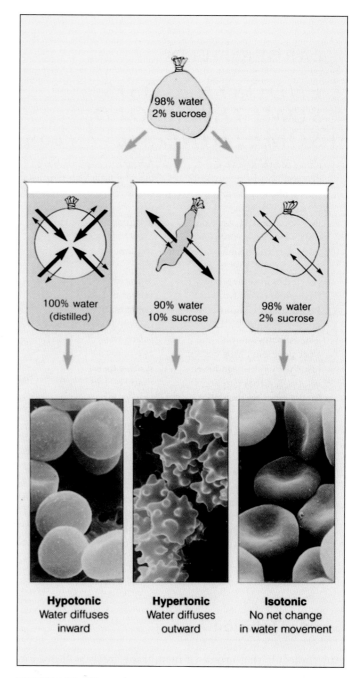

FIGURE 4.25   Osmosis.

(Art from Biology: The Unity and Diversity of Life, 6th ed., by C. Starr & R. Taggart, Brooks/Cole, 1992. All rights reserved.)

Such a process is necessary to bring essential nutrients into the cell when the concentrations outside are low. The movement of molecules against a concentration gradient requires energy. Energy can be supplied in the form of ATP or a concentration gradient of ions, usually protons (Chapter 5). This process is called active trans-

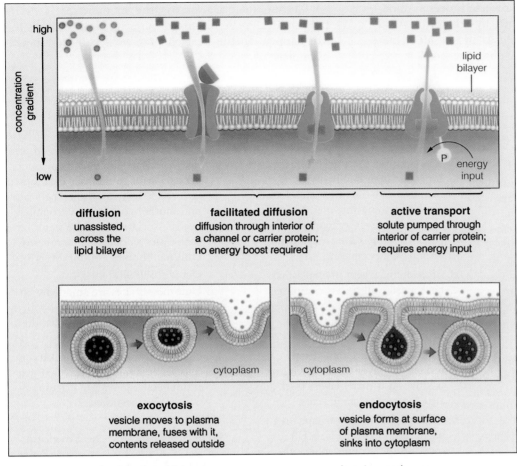

FIGURE 4.26   Mechanisms by which substances move across cytoplasmic membranes.

(Top art redrawn from Molecular Biology of the Cell, 2nd ed., by B. Alberts, et al. Copyright ©1989. Reproduced by permission of Rontledge, Inc., a part of the Taylor & Francis Group. Below art by Leonard Morgan from Biology: Concepts and Applications, 2nd ed. by C. Starr, Brooks/Cole, 1994. All rights reserved.)

port. Most nutrients, including sugars, amino acids, and vitamins, enter prokaryotes by **active transport.**

Group Translocation

Group translocation is a variation of active transport that occurs only in bacterial cells and only with certain molecules. In group translocation a molecule is transported into the cell and at the same time it is chemically changed into a slightly different molecule. The latter process requires energy. But the chemical modification prevents the molecule from leaving the cell. In a sense, group translocation is a trapping mechanism. For example, bacteria take up the nutrient glucose by group translocation. A carrier protein moves glucose molecules from the environment into the cell. Inside the cell it is **phosphorylated** (a phosphate group is added to it) so it

cannot leave. The phosphorylated form is used to fuel the cell's metabolic reactions.

Engulfment

Some eukaryotes can also move molecules across the cytoplasmic membrane by engulfing them (**Figure 4.26b**). Only eukaryotes without cell walls can do this. In engulfment the cytoplasmic membrane folds in on itself and wholly surrounds a foreign particle, bringing it into the cell. Inside it releases the particle as a separate membrane-bound structure called a **vacuole.** When solid material is engulfed, the process is called **endocytosis.** If a cell is taken in by certain types of white blood cells, the process is called **phagocytosis** ("cell eating"). If liquid material is taken in, the process is called **pinocytosis** ("cell drinking"). **Exocytosis** refers to the expulsion of material, such as waste, by the reverse process. Amoebae take in most of their nutrients by endocytosis and pinocytosis (Chapter 12). White cells in our blood destroy invading microorganisms by phagocytosis.

Getting materials across the membranes is essential for life. We've discussed here the kinds of ways this can happen. In the next chapter we'll consider specific examples.

SUMMARY

An Infectious Examination (p. 94)

1. This case illustrates how the unusual structure (possessing an outer waxy layer) of Mycobacterium tuberculosis affects its function (slow growth and acid fastness) and drug sensitivity (sensitivity to isoniazid, which inhibits synthesis of components of the waxy layer).

Structure and Function (pp. 84-85)

2. The cell is best understood in terms of structure and function, which are linked.

The Prokaryotic Cell (pp. 86-89)

3. Prokaryotic cells have an undifferentiated, grainy interior with no membrane-bounded organelles. They have a nucleoid (a mass of DNA), rather than a membrane-bound nucleus.

4. Prokaryotic cells are usually much smaller than eukaryotic cells.

Structure of the Bacterial Cell (pp. 87-98)

5. The prokaryotic cell has three basic parts: the envelope (the outside surface of the cell), structures outside the envelope, and the cytoplasm (which fills the interior and houses organelles).

The Envelope (pp. 87-93)

6. The outer membrane is a bilayer composed of phospholipid on the inner surface and mostly lipopolysaccharide (LPS) on the outer surface. LPS is not found in any other cell structure. Only Gram-negative bacteria have an outer membrane.

7. The periplasm is a gelatinous organelle bounded by two membranes—the outer membrane and the cytoplasmic membrane. It contains essential enzymes and other proteins. Only Gram-negative bacteria have a periplasm.

8. All bacteria except mycoplasmas have a cell wall that is composed of murein, which is a type of peptidoglycan. It is usually called peptidoglycan.

9. The cell wall in Gram-negative bacteria is one layer thick. In Gram-positives it is many layers thick and also contains teichoic acids.

10. The cell wall confers shape on a bacterium. Most bacteria fall into one of three groups, spherical (cocci), rod-shaped (bacilli), or spiral-shaped (spirilla). There are also intermediate shapes (for example, coccobacilli and vibrios).

11. The cell wall also contains a cell's turgor pressure, the internal osmotic pressure from its contents. A bacterial cell's turgor is extremely high because its contents are concentrated.

12. Mycoplasmas lack a cell wall. They lack turgor pressure because they actively pump sodium ions out of the cell.

13. L forms are strains of bacteria that have lost the ability to form walls but can still reproduce if provided osmotic protection.

14. The membrane that encloses the cytoplasm is called the cytoplasmic membrane. Cytoplasmic membranes contain the cytoplasm and regulate what enters and leaves the cell.

15. All membranes, including the cytoplasmic membrane, that have the same structural and chemical characteristics are called unit membranes.

16. Unit membranes are composed primarily of phospholipids that spontaneously arrange themselves into two layers. Their hydrophobic tails face one another and their hydrophilic heads associated with the water on both outer sides of the membrane. Unit membranes are thus described as a phospholipid bilayer.

17. Unit membranes are studded with proteins that can move, so the membrane structure as a whole fits the fluid mosaic model.

18. Unlike eukaryotic cells, prokaryotic cells also contain proteins that generate metabolic energy by respiration or photosynthesis.

Structures Outside the Envelope

Appendages (pp. 93-96)

19. Some bacterial species have appendages (structures attached to the envelope and extending beyond the cell).

20. Pili are straight, usually short, hairlike appendages that attach bacteria to other objects including other cells. The sex pilus attaches the mating pair of bacteria.

21. Flagella have a complex three-part structure that allows them to rotate and propel the cell toward favorable environments and away from harmful ones (a behavior called taxis). In chemotaxis, bacteria swim toward nutrients and away from toxic materials.

22. There are four types of flagellar arrangements: monotrichous, amphitrichous, lophotrichous, and peritrichous.

23. The spirochetes have axial filaments, bundles of flagella wrapped around the cell body that propel

the cell like a corkscrew through viscous environments.

Capsule (pp. 95-96)

24. The capsule is the slimy or gummy outermost layer of the cell. It is also called a slime layer or glycocalyx.

Cytoplasm (pp. 96-97)

25. The cytoplasm is a matrix composed primarily of water and protein where the cell's vital chemical reactions (metabolism) take place.

26. The nucleoid or nuclear region is a mass of DNA usually arranged in a single circular molecule, the bacterial chromosome. Some bacteria also have smaller circular DNA molecules called plasmids that encode specialized, nonessential functions.

27. Bacterial cytoplasm is packed with ribosomes, small structures that manufacture proteins. Prokaryotic ribosomes (called 70S) are smaller than eukaryotic ribosomes (called 80S).

28. Inclusions are visible structures in the cell other than the nuclear region and ribosomes. Examples are storage granules (which hold the cell's reserve supplies of nutrients), gas vacuoles (gas-filled regions), or magnetosomes (iron-containing structures).

Endospores (p. 97)

29. Endospores are extremely hardy, resting (nongrowing) structures that some bacteria—principally Gram-positives—produce through the process of sporulation when nutrients are exhausted. When favorable conditions return, endospores germinate to produce new vegetative cells, which grow and reproduce.

30. Sporogenesis, the complete cycle of sporulation and germination, allows survival under unfavorable conditions.

Structure of the Archaeon Cell (p. 98)

31. Archaeon cell walls are composed of either protein or a peptidoglycan called pseudomurein, not found in any other living things. Some archaea do not have cell walls.

32. The bonds in the cytoplasmic membrane of archaea cells are ether linkages, which are stronger than the ester bonds found in bacteria.

The Eukaryotic Cell (pp. 98-105)

33. Cells of algae, fungi, and protozoa are eukaryotic. So are human and plant cells.

Structure (pp. 99-105)

34. Many membrane-bound organelles help keep different chemical activities separate and provide systems of transportation in the more complex eukaryotic cell.

Appendages (pp. 99-100)

35. Eukaryotic cells do not have pili or axial filaments, but some have flagella and smaller versions called cilia that also function for locomotion.

36. Some human cells have cilia—projections whose function is the movement of materials rather than locomotion. Two groups of protozoa—the flagellates and the ciliates—have appendages.

Cell Wall (pp. 99-100)

37. Algae and fungi typically have a cell wall. Some protozoa have a pellicle, which functions much as a cell wall, though it is more flexible and skin-like. Human cells never have walls. The cell walls of plants and of most algae and some fungi consist of the polysaccharide cellulose.

Cytoplasmic Membrane (p. 100)

38. The cytoplasmic membranes of prokaryotes and eukaryotes both are unit membranes. Both also serve to regulate what enters and leaves the cell.

Cytoplasm and Its Contents (pp. 100-105)

39. The cytoskeleton is a lattice network composed of three types of protein structures that allows the cytoplasm in larger eukaryotic cells to move in an orderly way.

40. The eukaryotic nucleus is surrounded by a nuclear envelope containing nucleoplasm and nucleoli. Nucleoli are dense masses of RNA and protein that manufacture ribosomes.

41. Eukaryotic DNA is associated with proteins called histones that form a structure called a nucleosome. Nucleosomes form paired structures called chromosomes, which carry a cell's genes.

42. Eukaryotic cells undergo two types of cell division, mitosis (for growth) and meiosis (for reproduction).

43. The cytomembrane system sorts, organizes, and packages the different molecules that the cell produces. It consists of the endoplasmic reticulum (ER) and the Golgi apparatus.

44. Some regions of the ER are smooth (SER) and others are rough (RER). The SER is the major site of phospholipid synthesis, creating the components for the cell's membranes. The RER appears rough (grainy) because of the ribosomes (the major site of protein synthesis) on its surface.

45. The Golgi apparatus is made up of stacks of flattened membranes that modify and concentrate phospholipids and proteins. They also produce lysosomes, powerful packets of enzymes that destroy microbial cells captured by phagocytes.

46. The proteins involved in generating ATP are found in two types of double-membrane–bound organelles, mitochondria (respiration) and chloroplasts (photosynthesis). Plants and nearly all algae have chloroplasts. Animals (including humans), fungi, and some protozoa have mitochondria.

Membrane Function
(pp. 106-110)

47. The cytoplasmic membrane is semi-permeable: It allows certain molecules to cross freely but not others.

48. By simple diffusion, molecules flow into a cell along a concentration gradient until the molecules reach equilibrium.

49. The diffusion of water is referred to as osmosis. As water flows into a closed space, it increases the internal pressure, or osmotic pressure. In a cell it is called turgor pressure.

50. A cell's external environment may be hypertonic (higher concentration of solutes than the cell's interior, which may cause plasmolysis), hypotonic (lower concentration of solutes than the cell's interior, which can cause the cell to lyse), or isotonic (same concentration of solutes outside and inside the cell, so there is no osmotic effect).

51. In facilitated diffusion, carrier proteins bind a compound that cannot diffuse across the membrane and release it on the other side. Facilitated diffusion transports molecules only from regions of higher concentrations to regions of lower concentration, so no expenditure of energy is required.

52. In active transport, carrier proteins transport molecules from regions of lower concentration to regions of higher concentration. Energy is required.

53. Group translocation occurs only in prokaryotes and only with certain molecules. At the same time a molecule is transported into the cell, it is chemically changed to prevent it from leaving the cell. Energy is required.

54. In engulfment the cytoplasmic membrane folds in on itself, wholly surrounds a foreign particle, brings it into the cell, and releases it in a membrane-bound structure called a vacuole. Only eukaryotes without cell walls can engulf.

55. In endocytosis, solid material is engulfed; in phagocytosis a whole cell is taken in. In pinocytosis, liquid material is taken in. Exocytosis is expelling material from a cell by the reverse process.

REVIEW QUESTIONS

An Infectious Examination

1. Which structural property of *Mycobacterium tuberculosis* causes it to be slow-growing, acid fast, and sensitive to isoniazid? Explain the reason for all three of these consequences.

The Prokaryotic Cell

2. What are three major differences between prokaryotic and eukaryotic cells?

3. What are cellular appendages? Name the three types of prokaryotic appendages.

4. What are the functions of pili and flagella?

5. What are the major parts of a prokaryotic flagellum? What makes it turn? What are the four types of arrangement of flagella?

6. What does taxis mean? How does chemotaxis in *Escherichia coli* work?

7. How does the structure of axial filaments suit their function?

8. What are the parts of the prokaryotic cell envelope? Which parts are found only in Gram-negative prokaryotic cells?

9. How are structure and function related in the prokaryotic capsule?

10. What is the structure of the outer membrane? What is its function? What are porins? Why is lipid A called an endotoxin?

11. What is the structure and function of periplasm?

12. Do all bacteria have a cell wall? Explain.

13. What is the chief component of all bacterial cell walls? What is its chemical structure? What are the two major functions of prokaryotic cell walls? What is the difference between murein and peptidoglycan?

14. What are the differences among the cell walls in Gram-positive bacteria, Gram-negative bacteria, and mycoplasmas?

15. What are the three most commonly found shapes of bacteria?

16. How does the lysozyme experiment demonstrate that the cell wall confers shape on a prokaryotic cell?

17. What is turgor pressure? Why is a bacterial cell's turgor pressure so high? How do mycoplasmas survive without a wall?

18. What are the differences among L forms, protoplasts, and spheroplasts?

19. Describe the structure of the cytoplasmic membrane.

20. What are permeases?

21. What two functions do the cytoplasmic membranes of all prokaryotic and eukaryotic cells share? What function does a prokaryotic cytoplasmic membrane have that a eukaryotic cytoplasmic membrane lacks?

22. What is the structure and function of prokaryotic cytoplasm?

23. What is a nucleoid? What is the bacterial chromosome? What is a plasmid?

24. What is the function of ribosomes? What is the significance of the 70S-80S difference in ribosomes?

25. Define inclusion body, and give some examples of inclusion.

26. What are endospores? Describe sporogenesis.

27. Name two ways the archaeon cell differs from the bacterial cell.

The Eukaryotic Cell

28. Give some examples of eukaryotic cells. How are eukaryotic organelles different from prokaryotic organelles?

29. What types of appendages do eukaryotic cells have? How are eukaryotic flagella different from prokaryotic flagella? How are they the same?

30. Which eukaryotic microorganisms have cell walls? Do human cells have cell walls? What is a pellicle?

31. What is the main difference between cell walls in prokaryotes and cell walls in eukaryotes (when they have them)?

32. How are the cytoplasmic membranes of prokaryotic and eukaryotic cells the same? Name two important differences.

33. What is the cytoskeleton? Give some examples of how the cytoskeleton functions.

34. What are the parts of a eukaryotic nucleus?

35. Compare and contrast prokaryotic and eukaryotic DNA.

36. Sketch and explain mitosis.

37. What is the difference between mitosis and meiosis?

38. What is the cytomembrane system in a eukaryotic cell?

39. Describe the endoplasmic reticulum (ER). Why is some ER called smooth (SER) and some rough (RER)?

40. What is the function of SER? Of RER? How do vesicles function in the cytomembrane system?

41. Describe the Golgi apparatus and how it functions. What are lysosomes?

42. How are mitochondria and chloroplasts structurally alike? How are they functionally alike?

43. How do biologists explain the similarity between mitochondria and chloroplasts and prokaryotic cells?

Membrane Function

44. Why are cytoplasmic (unit) membranes described as semipermeable? How does the chemical structure of unit membranes determine which molecules can pass and which cannot?

45. Explain simple diffusion.

46. What is osmosis? Define osmotic strength. Define turgor pressure. Give examples of how a cell's external environment affects osmotic movement and the integrity of the cell.

47. How is facilitated diffusion like simple diffusion and how is it different?

48. What is active transport?

49. Describe group translocation.

50. What is engulfment? Define endocytosis, phagocytosis, pinocytosis, and exocytosis.

51. Which of these processes require energy and which do not? Which occur in both prokaryotic and eukaryotic cells? In only prokaryotic? In only eukaryotic?
    a. simple diffusion   d. group translocation
    b. active transport   e. facilitated diffusion
    c. osmosis   f. engulfment

CORRELATION QUESTIONS

1. Why are prokaryotes incapable of engulfment? Explain.

2. How could you distinguish between a prokaryotic and a eukaryotic flagellum just by watching it move?

3. If you had a scanning electron micrograph of a single pilus and a single bacterial flagellum, how would you know which was which? Explain.

4. Do you think a cell would have both flagella and gas vesicles? Explain.

5. Could a mycoplasma infection be treated with penicillin? Explain

6. What sort of evidence would you want to decide whether mitochondria are captured bacteria or captured archaea?

ESSAY QUESTIONS

1. Discuss the structural changes that had to occur for eukaryotic cells to become so large.

2. Discuss the evidence that supports the endosymbiont theory of the origin of mitochondria.

SUGGESTED READING

Fawcett, D. 1981. *The cell: An atlas of fine structure.* New York: W. H. Freeman.

Krawiec, S., and Riley, M. 1990. Organization of the bacterial chromosome. *Microbiological Reviews* 54:502–39.

Macnab, R. M. 1996. Flagella and motility. In Escherichia coli *and* Salmonella. F. C. Neidhardt, ed. Washington, D.C.: ASM Press

Nanninga, N. 2001. Cytokinesis in prokaryotes and eukaryotes: common principles and different solutions. *Microbiology and Molecular Biology Reviews* 65:319–33.

Nikaido, H. 1996. Outer membrane. In Escherichia coli *and* Salmonella. F. C. Neidhardt, ed. Washington, D.C.: ASM Press

Park, J. T. 1996. The Murein Sacculus. In Escherichia coli *and* Salmonella. F. C. Neidhardt, ed. Washington, D.C.: ASM Press

Rothman, J. 1985. The compartmental organization of the Golgi apparatus. *Scientific American* 244:57–67.

For additional readings, go to InfoTrac College Edition, your online research library at: http://www.infotrac.thomsonlearning.com

FIVE

Metabolism of Microorganisms

© Carolina Biological Supply/Visuals Unlimited

CHAPTER OUTLINE

LEARNING GOALS

To understand:

- *The function and importance of metabolism*

- *How the five sequential steps in metabolism—bringing nutrients into the cell, catabolic reactions, biosynthesis, polymerization, and assembly—synthesize all the cell's components*

- *How ATP is formed through substrate-level phosphorylation and chemiosmosis and the role it plays in metabolism*

- *How reducing power is formed and the role it plays in metabolism*

- *How glycolysis, the TCA cycle, and the pentose phosphate cycle proceed and how they form the 12 precursor metabolites, ATP, and reducing power*

- *How anaerobic metabolism proceeds by way of anaerobic respiration or fermentation*

- *How photoautotrophs carry out metabolism through light-driven processes*

- *How chemoautotrophs carry out metabolism by oxidizing inorganic compounds*

- *How metabolism is regulated so the proper amounts of products are made efficiently*

A Good Bug Gone Bad

L. B. was a 5-year-old boy whose parents brought him to see his pediatrician because he had been sick for 2 weeks. They were becoming more and more worried and wanted "lab tests" to find out what was wrong. The first sign of L. B.'s illness had been diarrhea. For several days the child's stools had just been frequent and watery. Then L. B. developed severe abdominal pain as well. Finally his parents noticed that the stools were tinged with blood. Because L. B. was able to drink and didn't have a fever, the family kept him at home and waited for him to improve. His diarrhea did improve markedly after about a week. But L. B. still didn't seem to be getting well. He looked tired and pale; his appetite didn't return.

The doctor who examined L. B. found little wrong with him. He was no longer having diarrhea; he was not dehydrated; he no longer had abdominal pain. But somehow he looked unwell. His doctor's and his parents' suspicions were confirmed an hour later when the laboratory results came back. L. B.'s complete blood cell count showed significant anemia (low red blood cell count). Direct microscopic examination of his red blood cells revealed that many were irregular and damaged. These results suggested that L. B.'s red blood cells were being destroyed, a condition called hemolytic (blood-destroying) anemia. L. B.'s blood also contained abnormally low number of platelets (blood-clotting cell fragments). Most worrisome of all were indications that L. B.'s kidneys were not functioning normally (high levels of creatinine and urea nitrogen were present in his blood).

The clinical picture L. B. presented is diagnostic of hemolytic uremic syndrome (HUS). This **syndrome** (complex of symptoms) is usually the result of an infection by one particular strain of *Escherichia coli* designated O15:H7. To confirm his diagnosis, L. B's physician sent a stool specimen to the microbiology laboratory.

L. B.'s physician admitted him to the hospital, because the most dangerous complication of HUS, kidney failure, can cause death, and L. B.'s laboratory results indicated that he was already in a serious state of kidney impairment. He could require renal dialysis. Two days later the microbiology laboratory reported that L. B.'s stool sample grew a culture of *E. coli* O157:H7, thus confirming his physician's diagnosis.

Fortunately, L. B.'s illness did not progress to the point where he required dialysis. After several days in the hospital, both his damaged kidneys and his blood abnormalities began to return to normal. He was discharged from the hospital with the expectation that he would make a full recovery from his potentially life-threatening infection.

Case Connections

- As we'll see in this chapter, *E. coli* has been studied so intensively that its metabolism has become the paradigm to which other microorganisms are compared. One of the main reason for this focus of attention is that most strains of *E. coli* do not cause disease; they are completely safe to work with.
- How could the O157:H7 strain have become so dangerous?
- We'll discuss the toxin that makes strain O157:H7 so dangerous in Chapter 23 and how a harmless strain probably acquired the genes to make it in Chapter 6.

METABOLISM: AN OVERVIEW

Metabolism is the acknowledged tough topic for beginning students of microbiology. The traditional approach has been to dissect metabolism into its various components—enzymes, energy production, carbohydrate metabolism, lipid metabolism, and so on—and discuss them sequentially. Often even the best students are left bewildered (and certainly disinterested) by this large dose of seemingly unrelated facts.

Without doubt metabolism is complex, but it's also completely logical. It's the step-by step process by which cells take nutrients from their environment and convert them into sufficient cellular components to double their mass and then become two cells. Microbial cells do little else. We'll consider metabolism here in this process-oriented way. We'll discuss the various steps of making a new cell—taking nutrients into the cell, processing some of them to conserve metabolic energy and produce starting materials for biosyntheses, biosynthesis of cellular building blocks,

SHARPER FOCUS

INTRODUCING ESCHERICHIA COLI

Without doubt *Escherichia coli* is the most thoroughly understood cellular organism. Why is that? First, *E. coli* is extremely convenient to study in the laboratory. It grows rapidly as single cells in simple media. Most strains do not cause disease, so they can be studied without taking special precautions. Furthermore, we know so much about *E. coli* that it is easier to learn more because it's easier to experiment on an organism if its basic properties are already known. Not surprisingly, the biotechnology industry uses *E. coli* extensively. Many biotechnology products—including insulin and human growth hormone—are made by inserting human genes into strains of *E. coli* and then harvesting the hormones from the cultures.

E. coli is also an old friend. We've shared a long and intimate history. *E. coli* probably first appeared between 120 and 160 million years ago—just about the time mammals did. *E. coli* typically inhabits the colon of mammals, and it probably evolved there. It was living with us about a million years ago when we emerged as humans.

E. coli has two habitats—the colon and the external environment. Life in the colon offers advantages and disadvantages. The temperature is a constant 37°C and nutrients are abundant, but the environment is anaerobic and contains some toxic compounds. *E. coli* prospers in this environment, but it can also grow in the very different environment it encounters when it leaves the colon in feces. This external environment is usually cooler, poorer in nutrients, and aerobic.

The metabolism of *E. coli* reflects these two habitats. Because it is a facultative anaerobe, it can grow in either the presence or absence of air. It has the relatively rare ability to use lactose (milk sugar) to support growth. Lactose is produced only by mammals and is often found in their intestines. Unlike most microorganisms, *E. coli* has a high tolerance for bile salts, which the liver produces and empties into the intestinal tract. *E. coli* can also rapidly shift its metabolism to respond optimally to nutritional and physical changes in both the colon and the external environment.

As is the case with all microorganisms, there are many strains (naturally occurring variants) of *E. coli*. Some strains can cause human diseases (Chapter 23). Traveler's diarrhea, infant diarrhea, infections of the urinary tract, and recently infections by a deadly strain, H157:O7, have attracted widespread public attention. But strains used for laboratory studies on basic biology and by industry are essentially harmless.

In many countries, including the United States, health departments routinely test drinking water for *E. coli*. Because *E. coli* is present in the colons of mammals, finding *E. coli* in a water supply is presumptive evidence that it has been contaminated by feces and is dangerous to drink.

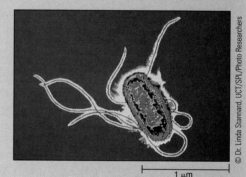

Escherichia coli shown in an image from a transmission electron microscope, with computer-enhanced color.

© Dr. Linda Stannard, UCT/SPL/Photo Researchers

---

polymerization of building blocks into macromolecules, and assembly of macromolecules into cellular components.

Let's begin our discussion of metabolism by considering what it does for the cell in which it occurs.

As soon as a new microbial cell is formed by cell division, it begins to synthesize more cellular materials and assemble them into organelles. When the cell has doubled its mass and all its components, it divides again. This process of synthesis and division continues as long as the environment is favorable. The steps of this process are or-

dinary chemical reactions, but because they occur in living cells they are called biochemical reactions. Almost all biochemical reactions are catalyzed by enzymes (Chapter 2). Collectively, all the biochemical reactions that take place in a cell are called its **metabolism.**

The biochemical reactions that make up the metabolism of most living things are strikingly similar. But there are some differences, depending on which organism we are talking about and the conditions under which it is growing. We'll start our discussion of metabolism by consider-

TABLE 5.1 Metabolic Tasks Required to Double Cell Mass

Metabolic Task	Function	Number of reactions[a]
1. Bringing nutrients into the cell	To transport nutrients across the cytoplasmic membrane and concentrate them in the cytoplasm	250
2. Catabolism	To process the major nutrient and produce the 12 precursor metabolites, ATP, and reducing power	406
3. Biosynthesis	To synthesize all necessary small molecules, including building blocks for macromolecules from precursor metabolites	438
4. Polymerization	To link together building blocks, forming macromolecules	482
5. Assembly	To assemble macromolecules into organelles	7

[a]Number of identified enzyme-catalyzed reactions in *E. coli* out of a total of 1925. In addition, 37 are required for cell division. Information from Riley M., and Labedan B., Chapter 116 in Escherichia coli and Salmonella. F. C. Neidhardt, ed. Washington, D.C.: ASM Press. *E. coli* has the genetic capacity for 4392 genes.

ing *Escherichia coli* growing aerobically (in the presence of air). Then we'll consider the metabolic variations that occur during anaerobic (in the absence of air) growth and those that occur in other microorganisms.

We'll focus our attention on reproduction of a cell. We do so because the major part of microbial metabolism is dedicated to reproduction—making more cellular components. This is in startling contrast to human beings and other animals who devote most of their metabolic effort to behavioral activity and relatively little to producing new cells.

Many biochemical reactions participate in cellular reproduction—about 2000 in *E. coli*—but only 5 metabolic tasks must be fulfilled for a cell to synthesize all the components of a cell (**Table 5.1**). All microorganisms, indeed all cells, carry out these same tasks, with minor differences in detail. The five metabolic tasks are:

1. Bringing Nutrients into the Cell: Cells obtain nutrients from their environment, but they must transport them across their cell membrane before they can enter metabolism. Usually cells concentrate these nutrients within their cytoplasm. Because microorganisms can use many different nutrients, many different biochemical reactions are needed to bring all of them into the cell.

2. Catabolism: The second metabolic task, called catabolism, is to convert nutrients into a group of organic compounds, which serve as starting points to synthesize all other cellular components. Remarkably, only 12 such starter compounds, called **precursor metabolites,** are needed. In addition, catabolism produces adenosine triphosphate (ATP, a compound that

stores metabolic energy) and reducing power (compounds that participate in various essential reductive biochemical reactions).

3. Biosynthesis: The third task, called biosynthesis, is to make all the small molecules the cell needs (including the building blocks of macromolecules) from the precursor metabolites. Some of these biosynthesis reactions use ATP and reducing power made by catabolism.

4. Polymerization: Next the building blocks, called monomers, produced by biosynthesis are chemically hooked together (polymerized) to produce the cell's macromolecules, including proteins, RNA, DNA, polysaccharides, and peptidoglycan.

5. Assembly: Finally some macromolecules are assembled to make cellular organelles, including the cell's wall, membranes, ribosomes, flagella, and pili.

As you can see, a microbial cell is like a factory that manufactures more factories identical to itself (**Figure 5.1**). The product of this factory in chemical terms is shown in **Table 5.2.** The arrangement of these chemicals is shown in **Figure 5.2.** Like any factory, the cell needs a plan to direct its operations, raw materials to be processed, and a driving force such as electrical power to make it work. The cell's plan is stored in the form of DNA. We'll discuss how it is stored and how it directs metabolism in Chapter 6. Here we'll discuss the raw materials, and how they are processed into a new cell. We'll also discuss how driving forces (ATP and reducing power) are generated and harnessed.

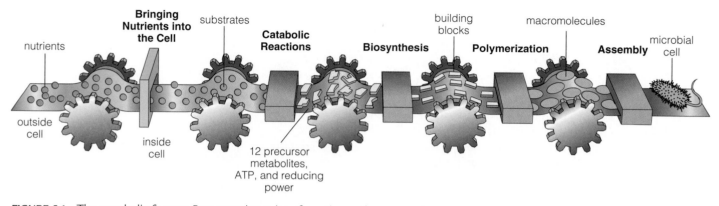

FIGURE 5.1 The metabolic factory. By a stepwise series of reaction pathways, nutrients are taken into the cell and converted into cell constituents.

TABLE 5.2 The Chemical Composition of *Escherichia coli*

Component	Percent of Total Dry Weight[a]	Number of Kinds of Molecules
Protein	55.0	4272
RNA	20.5	120[b]
DNA	3.1	1
Lipid	9.1	4
Lipopolysaccharides[c]	3.4	1
Peptidoglycan	2.5	1
Glycogen	2.5	1
Total macromolecules	96.1	
Small molecules[d]	2.9	
Ions	1.0	
TOTAL	100	

[a] Live cells of *Escherichia coli* are about 70 percent water. This makes up the other 30 percent.
[b] Not counting mRNAs.
[c] The combination of lipid and polysaccharide that makes up a part of the cell's outer membrane.
[d] Includes building blocks, metabolic intermediates, and vitamins.

AEROBIC METABOLISM

Now let's first look at how these metabolic tasks are accomplished by *Escherichia coli* when it's growing aerobically.

Bringing Nutrients into the Cell

The raw materials of metabolism are the nutrients present in the cell's environment. These must be brought across the cell envelope, which bars the passage of most molecules, in order to enter metabolism (**Figure 5.3**). (Recall from Chapter 4 the structure of the prokaryotic cell envelope and the ways molecules pass across cell membranes.)

Being Gram-negative, *E. coli*'s outer membrane presents the first barrier to the entry of nutrients. Because the outer membrane prevents the passage of both hydrophilic and hydrophobic molecules, all nutrients pass through the tiny water-filled pores in the outer membrane formed by proteins called **porins.** Molecules small enough to fit through these pores pass through them by simple diffusion. This means that:

1. Only molecules small enough to fit through a pore can cross the outer membrane. They can't be larger than a trisaccharide (three simple sugars joined together).

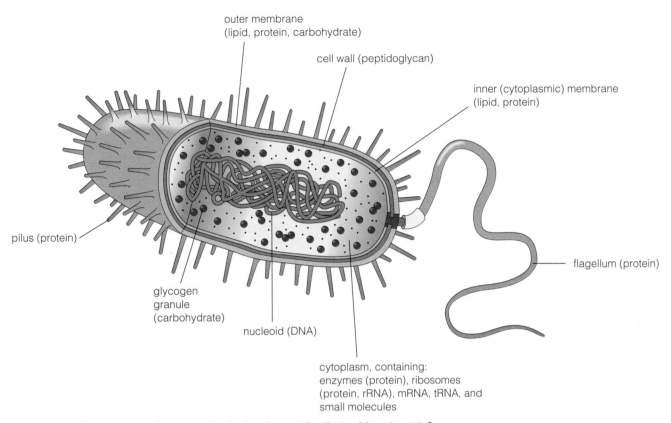

outer membrane
(lipid, protein, carbohydrate)

cell wall (peptidoglycan)

inner (cytoplasmic) membrane
(lipid, protein)

pilus (protein)

flagellum (protein)

glycogen
granule
(carbohydrate)

nucleoid (DNA)

cytoplasm, containing:
enzymes (protein), ribosomes
(protein, rRNA), mRNA, tRNA, and
small molecules

FIGURE 5.2   Arrangement of macromolecules in a bacterial cell. *E. coli* has 6 to 10 flagella rather than the single one shown here.

2. A nutrient can cross the outer membrane only if its concentration on the outside of the outer membrane exceeds its concentration on the inside. Porins cannot concentrate nutrients.

After crossing the outer membrane, a nutrient finds itself in the periplasm. To get into the cell, it must then cross the peptidoglycan cell wall and the cytoplasmic membrane. The cell wall is a loose molecular mesh, which presents no barrier to molecules small enough to traverse the outer membrane. The phospholipid matrix of the cytoplasmic membrane, on the other hand, is an absolute barrier to hydrophilic nutrients, although hydrophobic molecules pass readily through it. But proteins that extend all the way through the cytoplasmic membrane can take hydrophilic nutrients across it as well. These proteins are doors through an impassable wall. Called transporters (or permeases, facilitators, or carriers), these proteins bind to a nutrient in the periplasm, carry it through the membrane, and release it in the cytoplasm. Such carriers act like enzymes that catalyze the reaction:

nutrient in the periplasm ⟶ nutrient inside the cell

Like all enzymes, transporters are highly specific. With a few exceptions, one transporter can bring only a single compound into a cell. Therefore *E. coli* has many different transporters, each of which brings one or a few kinds of nutrients into the cell.

A few nutrients cross the cell membrane by transporter-mediated **facilitated diffusion.** The concentration of these nutrients inside the cell is slightly less than it is outside. Most nutrients, however, enter by **active transport** through the action of transporters that increase the concentration of nutrients inside the cell. They *pump* nutrients into the cell. Because pumping anything requires energy, active transport expends metabolic energy. Transporters use ATP or, as we'll see later in this chapter, a proton gradient (or gradient of some other ion) to pump nutrients into the cell. These pumps are highly effective. They can create and maintain nutrient concentrations in the cell that are a thousand times higher than their concentrations outside the cell. Another energy-requiring process that concentrates nutrients inside the cell is **group translocation** (Chapter 4). Group translocation chemically changes the nutrient as it is concentrated. In essence, the nutrient enters the cell by facilitated diffusion. Then it is immediately changed so it can't diffuse out again. For example, by

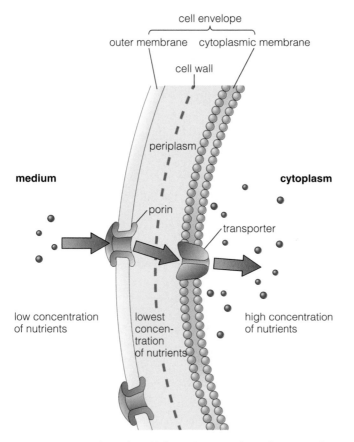

cell envelope

outer membrane   cytoplasmic membrane

cell wall

periplasm

medium   **cytoplasm**

porin

transporter

low concentration   lowest   high concentration
of nutrients   concen-   of nutrients
                           tration
                           of nutrients

FIGURE 5.3   Pathway by which nutrients are brought across the cell envelope and into the cytoplasm.

group translocation glucose diffuses into *E. coli* and is trapped inside in the form of glucose-6-phosphate. Energy is expended in the trapping step, in converting glucose to glucose-6-phosphate.

To summarize, bringing a nutrient into a cell means taking it across the cell envelope. The envelope of a Gram-negative bacterial cell such as *E. coli* consists of three layers: the outer membrane, the wall, and the cytoplasmic membrane. Each layer is crossed by different mechanisms. Nutrients diffuse through holes in the outer membrane created by transmembrane proteins called porins. Then nutrients diffuse without help through the loose molecule meshwork of the cell wall. Finally nutrients are carried across the cytoplasmic membrane by protein transporters. There are many different kinds of transporters because each can bring in only one or a few different nutrients. A few transporters let a nutrient flow into the cell by facilitated diffusion. Most bring the nutrient in by active transport, driven by metabolic energy, either as ATP or an ion gradient. Some by group translocation let a nutrient flow in and then chemically change it so it can't flow back out.

Catabolism

Escherichia coli, like all organisms, needs many different nutrients as raw materials to build a new cell. For example, it needs sources of nitrogen, sulfur, and phosphorous. The metabolism of most these nutrients is relatively simple. They are modified only slightly, by one or a few reactions. Then they are incorporated into components of the cell.

But the major nutrient, the source of carbon and energy, is quite a different story. It undergoes a many chemical changes on the route to becoming part of the cell. That's because carbon is the structural unit of all parts of the cell. Glucose is a commonly used source of carbon and energy for the growth of *E. coli*.

The first set of reactions in the series that a carbon and energy source such as glucose undergoes is called catabolism. Catabolic reactions supply *E. coli* with what it needs to synthesize cell components: the 12 precursor metabolites, ATP, and reducing power (**Figure 5.4**). The precursor metabolites are the starting materials for making all the cell's biochemical parts. ATP and reducing power are the driving forces for these reactions. There are many different catabolic reactions, but they are organized into just a few catabolic pathways. These pathways are sequences of biochemical chemical reactions, each catalyzed by a different enzyme that converts substrate molecules, such as glucose, into the three things it needs. Compounds that are starting points of a pathway or reactants of enzymes are called **substrates.** Compounds formed by one reaction in catabolic pathways and used by subsequent reactions are called **metabolic intermediates.** Some of these are the precursor metabolites. In this section we'll consider how *E. coli* growing aerobically at the expense of glucose as a carbon and energy source generates precursor metabolites, ATP, and reducing power.

Precursor Metabolites. *E. coli*'s various catabolic reactions produce all 12 of the precursor metabolites it and all cells need to grow (**Table 5.3**). These basic building materials for metabolism are metabolic intermediates of catabolic pathways. For example, glucose-6-phosphate is one precursor metabolite. It is produced from glucose by phosphorylation. Then a subsequent reaction converts leftover glucose-6-phosphate that isn't used as a precursor metabolite into a second precursor metabolite, fructose-6-phosphate. About 25 chemical reactions in *E. coli* are required to make the 12 precursor metabolites from glucose.

No single catabolic pathway produces all 12 precursor metabolites. A minimum of three different pathways are required: (1) **glycolysis,** which produces six precursor metabolites; (2) the **tricarboxylic acid (TCA) cycle,** which produces four more; and (3) the **pentose phosphate pathway,** which produces the final two. We'll look at these three essential pathways later in this chapter.

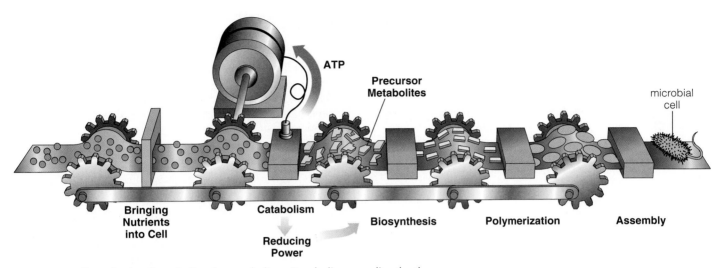

FIGURE 5.4 Central role of catabolism in metabolism. Catabolism supplies the three essential metabolic requirements: precursor metabolites for biosynthesis; reducing power, used largely for biosynthesis; and ATP, which drives all steps of the process.

Reducing Power. Many steps in a catabolic pathway involve **oxidation** (loss of electrons) from a metabolic intermediate; a few involve **reduction** (gain of electrons). Because only minuscule amounts of electrons can exist free in solution, oxidations and reductions are always linked. That is, as one compound loses electrons through oxidation, another compound gains them through reduction. As a result, the overall reaction, in which one compound is oxidized and the other is reduced, is called an oxidation-reduction reaction, often abbreviated as **redox reaction.** Redox reactions are always balanced. That is, the component being reduced uses all the electrons that the oxidized compound loses.

The term *oxidation* implies the addition of molecular oxygen. In inorganic chemical reactions, such as the oxidation of metallic iron to iron oxide (rust), oxygen gas does participate. The familiar reaction

$$\underset{\text{Metallic Iron}}{4\,Fe} + \underset{\text{Oxygen Gas}}{3\,O_2} \longrightarrow \underset{\text{Iron Oxide (Rust)}}{2\,Fe_2O_3}$$

illustrates the basic principles of redox reactions. Each iron atom oxidized loses three electrons (e^-):

$$Fe \longrightarrow Fe^{3+} + 3\,e^-$$

Each oxygen molecule is reduced by acquiring four electrons:

$$O_2 + 4\,e^- \longrightarrow O^{2-}$$

TABLE 5.3 The 12 Precursor Metabolites from Which All Cell Structures Are Made

Precursor Metabolite	Catabolic Pathway That Leads to Its Synthesis
Glucose-6-phosphate	Glycolysis
Fructose-6-phosphate	Glycolysis
Triose phosphate	Glycolysis
3-Phosphoglycerate	Glycolysis
Phosphoenolpyruvate	Glycolysis
Pyruvate	Glycolysis
Acetyl CoA	TCA cycle
Alpha-ketoglutarate	TCA cycle
Succinyl CoA	TCA cycle
Oxaloacetate	TCA cycle
Ribose-5-phosphate	Pentose phosphate
Erythrose-4-phosphate	Pentose phosphate

A total of 4 atoms of iron produce 12 electrons that are used by 3 molecules of oxygen, producing the balanced equation.

$$\underset{\text{Metallic Iron}}{4\,Fe} + \underset{\text{Oxygen Gas}}{3\,O_2} \longrightarrow \underset{\text{Iron Oxide (Rust)}}{2\,Fe_2O_3}$$

LARGER FIELD

REDUCING REDUCTIONISM

By the 1960s biochemistry had developed a good set of scientific tools for studying metabolism. They were based on reductionism, the philosophical approach that says that even highly complex systems (such as metabolism) can be understood by studying each component part and then collating the accumulated bits of information to produce a picture of the whole system. In the case of biochemical reductionism, the methods were well established: Break the cell, purify and characterize the individual enzymes, and analyze how the reactions fit together to make metabolic sense. If the product of one enzyme-catalyzed reaction is the substrate for another, the two reactions must function sequentially in a metabolic pathway. If the end product of a pathway is the starting material for another, they interconnect. It seemed only a matter of time until biochemists would be able to reconstruct the entire series of reactions that convert the raw materials in a medium into a microbial cell.

Biochemical reductionism is a powerful approach, and it worked very well. It identified most metabolic reactions, including those that generate ATP by substrate-level phosphorylation. But it failed completely when applied to formation of ATP by electron transport chains. When the cell was broken, electron transport continued, but formation of ATP stopped completely. Researchers speculated that perhaps the ATP-forming reactions were extremely **labile** (unstable), so they sought gentler ways to break the cell. But even the gentlest yielded no success.

In 1961 Peter Mitchell, a Scottish biochemist, took a different line of reasoning. Because components of the electron transport chain are embedded in the cell membrane, he speculated that ATP formation by electron transport occurred only when the cell membrane was intact. If this were true, then perhaps electron transport built a concentration gradient of protons across the cell membrane and the gradient drove protons back into the cell through a channel generating ATP from ADP. Scientists were slow to accept Mitchell's radical proposal (which he called **chemiosmosis**). But even the most skeptical were convinced when an intact but empty cell membrane was shown to convert ADP to ATP if acid was added to the suspending medium to establish an artificial proton gradient across the membrane. The experiment showed that the intact cell membrane alone could make ATP from ADP if provided with a proton gradient. Mitchell was awarded a Nobel Prize in 1978 for his remarkable achievement.

However, metabolic oxidations, including those that occur in the catabolic pathways, are different from inorganic oxidations in two ways:

1. In metabolic oxidations, oxygen gas seldom participates in the reaction.

2. In metabolic oxidations, electrons are sometimes transferred by themselves, but usually a proton is transferred along with the electron. This combination of a proton and an electron is a hydrogen atom:

$$H^+ \quad + \quad e^- \quad \longrightarrow \quad H$$
Proton   Electron   Hydrogen Atom

Oxidations in which protons are removed together with electrons are called **dehydrogenation reactions.** Reductions in which they are added together are called **hydrogenation reactions.**

In metabolic redox reactions, hydrogen atoms are not transferred directly from one metabolic intermediate to another. Instead, hydrogen atoms are usually transferred to or from one of two pyridine nucleotides, NAD or NADP (designated NAD[P] when the distinction between them is not important). Either of these two molecules in its reduced form stores the cell's reserves of hydrogen atoms, or its **reducing power** (**Figure 5.5**). Reduced molecules of NAD(P) (designated NAD[P]H) are needed to reduce metabolic intermediates and thereby drive subsequent steps in cellular synthesis. Reducing power can also be used to generate ATP. And often the reverse is possible: ATP can be used to generate reducing power (see ATP: Stored Energy). *The cell's stores of chemical energy (ATP) and reducing power are interchangeable.*

Reducing power is generated in the many dehydrogenation reactions that occur in catabolic pathways. In these reactions, electrons removed in the form of hydrogen atoms are transferred to a molecule of NAD(P). For example, in the first reaction of the pentose phosphate catabolic pathway, glucose-6-phosphate is oxidized to 6-phosphogluconolactone and NADP is reduced.

$$\text{glucose-6-phosphate} + NADP^+ \longrightarrow$$
$$\text{6-phosphogluconolactone} + NADPH + H^+$$

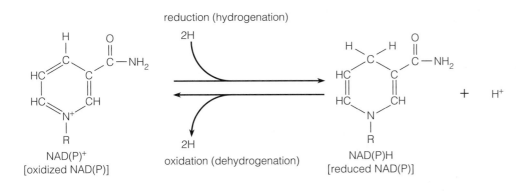

reduction (hydrogenation)

2H

oxidation (dehydrogenation)

2H

NAD(P)⁺
[oxidized NAD(P)]

NAD(P)H
[reduced NAD(P)]

$+$ H^+

FIGURE 5.5 Reducing power. When the oxidized form of the pyridine portion of pyridine nucleotide (R represents the nucleotide portion) accepts two hydrogen atoms from a catabolic reaction, it is converted to a reduced form. The reduced form, in turn, can donate two hydrogen atoms to other reactions and be converted back to the oxidized form.

Note that the oxidized form of NAD(P) is designated as having a single positive charge, NAD(P)⁺, and the reduced form as having a single attached hydrogen atom, NAD(P)H. In spite of this, reduction of NAD(P)⁺ or oxidation of NAD(P)H always involves the exchange of two hydrogen atoms.

$$NAD(P)^+ + 2H \longleftrightarrow NAD(P)H + H^+$$

Some of the NAD(P)H made during catabolism is reoxidized during biosynthesis. For example, one step in the biosynthesis of the amino acids tyrosine and phenylalanine involves the reduction of dehydroshikimate to shikimate at the expense of reduced NADP.

$$\text{dehydroshikimate} + NADPH + H^+ \longrightarrow \text{shikimate} + NADPH^+$$

But reducing power is not by itself enough to drive metabolic reactions. Stored metabolic energy is also needed.

ATP: Stored Energy. The principal compound that stores chemical energy in *E. coli* and all other cells is adenosine triphosphate, or ATP. The energy-storage capabilities of ATP depend on the two bonds that join the three phosphate groups in the molecule. To understand how these bonds store energy, we'll briefly review the role of energy in breaking and forming chemical bonds.

A chemical bond holds two atoms together. So it takes energy to break any chemical bond. But when reactants are activated (Chapter 2), chemical bonds form spontaneously, releasing energy. In any chemical reaction some bonds are broken and others are formed. Chemical reactions proceed only if the net energy released by forming new bonds exceeds that required to break existing bonds (Chapter 2). The energy difference between old bonds and new ones, which determines whether a reaction can occur, is called free energy. The key here is energy difference: More unstable reactant bonds and more stable product bonds drive a reaction.

The bonds that join the three phosphate groups to ATP are highly unstable. They are easy to break and there-

fore highly reactive bonds. Because of their extraordinary reactivity, they are called high-energy bonds and designated by the special bond symbol ~ (**Figure 5.6**). Because of the reactivity of these bonds, the third and sometimes the second and third phosphate groups of ATP are readily donated to other compounds. The compounds that receive a phosphate group from ATP are termed phosphorylated compounds. They participate in chemical reactions that would not occur if the reactants were unphosphorylated.

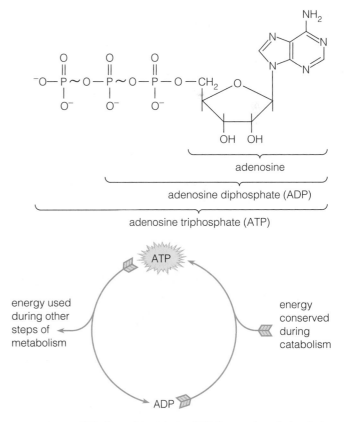

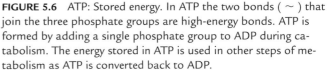

FIGURE 5.6 ATP: Stored energy. In ATP the two bonds (~) that join the three phosphate groups are high-energy bonds. ATP is formed by adding a single phosphate group to ADP during catabolism. The energy stored in ATP is used in other steps of metabolism as ATP is converted back to ADP.

This is why we say that the energy "stored" in the bonds of ATP is used to drive other metabolic reactions.

ATP Formation. But how are high-energy phosphate bonds formed in the first place? ATP is always formed by adding a single phosphate group to adenosine diphosphate (ADP). Some ADP is made from precursor metabolites by biosynthesis. The rest is made from ATP when it gives up a phosphate to drive some other metabolic reaction. For example, in one step of the biosynthesis of the amino acid proline, glutamate is phosphorylated by ATP, yielding glutamylphosphate and ADP.

glutamate + ATP ⟶ glutamylphosphate + ADP

ATP can be regenerated from ADP in two different ways: substrate-level phosphorylation and chemiosmosis.

Substrate-Level Phosphorylation. In substrate-level phosphorylation, ADP obtains a phosphate group attached to a metabolic intermediate by a high-energy bond. The high reactivity of the bond in the metabolic intermediate enables the phosphate group to be transferred to ADP, converting it to ATP (**Figure 5.7**). Most phosphate-containing metabolic intermediates, however, do not contain high-energy bonds and cannot be used to generate ATP.

Chemiosmosis. Chemiosmosis forms ATP from ADP by means of an enzyme called ATPase. This remarkable enzyme catalyzes the conversion of ADP to ATP as a result of a series of chemical events that occur in and around a membrane. In prokaryotes it is the cell membrane. In eukaryotes it is the mitochondrial membrane (**Figure 5.8**).

The energy for the chemiosmotic formation of ATP is a concentration gradient formed across the membrane. During metabolism, certain ions—usually protons, but sodium ions in some cells—are actively transported out of the cell, and so their concentration outside the cell membrane exceeds their concentration inside the cell. Because any concentration gradient naturally tends to equalize itself, the proton gradient across the cell membrane of prokaryotes (high outside, low inside) tends to force protons back into the cell (Chapter 4).

The proton gradient across a membrane constitutes a chemical energy potential, much as water stored in an elevated tank constitutes a mechanical energy potential. When water flows out of the tank by gravity, it can be passed through a waterwheel and be made to do work, such as generating electricity. When protons flow down a concentration gradient and across the membrane, they pass through a channel in the membrane that contains an ATPase enzyme (the only available channel through the

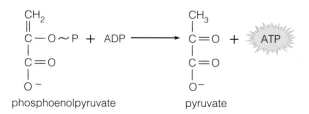

FIGURE 5.7 Substrate level phosphorylation. In this example a phosphate group joined to phosphoenolpyruvate by a high-energy bond (∼) is transferred to ADP, producing pyruvate and ATP.

otherwise proton-impermeable membrane) and do the work of generating ATP from ADP.

But creating a proton gradient, like storing water in an elevated tank, requires energy. How does *E. coli* accomplish this task chemically? The process begins when two hydrogen atoms (each consisting of a proton and an electron), usually from NADH, are transferred to one of a series of compounds embedded in the cell membrane. These compounds, collectively called an electron transport chain, perform a sequential series of oxidation-reduction reactions as they transfer the reducing power from one molecule to another. Electron transport chains differ from one organism to another. But all contain certain compounds, such as quinones, that accept only hydrogen atoms and other compounds, such as cytochromes, that accept only electrons.

As reducing power flows through the electron transport chain, the energy released is used to push protons out of the cell. At certain points in the chain, components act as pumps that push protons out of the cell. At other points in the chain, protons are moved out as hydrogen atoms are separated into protons and electrons. That happens when a hydrogen-carrying component of the electron transport chain transfers its reducing power to an electron-accepting component. During this transfer, protons are released on the outer side of the membrane, contributing to the proton gradient as the electrons are transferred to the electron-accepting component of the chain.

Eventually the chain ends. The compound at the end of an electron chain is called the **terminal electron acceptor.** In the case of aerobic respiration, it is oxygen. By accepting electrons from the electron transport chain, oxygen is reduced to water.

$$1/2\ O_2 + 2\ e^- + 2H^+ \longrightarrow H_2O$$

Thus the immediate consequences of reducing power flowing down such an electron transport chain are threefold: A proton gradient is established, oxygen is consumed, and water is formed. The net result is generation of ATP.

The composition of electron chains varies depending on which metabolic intermediate is donating reducing power to it and how much oxygen is available to the cell.

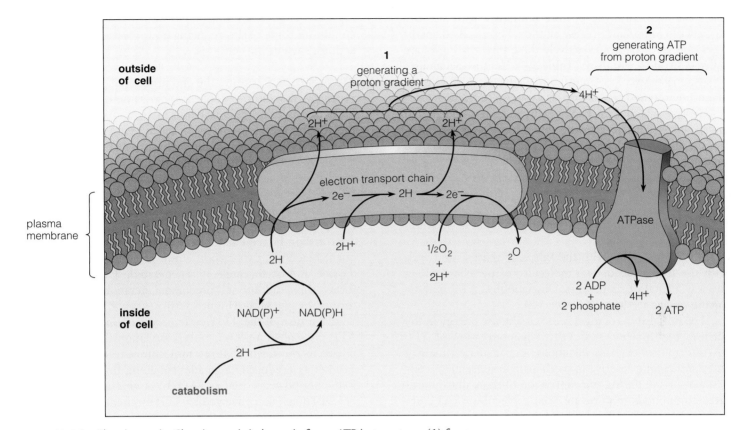

FIGURE 5.8 Chemiosmosis. Chemiosmosis in bacteria forms ATP in two steps: (1) forming a proton gradient as reducing power flows down an electron transport chain and (2) using the energy stored in the gradient to generate ATP as protons flow back into the cell through ATPase.

When NADH donates a pair of hydrogen atoms to the *E. coli* electron transport chain in the presence of adequate oxygen, four protons are pushed out of the cell. That's enough to form two molecules of ATP as they flow back into the cell through the membrane-bound ATPase.

Chemiosmotic synthesis of ATP occurs in most cellular organisms, but its efficiency varies greatly. Eukaryotic organisms generate about three molecules of ATP per pair of electrons, whereas some bacteria generate only one. Even *E. coli* makes only one ATP molecule per hydrogen pair if the concentration of oxygen in its environment is low.

As we've discussed, chemiosmosis usually converts reducing power in the form of NADH into ATP. But some organisms do the reverse. They use ATP to generate reducing power by **reverse electron flow.** That is, they force electrons/hydrogen atoms back through an electron transport chain and reduce NAD(P)⁺ to NAD(P)H. Thus the two driving forces of metabolism, reducing power (NAD[P]H) and stored energy (ATP), are interconvertible. Similarly, ATP and a proton gradient are interconvertible.

We've seen how protons flowing into the cell through the membrane-bound ATPase make ATP from ADP. But

ATPase can also hydrolyze ATP to ADP and use the energy to force protons out of the cell, establishing a proton gradient. The capacity to interconvert a proton gradient and ATP is essential because both forms of energy are needed by the cell. ATP is needed to drive many steps in metabolism. Proton gradients are needed for other purposes. They drive entry mechanisms and turn flagella for motility. In summary, by chemiosmosis, reducing power, a transmembrane proton gradient, and ATP are all interchangeable.

reducing power ⟷ proton gradient ⟷ ATP

Now that we have discussed how ATP and reducing power are generated in certain reactions, let's look at the overall flow of catabolism. That is, how groups of reactions are organized into catabolic pathways.

Catabolic Pathways. Catabolic pathways are the enzyme-catalyzed reaction sequences that transform substrate molecules into precursor metabolites, ATP, and reducing power. *E. coli* has catabolic pathways that metabolize many different substrates. But three pathways—glycolysis, the TCA cycle, and the pentose phosphate pathway—are particularly important because all the other metabolic pathways flow into them. Collectively these three pathways are called **central metabolism** (Table 5.4). Central metabolism is like a major river system. All other catabolic pathways are like tributaries and creeks that flow into these rivers.

Central metabolism begins with the sugar glucose. It ends, during aerobic growth, with precursor metabolites, carbon dioxide, and water. The three pathways of central metabolism are widely distributed in nature. They are found in organisms as diverse as *E. coli* and human beings, because all cellular organisms need the products of these reactions.

Glycolysis. The metabolic pathway of glycolysis is shown in **Figure 5.9.**

Glycolysis begins with the substrate glucose and ends with the formation of two molecules of pyruvate. Along the way a total of six precursor metabolites, ATP, and reducing power are formed (Table 5.4).

The initial steps of glycolysis actually use energy in the form of two molecules of ATP. This conversion of ATP to ADP phosphorylates metabolic intermediates, making them sufficiently reactive to participate in subsequent conversions. Then during the conversion of these phosphorylated compounds to pyruvate, four molecules of ATP are formed by substrate level phosphorylation and two molecules of NAD^+ are reduced to NADH. The net yield of ATP by substrate level phosphorylation is two (four produced minus two used). The total yield of ATP can be as high as six if the two molecules of NADH donate their four hydrogen atoms to an electron transport chain, yielding four ATPs by chemiosmosis. But under most growth conditions some of this reducing power is used in other reactions, such as biosynthesis.

The TCA Cycle. Pyruvate, the end product of glycolysis, is a precursor metabolite (Table 5.4), so some of it is used in biosynthesis. The rest is oxidized to another precursor metabolite, acetyl coenzyme A (acetyl CoA). Most acetyl CoA enters a cyclic catabolic pathway called the TCA cycle or the Krebs cycle. The TCA cycle forms more precursor metabolites, ATP by substrate-level phosphorylation, NAD(P)H, and carbon dioxide (**Figure 5.10**).

Acetyl CoA enters the TCA cycle by combining with the four-carbon precursor metabolite oxaloacetate to form a six-carbon intermediate, citrate. In a series of six subsequent reactions, two carbon atoms are released as carbon dioxide and oxaloacetate is regenerated. The TCA cycle forms three additional precursor metabolites to those formed by glycolysis (Figure 5.10).

Each turn of the TCA cycle produces one molecule of ATP by substrate-level phosphorylation and reducing power in the form of three molecules of NAD(P)H and another as a reduced flavin adenine dinucleotide ($FADH_2$). Reducing power stored as $FADH_2$, as well as NADH and NADPH, can be converted to ATP by chemiosmosis.

Pentose Phosphate Pathway. The key metabolic intermediates and reactions of the pentose phosphate pathway are shown in **Figure 5.11.**

At first glance the pentose phosphate pathway seems unnecessary. It begins with one intermediate of glycolysis,

TABLE 5.4 Principal Catabolic Pathways

Pathway	Number of Precursor Metabolites Formed[a]	Net Yield of ATP[b]	Yield of Reducing Power	Carbon-containing End Products
Glycolytic pathway	6	$(4 - 2) = 2$	2 NADH	2 pyruvate
TCA cycle[c]	4	1	1 NADPH	3 CO_2
			3 NADH	
			1 $FADH_2$	
Pentose phosphate pathway	2[d]	−1	2 NADPH	3 CO_2
				1 triose phosphate

[a] When a precursor metabolite molecule is used for biosynthesis, the yield of ATP, reducing power, and end products formed further down the pathway are reduced.

[b] The number of ATP molecules produced by substrate level phosphorylation minus those utilized. In some cases, reducing power is used to generate additional ATP.

[c] Yields calculated per molecule of pyruvic acid metabolized.

[d] Three others (glucose-6-phosphate, glyceraldehyde-3-phosphate, and fructose-6-phosphate) produced by the glycolytic pathway are also produced here.

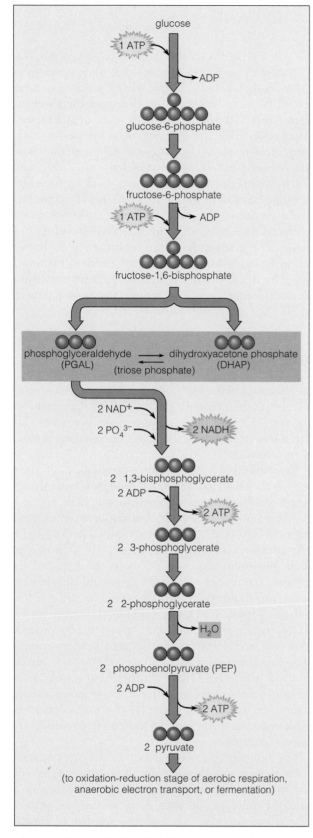

FIGURE 5.9 Glycolysis. Glycolysis is a pathway of central metabolism that converts a molecule of glucose into two molecules of pyruvate with a net yield of two molecules of ATP and two molecules of NADH, along with six precursor metabolites (shown in colored boxes).

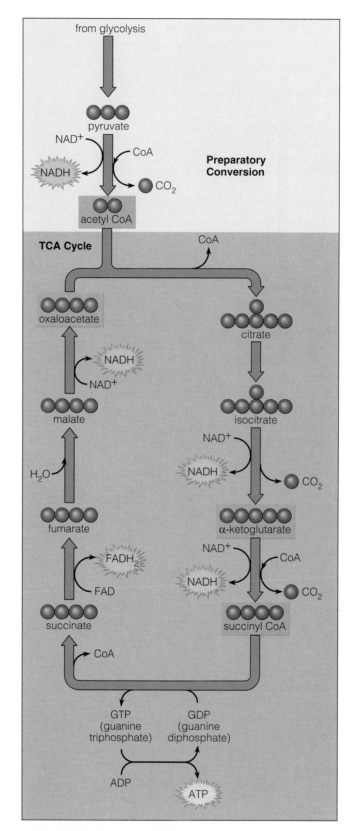

FIGURE 5.10 The TCA cycle. The tricarboxylic acid TCA cycle converts pyruvate into CO_2, reducing power, ATP (by substrate-level phosphorylation), and four precursor metabolites, shown in colored boxes. ($FADH_2$ is a carrier of reducing power capable of converting $NADH^+$ into NADH.)

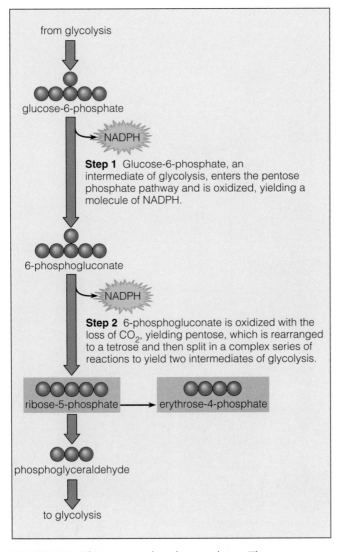

from glycolysis

glucose-6-phosphate

NADPH

Step 1 Glucose-6-phosphate, an intermediate of glycolysis, enters the pentose phosphate pathway and is oxidized, yielding a molecule of NADPH.

6-phosphogluconate

NADPH

Step 2 6-phosphogluconate is oxidized with the loss of CO_2, yielding pentose, which is rearranged to a tetrose and then split in a complex series of reactions to yield two intermediates of glycolysis.

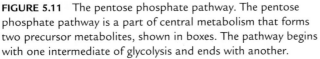

ribose-5-phosphate → erythrose-4-phosphate

phosphoglyceraldehyde

to glycolysis

FIGURE 5.11 The pentose phosphate pathway. The pentose phosphate pathway is a part of central metabolism that forms two precursor metabolites, shown in boxes. The pathway begins with one intermediate of glycolysis and ends with another.

new cell. By three catabolic pathways—glycolysis, TCA, and pentose phosphate—the cell gets all it needs to proceed with metabolism. These are 12 precursor metabolites, ATP, and reducing power. Precursor metabolites are intermediates of these pathways that are diverted to biosynthesis. ATP is generated by substrate-level phosphorylation in two of the pathways. All the pathways generate reducing power, some of which is used to generate ATP by chemiosmosis. The rest is used to drive other metabolic steps. A proton gradient across the cytoplasmic membrane, formed by passage of reducing power down an electron transport chain, is an intermediate of ATP formation. The gradient itself can do some metabolic work, such as bringing nutrients into the cell and turning flagella. The whole chemiosmotic system is reversible under some conditions: reducing power, the proton gradient, and ATP are interconvertible.

Biosynthesis

The next stage of the metabolic factory, called biosynthesis, uses the three products of catabolism—precursor metabolites, ATP, and reducing power—to construct all the small molecules it needs, including the building blocks for macromolecules.

Biosynthesis is a major part of metabolism. Of the approximately 2604 enzymes *E. coli* produces, more than 400 (almost a fifth) are used for biosynthesis (Table 5.1).

Like catabolic reactions, biosynthesis reactions are organized into reaction sequences, called pathways. Each pathway brings about the stepwise conversion of a precursor metabolite into a small molecule. Some of these pathways are branched. They use several different precursor metabolites or produce several different end products (**Figure 5.12**). For example, six amino acids (aspartate, asparagine, isoleucine, lysine, methionine, and threonine) are synthesized from two precursor metabolites (oxaloacetate and pyruvate) through a multiple branched pathway. Other pathways are unbranched. They produce a single small molecule from a single precursor metabolite. For example, the amino acid histidine is synthesized through an unbranched pathway from ribose-5-phosphate.

The biosynthetic pathways of *E. coli* are remarkably similar to biosynthetic pathways in all other organisms—bacteria, archaea, eukaryotic microorganisms, plants, and animals. Differences do arise, however, because some organisms cannot make the enzymes needed to complete certain pathways. But almost all organisms that can make a particular end product use the same pathway.

Essential small molecules (in addition to building blocks of macromolecules) include **coenzymes.** These compounds

glucose-6-phosphate, and ends with another, phosphoglyceraldehyde. It acts as an alternate route between these two points in glycolysis. But its role in catabolism is critically important because it forms two precursor metabolites that cannot be formed by either glycolysis or the TCA cycle. The pentose phosphate pathway produces no ATP by substrate-level phosphorylation, but it does form two molecules of NADPH.

The word *catabolism*, based on Greek roots, means "breaking down." It was chosen because during catabolism some substrates, such as glucose, are broken down into smaller molecules, the precursor metabolites, and finally carbon dioxide. But as we've seen, catabolism is actually the first step in building up the components of a

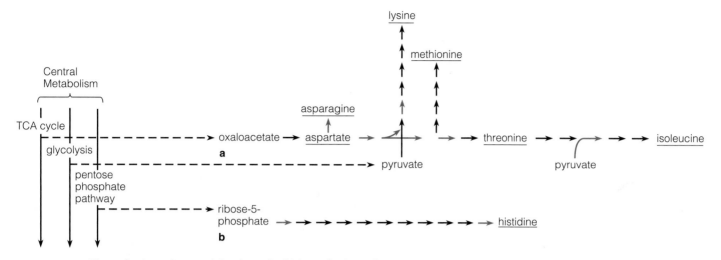

FIGURE 5.12 Biosynthetic pathways. (a) A branched biosynthetic pathway converts two precursor metabolites (pyruvate and oxaloacetate from glycolysis and the TCA cycle, respectively) into 6 amino acids (underlined) by 22 enzyme-catalyzed reactions (arrows). The pathway uses three molecules of ATP (red arrows) and four molecules of NADPH (green arrows). (b) An unbranched biosynthetic pathway converts a single precursor metabolite (ribose-5-phosphate from the pentose phosphate pathway) to a single amino acid (histidine) by 11 reactions. This pathway uses one ATP and one NADPH.

act together with enzymes to catalyze metabolic reactions. Organisms that cannot synthesize their own coenzymes grow only if small amounts of them or their precursors are supplied from an external source. The ready-made molecules that act as coenzymes or their precursors are called **vitamins.**

Organisms that cannot make a given essential small molecule grow only if that molecule is provided ready-made. Microorganisms must have it in their medium, animals in their diet. *E. coli* has a complete set of biosynthetic pathways. It can make all 20 amino acids needed to build proteins, and all its coenzymes, so it can grow in a medium that contains no amino acids or vitamins. Human beings, on the other hand, are unable to make nine amino acids and many coenzymes, so we must have these amino acids and many vitamins in our diets.

The principal driving force that fuels biosynthesis is reducing power, stored mostly in the form of NADPH. Reducing power is required because some biosynthetic reactions are reductions. Some ATP is also required, but the major expenditure of ATP in the making of a new cell occurs in the next step—the polymerization of building blocks into macromolecules.

To recapitulate, biosynthetic pathways flow out from central metabolism rather like minor catabolic pathways flow into it. These biosynthetic pathways make all the small molecules the cell needs.

Polymerization

In the next step of the metabolic pathway, called polymerization, the molecular building blocks made by biosynthesis are joined together to form macromolecules.[1] The major cellular polymerization reactions lead to the synthesis of DNA, RNA, proteins, polysaccharides, and peptidoglycan (**Table 5.5**).

For most macromolecules to be active, their building blocks must be joined in a specific order. The distinctive primary structure of a protein, for example, is determined by the order in which its particular amino acids are joined together (Chapter 2). *E. coli* must produce the approximately 2000 proteins, each with a distinct primary structure.

The ordering of polymerization reactions is determined by the information stored in the organism's DNA. Some polymerization reactions—those that build DNA, RNA, and protein—are directly determined by the physical structure of DNA. The ordering of building blocks in these macromolecules is accomplished as information flows from DNA to RNA and then to protein synthesis. We'll discuss how this occurs in Chapter 6.

---

[1]We use polymerization reactions to make plastics. In these processes the starting materials are usually called **monomers;** the products are called **polymers.** For example, nylon is the polymer of dicarboxylic acid and diamine monomers. The terms *monomer* and *polymer* are also used sometimes in place of *building block* and *macromolecule.*

TABLE 5.5 The Building Blocks Required to Synthesize a Cell's Macromolecules

Macromolecules	Building Blocks
Proteins	20 amino acids
Nucleic acids	Nucleotides consisting of:
RNA	Adenine, guanine, cytosine, uracil, phosphate, ribose
DNA	Adenine, guanine, cytosine, thymine, phosphate, deoxyribose
Polysaccharides	Sugars
Peptidoglycan	*N*-acetylmuramic acid, *N*-acetylglucosamine, 5 amino acids[a]
Lipids	Because lipids are defined only by physical properties, they are made of many different building blocks

[a] The number of amino acids in peptidoglycan varies by species.

Other polymerization reactions are indirectly determined by the information contained in DNA. For example, the polymerization reactions that build molecules such as polysaccharides and peptidoglycan are indirectly determined. In these cases, building blocks are ordered by the enzymes that catalyze the polymerization reactions. Because enzymes are proteins and protein structure is determined directly by DNA, the polymerization reactions that they catalyze are indirectly determined by DNA.

Ordering of building blocks is crucial in biological polymerizations, and so is the formation of the chemical bonds that join them to create macromolecules. Joining building blocks directly to form macromolecules is usually energetically impossible. For polymerization to occur the building block must first be activated by expending chemical energy in the form of ATP. In some cases ATP reacts directly with the building block, thereby supplying sufficient energy for polymerization to proceed.

The synthesis of glycogen, a metabolic reserve product stored by some microorganisms, including *E. coli*, requires a somewhat more complicated activation of the building block (**Figure 5.13**). Glucose, the building block of glycogen, is first phosphorylated by ATP to form glucose-1-phosphate and is then it is converted by another molecule of ATP to the activated building block, ADP-glucose. ADP-glucose is polymerized into glycogen by the enzyme, glycogen synthase.

Glycogen synthesis thus illustrates the basic principles of polymerization reactions: (1) The building block is activated by an ATP-utilizing reaction or a short pathway, and (2) activated building blocks enter into the enzyme-catalyzed polymerization, releasing the activating group, in this case ADP (Figure 5.13).

Assembly

In the final step of the metabolic factory, macromolecules made by polymerization are assembled into cellular structures. Assembly may occur spontaneously or as a result of

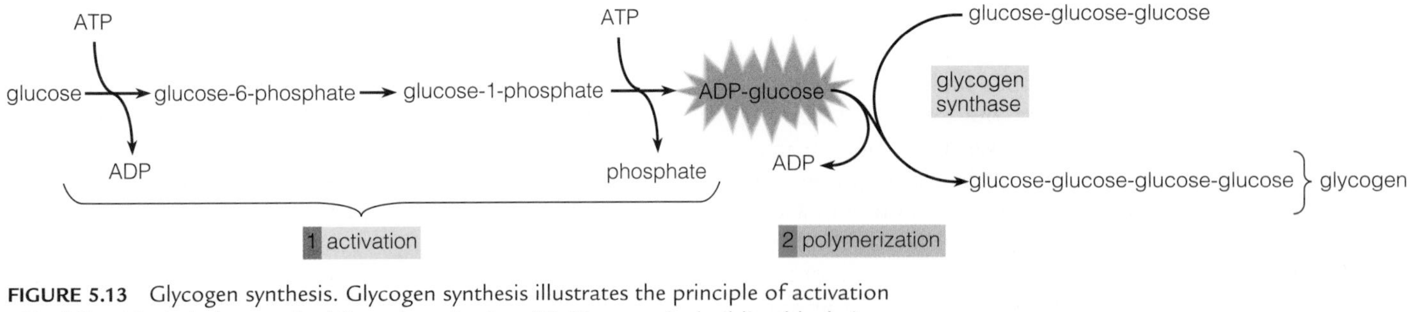

FIGURE 5.13 Glycogen synthesis. Glycogen synthesis illustrates the principle of activation of building blocks being required for polymerization. (1) Glucose, the building block, is activated in a pathway using two molecules of ATP to form ADP-glucose. (2) It is then polymerized. In the step shown here, one glucose unit is added to the glycogen chain as an ADP is released. (Although ATP-glucose is the activated form that *E. coli* and most bacteria use make glycogen, other organisms, including humans, use UTP-glucose.)

SHARPER FOCUS

NOT JUST PROTEINS ANYMORE

As biology advances, unifying principles emerge that simplify and pull together the masses of facts. Sometimes new principles force biologists to give up long-held convictions. For example, the belief that all enzymes are proteins and only they catalyze metabolic reactions was shattered in 1987 by T. R. Cech at the University of Colorado. Cech showed that an RNA molecule had catalytic properties. It could cut a long strand of RNA in two places and rejoin two of the resulting three RNA pieces, forming a shorter RNA molecule with a missing midsection. Such RNA processing is an essential step in gene expression of eukaryotes. Soon other RNA molecules were found that had similar catalytic activities. They were named ribozymes, short for ribonucleic acid (RNA) enzymes. However, known ribozyme-catalyzed reactions all involved cutting and rejoining other RNA molecules. Maybe ribozymes were able to catalyze only this one highly specialized kind of reaction.

Not at all. In 1992 H. F. Noller and his colleagues at the University of California, Santa Cruz, found that a ribozyme catalyzes what is probably the cell's most important metabolic reaction, the formation of peptide bonds that link amino acids to form proteins. The ribozyme itself is a portion of the RNA component of ribosomes.

The discovery of ribozymes has exciting implications about the origin of life. The fact that ribozymes can form the bonds that make RNA means RNA can reproduce itself. And ribozymes form the bonds that make protein. Thus RNA can reproduce itself and make protein! An RNA molecule with both these properties would have the fundamental characteristics of life.

reactions catalyzed by enzymes. Spontaneous assembly, also called self-assembly, is an intrinsic property of certain proteins. For example, flagellin, the molecular component of flagella on bacteria such as *E. coli*, self-assembles. The test of self-assembly is whether or not it will proceed spontaneously in vitro (outside the living organism). Under proper conditions of flagellin concentration and pH, and with the addition of a primer, a solution of flagellin will assemble spontaneously to form a flagellum with a characteristic width and helical shape. Even much more complex cellular structures such as ribosomes pass the in vitro test of self-assembly. Ribosomes, which are composed of 54 different proteins and 3 different RNA molecules, can assemble from purified components under carefully controlled laboratory conditions.

Formation of the bacterial cell wall is an example of assembly that is catalyzed by enzymes. Short units of peptidoglycan are released into the periplasm, where they are assembled into an intact wall by enzyme-catalyzed reactions. These wall-building enzymes are called penicillin-binding proteins because the antibiotic penicillin binds to them (Chapter 21). The particular **penicillin-binding proteins** that a cell contains determine the shape of the bacterial cell wall and thus the shape of the cell (Chapter 4).

A prerequisite of assembly and, for that matter, most cellular functions is **protein folding** (formation of tertiary structure; Chapter 2). How proteins attain the proper structure out of the many possible ones has been a topic of intense investigation over the past few years. Remarkably, the process has been found to be guided by other proteins called **chaperonins** (named by analogy to a human chaperon). Chaperonins bind to unfolded, nascent, misfolded proteins. They release them again only after they have become properly folded.

Assembly is the last step of the metabolic factory. The cell is ready to divide.

ANAEROBIC METABOLISM

The story we have just completed of how *E. coli* builds all the components of another cell identical to itself when growing aerobically is typical of the metabolism of many microorganisms. But many microorganisms can grow in the absence of oxygen. Now let's look at this slightly different situation—anaerobic metabolism. *E. coli* and other facultative anaerobes are capable of both aerobic and anaerobic metabolism; strict anaerobes rely exclusively on anaerobic metabolism.

The critical difference between aerobic and anaerobic metabolism lies in how ATP is generated. In aerobic metabolism, *E. coli* makes most of its ATP by aerobic respiration, producing a proton gradient by an electron transport chain with oxygen as its terminal electron acceptor. In the absence of oxygen the electron transport chain cannot function in this way. Thus aerobic respiration is impossible.

There are two ways that cells can make ATP from organic nutrients in the absence of oxygen. One, called **anaerobic respiration,** uses an electron transport chain with a compound other than oxygen as the terminal electron acceptor. The second, called **fermentation,** depends entirely on substrate-level phosphorylation.

Anaerobic Respiration

Aerobic and anaerobic respiration are similar processes. Both depend on an electron transport chain to form a proton gradient that can be used to generate ATP. The chains themselves and their yield of ATP differ somewhat. But the critical difference is the terminal electron acceptor. In aerobic respiration, oxygen accepts electrons and is reduced to water. In anaerobic respiration, another compound is reduced by accepting these electrons. Compounds that can act as a terminal electron acceptor in anaerobic respiration include sulfate, nitrate, fumarate, and trimethylamine oxide (**Table 5.6**). *E. coli*, for example, can use nitrate, fumarate, or trimethylamine oxide as an electron acceptor if oxygen is not available.

The reduction of some of the terminal electron acceptors of anaerobic respiration play critical roles in the cycles of matter that are essential to life (Chapter 28); others play less global roles.

Fermentation

Fermentation is a form of anaerobic metabolism in which all ATP is generated by substrate-level phosphorylation.

Fermentation generates fewer molecules of ATP per molecule of substrate than do aerobic and anaerobic respiration. For example, *E. coli* derives about 28 ATP molecules from glucose by aerobic respiration but only about 3 by fermentation. Because many molecules of substrate must be metabolized to supply a cell's ATP requirements, microorganisms can grow by means of fermentation only if substrate is abundant.

There is a striking difference between fermentation and respiration with respect to generating reducing power. As we have seen, oxidation and reduction must be balanced in any cell, regardless of its form of metabolism, because electrons removed from one compound as it is oxidized must be accepted by another compound, which is reduced. During respiration, a great deal of reducing power is consumed by reducing oxygen or some other terminal electron acceptor. Reducing power cannot be used this way in fermentation. Because organisms growing by fermentation do not reduce oxygen or any other terminal electron acceptor, the metabolic reactions of fermentation must use almost as much reducing power as they form. (The rest is needed for other steps of metabolism.) This situation is possible only if substrate enters fermentation pathways at an intermediate state of oxidation—that is, neither highly oxidized nor highly reduced. Sugars, which are at an intermediate oxidation state, are therefore almost the only substrates that can be used in fermentation.

To illustrate the principles of fermentation, let's look at one specific example—lactic acid fermentation (**Figure 5.14**). This type of fermentation is carried out by lactic acid bacteria, the organisms that cause milk to sour and are used to produce acidic dairy products such as yogurt and buttermilk. When deprived of oxygen, muscle tissue of animals, including humans, also carries out lactic acid fermentation—by the same pathway. Lactic acid fermentation proceeds through the glycolysis pathway, which we discussed earlier in the chapter. One molecule of glucose is metabolized to produce two molecules of pyruvate, two molecules of ATP, and two of NADH. In order to reoxi-

---

TABLE 5.6 Some Terminal Electron Acceptors of Bacterial Electron Transport Chains

Type of Respiration	Terminal Electron Acceptor	Reduced Product
Aerobic respiration	Oxygen (O_2)	Water (H_2O)
Anaerobic respiration		
Sulfate reduction	Sulfate (SO_4^{2-})	Hydrogen sulfide (H_2S)
Nitrate reduction	Nitrate (NO_3^-)	Nitrite (NO_2^-)
Fumarate reduction	Fumarate (HOOC—CH=CH—COOH)	Succinate (HOOO—CH_2—CH_2—COOH)
Denitrification	Nitrate (NO_3^-)	Nitrogen gas (N_2)
Trimethylamine oxide reduction	Trimethylamine oxide	Trimethylamine

LARGER FIELD

THE NOSE KNOWS

It's easy to tell if an ocean fish is fresh—just smell it. Completely fresh fish is almost odorless, but if stored unfrozen for even a short time, it takes on a characteristic fishy smell that gets stronger. On the other hand, beef, lamb, and chicken don't smell bad until they are almost completely spoiled. What is it about fish that makes it smell so bad so fast? The answer is trimethylamine oxide and anaerobic respiration. The flesh of marine fish contains trimethylamine oxide, an almost odorless compound that many marine species of bacteria can use as a terminal electron acceptor in anaerobic respiration. In doing so, the bacteria reduce trimethylamine oxide to trimethylamine, a compound with an intensely fishy odor. Only a tiny amount of bacterial growth by anaerobic respiration forms enough trimethylamine for us to detect it by smell. If fish doesn't smell fishy, you can be sure it contains very few bacteria. It is either very fresh or was freshly frozen.

dize the two molecules of NADH (and thus allow fermentation to continue), pyruvate is reduced to lactic acid. In this way almost as much reducing power is oxidized as is formed. An organism cannot multiply using this fermentative pathway alone because the pathway does not generate all 12 precursor metabolites. They are formed by certain reactions of the pentose phosphate pathway and the TCA cycle. But the overwhelming majority of substrate is metabolized through the fermentative pathway to meet the cell's requirement for ATP. Huge amounts of the fermentative end product, lactic acid, result, which gives the products of this fermentation their characteristic acidic taste.

There are many different kinds of fermentation, and each is characteristic of a particular kind or group of microorganisms (**Table 5.7**). Most fermentations metabolize glucose to pyruvate by means of glycolysis, but they differ in how pyruvate is reduced and NADH reoxidized. In lactic acid fermentation, pyruvate is reduced directly to lactic acid. In alcoholic fermentation, which is typical of yeast, pyruvate is converted to CO_2 and ethanol. In the mixed acid fermentation, which is characteristic of *E. coli* and related bacteria, pyruvate is converted to at least six different end products.

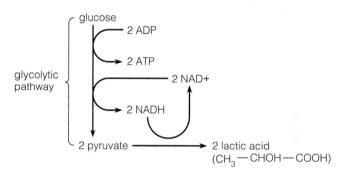

FIGURE 5.14 Fermentation. The lactic acid fermentation illustrates the principles of this form of anaerobic metabolism: ATP is formed only by substrate-level phosphorylation; NAD+ is reduced in one part of the pathway to NADH, which in another part is oxidized back to NAD+.

TABLE 5.7 Some Types of Fermentation

Fermentation	Some Organisms That Perform It	End Products
Alcoholic	Yeasts	Ethanol and CO_2
Lactic acid	Lactic acid bacteria	Lactic acid
Mixed acid	*Escherichia coli*	Lactic acid, acetic acid, formic acid, succinic acid, H_2, CO_2, ethanol
Butanediol	*Enterobacter aerogenes*	Lactic acid, acetic acid, formic acid, H_2, CO_2, 2,3-butanediol

NUTRITIONAL CLASSES OF MICROORGANIMS

As we've seen, *E. coli* must be supplied with at least one organic compound to make precursor metabolites, ATP, and reducing power. Not all microorganisms share this requirement. Some can make their precursor metabolites from an inorganic compound and carbon dioxide (CO_2). Some can generate ATP and reducing power from light energy; others can generate them by oxidizing inorganic compounds, such as iron. Now let's examine these various classes of microorganisms.

Microbiologists classify microorganisms into four nutritional classes, depending on (1) the source of carbon atoms they use to make precursor metabolites and (2) how they generate ATP and reducing power.

The class of organisms, which includes *E. coli*, that must be supplied with organic compounds as a source of carbon atoms from which to make precursor metabolites are called **heterotrophs** (different feeders). The class that are capable of using CO_2 as a source of all their carbon atoms are called **autotrophs** (self-feeders). Organisms, such as *E. coli*, that generate ATP and reducing power from chemical reactions are called **chemotrophs** (chemical feeders). Those that generate ATP and reducing power from light energy are called **phototrophs** (light feeders). All four possible combinations of generating precursor metabolites and ATP exist among microorganisms. There are **chemoautotrophs, chemoheterotrophs, photoautotrophs,** and **photoheterotrophs (Table 5.8)**.

First we'll look at how autotrophs (both photo- and chemo-) make precursor metabolites. Then we'll consider how chemoautotrophs and photoautotrophs generate ATP and reducing power.

Formation of Precursor Metabolites by Autotrophs

Autotrophs make precursor metabolites from CO_2. In other words, autotrophic organisms (certain microorganisms and plants) create organic molecules from an inorganic atmospheric gas. Not only do the organic molecules they create enable the autotrophs to synthesize their own cellular constituents; they supply organic nutrients for all heterotrophs. For example, plants, as autotrophs, use CO_2 to create organic molecules and grow, and animals eat plants or other animals that have eaten plants and catabolize the organic molecules to grow.

To synthesize precursor metabolites from CO_2, autotrophs have special metabolic pathways not found in heterotrophs. The most widespread of these pathways is the Calvin-Benson cycle found in plants and many autotrophic microorganisms (**Figure 5.15**). The critical step in this cycle is the incorporation of CO_2 into a preexisting organic molecule. As a consequence, the gaseous CO_2 becomes part of that organic molecule and becomes organic itself. The preexisting organic molecule is a five-carbon atom, phosphorylated sugar, ribulosebisphosphate. The product is a six-carbon atom sugar that splits immediately into two molecules of 3-phosphoglyceraldehyde. The remainder of the Calvin-Benson cycle regenerates another molecule of ribulosebisphosphate that can react with another molecule of CO_2. Each turn of the cycle converts one molecule of CO_2 into organic form. The Calvin-Benson cycle produces only two precursor metabolites. The other 10 precursor metabolites are made from intermediates of the Calvin-Benson cycle, which feed into slightly modified glycolysis, TCA, and pentose phosphate pathways. Thus the Calvin-Benson cycle together with the reactions of central metabolism can synthesize all 12 precursor metabolites from CO_2.

Unlike *E. coli*'s catabolic pathways, the Calvin-Benson cycle does not generate ATP or reducing power. Instead it uses them in large quantities: three molecules of ATP and two of NADPH for each molecule of CO_2 **fixed** (converted into nongaseous form). Thus autotrophs must generate ATP and reducing power by other means: from sunlight in the case of photoautotrophs or by the oxidation of inorganic compounds in the case of chemoautotrophs.

Various prokaryotes have other pathways that act like the Calvin-Benson cycle to fix CO_2 and connect with central metabolism.

TABLE 5.8 Nutritional Classes of Microorganisms

Source of Carbon Atoms	Source of Energy (ATP)	
	Chemical Reactions	Light Energy
Organic compounds	Chemoheterotrophs	Photoheterotrophs
CO_2	Chemoautotrophs	Photoautotrophs

Formation of ATP and Reducing Power by Photoautotrophs

Photoautotrophs obtain ATP and reducing power by one or the other of two types of photosynthesis: **oxygenic** (oxygen-producing) photosynthesis or **anoxygenic** (not oxygen-producing) photosynthesis. Certain bacteria (purple and green bacteria) carry out anoxygenic photosynthesis; other bacteria (cyanobacteria), plants, and algae carry out oxygenic photosynthesis. Both these processes depend on chlorophylls, which have the capacity to convert light energy into chemical energy.

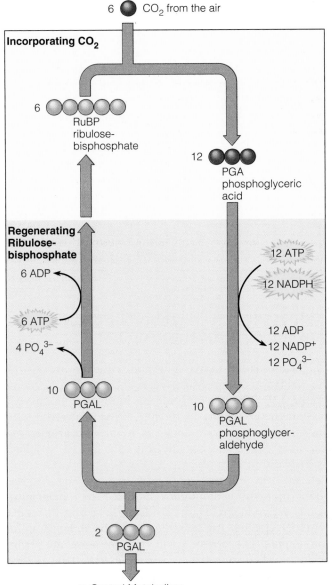

FIGURE 5.15 The Calvin-Benson cycle. The Calvin-Benson cycle is the way most autotrophs make precursor metabolites from CO_2.

Anoxygenic Photosynthesis. In anoxygenic photosynthesis (**Figure 5.16**), light energy ejects activated electrons from chlorophyll that flow through an electron transport chain. At the end of the chain, the electrons rejoin chlorophyll in its unactivated, or ground, state. This flow of electrons creates a proton gradient across the photosynthetic membrane in which the chain is embedded by mechanisms similar to those in the respiratory electron transport chain. Like heterotrophs, photoautotrophs use this proton gradient to generate ATP by passing protons through a transmembrane ATPase. Chlorophyll serves both as the electron donor to the electron transport chain and the chain's terminal electron acceptor. This cyclic electron flow that generates ATP is called **cyclic photophosphorylation.** Cyclic photophosphorylation cannot generate reducing power. That essential task must be accomplished by another means. Bacteria that carry out anoxygenic photosynthesis must be supplied with a reduced compound to generate reducing power in the form of NAD(P)H. For this purpose, some photoautotrophic bacteria use hydrogen gas, which is a sufficiently powerful reducing agent to reduce $NAD(P)^+$ directly. Others, which use less powerful reducing agents such as H_2S, depend on reverse electron flow to generate reducing power.

Certain bacteria that can use organic compounds as sources of electrons for anoxygenic photosynthesis also use them as sources of precursor metabolites. These phototrophs are photoheterotrophs; all others are photoautotrophs.

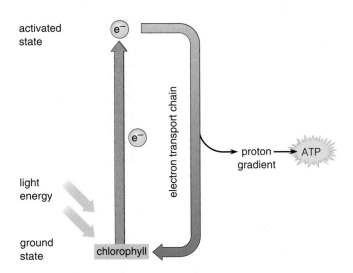

FIGURE 5.16 Anoxygenic photosynthesis. In anoxygenic photosynthesis, light energy strikes chlorophyll, ejecting an electron and raising it to an activated state. It has sufficient energy in its activated state to drive it through an electron transport chain back to chlorophyll at the ground (unactivated) state of energy. The flow through the electron transport chain forms a proton gradient that generates ATP.

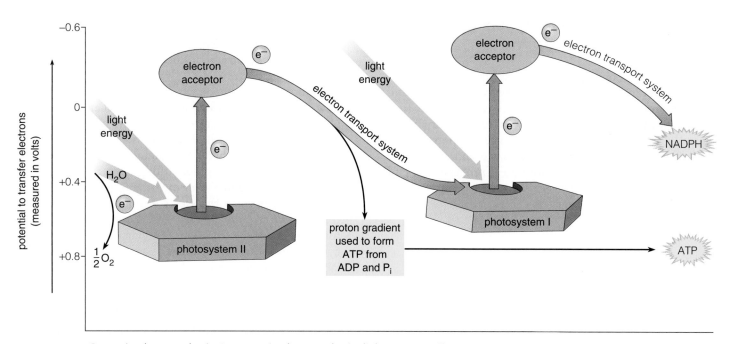

FIGURE 5.17 Oxygenic photosynthesis. In oxygenic photosynthesis, light energy striking chlorophyll in photosystem II, ejects an electron, and raises it to an activated state. Then the electron flows through an electron transport chain to join chlorophyll in photosystem I, where light energy raises it to a higher-energy activated state. From there it flows through an electron transport chain to NADP, reducing it to NADPH. The original source of electrons is water, which is split by light energy yielding an electron and O_2.

Oxygenic Photosynthesis. Green plants and cyanobacteria generate both their ATP and reducing power from light energy by the process of oxygenic photosynthesis (**Figure 5.17**). Chlorophyll molecules that participate in oxygenic photosynthesis occur in two different protein complexes, termed photosystem I and photosystem II. First light energy ejects an electron from a chlorophyll molecule in photosystem II, which then flows down an electron transport chain and generates a proton gradient, just as it does in cyclic photophosphorylation. The difference occurs at the end of the chain. Instead of completing the cycle by rejoining the original chlorophyll, the electron joins chlorophyll in photosystem I and light energy raises it to a higher activated state. The ejected electron flows through a second electron transport chain and then joins with a proton and is transferred to $NADP^+$, forming NADPH. Thus oxygenic photosynthesis produces both reducing power, in the form of NADPH, and ATP from the proton gradient.

Because oxygenic photosynthesis does not return the electron to a chlorophyll molecule, a source of electrons other than chlorophyll is required. Water is a source of electrons for oxygenic photosynthesis. It is photosplit to produce two electrons, two protons, and a half molecule of oxygen gas. Thus water is the source of oxygen that characterizes the process and electrons that feed into photosystem I.

Formation of ATP and Reducing Power by Chemoautotrophs

Chemoautotrophs derive ATP and reducing power from the oxidation of one of several inorganic compounds (**Table 5.9**). They accomplish this in much the same way as chemoheterotrophs do. They remove an electron from the inorganic substrate and pass it through an electron transport chain, thereby generating a proton gradient that is capable of producing ATP when the protons flow back into the cell through a membrane-bound ATPase. Some inorganic compounds oxidized by chemoautotrophs are reducing agents sufficiently strong to reduce NAD(P) directly. An example would be reduced sulfur compounds. But others—iron, for one—are not. Iron-oxidizing chemoautotrophs reduce NAD(P) by reverse electron transport. In reverse electron transport, the proton gradient drives electrons back through the electron chain, reducing NAD(P).

TABLE 5.9 Certain Classes of Chemoautotrophs[a] and Reactions They Use to Generate ATP

Class of Chemoautotroph	Substance Oxidized	Product of Oxidation
Hydrogen bacteria	Hydrogen gas (H_2)	Protons (H^+)
Sulfur bacteria	Sulfide (H_2S), sulfur (S), or sulfite (SO_3^{2-})	Sulfate (SO_4^{2-})
Iron bacteria	Ferrous ion (Fe^{2+})	Ferric ion (Fe^{3+})
Ammonia oxidizers	Ammonium ion (NH_4^+)	Nitrite ion (NO_2^-)
Nitrite oxidizers	Nitrite ion (NO_2^-)	Nitrate ion (NO_3^-)

[a] Also called chemolithotrophs.

REGULATION OF METABOLISM

Throughout this chapter we compared metabolism to a factory and traced how the various cellular assembly lines produce all the components necessary to make a cell. But how are these assembly lines regulated and coordinated? How does the cell ensure that assembly lines producing components needed in large quantities will run faster than those producing components needed in small quantities? How are the various lines geared up or down as growth accelerates or slows? How is one line selectively shut down if its product becomes available from an outside source? The answer to all these questions is metabolic regulation.

Purpose

Metabolic regulation ensures that each of the cell's different metabolic pathways operates at a rate that will supply the cell with the optimal amount of its end product. This regulatory network is complex, because different metabolic pathways operate at vastly different rates. Moreover, these rates can change suddenly and dramatically. Consider, for example, the need for the 20 amino acids that are incorporated into proteins. On the average, proteins in an *E. coli* cell contain 11 times as much glycine as tryptophan into its proteins. So, on the average, the biosynthesis pathway leading to glycine must operate 11 times faster than the one leading to tryptophan. And these basic rates must change in response to changing demands, both within the cell and in the environment. For example, if *E. coli* were provided with glycine in its growth medium, it would stop making glycine but would continue to make tryptophan and the other 18 amino acids at full rate. Adequate but not excessive amounts of building blocks must be synthesized under all conditions. Metabolic regulation makes this possible.

Metabolic regulation also increases the cell's efficiency in other ways. For example, cells usually synthesize enzymes for using substrates only when the substrate is available. Cells synthesize only the enzyme to use the best substrate—the one that supports fastest growth—when several substrates are available. Furthermore, cells synthesize only enough ribosomes to meet the demand for protein synthesis under a particular set of conditions (**Table 5.10**).

Types

There are two major types of metabolic regulation. The first type regulates the amount of an enzyme or any other protein that is synthesized. This type, called **regulation of gene expression,** is discussed in Chapter 6. The second type regulates the *activity* of an enzyme or a protein once it has been synthesized. This type of metabolic regulation, which is usually done by **allosteric proteins,** is discussed here.

Allosteric means "different site." All enzymes have a catalytic site where reactants—also known as substrates—are positioned properly to combine with one another (Chapter 2). Allosteric enzymes, however, have a second site where small signal molecules called **effectors** bind to the enzyme and change its activity.

Effectors either increase or decrease the rate of the enzymatic reaction. Some effectors bind to an allosteric site and activate enzymes, whereas others bind to an allosteric site and inhibit enzymes. Both activation and inhibition occur because binding at the allosteric site changes the **conformation** (shape) of the enzyme. The change in conformation subtly alters the catalytic site, which in turn influences the rate at which the enzyme binds its substrate and catalyzes the reaction (**Figure 5.18**).

TABLE 5.10 Some Regulatory Mechanisms That Operate in *Escherichia coli*

Level of Operation	Effect of Regulatory Mechanism
Bringing nutrients into the cell	Produces tranporters to bring certain nutrients into the cell only if its substrate is available
Catabolic pathways	Produces enzymes to metabolize a particular nutrient only if it is available
	Produces only the enzyme to metabolize the better of two substrates if both are available
	Metabolizes substrates just as fast enough to provide optimal levels of precursor metabolites, ATP, and reducing power
Biosynthesis	If a building block is present in the medium, enzymes for catalyzing its synthesis are not made
	Biosynthetic pathways operate just fast enough to maintain optimal levels of building blocks
Polymerization	The number of ribosomes made is just sufficient to synthesize proteins at an optimal rate
	Chromosome is replicated just fast enough to supply daughter cells with a complete genome
Behavioral	Flagella are not made if nutritional conditions are ideal (no need to move)

End-Product Inhibition. Most biosynthetic pathways are allosterically regulated by a feedback mechanism called **end-product inhibition (Figure 5.19)**. The end product of the pathway is the allosteric effector. The end product (effector) binds to the enzyme that catalyzes the first reaction in that pathway and inhibits its catalytic activity. Inhibiting the first enzyme decreases the flow of materials through the entire pathway.

Consider the regulation of the biosynthetic pathway that produces a particular amino acid. If a large (excessive) amount of the amino acid is present in the cell, the first reaction of the pathway that produces it is slowed down. If a small (inadequate) amount is present, synthesis speeds up.

Allosteric Activation. Some metabolic processes are regulated by **allosteric activation.** This means that a rise in intracellular concentration of an effector activates an allosteric protein. Synthesis of the storage product, glycogen, is an example of a pathway regulated by allosteric activation (**Figure 5.20**). When intermediates of pathways such as glycolysis rise, indicating that more carbon is available than is needed for growth, enzymes leading to the synthesis of glycogen are activated. This enables the cell to store the excess carbon for later use.

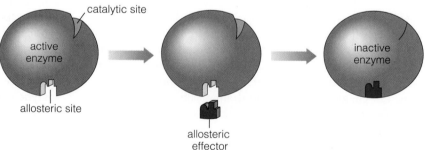

FIGURE 5.18 Allosteric enzymes. An allosteric enzyme has two active sites. One binds substrate to catalyze a reaction. The other binds a small molecule effector that changes the enzyme's activity. In this example the effector changes an active enzyme into an inactive enzyme.

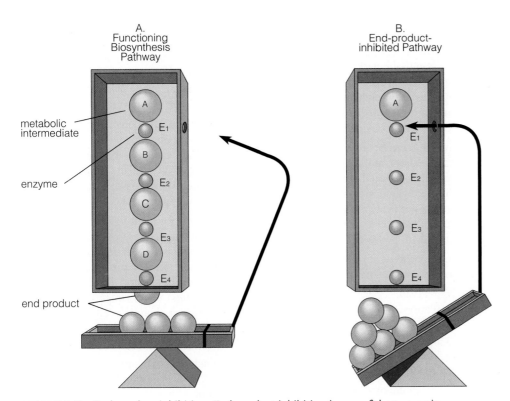

FIGURE 5.19 End-product inhibition. End-product inhibition is one of the two main types of metabolic regulation. When enough end product is produced, it shuts down or slows the process.

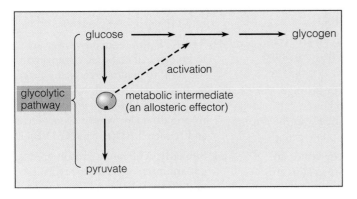

FIGURE 5.20 Allosteric activation of glycogen synthesis.

SUMMARY

Case Study: A Good Bug Gone Bad (p. 115)

1a. Most strains of *Escherichia coli*, being harmless, are excellent subjects for fundamental studies on metabolism, but some strains have gained the ability to be dangerous pathogens.

Metabolism: An Overview (pp. 115-117)

1. Metabolism is the total of all biochemical reactions that take place in a cell.

2. Because its primary function is reproduction, a microbial cell is like a factory that manufactures more factories.

3. The major product of metabolism, doubling all cellular components, can be divided into five tasks: bringing nutrients into the cell, catabolism, biosynthesis, polymerization, and assembly.

Aerobic Metabolism (pp. 118-131)

4. Aerobic metabolism of *E. coli* is a microbial paradigm.

Bringing Nutrients into the Cell (pp. 118-120)

5. Nutrients pass the outer membrane through pores formed by proteins called porins.

6. The cell wall is a porous structure that does not impede entry of nutrients.

7. Carrier proteins called transporters, permeases, facilitators, or carriers take nutrients across the cytoplasmic membrane.

8. Some transporters mediate facilitated diffusion; most mediate active transport, concentrating nutrients in the cytoplasm at the expense of adenosine triphosphate (ATP) or the proton gradient.

9. Some transporters mediate group translocation.

Catabolism (pp. 120-128)

10. Catabolic reactions are organized into catabolic pathways, sequences of chemical reactions, each catalyzed by a different enzyme, that convert substrates to a series of compounds (called metabolic intermediates).

11. Substrates are converted into the 12 precursor metabolites through enzyme-catalyzed reactions in three different catabolic pathways: the glycolysis pathway, the tricarboxylic acid (TCA) cycle, and the pentose phosphate pathway.

12. In the process of converting substrate to precursor metabolites, NAD(P) is reduced and ATP is formed.

13. Oxidations in which protons are removed along with electrons are dehydrogenation reactions. Reductions in which they are added together are hydrogenation reactions.

14. In metabolic oxidations and reductions, hydrogen atoms are not transferred directly from one metabolic intermediate to another. Instead they are transferred to or from NAD or NADP. Thus NAD(P) stores the cell's reserves of hydrogen, or reducing power, as NAD(P)H.

15. In biosynthesis, many hydrogenation reactions occur that reoxidize NAD(P)H.

16. ATP derives its energy-storage capabilities from the two bonds, called high-energy bonds, that join the last two of its three phosphate groups.

17. When the high-energy bonds are broken, the phosphate groups are readily donated to other compounds. Compounds that receive ATP phosphate groups are said to be phosphorylated. They participate in metabolic reactions that would not occur if reactants were unphosphorylated.

18. ATP is formed by adding a single phosphate group to adenosine diphosphate (ADP). Some ADP is formed as a product of phosphorylation reactions.

19. ADP can be converted to ATP through substrate-level phosphoryla-

tion. A phosphate bond that is already part of a metabolic intermediate becomes highly reactive during certain catabolic reactions, which enables the phosphate group to be transferred to ADP, converting it to ATP.

20. ADP can also be converted to ATP through chemiosmosis. The enzyme ATPase catalyzes the conversion of ADP to ATP in and around a membrane (in prokaryotes, it is the cytoplasmic membrane).

21. Energy for chemiosmosis comes from a concentration gradient (usually of protons) formed across the membrane. The proton gradient is created by passing electrons or hydrogen atoms through an electron transport chain that ends, in the case of aerobic respiration, with oxygen as a terminal electron acceptor.

22. Each pair of protons that reenters the cell through ATPase converts approximately one molecule of ADP to ATP. Each pair of electrons that passes through the electron transport chain adds two protons to the gradient and therefore generates about two molecules of ATP. Eukaryotic organisms generate three molecules of ATP per pair of electrons.

23. The ATP and reducing power are interconvertible. Some organisms use ATP to generate reducing power by reverse electron flow. That is, they force electrons/hydrogen back through the electron transport chain and reduce NAD(P)$^+$ to NAD(P)H.

24. Similarly, ATP and a proton gradient are interconvertible. Hydrolysis of ATP to ADP by ATPase forces protons out of the cell, establishing a proton gradient.

25. The three most important catabolic pathways for all cellular organisms—glycolysis, the TCA cycle, and the pentose phosphate cycle—are collectively called central metabolism.

26. Glycolysis begins with glucose, forms six precursor metabolites as intermediates, and produces two molecules of pyruvate. The net yield by substrate-level phosphorylation is

two (because two ATP molecules are spent at the beginning). The two molecules of NADH can donate two pairs of hydrogen atoms to an electron transport chain, yielding four ATPs by chemiosmosis.

27. Each turn of the TCA cycle forms four precursor metabolites as intermediates, one molecule of ATP, two molecules of NAD(P)H, and one of another reduced carrier called $FADH_2$.

28. The pentose phosphate pathway produces as intermediates the last 2 of the 12 precursor metabolites. It also produces two molecules of NADPH. No ATP is directly produced, but the NADPH may generate four molecules of ATP through chemiosmosis.

Biosynthesis (pp. 128-129)

29. Biosynthesis reactions are organized into biosynthesis pathways, enzyme-catalyzed reaction sequences that convert precursor metabolites into the cell's small molecules, including coenzymes and building blocks for macromolecules.

30. Some biosynthetic pathways are branched, fed by several different precursor metabolites or producing several different building blocks. Others are unbranched, producing a single building block from a single precursor metabolite.

31. Coenzymes are compounds that act with enzymes in certain metabolic reactions. Ready-made molecules that act as coenzymes or their precursors are called vitamins.

Polymerization (pp. 129-130)

32. The major polymerization reactions in cells are DNA replication, RNA synthesis, protein synthesis, polysaccharide synthesis, and peptidoglycan synthesis.

33. Ordering in polymerization is directed by the organism's DNA, either directly (DNA, RNA, and protein polymerization) or indirectly through enzymes (polysaccharides and peptidoglycan).

34. Glycogen synthesis illustrates the two basic principles of polymerization reactions: (1) The building block is activated by an ATP-utilizing reaction, and (2) activated building blocks enter into the enzyme-catalyzed polymerization, releasing the activating group to be used again.

Assembly (pp. 130-131)

35. Assembly of macromolecules into cellular structures may occur spontaneously (self-assembly) or as a result of enzyme-catalyzed reactions.

36. Flagella in bacteria and ribosomes are self-assembled, whereas the bacterial cell wall is assembled through enzyme-catalyzed reactions.

Anaerobic Metabolism (pp. 131-133)

37. In the absence of oxygen, aerobic respiration cannot occur.

38. In the absence of oxygen, nonphotosynthetic cells make ATP by anaerobic respiration or fermentation.

Anaerobic Respiration (p. 132)

39. Because anaerobic respiration, like aerobic, depends on an electron transport chain, ATP is generated by chemiosmosis.

40. The terminal electron acceptor for anaerobic respiration can be nitrate, sulfate, fumarate, or trimethylamine oxide.

Fermentation (pp. 132-133)

41. In fermentation, all ATP is generated by substrate-level phosphorylation.

42. With a given amount of substrate, fermentation generates the least amount of ATP.

43. Sugars are almost the only substrates that can be used in fermentation because they are at an intermediate oxidation state. There are many different kinds of fermentation, such as lactic acid fermentation, which is carried out by lactic acid bacteria.

Nutritional Classes of Microorganisms (pp. 134-136)

44. Autotrophs obtain their carbon from atmospheric carbon dioxide. Heterotrophs obtain carbon from organic compounds. Chemotrophs generate ATP and reducing power through chemical reactions. Phototrophs generate ATP and reducing power from light energy.

45. There are four nutritional classes of microorganisms: chemoautotrophs, chemoheterotrophs, photoautotrophs, and photoheterotrophs.

46. The most widespread pathway for forming precursor metabolites in autotrophs is the Calvin-Benson cycle. Autotrophic microorganisms obtain energy from sunlight (photoautotrophs) or by oxidizing inorganic compounds (chemoautotrophs).

47. Photoautotrophs obtain ATP and reducing power for the Calvin-Benson cycle, as well as for the other steps in metabolism, from anoxygenic photosynthesis or oxygenic photosynthesis.

48. Both processes involve chlorophyll and an electron transport chain that creates a proton gradient. Anoxygenic photosynthesis produces ATP exclusively, whereas oxygenic photophosphorylation produces both ATP and reducing power. Bacteria that mediate anoxygenic photosynthesis generate reducing power by reverse electron flow or by oxidation of a reduced sulfur or organic compound.

49. Chemoautotrophs derive ATP and reducing power by oxidizing an inorganic compound. Chemoautotrophs that use sulfur compounds can reduce NAD(P) directly, but iron-oxidizing chemoautotrophs cannot. They use reverse electron transport: The proton gradient drives electrons back through the electron chain, reducing NAD(P).

Regulation of Metabolism (pp. 137-139)

50. Metabolic regulation ensures that pathways operate at a rate that will supply the optimal amount of their end products and that cells synthesize only the enzyme that will use the best substrates available.

51. Regulation of gene expression controls the amount of enzyme that is produced. Allosteric enzymes regulate the activity of an enzyme once it has been produced. They have special sites to which effectors (small molecules) bind and increase or decrease enzyme activity.

52. Most biosynthetic pathways are regulated through end-product inhibition. When sufficient end product has been produced, an effector binds to the enzyme that catalyzes the first reaction in the pathway, thereby inhibiting the flow of materials through the entire pathway.

53. Allosteric activation refers to stimulation of the activity of an allosteric enzyme by a rise in intracellular concentration of an effector. The glycogen synthesis pathway is regulated by allosteric activation.

REVIEW QUESTIONS

Case Study: A Good Bug Gone Bad

1a. How would you diagnose hemolytic uremic syndrome (HUS)?

Metabolism: An Overview

1. Define metabolism.

2. Why is *E. coli* such a well-understood organism?

3. What is *E. coli*'s dual habitat and how does its metabolism reflect it?

4. How is cellular metabolism like a factory that makes more factories?

5. In what form does a cell store energy, and how is the energy released?

6. What is the source of a cell's reducing power? In what forms does a cell store reducing power?

7. What are the five sequential steps in the metabolic assembly line?

8. Define these terms: catabolic reactions, precursor metabolites, polymerization.

9. At which stage of the metabolic process are ATP and reducing power largely formed?

10. At which stage of the metabolic process is most of the ATP consumed? At which stage is most of the cell's stored reducing power required?

Aerobic Metabolism

11. Define aerobic metabolism.

12. Explain the entry mechanisms by which raw materials enter a cell. What role do transporters and ATP play?

13. Define substrate, metabolic intermediates, and precursor metabolites. How do they fit together in catabolic pathways?

14. Name the three primary catabolic pathways, and tell how many precursor metabolites each produces.

15. What is an oxidation-reduction reaction? How does metabolic oxidation differ from most nonbiological oxidations?

16. How are the two types of metabolic oxidations—dehydrogenation and hydrogenation reactions—different? Where are the cell's reserves of hydrogen, or reducing power, stored?

17. How does ATP store energy?

18. Why is phosphorylation an important part of metabolism?

19. How is ATP formed?

20. Explain substrate-level phosphorylation and tell why it is important.

21. What is chemiosmosis? Where does the energy for chemiosmosis come from?

22. Explain aerobic respiration and the role of the electron transport chain. What is the terminal electron acceptor in aerobic respiration?

23. Which parts of chemiosmosis are interconvertible? Why is interconvertibility important in metabolism?

24. How does glycolysis proceed? What is the total yield of ATP?

25. How does the TCA cycle proceed and what are its products?

26. Why are the two precursor metabolites made by the pentose phosphate pathway so important? How does the pentose phosphate pathway proceed?

27. Describe the two types of biosynthetic pathways. What roles do enzymes and coenzymes play in biosynthesis? What are vitamins?

28. What are the major polymerization reactions in a cell?

29. What role does DNA play in polymerization?

30. How does glycogen synthesis illustrate the two basic principles of polymerization reactions?

31. What are the two ways assembly occurs?

Anaerobic Metabolism

32. How do the catabolic reactions of aerobic and anaerobic metabolism differ?

33. What are the two ways nonphotosynthetic cells can make ATP in the absence of oxygen?

34. How are aerobic and anaerobic respiration the same? How are they different?

35. How is ATP generated in fermentation? Why are sugars virtually the only substrate for fermentation?

36. Compare the efficiency of generating ATP through aerobic respiration, anaerobic respiration, and fermentation.

37. Explain how the principles of fermentation are illustrated by lactic acid fermentation.

Nutritional Classes of Microorganisms

38. On what two factors does nutritional class depend? Define the fol-

lowing: chemoautotroph, chemoheterotroph, photoautotroph, and photoheterotroph.

39. Why are autotrophs so important in the chain of life? Describe the Calvin-Benson cycle.

40. Photoautotrophs obtain ATP and reducing power for metabolism from either of two light-driven processes—anoxygenic photosynthesis and oxygenic photosynthesis. How does each proceed? Which produces ATP exclusively?

41. How do chemoautotrophs form ATP and reducing power? Under what circumstances is reverse electron transport necessary?

Regulation of Metabolism

42. How does metabolic regulation increase a cell's efficiency?

43. What are the two major types of metabolic regulation? Explain how allosteric enzymes regulate metabolism.

44. What is end-product inhibition? Explain allosteric activation.

CORRELATION QUESTIONS

1. If two cultures of *E. coli* were growing in media of the same composition, one exposed to air and one anaerobi-

cally, which would have the larger biomass when growth ceased? Why?

2. Why is the concentration of certain nutrients in the periplasm less than that in the medium?

3. If tryptophan is added to a culture of *E. coli*, further synthesis of the amino acid stops immediately. Is that a consequence of regulation of gene expression or feedback inhibition? Why?

4. Which is greater? The requirement of reducing power for growth of a chemoheterotroph or a chemoautotroph? Why?

5. Certain specialized cells of cyanobacteria, called heterocysts, lack photosystem II. They carry out photosynthesis. Do they produce oxygen gas? Why?

6. Some bacteria that ferment are incapable of aerobic or anaerobic respiration. They do maintain a concentration gradient of protons across their cytoplasmic membrane. How can they accomplish this? Do you think their cytoplasmic membrane contains an electron transport chain? An ATPase? Why?

ESSAY QUESTIONS

1. Discuss this statement: The existence of extensive regulatory mecha-

nisms among microorganisms is proof that competition in a natural environment is intense.

2. Discuss this statement: Anaerobic respiration and fermentation are suited to different anaerobic environments.

SUGGESTED READINGS

Caldwell, D. R. 1999. *Miccrobial physiology and metabolism.* Star Publishing Co.

Gottschalk, G. 1979. *Bacterial metabolism.* New York: Springer-Verlag.

Harold, F. M. 1986. *The vital force: A study of bioenergetics.* New York: W. H. Freeman.

Moat, A. G. and Foster, J. W. 1995. *Microbial physiology.* 3d ed. New York: Wiley-Liss.

Neidhardt, F. C., Ingraham, J. L., and Schaechter, M. 1990. *Physiology of the bacterial cell.* Sunderland, Mass.: Sinauer.

Neidhardt, F. C., ed. 1996. Escherichia coli *and* Salmonella typhimurium: *Cellular and molecular biology.* Washington, D.C.: ASM Press.

For additional readings, go to InfoTrac College Edition, your online research library at: http://www.infotrac.thomsonlearning.com

SIX

The Genetics of Microorganisms

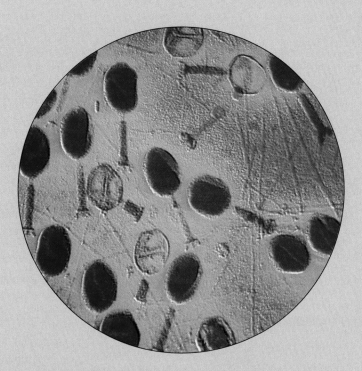

CHAPTER OUTLINE

LEARNING GOALS

To *understand*:

- *The structure of DNA and how it is replicated*
- *The two steps in gene expression: transcription and translation*
- *The ways that gene expression is regulated at the levels of transcription and translation*
- *The microbial genome and the difference between genotype and phenotype*
- *How the microbial genome changes by mutation*
- *How genetic information is transferred between prokaryotes*
- *through transformation, conjugation, and transduction*
- *How genetic information is exchanged between eukaryotic microorganisms*
- *How genetic change spreads through a population of bacteria*

Death by Mutation

B. B., a 20-year-old woman who used heroin intravenously, was admitted to a hospital because of a persistent fever with pain and swelling in her right leg. Her history of intravenous (IV) drug abuse suggested **endocarditis,** a life-threatening infection of the heart's lining (Chapter 27). The diagnosis was confirmed when the yeast *Candida parapsilosis* was found in her blood. Despite the odds, B. B.'s doctors hoped B. B. might be cured because *C. parapsilosis* is highly sensitive to the drug 5-flucytosine (5-FC). 5-FC is toxic to many fungi, including *C. parapsilosis,* because it is converted to a compound, 5-fluorouracil (5-FU), which becomes incorporated

in place of uracil into the RNA of the fungus. Such incorporation disrupts gene expression, killing the fungus. But 5-FC is harmless to humans because we lack the enzyme (cytosine deaminase) that converts 5-FC to 5-FU.

B. B. was treated with 5-FC and improved so dramatically that after only 4 days her fever was gone. *C. parapsilosis* could no longer be isolated from her blood. Sadly, however, despite continuing treatment with 5-FC, her infection reappeared a month later. She died within days. An autopsy revealed that her heart was heavily overgrown with *C. parapsilosis,* but this time the strain was resistant to 5-FC. Thus her treatment

had failed because a mutation had occurred in the fungus. The mutation destroyed cytosine deaminase, rendering the fungus, like humans, resistant to 5-FC. Unfortunately, mutations conferring drug resistance are common and present enormous problems for controlling infectious diseases.

Case Connections
- Consider while reading this chapter how a single mutation could lead to the entire culture's becoming resistant to 5-FC.
- Also consider why incorporating 5-FU into a cell's RNA would disrupt gene expression.

STRUCTURE AND FUNCTION OF GENETIC MATERIAL

In the last chapter we spoke of the plan that directs metabolism. Now we'll examine that plan—how it's written in the form of DNA, how it directs metabolism, how it's passed on to progeny cells, and how it occasionally changes. These topics are the essence of **genetics.**

The science of genetics began before anyone had the slightest idea about DNA. At first, genetics dealt exclusively with inheritance. It studied how certain characteristics such as blond hair or brown eyes were passed on to the next generation. The fundamental principles of inheritance were discovered in 1865 by Gregor Mendel, an Austrian monk. But only in the first part of the twentieth century did other scientists become aware of his results, and only in the 1940s were these principles applied to microorganisms. Then in 1953 James Watson and Frances Crick discovered the structure of DNA, and progress suddenly became explosive. Genetics could be studied biochemically.

The principles of genetics are the same in all organisms—bacteria, archaea, and eukaryotes. Only the

details differ. We'll begin by examining the structure of DNA because all genetics depends on it.

The Structure of DNA

DNA molecules are very long. Stretched-out, the DNA molecule that makes up a bacterial chromosome would measure about 1 mm—a thousand times longer than the cell itself. This huge molecule is a string of deoxyribonucleotide (**nucleotide** for short) building blocks hooked together, much like pearls are joined to form a necklace (Chapter 2). Each individual nucleotide consists of three smaller components: a molecule of deoxyribose, a molecule of phosphate, and a **nucleoside base** (usually referred to as a base). Because there are four kinds of nucleoside bases—adenine (A), guanine (G), cytosine (C), and thymine (T)—there are four kinds of deoxyribonucleotides. Nucleotides are **polymerized** (linked together) in such a way that the deoxyribose and phosphate portions form the long strand and the nucleoside bases (Gs, Cs, As, and Ts) stick out from it. This necklacelike structure is a single strand of DNA (**ssDNA**). DNA occurs as a single

FIGURE 6.1   Structure of DNA. Two strands (dark blue) of repeating deoxyribose and phosphate units form a double helix. The strands are held together by pairs of bases (light blue), which form two hydrogen bonds (dotted lines) between adenine (A) and thymine (T) and three between guanine (G) and cytosine (C). The two strands are antiparallel. Note that phosphate groups link deoxyribose units between the 3 prime (3′) of one deoxyribose unit and the 5 prime (5′) of the next. One strand runs in the 5′ to 3′ direction; the other runs in the 3′ to 5′ direction.

(Art by Margaret Gerrity.)

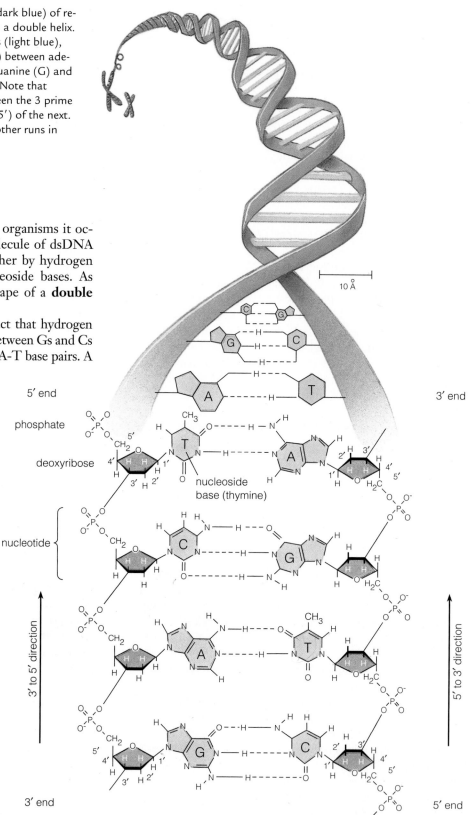

strand in some viruses, but in all cellular organisms it occurs as a double strand (**dsDNA**). A molecule of dsDNA consists of two single strands held together by hydrogen bonds that form between pairs of nucleoside bases. As these bonds form, DNA assumes the shape of a **double helix** (**Figure 6.1**).

The key to DNA's properties is the fact that hydrogen bonds form only between certain bases—between Gs and Cs and between As and Ts, forming G-C and A-T base pairs. A G in one strand is always opposite a C in the other. The same is true for A and T. As a result, the numbers of Gs and Cs in any molecule of dsDNA (from any organism) are always equal. So are the numbers of As and Ts. In contrast, the proportion of G-C to A-T base pairs can vary greatly among organisms. In some organisms the G-C pairs make up only 25 percent of the total. In others they are almost 80 percent (**Table 6.1**). As we'll see, the four bases in DNA are the letters in which the genetic language is written.

The critically important feature of DNA structure, then, is the way the two single strands are held together—by hydrogen bonds between As and Ts and between Gs and Cs.

Reactions of DNA

To fulfill its biological role, DNA enters into two kinds of reactions: replication and gene expression.

- **Replication** is the process by which a DNA molecule is precisely duplicated, so accurate copies can be passed on to progeny cells.

TABLE 6.1   Range of Percent of G-C Pairs
in the DNA of Organisms

Organisms	Range of Percent of G-C Pairs in DNA
Prokaryotes	
Archaea	27 to 61
Bacteria	25 to 78
Eukaryotes	
Fungi	22 to 60
Algae	37 to 68
Protozoa	23 to 67
Plants	33 to 48
Animals	
Invertebrates	32 to 50
Vertebrates	36 to 43

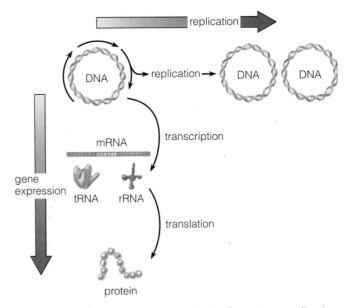

FIGURE 6.2   DNA enters into two kinds of reactions, replication and gene expression.

- **Gene expression** is the process by which the information stored in DNA is used to tell the cell what to do. The process determines which kinds of RNA and protein molecules are made. Then these molecules (mostly the proteins) do all the rest. They catalyze all the cell's metabolic reactions and make up all the cell's structures. The two steps of gene expression are called **transcription** and **translation** (**Figure 6.2**).

DNA $\longrightarrow$ RNA $\longrightarrow$ proteins
Transcription · Translation

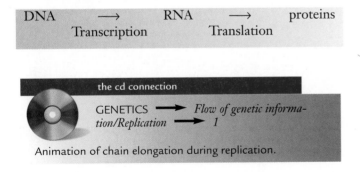

the cd connection

GENETICS $\longrightarrow$ *Flow of genetic information/Replication* $\longrightarrow$ *1*

Animation of chain elongation during replication.

Replication of DNA

During replication, deoxyribonucleotides are polymerized to form two molecules of DNA that are just like the one already in the cell. The primary challenge is how to make precise copies of the DNA. An additional challenge is marshalling sufficient chemical energy to join the deoxyribonucleotides together. In this section we'll examine how the cell meets these two challenges.

Replication: The Synthesis of DNA. The key to making an accurate copy of a cell's DNA is the specific pairing that occurs between nucleoside bases: A with T and G with C, the linkages that hold the two single strands of the double helix together.

The first step in replication is separating the double into two single strands. Separation exposes unpaired bases that can pair with the free nucleotides, thereby lining them up properly. The lineup of free nucleotides is **complementary** to the lineup of bases in the DNA strand, meaning a G-containing free nucleotide lines up opposite a C, a T opposite an A, and so on.

The nucleotides that are present in DNA contain only a single phosphate group (they are called **nucleoside monophosphates**). But the free nucleotides contain three phosphate groups (they are called **nucleoside triphosphates**). They are the activated forms of DNA building blocks (Chapter 5). As these activated nucleotides are polymerized to make a new strand of DNA, a **pyrophosphate** (two phosphates joined together) is split off each one of them. In this way a complementary copy of each single strand is made. The result is that one double-stranded DNA molecule is precisely copied, forming two double-stranded DNA molecules identical to it.

At any one time, only a small region of the parent double-stranded DNA molecule is separated into single strands (**Figure 6.3**), initially forming what looks like a bubble in the dsDNA molecule. Then as activated nucleotides polymerize within the bubble, dsDNA at its two ends continue to separate. The points of separation are

SHARPER FOCUS

SERENDIPITY AND SCIENCE

In a Persian fairy tale, three princes of Serendip are always discovering wonderful things by accident. Their adventures inspired the English word *serendipity.* Serendipity plays an important role in all aspects of our lives, and science is no exception. Setting out in one direction to answer a relatively routine question, scientists sometimes obtain results that lead to wonderful discoveries in a completely different direction. Serendipity is not just luck. The scientist must be astute enough to recognize the potential of unexpected experimental results. As Louis Pasteur said, "Chance favors the prepared mind."

An outstanding example of serendipity occurred in 1928, when Frederick Griffith, a British microbiologist, set out to investigate the connection between the ability of *Streptococcus pneumoniae* to produce capsules and its ability to cause disease. His results led to a major discovery that genetic exchange occurs among bacteria and eventually to the even more important one that genetic material is composed of DNA.

Griffith worked with two strains of pneumococci, as shown in the figure. One, called smooth (S) because it forms glistening colonies, produces lethal pneumonia when injected into mice. The other, called rough (R), forms nonglistening colonies and is harmless to mice because, when injected into mice, its cells are quickly killed. Smooth cells are surrounded by a thick protective capsule; rough cells have no capsule. By mutation, smooth cultures constantly produce a few rough cells. Griffith wondered how the presence of a capsule could be related to lethality.

Griffith found that rough cells and heat-killed smooth cells killed mice when injected together (ex-periment 4), but neither killed mice when injected separately (experiments 1 and 3). Could it be that all smooth cells needed to make them lethal was a bit of capsular material? Griffith answered this question by doing a critical experiment. He isolated pneumococci from the dead mice that had received rough and heat-killed smooth cells. He found that the pneumococci were smooth, but in most of their characteristics they resembled the rough strain. In other words, they had some properties of each strain he had injected: Genetic exchange had taken place between the heat-killed cells and the live cells. This was the serendipitous moment,

and Griffith grasped it. He astutely reasoned that some substance released from the heat-killed smooth cells had permanently changed the live rough cells, allowing them to survive and multiply in the mouse. The change became known as **transformation,** and the substance that caused it was called the **transforming principle.**

The chemical nature of the transforming principle was finally identified in 1944 by three American microbiologists, Oswald T. Avery, Colin MacLeod, and Maclyn McCarthy. They purified the material capable of causing genetic change and showed it to be exclusively DNA, establishing for the first time the chemical nature of genetic material.

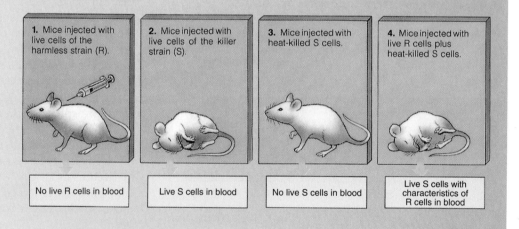

1. Mice injected with live cells of the harmless strain (R).

No live R cells in blood

2. Mice injected with live cells of the killer strain (S).

Live S cells in blood

3. Mice injected with heat-killed S cells.

No live S cells in blood

4. Mice injected with live R cells plus heat-killed S cells.

Live S cells with characteristics of R cells in blood

called **replication forks** or **replisomes.** At each of these forks, separation is followed by pairing of exposed bases with nucleoside triphosphates and polymerization. Thus the two forks move away from each other in opposite directions. As we'll see later, a package of enzymes is associated with replication forks.

The Mechanics of Replication. The two new strands of DNA that form at a replication fork are complementary to the two old strands. The old strands act as **templates** for forming the new strands, as a mold is used to form a statue. Together with their templates the two new strands formed at each replication fork become two new double

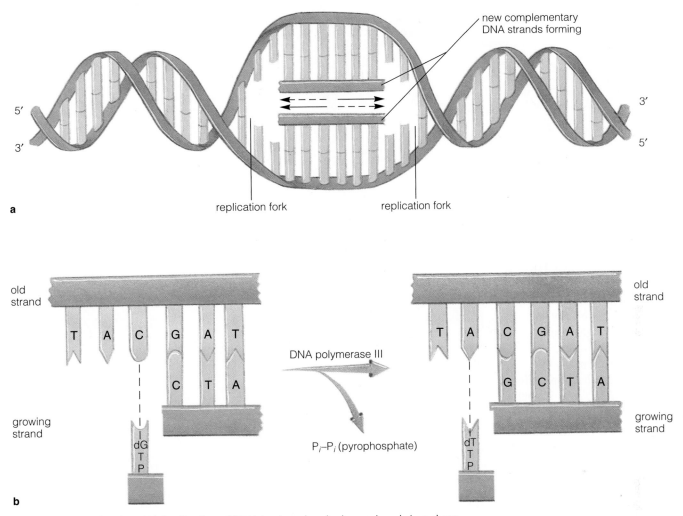

FIGURE 6.3   Replication. (a) Replication of DNA begins when hydrogen bonds in a short stretch break to form a bubble and two replication forks. (b) New strands grow within the bubble as nucleoside triphosphates (dGTP) pair with exposed bases (cytosine [C] here) on the old strand. Bases are added to the growing strand when DNA polymerase III releases pyrophosphate. The next nucleoside triphosphate (dTTP) pairs with the next exposed base (adenosine [A] here).

(Art by Margaret Gerrity.)

helices, each one identical to the original double helix. Because the new double helices are composed of one new and one old or conserved strand, the process is called **semi-conservative replication.**

The bubble that initiates replication of the circular bacterial chromosome forms at a genetically specified point on the chromosome called the **origin.** Then the two replication forks created by the bubble travel simultaneously in opposite directions around the circular *Escherichia coli* chromosome. Eventually they meet halfway around the chromosome, at a point called the **terminus.** Then the two completed chromosomes separate, ready to be distributed to daughter cells (**Figure 6.4**).

The replication fork is biochemically complex (**Figure 6.5**). It contains several enzymes that form a loose complex called the **replication apparatus,** which performs three functions:

- Separating the stands of the old dsDNA
- Synthesizing two new single strands
- Moving the replication fork

DNA helicase, one of the enzymes in the replication apparatus, performs two of these functions. It moves the fork by unwinding and separating the strands of the old double helix. Another protein, **single-strand binding protein (SSBP),** binds to the exposed single strands of DNA and

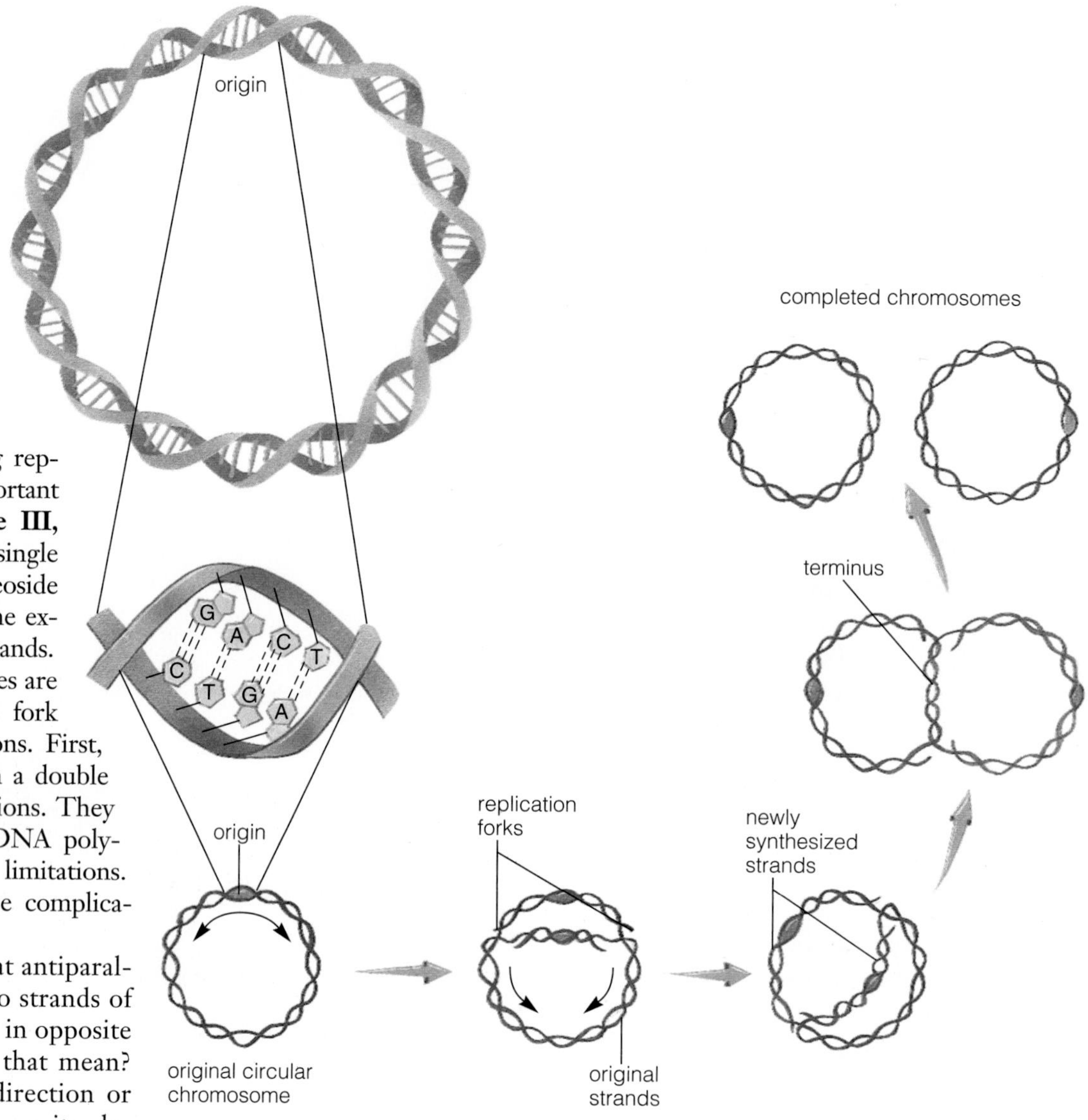
origin
G
A
C
T
C
T
G
A
origin
original circular chromosome
replication forks
original strands
newly synthesized strands
terminus
completed chromosomes

6.4 Replication of the bacterial chromosome. Replication begins when a bubble forms at the origin. This creates two replication forks at which polymerization occurs. The forks travel in opposite directions until they meet at the terminus. Then the two chromosomes separate.
(Art by Margaret Gerrity.)

keeps them separate during replication. A critically important enzyme, **DNA polymerase III,** synthesizes the two new single strands by joining the nucleoside triphosphates paired with the exposed bases on the single strands.

Three additional enzymes are needed in the replication fork because of two complications. First, the two strands of DNA in a double helix run in opposite directions. They are **antiparallel.** Second, DNA polymerase III has two peculiar limitations. Let's examine each of these complications separately.

First we'll consider what antiparallel means. It means the two strands of DNA in a double helix run in opposite directions. But what does that mean? Any strand of DNA has direction or **polarity** because of the way its deoxyribose units are joined together. Look again at Figure 6.1. Note that phosphate groups link deoxyribose units between the third, designated 3′ (pronounced 3 prime), carbon atom, of one and the fifth, the 5′ (5 prime) carbon atom, of the next. Therefore any strand of DNA has a free 5′ carbon (usually with an attached phosphate group) at one end and a free 3′ carbon (with its attached OH group) at the other. In other words, the two ends of a DNA strands have different orientations—as different as if they had a head at one end and a tail at the other. Different groups at the ends of strands means that there are different directions within the strand. There is a 5′ to 3′ direction and there is a 3′ to 5′ direction. That's what antiparallel means. In means the two single strands in any double helix are aligned in opposite directions. The 3′ end of one strand is next to the 5′ end of the other.

Now let's consider the two peculiarities of DNA polymerase III. First, it can lengthen a strand of DNA only in one direction—the 5′ to 3′ direction. (That's because it can add new nucleotides only to the 3′ OH end of an existing nucleic acid strand.) Second, DNA polymerase III can't start making a DNA strand from scratch. It can only lengthen an existing strand (with an exposed 3′ OH end). The strand that is being lengthened is called a **primer.** As we've already seen, a template is needed for DNA synthesis, so the primer must be paired to another strand of DNA.

Because DNA strands are antiparallel and because DNA polymerase III can move only in one direction, only one of the two strands at the replication fork can be extended continuously in the same direction that the replication fork is moving. This strand is called the **leading strand.** The other strand, called the **lagging strand,** must

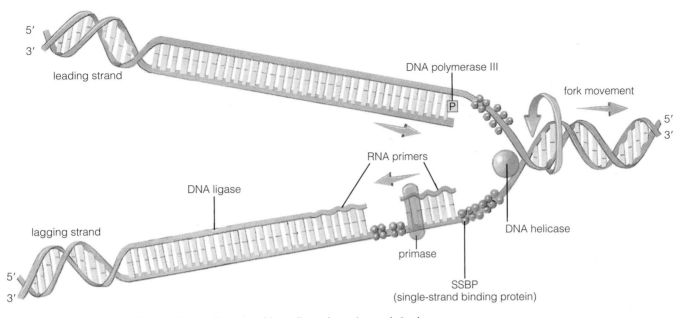

FIGURE 6.5 Replication of DNA. The leading strand is replicated continuously in the same direction the fork moves. The lagging strand is replicated in the opposite direction in sort bursts: Primase lays down a short RNA primer, which is extended by DNA polymerase (not shown); the primer is removed, and the gap is seal by DNA ligase.

(Art by Margaret Gerrity.)

be replicated in the opposite direction to the fork's movement. How is this possible? It's done by synthesizing short pieces of DNA as DNA polymerase III moves away from the fork. In this way many short pieces of the lagging strand are made and then joined together. To accomplish this discontinuous synthesis, three additional enzymes—**primase, DNA polymerase I, and DNA ligase**—are needed.

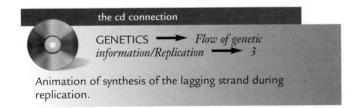

the cd connection

GENETICS ⟶ *Flow of genetic information/Replication* ⟶ *3*

Animation of synthesis of the lagging strand during replication.

First primase makes a short RNA primer. (Because primase is an RNA polymerase, it does not need a primer.) Then DNA polymerase III adds DNA to the RNA primer as it moves away from the replication fork and continues until it runs into the 5′ end of a previous RNA primer. When this collision occurs, DNA polymerase I takes over. It destroys the RNA primer while replacing it with DNA. Finally the third additional enzyme, DNA ligase, seals the gap between the newly synthesized fragment of DNA and the continuous strand in front of it. The entire lagging strand is replicated in this discon-

tinuous manner: Multiple RNA primers are made, extended, and destroyed, and gaps are sealed.

Gene Expression

Now we'll turn our attention to gene expression, the process by which the cell's genetic plan tells the cell what to do.

Transcription. Gene expression starts with **transcription**. During transcription, the cell's genetic plan, contained in DNA, is rewritten in the form of RNA molecules. Three kinds of RNA molecules are made—messenger RNA (mRNA), transfer RNA (tRNA), and ribosomal RNA (rRNA). Each of these kinds of RNA plays a particular role in translation, as we'll discuss in the following section.

In many respects transcription resembles replication. In both processes, DNA is used as a template for making a complementary strand of nucleic acid. In the case of replication the complementary strand is DNA; in the case of transcription it is RNA. Recall from Chapter 2 that DNA and RNA are structurally similar. There are two differences: (1) RNA has uracil (U) in place of the thymine (T) found in DNA, and (2) RNA has ribose in place of the deoxyribose found in DNA. During transcription a bubble forms in the DNA double helix (just as it does during replication). Then transcription follows a different pathway.

Ribonucleoside triphosphates (instead of deoxyribonucleoside triphosphates) pair with the exposed bases. The enzyme, RNA polymerase (instead of DNA polymerase III), links the nucleotides together, forming a strand of RNA (**Figure 6.6**). Thus the order of the bases in the RNA strand is determined by the sequence of bases in DNA. This means the information content of the DNA is **transcribed** (meaning rewritten) in the form of RNA. During transcription, the U that RNA contains pairs with A. All other pairings are the same, as they are during the replication of DNA.

In spite of the similarities between transcription and replication, there are four important differences.

1. Only one strand of the double helix is transcribed (it is transcribed continuously).

2. The product of transcription is a single-stranded instead of a double-stranded molecule. (Sometimes, however, the RNA product may fold back on itself to form short double-stranded regions where sequences of bases happen to be complementary to one another.)

3. The RNA produced by transcription is much shorter than the bacterial chromosome produced by replication. Transcription copies only one or a few genes at a time.

4. A single enzyme, RNA polymerase (a different enzyme from the primase that participates in DNA replication), is solely responsible for transcription in prokaryotes (as opposed to the numerous enzymes needed for replication).

Although the individual products of transcription are relatively short, at some time during the cell's life almost its entire bacterial **genome** (its total complement of DNA) is transcribed. Transcription begins at many different sites, called **promoters,** on the genome. Promoters have a particular sequence of bases that signals RNA polymerase to bind there. Once begun, transcription continues until it reaches another sequence of bases called a **terminator.** Then the **transcript** (the RNA product of transcription) and RNA polymerase are released and the DNA bubble closes.

Translation. After the genetic information is passed from DNA to RNA by transcription, it is used to make protein by **translation.** During translation, amino acids are linked together in the proper order to make a specific protein macromolecule. The name *translation* refers to translation of information from the language of nucleic acids into the language of protein.

The Role of RNA. The three products of transcription—**messenger RNA (mRNA), ribosomal RNA (rRNA), and transfer RNA (tRNA)**—play different roles in trans-

lation. mRNA carries the information that determines the order of amino acids in a protein. tRNA is the actual translator between RNA and protein. rRNA is a component of ribosomes, where translation occurs.

Let's consider tRNA first. tRNA plays the critical role of using the information in mRNA to determine the sequence of amino acids in a protein. Like any translator, tRNA recognizes two languages—nucleic acid and protein. Each tRNA molecule has one site that binds to specific regions of mRNA (thereby recognizing its encoded information) and another site that binds to an amino acid (thereby determining where that amino acid will be placed in a protein). Transfer RNA also plays a second critical role. It activates the amino acid monomer. Once activated, the amino acid is energetically capable of being joined by a peptide bond to a growing protein chain.

Each kind of tRNA molecule can recognize one particular amino acid and one specific region of mRNA. The amino acid is chemically linked to one end of a tRNA molecule (**Figure 6.7**). This linkage (also called **amino acid activation**) occurs at the expense of a molecule of ATP. The result is an energetically activated complex. The other end of the tRNA molecule, the mRNA-recognizing end, consists of an **anticodon.** An anticodon is three adjacent bases on tRNA that pair with a complementary codon (three adjacent bases) on mRNA. Codon-anticodon pairing positions the tRNA-bound amino acid so it can be linked to other amino acids in the proper sequence to make a particular protein. Each kind of tRNA molecule recognizes a particular amino acid and a particular codon.

The tRNA molecule shown in Figure 6.7 is linked to the amino acid methionine at one end. At the other end its anticodon (UAC) pairs specifically with a complementary mRNA codon (AUG). As a result of the action of tRNA, the sequence of bases in the mRNA molecule determines the sequence of amino acids in the protein molecule made from it.

The correspondence between codons in mRNA and amino acids is called the **genetic code (Figure 6.8).** There are four nucleic acid bases and therefore 64 (4³) possible three-base codons. But there are only 20 amino acids. As a result, the genetic code is **redundant:** Most amino acids are encoded by more than a single codon. Three codons, called **nonsense codons,** do not correspond to the anticodon of any tRNA molecules. They do not specify any amino acid. Instead they stop translation. They are located at the end of a gene. Each of the other 61 codons specifies one particular amino acid.

The Mechanics of Translation. Translation occurs on a ribosome, an organelle composed of protein and rRNA (Chapter 4). The ribosome moves down a molecule of mRNA,

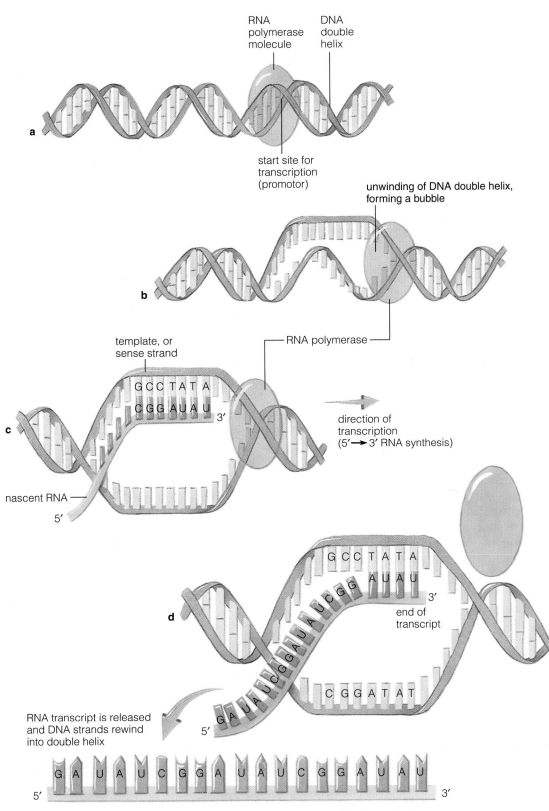

FIGURE 6.6 Transcription. (a) Transcription begins when a molecule of RNA polymerase binds to the DNA double helix at the start site. (b) The unwinding of the DNA helix forms a bubble. (c) Ribonucleoside triphosphates pair with the exposed bases and are polymerized into a nascent RNA molecule as the bubble moves in the direction of transcription. (d) When it reaches a terminator, the RNA transcript is released.

(Art by Margaret Gerrity.)

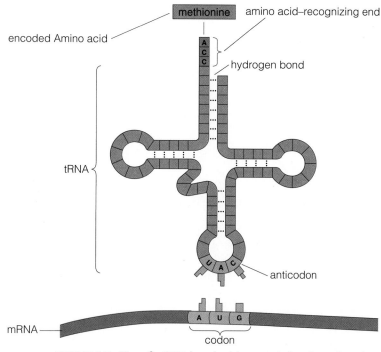

FIGURE 6.7   Transfer RNA has double-stranded regions that give it a cloverleaf-like shape. The amino acid–recognizing end (CCA) of the tRNA molecule becomes linked to an amino acid (in this case methionine). The anticodon end pairs with the codon on the messenger mRNA molecule (in this case AUG, the codon encoding methionine).

First Base	Second Base				Third Base
	U	C	A	G	
U	phenylalanine	serine	tyrosine	cysteine	U
U	phenylalanine	serine	tyrosine	cysteine	C
U	leucine	serine	**stop**	**stop**	A
U	leucine	serine	**stop**	**tryptophan**	G
C	leucine	proline	histidine	arginine	U
C	leucine	proline	histidine	arginine	C
C	leucine	proline	glutamine	arginine	A
C	leucine	proline	glutamine	arginine	G
A	isoleucine	threonine	asparagine	serine	U
A	isoleucine	threonine	asparagine	serine	C
A	isoleucine	threonine	lysine	arginine	A
A	(start) **methionine**	threonine	lysine	arginine	G
G	valine	alanine	aspartate	glycine	U
G	valine	alanine	aspartate	glycine	C
G	valine	alanine	glutamate	glycine	A
G	valine	alanine	glutamate	glycine	G

FIGURE 6.8   The genetic code tells which amino acids are encoded by codons on mRNA molecules. For example, tryptophan is encoded by UGG. The code is redundant. For example, six codons encode leucine. Three codons (UAA, UGA, and UAG) are stop, or nonsense, codons. The start codon is usually an AUG methionine.

exposing successive codons in a region of the ribosome called **the A site** (the amino acid binding site). Then an amino acid–bearing tRNA molecule pairs with the exposed codon. This pairing positions the amino acid so it can be polymerized onto the end of a growing protein that sits in the ribosome's P site (the peptide-binding site). Then the ribosome moves one codon down the mRNA molecule. The process repeats until a complete protein is made (**Figure 6.9**).

A protein produced by translation corresponds to a single gene. But most mRNA molecules are larger than a gene. A ribosome, therefore, must initiate translation within the mRNA molecule precisely at the beginning of a gene. It must stop translation precisely at the end of the gene. Two molecular signals direct the ribosome to the first codon in a gene. One signal is a codon itself. This **start codon** is usually an AUG codon, which codes for methionine. But there are AUG codon within genes, as well as at the beginning of genes (refer again to Figure 6.8). The start codon is distinguished from these other AUG codons by its closeness to a sequence of bases called the **Shine-Dalgarno sequence** or the **ribosome-binding site.** This sequence is complementary to a sequence of

bases in the ribosome's RNA. Presumably it grabs the ribosome and directs it toward the correct AUG codon.

Thus synthesis of a protein molecule begins when a ribosome binds to the start codon on a molecule of mRNA. Then the ribosome moves down the mRNA molecules one codon at a time. As a new codon is exposed, the anticodon of a complementary tRNA molecule (with an attached amino acid) pairs with the codon. Positioned in this way, the amino acid is attached to the growing new protein. This process continues until the ribosome reaches one of the three nonsense codons. Then the completed protein is released from the ribosome.

The two steps of gene expression—transcription and translation—occur simultaneously in bacteria. At any one time about 20 ribosomes are translating an average mRNA

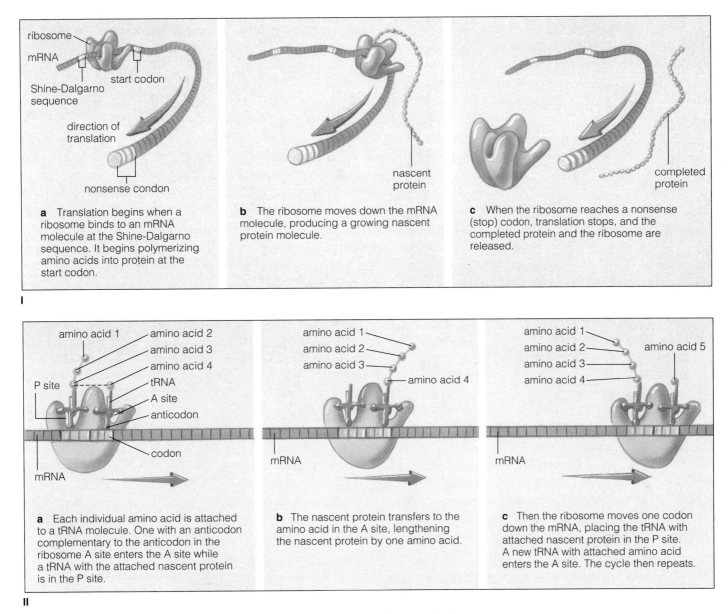

a Translation begins when a ribosome binds to an mRNA molecule at the Shine-Dalgarno sequence. It begins polymerizing amino acids into protein at the start codon.

b The ribosome moves down the mRNA molecule, producing a growing nascent protein molecule.

c When the ribosome reaches a nonsense (stop) codon, translation stops, and the completed protein and the ribosome are released.

a Each individual amino acid is attached to a tRNA molecule. One with an anticodon complementary to the anticodon in the ribosome A site enters the A site while a tRNA with the attached nascent protein is in the P site.

b The nascent protein transfers to the amino acid in the A site, lengthening the nascent protein by one amino acid.

c Then the ribosome moves one codon down the mRNA, placing the tRNA with attached nascent protein in the P site. A new tRNA with attached amino acid enters the A site. The cycle then repeats.

FIGURE 6.9 Translation. (I) Translation of a region of mRNA corresponding to a single gene into a protein product. (II) The steps by which the growing nascent protein molecule is lengthened by a single amino acid.

(Art by Margaret Gerrity.)

molecule while it is still being made. An mRNA molecule with attached ribosomes and nascent proteins is called a **polysome** (**Figure 6.10**).

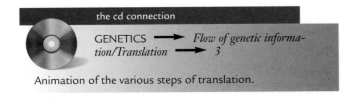

the cd connection

GENETICS ⟶ *Flow of genetic information/Translation* ⟶ *3*

Animation of the various steps of translation.

REGULATION OF GENE EXPRESSION

Microbial cells use substrates and synthesize macromolecules just fast enough to meet their needs. Making them too fast would be wasteful. Making them too slowly would starve the cell of essential components. Either extreme would slow the rate of growth and make the microorganism less able to compete with other organisms in its envi-

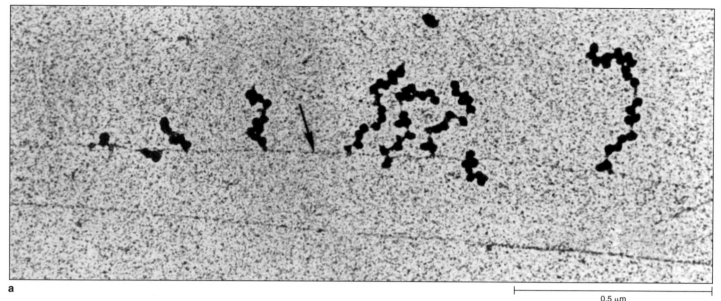

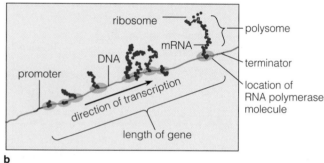

FIGURE 6.10   Simultaneous transcription and translation. (a) This transmission electron micrograph shows a gene being transcribed from left to right. The arrow points to the DNA molecule at the midpoint of the gene. (b) This sketch interprets the micrograph. RNA polymerase molecules are not visible in micrograph; neither are the nascent protein molecules that would be attached to each ribosome. Because a ribosome attaches to the end of an mRNA molecule and moves toward the DNA, protein molecules would lengthen as the ribosome approaches the DNA.

ronment. Cells can operate with such remarkable efficiency because their metabolism is precisely regulated.

Recall from Chapter 5 that metabolism is regulated in two ways: by changing the activity of certain enzymes (as discussed in Chapter 5) and by changing the rate of making proteins, including enzymes. This second type of regulation—which we'll discuss now—affects gene expression.

Regulation of gene expression increases a cell's metabolic efficiency in many ways. For example, some regulatory mechanisms allow the cell to make enzymes only when they are needed. On this basis enzymes are classified as follows:

- **Inducible enzymes** are made only when their substrates are present.
- **Repressible enzymes** are produced only when their product is needed.
- **Constitutive enzymes** are always produced because they are always needed.

The molecular mechanisms that cause enzymes to be induced or repressed are varied and complex, but they all fall into one or another of two broad classes. They either regulate transcription or they regulate translation. Let's look a little more closely at a few specific examples.

Regulation of Transcription

Most regulatory mechanisms act on transcription, usually by changing the activity of RNA polymerase. They do this by changing the frequency, rather than the rate, at which a particular gene is transcribed. When transcripts are made less frequently, less protein is made.

There are two principal mechanisms of regulating transcription: (1) by producing a regulatory protein that binds to DNA near the promoter of the gene that's being regulated and (2) by interrupting transcription after it has started. We'll examine examples of both mechanisms.

Regulatory Proteins That Bind to DNA. These proteins bind to specific regions of DNA, called **operators,** that are close to a gene's promoter (where RNA polymerase normally binds). Depending on what kind of protein is bound to the operator, transcription of the gene becomes more or less probable.

LARGER FIELD

JANNINE ZEIG

In the late 1970s Jannine Zeig, a graduate student at the University of California, San Diego, was doing research on the bacterium *Salmonella typhimurium*. Specifically, she was trying to explain phase variation, periodic changes in the type of flagella made by *S. typhimurium* and many other bacterial pathogens.

Phase variation protects *S. typhimurium* from the host's immune system because by the time it is able to detect the kind of flagella that the invading *S. typhimurium* cells produce, some of them start to produce a second kind that the immune system cannot detect. Later some of these bacteria again start to produce the first

kind of flagella. Zeig knew that the change was not a mutation because it occurred too often. Mutations in any particular gene occur in about one out of a hundred million cells. Phase variation occurs in about one out of a thousand. After extensive experimentation, Zeig found that an enzyme caused a small piece of DNA (con-

taining a promoter) in the *S. typhimurium* chromosome to invert. In one position the DNA transcribes genes encoding one kind of flagella. In the other position the DNA causes the other type to be encoded. Zeig discovered a new basis for phase variation. It has since been found that gene switching occurs in many bacteria.

Most regulatory proteins that bind to DNA are allosteric proteins. Like all allosteric proteins, their activity changes depending on whether or not an effector is bound to them (Chapter 5). We'll consider the lactose **operon** (an operon is a set of genes that is regulated and transcribed together) as an example.

The Lactose Operon. The lactose (*lac*) operon encodes a set of inducible enzymes that confer the ability to use **lactose** (milk sugar) as a growth substrate. Both the enzymes needed to metabolize lactose are encoded by the lac operon (**Figure 6.11**). One of these enzymes, **galactoside permease,** brings the disaccharide, lactose, into the cell, and the other, **β-galactosidase,** splits it into its component monosaccharides (glucose and galactose). Galactoside permease is encoded by the *lacY* gene and β-galactosidase by the *lacZ* gene. In addition to these two genes, the *lac* operon contains a promoter-operator region and the *lacA* gene, which encodes **galactoside transacetylase,** an enzyme with unknown function.

The DNA-binding protein, called the *lac* **repressor,** which regulates expression of the *lac* operon, is encoded by a gene called *lacI,* which lies close to but outside the *lac* operon. It has its own promoter and is expressed constitutively.

In the absence of lactose, the *lac* repressor binds to the *lac* operator and prevents transcription of the *lac* operon. When lactose is present, a small amount of it is converted to an effector, called **allolactose,** which binds to the *lac* repressor, changing it so that it no longer binds to the *lac* operator. Thus only when lactose is present are the genes of the *lac* operon transcribed and expressed. The result is

the *lac* operon is expressed only when its products are needed—when lactose is available.

Note that the effector, allolactose, has a negative effect of the *lac* repressor: It prevents the *lac* repressor from binding to the operator. Note also that when the *lac* repressor binds to the operator, it too has a negative effect of transcription: It prevents transcription. Other effectors have positive effects on binding proteins, and other binding proteins have positive effects on transcription. This variability permits a huge variety of regulatory mechanisms. Almost all you can imagine actually exist.

Now we'll examine **attenuation,** an example of the second method of regulating transcription, namely, interrupting transcription after it has started.

Attenuation. Attenuation is a mechanism that regulates transcription without the participation of a DNA-binding protein. Instead, transcription of a partially completed mRNA molecule either stops or proceeds, depending on whether the gene product is needed.

Attenuation is usually used to regulate production of enzymes that participate in biosynthesis of an amino acid. When the intracellular concentration of the amino acid rises higher than the cell needs, attenuation decreases transcription. The consequence is less enzyme and eventually less amino acid. When the concentration of the amino acid falls to an inadequate level, transcription and enzyme synthesis begin again.

Let's look at a typical attenuation-regulated operon, the histidine operon of *Escherichia coli*. It encodes the enzymes needed to synthesize the amino acid histidine. This operon (**Figure 6.12**) begins with a gene that encodes a

FIGURE 6.11   Regulation of transcription through the *lac* operon. (a) The lactose operon has three genes, *lacZ, lacY,* and *lacA,* which are regulated by the nearby regulatory gene, *lacI.* (b) When lactose is absent, the *lac* repressor binds to the operator, stopping transcription. (c) But when lactose is present, some of it is converted to allolactose, which binds to the repressor. This changes the repressor so it can no longer bind to the operator. Then the genes are expressed.

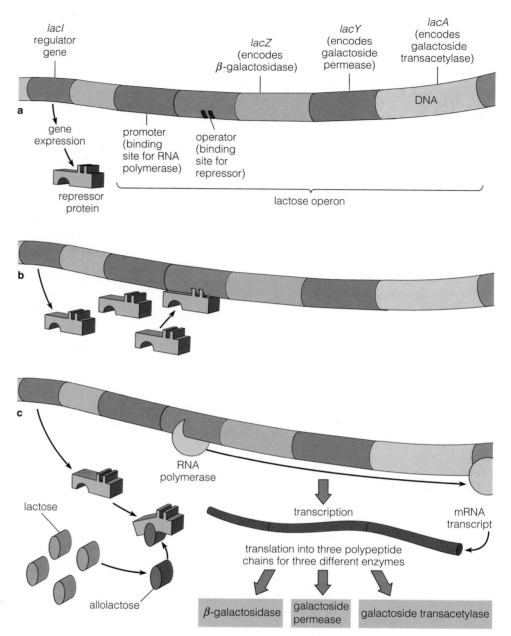

leader protein. A leader protein is only a test protein. Its sole function is to determine whether the cell has an adequate supply of histidine. The leader protein is well suited to its job, because it contains seven histidine units in a row! As a result, its rate of translation is highly sensitive to intracellular concentrations of histidine. When the supply of histidine is adequate, translation of the leader protein occurs rapidly, so that the first translating ribosome is only slightly behind RNA polymerase. As a result, only a short stretch of free mRNA separates them. But this short stretch is highly active: It folds back on itself, creating an **attenuator loop** which displaces the RNA polymerase molecule ahead of it, stopping transcription and thereby stopping expression of the histidine operon.

When the supply of intracellular histidine falls, the situation changes. Then the leading ribosome slows or stops when it reaches the seven-histidine stretch, while RNA polymerase proceeds ahead, exposing an extended stretch of mRNA between them. This extended piece of free RNA folds back on itself, producing a new loop, an **antiterminator loop.** When the antiterminator loop forms, the attenuator loop can't. RNA polymerase is not displaced. It proceeds to transcribe the rest of the operon. Histidine enzymes are made.

Thus attenuation helps adjust the supply of histidine to the cell's need for it. When more histidine is needed, the cell makes more of the enzymes that synthesize it. When excess histidine is present, lesser amounts of these enzymes are made. Histidine also regulates the activity of one of

these enzymes by end-product inhibition (Chapter 5). Together, attenuation and end-product inhibition set the intracellular concentration of histidine precisely at its optimal level. Attenuation spares the cell from making enzymes that are not needed.

Regulation of Translation

Synthesis of a few proteins is regulated at the level of translation rather than transcription. An example is synthesis of **ribosomal proteins** (the protein components of ribosomes) (**Figure 6.13**).

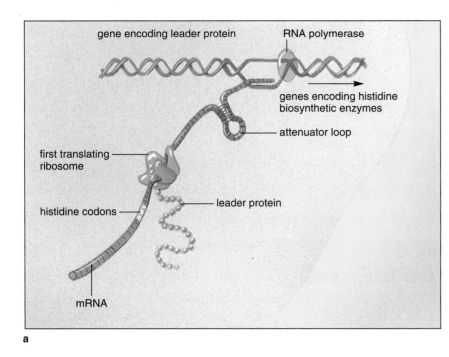

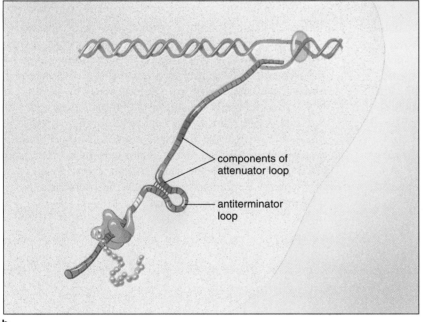

FIGURE 6.12 Attenuation. Regulation of the histidine operon in *Escherichia coli* is an example of attenuation. (a) When the histidine supply is adequate, the first ribosome closely follows the transcribing RNA polymerase molecule but is far enough behind for an attenuator loop form to form, stopping RNA polymerase so the genes encoding the enzymes that synthesize histidine are not transcribed. (b) When the histidine supply is inadequate, the ribosome slows or stops as it encounters histidine codons. This allows an antiterminator loop to form, which uses half the attenuator loop, preventing it from forming. Thus transcription continues and the histidine biosynthetic genes are expressed.

(Art by Margaret Gerrity.)

Further synthesis of the protein decreases until the excess of free molecules is depleted. This regulatory system maintains the concentration of free ribosomal proteins at an optimal level for synthesis of ribosomes.

Global Regulation

So far we have discussed how expression of a gene or an operon operates in response to a specific metabolic signal. A single compound such as lactose or histidine acts as a metabolic signal. But gene expression by microorganisms is also regulated by more general signals, such as the availability of any source of carbon or the shortage of any amino acid. This kind of regulation is called **global regulation** to reflect its broad impact (**Table 6.2**). We'll consider one example of global regulation, catabolite repression.

Catabolite repression regulates expression of many genes and operons in response to the availability of carbon sources for growth. The *lac* operon is one such set of genes that responds to catabolite repression. We've seen that the *lac* operon is expressed only in the presence of lactose. Catabolite repression adds an additional metabolic advantage. It saves the cell from expressing the *lac* operon if a "better" substrate (one that supports faster growth) than lactose is also available. If, for example, a mixture of lactose and glucose is available to *E. coli*, it uses the better substrate, glucose, first. In essence, catabolite repression allows the cell to make a choice among substrates.

Catabolite repression is mediated by two compounds: a small molecule, cyclic adenosine monophosphate (AMP), and a regulatory protein called catabolite activating protein

The genes encoding ribosomal proteins are grouped into operons. Each operon contains one gene that encodes a protein with two functions. The first function is the same as the function of all ribosomal proteins: to be incorporated into a ribosome. The second function is regulatory: to bind to its own mRNA and inhibit translation. When the supply of this protein is not excessive, all the molecules are incorporated into ribosomes. When the supply becomes excessive, the unused protein molecules bind to their encoding mRNA molecules and inhibit translation.

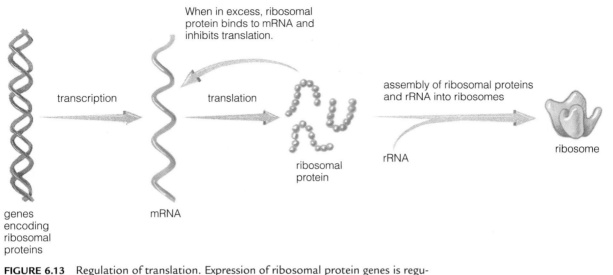

When in excess, ribosomal protein binds to mRNA and inhibits translation.

transcription

translation

ribosomal protein

assembly of ribosomal proteins and rRNA into ribosomes

rRNA

ribosome

genes encoding ribosomal proteins

mRNA

FIGURE 6.13   Regulation of translation. Expression of ribosomal protein genes is regulated at the level of translation. When free ribosomal proteins accumulate because they are not being used to make ribosomes, they bind to mRNA, preventing further translation. (Art by Margaret Gerrity.)

(CAP). Cyclic AMP serves only as a metabolic signal. It signals the rate at which catabolic reactions are proceeding. In the presence of a good substrate, such as glucose, catabolic reactions proceed rapidly and suppress the synthesis of cyclic AMP. If only a poor substrate (for example, dicarboxylic acid succinate) is available, catabolic reactions proceed slowly and cyclic AMP accumulates. Cyclic AMP combines with CAP and binds to the regulatory region of the *lac* operon (and other genes and operons controlled by catabolite repression) near the place where the *lac* repressor binds. But instead of preventing transcription, the CAP–cyclic AMP complex stimulates transcription. In fact, very little transcription of the *lac* operon occurs unless it is stimulated by CAP–cyclic AMP.

Two-component Regulatory Systems

Many regulatory systems in prokaryotes are composed of two rather than a single regulatory protein. One of these proteins, called a **sensor,** detects an environmental signal and transmits it to another protein, called a **response regulator,** that alters gene expression, often by binding to DNA and changing transcription. Sensors are usually located in or near the cytoplasmic membrane, where they can detect changes in the environment. They chemically transmit this information to a response regulator that is located in the cytoplasm, where it can exert its effect on DNA or some other target. Usually the chemical signal is transmitted by phosphorylation.

TABLE 6.2 Examples of Global Regulation that Function in *Escherichia coli*

Regulatory System	Function
Catabolite repression	Prevents expression of genes encoding catabolism of carbon sources when a better carbon source is available
Nitrogen regulation	Prevents expression of genes encoding catabolism of nitrogen sources when a better nitrogen source is available
Phosphorus regulation	Prevents expression of genes encoding catabolism of a phosphorus source when a better phosphorus source is available
Stringent response	Prevents synthesis of ribosomal RNA when any amino control acid is lacking
Heat shock	Stimulates synthesis of proteins that enable cells to withstand damage when temperature is raised

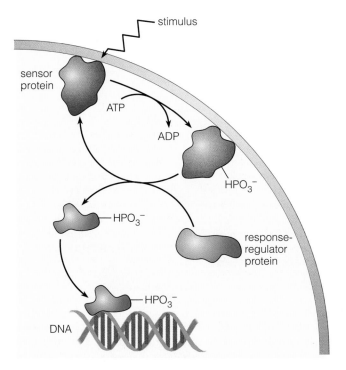

FIGURE 6.14 Two-component regulatory systems. The sensor protein, often associated with the cell surface, becomes phosphorylated when subject to a stimulus. The phosphate group is then passed to a response regulator protein, changing it so it reacts with DNA, modifying gene expression.

Regulation of chemotaxis of *Escherichia coli* (Chapter 4) illustrates the principles of two-component regulation (**Figure 6.14**). When signaled by a change in concentration of a chemotactic substance in the cell's environment, the sensor protein (CheA[1]) catalyzes its own phosphorylation at the expense of ATP. Then CheA passes its phosphate group to the response regulator, CheY. Phosphorylated CheY interacts with the motor that turns the flagella, causing it to reverse direction.

CHANGES IN A CELL'S GENETIC INFORMATION

Over time a microbial cell's genetic plan, written in its complement of DNA, changes. Chemical changes called **mutations** occur in the DNA, and changes are introduced into it by genetic exchange with related microorganisms. Before discussing these two mechanisms of genetic change, let's examine how DNA is stored within a cell. The sum total of all a cell's DNA is called its **genome.**

The Genome

The vast majority of a prokaryotic cell's DNA is contained in its chromosome. Most prokaryotes have a single circular chromosome. A few have a linear chromosome, and some have more than one chromosome. In addition, most bacteria have small DNA structures called **plasmids.** In contrast, most of a eukaryotic cell's genome is contained in pairs of chromosomes.

The single circular chromosome in *Escherichia coli* contains about 4288 genes. This is typical, though some bacteria contain about three times as many genes and some less than one-fourth as many. However, all prokaryotic cells contain much less DNA and therefore fewer genes than eukaryotic cells. We humans have about 30,000 genes, only about 7 times more than *E. coli.* Prokaryotic chromosomes carry all the genes that are essential for a cell's growth and survival. Plasmids carry nonessential genes.

Plasmids. Most bacteria contain plasmids. They are 20 to 50 times smaller than a chromosome, but their structure is much the same. Bacteria with circular chromosomes have circular plasmids. Those with linear chromosomes have linear plasmids. Plasmids are replicated the same way chromosomes are, and copies are passed on to daughter cells. Plasmids differ from chromosomes by encoding only functions that are not essential for growth and reproduction. Rarely a cell will fail to pass a plasmid to one of its daughters. When that happens, the progeny cell may be at a competitive disadvantage, but it does not die.

Functions that plasmids encode are beneficial in special environments. For example, some plasmids encode enzymes that make a bacterium resistant to antibiotics. These plasmids are called **R factors** (resistance factors). Some R factors confer resistance to only a single antibiotic. Others confer resistance to several antibiotics. Strains that carry these R factors have **multiple drug resistance.** Over the last few decades, R factors have proliferated to the point that some infections are difficult to cure with antibiotics. Both chromosome and plasmid-type antibiotic resistance are discussed in detail in Chapter 21. Plasmids also encode a number of other nonessential functions, including the ability to make certain disease-causing toxins. When these strains lose their toxin-encoding plasmids, they become essentially harmless (see Larger Field: Pasteur Revisited).

Some plasmids, called conjugative plasmids, have the ability to transfer a copy of themselves to another bacterial cell by the process of conjugation. If a plasmid confers a significant benefit on the cell, conjugation can spread

---

[1] You will notice that genes (for example, *lacZ*) are designated in italics with the first letter lowercase. Protein products of genes (for example, LacZ or CheA) are designated in roman type with the first letter capitalized.

plasmids quickly throughout a population of bacteria. (We'll discuss conjugation later in the chapter.)

Genotype and Phenotype.

The genome of any organism can be considered from two different points of view, **genotype** and **phenotype**. Genotype describes the cell's genetic plan. Phenotype describes the effect of the plan on the cell's appearance and function. If a cell's DNA changes, its genotype changes. Its phenotype might or might not change as well. For example, if a change occurred in one base pair in the gene encoding the enzyme β-galactosidase in *Escherichia coli*, the cell's genotype would change. If the change had no effect on the activity of β-galactosidase, the cell's phenotype would remain the same. But if the change inactivated β-galactosidase, the cell's phenotype would be different: It would be unable to metabolize lactose. We're used to judging the phenotypes of humans—what color their eyes or hair are, whether they're tall or short, and whether they have an inheritable disease, such as cystic fibrosis. Now modern genetics is learning what genotypes cause our various phenotypes.

Now we'll turn to one way a cell's genome can change—mutation.

Mutations

A mutation is any chemical change in a cell's DNA. Many different kinds of changes can occur (**Figure 6.15**). A **base substitution mutation** changes a single pair of bases to a different pair. A **deletion mutation** removes a segment of DNA. An **inversion mutation** reverses the order of a segment of DNA. A **transposition mutation** moves a segment of DNA to a different position on the genome. A **duplication mutation** adds an identical new segment of DNA next to the original one.

Incidence of Mutations.

Every time the chromosome is replicated, mistakes can occur, causing mutations, so mutations increase with the number of replications, not with time alone. As a consequence, **mutation rate** is calculated as the number of mutations per cell generation. Mutations that occur in the natural course of microbial growth are called **spontaneous mutations.** Those caused by intentional chemical, physical, or biological treatments are called **induced mutations.**

Spontaneous mutations are rare. Only about one cell in a hundred million (10^8) carries a mutation in any particular gene. This frequency is equivalent to about two persons being affected in the population of the United States. Such a rare event might seem inconsequential, but it is not, because microbial numbers are so huge. Most full-grown bacterial cultures contain about 109 cells per milliliter (four times as many as there are people in the United States). Each milliliter of such a culture contains about 10 cells with mutations in any particular gene. An average bacterial chromosome contains about 4000 genes (*E. coli* contains 4288). So each milliliter of culture contains about 40,000 mutations that weren't present when the culture started growing from one or two cells.

The vast majority of mutations either have no detectable impact or are only slightly detrimental, but some are lethal. A mutation that damages the cell is quickly lost from the population because cells that carry it are outgrown by the rest of the population. In rare cases a mutation benefits the cell. For example, a mutation might allow a cell to grow faster in a particular medium or resist a toxic agent. The **progeny** (succeeding generations of daughter cells) of a cell with such a mutation would soon dominate the population.

Induced Mutations.

The frequency of mutations is increased when a culture is treated with a **mutagen** (an agent

FIGURE 6.15 The five types of mutations are shown here. Base substitution changes a single base pair. Deletion removes base pairs. Inversion reverses the order of a segment of base pairs. Transposition moves base pairs from one location to another. Duplication replicates a segment of base pairs.

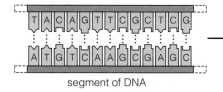

segment of DNA

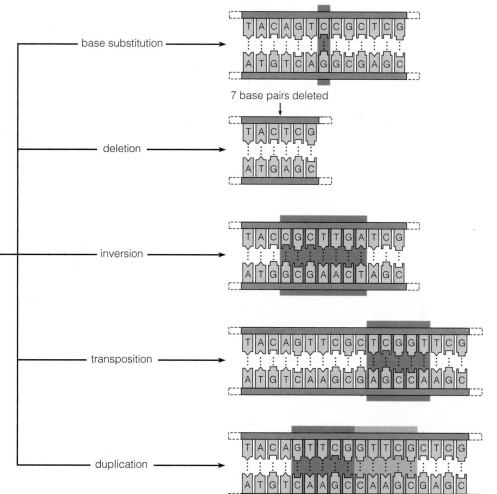

that induce mutations). Mutagens may be chemical, physical, or biological.

Chemical Mutagens. Chemical mutagens change DNA in various ways. Some are almost unbelievably powerful. For example, **nitrosoguanidine** can create a mutation in almost every surviving cell of a bacterial culture treated with this chemical mutagen.

Some chemical mutagens react with a component of DNA and change it. For example, hydroxylamine reacts specifically with cytosine in DNA, converting it to hydroxylaminocytosine. Unlike cytosine, which pairs only with guanine, hydroxylaminocytosine pairs with adenine as well. So following replication of a region of DNA where a cytosine has been converted to a hydroxylaminocytosine, the original C-G pair of bases is converted to a T-A pair.

Other chemical mutagens—for example, 5-bromouracil (5BU)—act by being incorporated into DNA (**Figure 6.16**). Usually, 5BU, like its structural analogue, thymine, pairs with adenine. Unlike thymine, however, 5BU often pairs with guanine as well. So incorporation of 5BU into DNA sooner or later changes an A-T pair of bases to a G-C pair.

Physical Mutagens. Certain physical agents, including ultraviolet (UV) light, x rays, gamma radiation, and decay of radioactive elements, also cause mutations. Even heat is slightly mutagenic.

UV light is highly destructive to DNA. All organisms have mechanisms to repair UV-damaged DNA, but sometimes during the process mutations occur (**Figure 6.17**). UV light stimulates adjacent pyrimidine bases, usually thymines, in DNA to react with one another, becoming linked to form a thymine **dimer** (a dimer is a chemical compound composed of two identical parts). This damage activates several repair systems, one of which allows DNA replication across the damaged region at the cost of accuracy. The resulting errors are mutations.

In contrast to UV light, x rays, gamma rays, decay of radioactive elements, and heat cause mutations by changing the chemical structure of DNA. X rays and gamma rays are **ionizing radiations** (Chapter 9). That is, they expel electrons from atoms, producing ions. The electrons react with other components of DNA to produce altered structures. Some reactions fragment the DNA backbone. Others cause mistakes to be made during subsequent replication.

Biological Mutagens. Microorganisms, and probably most higher organisms as well, carry bits of DNA within their genomes that are mutagenic. These sequences of DNA cause mutations by **transposing,** or moving themselves from one part of the genome to another. Because they

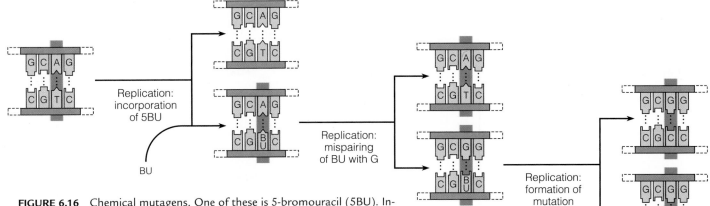

FIGURE 6.16 Chemical mutagens. One of these is 5-bromouracil (5BU). Incorporating 5-bromouracil into DNA in place of thymine (T) can change an A-T pair to a G-C pair. This occurs because on replication BU pairs with G, causing it to be inserted. The next replication generates a G-C pair.

move from place to place within a genome, they are called **transposable elements,** or **jumping genes.**

Two types of transposable elements exist, **insertion sequences** and **transposons.** Both are short regions of DNA, one to a dozen genes in length. Both encode enzymes that transpose the element. They do this by simultaneously replicating the element and moving it to another place on the genome. The original copy of the element is left where it was. Insertion sequences encode only the ability to transpose. Transposons carry other genes as well.

When the new copy of a transposable element is inserted within a gene, a mutation has occurred because the gene's function is destroyed. You might think that transposable elements would be a biological disaster, moving around the genome and destroying it. If they moved too often, they certainly would be disastrous. But transposable elements also encode regulatory proteins that inhibit their transposition. When a transposable element is first introduced into a cell, the inhibitory protein has not yet been produced, so the element transposes readily. Later, after the regulatory protein has been made, further transposition is rare.

Some transposable elements benefit the cell by bringing it useful genes. For example, some transposons carry genes that destroy antibiotics. They protect the cell from being killed by antibiotics. But what benefits do insertion sequences, that encode only their own ability to move, confer? Why have they been perpetuated during evolution? Some biologists speculate that they are examples of "selfish DNA." They evolved to perpetuate themselves, not to benefit the host cell.

The Consequences of Mutations. Two factors determine how serious a mutation will be for a cell: (1) how much damage the mutation does to the gene product and (2) how important the gene product is to the cell.

Damage to the Gene Product. All mutations change the coding properties of DNA, but their effects on the gene products can vary greatly (**Table 6.3**). If, for example, the mutation is a base substitution that changes a codon to another that encodes the same amino acid (Figure 6.8), the gene product is completely unchanged. Similarly, mutations that duplicate genes or operons might increase the quantity of gene products but usually have no effect on the cell's phenotype. A base substitution resulting in **missense mutation** (one that changes a codon to another that specifies a different amino acid) may alter the gene product only slightly if the new amino acid is chemically similar to the original one or profoundly if it is quite different.

If a bit of DNA is deleted, the gene product may be completely inactivated, and a **nonsense mutation** (one that changes a codon to a nonsense codon), by stopping translation before the protein product is completed, usually destroys its activity. Transposition of a piece of DNA into a gene usually also destroys the gene product as well.

Surprisingly, the vast majority of mutations cause no change in the cell's phenotype that can be detected in the laboratory. They might, however, cause subtle changes that affect the cell's long-term competitiveness in nature.

Importance of Mutated Gene Product. Some gene products are essential to a cell's survival regardless of the conditions of cultivation. Others are needed under only certain conditions. For example, if DNA polymerase III is destroyed by mutation, the cell cannot make more DNA, and it cannot multiply. This is called a **lethal mutation.** On the other hand, if an enzyme in a biosynthetic pathway is destroyed, the effect is less drastic. The mutant strain, called an **auxotroph,** can grow normally if the product of the pathway is present in the medium (Chapter 5). For example, we humans are multiple auxotrophs. A number of our biosyn-

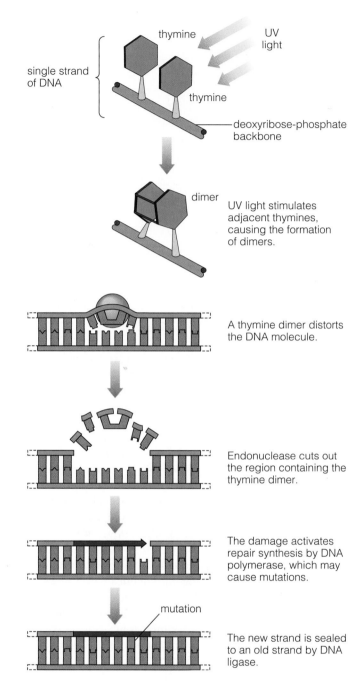

single strand of DNA

thymine

UV light

thymine

deoxyribose-phosphate backbone

dimer

UV light stimulates adjacent thymines, causing the formation of dimers.

A thymine dimer distorts the DNA molecule.

Endonuclease cuts out the region containing the thymine dimer.

The damage activates repair synthesis by DNA polymerase, which may cause mutations.

mutation

The new strand is sealed to an old strand by DNA ligase.

FIGURE 6.17 Physical mutagens. Mutagenic effect of UV irradiation.

thetic pathways (including those that make several amino acids and vitamins) are defective. But we do quite well because we obtain these amino acids and vitamins in our diet. Similarly, if a mutation destroys an enzyme in a pathway by which a particular carbon or nitrogen source is metabolized, the mutant strain will grow normally if another carbon or nitrogen source is provided. Such mutants are called **carbon source** or **nitrogen source mutants.**

Sometimes a mutation renders the gene product nonfunctional only under certain conditions. These are called **conditionally expressed mutations. Temperature-sensitive mutants** are an example. They make the gene products more sensitive to extremes of temperature. **Osmotic remedial mutants** make the gene product more sensitive to the concentration of salts in the medium. One type of temperature-sensitive mutant, a **heat-sensitive mutant,** might be unable to grow or carry out a certain function at 42°C but be perfectly able to perform that function at 30°C. The opposite might be true for the other type of temperature-sensitive mutant, a **cold-sensitive mutant.** An osmotic remedial mutant might be quite normal in a high-salt medium but completely unable to perform a certain function in a low-salt medium.

Conditionally expressed mutations are invaluable. They allow research microbiologists to study defects in essential genes. A temperature-sensitive mutant with a defect in an essential gene, for example, can be kept alive for study by growing it at the temperature at which the gene product is functional.

Selecting and Identifying Mutants

Mutant strains are valuable research tools. Let's consider how mutant strains can be isolated and identified.

The number of ways to select and identify mutant strains of microorganisms is limited only by the imagination of the researcher. The kinds of ways can be grouped into four general categories: **direct selection, indirect selection** or **counterselection, brute strength,** and **site-directed mutagenesis.** These procedures are usually applied to a culture that has been **mutagenized** (treated with a mutagen) to increase its complement of all mutations, including the one being sought.

Direct Selection. Direct selection means creating conditions that favor growth of the desired mutant strain. For example, penicillin-resistant mutants can be isolated by spreading a culture on a plate containing penicillin. Because only the penicillin-resistant mutant cells will grow, each colony that develops is a penicillin-resistant mutant clone.

Indirect Selection. Indirect selection, or counterselection, is the opposite of direct selection. First, conditions are created that prevent growth of the desired mutant strain. Then a condition is imposed that kills growing cells. Because the desired mutant cells cannot grow, they survive the lethal treatment. After counterselection the desired mutants makes up a larger fraction of the surviving population. They are thus easier to isolate.

TABLE 6.3 Consequences of Various Mutations

Probable Consequence	Mutation	Change
None	Base substitution	Changes a codon to another that encodes the same amino acid
	Duplication	Duplicates a gene or an operon
Moderate	Base substitution, missense	Changes a codon to one encoding a similar amino acid
Complete loss of activity	Base substitution, nonsense	Changes a codon to a nonsense codon
	Deletion	Removes DNA
	Transposition	Interrupts gene with transposed fragment
	Inversion	Inactivates genes in which inversion ends

Counterselection is used to isolate auxotrophic mutant strains. For example, an auxotrophic mutant that requires the amino acid histidine can be isolated using penicillin to counterselect. A mutagenized population is inoculated into a medium that lacks histidine. Then penicillin is added. Penicillin kills the parental cells because they can grow. The desired histidine auxotrophs survive because they cannot grow in a medium lacking histidine. Other agents, including radioactive substrates and certain toxic chemicals, kill only growing cells They too can be used for counterselection.

Brute Strength. Brute strength is the term applied to examining large numbers of clones one by one to find the desired mutant strain. There are many ways to examine. A researcher could test to see if a clone produces a particular enzyme or if it is unable to grow in a particular environment. Following treatment with a powerful mutagen, a researcher must, on the average, examine about 10,000 surviving clones to find a particular desired mutant strain.

One simplified brute-strength examination technique is replica plating (**Figure 6.18**). Replica plating allows the growth properties of many clones to be examined in a single operation. For example, replica plating can be used to detect a histidine auxotroph by testing for inability to grow on a medium lacking histidine. Replica plating is an easy way to transfer large numbers of colonies and keep track of them.

Site-directed Mutagenesis. Site-directed mutagenesis is a product of recombinant DNA technology (Chapter 7). Although these procedures are a form of chemical mutagenesis, they also involve selection because they allow the researcher to mutate one particular gene.

Uses of Mutant Strains

Mutant strains have answered many important biological questions. For example, they were used to determine the biosynthesis pathways of amino acids from precursor metabolites (Chapter 5). Auxotrophic mutants were isolated using the techniques described earlier. Such mutants lack one enzyme in the pathway. Substrates of missing enzymes sometimes are released into the medium, and metabolic intermediates beyond the missing step sometimes satisfy the mutant strain's auxotrophic requirements. In this way metabolic intermediates were identified

Mutant strains of microorganisms have been put to many practical uses. One such use is a test developed by Bruce Ames, a microbiologist at the University of California, Berkeley. His test (the **Ames Test**) determines whether a particular chemical might be **carcinogenic** (cancer causing). Tests for carcinogenic chemicals are traditionally done by administering the test chemical to animals and observing whether they develop cancer. But, knowing that most carcinogenic chemicals are mutagens, Ames was able to develop a simpler, faster, and reliable method using histidine auxotrophic mutants (**Figure 6.19**). The Ames test determines, by direct selection, the ability of the mutagen to cause **reversions** (mutations that reverse the original mutation). In this case the original mutant is a histidine auxotroph that is unable to grow in the absence of added histidine. Mutations that cause reversions permit the cells that harbor them to grow and produce colonies in the absence of added histidine. If the test chemical is mutagenic, it will cause many reversion mutations and many colonies will develop on plates lacking histidine. To increase the value of the test, the chemical to be evaluated is first mixed with macerated (mashed) rat liver, a rich source of enzymes that convert some harmless chemicals into carcinogens. If a

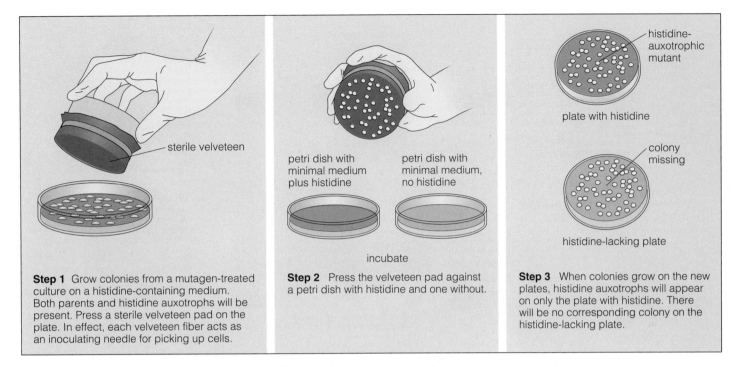

histidine-
auxotrophic
mutant

plate with histidine

colony
missing

histidine-lacking plate

sterile velveteen

petri dish with
minimal medium
plus histidine

petri dish with
minimal medium,
no histidine

incubate

Step 1 Grow colonies from a mutagen-treated culture on a histidine-containing medium. Both parents and histidine auxotrophs will be present. Press a sterile velveteen pad on the plate. In effect, each velveteen fiber acts as an inoculating needle for picking up cells.

Step 2 Press the velveteen pad against a petri dish with histidine and one without.

Step 3 When colonies grow on the new plates, histidine auxotrophs will appear on only the plate with histidine. There will be no corresponding colony on the histidine-lacking plate.

FIGURE 6.18 Replica plating is a technique for identifying and isolating mutant strains. This example shows how a colony of histidine auxotrophic mutant cells would be identified.

chemical is negative in the Ames test, it is probably harmless. If it is positive, it is certainly mutagenic and probably also carcinogenic. About 90 percent of the chemicals that test mutagenic in the Ames test are also shown to be carcinogenic in follow-up animal tests.

Genetic Exchange Among Bacteria

Genetic exchange (the transfer of genes from one cell to another) among bacteria is unlike the genetic exchange that occurs during sexual reproduction of eukaryotes. In most eukaryotes, genetic exchange is an essential part of the organism's life cycle. When it occurs, a complete set of chromosomes from two individuals come together in the same cell, called a **zygote**. In contrast, genetic exchange is not an essential step of the life cycle of bacteria, and when it occurs, only a portion of the DNA of one cell (the **donor cell**) is transferred to the other (the **recipient cell**). The result is a **merozygote** (partial zygote). Transfer travels one way. The recipient transfers no DNA back to the donor.

Usually the piece of bacterial DNA donated during genetic exchange is not a **replicon** (it cannot replicate itself by itself). The donated genes can be replicated and become a permanent part of the recipient cell's genome only if they are incorporated into one of the cell's own replicons—the

chromosome or a plasmid. Incorporation of donated genes into a replicon almost always involves **recombination,** or **crossing over** (the breakage and rejoining of pieces of DNA; **Figure 6.20**). Recombination is an essential feature of genetic transfer of chromosomal genes among prokaryotes. It is not an essential feature of genetic plasmid transfer. If the entire plasmid is transferred, it need not recombine with DNA in the recipient cell to become a part of its genome.

There are three forms of genetic exchange in bacteria: **transformation, conjugation,** and **transduction.** Although all three processes allow DNA to leave one bacterium and enter another, the mechanisms are different.

Transformation. During transformation, DNA leaves one cell and exists for a time in the extracellular environment. Then it is taken into another cell, where it may become incorporated into the genome. Transformation can be natural or artificial.

Natural Transformation. Some bacterial species are genetically programmed to take up DNA from their environment. It is a somewhat complex process. For example, *Streptococcus pneumoniae*, a bacterium that causes pneumonia in humans, has receptors on its cell surface that bind DNA free in the extracellular environment. Then surface enzymes called **nucleases** cut the bound DNA into

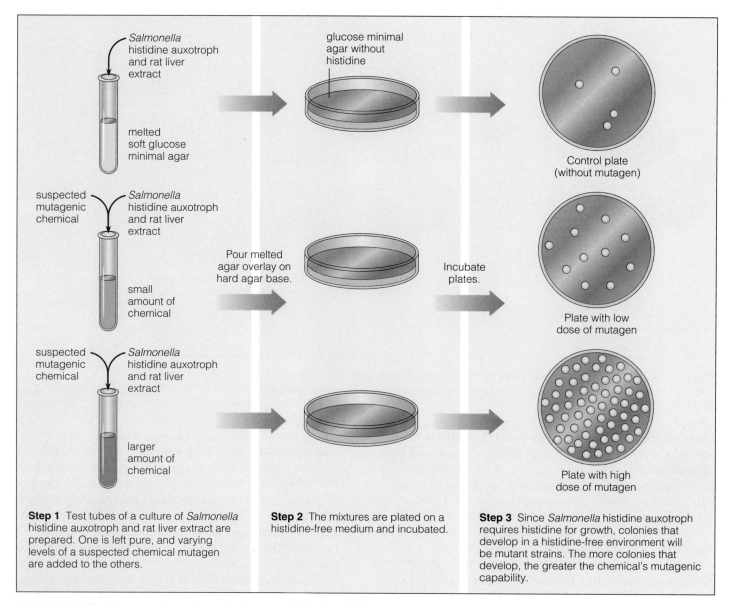

Salmonella histidine auxotroph and rat liver extract

melted soft glucose minimal agar

glucose minimal agar without histidine

Control plate (without mutagen)

suspected mutagenic chemical

Salmonella histidine auxotroph and rat liver extract

small amount of chemical

Pour melted agar overlay on hard agar base.

Incubate plates.

Plate with low dose of mutagen

suspected mutagenic chemical

Salmonella histidine auxotroph and rat liver extract

larger amount of chemical

Plate with high dose of mutagen

Step 1 Test tubes of a culture of *Salmonella* histidine auxotroph and rat liver extract are prepared. One is left pure, and varying levels of a suspected chemical mutagen are added to the others.

Step 2 The mixtures are plated on a histidine-free medium and incubated.

Step 3 Since *Salmonella* histidine auxotroph requires histidine for growth, colonies that develop in a histidine-free environment will be mutant strains. The more colonies that develop, the greater the chemical's mutagenic capability.

FIGURE 6.19   The Ames test determines whether a particular chemical is mutagenic and therefore probably carcinogenic.

fragments. One strand of each fragment is destroyed by another nuclease. The other strand enters the recipient *S. pneumoniae* cell. Inside the cell, the donated DNA is coated with a special protein that protects it from intracellular nucleases. Finally the newly absorbed DNA contacts the portion of the resident genome where matching genes are located, called the **homologous region.** Then by a series of enzyme-catalyzed reactions the incoming DNA replaces the homologous region. About 12 separate genes are needed to encode the various steps of natural transformation of *S. pneumoniae.*

The details of natural transformation vary somewhat among species, but in all cases the DNA is cut into small

pieces before it enters the recipient cell. As a result, the linear fragments of transforming DNA must be incorporated into a resident replicon in order to be copied and passed on to daughter cells. Chromosomal DNA that enters a cell by transformation can become part of the resident chromosome. But plasmid DNA can be incorporated into a functioning replicon only if the recipient cell happens to carry an identical plasmid. Consequently, plasmids are rarely transferred from one cell to another by natural transformation.

Natural transformation almost certainly evolved to make genetic exchange possible. It remains a mystery why only a few bacterial species are capable of it.

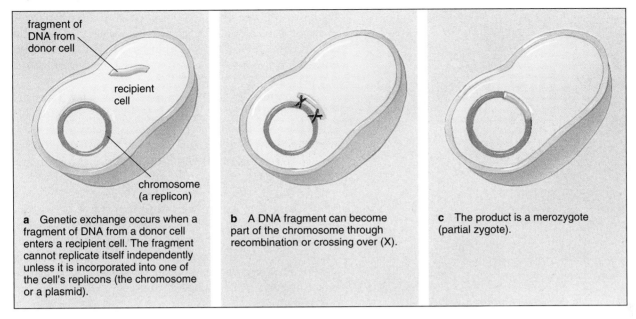

FIGURE 6.20   Genetic exchange in bacteria.

(Art by Margaret Gerrity.)

Artificial Transformation. In 1972 researchers discovered how to alter the envelope of *Escherichia coli* so DNA could cross it. Based on this work, other researchers went on to develop elaborate laboratory techniques called **artificial transformation** to introduce DNA into *E. coli* cells. These processes sometimes involve chilling cells and treating them with a strong solution of calcium chloride. Then plasmid DNA is added to the suspension. These intact plasmids can replicate by themselves because they are not cut as they enter the cell. Modifications of this procedure and totally different ones have since been developed. It is now possible to introduce intact plasmids into almost any bacterium, as well as into many eukaryotic microorganisms, plants, and animals. The development of artificial transformation was essential to the development of recombinant DNA technology, sometimes called genetic engineering (Chapter 7).

Conjugation.
Conjugation is carried out by **conjugative plasmids** (plasmids able to transfer themselves to another cell). The best-studied conjugative plasmid is the F plasmid, which occurs in *Escherichia coli* and closely related bacteria. Conjugation requires many genes. The F plasmid encodes at least 13 genes. Cells that carry an F plasmid are designated F+ and those that lack it F−. One of the genes on the F plasmid encodes a special pilus (Chapter 4), called a **sex pilus** or the F pilus (**Figure 6.21**). The F pilus allows the F+ cell to attach itself to an F− cell. Then attachment triggers a series of events that result in the transfer of the plasmid from the F+ to the F− cell. First the pilus retracts, bringing the two cells in direct contact. Then the plasmid DNA is **nicked** (one strand of the double helix is broken). A special kind of replication, called **rolling circle replication,** begins at the site of the nick, producing a linear single strand of plasmid DNA that enters the F− cell. Within the F− cell, the plasmid DNA forms a circle and duplicates. As a result of the plasmid transfer, the F− cell becomes F+ and thereby becomes capable of transferring a copy of the F plasmid to yet another F− cell. The original donor cell also remains F+ because it retains one copy of the conjugative plasmid.

Many other plasmids, including R plasmids, are transferred by similar mechanisms. Some of these are termed promiscuous because cells that carry them mate with almost any other bacterial cell. These plasmids are transferred between almost all species of Gram-negative bacteria. In contrast, the F plasmid is transferred only between strains of *E. coli* and closely related species.

Because conjugative plasmids and the chromosome sometimes share regions of homology, they can undergo recombination. If recombination does occur, the plasmid becomes integrated into the chromosome. Then the cell becomes an **Hfr** (high frequency of recombination) cell. Even when integrated into the chromosome, the plasmid can still initiate conjugation and bring some of the chromosome along with it when it transfers to an F− cell. Usually only a portion of the chromosome is transferred because the attachment between donor and recipient cells is fragile. Most cell pairs break apart long before the 100 minutes required for the entire chromosome of *E. coli* to be transferred from an Hfr cell to an F− cell. The F plasmid

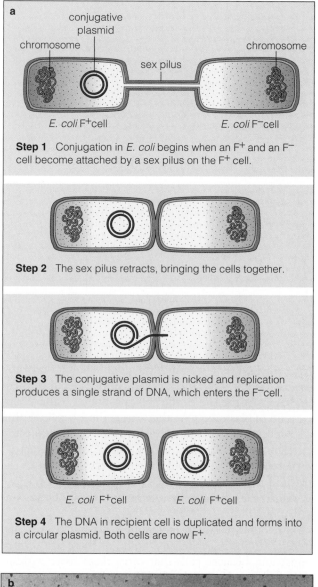

Step 1   Conjugation in *E. coli* begins when an F⁺ and an F⁻ cell become attached by a sex pilus on the F⁺ cell.

Step 2   The sex pilus retracts, bringing the cells together.

Step 3   The conjugative plasmid is nicked and replication produces a single strand of DNA, which enters the F⁻cell.

Step 4   The DNA in recipient cell is duplicated and forms into a circular plasmid. Both cells are now F⁺.

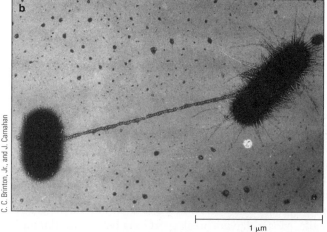

FIGURE 6.21   Conjugation. (a) Steps in the conjugative transfer of the F plasmid between two *Escherichia coli* cells. (b) Electron micrograph of two *E. coli* cells at Step 1 of the process.

begins to transfer itself at a point within the F plasmid. As a result, part of the F plasmid is transferred before the chromosome, and the rest is transferred after. Because the entire chromosome is rarely transferred, neither is the entire plasmid.

Any population of F⁺ bacteria includes a few Hfr cells. These Hfr cells can be purified, yielding pure cultures of Hfr cells. Such cultures have been extremely useful in studying bacterial genetics. They were used to **map** (locate the relative position of) genes on bacterial chromosomes (**Figure 6.22**).

Transduction. As we have seen, transfer of chromosomal genes by conjugation is an accident that occurs when a plasmid happens to integrate into the chromosome. Transfer of chromosomal genes during transduction is also an accident. It occurs when some of the viruses that infect bacteria, called **bacteriophages** or **phages,** reproduce themselves.

To understand transduction, we need to know something about how phages reproduce. We'll discuss phage reproduction in detail in Chapter 13. Here we'll consider just enough background to understand transduction. Based on their pattern of reproduction, there are two kinds of phages and hence two kinds of transduction. Some **virulent phages** (phages that always kill their host) bring about generalized transduction. Some **temperate phages** (phages that can be carried passively within their host without harming it) can bring specialized transduction, generalized transduction, or both.

Virulent Phages and Generalized Transduction. Virulent phages infect bacteria by attaching themselves to the surface of the victim cell and injecting their DNA into it (**Figure 6.23**). The phage DNA directs the cell to make phage components—DNA and protein. Then the components assemble into mature phage particles, which are released from the cell, ready to infect other cells.

Rarely—about once in 100,000 times—a mistake occurs in the final stages of assembly of the phage particle. Bacterial DNA instead of phage DNA becomes incorporated into the phage. This abnormal phage particle, called a **transducing particle,** can then attach to another cell as though it were to infect it. But instead of injecting phage DNA, it injects bacterial DNA into it. The result is genetic exchange. DNA from the bacterial cell where the transducing particle was formed is introduced into another cell. The amount of DNA exchanged by transduction is about the same as the amount in a normal bacteriophage. That's equivalent to a few percent of the bacterial chromosome. Transduction mediated by virulent phages is called **generalized transduction** because it transfers any portion of the bacterial chromosome from one cell to another.

SHARPER FOCUS

THE GENETICS OF ANTIBIOTIC RESISTANCE

Microbiologists have learned about the differences between prokaryotic and eukaryotic cells by studying the targets of antibiotic action (Chapter 4). For an antibiotic to be effective, the host eukaryotic cell must lack these targets or have different forms of them. The genetics of antibiotic resistance tells us even more about antibiotic action and prokaryotic cells.

Antibiotic-resistant strains of bacteria first appeared soon after antibiotics came into common use. How could bacteria change genetically to become resistant to an antibiotic that strikes at a vital cellular function? Subsequent research revealed three answers.

1. The target in the bacterial cell changes so that it remains capable of fulfilling its vital cellular function but is no longer sensitive to the antibiotic.
2. The bacterium becomes able to keep the antibiotic out of the cell or to pump it out of the cell after it enters.
3. The bacterium becomes able to destroy the antibiotic chemically.

Knowing the basis for antibiotic resistance can be the key to developing new antibiotics that are effective against resistant strains. Moreover, understanding the mechanisms of resistance has greatly benefited genetic research on prokaryotic cells. Mutations that change the target lie in genes that encode the target. Many such genes were first identified by antibiotic resistance. For example, the gene encoding RNA polymerase was first identified by a mutation conferring resistance to the antibiotic rifampicin (RNA polymerase is its target). Also, different genes encoding proteins in ribosomes were identified by mutations to the antibiotics streptomycin, erythromycin, kanamycin, and spectinomycin, among others.

Genetic changes that enable a cell to inactivate an antibiotic are almost always carried on plasmids. They encode reactions that add active groups (generally acetyl, phosphate, or adenyl groups) to the antibiotic or split it by hydrolysis. But where do these genes come from? What role did they play before antibiotics came into widespread use? They might have evolved to inactivate other toxic agents. Or they might be altered forms of genes that fulfill some completely different metabolic role. This is the case with the penicillinases, the enzymes some bacteria make that destroy penicillin. Penicillinases are altered forms of the enzymes that polymerize peptidoglycan monomers into bacterial cell walls.

Temperate Phages and Specialized Transduction. Temperate phages have two life cycles. One, the **lytic cycle,** is like the life cycle of virulent phages. The other life cycle, called the **lysogenic cycle,** does not kill the cell. Instead the phage DNA becomes quiescent. In this quiescent state the phage DNA is called a **prophage.** The prophages of some temperate phages exist as plasmids. Others become incorporated in the host cell's chromosome. The latter kind mediate specialized transduction. When the prophages occasionally become reactivated, they enter a lytic cycle that produces phage particles. Specialized transducing particles are formed when a mistake is made during reactivation. Instead of only the prophage leaving the chromosome, a few bacterial genes leave as well and become incorporated in a phage particle called a specialized **transducing particle.** When such a specialized transducing particle attaches to another bacterial host and injects its DNA, it injects the bacterial genes, as well the phage genes. Thus genetic exchange occurs between the bacterial cell in which the specialized transducing particle was formed and the one it later infects. The process is called **specialized transduction** because prophages are inserted only at a specific site on the bacterial chromosome. Only those bacterial genes adjacent to this site can be transferred by specialized transduction.

Genetic Exchange Among Eukaryotic Microorganisms

Genetic exchange between eukaryotic microorganisms is similar to genetic exchange between plants and animals. Following meiosis of a diploid cell, haploid gametes develop and fuse to form a new diploid cell (Chapter 5).

As an example, let's look at how this occurs in the life cycle of baker's yeast, *Saccharomyces cerevisiae* (**Figure 6.24**). A culture of baker's yeast is composed largely of diploid cells. These cells grow and reproduce asexually as long as adequate nutrients are available. When conditions for growth become less favorable, some cells undergo meiosis. The products of meiosis develop into thickened **ascospores** (spores enclosed within a saclike structure called an

Gene(s)	Function
thr	A cluster of genes encoding enzymes in pathway of biosynthesis of the amino acid threonine
leu	A cluster of genes encoding enzymes in pathway of biosynthesis of the amino acid leucine
pan	A cluster of genes encoding enzymes in pathway of biosynthesis of the vitamin pantothenic acid
metD	A gene encoding an enzyme in pathway of biosynthesis of the amino acid methionine
proA	A gene encoding an enzyme in pathway of biosynthesis of the amino acid proline
lac	The *lac* operon, a cluster of genes encoding enzymes in pathway of catabolism of the sugar lactose
tsx	A gene encoding a protein in the outer membrane to which the bacteriophage T6 attaches

Function of some of the Genes between 0 and 10 Units on the *Escherichia coli* Map

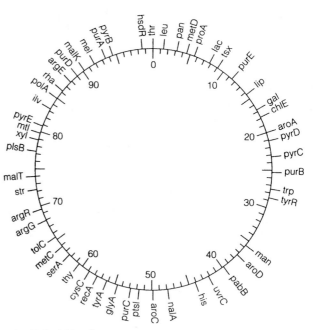

FIGURE 6.22   The map microbial geneticists use to locate the genes on the *Escherichia coli* chromosome is a circle divided into 100 units. The table names some of the genes between 0 and 10 units on the *E. coli* map and indicates their functions.

ascus). When conditions become favorable again, the ascospores germinate, producing haploid cells that act as gametes. Usually a gamete divides a few times before it fuses with another gamete. All gametes belong to one of two mating types, a or α. Because a gamete can fuse only with an opposite mating type, the probability of genetic exchange is high. A gamete cannot fuse with any neighboring cells, which would probably belong to the same mating type and be genetically identical. When gametes fuse, a diploid vegetative cell is formed and asexual reproduction again ensues.

Population Dynamics

A population of microorganisms is constantly undergoing genetic change. Mutations occur. Favorable ones are retained. Unfavorable ones are lost. If the culture is capable of transformation, conjugation, or transduction and the conditions are favorable, genetic exchange takes place.

Genetic change in a few cells can change an entire population of bacteria. This can be seen most dramatically in a hospital setting, where antibiotic resistance can spread rapidly among bacteria. Certain genes encode antibiotic resistance, allowing bacteria to survive in the presence of otherwise lethal antibiotics. These genes are often carried in transposons, and frequently the transposons are carried on conjugative plasmids that can move from one species of bacteria to another. The transposon can move to the chromosome or to another plasmid that the cells contains. In this way antibiotic resistance can spread through a bacterial population, almost like an infectious disease. In a hospital environment, where antibiotics are always present, strains of bacteria that carry genes for antibiotic resistance have a clear advantage over strains that do not. As a result, a population of bacteria can be converted from antibiotic sensitivity to antibiotic resistance in a short time. The stage is set for a **nosocomial infection** (a hospital-acquired infection) that is not treatable with antibiotics (Chapter 20).

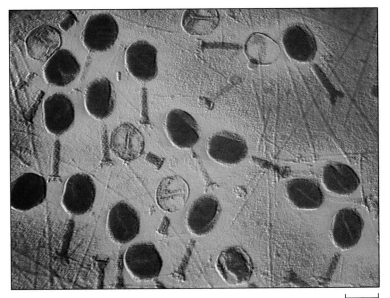

100 nm

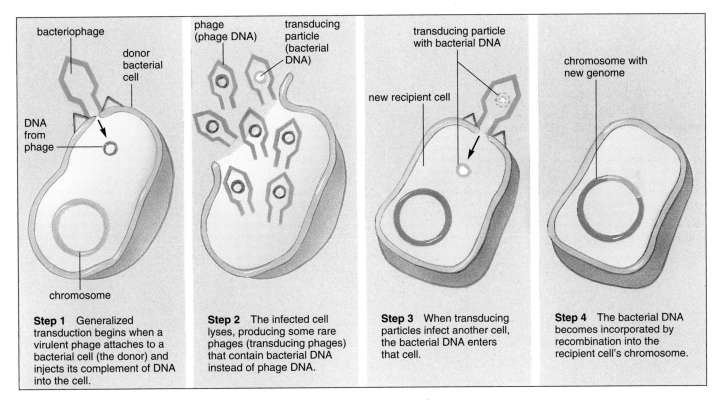

Step 1 Generalized transduction begins when a virulent phage attaches to a bacterial cell (the donor) and injects its complement of DNA into the cell.

Step 2 The infected cell lyses, producing some rare phages (transducing phages) that contain bacterial DNA instead of phage DNA.

Step 3 When transducing particles infect another cell, the bacterial DNA enters that cell.

Step 4 The bacterial DNA becomes incorporated by recombination into the recipient cell's chromosome.

FIGURE 6.23 Generalized transduction. The photo shows a mixture of phages and transducing particles; they cannot be distinguished by appearance.

(Art by Margaret Gerrity.)

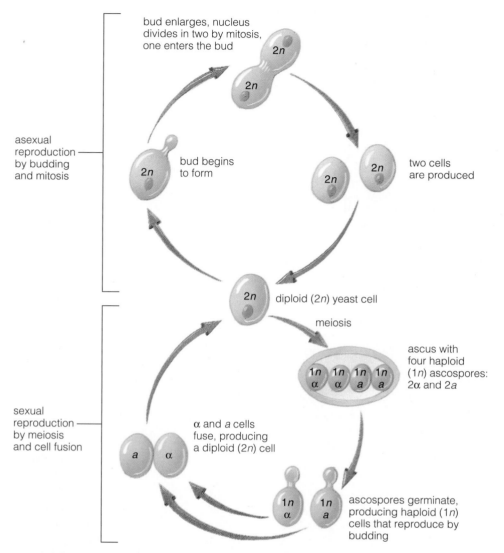

asexual
reproduction
by budding
and mitosis

bud enlarges, nucleus
divides in two by mitosis,
one enters the bud

2n

2n

2n

bud begins
to form

2n

2n

two cells
are produced

2n   diploid (2n) yeast cell

meiosis

1n   1n   1n   1n
α   α   a   a

ascus with
four haploid
(1n) ascospores:
2α and 2a

sexual
reproduction
by meiosis
and cell fusion

α and a cells
fuse, producing
a diploid (2n) cell

a   α

1n
α

1n
a

ascospores germinate,
producing haploid (1n)
cells that reproduce by
budding

FIGURE 6.24   Life cycle of baker's yeast.

(Art by Margaret Gerrity.)

SUMMARY

Case History: Death By Mutation (p. 145)

1a. Natural mutations occur constantly as microorganisms replicate. Some of these confer resistance to antimicrobial agents, such as 5-flucytosine, as the mutation that inactivated cytosine deaminase did.

Structure and Function of Genetic Material (pp. 145-146)

1. Genetics is the science that studies heredity.

The Structure of DNA (pp. 145-146)

2. DNA is composed of the sugar deoxyribose, phosphate, and a nucleoside base. The base may be adenine (A), guanine (G), cytosine (C), or thymine (T).

3. Deoxyribose and phosphate form long strands that wrap around a central core of bases, forming a double helix. Hydrogen bonds that form between G-C and A-T base pairs hold the two strands together.

Reactions of DNA (pp. 145-147)

4. DNA enters into two kinds of reactions: replication and gene expression.

Replication of DNA (pp. 147-151)

5. Replication is making a copy of a DNA molecule from deoxyribonucleotides.

6. Replication begins by breaking the A-T and G-C bonds within a short stretch of DNA, forming a bubble and exposing bases to pair with nucleoside triphosphates.

7. Each newly polymerized strand of DNA is complementary to and has bases that pair with the template (original strand). Each new double helix is composed of one new and one conserved strand. Thus the process is called semiconservative replication.

8. Replication begins at a genetically specified point on the chromosome called the origin.

9. Replication forks travel simultaneously in opposite directions around the chromosome. When they meet at a point called the terminus, the two completed chromosomes separate.

10. The replication forks contain the replication apparatus.

11. The two strands of a DNA double helix are antiparallel; one strand runs from a 3′ carbon to a 5′ carbon, and the other runs from a 5′ to a 3′. Because DNA polymerase III can synthesize a continuous strand of DNA in only one direction, there is continuous synthesis on the leading strand and discontinuous synthesis on the lagging strand.

Gene Expression (pp. 151-155)

12. Gene expression consists of two steps, transcription and translation.

13. Transcription is the polymerization of ribonucleotide building blocks into a molecule of RNA—messenger RNA (mRNA), transfer RNA (tRNA), or ribosomal RNA (rRNA).

14. Transcription begins when ribonucleoside triphosphates pair with the exposed bases on an exposed strand of DNA. The information in DNA is transcribed (rewritten): C pairs with G and A pairs with U (uracil).

15. Transcription begins near a site on the genome called a promoter. Here RNA polymerase separates the two strands, forming a bubble in the DNA, and uses one of them as a template. Polymerization proceeds in the 5′ to 3′ direction. It stops at a place called the terminator. The transcript is released, and the DNA bubble closes.

16. Translation is the polymerization of amino acids into a protein.

17. All three products of transcription participate in translation: mRNA carries the information that determines the order of amino acids in the protein; rRNA is a component of ribosomes where translation takes place; and tRNA does the actual translating. Each tRNA molecule has a site that binds to mRNA and a site that binds to an amino acid.

18. The nucleic acid–recognizing end of the molecule consists of an anticodon that pairs with a codon on mRNA.

19. The specificity of codon-anticodon pairing determines the sequence of amino acids in a protein. The correspondence between codon and amino acid is called the genetic code. Any of the three nonsense codons stops translation at the end of a gene.

20. As a ribosome moves down a molecule of mRNA, it exposes successive codons in a region called the A site. An amino acid–bearing tRNA molecule attaches at each codon, positioning the amino acid to be polymerized onto a peptide at the ribosome's P site. Then the ribosome moves one codon down and the process repeats.

Regulation of Gene Expression (pp. 155-161)

21. Gene expression is usually regulated by increasing or decreasing the rate of transcription or translation.

22. Inducible enzymes are produced only when their substrate is abundant. Repressible enzymes are produced only when their product is scarce. Constitutive enzymes are always produced.

23. Transcription is often regulated by regulatory proteins that bind to DNA and change its interaction with RNA polymerase. When fewer transcripts are made, less protein is made.

24. An operon is a set of genes that is regulated and transcribed together. The *lac* operon in *Escherichia coli*, which encodes the ability to use lactose as a substrate, is regulated by a repressor.

25. Attenuation regulates transcription in bacteria by sensing relative rates of transcription and translation. Regulation of the histidine operon is an example. When the supply of histidine

is adequate, translation of the leader protein occurs quickly. The small stretch of mRNA between RNA polymerase and the first ribosome forms an attenuator loop that prevents transcription of the subsequent genes in the operon. When the supply of histidine is low, translation of the leader protein is slow. The larger stretch of mRNA forms an antiterminator loop that prevents formation of the attenuator loop. Transcription of subsequent operon genes proceeds.

26. Translation of mRNA-encoding ribosomal proteins is regulated by a protein with two functions. One is to be incorporated into a ribosome. The second is to inhibit translation. When the supply of ribosomal protein is just right, all the molecules are incorporated into ribosomes. When there is too much free ribosomal protein, the second function is activated.

27. Catabolite repression is global regulation of gene expression in response to the availability of carbon. For example, catabolite repression inhibits the *lac* operon if a better substrate than lactose is available.

28. Many responses of bacteria to environmental signals are mediated by two-component regulatory systems composed of a sensor and a response regulator.

Changes in a Cell's Genetic Information (pp. 161-174)

The Genome (pp. 161-162)

29. The genome is the sum total of a cell's genetic information (DNA).

30. The genome can change by mutation or by genetic exchange.

31. Most of a prokaryote's genes are in its chromosome. Some are in small circular DNA molecules called plasmids. Most prokaryotes have a single circular chromosome, but a few have more than one circular chromosome and some have linear chromosomes. Most plasmids are circular. A few are linear. Plasmids encode nonessential features. R fac-

tors are plasmids that encode resistance to antibiotics.

32. Most eukaryotes do not have plasmids; most of their genome is in chromosome pairs.

33. Genotype is the genes that a cell contains. Phenotype is the outward expression of a cell's genes.

Mutations (pp. 162-165)

34. A base substitution mutation is a change in a single pair of bases to a different pair. A deletion mutation is total removal of a segment of DNA. An inversion mutation is the inversion of a segment of DNA. A transposition mutation is the movement of a segment of DNA to a different position on the genome. A duplication mutation is the addition of a new segment of DNA.

35. Every time the chromosome is replicated, mistakes can occur and mutations result. Mutation rate is number of mutations per cell per generation.

36. Spontaneous mutations are relatively rare, but their impact is great. Induced mutations are caused by chemical, physical, or biological mutagens.

37. Chemical mutagens may change a component of DNA or become incorporated into the DNA. Physical mutagens may fragment the DNA backbone or cause mistakes in replication. Biological mutations are sequences of DNA that themselves cause mutations by transposing.

38. Many mutations do not change the cell's phenotype. A missense mutation can be serious if the new amino acid that is encoded is not similar to the old. A nonsense mutation (one that changes a codon to a nonsense codon and stops translation) usually inactivates the gene product.

39. Lethal mutation results in destruction of an essential gene product. An auxotroph has a defect in a biosynthesis pathway that confers an additional nutritional requirement. Conditionally expressed mutations render a gene product nonfunctional only in certain environments.

Uses of Mutant Strains (pp. 166-167)

40. To isolate a mutant strain by direct selection, conditions are created that foster the growth of only the mutant strain.

41. In indirect selection (counterselection), conditions are created to prevent the growth of the desired mutant strain. Then growing cells are killed.

42. In the brute strength technique, large numbers of clones are examined one by one to find the desired mutant strain. Replica plating is a simplified brute strength technique.

43. By site-directed mutagenesis a researcher can introduce a particular mutation into a specific gene.

44. The Ames test determines whether a particular chemical is mutagenic and therefore potentially carcinogenic.

Genetic Exchange Among Bacteria (pp. 167-171)

45. Genetic exchange among bacteria is not an essential part of their life cycle. When it occurs, only a portion of the genome of the donor cell transfers to the recipient cell. A merozygote is produced.

46. A piece of transferred DNA must be incorporated into a replicon (the chromosome or a plasmid) before it is a permanent part of the recipient cell's genome. Incorporation occurs by recombination, or crossing over. A plasmid transferred intact to a recipient cell does not require recombination to become part of the recipient cell's genome.

47. Some bacteria are able to undergo natural transformation. DNA leaves the donor cell. Later it is taken up by the recipient cell and incorporated into its genome.

48. In artificial transformation, the bacterial cells are treated in the laboratory to make them able to take up DNA from their environment. Recombinant DNA technology (genetic engineering) depends on artificial transformation.

49. Conjugation is genetic transfer carried out by conjugative plasmids. The best-studied conjugative plasmid is the F plasmid in *Escherichia coli*. Conjugation begins when an F$^+$ cell encodes a sex pilus that attaches to an F$^-$ cell. The F plasmid is nicked and begins producing a single strand of plasmid DNA that enters the recipient cell. The recipient becomes F$^+$ and is now capable of transferring DNA to still another recipient.

50. Transduction is mediated by bacteriophages. When a phage particle is assembled that contains the bacterial DNA, it can be introduced into a new cell where genetic exchange occurs.

51. Specialized transduction occurs during infection by a temperate phage. A prophage leaves the cell and carries bacterial genes, as well as phage genes; it injects them into a new bacterium.

Genetic Exchange Among Eukaryotic Microorganisms (pp. 171-172)

52. Genetic exchange among microorganisms is similar to genetic exchange in other eukaryotes (plants and animals). A diploid cell undergoes meiosis; haploid gametes develop; they fuse to form a zygote or new diploid cell.

53. Baker's yeast, *Saccharomyces cerevisiae*, is a fungus that undergoes both asexual (budding and mitosis) and sexual (meiosis) reproduction.

Population Dynamics (pp. 172-175)

54. Mutagenic resistance to antibiotics can cause nosocomial (hospital-acquired) infections.

REVIEW QUESTIONS

Case History: Death By Mutation

1. Explain how a mutation can cause a microorganism to become resistant to 5-flucytosine.

Structure and Function of Genetic Material

2. Define genetics.

3. What are the building blocks of DNA? Name the bases and tell which pair with which.

4. What is DNA's function? What is a gene?

5. What are the two kinds of reactions DNA enters into? What is transcription? What is translation?

6. Explain this statement: DNA is the master plan of metabolism, and enzymes are the chief mediators.

7. What is replication?

8. How does the structure of DNA lead to the correct ordering of bases during replication?

9. Why is ATP necessary for replication? Where does it enter the replication process? What role does DNA polymerase III play?

10. Explain the mechanics of replication. In your explanation, use and define these terms: template, complementary strand, semiconservative replication, replication fork, origin, terminus.

11. What is the replication apparatus and how does it function?

12. Explain this statement: The two strands of any double helix are antiparallel.

13. Why is a primer important for the function of DNA polymerase III? What are the leading and lagging strands in replication? What roles do primase, DNA polymerase I, and DNA ligase play in the final phase of replication?

14. Define transcription.

15. How is the ordering of building blocks determined in transcription? How are replication and transcription similar? How are they different?

16. Describe how transcription occurs. Use and define these terms: promoters, terminator, transcript.

17. What are the three RNA products of transcription?

18. Explain this statement: Translation changes the information in the metabolic master plan from the language of nucleic acids into the language of proteins.

19. Briefly, what is the function of each of the three RNA products of transcription in translation?

20. Explain how tRNA functions in translation. Use and define these terms: activation, anticodon, codon.

21. What is the genetic code? Why is the genetic code called "redundant"? What are nonsense codons?

22. Describe the mechanics of translation. Use and define these terms: A site, P site, start codon, Shine-Dalgarno sequence/ribosome binding site, polysome.

Regulation of Gene Expression

23. What is metabolic regulation and why is it important?

24. What are two ways to regulate metabolism?

25. What are the usual roles of inducible enzymes and of repressible enzymes? Give an example of an enzyme produced constitutively.

26. What role do allosteric regulatory proteins play in regulating transcription?

27. Define operon. What does the *lac* operon encode? How does the *lac* operon function in the absence of lactose? How does it function in the presence of lactose?

28. What is attenuation?

29. Explain how expression of the histidine operon is regulated. In your explanation, use and define these terms: leader protein, attenuator loop, and antiterminator loop.

30. How is the translation of ribosomal proteins regulated?

31. How is global regulation different from regulation of the *lac* operon, for example, or different from attenuation?

32. Explain catabolite repression as an example of global regulation. What roles do cyclic AMP and CAP play?

33. What are the components of a two-component regulatory system?

Changes in a Cell's Genetic Information

34. Name two ways that changes can occur in a cell's genetic information.

35. What constitutes the genome in eukaryotes? In prokaryotes?

36. Describe the structure and function of plasmids.

37. What are R factors? Why are they medically significant?

38. What are conjugative plasmids?

39. Define genotype. Define phenotype.

40. Explain these types of mutations: base substitution, deletion, inversion, transposition, duplication.

41. Why is the rate of mutation calculated on the basis of replication rather than time?

42. What are spontaneous mutations? Why do they have a significant impact even though they occur relatively rarely?

43. What are induced mutations? What is a mutagen?

44. Give an example of a chemical mutagen. How does ultraviolet light act as a mutagen?

45. Why are biological mutagens called transposable elements, or jumping genes? What are transposons, and how do they benefit a cell? What are insertion sequences, and how do they function?

46. What two factors determine how serious a mutation will be for the phenotype of a cell? What is a missense mutation? A nonsense mutation?

47. What is a lethal mutation? An auxotroph?

48. What are conditionally expressed mutations? Give an example.

49. Explain these methods of isolating mutant strains: direct selection, indirect selection (counterselection), brute strength, replica plating.

50. How is the Ames test performed?

51. Name the three ways genetic exchange occurs in bacteria. Compare and contrast genetic exchange in prokaryotes and eukaryotes.

52. What is transformation? Describe natural transformation. Describe artificial transformation.

53. Describe conjugation in *Escherichia coli*. What is an Hfr cell?

54. What are bacteriophages (phages)? Describe generalized transduction. Describe specialized transduction. Use the terms *lytic cycle* and *lysogenic cycle* in your explanation.

55. Describe the forms of genetic exchange in baker's yeast as examples of genetic exchange in eukaryotes.

56. Discuss an example of how genetic change in a population of bacteria can lead to serious hospital-acquired (nosocomial) infections.

CORRELATION QUESTIONS

1. Do you think that primase would be able to substitute for RNA polymerase in gene expression? Why?

2. Do you think that replication would be more or less accurate than gene expression? Why?

3. You know that when cultures of two bacterial strains are mixed, genetic exchange occurs. What experiment would you do to determine whether the exchange takes place by transformation or conjugation?

4. Frame shift mutations often cause nonsense mutations to appear in the same gene. How could this happen?

5. Which is more energetically efficient—regulation that acts at the level of transcription or that acts at the level of translation? Why?

6. An early name for one type of RNA was adapter RNA. Do you think this name was used for mRNA, tRNA, or rRNA? Why?

ESSAY QUESTIONS

1. Mutation rate is regulated by metabolic processes that occur within a cell. Discuss why evolution has led to a low but detectable rate instead of a higher or lower one.

2. A mutation occurs in the chromosome of one cell of a bacterium that is infecting a patient in a hospital. The mutation makes the cell resistance to the particular antibiotic that is being used to treat the patient. Discuss how this resistance might spread to other bacterial cells in the patient and other bacteria in the hospital.

SUGGESTED READINGS

Ames, B. W. 1979. Identifying environmental chemicals causing mutations and cancer. *Science* 204:587.

Drake, J. W. 1991. Spontaneous mutation. *Annual Reviews of Genetics* 25:125–46.

Freifelder, D. 1987. *Microbial genetics*. Boston: Jones and Bartlett.

Hoch, J. A., and T. J. Silhavy. 1995. *Two-component signal transduction*. Washington, D.C.: ASM Press.

● For additional readings, go to InfoTrac College Edition, your online research library at: http://www.infotrac.thomsonlearning.com

SEVEN

Recombinant DNA Technology and Genomics

Courtesy Robert Hammer, Howard Hughes Medical Institute, Dallas

CHAPTER OUTLINE

LEARNING GOALS

To understand:

- *The nature of recombinant DNA technology and the potential it has to affect every aspect of our lives*

- *The fundamental tool of recombinant DNA technology—gene cloning—and the five steps involved: obtaining DNA, splicing genes into a cloning vector, putting recombinant DNA into a host cell, testing, and propagating*

- *The methods of finding the right gene for gene cloning*

- *The uses of* Escherichia coli *as a host cell for recombinant DNA*

- *Current applications of recombinant DNA technology in medicine, industry, agriculture, and criminal investigation and applications we can expect in the future*

- *The methods of sequencing*

- *The methods and information gained by genomics*

- *The methods and information gained by microarray technology*

Practical Genomics

R.G., an eighteen-year-old college student, went to the student health center the day before the campus was closing for winter break. R. G. was tired from late-night studying for finals but otherwise had been well until one day earlier when he developed a fever of 102°F and a very sore throat. He had no respiratory symptoms—no runny nose, cough, or congestion. But his throat was so painful he could barely eat or drink. His concerned roommates urged him to see a doctor.

On arrival at the student health center, R. G.'s temperature was still over 102°F and he looked ill. The examining doctor noted that his tonsils were enlarged and coated with patches of blood and pus. Although his neck was not stiff, numerous lymph nodes on both sides of his neck were enlarged to the size of grapes and quite tender to touch. His voice was muffled. His mouth was slightly dry. He had drunk only a small amount of water since the night before. R. C. told the doctor that he must travel home the next day.

The physician knew immediately that R. G. had an **acute pharyngitis** (throat infection), but what was its cause? A sudden, severe sore throat such as R. G.'s could be **strep throat** (caused by the bacteria *Streptococcus pyogenes*) or the beginning of **infectious mononucleosis** (caused by the Epstein Barr virus). If it were strep throat, 24 hours' treatment with an oral penicillin should virtually cure R. G.'s infection and allow him to travel home comfortably. But if it were the beginning of infectious mononucleosis ("mono"), giving penicillin would be of no help at all and might even cause R. G. to break out in a **mononucleosis rash.** Someone with mono and as sick as R. G. should be treated immediately with steroid medication to shrink the swollen tonsils in order to allow him to drink and avoid dehydration. With mono, R. G. certainly could not expect to be well enough to travel the next day.

Detecting *Streptococcus pyogenes* by throat culture takes 48 to 72 hours, the time necessary for colonies of the bacterium to develop on test plates. A blood test is available to detect mono, but it's not reliable until ten days after the patient experiences his first symptoms.

Fortunately for R. G., the student health center was able to perform a new rapid diagnostic test, employing a DNA probe, to detect *Streptococcus pyogenes*. The doctor used a cotton swab to obtain a sample from his throat and asked him to wait for the result. The test established *Streptococcus pyogenes* was present in R. G.'s throat, confirming that he had strep throat. R. G. received a prescription for penicillin and by the next day felt much better, making the trip home without difficulty.

Case Connections

- The test that allowed R. G.'s physician to rapidly diagnose and thereby treat his strep throat depends on the precision of hybridization (as discussed in Chapter 6), in this case between a bit of DNA used as probe and DNA that is unique to *Streptococcus pyogenes*.
- Knowledge of which DNA can be used as such a probe is a product of genomics, the rapidly developing field of study that is having a profound impact on all aspects of microbiology and will be discussed in this chapter.

RECOMBINANT DNA TECHNOLOGY AND GENOMICS

You might wonder why a book on microbiology includes a chapter on recombinant DNA technology and genomics. It does because these powerful methods are the products of past microbiological studies and exert a major impact on current studies. Much of modern microbiological research, both basic and applied, depends on recombinant DNA technology and genomics. We'll consider these two related topics sequentially.

Recombinant DNA Technology

What is recombinant DNA technology? It's easy to appreciate but hard to define. It's not a single technique or a procedure. Instead, it's a vast collection of different procedures. Many of them involve taking DNA from a cell, manipulating it in vitro, and putting it into another cell. Often the recipient cell belongs to a different species. The fundamental step to all these procedures is the one that gives the technology its name, recombinant: Genes from different organisms are recombined into a single molecule.

LARGER FIELD

COHEN, BOYER, AND CORNED BEEF

Beginnings in science are hard to pinpoint. But no one can argue that recombinant DNA technology didn't begin in 1972 in a Waikiki delicatessen over corned beef sandwiches.

Stanley Cohen and Herbert Boyer were attending a scientific conference in Honolulu. Cohen, who was studying bacterial plasmids, was struck by their flexibility. They replicate independently of the bacterial chromosome and can be moved readily from one bacterial cell to another even in the form of a pure solution of DNA. Cohen reasoned that he could move genes from any source just as readily if he could put them in a plasmid. In other words, he could clone a gene. But how to do it? At the conference, Cohen attended Boyer's talk on restriction enzymes and came up with an idea.

The two scientists met for dinner and decided to collaborate. Boyer's laboratory at the University of California in San Francisco was only an hour's drive from Cohen's at Stanford University. The project went quickly—and successfully. Cohen called the recombinant DNA (the plasmid and foreign gene) a chimera, after the mythological fire-breathing beast with a lion's head, a goat's body, and a serpent's tail. The age of recombinant DNA technology was born.

Recombination is a technical term geneticists use to describe the processes of forming a new combination of genes by any means, natural or artificial. For example, recombination occurs during sexual reproduction. All offspring acquire their genes from their parents. But each offspring has a different combination of genes from those found in either of its parents. Most such recombination in eukaryotes occurs because an offspring receives half its chromosomes (and the genes they carry) from each parent. The result is a recombination of parental genes.

Sometimes (usually during meiosis in eukaryotes and in all types of genetic exchange between prokaryotes) recombination has a different basis: Genes on the same chromosome or DNA molecule recombine by crossing over (Chapter 6). In essence, chromosomes or DNA molecules break, and then a fragment of one joins to a fragment of another. The result is a recombinant molecule, part of which is derived from one chromosome (or DNA molecule) and part from another. Such recombinant DNA molecules often form in living cells as a normal part of their development. But this recombination occurs only between chromosomes or DNA molecules that are **homologous** (similar enough for a single strand of one DNA molecule to form a hydrogen-bonded double strand with a single strand of the other DNA molecule, with only a few mismatched base pairs).

Recombinant DNA technology also produces recombinant DNA molecules, but with two important differences. First, recombination occurs in vitro, not in the cell. Second, the DNA molecules that join to form a recombinant molecule do not have to be homologous. They don't even have to be similar. Pieces of DNA from distantly related organisms—humans and bacteria, for example—can be joined into a single molecule.

In this chapter we'll first discuss gene cloning because it's the fundamental tool of recombinant DNA technology. Then we'll look at how recombinant DNA technology is applied in basic and applied science. Finally we'll consider the new but rapidly expanding field of genomics. We'll discuss specific uses of recombinant DNA technology and genomics throughout the rest of this book, and we'll discuss their impact on biotechnology in Chapter 29.

Gene Cloning

Gene cloning is the process of obtaining a large number of copies of a gene. It's called cloning because the copies are derived from a single copy of the gene. Recall that a colony on a petri dish is a clone of cells derived from a single cell (Chapter 3). The cell multiplies to become a clone.

To obtain a gene clone, the gene is introduced into a cell on a piece of DNA that the cell can replicate. Then the cell is allowed to multiply to become a clone of cells. Each of the cells in the clone has one or more copies of the gene. So collectively, all these copies of the gene are a clone of genes. Gene cloning is the cornerstone of recombinant DNA technology because it is the way to produce the massive numbers of gene copies needed for other recombinant DNA procedures.

Gene cloning involves five steps (**Figure 7.1**):

1. Obtaining a piece of DNA that carries the gene to be cloned

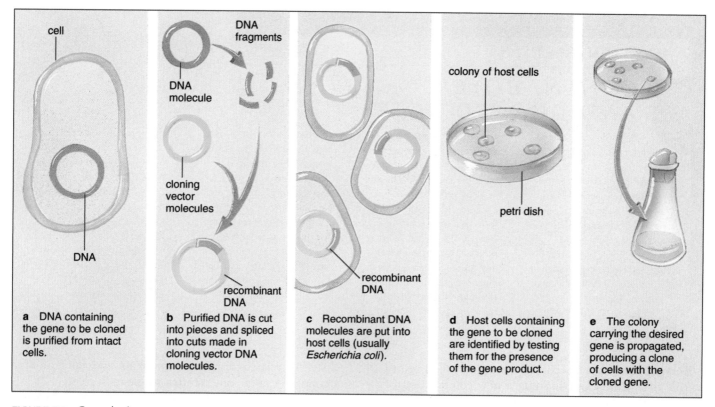

a DNA containing the gene to be cloned is purified from intact cells.

b Purified DNA is cut into pieces and spliced into cuts made in cloning vector DNA molecules.

c Recombinant DNA molecules are put into host cells (usually *Escherichia coli*).

d Host cells containing the gene to be cloned are identified by testing them for the presence of the gene product.

e The colony carrying the desired gene is propagated, producing a clone of cells with the cloned gene.

FIGURE 7.1　Gene cloning.
(Art by Margaret Gerrity.)

2. Splicing that DNA into a **cloning vector** (a DNA molecule that a host cell will replicate)

3. Putting the **recombinant DNA** (in this case, the cloning vector with the desired gene spliced into it) into an appropriate host cell

4. Testing to ensure that the gene has actually been put into the host cell

5. Propagating the host cell to produce a clone of cells that carries the clone of genes

Let's look at each of these steps individually.

1. Obtaining DNA. DNA makes up only a small fraction of a cell. It constitutes about 3 percent of the dry weight of a bacterial cell and an even smaller percentage of a eukaryotic cell. In spite of being a minor cellular component, DNA is easy to separate in pure form because it has unusual chemical and physical properties. First, DNA is an extremely large molecule. For example, the single DNA molecule that makes up the chromosome of *Escherichia coli* is a thousand times longer than the cell itself. Second, DNA is highly resistant to chemical and physical treatments that rapidly destroy other cell components. Finally, DNA is **denser** (heavier per unit volume) than most other cell components.

These special properties are used to purify DNA from a **cell extract** (the liquid content of ruptured cells; **Figure 7.2**). Because of its large size, DNA forms ropelike aggregates when alcohol is added to a cell extract. These aggregates can then be separated from the rest of the extract by wrapping them around a glass rod and pulling it out of the liquid. DNA's chemical resistance allows it to survive treatment with phenol, which denatures and precipitates proteins, and its high density allows it to be separated from the other less dense cell components by centrifugation. The first step in cloning a bacterial gene is purifying DNA, using one or a combination of such treatments.

Obtaining DNA to clone a prokaryotic gene is done simply by purifying it from other components of the bacterial cell. But obtaining DNA to clone a eukaryotic gene is more complicated because eukaryotic genes contain regions called introns. **Introns** are noncoding regions that enzymes in the eukaryotic cell cut out after the gene has been transcribed into mRNA (**Figure 7.3**). Because bacterial genes don't have introns, bacterial cells don't have the enzymes to cut introns out of mRNA. Thus intron-free DNA must be used to clone eukaryotic genes into bacteria.

SHARPER FOCUS

CLONING DINOSAURS

The movie *Jurassic Park* is about a theme park populated by cloned dinosaurs and the disaster that results. The basic premise is that DNA from dinosaur bones and the guts of ancient insects (whose last meal before they were preserved in amber was dinosaur blood) was cloned and then used to produce a dinosaur.

How plausible is this scenario? The second step—producing a dinosaur only from DNA without a dinosaur cell to nurture it—stretches credulity to the limit. No one knows enough about developmental genetics even to consider creating a living organism from cloned DNA alone. But what about the first step—cloning DNA from dinosaurs? If even a minuscule amount of DNA were found intact, it could be amplified (by polymerase chain reaction) and cloned. So the question becomes, could any DNA survive the 60 million years since dinosaurs became extinct?

Most scientists think it could not. They cite biochemical studies on the rate of decomposition of DNA that indicate DNA lasts only 40,000 to 50,000 years. Other scientists, however, point out that were DNA completely protected from water and oxygen, it would last much longer. This contention is supported by experimental evidence. Convincing experiments show that bits and pieces of bacterial DNA have survived in rock salt for 425 million years. Similar water-free, oxygen-free conditions might exist inside some dinosaur bones. In fact, DNA has been obtained by amplification of samples from dinosaur bones. The only question is whether it comes from a dinosaur or some other source. It might be from a human who touched the bone or from a fungus that grew on it. In this situation the enormous amplifying power of PCR leads to confusion rather than resolution. We simply do not know if any dinosaur DNA has survived.

Cloning particular genes and inserting them in another living cell is another story. Although scientists cannot now (nor probably ever will be able to) produce a living organism from the cloned DNA of its genome, they're readily able to produce particular protein products from particular cloned genes.

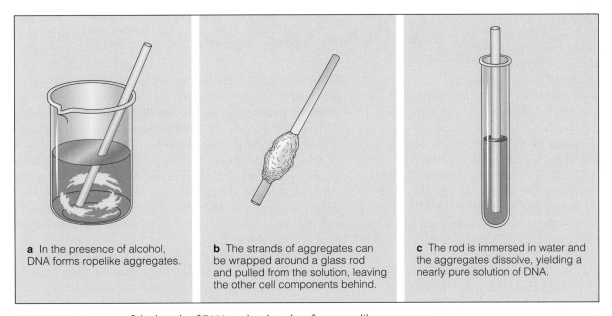

a In the presence of alcohol, DNA forms ropelike aggregates.

b The strands of aggregates can be wrapped around a glass rod and pulled from the solution, leaving the other cell components behind.

c The rod is immersed in water and the aggregates dissolve, yielding a nearly pure solution of DNA.

FIGURE 7.2 Because of the length of DNA molecules, they form ropelike aggregates when precipitated, providing a simple easy means of purification.

FIGURE 7.3   To obtain intron-free copies of eukaryotic genes, cDNA is made from RNA transcripts of the gene.

(Art by Margaret Gerrity.)

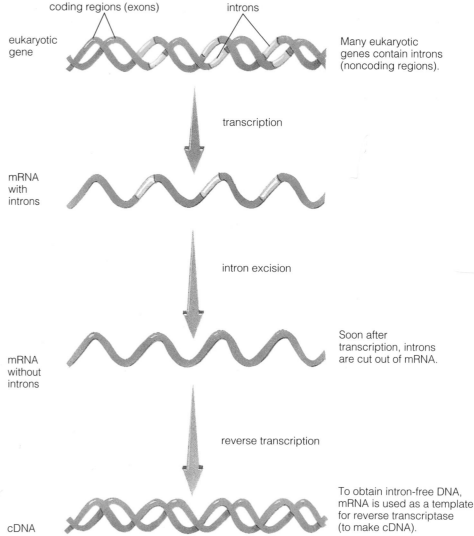

coding regions (exons)   introns

eukaryotic gene

Many eukaryotic genes contain introns (noncoding regions).

transcription

mRNA with introns

intron excision

mRNA without introns

Soon after transcription, introns are cut out of mRNA.

reverse transcription

cDNA

To obtain intron-free DNA, mRNA is used as a template for reverse transcriptase (to make cDNA).

To obtain intron-free DNA, the gene cloner purifies mRNA instead of DNA from the eukaryotic cell extract. Then the gene cloner uses **reverse transcriptase** (an enzyme that uses RNA as a template to make a complementary strand of DNA; Chapter 13) to make intron-free DNA from the mRNA. This DNA has the coding properties of normal eukaryotic genes but lacks their introns.

2. Splicing Genes into a Cloning Vector. After DNA containing the desired gene is obtained, it is spliced into a cloning vector. This step is necessary because only certain DNA molecules are replicated in cells. Cloning vectors are DNA molecules that are replicated because they contain a region called an **origin of replication** (Chapter 6).

Good cloning vectors need other properties in addition to an origin of replication. They must be relatively small so that their replication does not unduly tax the host cell's metabolic capacity. Also, they must carry other genes that identify host cells that contain the vector. For example, most cloning vectors carry genes encoding antibiotic resistance. Cells containing such a vector can be easily identified because that can multiply in antibiotic-containing media.

Plasmids and the genomes of certain viruses are used as cloning vectors. Almost all these cloning vectors are circular molecules of DNA. Inserting the gene to be cloned into the cloning vector involves two steps: (1) cutting the cloning vector, and (2) sealing the fragment of DNA with the gene to be cloned into the cut (a process called **ligation**). The result is a larger circular DNA molecule.

Cutting DNA with Restriction Endonucleases. Cloning vehicles and DNA to be inserted into them are usually cut with enzymes called **restriction endonucleases.** These enzymes have particularly useful properties for cloning. Before we discuss how restriction endonucleases are used to clone genes, let's consider why they don't cut up the DNA in the bacteria that produce them. The answer is that bacteria protect their own DNA from their own restriction endonucleases by **modifying** it. They modify their DNA by adding methyl groups to certain bases within the sequences where restriction endonucleases cut (**Table 7.1**). In contrast, foreign DNA, such as DNA injected by an infecting virus, is not modified. It is rapidly cut and thereby destroyed (see Larger Field: Werner Arber, Chapter 13). Most microbiologists believe that restriction endonucleases evolved to protect bacteria from viral attack.

Restriction endonucleases recognize specific base sequences and cut the DNA within or near them. Useful cloning vectors have only a single site at which specific endonucleases cut. There are many kinds of restriction endonucleases. Most bacterial species produce their own particular type. They are named for the species that produces them, using the initial of the genus and the two first

TABLE 7.1 Properties of Some Restriction Endonucleases

Restriction Endonuclease	Source (Bacterial Species)	Target Site (Cuts at Arrow)	Characteristics	
			Recognizes (No. Base Pairs)	Product
Eco RI	*Escherichia coli* R13	↓ G-A-A-T-T-C C-T-T-A-A-G ↑	6	4-base-long sticky ends
Hha I	*Haemophilus haemolyticus*	↓ G-C-G-C C-G-C-G ↑	4	2-base-long sticky ends
Sma I	*Serratia marcescens*	↓ C-C-C-G-G-G G-G-G-C-C-C ↑	6	Blunt ends
Hae III	*Haemophilus aegyptius*	↓ G-G-C-C C-C-G-G ↑	4	Blunt ends

letters of the species. For example, a restriction enzyme from *Escherichia coli* is designated *Eco*. If a species produces more than one enzyme, they are distinguished by Roman numerals. Also, if enzymes are encoded on an R factor (Chapter 6), they are designated R. Accordingly, one of the most commonly used restriction endonucleases is named *Eco* RI. It is one of several restriction enzymes produced by *E. coli*, and it is encoded by an R factor. Another useful restriction endonuclease—one obtained from the highly thermophilic bacterium *Thermus aquaticus*—is named *Taq* I.

Properties of Restriction Endonucleases. Members of a certain class of restriction endonuclease (called type II) are used for cloning. Type II restriction endonucleases cut asymmetrically within base sequences that are four, five, six, or seven base pairs long and that have an axis of **rotational symmetry.** That means the base sequences read the same way on one strand as they do in the opposite direction on the other strand. For example, the sequence

—A-G-A-T-C-T—
—T-C-T-A-G-A—

has an axis of rotational symmetry because it reads the same backward and forward. Such sequences are termed **palindromes** because they resemble word and phrase palindromes ("Otto" and "Madam I'm Adam").

Sequences such as

—A-G-C-C-G-A—
—T-C-G-G-C-T—

are also termed palindromes, though they occur within a single strand and do not have an axis of rotational symmetry.

Many type II restriction endonucleases cut asymmetrically within the axis of rotational symmetry. Thus they produce short, complementary, single-strand regions called **cohesive ends,** or sticky ends, because they readily pair with one another. For example, *Eco* RI cuts each strand between G and A (see arrows) in the sequence

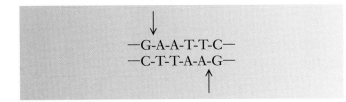

producing

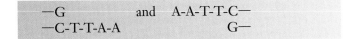

DNA from any organism—microorganism, plant, or animal—will produce the same sticky ends (A-A-T-T— and —T-T-A-A) when cut by Eco RI. When DNA molecules cut with this kind of restriction endonuclease are mixed and held at a low temperature (near the freezing point of water), the sticky ends pair and **anneal** (join together). They anneal because hydrogen bonds form between complementary bases. Sticky ends cut with *Eco* RI would anneal to form

—G A-A-T-T-C—
—C-T-T-A-A G—

The ends anneal, but gaps (missing phosphodiester bonds) exist between the G and A on both strands.

Thus, using a restriction endonuclease, any two pieces of DNA can be joined together. It is necessary only to cut both of them with the same restriction endonuclease, mix them, and allow them to anneal.

Following the Reaction. Restriction endonucleases catalyze a precise chemical reaction—cutting DNA at all copies of the sequences that a particular restriction endonuclease recognizes. How do we follow such a reaction? That is, how do we determine when all the susceptible sites have been cut? Usually gel electrophoresis is used: Samples of the reaction mixture are placed on a slab of gel (usually made of agarose, a derivative of agar, or polyacrylamide, a synthetic polymer similar to the polymer used to make acrylic fabric). Then the gel is exposed to an electrical field (**Figure 7.4**). Because DNA is negatively charged, it moves toward the positive pole of the field. Different DNA becomes separated on the gel because small molecules move faster than large ones. The location on the gel of the various-sized pieces of DNA is revealed by staining the gel with ethidium bromide (or a similar compound) and illuminating the gel with ultraviolet light. Ethidium bromide binds tightly to DNA and fluoresces. The locations of the DNA molecules appear as red-orange lines or bands on a light pink background. When cutting by restriction endonuclease is complete, no new (smaller) bands appear. When cut with a particular restriction endonuclease, each DNA molecule produces a characteristic pattern of bands after gel electrophoresis. This occurs because each DNA molecule has a definite number and definite location of sites at which the restriction endonuclease cuts.

Because specific patterns of bands are produced, cutting DNA samples with a restriction endonuclease and separating the fragments by electrophoresis determines if two samples of DNA are different or—with high probability—identical. Comparing gel patterns of DNA cut by restriction endonuclease is the way DNA samples are analyzed in criminal investigations. If a suspect's DNA pattern (called **restriction fragment length polymorphism** or **RFLP**) differs from that of an incriminating sample recovered at the crime scene, that individual may be ruled out as a suspect.

Ligation. After the sticky ends produced by a restriction endonuclease anneal, the gaps that remain at sites where the original cuts were made must be sealed. Otherwise the molecule would easily fall apart when the temperature is raised. The gaps are sealed, or ligated, by forming a covalent bond between the adjacent bases through the action of the enzyme called **DNA ligase** (Chapter 6). Using our earlier example, the ligated structure would then become

—G-A-A-T-T-C—
—C-T-T-A-A-G—

a   b   c

FIGURE 7.4   Gel electrophoresis is a way to separate fragments of DNA. (a) Each sample to be examined is placed in one of the wells in the top of a slab of gel held between two pieces of glass (the red color is due to a dye added to indicate location). (b) An electric field (positive on bottom, negative on top) is applied to the gel causing the negatively charged fragments of DNA to move down the gel. Small ones move faster than larger ones. (c) The gel is strained with a fluorescent dye and ethidium bromide and illuminated with ultraviolet light. Regions to which DNA fragments have moved appear as orange bands.

All organisms produce DNA ligases, but the one most commonly used in cloning is T4 ligase. T4 ligase is an enzyme formed by *Escherichia coli* when it is infected by bacteriophage T4.

Blunt-End Ligation. Although annealed sticky ends facilitate ligation, they are not absolutely necessary. Pieces of DNA without an extending single strand, which are termed **blunt-ended,** can also be ligated. The process is called **blunt-end ligation.** Blunt-end ligation is less efficient than ligation of gaps in annealed sticky ends because blunt ends form a weak joint that readily breaks. Nevertheless, blunt-end ligation has certain advantages and is often used in cloning.

One advantage of blunt-end ligation is that the DNA fragment bearing the cloned gene and the vector need not be cut by the same restriction endonuclease. Another advantage is that the DNA to be cloned can be cut mechanically by shearing. In shearing, the solution of DNA is vigorously stirred or rapidly drawn through a small orifice such as the opening of a pipette. Such treatments break a DNA molecule near its midpoint and then break the two fragments near their midpoints again. To ligate sheared DNA, we have to ensure that the ends are blunt by treating the solution of DNA fragments with an enzyme that removes any single strands on the ends of the molecule. Then the blunt-ended DNA fragment can be ligated into a vector that has also been cut to generate blunt ends. Some restriction endonucleases—for example, *Sma* I, from *Serratia marcescens*—form blunt ends directly. *Sma* I cuts the DNA sequence

$$\downarrow$$
$$-C\text{-}C\text{-}C\text{-}G\text{-}G\text{-}G-$$
$$-G\text{-}G\text{-}G\text{-}C\text{-}C\text{-}C-$$
$$\uparrow$$

between G and C, forming two blunt ends,

$$-C\text{-}C\text{-}C \quad \text{and} \quad G\text{-}G\text{-}G-$$
$$-G\text{-}G\text{-}G \qquad\qquad C\text{-}C\text{-}C-$$

Blunt ends are ligated by the same enzymes (usually T4 ligase) and under same conditions used to ligate gaps in sticky ends.

3. Putting Recombinant DNA into a Host Cell.

Ligation completes the construction of a recombinant DNA molecule. Then it has to be replicated to obtain enough to use. It's possible to replicate DNA in vitro using purified enzymes, but the only way to replicate recombinant DNA quickly, inexpensively, and in large amounts is to insert it into a living cell. A number of procedures are available to put DNA molecules into a host cell. They include transformation, transfection, microinjection, and electroporation.

Transformation is the process by which an intact cell takes up DNA from solution (Chapter 6). Some bacteria have an innate ability to be transformed by DNA from solution, a process called natural transformation. Most bacteria, however, can be transformed only after they have been specially treated, a process called artificial transformation. For example, *Escherichia coli* will take up DNA from solution in its environment after it has been exposed to high concentrations of calcium chloride (CaCl$_2$), chilled to 0°C, and suddenly heated to 42°C. Other treatments have been developed to transform other bacterial species, as well as plant and animal cells that do not undergo natural transformation.

If DNA from a virus is used as the cloning vector, the process is called **transfection.** A fundamental property of viruses is their ability to introduce their genome into a host cell (Chapter 13). So if recombinant DNA is incorporated into a virus particle, it can be inserted into a host cell the same way the virus inserts its own DNA. For example, when DNA from the bacteriophage lambda (λ) is used as the cloning vector, it can be incorporated into an infectious viral particle by adding purified viral proteins to a solution of the recombinant DNA (**Figure 7.5**). Then the artificially constructed virus can insert the recombinant molecule into a susceptible *E. coli* cell.

DNA can be inserted into some animal cells directly by **microinjection** (**Figure 7.6**). The cell is held by a slight vacuum to the end of a holding pipette, and DNA is injected into it from a tiny (about 1.5 mm in diameter) pipette. Microinjection cannot be used with bacterial cells because they are too small to be injected even by such a tiny pipette.

DNA can be introduced into many bacterial, animal, and plant cells by **electroporation.** In this process the cells are suspended in a solution of DNA and exposed to high-voltage electrical impulses. The electrical impulses destabilize the cytoplasmic membrane, making it possible for the DNA in solution to enter the cell.

4. Finding the Right Gene.

The basic step in cloning is cutting a cloning vector and splicing into the cut a fragment of DNA carrying the gene to be cloned. But how do we know this fragment carries the gene we want to clone? There are two ways—prior purification and subsequent identification.

Prior Purification. In prior purification, a DNA fragment carrying the gene to be cloned is purified and spliced into the cloning vector. To purify a gene or mRNA made from it, we have to be able to distinguish it from other similar molecules. We can do this by taking advantage of the specific hy-

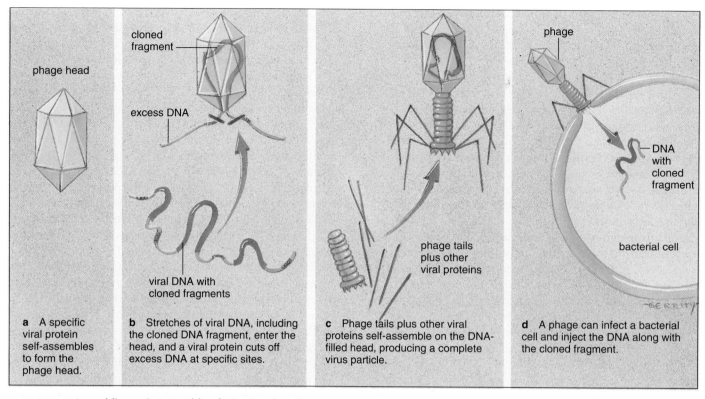

a A specific viral protein self-assembles to form the phage head.

b Stretches of viral DNA, including the cloned DNA fragment, enter the head, and a viral protein cuts off excess DNA at specific sites.

c Phage tails plus other viral proteins self-assemble on the DNA-filled head, producing a complete virus particle.

d A phage can infect a bacterial cell and inject the DNA along with the cloned fragment.

FIGURE 7.5   Assembling a virus capable of injecting cloned DNA into a bacterium.
(Art by Margaret Gerrity.)

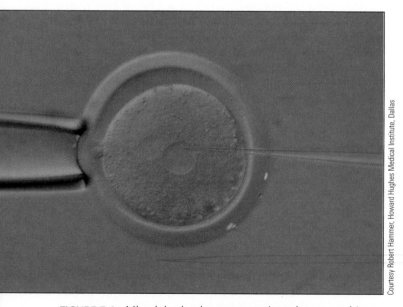

Courtesy Robert Hammer, Howard Hughes Medical Institute, Dallas

FIGURE 7.6   Microinjection is one way to introduce recombinant DNA into an animal host cell.

drogen bonding that occurs between complementary single strands of DNA or between a strand of DNA and an mRNA made from it. Pairing between such complementary molecules occurs spontaneously—a process called hybridization (Chapter 6). So to identify a particular gene or its mRNA product, we need only a complementary DNA strand appropriately tagged (usually with a radioactive isotope). A short DNA molecule that is complementary to its corresponding mRNA is called a **probe.**

But how do we obtain an appropriate probe? As an example, let's look at how scientists cloned the gene encoding human growth hormone (hGH)—the hormone produced by the pituitary gland that stimulates our growth and, when in short supply, causes dwarfism in children (**Figure 7.7**). hGH had been studied for a long time, so the scientists knew the sequence of amino acids that forms it. With this information and knowledge of the genetic code (Chapter 6), they deduced the possible sequences of the encoding DNA. More than one DNA sequence can encode the same amino acid sequence because most amino acids are encoded by more than one codon.

The scientists then synthesized in radioactive form all possible DNA fragments corresponding to a sequence of six amino acids in the protein. (Most modern research laboratories have machines that synthesize short pieces of DNA.) Such a DNA fragment (18 bases long [6 amino acids $\times$ 3 bases per codon]) is an adequate probe to identify the hGH-encoding mRNA molecule. They obtained mRNA from extracts of cadaver pituitary glands. Using the probes, they isolated the mRNA. Then, using reverse

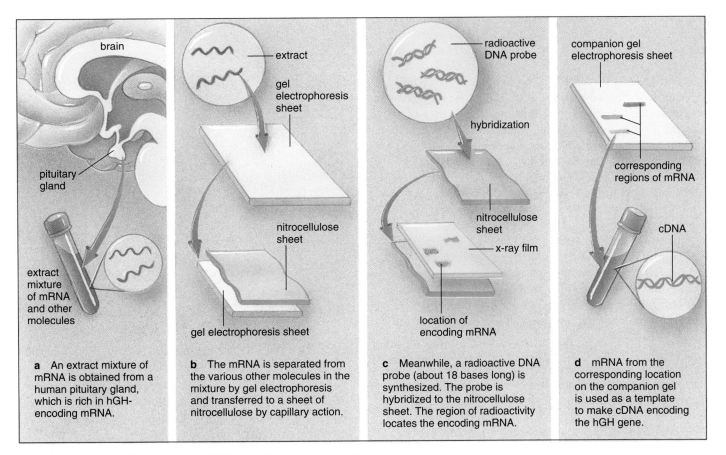

FIGURE 7.7   A method for obtaining cDNA-encoding human growth hormone.

(Art by Margaret Gerrity.)

a An extract mixture of mRNA is obtained from a human pituitary gland, which is rich in hGH-encoding mRNA.

b The mRNA is separated from the various other molecules in the mixture by gel electrophoresis and transferred to a sheet of nitrocellulose by capillary action.

c Meanwhile, a radioactive DNA probe (about 18 bases long) is synthesized. The probe is hybridized to the nitrocellulose sheet. The region of radioactivity locates the encoding mRNA.

d mRNA from the corresponding location on the companion gel is used as a template to make cDNA encoding the hGH gene.

transcriptase, they synthesized **cDNA** (DNA made from RNA), which they cloned into an appropriate vector.

Subsequent Identification. In subsequent identification, random fragments of DNA are spliced into the cloning vector and inserted into cells that are then plated to develop into colonies. Colonies that produce the product of the gene to be cloned are selected because they must carry that gene.

Scientists usually do not purify prokaryote genes before cloning because prokaryotes have relatively small genomes. Instead the entire genome is cut into pieces, each of which is spliced into molecules of the cloning vector and then transformed into bacterial cells. By this process, called **shotgun cloning**, they produce a **gene bank.** A gene bank is a set of bacterial clones, each of which carries a small clone of DNA. Altogether these clones carry the entire genome. Cloned DNA fragments are commonly about 30 kilobases (30,000 base pairs) long, so a complete gene bank of an ordinary-sized bacterium could be carried in 150 bacterial clones (4500 [genome size of E. coli] ÷ 30). Once made, a gene bank is saved and used later as a source of cloned genes. Then the scientist has to search the bank to find the gene he or she wants. This can be done in a number of different ways.

The gene may be expressed in the host cell. Expression occurs if the host cell's gene expression machinery, including transcription and translation machinery, is similar to the machinery of the cell from which the cloned DNA was obtained. In that case, it's possible to look for the product of the cloned gene. For example, the host cell might gain the characteristic encoded by the gene, such as an ability to synthesize a particular amino acid or break down a particular substrate. Even if the gene is not expressed by the host, an appropriate DNA probe can be made and used as we described for purifying the eukaryotic gene prior to cloning. Then, by hybridization, all the bacterial clones in the gene bank can be tested to find which one carries the gene we want.

Hosts for Recombinant DNA

Escherichia coli was the first organism used as a host for recombinant DNA. It's still the most commonly used. But with recent advances—particularly in transformation and other ways of putting recombinant DNA into cells—other microorganisms, insects, plants, and animals are also now being used as hosts. The organism selected depends on the reason for cloning. In basic biological studies on genes and their products, *E. coli* is almost always used. It is easy to cultivate, it grows rapidly, and we know more about it than any other organism.

Along with the advantages of *E. coli*, however, come some difficulties. For example, *E. coli* was the host used to make the first medically useful proteins, human growth hormone and insulin (for treating diabetes). But these proteins did not fold properly when made in *E. coli*. Instead, they **precipitated** (solidified) within the cell, forming inclusion bodies (**Figure 7.8**). To make biologically active protein from inclusion bodies, they had to be **renatured** (properly refolded). This is a relatively easy procedure for small proteins such as insulin and human growth hormone, but it is a major challenge for large proteins. Recently it has become possible to produce active proteins within *E. coli* by tagging them so they are exported to the periplasm and stimulating the formation of other proteins (**disulfide bond isomerases** and **chaperonins**) that aid folding.

E. coli also recognizes some human proteins as foreign and destroys them with proteases (protein-destroying enzymes). Finally, *E. coli* is unable to modify proteins as eukaryotic cells normally do. For example, human cells **glycosylate** (add sugar molecules to) many of their proteins. (Glycosylation usually does not change a protein's activity, but it does change its solubility, an important characteristic of some medically useful proteins.)

Researchers have tried using other microorganisms, both bacteria and fungi, as hosts, but none is superior to *E. coli*. However, cultured animal cells overcame two of *E. coli*'s drawbacks. Animal cells don't destroy human proteins as readily, and proteins often fold properly within them. Also, animal cells can glycosylate. At the same time, using animal cells is expensive. They grow more slowly, produce less protein from recombinant DNA, and require more expensive media than do bacteria. Chinese hamster ovary (CHO) cells are being used to produce medically useful proteins, including tissue plasminogen activator (tPA, used to treat heart attacks) and human DNase (used to treat symptoms of cystic fibrosis).

For further discussion of what makes a good host and how biotechnology industries are using intact insects and other hosts, see Sharper Focus: A Good Host in Chapter 29.

Applications of Recombinant DNA technology

Cloning is fundamental to recombinant DNA technology. Its applications are many. In the years to come, basic biology, medicine, agriculture, industry, and crim-

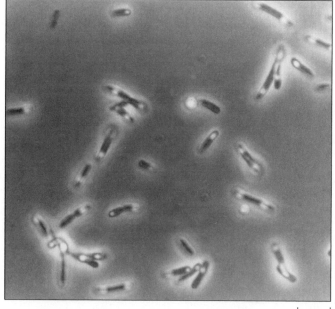

Courtesy Norman Lin, Genetech

⊢ 1 μm

FIGURE 7.8   Inclusion bodies (bright regions near the end of the cells) of human growth hormone in *Escherichia coli* cells.

inal investigation will all benefit (**Table 7.2**). Most applications fall into one of four classes—producing DNA, making proteins, amplifying genes, and engineering organisms.

Producing DNA.   Cloned genes can be produced in massive amounts by cultures of host cells. For example, there are 20 to 40 copies of certain *E. coli* plasmids (pBR type) per cell. When used as a cloning vehicle, an *E. coli* cell can produce as much cloned DNA as is present in all the rest of its genome. Such a culture can produce about 50 mg of cloned DNA per liter, a huge amount when you consider that most recombinant DNA procedures require only micrograms or nanograms. The cloned DNA that is produced is used in other recombinant DNA procedures, including use as a probe to detect similar DNA. DNA probes are used increasingly to identify microorganisms (see the beginning of Chapter 10).

Cloned DNA is also used as the starting material for the chemical reactions used to determine the sequence of bases in DNA. We'll discuss this procedure (called **sequencing**) later in this chapter (see Genomics).

Making Proteins.   We have already discussed how organisms carrying cloned genes are used to make medically useful proteins. This industry already exceeds several billion dollars a year. Industrial enzymes are also made from cloned DNA; the process is discussed in detail in Chapter 29.

LARGER FIELD

TREE TESTIMONY

Law enforcement has been quick to exploit the power of recombinant DNA technology. DNA from bits of dried human blood, semen, or tissue collected at the scene of a crime can be compared with DNA of a suspect to provide power-ful proof of guilt or innocence. Recently an Arizona judge, Susan Bolton, extended the forensic authority of DNA analysis when she admitted analysis of plant DNA as evidence in a murder trial. The body of a woman was found in the Arizona desert lying near a Palo Verde tree, a leafless tree typical of that region. Seeds from a Palo Verde tree were found in the back of a suspect's truck, but who could say from exactly which tree they had come? Recombinant DNA technology provided the answer. The pattern of DNA obtained by PCR from the seeds precisely matched the pattern obtained from DNA from the tree at the crime scene but from no others tested. The seeds almost certainly came from that particular tree.

TABLE 7.2 Some Applications of Recombinant DNA Technology

Field	Application	Importance
Basic biology	DNA sequencing	Answers questions about gene structure, gene function, and relatedness of genes and organisms
	Directed mutagenesis	Answers questions about gene function
Medicine	Therapeutic proteins	Makes human proteins for treating diseases such as diabetes, pituitary dwarfism, hemophilia
	Gene therapy	Treats genetic diseases such as cystic fibrosis
	Improved vaccines	Produces more effective vaccines with fewer side effects
	Diagnosis	Allows rapid, accurate diagnosis of infections and other diseases
	Veterinary medicine	Better diagnosis, prevention, and treatment of disease
Industry	Altering microorganisms	Improved production of antibiotics, amino acids, vitamins, and enzymes; also improved disposal of waste, including persistent toxic chemicals
Agriculture	Altering plants	More rapid breeding of disease-resistant and improved plants (e.g., tomatoes that stay fresh-tasting longer)
	Altering farm animals	More rapid development of superior breeds
Criminal	DNA fingerprinting	Can determine if a biological sample such as investigation blood, semen, or tissue is from a particular person

Amplifying Genes. An amazing procedure called the polymerase chain reaction (PCR) was developed in 1985. PCR makes it possible to zero in on a single gene in a complex mixture of DNA and multiply it. The product of PCR is an essentially pure solution of that gene or bit of DNA. PCR is so powerful it can amplify (increase) the amount of the gene more than a million times in a few hours. Because the amount of the other DNA in the sample is unchanged, the product of PCR is an almost pure solution of the multiplied gene.

PCR is catalyzed by DNA polymerase and depends on this enzyme's unusual properties. Recall from our discussion of replication in Chapter 6 that DNA polymerase requires single-stranded DNA to serve as a template and a free 3′ OH end of DNA to serve as a primer. To perform PCR, the DNA to be analyzed is mixed with DNA polymerase, two short primers, and the four deoxyribonucleoside triphosphates (dATP, dGTP, dCTP, and dTTP) that are the building blocks of DNA. Then the mixture is repeatedly

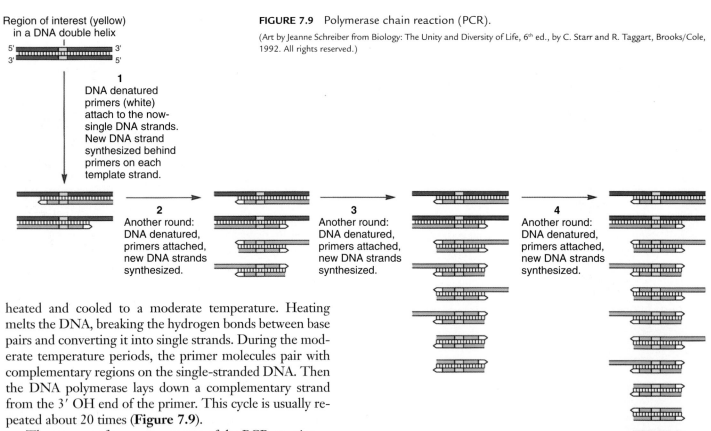

FIGURE 7.9   Polymerase chain reaction (PCR).

heated and cooled to a moderate temperature. Heating melts the DNA, breaking the hydrogen bonds between base pairs and converting it into single strands. During the moderate temperature periods, the primer molecules pair with complementary regions on the single-stranded DNA. Then the DNA polymerase lays down a complementary strand from the 3′ OH end of the primer. This cycle is usually repeated about 20 times (**Figure 7.9**).

The sources of two components of the PCR reaction—DNA polymerase and the primer molecule—are critical. The DNA polymerase is obtained from a thermophilic (heat-loving) bacterium, *Thermus aquaticus*, because this enzyme (*Taq* polymerase) can withstand repeated cycles of heating. The two primer molecules must be complementary to regions that border the gene to be amplified. The required primer molecules are synthesized chemically. Of course, to know what primer to make, you have to know the sequence of bases in the region bordering the gene. This information is usually available for a gene of interest.

Now let's consider the first heat-cool cycle of PCR. The heating step melts the DNA in the sample, converting it to single strands. On cooling, the primers pair with complementary regions of DNA in the sample. Then taq polymerase lays down a complementary strand in the 3′ direction. The result of this single heating and cooling cycle is to double the number of copies of the gene we're interested in. Then each subsequent cycle redoubles the number of copies. Twenty such cycles increase the number of copies about a millionfold ($2^{20} \cong 1,050,000$). Note that after about three cycles almost all the new DNA is no longer than the region bracketed by the primers. Note also that large amounts of the primers must be added to the reaction because the requirement for them doubles after each cycle.

The applications of PCR are almost limitless. In research laboratories it is a tool for studying genes. Forensic experts use it in criminal investigations, to identify indi-

viduals by examining samples of blood, hair, or semen (all contain small amounts of DNA). PCR has even been used to analyze the genes of ancient mummies.

PCR is revolutionizing medical diagnosis. Diagnosis often involves identifying small quantities of a specific biological substance—a task for which PCR is ideally suited. PCR can prevent the transmission of acquired immunodeficiency syndrome (AIDS) through blood transfusions (Chapter 27). Most contaminated blood can be identified because an infected person produces antibodies against the virus, but in some cases antibodies are absent. PCR, however, makes it possible to search for the genes of the virus in a sample of the patient's blood. PCR is sensitive enough to detect one gene from a single virus in a sample of blood. Because of its extraordinary power, PCR may soon become the diagnostic standard not only for AIDS but also for many other infectious and noninfectious diseases.

SHARPER FOCUS

CLINICAL NOTES: FIGHTING TB—NEW HOPE

Tuberculosis (TB) was the scourge of the nineteenth century, but with chemotherapeutic drugs, the incidence declined in developed countries during the twentieth century to the point that many public health officials believed TB might be the next infectious disease to be eradicated worldwide (Chapter 22). But the picture changed ominously in the mid-1980s. Strains of *Mycobacterium tuberculosis* appeared that were resistant to TB drugs, including isoniazid, which had been the cornerstone of tuberculosis treatment since the 1950s.

Obviously drugs to replace isoniazid had to be found quickly, but scientists did not even know at the time how isoniazid worked (see the beginning of Chapter 6). How did isoniazid stop the growth of sensitive *M. tuberculosis* strains, and what genetic changes caused strains to become resistant? Answers to these questions would come very slowly because *M. tuberculosis* grows at an exasperatingly slow pace. It takes between 2 and 8 weeks to isolate a strain of *M. tuberculosis* from a clinical sample and then from 1 to 13 weeks to determine its sensitivity to isoniazid.

But recently a team of microbiologists in London and Paris found a way to speed the research using recombinant DNA technology. They would transfer isoniazid-resistant genes from slowly growing *M. tuberculosis* to a closely related nonpathogenic species, *M. smegmatis,* which grows rapidly. The first step was to isolate an isoniazid-resistant strain of *M. smegmatis.* Then they transferred individual clones from a gene bank of an isoniazid-susceptible strain of *M. tuberculosis* into it. The clone that made *M. smegmatis* sensitive again must carry the gene in *M. tuberculosis* that makes it susceptible to isoniazid and the one that mutated to resistance.

The culprit gene was a surprise. It encoded an enzyme, catalase-peroxidase, that does two things. It breaks down hydrogen peroxide (H_2O_2) into water (H_2O) and oxygen gas (O_2). It also uses hydrogen peroxide to oxidize certain organic compounds. How could one of these reactions make *M. tuberculosis* sensitive to isoniazid? Did one of them convert isoniazid into an active form? If so, maybe the active product could be identified and used to treat patients infected with isoniazid-resistant strains of *M. tuberculosis.* Increasingly, recombinant DNA technology is providing hope and solutions in clinical research.

Engineering Organisms. As we've seen, recombinant DNA technology can produce large quantities of specific genes. In addition to using them for analysis, cloned genes can be inserted into organisms to change them genetically. A major promise of this capacity is **gene therapy** (introducing good genes to replace or overcome the effects of disease-causing genes). In similar ways, microorganisms, plants, and animals can be genetically altered or engineered for specific purposes. Microorganisms can be engineered to produce greater quantities of their products, such as antibiotics, amino acids, or enzymes, or they can be engineered to break down toxic or polluting compounds in our environment (Chapter 28). Plants can be engineered to produce more or better crops or to resist disease or insect damage. Animals can be engineered for similar purposes. The physicist Freeman Dyson predicts that recombinant DNA technology will benefit humanity more than the industrial revolution did.

No matter how vast their potential for good, hazards come with all powerful new technologies, and humans have intrinsic fears of the unknown. You'll continue to read and hear about the ongoing tensions between the power of DNA technology to benefit humanity and concerns that that this new capacity will be misused.

GENOMICS

Genomics is the study of an organism as revealed by the sequence of bases in its DNA. Of course, only after an organism's genome has been **sequenced** is genomics possible. So we'll begin by examining how sequencing is accomplished.

Sequencing

Like PCR, genome sequencing utilizes the enzyme DNA polymerase to duplicate single-stranded DNA in an in vitro reaction. In addition to the normal substrates for the enzyme—namely, a single-stranded DNA template, a primer, and the four deoxynucleoside triphosphates—small amounts of an **analogue** (a compound with slightly altered structure) of one of the four nucleoside triphosphates, called a **dideoxynucleoside triphosphate,** is added to the reaction mixture (**Figure 7.10a**). Dideoxynucleoside triphosphates have two special properties that make sequencing possible: (1) Each dideoxynucleoside triphosphate is sufficiently similar to its naturally occurring analogue to become incorporated in its place into a growing DNA chain, but (2) because dideoxynucleoside triphosphates have a second deoxy group, no subsequent nucleotide can be attached to them. So when incorporated, a dideoxynucleotide terminates growth of a DNA chain. If, for example, dideoxythymidine triphosphate (ddTTP) is added to the reaction mixture, a set of molecules will be synthesized, all of which terminate with a T (**Figure 7.10b**). If, on the other hand, ddTTG or ddTTA, or ddTTC is added, the set of molecules synthesized will terminate with a G, A, or C, respectively.

Carrying out one reaction in the presence of each of the four dideoxynucleoside triphosphates and then determining the length of the products of these reactions reveals sequence of bases in the DNA template. If the various reaction products are separated by gel electrophoreses, the sequence of bases can be read directly off the gel (**Figure 7.10c**).

The approach we have just described (sometimes called **hand sequencing**) can be done in most laboratories without extensive special equipment, but the process is slow and laborious. Using it, sequencing a DNA molecule that is 20,000 to 50,000 bases long is a year's work for an individual. But modern high-speed sequencing machines, which became available in the late 1980s, can accomplish the same task in about an hour.

These fully automated machines, which require only about 15 minutes of human attention per day, employ the same principles as hand sequencing, with two significant differences:

1. The products of the reactions are further modified by incorporating a fluorescent dye into the DNA chain—a different color (red, blue, yellow, or green) for the reaction done in the presence of each of the four dideoxynucleoside triphosphate reactions.

2. The products of the four reactions are then mixed and separated by electrophoreses (on a slab gel or gel packed into fine capillary tubes). A laser beam scans the effluent from the electrophoreses gel, causing the DNA to fluoresce, and a detector records the color. The sequence of colors that corresponds to the sequence of bases in the DNA is fed directly into a computer for further processing.

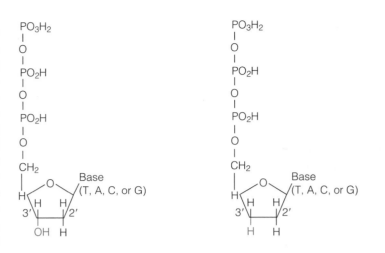

2'-deoxynucleoside triphosphate   2',3'-deoxynucleoside triphosphate

a

b

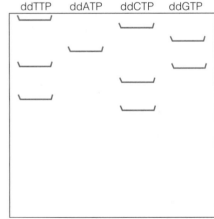

c

FIGURE 7.10   Sequencing DNA. (a) Comparison of the structure of the naturally occurring 2'-deoxynucleoside triphosphate and the analogue 2', 3'-dideoxynucleoside triphosphate. (b) Sets of DNA molecules made from the same template in the presence of each the four dideoxynucleoside triphosphates. (c) Pattern of DNA bands obtained by separation of the DNA molecules shown in (b). By reading from bottom of the gel upward, the sequence (-C-T-C-G-T-A-G-C-T-) complementary to the template is obtained.

TABLE 7.3 Range of Sizes of the Genomes of Some Organisms

Microorganism	Domain	Genome Size	Date Sequenced	Comments
Mycoplasma genitalium	Bacteria	0.58 million	1995	Human pathogen
Methanococcus jannaschii	Archaea	1.66 million	1996	Methane-forming bacterium
Synechocystis sp.	Bacteria	3.57 million	1996	Cyanobacterium
Escherichia coli	Bacteria	4.60 million	1997	Most thoroughly studied bacterium
Saccharomyces cerevisiae	Eukaryotes	13 million	1997	Yeast
Homo sapiens	Eukaryotes	3 billion	2001	Human being
Newt	Eukaryotes	ca. 15 billion	Not sequenced	Amphibian
Lily	Eukaryotes	ca. 100 billion	Not sequenced	Flowering plant

[a]Base pairs.

Assembling

Even sophisticated machines can only separate about 500 DNA molecules in a single electrophoresis run. As a result, only a bit of DNA 500 bases long can be sequenced at a time. These short sequences then have to be assembled in the proper order to reveal the sequence of bases in an organism's genome, which might be millions or billions of base pairs long (**Table 7.3**). Assembling can be accomplished by one of two methods: (1) The relative locations of the cloned bits DNA can be mapped (see Chapter 6) prior to sequencing and then lined up as the map indicates. (2) Random bits of DNA (about sevenfold more than would cover the entire genome) can be sequenced and then arranged in proper order by a computer searching for overlapping ends. But after a genome has been sequenced, what can it tell us?

Annotation

A completely sequenced genome is simply a long list of As, Ts, Gs, and Cs. The process of converting the list into useful information is termed **annotation.** The first step of annotation is asking a computer to search the sequenced genome for base sequences that signal the beginning and ending of genes (Chapter 6). Presumed genes, thus identified, are called **ORFs** (open reading frames). The next step, determining gene products that the various ORFs encode and what metabolic role they play, is also accomplished by a computer. The computer compares the sequence of each ORF with the sequences of genes with known function. This information is available in various databases, such as GenBank, that are accessible on the Internet. At present, by such comparisons the functions of between 20 and 70 percent of the ORFs in a newly sequenced organism can be identified, and this fraction will undoubtedly rise as the information content of gene databases increases. Even if the precise function cannot be identified, good guesses can be made by comparing parts of an ORF with known genes. If, for example, part of the ORF is similar to a part of a known gene that binds to adenosine triphosphate (ATP), it would be a good guess that one of the gene product's substrates in ATP—that the ORF encodes a kinase (Chapter 2).

As an example of what annotation can reveal about a newly sequenced genome, let's consider what was learned about the archaeon *Halobacterium* sp. NRC-1, which was sequenced in October 2000. These finding are summarized in **Table 7.4.**

TABLE 7.4 Some Information Learned from the Genomic Analysis of the Archaeon *Halobacterium* sp. NRC-1.[a]

Topic	Results
Genome size	2,571,010 base pairs
Replicons	3 chromosomes and 2 small replicons
Genes	
Total	2682
RNA-encoding	52
With matches in databases	1658
With known or predicted function	1067

[a]Ng, W. V, et al. 2000. Genome sequence of *Halobacterium* species NRC-1. *Proceedings of the National Academy of Science* 97:12176–81.

Benefits of Genomics

Genomics is a young field. At present, it's full impact on microbiology and our lives can only be imagined, but certainly it will be profound. Already genomics has lead to **proteomics** (the study of an organism as revealed by the complete set of proteins it contains). Genomics will simplify and make more focused the discovery of new, nontoxic antimicrobial drugs, which we so desperately need. For example, the genome of a pathogen can be screened

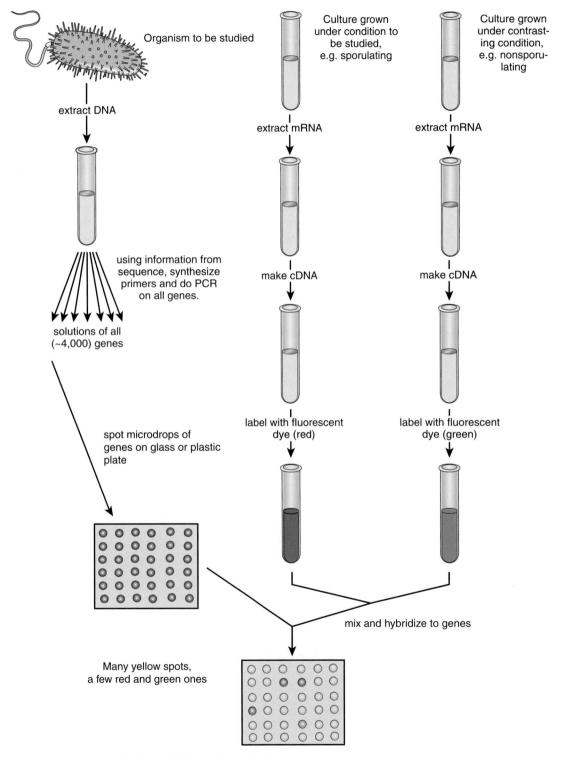

FIGURE 7.11   General scheme of microarray technology.

for possible targets of drug action by identifying genes and functions that we lack (Chapter 21). Certainly, important new insight will be gained concerning the relatedness of organisms and the pathways of their evolution (Chapter 10).

One highly productive benefit of genomics is the development of **microarray technology,** a means of determining which of an organism's genes are expressed under various conditions.

Microarray technology. Microarray technology allows researchers to answer questions about what a gene does even if the gene product is unknown. It does so by showing when various genes are transcribed, thereby revealing their function. For example, if a certain gene is expressed only when an organism is forming spores, one could conclude that that gene encodes one of the steps in the process of sporulation.

Microarray technology is conceptually simple but technically demanding (**Figure 7.11**). Copies of each of an organism's genes are made by PCR. A machine then makes

an array of miniature drops (about a thousand in a square centimeter), each containing one of the organisms genes. Then mRNA is extracted from cells, converted to DNA by the action of reverse transcriptase (Figure 7.3), labeled with a fluorescent dye, and hybridized to the array of genes. Hybridization only occurs between genes and the DNA made from their corresponding mRNA. Usually mRNA is isolated from cells grown under two different conditions to be compared by microarray technology. One is labeled with a green fluorescent dye and the other with a red dye. A mixture of the two is hybridized to the microarray. In the example we chose, mRNA would be extracted from sporulating and nonsporulating cells; one (sporulating, for example) would be labeled red and the other labeled green. After hybridization the array of genes appears to be an array of colored spots (**Figure 7.12**). Most of the spots would be yellow, having hybridized to both red- and green-labeled DNA because most genes are expressed under both sporulating and nonsporulating conditions. Those few genes expressed only by sporulation would be red.

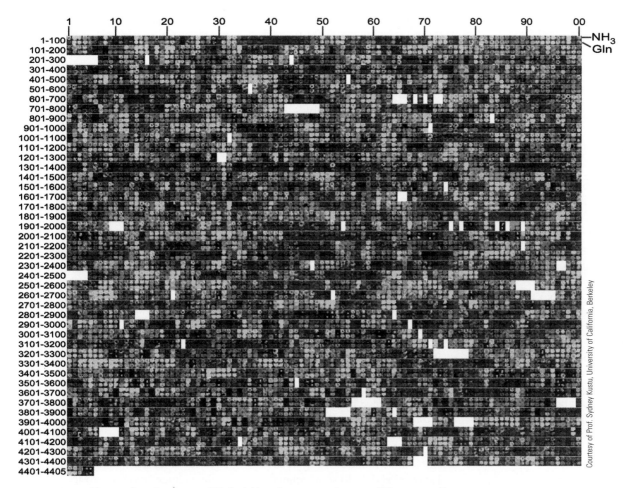

FIGURE 7.12   Photo of a microarray of *Escherichia coli* grown under two different conditions. Note the predominance of yellow spots.

Microarray technology is now being used extensively in microbiological research. It is dependant on and contributes to genomics. An organism's complete genome sequence must be known in order to synthesize a complete set of genes by PCR, and microarray technology can suggest the function of unknown genes in a newly sequenced organism. The uses of microarray technology will undoubtedly expand quickly to many fields, including diagnostic medicine.

SUMMARY

Recombinant DNA Technology and Genomics (pp. 180-193)

1. Recombinant DNA technology is a collection of procedures for manipulating DNA in vitro and putting it into a cell.

2. *Recombination* refers to producing a new combination of genes, whether naturally (as in sexual reproduction) or artificially (in vitro in recombinant DNA technology).

3. When genes on the same chromosome or DNA molecule recombine by crossing over, the result is a recombinant molecule. This happens often between homologous DNA molecules in living things. In recombinant DNA technology, the DNA molecules do not even have to be similar, let alone homologous.

Gene Cloning (pp. 181-189)

4. Gene cloning is the basic tool of DNA technology; it is the process of obtaining a set of identical copies (clones) of a gene.

5. Gene cloning involves five steps:
   a. Obtaining DNA that contains the gene to be cloned
   b. Splicing the piece of DNA containing the gene into a cloning vector (a DNA molecule that a cell will replicate)
   c. Putting the recombinant DNA molecule (the cloning vector with the desired gene spliced into it) into an appropriate host cell
   d. Testing to ensure that the desired gene has been inserted into the host cell
   e. Propagating the host cell to produce a clone of cells with the clone of genes

6. DNA is purified from a cell extract, the liquid content of ruptured cells. The large molecular size, chemical stability, and high density of DNA make it easy to purify

7. Good cloning vectors have an origin of replication; are relatively small, so their replication does not tax the host cell's metabolic capacity; and carry other genes that identify host cells containing the vector. Plasmids and the genomes of certain viruses are used as cloning vectors.

8. Cloning vehicles and DNA to be inserted into them are usually cut with enzymes called type II restriction endonucleases. The enzymes attack palindrome sequences. Most produce single-strand regions called sticky ends because they readily pair all the other sticky ends produced by the same enzyme. Sticky ends pair and anneal at low temperature. Restriction endonuclease reactions are monitored by gel electrophoresis.

9. Gaps that exist where the original cuts were made must be ligated (sealed) to stabilize the molecule. This is done with an enzyme called DNA ligase. Blunt-end ligation joins DNA molecules that lack sticky ends. They are more difficult to ligate, but the procedure has certain advantages: The DNA to be cloned can be cut mechanically by shearing.

10. Recombinant DNA molecules can be inserted into a living cell in several ways:
    a. Transformation: An intact bacterial cell takes up DNA from solution. Most bacteria must be specially treated, a process called artificial transformation.
    b. Transfection: A virus is used to inject DNA into the host cell.
    c. Microinjection: DNA can be injected directly into some animal cells.

d. Electroporation: Exposure to high-voltage electrical impulses renders the cell membranes of bacterial, plant, and animal permeable to DNA.

11. A short piece of DNA called a probe can be used to testing if a certain is present. The probe is complementary to the gene being cloned. Testing can also be done by subsequent identification: determining which colonies produce the products of the gene to be cloned.

12. Subsequent identification is usually done in cloning prokaryotes. The procedure is based on a process called shotgun cloning. A gene bank (a set of bacterial clones, each carrying a clone of DNA that collectively constitutes the entire genome) is made. Bacterial clones in the gene bank are tested by hybridization to determine which carries the desired gene.

13. *Escherichia coli* is the most common host for recombinant DNA because it is easy to maintain and cultivate, it grows rapidly, and it has been so thoroughly studied. However, some eukaryotic proteins do not fold correctly in *E. coli*. They form large insoluble aggregates called inclusion bodies. Also, *E. coli* destroys certain human proteins, perceiving them as foreign, and it cannot glycosylate proteins as eukaryotic cells normally do.

14. Cultured animal cells overcome the last two problems, but they grow more slowly, produce less protein, and require more expensive media. Chinese hamster ovary (CHO) cells are often used.

Applications of Recombinant DNA Technology (pp. 190-193)

15. Recombinant DNA technology produces huge amounts of DNA, used in sequencing.

16. Recombinant DNA technology produces medically useful proteins. It also produces industrial enzymes.

17. Polymerase chain reaction (PCR) can be used to amplify a single gene in a complex mixture of DNA and produce an essentially pure solution of it. PCR is used in research laboratories; in criminal investigations to match blood, hair, or sperm; and in many other fields. PCR is revolutionizing medical diagnosis.

18. Recombinant DNA technology allows us to genetically engineer microorganisms, plants, and animals. For example, plants can be engineered to produce better crops or to resist disease or insect damage.

Genomics (pp. 193–198)

19. Genomics is the study of an organism as revealed by the sequence of bases in its DNA.

Sequencing (pp. 193–194)

20. Sequencing employs dideoxynucleoside triphosphates in in vitro reactions to terminate at specific bases replication of bits of DNA. By determining the length of DNA synthesized in the presence of each of the four nucleoside triphosphates, the sequence of bases in the DNA molecule being replicated is revealed.

21. Hand sequencing does not require extensive special equipment but is slow and tedious.

22. High-speed, fully automated machines can sequence as many bases in an hour as an individual can in a year.

23. Because only about 500 bases can be sequenced in a single run, the information so gained must be assembled to determine the complete sequence of bases in a genome, which might be millions or billions of bases in length.

24. Assembling is accomplished either by mapping the relative location of bits of DNA to be sequenced or asking a computer to search for overlapping ends of sequenced bits of DNA.

Annotation (p. 195)

25. Annotation is the computer-based process of searching for presumptive genes, termed ORFs (open reading frames), and determining their function by comparing their sequences with those of known genes.

Benefits of Genomics (pp. 196–198)

26. Genomics has already told us much about little-studied microorganisms. It will aid the search for new antimicrobial drugs.

27. Microarray technology is a means of determining which of an organism's genes are being expressed under a given set of conditions.

REVIEW QUESTIONS

Recombinant DNA Technology

1. Explain this statement: Recombinant DNA technology is easy to appreciate but hard to define.

2. Explain the differences between recombinant molecules produced naturally in living things and those produced by recombinant DNA technology. How do the processes differ?

Gene Cloning

3. Define these terms: homologous DNA, gene cloning, clone, cloning vector.

4. What are the five steps in gene cloning?

5. How does the cloning of genes from prokaryotes and from eukaryotes differ?

6. What are the properties of a good cloning vector? How are cloning vectors cut? Define these terms: restriction endonuclease, palindrome sequence, sticky ends, anneal.

7. What is gel electrophoresis, and how is it used in gene cloning?

8. How is ligation done, and why is it necessary? What is blunt-end ligation, and what are its advantages?

9. Describe these ways of inserting recombinant DNA into a host cell: transformation, transfection, microinjection, and electroporation.

10. Discuss ways of testing to determine if a host cell that carries cloned genes carries a desired gene.

11. What roles do shotgun cloning and gene banks play in cloning genes from prokaryotes?

12. What are the characteristics of a good recombinant DNA host? What are the advantages and disadvantages of using *E. coli* as a host? What are the advantages and disadvantages of using CHO cells?

Applications of Recombinant DNA Technology

13. What are some of the medically important proteins produced by recombinant DNA technology?

14. What is PCR, and why is it considered such an amazing procedure? How is PCR being used today? Explain this statement: The potential applications of PCR are almost limitless.

15. What does it mean to engineer an organism? What is gene therapy? Discuss some other uses of genetic engineering.

Genomics

16. What is genomics?

Sequencing

17. What is sequencing?

18. What role do dideoxynucleoside triphosphates play in sequencing?

19. What role does electrophoresis play in sequencing?

20. How does an automated sequencing machine distinguish between terminal bases of DNA molecules that have been separated by electrophoresis?

Assembling

21. What does assembling mean as applied to sequencing?

22. What limitation of sequencing technology makes assembling necessary?

Annotation

23. What does annotation mean as applied to genomics?
24. What is an ORF?
25. How does genomics discover the function of an ORF?
26. What is proteomics?
27. What can learned by microarray technology?
28. Why is sequencing a necessary prerequisite of microarray technology?
29. Why are most spots on a microarray yellow?

CORRELATION QUESTIONS

1. If you found that an active preparation of *Eco* RI was unable to cut a particular sample of DNA, what would you conclude? Explain.
2. What changes in the PCR procedure would have to be made to use DNA ligase from *Escherichia coli* in place of the usually employed DNA ligase from *Thermus aquaticus*? Explain.
3. Would it be easier to anneal DNA fragments produced by the action of *Eco* RI or *Hha* I? Explain.
4. Using *Escherichia coli* as a host, would it be easier to clone a gene from *Bacillus subtilis* or the corresponding gene from *Saccharomyces cerevisiae*? Explain.
5. What will genomics allow microbiologists to do that they previously couldn't?
6. How could microarray technology be used in diagnostic medicine?

ESSAY QUESTIONS

1. Discuss the problems associated with using *Escherichia coli* to produce human proteins.
2. Discuss the possible hazards associated with recombinant DNA technology.

SUGGESTED READINGS

Bloom, B. 1992. Back to a frightening future. *Nature* 358:538–39.

DeWitt, P. E. 1994. The genetic revolution. *Time*, January 10, pp. 46–56.

Fish, S. A., T. J. Shepherd, T. J. McGenity, and W. D. Recovery of 16S ribosomal RNA gene fragments from ancient halite. *Nature* 417:432–436.

The Institute for Genomic Research. 2002. http://www.tigr.org.

Kyoto Encyclopedia of Genese and Genomes (KEGG). 2002. http://www.genome.ad.jp/kegg/kegg2.html.

Old, R. W., and S. B. Primrose. 1994. *Principles of gene manipulation.* 5th ed. Cambridge, England: Blackwell Science.

Persing, D. H., T. F. Smith, F. C. Tenover, and T. J. White. 1993. *Diagnostic molecular microbiology: Principles and applications.* New York: ASM Press.

Williams, R. C. 1991. *Molecular biology in clinical medicine.* New York: Elsevier Science.

For additional readings, go to InfoTrac College Edition, your online research library at: http://www.infotrac.thomsonlearning.com

EIGHT

The Growth of Microorganisms

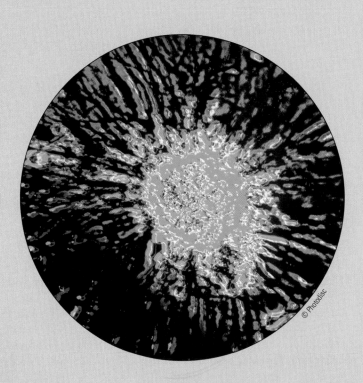

LEARNING GOALS

To understand:

- *How microorganisms grow*
- *The meaning of doubling time and exponential growth*
- *The phases of growth that microbial cultures pass through*
- *The methods that can be used to keep microorganisms growing continuously*

- *The kinds of nutrients microorganisms require for growth*
- *The environmental conditions that permit microbial growth: temperature, hydrostatic pressure, pH, and osmotic strength*
- *How to measure the size a microbial population using indirect methods—*

measuring turbidity, dry weight, or metabolic activity

- *How to measure the size of a microbial population by direct count, plate count, most probable number, and filtration*

It Looked Like an Epidemic

In 1983, thousands of migratory birds died at Kesterton National Wildlife Refuge, a wetland area in the Central California flyway. Many of the surviving young birds were crippled. Some had twisted beaks and deformed wings. Later, about a third of the field mice, house mice, deer mice, and voles at Kesterton were found to be hermaphroditic: they had both male and female reproductive organs.

Investigation of this apparent devastating epidemic showed that microorganisms were not responsible. Instead it was caused by selenium, an element that closely resembles sulfur, and is an essential nutrient for all organisms, humans included. But when selenium is available in high concentrations, the consequences can be disastrous. Then enzymes that normally incorporate sulfur into cellular building blocks incorporate selenium instead. When these selenium-containing building blocks become part of the organism's macromolecules, they fail to function or function in a distorted way, causing death or deformation.

The water at Kesterton, which is fed from runoff irrigation from soils that contain more selenium than most, contains dangerously high levels of selenium, although even here selenium concentrations are in the micromolar range.

Although even these low concentrations of selenium are toxic and teratogenic (presumably for all animals), tiny amounts of selenium are required by all forms of life because it is a cofactor for certain enzymes.

Bacteria, however, have a special relationship with selenium. Certain bacteria benefit by using oxidized selenium (selenate), much as some bacteria use oxidized sulfur (sulfate) as a terminal electron acceptor in an anaerobic respiration (Chapter 5) as a means of generating ATP. And the enzymes of some bacteria, including *Escherichia coli*, require selenium in the form of **selenocysteine** to be active. Selenocysteine is an amino acid identical to cysteine but with a selenium atom in place of the sulfur atom in ordinary cysteine. *E. coli* contains one selenocysteine monomer in an enzyme, **formic dehydrogenase,** that participates in central metabolism. If the selenium atom is replaced by a sulfur atom, the enzyme becomes almost completely inactive. If selenium were to replace sulfur in the other cysteines in the enzyme or any other enzyme, presumably they too would become inactive.

How can selenium replace sulfur in this cysteine and only this one? It happens in a highly unusual way. Selenocysteine is made on a special kind of tRNA molecule that serves only this purpose, and its incorporation into protein is encoded by a nonsense codon (UGA). Why does only this one particular nonsense codon signal the incorporation of selenocysteine? Presumably, the structure of the mRNA surrounding this codon is unique in some way.

Selenium, which can be so devastatingly toxic to most organisms, is also needed by all organisms, but some bacteria have special needs for it.

Case Connections

- In this chapter we'll consider growth of microorganisms, including the nutrients and trace nutrients that make it possible. Selenium might be the most dramatic example of a trace element. Miniscule amounts (only a few thousand molecules) are required by a cell that needs it.
- Selenium illustrates, in exaggerated form, a principle we'll revisit in Chapter 9: the distinction between a compound's being benign, essential, or toxic often depends on its concentration.

POPULATIONS

In Chapters 5 and 6, while discussing the metabolism and genetics of microorganisms, we focused on the individual microbial cell and how it reproduces itself. In this chapter and the next, we'll turn our attention to populations of microbial cells and how their numbers increase through growth and decrease or disappear as a result of death.

To focus on populations we'll change our point of view. In studying the individual cell our approach was chemical. We discussed how the cell produces the molecules it needs to build a new cell and the molecules it uses to preserve and express its genetic information. In contrast, the approach to the study of populations is necessarily mathematical.

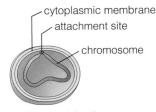

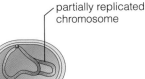

1 A bacterial cell (cutaway view) just after division.

2 The chromosome replicates as the cell grows.

3 The daughter chromosomes separate by a mechanism that remains unknown.

4 New membrane and wall material start growing through the cell midsection.

5 Membrane and wall material deposited at the cell midsection divide the cytoplasm in two.

a

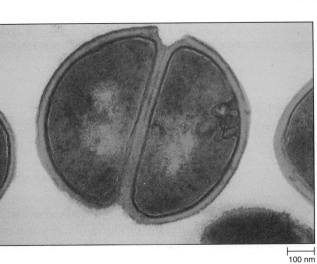

© George Musil/Visuals Unlimited

b   100 nm

FIGURE 8.1   Cell division by binary fission. (a) Growth and division of a rod-shaped cell. (b) Transmission electron micrograph of a dividing cell of *Staphylococcus aureus*.

isms (those that form long tubes) and microorganisms with complex life cycles follows more complicated rules.

Doubling Time and Growth Rate

Doubling time (formerly referred to as **generation time**) is the period required for cells in a microbial population to enlarge, divide, and produce two new cells for each one that existed before. Doubling time of all cells in a growing population is about the same. It doesn't change until nutrients become depleted or toxic metabolic products begin to accumulate. As conditions for growth become less favorable, doubling time increases. Eventually growth stops.

Doubling times vary depending on the species of microorganism and the growth conditions. For example, *Escherichia coli* grows with a doubling time of about 18 minutes in a rich laboratory medium at optimal temperature. But in the intestinal tract of vertebrates, where nutrients are less abundant, it grows with a doubling time of about 12 hours. A few bacteria can grow a little faster than *E. coli*, but most microorganisms, even under the most favorable conditions, grow more slowly, with doubling times of hours or days.

Doubling time tells us how fast a population is growing, but the relationship is inverse. That is, a low value of doubling time indicates rapid growth. A high value indicates slow growth. For this reason, **growth rate** (doubling times per hour) is often used to describe how fast a

Unless we mention otherwise, we'll use the term microbial growth to refer to the increase in numbers of a population, instead of the size of an individual cell.

THE WAY MICROORGANISMS GROW

Most bacteria elongate and divide by **binary fission** (cleavage near the midpoint to form two daughter cells of approximately equal size; **Figure 8.1**). Some unicellular microorganisms, including a few bacteria, replicate by **budding** (forming a bubblelike growth that enlarges and separates from the parent cell). Although we'll focus on growth of a bacterial population dividing by binary fission, the same principles apply to microorganisms that reproduce by budding (Chapter 5). The growth of filamentous microorgan-

# of Doublings	Time		Cells/milliliter		
	Min.	Hours	Numbers	Scientific Notation	Logarithm
	0	0	1000	10^3	3.0
1	20	0.33	2000	2×10^3	3.301
2	40	0.66	4000	4×10^3	3.602
3	60	1.00	8000	8×10^3	3.903
4	80	1.33	16,000	1.6×10^4	4.204
5	100	1.66	32,000	3.2×10^4	4.505
6	120	2.00	64,000	6.4×10^4	4.806
7	140	2.33	128,000	1.28×10^5	5.107
8	160	2.66	256,000	2.56×10^5	5.408
9	180	3.00	512,000	5.12×10^5	5.709
10	200	3.33	1,024,000	1.02×10^6	6.010

a

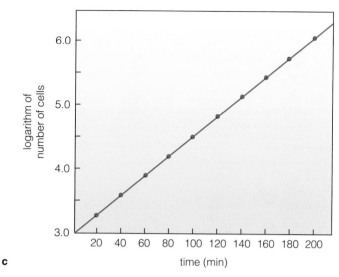

c

FIGURE 8.2 Exponential growth of a bacterial culture. (a) The number of cells per milliliter of culture growing exponentially with a doubling time of 20 minutes. (b) Plot of the number of cells against time. (c) Plot of the logarithm of the number of cells against time.

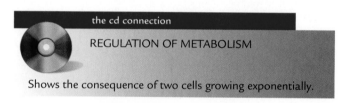

the cd connection

REGULATION OF METABOLISM

Shows the consequence of two cells growing exponentially.

This concept can be illustrated by a riddle about water lilies. Let's say that a water lily produces one new leaf each day for each leaf it had the day before. Starting with a single leaf, we see that it covers a 1-acre pond in 30 days. How long will it take to cover a 2-acre pond? The correct answer is 31 days, not 60, because the doubling time is 1 day. Under exponential growth conditions, the rate of producing new mass increases continuously. In the lily pond, one leaf was produced the first day. An acre-area of leaves was produced the thirty-first day.

Now let's look at a bacterial culture growing exponentially with a doubling time of 20 minutes (**Figure 8.2**). Starting with a population of a thousand cells per milliliter, the number of cells doubles every 20 minutes. After 10 doublings (less than 4 hours), the population has increased to more than a million cells per milliliter (**Figure 8.2a**). It's easier to get a feeling for microbial growth by looking at a graph instead of the actual numbers of cell. If we plot the number of cells against time (**Figure 8.2b**), we get an exponential curve. It gets steeper and steeper and finally shoots out of the range of the plot. A more useful way to view the data is to plot the logarithm of the number of cells against time (**Figure 8.2c**). That gives a straight line, which has advantages. For exam-

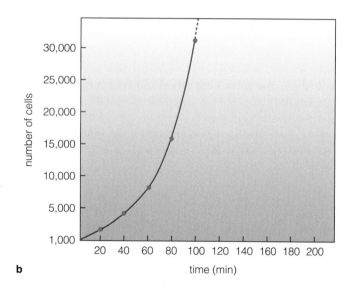

b

culture is growing. The growth rate of the culture of *E. coli* growing with a doubling time of 18 minutes would be 3.3 (60 minutes per hour/18 minutes per doubling time) doublings per hour.

Exponential Growth

As we've seen, the doubling time of a microbial population does not change as long as conditions are favorable. The result is **exponential growth,** which results in an almost explosive increase in numbers and mass of cells. During each doubling time, as many new cells and as much mass are produced as were produced cumulatively during all preceding doubling times.

ple, by plotting the logarithm of the number of cells in a population against time, we can test whether the population is undergoing exponential growth. If the plot is a straight line, we know it is. If it curves, we know it isn't. Moreover, the slope of the line is directly related to the growth rate of the population.

We might ask why this pattern of growth is called exponential. It's logical if you remember that with each doubling time the population increases by a factor of two. In other words, the total number of cells in the population is equal to two raised to an exponent. The exponent is the number of generations that have elapsed. Thus one cell (2^0) increases to two cells (2^1), then to four cells (2^2), eight cells (2^3), 16 cells (2^4), and so on. The general formula is exponential:

$$N = 2^n$$

where N is the number of cells in the culture; n is the number of doubling times that have passed.

Our discussion of doubling time might lead us to think that a growth curve would follow a stair-step pattern—that the numbers of cells in a growing culture would remain the same until the end of the doubling time. Then suddenly the number of cells would double. Such a culture in which all cells divide at the same time is called a **synchronous culture.** But microbial cultures, even those started from single cells, don't remain synchronous for long under ordinary circumstances. That's because individual cells in a culture don't all have exactly the same doubling time. Some divide sooner than others do. Doubling time of a culture is the average of the doubling times of all the cells in it.

Phases of Growth

Of course, exponential growth can't continue for long. The increasing number of cells uses nutrients and produces waste at ever increasing rates. Something has to happen. Usually an essential nutrient runs out. Sometimes toxic products stop further growth. Whatever the cause, growth rate slows and eventually stops. Then the culture is said to pass from the **exponential phase** (also called the **logarithmic** or **log phase**) of growth to the **stationary phase (Figure 8.3).**

Cells change as they pass from the exponential phase into the stationary phase. They become smaller, and they begin to synthesize components to help them survive when they can no longer grow. For example, as *Escherichia coli* enters the stationary phase, it synthesizes about 30 proteins not found in exponential-phase cells. It also changes the composition of some of the fatty acids in its membranes. In spite of these changes, stationary phase cultures begin to die after a day or so. The culture is said to enter the **death phase.** Dur-

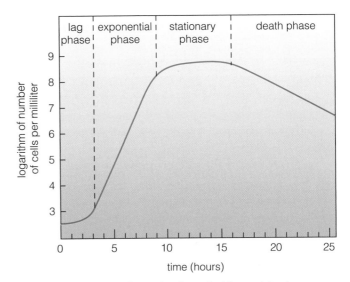

FIGURE 8.3   Phases of growth of a typical bacterial culture.

ing the death phase, cells of most microbial species die exponentially. (Note that the descending line during the death phase, like the ascending during the exponential phase, is a straight line; Figure 8.3). But the rate of death is much lower than the rate of increase of cells during the log phase.

Recently microbiologists have learned that the death phase is a highly dynamic state: Some cells die releasing nutrients that support the growth of other cells in the population. As a result, if culture tubes are sealed to prevent desiccation, live cells can be recovered from extremely old cultures, 35 years in the case of *Salmonella typhimurium*, for example.

When cells taken from stationary- or death-phase cultures are added to a fresh medium, they are unable to grow immediately. First the changes that occurred on entering the stationary phase and surviving the death phase must be reversed. This no-growth period is called the **lag phase.** The length of the lag phase depends, in general, on how long the **inoculum** (cells used to start the new culture) was in the stationary or death phase. If the cells were there only briefly, the ensuing lag phase may be as short as a few minutes. But if they hadn't grown for months, the ensuing lag-phase cells may last many hours. Changing medium or the environment can also cause cultures to enter a lag phase. If an exponential culture is moved from a **poor medium** (minimal supply of nutrients) to a **rich medium** (abundant supply of nutrients), no lag occurs. But a move from a rich to a poor medium usually results in a several-hour lag.

In summary, microbial cultures usually pass through four phases during their growth cycle: a lag phase, when they prepare to grow; an exponential phase, when cell numbers double at regular internals; a stationary phase, when growth ceases; and a death phase, when numbers of live cells decline.

LARGER FIELD

A WALKING CHEMOSTAT

The dream of industrial microbiology is to have microbes convert something abundant and cheap into a valuable product. Microbiologists might differ about the ideal product but not the ideal substrate. They all agree that cellulose—the major component of plant cell walls—is the world's most abundant organic compound and probably the cheapest. But only microorganisms use cellulose as a nutrient. Only they can convert it into something useful.

Robert Hungate, a leading American microbiologist, was asked if a commercially successful microbial conversion of cellulose would ever be discovered. He replied that it already had. It's the cow. The cow and other ruminant animals are mobile factories that use microorganisms to convert cellulose into valuable products—meat and milk. They gather their own cellulose as grass and other forage. They chew it into small bits, which flow into the **rumen** (the large chamber of a cow's four-chambered stomach). The rumen teems with bacteria and protozoa. They—not the cow itself—can use cellulose as a nutrient. The cow lives on the microbial cells and their fermentation products that come out of the rumen and enter the cow's intestines. In other words, the rumen is a fermenter that converts cellulose into food for the cow.

Hungate and his colleagues have shown that the cow's rumen functions much like a chemostat (**Figure 8.4**). The rumen, which acts like the culture vessel, is continuously supplied with cellulose and other nutrients. Microorganisms continuously grow in it. Some of its content continuously flows out into the cow's digestive system. Of course, cows don't eat continuously, but when they aren't eating, they supply a steady flow of cellulose-containing nutrients to the rumen by chewing their **cud** (regurgitated forage).

By feeding it almost continuously, cows and other ruminants have evolved to become highly efficient at maintaining an actively growing culture of microorganisms, which in turn feeds them.

© Morton Beebe/Corbis

Continuous Culture of Microorganisms

The four phases of growth we've just discussed occur when microorganisms are cultivated in **batch culture,** which means growth in a closed container—a test tube, a flask, or a fermenter—in a laboratory. No additional nutrients are supplied as the culture grows. No toxic end products are removed. Microorganisms rarely grow this way in nature. Under most natural conditions, nutrients continuously enter the cell's environment at low concentrations. Toxic products diffuse away or are used by other microorganisms. Under such conditions, populations grow at a low rate set by the concentration of the **limiting nutrient** (the scarcest essential nutrient available to them). It's possible to mimic these natural conditions in the laboratory. Under such conditions, cultures grow continuously for weeks or months. Three conditions must be fulfilled: (1) Maintain the concentration of one nutrient low enough to limit growth rate; (2) add nutrients, including the limiting nutrient, continuously; (3) remove cells and end products continuously.

These conditions can be fulfilled rather easily using an apparatus called the **chemostat** (**Figure 8.4**). A chemostat uses a metering pump to transfer fresh medium continuously from a reservoir to a growth vessel fitted with an overflow. Culture (cells and partially spent medium) leaves the growth vessel as rapidly as fresh medium enters. Under these conditions, the number of cells in the growth vessel doesn't change. The culture grows just fast enough to replace the cells lost through the overflow. Even if the rate of adding fresh medium is increased or decreased, the results are the same. Remarkably, the growth rate of the culture adjusts to match the rate of loss of cells through the overflow.

How can this happen? How is the culture able to grow just fast enough to replace itself? The concentration of the limiting nutrient is the key. It sets the rate of growth of the culture in the vessel to match the rate of loss of cells

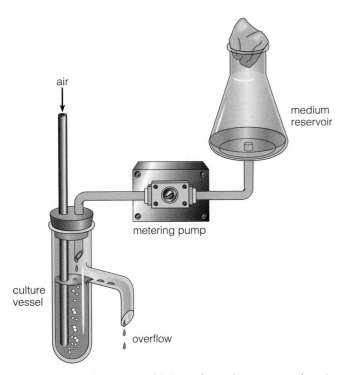

FIGURE 8.4   A chemostat. This is a schematic representation. An actual chemostat has many electronic controls.

through the overflow. The culture grows to a population that uses just enough limiting nutrient to set growth rate equal to **loss rate** (the rate of loss of cells through the overflow). The system adjusts itself. For example, if the growth rate exceeds loss rate, the population will increase. The larger population will use more limiting nutrient. The concentration of limiting nutrient in the growth vessel will drop. The growth rate will decrease until it just matches the loss rate. If the growth rate were to become less than the loss rate, the opposite set of events would occur. Thus in a chemostat, growth rate always equals loss rate and, as a result, the number of cells in the growth vessel remains constant.

Continuous culture has important industrial applications (Chapter 29). It is an efficient way to produce microbial cells or their products (such as antibiotics and vitamins) because a dense culture is constantly produced from the overflow of the growth vessel.

Growth of a Colony

Pure cultures of microorganisms that grow on a solid medium such as an agar-solidified medium in a petri dish form a mass of cells called a **colony.** Different microorganisms form colonies with different shapes and textures. Some colonies are so distinct that microbiologists can identify a microorganism just by glancing at the colony it forms.

Colonies are complex mixture of cells in different phases of growth. Cells in the center of the colony are in the stationary or death phase. They have stopped growing because they no longer have access to nutrients. At the same time, cells on the edge are in the exponential phase because they have direct contact with nutrients. No cells are in the lag phase. Once cells in the center of a colony stop growing, they can't start again until transferred to a fresh medium.

Location within a colony also determines a cell's access to oxygen. Cells growing on the surface of a colony exposed to air are fully aerobic, but cells embedded in the center of the colony are completely anaerobic because the surface cells remove all the oxygen from the air that diffuses into the colony. In a colony of a facultative anaerobe such as *Escherichia coli*, surface cells obtain energy by aerobic respiration at the same time those deeper in the colony are obtaining energy by fermentation or anaerobic respiration.

Biofilms

Laboratory cultures grown in liquid media usually consist of individual suspended cells. In contrast, the vast majority of microbial cells in natural liquid environments are found attached to solid surfaces in assemblages called **biofilms.** Biofilms consist of irregular layers, 10 to 200 μm thick, of microorganisms imbedded in the extracellular slime they excrete. Microbes within biofilms interact in complex ways, sometimes acting as **consortia** to break down complex nutrients or create mutually favorable environments. Biofilms play important roles in infections, natural environments, and industrial processes, as we'll discuss from time to time in subsequent chapters.

WHAT MICROORGANISMS NEED TO GROW

Let's turn our attention now from how a microbial population grows to what it needs in order to grow. Every species of microorganism has its own particular needs. Although the requirements vary greatly, they can all be placed in two general categories: nutritional and environmental. Let's look first at the nutritional requirements.

Nutrition

All living things need nutrients from the environment to build the molecules required to form new cells. We know that good nutrition improves the growth of all organisms. A well-nourished plant, animal, or human grows more rapidly than a poorly nourished one. For microorganisms, particularly bacteria, the effect of nutrition on growth rate is more dramatic. *Escherichia coli*, for example, grows up to 10 times faster in a rich medium than it does in a poor one. The poor medium meets all of *E. coli*'s essential needs, but the cell must synthesize more small molecules. The result is slower growth.

As we've just seen, growth rate depends on the *quality* of the medium. But **growth yield** (the maximum biomass a culture attains) depends on the *quantity* of nutrients in the medium. In other words, a poor medium will always support a slower growth rate but it might produce a greater yield. The yield of a culture is directly proportional to quantity of nutrient.

Now let's consider which nutrients a microorganism must have in order to grow. It must have a source of each of the elements of which it is built. The **major elements** are carbon, oxygen, nitrogen, phosphorus, and sulfur. It also must have **trace elements** (elements needed in small amounts) and **growth factors** (essential metabolites it is unable to make). We'll consider these nutritional needs one at a time in the following sections.

Major Elements

Carbon. Carbon is the structural basis of biochemicals. Autotrophs obtain their carbon from carbon dioxide (CO_2) in the atmosphere. Heterotrophs obtain theirs from organic compounds (Chapter 5).

Most heterotrophic microorganisms can use many different organic molecules as carbon sources. In fact, every naturally occurring organic compound is used as a carbon source by some microbial species. Most microbial organisms can use glucose, which plays a central role in metabolism. Consequently it is the most common carbon source in microbiological media. But other sugars, as well as polysaccharides, organic acids, alcohols, and amino acids, are often added to microbiological media.

Chemoheterotrophs also need an organic compound to serve as an **energy source** (a compound that can be metabolized to generate adenosine triphosphate [ATP]). Often the same compound serves both needs. For example, *E. coli* can use glucose both as a carbon and as an energy source—to build biochemicals and generate ATP (Chapter 5). How fast an organism can metabolize a particular carbon source often determines how fast it can grow. For example, *E. coli* grows twice as fast if it uses glucose instead of the amino acid lysine as a carbon source in an otherwise identical medium.

Oxygen. Oxygen is a major component of the molecules from which all organisms are built, but not all microorganisms require atmospheric oxygen. Most heterotrophic microorganisms obtain oxygen from the same molecule that serves as their carbon source. They also get some oxygen from water. In addition, most aerobic microorganisms have **oxy-**

genases, enzymes that add atmospheric oxygen to organic molecules, but this is a minor source of cellular oxygen.

In addition to needing oxygen as a nutrient, microorganisms that are capable of aerobic respiration use atmospheric oxygen to generate ATP (Chapter 5). Microorganisms that are dependent on aerobic respiration cannot grow in the absence of atmospheric oxygen. But in spite of being essential for some organisms, atmospheric oxygen is a highly reactive compound that is somewhat toxic to all organisms. Aging of humans, for example, is thought to be, in part, the cumulative consequence of long-term oxygen toxicity. Oxygen is particularly toxic to some enzymes. An extreme example is **nitrogenase,** the enzyme that enables nitrogen-fixing bacteria to use atmospheric nitrogen. Nitrogen-fixing bacteria that grow in the presence of air have evolved elaborate mechanisms to protect their nitrogenase from oxygen. Cyanobacteria (which produce oxygen, as well as live in the presence of air) accomplish this by segregating their nitrogenase into specialized cells, called **heterocysts,** that do not produce oxygen and are impermeable to it. Azotobacters (which grow in the presence of air but do not produce oxygen) use a different mechanism. They carry out aerobic respiration at a usually high rate. They use oxygen so rapidly at the cell surface (where respiratory enzymes are located) that the center of the cell (where nitrogenase is located) is anaerobic.

In addition to the toxicity of oxygen itself, organisms must deal with highly toxic compounds that are produced as unavoidable by-products of aerobic metabolism. Two such compounds, **hydrogen peroxide** (H_2O_2) and **superoxide** ($O_2 \bullet^-$), are direct by-products of aerobic metabolism. A third, **hydroxyl radical** ($OH\bullet$), is formed when hydrogen peroxide reacts with metal ions. (Compounds with an unpaired electron, symbolized by •, are called **free radicals.**) Living things that survive in the presence of air have evolved ways to destroy these compounds. All aerobes produce the enzyme **superoxide dismutase,** which destroys superoxide by converting it into oxygen and hydrogen peroxide.

$$2O_2 \bullet^- + 2H^+ \xrightarrow[\text{Dismutase}]{\text{Superoxide}} O_2 + H_2$$

The hydrogen peroxide produced by superoxide dismutase, along with that produced by metabolism, is destroyed by another enzyme, catalase, which converts hydrogen peroxide to oxygen and water.

$$2H_2O_2 \xrightarrow[\text{Catalase}]{} O_2 + 2H_2O$$

Hydrogen peroxide is also removed by **peroxidases** (enzymes that use hydrogen peroxide to oxidize certain organic compounds). No enzymes are available to destroy hydroxyl radical. It reacts immediately with any organic compound it contacts. Organisms that can tolerate exposure to oxygen contain superoxide dismutase, and almost all of them also contain catalase (**Table 8.1**). Organisms that lack both these enzymes quickly die in an oxygen-containing environment.

Nitrogen. Nitrogen makes up about 14 percent of the dry weight of most organisms, including microorganisms. It is a constituent of proteins and nucleic acids, as well as certain essential metabolites, so all living things require a source of nitrogen. But the form of nitrogen they can use depends on their metabolic capacity. Probably all microorganisms can use ammonia (NH_3). They incorporate this form directly into biosynthesis pathways. Most microorganisms can also use a variety of organic compounds as sources of nitrogen. Many can use amino acids. Some can use inorganic forms, including nitrate ion (NO_3^-). Some bacteria can **fix** (use) atmospheric nitrogen gas (N_2). These organisms prosper in natural environments where fixed nitrogen is the limiting nutrient.

Phosphorus. Phosphorus occurs in cells exclusively in the form of phosphate ion (PO_4^{3-}) or phosphate-containing organic compounds. It makes up about 3 percent of the dry weight of microorganisms. It is a constituent of nucleic acids, phospholipids, and certain essential metabolites. Phosphorus enters the cell as phosphate ion because most phosphate-containing organic compounds cannot pass through the cytoplasmic membrane. In spite of this, many microorganisms can use phosphate-containing organic compound as a source of phosphorus. They usually split off the phosphate from the compound outside

the cell or in the periplasm. They release phosphatases (phosphate-splitting enzymes) for this purpose. Then the phosphate ions enter the cell.

Sulfur. Sulfur is a minor constituent of cells. It amounts to about 1 percent of their dry weight. It is a component of two amino acids, a few species of transfer RNA, and certain essential metabolites. Although sulfur enters biosynthesis as sulfide (S^{2-}), the most common source of sulfur in laboratory media is sulfate ion (SO_4^{2-}). Sulfate is abundant in the ocean but relatively scarce in some lakes and soils. The major source of sulfur in many soils is organic sulfate-containing compounds that soil microorganisms take up and convert to sulfide.

Trace Elements. In addition to the major elements, small quantities of trace elements are required for growth of microorganisms. These include the metal ions shown in **Table 8.2**. Trace elements fulfill a variety of essential functions in the cell. Some serve as **cofactors** (ions an enzyme needs to be active) for certain enzymes. Others are constituents of **coenzymes** (organic molecules an enzyme needs to be active).

Availability of trace elements rarely limits microbial growth because they are required in such minute amounts. But it can happen. For example, availability of iron often limits microbial growth. Iron is abundant in nature, but it occurs predominantly in the form of iron (ferric) hydroxide, which is too highly insoluble to be available to microorganisms. Many bacteria acquire iron from ferric hydroxide by releasing iron-solubilizing compounds called **siderophores.**

TABLE 8.1 Relationship of Various Bacteria to Oxygen

| Microbial Class | Response to Oxygen | Presence of | | Example |
		Catalase	Superoxide Dismutase	
Obligate aerobes	Require oxygen	Present	Present	*Pseudomonas aeruginosa*
Facultative anaerobes	Can grow with or without oxygen	Present	Present	*Escherichia coli*
Microaerophiles	Grow best with low oxygen	Present	Present	*Campylobacter jejuni*
Aerotolerant anaerobes	Grow without oxygen, but not killed by it	Absent	Present	*Streptococcus pneumoniae*
Obligate anaerobes[a]	Killed by oxygen	Absent	Absent	*Methanococcus vannielli*

[a]Some contain small amounts of superoxide dismutase.

TABLE 8.2 Trace Elements in Microbial Cells

Element	Principal Ion Form	Function
Potassium	K^+	Maintains turgor pressure, cofactor for certain enzymes
Magnesium	Mg^{2+}	Cofactor for many enzymes
Calcium	Ca^{2+}	Cofactor for certain enzymes
Iron	Fe^{2+}/Fe^{3+}	In cytochromes[a] and certain enzymes
Manganese	Mn^{2+}	Cofactor for some enzymes
Molybdenum	Mo^{2+}	Present in coenzyme for several enzymes
Cobalt	Co^{2+}	Present in vitamin B_{12} and related coenzymes
Copper	Cu^{2+}	Present in several enzymes
Zinc	Zn^{2+}	Present in several enzymes

[a]Cytochromes are components of electron transport chains (Chapter 5).

Living tissue is a particularly poor source of available iron. Thus most pathogenic bacteria produce siderophores to take iron from host proteins that bind it tightly.

Growth Factors. As we discussed in Chapter 5, many microorganisms can synthesize all the building blocks and other small molecules (such as amino acids, nucleotides, sugars, fatty acids, and vitamins) they need to grow. They synthesize them from the sources of the elements we have just discussed. Microorganisms that cannot synthesize one of these compounds must have it supplied from the environment. Such essential compounds are called **growth factors.**

Requirements for growth factors vary widely among microorganisms. For example, *Escherichia coli* needs no growth factors. In contrast, *Leuconostoc citrovorum*, a lactic acid bacterium, must be provided with all 20 amino acids, several purines and pyrimidines, and 10 vitamins (Table 3.5).

If growth factors are present in the environment, most microorganisms use them even if they are able to make their own. That's why most bacteria grow faster in a rich medium, as we have discussed. Bacteria save metabolic capacity by synthesizing only those building blocks that are not available from the environment. The metabolic capacity they save allows them to grow faster. To accomplish this, they produce more ribosomes because when growing faster they have to make proteins faster. Such changes are dramatic. For example, when the bacterium *Salmonella typhimurium* is growing rapidly in a rich environment, each cell has more than 10 times as many ribosomes per cell as when it is growing slowly in a poor medium. This relationship between richness of a medium, growth rate, and number of ribosomes per cell is shown in **Table 8.3.**

The Nonnutrient Environment

Like ourselves, microorganisms do not live by food alone. They need a good environment. This includes a satisfactory temperature, hydrostatic pressure, pH, and osmotic strength.

TABLE 8.3 Effect of Medium Composition on Growth Rate and Ribosome Content of *Salmonella typhimurium*

Medium	Composition	Doubling Time (Minutes)	Number of Ribosomes per Cell
Lysine minimal	The amino acid lysine + salts	97	7000
Glucose minimal	Glucose + salts	50	17,000
20 amino acids	20 amino acids + salts	32	42,000
Brain-heart	An extract of beef brain and heart	21	83,000

SHARPER FOCUS

FIGHTING OFF A CHILL

Because microorganisms grow over a wide range of temperatures, they face environmental challenges that warm-blooded humans never do. One of the biggest is maintaining proper membrane fluidity. Unit membranes are composed of phospholipids, which, like most lipids, become more viscous at lower temperatures and more fluid at higher temperatures. To function properly, membranes must have the same (optimal) degree of fluidity over their growth temperature range.

How do they do this? They change the fatty acid composition of the phospholipids in their membranes according to their growth temperature.

The effect of fatty acid composition on the fluidity of fatty acid–containing lipids is dramatic. At room temperature, corn oil is more fluid than olive oil. In the refrigerator, olive oil solidifies but corn oil remains fluid. This difference results from corn oil being rich in **unsaturated fatty acids** (which contain double bonds) and **polyunsaturated fatty acids** (which contain more than one double bond), whereas olive oil has more **saturated fatty acids** (which contain no double bonds). By mixing corn oil and olive oil in the proper proportions, the same intermediate level of fluidity can be attained at room temperature, in the refrigerator, or at any intermediate temperature.

This same process occurs in microorganisms. As growth temperature declines, an increasingly high proportion of the fatty acids in their phospholipid membranes are unsaturated, so their fluidity is optimal at all growth temperatures. In *Escherichia coli* this precise adjustment is made quite simply. Fatty acids are synthesized by a branched pathway. One branch leads to saturated fatty acids, and the other to unsaturated fatty acids. One enzyme near the branch point, on the unsaturated side, is sensitive to higher temperature. As temperature rises, its activity declines and more fatty acids are made by the "saturated" branch. As temperature declines, the reverse happens. The result is constant membrane fluidity at all temperatures.

Temperature. Every species of microorganism grows over a range of temperature, from the **minimum temperature of growth** (the lowest that will support growth) to the **maximum temperature of growth** (the highest that will support growth). For most bacteria this range is approximately 40 Celsius degrees. Eukaryotic microorganisms have a narrower range. In between these temperature limits lies the **optimum temperature** (the temperature at which the microorganism grows most rapidly). Optimum temperature is usually only a few degrees below the maximum temperature. Growth rate increases from the minimum temperature to the optimal temperature because chemical reactions, including enzyme-catalyzed reactions, proceed more rapidly as temperature rises (**Figure 8.5**). Above the optimum temperature, growth rate declines rapidly because certain cellular components, especially proteins, are inactivated.

Microorganisms grow at extreme temperatures where most other living things cannot. In general, prokaryotes tolerate more extreme temperatures than do eukaryotes. No eukaryotes can grow at the boiling point of water. Some fungi, however, can grow at subfreezing temperatures. That's as low as bacteria can grow. Some algae grow near the freezing point of water. A type of red algae, called snow algae, grow as huge red patches on mountain snow fields.

By convention, microorganisms are divided into three major classes depending on the range of temperature over

which they can grow. **Thermophiles** (heat lovers) grow at high temperatures. **Mesophiles** (moderate temperature lovers) grow at intermediate temperatures. **Psychrophiles** (cold lovers) grow at cold temperatures.

Thermophilic prokaryotes grow above 50°C—a temperature at which water causes pain if you immerse

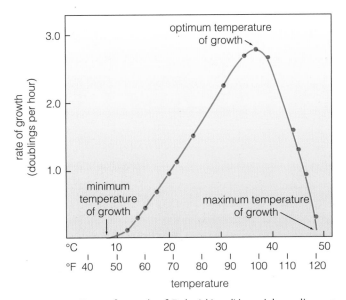

FIGURE 8.5   Rate of growth of *Escherichia coli* in a rich medium at various temperatures.

SHARPER FOCUS

THE RECORD TEMPERATURE

Until relatively recently, microbiologists were convinced that no vegetative microbial cells, even those of thermophiles, could survive exposure to 80°C. But now we know better. During the past 25 years, extremely heat-resistant microorganisms have been discovered. It was just a case of knowing where to look for them.

In 1972 Thomas Brock looked in hot springs in Yellowstone National Park. He found a remarkably heat-resistant bacterium, *Sulfolobus acidocaldarius*. It not only survives 80°C, it grows at 85°C. Karl Stetter looked in another hot environment—the hydrothermal vents on the ocean floor (see Sharper Focus: Deep-sea Hydrothermal

Vents, Chapter 28). He found an even more heat-resistant prokaryote, *Methanopyrus kandleri*. It grows at 110°C. That's 10° above the boiling point of water at sea level but not at the intense hydrostatic pressure where *Methanopyrus* grows. Eighty degrees Celsius is too cold for *Methanopyrus kandleri* to grow. Then in 1997 a new

record was set. Stetter found a microorganism he called *Pyroloblus fumarii* in a hydrothermal vent in the mid-Atlantic ridge near the Azores. It can grow at 113°C. Will new records be set? Probably, but there are limits. Holger Jannasch established that there is no evidence of life in water from hydrothermal vents at 250°C.

your hand. The extremely thermophilic *Pyrolobus fumarii* can grow at 113°C—13 Celsius degrees above the boiling point of pure water at sea level.

Mesophilic bacteria grow best near 37°C—the temperature of the human body. *Escherichia coli*, which grows naturally in the intestines of humans and other animals, is a mesophile. So are most disease-causing bacteria.

Psychrophilic (cold loving) bacteria grow at a significant rate at or below 5°C—the temperature of a properly functioning refrigerator. When milk spoils in a refrigerator, it often has the fruity odor of *Pseudomonas* spp. or the foul odor of *Achromobacter* spp. because these bacteria are

psychrophiles. In contrast, milk that sours at room temperature has the more pleasant, though sour, taste of yogurt or buttermilk, because at this temperature mesophilic lactic acid bacteria predominate.

Psychrophiles are subdivided into obligate and facultative groups. Thermophiles are subdivided into stenothermophiles, facultative thermophiles, and extreme thermophiles (**Table 8.4**). Facultative psychrophiles are also called **psychrotrophs** ("cold-growing") to indicate that they tolerate rather than benefit from decreasing temperature.

What determines the maximum or minimum temperature at which an organism is able to grow? Without doubt,

TABLE 8.4 Growth Responses to Temperature Among Various Groups of Bacteria

Class	Properties	Typical Environment
Psychrophiles (also called psychrotrophs)	Grow at appreciable rates below 5°C	
Obligate psychrophiles	Cannot grow at or above 20°C	Cold ocean water
Facultative psychrophiles	Can grow above 20°C	Soil and water
Mesophiles	Grow best at moderate temperature, around 37°C	Animals
Thermophiles	Grow above 50°C	
Facultative thermophiles	Can grow below 37°C	Soil
Stenothermophiles	Cannot grow below 37°C	Compost
Extreme thermophiles	Grow above 80°C (some above 100°C)	Hot springs

SHARPER FOCUS

WHY DOES CHAMPAGNE COME IN A HEAVY BOTTLE?

Most microorganisms can withstand high hydrostatic pressures. But yeasts are a striking exception. They stop growing and fermenting at only 8 atmospheres of pressure. Although no one knows why yeast is so sensitive to pressure, it is an advantage in winemaking. German winemakers exploit the pressure sensitivity of yeasts to slow the rate of the yeast fermentation that converts grape juice into wine. This prevents heat from building up and damaging the taste of wine. They carry out the fermentation in closed steel tanks and let the pressure from carbon dioxide production rise until it moderates fermentation rate.

The pressure sensitivity of yeasts is even more critical to making champagne. The carbon dioxide bubbles in champagne are formed by a second fermentation usually carried out in the bottle. A calculated amount of sugar is added to the wine to ensure that the proper amount of carbon dioxide is produced. But there is little risk that too much carbon dioxide will be produced and the bottle will explode. Fermentation stops when pressure in the bottle reaches 8 atmospheres, a pressure the heavy champagne bottle can withstand.

the maximum temperature of growth is determined by the heat stability of an organism's proteins. Proteins are usually the most heat-sensitive macromolecules in a cell, and growth cannot occur at temperatures that denature the cell's proteins (Chapter 2). Thermophiles grow at high temperatures because their proteins are exceptionally heat stable. Psychrophiles can grow only at low temperatures because some of their proteins are unusually heat sensitive. Heat stability reflects total protein structure. That is, a mutation that changes any amino acid in a heat-stable protein is likely to decrease its heat stability.

Causes of the minimum temperature of growth are more complex, but most have the same basis. Hydrophobic interactions within proteins become weaker as temperature decreases, causing the **conformation** (shape) of proteins to change slightly. The function of some proteins, including those that regulate metabolism, is particularly sensitive to such changes. At low temperatures, metabolic regulatory mechanisms become distorted and stop growth.

Hydrostatic Pressure. Hydrostatic pressure is pressure applied to a liquid. It is commonly measured in **atmospheres** (one atmosphere is 14.7 pounds per square inch, the pressure at the earth's surface). Ordinary bacteria, such as *Escherichia coli*, thrive at pressures as great as 300 atmospheres. Prokaryotes found in the deep ocean tolerate up to 1500 atmospheres, or 22,050 pounds per square inch, enough to crush all but the strongest steel vessels. In contrast, most yeasts cannot grow at pressures over 8 atmospheres, a low pressure.

Some prokaryotes, called **barophiles** (pressure lovers), benefit from high pressure. They grow more rapidly at pressures greater than 1 atmosphere. Certain barophiles, called **obligate barophiles,** can't grow unless the pressure on them exceeds 1 atmosphere.

Elevated pressure does not crush a microbial cell as it would a human being because water passes readily through the cell membrane. Thus the pressure inside and outside the cell is equalized. Increased pressure stops microbial growth not mechanically but biochemically. Molecular volume changes in the course of most chemical reactions. High pressure inhibits chemical reactions that undergo an increase in molecular volume. It favors chemical reactions that undergo a decrease in molecular volume. Barophiles exploit this principle.

pH. Most microorganisms grow over a range of pH (Chapter 2). In general, prokaryotes grow best at a slightly alkaline (basic) pH. Fungi grow best at a slightly acid pH, and protozoa and algae at a neutral pH. There are startling exceptions, however, especially among bacteria. **Acidophiles** (acid lovers) thrive in environments of extremely low pH. For example, some grow in the acid leachings of mine waste, where pH values are as low as 1.0—the acidity of sulfuric acid. **Alkaliphiles** (base lovers) thrive in environments of extremely high pH. Some grow in soda lakes, common in deserts of the American West. Here pH values are as high as 12.0—the alkalinity of hair remover.

Bacteria can grow over a wider pH range than their proteins can tolerate. They do this by adjusting their intracellular pH. By various mechanisms they pump hydrogen ions out of or into the cell. *Escherichia coli*, for example, can grow in environments that range from pH 5.0 to 8.0. But regardless of external pH, internal pH is maintained at fairly close to 7.6—the optimal value for its metabolism.

Osmotic Strength.

Osmotic strength is a measure of how much water is available (Chapter 4). All microorganisms need liquid water to grow. For this reason, they cannot grow at temperatures below the freezing point of their medium or above its boiling point. For the same reason they cannot grow at high osmotic strength, because it also deprives a cell of water. If the concentration of dissolved materials is high, the concentration of water is correspondingly low.

A medium of high osmotic strength presents microbial cells with a second problem. By osmosis it takes water out of the cell and robs it of its turgor pressure. Bacteria must maintain a positive turgor pressure because it provides the force they need to grow. It replaces the force provided by the cytoskeleton of eukaryotic cells.

If the concentration of solutes in the external environment increases, the bacterium reacts to maintain turgor pressure. To do so, it pumps potassium ions (K^+) and/or osmoprotectants, such as the amino acid proline, into the cell. It also synthesizes the disaccharide trehalose (see Larger Field: Bread in the Desert, Chapter 2). These solutes maintain a higher osmotic pressure inside the cell than the osmotic pressure outside. Eventually, however, this increase in solutes decreases the cell's available water and growth ceases.

When the solute concentration of the medium exceeds the concentration inside, the result is **plasmolysis** (Chapter 4). The volume of the cytoplasm shrinks. In bacteria the plasma membrane pulls away from the rigid wall (**Figure 8.6**). Eukaryotic microorganisms often buckle.

Although high osmotic strength prevents the growth of most bacteria, some species, called **halophiles** (salt lovers), can withstand extremely high salt concentrations. Red archaea called halobacteria flourish in water that is saturated with salt. They reach populations high enough to turn salt flats red. These bacteria not only tolerate a high intracellular concentration of salt, they depend on it for membrane stability. Placed in ordinary tap water, they lyse immediately.

MEASURING MICROBIAL GROWTH

Now let's turn our attention to how to measure microbial growth. Many methods are available. Some measure the mass of cells in the population. Others measure the number of cells. Still others measure some property, such as metabolic activity, that is related to the size of the population. We'll consider all three methods of measuring growth.

Measuring the Mass of Cells in a Population

The mass of cells in a population can be determined directly by weighing them—that is, determining their dry weight. It can also be determined indirectly by measuring the **turbidity** (cloudiness) of the culture and then relating turbidity to dry weight.

Direct Determination of Dry Weight.

To weigh the cells in a culture, they must be separated from the medium and dried. For accurate results, about 25 ml of a relatively dense culture is required. The cells are removed from the medium by filtration (Chapter 3) or **centrifugation** (using a centrifuge to rotate vessels called centrifuge tubes at a high rate of speed, generating a centrifugal force that pushes small particles, including microbial cells, to the bottom of the tube). Then any residual medium is removed by washing: The cells are resuspended in distilled water and filtered or centrifuged again. Then cells are dried in an oven at 105°C for about 24 hours and cooled in a **desiccator** (a chamber with low humidity that keeps materials dry). Finally the cells are weighed.

Measurement by dry weight is tedious, time-consuming, and insensitive. Nevertheless, it has certain uses. It must be used, for example, to construct a standard curve relating turbidity to cell mass.

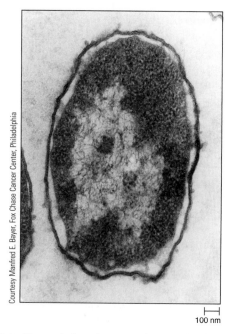

Courtesy Manfred E. Bayer, Fox Chase Cancer Center, Philadelphia

⊢————⊣
100 nm

FIGURE 8.6 Transmission electron micrograph of a plasmolyzed *Escherichia coli* cell. Note the empty space between the membrane-bounded cytoplasm and the cell wall.

Turbidity. The easiest and quickest way to estimate the dry weight of cells in a culture is to measure its turbidity. For this purpose an optical instrument called a **spectrophotometer** is generally used (**Figure 8.7a**). A spectrophotometer measures how much light a solution or a liquid culture of microbial cells transmits. As the mass of cells in the culture increases, turbidity increases, less light is transmitted through the culture, and the reading on the spectrophotometer is higher. Because the mass of a culture and number of cells are related, turbidity can also be used to determine cell number. But it is not a very sensitive measurement of numbers of bacterial cells. About 10 million average-size bacterial cells must be present per milliliter of liquid culture before the liquid becomes visibly turbid. A culture containing about a billion per milliliter is too turbid to see through (**Figure 8.7b**).

To convert spectrophotometer readings to mass or number of cells per milliliter, it's necessary to prepare a standard curve. Such a graph relates spectrophotometer readings to cell mass or number of one particular culture (**Figure 8.7c**).

Like all instruments, the spectrophotometer has advantages and disadvantages. Its measurements are rapid and reproducible. However, it can be used only on relatively dense cultures, and it cannot distinguish between dead and live cells. Nor can it be used with cells that tend to aggregate, because they rapidly settle out of suspension and cloudiness disappears.

Counting the Number of Cells in a Population

Microbial populations contain dead cells, as well as live ones. Some methods score both dead and live cells. They are called **total cell counts.** Methods that score only live cells are called **viable cell counts.**

Total Cell Count. Total cell counts can be done visually, using a microscope with a special microscope slide called a **counting chamber.** There are many styles of counting

FIGURE 8.7 Turbidity. (a) Components of a spectrophotometer, which measures turbidity in units of absorbance. (b) The clear tube on the left contains no bacterial cells. The turbid one on the right contains about a billion (10^9) cells per milliliter. (c) A standard curve relating absorbance to the number of a particular kind of bacterial cells per milliliter. Note that a reading of 1.6 absorbance units means the culture contains a hundred million (10^8) cells per milliliter.

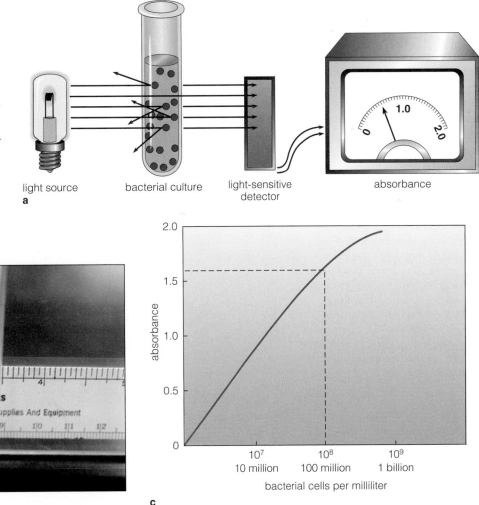

chambers. The **Petroff-Hauser** chamber, named for its developers, is perhaps the most commonly used in microbiology laboratories (**Figure 8.8**). All counting chambers have known depths and are marked on the bottom by squares of known dimensions. Thus each square that we see under the microscope marks off a known volume of liquid, for example, 1 microliter (0.001 ml). By counting the cells within several squares and averaging the results, we can calculate how many cells are present in 1 ml or any other volume of the culture. Because live and dead cells look much the same under a microscope, this method usually yields a total cell count. But vital stains (Chapter 3) can be used to distinguish live cells from dead ones. With these stains live or dead cells can be counted.

Total cell counts can also be done electronically, with an instrument called a **Coulter counter.** This instrument takes advantage of the fact that microbial cells do not conduct electricity as well as the medium in which they are suspended. A measured volume of the culture to be counted is placed in one chamber of the counter and drawn through a small pore (about 15 μm in diameter) into another chamber by a vacuum pump. The pore is part of an electric circuit, so each time a cell passes through, the conductivity of the circuit decreases. Drops in conductivity and hence number of cells is tallied electronically by the counter.

The two principal methods of counting total number of cells in a population—microscopic and electronic counting—give similar results. But they are used in different situations. Microscopic counting requires no expensive equipment, but it is a slow, tedious process. In contrast, electronic counting requires expensive equipment but is extremely rapid. Electronic counting is accurate if cells are the only particles present. If the sample contains foreign particles, such as blood cells or bits of soil, microscopic counting is best. A person doing a microscopic count can discriminate between microbial cells and foreign particles. An electronic counter will count them as cells if they are approximately the same size.

Viable Cell Count. Counting viable cells depends on a live cell's ability to grow—to form a colony or develop into a turbid culture. Plate counts and filtration counts depend on colony formation. Most probable number counts depend on development into a turbid culture.

Plate Count. The technique for performing a plate count is the same as the spread and pour plate methods of culturing microorganisms described in Chapter 3. A sample of the culture is serially diluted, usually tenfold at each dilution, and then a small amount of each dilution is spread on a plate or mixed with melted medium and incubated under appropriate growth conditions. After visible colonies have formed, usually a day or so later, plates with clearly separated colonies are selected and counted. Choosing plates with between 30 and 300 colonies offers a good compromise between speed and accuracy. The number of colonies on a plate, along with the dilution of the sample that was spread on that plate, allows us to calculate the concentration of cells present in the original sample (**Figure 8.9**).

The advantage of the plate count method is its extreme sensitivity. Even a single live cell can be detected with the appropriate medium and incubation conditions. Moreover, a plate count does not require complicated equipment. On the negative side, doing plate counts is slow and tedious and not very accurate. Accuracy increases with the numbers of colonies counted because of **sampling error** (the

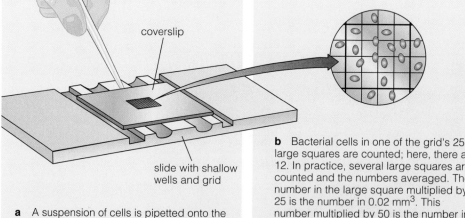

FIGURE 8.8   Total cell count using a Petroff-Hauser counting chamber. It consists of a slide with a shallow wells (0.02 mm deep). On the bottom of the slide, 1 mm² is inscribed with a grid dividing it into 25 large squares (outlined in heavy lines). Each large square is divided into 16 smaller squares. The volume over the entire grid is 0.02 (1/50) mm³. The volume over each large square is 1/25 of that amount.

coverslip

slide with shallow wells and grid

a   A suspension of cells is pipetted onto the slide at the edge of the coverslip, which fills the counting chamber by capillary action. In a few minutes the cells settle to the bottom and counting can begin.

b   Bacterial cells in one of the grid's 25 large squares are counted; here, there are 12. In practice, several large squares are counted and the numbers averaged. The number in the large square multiplied by 25 is the number in 0.02 mm³. This number multiplied by 50 is the number in 1 mm³. This number multiplied by 1000 is the number in a milliliter (a cubic centimeter). In this case:
12 × 25 × 50 × 1000 = 1.5 × 10⁷ cells/ml

inevitable inaccuracy, because all samples are not completely representative of the total population). Ninety-five percent of the time, the true number of viable cells does not differ from the number counted by more than twice the square root of the number of colonies counted.

Filtration Count. Filtration counts are a variation of plate counts. Instead of placing the cells directly on a plate as is done with plate counts, the cells are collected onto the surface of a filter (**Figure 8.10**), which is then placed on the plate. The number of colonies that develops is equal to the

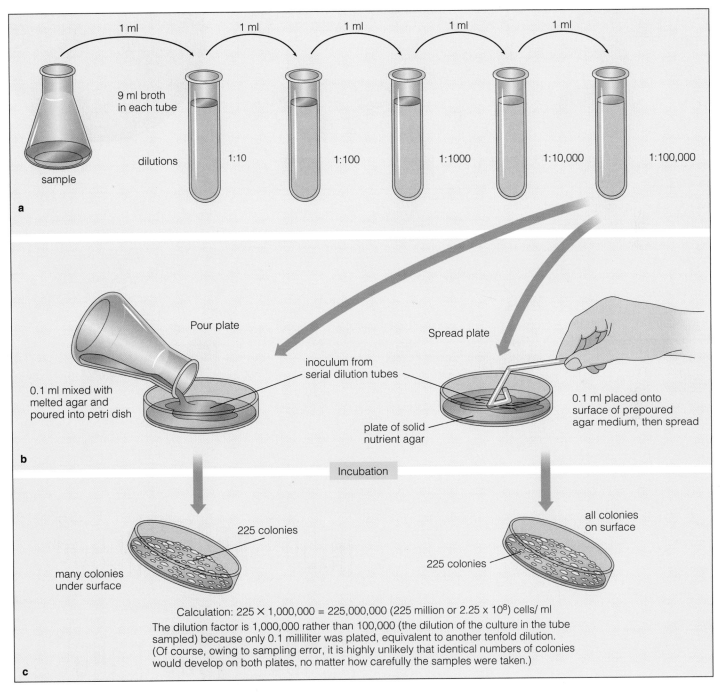

FIGURE 8.9 Example of a plate count of viable cells. (a) The sample is serially diluted to 1:100,000 (10⁻⁵). (b) Then 0.1 ml is added to a plate of nutrient agar either by spreading it on the surface (spread-plate method) or by pouring it with the medium (pour-plate method). (c) After incubation, the colonies that develop are counted.

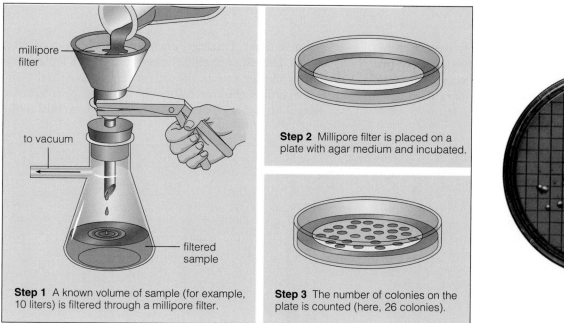

FIGURE 8.10 Counting viable cells by filtration.

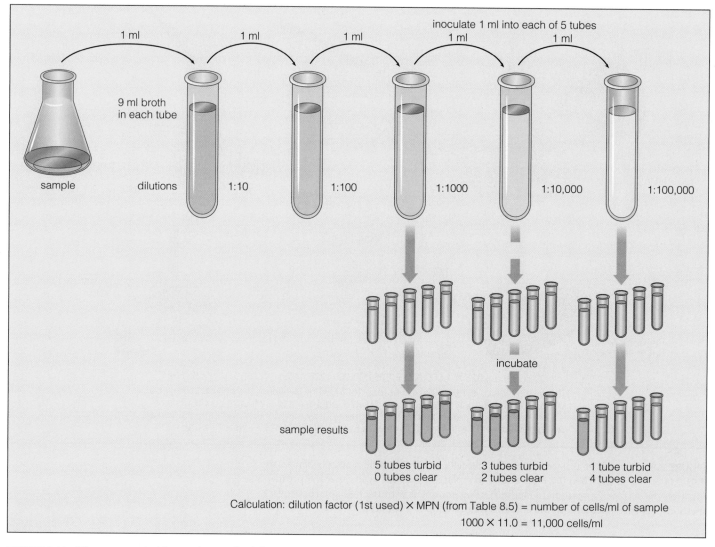

inoculate 1 ml into each of 5 tubes

1 ml   1 ml   1 ml   1 ml   1 ml

9 ml broth
in each tube

sample   dilutions   1:10   1:100   1:1000   1:10,000   1:100,000

incubate

sample results

5 tubes turbid   3 tubes turbid   1 tube turbid
0 tubes clear   2 tubes clear   4 tubes clear

Calculation: dilution factor (1st used) × MPN (from Table 8.5) = number of cells/ml of sample
1000 × 11.0 = 11,000 cells/ml

FIGURE 8.11 The most probable number method for counting viable cells.

number of viable cells the volume of liquid passed through the filter. Filter counts are valuable because they are so sensitive. By filtering a large volume of liquid, a small microbial population can be detected and counted. In the same way, the method can be used to count microorganisms in a large volume of air.

Most Probable Number. The most probable number (MPN) method is a statistical way to estimate number of viable cells in a sample. MPN involves making a series of tenfold dilutions of the culture sample in a liquid medium suitable for the growth of that organism (**Figure 8.11**). Then samples from these test tubes are incubated. After enough time has passed to allow for microbial growth, the test tubes are examined. Tubes that received one or more microbial cells from the sample become cloudy. Tubes that didn't receive any cells remain clear. As the di-

lution factor increases, a point is reached at which some tubes contain only a single organism and others contain none. By determining the probability that tubes did not receive cells, the number of microorganisms that were most probably present in the original sample can be determined using a statistically derived table. **Table 8.5** presents an MPN table.

Accuracy of the most probable number increases with the number of tubes used, but five tubes per dilution is regarded as a practical compromise between accuracy and economy. The MPN method is usually used to count particular microorganisms in a mixed culture. For example, it can be used to test for contamination in drinking water by determining the number of bacteria that can grow in a medium with lactose as a carbon source. Such bacteria are probably *Escherichia coli* from contaminating sewage (Chapter 28).

TABLE 8.5 Most Probable Numbers

Numbers of Turbid Tubes Inoculated from Three Successive Dilutions	MPN	Numbers of Turbid Tubes Inoculated from Three Successive Dilutions	MPN
0 1 0	0.18	5 0 0	2.3
1 0 0	0.20	5 0 1	3.1
1 1 0	0.40	5 1 0	3.3
2 0 0	0.45	5 1 1	4.6
2 0 1	0.68	5 2 0	4.9
2 1 0	0.68	5 2 1	7.0
2 2 0	0.93	5 2 2	9.5
3 0 0	0.78	5 3 0	7.9
3 0 1	1.1	5 3 1	11.0
3 1 0	1.1	5 3 2	14.0
3 2 0	1.4	5 4 0	13.0
4 0 0	1.3	5 4 1	17.0
4 0 1	1.7	5 4 2	22.0
4 1 0	1.7	5 4 3	28.0
4 1 1	2.1	5 5 0	24.0
4 2 0	2.2	5 5 1	35.0
4 2 1	2.6	5 5 2	54.0
4 3 0	2.7	5 5 3	92.0
		5 5 4	160.0

Note: Values of the most probable number (MPN) of viable cells in the first dilution of three successive tenfold dilutions used to inoculate five tubes (also see **Figure 8.11**). This table can be used for any unicellular microorganism as long as five tubes and tenfold dilutions are used. In **Figure 8.11** the number of turbid tubes from three successive dilutions was 5, 3, 1. From this table we see that the most probable number of viable cells in the sample (1 ml) from the first dilution (1:1000) added to each tube is 11.0. Therefore the number of viable cells in the original sample is 11,000 per milliliter (1000 + 11.0 = 11,000).

TABLE 8.6 Methods of Measuring Mass and Numbers of Microorganisms

Method	Measurement	Advantages/Limitations
Measuring Mass		
Dry weight	Mass of cells	Fundamental measurement/time-consuming, not sensitive
Turbidity	Mass or number of cells	Easy, rapid/not highly sensitive
Counting Numbers of Cells		
Microscopic count	Total number of cells	Allows human judgment about what is counted/time-consuming
Electronic counting	Total number of cells	Rapid, accurate/other particles interfere, expensive
Plate count	Viable cells	Simple, inexpensive/time-consuming
Filtration count	Viable cells	Highly sensitive/can't be used with dense cultures
Most probable number	Viable cells	Allows one physiological type to be counted in a mixture of others/time- and media-consuming

Measuring Metabolic Activity

There are several ways that metabolic activity can be used to indirectly measure a quantity of microbial cells. The rate of formation of metabolic products, such as gases or acids, that a culture produces reflects the mass of cells present. Also, the rate of utilization of a substrate, such as oxygen or glucose, reflects cell mass.

Finally, the rate of reduction of certain dyes can be used to estimate microbial mass. For example, methylene blue becomes colorless when reduced by components of microbial electron transport chains. Consequently, in the absence of oxygen, which oxidizes methylene blue, the rate of decolorization of this dye is an index of microbial mass. Rate of dye reduction is highly indirect and therefore not an accurate measurement of microbial mass. However, the technique can be used with complex materials, such as soil or milk.

The various ways of measuring numbers of microorganisms are summarized in **Table 8.6.**

SUMMARY

Populations (pp. 202–203)

1. Microbial growth refers to growth of a population—an increase in the number of cells.

The Way Microorganisms Grow (pp. 203–207)

2. Doubling time is the period required for cells in a microbial population to grow, divide, and produce two new cells for each one that existed before.

3. Growth rate is usually measured as doubling times per hour.

4. Exponential growth occurs when doubling time remains constant. The formula is $N = 2^n$, where N is the number of cells in the culture after n doubling times have passed.

5. Microbial growth is conveniently graphed on a logarithmic scale.

6. A microbial culture typically passes through four distinct phases of growth: lag phase, exponential phase, stationary phase, and death phase.

7. Lag phase, which follows inoculation of stationary-phase or death-phase cells into a fresh culture medium, is a no-growth period, but there is considerable metabolic activity as cells prepare to grow.

8. Exponential phase does not continue indefinitely because nutrients become depleted or toxic products accumulate.

9. No net increase of mass occurs during the stationary phase, but cells become smaller and synthesize components to help them survive non-growth periods.

10. During the death phase, cells die exponentially but at a much lower rate than they grow during the log phase. Death usually occurs because a cell has depleted its intracellular reserve of ATP.

11. Under most natural conditions, nutrients continuously enter a cell's en-

vironment at low concentrations. Growth rate is set by the concentration of the limiting nutrient.

12. In the laboratory, continuous culture is achieved with a chemostat. The concentration of the limiting nutrient in the growth vessel sets the growth rate.

13. Cells in a colony are in different phases of growth depending on their location. On the surface they are aerobic; those in the center are anaerobic.

What Microorganisms Need to Grow (pp. 207–214)

Nutrition (pp. 207–210)

14. Nutrients are chemicals from the environment that a cell needs to grow.

15. All microorganisms require a source of carbon to grow because carbon is the structural basis of biochemicals.

16. Oxygen plays a complex role in microbial metabolism—as nutrient, electron acceptor, and generator of toxic by-products. Microorganisms capable of aerobic respiration use oxygen gas to generate energy.

17. Toxic oxygen-containing compounds are by-products of aerobic metabolism. They include hydrogen peroxide and the free radicals superoxide and hydroxyl radical.

18. All living things require some form of nitrogen (a constituent of proteins and nucleic acids, as well as certain essential metabolites), phosphorus (a constituent of nucleic acids, phospholipids, and certain essential metabolites), and sulfur (a component of two amino acids, a few types of tRNA, and certain essential metabolites).

19. Trace elements, such as potassium, iron, and zinc, are nutrients that microorganisms need to grow. Some trace elements act as cofactors.

20. Growth factors are essential small organic molecules. If a microorganism cannot synthesize them, they must be supplied from the environment. Possible growth factors include amino acids, purines, pyrimidines, fatty acids, or vitamins.

21. Nutritional diversity refers primarily to the different requirements among microorganisms for organic growth factors.

22. Microorganisms exploit a rich nutrient environments by using available growth factors even if they are capable of making them.

The Nonnutritive Environment (pp. 210–214)

23. Every species of microorganism grows over a range of temperatures, from a minimum temperature of growth to a maximum temperature of growth. A species' optimal temperature is that at which it grows most rapidly.

24. Thermophiles can grow at temperatures above 50°C. Mesophiles grow best near 37°C, the temperature of the human body. Psychrophiles can grow at or below 5°C, refrigerator temperature.

25. The heat stability of a microorganism's proteins determines the maximal temperature at which it can grow. Minimum temperature is determined by weakening of hydrophobic interactions of proteins.

26. Hydrostatic pressure is pressure applied to a liquid. Most microorganisms tolerate high hydrostatic pressures. Barophiles grow better (facultative) or only (obligate) at pressures greater than 1 atmosphere.

27. In general, bacteria grow best at a slightly alkaline pH, and fungi at a slightly acid pH. Acidophiles thrive in low-pH environments. Alkaliphiles thrive in high-pH environments.

28. Bacteria maintain a positive turgor pressure. Their internal cell osmotic strength is greater than the osmotic strength of their external environment. Certain bacterial species, the halophiles, require high external osmotic strength.

29. Plasmolysis occurs when a high external osmotic strength draws water out of the cell, shrinking the cytoplasm.

Measuring Microbial Growth (pp. 214–220)

30. Some methods measure cell mass. Others measure cell numbers.

31. Determining dry weight depends on separating cells from their medium, drying them, and weighing them.

32. Turbidity is measured with a spectrophotometer. The spectrophotometer measures how much light a liquid culture of microbial cells transmits. A standard curve is used to relate spectrophotometer readings to cell mass or numbers in a particular culture.

33. A microscope count is done with a special microscope slide called a counting chamber, such as the Petroff-Hauser chamber. Based on the depth of the well and the grid, the number of cells in a set volume of the sample can be computed.

34. An electronic count is done with a Coulter counter. Electronic counting is possible because microbial cells do not conduct electricity as well as their medium.

35. In a plate count a sample of the culture is diluted tenfold or a hundredfold several times, and then a small amount is spread or poured on a plate and incubated. Plates with between 30 and 300 colonies are selected and counted.

36. Filtration is done prior to measuring small numbers of microorganisms in large volumes of liquid or gas.

37. The most probable number (MPN) method is based on the principle that a single live cell can develop into a turbid culture. After several tenfold dilutions, test tubes are incubated. Tubes that become cloudy received one or more cells. By determining the average dilution at which tubes did not receive cells, the number of microorganisms that were most probably present in the original sample can be computed using an MPN table.

38. There are several ways metabolic activity can be used to measure numbers: the rate of formation of metabolic products, such as gases or acids, that a culture produces; rate of utilization of a substrate, such as oxygen or glucose; and rate of reduction of certain dyes.

REVIEW QUESTIONS

Populations

1. To what does the term microbial growth refer?

The Way Microorganisms Grow

2. What is doubling time? Why is growth rate often used instead of doubling time?

3. Explain this statement: Microorganisms grow exponentially.

4. Name, in sequence, the phases of growth in a culture. Explain each one.

5. How do you maintain a continuous culture in the laboratory?

6. Under natural conditions, what usually sets the growth rate?

7. Define the term colony. How does location in a colony affect a cell's phase of growth? How does location affect its access to oxygen?

What Microorganisms Need to Grow

8. What are nutrients? Explain the function of each of these nutrients: carbon, oxygen, nitrogen, sulfur, trace elements, organic growth factors.

9. Explain this statement: Oxygen plays a complex role in microbial metabolism. Use the term free radical in your explanation.

10. To what does nutritional diversity refer?

11. What is a rich growth environment? How do microorganisms exploit a rich growth environment?

12. Name the significant physical factors in a microorganism's nonnutrient environment. Name the significant chemical factors.

13. Explain how different species grow over a range of temperatures. Use these terms in your discussion: minimum temperature of growth, maximum temperature of growth, optimum temperature. Explain what determines the maximum and minimum temperatures.

14. Define these terms: thermophile, mesophile, psychrophiles, psychrotroph.

15. What is hydrostatic pressure? Why can microorganisms withstand high hydrostatic pressures? How is hydrostatic pressure harmful to microorganisms?

16. What pH levels do bacteria prefer? Fungi? What are acidophiles and alkaliphiles? If metabolic functions require a relatively narrow pH, how can bacteria exist in environments with a wide range of pH values?

17. How are halophiles unlike most other bacteria? How do most bacteria maintain turgor pressure? What is plasmolysis?

Measuring Microbial Growth

18. What is the difference between a total count and a viable count?

19. Tell what each of the following methods measure. Describe the procedure, and give its advantages and disadvantages.
    a. turbidity
    b. dry weight
    c. metabolic activity
    d. direct microscopic
    e. direct electronic count
    f. plate count
    g. most probable number (MPN)
    h. filtration

CORRELATION QUESTIONS

1. If you wanted to know how many dead cells were in a culture, what would you do?

2. If you wanted to make cells in a chemostat grow more rapidly, what would you do?

3. How would you determine how many cells of nitrogen-fixing bacteria were present in a gram of soil? Explain.

4. How would you determine whether a slowly growing culture was just coming out of the lag phase or was in the exponential phase. Explain.

5. If you transferred cells to a fresh medium and they began to grow without a lag, what would you conclude?

6. If you diluted a culture a thousandfold, plated 0.1 ml of it, and 430 colonies developed, how many viable cells per milliliter would the culture contain?

ESSAY QUESTIONS

1. Discuss the advantages and disadvantages of using continuous culture to make antibiotics.

2. Discuss the effect of richness of a medium and concentration of a nutrient on growth rate.

SUGGESTED READINGS

Ingraham, J. L., O. Maaløe, and F. C. Niedhardt. 1983. *Growth of the bacterial cell.* Sunderland, Mass.: Sinauer.

Kolter, R. 1992. Life and death in stationary phase. *ASM News* 58:75–79.

Marr, A. G. 1991. Growth rate of *Escherichia coli. Microbiological Reviews* 55:316–33.

Mynell, G. G., and E. Mynell. 1970. *Theory and practice in experimental bacteriology.* New York: Cambridge University Press.

• For additional readings, go to InfoTrac College Edition, your online research library at: http://www.infotrac.thomsonlearning.com

NINE

Controlling Microorganisms

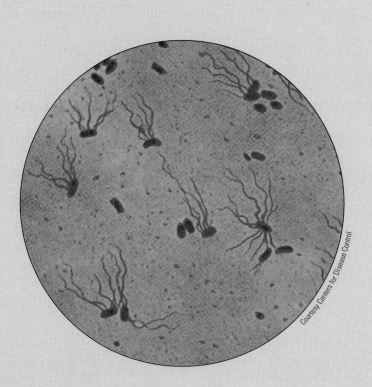

Courtesy Centers for Disease Control

CHAPTER OUTLINE

LEARNING GOALS

To understand:

- *The rate of microbial death and D-value*
- *The physical treatments used to control microorganisms*
- *The chemical treatments used to control microorganisms*
- *How we keep food from spoiling*

Four Hundred Thousand Patients

During April 1993, more than 400,000 people became ill in Milwaukee, Wisconsin. They suffered, to varying degrees, diarrhea, loose or watery stools, stomach cramps, upset stomach, and a slight fever. More than 4400 of them were sick enough to be hospitalized. Their illnesses were severe, lasting, on average, 12 days. At the peak of illness their average number of watery stools per day was 19. Their average weight loss was 10 pounds. Several immuno-compromised patients (those suffering from acquired immunodeficiency syndrome [AIDS] or undergoing chemo-therapy for cancer) died. The cause of this outbreak was quickly diagnosed as cryptosporidiosis, identified as such by the presence of characteristic micro-scopic (4 to 6 μm in diameter) oocysts of the protozoan *Cryptosporidium parvum* in their stools. There is no known cure for cryptosporidiosis, al-though some drugs, such as paromo-mycin, may reduce its symptoms.

The obvious source of these in-fections was water from the South Milwaukee waterworks plant because more than 50 percent of the people it served become ill. It is unlikely that *Cryptosporidium* had only recently en-tered the water supply, because it is widespread in the intestines of ani-mals. One would certainly think that something had gone wrong with the chlorination system at the plant. But that was not the case. Chlorination systems are designed to add suffi-cient chlorine to kill microorganisms in a water supply, and the South Milwaukee system was adding rec-ommended levels of the disinfectant. The problem is that the causative protozoan, *C. parvum*, produces thick-walled oocysts that are highly resistant to chlorine. They survive standard chlorine levels in water sup-plies, even exposure to full-strength household bleach for 2 hours. Why, then, was the water from the South Milwaukee waterworks safe before

the outbreak? Presumably the oocysts had been removed mechani-cally by flocculation and filtration. The outbreak was probably a result of subtle problems with these sys-tems. It's clear that chlorine, the dis-infectant that protects us against so many waterborne pathogens, can't protect us from *C. parvum*.

Case Connections
- As we'll see in this chapter, we humans throughout history have devised many effective ways to protect ourselves from the unde-sirable impact of some microor-ganisms. But as the Milwaukee outbreak of cryptosporidiosis illus-trates, they don't always work.
- Ridding water supplies of harm-ful microorganisms is a partic-ular challenge because the amount of material to be treated is so vast and single treatments are rarely effective against all microorganisms.

Microorganisms are vital links in our planet's web of life. We couldn't live without them. But they do make us sick. They spoil our food. They cause our possessions to rot and decay, and some of them produce foul odors. So it's not surprising that throughout history we humans have spent enormous efforts learning how to control microor-ganisms. In this chapter we'll consider what we've learned about controlling microorganisms by killing them or just by stopping their growth.

THE WAY MICROORGANISMS DIE

To control microorganisms, we're seldom interested in what happens to an individual microbial cell. Instead, we want to know how a particular treatment affects a popula-tion of microbial cells. Does it kill all the cells in the pop-ulation, or does it merely reduce their numbers? Does it stop their growth without killing them?

Some Useful Terms

A number of terms are used to describe the effects of var-ious treatments on microbial populations:

- **Sterilization** is a treatment that destroys all microbial life.
- **Disinfection** (or **sanitizing**) is a treatment that reduces the number of *pathogens* to a level at which they pose no danger of disease.
- **Decontamination** renders a surface that has been heavily exposed to microorganisms safe to handle.
- **Antisepsis** kills microorganisms on living tissue.

These useful terms overlap. For example, sterilizing an instrument also disinfects it or decontaminates it.

How can we know when microbial cells have been killed? For practical purposes, a microbial cell is considered to be dead if it fails to form a colony on a solid medium or to produce a turbid culture in a liquid medium that supports growth of live cells. Treatments that kill microorganisms are termed **microbiocidal.** Treatments that only inhibit microbial growth are termed **microbiostatic.** That is, treated cells appear to be dead, but they are able to grow again when the treatment is withdrawn. Refrigerating food is an example of a microbiostatic treatment. It preserves food by preventing microbial growth, not by killing microorganisms. The distinction between microbiostatic and microbiocidal treatments is not absolute. Some microbiostatic chemicals become microbiocidal if treatment is prolonged or if higher concentrations are used.

Now let's consider the question of how fast a microbiocidal treatment can kill a population of microorganisms.

Death Rate

As you might expect, not all cells in a population exposed to a microbiocidal treatment die at the same time. Curiously, they die the same way they grow—exponentially (Chapter 8). In other words, the time course of microbial death, like a graph of microbial growth, is a straight line when plotted on a logarithmic scale (**Figure 9.1**). This same pattern of exponential death occurs with exposure to most lethal agents, including heat, radiation, high osmotic strength, or chemicals.

Exponential death means that the number of surviving microbial cells continues to decrease by 1 logarithm (that is, tenfold) during each subsequent unit of time. In the particular example shown in Figure 9.1, the unit of time, called the **decimal reduction time** or **D-value,** is 60 minutes. To put it another way, D-value is the time in minutes it takes to kill 90 percent of the cells in a population. During subsequent D-values, 90 percent of the survivors will be killed. For thermal killing, the temperature of treatment is sometimes indicated by a subscript. For example, the D-value at 80°C is written $D_{80°C}$.

For heat treatments the D-value depends upon four factors: temperature, the type of microorganisms being treated, the physiological state of the microorganisms being treated, and the presence of other substances. (1) Temperature affects the rate because lethal chemical reactions, like all others, proceed more rapidly at higher temperatures. (2) Different types of microorganisms vary considerably in their resistance to killing. For example, the mycoplasmas (Chapter 4) are highly susceptible to treatment; endospores are extremely resistant. (3) Log-phase cells are in a physiological state that

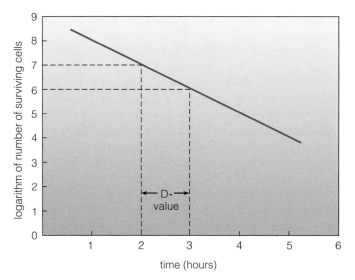

FIGURE 9.1   Microbial death. Note that the number of surviving cells in a microbial population declines 90 percent (tenfold or 1 logarithm value) during each D-value. (In this case, the D-value is 60 minutes.)

makes them generally more susceptible than stationary-phase cells to lethal treatments. Sometimes such differences are dramatic. For example, more than 99 percent of log-phase cells of *Escherichia coli* can be killed by cold shock (rapid chilling to 5°C or less), but no stationary-phase cells are killed by the same treatment. (4) The presence of other substances, especially proteins, protects microorganisms from many lethal treatments. Milk, for example, is quite protective. When added to a bacterial culture, it decreases killing by heat, cold, or chemicals because of the protein it contains. Microorganisms in vomit or feces are also hard to kill because of the protective proteins these materials contain.

Sterilization

D-time tells us how long it takes to kill 90 percent of the microbial population. But how can we calculate how long it will take to sterilize, to kill all cells in the population? The answer is like the answer to the mathematical riddle: How long will it take for a snail to reach its goal if each day it travels half the remaining distance? Of course, it would never reach its goal. But it would get very close. The same reasoning applies to the death of a bacterial culture. If the D-value remains the same throughout the period of treatment (it usually does), 90 percent of the surviving cells will die during each D time—but 10 percent will survive. However, what does it mean that 90 percent of the population will die when only one cell remains? Of course, with high probability, one D time later the material will be sterile—absolutely free of microorganisms. In other words, sterilization is a statistical concept. We must know three things to design a sterilization treatment: (1) the D-value of the treatment, (2) the number of cells present, and (3) the desired degree of certainty that no cells will be alive at the end of the treatment.

These calculations are vital in industry and research laboratories. For example, the canning industry must know how long food must be heated for it to be free of live cells. If the treatment is too short, endospores of the bacterium *Clostridium botulinum*, which causes botulism, may survive. Later they might germinate and produce a neurological toxin that could paralyze and kill whoever eats the canned food. On the other hand, excessive heating degrades the quality of the food. These calculations assure that materials are in fact sterilized—that with a high level of certainty, no microorganisms survive.

In addition to D-value, two other terms are useful to describe the susceptibility of a particular microorganism to killing by heat. **Thermal death point** (**TDP**) is the lowest temperature required to kill all microorganisms in a particular liquid suspension in 10 minutes. **Thermal death time** (**TDT**) is the minimal time required to kill all microorganisms in a particular liquid suspension at a given temperature.

In summary, killing microbial cells depends upon probability. We can't say with certainty when a particular cell will die. But we can determine when a microbiocidal treatment will kill a certain fraction of them. These results are expressed as D-time. Knowing D-time and how many cells are present, we can calculate how long it will take to sterilize populations of particular microorganisms. Thermal death point and thermal death time tell us the heat treatment required to sterilize a particular population.

PHYSICAL CONTROLS ON MICROORGANISMS

Now let's consider the various treatments used to control microorganisms. We've already discussed (in Chapter 3) sterilization treatments used in preparing a bacterial culture. Now we'll look at principles and applications of control measures in clinical medicine and industry (**Table 9.1**).

First we'll look at physical treatments: heat, cold, radiation, filtration, drying, and osmotic strength. Then we'll look at chemical methods.

Heat

Heat treatment is simple, inexpensive, and effective. It's the best method if the material being treated is not damaged by heat. Heat penetrates to kill microorganisms throughout the object. In contrast, some other sterilization treatments,

such as ultraviolet light and chemicals, sterilize only the surfaces they touch. The effectiveness of heat is illustrated by the fact that the National Aeronautics and Space Administration (NASA), after extensive and sophisticated research, chose this ancient method to sterilize the Viking spacecraft sent to Mars. (There are strict international standards to protect other planets from microbial contamination).

Heat sterilization can be dry or moist. Both kill by destroying proteins (Chapter 2). But dry heat kills largely by speeding oxidations, which inactivate proteins. Moist heat, on the other hand, denatures proteins by breaking the bonds that confer secondary and higher protein structure.

In a microbiological laboratory, dry-heat sterilization is commonly used in the form of flaming and hot-air ovens (Chapter 3). Quickly passing a pipette or the opening of a flask through a flame is considered reliable sterilization of the surface. Reliable hot-air sterilization requires severe treatments—171°C for 1 hour, 160°C for 2 hours, or 121°C for 16 hours. Large objects require even more time.

Moist heat in the microbiological laboratory is applied either by boiling or in autoclaves (Chapter 3). Moist heat is effective at a lower temperature than dry heat, and it penetrates more quickly. Boiling water kills most bacterial and fungal cells and inactivates many viruses in just a few minutes. However, vegetative cells of some thermophilic bacteria can withstand prolonged boiling. In fact, some grow vigorously at boiling temperature (100°C) because their proteins are not denatured. Endospores also survive boiling. The proteins in bacterial endospores resist inactivation by heat because the interior is dry and water cannot penetrate their thick walls. The autoclave is more effective than boiling because it uses pressure to raise the temperature considerably above that of boiling water. At a pressure of 15 pounds per square inch (normal operation), the temperature in an autoclave reaches the endospore-killing temperature of 121°C.

Cold

Low temperature by itself does not kill most microorganisms. In fact, cold storage is often used to preserve them for limited periods. If held at a temperature below their growth range, most microorganisms die slowly; and the lower the temperature, the more slowly they die. An exception to this rule is the phenomenon of **cold shock**—when a growing culture is suddenly chilled, many microorganisms are killed immediately. In spite of this, cold does not sterilize.

Although cold does not kill microbial cells, it is an effective and widely used microbiostatic treatment. Refrigerators (usually set at 5°C) preserve food because they stop the growth of most species of microorganisms, including most disease-causing microorganisms. Psychrophiles do

TABLE 9.1 Treatments to Control Microorganisms

Treatment	Effect	Mode of Action	Uses
Physical Methods			
Heat			
Dry heat	Sterilizes	Denatures protein	In the laboratory, used to sterilize dry materials that can withstand high temperature and any materials damaged by moisture.
Moist heat	Sterilizes	Denatures protein	In the laboratory, used to sterilize liquids and material easily charred. Used in food canning.
Pasteurization	Kills certain microorganisms	Denatures protein	Eliminates pathogens and slows spoilage of milk and dairy products, wine, beer. (Canned evaporated or condensed milk is sterilized.)
Cold	Slows or stops microbial growth	Slows chemical reactions	Preserves perishable materials, including food and microorganisms.
Radiation			
UV light	Sterilizes	Damages DNA	In the laboratory, sterilizes surfaces.
X rays and gamma rays	Sterilizes	Strips electrons from atoms	Used to sterilize plastic equipment and surface of fresh fruits and vegetables.
Filtration	Removes cellular microorganisms	Physically removes cells	In the laboratory, used with media, antibiotics, and other heat-sensitive materials. Used to preserve certain beverages.
Drying			
By evaporation or heat	Kills many microorganisms	Distorts membranes	Used to preserve foods.
By sublimation (freeze-drying)	Stops microbial growth	Stops most chemical reactions	In the laboratory, used to preserve cultures, proteins, blood. Used to make instant coffee, lightweight food.
Chemicals			
Phenols	Kills most microorganisms	Denatures proteins	Germicides.
Phenolics	Kills most microorganisms	Denatures proteins and disrupts plasma membrane	Disinfectants, antiseptics.
Alcohols	Kills most microorganisms	Denatures proteins and disrupts plasma membrane	Disinfects surfaces, including skin and thermometers.
Halogens	Kills microorganisms	Oxidizes vital biochemicals	Disinfects surfaces, including skin and water.
Hydrogen peroxide	Kills many microorganisms	Oxidizes vital biochemicals	Mild skin disinfectant.
Heavy metals	Kills many microorganisms	Reacts with sulfhydryl groups of proteins	Skin disinfectants.
Surfactants			
Soap, detergent	Washes away microorganisms	Physically removes microbes	Disinfects surfaces, including skin, bench tops.
Quaternary ammonium salts	Kills microorganisms	Disrupts membranes	Widely used sterilizing agents.
Alkylating agents			
Formaldehyde and glutaraldehyde	Kills microorganisms	Inactivates enzymes by adding alkyl groups	Preserves tissues, prepares vaccine, sterilizes surgical instruments.
Ethylene oxide	Kills microorganisms	Inactivates enzymes by adding alkyl groups	Gas used to sterilize heat-sensitive materials and unwieldy objects in hospitals.

grow slowly in a refrigerator. But most are not a health risk, although they eventually cause food to spoil. There are exceptions. *Listeria monocytogenes*, which causes listeriosis, a foodborne disease of the nervous system (Chapter 24), is a psychrophile.

The act of freezing a bacterial culture kills some cells, but the survivors remain alive for long periods in the frozen state. Because bacteria are too small for ice crystals to form within them, they are not killed by the mechanical destruction of essential intracellular structures, as eukaryotic cells are. Instead, bacteria are killed by the high osmotic strength that develops as water in the external environment freezes.

Bacterial cultures can be preserved by rapid freezing, sometimes with the addition of dimethylsulfoxide (DMSO), glycerol, or milk to protect proteins from denaturation. Such treatments minimize the number of cells killed by freezing (Chapter 3).

Radiation

Some forms of electromagnetic radiation (Chapter 3) kill living things, including microorganisms. In general, visible light and longer wavelength radiation are not significantly lethal. High doses of radio waves will kill microorganisms, but they do so by heating in much the same way a microwave oven does. Very intense visible light can kill in the presence of oxygen and pigments called **photosensitizers.** For example, intense light in the presence of methylene blue converts oxygen to a highly lethal form called singlet-state oxygen. However, this simple method of killing microorganisms has not yet found practical applications.

The types of radiation that kill bacteria directly are all of shorter wavelength than visible light. They are ultraviolet (UV) light (radiation with a wavelength of 10 to 400 nm) and ionizing radiation (with wavelength extending down to about 0.001 nm).

The UV wavelength that is most lethal to microorganisms is about 265 nm, because UV light kills principally by damaging DNA, which absorbs maximally at this wavelength. Because sunlight is rich in UV light, microorganisms have evolved various DNA repair mechanisms to correct the damage done by UV. One of these mechanisms is **photoreactivation.** Photoreactivation depends upon an enzyme that is active in the presence of visible light between 420 and 540 nm. Therefore controlled experiments on the killing effects of UV must be done in a room with dim light.

Mercury vapor lamps, called **germicidal lamps,** that emit maximally at 253.7 nm are commercially available. They are simple to use and leave no toxic residue. As soon as the lamp is extinguished, the UV light disappears. However, UV light kills microorganisms only on surfaces. It cannot even penetrate glass. UV light does present a health

hazard. If it reaches the eyes, either directly or by reflection, it damages the cornea. Moreover, the shorter UV wavelengths that kill microorganisms and damage the eye are not visible to humans. Damage can occur without an individual being aware of exposure. There are only a few practical applications of UV light. Some microbiological laboratories turn on UV lights at night, when no one is present. Such treatment decreases the microbial population and thereby minimizes the possibility of contaminating pure cultures that are being studied. Drying clothes in sunlight is a simple application of UV sterilization. The high UV content kills the microorganisms directly exposed to the light.

Two forms of ionizing radiation are used to kill microorganisms: x-rays (wavelengths from 0.1 to 400 nm) and gamma rays (wavelengths from 0.001 to 1 nm). Both forms cause ionization by stripping electrons from atoms composing the cell's essential macromolecules. Ionizing radiation also acts indirectly by ionizing water resulting in the formation of free radicals of its hydroxyl ($-OH$) and hydrogen (H) constituents. These highly reactive agents damage the cell's macromolecules, especially its DNA. The result is usually cell death. But ionizing radiation is rarely used in microbiological laboratories because it is so technically complex. It does, however, have considerable commercial application. It is used to sterilize plastic containers such as petri dishes and bottles used in microbiological laboratories. The radioactive isotope of cobalt, ^{60}C, used as a source of gamma radiation, has great potential as a safe and cheap way to preserve food. But even though it has been approved by the FDA, public suspicion of irradiated food has prevented significant use in the United States.

Filtration

Microorganisms other than viruses can be removed from liquids by filtration (Chapter 3). Thus filtration does not, in a strict sense, sterilize. But it is extensively used in microbiological laboratories to remove cellular microorganisms from certain media, vitamin solutions, and antibiotics that are heat sensitive. In industry, filtration is replacing pasteurization in some cases because filtration causes even less damage than the slight amount caused by pasteurization. Some sweet wines are now preserved with filtration, as are beers called "draft beer" or "cold-filtered beer."

Drying

Drying (removal of water) can be accomplished through evaporation or sublimation (Chapter 3). Drying by evaporation is almost never used in the laboratory because it causes chemical changes, but it is widely used in the food industry.

Lyophilization, or freeze-drying, removes water by **sublimation** (direct conversion from a solid state to a gaseous state). Materials to be lyophilized are frozen and placed in a chamber to which a partial vacuum is applied. Sublimation avoids the chemical changes caused by heat drying (Chapter 3). As a result, this method is often used to preserve perishable materials such as proteins, blood products, and reference cultures of microorganisms.

Osmotic Strength

High concentrations of salt or sugar are used to preserve certain foods for the same reason drying is: Microorganisms cannot grow without water. The high osmotic strength of salt and sugar solutions can also damage cells by plasmolysis. Osmotic strength is rarely used to control microbial growth in the microbiological laboratory because once added, solutes (such as salt) cannot be easily removed.

CHEMICAL CONTROLS ON MICROORGANISMS

Now let's turn to the chemical treatments for controlling microorganisms. Chemicals, including antibiotics, used to treat disease are called **chemotherapeutic agents.** We'll discuss them in Chapter 21. Here we'll discuss all other antimicrobial chemical agents. Chemicals that kill microorganisms are called **germicides.** Those that inhibit microbial growth are called **germistats.** Germicides used on inanimate objects are also called **disinfectants.** When applied to living tissue, germicides are called **antiseptics.** Depending upon how it is used, the same germicide can be used to disinfect or cause antisepsis. A number of other terms are used to describe compounds that kill or stop growth of various organisms: **biocide, bactericide, fungicide, virucide, biostat, bacteriostat, fungistat,** and so on. Their meanings are self evident.

The same germicide may be formulated differently by the manufacturer for particular uses—for example, as a disinfectant or an antiseptic. To register a formulation with the Environmental Protection Agency (EPA), the manufacturer must supply test results that establish its effectiveness and safety. Also, directions for use must be provided on the product label.

Many different compounds are used as germicides, and the number continues to increase. At least 1200 germicides made by 330 manufacturers are sold in the United States.

In this section we'll discuss how to select one of these germicides for a particular use. Then we'll look at how germicides are tested for effectiveness. Finally, we'll survey the various types of germicides.

Selecting a Germicide

Choosing the appropriate germicide depends upon answers to these questions:

- Will it damage the tissue or object being treated?
- Will it control the target microorganism(s)?
- What is the purpose of the treatment?

The first question is answered by information provided by the manufacturer. The second comes from standard classifications for germicides. Every registered germicide is classified as having high, intermediate, or low germicidal activity. Microorganisms are then grouped according to their sensitivity to these three classes (**Table 9.2**). As for the third question, more powerful germicides or higher concentrations must be used for sterilization than for disinfection. For example, a 3- to 6-percent solution of hydrogen peroxide is used for disinfection, but a 6- to 30-percent solution is needed for sterilization. If large amounts of protective substances, such as blood or feces, are present, a more powerful germicide is required. The price of a germicide may also be a consideration.

TABLE 9.2 Susceptibility of Some Types of Microorganisms to Germicides

Microorganism	Germicidal Susceptibility		
	High	Intermediate	Low
Bacteria			
Endospores	Killed	Not killed	Not killed
Vegetative cells[a]	Killed	Killed	Killed
Mycobacterium tuberculosis[b]	Killed	Killed	Not killed
Fungi	Killed	Killed	Some killed
Viruses			
Nonlipid and small	Killed	Some killed	Some killed
Lipid and medium-size	Killed	Killed	Killed

Source: After M. S. Favero and W. W. Bond, 1991, Sterilization, disinfection, and antisepsis. In *Manual of Clinical Microbiology*, 5th ed., ed. A. Balows (Washington, D.C.: American Society of Microbiology).
[a] Vegetative cells of most bacteria.
[b] Owing to its resistance and special importance, *M. tuberculosis* is considered separately.

FIGURE 9.2 Paper disc method of testing germicides. These petri dishes compare the effectiveness of four germicides (*quat* indicates Zephiran) against three bacterial species. Note that *Staphylococcus aureus* is more sensitive than *Pseudomonas aeruginosa* to all four germicides.

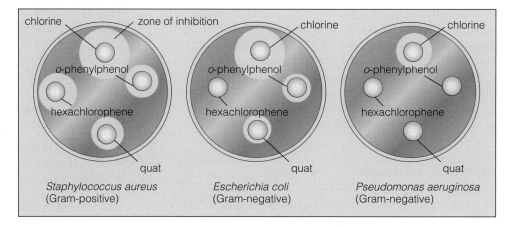

Testing Germicides

Germicides are tested by comparing their effectiveness to phenol, a traditional germicide. (Recall that Lister used phenol to treat James Greenlees; Chapter 1.) First the test germicide and phenol are serially diluted. Then a standard volume of a culture of a test bacterium is added (usually the intestinal pathogen *Salmonella typhi* or *Staphylococcus aureus*, a bacterium that infects wounds). After 5 and 10 minutes, various dilutions are tested to see if any cells survive. The samples are inoculated into a germicide-free medium to see if it becomes turbid after 48 hours' incubation. The highest dilution of the germicide that kills in 10 but not 5 minutes is the end point. The ratio of end points is the **phenol coefficient.** For example, the end point of germicide is a dilution of 1:1000 and phenol is 1:100; the germicide's phenol coefficient is 10 (1000 ÷ 100 = 10). In other words, it is 10 times more powerful than phenol.

The effectiveness of a germicide against various microorganisms can be determined by the **paper disc method** or the **use-dilution test.** In the paper disc method, a filter-paper disc impregnated with the test germicide is placed on the surface of an agar medium in a petri dish seeded with the test organism. After incubation, uniform bacterial growth develops with a clear, bacteria-free zone of inhibition around the disc. The size of the zone of inhibition is a useful index of the bacterium's sensitivity to the germicide (**Figure 9.2**).

In the use-dilution test the microorganism is added to dilutions of the germicide, much as is done in determining a phenol coefficient. But rather than sampling the tubes, the tubes are incubated. The highest dilution of the germicide that remains clear after incubation indicates the effectiveness of the germicide.

Classes of Germicides

There are too many germicides to discuss them individually. Instead we'll consider the six major classes: phenol and phenolics, alcohols, halogens and hydrogen peroxide, heavy metals, surfactants, and alkylating agents.

Phenol and Phenolics. Phenol, which Lister called carbolic acid, has been used since even before his time. Phenolics resemble phenol in having hydroxyl (—OH) groups attached to a benzene ring (**Figure 9.3**). These compounds kill microorganisms by denaturing vital cellular proteins, including enzymes. Phenolics (but not phenol) also act on lipids. The first noticeable effect of phenolics on microbial cells is disruption of the cytoplasmic membrane. Phenol itself is not often used clinically today because it is so powerful. When applied to skin it denatures enough protein to turn the area white. But some throat sprays do contain low concentrations of phenol.

Cresols are a common ingredients of household and hospital disinfectants because they remain active even in the presence of blood and feces. Paracresol is the most commonly used representative of the class. Its most familiar brand name is probably Lysol.

Hexachlorophene, a chlorinated phenolic, is very effective as an antiseptic. It persists on skin, providing long-lasting antimicrobial protection. It was once widely used

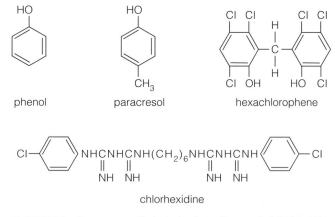

FIGURE 9.3 Structures of phenol, phenolics, and chlorhexidine. Paracresol and hexachlorophene are phenolics. Chlorhexidine is not a phenolic, but it has many of the same properties as hexachlorophene.

in soaps and lotions, particularly in hospitals to control staphylococcal infections of newborns. But unfortunately, it proved to be absorbed through the skin. It was associated with substantially increased risk of brain damage for premature babies. Now it is restricted to prescription use for difficult skin infections.

Hexachlorophene has been replaced by chlorhexidine for most hospital uses. Chlorhexidine is not a phenolic and is less toxic, but it has many of the desirable properties of hexachlorophene. Like the phenolics, chlorhexidine acts first to disrupt the cytoplasmic membrane.

Alcohols.

Alcohols (compounds with a hydroxyl [—OH] group) also kill by disrupting the lipids in cell membranes and denaturing proteins. Two alcohols, ethanol and isopropanol, are widely used as skin antiseptics (**Figure 9.4**). They are routinely used to disinfect sites for injection or for drawing blood. They act two ways: they dissolve lipids on the skin, exposing microorganisms; then they kill them. Oddly, a 50- to 70-percent solution of ethanol in water is more effective than the pure compound. Such solutions are used to disinfect thermometers. Isopropanol is somewhat more effective and only slightly more toxic. Alcohols do not kill endospores.

Halogens and Hydrogen Peroxide.

The halogens iodine and chlorine as well as hydrogen peroxide are oxidizing agents. They inactivate proteins including enzymes by oxidizing their functional groups. Iodine is used as an antiseptic, and chlorine is used as a disinfectant.

Iodine is used as an aqueous solution and also as a **tincture** (an alcohol solution). Iodine, like alcohols, disinfects the skin without damaging tissue. Iodophors (mixtures of iodine and detergent-like surfactants) are more powerful but safe germicides. They are used to disinfect skin before surgery and to sterilize dairy equipment.

Chlorine is used in many forms—chlorine, hypochlorites, inorganic chloramines, and organic chloramines. The activity of all these compounds depends upon their ability to generate chlorine (Cl_2). The microbiocidal activity of chlorine is markedly decreased by organic matter. So chlorine is used mainly to treat substances like water. But even in relatively pure water such as municipal water supplies (Chapter 28) or swimming pools extra chlorine must be added to overcome the presence of organic material. The **free chlorine** (that not bound to organic materials) is the

form that kills microorganisms. Free chlorine is remarkably lethal. A concentration of 0.6 to 1.0 part per million free chlorine kills almost all microorganisms in 15 to 30 seconds. *Cryptosporidium* is a notable exception (see Case History). Chlorine is also a widely used disinfectant in the dairy industry, the food industry, and hospitals.

Hydrogen peroxide (H_2O_2) is not a halogen. We include it here because, being an oxidizing agent, it kills microorganisms the way halogens do. It's used in a 3-percent solution as a weak antiseptic for cleaning wounds. It's also used to disinfect medical instruments and soft contact lenses because it does minimal damage to these fragile objects. When hydrogen peroxide comes in contact with tissue, it bubbles vigorously because all aerobes, including humans, contain the enzyme catalase, which decomposes hydrogen peroxide into oxygen and water. Hydrogen peroxide kills before it is destroyed by catalase or a similar enzyme, peroxidase. Some bacteria—for example, species of *Staphylococcus*—are relatively resistant to hydrogen peroxide because they produce large amounts of peroxidase.

Heavy Metals.

Salts (Chapter 2) of heavy metals react with the sulfhydryl (—SH) groups of proteins, thereby "poisoning" enzymes and killing microbial cells. Salts of mercury and silver have been used in medicine for years. Mercuric chloride was once widely used as an antiseptic but not today because it is highly toxic. Instead, mercury-containing organic compounds such as Merthiolate and Mercurochrome are used. They are less toxic.

Silver salts and colloidal silver (finely dispersed silver particles) were once widely used antiseptics. Once state laws uniformly required that a solution of silver nitrate be applied to the eyes of newborns to prevent ophthalmic gonorrhea and the blindness that results. Today, however, the trend is away from using silver nitrate and toward using antibiotics. Silver salts are used less today because they irritate skin. However, they continue to be helpful in protecting burned skin from infection.

Surfactants.

Surfactants are compounds that have hydrophilic (water-loving) and hydrophobic (water-fearing) parts (Chapter 2). They penetrate oily substances in water, break them into small droplets, and coat them. The hydrophobic ends of the surfactant stick into the droplets and the hydrophilic ends penetrate the water. The result is an **emulsion** (a fine suspension of oily droplets in water). Soaps and detergents are surfactants. They clean by emulsifying oily material so it can be washed away with water. Skin is covered with oily material in which microbial cells and viruses are imbedded (Chapter 26). Soaps and detergents remove this layer and the microorganisms it contains. As a result, frequent hand washing effectively controls the spread of many infections. It is

FIGURE 9.4 Structures of ethanol and isopropanol.

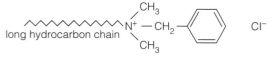

FIGURE 9.5 Structure of the quaternary ammonium salt benzalkonium chloride (commercially known as Zephiran). Note that four different groups are attached to a charged nitrogen atom (N^+).

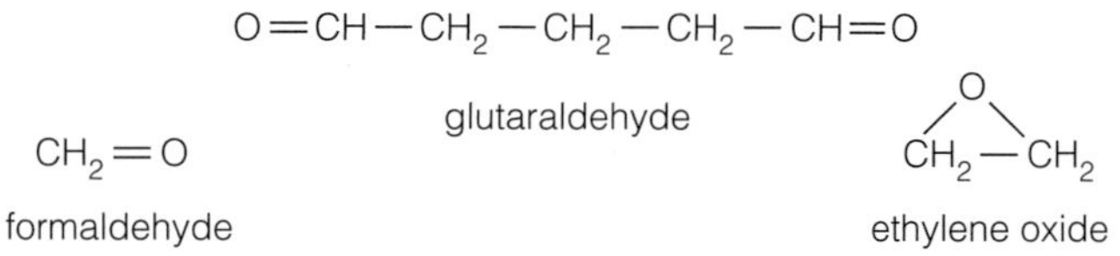

$$O = CH - CH_2 - CH_2 - CH_2 - CH = O$$
glutaraldehyde

$$CH_2 = O$$
formaldehyde

ethylene oxide

FIGURE 9.6 Structures of commonly used alkylating agents.

probably the best way to avoid colds (Chapter 22). Soaps with antibacterial agents are even more effective.

Some surfactants are germicides, as well as detergents. They kill microorganisms by penetrating and destroying their cytoplasmic membranes. Quaternary ammonium salts are the major class of surfactant germicides. Their hydrophilic group is a charged nitrogen atom with four (hence "quaternary") hydrophobic organic groups attached to it (**Figure 9.5**). These compounds kill all classes of cellular microorganisms and viruses that have membranes, although they are less effective against Gram-negative bacteria because of their protective hydrophilic outer membranes. Because quaternary ammonium salts are generally effective and nontoxic, they are widely used in the home, industry, laboratories, and hospitals. Two popular quaternary ammonium salts are Cepacol (cetylpyridinium chloride) and Zephiran (benzalkonium chloride). Some mouthwashes contain quaternary ammonium salts.

Alkylating Agents. Formaldehyde, glutaraldehyde, and ethylene oxide are called alkylating agents because they attach short chains of carbon atoms (alkyl groups) to proteins. They inactivate enzymes and kill cells (**Figure 9.6**).

Formalin, a 37-percent solution of formaldehyde, is used to preserve tissues and to embalm. It kills all microorganisms, including endospores. Lower concentrations of formaldehyde—0.2 to 0.4 percent—are used to inactivate microorganisms for killed vaccines (Chapter 20). Glutaraldehyde can be used to sterilize surgical instruments.

Another alkylating agent, ethylene oxide, has the special advantage of being a gas and therefore not remaining on an object after it's been sterilized. However, ethylene oxide is extremely toxic to humans, so it must be used in a sealed chamber. Ethylene oxide kills all bacteria, including endospores. It's used to sterilize materials that would be destroyed by heat, such as plastic medical equipment

TABLE 9.3 Uses of Some Common Germicides

Action	Example	Concentration
Sterilization	Ethylene oxide	450–500 mg/liter
	Glutaraldehyde	Variable
	Hydrogen peroxide	6–30%
	Formaldehyde	6–8%
	Chlorine dioxide	Variable
	Peracetic acid	Variable
Disinfection	Glutaraldehyde	Variable
	Hydrogen peroxide	3–6%
	Formaldehyde	1–8%
	Chlorine dioxide	Variable
	Peracetic acid	Variable
	Chlorine compounds	500–5000 mg/liter
	Alcohols (ethanol, isopropanol)	70%
	Phenolic compounds	0.5–3%
	Iodophor compounds	30–50 mg free iodine/liter
	Quaternary ammonium compounds	0.1–0.2%
Antisepsis	Alcohols (ethanol, isopropanol)	70%
	Iodophors	1–2 mg free iodine/liter
	Chlorhexidine	0.75–4.0%
	Hexachlorophene	1–3%
	Parachlorometaxylenol	0.5–4.0%

and animal feed. It's also used in hospitals to sterilize unwieldy objects such as mattresses and telephones.

Table 9.3 summarizes the antimicrobial chemical agents discussed here.

PRESERVING FOOD

Most foods are chemically quite stable. They spoil only when microorganisms grow on them. Preserving food, therefore, is largely a matter of eliminating microorganisms or slowing their growth. Effective and safe food preservation is a relatively recent human accomplishment. Even during the nineteenth century, most farm families in the United States were malnourished during the winter months because they could not preserve the food necessary for a well-balanced diet.

It this section we'll consider the major environmental factors humans manipulate to preserve food: temperature, pH, availability of water, and addition of chemicals.

Temperature

Temperature is the environmental factor most often used to preserve food. Low temperature slows or completely arrests the growth of microorganisms, and high temperature kills them.

The temperature inside a refrigerator (about 5°C) is low enough to stop the growth of many microorganisms, including most pathogens. Psychrophilic microorganisms are the exception. They continue to grow, some relatively rapidly, in refrigerators. When food does eventually spoil in a refrigerator, the psychrophilic population is responsible.

In contrast, almost no microorganisms can grow at the temperature inside a deep freeze (about −10°C), and those that can grow extremely slowly. As a result, food does not spoil in a properly functioning deep freeze. However, the quality of food, particularly vegetables, gradually deteriorates over time because of enzymes in the food itself. To minimize this deterioration, most vegetables are blanched (surface-heated by brief exposure to boiling water) before they are frozen. Such treatment inactivates many of the destructive enzymes in vegetables.

Canning.
Canning is the oldest and most widespread method of preserving food by temperature. At the beginning of the nineteenth century in France, Nicolas Appert discovered that food could be kept almost indefinitely if enclosed in an airtight container and heat-sterilized. Modern heat treatments used to can foods are designed to kill bacterial endospores because they are the most heat-resistant forms of life and they can germinate within canned food. Some, such as the spores of *Clostridium botulinum*, can germinate to produce a lethal toxin. Although toxin-containing food looks, smells, and (presumably) tastes normal, even a minute amount of it can be deadly. Canning procedures are designed to eliminate, with a high degree of assurance, all spores of *C. botulinum*.

As we discussed earlier in this chapter, D-value is a heat treatment that kills 90 percent of a particular microbial population. For canning, a 12D treatment for *C. botulinum* spores is usually applied. That means that the treatment reduces the population of *C. botulinum* by 12 log values. Such a treatment would decrease a population of 10^{12} *C.*

botulinum spores—a population many times denser than ever encountered in food—to a single viable spore.

Two factors—time and temperature—determine safe heat treatments for canning. If the temperature is higher, the time can be shorter. **High-temperature, short-time** (HTST) treatments hold great promise for improving the quality of canned foods. They are as effective as conventional treatments for killing microorganisms but do not give the food an overcooked taste.

Pasteurization.
Pasteurization is special heat treatment used to control microorganisms in milk, dairy products, wine, and beer. Pasteurization temperatures are too low to sterilize, but the treatment does kill most pathogens. It also decreases the total number of microbial cells present and thereby delays spoilage. Moreover, it causes minimal damage to the product.

Louis Pasteur developed the process to kill lactic acid bacteria in wine to keep it from spoiling. Pasteurization also extends the shelf life of milk, but its important function is to eliminate disease-causing bacteria in milk. These include *Mycobacterium tuberculosis*, which causes tuberculosis; species of *Brucella*, which cause brucellosis (a serious systemic disease); *Salmonella* spp., which cause intestinal disease; rickettsiae, which cause Q fever; and *Listeria* spp., which cause food poisoning. The standard treatment for pasteurizing milk in the United States is heating to 63°C for 30 minutes or 72°C for 15 seconds (HTST), which causes less flavor change. Milk can also be sterilized by **ultra–high temperature** (UHT) treatments so that it can be kept without refrigeration. UHT sterilized milk is popular in Europe, but not in the United States because it has a slightly different taste. Canned evaporated or condensed milk is sterilized, not pasteurized.

pH

As our earlier discussion of lactic acid bacteria pointed out, acidity—low pH—prevents the growth of most microorganisms, especially in an anaerobic environment. Adding vinegar to foods accomplishes much the same result. For example, some cucumber pickles are made by a lactic acid fermentation, while others are made by adding vinegar. Low pH also increases the effectiveness of heat treatments. Acidic foods such as tomatoes and fruits can be canned safely merely by boiling.

Water

Drying and salting food do not sterilize but preserve food by making it unable to support microbial growth for lack of water. Both treatments have been used for centuries to

Courtesy Stan Lester, Lester Farms, Winters, California

FIGURE 9.7 Field of apricots drying in the sun near Winters, California.

TABLE 9.4 Some Chemicals Used to Preserve Food

Chemical	Use and Activity
Calcium propionate	Antifungal agent added to bread to prevent mildew and *Bacillus* spp. that cause "ropy bread"
Sorbic acid (potassium sorbate)	Antifungal agent added to acid foods such as soft drinks, salad dressings, and cheeses
Sodium benzoate	Antifungal agent added to acid foods such as soft drinks, salad dressings, and cheeses
Sodium nitrate (nitrite)	By anaerobic respiration, nitrate is reduced to nitrite, an antibacterial agent that prevents germination of *Clostridium botulinum* spores when added to bacon, ham, and hot dogs

preserve foods. Drying is used most often for fruits and meat. Many fruits—apricots, peaches, prunes, pears, and grapes—are dried commercially in the sun. Most are first exposed to sulfur dioxide (SO_2) supplied as compressed gas or as the fumes of burning sulfur, but that step of the process is not essential for preservation. It's used principally to prevent the fruit from darkening in the sun (**Figure 9.7**). Other foods—for example, yeast and some prunes—are dried by heating. Smoking meat dries it and also adds chemicals that enhance flavor and inhibit microbial growth.

In freeze-drying (lyophilizing), food is frozen and then exposed to a vacuum to dehydrate it. The heat of sublimation keeps the food frozen until it is dry. Because the drying occurs at low temperature, the food suffers very little loss of quality. Unfortunately, freeze-drying is energy intensive and therefore costly, which restricts its use. Coffee is freeze-dried, as are specialty meals carried by backpackers.

Salting is most often used for fish and meat. Salt pork, a staple of nineteenth-century sailors and the American frontier, would last through an entire winter because the microorganisms that normally cause meat to spoil cannot withstand the high osmotic strength of a concentrated salt solution (Chapter 2). Today salt is still used to preserve pickles, olives, and meats. Similarly, jellies and jams are preserved by sugar, which also increases osmotic strength.

Chemicals

Various chemical preservatives are added to commercially prepared foods (**Table 9.4**). They are safe (the Food and Drug Administration regulates their use), and some are remarkably effective. For example, calcium propionate is now routinely added to bread. Before its use, bread would grow moldy in a week or so. Now it lasts almost indefinitely.

SUMMARY

Case History: Four Hundred Thousand Patients (p. 224)

1a. There are many effective ways to protect ourselves from the undesirable microorganisms. But standard methods don't work against some, such as *Cryptosporidium parvum*, that are particularly hard to kill.

The Way Microorganisms Die (pp. 224–226)

1. Sterilization completely eliminates all microorganisms, including viruses. Disinfection (sanitation) reduces the number of pathogens to a safe level. Decontamination renders an instrument or surface that has been heavily exposed to microorganisms safe. Antisepsis treats living tissue to destroy or inhibit microbial growth.

2. Microbial death is judged as the state when a cell can no longer divide to form new cells.

3. Microbiostatic treatments inhibit (prevent) growth. Microbiocidal treatments kill microbial cells. Some microbiostatic treatments become microbiocidal if prolonged.

4. A graph of microbial death against time of treatment is a straight line when plotted on a logarithmic scale.

5. Rates of microbial death from any particular treatment can be described by decimal time (D-value). D-value is the time in minutes it takes to kill 90 percent of the cells in a population.

6. The value of D depends upon temperature, the type of microorganism, its physiological state, and the presence of other substances that might be protective.

7. To design a sterilization treatment you must know the D-value, the number of cells to be eliminated, and the desired degree of certainty that no cells will be left alive.

8. Thermal death point (TDP) is the lowest temperature required to kill all microorganisms in a particular liquid suspension in 10 minutes. Thermal death time (TDT) is the minimal time required to kill all microorganisms in a particular liquid suspension at a given temperature.

Physical Controls on Microorganisms (pp. 226–229)

9. Heat penetrates to kill microorganisms throughout an object. Heat treatments may be dry (direct flaming, hot-air oven) or moist (boiling or autoclaving). Heat kills by denaturing protein.

10. Low temperature is a microbiostatic treatment.

11. Freezing kills most bacteria, but survivors remain alive for long periods in a frozen state. Rapid freezing is a way to preserve bacterial cultures.

12. Ultraviolet (UV) light and ionizing radiation (x-rays and gamma rays) kill microorganisms. UV kills by damaging DNA. Ionizing radiation kills by stripping electrons from atoms. Only exposed surfaces are sterilized.

13. Filtration eliminates microbes by physically removing them. It does not sterilize because viruses pass through filters.

14. Drying controls microbes by removing an essential nutrient, water.

15. Increasing osmotic strength by adding salt or sugar is used to preserve food.

Chemical Controls on Microorganisms (pp. 229–232)

16. Chemicals that kill microorganisms are called germicides. Germicides used on inanimate objects are also called disinfectants. Germicides applied to living tissue are called antiseptics. Germistats are chemicals that inhibit microbial growth.

17. To choose a germicide, ask if it will damage the object being treated, if it will control the target microorganism, and if the purpose is to control or eliminate all microorganisms.

18. Germicides are classified as having high, intermediate, or low germicidal activity. The presence of substances, such as blood or feces, diminishes the effectiveness of germicides.

19. Germicides are tested by comparing their effectiveness with phenol. The highest dilution of the germicide that kills in 10 but not 5 minutes is the end point. The ratio of end points is the phenol coefficient.

20. The effectiveness of a germicide against a specific microorganism can be determined by the paper disc method or the use-dilution test. The size of a bacteria-free zone around a disc impregnated with the germicide indicate its effectiveness. In the use-dilution test, the test organism is added to dilutions of the germicide and the tubes are incubated. The highest dilution of the germicide that remains clear indicates effectiveness.

21. Phenol is a compound with a hydroxyl group attached to a benzene ring. Phenolics are chemical variants of phenol. Both act by denaturing protein. Phenolics also act on lipids. Hexachlorophene is a phenolic once used as an antiseptic but now replaced with chlorhexidine, which is less toxic to humans.

22. Alcohols are compounds that contain a hydroxyl group. They kill microorganisms by denaturing proteins and disrupting the cytoplasmic membrane. They do not kill endospores. Ethanol and isopropanol are alcohols commonly used as clinical disinfectants.

23. Halogens are oxidizing agents. They inactivate enzymes by oxidizing certain functional groups. Iodine is an antiseptic and chlorine is a disinfectant.

24. Hydrogen peroxide is not a halogen, but it is an oxidizing agent. It is used as a weak antiseptic for cleaning wounds and to disinfect fragile medical instruments and contact lenses.

25. Salts of heavy metals react with the sulfhydryl groups of proteins to kill microorganisms. Merthiolate and Mercurochrome, organic compounds that contain mercury, are skin disinfectants.

26. Surfactants are compounds with hydrophilic and hydrophobic parts that penetrate oily substances in water and form an emulsion. Soaps and detergents are surfactants that control microorganisms by washing them away.

27. Quaternary ammonium salts are powerful surfactant germicides. They kill all classes of cellular microorganisms and viruses that have membranes. Higher concentrations are required to kill Gram-negative bacteria.

28. Alkylating agents attach short chains of carbon atoms to proteins, which kills the cell. Formaldehyde, formalin, and glutaraldehyde are examples.

29. Ethylene oxide is gas alkylating agent used sterilize heat-sensitive materials and unwieldy objects. It is extremely toxic to humans.

Preserving Food (pp. 232–234)

30. Temperature is often used to preserve food. Two factors—time and temperature—determine safe heat treatments for canning.

31. Refrigeration (about 5°C) is low enough to stop the growth of most microorganisms. Psychrophilic microorganisms are the exception.

32. Pasteurization is a heat treatment that controls microorganisms without sterilization and with minimal effect on the food.

33. Low pH prevents the growth of most microorganisms. Adding vinegar is one method of lowering pH.

34. Drying and salting food have been used for centuries to preserve meat, fish, and fruit. Both are means of removing water, which is essential for microorganisms.

35. Chemical preservatives are routinely used. For example, calcium propionate is routinely added to bread.

REVIEW QUESTIONS

Case History: Four Hundred Thousand Patients

1a. What properties of *Cryptosporidium parvum* make it particularly difficult to remove from water supplies?

The Way Microorganisms Die

1. Define these terms: sterilization, disinfection, sanitation, decontamination, antisepsis.

2. What is the usual method of determining whether a cell is alive or dead?

3. How do the number of live cells in a microbial population decrease with time of exposure to a lethal treatment. Use the term D-value in your explanation.

4. Define these terms: microbiostatic, microbiocidal, sterilization, thermal death point, thermal death time.

Physical Controls on Microorganisms

5. How does heat kill? What is the advantage of heat over most other sterilization treatments?

6. Explain this statement: Cold is a microbiostatic treatment.

7. How does freezing kill bacteria?

8. What two forms of radiation are used to sterilize? How do they act?

9. How does filtration control microorganisms? Does filtration sterilize? Explain.

10. How does high osmotic strength control microorganisms?

11. Give an industrial use for each of these physical antimicrobial treatments: heat, cold, radiation, filtration, drying, osmotic strength.

Chemical Controls on Microorganisms

12. What guidelines should be observed in choosing a germicide?

13. How do you determine a phenol coefficient?

14. Explain the paper disc method and the use-dilution test for determining the effectiveness of a germicide.

15. How are germicides classified?

16. For each of the following, tell how the germicide acts chemically to kill, and give an example of its use:
    a. phenol   e. hydrogen peroxide
    b. phenolics   f. heavy metals
    c. alcohols   g. surfactants
    d. halogens   h. alkylating agents

Preserving Food

17. Why does food spoil despite refrigeration?

18. What is the benefit of pasteurization?

19. What determines safe heat treatments for canning?

20. Explain why drying and salting food are effective preservation methods.

CORRELATION QUESTIONS

1. How long would it take to reduce of population of a billion bacteria to a thousand by a treatment with a D-value of 20 minutes?

2. Why are Gram-negative bacteria more resistant to quaternary ammonium compounds but not to heat?

3. How can freezing both protect and kill microorganisms?

4. What keeps pickles from spoiling?

5. Is desiccation the only antimicrobial effect of sun drying? Explain.

6. Do you think washing in soap and water or treatment with hydrogen peroxide is a better way to remove microorganisms? Explain.

ESSAY QUESTIONS

1. Discuss the benefits of pasteurization.
2. Discuss the basis for the public's objection to radiating food.

SUGGESTED READINGS

Block, S. C. 2000. *Disinfection, sterilization, and preservation*. 5th ed. Baltimore: Lea and Febiger.

Doyle, M. P., L. R. Beuchat, and T. J. Montville. 2001. *Food microbiology: fundamental and frontiers*. 2d ed. Washington, D.C.: ASM Press.

Mynell, G. C., and E. Mynell. 1970. *Theory and practice in experimental bacteriology*. New York: Cambridge University Press.

For additional readings, go to InfoTrac College Edition, your online research library at: http://www.infotrac.thomsonlearning.com

TEN

Classification

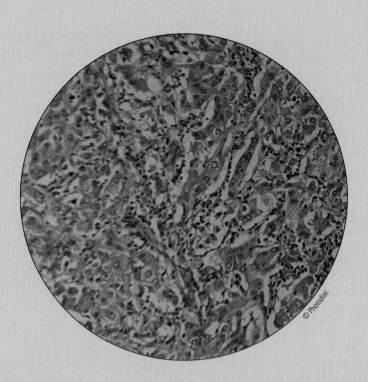

© Photodisc

CHAPTER OUTLINE

LEARNING GOALS

To understand:

- *The principles by which organisms are classified*
- *The difference between artificial and natural systems of classification*
- *The concept of species and how it applies to microorganisms*

- *How microbial species are named*
- *The evolutionary relationship among microorganisms, plants, and animals*

- *The methods used to classify microorganisms*
- *The methods used to identify microorganisms*

SHARPER FOCUS

THE UNIVERSAL TREE OF LIFE

Most of us share a curiosity about our family tree—who our ancestors were and who we're related to. This curiosity usually extends beyond family, to other species and how they're related to one another.

Since ancient times humans have assigned similar organisms to groups and named them. But the idea of the various groups being related had to wait until the concept of evolution was established. Then biologists began to study **pylogeny** (the evolutionary relationship of organisms). Through intensive studies on fossils, they learned about the phylogeny of plants and animals. Their principal tools were the many excellent fossils that were found and dated. They revealed the sequence of appearance of various groups of plants and animals and the evolutionary changes that occurred within them. The complex **morphology** (shape and structure) of plants and animals made their relationship to fossils and each other almost self-evident. By the middle of the twentieth century the general outlines of the family trees of plants and animals were clearly established.

Microorganisms, however, were left out because until relatively recently no microbial fossils were found. And there was another major problem. The shapes of most microorganisms, notably prokaryotes, are too simple and too similar to one another even to suggest relatedness among existing groups. Most microbiologists despaired that a family tree for microorganisms would ever be known. But all that began to change in the late 1960s, when new tools became available that made it possible to determine the sequence of monomers in macromolecules. Biologists realized that such sequences constituted a huge reservoir of information that could be applied to studies on phylogeny. Comparing sequences of **homologous molecules** (those with the same function) in various organisms did reveal the relatedness of the organisms. The number of differences in the sequence of monomers between homologous macromolecules in pairs of organisms is a quantitative measure of evolutionary distance between them. The most fruitful molecules to compare proved to be the

small subunit (SSU) of ribosomal RNA (Chapter 5).

Scientists soon determined with increasing rapidity the sequence of bases in SSU RNA from many species. (At this writing the compete sequence of SSU RNA molecules from more than 20,000 prokaryotic strains has been determined.) Such information lead to a remarkable achievement—construction of a **universal tree of life,** which shows the relationships among all organisms.

The universal tree of life confirms some things

we already suspected. For example, animals, plants, and Gram-positive bacteria comprise distinct, separate groups. But there are surprises. For example, the protozoa are not a distinct group. The ciliate group of protozoa are more closely related to animals than they are to flagellate group of protozoa. It was also a surprise to learn how closely related we are to fungi. The tree has three major branches, called **domains**—the bacteria, the archaea, and the eucarya (eukaryotes).

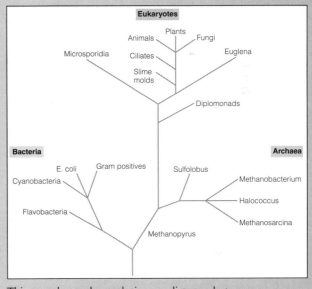

This tree shows the evolutionary distance between representatives of the three major groups of cellular organisms—bacteria, archaea, and eukaryotes.

In this chapter we'll consider how microorganisms are classified. We'll start with the general principles used to guide all classifications and then go on to methods used to deduce relatedness of microorganisms. Finally we'll examine methods used to identify microorganisms.

PRINCIPLES OF BIOLOGICAL CLASSIFICATION

Biologists are faced with the seemingly infinite variety of life. How can they make sense out of such vast diversity? The instinctive human and scientific response is to classify—to group similar things together. Then by studying one member of a group, we can hope to know something about all members. As knowledge expands, we can organize the groups by collecting similar groups into larger assemblages. A classification system based on this scheme of assigning individuals to groups and assigning these groups to progressively larger and more inclusive assemblages is called a **hierarchical scheme of classification.**

Biological classification is called **taxonomy.** It has fascinated humans for a long time. The Greek philosopher Aristotle made the first recorded attempt to classify all living things in the fourth century B.C. Others presented different schemes. But it wasn't until the eighteenth century that a Swedish biologist named Carolus Linnaeus devised a scheme

that was widely accepted and persisted. The Linnaean scheme remains the basis for biological classification today. We continue to name organisms the way he did, and we continue to use most of the hierarchical groups that he did.

Linnaeus's hierarchical sequence proceeded this way:

species

genus (*pl.*, genera)

family

class

order

phylum (*sing.*, phyla) or division

kingdom

One modern scheme groups kingdoms into domains (**Figure 10.1**).

Species is the lowest **taxon** (*pl.*, taxa), or group. All members of a species are very much alike. Progressively higher taxa accommodate progressively greater diversity of individual members. The highest taxa, such as kingdoms and domains, contain organisms that are quite different from one another. For example, a coral and an elephant are both members of the animal kingdom. Although highly dissimilar, they are more closely related to one another than either is to any member of the plant kingdom.

Now let's examine how species are named.

Taxon	Corn	Housefly	Human	Microbe
Domain	Eucarya	Eucarya	Eucarya	Bacteria
Kingdom	Plantae	Animalia	Animalia	Monera
Phylum	Anthophyta	Arthropoda	Chordata	Proteobacteria
Class	Monocotyledonae	Insecta	Mammalia	Gamma proteobacteria
Order	Commelinales	Diptera	Primates	Enterobacteriales
Family	Poceae	Muscidae	Hominidae	Enterobacteriaceae
Genus	*Zea*	*Musca*	*Homo*	*Escherichia*
Species	*mays*	*domestica*	*sapiens*	*coli*

Grant Heilman Photography   Runk/Schoenberger/Grant Heilman Photography   Dr. Charles Henneghien/Bruce Coleman Ltd.   Scott Camazine/Sharon Bilotta Best/Photo Researchers

FIGURE 10.1 Biological classification.

NO JOB TOO BIG

Carolus Linnaeus (1707–1778) was ambitious, hardworking, and had a passion for classification. He organized the biological information of his time and set up a system of nomenclature that serves as the basis for all modern systems.

Linnaeus set his lifetime goal as a youth. He wanted to classify what he considered to be the three kingdoms of nature: plants, animals, and minerals. He went to medical school in Sweden and then traveled extensively—to Lapland, Germany, England, France, and Holland. At age 28 he published three monumental books: one was a system of classifying plants, animals, and minerals; the second compared and analyzed all existing schemes for classifying plants; and the third described all known plant genera (935 at the time). He practiced medicine for a while, and then became a professor at the University of Uppsala. There he made his major contributions—publishing books on the binomial designation of species and the classification of species into hierarchical schemes. He even classified human diseases by genera, families, and orders. Carolus Linnaeus's impact on biology was immediate; it continues today.

Scientific Nomenclature

Linnaeus introduced the binomial nomenclature that we use today. Each organism is designated by two names. The first name is the organism's **genus** (*pl.*, **genera**) designation and the second its **specific epithet.** Together they constitute the name of a species. Species names are always Latinized and underlined or written in italics. The genus designation is capitalized, but not the specific epithet. Thus the name of the best-studied bacterium is written *Escherichia coli*, and that for humans, *Homo sapiens*. The genus designation can be replaced with an initial if the complete name has been used recently. The bacterium becomes *E. coli* and humans, *H. sapiens*. All cellular organisms are named this way, but viruses are not, as you will see later in this chapter.

Most genus and species names are chosen to tell us something about the organism—its appearance, its habitat, a characteristic property, or a scientist associated with it. For example, *Escherichia coli* is named for microbiologist Theodore Escherich and its usual habitat, the colon. *Staphylococcus aureus* is named for the characteristic way its cells aggregate to resemble a bunch of grapes (*staphyle* means "cluster") and its yellow colonies (*aureus* means "golden"). Some names are more obscure. For example, *Shigella etousae* is named for the Japanese microbiologist Kiyoshi Shiga and the European Theater of Operations of the United States Army (the terminal *e* gives the proper Latin ending) where it was first isolated.

Artificial and Natural Systems of Classification

The type of system Linnaeus used to group species into higher taxa is what biologists now call **artificial,** grouping species solely on the basis of visible similarities. Linnaeus and other biologists of the time believed that species were unchanging and created as such. (The concept of one species being related to another through a common ancestor had not yet entered biological thought.) Modern systems of classification strive to be **phylogenetic,** that is, to reflect evolutionary relatedness. Such systems are also called **natural.** Linnaeus's scheme couldn't be natural because he devised it about a hundred years before Charles Darwin developed his theory of evolution. With an understanding of evolution, the meaning and importance of biological classification vastly changed. Evolution explains how organisms are related to one another. Modern classification schemes incorporate this information.

But how do taxonomists get information about evolution? Much of it comes from studying the fossil record.

The Fossil Record

Fossils are the mineralized remains of organisms that lived in past ages. Many fossils reveal sufficient structural detail to identify and classify the organism that formed them (**Figure 10.2a**). In this way nearly 200,000 extinct species have been identified. Biologist know when these

FIGURE 10.2   The fossil record. (a) A 50-million-year-old fossil of a bat reveals enough detail to identify its species. (b) The oldest known fossil, a 3.5-billion-year-old filamentous microorganism that superficially resembles modern cyanobacteria but cannot be identified.

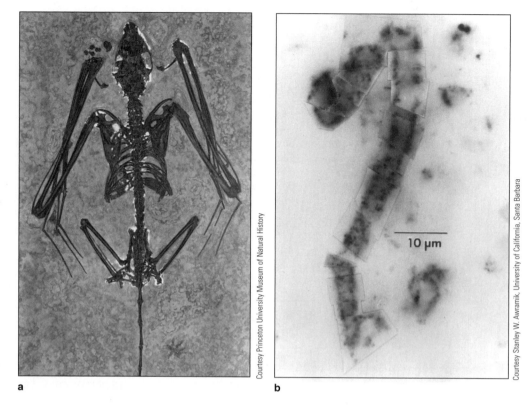

a

b

Courtesy Princeton University Museum of Natural History

Courtesy Stanley W. Awramik, University of California, Santa Barbara

10 µm

species lived because geologists can date the rocks in which their fossils were embedded. Such information provides a record of when species appeared and when they became extinct, thus enabling biologists to trace the sequence of evolution. But this rich record is largely limited to the past 600 million years. Before that time, the fossil record is sparse. Some older fossils do exist, but they are too few and lack sufficient detail to tell us much about the pattern of evolution during most of the 3.5 billion years that life has existed on Earth. The fossil record, therefore, tells us a lot about the evolution of plants and animals but almost nothing about the evolution of microorganisms.

Within the past four decades fossils of microorganisms have been discovered (**Figure 10.2b**). These were found by using electron microscopy to examine thin slices of ancient rocks that contained **stromatolites,** the fossilized remains of phototrophic prokaryotes that grew as **microbial mats** (layered masses of cells). Stromatolites continued to form up to relatively recent times and are relatively common in lagoons and hot springs (**Figure 10.3**). Microbial fossils in ancient stromatolites don't contain sufficient detail to identify species or reveal phylogenetic relationships. But they do establish how long microorganisms have been on Earth. The oldest known microbial fossils are about 3.5 billion years old. That's almost as old as the oldest rocks (3.8 billion years) and only about 1 billion years younger than Earth itself.

Now let's turn to the most vexing problem of microbial taxonomy: What constitutes a species?

The Concept of Species

Individual members of a species are not identical. The key question, then, is: How different can individuals be and still belong to the same species? In other words, what is the definition of species?

Before evolution was recognized, a species was considered to be an individual act of creation. The theory of evolution changed this definition. Many different definitions developed. Possibly the clearest and most useful is Ernst Mayr's **biological species concept,** stated in 1942: A species is a group of organisms that share a common pool of genes (that is, a group of organisms that can interbreed). New species appear when some members of an existing species change or become geographically isolated, so they no longer breed with the rest of the group.

This definition of species applies to eukaryotes, including most algae, fungi, and protozoa, but what about prokaryotes?

Prokaryotic Species. The biological species concept cannot be applied to prokaryotes, because sexual exchange of genes is not an essential part of their life cycle. Groups of bacteria need not share a common gene pool. Genetic exchange among bacteria does occur in nature, but it's sporadic. Worse yet, it sometimes occurs between distantly related organisms (Chapter 6).

We might ask, then, if the concept of species applies at all to prokaryotes, which reproduce almost exclusively by asexual means. We might expect that, as a result of mutations that occur during asexual reproduction, each individual bacterium would all be just slightly different, forming a continuous variation of types. Indeed, until the end of the eighteenth century, biologists considered all prokaryotes to

LOOKS CAN BE DECEIVING

The adage "Things are not always what they appear to be" certainly applies to *Pneumocystis carinii*—and in more ways than one. Until recently, this pathogen was known for causing mild, often asymptomatic respira-

tory infections (Chapter 22). But today it is known for the deadly pneumonia it causes in patients with acquired immunodeficiency syndrome (AIDS), whose weakened immune systems cannot fight off *P. carinii* (Chapter 27).

From its morphology and life cycle, *P. carinii* appeared to be a protozoan closely related to *Plasmodium*, the protozoan that causes malaria (Chapter 27). Sequencing the bases in its rRNA revealed that the sequence is

90 percent identical to the sequence in certain fungi and only 40 percent identical to the sequence of *Plasmodium* and related protozoa. These results show that *P. carinii* is a fungus, not a protozoan (see Figure 22.14).

belong to a single highly variable species. But this proved not to be the case. Groups of prokaryotes with similar properties do indeed exist. The term *species* is applied to these groups. Thus a species of prokaryotes is defined only by the similarities of its members, not by its capacity to interbreed.

Strains. Microorganisms that belong to the same species can be subdivided into **strains,** clones (Chapter 3) that are presumed or known to be genetically different. For example, if two clones of *Escherichia coli* are isolated at different places or at different times, they are designated as different strains. Or if a strain of *E. coli* is altered by mutation or genetic exchange, the product will be a new strain. Often strains are designated by capital letters followed by numbers. For example, K12 and ML30 are well-known strains of *E. coli*.

Microbiology laboratories accumulate large numbers of strains. A microbial genetics laboratory might have several thousand strains of the same species, each carrying a particular mutation or combination of mutations. An infectious disease laboratory would have strains of pathogenic bacteria isolated from different patients or outbreaks of disease. Identifying strains is an important step in tracing the spread of disease (Chapter 20). People suffering from disease caused by the same bacterial strain probably acquired it from the same source.

Viral Species. Classifying viruses presents problems as great if not greater than classifying prokaryotes because we have little or no evolutionary history to guide us. The International Committee on Taxonomy of Viruses (ICTV) has taken on this responsibility. The ICTV doesn't use a Linnaean scheme. Rather than Latinized names, viral species are given ordinary English names, which are not translated in most countries. Thus the virus that causes mumps is called the **mumps virus.** Most other virus names are similarly simple and direct. Viral species are grouped into genera

FIGURE 10.3 Stromatolites in shallow sea water in western Australia. These are 1000 to 2000 years old.

FIGURE 10.4   Evolution of the major schemes for classifying organisms.

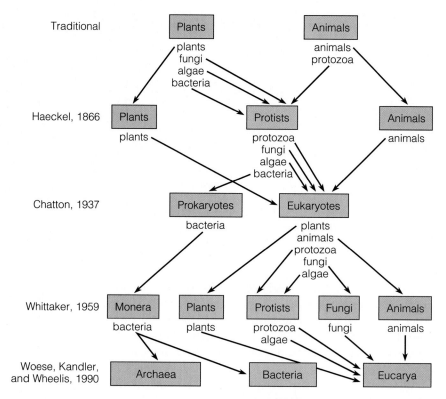

and genera into families. But no higher taxa are used to classify viruses (Chapter 13).

Viral species are also subdivided into strains. Some of them show important differences. For example, the Bangladesh strain of smallpox virus killed about 40 percent of the people it infected. Other strains of smallpox virus typically killed only a few percent of those they infected.

MICROORGANISMS AND HIGHER LEVELS OF CLASSIFICATION

Now that we have a workable definition of species that applies to prokaryotes and other microorganisms, let's consider how microorganisms fit into the upper hierarchical levels of biological classification. The earliest schemes assigned microorganisms to one or the other of the two major categories of living things—plants and animals (**Figure 10.4**). Protozoa, because they are largely motile, were assigned to the animal kingdom. Algae, because they are photosynthetic, and fungi, because they are nonmotile, were assigned to the plant kingdom. Bacteria were also put in the plant kingdom because there was no more logical place for them. The archaea weren't known at the time.

In the nineteenth century, scientists began to realize how unrealistic these assignments were. Clearly microorganisms are neither plants nor animals. Nearly every group has representatives with plant-like properties and others with animal-like properties. Some bacteria are photosynthetic, but many are motile. All algae are photosynthetic, but some are also motile. Some fungi are motile, but none is photosynthetic.

The obvious solution was to admit that microorganisms are neither plants nor animals. Thus the German biologist Ernst Haeckel proposed in 1866 that living things be divided into three kingdoms—plants, animals, and microorganisms, which he called **protists.** This scheme served biology well for nearly 60 years. In 1892 Dmitri Iwanowski discovered viruses. Where did they fit? To this day, viruses are not included in any classification scheme.

By the 1930s there was reason to revise Haeckel's classification scheme. Using the new electron microscope, scientists discovered that there were two types of cells: pro-

karyotic or eukaryotic. A French microbiologist, Edouard Chatton, considered this difference so fundamental that in 1937 he proposed all living things be divided into two groups, eukaryotes and prokaryotes, depending on the type of their cells. Under Chatton's scheme, bacteria were separated from other groups of microorganisms, as well as from plants and animals. In the 1940s two American microbiologists, Roger Y. Stanier and Cornelius B. Van Niel, marshaled extensive evidence to justify the logic of this scheme, making it generally accepted by biologists.

In 1959 Robert H. Whittaker, an American taxonomist, proposed that fungi differ sufficiently from other organisms to justify calling them a separate kingdom. This led to the five-kingdom scheme of classification—animals, plants, fungi, protists (containing protozoa and algae), and monera (bacteria).

As we've seen, biological classification reflects the state of our knowledge. It changes as knowledge expands. Science began to acquire a rich and completely new kind of information in the late 1960s when it became possible to determine the sequences of monomers in macromolecules. This flow of new information has continued at an ever increasing rate.

By the 1970s molecular biologists realized that prokaryotes consist of two different and unrelated groups, as distantly related to each other as they are to eukaryotes. To accommodate this new information three microbiologists, C. Woese, O. Kandler, and M. L. Wheelis, introduced a new classification scheme in 1990. They proposed that all

TABLE 10.1 The Three-Domain Scheme of Biological Classification

Domain	Eucarya	Bacteria	Archaea
Biological groups	Plants, animals, protozoa, algae, fungi	Bacteria	Archaea
Cell type	Eukaryotic	Prokaryotic	Prokaryotic
Contains microorganisms	Yes (some groups)	Yes (all)	Yes (all)

organisms be divided into three major groups called domains: the **Eucarya** (containing all eukaryotes), the **Bacteria** (containing most familiar prokaryotes), and the **Archaea** (originally called **archaebacteria,** containing prokaryotes that live mostly in extreme environments). This scheme is currently accepted by most biologists (**Table 10.1**).

Classification will undoubtedly continue to change. Now let's examine how biologists classify.

METHODS OF MICROBIAL CLASSIFICATION

Out of the early taxonomic studies on eukaryotes came the concept that some **characters** (observable properties) of organisms are more important than others in taxonomy. For example, flowers are more important than leaves in classifying plants. Means of reproduction is particularly important in classifying animals. Determining the most important characters proved very useful for classifying eukaryotes, but such properties weren't discovered in prokaryotes. An alternative approach, called **numerical taxonomy,** was applied to these hard-to-classify organisms.

Numerical Taxonomy

Numerical taxonomy follows one simple principle: It gives equal weight to all characters. Numerical taxonomy compensates for not knowing the important characters by measuring many characters of many strains. Then the taxonomist calculates the percentage of characters that various pairs of strains share. In this way relatedness of all pairs of strains is reduced to a single number, the **similarity coefficient (S_J).**

$$S_j = \frac{\text{The Number of Characters That Two Organisms Share}}{\text{Total Number of Characters Measured}}$$

With this number and the help of a computer, it's possible to construct a **dendogram** (a diagram that illustrates re-

lationships; **Figure 10.5**). A dendogram groups organisms with the highest similarity coefficients. Lines that divide the graph along different similarity coefficients indicate various levels of relatedness. For example, in Figure 10.5, a similarity coefficient of 0.85 might define a species and 0.65 might define a genus. These values vary depending on the organisms being studied and the taxonomists applying them.

Now let's turn our attention to the traditional characters microbiologists employed to classify microorganisms by methods such as numerical taxonomy. Later we'll consider modern molecular methods (**Table 10.2**). Many traditional characters are still used to identify microbial species.

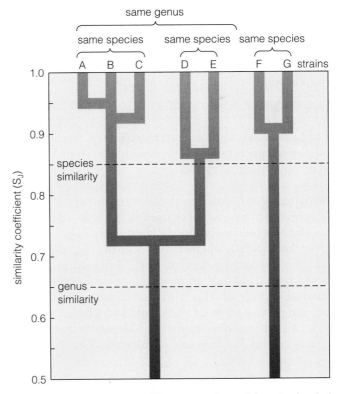

FIGURE 10.5 Dendogram relating seven bacterial strains by similarity coefficient. Setting species limits at 0.85 and genus limits at 0.65 places strains A, B, and C in one species, D and E in another, and F and G in a third. Strains A, B, C, D, and E are in one genus and F and G in another.

TABLE 10.2 Characters Used to Classify and Identify Prokaryotes

Character	Example	Comments
Traditional		
Morphology	Cell shape and arrangement; flagella presence and arrangement; Gram stain	Used to identify genera and species; Gram stain divides bacteria into two groups
Biochemistry and physiology	Conditions that permit growth (pH, temperature, osmotic strength, carbon sources utilized, products of fermentation [acids and gas])	Used to identify species, genera, and higher groups
Serology	Agglutination tests, fluorescent antibodies	Used largely to identify strains and species
Phage typing	Susceptibility to a group of bacteriophages	Used largely to identify strains
Genome Comparisons		
% G + C	Melting point or density of DNA	Values similar in related groups
Hybridization	Amount of annealing between single-stranded DNA from two sources	Whole genome used to define species, probes used to identify species
Sequences of bases in DNA	Complete genome, genes encoding rRNA, genes encoding proteins	Used to determine relatedness of all cellular organisms

Traditional Characters Used to Classify Prokaryotes

Traditionally, taxonomists used morphology, biochemistry and physiology, serology, and phage typing to make decisions about bacterial species.

Morphology. Looking at prokaryotes through a microscope reveals several useful characters: the shape of individual cells; how cells are grouped (singly or in chains, clusters, or regular packets); whether they bear flagella; and, if so, how they are arranged on the cell. Looking at Gram-stained bacteria reveals another significant morphological character. It tells whether a particular bacterium does or does not possess an outer membrane and whether it has a thick or thin peptidoglycan wall. This simple test distinguishes two major divisions of bacteria—Gram-positive and Gram-negative.

Biochemistry and Physiology. Many of the biochemical and physiological characters used to classify prokaryotes are based on conditions that support growth. Do they grow aerobically, anaerobically, or both? What incubation temperature is most favorable? Over what range of pH do

they grow? Are they able to withstand high osmotic strength? What end products and enzymes do they form?

Carbon sources that support growth are a particularly useful set of characters because most bacteria can use many different carbon sources. Whether or not a bacterium can use each particular carbon source becomes a character that can be used to calculate a similarity coefficient. In fact, it's possible to identify many species of microorganisms just by determining which carbon sources they can utilize for growth. Modern techniques make this a relatively easy task. A plastic plate with 96 depressions is prepared, each containing a colorless form of a **tetrazolium dye** (an oxidation/reduction indicator that is colorless when oxidized and colored when reduced) and a different carbon source. A suspension of microbial cells is added. If the organism can use the carbon source, the dye is rapidly reduced and becomes colored. The color pattern on the plate is read by a machine, and results are fed directly into a computer. In a few minutes, the computer displays the name of the microorganism (**Figure 10.6**).

Fermentation properties of bacteria are also useful characters. For example, are acids and/or gases produced during fermentation of a particular carbon source? Each of these products can be easily detected. Gases can be trapped in an

LARGER FIELD

KITS COME TO THE MICRO LAB

It has been said that taxonomy is an art, but identification is a science. Identification is also a rapidly evolving technology. Until about 15 years ago microbiologists made many kinds of media and used pipettes, inoculating loops, and racks of test tubes to identify microorganisms in clinical samples. Today microbiologists often use commercial kits.

Kits usually have small plastic chambers (in place of test tubes) filled with test media (eliminating most media preparation and pipetting). A system for simultaneous inoculation is provided (saving the time of multiple transfers with an inoculating needle). The test media contain indicators and sometimes films to collect gas. After inoculation and appropriate incubation (18 to 24 hours), the technician records color changes and gas accumulation and compares the results with charts. The charts relate patterns of results to specific microorganisms, thus completing the identification.

Some kits are designed to identify members of specific microbial groups, such as enteric bacteria (those related to *Escherichia coli*). One such kit provides a plastic tube divided into 12 chambers. All the chambers can be inoculated simultaneously from a single colony by touching the colony with a wire and drawing the wire up through the tube. Color changes and gas accumulation in the 12 chambers provide 15 test results, enough to identify most enteric bacteria. Color changes indicate whether the bacteria use certain substrates and produce specific metabolic products and enzymes. A wealth of information comes from one small plastic tube.

Some kits are designed to give especially rapid results. They use heavier inocula and depend on the activity of enzymes already present in the cell mass. It's not necessary to wait overnight—let alone a longer period—for growth. Positive or negative tests for one particular organism are also commercially available. They are usually based on immunological or molecular reactions.

A technician observing results from a test kit and entering them into a computer.

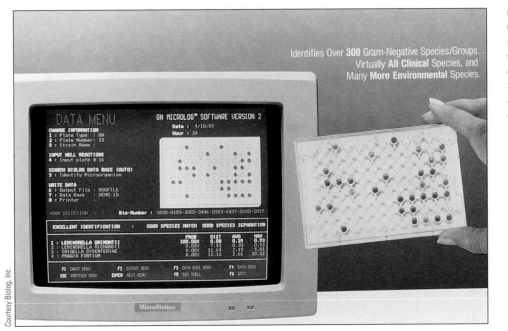

FIGURE 10.6   Identifying bacteria by their use of carbon sources. A suspension of the unknown bacterium is added to 96 wells in a plastic dish. Each well contains a different carbon source and an oxidation/reduction indicator. If the carbon source is metabolized, the indicator is reduced and becomes colored. A machine detects the pattern of colors and feeds the information to a computer, which displays the name of the bacterium.

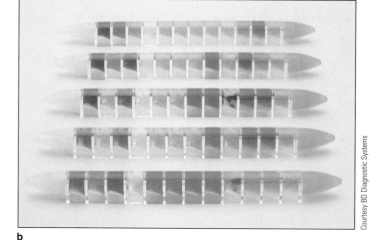

a   **b**

FIGURE 10.7   Fermentation properties of bacteria. (a) Laboratory methods. The uninoc-
ulated tube on the left is clear and red. Acid production changes the color of the pH indicator in the two inoculated tubes to yellow. Gas produced in the right-hand tube accumulates in the inverted tube. (b) Kit for determining fermentation reactions.

inverted tube or in an agar-solidified medium. Acids can be
detected because they lower the pH of the medium. The
lowered pH causes an indicator in growth medium to change
color (**Figure 10.7a**). Now some of these test can be done
easily and quickly using commercial kits (**Figure 10.7b**).

Serology. Another way to classify and identify microor-
ganisms is through **serology** (the science that studies
serum [*pl.*, sera], the noncellular fraction of blood; Chapter 19). Serum contains **antibodies** (protein molecules
made in response to infection; Chapter 17). Antibodies are
highly specific. They target specific microbes. Thus antibodies can distinguish between closely related microorganisms and even between strains of the same species. Sera
that inactivate particular bacteria are called **antisera.**

Antisera can be used in a number of ways. For example, they can be used in the slide agglutination test to
identify particular species or strains. Fluorescent-labeled
antibodies prepared from antisera are also used in a similar but more sensitive procedure. These and other antis-
era-based methods are described in Chapter 19.

Phage Typing. **Bacteriophages,** also called **phages**
(viruses that attack bacteria; Chapter 13), can be used to
classify bacteria. Determining the pattern of bacterial
strains attacked by a set of bacteriophages is called **phage
typing.** The diversity of bacterial strains that one bacteriophage can attack (termed the **host range**) is quite
narrow. Because only closely related bacterial strains are
attacked by the same phages, phage typing is usually restricted to classifying strains within species.

Phage typing is done by spreading a thin layer of the
bacterial strain on the surface of an agar plate and then

placing small drops of suspensions of various phages on it.
If the bacterial cells are susceptible to the phages in one or
more of the drops, they will lyse, producing a clear zone in
the **lawn** (confluent layer of bacterial growth) that forms
on incubation (**Figure 10.8**).

Strains with identical phage types are probably identical. Strains with similar phage types are closely related. Because phage typing can identify specific strains of bacteria,
it is used to determine if a cluster of bacterial infections is
caused by the same strain. Such information is useful
for determining the source and spread of a **nosocomial**
(hospital-acquired) infection (Chapter 20).

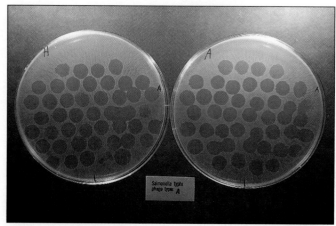

FIGURE 10.8   Petri dish used for phage typing. Suspensions of
various phages are dropped on a lawn of bacteria in an array that
identifies the phages. Clearing at the area where a phage was
placed indicates susceptibility of the bacterial strain to that particular phage. The pattern of clearing shown here identifies the strain
of *Salmonella typhi* being tested as belonging to phage type A.

GOOD QUESTION

The mole percentage of guanine plus cytosine (% G + C) in the DNA of various species has a considerable range. Some ciliate protozoa have a % G + C as low as 21 and some bacteria as high as 79. But in any one species the % G + C is essentially the same throughout the genome. If it varies among species, why doesn't it vary within a genome?

The answer is that, over a wide range, one value of % G + C is as good as another because the genetic code is redundant. For example, the UUC and UUA codons both specify the same amino acid, leucine. Frequent use of the UUC codon in place of the UUA codon raises the % G + C without affecting the protein being encoded. But the protein synthesizing machinery evolves to match the DNA's % G + C. In organisms with high % G + C, more tRNA molecules recognize UUC than recognize UUA. For this reason, % G + C tends not to change throughout one organism's genome.

This knowledge is important to biotechnology because when genes from high % G + C organisms are cloned into low % G + C organisms and vice versa, the genes are poorly expressed (protein is made slowly). The host cell's protein synthesizing machinery is ill suited to express DNA with a different % G + C than its own.

Now let's turn to the methods that dominate classification today—namely, comparing genomes.

Comparing Genomes

Taxonomists have long known that the key to taxonomy lies in the organism's genome. The traditional methods depend on observing expression of only a small portion of the genome. Most phage typing and serology methods, for example, depend on the product of a single gene. Therefore they examine only a minute fraction of the organism's genome. In the 1960s taxonomists began to develop methods to examine greater portions of the genome. In this section we'll consider the progress that has been made and its impact on classification. We will start with percent G + C and hybridization, which are highly indirect methods of evaluating an organism's genome. Then we will proceed to the most direct method possible—**sequencing DNA** (determining the sequence of bases in the DNA of an organism's genome).

Percent G + C. In the 1960s it became possible to determine the fraction of total base pairs in DNA (Chapter 6) that are guanine-cytosine (G-C) pairs, as opposed to adenine-thymine (A-T) pairs. This fraction varies widely among the DNA from various organisms. By convention, the fraction of G-C pairs is expressed as **mole percent guanine plus cytosine** or **% G + C.** Thus if % G + C of a sample of DNA is 40, 40 percent of its base pairs are G-C pairs and the other 60 percent are A-T pairs.

The % G + C of DNA affects its physical properties. Because three hydrogen bonds join G-C pairs and only two join A-T pairs, higher values of % G + C mean the two strands of DNA are joined by more hydrogen bonds. This increased bonding holds the strands more tightly together, making them more difficult to **melt** (separate by heating). Thus the **melting point** (temperature at which the strands separate) of a sample of DNA can be used to determine its % G + C. Measurements of density can also be used, because the density of a DNA sample is proportional to its % G + C.

The % G + C values can indicate the relatedness of organisms. These values vary enormously and even more widely in microorganisms than in plants and animals (**Table 10.3**). Still, % G + C values are useful because

TABLE 10.3 The Range of % G + C in DNA from Various Organisms

Organisms	Range of % G + C
Microbial Groups	
Prokaryotes	
Bacteria	25 to 78
Archaea	27 to 62
Eukaryotes	
Algae	37 to 68
Protozoa	21 to 65
Fungi	22 to 62
Nonmicrobial Groups	
Plants	33 to 48
Animals	32 to 50

they vary less within a species than within a genus. Thus closely related organisms have similar % G + C values. However, similar % G + C values do not prove that two organisms are closely related, because % G + C is unrelated to the coding properties of DNA. For example, humans and the bacterium *Bacillus subtilis* have almost identical % G + C values.

Better methods of examining the genome soon became available. The next major advance was DNA hybridization.

DNA Hybridization.

DNA hybridization (**Figure 10.9**) depends on the properties of the hydrogen bonds that hold stands of DNA together (Chapter 2). First, mild heating breaks these bonds, causing the DNA to "melt" and allowing the two strands of the double helix to separate. Second, on cooling, the hydrogen bonds spontaneously reform, causing the two single strands to **anneal** (be rejoined). In DNA hybridization, single strands of DNA from different sources are mixed and they anneal in regions where the sequences are the same or similar, creating a hybrid molecule. Therefore the extent of annealing is a quantitative index of the similarity of base sequences in the DNA from the two sources.

Total Genome Hybridization. Taxonomists hybridize the total genome of organisms to determine their relatedness. The procedure is relatively simple and precise, but it has drawbacks. First, for annealing to occur, the two samples of DNA—and hence the organisms from which they were obtained—must be similar. As a general rule, total genome hybridization gives useful information only about organisms in the same genus. Second, such hybridization tells how closely one organism is related to one other organism. To determine the relatedness to a group of organisms, many hybridizations must be made.

In spite of these difficulties, total genome hybridization has become the accepted standard for determining the limits of a prokaryotic species. The currently accepted limits of bacterial species are 70 percent homology or less than 5°C lowering of DNA melting point as determined by hybridization.

Hybridization has also proven to be extremely useful for identifying particular microorganisms using short pieces of tagged DNA called **probes.**

Probes. DNA probes are chosen to hybridize only with the DNA of specific organisms (**Figure 10.10**). Probes made by recombinant DNA technology (Chapter 7) are tagged with a fluorescent dye, a chemiluminescent molecule, or radioactive atoms so hybridization can be detected easily. The method can be used to identify colonies on a petri dish by lysing cells and adding the probe. If the probe hybridizes, the colony is identified.

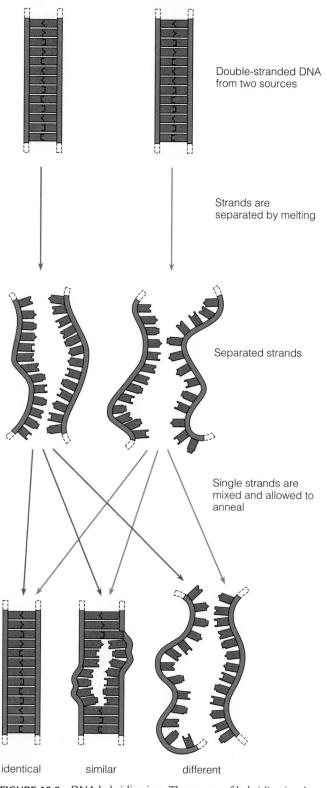

Double-stranded DNA from two sources

Strands are separated by melting

Separated strands

Single strands are mixed and allowed to anneal

identical   similar   different

FIGURE 10.9   DNA hybridization. The extent of hybridization between complete genomes from pairs of prokaryotes is used to define species. Ability of short pieces DNA (called probes) to hybridize to an organism's DNA is used to identify species.

LARGER FIELD

ESCHERICHIA COLI'S *FAMILY TREE*

The trees that scientists have developed from sequencing genes encoding SSU rRNA tell us which organisms evolved most recently from a common ancestor. But they don't tell us when. Until recently, evolutionary events could be dated only by the fossil record. Fossils are dated by the age of the rocks they are found in. Evolutionary events are deduced by comparing fossils of different ages with organisms alive today.

But could the sequence of bases in rRNA tell us dates in evolutionary history? It could if evolutionary changes occur at a constant rate. Then changes in rRNA would be a molecular clock ticking off evolutionary time and recording it. To test the possibility that changes in rRNA occur at a constant rate and, if so, set the molecular clock, some key evolutionary events had to be dated. Some were dated by the fossil record and other older events by relating them to geological change. For example, the time that oxygen appeared in the earth's atmosphere (which is recorded by changes in rocks) must have been when aerobes evolved.

Applying these tests showed that the rate of change in the sequence of rRNA is a pretty good molecular clock. By setting the molecular clock this way, base differences between a pair of organisms can be converted into years because they had a common ancestor. The method applies even to microorganisms, providing some fascinating insights. For example, it establishes that *Escherichia coli* separated from its close relative, *Salmonella typhimurium*, about 120 million years ago, the same time mammals first appeared. In short, *E. coli* evolved with mammals. It is well adapted to grow in the intestines of warm-blooded animals. It is resistant to bile, a fat-emulsifying liquid produced by the liver and toxic to many microorganisms. Finally, it can use lactose, a sugar produced only by mammals, as a carbon source in milk. The capacity to use lactose is relatively rare in bacteria, but in view of *E. coli*'s evolutionary history, we should not be surprised.

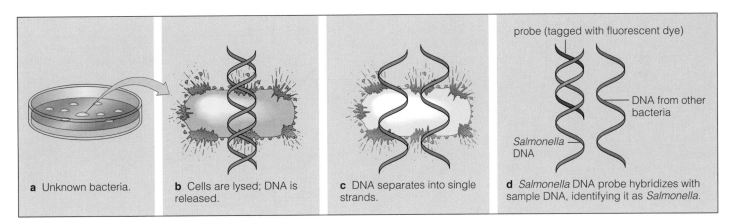

a Unknown bacteria.

b Cells are lysed; DNA is released.

c DNA separates into single strands.

d *Salmonella* DNA probe hybridizes with sample DNA, identifying it as *Salmonella*.

probe (tagged with fluorescent dye)

DNA from other bacteria

Salmonella DNA

FIGURE 10.10 Using a DNA probe to identify bacteria.

Increasing the Sensitivity of Probes. Until quite recently, probes weren't used routinely to identify microorganisms in clinical samples because they were too insensitive to detect the few microorganisms they usually contain. That changed when highly sensitive, rapid probe methods were developed (**Figure 10.11**). Increased sensitivity is gained by amplifying the amount of DNA (usually the DNA [rDNA] that encodes ribosomal RNA [rRNA]) that's present in the clinical sample. Rather than depending on the few copies of rDNA that each cell contains, the method starts with the several thousand copies of rRNA that are present. Then using a mixture of enzymes and primers, the number of copies of rRNA are increased about a billionfold and converted into DNA, to which the probe binds. This method is sensitive enough to detect a single microbial cell or viral particle. The method is somewhat reminiscent of the polymerase chain reaction (PCR) we discussed in Chapter 7, but there are distinct differences. Because reverse transcriptase

FIGURE 10.11 Method for amplifying ribosomal RNA and converting it to DNA for use as a probe target.

(Courtesy Craig S. Hill, Gen-Probe Transcription-Mediated Amplification: System Principles, Gen-Probe Incorporated, 10210 Genetic Center Drive, San Diego, CA, 92121.)

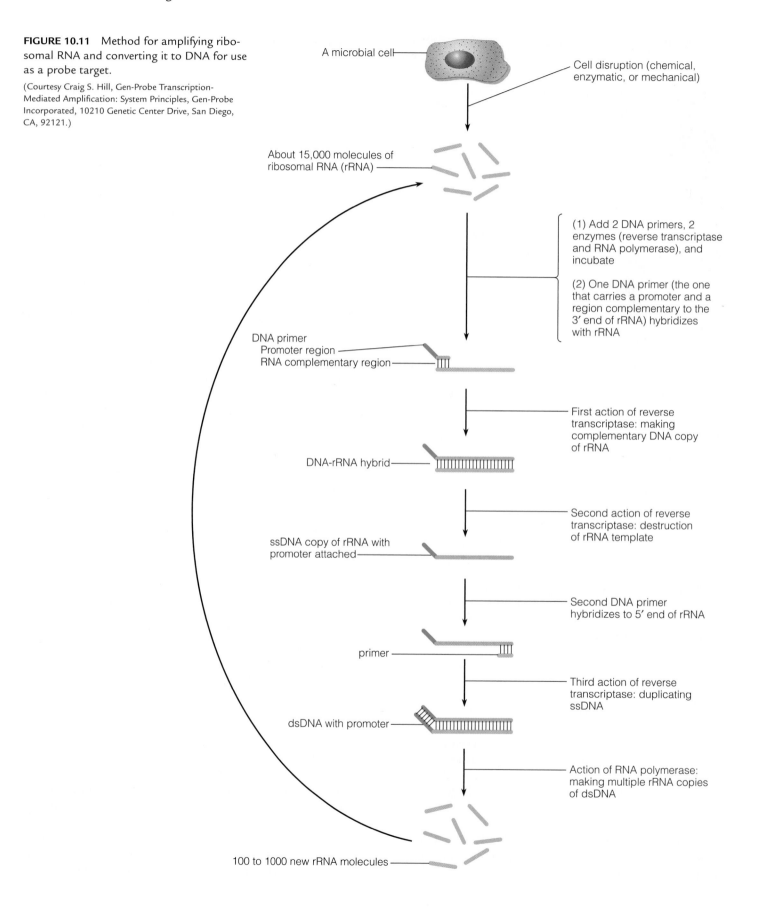

A microbial cell

Cell disruption (chemical, enzymatic, or mechanical)

About 15,000 molecules of ribosomal RNA (rRNA)

(1) Add 2 DNA primers, 2 enzymes (reverse transcriptase and RNA polymerase), and incubate

(2) One DNA primer (the one that carries a promoter and a region complementary to the 3' end of rRNA) hybridizes with rRNA

DNA primer
Promoter region
RNA complementary region

First action of reverse transcriptase: making complementary DNA copy of rRNA

DNA-rRNA hybrid

Second action of reverse transcriptase: destruction of rRNA template

ssDNA copy of rRNA with promoter attached

Second DNA primer hybridizes to 5' end of rRNA

primer

Third action of reverse transcriptase: duplicating ssDNA

dsDNA with promoter

Action of RNA polymerase: making multiple rRNA copies of dsDNA

100 to 1000 new rRNA molecules

(Chapter 13)—which regenerates ssDNA—is used, there is no need to repeatedly heat the sample in order to melt dsDNA. Because RNA polymerase is used, 100 to 1000 RNA copies are made from each molecule of dsDNA.

This amplified probe method is rapidly gaining favor in clinical laboratories, replacing well-established methods, such as culturing to detect gonococcal and group A streptococcal infections.

Now let's turn to the most powerful method for determining relatedness: determining the sequence of monomers in macromolecules.

Sequencing Macromolecules.

Since the mid-1960s, scientists realized that the sequence of monomers in macromolecules (amino acids in proteins or nucleotides in RNA or DNA) were rich sources of information. The U.S. chemist Linus Pauling referred to macromolecules as "documents of evolutionary history." At their beginning these studies relied on determining the sequence of amino acids in particular proteins, for example, the enzyme cytochrome C. But this approach did not proceed very far because determining the sequence of amino acids in proteins is laborious. In the late 1970s it became feasible to determine the sequence of nucleotides in ribosomal RNA; the results established relationships between distantly related organisms. By the mid-1980s it became practical to deduce the complete sequence nucleotides in ribosomal RNA by sequencing their encoding RNA (rDNA). These studies revealed the relationships between all cellular organisms (see box, The Universal Tree of Life) and became the foundation of modern taxonomy. Ribosomal RNA (rRNA) is generally agreed to be the best one. Now let's consider why this is true.

Ribosomal RNA Genes. The sequence of bases in rRNA has been remarkably successful in revealing relationships among all organisms—bacteria, archaea, and eukaryotes. In fact, studies on the sequences of bases in rRNA first revealed that the archaea are a separate major biological group, as different from bacteria as they are from eukaryotes. For the most part the small ribosomal subunit (SSU rRNA) has been sequenced. Such sequences for thousands of organisms are now known.

How can SSU rRNA show relationships over great evolutionary distances? First, all cellular organisms have this gene product. Second, during evolution, rRNA changes much more slowly than most macromolecules. Because the structure of the ribosome cannot tolerate much change and still remain functional, ribosomal RNA is highly conserved. Therefore the rRNA of even distantly related organisms is similar enough so we can count the number of differences between them and thus calculate the evolutionary distance between them. Using the DNA sequences of SSU rRNA from different organisms to determine relatedness between

organisms is based on the assumption that the molecules in all organisms have a common origin: They were originally identical. As one species evolves into another, the DNA sequences encoding rRNA change, but it continues to serve the same function. The similarity of DNA sequences encoding a molecule such as SSU rRNA is therefore a direct measure of the relatedness of the organisms that produced those molecules. Conversely, the number of differences in their DNA sequences is a measure of evolutionary distance between a pair of organisms.

Taxonomists have attempted to extract even more information from comparisons of DNA sequences by assuming that changes occur at a constant known rate. This assumption allows them to calculate the time in the past that two organisms evolved from a common ancestor. Over the past decade and a half, studies on the sequence of bases in rRNA have revolutionized our understanding of the relationships among microorganisms, as we'll see in the next three chapters.

Protein-encoding Genes. The relatedness of organisms can also be determined by comparing the sequence of bases in genes that encode proteins. This is possible because every organism has a few proteins that serve the same function in all organisms. Sequencing the genes that encode these proteins gives information much like that gained from sequencing SSU rRNA. For the most part this information supports the conclusions drawn from studies on SSU rRNA.

Characters Used to Classify Viruses

The primary character used to classify viruses is the kind of nucleic acid that constitutes the virus genome—whether it is DNA or RNA, single stranded or double stranded. Other characters are appearance in an electron micrograph, whether the virus is surrounded by a membrane, host range, and serology. For example, the rhabdoviruses, which include the rabies virus, are composed of single-stranded RNA. They are bullet shaped and surrounded by a membrane. They attack many animal species, and serum from an animal that survives an infection will protect other animals. In Chapter 13 we'll discuss these characters in detail.

Dichotomous Keys

Taxonomists sometimes construct **dichotomous keys,** which consist of a series of questions that offer only two alternatives, to simplify identification of unknown organisms. Dichotomous keys are arranged somewhat like a game of Twenty Questions. First, general questions are asked to divide organisms into two large groups. Then

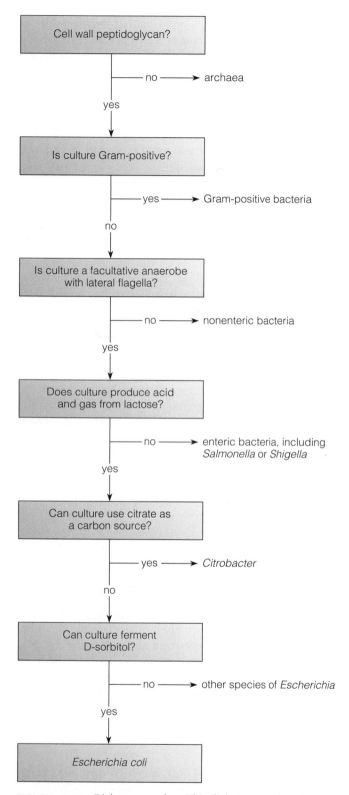

more specific questions divide organisms into smaller groups. Finally, highly specific questions distinguish between similar species. **Figure 10.12** gives an example of a dichotomous key that divides prokaryotes into smaller groups and eventually identifies *Escherichia coli*.

Dichotomous keys are helpful but not perfect. Relying on only a small number of characters can lead to incorrect or equivocal conclusions because rarely is any single character the same in all members of a species. For example, the last step of the key in Figure 10.12 to distinguish *E. coli* from other species of *Escherichia* depends on a character that is present in 94 percent of *E. coli* strains. In other words, 6 percent of the time the key will give an incorrect answer.

FIGURE 10.12   Dichotomous key. This dichotomous key divides prokaryotes into smaller groups. The answers shown lead to identifying an unknown prokaryote as *Escherichia coli*.

SUMMARY

Principles of Biological Classification (pp. 240–244)

1. Taxonomy is the science of classifying organisms. Biological taxonomists use a hierarchical scheme of classification: They collect individuals into groups and groups into progressively more inclusive and broader groups.

2. Biologists use the hierarchical classification scheme devised by Linnaeus: species, genera, families, classes, orders, phyla or divisions, and kingdoms. Kingdoms may be grouped into domains.

Scientific Nomenclature (p. 241)

3. Linnaean classification uses binomial nomenclature. The first name is the organism's genus, and the second is its specific epithet. Together the two constitute the species name.

Artificial and Natural Systems of Classification (p. 241)

4. Linnaeus's system was an artificial system of classification because it grouped organisms on the basis of visible similarities. A natural scheme of classification is based on phylogenetic (evolutionary) relatedness.

The Fossil Record (pp. 241–242)

5. Most natural systems of classification depend on information gained from the fossil record. Only recently have scientists found microbial fossils. Stromatolites are fossilized photosynthetic prokaryotes that grew as microbial mats. These fossils establish that microorganisms have been on the earth for about 3.5 billion years.

The Concept of Species (pp. 242–244)

6. A eukaryotic species is a group of organisms that share a common pool of genes. A bacterial species is defined by the similarities of its members.

7. Clones that are presumed or known to be genetically different are called strains.

8. Viral species are grouped into genera and genera into families. No higher taxa are used. They are given ordinary English names. Viral species are also divided into strains.

Microorganisms and Higher Levels of Classification (pp. 244–245)

9. Initially microorganisms were grouped with animals or plants. In 1866 Ernst Haeckel proposed three kingdoms—plants, animals, and microorganisms (or protists).

10. In 1937 Chatton divided all living things into eukaryotes and prokaryotes.

11. In 1969 Whittaker proposed the five-kingdom scheme—animals, plants, fungi, protists, and monera.

12. In 1990 Woese, Kandler, and Wheelis proposed three domains—archaea, bacteria, and eucarya (eukaryotes).

Methods of Microbial Classification (pp. 245–254)

13. Some characters are more important than others in classification of eukaryotes.

Numerical Taxonomy (pp. 245–246)

14. Numerical taxonomy gives equal weight to all characters (properties) of organisms. It aims to express the evolutionary distance between organisms in a number—the similarity coefficient (S_J).

15. Charts called dendograms group organisms with the highest similarity coefficients.

Traditional Characters Used to Classify Prokaryotes (pp. 246–249)

16. The shape and arrangement of bacterial cells and flagella are useful characters. The Gram stain divides bacteria into two groups, those with and those without an outer membrane.

17. Some of the biochemical and physiological characters used to classify bacteria are based on conditions that support growth.

18. Serology (the study of the properties of serum) is another traditional method of classifying bacteria. Antibodies contained in serum are highly specific and thereby distinguish between closely related strains and species.

19. Phage typing can be used to identify strains of bacteria.

Comparing Genomes (pp. 249–253)

20. The mole percent guanine plus cytosine (% G + C) expresses the percentage of total base pairs in DNA that are G-C, as opposed to A-T. Higher % G + C confers a higher DNA melting point and density. Closely related organisms have similar % G + C values.

21. DNA hybridization involves melting DNA strands from separate sources, mixing them, and cooling, so the two strands anneal, creating a hybrid. The extent of annealing indicates the extent of similarity in DNA bases. Sometimes a probe (a specific short piece of DNA) is used.

22. The ultimate tool of taxonomy is DNA sequencing—determining the sequence of bases in DNA.

23. Sequencing bases in genes that encode rRNA reveals relationships between all organisms.

24. Sequencing bases in genes that encode certain proteins also reveals relationships between distantly related organisms.

Characters Used to Classify Viruses (p. 253)

25. Viruses are classified according to the kind of nucleic acid that constitutes their genome—DNA or RNA, single or double stranded.

26. Viruses are also classified on the basis of morphology and host range.

Dichotomous Keys (pp. 253–254)

27. Dichotomous keys are a series of questions about characters that offer only two alternatives.

28. Dichotomous keys can lead to incorrect conclusions because a single character is rarely the same in all members of a species.

REVIEW QUESTIONS

Principles of Biological Classification

1. Define these terms: classification, hierarchical scheme of classification, taxonomy.

2. Arrange these taxa from the smallest group to the largest:
   a. order   e. family
   b. kingdom   f. species
   c. genus   g. class
   d. phylum or division h. domain

3. How is species named? Give an example.

4. What is the difference between an artificial and a natural system of biological classification? What is the fossil record? What is the significance of stromatolites?

5. How are eukaryotic species defined? How are bacterial species defined?

6. What is a strain?

7. On what basis are viral species defined?

Microorganisms and Higher Levels of Classification

8. Explain this statement: Biological classification schemes change as we get new information. Use schemes developed by Haeckel, Chatton, Whittaker, and Woese, Kandler, and Wheelis as examples.

9. Why are viruses not included in biological classification schemes?

Methods of Microbial Classification

10. How is an organism classified according to numerical taxonomy?

11. Explain how each of these traditional characters is used to classify bacteria:
    a. morphology   c. serology
    b. biochemistry   d. phage typing
       and physiology

12. Explain how each of these characters is used to classify bacteria:
    a. % G + C
    b. DNA hybridization
    c. sequencing DNA bases

13. Describe the characters used to identify viruses.

14. What is a dichotomous key? What are its advantages and disadvantages?

CORRELATION QUESTIONS

1. The word *strain* is used to describe a subdivision of a bacteria species. What word would you use to describe an equivalent subdivision of the human species? Explain.

2. What is the inherent weakness of a dichotomous key?

3. Why was the rate of evolution of eukaryotes known before anything was known about the rate of evolution of prokaryotes?

4. Do you think fossils of prokaryotes will be found that are significantly older than those we now have? Explain.

5. Explain why it is that bacteria exchange genes but bacterial species do not necessarily share a gene pool.

6. Why have biochemical and physiological characters been used more often to classify bacteria than to classify plants?

ESSAY QUESTIONS

1. Discuss what properties a character must have to be useful in revealing relationships among organisms in different kingdoms.

2. Marshal evidence supporting the theory that all living things are related.

SUGGESTED READINGS

Doolittle, R. F. 1998. Microbial genomes opened up. *Nature* 392:339–42.

Margulis, L., and K. V. Schwartz. 1988. Five kingdoms: An illustrated guide to the phyla of life on earth. San Francisco: W. H. Freeman.

Rosselló-Mora, R., and R. Amann. 2000. The species concept for prokaryotes. *FEMS Microbiology Reviews* 25:39–57.

Sneath, P. H. A., and R. R. Sokal. 1973. Numerical taxonomy: The principles and practice of numerical classification. San Francisco: W. H. Freeman.

Woese, C. R. 1987. Bacterial evolution. *Microbiological Reviews* 51:221–71.

Woese, C. R., O. Kandler, and M. L. Wheelis. 1990. Towards a natural system of organisms: Proposal for the domains archaea, bacteria, and eukarya. *Proceedings of the National Academy of Science* 87:4576–79.

For additional readings, go to InfoTrac College Edition, your online research library at: http://www.infotrac.thomsonlearning.com

ELEVEN

The Prokaryotes

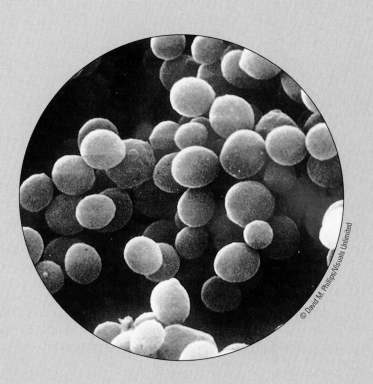

© David M. Phillips/Visuals Unlimited

CHAPTER OUTLINE

LEARNING GOALS

To understand:

- *Natural schemes of classifying prokaryotes*
- *The* Bergey's Manual *scheme of classifying prokaryotes*

To become familiar with:

- *The better-known prokaryotic genera and species in* Bergey's Manual
- *The properties and activities of these organisms*

- *The impact of these organisms on our lives and our environment*

Growing a Variety of Bacteria

I. B., a 2-year-old female with a history of repeated urinary tract infections, was taken to her pediatrician with a temperature of 101°F. X-ray studies done when she was a baby had revealed **vesico-ureteral reflux,** a condition in which some urine flows backward toward the kidneys, rendering patients prone to repeated infection by a variety of bacteria. The causes of these infections represent a sampling of bacteria in the patient's environment. I. B. had suffered several such infections, one sufficiently serious to require hospitalization. To prevent such infections, I. B. took daily small doses of an antibiotic. But the infections continued, and I. B.'s parents promptly took her to her pediatrician whenever she developed a fever, even a mild one.

In spite of her elevated temperature I. B. looked well, was playful, and showed no signs of a cold or flu-like illness. Because I. B. was just learning to talk, her pediatrician couldn't ask about the usual symptom of urinary infection: Did she suffer burning on urination? The only way to know if she had another urinary tract infection was to obtain a sample of urine, analyze it, and culture it.

Fortunately I. B. was able to urinate into a cup. Microscopic observation of her urine in the clinical laboratory strongly suggested an infection: Her urine contained a few bacterial cells and many white blood cells. Her pediatrician wanted to begin treatment immediately, before she became more seriously ill. But although he was sure that this infection was caused by a bacterium (urinary tract infections of this nature are always bacterial), he had no way to guess which bacterium it might be. *Escherichia coli* is the most common, but *Klebsiella*, *Proteus*, or *Enterobacter* can also be the culprit. And even if her physician had known which bacterium it was, he couldn't be sure which antibiotic would provide the best treatment.

To answer these important questions, I. B.'s doctor sent a sample of her urine to the microbiology laboratory for culture and sensitivity testing. Because the bacteria had to be cultivated in the laboratory, it would be 2 days before he had the results of these tests. In the meantime, he had to make a guess about which antibiotic was statistically most likely to cure her infection based on information obtained from past cases in his community. He prescribed an oral dose of cephalexin three times a day and asked I. B.'s parents to watch her closely and call if her fever went higher or if she seemed more listless or reluctant to eat or drink.

Two days later I. B.'s physician called the microbiology laboratory to ask a technician to look at the petri plates and tell him as soon as possible about the antibiotic sensitivity of the infecting organism. The result of the culture was *Proteus mirabilis*, which was sensitive to cephalexin. By the time the physician called to tell I. B.'s family the result of her culture, the child had already recovered completely from her infection. She had no fever; life had returned to normal.

Case Connections

- In this case we'll learn about the various candidate bacteria for causing I. B.'s urinary tract infection, their characteristics, and how they are identified.
- We see that all the candidate bacteria belong to the bacterial phylum Proteobacteria and the class Gammaproteobacteria and what such a relationship implies.

PROKARYOTIC TAXONOMY

Results of DNA sequencing have completely changed prokaryotic taxonomy. Such information has allowed taxonomists to construct natural classification schemes that most probably show the true relationships among all organisms, including prokaryotes. The most complete and widely accepted of these natural schemes of classification is based on the sequence of bases in 16S ribosomal RNA. Such sequences from thousands of prokaryotes are now known. This wealth of data has allowed taxonomists to construct a genealogical tree of prokaryotes (**Figure 11.1**). The tree shows that prokaryotes can be divided into two domains—Archaea and Bacteria. The domain Archaea can be subdivided into 2 phyla and the domain Bacteria into 23 phyla (only 8 are shown).

Microbiologists have hoped for such a genealogical tree of prokaryotes since the time of Darwin. Having it has been a major boon to research microbiologists, who have used it for the past decade and a half to correlate their results and

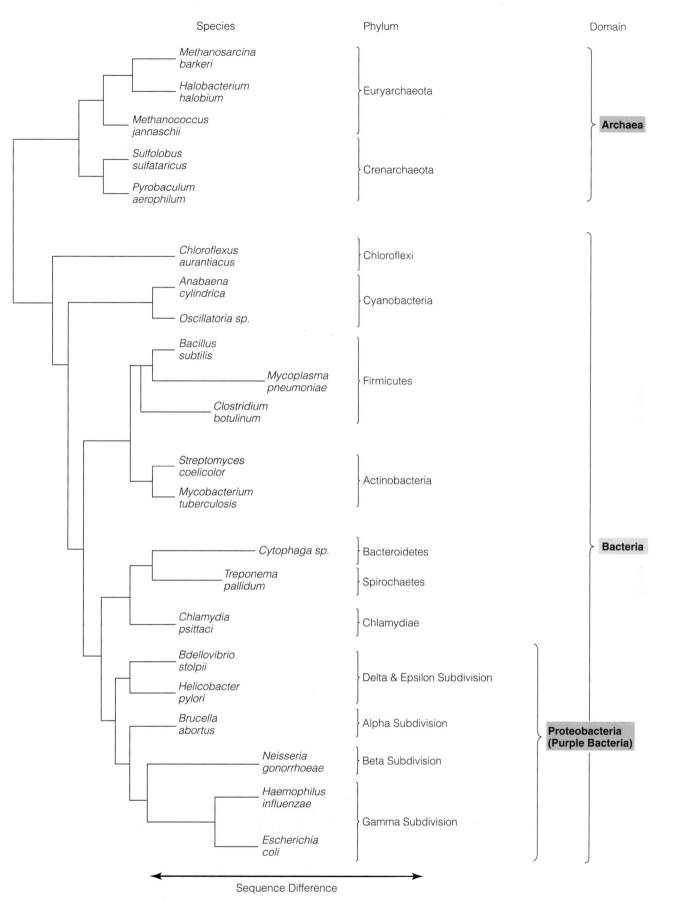

FIGURE 11.1 The genealogical tree of prokaryotes. The position of species on the tree reflects differences in the sequences of their rRNA. Vertical lines join species or groups of species at the level when there were no differences—that is, when now-existing species shared a common ancestor. For example, the line joining all species of Archaea to all species of Bacteria is at the base of the tree (vertical line, far to the left). The position of this line indicates that the 16S RNA sequences of all members of one of the domains are quite different from the sequences of all members of the domain. Thus the two groups are only distantly related. In contrast, we see within the domain Bacteria that *Bacillus subtilis* and *Mycoplasma pneumoniae* are closely related. *M. pneumoniae* is farther from the closest vertical line than *B. subtilis* is, indicating that *M. pneumoniae* has evolved more than *B. subtilis* since they shared a common ancestor.

make predictions. But only in 2001 was this information incorporated into **Bergey's Manual,**[1] the scheme that most microbiologists use on a daily basis. *Bergey's Manual* also contains the authoritative listing of recognized prokaryotic species.

In this chapter we'll consider some clinically and ecologically important groups of prokaryotes, as well as certain groups of prokaryotes that have unusual properties and are not discussed elsewhere in this text. We'll use the latest *Bergey's Manual* as a framework for our discussion.

THE *BERGEY'S MANUAL* SCHEME OF PROKARYOTIC TAXONOMY

Two (Archaea and Bacteria) of the biological world's three domains share the same prokaryotic cell structure. Nevertheless, they differ from each other and members of the third domain (Eucarya or eukaryotes) in many respects (**Table 11.1**), and they are not closely related to each other. In fact, Archaea are somewhat more closely related to eukaryotes than they are to the Bacteria. *Bergey's Manual* divides the Archaea into 2 phyla and the Bacteria into 23 (**Table 11.2**). We will discuss some representative species in this chapter.

DOMAIN ARCHAEA

The domain Archaea consists of two phyla, the Crenarchaeota and the Euryarchaeota.

Al Crenarchaeota

The Crenarchaeota is a relatively small phylum consisting of a single class (Thermoprotei). Representatives are morphologically diverse: They can be rods, cocci, filaments, or disc shaped. They are also metabolically diverse: aerobic, facultatively anaerobic, strictly anaerobic, chemolithoautotrophs, or heterotrophs. But they all share the property of being ex-

---

[1]*Bergey's Manual* is the term commonly used for either of two sets of volumes: *Bergey's Manual of Determinative Bacteriology* and *Bergey's Manual of Systematic Bacteriology.* The former has gone through nine editions, the first published in 1923, the ninth in 1994. The latter, which contains more detailed taxonomic and biological information, was first published in four volumes appearing sequentially between 1984 and 1989. The first of five volumes of the second edition was published in 2001.

tremely thermophilic, and most of them metabolize sulfur. The most highly thermophilic known organism, *Pyrolobus fumarii* (Chapter 8), belongs to this phylum. Members of one genus of the Crenarchaeota, *Sulfolobus*, have oddly irregular lobed cells (**Figure 11.2**).

A2 Euryarchaeota

The Euryarchaeota is a large phylum consisting of five classes, which include just about every conceivable cell shape, including spiral and triangular. One class, the Thermoplasmata, contains all the wall-less archaea. The phylum contains three important metabolic types, the methanogens, the extreme halophiles, and the thermoacidophiles, which we'll discuss in turn.

The Methanogens. The **methanogens** (methane formers) make **natural gas** (methane, CH_4) from hydrogen gas (H_2) and usually either carbon dioxide (CO_2) or acetate (CH_3-COO^-). This conversion occurs only in anaerobic environments where H_2, CO_2, and acetate are produced by other species of bacteria.

The methanogens include many genera, and they are everywhere. So if organic material is put in an anaerobic environment, methane will almost certainly form. This is particularly likely at the bottom of a body of water, where bacteria that metabolize organic compounds quickly create anaerobic conditions. Bubbles rising from a quiet pond usually contain methane formed by methanogens in the anaerobic mud at its bottom. Cows and other ruminants produce prodigious amounts of methane in their rumen, which they belch into the atmosphere. Some humans har-

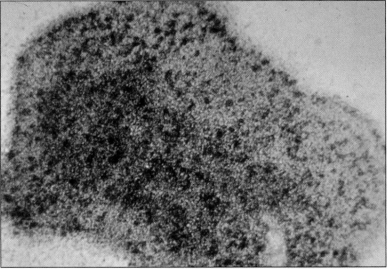

FIGURE 11.2 *Sulfolobus* (A1 Crenarchaeota). Note its irregular lobed shape.

TABLE 11.1 Major Distinctions Between Prokaryotes and Eukaryotes, and Between Archaea and Bacteria

Characteristic	Prokaryotes	Eukaryotes
Membrane-surrounded nucleus	Absent	Present
Usual size of smallest cell dimension	0.2 to 2.0 μm	2.0 μm or larger
Mitochondria	Absent	Usually present
Chloroplasts in phototrophs	Absent	Present
Gas vacuoles	Present in some	Never present
Golgi apparatus, lysosomes, microtubular system, endoplasmic reticulum	Absent	Usually present
Cytoplasmic streaming, pseudopodal movement, endocytosis, exocytosis	Absent	Often present
Location of ribosomes	Dispersed in cytoplasm	Attached to endoplasmic reticulum
Size of ribosomes	70S	80S (except in mitochondria and chloroplasts)
Diameter of flagella	~0.02 μm	~0.2 μm
Endospores	Present in some	Never present
Sterols in membranes	Absent (except mycoplasmas)	Common
Location of respiratory functions	Cytoplasmic membrane	Mitochondria
Number of chromosomes	Usually one	More than one

Characteristic	Archaea	Bacteria
Form methane as predominant end product	Some	None
Extremely thermophylic (growth at 110°C)	Some	None
Cell walls contain muramic acid	None	Most
Susceptible to β-lactam antibiotics	None	Many
Some genes contain introns	All	None
Membranes contain fatty acids that are ester-linked to glycerol	None	All
Contain rifampin-sensitive RNA polymerase	None	All

bor methanogens. Metabolism by methanogens is also put to practical use to produce fuel from sewage (Chapter 28).

The Extreme Halophiles. Many prokaryotic species are **halophiles** (salt lovers), but the halophilic archaea, such as *Halobacterium halobium*, set the record for tolerating high concentrations of salt. These organisms can't grow with less than 10 percent sodium chloride (NaCl), and they grow best with more than twice that amount. These bright-red archaea flourish even in saturated salt solutions.

The halophilic archaea are aerobes that generate most of their adenosine triphosphate (ATP) by respiration, but they can also carry out a primitive form of photosynthesis. Their cytoplasmic membranes contain patches called **purple membranes,** which contain the pigment **bacteriorhodopsin.** (Bacteriorhodopsin is similar to rhodopsin, the light-sensitive pigment in the eyes of vertebrates.) Bacteriorhodopsin, acting with a cofactor, **retinal,** converts light into chemical energy by using light energy to expel a proton from the cell, thus creating a proton gradient that generates ATP as the proton reenters the cell.

The Thermoacidophiles. The **thermoacidophiles** (heat and acid lovers) thrive in extremely hostile environ-

TABLE 11.2 Phyla of Prokaryotes

Phylum	Properties	Representative Species	Features (Chapter)
Domain Archaea			
A1 Crenarchaeota	Most are obligately thermophilic; most metabolize elemental sulfur	*Sulfolobus acidocaldarius*	Lives in acid-hot environments, such as hot springs in Yellowstone (11)
A2 Euryarchaeota	Contains methane-forming, extremely halophilic, wall-less, sulfate-reducing, and extremely thermophilic sulfur metabolizing species	*Metanobacterium formicicum*	Produces methane gas from hydrogen gas and carbon dioxide; found in fresh water sediments, marshy soils, and the rumen of cattle and sheep (11)
Domain Bacteria			
B1 Aquificae	Gram-negative, nonsporulating, thermophilic rods or filaments.	*Aquifex pyrophilus*	Occurs in marine hydrothermal vents, maximum temperature for growth 95°C, chemolithotroph
B2 Thermotogae	Gram-negative, nonsporulating rods with a sheathlike outer layer, or "toga"	*Thermotoga maritima*	Thermophilic, Gram-negative; found in deep sea hydrothermal vents, sulfurous hot springs, oil deposits
B3 Thermodesulfobacteria	Gram-negative, rod-shaped cells; outer membrane forms protrusions, thermophilic sulfate reducers	*Thermosulfobacterium commune*	Isolated from hot springs and oil deposits at incubation temperatures of 60° to 70°C
B4 Deinococcus-Thermus	Includes Gram-positive radiation-resistant cocci and rods, as well as Gram-negative thermophiles	*Deinococcus radiodurans* *Thermus aquaticus*	Naturally resistant to high levels of gamma and ultraviolet radiation Isolated from Yellowstone National Park; grows well at 70°C
B5 Chrysiogenetes	Represented by single species	*Chrysiogenes arsenatis*	Capable of anaerobic respiration using arsenate as a terminal electron acceptor
B6 Chloroflexi	Gram-negative, filamentous bacteria with gliding motility; some are anoxygenic phototrophs; others are chemoheterotrophs	*Chloroflexus aurantiacus*	Grows within microbial mats in hot springs
B7 Thermomicrobia	Represented by single species	*Thermobacterium roseum*	Thermophilic, aerobic, Chemoheterotroph; isolated from a hot spring in Yellowstone National Park
B8 Nitrospirae	Metabolically diverse group containing nitrifiers, sulfate reducers, and magnetotactic forms	*Nitrospira marina* *Candidatus*[a] *Magnetobacterium bavaricum*	Oxidizes nitrite to nitrate Magnetotactic bacterium isolated from a lake in Southern Germany
B9 Deferribacteres	Heterotrophs that respire anaerobically. Terminal electron acceptors include Fe3+, Mn4+, S, Co3+.	*Deferribacter thermophilus*	Thermophilic can use Fe^{3+}, Mn^{4+}, and nitrate as terminal electron acceptors
B10 Cyanobacteria	Gram-negative unicellular, colonial, or filamentous, oxygenic photosynthetic bacteria	*Nostoc punctiforme*	Forms nitrogen-fixing symbioses with plants
B11 Chlorobi	Gram-negative bacteria, anoxygenic photoheterotrophs ("Green sulfur bacteria")	*Chlorobium limicola*	Grows in stagnant fresh water containing hydrogen sulfide exposed to light
B12 Proteobacteria	Largest bacterial phylum, containing 384 genera and 1300 species; contains 5 classes: Alphaproteobacteria	 *Rickettsia rickettsii* *Caulobacter* spp. *Rhizobium* spp.	 Causes Rocky Mountain spotted fever (27) Prosthecate bacteria (11) Form symbiotic nitrogen-fixing nodules on leguminous plants (11, 28)
	Betaproteobacteria	*Bordetella pertussis* *Neisseria meningitidis*	Causes whooping cough (15, 22) Causes meningitis (25)
	Gammaproteobacteria	*Nitrosomonas* spp. *Zooglea ramigera* *Beggiatoa* spp. *Francisella tularensis* *Legionella pneumonophila*	Oxidize ammonia to nitrite (28) Active in sewage treatment (28) Oxidize hydrogen sulfide (11) Causes tularemia (27) Causes legionellosis (22)

TABLE 11.2 Phyla of Prokaryotes (continued)

Phylum	Properties	Representative Species	Features (Chapter)
B12 Proteobacteria (continued)	Deltaproteobacteria	*Pseudomonas aeruginosa*	Opportunistic pathogen (26)
		Azotobacter spp.	Free-living nitrogen fixers (11)
	Epsilonproteobacteria	*Vibrio cholerae*	
		Escherichia coli	Causes cholera (23)
			Some strains cause diarrhea (23)
		Salmonella typhi	Causes typhoid fever (23)
		Desulfovibrio spp.	Reduce sulfate to sulfide(11)
		Bdellovibrio bacteriovorus	Preys on other bacteria (3, 11)
		Campylobacter jejuni	Causes bloody diarrhea (23)
		Helicobacter pylori	Causes gastritis and peptic ulcer disease (23)
B13 Firmicutes	Contains low G + C Gram-positive bacteria and mycoplasmas	*Clostridium tetani*	Causes tetanus (25)
		Clostridium botulinum	Causes botulism (25)
		Mycoplasma pneumoniae	Causes primary atypical pneumonia (24)
		Bacillus anthracis	Causes anthrax (27)
		Listeria monocytogenes	Causes listeriosis (24)
		Staphylococcus aureus	Causes impetigo, boils, abscesses (26), toxic shock syndrome (24)
		Lactobacillus spp.	Lactic acid bacteria (29)
		Pediococcus spp.	Lactic acid bacteria (29)
		Oenococcus oeni	Mediates malolactic fermentation (29)
		Streptococcus pneumoniae	Causes pneumonia (22)
B14 Actinobacteria	Contains actinomycetes mycobacteria	*Corynebacterium diphtheriae*	Causes diphtheria (22).
		Mycobacterium tuberculosis	Causes tuberculosis (22)
		Nocardia spp.	Acid-fast bacterium (11)
		Propionibacterium acnes	Associated with acne (26)
		Streptomyces griseus	Produces streptomycin(21)
		Frankia spp.	Symbiotic nitrogen fixers (28)
		Bifidobacterium spp.	Microbiota of breastfed infants (14)
		Haemophilus vaginalis	Causes vaginitis (23)
B15 Planctomycetes	Gram-negative bacteria; some reproduce by budding, some have appendages	*Planctomyces bekefii*	Budding bacterium
B16 Chlamydiae	Nonmotile, obligatively parasitic coccoid bacteria that live within vacuoles in the cytoplasm of host cells	*Chlamydia trachomatis*	Causes chlamydia (24)
B17 Spirochaetes	Gram-negative, spiral-shaped, flexible bacteria; motile by periplasmic flagella	*Borrelia burgdorferi*	Causes Lyme disease (27)
		Treponema pallidum	Causes syphilis (24)
		Leptospira interrogans	Causes leptospirosis (24)
B18 Fibrobacteres	Gram-negative anaerobes associated with the digestive tract of herbivores	*Fibrobacter*	
B19 Acidobacteria	Gram-negative, aerobic, acid-tolerant heterotrophs and Gram-negative anaerobes	*Geothrix*	
B20 Bacteroidetes	Phenotypically diverse group of Gram-negative bacteria containing aerobic rods, anaerobic rods, gliding, sheathed, and curved bacteria	*Bacteroides gingivalis*	Associated with periodontal disease (23)
		Cytophaga spp.	Metabolize cellulose (11)
B21 Fusobacteria	Anaerobic, Gram-negative rods with heterotrophic metabolism	*Fusobacterium* spp.	Resident of the mouth (14)
B22 Verrucomicrobia	Gram-negative, mesophilic, heterotrophs; some produce prosthecae; some reproduce by budding	*Prosthecobacter fusiformis*	Nonmotile prosthecate bacterium
B23 Dictyoglomi	A single genus of Gram-negative, rod-shaped, extremely thermophilic, obligately anaerobic, heterotrophic bacteria	*Dictyoglomus*	

ªSpecies that have not been cultivated in pure culture are designated *Candidatus*.

ments. They grow at temperatures near the boiling point of water and pH values as acid as the contents of the human stomach. They are found, for example, in geothermal springs containing sulfuric acid.

Although all members of the group are adapted to these extreme environments, they are otherwise quite diverse. Members of one genus, *Thermoplasma*, lack a cell wall, thus resembling the mycoplasmas of bacterial domain Firmicutes.

DOMAIN BACTERIA

The 23 phyla of the Bacteria (Table 11.2) vary enormously in size and complexity. Some contain many representatives. Others, for example B3 Thermodesulfobacteria, are represented by only a single species. We'll discuss only the more important phyla.

B4 Deinococcus-Thermus

The phylum B4 Deinococcus-Thermus contains two notable bacteria. *Deinococcus radiodurans* is notable for its remarkable resistance to gamma radiation (up to 1.5 kGy) and ultraviolet light (up to 1500 J/sq m). It can be isolated almost in pure culture from soil exposed to high levels of gamma radiation. *Thermus aquaticus*, which can grow at the boiling point of water, is notable for producing *Taq* **polymerase,** an economically and scientifically valuable enzyme. *Taq* polymerase is used in the polymerase chain reaction (PCR) procedure (Chapter 7) because it is sufficiently heat resistant to survive the multiple heat treatments used to melt DNA.

B5 Chrysiogenetes

The phylum B5 Chrysiogenetes is represented by a single species, which has the unusual property of carrying out anaerobic respiration using the highly toxic substance arsenate as terminal electron acceptor.

B10 Cyanobacteria

Cyanobacteria are oxygenic photosynthetic bacteria. Because they are about the same size and shape as some algae, and they produce oxygen as algae do, and some of them are blue-green, they were once called the blue-green algae. But cyanobacteria are not algae. They are prokaryotes with an outer membrane and a thin peptidoglycan wall typical of Gram-negative bacteria.

Some cyanobacteria can fix atmospheric nitrogen. As a result, they have the simplest possible nutrition. They ob-

tain their carbon (as CO_2), as well as their nitrogen (as N_2), from the atmosphere. In addition to air, they need only a few minerals and a source of light. Cyanobacteria are major nitrogen fixers in nature (Chapter 28).

Nitrogen fixation and oxygenic photosynthesis might appear to be incompatible because all nitrogen-fixing enzymes are rapidly destroyed by oxygen (Chapter 28). Many filamentous cyanobacteria solve this problem by separating the two processes. They fix nitrogen in specialized cells called **heterocysts** (**Figure 11.3**). They carry out photosynthesis in all the other cells. Because heterocysts do not photosynthesize, they do not produce O_2 intracellularly and their thickened cell walls exclude extracellular O_2.

Cyanobacteria are tough. They resist desiccation and high temperatures, and they flourish in almost all environments—oceans, lakes, streams, and soil.

Cyanobacteria form many symbiotic associations with plants. They fix nitrogen, and the plant provides a protective environment (Chapter 28).

B11 Chlorobi

The phylum B11 Chlorobi, along with the phylum B20 Bacteroidetes, contains the anoxygenic phototrophic bacteria. They are found in anaerobic environments, but because they are photosynthetic, they must have light. As a result, these bacteria are found in deep, clear bodies of

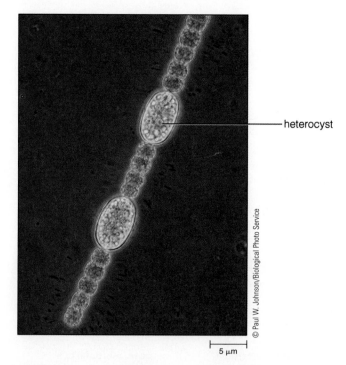

heterocyst

5 μm

© Paul W. Johnson/Biological Photo Service

FIGURE 11.3 Cyanobacteria. Note the striking difference between heterocysts and other cells.

LARGER FIELD

GIANT BACTERIA

Smallness is the hallmark of Prokaryotes (Chapter 1). It's the essence of their life strategy. Being small offers them many advantages, including ready access to dissolved nutrients in their environment and rapid movement of metabolites by diffusion to all parts of the cell's interior. Because of their small size, prokaryotes don't have or need protoplasmic streaming or elaborate intracellular compartments. The net benefit of such savings is rapid growth and successful competition with larger, more complex eukaryotic microbes.

terium, *Beggiatoa* sp., was found in the Gulf of Mexico. And in 1999 the current biovolume record holder, *Thiomargarita namibiensis,* was discovered on the ocean floor off Namibia, Africa. This spherical bacterium, 750 μm in diameter, forms chains of cells that become packed with sulfur granules, making them appear like a string of pearls on the black mud of the ocean floor. Equivalent-sized archaea have not been found. The largest archaeon yet discovered is *Staphylothermus marinus,* which is only about 15 μm in diameter. But the archaea probably hold the

bacteria must have been seen, but because of their large size they were assumed to be eukaryotes. Only with today's new technology (use of fluorescent-labeled probes; Chapter 10) has it become possible to determine unequivocally whether a single cell viewed in its natural environment is a prokaryote or a eukaryote.

2. How can metabolites inside such large prokaryotes get from where they are made to where they are needed fast enough to permit growth at a significant rate? That's a major challenge for a large prokaryote because diffusion occurs

rapidly only over short distances. For example, an oxygen molecule can travel by diffusion 1 μm (the typical length of an ordinary bacterial cell) in 0.001 second, but it takes an hour to diffuse 1 mm (the approximate length of giant bacteria).

The answer to this seeming dilemma is that functional distances within giant bacteria remain small because their cells are packed with inert material—sulfur granules, calcium carbonate, or vesicles. Thus the distances that metabolites must diffuse is small even in these huge prokaryotic cells.

the cd connection

MICROBIAL CLASSIFICATION ⟶
Determining relatedness/DNA probes

Shows a natural population of bacteria in which species have been identified by fluorescent DNA probes

During the past decade, however, the prokaryotes-are-small generalization was shattered by the discovery of several huge bacteria, none of which have yet been cultured in the laboratory. In 1993 *Epulopiscium fishelsoni* was discovered in the gut of tropical fish. This bacterium is about 600 μm long and 80 μm thick, easily visible to the naked eye and large enough for more than a million ordinary-sized bacteria (such as *Escherichia coli*) to fit inside. Then in 1996 an even larger bac-

record for the smallest prokaryotes. The disc-shaped archaeon *Thermodiscus* spp. are as small as 0.2 μm in diameter and only 0.1 μm thick. Thus the biovolume range of prokaryotes is immense, ranging from less than 0.01 μm³ to 200,000,000 μm³—more than 10 billion–fold.

The discovery of huge bacteria poses two fundamental questions:

1. Why weren't these bacteria discovered earlier? The probable answer is simple enough. Such

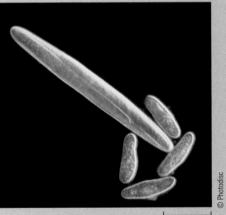

© Photodisc

The huge bacterium *Epulopiscium fishelsoni* is shown with four smaller cells of the protozoan *Paramecium* to show its relative size. An ordinary-sized bacterium such as *Escherichia coli* would be just barely visible at this magnification.

water. They also thrive near the surface of ponds or mud flats that are rich in organic matter. Such environments are made anaerobic by other microorganisms that use oxygen.

Chlorobium, a well-known genus of the group commonly called **green bacteria,** belongs to this phylum. By producing elemental sulfur as a by-product of their photosynthesis, these bacteria probably caused the huge deposits of elemental sulfur that occur in various parts of the world. Certain lakes in North Africa seem now to be in the process of forming new sulfur deposits. They contain large populations of anoxygenic phototrophs, and their bottoms are covered with elemental sulfur.

B12 Proteobacteria

The Proteobacteria, with more than 1300 species and 384 genera, constitutes the largest phylum of the domain Bacteria. We'll discuss each of its five classes—Alphaproteobacteria, Betaproteobacteria, Gammaproteobacteria, Deltaproteobacteria, and Epsilonproteobacteria—separately.

Alphaproteobacteria. The Alphaproteobacteria class contains several environmentally important genera. Members of one genus, *Acetobacter*, produce acetic acid, the acid component of vinegar. Only microbially produced vinegar can be sold legally in the United States and many other countries, so *Acetobacter* is widely cultivated for this purpose. However, some species of *Acetobacter* are a bane of the food industry. By converting alcohol to acetic acid, they cause beer and wine to sour. Because *Acetobacter* spp. are aerobes, beer and wine can be protected simply by excluding air.

Members of another genus, *Agrobacterium*, cause cancerlike diseases of plants. *Agrobacterium tumefaciens* causes crown gall (**Figure 11.4**) when it enters the plant through a wound at the **crown** (where the root and stem meet). Then some of its DNA, called T-DNA, on a plasmid called the **Ti** (tumor-inducing) plasmid becomes incorporated into a plant cell's chromosome. These genes transform the plant cell into a tumor cell. This cell multiplies, eventually producing a tumorlike growth, or gall, at the crown of the plant. The bacteria flourish inside the gall by causing the plant to produce **opines** (unusual amino acids), which *A. tumefaciens* uses as nutrients. In other words, *A. tumefaciens* subverts the metabolism of plant cells, causing them to supply food and a protective environment.

Plant molecular biologists use the Ti plasmid to insert new genes into the genome of plants. By recombinant DNA technology, almost any gene can be placed within the T-DNA region of the Ti plasmid. Then the new gene can enter the plant genome along with the T-DNA. The product is a **transgenic plant** (one that contains genes from another organism). The process holds great promise for

FIGURE 11.4   A gall on an oak tree caused by *Agrobacterium* (Alphaproteobacteria).

improving plant varieties and for using plants to make valuable products. For example, genes from humans or animals could enable plants to make medically useful proteins.

Two other genera in this group are important in the earth's nitrogen cycle (Chapter 28). *Rhizobium* and *Bradyrhizobium* are symbiotic nitrogen-fixing bacteria. They enter the roots of legumes such as beans, peas, clover, and alfalfa and produce tumorlike nodules where nitrogen is **fixed** (reduced from atmospheric nitrogen gas to ammonia). The relationship between bacterium and plant is symbiotic. The plant supplies nutrients and favorable conditions of low-oxygen supply. The bacterium supplies the plant with usable nitrogen.

Other Alphaproteobacteria have unusual shapes. Most bacteria are simple geometric shapes—spheres, rods, or helices. But there are exceptions, and some of them are included in this class.

Possibly the most unusually shaped bacteria are the appendaged, or **prosthecate,** bacteria found in aquatic environments. Prosthecates have one or several filamentous or conical extensions of the cell, called **prosthecae.** They give the cell a spidery appearance (**Figure 11.5**). Prosthecae are part of the cell: Their contents lie within the cytoplasmic membrane. Their physiological function is not established, but there are two possibilities. First, prosthecae may give these nonmotile aerobic cells better access to air by slowing their rate of settling and allowing even gentle currents to raise them. Second, prosthecae dramatically in-

Alphaproteobacteria also contains genera, such as *Beijerinckia*, that can fix nitrogen in a free-living state. The profound impact these bacteria have on global ecology is discussed in Chapter 28. Members of another genus of free-living nitrogen fixers, *Azospirillum*, which are helical, maintain the fertility of tropic soils. They live in close association with roots of grasses, where they fix nitrogen from the atmosphere, making it available to the plants (Chapter 28).

Betaproteobacteria. The Betaproteobacteria class contains two important genera of pathogens, *Bordetella* and *Neisseria*. *Bordetella pertussis* causes the childhood disease whooping cough (Chapter 22). *Neisseria gonorrhoeae* causes the sexually transmissible disease gonorrhea (Chapter 24). *Neisseria meningitidis* causes meningococcal meningitis, an infection of the **meninges** (membranes covering the brain and spinal cord; Chapter 25).

Some members of this class develop within **sheaths** (long, transparent polysaccharide tubes; **Figure 11.7**). As cells are pushed out by cell division, they form new sheath material to elongate the tube. When growth conditions become unfavorable, cells swim out, leaving an empty sheath behind. When swimming cells enter a favorable environment, they multiply, producing a new cell-packed sheath.

Sheathed bacteria are found in contaminated streams and sewage treatment ponds. Some are capable of oxidizing iron and manganese. In environments where these minerals are abundant, the sheaths become heavily encrusted with insoluble iron and manganese oxides. The encrustations form tufts, which are easily visible in polluted water. Tufts of *Sphaerotilus* spp. in sewage treatment plants sometimes become so profuse that they interfere with clarification of the treated water.

One species of Betaproteobacteria, *Aquaspirillum magnetotacticum*, contains magnetosomes (Chapter 4). These

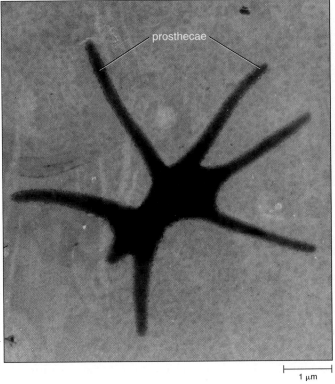

prosthecae

1 μm

FIGURE 11.5   The prosthecate bacterium *Ancalomicrobium adetum* (Alphaproteobacteria).

crease a cell's surface-to-volume ratio, thereby facilitating the entry of nutrients. This is particularly beneficial to these organisms because they live in extremely nutrient-poor environments. This second possibility is supported by the observation that prosthecae become longer as the growth medium becomes more dilute.

One prosthecate bacterium, *Caulobacter*, has an elaborate life cycle by bacterial standards (**Figure 11.6**). *Caulobacter* prospers in nutrient-poor aquatic environments, where it attaches itself to solid surfaces by a sticky pad called a holdfast.

Caulobacters have been intensively studied as a model system to learn how cells differentiate. Caulobacters are easy to find in nature. A microscope slide immersed in unsterilized tap water for several days will often have *Caulobacter* cells attached to it.

In certain species of budding bacteria—for example, *Hyphomicrobium*—budding occurs at the tips of the prosthecae. The result is a string of cells linked by prosthecae.

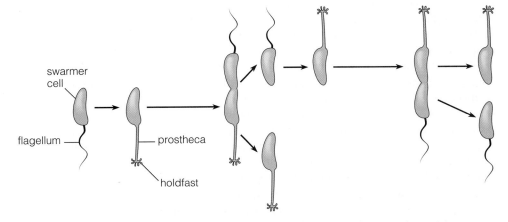

swarmer cell

flagellum

prostheca

holdfast

FIGURE 11.6   Life cycle of the prosthecate bacterium *Caulobacter* (Alphaproteobacteria).

FIGURE 11.7 Cells of the sheathed bacterium *Sphaerotilus* (Betaproteobacteria) within long transparent tubes.

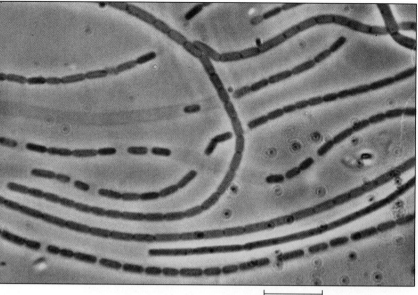

10 µm

inclusions allow these bacteria to follow magnetic lines of force toward the bottom of bodies of water, their optimal environment. These unusual organisms are found in both fresh and salt water.

Gammaproteobacteria.

The Gammaproteobacteria class contains 13 orders, including the Enterobacteriales, the Vibrionales, the Pasteurellales, and the Pseudomonales.

The Enterobacteriales. The Enterobacteriales, also commonly called the **enteric bacteria,** include human pathogens, most of which infect the digestive system. *Salmonella typhi* causes typhoid fever; *Shigella* spp. cause shigellosis, a form of dysentery. The enteric family also includes *Yersinia pestis,* which causes bubonic plague (Chapter 27), and the well-studied *Escherichia coli,* which has a few pathogenic strains (Chapter 23).

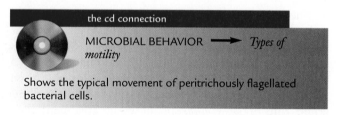

the cd connection

MICROBIAL BEHAVIOR ⟶ *Types of motility*

Shows the typical movement of peritrichously flagellated bacterial cells.

Enteric means "intestinal," but not all members of the enteric group are found in digestive tracts of animals. The enterics are linked by similar metabolism, not habitat. Some cause diseases of plants. *Erwinia amylovora,* for example, causes fire blight of pears and related fruits. Diseased plant leaves look as if they have been burned. *Erwinia carotovora* causes soft rots of many plants. For example, it turns carrots to mush.

Some members of the enteric group are motile by means of flagella. Others are nonmotile. All are rod shaped, and all are facultative anaerobes that ferment sugars if oxygen is not available. Some genera do this via a mixed-acid fermentation and others via a butanediol fermentation. Mixed-acid fermentation produces a mixture of organic acids, as well as gas (CO_2 and H_2). Butanediol fermentation produces **butanediol** (a four-carbon dihydroxy alcohol) and less acid. Which fermentation an enteric species carries out is a basis for subdividing the group (**Table 11.3**). It's easy to distinguish them. Mixed-

acid fermentation can be detected just by including an acid-base indicator—for example, neutral red—in the medium. If the indicator changes color, the organism carries out mixed-acid fermentation. Butanediol fermentation can be detected by a simple chemical test for the presence of **acetoin,** the immediate precursor to butanediol, in the growth medium.

Besides carrying out a mixed-acid fermentation, *Escherichia coli* exhibits three other distinctive properties: (1) It can ferment lactose (milk sugar). (2) It can't use citric acid as a carbon source. (3) It converts the amino acid tryptophan to indole. Indole gives tryptophan-containing cultures of *E. coli,* and therefore feces, their characteristic odor. These characteristics are the basis for the tests used to distinguish *E. coli* from other enteric bacteria (**Figure 11.8**). Such tests are commonly performed to test for sewage contamination in a water supply (Chapter 28).

The Vibrionales. The vibrios are curved rods. Most are motile by polar flagella and produce hydrogen and CO_2 by fermentation. The family includes *Vibrio cholerae,* the species that causes cholera (Chapter 23).

Two genera, *Vibrio* and *Photobacterium,* contain **luminescent** (light-emitting) species. Most, possibly all, luminescent bacteria are associated with marine life (**Figure 11.9**). Some of these associations are highly evolved. For example, the flashlight fish, *Photoblepharon palpebratus,* in the eastern Mediterranean Sea has a special light-generating organ below each eye that contains a dense culture of luminous bacteria. The organ secretes nutrients to support the bacteria, has a layer of reflective tis-

TABLE 11.3 Fermentation Patterns of Key Genera of Enteric Bacteria (Gammaproteobacteria)

Mixed Acid Fermentation		Butanediol Fermentation	
Produce H$_2$ and CO$_2$	No Gas Produced	Produce H$_2$ and CO$_2$	Produce only CO$_2$
Escherichia	Shigella	Enterobacter	Serratia
Proteus	Salmonella typhi		Erwinia
Salmonella (most spp.)	Yersinia		

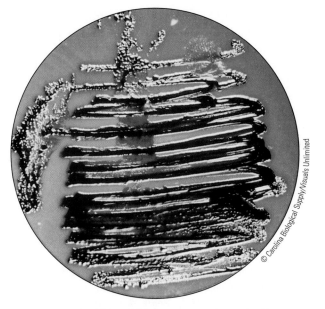

FIGURE 11.8   *Escherichia coli* (Gammaproteobacteria, Enterobacteriales). Distinctive black colonies with a metallic green sheen forms on a lactose-EMB plate. By fermenting lactose, it produces enough acid to cause the two dyes in the medium—eosin (E) and methylene blue (MB)—to interact, creating this unusual color.

sue under it to direct the light outward, and has a closeable flap to turn off the light. It probably functions as a defense mechanism at night. When the fish changes direction, it closes the flap. A pursuing predator is disoriented because the fish seems to disappear and then reappear somewhere else.

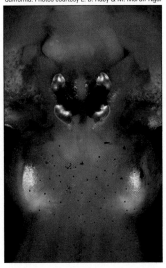

a   1 μm   b   c

FIGURE 11.9   *Vibrio* spp. (Gammaproeobacteria, Vibrionales) are curved rods. The species shown in (a) is *V. cholerae,* which causes cholera. (b) Like the fish *Photoblepharon,* the small (approximately 4 cm) sepiolid squid *Euprymna scolopes* is symbiotically bioluminescent. The squid is a common nocturnal inhabitant of the shallow waters above reef flats in the Hawaiian Islands. (c) A ventral dissection of *E. scolopes* revealing the light-emitting organ as a complex, bilobed structure (approximately 0.8 cm in length) located in the center of the mantle cavity. An adult animal maintains about 100 million symbiotic luminous bacteria of the species *Vibrio fischeri* in the central tissue of the organ.

The Pasteurellales. The Pasteurellales are small nonmotile rods. They include two genera of devastating pathogens—*Pasteurella* and *Haemophilus.* Some species of *Pasteurella* (named for Louis Pasteur) cause diseases of animals. Some *Haemophilus* species are obligate parasites that cause many human illnesses, from minor eye infections to potentially fatal meningitis (Chapter 25).

The Pseudomonales. The Pseudomonales include the genus *Pseudomonas,* with species that are critically important to soil ecology. All are aerobic rods with polar flagella that can metabolize many carbon sources. It is their ability to break down organic compounds that makes them interesting to environmentally conscious industries today. For example, species that degrade petroleum have been used to clean up oil spills. But some species are harmful. One species, *Pseudomonas aeruginosa,* commonly causes of infections in weakened hosts, especially burn victims and cystic fibrosis patients (Chapter 26). *P. aeruginosa* produces a blue-green pigment, and related species produce green or gold pigments. These fluorescent, water-soluble pigments are distinctive enough to identify colonies on sight (**Figure 11.10**).

Another genus of Gammaproteobacteria, *Zymobacter,* produces large quantities of ethanol as an end product of fermentation. Although alcohol production is common among yeasts, it's rare among bacteria. Yeasts are used to make most alcohol beverages. But pulque is an exception. This alcoholic drink from Mexico is the juice of the cactuslike agave plant, fermented by *Zymobacter.* Tequila is the liquor made by distilling pulque.

Other Pathogens. Other pathogenic Gammaproteobacteria include *Francisella tularensis,* which causes tularemia, a systemic infection spread primarily by infected rabbits (Chapter 27), and *Legionella pneumophila,* which causes legionellosis, a pneumonia (Chapter 22).

Commercially Important Species. One species of the genus *Xanthomonas* is cultivated for its capsule, which contains **xanthan gum,** a widely used thickener for foods and paints. Many labels on foods such as salad dressing, cottage cheese, and yogurt list xanthan gum as an ingredient.

Gliding Bacteria. Some Gammaproteobacteria are capable of an unusual form of motility, known as **gliding.** Other groups also contain some gliders. Those that form fruiting bodies are Deltaproteobacteria. Gliding cyanobacteria are in their own phylum. Gliders can move only when they are in contact with a solid surface. Some gliders move relatively rapidly, at rates as high as 600 μm per minute. In terms of body length, this rate for a 1-μm bacterial cell is equivalent to a 6-foot human moving about 40 miles per hour. In contrast, other gliders move 600 times more slowly, at 1 μm per minute.

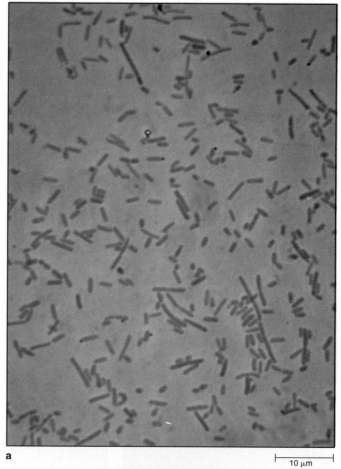

Courtesy Dennis Ohman, University of Tennessee

a   10 μm

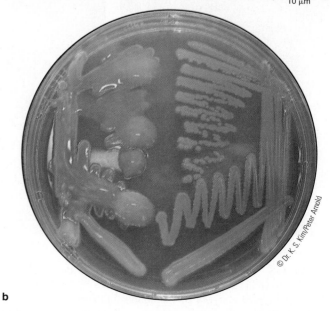

© Dr. K. S. Kim/Peter Arnold

b

FIGURE 11.10 *Pseudomonas aeruginosa* (Gammaproteobacteria, Pseudomonales). (a) Individual cells. (b) Typical blue-green colonies on a petri dish. The strain on the left (derived from the parent strain on the right) is a mutant that makes larger amounts of extracellular slime. Such slime-producing strains often infect the lungs of persons who have cystic fibrosis.

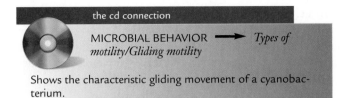

Microbiologists have intensively studied the mechanism of gliding motility, but it remains a mystery. A popular but unproved hypothesis is that these bacteria have a fluid outer membrane that flows back and forth in grooves in the peptidoglycan wall. When the moving outer membrane attaches to a solid surface, the cell glides.

Gliding motility, like all forms of motility, offers an organism a selective advantage in nature. Combined with tactic responses, it allows an organism to seek out favorable environments and avoid unfavorable ones. For example, *Beggiatoa*, which lives by oxidizing hydrogen sulfide (H_2S), glides to an environment where both H_2S and oxygen (O_2) are present.

Other members of this group attack other substrates. Some, such as *Cytophaga* (Bacteroidetes), break down cellulose in plants. They are found in almost all soils. *Lysobacter* spp. are microbial predators. They lyse and consume living algae.

Deltaproteobacteria.

The Deltaproteobacteria include a group of anaerobes that participate in the cyclic interchanges of sulfur-containing compounds in nature (Chapter 28). Through anaerobic respiration they reduce sulfate (SO_4^{2-}) and elemental sulfur to hydrogen sulfide gas, H_2S. The group contains rods, curved rods, spirals, cocci, and packets of cocci.

These bacteria live in mud flats, usually those bordering salt water or brackish water in a bay or estuary. Conditions there are ideal: The salt water is rich in sulfate, and the mud is anaerobic. The rotten egg odor and black color of mud flats signals the presence of these bacteria. The H_2S they produce is responsible for both. The mud is blackened by the metal sulfides that form when H_2S reacts with metal ions, especially iron, present in it.

Desulfovibrio spp. are responsible for the color and hence the name of the Black Sea. The H_2S these bacteria produce in its depths forms black metal sulfides in the water. *Desulfovibrio* and other sulfate- and sulfur-reducing bacteria also cause considerable economic loss. They corrode iron pipes buried in damp soil by converting metallic iron to iron sulfide and iron hydroxide.

Gliding Fruiting Bacteria. The gliding fruiting bacteria, also called myxobacteria, are soil organisms that undergo an unusual degree of differentiation for prokaryotes. They glide in masses over surfaces, consuming microorganisms in their path. When nutrients are depleted, they aggregate and differentiate into fruiting bodies that contain spores called myxospores. Fruiting bodies are easily visible to the naked eye. Some, such as those formed by species of *Myxococcus*, are simple dome-shaped structures. Others, such as those produced by species of *Stigmatella*, are treelike, branched, and quite beautiful (**Figure 11.11**). The surface of animal dung in a damp area is a good place to find myxobacteria because it contains large numbers of bacteria, the food of myxobacteria.

The Epsilonproteobacteria.

This group contains several interesting and important organisms. *Campylobacter jejuni* is a major cause of diarrheal illness in the United States (Chapter 23). *Campylobacter fetus* causes a sporadic abortion in cattle and sheep, but rarely infects humans. The closely related *Helicobacter pylori* causes gastric ulcers in humans.

B13 Firmicutes

The Firmicutes are divided into three classes: the Clostridia, the Bacilli, and the Mollicutes.

Clostridia and Bacilli.

The classes Clostridia and Bacilli contain two genera of Gram-positive rod-shaped bacteria, *Clostridium* and *Bacillus*, that form endospores. They are widespread in soil and water and some are important clinically. They differ in their response to oxygen: All *Clostridium*

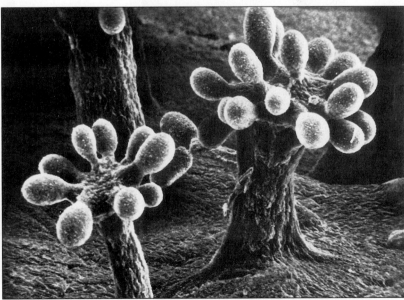

100 μm

Courtesy Patricia Grilione and J. Pangborn, San Jose State University

FIGURE 11.11 Fruiting bodies of the gliding fruiting bacterium *Stigmatella* (Deltaproteobacteria, gliding fruiting bacteria).

spp. are strict anaerobes; all *Bacillus* spp. are capable of aerobic respiration. Some Bacilli are facultative anaerobes.

Because endospores are the most heat-resistant biological structures, heat sterilization procedures are designed to eliminate spore formers. When they are killed by a certain treatment, all other microorganisms are certainly dead (Chapter 9).

Often endospores have a greater diameter than the cells in which they are formed, so they appear as bulges. Their location within the mother cell is genetically determined. Thus it can be used to distinguish species (**Figure 11.12**). Certain species of *Bacillus* produce protein crystals near their spores that are highly toxic to insects (see Figure 29.7). The protein produced by *B. thuringiensis*, sometimes called Bt toxin, is commercially available for garden and agricultural use.

Both *Bacillus* and *Clostridium* contain pathogenic species. *B. anthracis* causes anthrax, a disease of animals that also infects humans and is a feared weapon of biological warfare (Chapter 27). *Clostridium tetani* causes tetanus (Chapter 25). *Clostridium botulinum* causes botulism (Chapter 25). *Clostridium perfringens* causes gas gangrene (Chapter 26) and, occasionally, food poisoning (Chapter 23). *Clostridium difficile* causes diarrhea (Chapter 23).

A group of Gram-positive cocci are distributed between the classes Clostridia and Bacilli. All are spherical or nearly so, but they vary considerably in size. Members of some genera are identifiable by the way cells are attached to one another: in packets, chains, or grapelike clusters (**Fig-**

ure 11.13). These arrangements reflect patterns of cell division and the fact that cells stick together. *Sarcina* cells, for example, are arranged in cubical packets because cell division alternates regularly among the three perpendicular planes.

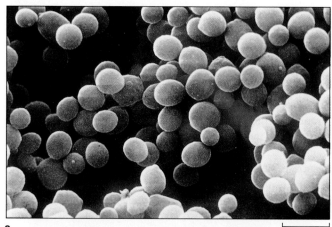

a

1 µm

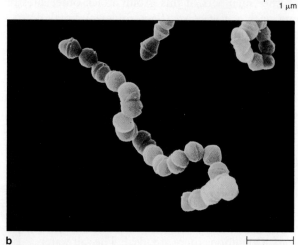

b

1 µm

c

0.5 µm

FIGURE 11.13 The Gram-positive cocci (Firmicutes). (a) Grapelike clusters of *Staphylococcus* cells. (b) Long chain of *Streptococcus* cells. (c) Cubical packets of *Sarcina* cells.

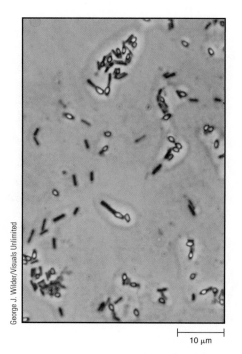

10 µm

FIGURE 11.12 *Bacillus* cells with endospores (B13 Firmicutes). Note how these phase-bright endospores swell the cell at one end.

Streptococcus spp. resemble a string of beads because division always occurs in the same plane. Some of these strings, such as those of *Streptococcus pneumoniae*, are only two cells long. They are called **diplococci.** Species of *Staphylococcus* have no regular plane of division. They form grapelike clusters.

The various Gram-positive cocci differ physiologically and by habitat. *Staphylococcus* spp. are facultative anaerobes that inhabit human skin. They ferment sugars, producing lactic acid as an end product. Many of these species produce carotenoid pigments, which color their colonies yellow or orange. *Staphylococcus aureus* is a major human pathogen. It can infect almost any tissue in the body, often the skin (Chapter 26). It often causes **nosocomial** (hospital-acquired) infections (Chapter 20).

Four genera of this group—*Streptococcus*, *Leuconostoc*, *Pediococcus*, and *Lactobacillus*, constitute the lactic acid bacteria. They share two distinctive physiological properties: They ferment sugars to produce lactic acid, and they are **aerotolerant anaerobes** (they grow fermentatively with or without oxygen).

Being resistant to the quantities of acid they produce, they are often the sole survivors in environments where they have grown. This property has been used by humans since the beginning of civilization to make and preserve foods (Chapter 29).

The genus *Streptococcus* contains several human pathogens. *Streptococcus pyogenes* causes several diseases, including strep throat (Chapter 22), rheumatic fever (Chapter 22), and endocarditis (Chapter 27). *Streptococcus pneumoniae* causes a life-threatening pneumonia (Chapter 22). *Streptococcus mutans* causes dental caries (Chapter 23).

An aerobic vibrioid member of the Bacilli, *Bdellovibrio bacteriovorus*, preys on other bacteria. It is so small (0.3 mm × 1.4 mm) and moves so fast that it is easily missed in the field of a light microscope (see box, Just by Looking, in Chapter 3). It moves about 100 cell lengths per second, equivalent to a person moving at about 400 miles per hour. When *Bdellovibrio* strikes its bacterial prey, it bores through the outer membrane. Then, in the host's periplasm, it grows in length, using host cytoplasm for nutrients (**Figure 11.14**). Three to four hours later, the long cell divides and the host cell lyses, releasing three or four *Bdellovibrio* cells. Various strains are widespread in soil, fresh water, and marine environments. They consume large numbers of bacteria.

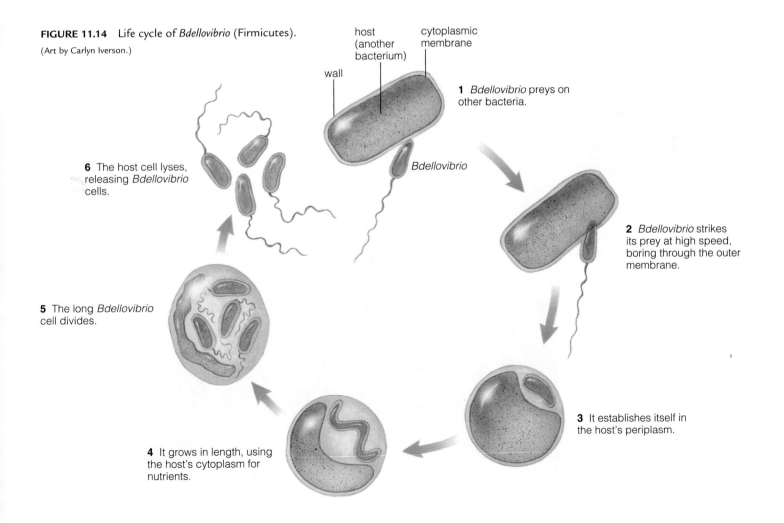

FIGURE 11.14   Life cycle of *Bdellovibrio* (Firmicutes). (Art by Carlyn Iverson.)

host (another bacterium)

cytoplasmic membrane

wall

1 *Bdellovibrio* preys on other bacteria.

Bdellovibrio

6 The host cell lyses, releasing *Bdellovibrio* cells.

5 The long *Bdellovibrio* cell divides.

4 It grows in length, using the host's cytoplasm for nutrients.

3 It establishes itself in the host's periplasm.

2 *Bdellovibrio* strikes its prey at high speed, boring through the outer membrane.

Mollicutes. The terms *Mollicutes* and *mycoplasmas* are synonyms. These bacteria were traditionally treated as a distinct group because they lack a cell wall. Now molecular studies show that they are closely related to Gram-positive bacteria. Mycoplasmas share a number of properties: All are parasites of humans, animals, or plants; almost all are **obligate fermenters** (they ferment in the presence of oxygen); their colonies have a distinctive fried egg appearance (**Figure 11.15**).

Without a cell wall, how do the mycoplasmas keep from bursting? There are two reasons. First, unlike walled prokaryotes, mycoplasmas have sterols in their cytoplasmic membrane. These lipids strengthen the membrane slightly. But more importantly, mycoplasmas maintain their cytoplasm at nearly the same osmotic pressure as their external environment by pumping sodium ions (Na^+) out of the cell. Blocking this mechanism, by depriving mycoplasmas of an energy source, causes them to swell and lyse.

Mycoplasmas come in a variety of shapes, from long spirals to perfectly round cocci. The layer of carbohydrate located outside the cytoplasmic membrane is probably responsible for maintaining such shapes. But when growth conditions are less than ideal, mycoplasma cells become distorted, forming long strands that resemble fungi (thus accounting for their name; *myco* means "fungus"). Being wall-less, mycoplasmas can squeeze through small holes, even the pores in membrane filters used to sterilize liquids. As a result, tissue cultures of animal cells cannot be reliably sterilized by filtration; antibiotics, usually penicillin and streptomycin, are added to the media to suppress growth of contaminating mycoplasmas.

Mycoplasma pneumoniae causes a mild type of pneumonia called primary atypical pneumonia (Chapter 22). Many mycoplasmas cause diseases of animals, such as **rhinitis** (inflammation of the nose) in chickens and turkeys. Another mycoplasma, *Ureaplasma urealyticum*, is a harmless inhabitant of the vaginal tract of 60 percent of normal women. Occasionally it enters the bloodstream during delivery, causing a mild postpartum fever.

(vertical credit) © Michael Gabridge/Visuals Unlimited

FIGURE 11.15 *Mycoplasma* colonies (Firmicutes, Mollicutes). Note their distinctive fried-egg appearance.

B14 Actinobacteria

The Actinobacteria contains oddly shaped bacteria, acid-fast bacteria, and spore-forming bacteria.

Oddly Shaped Bacteria. *Arthrobacter* spp. change shape during growth (**Figure 11.16**). Abundant in soil, these aerobes are responsible for much of the mineralization of organic material that occurs there. *Arthrobacter* cells exhibit a

FIGURE 11.16 Changes in the morphology of *Arthrobacter* (Actinobacteria).

(a, b, c) Courtesy H. Veldkamp, G. Vanden Berg, and LPTM Zevenenhuizen, "Glutamic acid production by *Arthrobacter globiformis*," *Antonie van Leeuwenhoek* 29:35–51, 1963

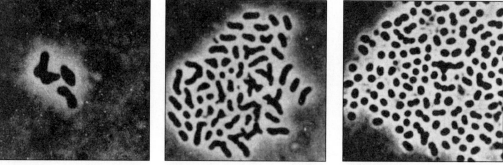

a   b   c   5 μm

LARGER FIELD

CAN PROKARYOTES GROW ANYWHERE? WHAT ABOUT MARS?

Some members of the phylum Cyanobacteria need no organic nutrients to live; inorganic ions and light are sufficient. Chemolithotrophs (such as members of the Crenarcheota) don't even need light. They derive energy by oxidizing inorganic compounds. And over broad ranges, temperature is not a problem. Some prokaryotes can grow at temperatures higher than the boiling point of water, others at temperatures lower than its freezing point. Some bacteria thrive in the crushing pressures of the deepest oceans, and others at a pH as low as 1 or as high as 12. Some bacteria can even survive high doses of radiation or starvation for hundreds of years. With such wide-ranging capabilities, it's not surprising that prokaryotes can survive almost everywhere on Earth. But what about other planets?

The reality of space travel has made this a critically important question in two regards. First, extraterrestrial microorganisms

might wreak ecological havoc on Earth or start a deadly epidemic. Conversely, microorganisms from Earth might materially alter other planets. Certainly their presence would confound future scientific investigations because scientists could not know if they were native to the planet. For these reasons, the United States and other nations engaged in space exploration agreed in the 1970s to minimize the possibility of contaminating other planets with terrestrial microorganisms. By international agreement, the highest acceptable probability of contaminating a planet is to be one chance in a thousand. This limit is a compromise between safety and practicality.

Many factors affect the probability of contaminating other planets. They include the number of microorganisms on spacecrafts, the likelihood that they can survive the trip, and the number of space missions to the planet. But

merely introducing microorganisms to a planet does not permanently contaminate it. Eventually the microorganisms will die out—unless they multiply. The crucial factor is the probability of a terrestrial microorganism's being able to multiply there and thereby persist indefinitely. Microbiologists estimate the probability of microbial growth by comparing the known or suspected general conditions on planets with the requirements for microbial growth. Of the planets in our solar system, our neighbor Mars is by far the most hospitable. It has the elements and temperature necessary for life and, most importantly, it has water. But the Martian environment is severe. The temperature rarely rises above the freezing point of water. Most regions are extremely dry. Also, the planet is exposed to intense ultraviolet radiation.

In 1977 the United States sent the Viking space probe to Mars with

two landers to test for the presence of life. At that time, the probability of a terrestrial organism growing on Mars was considered high enough to require that the landers be heat sterilized. The landers tested a few square meters of surface and found no evidence of microorganisms.

Data collected by the landers also revealed how truly hostile the Martian environment is. Powerful unknown oxidants that are probably lethal to terrestrial microorganisms were found on the surface. Also, the low temperature combined with tiny amounts of water suggests that liquid water—an absolute requirement for life as we know it—might not exist on Mars. The chance of microorganisms growing on Mars is now considered so low that the next planned space probe probably will not be sterilized before launch. Current scientific opinion says Earth is the only place life exists in our solar system.

curious sort of movement called **snapping post-fission movement.** Immediately after the cross wall forms between two daughter cells, they appear to snap apart, forming a V-shaped pair of cells (**Figure 11.17**).

A closely related genus, *Bifidobacterium*, has irregular, often swollen and branched cells. It flourishes in the intestinal tract of breastfed babies because the *N*-acetylglucosamine it needs for growth is abundant in breast milk (Chapter 14).

Bifidobacteria are anaerobes that ferment lactose to lactic and acetic acids. The large amount of acid they produce suppresses the growth of many diarrhea-causing pathogens.

The human pathogens that belong to this group include *Propionibacterium acnes*, which causes the skin infection acne (Chapter 26); *Corynebacterium diphtheriae*, which causes the respiratory disease diphtheria (Chapter 22); and *Actinomyces israelii*, which causes abscesses, usually of the head and

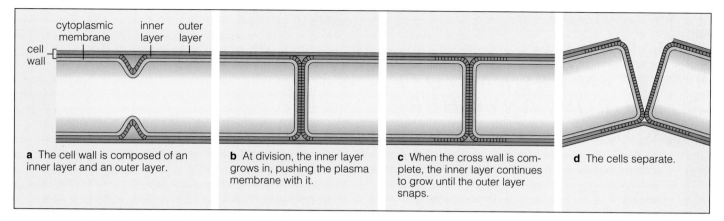

FIGURE 11.17 Snapping post-fission movement of *Arthrobacter* and some other members of Actinobacteria.

neck (Chapter 26). Corynebacteria exhibit snapping fission. Actinomyces have irregularly shaped branching cells.

Acid-fast Bacteria. Members of the mycobacterial and nocardioform groups have an unusual envelope structure, within which the peptidoglycan wall is attached to a polysaccharide layer, which in turn is attached to **mycolic acids** (branched chains of fatty acids with hydroxyl [—OH] groups).

The mycolic acids of two genera, *Mycobacterium* and *Nocardia*, have especially long carbon chains. They form a waxy outer surface that protects against hostile environments (see the beginning of Chapter 4). It also affects staining because it is a barrier to many compounds, including most dyes. To stain them, cells must be heated to about 75°C. Then they become difficult to destain, even with acid. These unusual characteristics allow *Mycobacterium* and *Nocardia* to be identified by the acid-fast stain procedure, so they are known as **acid-fast bacteria** (Chapter 3).

Mycobacterium and *Nocardia* are closely related. They can be distinguished by the shape of their cells. *Mycobacterium* cells are mainly rods that may be branched. *Nocardia* cells are long and usually branched. *Nocardia* are abundant in the soil, but several species are pathogenic to humans. They cause acute and chronic infections that can spread throughout the body. Nineteen species of mycobacteria are pathogenic. One, *Mycobacterium tuberculosis*, causes tuberculosis, which is now undergoing a resurgence in the United States (Chapter 22). Another species, *Mycobacterium leprae*, causes **Hansen's disease,** or **leprosy,** a chronic disease characterized by loss of sensation and deformed extremities (Chapter 26).

Diagnosis and basic research on mycobacterial diseases is hampered by the difficulty of culturing pathogenic mycobacteria. They grow very slowly because their waxy surface impedes entry of nutrients. Even those called rapid-growers require several days' incubation before a colony can be seen. Disease-causing mycobacteria, including *M. tuberculosis*, require 2 to 6 weeks of incubation. *M. leprae* has not yet been cultivated in vitro.

Colonies of *M. tuberculosis* have a waxy, warty appearance. In liquid culture, cells aggregate into characteristic ropelike structures (**Figure 11.18**).

Spore-forming Bacteria. The spore-forming Actinobacteria are known as **actinomycetes.** Their abundant presence in soil accounts for much of the breakdown of organic matter that occurs there. But the greatest impact actinomycetes have had on our lives comes from the ability of some of them to produce antibiotics. Most antibiotics in current use are produced by species of *Streptomyces* (Chapter 21).

Members of this genus are abundant in most soils. In fact, the pleasant odor of freshly turned soil comes from the volatile compounds these bacteria produce. On an agar plate inoculated with soil, numerous *Streptomyces* colonies develop. The colonies are easy to recognize by their pastel colors, characteristic texture (a hard mass that extends into the agar), and soil-like odor. Some of the filaments grow into the agar. The rest extend up from the agar and form chains of spores at the tips (**Figure 11.19**). The spores give the colony its characteristic pastel color.

Although their general growth pattern is funguslike, actinomycetes are distinctly prokaryotic in size and structure. In cross section an actinomycete mycelium looks like a typical Gram-positive bacterial cell.

Members of one group of spore-forming Actinobacteria form spores within a **multilocular** (many-chambered) sporangium at the end of a filament (**Figure 11.20**). The group contains three genera. *Geodermatophilus* is found in soil. *Dermatophilus* infects animal and sometimes hu-

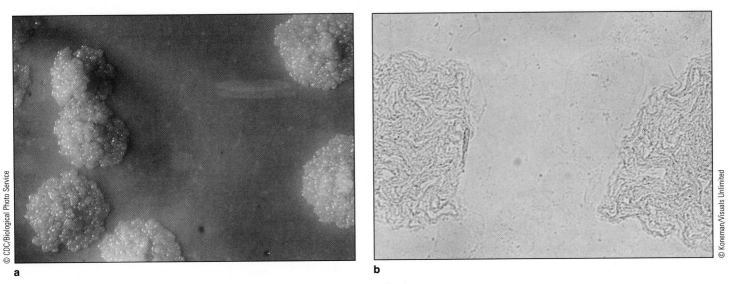

FIGURE 11.18 *Mycobacteria tuberculosis* (Actinobacteria). (a) Waxy, warty colonies. (b) Ropelike aggregates in a liquid medium.

man skin (Chapter 26). *Frankia* fixes nitrogen. *Frankia* spp. are ecologically important. They live in nodules on the roots of 17 different genera of plants, including alder trees. When infected with *Frankia*, alders can grow in the complete absence of fixed nitrogen. *Frankia* supplies all they need.

B16 Chlamydiae

Two groups of Gram-negative bacteria—the rickettsiae and the chlamydiae—were once thought to be viruses because they are so small. These two groups are not closely related. The rickettsiae are in the Gammaproteobacteria class.

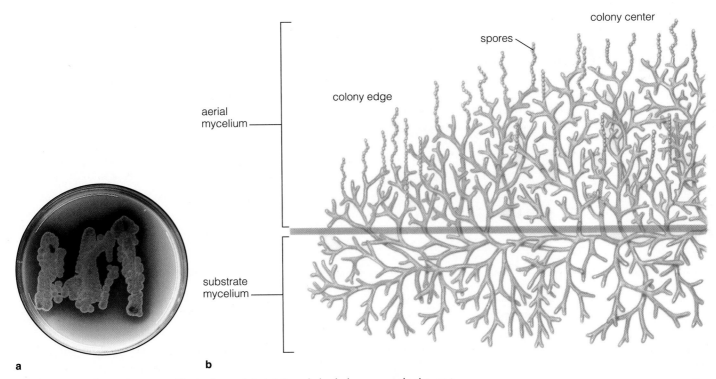

FIGURE 11.19 *Streptomyces* spp. (Actinobacteria). (a) Pastel-shaded orange colonies on an agar plate. (b) Cross section sketch of a colony.

(Art by Carlyn Iverson.)

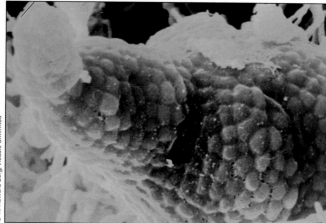

FIGURE 11.20   Multilocular sporangium of *Frankia* (Actinobacteria).

Chlamydiae range from 0.2 to 0.7 mm in diameter. Rickettsiae are only slightly larger at 0.3 to 1.0 mm. Otherwise, these look like other nonmotile Gram-negative rods or coccobacilli. Most species are obligate intracellular parasites. They can't be cultivated outside a living host cell.

Rickettsiae and chlamydiae cause serious human infections (**Table 11.4**). In general, rickettsial pathogens are carried from one human host to another by arthropods (such as ticks, lice, and mites). Chlamydiae are spread directly from one infected human to another. Not all *Chlamydia* spp. are pathogens. Some are harmless intracellular symbionts that may even confer an advantage to their host cells.

Rickettsiae reproduce by simple division, but not chlamydiae. They have a more complicated reproductive cycle that alternates between two cell types: the reticulate body (or vegetative cell) and the elementary body (or chlamydiospore). Reticulate bodies multiply, and elementary bodies spread the infection (**Figure 11.21**).

Rickettsiae and chlamydiae can multiply only within a living host cell because they are energy parasites. Instead of making ATP, they take it from their host by exchanging their ADP for the host's ATP.

B17 Spirochaetes

The phylum Spirochaetes is composed of Gram-negative bacteria distinguished by their unusual shape and unusual mechanism for motility. Spirochetes are long and helical—they look like a corkscrew or coil (**Figure 11.22**). Their flagella are bundled into two axial filaments (Chapter 4), which account for the spirochetes' unusual motility. As the endoflagella turn, they cause the helical ridge they form on the outer membrane to move down the cell and propel it through the medium. Spirochetes move so rapidly that when viewed under the microscope they seem to disappear from one place and suddenly reappear in another. They are motile in viscous environments that immobilize ordinary flagellated bacteria. Thus spirochetes are often found in thick mud and on mucous membranes of animals. They can also move on the surface of solid media. By changing the direction of rotation of their flagella, they twitch and writhe, moving like an inchworm along the surface.

Spirochetes are metabolically diverse. Some are aerobes. Others are facultative anaerobes. Still others are an-

TABLE 1.4   Some Diseases Caused by Rickettsiae and Chlamydiae

Organism	Disease	Chapter	Comments
Rickettsia prowazekii	Epidemic typhus	27	Causes epidemics that rage during war and times of social disaster; transmitted by body lice
Rickettsia typhi	Endemic typhus	27	Transmitted by rat fleas
Rickettsia rickettsii	Rocky Mountain spotted fever	27	Named for area where it was first identified
Coxiella burnetii	Q fever	22	Designated Q for question
Chlamydia trachomatis	Trachoma	26	Most common cause of blindness worldwide
	Nongonococcal urethritis	24	Sexually transmissible
Chlamydia psittaci	Human psittacosis or Parrot fever	22	Primarily a bird pathogen but can cause pneumonia in humans
Chlamydia pneumonia	Pneumonia	22	Causes pneumonia and is associated with coronary artery disease

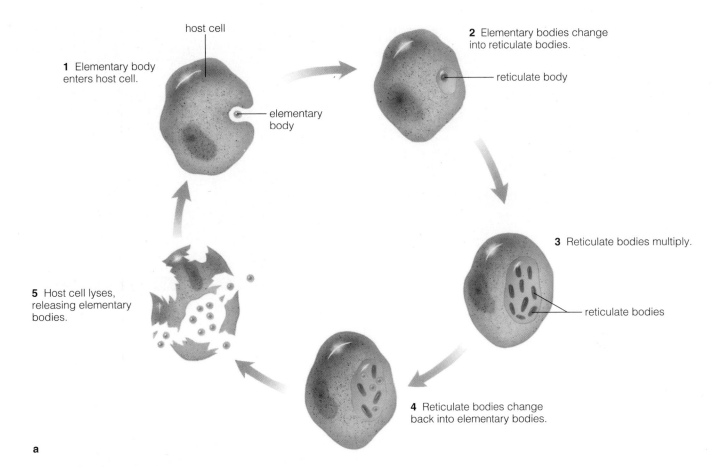

1 Elementary body enters host cell.

host cell

elementary body

2 Elementary bodies change into reticulate bodies.

reticulate body

3 Reticulate bodies multiply.

reticulate bodies

4 Reticulate bodies change back into elementary bodies.

5 Host cell lyses, releasing elementary bodies.

a

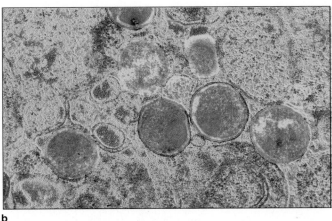

b

FIGURE 11.21 *Chlamydia trachomatis* (Chlamydiae). (a) Life cycle. (b) Reticulate bodies within a human cell.

(Art by Carlyn Iverson.)

aerobes. Some spirochetes live harmlessly in our mouths (Chapter 14), but others cause devastating human diseases. *Treponema pallidum* causes the sexually transmissible disease syphilis (Chapter 24); *Leptospira interrogans* causes leptospirosis, a kidney infection of animals and humans (Chapter 24). Some species of *Borrelia* cause relapsing fever,

a debilitating systemic infection (Chapter 27). Another species, *Borrelia burgdorferi*, causes Lyme disease, a systemic disease that was first recognized in the 1970s (Chapter 27).

B20 Bacteroidetes

The phylum Bacteroidetes contains the purple nonsulfur bacteria genus *Bacteroides*.

Purple Nonsulfur Bacteria. Those nonoxygenic photosynthetic bacteria that use organic compounds as a source of electrons were traditionally known as **purple nonsulfur bacteria.** Often these bacteria are highly colored—red, orange, purple, and bright green—because of the chlorophyll and other photosynthetic pigments they contain. Water samples from deep regions of lakes with an abundance of these organisms are also intensely colored.

Bacteroides. Species of *Bacteroides* are among the most abundant microorganisms in the mouth and intestinal tract of animals, including humans. For example, there are 10 billion to 100 billion (10^{10} to 10^{11}) cells of *Bacteroides* spp. per gram of intestinal contents. These organisms are

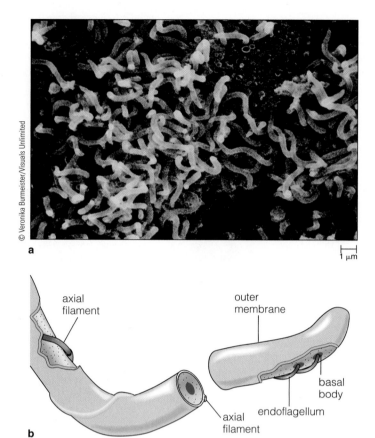

FIGURE 11.22   The Spirochaetes. (a) A mass of cells. (b) The structure and location of the axial filament.

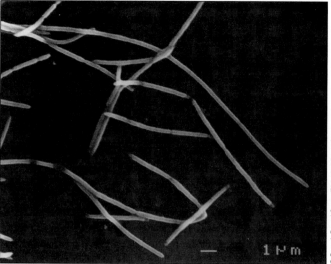

FIGURE 11.23   Cells of *Fusobacterium nucleatum* (Fusobacteria). Note their pointed ends.

usually harmless, but occasionally they do cause disease. For example, *Bacteroides* spp. cause infections of the abdominal and pelvic cavities of weakened hosts.

B21 Fusobacteria

Fusobacterium periodonticum causes dental abscesses. *Fusobacterium nucleatum* can infect the respiratory system (**Figure 11.23**).

SUMMARY

Case History: Growing a Variety of Bacteria (p. 258)

1. Urinary tract infections can be caused by a variety of bacteria, which must be identified by their properties, not the symptoms they cause.

Prokaryotic Taxonomy (pp. 258–260)

2. From studies on the sequences of 16S RNA, microbiologists have learned a great deal about the evolutionary relatedness of prokaryotes.

3. In 2001 a natural scheme of classification scheme was incorporated into *Bergey's Manual.*

The Bergey's Manual Scheme of Prokaryotic Taxonomy (p. 260)

4. *Bergey's Manual* divides prokaryotes into two domains: Archaea and Bacteria.

Domain Archaea (pp. 260–264)

5. The domain Archaea consists of two phyla, the Crenarchaeota and the Euryarchaeota.

Crenarchaeota (p. 260)

6. Crenarchaeota consists of a single class. All its members are thermophilic and most metabolize sulfur.

Euryarchaeota (pp. 260–264)

7. The Euryarchaeota contains three important metabolic types: the methanogens, the extreme halophiles, and the thermoacidophiles

8. The methanogens are strict anaerobes that make natural gas (methane).

9. The extreme halophiles are aerobes that require at least 10 percent salt to grow; they are capable of a primitive form of photosynthesis.

10. The thermoacidophiles thrive in extremely hostile environments, such as acidic geothermal springs.

Domain Bacteria (pp. 264–280)

11. The domain Bacteria contains 23 phyla.

B4 Deinococcus-Thermus

12. B4 Deinococcus-Thermus phylum contains two notable species: *Deinococcus radiodurans*, which is remarkably resistant to radiation, and *Thermus aquaticus*, which is the source of a heat-resistant *Taq* polymerase used in the polymerase chain reaction.

B5 Chrysiogenetes

13. B5 Chrysiogenetes phylum contains a single species, which uses arsenate as a terminal electron acceptor for anaerobic respiration.

B10 Cyanobacteria

14. Cyanobacteria carry our oxygenic photosynthesis, and some fix nitrogen in specialized cells called heterocysts.

15. Cyanobacteria, which are the size and shape of some algae, were once called blue-green algae.

B11 Chlorobi

16. B11 Chlorobi phylum contains anoxygenic phototrophic bacteria that produce elemental sulfur as a by-product of photosynthesis.

B12 Proteobacteria

17. B12 Proteobacteria phylum, with 1300 species, is the largest phylum of the domain Bacteria.

18. The class Alphaproteobacteria contains the genus *Acetobacter*; its species produce acetic acid, which is used in the manufacture of vinegar. *Agrobacterium tumefaciens* causes crown gall of plants and is used by molecular biologists to insert foreign genes into plants. *Rhizobium* and *Bradyrhizobium* spp. form symbioses with legumes to fix nitrogen. The class also contains prosthecate bacteria and free-living nitrogen fixers such as *Beijerinckia* and *Azospirillum* spp.

19. The class Betaproteobacteria contains two important genera of pathogens, *Bordetella* and *Neisseria*, with species that cause whooping cough and gonorrhea. Some Beta-proteobacteria, such as *Sphaerotilus* spp., live within sheaths. One species, *Aquaspirillum magnetotacticum*, is magnetotactic.

20. The Gammaproteobacteria is a large class with 13 orders, including the Enterobacteriales, the Vibrionales, the Pasteurellales, and the Pseudomonales.

21. The Enterobacteriales, also called enteric bacteria, contains *Escherichia coli* and pathogens such as *Salmonella typhi*, which causes typhoid fever, and *Yersinia pestis*, which causes bubonic plague. It also contains plant pathogens, including *Erwinia* spp.

22. The Vibrionales contains *Vibrio cholerae*, which causes cholera. Some species of *Vibrio* are luminescent, as are species of *Photobacterium*.

23. The Pasteurellales includes two genera of devastating pathogens, *Pasteurella* and *Haemophilus*.

24. The Pseudomonales are aerobic, motile rod-shaped bacteria that are critically important to soil ecology. One species, *Pseudomonas aeruginosa*, commonly causes infections in weakened hosts.

25. Also included in the Gammaproteobacteria are *Zymobacter*, which produces alcohol; *Francisella tularensis*, which causes tularemia; and *Legionella pneumophila*, which causes legionellosis.

26. Some Gammaproteobacteria are capable of gliding motility.

27. The Deltaproteobacteria includes *Desulfovibrio* spp. that produce hydrogen sulfide from sulfate by anaerobic respiration. Gliding fruiting bacteria are also included.

28. The Epsilonproteobacteria includes two important pathogens, *Campylobacter jejuni*, which causes diarrhea, and *Helicobacter pylori*, which causes gastric ulcers.

B13 Firmicutes

29. The Firmicutes are divided into three classes: the Clostridia, the Bacilli, and the Mollicutes.

30. The Clostridia and Bacilli contain the genera *Clostridium* and *Bacillus*, which are Gram-positive, endospore-forming bacteria. They contain pathogenic species, including *Bacillus anthracis*, which causes anthrax; *Clostridium tetani*, which causes tetanus; and *Clostridium botulinum*, which causes botulism. These classes also contain the non–spore-forming cocci genera *Streptococcus*, *Sarcina*, and *Staphylococcus*, along with *Bdellovibrio*, which preys on other bacteria.

31. The Mollicutes, another term for mycoplasmas, are wall-less bacteria, one species of which, *Mycoplasma pneumonia*, causes primary atypical pneumonia.

B14 Actinobacteria

32. The Actinobacteria contains oddly shaped bacteria, acid-fast bacteria, and spore-forming bacteria.

33. *Arthrobacter* spp. are oddly shaped soil bacteria that exhibit snapping post-fission movement. *Bifidobacterium* spp. are irregular, branched, swollen cells that flourish in the intestinal tract of breastfed babies. *Corynebacterium diphtheriae* cause the respiratory disease diphtheria.

34. Acid-fast bacteria, named for their positive reaction to the acid-fast stain, include the genera *Mycobacterium* and *Nocardia*. *Mycobacterium tuberculosis* causes tuberculosis. *Mycobacterium leprae* causes Hansen's disease, or leprosy.

35. The spore-forming Actinobacteria are known as actinomycetes. One genus, *Streptomyces*, which is abundant in the soil, contains species that produce antibiotics. *Frankia* spp., which produces spores within a multilocular sporangium, fixes nitrogen with nodules of many plant species.

B16 Chlamydiae

36. Chlamydiae, along with the rickettsiae (which are in the Gammaproteobacteria), are tiny Gram-negative bacteria, many of which cause serious human infections.

37. Chlamydiae have a complex life cycle that alternates between reticulate bodies that multiply and elementary bodies that spread the infection.

B17 Spirochaetes

38. Spirochaetes are Gram-negative, corkscrew-shaped shaped cells that move by the turning of flagella trapped within their periplasm.

39. *Treponema pallidum*, which causes syphilis, *Leptospira interrogans*, which causes leptospirosis, and *Borrelia burgdorferi*, which causes Lyme disease, are Spirochaetes.

B20 Bacteroidetes

40. The phylum Bacteroidetes contains the purple nonsulfur bacteria and the genus *Bacteroides*.

41. The purple nonsulfur bacteria are nonoxygenic photosynthetic bacteria that use organic compounds as a source of electrons.

42. Species of *Bacteroides* are among the most abundant microorganisms in the mouth and intestinal tract of animals.

REVIEW QUESTIONS

Case History: Growing a Variety of Bacteria

1. What property did the all the bacteria suspected of causing I. B.'s urinary tract infection share?

Prokaryotic Taxonomy

2. What is the basis for a natural scheme of classifying prokaryotes?

3. How does the genealogical tree of prokaryotes show which organisms are most closely related?

4. What kind of classification scheme does *Bergey's Manual* use and why?

5. What kinds of information do you find in *Bergey's Manual*?

The Bergey's Manual Scheme of Bacterial Taxonomy

6. What are the major divisions of *Bergey's Manual*?

7. What are the major divisions of the Archaea? What are their properties?

8. On what basis are bacteria subdivided into phyla? Are there examples of physiologically related bacteria being in different phyla?

9. B4 Deinococcus-Thermus: What notable bacteria belong to this group?

10. B10 Cyanobacteria: Why were these bacteria once called blue-green algae? How do they obtain energy? How do they protect their nitrogen-fixing system?

11. B11 Chlorobi: How do these bacteria obtain energy? Why are they found in deep, clear bodies of water?

12. Alphaproteobacteria: Describe the process by which *Agrobacterium tumefaciens*, a member of this group, causes crown gall disease in plants. How are xanthan gum and acetic acid, two products of this group, used industrially? What is the significance of the symbiotic relationship *Rhizobium* and *Bradyrhizobium* have with legumes?

13. Betaproteobacteria: Give an example of a disease caused by a member of this group.

14. Gammaproteobacteria: Name some human diseases caused by members of this group. What four properties of *Escherichia coli* distinguish it from other enterics?

15. Deltaproteobacteria: How do members of this group get their energy by metabolizing sulfate, and why is this ecologically significant? Where are these bacteria found?

16. Epsilonproteobacteria: What diseases do these bacteria cause?

17. B13 Firmicutes: One members of this group is *Bdellovibrio bacteriovorus*, which preys on other bacteria. What does it mean that *B. bacteriovorus* is host specific?

18. What characteristic unites the mollicutes? What does it mean that they are obligate fermenters? How do you account for their being called mycoplasmas?

19. Tell how cells in these genera are arranged: *Sarcina*, *Streptococcus*, *Staphylococcus*. Some members of this group are called lactic acid bacteria. What does that mean?

20. What are endospores? Why is survival of *Bacillus* and *Clostridium* used as an index of sterilization? Name some of the serious human diseases these two genera cause.

21. B14 Actinobacteria: What are some of the irregular shapes included in this group? Some members of this group, including *Arthrobacter*, exhibit snapping post-fission movement. Describe this. *Bifidobacterium* is also a member of this group. Why is it important?

22. Acid-fast bacteria: What is unusual about the wall structure of these groups? Name some of the human diseases caused by mycobacteria.

23. B16 Chlamydiae: Why were these bacteria and the rickettsiae once thought to be viruses? What does it mean that most are obligate intracellular parasites? Name some of the human infections they cause. Describe chlamydial reproduction.

24. B17 Spirochaetes: Describe how spirochetes move. What is their shape? Name some diseases they cause.

25. B20 Bacteroidetes: What are the properties of purple nonsulfur bacteria? What importance do members of the genus *Bacteroides* have to human health?

CORRELATION QUESTIONS

1. Is the title of the latest *Bergey's Manual* inconsistent with its contents? Explain.

2. Why can't a Gram-positive bacterium have an axial filament?

3. Many bacteria make acid by fermentation. Which one makes it by respiration?

4. Would you expect an iron pipe buried in a mud flat to last longer than one buried in well-aerated soil? Why?

5. If you wanted to find a microorganism that produced an antibiotic, would you look for it in a mud flat or in well-aerated soil? Why?

6. After a biological warfare attack that spread anthrax, would boiled water be safe? Why?

ESSAY QUESTIONS

1. Discuss the advantages and disadvantages of the *Bergey's Manual* approach to classification.

2. Discuss selective pressures that led some bacteria to become luminescent.

SUGGESTED READINGS

Balows, A., Trüper, H., Dworkin, M., Harder, W., and Schleifer, K-H. 1991. *The prokaryotes: A handbook on the biology of bacteria: Ecophysiology, isolation, identification, applications.* 2d ed. New York: Springer-Verlag.

Boone, D. R., and R. W. Castenholz. 2001. *Bergey's manual of systematic bacteriology.* 2d ed. New York: Springer-Verlag.

Schulz, H., and B. B. Jørgensen. 2001. *Annual review of microbiology: Big bacteria.* Palo Alto, Calif.: Annual Reviews, Inc.

For additional readings, go to InfoTrac College Edition, your online research library at: http://www.infotrac.thomsonlearning.com

TWELVE

Eukaryotic Microorganisms, Helminths, and Arthropod Vectors

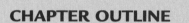

© Cath Ellis, Dept of Zoology, University of Hull/SPL/Photo Researchers

CHAPTER OUTLINE

LEARNING GOALS

To understand:

- *The characteristics, ecological roles, and classification of the major groups of eukaryotic microorganisms—the fungi, algae, and protozoa*
- *The nature of lichens*
- *The properties of slime molds, a small, distinct group of eukaryotic microorganisms*

- *The phylogenetic relationships among various groups of eukaryotic microorganisms*
- *The major groups of helminths—their appearance, their life cycles, and the diseases they cause*

- *The characteristics of certain arthropods, including insects, ticks, and mites, that cause disease or act as vectors or reservoirs*

A Fatal Fungus

A H., a 32-year-old white male, had always enjoyed good health, but gradually he began to suspect something was wrong. At first he noticed unusually frequent, severe headaches. As weeks and months went by, he developed fever, fatigue, and occasional morning vomiting. Then one day he realized that he wasn't thinking clearly. He became alarmed and called his doctor. His doctor did a thorough medical evaluation. A. H.'s symptoms of headache, confusion, and morning vomiting strongly suggested an illness that was affecting his brain. His fever made his physician suspect infection, but A. H.'s history did not suggest a bacterial or viral cause.

Patients with bacterial (see Case History: An Infectious Examination, Chapter 4) or viral infections of the central nervous system become very ill quite suddenly. A. H.'s illness had come on slowly and gradually. His physician had to consider the eukaryotic microorganisms, neither bacteria nor viruses, which can cause chronic **meningitis** (infection of the meninges, the membranes surrounding the brain and spinal column). His physician performed a **spinal tap** (drew a sample of spinal fluid by inserting a needle between vertebrae in lower back). That provided the diagnosis. A portion of the spinal fluid was negatively stained with India ink. It revealed many perfectly round cells. Each was surrounded by a capsule several times larger than the cell itself. The laboratory technologist identified these encapsulated cells as being the fungus *Cryptococcus neoformans*. This identification was confirmed later by culturing the organism. *C. neoformans* causes a life-threatening infection of the membranes surrounding the brain and spinal cord (Chapter 25). It is extremely rare in healthy people and always difficult to treat. But it's rather common in people with acquired immunodeficiency syndrome (AIDS). Further testing established, unfortunately, that A. H. did have AIDS. He died 4 months later despite aggressive medical treatment.

It's no coincidence that many potentially fatal infections are caused by capsule-forming microorganisms. A capsule greatly enhances a microbe's virulence (Chapter 15). Bacterial meningitis, pneumonia, and epiglottiditis are other potentially fatal infections caused by encapsulated microbes.

Case Connections

- In this chapter we'll survey eukaryotic microorganisms, including fungi, such as *Cryptococcus neoformans*, that infect humans and animals, as well as others that infect plants.
- We'll examine the properties of these organisms that make them difficult to treat with antimicrobial drugs.

EUKARYOTIC MICROORGANISMS

In Chapter 11 we discussed how molecular studies (Chapter 10) have changed completely the taxonomy of prokaryotes by revealing the true relationships among these microorganisms. We based our discussion of prokaryotes on the phylogenetic scheme presented the latest edition of *Bergey's Manual.* We'll take a different approach here. We won't use a phylogenic scheme to organize the eukaryotic microorganisms. Instead we'll use the traditional grouping based on easily observable properties.

We'll consider the eukaryotic microorganisms as consisting of four groups: fungi, algae, protozoa, and slime molds. In the five-kingdom scheme of classification (Chapter 10), the eukaryotic microorganisms are divided between two kingdoms—the Fungi and the Protista. The Protista includes protozoa, unicellular algae, and slime molds. For simplicity, we'll consider the fungi, algae, lichens (symbiotic associations of fungi and algae or cyanobacteria), protozoa, and slime molds in that order, without special regard for their kingdom classification. We'll mention actual phylogenetic relationships as we proceed and summarize them at the end of the chapter.

In this chapter we'll also consider **helminths** (worms) and certain **arthropods** (animals with jointed legs, such as crabs, lobsters, insects, and spiders, mites and ticks). Helminths and arthropods are not microorganisms by anyone's definition, but we'll consider them because some members of both groups cause infectious diseases and some arthropods transmit them (Chapter 15).

SHARPER FOCUS

PARASITOLOGY: A CAN OF WORMS?

Parasitology is the branch of clinical medicine that studies diseases caused by protozoa and helminths. Of course, pathogenic bacteria, viruses, and fungi are also parasites. So why should protozoal and worm infections be set apart from the others to make parasitology a separate science? There are two reasons.

First, although protozoal and helminthic infections are only moderately important in temperate countries such as the United States, they are immensely important in trop-

ical countries. About 1 billion people are infected by the worm *Ascaris,* 600 million by malarial protozoa, and 300 million by larval roundworms called filaria. Almost all these people live in tropical countries of the developing world. That's why parasitology is sometimes called tropical medicine. But the situation is beginning to change. Increasingly these diseases are a threat to persons living in the United States. They are becoming more common in this country, because more infected people are moving to the

United States and because immune deficiencies such as acquired immunodeficiency syndrome (AIDS) are increasing, and they make people more susceptible to parasitic diseases.

Protozoal and helminthic parasites are complex organisms. We still don't know enough about them to control them adequately. For example, we know that protozoa and worms activate our immune responses. But we don't understand why the immune system is usually unable to rid the body of them. Typically, parasitic infections

last for years. Their persistent debilitating impact makes these infections especially tragic. They lower the quality and the length of life of millions of people. Their social and economic impact is incalculable.

Parasitic infections are set apart for a second reason: They tend to have elaborate life cycles. They pass through multiple hosts. They change their body type as they develop. Creatures as diverse as mosquitoes, snails, and cattle are essential alternate hosts for some parasites that infect humans.

Fungi

The fungi are a large and diverse group. With about 70,000 named species, they constitute a kingdom that includes single-celled organisms called **yeasts;** filamentous organisms called **molds;** and organisms that aggregate to form fleshy, macroscopic, fruiting structures called **mushrooms, puffballs,** or **shelf fungi,** depending on their shape (**Figure 12.1;** see also Figure 1.5b). The branch of microbiology that studies this group of organisms is called **mycology** (*mykes* is Greek for "mushroom").

Fungi are **heterotrophic** (they use organic compounds as a carbon source), **nonphototrophic** (they do not use light as an energy source), and **absorptive** (they take up nutrients in solution).

Most fungi are **saprophytes** (they obtain nutrients by decomposing dead organic matter), but some are pathogens of plants and animals, including humans. Although they cause most plant diseases, they cause only a small number of human diseases. Most of these are not serious, but all are hard to treat and some are lethal.

We'll consider the fungi as two groups: the **higher fungi** and the **lower fungi.** But first let's look at some of the general properties of both groups.

Morphology. All fungi, with the exception of **yeasts,** grow as an interconnected system of branched, tubelike filaments. The individual filaments are called **hyphae.** The entire interconnected set of such hyphae, which contains a multinucleate mass of cytoplasm, is called a **mycelium.** The mycelia of lower fungi are called **coenocytic** (from Greek, meaning "common cell") because they are completely undivided by cross walls. Mycelia of the higher fungi appear to be divided by structures called **septa,** but these septa are incomplete. They have a central opening through which cytoplasm can flow freely. In this sense, then, the **thallus** (body) of all fungi can be viewed as being a single cell. A yeast is and looks like an ordinary single cell. The filamentous higher and lower fungi look like they are multicellular, but they aren't because their multinucleate mass is undivided by cross walls (**Figure 12.2**).

Fungi are often characterized by the overall appearance of their thalli. They are called yeasts, molds, or fleshy fungi, depending on how their component cells or hyphae are arranged. These terms have no taxonomic significance.

Yeasts are single cells, mostly oval shaped, that reproduce by **budding.** In budding a bubble forms on the cell surface, grows, pinches off, and usually separates. Some-

FIGURE 12.1 Fungi. (a) A shelf fungi growing from a tree trunk and (b) Grapes infected with the mold *Plasmospora viticola,* which causes powdery mildew.

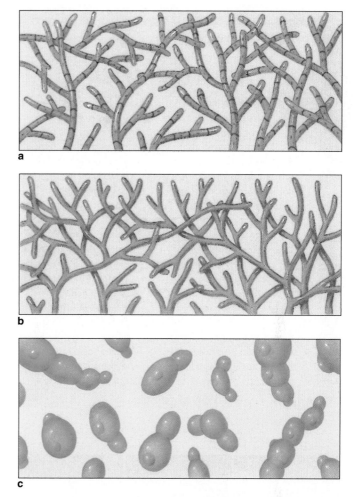

FIGURE 12.2 Fungal thalli. (a) Mycelium of a higher fungus with septa (incomplete cross walls) at regular intervals. (b) Coenocytic mycelium of a lower fungus. (d) Pseudohyphae typical of some yeasts.

(Art by Carlyn Iverson.)

times buds remain attached to the mother cell, forming a chain of cells called a **pseudohypha.** A few yeasts divide by forming cross walls. On solid media, yeasts form round, pasty, or mucoid colonies with a pleasant odor.

Molds are highly branched and loosely intertwined hyphae that form an open mycelium. The mycelia of molds form woolly growth on damp and decaying material in nature and fuzzy spreading colonies on solid media in culture (**Figure 12.3**).

Fleshy fungi (those that form the fruiting structures called, for instance, mushrooms, puffballs, or shelf fungi) consist of tightly intertwined hyphae. Sexual spores form within these structures. Some of these fruiting structures appear quickly—sometimes overnight—because they are formed from an extensive, usually underground, moldlike mycelium. The fruiting structure forms as cytoplasm flows into it from the underground mycelium.

Fungal hyphae are remarkably tough. They are composed of cellulose (Chapter 2), **chitin** (a polymer of acetyl-glucosamine, also found in the hard exoskeleton of crabs, lobsters, and some insects), or a combination of the two.

Ecology. Fungi survive and prosper in nearly every environment on Earth, both aquatic and terrestrial. They grow in highly acidic environments, such as acidic fruit, and in the presence of high concentrations of salts, sugars, and other nutrients, such as on the surface of home-canned jams and jellies. They grow at temperatures well below the freezing point of water (below those that support the growth of bacteria), but they cannot tolerate temperatures as high as bacteria do. Fungi can grow with minuscule concentrations of nutrients. In the laboratory, for example, they often develop in "nutrient-free" reagents, by getting their nutrients from airborne organic compounds that dissolve in these liquids.

SHARPER FOCUS

GERALD FINK

In 1992, at the Whitehead Institute in Boston, Gerald Fink discovered that *Saccharomyces cerevisiae* exhibited dimorphism—it could switch from being yeastlike (unicellular) to being mycelial. The discovery came 132 years after Louis Pasteur discovered the phenomenon and in a microorganism that has been intensively studied by hundreds of microbiologists (baker's yeast is the *Escherichia coli* of eukaryotic microbiology) and used by bakers and brewers for thousands of years.

How could dimorphism in this organism be overlooked all this time by so many people? Fink himself was suspicious of his discovery, believing the mycelial form might be a contaminant blown in from the ventilation system. He became convinced only when one of his graduate students showed that the mycelial form could mate with unicellular forms. Fink discovered dimorphism in baker's yeast in the course of studying how *S. cerevisiae* responds to the harsh conditions it undoubtedly encounters in nature. He

found that near-starvation for a nitrogen source triggers the change. Fink speculates that by providing a way to forage for food, dimorphism offers a selective advantage. Being nonmotile, baker's yeast cannot swim to a new source of nutrients, and budding cells form a compact colony on a surface. Filaments, however, spread over a surface, enabling the organism to obtain nutrients from a larger area. Many mycologists now believe that dimorphism is the rule among the fungi. It's just a case of finding

the conditions that trigger the change.

Finding that baker's yeast is dimorphic has significance beyond learning more about this one organism. The ability of most fungal pathogens to infect humans depends on their being dimorphic. But most pathogenic fungi have not been intensively studied, and we badly need antifungal drugs. A drug that stops dimorphic change might cure fungal infection and not be toxic to the human host.

pine needles

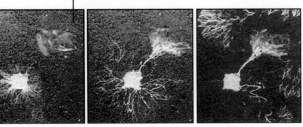

Courtesy Alan D. M. Rayner, University of Bath, and the New Phytologist Trust

FIGURE 12.3 A mold seeking nutrients. Note how the mold senses the presence of nutrients (in the pine needles), grows toward it, and covers it with mycelium.

Most fungi are aerobes. A few species are anaerobes, and some species, including many species of yeast, are facultative anaerobes.

Along with bacteria, fungi are the prime decomposers of organic materials, including food. They convert dead plant and animal materials into forms that plants can use (Chapter 28). Some fungi are even better than bacteria at decomposing decay-resistant organic materials such as **lignin** (the dark-colored component of wood). Fungi such as *Polystictus* spp. and *Armillaria* spp. cause **white rot** of wood by attacking the dark-colored lignin and leaving white cellulose. Other fungi (and bacteria as well) cause **brown rot** by attacking cellulose and leaving lignin. They convert wood

into a pulpy reddish-brown mass. Other fungi, such as *Pleurotus* spp. and *Polyporus* spp., attack both cellulose and lignin. Fungi can break down just about anything organic—including clothing, paper, leather, and paint. Some fungi even grow in jet fuel, causing clogged fuel lines. Fungi are abundant in garden compost piles, where they convert plant materials and other organic waste into rich, friable soil.

Reproduction. Fungi are classified by how they reproduce, sexually or asexually.

Asexual reproduction occurs by elongation and fragmentation of hyphae, by budding or division of yeast cells, or by production of asexual spores. These specialized cells are called asexual because their formation does not require fusion of gametes. These spores are adapted for dispersal. They travel to a favorable environment and germinate there to produce new thalli. Different fungi produce one or the other of two general types of asexual spores—**sporangiospores** or **conidiospores** (also called **conidia**). Sporangiospores are produced *within* a spore-containing structure called a **sporangium**. Conidia are borne naked *on* the tips of specialized hyphae called **conidiophores.**

Sexual reproduction involves the production of sexual spores. These form following sexual fusion of gametes. Usually, meiosis also occurs before the spores form, but it can occur after. **Homothallic** fungi produce both male and

TABLE 12.1 Properties of Major Groups of Fungi

Class	Distinguishing Characteristics	Examples
The Lower Fungi	Mycelia (if present) are coenocytic and lack septa.	
Chytridiomycetes	Water molds with uniflagellated sporangiospores and gametes.	*Allomyces* (seen as fuzzy growth in tropical ponds and streams)
Oomycetes	Water molds with biflagellate sporangiospores and male gametes, nonmotile female gametes and sexual spores.	*Saprolegnia* (seen as fuzzy growth on fish in a stream)
Zygomycetes	Terrestrial fungi with nonmotile sporangiospores and zygospores. Several genera parasitize weakened human patients.	*Rhizopus* (black bread mold)
The Higher Fungi	Mycelia (if present) are septate; all species are terrestrial; conidia and gametes (if present) are nonmotile.	
Ascomycetes	Ascospores formed in a sac (ascus); some parasitic on plants or animals.	*Peziza* (an orange cup fungus seen on soil in damp woods)
Basidiomycetes	Basidiospores formed on a club-shaped basidium; some as parasitic on plants or animals.	*Amanita* (red or white mushrooms seen in forests)
Deuteromycetes (Fungi Imperfecti)	Higher fungi that lack sexual spores; some are parasitic on plants or animals.	*Penicillum* (blue or green mold seen on fruit; used to make penicillins)
Yeasts	Unicellular, oval cells, mostly Ascomycetes and Deuteromycetes but some Basidiomycetes and Zygomycetes; some parasitic on plants or animals.	*Saccharomyces* cerevisiae (yeast used in baking, brewing, and wine making)

female gametes on the same thallus. **Heterothallic** fungi produce them on different thalli. Usually, sexual spores are more resistant than asexual spores and vegetative cells to heat, desiccation, and other hostile conditions.

The type of spores fungi produce is a major basis for subdividing the two major groups—lower fungi and higher fungi—into classes. The major groups themselves are distinguished by hyphal structure (**Table 12.1**).

The Lower Fungi. The lower fungi are all coenocytic. According to the structure of their spores and gametes, they are divided into five classes. The most important of these are the **Chytridiomycetes,** the **Oomycetes,** and the **Zygomycetes.** All members of these groups produce sporangiospores.

Chytridiomycetes. The Chytridiomycetes are water molds. They grow as fuzzy, whitish masses on organic materials in streams or ponds. Within a few days after being immersed in water, a piece of fruit or a dead fish, for example, becomes covered with a growth of Chytridiomycetes and other water molds. Chytridiomycetes are distinguished by producing sporangiospores and gametes that are motile by means of single posterior flagella. Some Chytridiomycetes

are unicellular. Some classification schemes place them among the Protista. *Allomyces* is a well-studied Chytridiomycete. It grows in tropical ponds and streams, producing beautiful bright orange sporangia.

Oomycetes. The Oomycetes are also water molds. They also produce motile sporangiospores, but their sporangiospores are **biflagellate** (two flagella) and their sexual spores are not motile. They form after nonmotile female gametes fuse with motile or nonmotile male gametes. In spite of having motile sporangiospores, these organisms do not have to live immersed in water. The thin film of water on damp soil particles is adequate for them to complete their life cycle. Some Oomycetes infect plants. Very few infect humans. Some species, such as those belonging to the genus *Saprolegnia*, that live in streams and ponds are parasitic on fish (**Figure 12.4**).

Zygomycetes. The Zygomycetes are completely terrestrial. They can complete their life cycle without liquid water because they don't produce swimming cells. The black mold *Rhizopus nigricans*, which develops on stale bread and other cereal foods, has a life cycle that's typical of Zygomycetes (**Figure 12.5**). It grows as a haploid mycelium that spreads

© Heather Angel/Biofotos

FIGURE 12.4 The Oomycete *Saprolegnia* parasitizing a fish. Note the peeled skin and the fuzzy growth on the tail.

by producing airborne sporangiospores. These haploid mycelia belong to one or the other of two mating types. They are designated plus (+) and minus (−) instead of male and female because they can't be distinguished by appearance. When a plus and minus strain grow close to one another, some ordinary looking hyphae grow toward each other and act as gametes. They fuse and produce a diploid zygospore which can withstand harsh conditions. When favorable conditions return, the zygospore germinates, undergoes meiosis, and develops into a sporangium. Half its sporangiospores germinate to produce plus mycelia. The other half produce minus mycelia.

Several genera of Zygomycetes, including *Rhizopus*, contain species that cause diseases called zygomycoses. They attack individuals who are immunologically weakened.

The Higher Fungi. The higher fungi are not coenocytic. Their hyphae have regularly spaced septa (perforated by a central pore). There are two groups—Ascomycetes and Basidiomycetes—that are distinguished by the kind of sexual spores they form. A third group—Deuteromycetes—consists of those that fail to form sexual spores.

Ascomycetes. The sexual sores of Ascomycetes are called **ascospores.** They form after hyphal tips on the same or different thalli (**homothallic** or **heterothallic,** respectively) and the nuclei they contain fuse. Then the diploid nucleus within the resulting zygote undergoes meiosis. Sometimes the four nuclear products divide again before these haploid nuclei develop into ascospores. The cell that contains them becomes a saclike structure called an **ascus** (**Figure 12.6a**). When released from the ascus, ascospores germinate to produce a haploid mycelium which can reproduce asexually by forming conidia.

The major difference among species of Ascomycetes is the way that asci are arranged. Some species produce individual asci. They are moldlike throughout their life cycle. Other species produce arrays of asci arranged in sometimes elaborate structures called **ascocarps.** The shapes of ascocarps are quite varied. Some, like those of *Peziza* spp., are cup shaped (**Figure 12.6b**). These are the bright orange or red structures seen on forest floors. Other species, such as *Morchella* spp. (morels), are mushrooms, with globular heads on stalks. Morels are prized for their delicious taste.

Basidiomycetes. The sexual spores of Basidiomycetes are called **basidiospores.** In most species they form in structures called **basidiocarps** (commonly called mushrooms, puffballs, or shelf fungi), but some don't form basidiocarps. They are molds, and a few are yeasts. Basidiospore formation by mycelial Basidiomycetes passes through a biologically unique phase called a **dikaryon** (**Figure 12.7**). A dikaryon is a mycelium that contains two different kinds of nuclei—one from each of the two parents. Dikaryons form when the hyphae from two mating types fuse. Then the cytoplasms from the two mycelia mix, but the nuclei do not fuse. There are other organisms that are briefly dikaryotic, but only basidiomycetes exist as dikaryons throughout most of their life cycle. The dikaryon phase of mushroom- and puffball-forming Basidiomycetes develops underground. When conditions are right, much of the cytoplasm from the mycelium flows to the soil surface, where it quickly forms the basidiocarp, the mushroom or puffball. Within the basidiocarp, pairs of nuclei within the same hypha fuse. The resulting diploid nucleus then undergoes meiosis, producing four nuclei. Each develops into a basidiospore. The four of them sit on a single club-shaped cell called a **basidium.** Masses of these basidia line the gills of gilled mushrooms, the pores of pored mushrooms, or the inside of puffballs. When mature, basidiospores are shot from the basidium into the space between the gills or the openings of pores. Thus they fall free of the mushroom and land on the ground, where they germinate.

Deuteromycetes. Deuteromycetes do not produce sexual spores. They only produce conidia, and they grow as molds or yeasts. Deuteromycetes are also called Fungi Imperfecti because nineteenth-century mycologists considered fungi that lacked a sexual stage to be imperfect.

Deuteromycete is a classification of convenience. A higher fungus that does not form sexual spores cannot be assigned to either the Ascomycetes or the Basidiomycetes, so it is assigned, sometimes temporarily, to the Deuteromycetes. Maybe it can't form sexual spores because its opposite mating type is not available. Maybe it has lost the capacity to form sexual spores. If it later forms sexual spores, it is reclassified as an Ascomycete or Basidiomycete (**Table 12.2**).

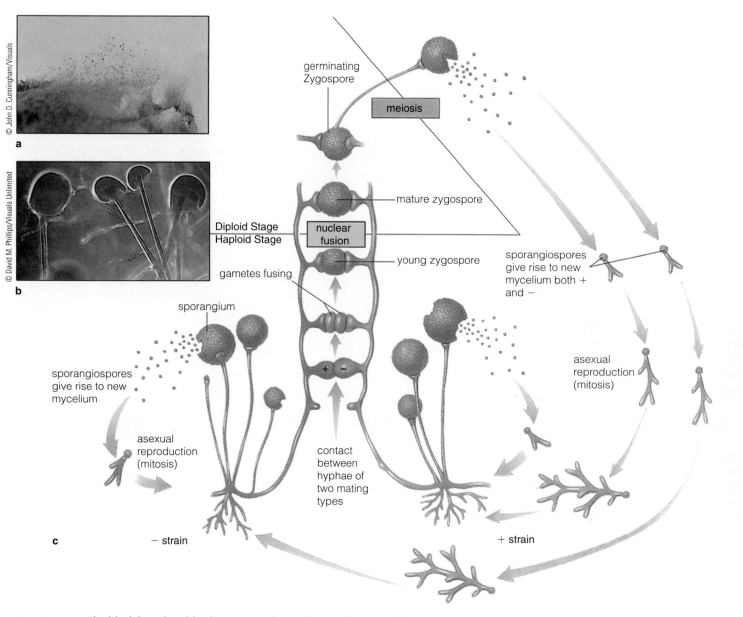

© John D. Cunningham/Visuals

© David M. Phillips/Visuals Unlimited

FIGURE 12.5   The black bread mold, *Rhizopus*. (a) Photo of the surface of a piece of bread covered with *Rhizopus*. Note the abundance of black sporangia. (b) Close-up of four sporangia packed with black sporangiospores. (c) Life cycle of *Rhizopus*.

(Art by Carlyn Iverson.)

Because they lack sexual spores, Deuteromycetes are identified principally by the shape and arrangement of their conidia (**Figure 12.8**). For example, the well-known imperfect genus *Penicillium* is distinguished by forming long chains of conidia on branching conidiophores, creating a brushlike structure (*penicillus* means "brush" in Latin). The closely related genus *Aspergillus* forms long chains of conidia on a globelike conidiophore.

Penicillium is well known because some of its species produce the antibiotic penicillin. *Aspergillus* is known for causing aspergillosis, a disease of animals, including humans. Most fungi that cause human disease—including *Blastomyces*, *Cryptococcus*, *Histoplasma*, and *Candida*—are Deuteromycetes.

Yeasts. Yeast is a descriptive term, not a taxonomic one. It simply means a fungus the grows as single cells. Originally only strains of *Saccharomyces cerevisiae* used to make bread, beer, and wine were referred to as yeasts. This facultative anaerobe (Chapter 5) was cultivated by humans because it

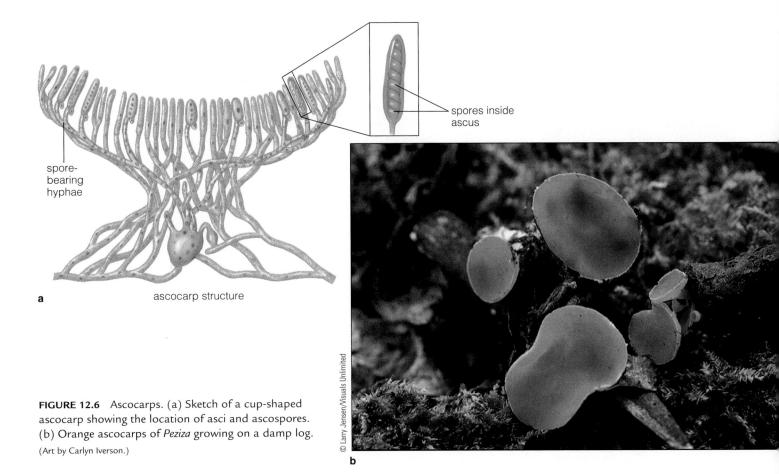

spores inside
ascus

spore-
bearing
hyphae

a   ascocarp structure

© Larry Jensen/Visuals Unlimited

FIGURE 12.6   Ascocarps. (a) Sketch of a cup-shaped ascocarp showing the location of asci and ascospores. (b) Orange ascocarps of *Peziza* growing on a damp log. (Art by Carlyn Iverson.)

b

ferments sugars to ethanol and carbon dioxide. Carbon dioxide gas, which makes bread dough rise and fermenting beer or wine appear to boil, gives yeasts their name. It comes from the Greek word *zestos*, meaning "to boil."

TABLE 12.2 How Deuteromycetes Are Reclassified if Sexual Spores Form		
Deuteromycete Genus (No Sexual Spores)	Reclassification Sexual Spores Form	
	Class	Genus
Aspergillus	Ascomycete	*Eurotium*
Blastomyces	Ascomycete	*Ajellomyces*
Candida	Ascomycete	*Pichia*
Cryptococcus	Basidiomycete	*Filobasidiella*
Histoplasma	Ascomycete	*Gymnoascus*
Penicillium	Ascomycete	*Talaromyces*
Trichophyton	Ascomycete	*Arthoderma*

Today microbiologists use the term yeast for any of the hundreds of species of nonfilamentous, single-celled, round or oval-shaped fungi. Most yeasts are Ascomycetes or Deuteromycetes, some are Basidiomycetes, and a few are Zygomycetes. Yeasts, which include aerobes, as well as facultative anaerobes, are widely distributed in nature. They are found on leaves, fruits, and cured meats such as bacon and ham, in the nectar of flowers, in soil, and on our bodies (Chapter 14). Most yeasts multiply by budding. A few divide by fission.

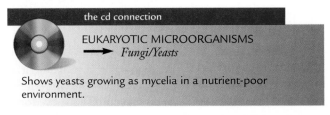

the cd connection

EUKARYOTIC MICROORGANISMS
→ *Fungi/Yeasts*

Shows yeasts growing as mycelia in a nutrient-poor environment.

Dimorphism. Some fungi switch between growing as a single-celled yeast and as a mycelium. This switching, called **dimorphism** (from Greek, meaning "two shapes"), has fascinated microbiologists since Louis Pasteur discovered it in 1860. He noticed that the mycelial Zygomycete *Mucor rouxii*

FIGURE 12.7 A gilled mushroom (Basidiomycete). (a) Life cycle. The gills on the lower side of the cap of this mushroom bear swollen club-shaped cells called basidia (b), some of which can be seen to bear basidiospores. When first formed, the basidia are dikaryons with one nucleus of each mating type. At fertilization the two nuclei fuse within the basidium, initiating the brief diploid stage of the life cycle. Soon the diploid nuclei undergo meiosis, producing four haploid nuclei, which migrate to the tip of the basidium, where basidiospores form. The basidiospores are shot off the gills and fall to the ground, where half germinate to become mycelia of one mating type; the other half become mycelia of the other mating type. After a brief period of growth, the mycelium of one mating type fuses with a mycelium of the other mating type. Because the nuclei do not fuse, the resulting mycelium is a dikaryon; it contains two kinds of nuclei. The dikaryon form undergoes extensive growth underground. Finally the massive underground mycelium aggregates to form an above-ground mushroom.

(Art from Biology: The Unity and Diversity of Life, 6th ed., by C. Starr and R. Taggart, Brooks/Cole, 1992. All rights reserved.)

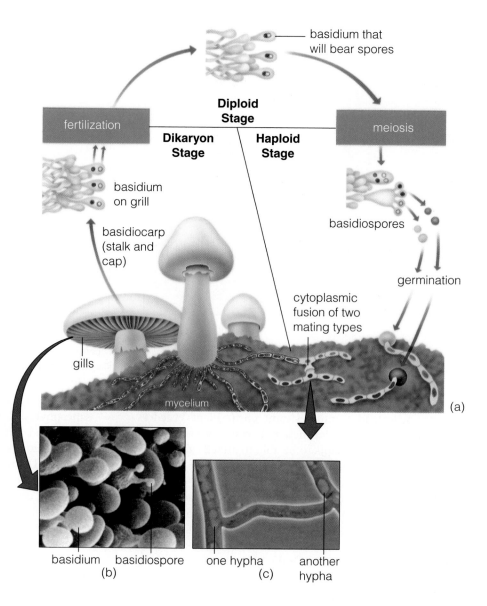

became yeastlike if deprived of oxygen. Later other mycelial fungi were found to become yeastlike at higher temperatures.

A capacity for dimorphism is critically important for pathogenic fungi such as *Histoplasma capsulatum*, *Blastomyces dermatitidis*, *Coccidioides immitis*, and *Candida albicans*, that can cause systemic infections (Chapter 27). They are mycelial outside the host and yeasts inside it. Being single celled inside allows them to spread throughout the body in the bloodstream. Being mycelia outside allows them to spread and compete more effectively for scarce nutrients. The higher temperature and higher concentrations of nutrients inside the body trigger the shift of most pathogenic fungi. Dimorphism is probably a more common property of fungi than we now recognize. Only recently was baker's yeast found to be dimorphic. It's probably just a case of finding the right conditions for switching.

Plant Disease. Bacteria and viruses cause plant diseases, but fungi are probably the major cause (**Table 12.3**). Two genera of Oomycetes, *Pythium* and *Phytophthora*, are particularly devastating. *Pythium* spp. attacks young plants near the ground, blackening them, causing them to collapse, and

killing them. The disease process is called **damping off**. *Phytophthora infestans* turns potato tubers into a dark slime. It caused starvation and massive emigration from Ireland during the nineteenth century (Chapter 1). A recently discovered species, *Phytophthora ramorum*, causes sudden oak death, which infects several species of oak trees and other plants and threatens to eliminate tan oaks in California.

Higher fungi cause apple scab, which disfigures apples (**Figure 12.9**); corn smut, which turns ears of corn into swollen, black, powdery masses; and wheat rust, which forms rusty red spots on the plant. Corn smut and wheat rust have destroyed vast quantities of these cereal grains. Another fungus causes Dutch elm disease, which has virtually eliminated that graceful shade tree from the United States.

Many plant diseases are successfully controlled by applying antifungal sprays or dusts. Others are controlled by

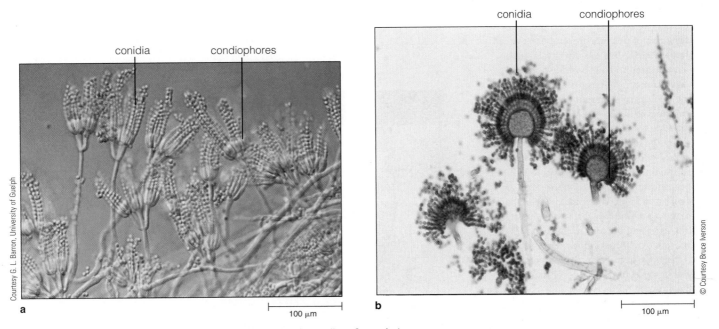

FIGURE 12.8   Conidia of Deuteromycetes. (a) Conidia of *Penicillium* form chains on branching conidiophores. (b) Conidia of *Aspergillus* form chains on globelike conidiophores.

breeding fungus-resistant plant varieties, such as rust-resistant wheat and smut-resistant corn. Plant breeders must continuously develop new rust-resistant varieties of wheat as the fungus adapts to be able to attack existing varieties. It's an unending race between fungus and breeder.

Human Health. Fungi do not play the dominate role in human health that they do in agriculture. Of the 70,000 named fungal species, fewer than a hundred cause human disease. Fungal infections, called **mycoses,** are discussed in Chapters 22 through 27 (**Table 12.4**). In general, these infections are not highly contagious. Humans usually ac-

quire them from nature, where the organisms exist as saprophytes. But when fungi do cause disease, they can be deadly and difficult to treat because many fungal pathogens infect tissues such as skin that are difficult for drugs to enter. Treatment itself can be dangerous. Some antifungal agents are highly toxic to humans.

Fungi threaten us in another way. Some of them produce toxins that are hallucinogenic or poisonous. For example, **muscarin,** produced by the mushroom *Amanita muscaria*, is highly hallucinogenic. It is consumed as part of the religious rites of peoples native to the northern rim of the Pacific Ocean. The toxins **phalloidin** and **amanitin,**

TABLE 12.3   Some Important Plant Diseases Caused by Fungi

Disease	Fungus and Class	Impact
Powdery mildew	*Plasmopara viticola* (Oomycete)	Farmers must spray several times a year to protect grapes from this infection, which destroys fruit
Potato blight	*Phytophthora infestans* (Oomycete)	Caused famine in Ireland (1845–1860)
Wheat rust	*Puccinia graminis* (Basidiomycete)	Sporadic epidemic that costs North American farmers billions of dollars
Corn smut	*Ustilago mayidis* (Basidiomycete)	Major destroyer of corn, turns ears black
Dutch elm disease	*Ceratocystis ulmi* (Ascomycete)	Almost eliminated elm trees from North America
Apple scab	*Venturia inaequalis* (Ascomycete)	Disfigures apple fruit

© Eric Crichton/Bruce Coleman Ltd.

FIGURE 12.9 Apples infected with the Ascomycete *Venturia inaequalis,* which causes apple scab.

TABLE 12.4 Some Higher Fungi That Infect Humans

Fungus	Infection (Chapter)
Blastomyces dermatitidis	Blastomycosis (22)
Coccidioides immitis	San Joaquin Valley fever (22)
Histoplasm capsulatum	Histoplasmosis (22)
Pneumocystis carinii	*Pneumocystis* pneumonia (22)
Candida albicans	Thrush, vaginitis, candidiasis (24, 16)
Cryptococcus neoformans	Cryptococcosis (25)
Dermatophytes (*Trichophyton, Microsporum, Epidermophyton*)	Tinea (ringworm, athlete's foot) (26)

produced by the closely related *A. phalloides* (also known as the death cap), are highly poisonous. Ingesting even a small quantity of this mushroom can cause death from irreversible liver failure. But another closely related species, *A. caesarae* (Caesar's mushroom), is a prized delicacy. So it's a good idea not to collect and eat mushrooms unless you know them well.

Some toxins that fungi produce threaten our food supply. For example, *Claviceps purpurea,* a mold that grows on rye (rarely in other grains), produces highly toxic **ergot.** Once a major killer (see Larger Field: St. Anthony's Fire), it is well controlled now. But **aflatoxin,** produced by certain species of *Aspergillus,* principally *A. flavus,* is still a serious problem in parts of Asia and Africa. These species grow in many kinds of plant materials. Crops such as peanuts and cereal grains, if not properly dried, can contain enough aflatoxin to cause severe liver damage and death. Even low levels are dangerous because they may be carcinogenic. In the United States, rigid standards of food processing and testing have effectively eliminated aflatoxin from our food supply. Another toxin-producing fungus, *Stachybotrys chartarum,* sometimes called black mold, is responsible for "sick building syndrome" (see Sharper Focus: Sick Horses, Sick Buildings).

Other fungi are on our side in the battle against disease. The Deuteromycete species *Penicillium notatum* and *P. chrysogenum* produce penicillins. *Cephalosporium* spp. produce a closely related family of antibiotics, the cephalosporins. Even

deadly ergot can be beneficial. Given in small amounts, it mitigates migraine headaches and aids childbirth.

Algae

Algae are phylogenetically diverse but metabolically unified. They all generate adenosine triphosphate (ATP) by oxygenic photosynthesis using the same kind of chlorophyll (chlorophyll a) that plants use. And they all make their precursor metabolites from carbon dioxide (CO_2) through the Calvin-Benson cycle (Chapter 5). But their structure and life cycles are quite varied. Some are microscopic single-celled organisms. At the other extreme, some, commonly called seaweeds, are huge multicelled structures up to 50 meters long. But all algae, even the large ones, lack the tissue differentiation that's typical of higher plants. All the cells of algae are similar. Algae also differ from plants in the way they reproduce. They do not produce **embryos** (miniature organisms in a seed). Instead, they produce single-celled asexual spores and gametes. Most of them are motile by means of flagella.

Ecology and Uses. Algae are mostly aquatic organisms. A few, however, can live in moist terrestrial environments such as the surface of soil, rocks, and tree trunks in damp areas. Algae live in salt water, as well as freshwater. Their metabolic activities profoundly affect both environments.

LARGER FIELD

ST. ANTHONY'S FIRE

In August 1951, 300 inhabitants of Point-Saint-Esprit, a small village in France, became ill and 5 died of a bizarre illness. People hallucinated that they were being chased by tigers or saw death walk into the room. They suffered convulsions. They jumped from rooftops. The streets were filled with screaming people. Not just humans were affected. A cat writhed, twisted, and tried to climb a wall. A dog leaped in the air, snapping viciously, and crushed rocks with its teeth until blood dripped from its mouth. Ducks strutted like penguins, flapped their wings, quacked to a crescendo, and died.

Laboratory tests on the victims were too late to detect the cause, but experts agreed that this was an outbreak of St. Anthony's fire, or ergot poisoning. It was almost certainly caused by illegally distributed ergot-contaminated rye flour that was made into bread at the village bakery. It was verified that the sick animals had also eaten the contaminated bread. John J. Fuller,

a science writer who published a definitive account of the episode, concluded, "There is one and only one cause of the tragedy: some form of ergot, and that form has logically got to be akin to LSD."

Ergot poisoning is caused by the fungus *Claviceps purpurea,* an Ascomycete that infects rye, a major cereal grain. The infection does little damage to the plant. Its body and most of the grain it produces look normal—only a few grains are replaced by a hardened purple-black mass of *Claviceps* mycelium called ergot. But when even the tiniest amount of ergot is consumed by an animal or human, it can cause disorientation, hallucinations, gangrene, abortion, or death.

Ergot poisoning was known to the Assyrians as early as 600 B.C. There were few occurrences during Roman times because Romans didn't care for rye. But during the Middle Ages, when Roman influence ended, rye became popular throughout Western Europe and major out-

breaks of ergot poisoning occurred. Entire villages suffered hallucinations. In 1994, 40,000 people died in the Limoges district of France. About that time ergot poisoning became known as St. Anthony's fire because the Pope authorized the Order of St. Anthony to treat its victims, some of whom complained of fiery pain. At first, theories of what caused St. Anthony's fire were as bizarre as the symptoms; they included witchcraft and poisonous air. Gradually, however, people came to associate the symptoms with infected rye, and ergot poisoning became rare.

What of the physiological symptoms, abortion and gangrene? Ergot contains an alkaloid, ergotamine, that makes smooth muscles contract. Contractions can be severe enough to induce abortion and restrict blood supply, so that gangrene results. Today purified ergotamine is used clinically to induce birth, control postpartum bleeding, and control migraine headaches.

The psychological symptoms of ergotism are not as well understood. Ergotamine does not cause hallucinations, and it is not mind altering. It is, however, structurally similar to lysergic acid diethylamide (LSD), an extremely powerful hallucinogen. A minuscule amount of LSD causes pigeons to "strut like penguins" and people to jump from high places. The threshold dose of LSD is 25 μg, and about 100 μg produces a reaction resembling a psychotic state. For comparison, an ordinary aspirin tablet is 300,000 μg. What, then, is the connection between ergot poisoning and its LSD-like psychological effects? In damp climates, other molds grow on ergot, and circumstantial evidence suggests that they convert a small amount of ergotamine to LSD or a similar mind-altering chemical. The outbreak in Point-Saint-Esprit was relatively small; St. Anthony's fire must have caused mass chaos and terror during the Middle Ages.

They are major fixers of CO_2. It is estimated that 80 percent of Earth's photosynthesis is carried out by the ocean's **phytoplankton** (microscopic, floating algal and cyanobacterial species; *planktos* means "drifting" in Greek). In turn phytoplankton nourish the ocean. They are the base of a food chain leading to fish and other aquatic animals.

Indirectly, algae have great impact on our lives. Without them the planet would not be habitable. But they have less direct impact. They do not cause infectious disease. Because they are photosynthetic, they cannot live inside our bodies.

The colorless alga *Prototheca* is a notable exception. It is associated with **bursitis** (an inflammation of the joints).

Certain algae have commercial value. For example, some marine algae (seaweed) are eaten as food in Asian countries. And **kelp,** the huge brown seaweed that grows profusely off the Pacific Coast (Chapter 1), is harvested to obtain alginic acid (algin), a polysaccharide that's used extensively as a thickener in ice cream, salad dressings, prepared sauces, and other foods. The red alga *Chondrus crispus,* also known as Irish moss, is the source of another

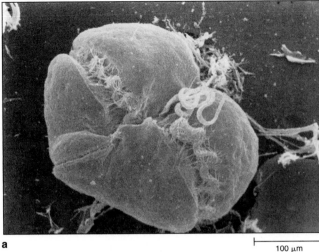

FIGURE 12.10   Dinoflagellates. (a) A *Gymnodinium breve* cell. (b) An estuary filled with fish killed by red tide consisting of large numbers of these of these cells.

polysaccharide thickener, called carrageenan. It's used in milk products, including ice cream, custards, and evaporated milk. Other red algae, species of *Gelidium* and *Gracilaria*, that grow mainly in the western Pacific Ocean are the total source of agar, which microbiologists use to solidify media. The shells of diatoms (an algal group), which occur in huge deposits called diatomaceous earth in regions that were once the floors of ancient seas, are used extensively in industry for polishing and insulating, as well as an additive to speed the rate of filtering liquids.

In spite of their not being infectious, a few algae are a threat to animal and human health. Certain dinoflagellates,

such as *Gymnodinium* and *Gonyaulax*, produce a potent neurotoxin and a red pigment. They sometimes proliferate in shallow water during spring and summer, forming **blooms** (dense accumulations of cells) called **red tides.** Shellfish that feed on them are not harmed, but fish and humans who eat the shellfish can suffer life-threatening paralysis (**Figure 12.10**).

Classification. Algae are divided into six groups according to the form of their thalli (whether they are unicellular, coenocytic, filamentous, or plantlike), the structure of their walls, and the pigments they produce (**Table 12.5**).

TABLE 12.5 Properties of Major Groups of Algae

Phylum	Thallus Structure	Wall Composition	Pigments Chlorophylls	Others
Euglenophyta: euglenoids	Unicellular	No wall	a, b	
Pyrrophyta: dinoflagellates	Unicellular	Cellulose	a, c	Carotenoids
Chrysophyta: diatoms	Unicellular, coenocytic, filamentous	Silica	a, c	Carotenoids
Chlorophyta: green algae	Unicellular, coenocytic, filamentous, or plantlike	Cellulose	a, b	
Phaeophyta: brown algae	Plantlike	Cellulose and algin	a, c	Carotenoids
Rhodophyta: red algae	Plantlike	Cellulose	a	Phycobilins

Algal pigments participate in photosynthesis. The chlorophylls (a, b, and c) mediate the basic reactions of converting light energy into chemical energy (Chapter 5). The accessory pigments, such as carotenoids (which give carrots their characteristic color) or phycobilins (which are structurally similar to cytochromes; Chapter 5), collect light energy and pass it on to the chlorophylls. The pigments an alga contains dictate its habitat. It grows best in an environment rich in the wavelengths of light that its pigments absorb.

Pigments are also the basis for the common names of three groups—the green algae, the brown algae, and the red algae. The other three groups—the euglenoids, the dinoflagellates, and the diatoms—are distinguished by morphology. Euglenoids are single motile cells with two flagella of unequal length. Dinoflagellates are single cells, usually with two flagella, one wrapped in a groove around the middle of the cell and the other extending from the groove. Diatoms are single cells enclosed in an elaborately sculptured, rigid silica shell that resembles the two parts of a nested petri dish (**Figure 12.11**).

---

the cd connection

EUKARYOTIC MICROORGANISMS
→ *Protists/Flagellate protozoa*

Shows swimming euglenoid with red eye spot.

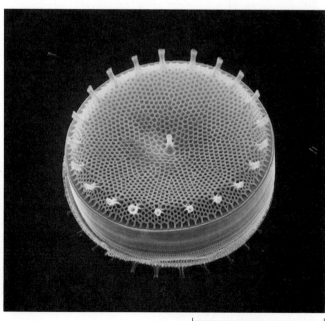

Courtesy Greta A. Fryxell, Texas A & M University

|———— 10 μm ————|

FIGURE 12.11   A diatom. Note the typical intricately sculptured silica shell.

Phycology.

Phycology, the study of algae and the province of specialists called **algologists,** is largely a natural science. But certain algae are studied experimentally in culture. Notable among these are *Chlamydomonas* (a unicellular green alga), *Euglena* (a euglenoid), and *Navicula pelliculosa* (a diatom). Each holds special interest.

Chlamydomonas, which swims actively by means of a posterior flagellum, can be propagated easily in the laboratory. It is an ideal model organism for studies on genetics, photosynthesis, and herbicide action. Herbicides that inhibit photosynthesis kill *Chlamydomonas,* as well as plants, but it is easier to study cultures of *Chlamydomonas* than fields of plants.

Euglena are motile single-celled organisms (**Figure 12.12**). They have a bright red eye spot that senses direction and intensity of light, allowing them to swim to bright regions where they can photosynthesize more rapidly. Studies on this response mechanism reveal how eukaryotes sense and respond to visible light.

The diatom *Navicula pelliculosa* is a model organism for studies on silicon metabolism, an unusual biological process. *Navicula* uses silicon to make its silica shells and to facilitate certain metabolic reactions.

Lichens

Lichens are not organisms. They are **mutualistic associations** (both partners benefit) between a fungus (usually an Ascomycete) and a phototroph (either a green alga or a cyanobacterium). These various associations are so intimate that they appear to be separate species. More than 16,000 such "species" have been described.

Different lichens vary enormously (**Figure 12.13**), but in all of them the fungus is the dominant partner. It makes up most of the mass and gives the thallus its shape. It may be upright (fruticose), leaflike and flat (foliose), or encrusted and flat (crustose). Usually some fungal mycelia extend beyond the thallus to form **rhizoids** (rootlike structures).

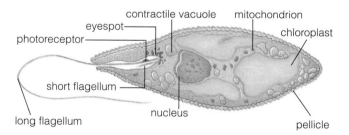

contractile vacuole   mitochondrion
eyespot
photoreceptor   chloroplast

short flagellum

long flagellum   nucleus   pellicle

FIGURE 12.12   Sketch of a euglenoid alga showing its complex structure.

(Art by Palay/Beaubois from Biology: The Unity and Diversity of Life, 6th ed., by C. Starr and R. Taggart, Brooks/Cole, 1992. All rights reserved.)

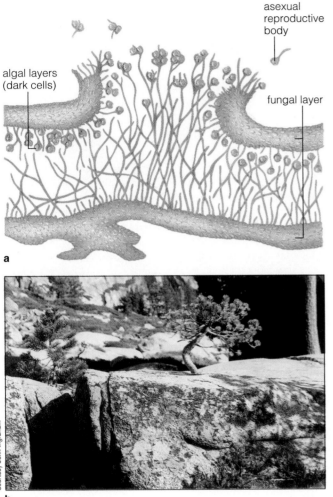

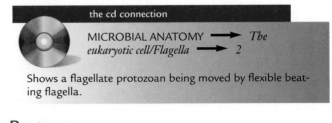

FIGURE 12.13 Lichens. (a) A cross section through the thallus of a typical lichen showing its two layers of fungal mycelium with algal cells in between. (b) Black, yellow, orange, and white foliose lichens growing on granite in the Sierra Nevada mountains. (c) *Usnea* (old man's beard) looks like moss hanging from the limbs of a tree.

(Art by Carlyn Iverson; adapted from Biology of Plants, by Peter H. Raven, Ray F. Evert, and Susan E. Eichhorn 1971, 1975, 1976, 1981, 1986, 1992, 1999 by W.H. Freeman and Company/Worth Publishers. Used with permission.)

These attach the lichen to a tree, a rock, or some other solid surface. Some lichens that hang from trees look like mosses. Others look like thin, colored patches on rocks. All of them grow where nutrients are scarce and few other organisms can survive. The manna that the Bible describes as food miraculously supplied to the Israelites in the desert wilderness is thought to have been lichens. Litmus, a dye used as a pH indicator, is extracted from the lichen *Roccella tinctoria*. Most lichens grow very slowly, only about 0.1 to 4 cm per year, but they are very long lived. Some in the Arctic are estimated to be 1000 to 4000 years old.

A lichen is an intimate association but not necessarily a permanent one. The two partners can be teased apart and cultivated separately. By itself, the fungus grows into a shape quite different from its shape in the lichen. In contrast, the alga, which usually exists as individual cells in the lichen, continues to do so. It's possible to re-form a lichen from its constituent alga and phototroph by growing them together and restricting their supply of nutrients. A reconstituted lichen looks just like a natural one.

The benefits the fungus derives from being part of a lichen are obvious. It's nourished by products of the phototroph's photosynthesis. If the phototroph is a cyanobacterium, it might provide fixed nitrogen as well. The benefits to the phototrophs are less clear. At the least, the association provides them with a home that protects them from desiccation and extremes of temperature and light.

the cd connection

MICROBIAL ANATOMY ⟶ *The eukaryotic cell/Flagella* ⟶ *2*

Shows a flagellate protozoan being moved by flexible beating flagella.

Protozoa

Protozoa are nonphotosynthetic, unicellular eukaryotes that cause some of the most devastating human diseases, including malaria (Chapter 27), leishmaniasis, and

SHARPER FOCUS

NOT GONE AND NOT FORGOTTEN

The mitochondria (and chloroplasts) in eukaryotes are remnants of prokaryotic cells engulfed by some primitive eukaryote. Microbiologists have suspected this for decades, and now the evidence is overwhelming. The biochemical and morphological structure of these organelles is clearly prokaryotic. Also, antibi-

otics active against prokaryotes inhibit much of their metabolism without affecting other parts of eukaryotes. Certainly, a primitive eukaryote engulfed the prokaryotic precursor of mitochondria. But what about the mitochondria-less eukaryotes that exist today? Did their ancestors never have mitochondria? Or did they once have

them only to lose them later? Now the answer is known.

Mitochondria-less eukaryotes that persist today include *Giardia* spp. and a closely related group, the Microsporidia. Microbiologists have known for some time that these protozoa lack mitochondria and consequently are anaerobes. They thought these

organisms lost their mitochondria because they were unnecessary in anaerobic environments. Now that suspicion is supported by strong evidence. These organisms have genes that certainly came from mitochondria. So mitochondria must have been in their ancestral cells. They left incriminating mitochondrial footprints behind.

African sleeping sickness (Chapter 25). Malaria alone affects more people than any bacterial disease.

Protozoa are the pinnacle of unicellular complexity. Some have organelles that are almost as complex as the organs of multicellular organisms. For example, *Tetrahymena* has a mouthlike **cytostome** that takes in particulate food. It has cilia that sweep food into the cytostome, as well as move the cell. It has food vacuoles that digest the food. It has an anuslike **cytoproct,** through which undigested food is expelled. In addition, it has a contractile vacuole that keeps the cell from bursting from turgor pressure by expelling the water that constantly flows into the cell by osmosis. All this exists within a single cell.

The protozoa are divided traditionally into four groups on the basis of their means of locomotion (**Table 12.6**). The flagellates move by means of one or more flagella. The amoeboids move by extending pseudopods, long lobes that form on the cell (Chapter 4). The sporozoa are nonmotile, and the ciliates move by means of many cilia. It shouldn't be surprising that the groups of protozoa are so diverse because they are only distantly related. For example, the ciliates are more closely related to the brown algae than to the flagellates.

Flagellate Protozoa. The flagellate protozoa (also called Mastigophora) resemble (and are related to) the euglenoid algae, differing principally by not being photosyn-

TABLE 12.6 Major Groups of Protozoa

Group	Means of Motility	Parasitic Species	Disease (Chapter)
Mastigophora (flagellates)	One or more flagella	*Trypanosoma brucei*	Sleeping sickness (25)
		Trypanosoma cruzi	Chagas' disease (27)
		Giardia lamblia	Giardiasis (23)
		Trichomonas vaginalis	Trichomoniasis (24)
Sarcodina (amoeboids)	Pseudopods	*Entamoeba histolytica*	Amoebic dysentery (23)
		Naegleria fowleri	Primary amoebic encephalitis (25)
Sporozoa	Nonmotile	*Plasmodium*	Malaria (27)
		Toxoplasma gondii	Toxoplasmosis (27)
Ciliophora (ciliates)	Numerous cilia	*Balantidium coli*	Balantidiasis (23)

thetic. Their flagella are unlike the flagella of bacteria (Chapter 4). Instead of turning, they beat with a whiplike motion. Usually there are two of them. They can be either anterior or posterior. That is, they either push or pull the cell along. In some species the flagella lie close to the cell surface and are covered by an expanded membrane called an **undulating membrane.** As the flagella moves, the membrane undulates propelling the cell.

Members of one group of flagellate protozoa, the trypanosomes (including the species that cause African sleeping sickness and leishmaniasis), have a leaflike appearance, a single flagellum, and an undulating membrane at one stage of their life cycle (**Figure 12.14a**). Members of another group, the diplomonads, have a double body containing two sets of most organelles, giving the cell the appearance of a human face (**Figure 12.14b**). Diplomonads have two to six flagella. One diplomonad, *Giardia lamblia,* causes a common waterborne dysentery (Chapter 23).

Amoeboid Protozoa.

Amoeboid protozoa (also called Sarcodina) are found just about everywhere—in marine environments, freshwater, soil (including desert), and in association with animals. This is a huge and phylogenetically diverse group of microorganisms, with 48,000 described species. All members, however, have one common characteristic: At some stage of their life cycle they produce pseudopods.

the cd connection

MICROBIAL BEHAVIOR ⟶ *Types of motility/Amoeboid motility*

Shows an amoeba extending its pseudopods.

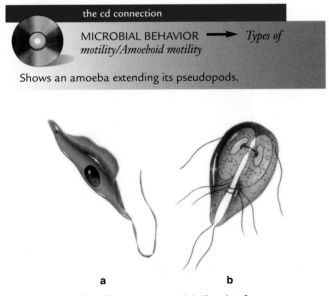

a   **b**

FIGURE 12.14 Flagellate protozoa. (a) Sketch of a trypanosome showing its leaflike appearance, single flagellum, and undulating membrane. (b) Sketch of a diplomonad showing its double body containing two sets of most organelles and eight flagella. Note its resemblance to a human face.

(Art by Carlyn Iverson.)

Pseudopods are organelles of locomotion and feeding. They extend to pull the cell along a solid surface and they engulf bits of food (**Figure 12.15**). The protein actin (related to the protein in the muscles of animals) is the driving force. It aggregates to form microtubules. These push against the cell membrane, causing bulges that the cytoplasm streams into, forming a pseudopod. When the actin molecules disaggregate, the pseudopod retracts, pulling the cell along.

Very few amoeboids cause human disease. But one species, *Entamoeba histolytica,* can cause a serious dysentery (Chapter 23).

Sporozoa.

In contrast to the generally benign amoeboid protozoa, all sporozoa are parasitic. The group constitutes more than 300 genera and 4000 species.

Some sporozoa have an elaborate life cycle during which they assume different forms and require different host. For example, the actively multiplying form of *Plasmodium* spp., which causes malaria, is called a **trophozoite.** The form that is infectious to animals or humans is called a **sporozoite.** And the form that enters and infects red blood cells is called a **merozoite** (**Figure 12.16**). *Plasmodium* spp. must enter the body of a female *Anopheles* mosquito, as well as an animal, to complete its life cycle (see Figure 27.14).

Toxoplasma gondii, which causes toxoplasmosis, a systemic disease of animals, particularly cats, and humans, has a less complicated life cycle (Chapter 27). Humans acquire the disease by consuming cysts in the meat of infected animals or by ingesting material contaminated by cat feces containing *Toxoplasma* cells.

M. Abbey/Visuals Unlimited

FIGURE 12.15 Micrograph of the amoeboid protozoa, *Amoeba proteus.* Note the three large and several small pseudopods.

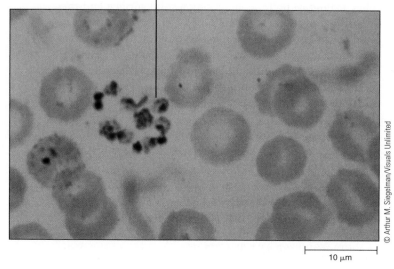

merozoite

10 µm

© Arthur M. Siegelman/Visuals Unlimited

FIGURE 12.16 Micrograph of red blood cells from a person suffering from malaria caused by *Plasmodium vivax*. Note that two cells contain the merozoite stage of the parasite.

Ciliate Protozoa. The ciliate protozoa (Ciliophora) often have an odd and distinctive nuclear pattern. They have two different nuclei, a larger **macronucleus** and a smaller **micronucleus.** The macronucleus is polyploid (has several copies of its complement of chromosomes). It directs vegetative growth and cell division. The micronucleus participates in sexual reproduction. Strains that lack a micronucleus cannot reproduce sexually, but they can grow and divide indefinitely.

Cilia are organs of locomotion. They have the same structure as flagella, only they are smaller (Chapter 4). By coordinated beating they can move the cell rapidly, turn it sharply, and stop it abruptly. Ciliates need these abilities because they prey on bacteria, fungi, and other protozoa. Predation is in some cases remarkably specific. For example, *Didinium nasutum* attacks only species of *Paramecium* (another ciliate), which it swallows whole (**Figure 12.17**). This is quite a feat because *Didinium* is not much larger than *Para-*

mecium. It enlarges its cytostome enormously and doesn't eat again for at least 2 hours.

the cd connection

EUKARYOTIC MICROORGANISMS
→ *Protists/Ciliate protozoa*

Shows protozoan using its cilia in feeding.

Cilia are also accessory feeding organs. They sweep small bits of food into the cytostome. For some nonmotile stalked species that's all they do.

Many ciliates are associated with animals. The rumen of cattle and other grazers teems with ciliates that digest cellulose and in turn become food for the animal (see the beginning of Chapter 8). A few ciliates are animal parasites. But only one, *Balantidium coli*, is a human pathogen. It causes balantidiasis, a severe diarrheal infection of the large intestine (Chapter 23).

The Slime Molds

In spite of their name, slime molds are not fungi. This small group of about 700 species is divided into two groups, the true slime molds and the cellular slime molds. None is a pathogen or is put to any industrial use.

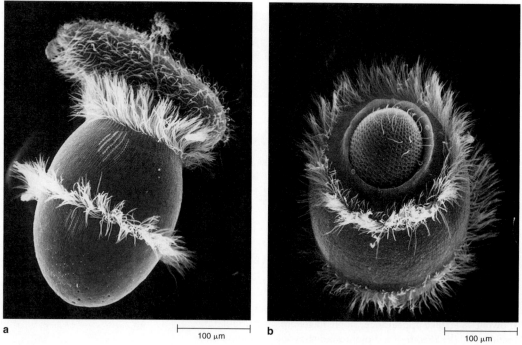

a   100 µm   b   100 µm

(a, b) Gary Grimes and Steven L'Hernault, Hofstra University

FIGURE 12.17 Ciliate protozoa. (a) Scanning electron micrograph showing one ciliate protozoan, *Didinium,* that has captured another, *Paramecium.* (b) Scanning electron micrograph of a *Didinium* in the process of swallowing a *Paramecium.*

SICK HORSES, SICK BUILDINGS

Recently we've all read and heard increasingly about "sick buildings" and the dangers coming from molds growing in them. Many people in the United States are profoundly worried that these molds will make them sick or cause them economic loss. The insurance industry estimates that $600 million in mold-related claims will be filed during 2003.

What's the basis for this concern? Molds are everywhere, and many benefit us in countless ways, including making some of our most valuable antibiotics, delicious cheese, and soy sauce. One concern is allergy. It's estimated that 10 percent of the population is allergic to mold. The other concern is more specific. A black mold,

Stachybotrys chartarum, that can grow on damp cellulose, such as the surface of wall board, produces about a dozen mycotoxins.

Although concern about *S. chartarum* in the United States is limited to the past decade or so, it has a longer disease-causing history in Eastern Europe. During the early 1930s a new disease of horses appeared in the Ukraine. They suffered a variety of symptoms, including irritation of mucous membranes, nervous disorders, hemorrhage, and death. Russian scientists established that the disease was caused by the *S. chartarum* that grew on the hay and grain fed to the animals. The infestations were severe, turning the feed intensely black. Such disease outbreaks

have not been reported in North America. In the late 1930s farm workers in Russia showed similar but milder symptoms 2 to 3 days after handling *S. chartarum*-infested hay or feed.

But the possibility of *S. chartarum*'s causing disease by growing inside buildings was not recognized until 1986, when over a 5-year period a family in Chicago complained of headaches, dermatitis, and general malaise. Their house had damp walls that favored mold growth. Air sampling of the house revealed the presence of mycotoxins that *S. chartarum* was known to produce. When the dampness was eliminated, mold growth stopped and the symptoms of the family members disappeared.

Since 1986 many cases of black mold–associated disease symptoms, including pulmonary hemorrhage in infants, have been reported. Because of the number of mycotoxins produced by *S. chartarum* and the broad spectrum of symptoms associated with them, it is usually difficult to associate unequivocally a particular illness with *S. chartarum*.

But the disease scenario is clear. Dampness within a building or the aftermath of a flood creates conditions in which the black mold *S. chartarum* can flourish and produce its spectrum of mycotoxins. The result can be a "sick building." People who live or work there can suffer symptoms.

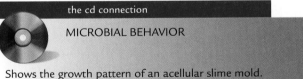

the cd connection

MICROBIAL BEHAVIOR

Shows the growth pattern of an acellular slime mold.

True Slime Molds. The true slime molds (also called Myxogastria) are phylogenetically related to the Oomycetes, but they don't resemble them in appearance. They look like a slimy, veined mass (**Figure 12.18a**). This sometimes colorful structure, called a **plasmodium,** is commonly seen on damp decaying logs or stumps. It is simply multinucleate cytoplasm. It flows along consuming microorganisms and bits of plant material in its path. As long as nutrients are available, it continues to expand, sometimes attaining a mass of half a pound or so. When it reaches a relatively dry spot, it devel-

ops fruiting bodies. Meiosis occurs within these raised structures and spores form. The spores, which are haploid, can withstand prolonged starvation, and when favorable conditions return, they germinate, producing amoeboid cells that feed on bacteria. The amoeboid cells can fuse, or, if water is present, they develop into flagellated cells that fuse. The product is a zygote that develops into a new plasmodium.

Cellular Slime Molds. The cellular slime molds (also called Acrasieae) are phylogenetically related to the red algae, but they produce cells that resemble amoeboid protozoa. One species, *Dictyostelium discoideum,* is a favorite of microbiologists who study morphogenesis because it undergoes startling morphological changes during its life cycle. Its amoeboid cells (which have a single nucleus) feed mainly on bacteria, by phagocytosis. When bacteria run out, the cells aggregate to form a mass of cells, called a **grex**

© Edward S. Ross

a

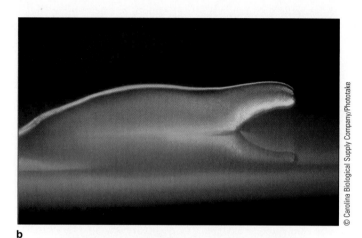

© Carolina Biological Supply Company/Phototake

b

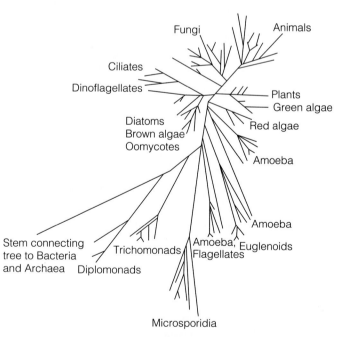

From "Morphogen Hunting in Dictyostelium," Robert R. Kay, Mary Berks, and David Traynor, Development 1989 Supplement, 81–90. The Company of Biologists 1989

c

FIGURE 12.18   Slime molds. (a) Plasmodium of a true slime mold, *Physarum,* on a decaying log. (b) Slug-shaped grex of the cellular slime mold, *Dictyostelium discoideum.* (c) Fruiting body of *D. discoidium.*

(**Figure 12.18b**). This sluglike structure moves toward light. When it reaches a bright area of light, it stops and differentiates into a fruiting body—a stalk with a bulbous sac filled with spores (**Figure 12.18c**). Then, under favorable conditions, the spores germinate, forming more amoeboid cells.

Relatedness of Eukaryotic Microbes

Figure 12.19 is a tree that shows the relatedness of the eukaryotes. It's apparent that the groupings—fungi, algae, and protozoa—that are commonly used to divide eukaryotic microbes are not all made up of closely related organisms. Most organisms called fungi, with the exception of Oomycetes, are closely related to each other and to animals. The various groups of algae—green, red, brown and Euglenoids—are not closely related. And the protozoa are even more disperse. Even various groups of amoeba are not closely related.

Helminths

Helminths are not microorganisms. They are macroscopic animals commonly called worms. We include these creatures in this microbiology text because some of them cause parasitic diseases (**Table 12.7**). In later chapters we'll consider

FIGURE 12.19   Relatedness of various eukaryotes. This tree, based on the sequences of bases in the small subunit of their ribosomal RNA, indicates relatedness by the length of the lines that connect various groups—the shorter the line, the closer the relationship. For example, fungi and animals are relatively closely related, as are plants and green algae, whereas animals and diplomonads are only distantly related.

TABLE 12.7 Major Groups of Helminths

Group	Principal Characteristics	Examples
Platyhelminthes: flatworms		
Cestoda: tapeworms	Scolex; flattened, segmented body	*Taenia saginata* (beef tapeworm) *Echinococcus granulosus* (dog tapeworm)
Trematoda: flukes	Shaped like flattened, pointed ovals; unsegmented	*Paragonimus westermani* (lung fluke) *Schistosoma* spp. (cause schistosomiasis)
Nemathelminthes: roundworms	Long cylindrical, unsegmented bodies, pointed bodies, pointed ends	*Trichuris trichiura* (whipworm) *Necator americanus* (hookworm) *Trichinella spiralis* (causes trichinosis) *Wuchereria bancrofti* (causes elephantiasis) *Enterobius vermicularis* (pinworm) *Ascaris lumbricoides* (causes ascariasis) *Onchocerca volvulus* (causes river blindness) *Loa loa* (causes loaiasis)

the infections they cause. Here we'll discuss their biology. Not all helminths are human parasites. Some parasitize other animals. Some parasitize plants, but most are free living.

The helminths include two phyla—the flatworms and the roundworms. We'll consider them separately.

Flatworms. Flatworms (also called Platyhelminthes) are named for the flattened bodies of the adult worms. They have a head and bilaterally symmetrical bodies, as well as specialized organ systems, including a nervous system, an excretory system, and a reproductive system. Most species have a digestive tract with a single opening through which food enters and waste leaves. Some parasitic species have no digestive tract at all. They simply absorb nutrients through their outer covering.

We'll consider two groups of flatworms that infect humans: tapeworms and flukes.

Tapeworms. Tapeworms (also called Cestoda) are huge animals that live in the intestines of animals including humans (**Figure 12.20**). They can grow to be 30 feet long! Their flat, segmented bodies consist of three parts: a **scolex** with hooks

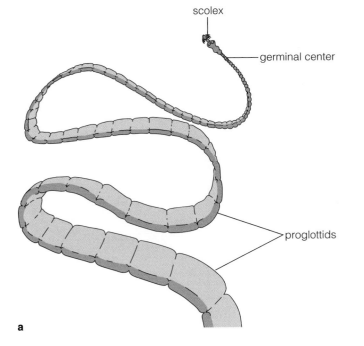

scolex

germinal center

proglottids

a

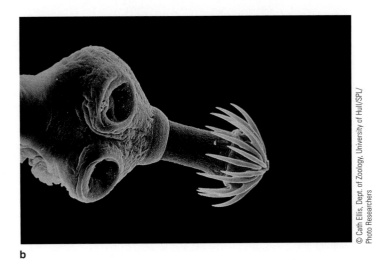

FIGURE 12.20 Tapeworm. (a) Sketch showing the principal components. (b) Micrograph of the scolex with sharp hooks.

b

or suckers that attaches the worm to the intestinal lining, a necklike **germinal center** where new segments are formed, and the segments (also called **proglottids**) themselves. Progressing from the germinal center down the worm traces the life history of a proglottid. Below the germinal center, it grows in size, and by about halfway down the worm it becomes sexually mature. Then fertilization occurs within the segment. (Being **hermaphroditic,** each segment has both male and female reproductive organs.) Further down, the larger and older segments become packed with fertilized eggs that are shed in the feces. Proglottids are egg-making machines. They have no digestive or excretory system. They just absorb nutrients, excrete wastes, and make eggs.

The life cycles of some tapeworms pass through two forms: larval and adult. Larval forms generally develop in muscle, brain, eye, liver, or heart tissue. They displace surrounding tissue, causing serious or even life-threatening damage.

People acquire tapeworm infections in various ways, depending on how they fit into the worm's life cycle. To illustrate, let's consider two different tapeworm infections—

Taenia saginata, the beef tapeworm, in which humans are parasitized by the adult form, and *Echinococcus granulosus,* in which humans harbor the larval form.

Cattle can carry *Taenia saginata* larvae, which become encysted in their muscle as **cysticerci.** People become infected by eating undercooked infected meat (**Figure 12.21**). The larvae hatch in their intestine and develop into a tapeworm in their intestines. Proglottids containing eggs are excreted in their feces. If untreated human waste is deposited on grazing land and cattle consume the eggs, they hatch in the cow's intestine. Larvae penetrate the intestinal wall and encyst in the cow's muscle, ready to infect another human.

In the case of *Echinococcus granulosus* the roles of human and animal are reversed. Dogs carry the adult. Grazing animals and humans carry the larval form. The dog comes off best. The adult tapeworm it harbors is relatively harmless. But the consequences for other hosts are more serious. The larvae damage the muscles of grazing animals and various tissues, including the liver and lungs, of humans.

The most important tapeworm infections of humans are discussed in Chapter 23.

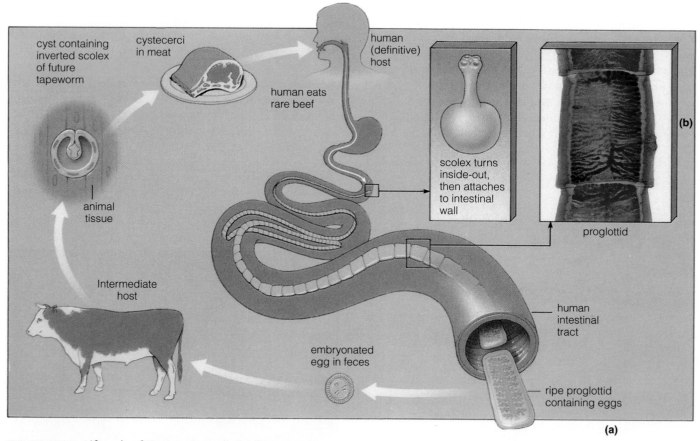

FIGURE 12.21   Life cycle of *Taenia saginata,* the beef tapeworm.

(Art by Keith Kasnot from Biology: The Unity and Diversity of Life, 6th ed., by C. Starr and R. Taggart, Brooks/Cole, 1992. All rights reserved.)

Flukes. Like tapeworms, flukes (also called Trematoda) have flattened bodies, but they are not segmented. Adults are shaped like flat, pointed ovals. They are small but deadly. Depending on species, they range in size from a few millimeters to a few centimeters. They have a ventral sucker to attach firmly to their host.

Flukes have complex life cycles that pass through larval and adult forms. Humans harbor the adults, usually in specific tissues. For example, some species parasitize the lung. Others parasitize the liver. Still others live in blood vessels. Aquatic animals or plants, called intermediate hosts, harbor the larvae. All flukes require at least one intermediate host—a snail or a clam. Some require a second—a fish, a crab, or some water plant.

To illustrate a fluke's complex life cycle, let's briefly consider *Paragonimus westermani*, the lung fluke (for a detailed life cycle of the liver fluke, see Figure 23.20). Adult flukes develop in the human lung, causing fever and a cough that produces blood-tinged sputum. Eventually lung damage may be so extensive that breathing is impaired, and the patient may die. The hermaphroditic fluke produces eggs that are coughed up, swallowed, and passed in the feces. If raw sewage is dumped in lakes or streams, the eggs hatch and release free-swimming larvae called **miracidia.** These penetrate a snail and multiply there, producing numerous **redia.** Each of these develops into a different free-swimming form called a **cercaria.** They penetrate crabs or crayfish, forming cysts called metacercariae in them. If a person eats one of these crabs or crayfish without cooking them completely, larvae hatch from the cysts. They burrow through the human intestine and travel to the lung, where they mature.

Paragonimus westermani are acquired in Asia, South America, and occasionally Africa. Other fluke infections are discussed elsewhere: the liver fluke in Chapter 23 and the blood fluke in Chapter 27.

Roundworms. Roundworms (also called Nemathelminthes or nematodes) have cylindrically shaped bodies. These worms are bilaterally symmetrical and usually tapered at both ends. They are more highly developed than flatworms. They have a complete digestive system (with mouth and anus), a well-developed reproductive system (with separate male and female individuals), and even a primitive nervous system. They inhabit almost every environment. About 30 species are parasitic on humans.

The life cycles of roundworms also pass through larval and adult forms, sometimes in two hosts. The life cycle of *Trichinella spiralis*, the causative agent of trichinosis, illustrates the general principles. The larvae form cysts in the muscles of animals (**Figure 12.22**). When meat-eating animals ingest these cysts, larvae hatch from them in the animal's small intestine. There they develop into

adult worms. These adults produce even more larvae—approximately 1500 from each fertilized female. Some larvae burrow through the intestinal wall and enter the bloodstream. When they reach muscle, they form cysts. Trichinosis used to be relatively common when pigs were fed raw garbage. The pigs became infected from eating infected meat in the garbage. Humans became infected by eating pork that wasn't thoroughly cooked. It's no longer legal to feed uncooked garbage to pigs, and trichinosis has become quite rare in the United States.

The life cycles of other roundworms are discussed elsewhere in connection with the infections they cause: the whipworm, *Trichuris trichiura; Ascaris lumbricoides;* and the hookworm, *Necator americanus,* in Chapter 23. *Wuchereria bancrofti,* the roundworm that causes elephantiasis, is discussed in Chapter 27.

Arthropod Vectors

Arthropods certainly aren't microorganisms. They are invertebrate animals with jointed legs (*arthro* means "joint," *pod* means "foot"). But we will consider them briefly because they play an important role in human disease (Chapters 15 and 20). They are **vectors** (they transmit disease between hosts), and they serve as **reservoirs of infection** (where pathogens can survive between human hosts).

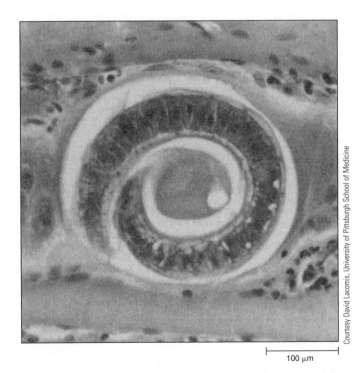

FIGURE 12.22   Larvae of the roundworm *Trichinella spiralis,* which causes trichinosis.

Courtesy David Lacomis, University of Pittsburgh School of Medicine

100 µm

The Organisms. Arthropods are a diverse group, ranging from centipedes, honeybees, and spiders to crabs and lobsters. More than a million species of arthropods have been described, and new ones are still being discovered. Of the three main arthropod groups—chelicerates, crustaceans, and insects—only two are important in human disease: the chelicerates (ticks and mites) and the insects (mosquitoes, flies, lice, fleas, and bugs). Some of these are shown in **Figure 12.23.**

Ticks and mites (also called arachnids) hatch from eggs as six-legged larvae and change into eight-legged adults with two body segments. Although ticks and mites are closely related, they differ in appearance. Ticks tend to be larger and to have less hair. Ticks and some mites as well can suck their host's blood. Some ticks can pass infectious microorganisms into their eggs, a process called **transovarial transmission.** These ticks are reservoirs for disease-causing microorganisms (Chapter 15).

Insects are a huge group. More than 800,000 species have been described. All have three body segments—a head, thorax, and abdomen. Six legs are attached to the thorax. Most insects also have two pairs of wings. Some insect species are of great medical importance. Female mosquitoes and many flies can puncture the skin and suck blood. Lice (which are wingless) can chew tissues of the body's surface, as well as suck blood. They spend their lives on the body of an animal or a human. Fleas, jumping insects without wings, also live on the bodies of animals or humans and suck their blood. Some true bugs, notably the reduviid bug found in North and South America, do the same.

Arthropods and Human Health. Arthropods are vectors for viruses, bacteria, protozoa, and helminths that cause human diseases (**Table 12.8**). They do this either mechanically or biologically.

A mechanical vector picks up pathogens on its body and carries them to another place. For example, when a housefly lands on decaying matter such as garbage or feces, it can pick up disease-causing microorganisms on its feet.

Later when they land on people or their food, they spread disease. Mechanical vectors transmit disease, but they are never essential to perpetuating a disease.

A biological vector, on the other hand, is an essential link in the transmission of a disease because the arthropod is an essential stage in the microorganism's life cycle. For example, the protozoan *Plasmodium*, which causes malaria, must have a mosquito host as well as a human host to survive (Chapter 27). *Plasmodium* spp. are passed back and forth from mosquitoes to humans when the mosquito takes a blood meal. Because the mosquito is an essential link in the chain of infection, mosquito abatement is an effective way to control malaria. The same applies to biological vectors of other diseases.

Arthropods do not have to carry disease agents to affect our health. Some are themselves infective. Their infections, called **infestations,** are usually not serious. For example, a microscopic mite (*Sarcoptes scabiei*) that lives in the skin's outer layers causes the itchy skin condition called **scabies** (Chapter 26). The head louse (*Pediculus capitis*), which lives on hairs in the scalp, and the pubic or crab louse (*Phthirus pubis*), which lives on hairs in the genital area, can cause significant skin irritation (Chapter 26). However, the body louse (*Pediculus corporis*), which lives in unwashed clothing, does spread microbial disease, including typhus.

Still other arthropods threaten human health in different ways. Some biting arthropods inject a venom. These include the black widow spider, whose bite can cause painful or even fatal muscle contractions, and the brown recluse spider, whose bite can kill nearby skin and muscle. Certain ticks can also cause a severe neurological syndrome called tick paralysis if they remain attached for several days. This rare disorder can be cured by removing the tick.

Honeybee and wasp stings produce a relatively harmless venom, but in rare instances they cause sudden death by triggering a severe reaction of the immune system (Chapter 18).

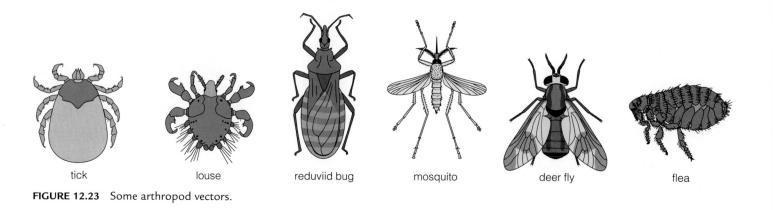

tick   louse   reduviid bug   mosquito   deer fly   flea

FIGURE 12.23 Some arthropod vectors.

TABLE 12.8 Some Microbial Infections Transmitted by Arthropod Vectors

Microbial Pathogen	Arthropod Vector	Human Disease (Chapter)
Viruses		
Encephalitis virus (Eastern, Western, St. Louis, Japanese B)	*Culex* mosquito	Viral encephalitis (25)
Yellow fever virus	*Aedes* mosquito	Yellow fever (27)
Dengue fever virus	*Aedes* mosquito	Dengue fever (27)
Bacteria		
Rickettsia rickettsii	*Dermancentor* tick	Rocky Mountain spotted fever (27)
Rickettsia prowazekii	*Pediculus* body louse	Epidemic typhus (27)
Rickettsie typhi	*Xenopsylla* flea	Endemic (murine) typhus (27)
Francisella tularensis	Ticks, deer flies, mosquitoes	Tularemia (27)
Yersinia pestis	*Xenopsylla* flea	Plague (27)
Borrelia burgdorferi	*Ixodes* tick	Lyme disease (27)
Borrelia spp.	*Ornithodorus* tick	Relapsing fever (27)
Protozoa		
Plasmodium spp.	*Anopheles* mosquito	Malaria (27)
Trypanosoma cruzi	*Triatoma* reduviid bug	Chagas' disease (American trypanosomiasis) (27)
Trypanosoma brucei, gambiense, and rhodesiense	*Glossina* tsetse fly	African sleeping sickness (African trypanosomiasis) (25)
Helminths		
Onchocerca volvulus	*Simulium* black fly	River blindness (26)
Loa loa	*Chrysops* mango fly	Loaiasis (26)
Wuchereria bancrofti	Mosquitoes	Filariasis (elephantiasis) (27)

SUMMARY

Case History: A Fatal Fungus (p. 285)

1a. Eukaryotic microorganisms can cause deadly infections that are hard to treat.

Eukaryotic Microorganisms (pp. 285–309)

1. We consider four groups of eukaryotic microorganisms: fungi, algae, protozoa, and slime molds.

Fungi (pp. 286–295)

2. Fungi include single-celled organisms (yeasts), filamentous organisms (molds), and organisms that form fleshy fruiting structures (mushrooms, puffballs, and shelf fungi).

3. All fungi are eukaryotic, heterotrophic, nonphototrophic, and absorptive. Most fungi are saprophytes; some are parasites.

4. Mycelial fungi contain a multinucleate mass of cytoplasm enclosed within a system of tubelike filaments called hyphae. The mycelia of lower fungi are coenocytic. The mycelia of higher fungi have septa, incomplete cross walls. The reproductive structures in fungi are called spores.

5. The body of a fungus is called the thallus. Fungi that have fleshy fruiting structures have a two-part thallus, an extensive mycelium underground and a solid mass containing sexual spores above ground (such as a mushroom).

6. Most yeasts reproduce by budding. A bubble forms on the cell surface, grows, pinches off, and separates. Pseudohyphae form when buds remain attached to the mother cell in long chains.

7. Most fungi are aerobes. Many yeasts are facultative anaerobes. Fungi can break down most organic matter.

8. Fungi are classified by how they reproduce. Asexual reproduction occurs by elongation of hyphae, dividing or budding of single cells, or the production of asexual spores. Sporangiospores are produced within a special structure called a sporangium, and conidiospores, or conidia, are produced naked on tips of special hyphae called conidiophores. Sexual reproduction occurs by producing sexual spores.

9. Important lower fungi include Chytridiomycetes and Oomycetes (water molds) and the Zygomycetes (terrestrial).

10. The higher fungi are divided into three groups. Ascomycetes form ascospores within a sac called the ascus. Basidiomycetes form basidiospores on basidia, club-shaped cells; some have a dikaryon stage. Deuteromycetes do not form sexual spores; they are also called Fungi Imperfecti.

11. Yeast is a descriptive term for any single-celled fungus.

12. Dimorphic fungi switch between a single-celled yeast phase and a mycelial phase.

13. Fungi are the major cause of infectious disease in plants. Only a few fungi are pathogenic for humans, but fungal infections can be deadly and difficult to treat. Fungi produce penicillins and the cephalosporins.

Algae (pp. 295–298)

14. Metabolically, algae resemble higher plants, but they reproduce as fungi do, producing asexual spores and gametes that fuse to form zygotes.

15. Most algae are aquatic organisms. They are major fixers of CO_2; phytoplankton carry out about 80 percent of the earth's photosynthesis.

16. Algae do not cause disease because they are photosynthetic and cannot live inside the body. An exception is *Prototheca*, which has lost its ability to photosynthesize.

17. Alginic acid, derived from kelp, and carrageenan from the red alga *Chondrus crispus* are food thickeners. Other red algae are the source of agar.

18. Algae are divided into six groups. Red algae and brown algae are plantlike and multicellular. Green algae range from unicellular to filamentous to plantlike and multicellular. Euglenoids are single motile cells with two flagella of unequal length. Dinoflagellates are single cells, usually with two flagella, one wrapped around the cell and the other extending. Diatoms are single cells enclosed in a silica shell.

Lichens (pp. 298–299)

19. Lichens are a mutualistic association between a fungus (usually an Ascomycete) and a green alga or a cyanobacterium. The fungus makes up most of the mass and gives the lichen its shape.

20. Lichens grow under harsh conditions and very slowly.

21. Both partners can be cultivated separately. The alga continues growing as individual cells, but the fungus assumes a quite different shape.

Protozoa (pp. 299–302)

22. Protozoa are nonphotosynthetic, unicellular eukaryotes. They have elaborate organelles and are the pinnacle of unicellular complexity.

23. Protozoa reproduce asexually by cell division and sexually by conjugation.

24. Some protozoa produce cysts.

25. Protozoa are divided into four groups on the basis of locomotion. The flagellates (Mastigophora) move by means of flagella. The amoeboids (Sarcodina) move by extending pseudopods. The sporozoa (Sporozoa) are nonmotile, and the ciliates (Ciliophora) move by means of cilia.

26. Protozoa cause such serious diseases as malaria, amoebic dysentery, and African sleeping sickness.

The Slime Molds (pp. 302–304)

27. The true slime molds (Myxogastria) are multinucleate at one stage of their life cycle. They are commonly seen on decaying logs as a slimy mass called a plasmodium. When a plasmodium reaches a relatively dry region, it develops fruiting bodies, which contain spores.

28. The cellular slime molds (Acrasieae) produce cells that resemble amoeboid protozoa.

Relatedness of Eukaryotic Microbes (p. 304)

29. The groupings—fungi, algae, and protozoa—are not all made up of closely related organisms.

Helminths (pp. 304–307)

30. Helminths are animals, but they are studied in microbiology because some cause infectious diseases.

31. Flatworms (Platyhelminthes) include two main groups of human parasites, tapeworms (Cestoda) and flukes (Trematoda). Roundworms (Nemathelminthes) include numerous parasitic species.

32. Tapeworms have segmented bodies (segments are called proglottids) and life cycles with larval, as well as adult forms. Adult cestodes attach to the intestines of the host with the scolex; eggs are passed in the stool. Larval tapeworms generally develop in tissue (such as heart or muscle).

33. Humans can acquire tapeworm infections by eating undercooked meat containing cysticerci (encysted larvae).

34. Flukes are not segmented; adults are shaped like flat, pointed ovals. Some fluke species parasitize human tissue, such as the liver or lung, whereas others live in blood vessels.

35. Flukes have complex life cycles. Parasitic human species are limited to areas where the proper intermediate hosts are found and where untreated human sewage enters water.

36. Roundworms are more highly developed than flatworms. Life cycles can be simple or complex.

37. People acquire roundworm infections in various ways. They can ingest eggs in fecal-contaminated material (whipworm and *Ascaris lumbricoides* infections), ingest the worm's larvae (trichinosis), become infected by walking barefoot on infected ground (hookworm), or have free-living filariform larvae penetrate their skin.

Arthropod Vectors (pp. 307–309)

38. Arthropods are invertebrate animals with jointed legs. They are studied in microbiology because some are vectors and some are also reservoirs of infection.

39. The most medically important arthropods are certain chelicerates (ticks and mites) and insects (mosquitoes, flies, lice, fleas, and bugs).

40. Ticks that pass infectious organisms into their eggs (transovarial transmission) are also reservoirs.

41. A mechanical vector picks up a pathogen on its body and carries it from one place to another.

42. A biological vector is an essential link in the transmission of a disease.

43. Some biting arthropods inject a venom that can be painful or even fatal.

REVIEW QUESTIONS

Case History: A Fatal Fungus

1a. What critical observation led A. H.'s physician to suspect that his infection was caused by a eukaryotic microorganism, even before he performed the spinal tap?

Fungi

1. What three characteristics define all fungi? Distinguish among yeasts, molds, and mushrooms.

2. Describe the typical structure of fungi. How do higher fungi differ from lower fungi?

3. How do yeasts reproduce?

4. Discuss how lower fungi reproduce and give some examples of lower fungi.

5. Discuss how each of the following types of higher fungi reproduces:
   a. Ascomycetes
   b. Basidiomycetes
   c. Deuteromycetes

6. What are dimorphic fungi?

7. Give some examples of how fungi benefit us and how they harm us.

Algae

8. How are algae like higher plants and how are they like fungi? What are their chief characteristics?

9. Why can't most algae cause diseases?

10. Name the six types of algae and distinguish between them.

Lichens

11. What are lichens? Where are they found?

12. What happens when you tease apart the thallus of a lichen?

Protozoa

13. What are protozoa? How do they reproduce?

14. On what basis are protozoa classified? Name the four types.

15. Give some examples of diseases caused by protozoa.

The Slime Molds

16. Distinguish between true slime molds and cellular slime molds.

17. What is a plasmodium? What is a fruiting body?

Relatedness of Eukaryotic Microbes

18. Is the term protozoa valid phylogenetically? Why?

Helminths

19. What are helminths and why do microbiologists study them?

20. Distinguish among the following:
   a. Platyhelminthes
   b. Cestoda
   c. Trematoda
   d. Nemathelminthes

21. How do humans acquire a tapeworm infection? What are the symptoms?

22. Describe the life cycle of a fluke.

23. How do humans get trichinosis?
   a. whipworm
   b. trichinosis
   c. hookworm

Arthropod Vectors

24. What are arthropods and why do microbiologists study them?

25. What are the most medically important arthropods?

26. Distinguish between a mechanical vector and a biological vector and give an example of each.

CORRELATION QUESTIONS

1. How would you distinguish between an Oomycete and a Zygomycete?

2. Could a Deuteromycete be a mushroom? Explain. Could it be a yeast?

3. How would you distinguish a colorless alga from a fungus?

4. *Pneumocystis carinii*, a notorious killer of persons with acquired immunodeficiency syndrome (AIDS), was long thought to be a protozoan, but molecular studies showed it to be a fungus. What do these facts tell you about the properties of *P. carinii*?

5. Do you think it would be possible for a member of either group of slime molds to be a human pathogen? Explain.

6. How would you determine whether an adult arthropod was a tick or a mite?

ESSAY QUESTIONS

1. Discuss the phenomenon of heteromorphism from the standpoints of survival value and triggering mechanisms.

2. Discuss the difficulties, as well as the unique opportunities, available for controlling helminth infections.

SUGGESTED READINGS

Alexopoulos, C. J., C. W. Mims, and M. Blackwell. 1996. *Introductory mycology.* 4th ed. New York: John Wiley and Sons.

Garcia, L. S., and D. A. Bruckner. 1993. *Diagnostic medical parasitology*. 2d ed. Washington, D.C.: American Society for Microbiology.

Jeffrey, H. C., and R. M. Leach. 1975. *Atlas of medical helminthology and protozoology*. New York: Churchill Livingstone.

Large, E. C. 1940. *The advance of the fungi*. New York: Henry Holt.

Lee, J. J., S. H. Hutner, and E. C. Bovee. 1985. *An illustrated guide to the protozoa*. Lawrence, Kan.: Society of Protozoologists.

Sleigh, M. 1989. *Protozoa and other protists*. New York: Hodder and Stoughton.

Sogin, M. L., H. G. Morrison, G. Hinkle, and J. D. Silberman. 1996. Ancestral relationships of the major eukaryotic lineages. *Microbiología SEM* 12:17–28.

Sze, P. 1986. *A biology of the algae*. Dubuque, Iowa: William C. Brown.

Tanner, J. R. 1987. St. Anthony's fire, then and now: A case report and historical review. *The Canadian Journal of Surgery* 30:291–93.

For additional readings, go to InfoTrac College Edition, your online research library at: http://www.infotrac.thomsonlearning.com

THIRTEEN

The Viruses

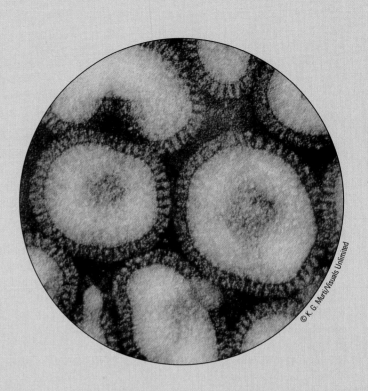

© K. G. Murti/Visuals Unlimited

CHAPTER OUTLINE

LEARNING GOALS

To understand:

- *The nature of viruses—their host range, size, structure, and life cycle—and how they are classified and named*

- *The life cycles of virulent and temperate bacteriophages*

- *The general properties of animal viruses and, in more detail, the properties of retroviruses, influenza viruses, and tumor viruses*

- *The general properties of plant viruses*

- *The nature of viroids and prions*

A Double Hit

V. O., a 2-month-old female born in October, was brought to her pediatrician for well baby care. Although she had gained weight well and seemed healthy, her mother worried because an older sibling had developed complications from congenital heart disease during her first few months of life.

Two weeks later, in late December, V. O. developed a fever and labored breathing. Thick nasal secretions made it difficult to keep her airway clear. She was too tired to feed. When her mother brought V. O. to her pediatrician, she was in significant respiratory distress and her oxygen saturation was low. V. O. was admitted to the hospital to be given oxygen and intravenous fluid. Because of V. O.'s age, clinical presentation, and the time of year, infection with respiratory syncytial virus (RSV) was suspected. A sample of her nasal discharge, taken with a swab, was sent to the microbiology laboratory for rapid diagnosis. The immunological test confirmed that V. O. was infected with RSV. V. O. improved rapidly. She was discharged from the hospital several days later. Her mother was reassured that RSV infections are common during the winter among healthy infants. V. O.'s illness did not indicate any serious underlying health problem.

But 3 weeks later V. O. was back in the doctor's office again, this time with a fever of 103.5°F. She was too listless to eat. She would not smile or make eye contact with her parents. Her heart rate and respiratory rate were elevated. She looked ill. Her doctor ordered a blood cell count and obtained blood and urine for bacterial cultures. Because an outbreak of influenza A was known to be occurring in early January, her doctor sent material obtained by a nasal swab for rapid diagnosis of influenza A. The baby was so ill that she received an injection of antibiotics while the laboratory tests were pending. The next morning the laboratory reported that the test for influenza A was positive. V. O. had yet another viral infection. Antibiotics were discontinued. She was given medications to control her fever, and she recovered within a week.

Although V. O. looked and was perfectly healthy after her recovery, her mother worried because V. O. had been seriously ill twice during her first 4 months of life. But because definitive laboratory tests had been done, V. O.'s pediatrician was able to reassure the family that viruses were the cause of both illnesses. Although V. O. was unlucky to have contracted both these serious viral infections so early in her life, she was a healthy baby with a good prognosis. As expected, V. O. continued to grow and develop normally and had no further problems.

Case Connections

- As V. O.'s illnesses illustrate, viral infections are common and they can be severe, but they are usually self-limiting.
- Both of the viruses that infected V. O. contain minus-strand RNA. In this chapter we'll explore how such viruses can replicate and reproduce this unusual form of genetic material.
- In Chapter 22 we'll explore further the diseases caused by RSV and influenza A virus.
- Why did V. O.'s pediatrician discontinue administering antibiotics when she learned that V. O. suffered from an infection by RSV? If the answer to this important question isn't clear now, it should be when you've finish reading this chapter.

THE ULTIMATE PARASITES

Viruses are parasites, but they are not composed of cells. They are bits of genetic information (nucleic acid) packaged inside a protein coat. When they insert their package of genetic information into a host cell, it directs the cell's metabolic machinery to make more virus. Thus making viruses is a shared process: The virus supplies the information; the host cell supplies the metabolic machinery (Chapter 5). In some cases only a small portion of a cell's metabolic capacity is diverted to make viruses, and the host cell is almost unaffected. In other cases all host cell functions are diverted, and the cell dies.

Viruses can attack all cellular organisms. A particular virus usually parasitizes only a small group of closely related organisms. But there are many different kinds of viruses. All prokaryotes, eukaryotic microorganisms, plants, and animals, including humans, are vulnerable to attack by many different viruses. Each of us has been infected by viruses many times during our lives. Even when we feel completely healthy, we harbor numerous viruses that are in a quiescent state.

The Discovery of Viruses

Cellular microorganisms can be removed from a liquid by passing it through a filter. For well over a hundred years, unglazed porcelain filters were used. Investigators assumed that such filtration would sterilize all liquids. In 1892, however, a young Russian scientist named Dmitri Iwanowski reported that filtration did not sterilize tobacco plant extracts. The liquid from diseased plants that came through the filter was still able to cause mosaic disease (**Figure 13.1**). Using his light microscope, he couldn't see any microbial cells in the liquid, and he was unable to cultivate microorganisms from it using standard methods. He reported to the St. Petersburg Academy of Science that he had discovered a new kind of infectious agent that was smaller even than bacteria. But he did express some concern that the filter was faulty.

Six years later the well-known Dutch microbiologist Martinus W. Beijerinck made the same observation, but he had no doubts about his methods. He reported that the infectious agent of tobacco mosaic disease was an organism small enough to pass through the pores in the filter. He called the organism a **filterable virus** (*virus* is Latin for "poison" or "slime"). Beijerinck predicted that other filterable viruses would be discovered. They soon were. In the next several years other filterable viruses were discovered. Some caused plant diseases. Others caused animal diseases. Gradually the modifier filterable was dropped. Now we simply say viruses.

In 1935 Wendell M. Stanley, a U.S. biochemist, purified and crystallized tobacco mosaic virus. His discovery fascinated the scientific community, and the public as well. Causing infectious disease is a biological capacity. Crystallization is associated with purified chemicals. Only with the development of the electron microscope were viruses finally identified as objects with definite sizes and shapes. But are viruses alive? Beijerinck had speculated that they might be "self-supporting molecules."

Are Viruses Alive?

Deciding whether something is living usually isn't much of a problem. Most living things have a number of properties that inanimate objects lack. They reproduce and carry on metabolism. They are organized as cells. They contain enzymes, nucleic acids, carbohydrates, and lipids. They evolve and adapt to changing environments. Something that has all these properties is certainly alive. But what about something that has most but not all of them?

Viruses reproduce. They contain some of the same macromolecules found in cellular organisms. They evolve and adapt to a changing environment. In these respects they are like other living things. But viruses are not cells, and they lack any significant metabolism of their own. A purified preparation of **virions** (intact, nonreplicating virus particles) shows no obvious signs of life.

Most biologists consider viruses to be living organisms, albeit very simple ones. They resemble other living things in many respects and appear to have evolved from them. Still, some biologists insist that only cellular organisms are alive. Issues such as this one stimulate great discussion and passion, but scientifically they are unimportant. Regardless of whether or not we might think viruses are living organisms, we would study them the same way. In this textbook we'll consider viruses as one group of microorganisms.

Now let's consider how microbiologists classify the myriad viruses that have been discovered.

FIGURE 13.1 Tobacco mosaic virus. (a) The mottled lesions on a tobacco leaf that give tobacco mosaic disease its name. (b) Tobacco mosaic virions.

Kenneth M. Corbett

a

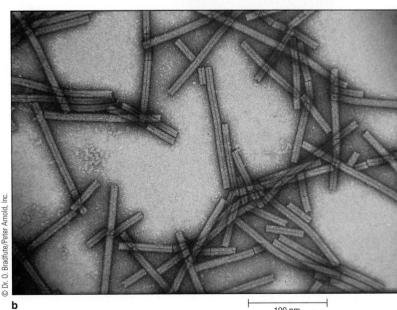

© Dr. O. Bradfute/Peter Arnold, Inc.

b

100 nm

CLASSIFICATION OF VIRUSES

Viruses are classified according to (1) host range, (2) size, (3) structure, and (4) life cycle.

Host Range

The primary classification of viruses is based on **host range** (the spectrum of organisms a particular virus attacks). There are animal viruses, plant viruses, viruses of eukaryotic microorganisms, and bacterial viruses. Bacterial viruses are called **bacteriophages** or simply **phages** (*phage* is from the Greek for "eat") because they usually **lyse** (break down and destroy) bacterial cells.

A virus usually attacks only a single species and sometimes only certain members of a species. Some viruses attack only particular types of cells within a plant or animal.

Host range is determined largely by the presence of appropriate **receptors** (usually proteins) on the cell's surface. Each type of virus attaches to a particular type of receptor, so each type of virus can attack only those cells that have a receptor for it.

Size

Smallness and structural simplicity are the notable features of viruses. Some viruses are as small as 25 nm. Others are as large as 300 nm. They range from about one-tenth to one-third the size of a small bacterial cell (**Figure 13.2**). Viruses lack most cellular structures, including cytoplasm, ribosomes, and a nucleus or nucleoid. Most are merely a piece of nucleic acid wrapped in a protein coat. Viruses are also genetically simple. Even one of the largest viruses, T4, contains only 77 genes (about fiftyfold fewer than *Escherichia coli*). The smallest viruses, such as Qβ and MS2, contain only three genes, yet both are powerful pathogens.

Structure

The basic structure of a virion is a nucleic acid core surrounded by a protein coat called a **capsid**. In addition, there may be a few enzyme molecules, additional structural proteins, and a surrounding membrane, which is called the **envelope** (**Figure 13.3**). Often other proteins form structures, called **spikes,** that protrude from the envelope. Viruses that lack an envelope are called **naked viruses.**

Nucleic Acid. All cellular organisms store their genetic information in the form of double-stranded (ds) DNA. During gene expression, they transcribe their genetic information into RNA, but they never store it in RNA. In contrast, various viruses store their genetic information in all forms of nucleic acid—dsDNA, single-stranded (ss) DNA, dsRNA, and ssRNA—but each virus uses only a single form. Viruses that use ssDNA convert it into usable double-stranded form only after it enters its host. Viruses that store information in RNA short circuit the flow of information from DNA to RNA to protein (Chapter 6).

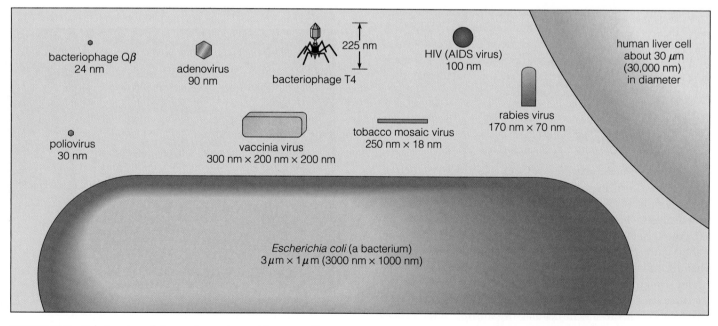

FIGURE 13.2   Relative size of viruses, bacteria, and human cells.

FIGURE 13.3   Basic structure of naked and enveloped virions.

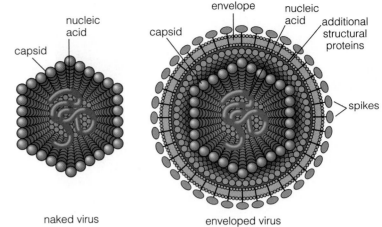

naked virus   enveloped virus

Some of these viruses use ssRNA that is translated directly by the host ribosomes. Such RNA, called the **plus strand,** is in fact messenger RNA (mRNA). Other viruses store the complementary ssRNA called the **minus strand.** Minus-strand RNA is converted into mRNA (a plus strand) after it enters the host cell.

Some viral nucleic acid molecules are circular, and some are linear. Most viruses have a single nucleic acid molecule, but a few have more than one.

In summary, viruses can store their genetic information in five different types of nucleic acid: dsDNA, ssDNA, dsRNA, or ssRNA (either a plus or a minus strand). But no known virus contains more than one kind of nucleic acid.

Viral Capsids. The capsid, which surrounds the nucleic acid core of a virus (Figure 13.3), is made up of subunits called **capsomeres,** which in turn are composed of one or more different proteins. Capsids come in three basic shapes—helical, polyhedral, and complex (**Figure 13.4**). The particular shape depends upon how the capsomeres are arranged and how many kinds there are. Some virus capsids are built from only one type of capsomere. Others are built from several.

The capsomeres of **helical viruses** fit together as a spiral to form a rod-shaped structure. Tobacco mosaic virus is an example of such a helical virus.

The capsomeres of **polyhedral viruses** are usually arranged in equilateral triangles that fit together to form a structure resembling a geodesic dome (**Figure 13.5**). The most common polyhedral virus structure has 20 triangular surfaces; it is **icosahedral** (20-sided). Viruses with this many faces appear almost spherical.

Complex viruses are combination structures with a helical portion, called a tail, attached to a polyhedral portion, called a head. Many bacteriophages are complex, and some have more structures that just head and tail; they include a tail sheath, a plate, pins, and tail fibers. The plate, pins, and tail fibers help the virion attach to a host cell. The tail sheath participates in injecting viral DNA into the host cell.

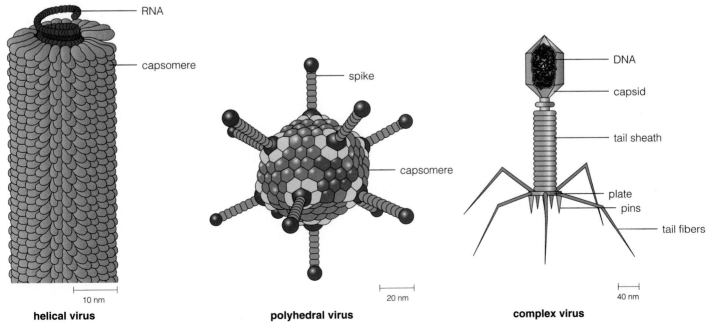

helical virus   polyhedral virus   complex virus

FIGURE 13.4   Three basic shapes of viral capsids.

Courtesy John L. Ingraham

FIGURE 13.5 Virologists got the idea of how capsomeres fit together to form the triangular faces of icosahedral and other polyhedral viruses from the way the U.S. engineer Buckminster Fuller used triangles to create geodesic domes.

Viral Envelopes. The membrane that surrounds enveloped viruses is actually a piece of the host cell's own cytoplasmic membrane. A virion acquires its envelope as it emerges from its host cell. Like the cytoplasmic membranes from which they were derived, viral envelopes contain both phospholipids and proteins. The phospholipid portion of a viral envelope comes from the host, but the proteins are encoded in the viral genome. Some of these proteins are **glycoproteins** (proteins that have sugar molecules attached to them). Enveloped viruses are particularly sensitive to nonpolar solvents such as chloroform and ether because these solvents destroy the virus's membrane and thereby its ability to infect. Naked viruses are relatively resistant to nonpolar solvents. In practice, a quick and easy way to determine if a virus is enveloped or not is to test whether it is inactivated by the nonpolar solvent, ether. Enveloped viruses are inactivated; naked viruses are not.

Life Cycle

All viruses have the same basic life cycle, which is radically different from that of any cellular organism. Outside a host cell, viruses exist only in the form of virions, which cannot replicate. To replicate, a virus must infect a host cell. To do so, the virion attaches itself to the cell, a process called **adsorption.** Then the viral genome enters the host cell, a process called **penetration.** During penetration, some types of viruses open and disassemble so that only the nucleic acid enters the host cell. The process of removing the capsid and envelope is called **uncoating.** In other cases

the entire virion is taken into the host cell, and uncoating occurs later. After uncoating, viral nucleic acid directs the cell to make viral components (nucleic acid and protein). Intact new virions reappear only as these components are reassembled in the process of **maturation.** Soon afterward the virus particles exit the infected cell during the process of **release,** often—but not always—killing the cell. Later in this chapter, we'll look at the replication of various types of viruses in more detail.

Now we'll consider the taxonomy of viruses.

Taxonomy

At first, viruses were named according to their host range or the organ system they affected. By the 1960s, however, virologists realized that some viruses can affect more than a single organ system or host. Classifying a virus as a brain virus or a skin virus or one that infects a particular organism was no longer adequate.

A new classification system was needed. A committee, later named the International Committee on Taxonomy of Viruses (ICTV), set about that task at the International Congress of Microbiology in Moscow in 1966. The system they proposed is not the traditional Linnaean scheme used for all cellular organisms. The ICTV scheme has only three hierarchical levels—family (including some subfamilies), genus, and species. Family names all end in viridae. Genus names end in virus. Species names are English words.

Using the ICTV system, the virus that causes acquired immunodeficiency virus (AIDS) is classified as:

Family: Retroviridae

Genus: Lentivirus

Species: human immunodeficiency virus (HIV)

Family names are often converted into English. Thus the Retroviridae are called retroviruses. This practice obscures the distinction between family and genus.

The names of viral families generally indicate something about their members. For example, the family of hepatitis viruses that contain DNA is the Hepadnaviridae (*hepa* [Greek, meaning "liver"] + DNA). The family containing smallpox, monkeypox, and cowpox viruses is the Poxviridae. The family with viruses that contain two molecules of RNA is the Birnaviridae (bi + RNA).

A number of human diseases discussed in Chapters 22 through 27 are caused by viruses. The properties of the families to which they belong are summarized in **Table 13.1.**

Now we'll survey the various types of viruses.

TABLE 13.1 Animal Viruses Discussed in Part IV

Family	Properties	Virus (Species or Genus)	Diseases
DNA Viruses			
Herpesviridae	Enveloped dsDNA	Herpes simplex types 1 and 2 (HSV-1, HSV-2)	Cold sores, genital herpes
		Varicella zoster virus	Chickenpox/shingles
		Epstein-Barr virus (EBV)	Mononucleosis, Burkitt's lymphoma
Poxviridae	Enveloped dsDNA	Smallpox virus	Smallpox (variola)
Hepadnaviridae	Enveloped dsDNA	Hepatitis B virus	Hepatitis B
Papovaviridae	Naked dsDNA	Human papillomaviruses	Warts
Adenoviridae	Naked dsDNA	Human adenovirus	Respiratory disease
			Enteric diseases
			Infectious pinkeye
RNA Viruses			
Retroviridae	Enveloped plus-strand RNA	Human immunodeficiency viruses (HIV-1 and HIV-2)	AIDS
		Human T-cell leukemia viruses (HTLV-1 and HTLV-2)	T-cell leukemia
Togaviridae	Enveloped plus-strand RNA	Alphavirus	Some forms of encephalitis
		Rubella virus	Rubella (German measles)
Flaviviridae	Enveloped plus-strand RNA	Yellow fever virus	Yellow fever
		Dengue fever virus	Dengue fever
		Hepatitis C virus	Hepatitis C

(continued)

TABLE 13.1 Animal Viruses Discussed in Part IV (continued)

Family	Properties	Virus (Species or Genus)	Diseases
RNA Viruses			
Coronaviridae	Enveloped plus-strand RNA	Coronavirus	Upper-respiratory infections
Picornaviridae	Naked plus-strand RNA	Enteroviruses (includes poliovirus, coxsackie-virus, echovirus)	Polio, myocarditis, pericarditis
			Gastroenteritis, meningoencephalitis
		Rhinovirus	Common cold
		Hepatitis A virus	Hepatitis A
Calciviridae	Naked plus-strand RNA	Norwalk agents	Gastroenteritis
Orthomyxoviridae	Enveloped minus-strand RNA	Influenza virus	Influenza
Rhabdoviridae	Enveloped minus-strand RNA	Rabies virus	Rabies
Paramyxoviridae	Enveloped minus-strand RNA	Mumps virus	Mumps
		Measles virus	Measles (rubeola)
		Parainfluenza virus	Croup
		Respiratory syncytial virus	Bronchiolitis
Bunyaviridae	Enveloped minus-strand RNA	Hantavirus	Hantavirus-associated respiratory distress syndrome
Reoviridae	Naked dsRNA	Rotavirus	Infant diarrhea

LARGER FIELD

WERNER ARBER

The history of science is filled with stories of basic research leading to surprising practical applications. In the 1960s a Swiss microbiologist, Werner Arber, was working on an esoteric problem of phage replication called host-induced modification. Arber's work laid the groundwork for what promises to be the most important scientific tool of the twenty-first century—genetic engineering.

In host-induced modification, phage λ (and other phages) changes its properties depending upon the strain of *Escherichia coli* on which it is grown. When phage λ is grown on strain B, each λ virion forms a plaque when plated on a lawn of strain B, but only 1 in 10,000 virions forms a plaque on a lawn of strain K. The opposite occurs with λ virions grown on strain K—all form plaques on a lawn of strain K but only 1 in 10,000 forms a plaque on strain B.

On investigating, Arber found that DNA from strain-B–grown phage λ was rapidly degraded when it entered a strain-K cell but not when it entered a strain-B cell. On the other hand, K-grown phage λ was rapidly degraded in strain-B cells but not in strain-K cells. Apparently these two strains of *E. coli* (and other strains as well) treated λ DNA like their own if it was made in a cell of their own strain but like foreign DNA if it was made in a cell of a different strain. They could recognize foreign DNA and destroy it.

But how could a bacterium distinguish between its own and foreign DNA? Arber found that each strain of bacteria produces its own **restriction endonuclease** that recognizes a particular sequence of four to six bases and cuts DNA there (Chapter 7). Then other enzymes destroy the DNA beginning at the site of the cut. To avoid destroying its own DNA, each strain of bacteria also has a **modification enzyme** that adds a methyl group to one of the bases within the sequence that the restriction endonuclease recognizes. The restriction endonuclease cannot cut the methylated DNA. Thus DNA in strain-K-grown λ, for example, is modified so that it cannot be cut by the K-strain restriction endonuclease, but it can be cut by B-strain endonuclease.

Arber's studies on an obscure biological phenomenon led to an application that is now touching all our lives. Restriction endonucleases are the fundamental tools of recombinant DNA technology (Chapters 7 and 29), which is the basis of genetic engineering.

BACTERIOPHAGES

Bacteriophages were discovered in 1915. In the 1940s a small group of microbiologists, led by two Americans, Max Delbrück and Salvador Luria, initiated studies to focus on seven bacteriophages (called T1 through T7) that attack *Escherichia coli*. Their achievements were groundbreaking for two reasons. First, they learned how to grow and count bacteriophages. Second, they discovered how bacteriophages replicated. These results with bacteriophages proved to be applicable to all other types of viruses. For that reason, we'll start by examining bacteriophages.

Phage Counts and Phage Growth

The methods used to count and study bacteriophages (or other viruses) are based on their mode of infecting and destroying host cells. Plaque counts are used to enumerate phages. The one-step growth experiment reveals how they replicate.

Plaque Count. To determine how many bacteriophages are present in a sample, we add a drop of a culture of the host bacterium to the virus sample. Then we spread the virus-bacteria mixture (or pour it in melted agar) over the surface of agar-solidified medium in a petri dish. After an appropriate incubation period—from a few hours up to a day or so—the surface of the petri plate becomes covered with a **lawn** (confluent growth) of host bacterial cells. Within the lawn are clear circular **plaques** (regions that contain only a few or no bacteria [**Figure 13.6**]). In most cases one plaque develops for each phage virion that was present in the suspension.

The plaque is a result of an epidemic started by a single virion. That virion infected a single host bacterial cell, that produced many more virions. These virions infected surrounding cells, producing still more virions. The epidemic continued until the bacteria stopped growing, preventing further phage development. By that time most of the cells in the region surrounding the original site of infection had been eliminated, producing a clear or cloudy plaque.

The number of plaques that develop on the plate tells us how many viruses were present in the sample we plated.

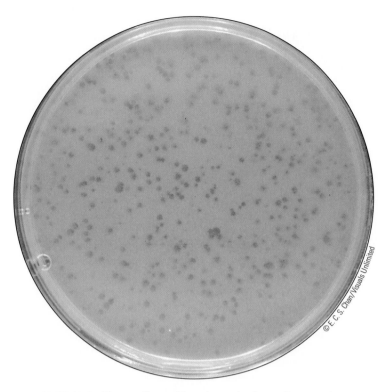

FIGURE 13.6   Plaques formed by a bacteriophage. Because these plaques were formed by a temperate phage (phage λ), they are slightly cloudy.

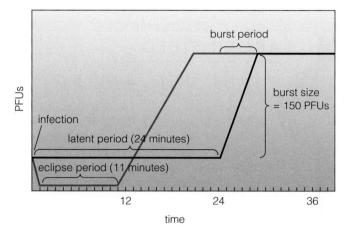

FIGURE 13.7   A one-step growth experiment using phage T4. Black curve: total PFUs (virions plus infected cells). Red curve: virions only.

by doing plaque counts on samples taken from it. Plotting the number of PFUs in the sample against the time at which the sample was taken produces the one-step growth curve (**Figure 13.7**).

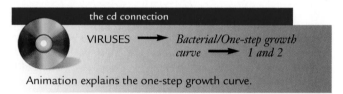

Animation explains the one-step growth curve.

Usually, however, it is necessary to dilute a virus sample before we plate it. Otherwise the plaques would be confluent and uncountable. Then we can calculate the number of bacteriophages present in the original sample, the same way we would calculate the results of a viable count of bacterial cells (Chapter 8): multiply the number of plaques times the dilution factor. This procedure is termed a **plaque count.** A virion causes a plaque to form, but so does an infected bacterial cell, because it's destined to produce virions. Both are present in any culture infected with bacteriophages. Virions together with infected cells are called **plaque-forming units** (PFUs).

The One-Step Growth Curve. Being able to do plaque counts allowed microbiologists to do quantitative experiments with phages, including the one-step growth curve experiment. This classic experiment tells us a great deal about viral replication. The one-step growth curve experiment starts by adding bacteriophage virions to susceptible bacterial cells. After a few minutes (during which the virions become attached to bacterial cells), the mixture is diluted at least fiftyfold. Such dilution makes further contact between virions and host cells unlikely. Then, at intervals, the number of PFUs in the infected culture is determined

The phage one-step growth curve looks quite unlike the growth curve of cellular organisms. The number of cellular organisms increases steadily with time. In contrast, the number of PFUs in the infected culture does not increase at all for a period of time following infection. This period of no increase is termed the **latent period.** Then (in the case of those phages that cause their hosts to burst) the **burst period** occurs: The number of PFUs abruptly increases several-hundred-fold. The burst occurs in a single step, giving the experiment its name. Although no new PFUs are made during the latent period, a lot is happening. The host cell, under the direction of viral nucleic acid, is producing all the molecular components needed to form new virions. Just before and during the burst period, the components are assembled into virions that are released from the cell as it bursts. In the process the cell is destroyed. The fold increase in PFUs that occurs during the burst period is the **burst size.** It represents the average number of new virions released from each infected cell.

By counting the number of virions (instead of total PFUs) in the infected culture, we can learn more about viral replication. To count only virions, each sample taken from the infected culture is first treated with chloroform. Chloroform kills the cells in the sample, including infected ones. Chloroform does not harm intact phage virions (they are not en-

LARGER FIELD

HOW SAFE IS SAFE?

We believe that our municipal water supplies are safe. Filtration systems chemically treat water with chlorine and ozone to kill pathogenic bacteria and parasites. Cholera, typhoid fever, and shigellosis have virtually disappeared in developed countries. But what about viruses? They aren't removed by filtration, and they might survive chemical treatment. Moreover, small numbers can cause disease. It takes hundreds or thousands of bacterial *Salmonella* cells to cause enteritis, but only a few echovirus virions.

Even if there are virions in municipal drinking water supplies, do they cause disease? Most microbiologists and public health experts believe they do not. But in 1991 a Canadian microbiologist, Pierre Payment, studied 2400 people in suburban Montreal. Half of them used their regular tap water and half were supplied with water that had been specially processed to also remove viruses. Over the next 18 months he monitored the health of both groups. The results were surprising. Montrealers using regular tap water had a 30 to 35 percent greater chance of getting gastroenteritis than did the people drinking virus-free water. Something in the tap water was causing 20 extra illnesses for every 100 people each year.

The illnesses were relatively mild (mostly 1 day at home in bed) and the incidence was small (one extra case of gastroenteritis per family). But applying this incidence to the U.S. population, for example, amounts to millions of extra cases of gastroenteritis per year, with an economic loss of several billion dollars. If enteritis-causing viruses are indeed present in Montreal's water, the numbers must be extremely low. One virus per thousand liters of water would be enough to account for the incidence of illness Payment found in Montreal.

veloped). After infection the number of virions goes to zero. It stays there until the burst period begins. This result tells us that after infection the virion ceases to exist as an entity. It disassembles. New virions appear as they are assembled just before and during the burst period. The period when no intact virions are present is called the **eclipse period.**

In summary, in the one-step growth curve experiment infection of a bacterial culture is synchronized. At any time during the experiment each cell is at the same stage of infection. So by examining the whole culture we learn what is happening in each individual cell. From the experiment we learn when the phage virion is disassembled, when progeny phage virions are assembled, and how many virions are produced by each infected bacterial cell.

Replication Pathways

In broad outline the replication cycle of all phages is the same. There are variations, however, such as in how the viral genome enters the host cell and how the mature virions leave. In addition, there are different developmental pathways. Some phages enter the **lytic pathway,** leading directly to the production of more virions. Other phages can enter the **lysogenic pathway,** leading to a prolonged quiescent state, termed **lysogeny.** Phages that follow only the lytic pathway are termed **virulent;** phages that can follow either pathway are termed **temperate.**

Virulent Phages. The steps in the replication cycle of a typical virulent bacteriophage such as T4 are described below and summarized in **Figure 13.8.** T4 is a naked, complex, dsDNA bacteriophage.

First T4 attaches (adsorption) by its tail to a specific receptor site, the core region of a lipopolysaccharide molecule in the outer membrane of its host, *Escherichia coli*. Each kind of phage adsorbs to one specific kind of molecule on the cell surface. It might be a lipopolysaccharide or protein component of the outer membrane of a Gram-negative bacterium, the teichoic acid of a Gram-positive bacterium, a protein of the flagellum, or a pilus. Phages don't adsorb to particular structures. They adsorb to specific molecules. For example, some phages adsorb only to proteins at the end of F-pili. Others adsorb only to proteins that make up the bulk of the F-pili.

Next during the penetration or infection stage, the tail sheath of T4 contracts, forcing its dsDNA into the cell (Chapter 4). This initiates the eclipse period. An intact virion no longer exists. The protein component of the phage, which plays no further role in replication, remains attached to the cell's surface. Thus uncoating of T4 occurs as it enters the host cell.

During the latent period that follows infection, the phage DNA directs the host cell to synthesize viral components, including more viral DNA and protein components.

At the end of the latent period, the component parts are assembled into new virions, a process called **maturation.**

FIGURE 13.8   Life cycle of a virulent phage (T4).

(Art by Carlyn Iverson.)

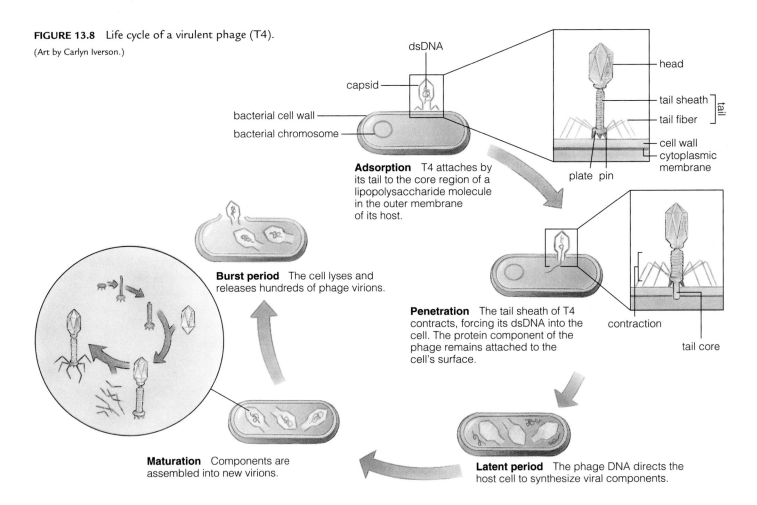

Adsorption   T4 attaches by its tail to the core region of a lipopolysaccharide molecule in the outer membrane of its host.

Penetration   The tail sheath of T4 contracts, forcing its dsDNA into the cell. The protein component of the phage remains attached to the cell's surface.

Latent period   The phage DNA directs the host cell to synthesize viral components.

Maturation   Components are assembled into new virions.

Burst period   The cell lyses and releases hundreds of phage virions.

Finally the intact virions are released during the burst period as the cell is destroyed. A phage-encoded lysozyme (Chapter 4) dissolves the cell wall, causing the cell to lyse and release the hundreds of phage virions inside it.

Virulent phage infections are so swift, deadly, and productive of phage virions that it is surprising that any bacterial cells in a phage-infected culture survive. But some always do. They are mutant cells that are resistant to phage attack. In most cases the specific receptor site on their surface is altered so phage virions can no longer bind to them. Such mutant cells are generally present at a frequency of about one in a hundred million (10^{-8}). An ordinary bacterial culture with more than a hundred million (10^8) cells in each milliliter will almost certainly contain cells that survive the infection.

Temperate Phages. The life cycle of the temperate phage lambda (λ), which also infects *Escherichia coli*, is shown in **Figure 13.9.** Like phage T4, phage λ is a dsDNA complex virion.

The initial stages of infection by λ and T4 are quite similar. The λ virion adsorbs by its tail to a particular receptor protein in the bacterium's outer membrane (specifically, a porin that brings maltose across the outer membrane). Next it injects its linear dsDNA into the host cell. The dsDNA has cohesive ends, which join together inside the host, forming a circle. Then λ takes either the lytic or the lysogenic pathway. The lytic pathway is almost identical to the one T4 always takes. It leads to lysis of the host cell and release of many virions. In contrast, the lysogenic pathway leads to a quiescent state. No phage components are synthesized. The host cell is not damaged. The circular phage genome simply recombines with the host cell's chromosome and becomes integrated into it at a specific location. In this integrated state the phage DNA is called a **prophage.** A repressor protein encoded by the prophage maintains it in a quiescent state. The prophage is replicated and distributed to progeny cells along with the rest of the bacterial chromosome. Such cells are termed **lysogenic** because they have the potential to lyse if the repressor is inactivated.

In the prophage state a few phage genes (in addition to the one encoding the repressor) are expressed. Sometimes they slightly change the host cell's phenotype. For example, colonies of lysogenic strains may have a slightly different appearance than colonies of phage-free strains. Another inter-

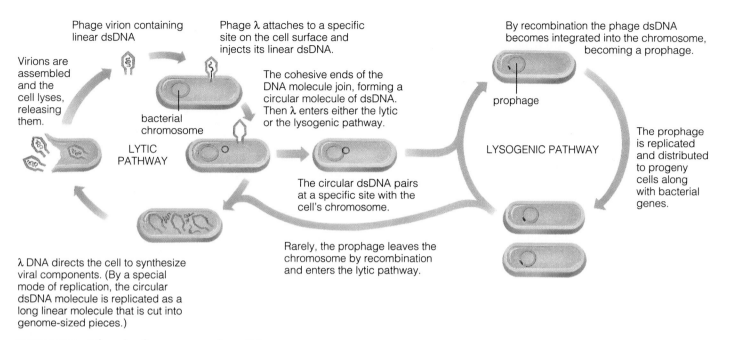

Phage virion containing linear dsDNA

Phage λ attaches to a specific site on the cell surface and injects its linear dsDNA.

Virions are assembled and the cell lyses, releasing them.

bacterial chromosome

LYTIC PATHWAY

The cohesive ends of the DNA molecule join, forming a circular molecule of dsDNA. Then λ enters either the lytic or the lysogenic pathway.

By recombination the phage dsDNA becomes integrated into the chromosome, becoming a prophage.

prophage

LYSOGENIC PATHWAY

The prophage is replicated and distributed to progeny cells along with bacterial genes.

The circular dsDNA pairs at a specific site with the cell's chromosome.

λ DNA directs the cell to synthesize viral components. (By a special mode of replication, the circular dsDNA molecule is replicated as a long linear molecule that is cut into genome-sized pieces.)

Rarely, the prophage leaves the chromosome by recombination and enters the lytic pathway.

FIGURE 13.9   Life cycle of a temperate phage (λ).

(Art by Carlyn Iverson.)

esting example of this phenomenon, termed **lysogenic conversion,** occurs in *Corynebacterium diphtheriae*, the bacterium that causes the respiratory disease diphtheria (Chapter 22). Only lysogenic strains of *C. diphtheriae* cause disease because the disease-causing toxin is encoded in a prophage.

A lysogenic bacterial cell can grow, divide, and produce lysogenic progeny almost indefinitely. In fact, most bacterial cultures probably carry one or more prophages. But on rare occasions (in one infected cell out of 10,000 to 100,000) the prophage leaves the bacterial chromosome and enters the lytic pathway. Soon that host cell lyses and mature virions are released into the medium.

The virions that are released do not affect other cells in the lysogenic culture. The same repressor protein that keeps the prophage in a quiescent state makes lysogenic cells immune to infection by the same virus.

Because of this immunity, temperate phages make plaques only when plated on a nonlysogenic strain. These plaques have a distinctive appearance. Instead of being clear, like the plaques formed by virulent phages, they are slightly cloudy. The plaques are visible because many cells are lysed by phages that enter the lytic pathway. The plaques are cloudy because some cells become lysogenic and survive. They grow within the plaque, making it cloudy.

In summary, temperate phages can live a double life. They can act like virulent phages and destroy their host. They can also integrate their DNA into the host's chromosome and exist there peacefully. When integrated, they

retain their deadly potential. In any lysogenic culture, cells are constantly lysing and releasing mature phage virions.

Phage Diversity. The two bacteriophages that we have discussed, T4 and λ, are structurally similar. They are large (about 225 nm long), complex, naked phages that contain dsDNA. But there are many other kinds of bacteriophages. Some bacteriophages, such as Qβ, are as small as 24 nm. Some phages are helical; others are polyhedral. A few phages are enveloped. All the types of nucleic acid—dsDNA, ssDNA, dsRNA, and ssRNA—are found among phages. The nucleic acid of most phages is a single molecule, but some phages contain several. For example, f6 contains three different molecules of dsRNA.

In summary, phages have very little impact on our lives. They are similar in structure, life cycle, and diversity to plant and animal viruses, but they are easier and less expensive to study. Phage studies have revealed all the basic principles of virology. They are the foundation for studies on animal and plant viruses.

ANIMAL VIRUSES

Viruses that infect animals are as diverse in size, shape, and structure as the phages that infect bacteria. For the most part, animal-virus life cycles resemble those of the bacteriophages, but there are some differences.

The methods of studying animal viruses are modeled after the methods used to study phages.

Cell Cultures

For a long time, research on animal viruses lagged behind research on bacteriophages. Bacteriophages can be cultivated by infecting a bacterial culture and counted by plaque assay. Both are simple procedures. In contrast, animal viruses had to be cultivated by infecting animals and counted by dilution end points (Chapter 8). To determine a dilution end point, many animals had to be infected. The process was time-consuming and expensive.

In the 1930s virologists made a significant advance. They discovered that some animal viruses could be grown in **embryonated chicken eggs** (fertile eggs that contain a developing embryo). An incubator with eggs can contain as many individuals as a roomful of mice or rats. To grow viruses, embryonated eggs are inoculated with virus; after the virus has multiplied, virus-containing fluids are withdrawn with a syringe. To determine dilution end points, embryonated eggs are inoculated with dilutions of a virus sample to discover the highest dilution that kills the embryo. Although time-consuming, this procedure is still used today to study viruses and to produce viruses for vaccines.

In the 1950s an even more important advance was made. Methods were developed to grow plant and animal cells much like bacteria are grown. In 1952 **cell culture,** as the set of methods is called, was applied to virus research. Cell culture made it possible to count animal viruses quickly and inexpensively by plaque count. A monolayer of animal cells (confluent growth, one layer thick) could be grown in a petri dish, and the cells could then be infected with a sample of an animal virus. A virion that kills the cultured cells causes a region of the cell layer to die, forming a plaque. Such plaques become easily visible when stained with a vital stain (one that stains only live cells). Even some viruses that don't kill usually change the appearance of infected cells sufficiently to produce countable plaquelike spots on the monolayer. Today many types of cells can be cultured, making it possible to study many types of viruses by plaque counting (**Table 13.2**).

For obvious reasons, cell culture was a particular boon to studies on viruses that infect only humans. For example, even deadly HIV can be cultivated safely in cultured human white blood cells.

Cells are cultured in ordinary plastic glassware—flasks, bottles, or petri dishes. The growth medium is necessarily quite complex. It usually contains serum from a fetal calf (**Figure 13.10**). The cells can be kept suspended by stirring, or they can be allowed to settle and grow as a monolayer. To subculture, monolayers can be separated into individual cells by treating them briefly with the proteolytic enzyme trypsin. Such treatment dissolves the proteins that hold the cells together without damaging the cells themselves.

Cultured cells lines (analogous to strains of bacteria) are categorized by their source, their chromosome number, and how long they continue to multiply in culture. **Primary cell cultures** are started from normal tissues taken directly from humans or other animals. After several subcultures, the cells that grow best come to dominate the culture. If these cells contains the same number of chromosomes as cells in the tissue from which it was derived, the line is called a **diploid cell line.** If they have a different number, it is called a **heteroploid cell line.**

Most diploid cell lines from normal tissue grow increasingly slowly after 20 to 30 subcultures, and they lose their ability to support viral replication. In contrast, some

---

TABLE 13.2 Cell Cultures Used to Culture Certain Animal Viruses

Virus Cultivated	Source of Cultured Tissue Used to Cultivate Virus	Cell Type	Culture Type
Human immunodeficiency virus	Human lymphocytes	White blood cells	Primary cell culture
Adenovirus	Human kidney	Human embryonic kidney	Primary cell culture
Rhinovirus, enterovirus	Human kidney, lung, foreskin	Human fetal fibroblast	Diploid cell line
Herpes simplex virus	Human cervical cancer	HeLa[a]	Continuous cell line
Influenza virus	Canine kidney	HDCK	Continuous cell line
Enterovirus	Monkey kidney	MGMK	Continuous cell line

[a]Continuous cell lines are usually designated by four letters.

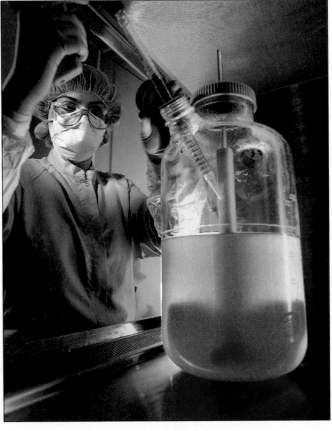

FIGURE 13.10   Cell culture. The colored liquid in this flask is the medium that supplies nutrients to the cultured cells.

cell lines, usually derived from cancerous tissue, grow indefinitely in culture. They are called **continuous cell lines.** Probably the most famous continuous cell line is the HeLa cell line (for Helen Lack, the donor). It has been cultivated continuously in various laboratories throughout the world since 1951. It was derived from human cervical cancer tissue. Cell culture has largely replaced laboratory animals to detect and identify animal viruses.

In spite of the dramatic advances made in research in the last 40 years, methods for diagnosing viral disease in most clinical practice have not changed much. Cell culture methods take too long for the results to be useful in planning treatment. Antibody-based methods (Chapter 17) do not work until the patient develops antibodies against the virus, which usually occurs after the patient has recovered. (AIDS is an exception—antibodies develop before the patient becomes seriously ill.) Therefore for the most part physicians do not culture viruses or look for antibodies. They still rely on symptoms to diagnose viral disease. In a few cases they use immunoassays (Chapter 19) if diagnosis is clinically important.

Development of cell culture was a major boon to research on animal viruses. It allowed virologists to deal with animal viruses almost as easily as they had with bacteriophages. It also enabled scientists to produce effective vaccines, including those that lead to the elimination of polio and the control of measles.

Replication

Animal viruses replicate much as phages do. The stages are adsorption, penetration, uncoating, viral synthesis, maturation, and release.

Adsorption. The plasma membrane is the outermost layer of animal cells. Proteins embedded in it act as **receptors** for a virus. During adsorption, **receptor-binding proteins** on the virus surface bind to these receptors on the surface of the host cell. Receptor-binding proteins are usually in the glycoprotein spikes of enveloped viruses or the capsid of naked viruses. Adsorption determines the tissue specificity of animal viruses because a virus attacks only those cells that have the proper receptors on their surfaces.

Penetration. Adsorption of a virion to its receptor triggers penetration. The intact virion or only its nucleic acid (with or without a few proteins) are brought into the cell.

Viruses penetrate animal cells in one of three ways (**Figure 13.11**):

1. The membranes of some enveloped viruses fuse with the cell's plasma membrane, emptying the rest of the virion inside the cell.

2. Other enveloped viruses enter the cell by phagocytosis (Chapter 4). In this case the host cell engulfs the entire virion, bringing all of it into the cell within a vesicle composed of the cell membrane.

3. Most naked viruses enter a cell as most bacteriophages do. The capsid adsorbs to the cell surface and remains there; the nucleic acid enters the cell.

Uncoating. Viral genes are not expressed (Chapter 6) until the virion is uncoated. The membranes of some enveloped viruses and the capsids of most naked viruses are removed in the process of penetration. Virions that enter the cell partially or completely intact are uncoated inside the cell by the cell's own hydrolytic enzymes, sometimes those in the cell's lysosomes (Chapter 4).

Viral Synthesis. Once uncoated, the virus's nucleic acid directs the host cell's metabolic machinery to synthesize components of more virions. The type of nucleic acid that the virion contains—either dsDNA, ssDNA, plus-strand RNA, minus-strand RNA, or dsRNA—determines how the process proceeds (**Table 13.3**).

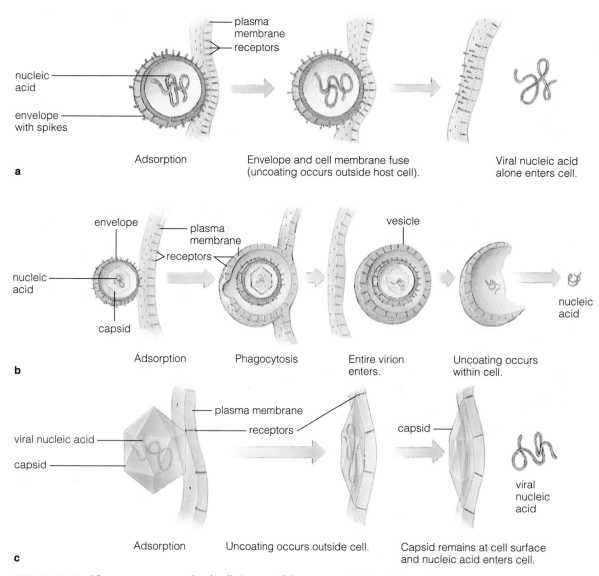

FIGURE 13.11   Viruses penetrate animal cells in one of three ways. (a) The viral envelope fuses with the cell's cytoplasmic membrane, uncoating the virus. The nucleic acid alone is emptied into the cell. (b) An enveloped virus is phagocytized. The entire virion enters the cell. (c) A naked virus adsorbs to the cell surface. The capsid remains outside. The nucleic acid enters the cell.

(Art by Carlyn Iverson.)

dsDNA Viruses. If the virion contains dsDNA, viral synthesis occurs much like the synthesis of any host-cell component (**Figure 13.12**). Viral genes are transcribed into messenger RNA (mRNA); mRNA is translated into viral proteins; and viral DNA is replicated. The host cell has all the enzymes needed to carry out these syntheses. But sometimes for special reasons the virus directs the cell to make new enzymes to catalyze certain steps. These new enzymes are made from viral genes. For example, poxviruses direct their host to make a new RNA polymerase. This new enzyme ensures that viral components are made at the time they are needed. Components that are needed right away are made from viral genes, called **early genes,** which can be transcribed early in the infection by the RNA polymerase that is already present in the host cell. The viral RNA polymerase is made from one of these early genes. **Late genes** can be transcribed only by viral RNA polymerase. Therefore the late genes are transcribed only after the viral enzyme has been made, and the products of late genes are made only when they are needed, late in the infection.

TABLE 13.3 Patterns of Animal Virus Replication

Nucleic Acid in Virion	Replication of Nucleic Acid	Transcription/Translation	Examples
dsDNA	Replication of DNA begins after some viral protein has been made.	dsDNA is transcribed by the same mechanisms as chromosomal DNA.	Herpesviruses, poxviruses
ssDNA	In host cell nucleus, ssDNA is converted to double-strand replicative form, which is copied and generates ssDNA.	Host-cell RNA polymerase transcribes double-stranded replicative-form DNA.	Parvoviruses
Plus-strand RNA	Virus-encoded RNA-dependent RNA polymerase makes minus-strand RNA, which serves as a template for more plus-strand.	Translation of plus-strand RNA begins immediately, forming capsid protein and RNA-dependent RNA polymerase.	Picornaviruses, togaviruses
Minus-strand RNA	RNA-dependent RNA polymerase makes plus-strand RNA, then more minus strand.	Transcription is catalyzed by RNA-dependent RNA polymerase in virion.	Orthomyxoviruses, rhabdoviruses
dsRNA	RNA is replicated by virus-encoded RNA polymerase.	dsRNA is converted to mRNA by RNA-dependent RNA polymerase.	Reoviruses

ssDNA Viruses. If the virion contains ssDNA, only one additional step is necessary. The ssDNA must be converted to dsDNA before viral synthesis starts. Then it proceeds the same way as synthesis from dsDNA does.

RNA Viruses. Viral synthesis by all three classes of RNA viruses (those with dsRNA, plus-strand RNA, or minus-strand RNA) is quite different from viral synthesis by DNA viruses. RNA viruses use the RNA they contain as a template to make more RNA for progeny virions. To do

this, RNA viruses direct their host cells to make an enzyme that is never found in uninfected cells. This enzyme, called **RNA-dependent RNA polymerase,** can make a complementary copy of a ssRNA template. RNA-dependent RNA polymerase plays a different role in the replication of each class of RNA viruses (**Figure 13.13**).

Plus-Strand RNA Viruses. Plus-strand RNA viruses use RNA-dependent RNA polymerase only to replicate their genomes; they don't need it for gene expression. Enzymes

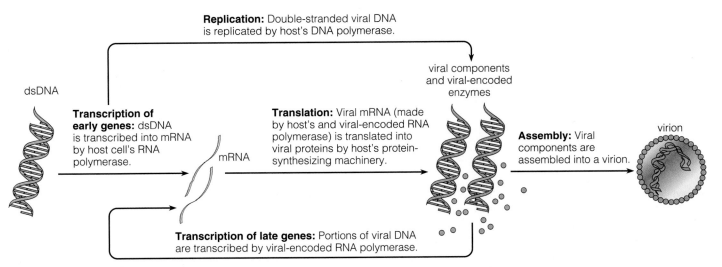

FIGURE 13.12 Replication and gene expression by dsDNA viruses. In the case of ssDNA virus synthesis, there is an additional step not shown here: A complementary strand to the ssDNA is made, converting it to dsDNA.

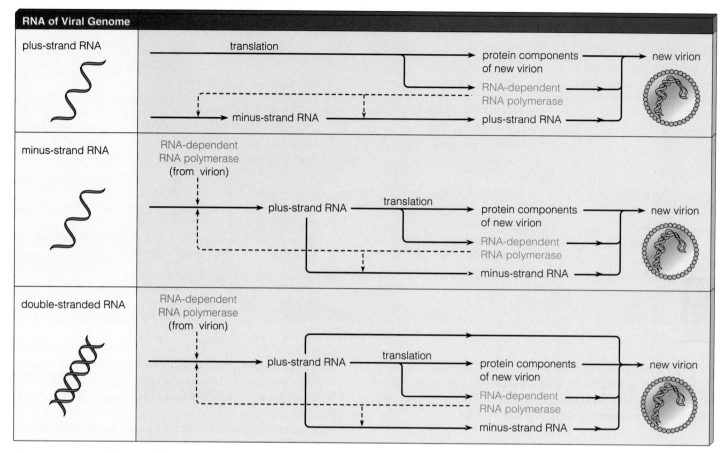

FIGURE 13.13 Replication of the genome of RNA viruses.

already in the host cell can translate the virion's plus-strand RNA into viral proteins. But the cell doesn't have the enzyme needed to make more plus-strand RNA for progeny virions. The viral-encoded RNA-dependent RNA polymerase does it in two steps. First the enzyme uses the plus-strand RNA as a template to make minus-strand RNA. The only purpose of this minus strand is to serve as a template for making more plus-strand RNA for progeny virions. The newly synthesized plus-strand RNA also speeds the rate of viral synthesis by acting as addition mRNA. Most plus-strand RNA viruses follow this route of viral synthesis. But retroviruses, including HIV (they too are plus-strand RNA viruses), follow a completely different route, which we'll discuss later (see Retroviruses).

Minus-Strand RNA Viruses. Minus-strand RNA viruses also use RNA-dependent RNA polymerase to replication their genome. But in addition they use it for gene expression. The minus-strand RNA can't participate directly in gene expression. It must first be transcribed into plus-strand RNA. RNA-dependent RNA polymerase does this. In other words, the only purpose of the minus strand in the virion is to serve as template for RNA-dependent RNA

polymerase to make plus-strand RNA. Then the plus strand enters the normal pathway of gene expression: It serves as mRNA to make viral proteins. The plus strand is also needed as a template for RNA-dependent RNA polymerase to make more minus-strand RNA for progeny viruses. Minus-strand RNA viruses are faced with an additional problem—a sort of catch-22. The genes encoded in the minus strand can't be expressed without RNA-dependent RNA polymerase, and a viral gene must be expressed in order to make RNA-dependent RNA polymerase. The solution for this seeming dilemma is to include a few molecules of RNA-dependent RNA polymerase in the virion. These few molecules allow enough viral gene expression to make enough RNA-dependent RNA polymerase to start viral synthesis at a significant rate. For this reason, the virions of all minus-strand RNA viruses contain a few molecules of RNA-dependent RNA polymerase.

dsRNA Viruses. Double-stranded RNA viruses need RNA-dependent RNA polymerase for the same two purposes as minus-strand RNA viruses do: to replicate their genome and to express it. The dsRNA genome is used as a template to make plus-strand RNA. This plus-strand RNA is used as mRNA to make viral proteins. It is also used as a template to make minus-strand RNA, which combines with plus strands to make dsRNA for progeny virions. RNA-dependent RNA polymerase must be present in double-stranded RNA viruses for the same reason it is in minus-strand RNA. In some dsRNA viruses, RNA-dependent RNA polymerase is part of the capsid.

In summary, viral synthesis proceeds differently depending upon what type of nucleic acid its virion contains, but all viruses share two common tasks: (1) They must make a new copy of their genetic information, and (2) they must use their genetic information to force the host cell to make viral proteins. To accomplish these tasks, some viruses need enzymes that their hosts lack. All RNA viruses need RNA-dependent RNA polymerase make a new copy of their genome. Some also need it to make viral proteins. Those (minus-strand and dsRNA viruses) that need it to make proteins have to have a few molecules of the enzyme their virion.

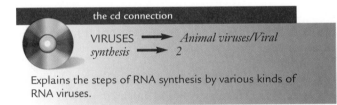

the cd connection

VIRUSES ⟶ *Animal viruses/Viral*
synthesis ⟶ *2*

Explains the steps of RNA synthesis by various kinds of RNA viruses.

Maturation. How capsids assemble around nucleic acid in animal viruses is not understood as thoroughly as maturation of phages. It occurs largely spontaneously. Some host cell proteins may help.

Release. The way that an animal virus is released from its host cell depends upon whether or not it has an envelope.

Enveloped Viruses. An enveloped virus acquires its protein-studded membrane as it leaves the host cell. Membrane proteins are among the components that viral genes direct the host cell to make. These become embedded in a localized region of the host-cell membrane where the virus is destined to leave the cell. As the virus emerges, it pushes out the plasma membrane (or the endoplasmic reticulum), forming a bud that encloses the virus. Then the bud pinches off behind the virion, resealing the host cell. As a result of leaving the cell this way, the virion is composed of host-cell, as well as viral, components. The contents of the virion and most of the proteins in its membrane are encoded by viral genes, but the lipid portion of the viral envelope is encoded by host-cell genes.

Nonenveloped Viruses. Nonenveloped animal viruses leave their host the way phage T4 does. Completely assembled virions accumulate in the cytoplasm; then the cell lyses, releasing them.

Latency

Sometimes animal viruses cause a latent infection, just as temperate bacteriophages do. Such latent viral infections are typical of DNA viruses belonging to the herpesvirus family. First these viruses cause a **primary infection** that produces symptoms of disease. Then the viruses become latent, and the patient is symptomless. Later the virus can be reactivated and cause other symptoms. One strain of herpes simplex virus (HSV-1) settles into a latent, symptomless infection of nerve cells that go to the lips and mouth (Chapter 26). Then, when the virus is reactivated by a fever, a cold, too much sun, or stress, painful blisters, or "cold sores," appear on the lips.

Another herpesvirus (varicella zoster) causes chickenpox (varicella) as a primary infection followed by a latent infection of nerve cells all over the body. Sometimes much later, the virus can be reactivated and cause a completely different disease called **shingles** (zoster), a painful blistering of a strip of skin.

Animal Viruses of Special Interest

Here we'll consider three animal viruses—retroviruses, influenza viruses, and tumor viruses—in more detail because each exhibits an unusual characteristic.

Retroviruses (Family Retroviridae). Retroviruses, which are probably the most thoroughly studied class of animal viruses, exhibit an unusual pattern of replication. Most members of this large viral family cause no symptoms at all in their animal hosts. But some cause malignant tumors and leukemias in animals and one, **immunodeficiency virus (HIV)**, causes the devastating human disease **AIDS** (Chapter 27). We'll discuss HIV as a representative retrovirus.

HIV. Retroviruses are plus-strand RNA viruses, but they replicate differently than all other RNA viruses. The family name *retro* (Latin, meaning "backward") refers to the unique, seemingly backward biochemical step in the replication cycle of these viruses. All cellular organisms use DNA as the template to make RNA (RNA viruses other than retroviruses use RNA); retroviruses do the reverse: They use RNA as a template to make DNA, employing a unique enzyme called **reverse transcriptase.** Only retrovirus-infected animal cells (and to a minute extent a few bacteria) produce this enzyme.

Superficially, an HIV virion resembles the virions of most enveloped RNA viruses (**Figure 13.14**). Its capsid is surrounded by a membrane with embedded protein spikes. Just under the membrane is a layer of structural protein called a **matrix.** The capsid, which is shaped like a **truncated** (cut-off) cone, lies inside the matrix. The unique

FIGURE 13.14   An HIV virion.

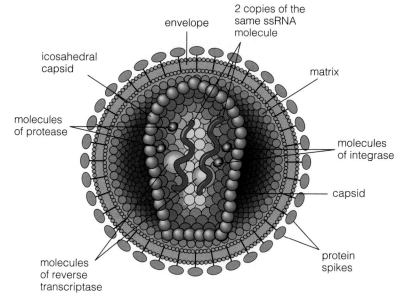

aspects of the retroviral virion are found in the **core** (the region inside the capsid). First, the core contains two copies of the same plus-strand RNA molecule. In other words, the virion is diploid (it's not clear what advantage being diploid offers the virus). Second, the core contains molecules of three enzymes: reverse transcriptase, integrase, and protease. (We'll discuss integrase and protease later.) As we've discussed, viruses for the most part use enzymes available in their host cell. They also direct the host to make additional enzymes that they need but the host normally lacks. If one of these enzymes is needed for gene expression, it must be included in the virion. Reverse transcriptase, is such an enzyme. HIV virions contain reverse transcriptase (and integrase, as we will see later) for the same reason that minus-strand RNA viruses contain RNA-dependent RNA polymerase.

The plus-strand RNA contained in an HIV virion should be able to serve immediately as mRNA within the host cell, but it does not. Instead it is used exclusively as a template for reverse transcriptase to make DNA. This process of reverse transcription occurs in three steps. All are catalyzed by reverse transcriptase (**Figure 13.15**). The final product is a molecule of dsDNA.

Enzymes in the host cell convert the linear DNA product of reverse transcriptase into circular form. Then integrase (the second enzyme present in the virion) integrates the circularized viral DNA into the host chromosome by a reaction similar to the one that integrates temperate bacteriophages' DNA into the bacterial chromosome. Unlike the genome of the temperate phage λ, however, the retroviral DNA does not always recombine at the same position on the host chromosome. The integrated DNA copy of the retroviral genome is called a **provirus** (after the prophage of a temperate phage).

As occurs with a prophage, the provirus can be replicated along with the host cell DNA (**Figure 13.16**). Proviral DNA can also be transcribed into mRNA and translated into viral proteins by the host cell. These proteins migrate to the cell surface, where, along with the mRNA, they assemble and bud through the cell membrane to produce virions. Here again HIV is unusual. The virion is not mature as it leaves the host cell. The core is not yet formed, and most viral proteins are connected together into long strands. At this stage the third virion protein, protease, acts. Within the maturing virion, it cuts the strands into proper units that assemble to form the core.

The special roles reverse transcriptase and protease play in the replication of HIV offer targets for drugs to control AIDS. Azidothymidine (AZT) and related anti-AIDS drugs inhibit reverse transcription. Protease inhibitor drugs act on protease.

Among the several unusual properties of HIV is its high rate of genetic change (**Table 13.4**). The genetic diversity of HIV within a single infected person can be greater than the worldwide diversity of influenza virus during an epidemic. This remarkable capacity for HIV to change may account for its transition from a symptomless resident of chimpanzees to a lethal pathogen of humans. Further changes, such as in mode of transmission, could render HIV even more dangerous.

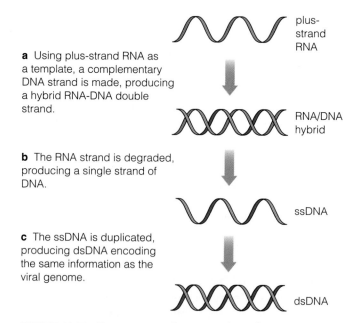

a Using plus-strand RNA as a template, a complementary DNA strand is made, producing a hybrid RNA-DNA double strand.

b The RNA strand is degraded, producing a single strand of DNA.

c The ssDNA is duplicated, producing dsDNA encoding the same information as the viral genome.

FIGURE 13.15 Reverse transcriptase catalyzes three reactions.

LARGER FIELD

GOOD TIMING

To enter a host cell, plant viruses must depend upon the wall's being breached by mechanical abrasion or the bite of an insect. That gets the virus into the first cell, but how does it spread to other cells to cause general infection? Plant viruses en-code special movement proteins that spread the virus from cell to cell through plasmodesmata, the thin protoplasmic bridges between adjoining plant cells. Tobacco mosaic virus, for example, encodes only four proteins, but one of them is a movement protein.

Plant virologists discovered movement proteins only in the last decade, but biotechnology is far enough along that commercial applications are already possible. By replacing the virus's cell-killing genes with genes for a protein of commercial value and keeping the movement protein, biotechnologists can create viruslike packages that spread from cell to cell throughout a plant, carrying the protein-encoding gene. This technique is already being used to make tricosanthin, an experimental AIDS drug, in a tobacco field.

AIDS, the disease, is considered in Chapter 27.

Influenza Viruses. The influenza viruses make up the family Orthomyxoviridae. This family is divided into types A, B, and C, groups that correspond roughly to genera. Type A is the most common. It has an extremely wide host range with strains that infect many animals, including humans, seals, pigs, and birds. Type A strains have been responsible for many epidemics and **pandemics** (worldwide epidemics). The most notorious was the great flu epidemic of 1918, at the end of World War I, which caused at least 20 million deaths, more than the war itself (Chapter 20). Types B and C, which infect only humans, don't cause pandemics. Type B causes outbreaks every 2 or 3 years. Type C causes subclinical infections or mild coldlike illnesses. All three types of flu viruses are similar in structure and mode of replication; their capsid proteins differ.

Flu viruses are enveloped minus-strand RNA viruses with two kinds of protein spikes in their membranes, **hemagglutinin (HA)** and **neuraminidase (NA)**. HA is the protein by which the virus attaches to its host

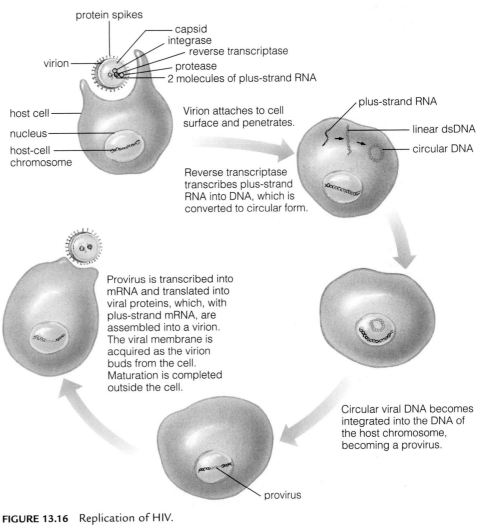

protein spikes
capsid
integrase
reverse transcriptase
virion
protease
2 molecules of plus-strand RNA
host cell
nucleus
host-cell chromosome

Virion attaches to cell surface and penetrates.

plus-strand RNA
linear dsDNA
circular DNA

Reverse transcriptase transcribes plus-strand RNA into DNA, which is converted to circular form.

Provirus is transcribed into mRNA and translated into viral proteins, which, with plus-strand mRNA, are assembled into a virion. The viral membrane is acquired as the virion buds from the cell. Maturation is completed outside the cell.

Circular viral DNA becomes integrated into the DNA of the host chromosome, becoming a provirus.

provirus

FIGURE 13.16   Replication of HIV.

(Art by Carlyn Iverson.)

TABLE 13.4 Unusual Properties of HIV (and Retroviruses)

Property	Comments
RNA genome is diploid	Significance is not understood.
Intact virion contains three enzymes: reverse transcriptase, integrase, and protease	Inhibitors of two of these enzymes are powerful anti-AIDS drugs.
Plus-strand genome serves as a template for making dsDNA (action of reverse transcriptase), not as mRNA	Certain anti-AIDS drugs, including AZT, act by inhibiting this reaction.
Maturation is completed within the intact virion when protease cuts strings of connected proteins into active units	Anti-AIDS drugs called protease inhibitors act by inhibiting this reaction.
Capacity for rapid genetic change	Possibly accounts for its having become a human pathogen; may foretell its gaining further pathogenic potential.

cell. (As its name implies, it can also attach to red blood cells, causing them to **agglutinate** [aggregate].) NA is an enzyme that breaks down **sialic acid** (a component of the host-cell membrane), thereby releasing the virion from the surface of its host cell.

Most viruses have a single characteristic shape, but the shape of the influenza virus is highly variable (**Figure 13.17**). Within a single strain of the virus, some virions are nearly round. Others are elongated and sometimes bent. Still others form long filaments.

Like virions of other minus-strand RNA viruses, influenza virions contain molecules of RNA-dependent RNA polymerase.

Many of the unusual properties of the influenza virus arise from its having a segmented genome. That is, its RNA exists in the virion as eight separate pieces, each of which is enclosed within a helical capsid. Each piece encodes a single protein, except for the smallest two, which encode two overlapping genes for two proteins. All eight helical capsids are packaged in a single larger capsid, which is surrounded by an envelope.

Once infected by a strain of influenza virus, humans become immune, but only to that particular strain (Chapter 27). Immunity depends upon the ability to inactivate the HA and NA on the virus surface. Most humans get influenza several times in their lives, because influenza constantly changes the HA and NA it presents. These changes occur suddenly and dramatically by **antigenic shift** or gradually and incrementally by **antigenic drift.**

The segmented genome of influenza virus is the basis for its ability to change by antigenic shift. This happens when two different influenza viruses infect the same cell. The progeny virions then contain some RNA molecules from each of the infecting virions. In other words, the

RNA molecules of the two infecting virions reassort in various ways among the progeny virions. The product is a virus that is significantly different from either of the original infecting strains. Antigenic shift sometimes produces highly dangerous strains that cause deadly pandemics. Mutational changes cause antigenic drift, small changes but sufficient to make yearly flu shots necessary. Influenza, the disease, is discussed in Chapter 22.

Next we'll consider certain capacities of some viruses, namely their ability to form tumors. Viruses that have this capacity are termed **tumor viruses.**

Tumor Viruses. One family of RNA viruses and four families of DNA viruses cause **tumors** (uncontrolled growth of new tissue) in animals and humans (**Table 13.5**). Before we discuss these viruses we'll briefly consider tumors and their cause.

Tumors. Most tumors are **benign** (not life threatening), but some are **invasive** (they grow into other tissue) and **malignant** (deadly). **Cancer** is another term for malignant tumor.

Much of what we know about the causes of tumors comes from studies on tumor viruses, particularly the RNA tumor viruses (all of which are retroviruses). In spite of the wealth of information that came from studies on tumor viruses, most human cancers are not a result of viral infection. Tumors result from unrestrained growth of cells, which occurs as a consequence of genetic change. The number of ways this can happen is limited. Most probably, changes in only about 40 genes can lead to development of tumors.

A gene that causes a tumor is called an **oncogene** (onco refers to tumor). Oncogenes are derived through mutation from normal mammalian genes called **proto-oncogenes.** In general, proto-oncogenes regulate normal

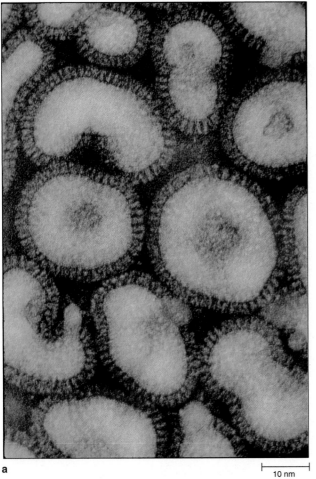

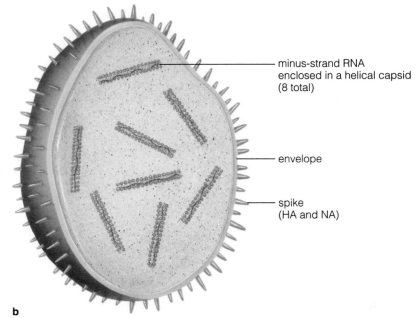

minus-strand RNA
enclosed in a helical capsid
(8 total)

envelope

spike
(HA and NA)

b

a

10 nm

FIGURE 13.17 Influenza virus. (a) Unlike most viruses, influenza virus particles vary greatly in shape, as this electron micrograph shows. (b) An influenza virus has a segmented genome—eight different minus-strand RNA molecules. Each RNA molecule is enclosed in a helical capsid. All are enclosed in a single large capsid and surrounded by an envelope with hemagglutinin or neuraminidase spikes.

(Art by Carlyn Iverson.)

© K. G. Murti/Visuals Unlimited

cellular processes. When they become oncogenes, however, normal regulation is distorted and uncontrolled cancerous growth may result.

A second type of genetic change can also result in formation of a tumor, namely loss or destruction of a gene that normally acts as a **growth suppressor** or a **tumor suppressor.** When these genes (examples are Rb and p53) are inactivated, uncontrolled growth occurs and tumors result. But how do viruses bring about these tumor-forming genetic changes? Now we'll examine how RNA viruses and then DNA viruses cause tumors to form.

RNA Tumor Viruses (Retroviruses). Certain RNA tumor virus are known always to cause cancer in animals but not in humans. A person infected by one of these viruses may develop a specific type of cancer, but it doesn't always happen (Chapter 15).

Retroviruses cause tumors to form by introducing oncogenes into their host; they do so in one of three ways. The first way is typified by Rous sarcoma virus (RSV), which causes slowly developing tumors in chickens. RSV contains an oncogene (designated *src*) as part of its normal genome. This oncogene is inserted into the host's genome when the virus becomes a provirus. That cell becomes a tumor cell.

The second way retroviruses can convert a normal cell to a tumor cell is typified by Moloney murine sarcoma virus (Mo-MSV). Such viruses are defective. They have fewer viral genes than normal viral parents because they have lost essential viral genes to make room for an oncogene, which they can insert into cells they infect. These defective retroviruses cannot replicate by themselves. They depend upon the presence of a normal retrovirus in the same cell to supply the vital functions lost from the genome of the defective virus.

The third was is typified by viruses such avian leukosis virus (ALV) that do not carry an oncogene. This class of viruses cause tumors at a low frequency depending upon location on the host cell genome where the proviral genome integrates. Integration can occur anywhere on the genome, usually with no effect. But when they happen to integrate next to a proto-oncogene, they stimulate its expression, causing it to act like an oncogene.

TABLE 13.5 Some Tumor Viruses

Virus	Comments
RNA Viruses	
Family: Retroviridae	
Rous sarcoma virus	This virus carries an oncogene, *src,* as an extra gene. Causes, with high frequency, slowly developing cancers in chickens.
Moloney murine sarcoma virus	This defective virus carries an oncogene, *mos,* which causes cancers in mice.
Avian leukosis virus (ALV)	ALV does not carry an oncogene but causes, at low frequency, cancers in birds (only when it integrates next to a proto-oncogene, *myc,* in the host).
DNA Viruses	
Family: Papovaviridae	
Papillomaviruses	These cause warts in humans, which can convert to cancer; cause natural cancers in animals.
Simian virus 40 (SV40)	SV40 is a monkey polyoma virus that causes leukemia in juvenile hamsters.
Family: Adenoviridae	
Adenoviruses	These viruses are highly oncogenic in animals; no human cancers have been proven to be associated with adenoviruses.
Family: Herpesviridae	
Epstein-Barr virus (EBV)	This virus is causally associated with cancers, including Burkitt's lymphoma in Africa and nasopharyngeal cancer in China and Southeast Asia.
Family: Hepadnaviridae	
Hepatitis B Virus (HBV)	Certain cancers of the liver, which are among the most common cancers, may be caused by this virus.

DNA Tumor Viruses. DNA tumor viruses cause tumors in animals at a very low frequency. Only 1 in 1 million to 10 million cells actually becomes a tumor cell. Hepatitis C virus is an example of such an virus. It is associated with liver cancer in humans.

One reason DNA viruses are inefficient at causing tumors is that DNA viruses lyse their host cells as part of their normal replicative cycle. Once dead, the cell cannot be transformed. DNA tumor viruses apparently transform when unfavorable conditions suppress their normal replication. In laboratory studies, irradiation with ultraviolet light, incubation at an abnormal temperature, and exposure to certain chemicals stimulate formation of tumor cells by DNA viruses.

DNA viruses also form tumors when introduced into an animal that does not normally serve as a host. For example, human adenoviruses (*adeno* is from the Latin, meaning "glands") are DNA viruses found in tonsil and adenoid glands. They cause mild respiratory infections in humans, but never tumors. They do, however, cause tumors when injected into newborn hamsters. Virologists speculate that the virus becomes oncogenic because it cannot undergo its normal replication cycle and kill the hamster cell.

Certain viral proteins seem to be the actual cause of tumors. Some interact with products of proto-oncogenes, making them oncogenic. Others interact with host proteins that normally suppress tumor formation, inhibiting their antitumor activity.

PLANT VIRUSES

Plant viruses are generally named for the disease they cause. For example, the virus that causes mosaic disease of tobacco plants is called tobacco mosaic virus (TMV). The virus that causes bushy stunt disease of tomato plants is called tomato bushy stunt virus.

Plant viruses can be studied the same way as animal viruses—using cell cultures.

Growth, Replication, and Control

Animal cells are surrounded only by membranes, but plant cells have thick cell walls. In addition, some plant tissues, including leaves and stems, are covered by layers of waxy material. A plant virus must break through these protective layers to enter its host cell. Some plant viruses enter through mechanical injuries—abrasions and nicks from animals or weather. Other plant viruses enter with an insect bite. Aphids, leafhoppers, and other sap-consuming insects all transmit plant viruses.

Considering the ways they enter their hosts, it's probably not surprising that plant viruses are never enveloped. But in other respects, plant viruses resemble animal viruses. Some are helical and some are polyhedral. Most contain RNA. Some contain dsDNA or ssDNA. Once inside the plant cell, viral replication proceeds much like phage and animal virus replication. Viral nucleic acid directs the host cell to make more viral components. These are assembled into virions and released.

Most viral diseases of plants are chronic degenerative diseases They merely slow plant growth and diminish crop yield. Still, their economic impact is enormous. There are two general ways to control plant viruses: (1) control the insects that spread the virus, and (2) develop virus-resistant and virus-free plant strains. Developing resistant plants is preferable because viruses are often spread in seeds or by grafting.

Ironically, virus-infected plants are often quite beautiful. In spite of their diminished crop, potato plants with leaf roll disease have thickened, healthy-looking leaves. Tulip plants with tulip break disease have beautifully variegated flowers (**Figure 13.18**).

Tobacco Mosaic Virus

Tobacco mosaic virus (TMV) was the first virus to be discovered. It was the first to be crystallized. It was also the first to be disassembled into RNA and protein and then reassembled into infectious virions. TMV is the most intensively studied of the plant viruses. It is the *Escherichia coli* of plant viruses.

TMV is named for the mottled mosaiclike patterns it causes on leaves of infected plants. It is a helical plus-strand RNA virus. When it infects a tobacco plant, it multiplies rapidly and accumulates in large quantities. Its abundance in infected plants attracted virologists to it. Many fundamental principles of plant virology came from studies on TMV.

Courtesy John L. Ingraham

FIGURE 13.18 Variegated tulips are the result of tulip break disease.

VIRUSES OF EUKARYOTIC MICROORGANISMS

Studies on viruses that infect eukaryotic microorganisms have lagged behind studies on other viruses. These viruses do not have the economic impact of plant and animal viruses, and they are not as easy to study as phages. In fact, until about 40 years ago microbiologists did not believe viruses infected eukaryotic microorganisms. Viruses in protozoa were not definitely identified until 1986. Now many viruses are known to infect these organisms. Viruses that infect fungi are called **mycoviruses.** They are mainly dsRNA viruses and may be enveloped. Known viruses that infect protozoa are RNA viruses. Most if not all are double stranded.

INFECTIOUS AGENTS THAT ARE SIMPLER THAN VIRUSES

Viruses are not the simplest agents that cause disease. There are two even simpler agents, viroids and prions.

Viroids

A **viroid** is a circular molecule of ssRNA without a capsid. Viroids cause several economically important diseases of plants, including citrus exocortis. One such disease, potato spindle tuber, causes the edible part of the potato (the tuber) to become long and pointed. It also stunts the growth of tomato plants (**Figure 13.19**). As few as 10 molecules of the viroid are enough to infect a potato plant. Many more virions are needed to cause most plant diseases.

A viroid is only about one-tenth the size of the smallest plant virus. Spindle tuber viroid is composed of only 359 nucleotides, about a third as many as make up an ordinary-size gene. How such a small molecule can cause disease remains a mystery. It seems unlikely that the viroid encodes a protein that causes disease. More probably it interacts directly with the host genome, changing the expression of the host genes and causing disease.

Viroids may be bits of mRNA from plant genes that have escaped during normal processing.

Prions

Prions (pronounced PREE-onz), meaning "proteinaceous infectious particles," are thought to cause about a dozen fatal degenerative disorders of the central nervous system of humans and other animals (Chapter 25). Prions are unique among infectious agents because they contain no nucleic acid. They are composed exclusively of protein (**Figure 13.20**).

The existence of an infectious agent composed only of protein seems to defy scientific dogma. How can a protein multiply and cause disease? These questions are just now being answered. First, a prion does not replicate itself from scratch. It causes another protein to change its shape and thereby become a prion. A prion protein (designated PrP^{sc}) is an altered form of a normal host protein (PrP). A prion has the capacity to change the secondary structure of a molecule of PrP so that it too becomes a prion. The secondary structure of PrP is largely alpha helices. A prion is largely beta sheets. When a prion enters a host, it forms more prions from the cell's supply of PrP. Prions accumulate, aggregate into fibrils, and cause disease.

Stanley Prusiner, who did most of the work on prions, was awarded the Nobel Prize in 1998. Some virologists, however, are still not convinced that prions alone cause these diseases. They suspect that a still-undiscovered virus participates. Prions cause diseases of animals and humans, including bovine spongiform encephalopathy (mad cow disease), which are discussed in Chapter 25.

THE ORIGIN OF VIRUSES

How did viruses originate? Because they are so extremely simple, we might think they are primitive and evolved early in the history of life on Earth. But they are obligate parasites. They could not have existed before their hosts evolved. Viruses must have evolved from more complex things.

Many scientists believe that viruses evolved from genes in the cells they infect. They may be "genes on the loose." A hint of this evolutionary origin comes from the properties of viroids. They seem to be pieces of plant genes. Maybe an excised gene could become capable of entering another cell and cause the cell to reproduce it. Such a gene

FIGURE 13.19 Growth stunting and leaf distortions in the tomato plant on the right are caused by the potato spindle tuber viroid.

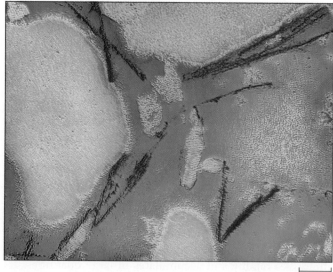

FIGURE 13.20 Prions are infectious agents that seem to be composed only of protein. Prions aggregate into the fibrils shown here.

would be a primitive virus. Further evolution might produce the highly successful and widespread infectious agents we call viruses. This line of reasoning suggests that viruses evolved late from cellular organisms and that viroids are primitive viruses.

New viruses continue to emerge. Some of these, such as hepatitis C virus (Chapter 23), have probably existed for a long time but have been discovered only recently. Others, such as dengue fever virus (Chapter 27), have spread into new host populations. Still others, such as HIV (Chapter 27) and a strain of hantavirus that causes pulmonary syndrome (Chapter 22), are probably genuinely new. They might have arisen by a combination of mutation, cross-species transfer, and changed human behavior.

SUMMARY

Case History: A Double Hit (p. 314)

1a. Viral infections are common; they can be severe, but they are usually self-limiting.

The Ultimate Parasites (pp. 314–315)

1. A virus is a package of nucleic acid wrapped in a protein coat. It is acellular and lacks a metabolism of its own.

2. Virions are intact, nonreplicating virus particles. Some of them have been crystallized.

3. In 1892 Dmitri Iwanowski found that filtering tobacco plant extracts did not remove the disease-causing agent. This observation led to the discovery of viruses. Viruses were not seen until the development of the electron microscope in the 1930s.

Classification of Viruses (pp. 316–320)

4. The primary classification of viruses is based on host range. Viruses that attack bacteria are called bacteriophages, or simply phages. Host range is determined principally by receptors on the cell surface to which virions attach.

5. The size of viruses ranges from 25 nm to more than 300 nm. Viruses are smaller than the smallest prokaryotic cell.

6. Viruses have a nucleic acid core, a protein coat, and sometimes a membrane.
   a. The nucleic acid core may contain double-stranded (ds) DNA, single-strand (ss) DNA, dsRNA, or ssRNA (either plus strand or minus strand). Plus strand is mRNA, translated directly by the host ribosome. Minus strand must be transcribed into mRNA (a plus strand) after it enters the host cell.
   b. The protein coat that surrounds the nucleic acid core of a virus is called a capsid. A capsid can be helical, polyhedral, or complex, depending upon the arrangement of its constituent protein molecules, called capsomeres. Complex viruses have a helical portion called a tail attached to a polyhedral portion called a head. They may have additional structures, such as a tail sheath, a plate, pins, and tail fibers.
   c. Viruses surrounded by a membrane are called enveloped viruses. Viral envelopes are pieces of the host cell's plasma membrane that the virion acquires as it emerges from its host cell. Viruses without envelopes are called naked viruses. Enveloped viruses have protruding proteins called spikes.

7. All viruses have similar life cycles. Outside a host cell they exist as non-replicating virions. The virion attaches itself to a host cell (adsorption). The genome enters the host cell (penetration). The process of removing the capsid and envelope (uncoating) can occur before or after penetration. The host cell manufactures viral components (viral synthesis). The components are assembled into intact new virions (maturation). Virus particles leave the infected cell (release).

8. Viral taxonomy does not follow the Linnaean scheme. It has only three levels—family, ending in viridae; genus, ending in virus; and species, English words.

Bacteriophages (pp. 321–325)

9. Bacteriophages are studied by plaque count and one-step growth curve. A plaque count determines the number of bacteriophages in a sample by counting the plaques on a lawn of bacteria. Virions and infected cells together are called plaque-forming units (PFUs).

10. In the one-step growth curve experiment, plaque counts are done on an infected culture to determine the number of PFUs present. The phage growth curve shows a latent period, a burst period, and an eclipse period. The burst size is the fold increase in PFUs that occurs during the burst period.

11. Bacteriophages may immediately enter the lytic pathway, leading to production of more virions, or they enter the lysogenic pathway, leading to a prolonged quiescent state termed lysogeny. Phages that follow only the lytic pathway are termed virulent. Phages that can follow either pathway are termed temperate.

12. Virulent phages, such as T4, lyse the host cell during release. Lambda (λ), a typical temperate phage, lyses the host cell if it takes the lytic pathway. If it takes the lysogenic pathway, no phage components are synthesized. Instead the phage genome becomes part of the host-cell chromosome, forming a prophage. The cell is termed lysogenic; it has the potential to cause lysis if the genes in the prophage become reactivated. Lysogenic cells are immune to further attack by the same virus.

Animal Viruses (pp. 325–326)

13. Animal viruses are often studied in cell cultures. A monolayer of cells is infected with virus, causing plaques that can be counted. Cells that dominate a culture are called a cell line. Cell lines that grow indefinitely in culture are called continuous cell lines.

14. There are three modes of viral penetration: The viral envelope fuses with the cell's plasma membrane; the host cell engulfs the virion (phagocytosis); or the virus adsorbs to the cell surface while the nucleic acid component enters.

15. Single-strand DNA must be converted to dsDNA, which is transcribed into mRNA and translated into viral proteins. RNA viruses encode RNA-dependent RNA polymerase, an enzyme that plays a different role in viral synthesis of each class of RNA virus.

16. In latent viral infections, such as those caused by the herpesvirus family, viral replication is arrested until reactivated.

17. Retroviruses use RNA as a template to make DNA. Influenza viruses undergo antigenic shift and antigenic drift. Some viruses are associated with human cancer.

Plant Viruses (pp. 336–337)

18. Plant viruses are usually named for the disease they cause. All plant viruses are naked.

19. Plant viruses must penetrate cell walls and sometimes protective waxy outer coverings on stems and leaves. Some plant viruses enter through mechanical injuries to plants. Others enter through an insect bite.

20. A plant virus—tobacco mosaic virus (TMV)—was the first to be discovered, the first to be crystallized, and the first to be disassembled into its component parts and then reassembled into infectious virions.

Viruses of Eukaryotic Microorganisms (p. 337)

21. The study of viruses that infect eukaryotic organisms is in its infancy. Viruses in protozoa were not definitely identified until 1986.

22. Viruses that infect fungi are called mycoviruses.

Infectious Agents That Are Simpler Than Viruses (pp. 337–338)

23. A viroid is a circular molecule of ssRNA without a capsid. Viroids cause several plant diseases. Viroids are about one-tenth the size of the smallest plant virus.

24. A prion is an infectious agent that seems to be composed only of protein. Prion-caused diseases affect the nervous system.

The Origin of Viruses (pp. 338–339)

25. Viruses may have evolved from genes in the cells they infect.

26. Some viruses, such as the hepatitis C virus, have existed for a long time but have been discovered only recently. Others, such as HIV, are probably new, arising from an interplay of mutation, cross-species transfer, and human behavior.

REVIEW QUESTIONS

Case History: A Double Hit

1a. Which application of microbiological knowledge gave V. O's parents the greatest comfort? Explain.

The Ultimate Parasites

1. Define these terms: virus, virion, acellular.

2. Briefly discuss the key events in the history of virology.

Classification of Viruses

3. On what four bases are viruses classified?

4. What is host range? How is host specificity determined?

5. What is the basic structure of a virus? Name the types of nucleic acid that may be found in viruses. What is the difference between plus-strand and minus-strand RNA?

6. Define these terms: capsid, capsomere, tail, head. What purpose is served by additional viral structures such as a tail sheath or plate?

7. What is the difference between an enveloped virus and a naked virus? What are spikes, and how do viruses acquire them?

8. Explain these steps in the viral life cycle: adsorption, penetration, uncoating, viral synthesis, maturation, release.

9. How is viral taxonomy different from bacterial?

Bacteriophages

10. Explain plaque count and the one-step growth curve. What is a PFU? What do eclipse and latent period mean?

11. Trace the two pathways bacteriophages may take on entering a host cell.

12. Explain these terms: virulent phage, temperate phage, prophage, lysogenic cell.

Animal Viruses

13. How are animal viruses studied? What is a continuous cell line?

14. How is animal virus replication the same as phage replication? How is it different?

15. What is a latent viral infection? Give an example.

16. What is a retrovirus?

17. Explain the difference between antigenic shift and antigenic drift.

18. Explain this statement: Some viruses are associated with human cancer.

Plant Viruses

19. How are plant viruses studied? How are they named?

20. Explain this statement: It is not surprising that all plant viruses are naked viruses.

21. Why do we speak of tobacco mosaic virus as the plant-world equivalent of *Escherichia coli* in the history of microbiology?

Viruses of Eukaryotic Microorganisms

22. Explain this statement: The study of viruses in eukaryotic microorganisms is in its infancy.

23. What are mycoviruses?

Infectious Agents That Are Simpler Than Viruses

24. What is a viroid? What kinds of diseases do viroids cause?

25. What is a prion? Why are prions so puzzling? What kinds of diseases do they cause?

The Origin of Viruses

26. Why are viruses probably an example of retrograde evolution?

27. Explain this statement: Viruses probably evolved from a gene on the loose.

28. How do new viruses arise?

CORRELATION QUESTIONS

1. If you did a one-step growth curve with an enveloped bacteriophage, would it be possible to determine the extent of the eclipse period? Why?

2. HIV and other retroviruses have higher rates of mutation than other viruses. At what step in their life cycle would you expect most of these mutations to occur? Explain your reasoning.

3. HIV virions contain plus-strand RNA but it does not serve as mRNA immediately after entering a host cell. Would you suspect that this is true simply from knowing the composition of an intact virion? How would you come to this conclusion?

4. Why do only some virions contain enzymes?

5. What changes in a virus must occur for it to be able to infect a new host? Explain.

6. How do viroids differ from RNA viruses?

ESSAY QUESTIONS

1. Discus your reasons for believing that viruses are or are not living things. Does the same reasoning apply to viroids? Prions?

2. Discuss possible reasons why plant viruses largely cause nonlethal diseases.

3. Discuss the possibility of HIV's being a tumor virus.

SUGGESTED READINGS

All the virology on the WWW: http://www.tulane.edu/~dmsander/garryfavweb.html.

Dimmock, N. J., and S. B. Primrose. 1994. *Introduction to modern virology.* 4th ed. Cambridge, U.K., Blackwell Scientific.

Fields, B. N., and D. M. Knipe. 1990. *Fields' virology.* 2d ed. New York: Raven Press.

Fraenkel-Conrat, H., P. C. Kimball, and J. A. Levy. 1988. *Virology.* 2d ed. Englewood Cliffs, N.J.: Prentice-Hall.

Levine, A. J. 1992. *Viruses.* New York: Scientific American Library.

Nature Insight. 2001. AIDS. *Nature* 963-1007.

Prusiner, S. 1995. The prion diseases. *Scientific American*, January.

Summers, W. C. 2001. Bacteriophage therapy. *Annual Reviews of Microbiology* 55:437–51.

Wang, A. L., and C. C. Wang. 1991. Viruses of protozoa. *Annual Reviews of Microbiology* 45:251–64.

For additional readings, go to InfoTrac College Edition, your online research library at: http://www.infotrac.thomsonlearning.com

FOURTEEN

Microorganisms and Human Health

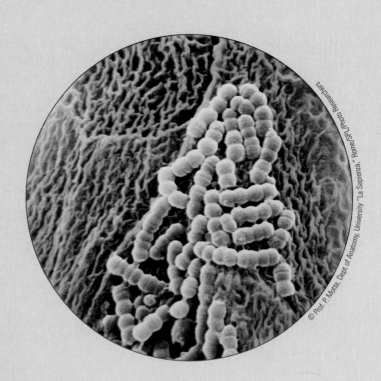

© Prof. P. Motta, Dept of Anatomy. University "La Sapienza." Rome/SPL/Photo Researchers

CHAPTER OUTLINE

LEARNING GOALS

To understand:

- *The nature of the normal human microbiota and how it can change*
- *The three types of human-microbe symbioses: commensalism, mutualism, and parasitism*
- *The human factors that determine the normal microbiota: structural defenses, mechanical defenses, and biochemical defenses*

- *The microbial factors that determine the normal microbiota: physical, nutritional, and special adaptations to life on living tissue*
- *The nature of the microbiota on the skin and conjunctivae; in the nasal cavity and nasopharynx; and in the mouth, intestinal tract, vagina, and urethra*

- *The impact of the normal microbial biota on its human host*
- *The disease-causing ability of microorganisms as it exists along a continuum, from highly virulent to almost harmless*

Avoiding a Serious Infection by Treating Someone Else

A few hours before birth, Baby Girl A faced the danger of life-threatening infection. As is the case with most babies, the protective fetal membranes that surrounded her suddenly broke. Soon, as she would pass through the birth canal, she would be exposed to the wide variety of microorganisms present in her mother's vagina. Prior to the moment when the membranes ruptured, Baby Girl A had existed in a completely microbe-free environment. When the first approximately 100 of her cells that arose from the union of egg and sperm burrowed into the lining of her mother's uterus, she was protected by her mother's tissues. Then, as the fetal membranes formed around her, she was provided the added protection of being surrounded by a microbe-impermeable sac.

Bursting of fetal membranes just before birth and thereby sudden exposure to the world of microorganisms is a normal and usually risk-free event. Most of the microorganisms Baby Girl A will encounter in the birth canal are harmless. Some of them will become intimately associated with her. They will live in and on her, becoming part of her **normal microbiota** (a characteristic population of microorganisms associated with humans) and will accompany her through life. Her normal microbiota will protect her by creating an environment that is unfavorable for the growth of **pathogens** (disease-causing microorganisms).

But Baby Girl A was at risk of encountering Group B streptococci because this pathogen had been present in her mother's vagina. Although Group B streptococci can cause a variety of serious diseases, Baby Girl A's mother was unaware of being colonized by this pathogen because she had no symptoms. Between 5 and 35 percent of pregnant women are similarly colonized by Group B streptococci and like Baby Girl A's mother do not show symptoms. But about 1 percent of infants born to such colonized women become infected and 5 to 8 percent of them die. The disease newborns suffer when infected with Group B streptococci at birth can strike quickly or be delayed. Either eventuality is dangerous. Early-onset disease, which begins within the first 6 days after birth, usually develops into meningitis (see Chapter 24, Case History: A Good Outcome to a Bad Scare). Late-onset disease typically begins at 3 to 4 weeks of age. It usually results in septicemia (Chapter 27).

Because Group B streptococci rarely cause symptoms in colonized women, the obstetrician caring for Baby Girl A's mother couldn't know whether or not she carried the pathogen without testing. But knowing the consequences to the baby could be severe, he took swab samples from her vagina and rectum during the thirty-sixth week of her pregnancy. These samples were examined by culturing for the presence of Group B streptococci. The culture results were reported as positive. She did carry Group B streptococci, and Baby Girl A would be exposed to it at birth. Baby Girl A's mother was given intravenously large doses of penicillin G. Because of this preventive treatment, Baby Girl A passed through a birth canal that was loaded with a variety of microorganisms (penicillin G kills only certain bacteria, including Group B streptococci), but not Group B streptococci. She began to acquire her normal microbiota. Baby Girl A was observed carefully in the nursery for any signs of infection. She showed none and went home with her mother at 2 days of age as an entirely healthy newborn.

By treating her mother with penicillin G, the Group B streptococci that could have been the source of a serious infection for Baby Girl A were eliminated.

Case Connections

- Because the antibiotic Baby Girl A's mother received was highly selective in its antimicrobial action, she was exposed at birth to an almost normal microbiota.
- As Baby Girl A's case suggests and as we'll learn in this chapter, normal microbiota is established quickly and helps prevent disease.
- As we'll see in this chapter and the next, the distinction between a microorganism's being a pathogen or a harmless member of the microbiota is sometimes indistinct. The Group B streptococci that caused her mother no detectable harm could have killed Baby Girl A.

NORMAL BIOTA

In this chapter we'll consider the relationships between microorganisms and healthy humans. For our purposes, we'll define health as a stable state in which all the body's organ systems function adequately. Disease is an unstable state. It ends either with recovery or death.

Throughout our evolution, we developed an intimate and complex relationship with microorganisms. Hundreds of species of bacteria and fungi live in and on our bodies—some permanently, others just temporarily. Altogether, the numbers are prodigious. Even when we are healthy, the number of microbial cells in or on our body is 10 times greater than the number of human cells we're made of.

The microorganisms that live with us stably are called our **normal biota.** They thrive and multiply because they are adapted to life on our bodies. Under most circumstances, they don't cause disease. Our normal biota inhabits only the surfaces of our body—both external and internal. The external surfaces, such as the skin, come in direct contact with microorganisms in the environment. The internal surfaces are exposed indirectly. The surface of our intestines, for example, is exposed to microorganisms that enter through the mouth.

External body surfaces that support a normal biota include the skin and the outer covering of the eye. Interior surfaces with a normal biota include the nose, mouth, intestinal tract, vagina, and urethra (**Figure 14.1**). Microorganisms found in any other tissues of the body—such as the brain, heart, muscle, or bone—are not normal biota. They can be expected to cause disease.

Resident Biota

The first microorganisms on the surfaces of Baby Girl A's body were those that just happened to land there. Other babies of similar age might have quite different microorganisms on them. But by the time Baby Girl A was 2 weeks old, the population of microorganisms on her body became strikingly similar to the population on all humans. This

population of microbes is called the **resident biota.** Each body surface has its own characteristic resident biota, made up of particular microbial species. The exact combination of microbial species and the size of their populations may vary somewhat, but the resident biota is characteristic and permanent.

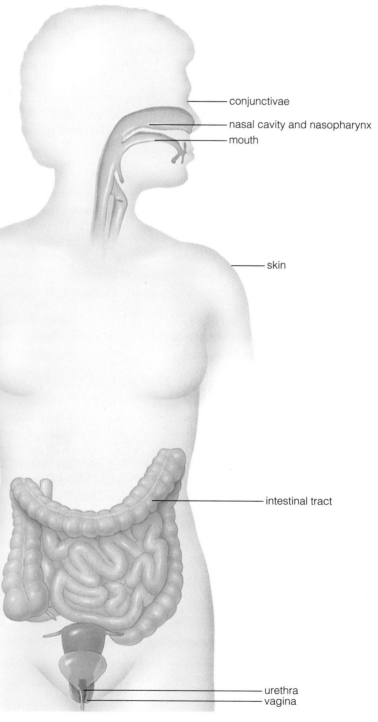

FIGURE 14.1   Sites of colonization by the normal biota.
(Art by Carlyn Iverson.)

LARGER FIELD

TORCH

Baby Girl A had no contact with microorganisms before her birth, but not all babies are so fortunate. An unborn baby's lifeline is the **placenta,** the organ that communicates with the mother's bloodstream. The placenta delivers oxygen and nutrients to the fetus and carries away its metabolic waste. But in about 2 percent of pregnancies, the placenta also provides access for unwelcome pathogens.

Transplacental infection occurs when microorganisms from an infected mother's bloodstream cross the placenta and infect the fetus. Such infections can interfere with the complex processes that shape a baby's organs and tissues. Even an infection so mild that the mother is unaware of any illness can kill the fetus. The most common infections that cause severe prenatal damage and malformation are abbreviated TORCH. The initials stand for **t**oxoplasmosis (Chapter 27), **o**thers (such as syphilis; Chapter 24), **r**ubella (Chapter 26), **c**ytomegalovirus (Chapter 24), and **h**erpes simplex (Chapter 24).

The stage of pregnancy at which infection occurs is critical. Tissues that are undergoing active development are the most likely to be harmed. For example, if rubella (German measles) infection occurs before the twelfth week of development, heart defects are likely. But if infection occurs between the twelfth and sixteenth week, the result is more likely to be deafness. Infections that are transmitted from mother to infant during pregnancy are discussed further in Chapter 24.

Transient Biota

Occasionally other microbial species come to rest on our body surfaces. But most are not well enough adapted to life on the human body to persist there indefinitely. They constitute a **transient biota.** Like the vast majority of all microbes, most members of a transient biota are harmless, but a few may be pathogens.

When transient microorganisms land on exposed areas of skin, particularly the hands, they can become loosely attached by oil or dirt. But they don't persist. The skin's resident biota can be reduced by washing or applying disinfectants, but it can't be eliminated. In contrast, a transient biota can be completely removed by careful washing.

Most hospital workers have a large transient microbiota of pathogens because they are exposed to them daily. These often include the pathogenic species *Staphylococcus aureus* (see Figure 11.10). Such pathogens are harmless to the healthy hospital worker but not to some patients—for example, those with lowered resistance, those with open wounds, the elderly, and infants. Even healthy babies, such as Baby Girl A, are extremely vulnerable to staphylococcal skin and eye infections. So transient biota can be dangerous. Good habits of hand washing are crucial in a hospital.

Opportunists

Opportunists are microorganisms that cause disease when the proper "opportunity" arises. Usually it is a breakdown in the immune system. Or it may be a result of medical treatment, such as use of **broad-spectrum antibiotics** (antibiotics that act against a wide variety of bacteria) or implantation of devices such as plastic catheters or artificial metal joints. Under such conditions, members of the normal biota can become **opportunistic pathogens.** For example, *Candida albicans*, a normal resident, can become an opportunist during antibiotic therapy and cause disease (see Figure 24.14). Some organisms such as *Pseudomonas aeruginosa*, which are never normal residents, are opportunists that cause disease only in people whose health is already weakened by other factors.

Changing Biota

Although the normal biota of the human body is relatively stable, it does change over time. As Baby Girl A develops, her body will change in ways that alter the environment of her microbial biota. Some established species will disappear. Some new species will become established.

For example, in about 6 months, when she gets teeth, the environment of her mouth will change dramatically. The teeth themselves provide new surfaces for microbial growth, especially of streptococcal species, and the anaerobic crevices between teeth and gums provide a favorable habitat for strict anaerobes, such as *Fusobacterium* spp. and *Bacteroides* spp.

Another change will occur when Baby Girl A is weaned. While she is being fed only breast milk, her intestinal biota will be composed almost exclusively (more than 90 percent) of one bacterium, *Bifidobacterium* (**Figure 14.2**). But when she begins to drink infant formula or

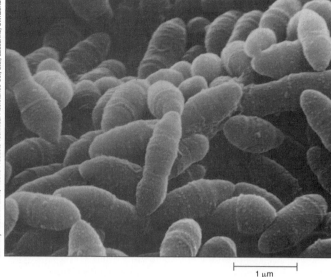

Courtesy Research Labs, Nestlé Products Technical Assistance Co., Ltd., Lausanne, Switzerland

1 µm

FIGURE 14.2 *Bifidobacterium,* the dominant bacterium in the intestines of breastfed infants.

eat solid food, her population of *Bifidobacterium* will be replaced by the widely varied intestinal microbiota typical of adults. It's thought that a carbohydrate in breast milk called **bifidus factor** favors the growth of *Bifidobacterium*.

Bifidobacterium promotes infant health. It ferments milk sugars to acetic and lactic acids (Chapter 5). The result is an acidic intestinal tract that's inhospitable to many pathogens, including those that cause diarrhea. Infant diarrhea is a common cause of illness and death in many parts of the world, so programs that encourage breastfeeding have a major effect on public health.

The microbiota of Baby Girl A's vagina will also change dramatically during her life. The female hormone **estrogen** is responsible for these changes. When she was born, estrogen that crossed the placenta from her mother encouraged the growth of lactobacilli, creating an acidic vaginal environment. Over the next 2 to 3 weeks, as the influence of maternal estrogen declines, lactobacilli nearly disappear. The pH becomes more alkaline and stays that way through infancy and girlhood. These conditions favor colonization by only a few microorganisms that don't com-

pete well against potential pathogens. Fortunately, prepubertal girls who are not sexually active are not exposed to many vaginal pathogens. However, occasional vaginal infection may occur, usually caused by normal intestinal inhabitants, such as *Escherichia coli*. Then during puberty, Baby Girl A will begin to make estrogen, which will stimulate her vaginal cells to produce glycogen. As these cells are sloughed during normal epithelial growth, the glycogen they contain becomes available to lactobacilli. They reestablish the acidic environment of the vagina that competes effectively against pathogens. This crucial protection will continue during Baby Girl A's sexually active and reproductive years. Then, as estrogen levels fall again after menopause, an alkaline vaginal environment with a sparse microbiota will return.

As we've seen, the normal biota varies over time. It also varies among individuals. We'll discuss individual variation later in this chapter.

SYMBIOSIS

Baby Girl A and members of her normal biota live in symbiotic relationships. **Symbiosis** refers to two kinds of organisms living together. Baby Girl A, whose body provides a habitat for its symbiotic partners, is the **host**. There are three kinds of symbioses—commensalism, mutualism, and parasitism. We'll discuss them separately (**Table 14.1**).

Commensalism

Commensalism is a symbiosis in which one partner benefits and the other neither benefits nor is harmed. That's the relationship that exists between humans and members of their normal microbiota. The microorganism benefits by being housed and fed. The human host is largely unaffected. Because neither partner is harmed, commensalisms tend to be long-lasting and stable. They are usually the product of long-term evolution.

As we've said, individual species of commensal microorganisms don't benefit their host. But collectively they

TABLE 14.1 Types of Symbioses

Type	Host	Symbiotic Partner	Relationship
Mutualism	Beneficial	Beneficial	Stable
Commensalism	Neither beneficial nor harmful	Beneficial	Stable
Parasitism	Harmful	Beneficial	Unstable

FIGURE 14.3 Microbial antagonism. This photo of a petri dish shows antagonism of *Serratia* (red colonies) against *Micrococcus* (confluent gold colonies).

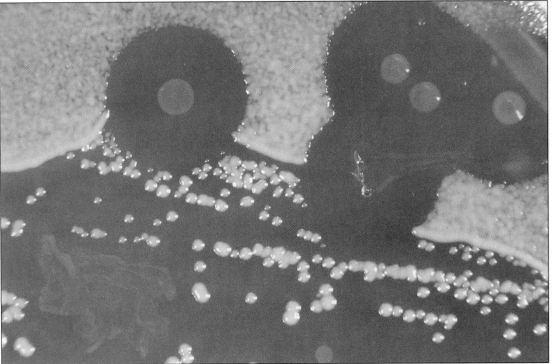

© Larry Jensen/Visuals Unlimited

do. Their most important benefit is competing with pathogens so they can't establish an infection. This kind of protective effect, in which one microorganism interferes with the growth of another, is called **microbial antagonism** (**Figure 14.3**).

Mutualism

Mutualism is a symbiosis in which both partners benefit. Usually neither can survive without the other. Like commensalism, mutualism is highly evolved and extremely stable. Ruminants (cows and similar cud-chewing mammals) have mutualistic relationships with microorganisms in their digestive tract that ferment cellulose. Microorganisms enable the ruminant to use cellulose as a primary source of food. The ruminant provides the microorganism with a nutrient-rich habitat (Chapter 8).

No true mutualism exists between humans and microorganisms because no single species is essential to human survival.

Parasitism

In **parasitism** the host is harmed and the parasite benefits. (Most microbial parasites are called **pathogens.** The term **parasite** is reserved for protozoa, worms, and insects.) Parasitism is usually an unstable symbiosis (exceptions are dis-

cussed in Chapter 15). Either the host dies or successfully eliminates the pathogen from its body. Such instability shows that the two organisms are poorly adapted to live together; they have evolved together for only a relatively short time.

But the distinction between parasitism and commensalism is not absolute. The commensals that make up the vast majority of the normal biota are potential pathogens under certain circumstances. These commensals are opportunists.

True parasitism is not rare. Aside from opportunistic infections, Baby Girl A will experience at least 150 different infections, including the common cold, over her lifetime.

FACTORS THAT DETERMINE THE NORMAL BIOTA

The microorganisms that inhabit the human body, like all microorganisms, survive because they are adapted to their environment. But what determines which microorganisms are found in a given location on the body? There are two factors: the environment and how microbes adapt to it.

Living Tissues as an Environment for Microorganisms

In Chapter 8 we considered the environmental factors that influence microbial growth: nutrients and non-nutrient features, such as temperature, pH, and oxygen supply.

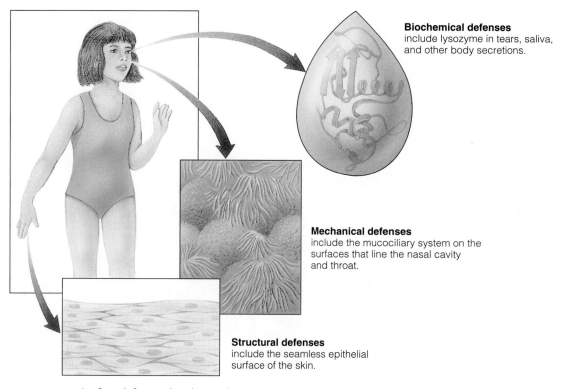

Biochemical defenses
include lysozyme in tears, saliva, and other body secretions.

Mechanical defenses
include the mucociliary system on the surfaces that line the nasal cavity and throat.

Structural defenses
include the seamless epithelial surface of the skin.

FIGURE 14.4 Surface defenses—biochemical, mechanical, and structural.
(Art by Carlyn Iverson.)

These are important on living surfaces. In addition, living surfaces can present structural, mechanical, and biochemical barriers to microbial growth (**Figure 14.4**).

These barriers are called **nonspecific surface defenses (Figure 14.5)** because they act against most or all microorganisms. They constitute our first line of defense against infection. We'll discuss them here and the body's second and third lines of defense in Chapters 16 and 17.

Structural Defenses. All the microbially colonized surfaces are **epithelial surfaces,** which are composed of cells that are tightly joined together to form a seamless and relatively impermeable barrier. If commensals were to cross this barrier, they would do significant damage to underlying tissues.

There are two main types of epithelial surfaces. The first type is skin. It covers all the body's exterior with the exception of the eye. Skin is relatively thick because it is composed of many layers of epithelial cells. The second type of surface is **mucous membrane.** It covers the surface of the eye (the **conjunctiva**) and all the body's interior surfaces. Compared with skin, mucous membranes are relatively thin. They provide less protection.

Epithelium also defends us in a second way. Because it grows, cells are constantly being sloughed off the surface and replaced by new ones from below. This process is a

relatively rapid. Surface cells and the microorganisms that cling to them are lost every few days or weeks. Sloughing controls the size of the microbial population and restricts it to those species that can multiply more rapidly than they are lost.

Mechanical Defenses. There are three kinds of mechanical defenses. One is the movements that occur along many body surfaces and eliminate most of the microbes that happen to land there. Some surfaces are moved by underlying muscles. Intestinal surfaces, for example, are moved almost constantly by the vigorous muscular activity that moves food and waste through the digestive system.

The epithelial surfaces that line the nasal cavity and throat are protected by a second kind of defense, called the **mucociliary system.** It's a combination of mucus production and ciliary movement. Cells on the surface produce a viscous mucus, forming a layer that entraps microorganisms. Cilia on epithelial cells constantly move this mucous layer toward the mouth, where it—along with the trapped microorganisms—is swallowed and eliminated.

A third kind of defense is the moving body fluids that bathe some surfaces and dislodge microbes. In the urethra, for example, the rapid flow of urine washes away most microorganisms. Tears wash microorganisms off the conjunctivae.

FIGURE 14.5 The body's three lines of defense against microorganisms. The first line (discussed in this chapter) is shown here being breached by a thorn. When it is overcome, the second and third lines are activated. They are discussed in Chapters 16 and 17.

(Art by Carlyn Iverson.)

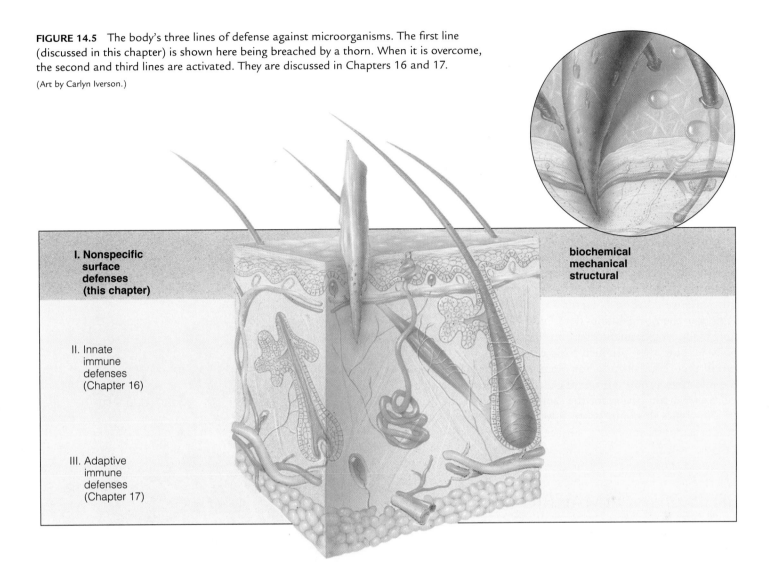

I. Nonspecific
surface
defenses
(this chapter)

II. Innate
immune
defenses
(Chapter 16)

III. Adaptive
immune
defenses
(Chapter 17)

biochemical
mechanical
structural

Biochemical Defenses. Biochemical defenses are chemical warfare against microorganisms. Some take place inside cells. For example, the skin cells produce enough of the protein keratin to fill the cells in the outermost layer of the skin, making it so dry that it is inhospitable to microorganisms.

Other biochemical defenses make body surfaces inhospitable by acidifying them. For example, the skin secretes fatty acids, and the stomach secretes hydrochloric acid.

Still other biochemical secretions have more specific antimicrobial effects. Bile, produced in the liver, is secreted into the intestines, where it kills many microorganisms by disrupting their cell envelopes. **Lysozyme,** an enzyme found in many secretions, including tears and saliva, breaks specific chemical bonds in peptidoglycan. Gram-positive bacteria are particularly vulnerable to lysozyme's lethal effect because their peptidoglycan wall is exposed at their cell surface.

Microbial Adaptations to Life on Body Surfaces

In spite of their defenses, some microorganisms are well adapted to live on body surfaces. Some species are adapted to survive the extreme dryness of skin. Others are able to tolerate the toxic chemicals the body produces. For example, *Escherichia coli* and the other organisms that colonize the human intestine are unaffected by bile, which is toxic to many microbes. And certain species of skin fungi can grow at the expense of the fatty acids that kill many other microorganisms.

Availability of nutrients also determines which microbes can survive. Within the human body, most nutrients are plentiful. Iron is the notable exception. The human body contains lots of iron. But because it is bound tightly to a variety of human proteins, it is not available to microorganisms. Thus an ability to take iron away from these proteins is critical for microorganisms that live on or in the human body.

But perhaps the most critical adaptation of members of the human microbiota is their ability to adhere to human cells. Otherwise they would be swept away by the body's mechanical defenses. Members of the normal biota attach themselves firmly to the epithelial surface by means of protein molecules called **adhesins.** Adhesins are located on the microbial cell surface or on pili. They bind to complementary sugar molecules, called **receptors,** on the surface of epithelial cells. Some strains of *Escherichia coli*, for example, have adhesins on their type 1 pili that bind to D-mannose sugar receptors on cells of the intestinal epithelium (**Figure 14.6**). Adhesins are remarkably specific. Most bind only to one kind of epithelial cell in one kind of animal. Receptors are also highly specific. For example, receptors in the human mouth vary from one tiny area to another. As we'll see in Chapter 15, pathogens adhere the same way the normal biota does.

Finally, some microorganisms persist on the human body because they produce toxic compounds that inhibit or kill their competitors. For example, certain bacteria in the intestines produce proteins called **bacteriocins,** which kill bacteria competing for the same space and nutrients.

Now let's become more specific and consider the properties of various body surfaces and which organisms live on them.

SITES OF NORMAL BIOTA

The skin, conjunctivae, nasal cavity, mouth, intestinal tract, vagina, and urethra all support a normal microbial biota. We'll consider each of these sites individually (**Table 14.2**).

The Skin

The skin is our largest organ. It makes up about 15 percent of our total body weight and has two functions: (1) It protects the body's vulnerable interior tissues from the outside environment, and (2) it helps regulate body temperature. We'll discuss the skin further in Chapter 26.

Microbial Environment. The predominant factor that limits microbial growth on the skin is dryness (**Figure 14.7**). Microorganisms are present on all normal skin, but they're numerous only in relatively moist areas. Such moisture as there is comes largely from the skin's own secretions. These include oil and perspiration. Additional moisture comes from mucous membrane surfaces where they contact the skin, near the nose, mouth, urethra, and anus.

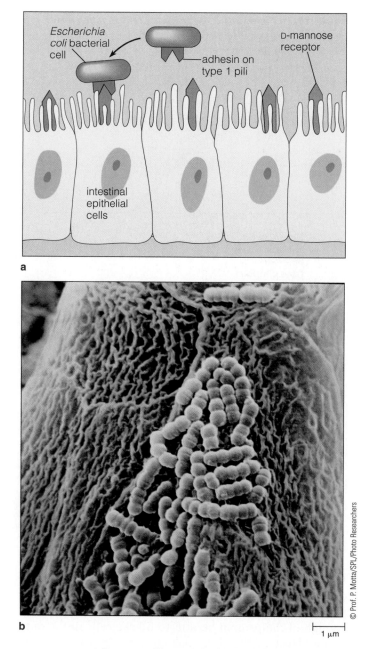

FIGURE 14.6 Adherence of bacteria to body surfaces. (a) Schematic sketch of how adhesins on the pili of *Escherichia coli* attach to D-mannose receptors on the surface of intestinal epithelial cells. (b) Bacteria (gray-green chains of cells) adhering to the tongue's surface (roughened pink background).

Skin secretions may have lethal, as well as beneficial, effects on microorganisms. For example, the oils on skin contain fatty acids which inhibit some microorganisms. In addition, perspiration contains **lysozyme,** an enzyme that kills some bacteria, and enough salt to slow or stop the growth of many microorganisms.

TABLE 14.2 Microorganisms Commonly Found on the Bodies of Healthy Human Beings

Organ	Microorganisms (Bacteria Unless Indicated as Fungus or Mite)
Skin	*Staphylococcus epidermidis* and *aureus*
	Corynebacterium spp.
	Propionibacterium acnes
	Pityrosporum ovale and *orbiculare* (fungus)
	Demodex folliculorum (mite)
Conjunctivae	*Staphylococcus* spp.
	Corynebacterium spp.
Nasal cavity and nasopharynx	*Staphylococcus epidermis* and *aureus*
	Streptococcus spp.
	Moraxella catarrhalis
	Lactobacillus spp.
	Corynebacterium spp.
	Haemophilus spp.
Mouth	*Streptococcus salivarius, mitis,* and alpha and gamma streptococci
	Staphylococcus epidermidis and *aureus*
	Moraxella catarrhalis
	Lactobacillus spp.
	Corynebacterium spp.
	Haemophilus spp.
	Bacerteroides spp.
	Fusobacterium spp.
	Enterobacterial spp.
	Candida albicans (fungus)
Intestinal tract	*Bacteroides* spp.
	Bifodobacterium spp.
	Clostridium spp.
	Escherichia coli
	Fusobacterium spp.
	Klebsiella spp.
	Proteus spp.
	Enterobacterial spp.
	Lactobacillus spp.
	Group D *Streptococcus*
	Staphylococcus aureus
Vagina	*Lactobacillus* spp.
	Clostridium spp.
	Bacteroides spp.
	Fusobacterium spp.
	Candida albicans (fungus)
Urethra	*Staphylococcus epidermidis*
	Enterococci

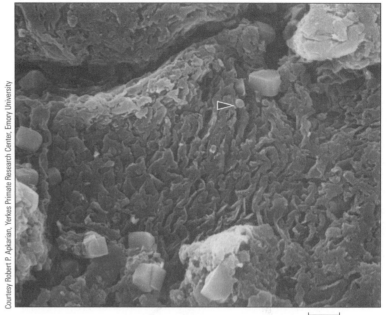

1 µm

FIGURE 14.7 Region of the human skin with an attached staphylococcal cell (arrow). The presence of salt crystals (small cubes) show how dry the surface is.

The importance of the skin as a defense against infection is dramatized by the consequences of losing large areas of it through burns. If a burn victim survives the initial crisis, the most probable cause of death is infection, often due to the opportunistic pathogen *Pseudomonas aeruginosa*. Even intensive care in a hospital burn unit, with antibiotic therapy and scrupulous attention to cleanliness, cannot protect reliably against infection.

Although the skin is a hostile environment, many microorganisms manage to survive there. Humans have up to 10,000 microorganisms on each square centimeter of dry skin, and up to a million on moist areas. About 90 percent of them can be removed by thorough washing, but they grow back to near-normal numbers within 8 hours.

Biota. Like all normal biota, the vast majority of the skin's biota are commensals. No species directly benefits the host. But collectively they do. They keep the skin healthy by occupying the space pathogens would need to become established and thereby excluding them. And they also produce pathogen-inhibiting fatty acids as they break down some of the skin's oils.

Several groups of microorganisms grow on the skin. They are *Staphylococcus* spp., diphtheroids, and fungi (**Figure 14.8**).

Staphylococcus spp. (*Bergey's Manual* Firmicutes; Chapter 11) are facultative anaerobes, so they can grow aerobically

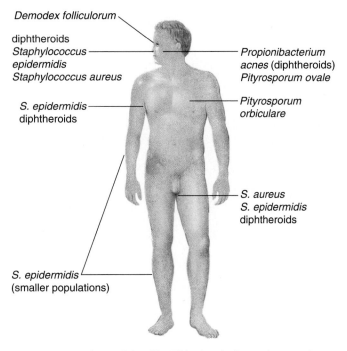

FIGURE 14.8 Biota of the skin. This sketch shows the prominent locations of various bacteria and yeasts, as well as the mite, *Demadex folliculorum.*

(Art by Lewis Calver from Biology: The Unity and Diversity of Life. 6th ed., by C. Starr and R. Taggart, Brooks/Cole, 1992. All rights reserved.)

on the skin's surface and anaerobically in its airless pores. They're hardy enough to withstand the skin's low moisture, its high salt, and its extremes of temperature. But they reach the highest populations in relatively moist areas such as under the arms, around the nose, and near the anus.

All *Staphylococcus* spp. have some disease-producing potential. But the severity varies. *Staphylococcus aureus* is the most pathogenic. Only 5 to 10 percent of humans carry it, usually around the anus (Chapter 26). Another species, *Staphylococcus epidermidis*, is found on everyone's skin. Some strains of it are nonpathogenic. Others are weakly pathogenic opportunists that cause problems for people with special vulnerabilities. For example, they are the most common cause of infections in artificially implanted devices, such as plastic catheters and artificial joints.

The diphtheroids are Gram-positive rods related to *Corynebacterium diphtheriae* (Actinobacteria; Chapter 11). Some are aerobes and others are anaerobes. Most are commensals. Certain of the anaerobes grow deep within hair follicles at the expense of **sebaceous** (fatty) secretions that are produced there. One of these species, *Propionibacterium acnes*, changes the sebaceous secretions into irritating fatty acids that initiate the inflammation called **acne** (Chapter 26). In a sense, *P. acnes* is an opportunistic pathogen because it causes disease primarily when sebaceous secre-

tions are abundant during adolescence. The aerobic diphtheroids grow nearer the skin surface.

Staphylococcus spp. and diphtheroids are on everyone's skin, but their relative abundance varies among individuals. *Staphylococcus* spp. predominate on the **axillary** (underarm) skin of some people. Diphtheroids predominate on others. "Coccal people" and "diphtheroid people" also differ in other respects. For example, coccal people have a much larger population of *Staphylococcus* spp. on their faces than diphtheroid people do.

Fungi, the third major group of microorganisms on the skin, are mostly species of *Pityrosporum*. They are yeasts that use fats as nutrients, so they are found where sebaceous secretions are plentiful. One species, *Pityrosporum ovale*, predominates on the face and scalp. Another, *Pityrosporum orbiculare*, predominates on the chest and back. Other yeasts on the skin include *Torulopsis glabrata* and *Candida albicans*. *Pityrosporum* spp. and *T. glabrata* are commensals. *C. albicans* is an opportunistic pathogen.

A microscopic mite, *Demodex folliculorum*, also inhabits our skin. It's the only arachnid normally present on the human body. It lives on the face, within hair follicles and in the openings of sebaceous glands. Baby Girl A won't acquire this mite during her infancy. But she, like all other humans, probably will acquire it by the time she reaches adulthood. It probably won't bother her, but it may cause an irritation of the eyelid margins called **blepharitis.**

Now let's consider the microbiology of our other external body surface, the conjunctiva.

The Conjunctiva

The conjunctiva (*pl.*, conjunctivae) is the mucous membrane that covers the exposed surface of the eye and the interior surface of the eyelid. It protects the surface of the eye.

Microbial Environment. The conjunctiva is a continuous surface that's relatively impermeable to microorganisms (**Figure 14.9**). It's protected mechanically and biochemically by tears (**Figure 14.9a**), which are produced in small amounts by a lacrimal gland that lies beneath each eyelid. Tears constantly bathe the conjunctivae. They are spread over its surface by movement of the eyelids, and they accumulate at the inner corner of the eye. Then they flow through the nasolacrimal duct into the nose. Tears defend the conjunctivae mechanically by their flushing action and biochemically by the lysozyme they contain.

Biota. The tear-washed conjunctiva is an inhospitable environment for microorganisms. Normally, few species are found there. Those that are come from the microbiota of the skin. *Staphylococcus* spp. are found in more than 90 per-

LARGER FIELD

GONOCOCCAL BLINDNESS

Until the 1920s, the most common cause of blindness in the United States was newborn infection with *Neisseria gonorrhoeae*. As a newborn passed through the

birth canal of a mother with gonorrhea, bacteria infected the infant's eyes. In 2 to 5 days an intense inflammation appeared, often causing permanent eye damage. Gonococcal

blindness was effectively eliminated in this country by making silver nitrate eyedrops a routine part of the treatment of newborns. In the 1980s most nurseries began using the

less irritating antibiotic erythromycin. Some type of treatment is required by law in most states (Chapter 26).

cent of people. Diphtheroids are found in about 80 percent (**Figure 14.9b, c**). But they're present in small numbers compared with the populations on skin. Most of them are commensals, but opportunistic pathogens may appear if the normal flow of tears is interrupted.

Now we'll consider the body's internal surfaces (**Figure 14.10a**).

The Nasal Cavity and Nasopharynx

The nasal cavity and the space behind it, called the nasopharynx, are the passages that deliver air to the lungs. Air that enters these passages contains dust, debris, and

microorganisms. The average adult inhales about 10,000 microorganisms each day. Baby Girl A's nasal cavity and nasopharynx were sterile at birth, but 24 hours later they contained many microorganisms.

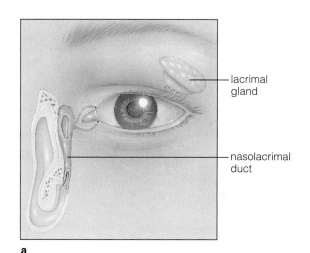

lacrimal gland

nasolacrimal duct

a

FIGURE 14.9 The conjunctivae. (a) Source and drainage of tears. (b and c) Bacterial biota: (b) Micrograph of *Staphylococcus epidermidis.* (c) Micrograph of *Corynebacterium pseudodiphtheriae,* a diphtheroid.

(Art by Carlyn Iverson.)

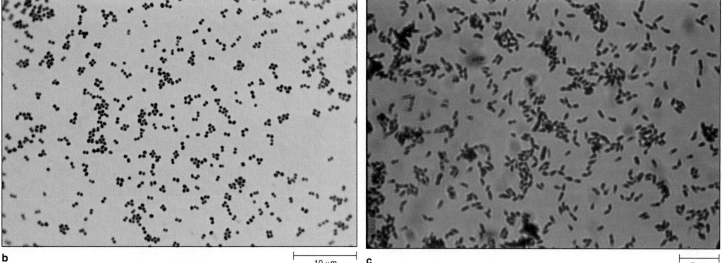

© Paul W. Johnson/Biological Photo Service

b

10 µm

c

© Martin M. Rotker

5 µm

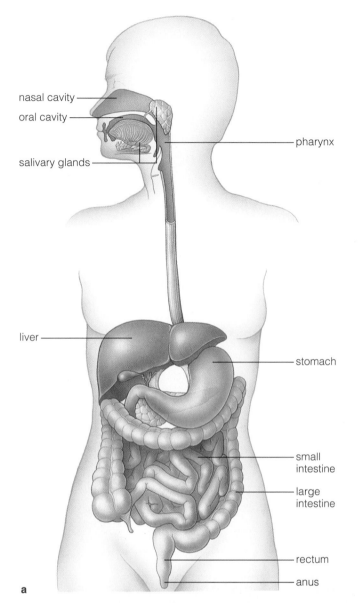

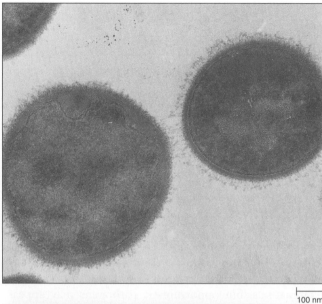

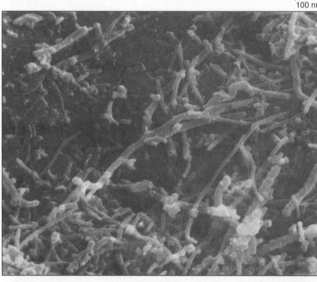

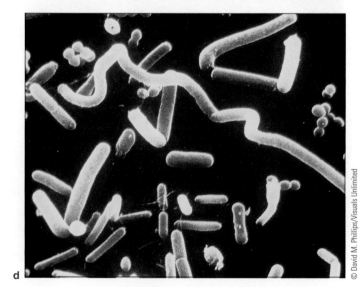

FIGURE 14.10   The body's internal surfaces. (a) Locations of the biota. (b) *Streptococcus salivarius.* (c) Plaque on the enamel of human tooth, the habitat of *Streptococcus mutans.* (d) Diverse types of intestinal bacteria.

(Art by Kevin Somerville from Biology: Concepts and Applications, 2nd ed., by C. Starr, Brooks/Cole, 1994. All rights reserved.)

Microbial Environment. The nasal cavity and nasopharynx is a warm and moist environment. Many microorganisms would do well there if they weren't removed by the mucociliary system described earlier.

If, for whatever reason, this system doesn't function normally, repeated and potentially life-threatening respiratory infections occur. For example, the system doesn't work well in persons with cystic fibrosis because they produce abnormally thick mucus that is difficult for the cilia to move. The mucociliary system can also be weakened by toxic substances, such as cigarette smoke, in the environment. If Baby Girl A's parents smoke, she will have a significantly increased risk of frequent respiratory infections. Because a functioning mucociliary system is so effective, the normal microbiota of the nasal cavity and nasopharynx has to be able to adhere tightly to stay there.

Biota. Surprisingly, the nasal cavity and nasopharynx are densely colonized. Many of the microorganisms found on the skin are also found in these passages. *Staphylococcus epidermidis* is in the nasal cavities of more than 90 percent of humans, and *Staphylococcus aureus* is present in most. Diphtheroids are in the nasal cavity of up to 80 percent of humans.

The nasal cavity and nasopharynx are also colonized by microorganisms not usually found on the skin. These include various species of *Streptococcus*, most of which are harmless commensals. The closely related *Lactobacillus* spp. are also commonly found in the nasal cavity. The nasopharynx often contains some Gram-negative pathogens. These included *Moraxella catarrhalis*, which is responsible for a significant number of middle ear infections in young children, and *Haemophilus influenzae*. Some these strains of *H. influenzae* are highly pathogenic (Chapter 22). Others are relatively harmless.

The Mouth

The mouth is readily and quickly colonized. Microbes entered Baby Girl A's mouth as she was passing through her mother's vagina. Then more from the skin around her mother's breast entered during her first meal. Huge numbers of various kinds of microbes will continue to enter during her first few months of life as she explores her environment—mouthing blankets, her own hands, and numerous other objects.

Microbial Environment. Many features of the mouth make it an ideal environment for microbes. It's moist, warm, and abundantly supplied with nutrients. But mechanical and biochemical factors limit the types of microorganisms that can survive there.

Talking, chewing, and swallowing dislodge microorganisms that are not able to adhere. And saliva, which constantly enters the mouth from the salivary glands and leaves by swallowing, flushes microorganisms out of the mouth. Lysozyme present in saliva also kills some of them. People who don't produce normal amounts of saliva don't have a normal mouth biota. Many of them suffer extensive tooth decay.

Biota. The human mouth is densely populated with more than 80 different species of microorganisms. Many of the bacteria belong to the genus *Streptococcus*. They will be the principal microorganisms in Baby Girl A's mouth until her teeth erupt. Thereafter, they will remain but in smaller numbers. Mouth streptococci include *Streptococcus salivarius* (**Figure 14.10b**), *Streptococcus mitis*, and many other species. There are even a few enterococci in the mouth, although most are in the intestines. The streptococci species are only on the surface of the teeth, the roof of the mouth, and the tongue.

Staphylococci are also widespread. *Staphylococcus epidermidis* is in the mouths of nearly all humans, and *Streptococcus aureus* is fairly common. *Moraxella catarrhalis*, lactobacilli, diphtheroids, and *Haemophilus influenzae* are nearly always present. Even anaerobes such as *Bacteroides* spp. and *Fusobacterium* spp. are common. And almost half of us harbor small populations of the yeast *Candida albicans*.

The complex microbial population that exists in the mouth is generally quite stable. Most are commensals that maintain a healthy mouth environment, but some do cause problems. One example, *Streptococcus mutans* is the primary cause of tooth decay (**Figure 14.10c**; Chapter 23). *Haemophilus influenzae*, as well as *Moraxella catarrhalis*, can infect the sinuses and the middle ear.

The Intestinal Tract

From the mouth, food and microorganisms move into the intestinal tract. They pass through the esophagus, the stomach, the small intestine, and finally the large intestine before undigested material is eliminated as feces (Chapter 23).

The first evacuations from Baby Girl A's digestive system, a sticky green material called **meconium,** contain no microorganisms. Only after the microorganisms that Baby Girl A takes into her mouth pass through her intestinal tract will microorganisms appear in her feces. Clinical studies on infants born with a blockage of the intestinal tract confirm that microorganisms enter the intestines only from the mouth. Newborn infants with bowel obstruction have a normal population of microorganisms above the blockage, but none below it.

Microbial Environment. The esophagus, stomach, and upper small intestine are too inhospitable to sustain a normal biota. For one thing, biochemical products of the digestive system inhibit microbial growth. Stomach acid lowers the pH in the stomach to 1. Bile salts destroy many organisms in the upper small intestine.

These biochemical factors limit organisms from the upper intestinal tract as well. But mechanical factors are more important. **Peristalsis** (the wavelike muscular contractions that move food through the digestive tract) and gastric churning are extremely powerful. Digestive contents flow past the epithelial surface so rapidly that most microorganisms are swept away. The overriding importance of mechanical defenses in limiting the biota of the upper intestinal tract is revealed by autopsies. Within hours after death, when intestinal movement has stopped but the chemical content has not changed, the upper small intestine acquires a large population of microorganisms.

The large intestine, in contrast, houses a huge and diverse microbiota. Some toxic biochemical agents, such as bile salts, are present, but intestinal movement is much slower. Therefore enormous microbial populations accumulate. Indeed, microorganisms make up about a third of the dry weight of the feces. People eliminate between a hundred billion (10^{11}) and a hundred trillion (10^{14}) bacteria in their feces every day.

Biota. Hundreds of species of bacteria coexist in the large intestine (**Figure 14.10d**), but by far the most abundant are the strict anaerobes. Four genera of anaerobes, *Bacteroides*, *Bifidobacterium*, *Fusobacterium*, and *Clostridium*, account for more than 90 percent of all the bacteria in the bowels.

Facultative anaerobes are present in smaller numbers. They include members of the Enterobacteriaceae, such as *Escherichia coli*, *Proteus* spp., *Enterobacter* spp., and *Klebsiella* spp. Gram-positive organisms include a group of streptococci called enterococci (*enteron* means "intestines" in Greek), *Staphylococcus aureus*, and *Lactobacillus* spp.

Microorganisms in the intestine interact in complex ways. Some produce bacteriocins that kill competing organisms. Others control the growth of their neighbors by producing acids. Still others change the oxidation state of the environment. All these changes limit the growth of competing microorganisms, including pathogens.

Genetic exchange occurs among the various members of the dense intestinal microbiota, sometimes with important consequences. For example, when plasmids that encode resistance to antibiotics are exchanged, a large portion of the intestinal population can become drug resistant in a short time.

The Vagina

The vagina lies at the outlet of the female reproductive tract (**Figure 14.11**). It receives semen during sexual intercourse and is the birth canal through which babies pass as they leave the womb and

enter the outside world. The vagina is the only site in the female reproductive tract that sustains a normal biota.

Microbial Environment. The vagina is a warm, moist, protected environment with few mechanical features. Only the scarcity of nutrients and low pH limit microbial growth.

Biota. The lactic acid bacteria that are responsible for the low pH of the vagina are its primary inhabitants, along with small populations of some other microbial species. These include aerobic and anaerobic streptococci, *Staphylococcus* spp., and anaerobes such as *Bacteroides*, *Clostridium*, and *Fusobacterium* spp. Many women also support a small population of the fungus *Candida albicans*, which can cause opportunistic vaginal infections.

The Urethra

The urethra lies at the very end of the urinary tract in both males and females. The outermost part of the urethra, near the point at which the mucous membranes of the urinary tract meet the skin, is the only part of the urinary tract that supports a microbiota. The upper part of the urethra, which joins the bladder, is normally free of microorganisms.

Microbial Environment. The structure of the urethra limits microbial growth because it is composed of tightly joined epithelial cells that resist attachment of microorganisms. In addition, the mechanical forces of urine flowing rapidly over the epithelial surface flush out microorganisms. The fluid that bathes the urethra doesn't provide any biochemical protection. In fact, urine is an excellent culture medium.

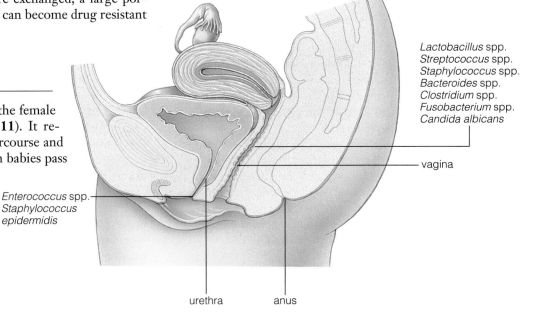

Lactobacillus spp.
Streptococcus spp.
Staphylococcus spp.
Bacteroides spp.
Clostridium spp.
Fusobacterium spp.
Candida albicans

vagina

Enterococcus spp.
Staphylococcus epidermidis

urethra   anus

FIGURE 14.11 The vagina and urethra and their microbiota.

LARGER FIELD

ONCE A HYDROGEN-PRODUCER, ALWAYS A HYDROGEN-PRODUCER

Curiously, not everyone produces the same type of intestinal gas. Although most of us are hydrogen-producers, a significant minority produces methane in their large intestine. The type of gas we produce depends upon our normal intestinal biota. Everyone is colonized by bacteria, such as *Escherichia coli,* that produce hydrogen

when they ferment sugars under the anaerobic conditions of the large intestine. But some people are also colonized by a type of archaea, the methanogens, that convert hydrogen and carbon dioxide into methane.

Whether we are a hydrogen-producer or a methane-producer is determined early in childhood,

when the intestinal biota becomes established, so the type of intestinal gas we produce is fixed for life. The type of gas we produce has not been associated with any clinical problem.

Although the type of gas we produce depends upon our microbiota, the quantity depends upon our diet. Most of the

sugars we eat are absorbed as nutrients in the small intestine, but some aren't. These sugars pass into the large intestine, where they are metabolized by intestinal bacteria. Beans contain two such nonabsorbable sugars, raffinose and stachyose. They account for our increased gas production after a meal with beans.

Biota. The normal biota of the urethra is scant. There are some enterococci and *Staphylococcus epidermidis* at the end of it.

IS NORMAL BIOTA HELPFUL OR HARMFUL?

Whether the normal biota is helpful or harmful has fascinated microbiologists for years. Many passionately argue one or the other position.

Conflicting Theories

Louis Pasteur believed that normal microbial biota is not only helpful, it's essential to higher animals. His associate, Elie Metchnikoff, took the opposite position. He believed that all microorganisms compete with humans for life-sustaining chemicals and therefore they are all harmful to varying extents.

Studies on germ-free animals proved Pasteur wrong. These animals are maintained germ-free by transferring them directly from their sterile prenatal environment to a sterile chamber. If they're fed sterilized food and handled with sterilized gloves, they remain germ-free throughout their lives (**Figure 14.12**). They're healthy. They produce healthy offspring. And they live as long as normally colonized animals. They do, however, differ from their natural

counterparts in subtle ways. For example, they acquire no immunity to microorganisms. But in a germ-free world they suffer no ill effects from not having a normal microbial biota.

Metchnikoff's theory is more difficult to prove or disprove, but there's no good evidence that our microbiota saps our strength in any way. In a natural environment, its benefits probably outweigh its harmful effects.

Courtesy, Allentown Caging Equipment Co., Inc.

FIGURE 14.12   A sterile chamber with two germ-free rats.

Harmful Effects

Outweighed or not, is there any evidence that the microbiota is harmful, even subtly? There's a little. For example, some intestinal microorganisms use vitamin C as a substrate for their growth and thereby destroy it. But there is no evidence that this leads to a vitamin deficiency.

Also, some chemicals are carcinogenic in normal animals but not in germ-free animals. These chemicals are converted to carcinogens by bacterial metabolism. However, germ-free and normal animals develop cancers at similar rates, so it's unlikely that microbially produced carcinogens are a significant cause of cancer in humans.

Beneficial Effects

In the real, germ-laden world, our normal microbiota benefits us in several ways. The most significant is by microbial antagonism. As we've seen, our normal biota produces antimicrobial compounds, competes for receptors on epithelial cells, and changes pH. These activities kill pathogens outright or inhibit their growth. Also, certain members of the intestinal microbiota inactivate certain toxins produced by pathogenic bacteria.

Our microbiota also stimulates our immune system. This stimulation is nonspecific. It doesn't provide immunity against any particular disease, but it does improve the entire defensive system. Normal animals are better than germ-free animals at mounting an immune defense against pathogens, even pathogens with which they've had no previous contact.

Finally, our intestinal microbiota does provide us with supplemental sources of certain vitamins. Two of these are vitamin K (important for blood clotting) and vitamin B_{12} (important in the production of red blood cells and normal functioning of the nervous system).

Loss of Normal Biota

Any change of our normal biota may have undesirable effects. Pathogens might establish themselves in the underpopulated regions and disease may result, or a normally harmless opportunist may proliferate and cause an infection. Let's look at how this can happen as a result of antibiotic therapy.

By the age of 2 to 3 weeks, Baby Girl A will have acquired most of the normal biota typical of an adult. Suppose that some time later she develops an infection of the middle ear. She will probably be treated with a broad-spectrum antibiotic to ensure that the causative bacterium is eliminated regardless of whether it is Gram-positive or Gram-negative. But, in addition to killing the bacteria infecting her ear, the antibiotic will kill many bacteria in other parts of her body. These microbial vacuums may be filled by growth of the yeast *Candida albicans*, which won't be affected by the antibiotic because it's a fungus. The result is likely to be a severe and persistent *Candida* diaper rash. Similar antibiotic therapy in adult women often causes *Candida* vaginitis, characterized by redness, itching, and a yeast-laden vaginal discharge. It can also cause *Candida* infections of the mouth, called thrush.

A Spectrum of Disease-causing Abilities

Categorizing microorganisms as pathogens, opportunists, or commensals is useful, but it implies unrealistically rigid distinctions. In reality, the distinctions are blurred because the host's health is critical in determining whether disease will occur. Therefore it's more useful to think of microorganisms as existing along a continuous spectrum of disease-causing potential (**Figure 14.13**). At one end of the spectrum are highly virulent pathogens that cause disease in almost all cases. Next are less virulent microorganisms that often cause disease. And then there are opportunists of greater or lesser degrees that cause disease in more or less weakened hosts. Most microorganisms never cause any disease, but any microbe that can sustain itself in the human body has the potential of harming the host if debilitation is extreme.

With our advanced medical technology, people with serious diseases and badly weakened defenses now survive much longer. Today in industrialized societies more deaths are caused by opportunistic infections than by "true" pathogens.

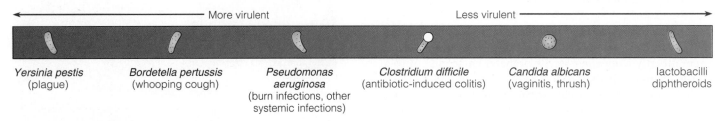

FIGURE 14.13   Location of some microorganisms on the spectrum of virulence.

SUMMARY

Case History: Avoiding a Serious Infection by Treating Someone Else (p. 343)

1. A baby begins to acquire its normal microbiota as it passes through the birth canal, but it can also be infected if pathogens are present, even though the mother shows no symptoms.

Normal Biota (pp. 344–346)

2. Health is a stable state in which the body's organ systems function adequately. Disease is an unstable state that ends in recovery or death.

3. The microorganisms we coexist with in a stable relationship constitute our normal biota. Normal biota inhabit only the surfaces of our bodies: the skin and the conjunctivae; the nasal cavity and nasopharynx; and the mouth, intestinal tract, vagina, and urethra.

4. Resident biota inhabit the human body throughout life. They can be temporarily reduced but never eliminated.

5. Transient biota inhabit the human body only under certain circumstances. They can be removed by careful cleansing.

6. Opportunists are microorganisms that cause disease when the host's normal defenses are weakened or when the normal biota is altered.

7. The normal biota changes over time.

Symbiosis (pp. 346–347)

8. Symbiosis is two different kinds of organisms living together. The larger one, which provides a habitat for its symbiotic partners, is the host.

9. Symbiotic relationships are classified according to whether they harm or benefit. In commensalism, one partner benefits (the commensal organism) and the other is neither benefited nor harmed (the host).

10. Most microbes making up the normal biota are commensals. Individ-
ual species do not benefit the human host, but they do collectively because of microbial antagonism.

11. In mutualism, both partners benefit. In parasitism, the host is harmed and the parasite benefits.

12. Bacterial, fungal, and viral parasites are called pathogens. The term parasite is reserved for protozoa, worms, and insects.

13. Many commensals are potential pathogens. They can become opportunists and cause infection.

Factors That Determine the Normal Biota (pp. 347–350)

14. Our microbiota survive because they are adapted to life on living tissue.

15. Structural, mechanical, and biochemical features of body surfaces constitute nonspecific surface defenses, the body's first line of defense against infection. They are nonspecific because they act against all pathogens.

16. Our structural defenses are our epithelial surfaces, the skin and the conjunctivae, and the interior surfaces, which are mucous membranes. The normal sloughing of epithelium as it grows is another structural defense. When the dead cells are lost, the microbes on them are also lost.

17. Mechanical defenses—movements—eliminate many transient microorganisms from body surfaces. Some surfaces move because of the action of underlying muscles. The mucociliary system protects by cilia moving a layer of mucus. Urine washes microorganisms out of the urethra. Tears wash microorganisms off the conjunctivae.

18. Biochemicals that inhibit microbial growth include keratin, which keeps the skin surface dry; stomach acid and fatty acids, which lower pH; and lysozyme and bile, which kill some microbes.

19. The normal biota are adapted to growth at a particular body site.

20. Adhesins are protein molecules that attach microbes to human cells. Ad-
hesins are very specific. Some microbiota produce bacteriocins, proteins that kill other microorganisms.

Sites of Normal Biota (pp. 350–357)

The Skin (pp. 350–352)

21. Skin biota are mainly *Staphylococcus* spp., diphtheroids, or fungi.

22. *Staphylococcus* spp., which are facultative anaerobes, include the pathogen *Staphylococcus aureus* and the opportunist *Staphylococcus epidermidis*.

23. Diphtheroids include the anaerobe *Propionibacterium acnes*, which causes acne.

24. Fungi on the skin include yeasts belonging to the genus *Pityrosporum*, which use fats as a substrate for growth and grow on oily areas on the face, scalp, chest, and back.

25. The mite *Demodex folliculorum* lives on the face, within hair follicles and in the openings to oil glands.

The Conjunctiva (pp. 352–353)

26. The conjunctivae are defended by their continuous and relatively impermeable surface and by tears.

27. Among the few species found at this site are *Staphylococcus* spp. and diphtheroids.

The Nasal Cavity and Nasopharynx (pp. 353–355)

28. Microbiota of the nasal cavity and nasopharynx are particularly well adapted to adhere because of the mucociliary system.

29. The same species that colonize the skin are found in these densely colonized sites. They include *Staphylococcus epidermidis*, *Staphylococcus aureus*, diphtheroids, *Lactobacillus* spp., and *Moraxella catarrhalis*.

The Mouth (p. 355)

30. The mouth is a warm, moist environment with abundant nutrients. It is densely populated with

microorganisms. Streptococci predominate before teeth erupt and thereafter are present in smaller numbers.

31. This complex microbial population is stable and consists largely of commensals. One exception is *Streptococcus mutans*, which causes tooth decay.

The Intestinal Tract (pp. 355–356)

32. The first intestinal tract evacuations in a newborn are germ-free.

33. The esophagus, stomach, and upper intestine are too inhospitable to sustain a normal biota. Peristalsis and stomach churning keep this region nearly germ-free.

34. In the lower intestine, intestinal movement is much less vigorous. A complex microbial community develops there.

35. The majority of species in the lower intestine are strict anaerobes belonging to the genera *Bacteroides*, *Bifidobacterium*, *Fusobacterium*, and *Clostridium*. Facultative anaerobes include the Enterobacteriaceae (such as *Escherichia coli*) and *Lactobacillus* spp.

The Vagina (p. 356)

36. The vagina is a warm, moist, protected environment. When influenced by estrogen, it becomes acidic with the growth of lactobacilli. When estrogen is not being produced, it is alkaline and more prone to infections.

37. Aerobic and anaerobic species colonize this site, including the fungus *Candida albicans*, which can cause opportunistic infections.

The Urethra (pp. 356–357)

38. Only the outermost part of the urethra, where the mucous membranes meet the skin, supports a microbiota.

39. The normal biota is scant, but usually enterococci and *Staphylococcus epidermidis* are present on the outer part of the urethra.

Is Normal Biota Helpful or Harmful? (pp. 357–358)

40. The benefits of normal biota probably outweigh possible harmful effects. Research has yielded no proof of even subtle harm.

41. The most significant beneficial effect of microbiota derives from microbial antagonism.

42. Microbiota also stimulate our immune system in a nonspecific way. Some intestinal bacteria provide supplemental sources of vitamins K and B_{12}.

43. If our relationship with microorganisms is altered, pathogens can establish themselves in underpopulated areas or normally harmless commensals can proliferate and cause opportunistic infections.

44. Categorizing microorganisms as harmful or harmless is not always clear-cut. The host's state of health often determines whether disease will occur. Microorganisms exist along a continuous spectrum of disease-causing potential.

45. Today, because of advanced medical technology, more people die from infection by opportunists than by true pathogens.

REVIEW QUESTIONS

Case History: Avoiding a Serious Infection by Treating Someone Else

1. Why was penicillin a wise choice of an antimicrobial drug with which to treat Baby Girl A's mother?

Normal Biota

2. What are normal biota and where are they found?

3. How do we acquire our normal biota?

4. Describe, as fully as you can, these three types of biota: resident biota, transient biota, and opportunists.

5. How do normal biota in the mouth and intestines change during infancy?

6. Describe the influence of estrogen on vaginal biota.

Normal Biota and Symbiosis

7. Define these terms: symbiosis, host. How are symbiotic relationships classified?

8. Explain the difference between commensalism, mutualism, and parasitism. Which are stable, highly evolved relationships and which are not? Explain.

9. Explain this statement: There are no true mutualistic relationships between humans and microorganisms.

10. What is microbial antagonism?

11. How do commensals benefit us? How can they harm us?

12. How are the terms parasite and pathogen used in clinical medicine?

Factors That Determine the Normal Biota

13. In addition to the physical and nutritional factors, what unique factors influence microbial growth on human tissue?

14. What are our nonspecific surface defenses?

15. Name our major structural defense against microbial invasion. What are the two types of epithelium and where are they found? Name two ways the skin defends against microbes.

16. What are mechanical defenses? Give some examples.

17. What are biochemical defenses? Give some examples.

18. Describe the ways microorganisms must adapt if they are to live on body surfaces. What are adhesins, and what role do they play?

19. What are bacteriocins, and why are they important?

Sites of Normal Biota

20. For each of the following sites, describe the environment and name

some of the representative microbiota found there:

a. skin
b. conjunctivae
c. nasal cavity and nasopharynx
d. mouth
e. intestinal tract
f. vagina
g. urethra

21. What is *Demodex folliculorum*? Does it harm us?

22. Why are the first evacuations in a newborn free of microorganisms?

23. Explain this statement: The intestinal tract is a complex microbial community.

24. Describe some of the complex interactions in the intestinal ecosystem.

Is Normal Biota Helpful or Harmful?

25. Are microbiota essential to human life? Are they inevitably harmful? How would most microbiologists answer the question of whether normal biota are helpful or harmful?

26. What are the major beneficial effects of normal biota?

27. Why is the loss of normal biota dangerous? Why does the loss of normal biota from broad-spectrum antibiotics affect the population of *Candida albicans*?

28. What is the single most important reason why categorizing microorganisms as commensals, opportunists, or pathogens is not always clear-cut?

29. Explain this statement: Microorganisms exist along a continuous spectrum of disease-causing potential.

CORRELATION QUESTIONS

1. Is lysozyme more lethal to *Staphylococcus aureus* or *Escherichia coli*? Why?

2. How does skin oil prolong and shorten the presence of a transient microbiota?

3. Microbiota used to be called microflora. Why is the change justified?

4. How does breastfeeding affect the incidence of infant diarrhea?

5. How is it that the normal microbiota is irrelevant in a germ-free environment but important in a normal one?

6. If you had to remove a germ-free animal from its sterile environment, what might you do to save its life?

ESSAY QUESTIONS

1. Discuss the human and microbial factors that lead to a continuing relationship between a microorganism and humans.

2. What have germ-free animal studies taught us about the relationship between humans and their normal biota?

SUGGESTED READINGS

Drasar, B. S., and P. A. Barrowk. 1985. *Intestinal microbiology.* Washington, D.C.: American Society for Microbiology.

Mackowiak, P. A. 1982. The normal microbial biota. *New England Journal of Medicine* 307:83.

Marsh, P., and M. Martin. 1984. *Oral microbiology.* 2d ed. Washington, D.C.: American Society for Microbiology.

Roth, R., and W. Jenner. 1988. Microbial ecology of the skin. *Annual Reviews of Microbiology* 42:441.

Van der Waaij, D. 1989. The ecology of the human intestine and its consequences for overgrowth by pathogens such as *Clostridium difficile. Annual Reviews of Microbiology* 43:69–87.

For additional readings, go to InfoTrac College Edition, your online research library at: http://www.infotrac.thomsonlearning.com

FIFTEEN

Microorganisms and Human Disease

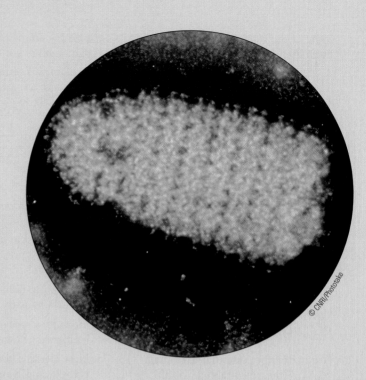

© CNRI/Phototake

CHAPTER OUTLINE

LEARNING GOALS

To understand how microbial pathogens that cause human infection:

- *Survive between hosts*
- *Come in contact with the body*
- *Adhere to body surfaces*

- *Penetrate into deeper tissues*
- *Avoid host defenses and multiply*

- *Cause disease*
- *Leave the host*

The Infection of Baby Girl B

Baby Girl B was a healthy 3 month old. She weighed nearly 13 pounds. She could smile, laugh, and roll over by herself. Her development was normal. But because of a minor illness, she had missed her first set of baby shots at her 2-month medical checkup.

One day Baby Girl B got a runny nose and began sneezing. Her mother called the doctor and was reassured that infants often catch colds and recover from them uneventfully. But Baby Girl B did not improve. After 10 days of cold symptoms, she developed a cough that produced thick, sticky mucus. She began to have fits of repeated coughing so intense that she couldn't catch her breath. After one of these fits or paroxysms, Baby Girl B would gasp so loudly it sounded like a "whoop." Often her coughing episodes ended in vomiting. Baby Girl B's mother took her to the doctor.

The pediatrician made a provisional diagnosis of pertussis, or whooping cough. It was confirmed when the Gram-negative bacterium *Bordetella pertussis* was cultured from the back of Baby Girl B's nose. Like most infants with pertussis, Baby Girl B's was seriously ill. The pediatrician immediately admitted Baby Girl B to a pediatric intensive care unit. Because her coughing paroxysms were now almost constant, she was unable to eat or rest. During her worst coughing spells, she couldn't get enough oxygen. Occasionally she even stopped breathing for short periods. Physiological abnormalities developed: Her blood contained an excess of lymphocytes (white blood cells); cells in her tissues became unusually sensitive to histamine (a chemical released at times of stress or injury that causes many physiological abnormalities); and she produced excessive amounts of insulin that interfered with normal regulation of levels of glucose in her blood.

Nurses frequently suctioned the sticky mucus that clogged Baby Girl B's nose and throat, and she received supplemental oxygen and intravenous fluids and nutrition. She was also treated with antibiotics. They prevented her from spreading the infection to others, but they didn't hasten her recovery. The supportive care she received allowed her to survive a month-long ordeal of almost continuous coughing.

By the time she was discharged from the hospital, Baby Girl B had lost considerable weight and was weak. She could no longer roll over or lift her head as well as she used to, and she continued to suffer occasional coughing paroxysms over the next few months. But eventually her recovery was complete. By her first birthday, Baby Girl B was chubby, playful, and trying to take her first wobbly steps.

Case Connection

■ As you read through this chapter, keep track of the seven capabilities *Bordetella pertussis* needed to cause Baby Girl B's infection.

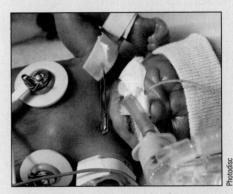

Infants with whooping cough often require supplemental oxygen in a pediatric intensive care unit.

THE SEVEN CAPABILITIES OF A PATHOGEN

In Chapter 14 we considered the hundreds of species of microorganisms that colonize our body surfaces and how they help us avoid infection. Now we'll consider how infection occurs—how pathogens are sometimes able to overcome these and other defenses to cause disease. We'll look at the process from the microbe's point of view. That is, we'll examine what the microbe must be able to do to cause an infection. Such disease-causing capabilities are called **virulence factors.** Throughout our discussion we'll return to Baby Girl B's particular infection and how *Bordetella pertussis* was able to cause it.

To be successful a pathogen must be able to do most if not all of these seven things:

1. Maintain a **reservoir** (a place to live before and after infection)

2. Leave its reservoir and enter a host

3. Adhere to the surface of the host

4. Invade the body of the host

5. Evade the body's defenses

6. Multiply within the body

7. Leave the body and return to its reservoir or enter a new host

While in the host's body, the pathogen causes disease. Sometimes its disease-causing ability helps it multiply or survive. Other times, it's irrelevant to the pathogen's welfare. We'll often use the terms infection and disease interchangeably. But there is a difference. Infection refers to the growth of a pathogen in the body. Disease is its consequence.

Now we'll consider the seven capabilities of a pathogen, one by one.

ONE: MAINTAINING A RESERVOIR

Every pathogen must have at least one reservoir. If all its reservoirs are eliminated, it too ceases to exist. The most common reservoirs for human pathogens are other humans, animals, and the environment. We'll examine each of these in turn.

Human Reservoirs

Humans are the only reservoirs of many pathogens, including those that cause pertussis, measles, gonorrhea, and the common cold. That's because most of these organisms are too fragile to survive in any other environment. They die quickly outside the human body.

There are two kinds of human reservoirs: (1) people who are sick, such as Baby Girl B, and (2) people who are apparently healthy. The apparently healthy people who are infected but have not yet developed symptoms are called **incubatory carriers.** Those who are able to harbor the pathogen for months or years and never become sick are called **chronic carriers.** People infected with human immunodeficiency virus (HIV) who haven't yet developed acquired immunodeficiency syndrome (AIDS) are examples of incubatory carriers. Typhoid Mary (Chapter 23) was a notorious example of a chronic carrier. She contracted typhoid and recovered from it. But the typhoid pathogen, *Salmonella typhi*, continued to multiply in her body and to be excreted in her feces for years. Other chronic carriers never become ill. For example, some people have *Streptococcus pyogenes* (which causes strep throat) in their throats for years without becoming sick.

Animal Reservoirs

Animals, both domestic and wild, are also reservoirs for certain human pathogens. They provide an environment quite like that of the human body. For some pathogens, any of several animals will do. For example, almost any mammal, including cats, dogs, skunks, and bats, can be a reservoir for rabies (Chapter 25). Humans become infected when they come into contact with an animal reservoir. For rabies, contact usually means an animal bite. More often, however, pathogens pass from their animal reservoir to human beings when people consume or handle contaminated animals or animal products. Or biting insects that have previously bitten an infected animal might pass them along.

Such a human disease caused by a pathogen with an animal reservoir is called a **zoonosis.** More than 150 zoonoses are known, but fewer than half of them are clinically important (**Table 15.1**). New zoonoses continue to be discovered. Lyme disease, for example, was discovered only in the late 1970s (Chapter 27), and Hantavirus pulmonary syndrome first surfaced in 1993 (Chapter 22). Wild deer and mice are the primary reservoir for Lyme disease, and at least eight rodent species, but principally the long-tailed deer mouse, are reservoirs for Hantavirus pulmonary syndrome.

Having an animal reservoir profoundly affects the pattern of a disease because new forms of a disease can develop in an animal reservoir and animal reservoirs can make diseases hard to control. Influenza and yellow fever, two viral diseases, illustrate these principles.

Influenza is a respiratory disease of humans. Closely related strains of the same virus infect birds, horses, and pigs. As these viral strains multiply in different animal hosts, different selective pressures favor different mutant forms of the virus. Then, when diverse strains happen to infect the same animal, genetic recombination can suddenly produce a "new" virus that might have enhanced virulence for humans.

Yellow fever illustrates how animal reservoirs complicate disease control. At one time, yellow fever ravaged North America, South America, and Africa. But the situation changed when Walter Reed (**Figure 15.1**) discovered that mosquitoes spread the disease. With this information, yellow fever was eradicated from Havana, Cuba, simply by controlling the mosquito population. The program worked because there wasn't a significant animal reservoir for yellow fever in Cuba. But the same approach failed in Panama, where the disease was slowing work on the Panama Canal. The approach failed because monkeys in the surrounding jungle constituted an immense animal reservoir. So instead of eradicating yellow fever in Panama, public health officials had to settle for controlling its spread.

TABLE 15.1 Some Important Zoonoses: Human Diseases with Animal Reservoirs

Disease	Microorganism	Animal Reservoir	Transmission from Reservoir	Chapter Reference
Bacterial				
Plague	*Yersinia pestis*	Rodents, including wild burrowing species and rats	Flea bite	27
Anthrax	*Bacillus anthracis*	Cattle, goats, horses, pigs, sheep	Inhaling spores from infected soil or animal products	27
Brucellosis	*Brucella* spp.	Cattle, goats, horses, pigs, sheep	Consuming dairy products from infected animals or direct contact with infected tissue	27
Salmonellosis	*Salmonella* spp.	Many, including poultry, turtles, rats	Eating contaminated food, drinking contaminated water	23
Tularemia	*Francisella tularensis*	Wild animals, especially rabbits	Insect bite or direct contact with infected tissue	27
Leptospirosis	*Leptospira interrogans*	Mammals, including wild species and domestic dogs and cats	Contact with infected urine or water contaminated by infected urine	24
Psittacosis	*Chlamydia psittaci*	Many kinds of birds	Inhaling the bacteria	22
Lyme disease	*Borrelia burgdorferi*	Wild mammals, especially deer and mice	Tick bite	27
Fungal				
Ringworm	*Epidermophyton* spp. *Microsporum* spp. *Trichophyton* spp.	Dogs and cats	Direct contact	26
Histoplasmosis	*Histoplasma capsulatum*	Many kinds of birds	Inhaling spores from feces	22
Protozoal				
Giardiasis	*Giardia lamblia*	Beaver, marmot, muskrat	Drinking water contaminated by feces	23
Toxoplasmosis	*Toxoplasma gondii*	Domestic cats, wild carnivores, birds, rodents	Eating undercooked meat from infected animals or contact with cat feces	27
Trypanosomiasis (African sleeping sickness)	*Trypanosoma* spp.	Wild game animals	Bite of tsetse fly	25
Viral				
Influenza	Influenza virus	Swine, birds, horses	Inhaling the virus	22
Yellow fever	Yellow fever virus	Monkeys	Mosquito bite	27
Rabies	Rabies virus	Most animals, especially carnivores	Animal bites	25
Viral encephalitis	Encephalitis virus spp.	Birds, horses	Mosquito bite	25
Helminthic				
Tapeworm	*Taenia* spp.	Cattle, pigs	Eating undercooked contaminated beef or pork	23
Trichinosis	*Trichinella spiralis*	Pigs	Eating undercooked contaminated pork	23

SHARPER FOCUS

KOCH'S POSTULATES

In the early 1880s Robert Koch was working with the bacterium *Mycobacterium tuberculosis*. The organism was of great interest because researchers suspected it caused the widespread and often-lethal infection tuberculosis. Koch made two important discoveries. He found a way of staining human tissue for microscopic examination that showed *M. tuberculosis* cells as thin blue rods on a brown background of human cells. He also found that *M. tuberculosis*—a slow-growing and highly fastidious bacterium—would grow on coagulated blood serum. With these tools, Koch set out to prove that *M. tuberculosis* caused tuberculosis. In the 1880s no connection between the two had been proved. For that matter, there was no proof that any particular microorganism caused a particular disease.

Koch began by examining tuberculosis patients for the presence of *M. tuberculosis* cells. He found the bacterium in every patient—blue rods against brown tissue. Then Koch cultured the tuberculosis cells on coagulated blood serum and isolated pure cultures of *M. tuberculosis*, which he injected into guinea pigs. They succumbed to tuberculosis. Unequivocally, *M. tuberculosis* caused tuberculosis.

By these experiments Koch provided absolute proof of the microbial **etiology**, or cause, of an important infectious disease of humans. Moreover, he enunciated a principle that is valuable today. Fulfilling **Koch's postulates,** as the listing of his three experiments became known, provides absolute proof that a particular microorganism causes a particular

disease. The postulates are as follows:

1. The causative microorganism must be present in every individual with the disease.
2. The causative microorganism must be isolated and grown in pure culture.
3. The pure culture must cause the disease when inoculated into an experimental animal.

Most modern textbooks add a fourth postulate:

4. The causative microorganism must be reisolated from the experimental animal and reidentified in pure culture.

Of course, Koch's postulates cannot be met if there is no way to culture the pathogen or if it infects only humans. Koch himself faced this dilemma later in his career when he studied

cholera. He discovered that a bacterium, *Vibrio cholerae,* was present in the intestines of all patients he examined, and he was able to culture the organism. But he could not find an experimental animal susceptible to the disease. The third postulate was ultimately fulfilled when a physician studying at Koch's institute accidentally swallowed cholera bacteria and developed the disease.

Koch's postulates are not the sole route to determining infectious etiology. There are many infections for which etiology is known but Koch's postulates have not been fulfilled. *Treponema pallidum,* for example, is the cause of syphilis; but because it has never been cultured, Koch's postulates cannot be met. No viral pathogens can be tested according to Koch's postulates because they reproduce only within

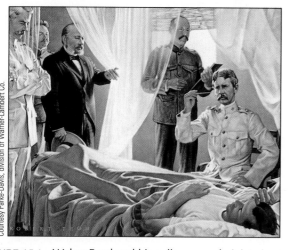

Courtesy Parke-Davis, division of Warner-Lambert Co.

FIGURE 15.1   Walter Reed and his colleagues administering to a soldier with yellow fever.

Environmental Reservoirs

Some pathogens are robust enough to survive for long periods in nonliving reservoirs or versatile enough to multiply in them. These pathogens can adapt to two quite different environments: the human body and environmental reservoirs such as soil, water, and house dust.

Soil is a reservoir for many pathogens, including *Clostridium tetani*, the bacterium that causes tetanus (Chapter 25). In the moist, warm environment of human tissues, *C. tetani* multiplies rapidly and produces a toxin that causes the deadly disease. In the soil, where conditions can be less than ideal, *C. tetani* survives for years by producing endospores (Chapter 4).

Water, usually contaminated drinking water, is a reservoir for many pathogens that infect the gastrointestinal tract. One of these, *Vibrio cholerae*, causes cholera. Another,

SHARPER FOCUS

KOCH'S POSTULATES (continued)

a living cell. To determine the etiological agent of a viral infection, we use Rivers' postulates, codified by T. M. Rivers in the 1930s. They are as follows:

1. The viral agent must be found either in the host's body fluids at the time of the disease or in the infected cells.

2. The viral agent obtained from the infected host must produce the disease in a healthy animal or plant or must induce production of antibodies (proteins produced by the host in response to infection).

3. Viral agents from the newly infected animal or plant must in turn transmit the disease to another host.

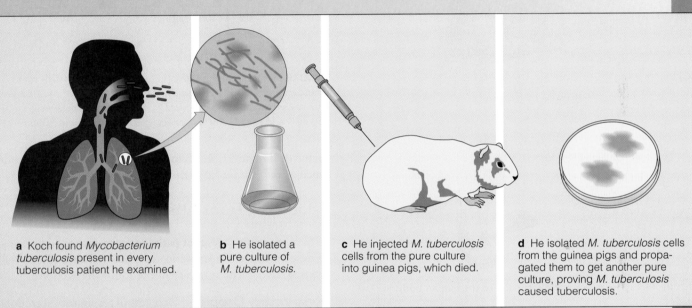

a Koch found *Mycobacterium tuberculosis* present in every tuberculosis patient he examined.

b He isolated a pure culture of *M. tuberculosis*.

c He injected *M. tuberculosis* cells from the pure culture into guinea pigs, which died.

d He isolated *M. tuberculosis* cells from the guinea pigs and propagated them to get another pure culture, proving *M. tuberculosis* caused tuberculosis.

If fulfilled, Koch's postulates provide absolute proof of the etiology, or cause, of an infection.

Salmonella typhi, causes typhoid fever. Both these pathogens adjust quickly to the environment of the human intestine. They can withstand the high concentration of acid in the stomach and the destructive effects of bile salts in the small intestine. Then, when they reenter a lake, stream, or ground water, their metabolism readjusts to that quite different environment with its lower temperature and scarcer nutrients.

TWO: GETTING TO AND ENTERING A HOST

A critical phase of any disease process is **transmission.** That's when a pathogen leaves its reservoir to enter the body of a host. Most pathogens are transmitted in a particular way, and they enter the body at a particular site called the **portal of entry.** For example, *Bordetella pertussis* is transmitted by respiratory droplets; its portal of entry is the nose. Yellow fever virus is transmitted by mosquitoes; its portal of entry is the bloodstream.

Portals of Entry

The most common portals of entry are the body's surfaces—the same sites that are normally colonized by members of our microbiota (Chapter 14). But these are not the only portals of entry for pathogens (**Figure 15.2**). We've just seen that a mosquito bite makes the bloodstream the portal of entry for yellow fever. The lungs also can be a portal of entry if a pathogen is inhaled directly into them. Tissues below the skin become a portal of entry for a person with an

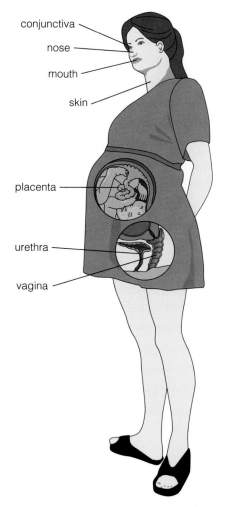

FIGURE 15.2   The major portals of entry to the human body.

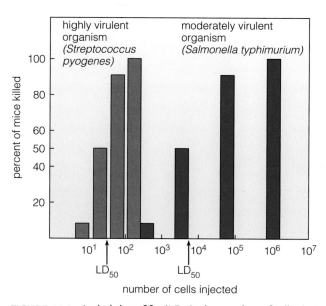

FIGURE 15.3   Lethal dose fifty (LD_{50}): the number of cells that must enter the body to cause the death of 50 percent of test animals. As the graph shows, the LD_{50} of *Streptococcus pyogenes* is about 50 cells and the LD_{50} of *Salmonella typhimurium* is about 8000 cells.

value of ID_{50} and LD_{50} than does a moderately virulent one such as *Salmonella typhimurium* (**Figure 15.3**).

Modes of Transmission

There are many ways that pathogens are carried to humans. These modes of transmission are listed in **Table 15.3.** We'll discuss each of them briefly in the following sections.

Respiratory Droplets. Droplets of respiratory secretions are formed and expelled when we cough, sneeze, laugh, or talk (**Figure 15.4**). They transmit many pathogens, including *Bordetella pertussis*, directly from one host to another. About a week before her first symptoms appeared, Baby Girl B must have inhaled *B. pertussis*–containing respiratory droplets that someone expelled. She must have been reasonably close to that person because tiny, warm respiratory droplets last only a brief time. They can travel only about a meter before they dry, killing the bacteria they contain.

Whooping cough is **communicable.** That means it can be transmitted from one person directly to another. In fact, whooping cough is highly communicable. More than 80 percent of people exposed to *B. pertussis* become infected. Children can contract whooping cough from someone they meet only briefly in a public place. That must have been the case with Baby Girl B because no one in her family or their circle of friends had the disease.

More human diseases are transmitted by respiratory droplets than by any other mode of transmission. And respiratory droplets are extremely efficient. One person exhaling infected droplets can transmit a respiratory

open wound, and the placenta may be a portal of entry into the body of the unborn baby if a pregnant woman has pathogens in her bloodstream. Some pathogens use different portals of entry under different circumstances. For example, the usual portals of entry for sexually transmissible pathogens are the urethra of males and the vagina of females, but they may become the throat or the rectum if these are points of sexual contact (**Table 15.2**).

The likelihood of infection depends upon how many of cells of the pathogen enter the portal—the more that do, the more likely infection becomes. This relationship between numbers of cells of the pathogen and likelihood of infection can be measured quantitatively using experimental animals. The result is expressed as the **ID_{50}** (infectious dose)—the number of microbial cells that must enter the body to establish an infection in 50 percent of the test animals. A similar measurement, the **LD_{50}** (lethal dose), is the number of microorganisms that must enter the body to cause the death of 50 percent of test animals. A highly virulent pathogen such as *Streptococcus pyogenes* has a lower

TABLE 15.2 Portals of Entry for Some Microbial Pathogens

Portal of Entry	Disease	Pathogen	Chapter Reference
Skin	Abscesses (through the blood)	Many organisms, including *Staphylococcus aureus*	26
	Tetanus (through wounds)	*Clostridium tetani*	25
	Plague (through insect bites)	*Yersinia pestis*	27
	AIDS (through injections)	HIV virus	27
Nose (primarily respiratory pathogens)	Pertussis	*Bordetella pertussis*	22
	Influenza	Influenza virus	22
	Common cold	Many different viruses	22
	Pneumococcal pneumonia	*Streptococcus pneumoniae*	22
	Measles	Measles virus	26
	Diphtheria	*Corynebacterium diphtheriae*	22
	Smallpox	Variola virus	26
	Chickenpox and shingles	Varicella zoster virus	26
	San Joaquin Valley fever	*Coccidioides immitis*	22
Conjunctivae	Ophthalmia neonatorum (gonococcal eye infection of the newborn)	*Neisseria gonorrhoeae*	24
	Trachoma	*Chlamydia trachomatis*	26
Mouth (primarily gastrointestinal pathogens)	Cholera	*Vibrio cholerae*	23
	Typhoid	*Salmonella typhi*	23
	Hepatitis A	Hepatitis A virus	23
	Salmonellosis	*Salmonella* spp.	23
	Poliomyelitis	Poliovirus	25
	Giardiasis	*Giardia lamblia*	23
Urethra	Urinary tract infections (males and females)	Many organisms, especially *Escherichia coli* and other enterobacteria	24
	Gonorrhea (males)	*Neisseria gonorrhoeae*	24
	Syphilis (males)	*Treponema pallidum*	24
	AIDS (males)	HIV virus	27
Vagina	Vaginitis	Many organisms, including *Gardnerella vaginalis*	24
	Gonorrhea (females)	*Neisseria gonorrhoeae*	24
	AIDS (females)	HIV virus	24
Placenta	Rubella (German measles)	Rubella virus	26
	Syphilis	*Treponema pallidum*	24
	AIDS	HIV virus	27

pathogen to every susceptible person in the immediate environment, so diseases transmitted by respiratory droplets spread rapidly where people live in crowded conditions or within the same household.

Fomites. Some pathogens are hardy enough to remain infectious on inanimate objects such as cups, towels, bedding, and handkerchiefs. Such objects, called **fomites,** can become vehicles for transmitting these pathogens. The

TABLE 15.3 Modes of Transmission

Mode of Transmission	Disease Examples (Chapter)
Respiratory droplets	Pertussis, pneumonia (22); measles (26)
Fomites	
Facial tissues, household surfaces	Common cold (22)
Eating utensils	Typhoid (23)
Contaminated needles	HIV infection (27)
Direct contact	Gonorrhea, herpesvirus infections, syphilis (24); AIDS (27)
Fecal-oral	Cholera, viral gastroenteritis, hepatitis A, giardiasis (23)
Vectors	
Mechanical	Cholera, typhoid (23)
Biological	Malaria, yellow fever (27)
Airborne	Tuberculosis, San Joaquin Valley fever (22)
Parenteral	Tetanus (25); gas gangrene (26)
By injection	Hepatitis B (23); HIV infection (27)

object that can serve as a fomite for a particular pathogen depends upon the pathogen's portal of entry. For example, eating utensils are likely fomites for pathogens of the intestinal tract whose portal of entry is the mouth. Contaminated hypodermic needles can be fomites for bloodborne pathogens such as those that cause AIDS or hepatitis B.

Fomites aren't a reservoir because pathogens don't survive on them very long. As a result, they usually transmit disease only among people who come into fairly close contact with one another.

The viruses that cause the common cold are an example. A person with a cold produces large amounts of mucus that contain huge numbers of virus particles. If the sufferer rubs his or her nose or eyes, the hands become heavily contaminated with virus particles. These particles can then be transferred to fomites, such as facial tissues, tabletops, or eating utensils. When an uninfected person touches one of these and then touches his or her own nose or eyes, the cold virus completes its journey to a new host. Disinfecting fomites effectively decreases this type of transmission. For example, using facial tissues that contain a virus-killing chemical decreases the spread of colds within a household. But the most effective way to decrease the spread of colds is frequent hand washing. It prevents the transfer of viruses to and from fomites.

Direct Body Contact. Some pathogens are spread by direct contact between the body of one person and another, such as by touching, kissing, or sexual intercourse.

Pathogens spread by direct body contact are too fragile to survive outside a human for even a brief period. To be transmitted, they must be deposited directly onto the body of a new host. Most such intimate contact occurs during sexual relations. Some pathogens are transmitted primarily or exclusively this way. Gonorrhea is an example of such a **sexually transmissible disease (STD)** (Chapter 24). STDs spread when the mucous membranes of an infected person come into direct contact with the mucous membranes of an uninfected person.

But transmission by direct body occurs by nonsexual means as well. For example, herpes simplex virus type 1, which usually infects the mouth or lips, is most commonly spread by kissing or exchange of saliva. Infants and toddlers

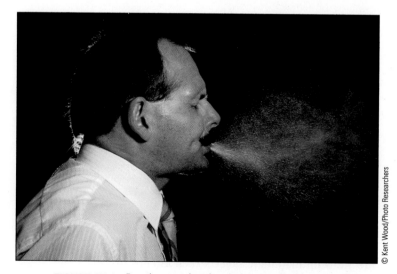

FIGURE 15.4 Respiratory droplets being expelled by a cough.

© Kent Wood/Photo Researchers

TABLE 15.4 Biological Vectors of Infection

Disease	Pathogen	Vector	Chapter Reference
Plague	*Yersinia pestis*	Flea	27
Malaria	*Plasmodium* spp.	Anopheles mosquito	27
Lyme disease	*Borrelia burgdorferi*	Tick	27
Yellow fever	Yellow fever virus	Aedes mosquito	27
Rocky Mountain spotted fever	*Rickettsia rickettsii*	Tick	27
Viral encephalitis	Various arboviruses	Mosquitoes	25
Epidemic typhus	*Rickettsia prowazekii*	Body louse	27
African trypanosomiasis (African sleeping sickness)	*Trypanosoma brucei*	Tsetse fly	25
American trypanosomiasis (Chagas' disease)	*Trypanosoma cruzi*	Reduviid bug	27
Filariasis (elephantiasis)	*Wuchereria bancrofti*	Mosquitoes	27

who mouth each other and share toys that they put into their mouths commonly exchange saliva. As a result, most cases of primary herpes simplex type 1 infections are seen in small children. The virus can also be transmitted from one place to another on a person's body. For example, children with oral herpes lesions who suck their fingers may develop typical blistering sores of herpes on their hands.

Another nonsexual form of disease transmission by direct contact is from mother to infant. This is called **vertical transmission,** which means transmission from one generation to the next. Vertical transmission can be **prenatal** if it occurs across the placenta before birth or **perinatal** if it occurs during passage through the birth canal or soon after birth. Many sexually transmissible infections are also transmitted from mother to infant. These include syphilis, gonorrhea, hepatitis B, and HIV infection.

The Fecal-Oral Route.

The **fecal-oral route** is between infected feces of one host to the mouth of a new one. This route overlaps other modes of transmission. It can occur directly by body contact or indirectly through water, food, fomites, or vectors. Most of the gastrointestinal pathogens that cause diarrhea are transmitted by the fecal-oral route.

Cholera illustrates the possible variations of the fecal-oral route. This deadly disease is caused by a particularly robust bacterium, *Vibrio cholerae*, which is transmitted in public water supplies that become contaminated with infected human feces. Most major epidemics of cholera spread this way (Chapter 20). Cholera can also spread within a household if one family member is infected. Each day the infected person passes several liters of watery feces,

each drop of which contains millions of cells of *V. cholerae*. Therefore traces of feces are certain to get on the hands of those who care for the sick person. Unless hand washing is scrupulous, *V. cholerae* will eventually be transmitted to the caretaker's mouth. If the caretaker also prepares the family's food, everyone in the household is at risk. Fomites may also play a role. For example, anyone who shares or touches fecally contaminated bedding is at risk of infection. Flies that land on feces or other infected material can act as vectors that carry the organisms to the food or bodies of uninfected people—perhaps the family next door. On a grander scale, cholera has been transmitted from one continent to another in fecally contaminated bilge water of ships (Chapter 23).

Arthropod Vectors.

Arthropods that act as vectors include flies, mosquitoes, ticks, fleas, and lice (Chapter 12). Some of these are **mechanical vectors.** That is, they act as living fomites by picking up pathogens on their bodies and carrying them from one place to another. The flies we just discussed were mechanical vectors of *Vibrio cholerae*. Mechanical vectors are never the only way a pathogen is transmitted: They aren't an essential link in the transmission of a disease. But **biological vectors** are. They transmit disease, and they also play a role in the pathogen's life cycle. Such pathogens develop and multiply in the body of the arthropod host, as well as in the body of the human host (**Table 15.4**). Biological vectors usually transmit infection by biting. Fleas that transmit plague, mosquitoes that transmit malaria, and the ticks that transmit Lyme disease are examples of biological vectors (**Figure 15.5**).

FIGURE 15.5 The deer tick that causes Lyme disease, shown on a person's index finger.

© Scott Camazine/Photo Researchers

Airborne Transmission. Airborne transmission occurs when pathogens that can survive being suspended in air are inhaled into the respiratory system. Unlike pathogens transmitted by respiratory droplets, these microorganisms can withstand prolonged drying. As a result, they can be transmitted across considerable distances (greater than a meter) between people who have had no close contact with one another. Airborne pathogens don't necessarily stay suspended in air. They might settle out on dust particles and become airborne again (and infectious) months or years later.

Pathogens that are transmitted this way include *Mycobacterium tuberculosis*, the bacterium that causes tuberculosis (Chapter 22). It can survive desiccation and remain infectious for months. Some fungal pathogens can remain viable in dry soil for years. They become airborne and infectious whenever dust forms. *Coccidioides immitis*, which causes San Joaquin Valley fever (Chapter 22), is such an organism. In the 1970s, during a major dust storm in California, *C. immitis* infected people who lived hundreds of miles away from the infected soil.

Parenteral Transmission. Parenteral transmission means the pathogen is deposited directly into a blood vessel, tissue below the skin, or mucous membranes. The viruses that cause hepatitis B and AIDS, the protozoan that causes malaria, and the helminth that causes elephantiasis can be transmitted this way.

Parenteral transmission can occur in several ways, for example when an arthropod vector pierces the skin by biting or when preexisting breaks in the skin or mucous membranes provide access for pathogens. It can also occur when a hypodermic needle or other sharp object penetrates the skin. Sharing of hypodermic needles by intravenous drug users provides opportunities for parenteral transmission of HIV and the hepatitis viruses. Transfusions of blood also provide such opportunities. Although blood products for transfusions are screened for certain pathogens, absolute safety cannot be ensured.

A special type of parenteral transmission occurs when people sustain significant wounds. Any small break in the skin can lead to a localized infection by normal skin biota, such as staphylococci. But deep wounds surrounded by dead tissue are an environment in which certain anaerobic pathogens thrive. These include *Clostridium tetani*, which causes tetanus, and *C. perfringens*, which causes gas gangrene (Chapter 26).

THREE: ADHERING TO A BODY SURFACE

When a pathogen comes to rest on a body surface, it's immediately confronted by defenses that could dislodge it (Chapter 14). Successful pathogens aren't swept away because they, like the normal biota, attach tightly to body surfaces by means of adhesins. These proteins bind to specific molecular receptors, usually on the surface of the pathogen's target cells (see Figure 14.6), and the presence or absence of receptors determines a pathogen's **tissue trophism** (which cells or tissues it attacks). The many kinds of adhesins that various pathogens produce are divided into two broad groups: those that are attached to pili and those that aren't. Some pathogens, probably most, produce several types of adhesions. For example, *Bordetella pertussis*, the organism that caused Baby Girl B's whooping cough, produces at least eight.

When the *B. pertussis* cells entered Baby Girl B's body, they probably attached to cilia on epithelial cells lining her respiratory tract by means of the adhesin called **filamentous hemagglutinin (FHA)**. FHA is a large protein that forms filamentous structures on the surface of *B. pertussis*. These structures bind to a molecule (sulfatide) on the membranes of ciliated human cells (**Figure 15.6**). This firm, bridgelike attachment keeps the bacterium from being swept out of the body by the mucociliary system. *B. pertussis* sticks to the respiratory epithelium and multiplies there.

FHA is a relatively unusual type of adhesin. The structures it forms on the cell surface aren't true pili. They're thin, wavy filaments. Most adhesins, including those on pathogenic strains of *Escherichia coli* and species of *Neisseria*, are components of pili. The presence of pili on these bacteria is essential for their virulence. Strains that can't make pili can't adhere to human tissue and therefore can't initiate infection. They become **avirulent** (harmless).

Some pathogenic bacteria produce different adhesins over time. Changing its repertoire of adhesins helps a pathogen in several ways. It helps a pathogen evade the body's immune defenses, and it allows the pathogen to attack more than one kind of host cell. For example, *Neisseria gonorrhoeae* (Chapter 24) can produce specific adhesins that bind to genital, rectal, pharyngeal, or conjunctival surfaces.

---

the cd connection

MICROBIAL ANATOMY ⟶ *The prokaryotic cell/The appendages* ⟶ *1*

Micrograph of *Escherichia coli* covered with type I pili.

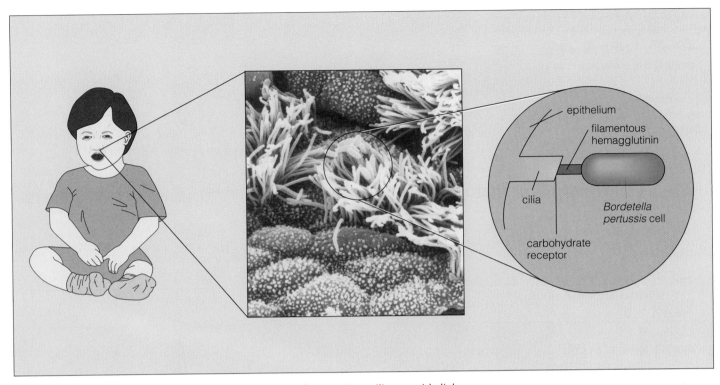

FIGURE 15.6 Filamentous hemagglutinin attaches *Bordetella pertussis* to cilia on epithelial cells that line the nose.

FOUR: INVADING THE BODY

After attaching to a body surface, some pathogens, termed **noninvasive,** stay where they attach. *Bordetella pertussis* is an example of a noninvasive pathogen. The bacteria that attached to the cilia on cells in Baby Girl B's air passages multiplied on the surfaces of these cells and caused her disease. *Streptococcus pneumoniae,* which causes a life-threatening pneumonia, is another example of a noninvasive pathogen.

Other pathogens are **invasive.** Some of these penetrate into host cells on their way to deeper tissues (**Figure 15.7**). Others remain within the cell they invade. Invasive pathogens attach to host cells by using a special class of adhesins called **invasins,** which trigger the host cell to take them in.

Some invasive pathogens deliberately trigger **phagocytes** to engulf them. The normal function of phagocytes is to engulf and destroy invading microbes. But these invasive pathogens have the ability to survive and multiply within the phagocyte's lethal vacuole (called a phagolysosome; Chapter 16) or to escape from it unharmed. For example, *Coxiella burnettii,* which causes Q fever (Chapter 22), thrives in the phagolysosome. So does *Mycobacterium tuberculosis.* The destructive enzymes and chemicals that kill most phagocytized microbes never get through the waxy cell wall that surrounds *M. tuberculosis* cells (Chapters 4 and 22). By taking up residence within normally lethal phagocytic cells, *M. tuberculosis* becomes distributed throughout the body and establishes a successful infection. *Listeria monocytogenes* takes another approach. It escapes from the hostile phagolysosome by dissolving the surrounding membrane and thereby enters the nutrient-rich cytoplasm.

Other invasive pathogens carry invasins that stimulate cells that are not normally phagocytic to become phagocytic and engulf them. For example, when *Salmonella* and *Shigella* spp. bind to the epithelial cells on the surface of the intestinal mucosa; then these cells engulf them.

Invasiveness is a survival mechanism. It allows a pathogen to enter a nutrient-rich environment, where it's protected from most of the host's defenses. Some invasive pathogens stay inside cells and multiply there. They're called **intracellular pathogens.** Others just pass through epithelial cells on their way to deeper tissues, such as the blood or lymphatic circulations.

Viruses enter cells by other means. We considered them in Chapter 13.

Most eukaryotic pathogens don't invade cells, but some do. For example, *Plasmodium* spp., the protozoan that cause malaria, invade red blood cells by mechanisms that are not well understood.

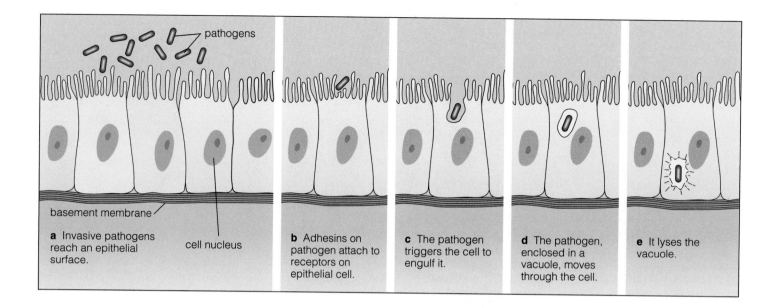

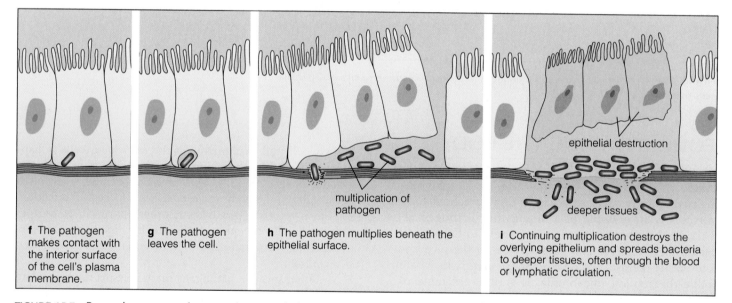

FIGURE 15.7   Route that some pathogens take to reach deeper tissues.

FIVE: EVADING THE BODY'S DEFENSES

So far we've discussed the ways pathogens reach the part of the human body where they are best adapted to multiply. But along the way and after they arrive, they're confronted with the host's arsenal of defenses. We'll discuss these defenses in Chapters 16 and 17. Here we'll look at how the pathogen avoids them.

Evading Phagocytosis

Phagocytosis is a major host defense against invading pathogens. In the previous section we considered how some pathogens survive it and others exploit it. Now we'll consider how many pathogens avoid it.

Capsules. Before a phagocyte can **phagocytize** (consume and destroy) an invading microorganism, it must first come in direct contact with the surface of the microbial cell. Some pathogens avoid such direct contact by producing a slippery mucoid capsule (**Figure 15.8**). *Bordetella pertussis* produces such a capsule, so the phagocytes defending Baby Girl B's airways were unable to eliminate these bacterial cells.

Extracellular capsules are vital to the **pathogenicity** (disease-causing ability) of many other bacteria. Stains that lose their capsules through mutation become **avirulent** (unable to cause disease). Strains that produce the thickest capsule are the most virulent. *Streptococcus pneumoniae* (Chapter 22) is an example of such a capsule-dependent pathogen. Almost all the bacteria and fungi that infect the central nervous system, including *Neisseria meningitidis*, *Haemophilus influenzae*, and *Cryptococcus neoformans*, must produce a capsule to survive within their host and cause disease (Chapter 25).

Surface Proteins. Some pathogens defend themselves against phagocytosis by producing specialized surface proteins. Like capsules, these antiphagocytic proteins interfere with direct contact between phagocyte and pathogen. *Streptococcus pyogenes*, the bacterium that causes strep throat, produces such a surface protein, called M protein, that projects beyond its cell surface (**Figure 15.9**). A similar surface protein, called protein II, protects *Neisseria gonorrhoeae*, the bacterium that causes gonorrhea.

Evading the Immune System

Microorganisms that survive phagocytes and the other nonspecific body defenses (Chapter 16) are then confronted by an even more sophisticated line of defense—the adaptive immune response (Chapter 17). But pathogens have evolved mechanisms to evade these defenses as well. The principal evasion mechanisms are antigenic variation, IgA proteases, and serum resistance. We'll discuss them in the following sections.

Antigenic Variation. The adaptive immune system recognizes molecules called **antigens** on the surface of microorganisms. Many kinds of molecules, including adhesins, function as antigens. Antigenic recognition allows the immune system, after a delay of a week or so, to direct highly specific defenses against the antigen-bearing invaders, leading to their inactivation or destruction. But some pathogens evade this defensive action by periodically changing their

surface antigens, a process called **antigenic variation.** By the time the immune system organizes its defensive response, the pathogen has changed its antigens and therefore has become immune to them. Many pathogens are capable of antigenic variation. *Neisseria gonorrhoeae* is particularly good at it. It has an unusually large repertoire of antigenic variation, as does the protozoan pathogen *Trypanosoma brucei*, which causes African sleeping sickness (Chapter 25).

IgA Proteases. The immune system reacts to the presence of antigens by making antibodies, the basis of the adaptive immune system's defensive response. Some pathogens defend themselves by attacking some of these antibodies with antibody-destroying enzymes, called **IgA proteases,** that destroys the IgA class of antibodies (Chapter 17). IgA antibodies accumulate on epithelial surfaces and prevent infection by interfering with the pathogen's ability to adhere

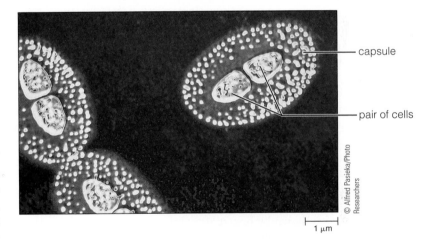

capsule

pair of cells

© Alfred Pasieka/Photo Researchers

1 μm

FIGURE 15.8 *Streptococcus pneumoniae* surrounded by a slippery mucoid capsule that helps it avoid phagocytosis.

From *Clinical Microbiological Review*, July 1989, p. 288. Reprinted by permission of the American Society for Microbiology. Photo courtesy Vincent A. Fischetti, Rockefeller University.

FIGURE 15.9 M protein on the surface of *Streptococcus pyogenes* cell that helps it avoid phagocytosis.

SHARPER FOCUS

CRITICAL CUES

When some pathogenic microorganisms encounter a human host, they pick up environmental cues that turn on their disease-causing capabilities. For some microbes the critical cue is a low concentration of available iron—an essential nutrient that is much scarcer inside the human body than elsewhere in the environment. When pathogenic members of the bacterial genus *Shigella*, for example, leave their watery reservoir and enter cells of the intestinal tract, they sense the drop in iron concentration and begin to produce a cell-damaging toxin. For other microbes the critical cue is body temperature. Intestinal pathogens of the genus *Yersinia*, for example, turn on disease-causing functions only at 37°C—the temperature inside the human body.

to that surface. *Neisseria gonorrhoeae* is one of the pathogens that makes an IgA protease. By removing IgA, *N. gonorrhoeae* can bind to the surface and remain infective. *Neisseria meningitidis* and most other pathogens that attack the central nervous system do the same thing.

Serum Resistance. Some microorganisms have a biochemical weapon called **serum resistance.** Serum resistance acts against the complement system, a group of host proteins with many defensive functions (Chapter 16), including forming a complex on the surface of a Gram-negative bacterium that makes holes in them, causing them to lyse. Bacteria with serum resistance have features on their surface that interfere with the formation of this destructive complex. This mode of protection is called serum resistance because complement is contained in human serum. The power of serum resistance is shown by how it affects the virulence of *Neisseria gonorrhoeae*: Strains of *Neisseria gonorrhoeae* that have serum resistance spread throughout the body and cause life-threatening disease. Strains that lack it remain localized in the genital tract and cause less serious disease.

Obtaining Iron

Let's assume that a pathogen has entered human tissue and successfully evaded the body's defenses. It's now faced with another major challenge—obtaining adequate amounts of iron in order to multiply. Human tissue is a rich source of all the nutrients the pathogen needs, except iron. There's lots of iron in human tissue, but it's not available to the pathogen because it's tightly bound to iron-transport proteins such as **transferrin, lactoferrin,** and **ferritin.** As a result, microorganisms that infect humans are constantly engaged in a struggle for iron.

Some pathogens obtain iron by producing and releasing their own iron-binding proteins, called **siderophores,** which take iron away from the iron-transport proteins by binding it more tightly. Then the pathogen takes up the siderophore with its bound iron. *Neisseria meningitidis,* a pathogen that causes meningitis, uses a different strategy. It produces receptors on its cell surface that bind directly to the human iron-transport proteins. Then it takes up these proteins, along with the iron atoms they contain (Chapter 25).

The amount of iron in our diet seems to affect the amount that is available to a pathogen. Studies suggest that giving infants large amounts of supplementary iron makes them more prone to infection. However, iron is relatively scarce in an infant's diet and iron deficiency is common during the first year of life, when iron requirements are high. Many physicians therefore recommend iron supplementation for all babies but prescribe large doses only if there is an indication of deficiency. Other clinical studies suggest that the high levels of free iron in the blood of patients with sickle cell anemia contribute to their unusual vulnerability to infection.

SIX: MULTIPLYING IN THE HOST

A pathogen in human tissue that survived the body's defenses and is able to obtain adequate iron begins to multiply. In so doing, it causes disease. We might ask how this benefits the pathogen. In some cases it's clear. For example, killing human cells makes more nutrients available to the pathogen. In other cases, possible benefits are obscure. Indeed, there might not be any.

In this section we'll discuss **pathogenesis**—how pathogens damage tissue and cause human illness. We'll consider the two most common forms of bacterial pathogenesis: production of **toxins** (poisonous products that damage human tissue) and the indirect damage they do by stimulating the body's defenses. **Hypersensitivity** (an exaggerated immune

SHARPER FOCUS

NO HARM, NO FOUL

Pathogens face a dilemma. To survive in their human host, they inevitably cause some damage. But if the damage is too great, the host will die, leaving them homeless. Ideally a pathogen should modulate its activities so that it can grow while damaging the host only slightly. Microbiologist John Mekalanos calls this survival strategy "no harm, no foul." Consider how this mechanism works with toxin-producing pathogens that kill host cells to obtain iron. When iron is scarce, the pathogen produces toxin, host cells die and release iron, and the local iron concentration rises, which signals the microbe to decrease toxin production. This feedback loop ensures that the pathogen gets enough iron to continue to grow, but host damage is minimized.

response) is another form of microbial pathogenesis. It causes damage done by systemic fungal and helminthic infections and by tuberculosis. It's discussed in Chapter 18. Viral pathogenesis, which is quite different from the pathogenesis of cellular organisms, is discussed later in this chapter.

Toxins

Many bacterial pathogens, both Gram-positive and Gram-negative, produce **exotoxins.** They're secreted proteins, usually enzymes, that kill host cells at extremely low concentrations. All Gram-negative bacteria, both pathogenic and nonpathogenic, contain another toxin called **endotoxin.** It's is the lipopolysaccharide component of their outer membranes. Endotoxin is much less toxic than exotoxins, although its toxicity varies somewhat among species (**Table 15.5**). In addition to these two kinds of toxins, some bacterial pathogens produce other proteins that are toxic but at much higher concentrations than exotoxins. We'll discuss all three classes of toxins in the following sections.

TABLE 15.5 Differences Between Exotoxins and Endotoxins

Characteristic	Exotoxin	Endotoxin
Chemical composition	Proteins, usually two components (A and B)	Lipid portion (lipid A) of lipopolysaccharide outer membrane
Produced by	Certain Gram-positives and Gram-negatives	All Gram-negatives
Location	Excreted outside cell	Part of outer membrane
Heat stability	Most are heat labile (sensitive); inactivated at 60–80°C (except staphylococcal enterotoxin)	Relatively heat stable; many withstand heating above 100°C
Toxicity	High toxicity	Low toxicity
Effect on host	Highly variable from one toxin to another	Similar for all endotoxins
Immune response	Active, antitoxins provide host immunity	Poor, no antibodies to endotoxin produced
Toxoid production	Chemically treated toxin (toxoid) used as vaccine	Toxoid cannot be made
Diseases	Many, including tetanus, botulism, diphtheria	Meningococcemia and overwhelming Gram-negative infections

TABLE 15.6 Some Types of Exotoxins

Type	Action	Toxin	Disease/Microbe	Clinical Effect
AB toxins	B subunit binds to host cell; A subunit exerts effect	Heat-stable *Escherichia coli* enterotoxin	Travelers diarrhea/ *E. coli*	Stimulates intestinal epithelium to secrete fluids, causing profuse diarrhea.
		Cholera toxin	Cholera/*Vibrio cholera*	Same as above.
		Pertussis toxin	Whooping cough/ *Bordetella pertussis*	Disrupts cellular regulation.
Proteolytic toxins	Split essential neuroproteins	Botulinum neurotoxin	*Botulism*/Clostridium *Botulinum*	Causes flaccid (limp) muscle paralysis.
		Tetanus neurotoxin	Tetanus/*Clostridium tetani*	Causes rigid paralysis. Symptoms differ from those of botulism because the toxin is released at a different site within the body.
Pore-forming toxins	Toxin inserts in host cell's cytoplasmic membrane, forming pore that leads to cell lysis	Hemolysin	Hemorrhagic colitis/ *E. coli*	Causes bloody diarrhea.
		Listeriolysin O	Listeriosis/*Listeria monocytogenes*	Causes meningitis, septicemia, or endocarditis.

Exotoxins. Exotoxins (so named because they're released outside the cell) play a central role in pathogenesis. For example, purified exotoxins alone can cause most of the clinical features of diseases such as cholera, tetanus, and diphtheria. There are many kinds of exotoxins (**Table 15.6**). We'll focus on just a few of them as examples.

Many exotoxins, called AB toxins, are composed of two protein subunits, designated A (active) and B (binding). The B subunit attaches the exotoxin to particular component of the host cell. In some cases these are structural components, but more often they're enzymes. Then the A subunit, which is an enzyme, alters that cellular component. For example, cholera toxin alters the enzyme adenyl cyclase, and diphtheria toxin inactivates a protein, called EF 2, that eukaryotes need to synthesize proteins.

Another class of exotoxins, called **proteolytic toxins,** act by attacking vital host proteins and splitting them. A third class, called **pore-forming toxins,** act by inserting themselves into the membranes of host cells, creating a pore that kills the cell.

The enzymatic activity of exotoxins accounts for their astounding potency. For example, as little as 130 μg (micrograms) of tetanus toxin—an amount about equal to the size of the period at the end of this sentence— can kill an adult. Because it's an enzyme, one molecule of the toxin can quickly destroy many molecules of its target.

Many exotoxins alter or inactivate their targets by adenosine diphosphate (ADP)-ribosylation. That is, they catalyze the following reaction (Chapter 5):

$$NAD^+ + \text{target protein} \longrightarrow$$
$$ADP—ribose—\text{target protein} + \text{nicotinamide}$$

Some exotoxins enter the bloodstream. Then they exert their harmful effects far from their site of production. This explains how pathogens such as *Bordetella pertussis*, which remains localized in the throat, can cause systemic disease that affects tissues throughout the body.

Pertussis toxin (Ptx), aided by other proteins, disrupts the host cell's **adenylate cyclase system,** which mediates intracellular communication. Ptx is an AB type toxin. The B component binds to the host cell, and the A component enters the cell where it does its damage. (**Figure 15.10**). Because it enters the circulation, many cells in the body are disrupted. This damage produces the nonrespiratory symptoms Baby Girl B suffered (oversensitivity to histamine, abnormally high levels of insulin, and a vast excess of lymphocytes). The toxin probably also accounts for the lingering effects of her illness. The weeks of disabling coughing occurred relatively late in the infection. By that time probably only a few microorganisms were still actively dividing in her respiratory tract. But toxin that had previously entered cells in the brain center where coughing is controlled continued to exert their effect.

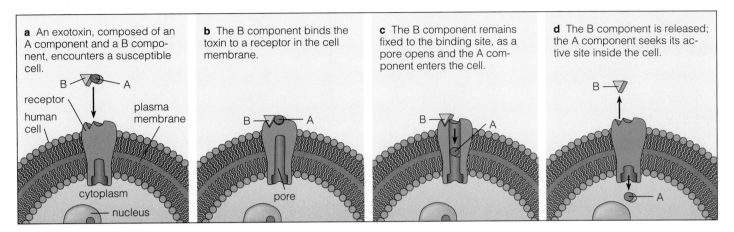

a An exotoxin, composed of an A component and a B component, encounters a susceptible cell.

b The B component binds the toxin to a receptor in the cell membrane.

c The B component remains fixed to the binding site, as a pore opens and the A component enters the cell.

d The B component is released; the A component seeks its active site inside the cell.

B A
receptor
human cell
plasma membrane
cytoplasm
pore
nucleus
A

FIGURE 15.10 Exotoxins. The action of some AB toxins.

The immune system can neutralize some exotoxins with antibodies called **antitoxins.** This capacity is exploited to make certain vaccines. The exotoxin is modified in the laboratory with heat or chemicals to produce a **toxoid.** Toxoid can no longer cause disease, but it can still stimulate the immune system to produce antitoxin. The vaccine against diphtheria, for example, is a toxoid (Chapter 20).

Curiously, diphtheria exotoxin is encoded by genes carried on a temperate bacteriophage (Chapter 13), not on the bacterium's chromosome. Strains of *Corynebacterium diphtheriae* that aren't infected by this bacteriophage don't produce toxin and aren't virulent.

It's not completely clear what value exotoxins offer the pathogens that produce them, but there are some hints. For example, the exotoxin *C. diphtheriae* produces kills cells in the throat. Because *C. diphtheriae* grows better than most other bacteria in this environment of dead and dying cells, the toxin probably gives *C. diphtheriae* a selective advantage over its competitors (Chapter 22).

Some disease-causing toxins are produced outside the body and taken in with contaminated food. They cause illnesses such as **botulism** (Chapter 25) and **staphylococcal food poisoning** (Chapter 23). These illnesses are not infections. They are poisonings.

Endotoxins. Endotoxins (so named because they're part of the cell) are the lipopolysaccharide (LPS) component of the outer membrane of Gram-negative bacteria (Chapter 4). Lipid A is the toxic portion of it. Because Lipid A is embedded in LPS, it's released and able to exert its toxic effect only if the bacterium is lysed. And during an infection, some cells of the pathogen are constantly being lysed, either by the body's own defenses or by treatment with certain antibiotics. Lipid A is toxic because it activates complement (Chapter 16) and stimulates the release of cytokines (Chapter 16). Cytokines are molecular signals that activate the

body's immune defenses. They become toxic if they're produced in too high concentrations. Therefore endotoxin toxicity requires two steps: (1) release from a Gram-negative bacterial cell by lysis and (2) stimulation of the body's immune defenses to produce excessive amounts of cytokines.

Because endotoxin stimulates release of many cytokines, it can exert an astonishing variety of toxic effects on the human body—fever, increased or decreased numbers of white blood cells, diarrhea, shock, extreme weakness, and death. For example, one cytokine, **interleukin-1,** causes fever. Another, **tumor necrosis factor,** causes **endotoxic or septic shock,** a life-threatening loss of fluid from the circulation.

Because endotoxins are not very potent, they generally play a minor role in Gram-negative infections, including Baby Girl B's whooping cough. But there are exceptions. For example, most of the clinical findings of **meningococcemia** (a condition in which *Neisseria meningitidis* invades the blood) are caused by endotoxin. And endotoxins become clinically significant when large numbers of dying Gram-negative bacteria are circulating in the bloodstream. This can happen during an overwhelming Gram-negative infection. Paradoxically, agents such as antibiotics or complement that kill Gram-negative bacteria may actually increase endotoxin-causing damage.

Other Toxic Proteins. Pathogenic bacteria produce other toxic proteins in addition to exotoxins and endotoxins. For example, *Bordetella pertussis* produces at least three toxic agents in addition to pertussis exotoxin. It produces **tracheal cytotoxin,** which is a small fragment of peptidoglycan that selectively kills ciliated respiratory cells and stimulates the release of cytokines that cause fever. It produces an **adenylate cyclase,** an enzyme that enters eukaryotic cells and potentiates the effect of pertussis exotoxin. And it produces **dermonecrotic toxin,** which

TABLE 15.7 Actions of Extracellular Enzymes

Type of Extracellular Enzyme	Example of Producing Pathogen	Action
Cytolysins (lyse cells)		
Hemolysin	*Staphylococcus aureus, Streptococcus pyogenes*	Lyses red blood cells
Leukocidin	*Staphylococcus aureus*	Lyses white blood cells
Lecithinase	*Clostridium perfringens*	Destroys cell membranes
Breaks Down Materials That Hold Cells Together		
Hyaluronidase	*Streptococcus pyogenes*	Breaks down hyaluronic acid, an essential component of connective tissue
Collagenase	*Clostridium perfringens*	Breaks down connective tissue
Protease	*Clostridium perfringens*	Destroys protein in muscle tissue
Disturbs Normal Blood Clotting		
Coagulase	*Staphylococcus aureus*	Causes plasma to coagulate
Streptokinase	*Streptococcus pyogenes*	Dissolves clots

causes skin lesions at low doses and death at high does in mice. Its role in whooping cough, if any, is unknown.

Certain bacterial pathogens produce toxic **extracellular enzymes** (secreted outside the cell; **Table 15.7**). Some of these lyse host cells. Others help spread the infection. Still others interfere with blood clotting. The first class includes **cytolysins,** which attack cell membranes; **hemolysins,** which lyse red blood cells; and **leukocidins,** which lyse white blood cells (**Figure 15.11**). The second class breaks down the materials that hold cells together to form tissues. **Hyaluronidase** degrades the polysaccharide hyaluronic acid that cements cells together in many different tissues. **Collagenase** degrades the protein collagen, a major structural component of connective tissue. The third

class includes **coagulases,** which split the serum protein **fibrinogen** to form **fibrin,** thus creating blood clots. It also includes **kinases** that split fibrin and dissolve clots. (One of these bacterial kinases, streptokinase, is used clinically to dissolve the blood clots that block coronary arteries during a heart attack. One of the authors received this. That's why you're reading this book.) Interfering with the blood clotting system might benefit microbes by forming protective clots around themselves or by escaping from clots the body produced.

Now we'll turn to another type of damage an infection causes—collateral damage (as the military would say) done by the body's defenses while fighting the infection.

Damage Caused by Host Responses

Sometimes the body's defenses against a pathogen disrupt normal body functions in ways that produce disease. A pathogen that damages the body in this way need not produce a toxin to be virulent.

Streptococcus pneumoniae, for example, produces no known toxins, but it does cause a deadly pneumonia (Chapter 22). When *S. pneumoniae* multiplies in the lungs, phagocytic white blood cells come to combat the infection, but the phagocytes are ineffective because *S. pneumoniae* is protected by a capsule. More and more phagocytes arrive to help. The dead phagocytes, along with some bacteria cells, accumulate in the lungs, impairing breathing and causing pneumonia.

A similar event potentiates the action of pertussis toxin. When it stops the beating of cilia on the surface of the res-

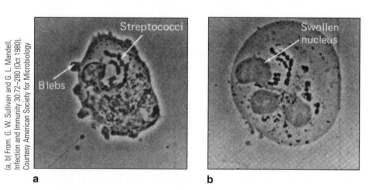

(a, b) From: G. W. Sullivan and G. L. Mandell, Infection and Immunity 30:72–280 (Oct 1980). Courtesy American Society for Microbiology

a   b

FIGURE 15.11 Effect of leukocidin on a phagocyte. (a) A phagocyte that has just ingested a chain of streptococcal cells. (b) A phagocyte after it has been damaged by leukocidin. It has stopped moving and is swollen. Soon most internal structures will disappear.

piratory epithelium surfaces, masses of bacteria and phagocytic white blood cells accumulate. They produced the sticky respiratory secretions that caused Baby Girl B to choke and cough. These defenses and their protective, as well as damaging, effects are further discussed in the section on Inflammation in Chapter 16.

Now let's look at the ways viruses damage host cells. They're quite different from the ways cellular organisms do.

Viral Pathogenesis

Recall from Chapter 13 that viruses depend completely on the metabolism of their host cell to reproduce themselves. Some viruses kill the cells they infect. Others damage them or have little impact. **Figure 15.12** summarizes the four possible outcomes of a viral infection.

Lytic infections kill the host cell by lysing it. The virus takes over so much of the cell's metabolic machinery, it can't carry on essential maintenance functions. If the cytoplasmic membrane disintegrates, the cell lyses immediately. If the membranes that surround lysosomes disintegrate, destructive enzymes are released into the cytoplasm,

which destroy the cell from within, a process termed **autolysis.** When an infected cell lyses, new viruses are released that infect more host cells.

In contrast, **persistent viral infections** can last for years. Without killing the cell, they produce new virus particles by budding out through the cell membrane. This process can cause little damage to the host.

In a **latent infection,** the virus lies dormant within the host cell. No new viral particles are produced. Latent infections can last for the rest of a person's life. No damage to the host occurs as long as the infection remains latent, but various stimuli may reactivate the virus, initiating a destructive lytic infection.

Some viral infections change human cells in ways that can be seen under the microscope. For example, they might form **inclusion bodies** (collections of viral components such as protein and nucleic acid, waiting to be assembled into new viral particles) or giant multinucleated cells. Rabies virus produces inclusion bodies called **Negri bodies** in infected nerve cells (**Figure 15.13**).

Viral infections can also affect cells in ways that aren't visible under the microscope but can have important consequences. For example, they might make new antigens on

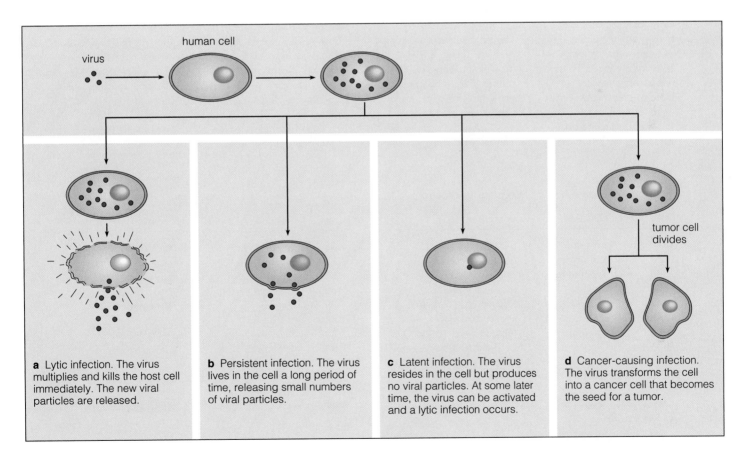

a Lytic infection. The virus multiplies and kills the host cell immediately. The new viral particles are released.

b Persistent infection. The virus lives in the cell a long period of time, releasing small numbers of viral particles.

c Latent infection. The virus resides in the cell but produces no viral particles. At some later time, the virus can be activated and a lytic infection occurs.

d Cancer-causing infection. The virus transforms the cell into a cancer cell that becomes the seed for a tumor.

FIGURE 15.12 Viral pathogenesis. The four possible outcomes of a viral infection.

LARGER FIELD

ANOTHER GOLDEN AGE?

The study of microbial pathogenesis—the relationship between pathogenic microorganisms and the diseases that they cause—began a little more than 100 years ago, when Robert Koch conclusively proved that a specific species of bacterium caused a specific human disease (see Sharper Focus: Koch's Postulates earlier in this chapter).

Koch's success, along with Louis Pasteur's at about the same time, initiated a period of intense research. This period—from the late 1800s through the early 1900s—became known as the golden age of medical bacteriology. Most of the major bacterial pathogens were isolated during this time. The techniques were fairly basic: Isolate the microorganism, grow it in pure culture, and examine human and microbial cells under the microscope.

Until relatively recently, most of our knowledge about microbial virulence and pathogenesis was based on the same techniques. Clinicians observed the manifestations of a disease. Pathologists examined diseased tissue, grossly and under the microscope. And microbiologists used Koch's postulates, or some of them, to prove that a given microorganism was the etiological agent of a given disease.

Thus basic microbiology explained which microorganisms caused which diseases but—except for identifying a few toxins—little more. The molecular and biochemical mechanisms of pathogenesis were a mystery.

Today the story is different. In the last 25 years, advances in molecular biology and recombinant DNA technology have provided microbiologists with new and amazingly powerful tools. The technological sophistication of biomedical research, including recombinant DNA technology (Chapter 7), has led to major breakthroughs in our understanding of microbial pathogenicity. Geneticists have also discovered ways in which disease-causing genes are regulated. In some organisms, such as *Bordetella pertussis,* genes that encode a set of interacting virulence factors are turned on or off as a group. All the *B. pertussis* virulence factors discussed in this chapter, with the exception of tracheal cytotoxin, are controlled by a single regulatory gene.

Knowing the molecular basis of a pathogenesis helps us develop new ways to prevent and cure illness. For one thing, we can develop better vaccines. For example, the traditional pertussis vaccine was made of whole pertussis cells—a crude preparation that provided good immunological protection against pertussis but also caused side effects such as fever, soreness at the vaccination site, and rare neurological complications, including seizures. An increased understanding of the key pertussis virulence factors allowed researchers to design a more highly purified vaccine that contains no bacterial cells, only the few antigens critical to protection against infection. This new **acellular vaccine,** which has few side effects, has now replaced the whole cell vaccine for infants and children.

The First Golden Age of Medical Microbiology

Date	Disease	Bacterium
1876	Anthrax[a]	*Bacillus anthracis*
1879	Gonorrhea	*Neisseria gonorrhoeae*
1880	Typhoid fever	*Salmonella typhi*
1880	Malaria	*Plasmodium* spp.
1881	Wound sepsis	*Staphylococcus aureus*
1882	Tuberculosis[a]	*Mycobacterium tuberculosis*
1883	Cholera[a]	*Vibrio cholerae*
1883–84	Diphtheria	*Corynebacterium diphtheriae*
1885	Tetanus	*Clostridium tetani*
1885	Diarrhea	*Escherichia coli*

Date	Disease	Bacterium
1886	Pneumonia	*Streptococcus pneumoniae*
1887	Meningitis	*Neisseria meningitidis*
1887	Brucellosis	*Brucella* spp.
1892	Gas gangrene	*Clostridium perfringens*
1894	Plague	*Yersinia pestis*
1896	Botulism	*Clostridium botulinum*
1898	Dysentery	*Shigella dysenteriae*
1905	Syphilis	*Treponema pallidum*
1906	Whooping cough	*Bordetella pertussis*
1909	Rocky Mountain spotted fever	*Rickettsia rickettsii*

[a]Koch discovered these.

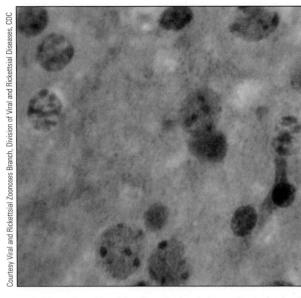

FIGURE 15.13 Negri bodies. Stained tissue from the brain of a dog with rabies.

the surface of host cells, setting them up as targets for destruction by the body's own immunological defenses (Chapter 17).

Finally, certain viruses establish latent infections in human cells that can later cause cancer. We discussed these in Chapter 13.

SEVEN: LEAVING THE BODY

The final link in any chain of infection occurs when the pathogen leaves the body of one host in order to reach another. The anatomical route through which a pathogen usually leaves the body of its host is called its **portal of exit.**

For most respiratory pathogens, the portal of exit is the same as the portal of entry—the nose. The pathogens that cause influenza, tuberculosis, strep throat, meningococcemia, and all types of pneumonia are expelled through respiratory secretions. Being a respiratory pathogen, *Bordetella pertussis* exited this way from Baby Girl B. After the bacterium reached a sufficient concentration in her respiratory tract, bacterial cells were expelled in respiratory droplets every time she coughed, sneezed, or breathed. The greatest concentration of microorganisms existed during the early coldlike phase of her illness (before she was admitted to the hos-

pital). At that stage, she might have transmitted the pathogen to other susceptible children or adults.

For most gastrointestinal pathogens, the portal of entry is the mouth and the portal of exit is the anus. Examples are the bacterial pathogens that cause cholera and typhoid, as well as the many viruses that cause diarrhea. Protozoal causes of diarrhea, including *Giardia lamblia* and *Entamoeba histolytica*, also exit the body through this portal.

Most sexually transmitted pathogens exit the body in the same way they entered—across the mucous membrane surfaces of the genital tract. The pathogens that cause gonorrhea, syphilis, genital herpes, and AIDS all exit this way most of the time.

Pathogens that are transmitted parenterally by arthropod vectors also exit the same way they entered—in a drop of blood. They include the pathogens that cause malaria, yellow fever, and viral encephalitis.

Most microbial pathogens are able to accomplish most of the seven things we've just discussed. We'll return to this list as we discuss the various microbial diseases in Chapters 22 through 27.

EMERGING AND REEMERGING INFECTIOUS DISEASES

Possibly, medical microbiology's greatest accomplishment was the discovery that a particular microorganism causes a particular disease (see Sharper Focus: Koch's Postulates and Larger Field: Another Golden Age?). But the relationships between microorganisms and diseases are not fixed. New infectious diseases suddenly emerge; others become less virulent, seem to disappear, and then sometimes reemerge. The reasons for the comings and goings of particular infectious diseases are variable. Some diseases that previously have gone undetected seem to emerge when they strike a particularly vulnerable population. Some pathogens cross a **species barrier** (become able to infect a different species of plant or animal) to cause devastating diseases of humans. Some harmless microorganisms become deadly pathogens by gaining toxin-encoding genes from another pathogen.

Most experts agree that the rate at which infectious diseases emerge and reemerge has increased in recent years. There are many reasons, but the most important are dramatic increases in human populations, rapid worldwide travel, and environmental changes.

Some emerging and reemerging infectious diseases are listed in **Table 15.8.**

TABLE 15.8 Some Emerging and Reemerging Infectious Diseases

Disease	Cause	Comments	Chapter Reference
Emerging Diseases			
New variant Creutzfeldt-Jacob disease	Prion	A variant of bovine spongiform encephalitis (BSE; mad cow disease) affecting humans, which was first described in the United Kingdom in the 1996. BSE emerged in the 1980s.	25
Ebola hemorrhagic fever	Ebola virus	First outbreaks occurred in Central Africa in 1976.	27
HIV/AIDS	Human immunodeficiency virus	The virus was first isolated in 1983.	27
Hepatitis C	Hepatitis C virus	The virus was first identified in 1989. Now up to 3 percent of the world's population is estimated to be infected.	23
Hantavirus pulmonary syndrome	Sin nombre virus	The virus was first isolated from cases of highly fatal respiratory disease in the Southwestern United States in 1993.	22
Lyme disease	*Borrelia burgdorferi*	This bacterium was detected and identified as being the cause of Lyme disease in 1982.	27
Legionellosis	*Legionella pneumophila*	This bacterium was first identified in 1977 as the cause of a severe pneumonia at a convention center in 1976.	22
Hemolytic uremic syndrome	*Escherichia coli* O157:H7	This bacterium was detected first in 1982. It probably is a strain of *E. coli* that acquired toxin-encoding genes from a strain of *Shigella*.	23
Reemerging Diseases			
Cholera	*Vibrio cholerae*	In 1991 a cholera epidemic of more than 390,000 cases reemerged in South America, where it had not been seen for more than a century.	23
Dengue fever	Dengue fever virus	After a long absence and following a relaxation of active mosquito control, a severe outbreak of dengue fever occurred in Cuba in 1991. Epidemics have since spread to Central and South America.	27
Diphtheria	*Corynebacterium diphtheriae*	An epidemic of diphtheria with more than 50,000 cases occurred in Russia in 1994–1995 following a decline in immunizations.	22
Meningococcal meningitis	*Neisseria meningitidis*	Since the mid-1990s, a new strain of *N. meningitidis* (serogroup A, clone III.1) has spread through sub-Saharan regions of Africa, causing epidemics on an unprecedented scale.	25

Source: World Health Organization: *Emerging and Re-emerging Infectious Diseases*, 1998, Geneva.

SUMMARY

The Seven Capabilities of a Pathogen (pp. 363–364)

1. A successful pathogen must be able to do most or all of the following:
   a. maintain a reservoir
   b. leave its reservoir and enter a host
   c. adhere to the surface of the host
   d. invade the body of the host
   e. evade the body's defenses
   f. multiply within the body
   g. leave the body and return to its reservoir or enter a new host

2. Pathogenesis refers to the process by which a microorganism causes disease. Infection is the growth of a disease-causing microorganism in the body.

One: Maintaining a Reservoir (pp. 364–367)

3. A disease reservoir is a place where pathogens are maintained between infections.

4. A human serving as a reservoir of infection is a carrier. An incubatory carrier is an apparently healthy individual who may be in the earliest, symptomless stages of the disease. A chronic carrier is someone who harbors a pathogen for months or years.

5. Most animal reservoirs provide an environment similar to that of the human host. A human disease caused by a pathogen that maintains an animal reservoir is called a zoonosis.

6. Animal reservoirs can profoundly affect the pattern of human disease, leading in some cases to a pandemic.

7. Some pathogens survive in water, soil, and house dust.

Two: Getting to and Entering a Host (pp. 367–372)

8. Disease transmission takes place when a pathogen leaves a reservoir and enters the body of a host.

9. The portal of entry for a pathogen is where it enters the host's body. The most common portals of entry are the same anatomical surfaces colonized by microbiota.

10. The ID_{50} (infectious dose) is the number of microorganisms that must enter the body to establish infection in 50 percent of test animals. The LD_{50} (lethal dose) is the number of microorganisms that must enter the body to cause death in 50 percent of test animals.

11. Modes of transmission are categorized as contact between humans (direct or indirect), vehicles (inanimate objects), and vectors (living transmitters, usually arthropods).

12. Droplets of respiratory secretions are transmitted directly from one host to another through sneezing, coughing, laughing, or speaking. More human diseases are transmitted this way than by any other.

13. Fomites are inanimate objects such as eating utensils, towels, bedding, and handkerchiefs that transmit disease. Hand washing can break the transmission cycle.

14. Direct body contact takes place by touching, kissing, or sexual intercourse. Sexually transmissible disease (STD) is spread by direct mucous membrane contact.

15. Vertical transmission is transmission of pathogens from mother to infant. It can be prenatal or perinatal (occurring in the birth canal or immediately after birth).

16. In the fecal-oral route, pathogens are transmitted from infected feces to the mouth of a new host. Fecal-oral transmission can be direct hand-to-hand or hand-to-mouth, by vehicles such as water, food, and fomites or by vectors.

17. Arthropods can be mechanical vectors, carrying pathogens on their bodies, or biological vectors, an essential link in the transmission of a disease because the microorganism spends part of its life cycle in the arthropod host.

18. Airborne transmission occurs when microorganisms that can survive in air are inhaled.

19. In parenteral transmission, pathogenic microorganisms are deposited directly into blood vessels or deep tissues. This occurs when a biological vector bites through the skin or when intravenous drug users share needles.

20. Deep wounds can allow anaerobic pathogens such as *Clostridium tetani* (which causes tetanus) to enter.

Three: Adhering to a Body Surface (pp. 372–373)

21. Pathogens—like the body's normal biota—adhere to a body surface by means of adhesins on their pili or surface.

Four: Invading the Body (pp. 373–374)

22. Most pathogens are invasive. They enter host cells or tissues. A few are noninvasive. They remain on the surface.

23. Invasive pathogens that enter host cells to live are called intracellular pathogens. The cell provides a nutrient-rich environment safe from body defenses.

24. In the 1880s Robert Koch proved the germ theory of disease, that a particular microorganism causes a particular infection.

25. We still use Koch's postulates in certain cases today to prove the cause of an infectious disease.
   a. The causative microorganism must be present in every individual with the disease.
   b. The causative microorganism must be isolated and grown in pure culture.
   c. The pure culture must cause the disease when inoculated into an experimental animal.
   d. The causative microorganism must be reisolated from the experimental animal and reidentified in pure culture.

26. We use Rivers' postulates to determine the etiology of viral infections.
   a. The viral agent must be found either in the host's body fluids

at the time of disease or in infected cells.

b. The viral agent obtained by the host must produce the disease in a healthy animal or plant or must produce antibodies.

c. Viral agents from the newly infected animal or plant must in turn transmit disease to another host.

Five: Evading the Body's Defenses (pp. 374–376)

27. Some pathogens have capsules that keep a phagocyte from establishing direct contact. Strains of *Streptococcus pneumoniae* without capsules are avirulent (harmless). Those with the thickest capsules are the most virulent.

28. Surface proteins on pathogens also keep phagocytes from establishing contact. *Streptococcus pyogenes*, which causes strep throat, produces M protein.

29. Some pathogens, such as *Mycobacterium tuberculosis*, survive inside the phagocyte.

30. Pathogens that evade nonspecific body defenses encounter the body's immune defenses. The immune system recognizes pathogens by means of markers on their surface called antigens.

31. Some pathogens change their surface antigens to avoid recognition (antigenic variation). Some attack antibodies directly with enzymes called IgA proteases. Some use serum resistance, features on the bacterial surface that interfere with the host's defensive complement system.

32. Some pathogens obtain iron by producing iron-binding compounds called siderophores.

Six: Multiplying in the Host (pp. 376–383)

33. The two most common forms of bacterial pathogenesis are production of toxins and damage caused by stimulation of the body's defenses.

34. Exotoxins are highly destructive proteins produced by both Gram-positive and Gram-negative bacteria. Most exotoxins are composed of two units, the A (active) unit and the B (binding) unit, and are highly specific.

35. Endotoxin is the lipopolysaccharide (LPS) component of the outer membrane of Gram-negative bacteria. It acts by stimulating human cells to secrete particular messenger proteins. For example, fever results when endotoxin stimulates white blood cells to secrete the protein interleukin.

36. Endotoxin is generally not very potent.

37. Some pathogens produce extracellular enzymes. There are three types: cytolysins attack cell membranes; hemolysins lyse red blood cells; leukocidins lyse leukocytes.

38. The body defenses that fight off pathogens can disrupt body function in ways that cause disease. *Streptococcus pneumoniae*, for example, multiplies in the lungs and summons great numbers of phagocytes. As dead cells of both kinds accumulate, normal gas exchange is impaired and breathing becomes difficult.

39. Some pathogens stimulate hypersensitivity, an exaggerated immune response that causes damage.

40. Some viral infections are cytocidal (they kill cells), whereas others are cytopathic (they damage but do not kill the cell).

41. Lytic infections kill the host cell by lysing it.

42. A persistent viral infection can last for years, producing new virus particles without killing the infected cell. In a latent viral infection, the virus lies dormant within the host cell, not producing new viral particles. Latent infections can last a lifetime and not be damaging unless the virus is reactivated.

43. Inclusion bodies are collections of viral components such as protein and nucleic acid.

44. Oncogenic viruses establish latent infections in human cells that transform the infected cells into cancer cells. Only a few virally caused cancers have been definitely established.

Seven: Leaving the Body (p. 383)

45. The anatomical route through which a pathogen leaves the body of its host is called its portal of exit.

46. For most respiratory pathogens, the portal of exit is the same as the portal of entry, the nose. For most gastrointestinal pathogens, the portal of exit is the anus. Most sexually transmissible diseases exit the same way they entered, through the genital mucous membranes. Pathogens transmitted parenterally by arthropod vectors exit the same way, in a small amount of blood.

REVIEW QUESTIONS

The Seven Capabilities of a Pathogen

1. What are the seven capabilities that most pathogens have?

2. Define pathogenesis and infection.

One: Maintaining a Reservoir

3. What is a disease reservoir? Name three disease reservoirs and describe the kinds of pathogens that exist in each.

4. Distinguish among a carrier, an incubatory carrier, and a chronic carrier.

5. What is a zoonosis? Give an example.

6. Give an example of how an animal reservoir can profoundly affect the pattern of human disease.

Two: Getting to and Entering a Host

7. Define disease transmission. What are the three major modes of transmission?

8. What is a portal of entry? Name the most common portals of entry for pathogens.

9. Explain ID$_{50}$ and LD$_{50}$.

10. Describe how a pathogen is transmitted by each of the following means:
    a. respiratory droplets
    b. fomites
    c. direct body contact
    d. the fecal-oral route
    e. arthropod vectors
    f. air
    g. parenteral means

11. Why are more diseases transmitted by respiratory droplets than by any other mode?

12. Compare and contrast the types of pathogens transmitted by respiratory droplets, fomites, direct body contact, and air.

13. What is an STD? Give an example. Are STDs communicable diseases? Explain.

14. Explain this statement: The fecal-oral route overlaps with other modes of transmission.

15. Name the two types of vertical transmission.

16. What is the difference between a mechanical vector and a biological vector?

17. Why are deep wounds more likely than surface breaks in the skin to lead to lethal infections?

Three: Adhering to a Body Surface

18. Why do pathogens face the same challenge as normal microbiota in maintaining a foothold on body surfaces?

19. Describe how filamentous hemagglutinin functions as an adhesin for *Bordetella pertussis*. Why is it not a typical adhesin?

Four: Invading the Body

20. What is the difference between an invasive pathogen and a noninvasive pathogen?

21. What is an intracellular pathogen? Why does it have an adaptive advantage?

22. Explain the process by which a bacterium invades the body's interior. How does a virus enter? Give some examples showing the variety of eukaryotic modes.

23. Define etiology.

24. What are Koch's postulates and what is their significance?

25. Can Koch's postulates be applied to all bacteria? Explain. Give an example.

26. What are Rivers' postulates?

Five: Evading the Body's Defenses

27. Define phagocytosis. What kinds of cells are phagocytes?

28. Describe two ways pathogens evade phagocytosis.

29. Instead of evading phagocytosis, some pathogens are readily consumed and then survive within the phagocyte. Explain how *Mycobacterium tuberculosis* does this.

30. Pathogens overcome the body's more sophisticated immune defenses by means of antigenic variation, IgA proteases, and serum resistance. Explain each of these.

31. Why must pathogens constantly compete for iron in the human body? Give an example of how some pathogens succeed.

Six: Multiplying in the Host

32. What are the two most common ways pathogens cause disease?

33. What are exotoxins? Describe their structure. What determines if a particular cell will produce an exotoxin?

34. In what ways are exotoxins very specific? Define neurotoxin and enterotoxin, and name a disease caused by each.

35. How is botulism toxin different from pertussis toxin? How are the diseases they cause different?

36. Compare and contrast endotoxin with exotoxin.

37. How does endotoxin act? What is interleukin 1? What is tumor necrosis factor?

38. What are extracellular enzymes? Name the three types. Give an example of each.

39. Explain this statement: The response that pathogens evoke in the body is a form of pathogenesis. Give an example.

40. What is the difference between cytocidal and cytopathic viral pathogenesis?

41. Classify each of the following as cytocidal or cytopathic and then explain the process: lytic infection, persistent infection, latent infection, cancer-causing infection.

42. Define these terms: autolysis, oncogenic virus, inclusion bodies, transformed cell.

Seven: Leaving the Body

43. Define portal of exit. Name the typical portals of exit for respiratory, gastrointestinal, sexually transmissible, and arthropod-transmitted pathogens.

CORRELATION QUESTIONS

1. How would you distinguish between an incubatory carrier and a chronic carrier of typhoid fever.

2. Would it be easier to eradicate a disease with an exclusively human reservoir or one with an animal reservoir? Explain.

3. Why are exotoxins more potent than endotoxins?

4. Assume there were two strains of the same bacterial pathogen that differ with respect to the tissue they infect. What molecular difference would you expect to find between the two strains? Explain.

5. Infection by *Bordetella pertussis* affects the throat and other parts on the body as well. Infection by *Streptococcus pneumoniae* affects the lungs

only. What is the basis for this difference? Explain.

6. How can being phagocytized be an advantage to a pathogen?

ESSAY QUESTIONS

1. Trace *Bordetella pertussis*—or another pathogen of your choice—through the seven challenges of infecting a human host.

2. From time to time, there is an outbreak of disease of unknown origin. In 1976, for example, large numbers of war veterans attending a convention in Philadelphia came down with a mysterious illness; 29 died. How would you have applied Koch's postulates to identify the causative microorganism for what we now call legionnaires' disease?

SUGGESTED READINGS

Finlay, B. B., and S. Falkow. 1997. Common themes in microbial pathogenicity revisited. *Microbiological Reviews* 61:136–69.

Flint, S. J., L. W. Enquist, R. M. Krug, V. R. Ranconiello, and A. M. Skalka, 1999. *Principles of virology—molecular biology, pathogenesis, and control.* Washington, D. C.: ASM Press.

Miller, V. L., J. B. Kaper, D. A. Portnoy, and R. Isberg. 1994. *Molecular genetics of bacterial pathogenesis.* Washington, D. C.: ASM Press.

Roth, J. A. , C. A. Bolin, K. A. Brogden, F. C. Minion, and M. J. Wannemuehler. 1995. *Virulence mechanisms of bacterial pathogens.* Washington, D. C.: ASM Press.

Salyers, A. A., and D. D. Whitt. 2001. *Bacterial pathogenesis.* 2d ed. Washington, D. C.: ASM Press.

Zur Hausen, H. 1991. Viruses in human cancers. *Science* 254:1167–72.

For additional readings, go to InfoTrac College Edition, your online research library at: http://www.infotrac.thomsonlearning.com

SIXTEEN

The Immune System: Innate Immunity

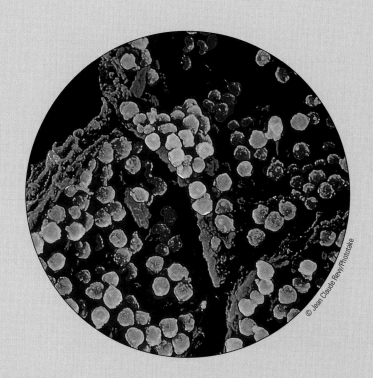

© Jean Claude Revy/Phototake

CHAPTER OUTLINE

LEARNING GOALS

To understand:

- *The three lines of defense against microbial infection*
- *The steps in the inflammatory response and how inflammation activates and coordinates the body's nonspecific defenses*

- *The different types of leukocytes (white blood cells) and how they contribute to the body's defenses*
- *The steps in phagocytosis and its central role in the body's nonspecific internal defenses*

- *The complement system, including the complement cascade and the classical, alternate, and terminal pathways*
- *Interferon, particularly its role in defending against viral infection*

T. L.'s Close Call: Part One

T. L., a college student, was backpacking in a remote wilderness region. While pitching a tent, he tripped and fell toward the branch of a dead tree. In an attempt to break his fall, he extended his arm and sustained a puncture wound to his right palm. Although the wound was painful and bled for a short time, it did not appear to be serious. He fell asleep that night unconcerned about his condition.

The next morning, T. L. noticed that the tissues immediately surrounding his wound were red, swollen, and warm. A round area about 1 inch in diameter clearly looked abnormal when compared with the rest of his hand. The affected part of T. L.'s hand was painful, especially when he touched or bumped it. That afternoon, after 8 hours of hiking, the sore hand was even more painful. The red area had doubled in size, and a thick yellow discharge oozed from the open wound. T. L. felt unusually tired, his body ached, and a brief chill made him aware of a developing fever. His companions helped to elevate his arm and apply warm compresses to his palm, hoping that he would feel well enough by the next day to continue their trip. (Continued in Chapter 17.)

Case Connection

■ T. L. experienced a number of symptoms as a result of his injury. Keep them in mind because they are consequences of the innate immune system's response to injury. Each of them will be considered in this chapter.

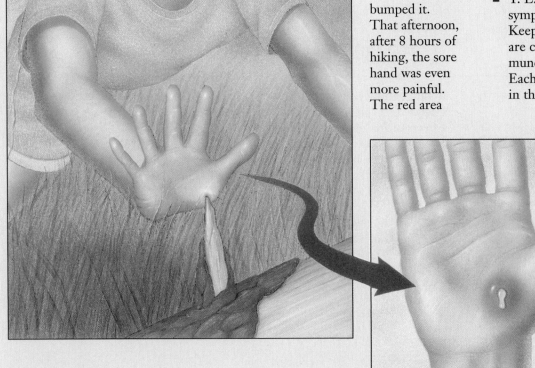

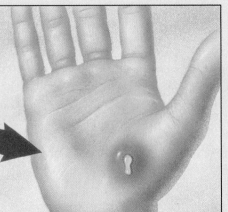

Puncture wounds can cause life-threatening infections.

(Art by Carlyn Iverson.)

THE BODY'S THREE LINES OF DEFENSE AGAINST INFECTION

All of us come into daily contact with a multitude of microorganisms. Most of them are harmless or even beneficial (Chapter 14). But a few, if they invade us, can cause disease and threaten our lives (Chapter 15).

We have three lines of defense against such invading microorganisms (**Figure 16.1**). The first consists of surface defenses—structural, mechanical, and biochemical—that prevent microorganisms from entering our bodies (Chapter 14). If this line is penetrated, as it was when T. L. suffered his puncture wound, our immune system swings into action. Almost immediately one part of it, our **innate immune system,** attacks the invaders, usually eliminating them. But if

SHARPER FOCUS

FEVER

Within hours of his accident, T. L. began to run a fever. Normally the human body is maintained at a relatively constant 37°C (98.6°F) by a regulatory center in the brain called the anterior hypothalamus. This region functions like a thermostat, registering body temperature and modifying the production or loss of heat to maintain temperature at the normal set point. Body temperature can be regulated by simple behavioral responses such as putting on a sweater or by complex physiological processes such as sweating (which dissipates heat) or shivering (which produces heat). Another physiological determinant of body temperature is the amount of blood flowing through surface skin vessels. When **vasodilation** enlarges blood vessels and shunts large amounts of blood through surface vessels, heat is lost from the body and the skin appears flushed. When **vasoconstriction** (the narrowing of blood vessels) shunts blood away from the surface, heat is conserved within the body and the skin feels cold and clammy.

During fever, the body's normal temperature set point is adjusted slightly upward. At the onset of fever, shivering and vasoconstriction elevate body temperature. While fever persists, an increased rate of metabolism maintains the elevated temperature. As fever resolves, vasodilation and sweating allow body temperature to return to normal.

What causes the abnormal elevation of the temperature set point? Certain microorganisms, including bacteria and viruses, produce substances that can cause fever. Collectively these substances are called **exogenous pyrogens** (**exogenous,** "from the outside"; **pyrogens,** "fever-generating"). One well-known exogenous pyrogen is lipid A, a component of endotoxin (Chapter 15). Exogenous pyrogens cause phagocytes to produce a protein called **endogenous pyrogen** (**endogenous,** "from the inside")—also known as interleukin-1 (see Chapter 15). Endogenous pyrogen stimulates the anterior hypothalamus to produce prostaglandins, causing the change in body temperature that we recognize as fever.

Some of the most uncomfortable sensations of illness—chills, shivering, sweating, and a feeling of being overheated—are attributable to fever. Antipyretic medications such as aspirin and acetaminophen reduce fever by interfering with the last step in the fever-generating process—the synthesis of prostaglandins. The ill person usually feels much better, although the medication has no effect on the underlying infection. Medication to control fever is an example of treating the symptom rather than the disease.

People sometimes worry that fever will harm them—that the brain or some other part of the body will be permanently damaged by becoming too hot. Greatly elevated body temperature can indeed be quite harmful. Temperatures above 41°C, or 105.8°F, can cause permanent tissue damage, and people rarely survive temperatures above 45°C, or 113°F. But such high temperatures virtually never occur as a result of fever. They usually occur from **heatstroke,** environmental conditions (such as high heat, high humidity, and exercise) that overwhelm the body's normal temperature-regulating mechanisms. It is exceedingly rare for fever to elevate body temperature above 41°C. If you have a high fever, you should get medical attention to rule out the possibility of a potentially serious infection, such as meningitis or pneumonia, but the fever itself is seldom dangerous.

Although fever is rarely harmful, is it useful? Does fever help fight infection? No experimental evidence definitely establishes that people with fever survive infection better than people without it, but several factors suggest that fever may serve a useful purpose. Certain defensive cells, including phagocytes and cells of the specific immune system, are more active at higher temperature. Human leukocytes display maximum phagocytic activity at 38° to 40°C. Fever also decreases the concentration of iron circulating in the blood, which may restrict microbial growth. Some microorganisms grow more slowly at temperatures above 37°C, and fever may help bring these infections under control. Finally, many lower vertebrates, including reptiles and fish, have elevated body temperatures when infected, suggesting that the elevation of body temperature during infection is an ancient evolutionary adaptation; fever in humans probably confers a selective advantage.

it, too, is overwhelmed, our third and last line of defense, the **adaptive immune system,** is activated. But unless we have been previously exposed to the invader, our adaptive immune system is activated only days or weeks later.

The two parts of our immune system, innate and adaptive, differ in the ways they recognize microorganisms, but then both systems use the same means to destroy them (**Table 16.1**). Until recently, the innate immune system

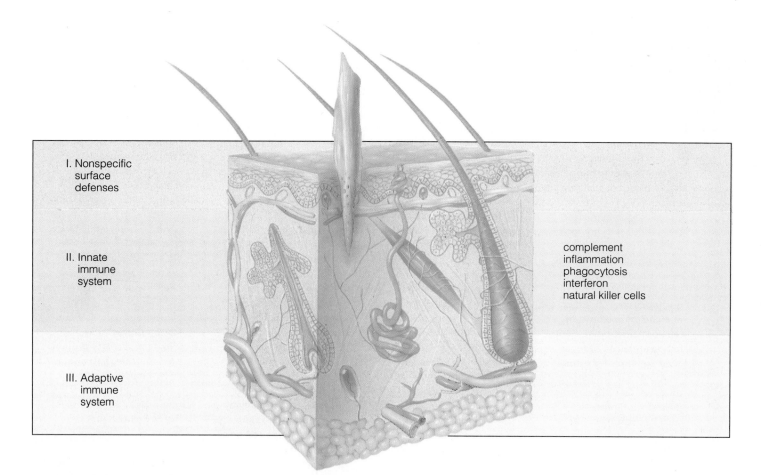

FIGURE 16.1   Lines of defense against infection. Although the three lines of defense are covered in separate chapters—14, 16, and 17—they are highly interrelated. The second line of defense includes five components of the innate immune system that immediately attack any invader that reaches the body's interior—complement, inflammation, phagocytosis, and, in the case of viruses, interferon and natural killer cells. These defenses often stop infection before symptoms appear.

(Art by Carlyn Iverson.)

TABLE 16.1   Characteristics of Innate and Adaptive Immunity

Characteristic	Innate Immunity	Adaptive Immunity
Speed of response	Rapid, begins to act within a few hours	Slow, takes days or weeks to act
Distribution in nature	Present in all animals	Present only in vertebrates
Variation over time	Invariant, an inherited capability that doesn't change over our lifetime	Changes with our exposure to microorganisms and foreign substances
Means of recognizing pathogens	Recognizes classes of molecules shared by groups of microorganisms (e.g., peptidoglycan)	Recognizes unique molecules on particular pathogens (e.g., cholera enterotoxin)
Numbers of molecules recognized	Only a few	An almost infinite number
Change during response	Constant	Improves during response

was referred to as the **nonspecific interior defenses,** the term immune system being reserved for what is now called the adaptive immune system. But the many interconnections between the two defenses make it clear that they are two parts of the same immune system. They evolved to work together, but the innate system evolved first. Animals other than vertebrates lack an adaptive immune system. They rely completely on their innate immune system and do quite well. They are highly resistant to microbial attack.

We will discuss the innate immune system in this chapter and the adaptive immune system in the next (Chapter 17).

THE INNATE IMMUNE SYSTEM

Both the innate and the adaptive parts of the immune system depend upon the actions of certain **leukocytes** (white blood cells) and proteins present in **plasma** (the noncellular fraction of blood). There are about 10 kinds of leukocytes, and more than that number of proteins. All leukocytes are formed in the bone marrow. At first they are much alike; then, as they differentiate and mature, they take on the various sizes, morphologies, and functions that distinguish them (**Table 16.2**).

The first response of the innate immune system to invading microorganisms is the activation of **complement,** a set of microbe-attacking proteins in blood, lymph, and extracellular fluids. The first effective attack on the invaders is carried out by **phagocytes,** a class of leukocytes that destroy pathogens by **phagocytosis.** We'll discuss complement first, then phagocytes, and finally how they interact.

Activation of Complement

When activated, complement plays several important roles in both the innate and adaptive immune responses. Activation of complement in the innate response proceeds through the **alternative pathway (Figure 16.2).**

The first step of the alternate pathway is the spontaneous **hydrolysis** (splitting with water) of a complement component designated **C3.** Such hydrolysis, which produces **C3(H$_2$O),** occurs constantly, but it speeds up dramatically in the presence of invading microorganisms. As shown in Figure 16.2, interactions between C3(H$_2$O) and two other proteins (**factors B and D**) lead to the splitting of C3 into two fragments designated, C3a and C3b. As we'll soon see, C3b plays several critical roles in the immune response. Most importantly, it catalyzes the splitting of more molecules of C3 to form more molecules

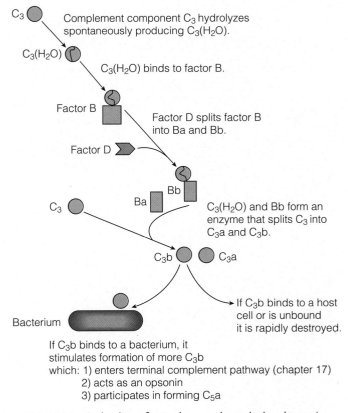

FIGURE 16.2   Activation of complement through the alternative pathway.

of C3b, *but only if invading microbes are present.* Normally the C3b molecules that form spontaneously are quickly destroyed (hydrolyzed). But when microbial cells are present, C3b is stabilized by binding to their surfaces. There it interacts with other complement proteins to form **C3 convertase,** a powerful enzyme that rapidly forms many more molecules of C3b. (C3b also binds to host cells, but they have the capacity to displace it, so C3 convertase doesn't form on host cells.)

Critical Roles of C3b.   C3b has three critical roles in the immune response:

1. C3 enters the **terminal complement pathway,** which leads to the formation of the **membrane-attack complex.** This complex kills microbial invaders by making holes in their membranes. We'll discuss the terminal complement pathway and the membrane-attack complex formed by it in Chapter 17.

2. C3 acts as an **opsonin,** molecules that facilitate phagocytosis, as we'll see in the next section.

3. C3 participates in forming C5a (Figure 16.2), which attracts phagocytes.

TABLE 16.2 Leukocytes: Their Properties and Roles in Immune Responses

Type	Class Name(s)	Role	Participate in Innate Immune System?	Other Properties
Macrophage	Phagocyte	Traps invading microorganisms by phagocytosis; kills them when phagosome fuses with lysosome initiates inflammation	Yes	Large cells located in tissues
Neutrophil	Phagocyte and granulocyte	Traps invading microorganisms by phagocytosis; kill them with toxic agents in its granules	Yes	Smaller than macrophages; not resident in tissues but enter them when infected
Monocyte		Circulates in blood and give rise to macrophages	Yes	
Eosinophil	Granulocyte, polymorphonuclear leukocyte	Protect against parasites such as helminthic worms		
Basophil	Granulocyte, polymorphonuclear leukocyte	Unknown		
Mast cell	Granulocyte	Protect against parasites such as helminthic worms		Located in connective tissues
Natural killer cell		Kills host cells that contain intracellular pathogens	Yes	

TABLE 16.2 Leukocytes: Their Properties and Roles in Immune Responses (continued)

Type	Class Name(s)	Role	Participate in Innate Immune System?	Other Properties
B cell	Small lymphocyte	When activated, differentiates into plasma cells		
Plasma cell	Lymphocyte	Makes antibodies		
T cell	Small lymphocyte	Some kill infected host cells; others regulate adaptive immune response		
Dendritic cell		Stimulates B cells to differentiate		Star-shaped cell located in many tissues, as well as blood

Phagocytes and Phagocytosis

When phagocytes recognize pathogens as invaders, they engulf and destroy them (**Figure 16.3**). Finally phagocytes expel the resulting bacterial debris. Minor invasions by microorganisms are relative common, so phagocytes are called on often, probably almost daily, particularly in areas of the body such as mouth tissues, which are bathed in microorganisms. Phagocytes are our principal means of eliminating pathogens, and they are usually successful, eliminating all invaders.

We have two kinds of phagocytes: **macrophages** and **neutrophils.** Macrophages are the long-lived, mature form of **monocytes** that have left the bloodstream and taken up residence in tissues. Neutrophils continue to circulate in the blood, but they too enter tissues that become infected. Neutrophils are most abundant white blood cells we have, but they are short lived. After entering infected tissue they live only a few hours.

Recognition. **Recognition,** the first and critical step of phagocytosis, is the ability of phagocytes to direct their killing power exclusively toward invading microorganisms, not other host cells. Phagocytes can do this because on their outer surface they bear proteins termed **cell-surface**

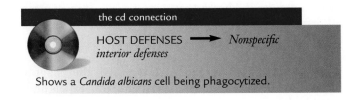

the cd connection

HOST DEFENSES ⟶ *Nonspecific interior defenses*

Shows a *Candida albicans* cell being phagocytized.

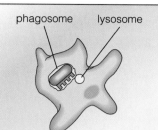
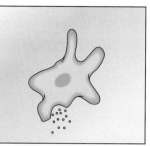

a **Recognition.** The phagocyte recognizes a bacterium as an invader and binds to it, aided by an opsonin (see Figure 16.4).

Engulfment. Attacking phagocyte extends pseudopods which surround the bacterium.

Destruction of engulfed bacterium. When tips of the phagocyte fuse, the bacterium is trapped within a membrane-bound phagosome. The bacterium is destroyed by chemicals within the lysosome, which fuses with the phagosome.

Expulsion. Undigested bacterial debris is expelled by the phagocyte.

FIGURE 16.3   Phagocytosis. (a) Steps in the process. (b) Scanning electron micrograph of a phagocyte (a macrophage) ingesting a yeast cell.

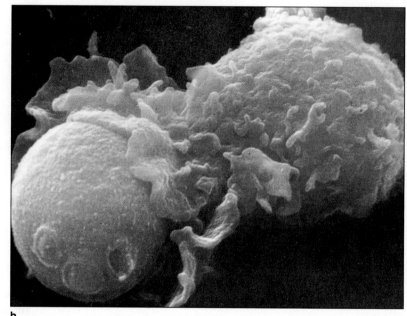

b

receptors. These proteins can bind to molecules on the surface of microbial cells that are not present on host cells. In this way phagocytes and (as we'll see later) protein components of the innate immune system are able to recognize groups of microorganisms (**Table 16.3**). For example, one cell-surface receptor recognizes Gram-negative pathogens by binding to lipopolysaccharide (LPS) on their outer surface. Another can bind to peptidoglycan, which is exposed on the outer surface of Gram-positive bacteria.

Attachment of a phagocyte to an invading microorganism is facilitated by C3b and other opsonins. They form a bridge between the phagocyte and its microbial prey (**Figure 16.4**).

Engulfment. Once attached to a bacterial cell, the attacking phagocyte extends **pseudopods** (armlike extensions) that engulf it. When the tips of the pseudopods meet, they fuse. This forms a **phagosome,** a membrane-bound vacuole within the phagocyte. The bacterial cell is trapped inside the phagosome.

Destruction of Engulfed Bacteria. Normally a phagosome quickly fuses with a **lysosome** (a membrane-bound package of lethal chemicals) in a macrophage (granules within neutrophils play a similar role), forming a **phagolysosome.** Then chemicals—including lysozyme, lactic acid, nitric oxide (NO), and other enzymes and **oxidants** (chemicals that oxidize vital biochemicals)—from the lysosome attack the bacterium. Working together, these agents kill

most bacteria in 30 minutes or less. Finally, other enzymes digest almost all the bacterial remains.

Although phagocytosis is extremely effective in destroying most pathogens, including those that infected T. L.'s hand, some pathogens have evolved ways to survive, and even multiply, within phagocytes (Chapter 15). One such bacterium is *Mycobacterium tuberculosis.* It must establish itself within phagocytes in order to cause tuberculosis (Chapters 15 and 22).

Expulsion. After a bacterium is digested, the phagolysosome fuses with the cell membrane and expels indigestible debris.

SHARPER FOCUS

WHEN PHAGOCYTES DON'T WORK

Chronic granuloma-tous disease (CGD) is an inherited disorder in which phagocytes are de-fective. The patient's phagocytes are normal looking; they can ingest mi-

croorganisms, but they can't kill them. Patients with CGD suffer greatly from this failure of phago-cytosis. Most develop seri-ous and recurrent bacterial infections before the age

of 2. They often develop multiple abscesses and large, infected lymph nodes that must be opened surgi-cally before they will heal. They suffer repeated episodes of pneumonia, of-

ten caused by pathogens that are not usually viru-lent. The average life ex-pectancy of a child with CGD is 10 years. Survival depends on phagocytosis.

TABLE 16.3 Some Components of Pathogens That Activate Innate Immunity

Component	Pathogen(s) Recognized	Comments
Lipopolysaccharide (LPS)	Gram-negative bacteria	Called endotoxin because of toxic aspects of innate immune response
Peptidoglycan	Gram-positive bacteria	Although all bacteria except mycoplasmas contain peptidoglycan, it is on outer surface of Gram-positives
Lipoteichoic acid	Gram-positive bacteria	Component of Gram-positive wall
Lipoarabinomannan	Mycobacteria	Component of mycobacterial wall
Bacterial DNA	Many bacteria pathogens	DNA sequences that contain unmethylated adjacent Cs and Gs
Lipopeptides	Most bacteria	Proteins that link wall and cytoplasmic membrane
Zymosan	Yeasts	Component of cell wall of yeasts

Inflammation

Soon after bacteria enter our tissues and begin to multiply, macrophages waiting there recognize them as invaders. Then, with the help of C3b, the macrophages phagocytose them. As they degrade these bacteria, cellular components are released, which stimulate the macrophage to release compounds termed **cytokines.** Cytokines signal other components of the immune system to respond to the invaders, organizing a coordinated defense (**Table 16.4**). All the cytokines released by macrophages exert local effect on the infected tissue, and some enter the bloodstream to exert systemic effects. Certain cytokines alert the adaptive immune system. Others coordinate and expand the antipathogen activities of the innate immune system. Among these is developing a state of **inflammation,** a condition in

which fluid accumulates within the affected tissues. Such accumulation causes swelling, redness, and pain—the symptoms T. L. experienced the morning after his accident. The cytokines that cause inflammation are termed **inflammatory mediators.** But they are not the only ones. Other inflammatory mediators are listed and described in **Table 16.5.** Some of these are released, like cytokines, by phagocytosing macrophages. Others are released by **mast cells** (Table 16.2). Mast cells are attracted to the site of inflammation by complement fragments C3a and C5a and stimulated to **degranulate** (break down their internal granules), thereby releasing more inflammatory mediators.

Bacterial infection is not the only cause of inflammation. Any event that damages tissue can cause inflammation, but the most common ones are injury and infection. (T. L. suffered both of them.) They produce **acute**

COMPLEMENT OUT OF CONTROL

Complement out of control can be nearly as serious as an inadequate complement defense. Hereditary angioedema, in which patients lack the ability to produce a normal C1 inhibitor, is one such disorder. With-

out this inhibitor, uncontrolled C1 activity leads to the splitting of C2 and the release of a protein fragment that dilates nearby blood vessels and causes tissue swelling. When C1 is activated by injury, vigorous exercise, or emotional

stress, swelling occurs very rapidly. In some areas, such as the arms or legs, edema may not be much of a problem, but swelling of the intestines can cause severe cramps, and swelling near the throat can cause death from suffoca-

tion. For some reason, patients with hereditary angioedema rarely experience problems until they are older children or teenagers. The attacks last 2 or 3 days when they do occur.

FIGURE 16.4   Opsonization. (a) Scanning electron micrograph of a phagocyte recognizing and adhering to several clumps of bacterial cells. (b) Schematic diagram of how an opsonin aids the process by forming a bridge between the two cells.

inflammation, the process we'll be discussing here. Another form of inflammation, called **chronic inflammation,** is produced by disordered immune responses or other disease processes. It's less dramatic than acute inflammation, but it's much more destructive. We'll discuss chronic inflammation in Chapter 18.

Effects of Inflammatory Mediators

The primary effects of inflammatory mediators are to: (1) stimulate production of other inflammatory mediators; (2) stimulate nerve endings, causing pain; (3) alter capillaries; and (4) attract and stimulate phagocytes. Now let's consider these last two effects in a little more detail.

Altering Capillaries. When capillaries dilate, the blood supply to inflamed tissues increases. The resulting greater flow of blood carries more infection-fighting weapons, including leukocytes and serum proteins to the threatened region. The increased blood flow also causes **erythema** (redness) and warming of this tissue. Erythema is usually the most obvious indication of inflammation of the skin, but warming is also a useful clinical sign. To diagnose inflammation, clinicians often touch the affected area and compare its temperature with that of normal skin.

The traditional home remedy for infection, warm compresses, also augments blood flow to the area. Such

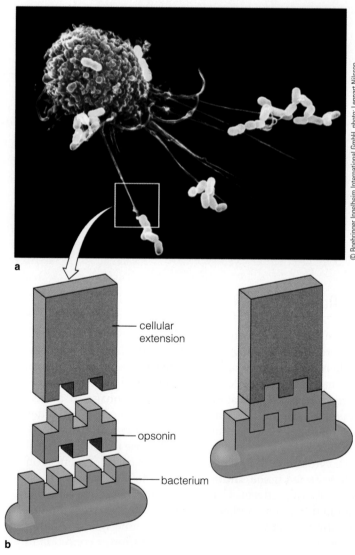

cellular extension

opsonin

bacterium

a

b

© Boehringer Ingelheim International GmbH, photo: Lennart Nilsson

TABLE 16.4 Action of Certain Cytokines Released by Macrophages

Cytokine	Local Effects[a]	Systemic Effects
Interleukin-1	Tissue destruction; increases access of other leukocytes	Fever
Interleukin-6	Stimulates adaptive immune response (antibody production; Chapter 17)	Fever
Interleukin-8	Chemotactic factor; attracts leukocytes including neutrophils to infected area	
Interleukin-12	Activates NK cells; also induces CD4 T cells to differentiate (Chapter 17)	
Tumor necrosis factor-alpha (TNF-α)	Increases permeability of blood capillaries in infected area	Fever; shock

[a]Most cytokines exert multiple effects; only the most relevant are listed here.

TABLE 16.5 Inflammatory Mediators[a]

Inflammatory Mediator	Source/Identity	Activity
Histamine	Degranulation of mast cells/a small molecule derived from the amino acid histidine	Causes vasodilation and increased blood vessel permeability; stimulates nerve endings, causing pain
Complement fragments	Formed during the activation of the complement system/small proteins	Cause vasodilation; activate and attracts leukocytes
Kinins	Formed as blood clots/a family of short peptides	Cause vasodilation and increased blood vessel permeability; stimulate nerve endings causing pain
Bacterial by-products	By-products of the breakdown of bacteria by phagocytes/for example, the short peptide formyl-methionyl-leucyl-phenylalanine	Attract leukocytes to damaged area and activates them
Prostaglandins	Formed in mast cells from arachidonic acid, a fatty acid component of the cell membrane/small molecules	Cause vasodilation and increased blood vessel permeability[b]
Leukotrienes	Formed in mast cells from arachidonic acid, a fatty acid component of the cell membrane/small molecules	Have the same activity as histidine but are 100 times more active

[a]Some cytokines also act as inflammatory mediators; see Table 16.4.
[b]Aspirin reduces inflammation by inhibiting the synthesis of prostaglandin.

warmth also enlarges blood vessels. Although they didn't stop T. L.'s infection, the warm compresses his friends applied to his hand probably helped.

The other effect of inflammatory mediators on capillaries, increased permeability, allows infection-fighting leukocytes and serum proteins to enter the affected tissue. Normally the endothelial cells that line blood vessels make near-leakproof contact with each other, but inflammatory mediators cause them to separate. And as we've noted, the accompanying fluid causes swelling

SHARPER FOCUS

BLOOD: A COMPLEX BODY TISSUE

Although it looks homogeneous, blood is in fact a complex body tissue. It's composed of many different types of cells and molecules, including water, and its composition changes with our state of health. Therefore it is important, clinically, to be able to detect these changes. One way to do this is to spin a sample of it in a centrifuge—an instrument that separates materials by density, size, and shape. Centrifuging a sample of blood is a routine part of diagnosing many illnesses. The procedure is quick and simple. The patient's finger is punctured with a small lance, and blood is collected in a tube that has been treated to prevent clotting. The tube is centrifuged for 5 minutes and examined.

The upper half of the tube is filled with clear liquid. This is plasma, the fluid component of blood. Plasma contains no cells, but it does contain many different proteins. These include complement (which is discussed in this chapter), **albumin** (which increases the osmotic strength of blood and helps keep it in the circulatory system), **globulins** (which include the antibodies), and **fibrinogen** (one of the proteins that allows blood to clot). Because plasma contains clotting proteins, it will clot if allowed to sit outside the body. The fluid left after the clot is removed from plasma is called **serum.** In addition to proteins, plasma also contains various ions, sugars, lipids, amino acids, hormones, vitamins, and dissolved gases. Many of the substances in plasma are being transported from one part of the body to another.

Below the plasma is a barely discernible band of whitish material called the **buffy coat.** This small but important component of blood is made up of leukocytes. **Platelets,** subcellular fragments that participate in blood clotting, are also found in the buffy coat.

At the bottom of the centrifuged tube is a long red column made up of **erythrocytes** (red blood cells). They contain the iron-containing pigment hemoglobin, which binds oxygen and gives blood its color. Erythrocytes transport oxygen and carbon dioxide between the lungs and other tissues. The erythrocyte volume, measured as a percentage of the total blood volume, is called the blood **hematocrit.** Normally it is around 45 percent.

Usually blood is centrifuged to get a quick reading of the patient's hematocrit. A decrease in erythrocyte volume, called **anemia,** is characteristic of many different disorders, including iron deficiency and certain deficiencies of vitamins, such as folate and vitamin B_{12}. But examination of a spun hematocrit tube can also provide clues to other diagnoses, often before the results of more complicated laboratory tests can be obtained. For example, the plasma of patients with jaundice is yellow, whereas the plasma of people with abnormally high levels of fat in their blood has a milky appearance. Patients with leukemia, whose leukocyte count is often very high, may have an unusually thick buffy coat layer. People suffering from dehydration have an unusually wide erythrocyte band and a diminished plasma layer.

and pain. By raising T. L.'s arm, his friends minimized the swelling of his inflamed hand and thereby decreased his pain.

Attracting and Stimulating Phagocytes. Phagocytes are normally inactive. They become activated when inflammatory mediators (and other agents) come in contact with their cell membrane. Then they become powerful destroyers of microbial cells. In addition to making phagocytes active, inflammatory mediators attract more of them to the site of the battle against infection. In other words, they are **chemotactic.** Phagocytes move toward them along a concentration gradient (Chapter 4). Inflammatory mediators attract

BLOOD: A COMPLEX BODY TISSUE (continued)

Components	Relative Amounts (% of plasma volume)	Functions
Plasma represents 50–60% of total volume		
Water	91–92%	Solvent
Plasma proteins (albumin, globulins, fibrinogen, etc.)	7–8%	Defense, clotting, lipid transport, roles in extracellular fluid volume, etc.
Ions, sugars, lipids, amino acids, hormones, vitamins, dissolved gases	1–2%	Roles in extracellular fluid volume, pH, etc.

Components	Relative Amounts (no. per μl)	Functions
Buffy coat cells represent <1% of total volume		
Leukocytes		
Neutrophils	3,000–6,750	Phagocytosis
Lymphocytes	1,000–2,700	Immunity
Monocytes (macrophages)	150–720	Phagocytosis
Eosinophils	100–360	Roles in inflammatory response, phagocytosis, and immunity
Basophils	25–90	Roles in inflammatory response
Platelets (cell fragments)	250,000–300,000	Roles in clotting

Components	Relative Amounts (no. per μl)	Functions
Erythrocytes represent 40–50% of total volume	4,500,000–5,500,000	O_2, CO_2 transport

plasma

buffy coat

erythrocytes

$$\frac{a}{b} \times 100 = \text{hematocrit \%}$$

b

a

material to seal tube and prevent blood from spilling

Span hematocrit tube and components of blood.

them in such a way that each type arrives at the correct time to fulfill its special role. The first to arrive consume and destroy invading bacteria. Those that arrive later remove debris and thereby help repair the damaged tissue.

Phagocytes travel through blood vessels to the site of infection. When they reach it, they migrate to the capillary wall and adhere to it, a process called **margination.** Then they squeeze through gaps in the blood vessel's endothelial lining, dissolve the basement membrane layer that normally surrounds all blood vessels, and enter tissues, a process called **diapedesis.** Together, chemotaxis, margination, and diapedesis bring the phagocytes to where they are needed.

Acute Inflammation

The pathway from inflammatory stimulus through inflammatory mediators and stimulated capillaries and phagocytes leads to acute inflammation of the sort that occurred in T. L.'s hand (**Figure 16.5**). As we've seen, inflammation can be painful, debilitating, and an effective way to fight infection.

Inflammation also damages human tissue. The phagocytes and complement that kill microorganism also kill human cells. As a result, the site of an infection such as T. L.'s becomes filled with a mixture of dead microorganisms, leukocytes, microorganisms, and tissue cells. This mixture is a yellow fluid called **pus.** It often oozes from a contaminated wound such as T. L.'s. Eventually this debris of battle is cleared away by a process called **inflammatory repair.**

Inflammatory Repair

The first step in repair is clean up. Macrophages consume and destroy the dead microorganisms, the dead and dying host cells, and any foreign particles that may have been introduced into the wound. If the inflammation was severe and large quantities of pus accumulated, repair may take a long time. As we'll see in Chapter 17, T. L.'s hand didn't return to normal for 2 weeks. Then after the debris has been cleared away, cells in the affected tissues regenerate and finally heal. In many cases, such as T. L.'s, repair is complete. Permanent damage may result, however, if tissue damage is extensive enough to cause scarring. And certain tissues, including nervous tissue, can't regenerate. That's why infection of the nervous system can lead to permanent brain damage.

Now we'll consider our innate immune system's defense against viral infection.

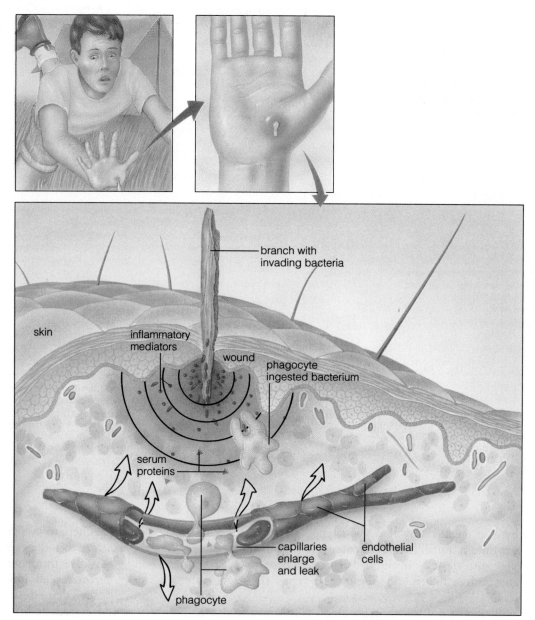

FIGURE 16.5 Acute inflammation. In T. L.'s case the stimulus was a penetrating wound caused by a bacteria-laden branch of a tree. Inflammatory mediators were produced at the site of the stimulus. They caused blood vessels to dilate and increase their permeability. They also attracted phagocytes to the site and activated them. Serum leaked into the damaged tissue. Phagocytes are seen leaving the capillary by diapedesis.

(Art by Carlyn Iverson.)

DEFENSE AGAINST VIRAL INFECTIONS

The innate immune system has two defenses against viral infections: interferons and the lymphocytes termed natural killer (NK) cells.

Interferons

Interferons are a group of small proteins that "interfere" with viral replication. These natural antiviral agents are startlingly potent. They impair viral replication at concentrations as low as 3×10^{-14} M—or approximately 1 molecule for every 2000 trillion water molecules.

There are three main classes of interferons, called alpha, beta, and gamma. **Alpha interferon** and **gamma interferon** are produced by a type of lymphocyte called T cells (Chapter 17). Beta interferon is produced by **fibroblasts** (a type of tissue cell). There are many different interferons—13 alphas, 5 betas, and an unknown number of gammas. Interferons are distinguished by chemistry and genetics, rather than function. In general, the most powerful antivirals belong to the alpha and beta groups. The most powerful immune regulators (Chapter 17) belong to the gamma group.

Cells that have been infected by a virus are stimulated to produce and release interferons. They are then taken up by healthy neighboring cells, causing them to produce antiviral proteins (AVPs). AVPs are enzymes that interfere with viral protein synthesis and thereby stem the spread of infection (**Figure 16.6**). Most types of viruses stimulate cells to produce interferon. So do some species of bacteria and protozoa, double-stranded RNA, and endotoxins.

Interferons are virus nonspecific. That is, the same interferon is active against many different viruses.

At one time, scientists hoped that interferon might be used to treat diseases, including viral disease and maybe even cancer. Enthusiasm soared when it became possible to produce large quantities of pure human interferon using recombinant DNA technology. However, the results were not spectacular. They were ineffective against many cancers and viral diseases, and they were highly toxic. They have, however, proven useful in treating certain rare leukemias (cancers of leukocytes) and some chronic viral infections, including hepatitis and herpes. In the future they might find other applications, either alone or in combination with other agents.

Interferons do more than defend against viral infection. They also help regulate many cell functions, including cell motility, cell division, activation of macrophages, and transplant tissue rejection.

Natural Killer Cells

Natural killer (NK) cells fight viral and other intracellular infections by killing infected host cells, thereby stopping multiplication of the pathogen. Natural killer cells are

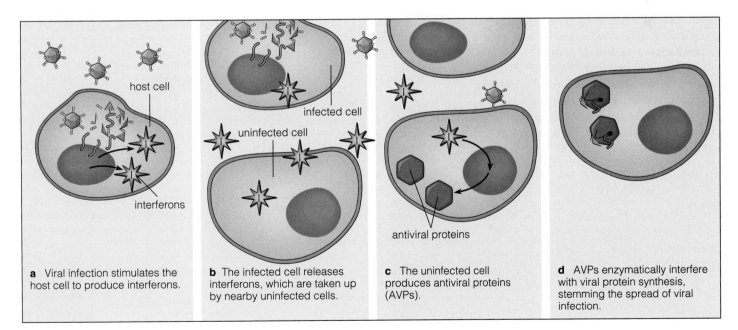

a Viral infection stimulates the host cell to produce interferons.

b The infected cell releases interferons, which are taken up by nearby uninfected cells.

c The uninfected cell produces antiviral proteins (AVPs).

d AVPs enzymatically interfere with viral protein synthesis, stemming the spread of viral infection.

FIGURE 16.6 Antiviral action of interferon.

FIGURE 16.7 How NK cells are able to identify virus-infected host cells and kill them.

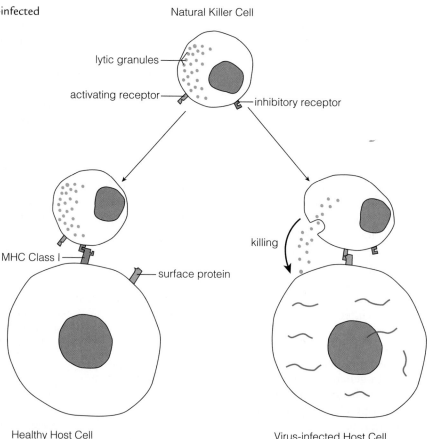

leukocytes (Table 16.2) that normally circulate in the blood. They migrate from the blood into infected tissues, as phagocytes do, in response to inflammatory cytokines. They are immediately able to kill infected cells, but their killing potential is stimulated twenty- to a hundredfold after being exposed to certain cytokines (notably interferon-alpha [IFN-α], interferon-beta [IFN-β], and interleukin-2 (IL-2) in the infected area. As they attack infected cell, NK cells also produce cytokines that stimulate macrophages.

We might ask how NK cells are able to discriminate between viral-infected and normal healthy host cells. Although the basis for an NK cell's being able to make this critical choice is not completely clear, it seems to depend upon the ways that viruses change the cells they infect. Many viruses cause the cells they infect to stop making certain surface proteins (major histocompatibility complex [MHC] class I; Chapter 17), and NK cells kill host cells that lack these proteins (**Figure 16.7**).

The power of NK cells to fight viral infections is proven by the consequences of their loss. Patients who lack NK cells (as a result of a genetic defect) suffer from repeated viral infections even if they have a normal adaptive immune system.

In this chapter we saw how T. L.'s innate immune system fought to control his serious wound infection. But it was overwhelmed by the infection. In the next chapter we'll consider the adaptive immune system and how it brought T. L.'s infection under control.

Natural Killer Cell

lytic granules

activating receptor

inhibitory receptor

MHC Class I

surface protein

killing

Healthy Host Cell
Carries MHC class I complex.
NK cell binds by its inhibitory receptor to this complex, which inactivates the activating receptor.

Consequence:
Host cell in unaffected.

Virus-infected Host Cell
Virus infection causes host cell to lose MHC class I complex. NK cell binds by its activating receptor to a surface protein, which stimulates NK cell to release its lytic granules.

Consequence:
Virus-infected cell is killed.

SUMMARY

The Body's Three Lines of Defense Against Infection (pp. 390–393)

1. We have three lines of defense against invading microorganisms: surface defenses, the innate immune system, and the adaptive immune system.

2. Our innate immune system attacks invading microorganisms immediately and usually eliminates them.

3. If we have not been exposed previously to an invading microorganism, our adaptive immune system attacks them, but only days or weeks after they enter the body.

4. All animals have an innate immune system. Only vertebrates have an adaptive immune system.

The Innate Immune System (pp. 393–402)

5. Both the innate and adaptive immune system depend on leukocytes and proteins in plasma.

6. There are about 10 kinds of leukocytes. All are formed in the bone marrow. They are distinguished by size, morphology, and function.

7. The first response of the innate immune system is activation of complement.

8. The first attack on invading microorganisms is phagocytosis.

Activation of Complement (pp. 393–395)

9. Activation of complement in the innate immune system occurs through the alternative pathway.

10. In a series of reactions, C3 is converted to C3b, but only if invading microorganisms are present.

11. Normally C3b is unstable, but when bound to microbial cells it interacts with other complement components to form C3 convertase, which catalyzes the formation of much more C3.

12. C3b has three critical roles: It enters the terminal complement pathway, which leads to the formation of the membrane attack complex; it acts as an opsonin; and it participates in forming C5a.

Phagocytes and Phagocytosis (pp. 395–397)

13. Phagocytes recognize, engulf, and destroy invading microorganisms. They expel undigested debris.

14. We have two kinds of phagocytes: macrophages and neutrophils.

15. Macrophages are long-lived mature forms of monocytes that are located in tissues.

16. Neutrophils circulate in the bloodstream. They enter infected tissues.

17. Cell-surface receptors on a phagocyte's outer surface bind to molecules on microbial cells that do not occur on host cells. These include lipopolysaccharide and peptidoglycan.

18. Phagocytes extend pseudopods, which enclose cells, forming a membrane-bound phagosome with the cell inside.

19. In macrophages the phagosome fuses with a lysosome (forming a phagolysosome); in neutrophils they fuse with intracellular granules. Both contribute chemicals that digest and kill most pathogens.

20. Some pathogens can survive in phagocytes.

21. Debris from digested bacteria is expelled from the phagocyte as the phagosome fuses with the cell membrane.

Inflammation (pp. 397–398)

22. When macrophages engulf invading microorganisms, they produce cytokines, proteins that signal and coordinate other components of the immune system.

23. Various cytokines cause tissue damage, attract leukocytes, stimulate the adaptive immune system, increase permeability of blood capillaries, and cause fever and shock.

24. Certain cytokines and other compounds (including histamine, complement fragments, kinins, bacterial by-products, prostaglandins, and leukotrienes) are inflammation mediators; they induce inflammation.

Effects of Inflammatory Mediators (pp. 398–401)

25. Inflammation is characterized by redness (erythema), warmth, and pain.

26. Inflammation mediators enlarge blood vessels and make them more permeable, allowing infection-fighting proteins and leukocytes to enter the affected area.

27. Phagocytes travel to the site of infection, adhere to the capillary wall (margination), and squeeze through it (diapedesis) to enter the infected tissue.

28. Pus is a mixture of dead microorganisms, leukocytes, and tissue cells.

Inflammatory Repair (p. 402)

29. Macrophages consume dead microorganisms, host cells, and foreign particles in a wound.

30. After the debris has been cleared, the affected tissues regenerate and finally heal.

31. Nerve tissues cannot regenerate.

Defense Against Viral Infections (pp. 403–404)

32. The innate system has two defenses against viral infections: interferons and natural killer cells.

Interferons (p. 403)

33. Interferons are startlingly potent.

34. There are three kinds of interferons: alpha, beta, and gamma.

35. Virus-infected cells produce interferons that cause neighboring cells to produce antiviral proteins (AVP), which interfere with viral protein synthesis.

36. Some bacteria, some protozoa, double-stranded RNA, and endotoxin stimulate cells to make interferons.

37. Interferons are useful in fighting certain leukemias.

38. Interferons also regulate cell functions, including motility, cell division, activation of macrophages, and transplant tissue rejection.

Natural Killer Cells (pp. 403–404)

39. Natural killer cells fight viral infections by killing infected host cells.

40. Natural killer cells are attracted to infected tissues and activated by cytokines.

41. Lethal compounds are released from intracellular lytic granules when natural killer cells bind to virus-infected cells.

42. Viral infection causes host cells to lose major histocompatibility complex (MHC) class I, allowing natural killer cells to bind.

43. Patients who lack natural killer cells suffer repeated viral infections even if their adaptive immune system is normal.

REVIEW QUESTIONS

The Body's Three Lines of Defense Against Infection

1. What are the body's three lines of defense? What are the key differences between the innate and adaptive immune systems?

2. Which animals have an innate immune system? An adaptive immune system?

3. What happens if a pathogen evades the third line of defense?

The Innate Immune System

4. What are leukocytes? What is plasma?

5. Where is complement located?

6. What is a phagocyte?

Activation of Complement

7. By which pathway is complement activated in the innate immune system?

8. What happens to C3b if no invading microorganisms are present?

9. What three roles does C3b play in the presence of invading microorganisms?

10. What does the terminal complement pathway lead to?

11. What does an opsonin do?

12. How is C5a formed? What role does it play in the innate immune system?

Phagocytes and Phagocytosis

13. What are the four steps of phagocytosis?

14. What are the two types of phagocytes?

15. From what type of leukocytes are macrophages derived?

16. Where are neutrophils normally located?

17. What do the cell-surface receptors on phagocytes recognize?

18. Why is the process of recognition particularly critical to the successful action of phagocytes?

19. Cite some examples of the proteins that various cell-surface receptors recognize.

20. What is a pseudopod? What is a phagosome?

21. What is the source of the phagosome's membrane?

22. What is a phagolysosome? Do neutrophils produce one? Why?

23. Are the contents of phagolysosomes able to kill all bacteria? Cite an example.

24. How does a phagocyte expel undigested bacterial debris?

Inflammation

25. What are cytokines?

26. In the process of inflammation, which cells are the first to release cytokines?

27. Name some cytokines and tell what they do.

28. What are some systemic effects of cytokines?

29. What are the symptoms of inflammation?

30. What is the role of mast cells in inflammation?

31. Name some inflammatory mediators and tell what they do.

32. What is the difference between acute and chronic inflammation?

Effects of Inflammatory Mediators

33. What are the three principal effects of inflammatory inhibitors?

34. What is erythema? What causes it?

35. How do warm compresses help fight and infection such as T. L.'s?

36. How is one way to diagnose inflammation?

37. Why did raising T. L.'s arm decrease his pain?

38. What does margination mean?

39. What does diapedesis mean? Why does it occur during inflammation?

40. What is pus composed of?

Inflammatory Repair

41. What events occur during inflammatory repair? Why is it a lengthy process?

Defense Against Viral Infections

42. What two defenses can the innate immune system against viral infection?

Interferons

43. What are interferons? How do they act? What stimulates their formation?

44. Are interferons active against any cancers? If so which?

45. What is the relationship between interferons and antiviral proteins?

Natural Killer Cells

46. What cells do NK cells kill?

47. Why don't NK cells kill healthy host cells?

48. What are the consequences of not having NK cells?

CORRELATION QUESTIONS

1. Why is the innate immune system called the second line of defense?

2. Could an uninfected wound lead to inflammation? How?

3. During inflammation, how are erythema and phagocytosis linked? Explain.

4. Why is complement activation the first response of the innate immune system to invading pathogens?

5. What effect do interferons have on viruses?

6. What is the common origin of all leukocytes?

ESSAY QUESTIONS

1. You are examining a patient who has recently sustained a minor injury that penetrated the skin. Discuss how you might know whether treatment is necessary, what it might be, and what favorable and unfavorable consequences might result.

2. People with advanced cases of diabetes often have changes in their small blood vessels that interfere with the normal circulation of blood, especially to their feet and toes. Discuss how this condition might affect vulnerability to infection.

SUGGESTED READINGS

Aderem, A., and R. J. Ulevich. 2000. Toll-like receptors in the induction of the innate immune response. *Nature* 406:782–84.

Kiester, E. 1984. A little fever is good for you. *Science* 84 (November): 168–73.

Nilsson, L. 1987. *The body victorious.* New York: Dell.

Parham, P. 2000. *The immune system.* New York: Garland Publishing.

For additional readings, go to InfoTrac College Edition, your online research library at:
http://www.infotrac.thomsonlearning.com

SEVENTEEN

The Immune System: Adaptive Immunity

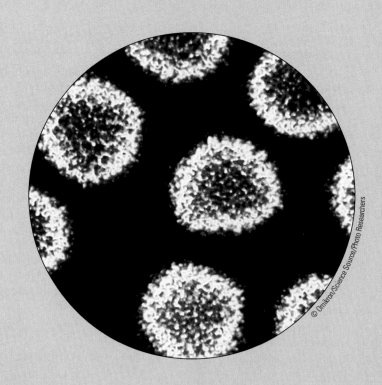

© Omikron/Science Source/Photo Researchers

CHAPTER OUTLINE

LEARNING GOALS

To understand:

- *The components and function of the adaptive immune system*
- *The humoral immune response: how a diversity of antibodies are made by B cells and how they defend us against pathogens*

- *The cell-mediated immune response: how T cells protect us against intracellular pathogens and how they regulate our immune system*
- *Immune tolerance: why our immune system does not attack our own cells and tissues*

- *The types of adaptive immunity: active, passive, naturally and artificially acquired*
- *How T. L.'s adaptive immune system saved him*

T. L.'s Close Call: Part Two

When T. L. awoke 2 days after sustaining a puncture wound to his palm, his condition was clearly deteriorating. The first thing he noticed was the spread of the area of redness and swelling. It now included his entire hand. Its margin was now above his wrist. In addition, faint red streaks had appeared along the inner aspect of his arm, and tender lumps were noticeable behind his elbow and under his arm. His fever seemed higher than the night before. He felt too weak to walk.

T. L. was clearly too sick to travel. His companions estimated that it would take several days to return with medical help. Reluctantly they decided to stay in camp and allow T. L. to rest. Over the next few days T. L. and his friends were relieved to find him gradually improving. The red area stopped advancing at his forearm. His hand remained painful and swollen, as did the lymph nodes at his infected elbow. But he felt less tired and his fever was gone. Four days later, when he was continuing to recover, the party turned back and took T. L. to the nearest emergency room.

Doctors diagnosed a presumed bacterial infection of T. L.'s hand and prescribed a broad-spectrum antibiotic that was likely to be effective against the unknown infecting organism. The medication could be taken by mouth, so T. L. was sent home with instructions to return if his condition failed to improve. The redness and swelling of his hand subsided gradually over the next 2 weeks. He had no further problems related to his infection.

Case Connection

■ T. L.'s symptoms continued because his second line of defense, his innate immune system, failed to stop his developing infection. In this chapter, we'll examine how his third line of defense, his adaptive immune system, was activated and how it did eliminate his infection.

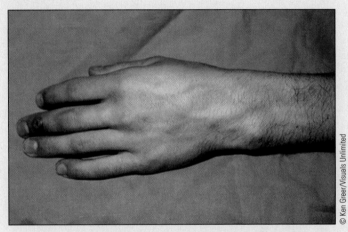

Red streaks indicate the development of lymphangitis as the infection progresses.

© Ken Greer/Visuals Unlimited

THE ADAPTIVE IMMUNE SYSTEM: AN OVERVIEW

T. L. was well on his way to recovery when he went to the emergency room. The antibiotic he received there undoubtedly shortened his illness and ensured his recovery, but he almost certainly would have recovered without it. T. L.'s own adaptive immune system had controlled his life-threatening infection. His first line of defense (Chapter 14) was overwhelmed when the branch of the tree penetrated his skin. His second line of defense, the innate immune system, (Chapter 16) was helpful. Its phagocytes and complement killed many of the invading bacterial cells, but it didn't control the infection. Then when his third line of defense, the adaptive immune system, was activated, his infection was controlled (**Figure 17.1**).

In this chapter we'll discuss the adaptive immune system and examine how it stopped T. L.'s infection. Of the three lines of defense, the adaptive immune system (on first exposure to a particular pathogen) is the slowest to respond. It is by far the most complicated, but it's also the most powerful. Its hallmark is specificity. Whereas the innate immune system identifies and attacks microbial invaders by recognizing molecules shared by groups of microorganisms, the adaptive immune system detects the specific molecular components of specific strains of microorganisms. It responds by inactivating or killing the microorganisms that bear them. The adaptive immune system is effective against all classes of microorganisms—prokaryotic, eukaryotic, and viral. It even acts against helminths.

FIGURE 17.1   The three lines of defense against infection.
(Art by Carlyn Iverson.)

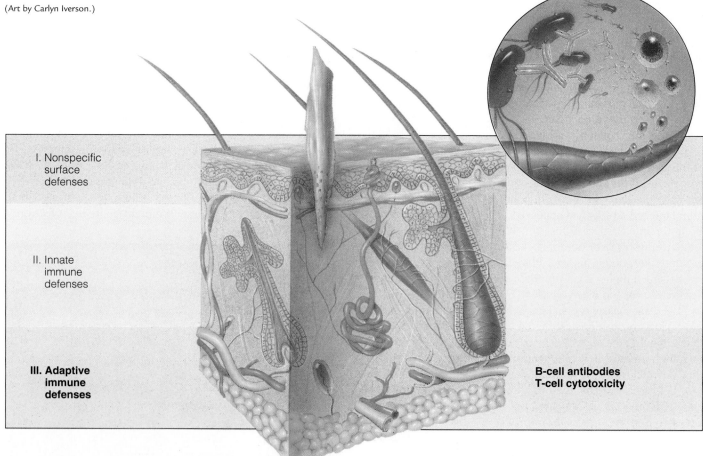

I. Nonspecific
surface
defenses

II. Innate
immune
defenses

**III. Adaptive
immune
defenses**

**B-cell antibodies
T-cell cytotoxicity**

Although our adaptive immune system is slow to act when we first encounter a particular pathogen, it responds quickly and powerfully on a second and subsequent exposures to the same organism. These later responses are so rapid that we are usually unaware of being exposed to potentially deadly pathogens. Humans have known from ancient times that surviving a particular infectious disease produces subsequent immunity to it. This knowledge lead to the development in the eighteenth century of treatments termed **vaccination,** which conferred immunity but avoided the dangers of a first infection.

Cells called lymphocytes are the central actors in the adaptive immune system. The other components are the lymphoid tissues where lymphocytes are formed, where they **differentiate** (mature and acquire special capabilities), where they are stored, and through which they are transported throughout the body.

First we'll examine the components (lymphocytes and lymphoid tissues) of the adaptive immune system. Then we'll consider what they do.

Lymphocytes

Lymphocytes (**Table 17.1**) are a type of leukocytes (Chapter 16, Table 16.2). They are smooth and round, lack visible granules, and are relatively small. There are three kinds of lymphocytes: B lymphocytes (B cells), T lymphocytes (T cells), and natural killer (NK) cells. B cells and T cells play central roles in the adaptive immune system. As we saw in Chapter 16, NK cells participate in the innate immune system. B cells and T cells are distinguished by function, not by appearance as the other leukocytes are.

Both B cells and T cells are triggered into defensive action against invading pathogens when they encounter **antigens** (certain foreign molecules that we lack). Because every microorganism has its own unique constellation of molecules that acts as antigens, B and T cells are able to attack all invading pathogens. The two types of lymphocytes respond differently. All B cells respond the same way to antigens: They produce defensive proteins called **antibodies.** T cells respond in either of two different ways. Some T cells called **cytotoxic T cells (CD8 T cells)** re-

TABLE 17.1 Types of Lymphocytes and Their Roles

Type of Lymphocyte	Function
B CELLS	Agents of antibody-mediated (humoral) immunity
Naïve B cells	Differentiated B cells that have not yet encountered a complementary antibody
Plasma cells	Fully differentiated B cells that mass produce antibodies to fight current infection
Memory B cells	Fully differentiated B cells produced during a previous infection that mount an accelerated and amplified response if reinfection with the same organism occurs
T CELLS	Agents of cell-mediated immunity and producers of lymphokines
CD4 T cells	Helper T cells
T_H1 cells	Secrete lymphokines that induce a cell-mediated immune response
T_H2 cells	Secrete lymphokines that regulate an antibody-mediated immune response
CD8 cells	Cytotoxic T cells: destroy infected cells, cancerous cells and transplanted cells
NATURAL KILLER CELLS	Components of the innate immune system; function with recognizing an antigen; target infected cells, cancerous cells, and transplanted cells

spond by killing those cells in our body that have been infected by a pathogen (and therefore bear its antigens). Other T cells called **helper T cells (CD4 T cells)** augment the antimicrobial activities of other lymphocytes: Some activate macrophages (Chapter 16) to become more efficient phagocytes; others stimulate B cells to produce more antibodies.

B cells and T cells confer different types of adaptive immunity, respectively termed humoral and cell-mediated immunity.

Humoral Immunity

Because B cells respond to antigens by differentiating and producing antibodies, which circulate in our blood, the type of immunity they confer is called **antibody-mediated immunity** or **humoral** (meaning fluid) **immunity.** Antibodies protect against invading pathogens in three ways: (1) They **neutralize** (inactivate) toxins or vital compounds on the surface of a pathogen by binding to them. (2) They bind to antigens of the surface of pathogens, which facilitates their destruction by phagocytes, the process called **opsonization.** (3) When bound to the surface of pathogens, they also activate **complement** (a set of pathogen-destroying proteins in serum; Chapter 16.) We'll discuss neutralization, opsonization, and activation of complement in greater detail later in this chapter.

Cell-mediated Immunity

T cells respond to antigens that appear on the surface of pathogen-infected cells. Because the T cell itself attacks infected cells, the type of immunity they confer is called **cell-mediated immunity.** Such destruction of our own cells might seem dangerously harmful, but it effectively protects us against intracellular pathogens (including viruses) by interrupting their cycle of infection.

Now let's examine the tissues of our body, called **lymphoid tissues,** in which lymphocytes are formed, undergo differentiation, and are stored.

Lymphoid Tissues

The tissues where lymphocytes are formed or differentiate are called **primary lymphoid tissues (Figure 17.2, Table 17.2).** They are the bone marrow and the thymus.

Bone marrow is the jellylike mass in the core of our large bones—vertebrae, ribs, sternum, long bones of the arms and legs, and pelvis. Bone marrow is composed primarily of fat, but it also contains the **stem cells** (rapidly dividing cells yielding progeny with the capacity to develop in a variety of cell types) that give rise to all our blood cells, including both B and T lymphocytes (Chapter 16). Once formed, B cells remain in the bone marrow to differentiate. But newly formed

FIGURE 17.2 Tissues of the immune system.

(Art from Biology: Concepts and Applications, 2nd ed., by C. Starr, Brooks/Cole, 1994. All rights reserved.)

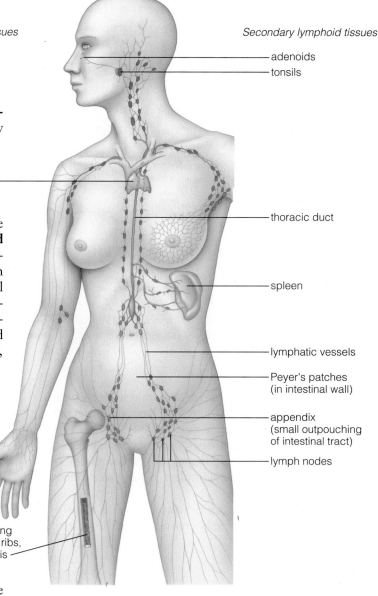

Primary lymphoid tissues

Secondary lymphoid tissues

adenoids
tonsils
thymus
thoracic duct
spleen
lymphatic vessels
Peyer's patches (in intestinal wall)
appendix (small outpouching of intestinal tract)
lymph nodes
bone marrow of long bones, vertebrae, ribs, sternum, and pelvis

T cells leave the bone marrow and migrate to the **thymus** (a gland located in front of the heart), where they differentiate.

When fully differentiated, B cells and T cells escape from their primary lymphoid tissue by entering the small blood vessels that pass through these tissues. Then they travel through the bloodstream to one of a group of storage sites called **secondary lymphoid tissues,** where they leave the bloodstream through specialized blood vessels. Most lymphocytes are stored in lymph nodes or the spleen. **Lymph nodes** are small kidney-shaped structures (**Figure 17.3**) scattered throughout the body. The **spleen** is a single fist-sized organ located high in the abdomen. Other lymphocytes are stored in the **tonsils** (a pair of bean-sized tissues in the throat), the **adenoids** (located in the nose), the **appendix** (an outpouching of the intestinal tract), and **Peyer's patches** (small pockets of lymphoid tissue in the intestinal wall). Secondary lymphoid tissues do more that just store lymphocytes. They are the sites where B and T lymphocytes interact and become activated to attack invading pathogens.

Lymphocytes constantly leave secondary lymphoid tissues and enter the **lymphatic circulation.** This circulation is a system of vessels that collects excess fluid, called **lymph,** from the body's tissues and returns it to the bloodstream. Lymph is formed continuously in our tissues as fluid from blood leaks across thin capillary walls. The many lymphatic vessels that comprise the lymphatic circulation merge at the large **thoracic duct,** which empties in the heart. Thus the lymphatic circulation and bloodstream are interconnected. Lymphocytes constantly circulate through both of them.

Lymphatic vessels and lymph nodes usually become swollen and inflamed during infections that activate the immune defenses. The red streaks that T. L. noticed on his arm were inflamed lymphatic vessels. The tender swellings behind his elbow and under his arm were lymph nodes that became enlarged as lymphocytes proliferated in them.

Before examining the details of antibody-mediated and cell-mediated immunity, let's consider the common feature of both: They interact with antigens, the molecules that direct the attack on pathogens.

Antigens

The adaptive immune system recognizes and then reacts to molecules called **antigens.** Many large molecules act as antigens to varying degrees. Almost all proteins are strong antigens. Polysaccharides are weaker antigens. In general, lipids and nucleic acids are almost devoid of antigenic activity. But these and other weak antigens become stronger when coupled to proteins.

Antigens on the surface of microorganisms play a particularly important role in the immune response because they are directly exposed to the immune system. Such surface antigens include molecules in the capsules, cell envelopes, and flagella of bacteria, as well as those

TABLE 17.2 Tissues and Vessels of the Immune System

Tissue	Location	Role in Adaptive Immune System
Primary Lymphoid Tissues		
Bone marrow	Within vertebrae, ribs, sternum, long bones, pelvis	Produces all blood cells, including leukocytes and lymphocytes; site of B-cell differentiation
Thymus	Gland that lies in front of the heart	Site of T-cell differentiation
Secondary Lymphoid Tissues		
Lymph nodes	Small kidney-shaped organs located on lymphatic vessels throughout the body	Mature lymphocytes are stored, interact with one another, and move between blood and lymph vessels here
Spleen	Upper abdomen	Mature lymphocytes stored and interact here
Tonsils	Pair of lymphoid tissues in the throat	Mature lymphocytes stored and interact here
Adenoids	Ring of lymphoid tissues in the nose	Mature lymphocytes stored and interact here
Appendix	Small outpouching of the intestinal tract	Mature lymphocytes stored and interact here
Peyers's patches	Small collections of lymphoid tissue within the intestinal wall	Mature lymphocytes stored and interact here
Lymphatic Circulation		
Lymph vessels (lymphatics)	Vessels throughout the body that collect lymph and return it to the bloodstream	Transports lymphocytes
Thoracic duct	Largest lymph vessel	Empties into heart to return lymph to the bloodstream

on the surfaces of fungal and helminthic cells and in the capsids or envelopes of viruses.

Our immune system, when functioning properly, does not recognize any of the body's own molecules as being antigens. This ability not to react to the body's own molecules is called **immune tolerance.** We'll discuss its basis later in this chapter. The consequence of immune tolerance is that the immune system, when operating properly, recognizes only foreign molecules as being antigens and does not attack our own tissues.

Lymphocytes are the components of the immune system that recognize antigens. They're capable of such recognition because they have **antigen receptors** (antigen-recognizing molecules) on their cell surface (**Figure 17.4**). The antigen receptors on B cells are tethered antibody molecules, called **surface immunoglobulins.** Those on T lymphocytes are called **T-cell antigen receptors.** Each lymphocyte can recognize only a small part (called an **epitope** or **antigenic determinant**) of a single antigen, but our immune system is able to recognize myriad epitopes and therefore myriad antigens. This is possible because we have a huge number of different lymphocytes. Each one is able to recognize one particular epitope. We'll discuss how it's possible to form so many different lymphocytes later in this chapter.

Differently shaped epitopes are recognized by lymphocyte receptors with complementary shapes. In other words, there is a shape-to-shape recognition (Figure 17.4). The two molecules fit together rather like two pieces of

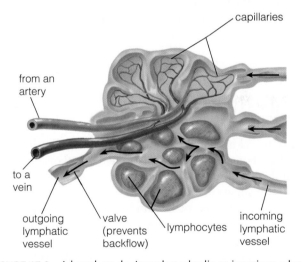

FIGURE 17.3 A lymph node. Lymph nodes lie at junctions where several incoming lymph vessels merge to form a single outgoing vessel. The node is packed with lymphocytes and macrophages. Lymphocytes arrive in arterial blood and pass through the capillary walls to enter storage areas. Lymphocytes at sites where incoming lymph vessels drain areas of infection swell due to multiplication of activated lymphocytes within them.

a jigsaw puzzle. The immune system's remarkable ability to recognize individual epitopes is the key to its specificity. Actual recognition occurs when an antigen molecule comes in direct contact with an antigen receptor on a lymphocyte. This activates either the humoral or

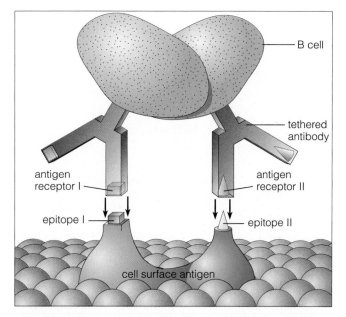

FIGURE 17.4 Antigen recognition by B cells. Sketch illustrates two B cells with tethered antibodies, each capable of recognizing a different epitope on an antigen.

cell-mediated immune response, depending upon whether the lymphocyte is a B cell or a T cell.

We'll consider the two adaptive immune responses (humoral and cell-mediated) separately, although, as we'll see, they interact.

THE HUMORAL IMMUNE RESPONSE

The humoral immune defense is mounted by B cells. Each of us has about 10 trillion (10^{13}) of them. As they differentiate in the bone marrow, they become **immunocompetent.** That means that each individual B cell acquires the capability to produce one particular kind of antibody attached to it that serves as its **antigen receptor.** Our 10 trillion B cells can produce about 100 million distinct antibodies. In other words, we have about 100 million different kinds of B cells, each of which has a different kind of antibody on its surface. That staggering number of different antibodies is enough so some of them are able to recognize any antigen that we're likely to encounter during our lifetimes.

Generating Antibody Diversity

How can the process of differentiation generate such an enormous diversity of B cells, enabling us to produce an antibody for every antigen? Each protein we produce (and

an antibody is a protein) is encoded by a separate gene (Chapter 6). But we have only about 30,000 genes in our genome, and we make 100 million antibodies. Something else intervenes.

B-cell diversity is generated by a process that involves a type of genetic recombination (Chapter 7; **Figure 17.5**). Instead of there being a separate gene in our genome to encode each antibody, individual B lymphocytes create, during differentiation, their own genes to encode a unique antibody. B cells do this by genetically cutting and rejoining gene fragments together in different combinations. Our genome contains a library of modular bits of DNA that can be pieced together to create complete antibody genes. As differentiation occurs, one copy of each necessary piece of information is chosen at random from the library. Because the number of possible combinations of these small bits of genetic information is enormous, the 100 million different antibody genes we possess genetically are produced from a relatively limited amount of DNA. A similar process occurs as T lymphocytes differentiate in the thymus, which we'll discuss later.

Each fully differentiated B cell, termed a **naïve B cell** before it encounters an antigen, has about 100,000 identical antibody molecules attached to its surface. When one of them binds to a matching antigen on the surface of an invading microbial cell, a virus, or some other foreign biological material, that B cell undergoes activation. During activation, the structure and class of antibody it produces change. We'll discuss the structure of antibodies and then the process of activation.

Structure of Antibodies

Antibodies are Y-shaped molecules composed of four polypeptide chains: two identical **light chains** (smaller ones) and two identical **heavy chains** (larger ones) (**Figure 17.6**). They are **glycoproteins** (carbohydrate groups are attached to the polypeptide chains). **Antigen binding sites** are located at each of the ends of the two arms of the Y. They and a section just below them are the **variable regions** which change so dramatically during differentiation (and as we'll see, also during activation), thereby generating the huge diversity of antigen binding sites found on antibodies. In contrast, the rest of the antibody is called the **constant region** because it changes to a lesser extent. In contrast to the variable region, which determines which antigen an antibody recognizes, the constant region determines what it might do. The constant region carries the sites for binding to phagocytes and to complement.

The plant enzyme papain splits antibodies near the point where the arms of the Y join the stem, producing two **Fab** (antibody-binding) fragments and one **Fc** (constant)

SHARPER FOCUS

RUNAWAY ANTIBODIES

For most of us, antibodies help protect against infection. But victims of a plasma cell cancer called **multiple myeloma** suffer runaway antibody production that can be fatal. In these patients one particular clone of plasma cells turns cancerous and begins to multiply out of control, producing immense quantities of antibody in the process. Because the cancerous transformation occurs within a single clone, the excess antibody molecules are of a single type—either IgG, IgA, IgD, or IgE. This enormous quantity of antibodies clogs the kidneys and eventually destroys them. Meanwhile, the cancerous plasma cells themselves become so numerous they destroy bone and crowd normal cells out of the marrow. Paradoxically, people with multiple myeloma often die from infection. The malignant plasma cells overwhelm normal lymphocytes, leading to a deficiency of normal antibodies.

FIGURE 17.5 Generating antibody diversity. Modular segments of DNA from various segments of our genome (v, j, and c) are selected at random. These recombine to create a unique combination of segments in the DNA of each differentiating B cell. These are transcribed, processed, translated, and assembled to produce a unique antibody.

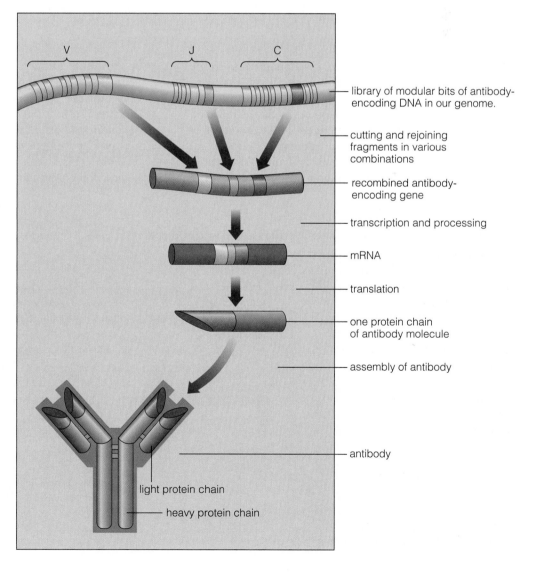

- library of modular bits of antibody-encoding DNA in our genome.
- cutting and rejoining fragments in various combinations
- recombined antibody-encoding gene
- transcription and processing
- mRNA
- translation
- one protein chain of antibody molecule
- assembly of antibody
- antibody
- light protein chain
- heavy protein chain

fragment. These terms (Fab and Fc) are used to describe parts of antibody molecules.

Groups of antibodies share similar constant regions of the heavy chain and have the same numbers of attached carbohydrate groups. On this basis they are divided into five classes.

The Five Classes of Antibodies.

All antibodies are called **immunoglobulins** because they are immunologically active and they are found in the globulin protein fraction of serum. The five different classes are named immunoglobulin (abbreviated Ig), followed by a letter—G, A, M, D, or E—to designate one of the classes. The properties of these various classes of antibodies are summarized in **Table 17.3.**

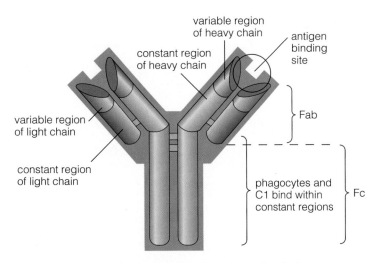

FIGURE 17.6 Antibody monomer. Locations of carbohydrate groups are not shown because their positions vary among monomers of the five classes of antibodies. Neither is the membrane-anchoring region, which is attached to the stem end of antibodies that are tethered to B cells.

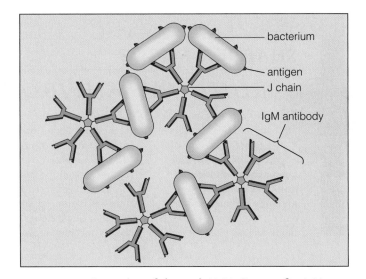

FIGURE 17.7 Formation of clumps by IgM. Because five IgM monomers are joined by a J chain, its 10 antigen-binding sites can bind to several antigens and form a clump.

All classes of antibodies occur as monomers (a single Y-shaped structure) on the surface of the B cell that produces them. Three classes (IgD, IgE, and IgG) are also secreted as monomers, but the monomers of classes IgM and IgA are joined together when secreted. Five monomers of IgM are joined together to form pentamers. IgA is secreted as a mixture of monomers and dimers. Because monomers have two antigen binding sites, they are said to

be **divalent.** Dimers are tetravalent and pentamers are decavalent. The high valence of IgM makes it particularly suitable for forming clumps with some antigens (**Figure 17.7**). Antigens can be monovalent or multivalent, depending upon how many epitopes they carry. Only multivalent antigens are recognized by antibodies.

During his illness, T. L. produced all five classes of antibodies. Some were indispensable to his recovery. Others had

TABLE 17.3 Properties of the Five Classes of Antibodies[a]

Class	Secreted Form	Abundance (%)	Location	Functions
IgA	Monomers and dimers	15 to 20	Monomer in blood; dimer in body secretions (e.g., saliva, milk, mucus)	Protects body's mucous membrane surfaces, such as those on lungs, intestines, and bladder, from pathogens, particularly viruses. Presence in colostrum defends newborns against gastrointestinal infections.
IgD	Monomers	Less than 1	On surface of B cells	Unknown.
IgE	Monomers	Less than 0.01	On surface of mast cells, eosinophils, and basophils	Controls worms by causing mast cells, eosinophils, and basophils to degranulate. Contributes to allergies.
IgG	Monomers	75	Blood and extracellular fluids	Acts as opsonin. Activates complement. Mother's IgG enters fetal circulation and protects newborns for several months.
IgM	Pentamers	5 to 10	Blood and extracellular fluids	Active during earliest phase of a primary immune response. Binds weakly to antigens, but its high valence enhances its effect. Activates complement.

[a]In addition to the variables listed here, the various classes of antibodies differ with respect to the length of the heavy-chain C region, the number of attached carbohydrate groups, and the number and location of disulfide bonds that join the various chains.

THE INDISPENSABLE MACROPHAGE

Body structures that we cannot live without always hold a special fascination for clinicians and researchers. Macrophages fall into this category. No one has been found to be without macrophages. Macrophages are essential not only for their voracious phagocytic activity—for which they were named—but also for their ability to secrete more than 100 different substances. Among macrophage secretions are cytokines, including interleukin-1, tumor necrosis factor, and interferons (when produced by macrophages and other mononuclear cells, these cytokines are also called monokines). Other macrophage secretions are enzymes, such as lysozyme; still others, such as prostaglandins and leukotrienes, mediate inflammation and intercellular communication. Macrophages even secrete components of the complement system.

limited usefulness during his particular struggle. IgG helped him the most. As an opsonin, it helped T. L.'s phagocytes recognize and kill the invading bacteria, and as an activator of the complement system, it produced more opsonins in the form of C3b. Moreover, it generated inflammatory mediators that attracted phagocytes to his infection and activated them.

Now let's consider the next step in the development of a B cell, activation.

B-Cell Activation

Before we are exposed for the first time to a particular pathogen, only a few of our B cells can recognize the antigens it bears. These naïve B cells are in a resting state, incapable of mounting a defense against an infection.

Naïve B cells produce membrane-bound IgM and IgD on their outer surface. When one of them binds to an antigen, activation is initiated and many changes occur. That B cell begins to multiply, to undergo further differentiation, and eventually produces quantities of soluble antibodies. The end result is a clone of cells that make an effective form of the same antibody. This process of selective B-cell stimulation is called **clonal selection (Figure 17.8)**.

At first the stimulated B cell produces large amounts of IgM and smaller amounts of IgD. The IgM begins to confer protective immunity; the IgD has no known function. It's logical that IgM should be the first type of antibody to be produced. The 10 antigen binding sites each molecule bears compensate in part for the still-imperfect state of its antigen binding site. The binding site improves during activation because the rate of mutation in the genes encoding the variable region of antibody increases about a millionfold. Some of the many mutations that result improve the antigen binding sites. Cells bearing improved sites are preferentially selected to develop into antibody-producing cells. This process is called **affinity** (binding) **maturation.**

Later, by recombination, **class switching** occurs. The variable region becomes attached to different constant regions. As a result, some cells in the clone switch from making IgM to making IgA, IgE, or IgG. But regardless of the class of antibodies produced, all cells in the clone produce the antibodies with the same binding sites, which was perfected during the process of activation.

Clonal selection produces a group of identical B cells capable of fighting an infection. Some members of this clone differentiate to become **effector cells;** these are **plasma cells** that produce huge quantities of antibodies to fight the current infection. Plasma cells are prodigious producers of antibodies. One cell can make 1000 antibody molecules per minute.

Plasma cells are relatively short lived; they disappear soon after the infection is controlled. Other cells in the clone become **memory B cells** (which we discuss later) held in reserve to fight future infections by the same pathogen.

In most cases, activation depends upon interaction between B cells and T cells that have been activated by the same antigens. (Some antigens, including polysaccharides, do not need the participation of T cells; this is discussed later.) Because contact with T cells is usually required, activation occurs to a large extent in secondary lymphoid tissues, such as lymph nodes, that have large concentrations of B and T cells. T cells stimulate and modulate activation of B cells by producing lymphokines.

Activating Factors: Lymphokines. Binding to an antigen starts the process of B-cell proliferation and differentiation, but in most cases it can't proceed very far

FIGURE 17.8 Clonal selection. (a) The myriad naïve B cells in our body are represented here by those numbered B_1 through B_8. The tethered antibody on one kind of them (B_2) can recognize and bind to an antigen on the invading bacterium's surface. (b) Antigen binding stimulates further differentiation leading to a clone of identical B cells that produce antibodies that bind more tightly to the antigen. Some of them differentiate cells to become plasma cells that produce large amounts of soluble antibody to fight the present infection. Others become memory cells, held in reserve to fight a future infection by the same pathogen.

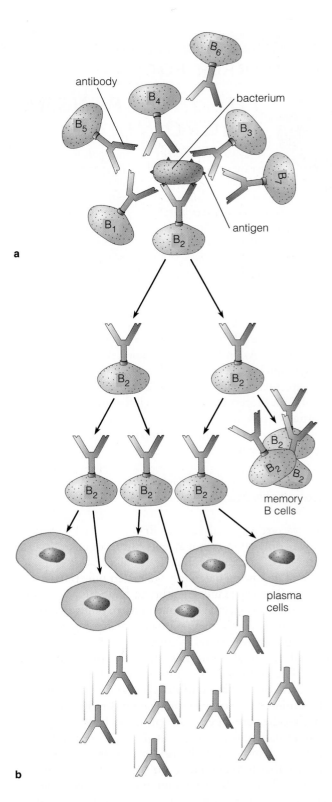

without the participation of lymphokines. Lymphokines are small signal proteins produced by lymphocytes. More than 40 of them have been discovered, many just in the past few years. Various lymphocytes, including macrophages, B cells, and T cells, produce lymphokines. We've already discussed the roles some of them play in inflammation by attracting cells and regulating their activity (Chapter 16). **Table 17.4** summarizes the origins and activities of a few lymphokines that play important roles in the adaptive immune response. One class of helper T cells, Th1 cells, is dedicated to producing lymphocytes that stimulate and regulate differentiation and antibody-production by B cells.

The activities of lymphokines are interrelated, and often they act sequentially. For example, macrophages (and other antigen-presenting cells; see later) secrete **interleukin-1,** which activates both B and Th1 cells. Activated Th1 cells in turn secrete other lymphokines that activate additional B cells and increase their defense capabilities in various ways. These include **interleukin-2** (which is itself a potent T-cell stimulator), **interleukin-4, interleukin-5, interleukin-6,** and interferon-gamma (see Chapter 16).

Some lymphokines are transferred from helper T cells to developing B cells by direct contact between a pair of cells. But this only occurs with B cells that have been activated by certain antigens. For example, many large antigens with many identical epitopes—for example, the polysaccharide capsule of pneumococci—can activate B cells without any help. These **T-independent antigens** stimulate a relatively weak immune response. Most antigens, however, are **T-dependent antigens.** The B cells they activate must come in contact with helper T cells in the secondary lymphoid organs. There they contact and receive lymphokines from helper T cells that have been activated by the same antigen.

Thus B-cell activation is an incremental process, fostered by many stimuli. One first essential component is antigen recognition itself. Another is secreted lymphokines. A third is lymphokines transferred by direct cell-to-cell contact.

In addition to participating in B-cell activation, lymphokines amplify the immune response in a variety of ways (Table 17.4). Interleukin-1 plays a role in fever production (Chapter 16) and has other actions as well.

TABLE 17.4 Properties of Some Important Lymphokines

Lymphokine	Source	Role in Immune Activation	Other Actions
Interleukin-1	Macrophages, other antigen-presenting cells, B cells	Activates B cells and T cells; stimulates them to produce other lymphokines	Inflammatory mediator, induces fever and sleep, stimulates synthesis of erythrocytes
Interleukin-2	T cells and NK cells	Activates CD8$^+$ T cells, growth factor for activated B cells	Activates monocytes
Interleukin-3	T cells		Stimulates production of all types of blood cells
Interleukin-4	T cells	Growth factor for activated B cells, activates T cells	Mast cells growth factor
Interleukin-5	T cells	Stimulates B cells to form a clone and to differentiate into plasma cells	Stimulates differentiation of eosinophils
Interleukin-6	T cells	Stimulates B cells to differentiate into plasma cells	Stimulates differentiation of thymus cells
Gamma interferon	T cells and NK cells	Stimulates B cells to differentiate into plasma cells	Activates macrophages; antiviral agent; activates CD8$^+$ T cells and NK cells
Tumor necrosis factor	T cells		Toxic to some tumors and virus-infected cells; inflammatory mediator; induces fever, sleep

Gamma interferon activates cytotoxic lymphocytes. A closely related group of cytokines includes tumor necrosis factor (actually a lymphokine), which is directly toxic to some tumor cells and virus-infected cells.

Reactivation: Immunological Memory

When an infection is over, plasma cells die and disappear. However, memory B cells continue to circulate within the body and retain their ability to recognize the same antigen if they encounter it again. They function in **immunological memory**, which greatly accelerates and amplifies the immune response when the same pathogen is encountered again.

Memory B cells that recognize a given antigen are far more numerous than the never-stimulated naïve B cells, so they are able to mount an immune response more quickly. Memory B cells differ from naïve lymphocytes in other ways too. They produce higher-affinity (and therefore more effective) antibodies that were developed during activation, and their plasma cells manufacture a different profile of antibody classes.

Production of antibody during a **primary immune response,** which occurs when a person first encounters a particular antigen, may take from 1 to 2 weeks to develop fully. But a **secondary immune response,** which is initiated by memory B cells, produces antibody within a few days. Moreover, a secondary response generates a much greater quantity of antibody (**Figure 17.9**).

Immunological memory has profound implications. Its rapid response can sometimes eliminate pathogens from the body before the clinical signs and symptoms of illness develop. For this reason, people who have contracted infections such as smallpox, measles, chickenpox, mumps, or German measles are immune to reinfection. For the same reason, **vaccines,** which stimulate the formation of memory lymphocytes without causing a clinically significant infection, can prevent infectious disease. (Vaccines are discussed in Chapter 20.)

Because T. L. began to recover from his infection within a few days, his immune response was probably the secondary type initiated by memory B cells. His immune system must have been exposed previously to the same (or very similar) bacterial antigens. T. L. and his friends made a serious error in judgment, however, when they decided to wait a day and see how his infection progressed, because if memory B cells had not been present to accelerate his immune response, his infection could have been fatal.

Now let's consider the ways that antibodies produced by B cells fight infections.

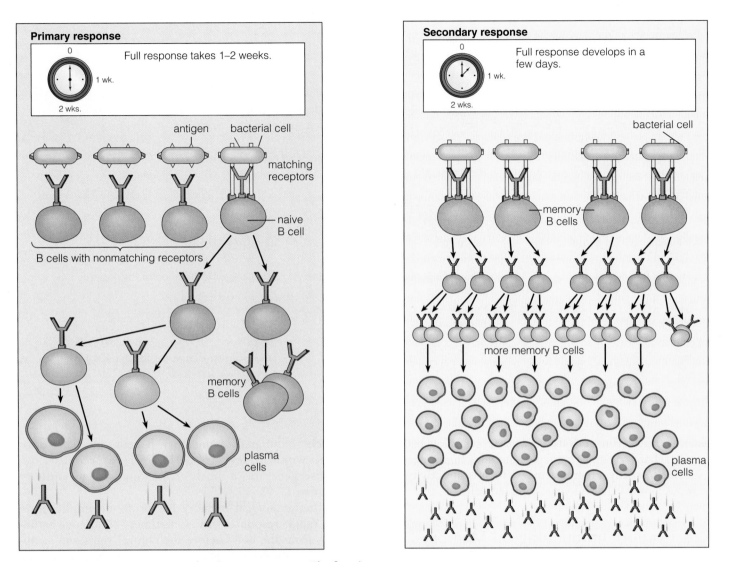

FIGURE 17.9 Primary versus secondary immune response. The first time a person encounters an antigen, virgin B cells mount a primary immune response by producing antibodies and some memory B cells. If a person encounters the same antigen again, the memory B cells initiate a secondary immune response, which is more rapid and effective than a primary response.

Action of Antibodies

Antibodies are not toxic to pathogens, they just bind tightly onto matching antigens to form an **antigen-antibody complex.** A uniquely shaped cavity in the antigen binding site of the antibody molecule is filled by an epitope of complementary shape on the antigen. The fit between antigen and antibody is somewhat variable. Some antigen-antibody matches are precise, but others are poorer. Poorer fit leads to weaker antigen-antibody interaction and eventually to a poorer immune response. As stated earlier, by binding to antigens, antibodies can exert three possible defenses: neutralization, opsonization, or activation of complement.

Neutralization. Antibodies can bind to and cover sites on the surface of a pathogen that are necessary for growth or infection. Some antibodies, called **antitoxins,** bind to toxins that act as antigens (Chapter 15). Such binding inactivates the toxin, rendering it harmless. Other antibodies neutralize viruses by binding to the antigens that attach the virus to the host cell. As a result the virus cannot attach to its host cell to initiate an infection.

Opsonization. As we learned in Chapter 16, phagocytes can recognize, ingest, and destroy some pathogens without the help of antibodies. But even in these cases the speed of the process is greatly enhanced if the pathogen is first

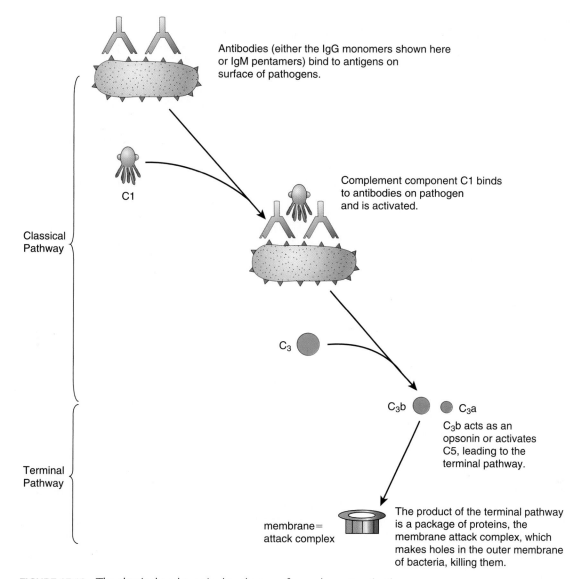

Antibodies (either the IgG monomers shown here or IgM pentamers) bind to antigens on surface of pathogens.

C1

Complement component C1 binds to antibodies on pathogen and is activated.

Classical Pathway

C₃

C₃b   C₃a

C₃b acts as an opsonin or activates C5, leading to the terminal pathway.

Terminal Pathway

membrane-attack complex

The product of the terminal pathway is a package of proteins, the membrane attack complex, which makes holes in the outer membrane of bacteria, killing them.

FIGURE 17.10   The classical and terminal pathways of complement activation.

opsonized (coated with bound antibodies). Engulfment is facilitated because phagocytes interact with the Fc region of some antibodies (IgG). Some pathogens, such as encapsulated strains of *Streptococcus pneumonia*, are highly resistant to attack by phagocytes unless they are opsonized.

Parasites, including protozoa and worms (although they are too large to be engulfed), are effectively attacked and destroyed by eosinophils (Chapter 16) if they have been opsonized by IgE antibodies.

Activation of Complement. In Chapter 16 we saw how complement can be activated in the absence of antibodies via the alternative pathway to produce an opsonin (C3b) and thereby facilitate phagocytosis. Now we will

examine the more powerful **classical pathway (Figure 17.10)**, which is activated by the Fc region of an antibody (IgM or IgG) when it is bound to an antigen. Unbound antibodies do not activate complement.

The classical pathway leads to the formation of C3b, which is a highly effective opsonin (its most important function). But C3b also leads to the formation of another complement fragment, C5b, which initiates the formation of a **membrane-attack complex.** A membrane-attack complex makes holes in the membranes of bacteria, killing them. In spite of its dramatic effect, the membrane-attack complex does not play a major role in immune defense. Its activity seems to be restricted to fighting infections caused by certain species of *Neisseria*.

SHARPER FOCUS

ANTI-IMMUNE TRICKERY: VARIOUS STRATEGIES

Our immune system evolved to protect us from all kinds of pathogens. But in parallel, many pathogens evolved ways to evade our immune defenses. Let's look at the ways that some pathogens evade our immune defenses.

Some viral pathogens elude our immune defenses by hiding vital proteins that act as antigens. All viruses must enter cells to infect them. First, specialized proteins on their surface must bind to receptor molecules on the surface of the host cell. These host-binding proteins on the virus's surface are especially vulnerable to antibody attack. Neutralizing antibodies bind to these viral proteins, preventing them from binding to host-cell receptors and thus preventing infection.

Rhinovirus-14, one of many viruses that cause the common cold, avoids such neutralizing anti-

bodies by hiding its host-cell binding protein at the bottom of a deep canyon on the surface of the viral particle. Its binding target—the host-cell receptor protein—is able reach into the canyon and establish the necessary link for viral attachment and infection. But antibody proteins are too large to enter the canyon and therefore cannot neutralize the binding protein. As a result, the rhinovirus is protected against antibody attack. The viruses that cause influenza, poliomyelitis, and acquired immunodeficiency syndrome (AIDS) also conceal their host-binding proteins on the floor of narrow surface canyons.

Other pathogens confuse immune defenses by changing their surface antigens. Trypanosomes, the protozoal pathogens that cause African sleeping sickness (Chapter 25), are masters at such anti-

genic switching. Mounting a primary immune defense takes time, so significant amounts of antibody are not produced for almost 2 weeks after the body detects a foreign antigen. When a trypanosome first enters a new human host, it displays one type of surface antigen that stimulates an immune response against it. But by the time significant quantities of antibody are produced, the surface antigen has changed, and the immune system can't recognize the new, differently shaped antigen. This cycle of immune response and antigen change can continue almost indefinitely because the trypanosome is capable of producing more than 1000 different antigens. In this way the trypanosome always remains one jump ahead of the immune system. As a result, our immune systems can't control the pathogen. Therefore

African sleeping sickness is a chronic, usually fatal infection. Some bacteria, including *Neisseria gonorrhoeae* and *Haemophilus influenzae,* also rely on antigenic switching to evade our immune system.

T-cell defenses, too, can be sabotaged. Adenovirus, which infects the human respiratory and gastrointestinal systems, is an example of such a saboteur. CD8 cells, the usual defenders against viral infections, can recognize viral antigens only if they are displayed in complex with class I MHC antigens on a host-cell surface. But adenoviruses produce proteins that bind to class I MHC antigens and trap them within the cell. In doing so, they prevent the MHC protein from transporting their antigens to the cell surface. Thus CD8 cells can't recognize adenovirus-infected cells. Such cells survive and produce many more virions.

SHARPER FOCUS

ANTI-IMMUNE TRICKERY: VARIOUS STRATEGIES (continued)

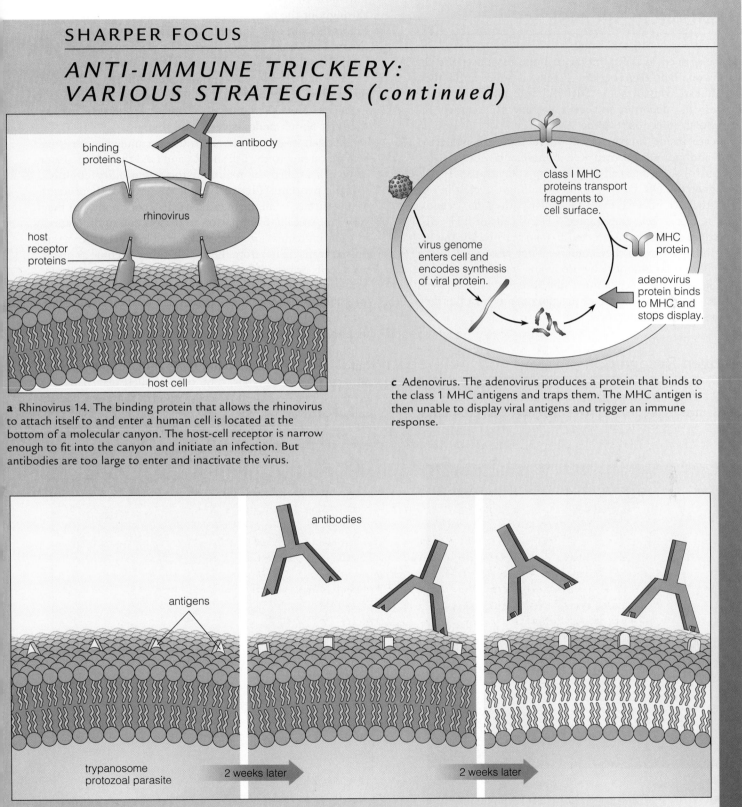

a Rhinovirus 14. The binding protein that allows the rhinovirus to attach itself to and enter a human cell is located at the bottom of a molecular canyon. The host-cell receptor is narrow enough to fit into the canyon and initiate an infection. But antibodies are too large to enter and inactivate the virus.

c Adenovirus. The adenovirus produces a protein that binds to the class 1 MHC antigens and traps them. The MHC antigen is then unable to display viral antigens and trigger an immune response.

b Trypanosomes. Antigens on the surface of trypanosomes change so frequently that the antibodies produced by a primary immune response cannot keep up. Consequently, untreated African sleeping sickness is a chronic and often fatal disease.

THE CELL-MEDIATED IMMUNE RESPONSE

As we've seen in the humoral immune response, B cells, with some help from certain T cells, produce antibodies, which can control some bacterial infections. But antibodies can't penetrate into cells, so they can't control intracellular infections. Some bacteria and all viruses do replicate inside human cells, where they are protected from antibodies. Cell-mediated immunity, however, is able to control such intracellular infections. Cytotoxic T cells (designated CD8 T cells) do so by killing the cells in our body that are infected by intracellular pathogens. In addition, helper T cells (designated CD4 T cells or TH cells) can stimulate bacteria-infected human cells to kill the intracellular pathogen they contain. They also regulate the overall immune response by producing lymphokines.

Like B cells, T cells must first interact with antigens before they are able to attack invading pathogens. But they do so somewhat differently.

Antigen Recognition

T cells don't recognize intact antigens the way B cells do. The antigen must first be processed by **antigen-presenting cells.**

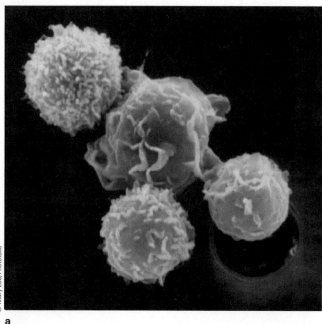

a

Antigen-presenting Cells. Antigen-presenting cells throughout the body ingest protein antigens (from microorganisms and other sources) they encounter and fragment them. Then they display pieces of the antigen (peptides) on their cell surface in a form that T cells recognize.

The most important antigen-presenting cells are the macrophages, but **dendritic cells,** named for their branching shape, are also important. They are found primarily in the skin, lymph nodes, spleen, and thymus.

Antigen fragments formed within an antigen-presenting cell are carried to the cell surface and held there by a group of proteins called the **major histocompatibility complex (MHC) proteins (Figure 17.11).** We have MHC proteins on the surface of all the nucleated cells in our body, but they vary dramatically from person to person due to differences in each individual's genetic makeup. They get their name (histo means tissue) from the fact that they become antigens when cells or tissues bearing them are given to another person. The other person's immune system responds by attacking them. Because of them, tissues can be transplanted (are compatible) only if the other person has identical or nearly identical MHC proteins, a very rare situation.

There are two classes of MHC proteins. One of them, called class II MHC proteins, holds antigen fragments on antigen presenting cells. The other, called class I MHC proteins, plays a role in T cell response to virus-infected cells, which we'll discuss later.

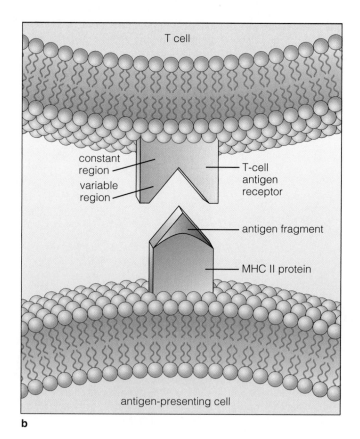

b

FIGURE 17.11 T-cell antigen recognition. (a) Antigen-presenting cell surrounded by three T cells. (b) T-cell antigen receptor recognizing an antigen fragment and class II MHC protein on an antigen-presenting cell.

Once the antigen is displayed, the antigen-presenting cells travel to lymph nodes or other lymphoid organs where large numbers of lymphocytes are concentrated. There the odds are increased that the antigen fragment will encounter a T cell with a matching receptor.

T-cell Antigen Receptors. In many respects, T-cell antigen receptors resemble the tethered antibody molecules on the surface of B cell that act as antigen receptors. Just as happens during B-cell differentiation, millions of genetically unique T cells are produced by recombination. Most of them have a unique antigen receptor protein with a unique antigen-binding specificity. The T-cell receptor also has a constant region that recognizes the MHC protein on the antigen-presenting cell.

Now let's turn our attention to how T cells become activated, the first step of the cell-mediated immune response.

T-Cell Activation

All T lymphocytes exist in a resting state until their antigenic receptors encounter and bind to a matching antigen fragment and MHC protein on an antigen-presenting cell. Such binding stimulates the T cell to divide and secrete lymphokines. The first cells to be activated in any immune response are the TH cells. These were the cells that initiated T. L.'s immune response. Activation is enhanced when the TH cell is stimulated by macrophage-produced interleukin-1. The stimulated TH cell stimulates itself still further by producing interleukin-2. It becomes more responsive to interleukin-2 by producing more receptors that recognize interleukin-2. Resting T cells have approximately 100 interleukin-2 receptors. Activated T cells have as many as 12,000. Interleukin-4 also helps activate TH cells.

The production of interleukin-2 by clones of activated TH cells also helps to activate cytotoxic T cells (TC) and other TH cells. When these cells recognize antigen fragments and class I MHC on antigen-presenting cells, they too respond to stimulation by interleukin-2. The combination of antigen recognition and interleukin-2 causes these cells to multiply, producing clones of activated cells that can participate in the cell-mediated immune response. This clonal selection is similar to B-cell clonal selection.

T-Cell Response

Now we'll examine how activated T cells kill virus-infected cells, signal bacteria-infected cells to kill their intracellular parasites, and regulate the intensifying of the immune response.

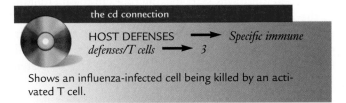

the cd connection

HOST DEFENSES ⟶ *Specific immune*
defenses/T cells ⟶ 3

Shows an influenza-infected cell being killed by an activated T cell.

Against Viruses. Human cells are able to signal that they have been infected by a virus. They do this by breaking down some viral proteins into peptide fragments. These fragments bind to class I MHC proteins, which take them to the cell surface and display them much like class II MHC proteins display antigen fragments on antigen-presenting cells (**Figure 17.12**).

The infected cell is then marked as a target for cytotoxic CD8 cells. The attack begins when a CD8 cell with a complementary antigenic receptor encounters and binds to the displayed viral antigen. The CD8 cell remains

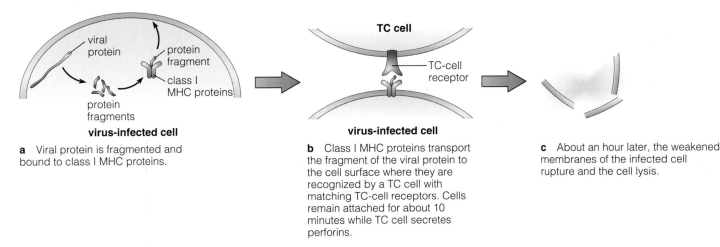

a Viral protein is fragmented and bound to class I MHC proteins.

b Class I MHC proteins transport the fragment of the viral protein to the cell surface where they are recognized by a TC cell with matching TC-cell receptors. Cells remain attached for about 10 minutes while TC cell secretes perforins.

c About an hour later, the weakened membranes of the infected cell rupture and the cell lysis.

FIGURE 17.12 Cytotoxicity. TC cells can destroy a virus-infected host cell.

TABLE 17.5 T-cell Response to Bacteria-infected and Virus-infected Cells

Property	Virus-infected Cell	Bacteria-infected Cell
Cellular location of antigen	Cytoplasm	Within phagosome
Transporting and displaying MHC proteins	Class I	Class II
T cells that recognize displayed antigen	Killer T cells, also called TC or CD8$^+$ T cells	Helper T cells, also called TH or CD4$^+$ T cells
Action of T cell	Kills the infected cell	Stimulates the infected cell to kill bacteria in its phagosome

attached for about 10 minutes. During this time, it secretes toxic proteins called **perforins** from its lymphocytic granules. These perforins start making holes in the membrane of the infected cell. About an hour later, the infected cell swells and lyses. The CD8 cell can then go on to kill many other infected cells. By killing infected cells, CD8 cells stop the production of new viral particles and thus stem the spread of infection.

Virus-infected cells are not the only targets for the lethal action of CD8 cells. Any host cell bearing a foreign antigen and class I MHC proteins is vulnerable. These include some cancer cells and cells in transplanted organs. CD8 cell killing is the basis of transplant rejection (Chapter 18).

Now let's turn to how T cells help to combat infections by bacteria that grow inside human cells.

Against Intracellular Bacteria.

Human cells, usually macrophages, can signal that they have been infected by bacteria and thereby get the help of TH cells. This signaling resembles the way virus-infected cells signal CD8 cells. But there are differences in the signal and the consequences. Antigens from the infecting bacteria are picked up by class II MHC proteins and taken to the cell surface. Receptors on TH cells bind to the antigen class II MHC. Then the bound TH cell releases lymphokines that stimulate the infected cell to kill the bacteria it contains. The infected cell survives.

The many differences between T-cell response to bacteria-infected and virus-infected cells (**Table 17.5**) depend upon the different locations and activities of class I and class II MHC proteins. Class I MHC proteins are in the cytoplasm where viral proteins are synthesized, so they pick up and display fragments of them. In contrast, class II MHC proteins can associate with intracellular membranes, so they pick up antigens from intracellular bacteria that multiply within membrane-bound phagosomes. Then, when displaying foreign antigens on the surface of the infected cell, class I MHC proteins bind to cell-killing CD8 cells. But class II MHC proteins bind to TH cells, which stimulate the cell to kill the invaders within its phagosome.

Immunoregulation.

Two classes of TH cells, designated TH1 and TH2, act as immunoregulatory cells. TH1 cells induce cell-mediated immunity. TH2 cells regulate antibody-mediated immunity.

T cells balance and regulate the immune response. They suppress it, as well as stimulate it. It appears that both CD8 cells and TH cells participate in suppression.

NATURAL KILLER CELLS

The third group of lymphocytes, natural killer cells (sometimes called non-B, non-T cells), makes up less than 3 percent of the total lymphocyte population. These cells are fundamentally different from B and T lymphocytes. Because they can function without recognizing an antigen, they are part of the innate immune system. We discussed them in Chapter 16. But natural killer cells lyse target human cells in much the same way CD8 cells do, by secreting perforins. Soon after an NK cell contacts a target, the target cell lyses. NK cells destroy cells infected by viruses and cells in transplanted organs. They are thought to be particularly effective against cancer cells (**Figure 17.13**).

Sometimes natural killer cells act as though they were part of the adaptive immune system because they destroy cells that have been coated by antibody. But their antibody-direct attack is nonspecific. They cannot distinguish among antibodies.

Now let's turn to one of the most important and most vexing questions of immunology. Why don't the body's own molecules act as antigens and stimulate a disastrous immune attack on its own tissues?

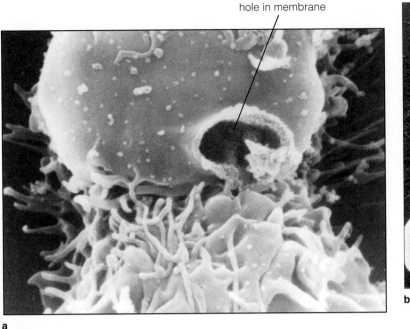

hole in membrane

Courtesy Gilla Kaplan, Rockefeller University

a

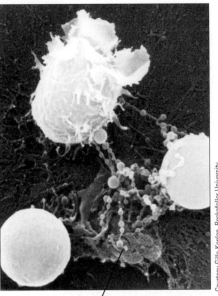

Courtesy Gilla Kaplan, Rockefeller University

b

remnants of lysed
cancer cell

FIGURE 17.13   Natural killer cells. These scanning electron micrographs show an NK cell destroying a tumor cell. (a) The large hole in the cancer cell was made by perforins secreted by the NK cell. (b) View about an hour later, after the cancer cell has been destroyed.

IMMUNE TOLERANCE

The immune system's lack of response to self antigens is called **immunological tolerance.** The ability to distinguish self antigens from foreign antigens means the immune system can distinguish self from nonself.

How this distinction is made is still an area of active research. One answer is a process called **clonal deletion.** The theory states that differentiating B cells in the bone marrow and T cells in the thymus are quite different from fully differentiated cells. Instead of being activated by antigens, they are killed by them. They undergo **apoptosis** (programmed cell death). And, of course, these immune cells differentiate in a sea of the body's own constituents (self antigens). But very few or no foreign antigens are present, so B cells or T cells that react with self antigens are eliminated. Only those that react with foreign antigens survive.

Still, some self-antigen–reacting immune cells survive the process of clonal deletion. But the body also has the ability to destroy mature immune cells that react with self antigens. Although the details aren't yet clear, the process seems to depend upon the requirement that T cells must simultaneously recognize self antigens (MHC proteins and at least one other, called B7) to recognize a foreign antigen.

When these mechanisms of immune tolerance fail, the result is **autoimmune disease** (Chapter 18). In effect, the body tries to destroy itself.

So far, we've discussed only immunity that results from naturally occurring foreign antigens, but there are other forms as well.

TYPES OF ADAPTIVE IMMUNITY

Immunity we have discussed so far is called **active immunity.** It's the product of an individual's own immune system that was acquired by infection. But immunity can also be acquired passively, and it can be artificially induced. **Passive immunity** is acquired by receiving antibodies that were produced elsewhere. Both active and passive immunity can be acquired naturally or artificially (**Table 17.6**).

Naturally acquired active immunity comes from infections encountered in daily life. **Artificially acquired active immunity** is stimulated by vaccines. We'll discuss vaccination (also called immunization) in Chapter 20.

Naturally acquired passive immunity refers to the antibodies transferred from mother to fetus across the placenta and from mother to the newborn in colostrum. **Artificially acquired passive immunity** refers to antibodies

TABLE 17.6 Types of Immunity

Type	Natural	Artificial
Active	B cells and T cells	B cells and T cells
	Immunity developed from a naturally acquired infection	Immunity developed in response to a vaccine
Passive	Antibodies only	Antibodies only
	Antibodies transferred from mother to child across the placenta and in breast milk	Antibodies from a donor are injected into an individual

formed by an animal or a human and administered to prevent or treat infection. We'll discuss this too in Chapter 20.

THE ROLE OF THE ADAPTIVE IMMUNE SYSTEM IN T. L.'S RECOVERY

Now let's return to the case of T. L. and see how what we've learned about immunology applies to it.

When T. L.'s skin was punctured, large numbers of bacteria were introduced directly into his subcutaneous tissues. The bacteria multiplied rapidly, overwhelming local inflammatory defenses of his innate immune system. Without help from his adaptive immune system, the infection would have spread throughout his body, probably killing him.

Because T. L. was infected by bacteria that didn't multiply inside his cells, B cells play the major role in his defense. But T cells had to be activated in order to activate the B cells. This began when antigen-presenting cells such as macrophages phagocytized some of the invading bacteria. They destroyed most of the bacterial cell and processed antigens into fragments. These were transported to the cell surface by class II MHC proteins and displayed. Very few responsive T cells were present near T. L.'s wound, so the likelihood that the displayed antigens would happen to encounter T lymphocytes with matching receptors at the site was extremely remote. But macrophages transported the processed antigens to T. L.'s lymph nodes, which housed millions of immunocompetent T cells with different antigen receptors.

Within the lymph nodes, antigen-presenting macrophages encountered TH cells with antigen receptors that matched the displayed bacterial antigens. Recognition occurred when the lymphocyte antigen receptors bound to the displayed antigens. This recognition stimulated the lymphocytes to enlarge and become more metabolically active. The TH cells were stimulated further by macrophage-produced

interleukin-1 molecules. TH cells in turn produced self-stimulating lymphokines, particularly interleukin-2, that caused cell division and formation of an active TH-cell clone.

At the same time, B cells recognized and bound to antigens on the surface of the invading bacterial. Because he had been exposed previously to bacteria with these same antigens, T. L. had a relatively large number of memory B cells with receptors that matched the particular bacterial antigens. This significantly accelerated his immune response.

Memory B cells took up and presented bacterial antigens, causing some B and TH cells to become coupled together. The close proximity of these two types of lymphocytes allowed the B cells to receive large quantities of activating and growth-promoting lymphokines from the TH cells. This stimulation reactivated the memory B cells, producing an army of active B cells that became plasma cells capable of producing large quantities of antibodies to fight the bacteria infecting T. L.'s hand. Because this was a secondary immune response, these plasma cells produced mainly IgG.

Millions of IgG molecules entered T. L.'s circulation. Wherever they encountered an invading bacterium, their variable region bound to antigens on the bacterial cell surface. At sites within their constant region, these same antibody molecules bound to complement or phagocytes, activating complement and causing opsonization. Opsonization made the invading bacteria easy prey for neutrophils and other phagocytes. Before antibody levels rose, the bacteria multiplied faster than T. L.'s phagocytes could eliminate them. But after antibody became plentiful, phagocytosis outpaced bacterial multiplication and T. L.'s immune system won the battle against the invading bacteria.

As T. L.'s infection subsided, suppresser T cells returned the activated lymphocytes to a resting state. Some memory lymphocytes, however, persisted in his body to defend him against yet another infection by the same type of bacteria.

SUMMARY

The Adaptive Immune System: An Overview (pp. 409–414)

1. When our first line of defense (Chapter 15) is breached and our innate immune system (Chapter 16) fails to control an infection, our adaptive immune system is activated.

2. Our adaptive immune system is powerful, highly specific, but slow to act on first encounter with a particular pathogen.

3. Our adaptive immune system is effective against prokaryotic and eukaryotic microbes, viruses, and helminths.

4. Our adaptive immune system responds quickly and more powerfully on subsequent exposures to a particular pathogen.

5. Vaccination confers immunity but avoids the dangers of a first infection.

Lymphocytes (pp. 410–411)

6. Lymphocytes are small, smooth, round leukocytes that lack granules.

7. There are three kinds of leukocytes: B cells, T cells, and natural killer (NK) cells.

8. B cells and T cells are triggered into defensive action when they encounter antigens, which are foreign molecules we lack.

9. B cells respond to antigen by producing antibodies.

10. Cytotoxic T (CD8) cells respond to antigens by killing our own pathogen-infected cells.

11. Helper (CD4) T cells respond by stimulating phagocytes and B cells.

Humoral Immunity (p. 411)

12. Humoral immunity depends upon the production of antibodies by B cells.

13. Antibodies protect against invading microbes by neutralization, opsonization, and activating complement.

Cell-mediated Immunity (p. 411)

14. Because T cells attack infected cells, the immunity they confer is termed cell mediated.

Lymphoid Tissues (pp. 411–412)

15. Primary lymphoid tissues are the bone marrow where all blood cells, including lymphocytes, are formed and where B cells differentiate; the thymus is where T cells differentiate.

16. Differentiated B and T cells travel to secondary lymphoid tissues, where they are stored and interact.

17. Secondary lymphoid tissues include lymph nodes, spleen, tonsils, adenoids, appendix, and Peyer's patches.

18. Lymph is formed in our tissues as fluid from blood leaks across capillary walls.

19. A system of lymph vessels collects lymph and returns it to the thoracic duct that empties into the heart.

20. Lymphocytes circulate through blood and lymph.

21. Lymphatic vessels and lymph nodes usually become swollen and inflamed during infections that activate an immune response.

Antigens (pp. 412–414)

22. The adaptive immune system recognizes and reacts to antigens.

23. Many macromolecules, including proteins and polysaccharides, can act as antigens.

24. Owing to immune tolerance, macromolecules in our body do not act as antigens.

25. Antigen receptors on B and T cells bind to epitopes (antigenic determinants) or antigens by shape-to-shape recognition.

The Humoral Immune Response (pp. 414–423)

26. Humoral immune defense is mounted by the fraction of our 10 trillion B cells that bear antigen receptors able to attach to a particular epitope.

Generating Antibody Diversity (p. 414)

27. The diversity of B cell receptors is sufficient to recognize epitopes on any pathogen we are likely to encounter in our lifetime.

28. B-cell diversity is generated by recombining bits of DNA in our genome to generate 100 million different antibody genes.

Structure of Antibodies (pp. 414–417)

29. Antibodies are Y-shaped molecules composed of four different polypeptide chains, two of which are identical light chains and two of which are identical heavy chains.

30. Antibodies are glycoproteins because carbohydrate groups are attached to them.

31. Antigen binding sites are located at the ends of the arms of the Ys, below which are the variable regions. The rest of the molecule is the constant region.

32. The constant region binds to and activates phagocytes and complement.

33. There are five classes of antibodies: IgA, IgD, IgE, IgG, and IgM.

34. IgD, IgE, and IgG are secreted as monomers, IgM as pentamers, and IgA as monomers or dimers.

B-Cell Activation (pp. 417–419)

35. Naïve B cells are fully differentiated B cells that have not yet encountered a complementary antigen; they bear membrane-bound IgM and IgD on their outer surface.

36. Clonal selection begins when naïve B cells bind to their complementary antibody: Then they multiply, produce soluble IgM and IgD, and undergo a high rate of mutation in their DNA-encoding region, resulting in affinity maturation.

37. Later, class switching occurs: Various activated B cells begin to make all classes of antibodies.

38. Some members of the clone become short-lived effector (plasma) cells, which make large quantities of antibody; others become long-lived memory cells, which fight future infections by the same pathogen.

39. Lymphocytes, including interleukins-1, -2, -4, -5, and -6, produced by TH1 cells, macrophages, B cells, and NK cells help B-cell activation stimulated by most antigens. Some T-independent antigens stimulate B-cell activation without help.

Reactivation: Immunological Memory (pp. 419–420)

40. The powerful and rapid secondary immune response initiated by memory B cells acts within a few days, sometimes suppressing all symptoms.

41. The action of vaccines depends upon their stimulating the formation of memory B cells.

Action of Antibodies (pp. 420–423)

42. Antibodies protect against infection by neutralizing pathogens, acting as opsonins, and activating complement via the classical pathway.

43. The classical pathway, which is activated by antigen-bound IgM or IgG, produces C3b, a powerful opsonin, and, via the terminal pathway, bacteria-killing membrane-attack complex.

The Cell-mediated Immune Response (pp. 424–426)

44. Cytotoxic (CD8) T cells, with the aid of helper (CD4 or TH cells) T cells, are responsible for cell-mediated immunity.

Antigen Recognition (pp. 424–425)

45. T cells recognize antigens that have been processed by antigen-presenting cells (macrophages and dendritic cells).

46. Antigen fragments formed by antigen-presenting cells are carried to the cell's surface and presented there by class II major histocompatibility complex (MHC) proteins.

47. T-cell antigen receptors resemble in diversity and activity tethered antibodies on B cells.

T-Cell Activation (p. 425)

48. Binding of T-cell receptors to antigen fragments on antigen-presenting cells initiates T-cell activation.

49. Activated TH cells produce lymphokines that activate cytotoxic T cells and other TH cells.

T-Cell Response (pp. 425–426)

50. Virus-infected human cells fragment viral proteins, which class I MHC proteins carry to their surface, signaling cytotoxic T cells to kill the cell by producing perforins.

51. Killing infected cells stops production of new viral particles and stems the spread of infection.

52. Class II MHC proteins in human cells infected by intracellular bacteria carry antigens from the pathogen to the cells surface, signaling TH cells to release lymphokines that stimulate the infected cell to kill the bacteria it contains.

Natural Killer Cells (pp. 426–427)

53. NK cells act principally in innate immunity, but they also kill antibody-coated pathogens. They cannot distinguish among antibodies.

Immune Tolerance (p. 427)

54. Immune tolerance is lack of response to the body's own constituents.

55. One basis of immune tolerance is clonal deletion: Differentiating immune cells are killed by antigens. Cells that react against the body's own constituents are eliminated.

56. The body also has mechanisms to destroy fully mature immune cells if they act against the body's constituents.

Types of Adaptive Immunity (pp. 427–428)

57. Immunity can be active (a product of an individual's own immune system) or passive (antibodies received from another individual). Both types can also be naturally or artificially acquired.

58. Naturally acquired active immunity results from infections. Artificially acquired active immunity comes from vaccines. Naturally acquire passive immunity refers to the antibodies a child receives from the mother across the placenta and in breast milk. Artificially acquired passive immunity consists of antibodies formed by an animal or human and administered to an individual.

REVIEW QUESTIONS

The Adaptive Immune System: An Overview

1. What is the adaptive immune system? Name the three kinds of lymphocytes.

2. What is humoral immunity? What else is it called?

3. What is cell-mediated immunity?

4. Name the primary lymphoid organs and describe their functions. Do the same for the secondary lymphoid organs.

5. Describe lymphatic circulation. Why are lymph nodes critical to immune-system functioning?

6. What is an antigen? An antibody?

The Humoral Immune Response

7. Explain the process of genetic recombination that leads to B-cell diversity.

8. How is an antigen-antibody complex formed?

9. Explain this statement: Unlike the lock-and-key fit between enzyme and substrate, antigen-antibody binding is variable.

10. What is clonal selection in B-cell activation? How do effector B cells differ from memory B cells?

11. How do TH cells function in B-cell activation?

12. What are lymphokines? Which lymphokines are involved in B-cell function? How are they involved?

13. What is immunological memory? What is the difference between a primary immune response and a secondary immune response?

14. What are plasma cells?

15. How does the linking activity of antigens protect us against infection?

16. Describe the structure of antibodies. Use these terms in your explanation—antibody monomer, heavy chains, light chains, antigen binding sites, variable regions, constant regions, valence, immunogenic.

17. Name the five classes of antibodies. For each, tell what percent of the total antibody population it constitutes and what its main functions and features are.

The Cell-mediated Immune Response

18. What kind of infections do T cells act against?

19. Describe how T-cell antigen recognition occurs.

20. What role do lymphokines play in T-cell activation?

21. How does a TC cell know that a human cell is infected by a virus?

22. How does a TH cell know that a human cell is infected by intracellular bacteria?

23. Describe how a TC cell kills.

24. Describe how TH cells stop intracellular bacterial infections.

25. What are MHC proteins?

Natural Killer Cells

26. Name the ways non-B, non-T cells are different from B and T cells.

27. How do NK cells kill?

28. Explain this statement: Non-B, non-T cells are more like nonspecific defenses than like specific defenses.

Immune Tolerance

29. What does immune tolerance mean? Why is it important?

30. What does clonal deletion mean? How does it occur?

31. What does apoptosis mean?

Types of Adaptive Immunity

32. What distinguishes active from passive immunity?

33. Define these types of immunity and give an example of each: naturally acquired active immunity, naturally acquired passive immunity, artificially acquired active immunity, artificially acquired passive immunity.

CORRELATION QUESTIONS

1. Could you distinguish a B cell from a T cell by looking at it under a microscope? Could you distinguish a resting T cell from an activated T cell? Explain.

2. How do we know that memory B cells participated in defending T. L. against his infection?

3. Distinguish between the roles played by B cells and T cells in fighting a viral infection.

4. What determines whether a TH cell or a TC cell will bind to an infected cell during an immune response?

5. Could IgG cause a soluble antigen to form visible clumps? Why?

6. Would passive immunity be able to rid the body of virus-infected cells? Explain.

ESSAY QUESTIONS

1. Discuss the interconnections between the immune system and the body's nonspecific defenses.

2. Discuss the value of the immune system's being able to make five kinds of antibodies.

SUGGESTED READINGS

Benjamini, E, G. Sunshine, and S. Lleskowitz. 1996. *Immunology: A short course.* New York: John Wiley and Sons.

Parham, P, 2000. *The immune system.* New York: Garland Publishing.

Scientific American. 1993. Life, death, and the immune system. Special issue, September.

For additional readings, go to InfoTrac College Edition, your online research library at: http://www.infotrac.thomsonlearning.com

EIGHTEEN

Immunological Disorders

© David Scharf/Peter Arnold, Inc.

CHAPTER OUTLINE

LEARNING GOALS

To understand:

- *How abnormal or misdirected immune responses, called hypersensitivities, can harm the body*

- *The basis of type I hypersensitivity and its relationship to allergy*

- *The basis of type II hypersensitivity and its relationship to blood transfusion and certain autoimmune diseases*

- *The basis of type III hypersensitivity and its relationship to certain autoimmune diseases*

- *The basis of type IV hypersensitivity and its relationship to tissue transplantation, contact dermatitis, and granulomatous infections*

- *Congenital and acquired immunodeficiency disorders*

- *How failure of immune surveillance might lead to cancer*

"It's Nothing—Just a Bee Sting"

Mr. C. was a 31-year-old operator of farm machinery. One day, while cultivating an almond orchard that was being pollinated by honeybees, he was stung on his hand. Within 5 minutes his hand was red and swollen. Having experienced similar reactions to bee stings, he was unconcerned and continued his work. But during the next 45 minutes, Mr. C. became alarmed. The redness and swelling spread up his forearm almost to his elbow. His entire body began to itch. Red, blotchy patches appeared on his face, chest, and left arm. His lips became swollen. He felt nauseated. The itching became particularly intense under his chin. When he called to a fellow worker for help, he was hoarse and could barely talk because of a dry, uncontrollable cough.

Mr. C. was rushed to a nearby emergency room and immediately given a subcutaneous injection of epinephrine (adrenaline). Within minutes he improved markedly. Fifteen minutes later he felt back to normal. But the emergency room doctor kept Mr. C. several hours for observation. She told him he had suffered an anaphylactic reaction to a bee sting allergy. Before discharging him, she gave him an injection of a long-acting adrenaline-like medication and a prescription for an antihistamine to take orally at home. She also gave him a prescription for an adrenaline syringe for him to carry and use at once if he suffered similar symptoms from another bee sting. Finally, considering his constant risk of being stung, she advised him to consult an allergist to see if he should undergo immunotherapy to prevent similar reactions.

Case Connections

- Our adaptive immune system (Chapter 17), which evolved to protect us from pathogenic microbes, can make us vulnerable to sudden death by exposure to nonmicrobial substances such as bee venom.
- Note the suddenness and intensity of Mr. C.'s reaction. We'll examine the reasons in this chapter.
- Is it significant that this was not the first time that Mr. C. had been stung? We'll see.

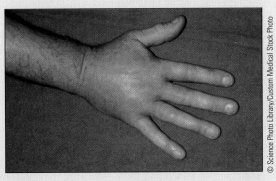

a A typical allergic reaction to a bee sting.

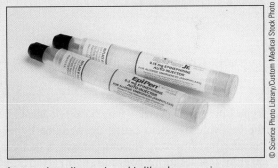

b An adrenaline syringe kit like the one given to Mr. C. by his physician. Such a kit usually contains an alcohol pad for cleaning the injection site, a syringe filled with epinephrine, antihistamine pills to extend the effect of the epinephrine, and a tourniquet to tie above the site of the sting to prevent the spread of venom.

IMMUNE SYSTEM MALFUNCTIONS

In Chapters 16 and 17 we saw how complex and powerful our immune system is. We probably shouldn't be surprised that such a complicated system could malfunction or even harm us. In this chapter we'll see how it sometimes does. We'll examine the resulting immunological disorders. There are two kinds of them: **hypersensitivities** (inappropriate immune responses that harm us) and **immunodeficiencies** (inadequate immune responses that fail to protect us). We'll discuss them sequentially. Then we'll consider more subtle immune inadequacies that might lead to cancer.

HYPERSENSITIVITY

There are four types of hypersensitivities. They differ with respect to the events that trigger them and the damage they cause (**Table 18.1**).

TABLE 18.1 Hypersensitivity

Type	Names	Examples	Reaction Time
Type I	Anaphylactic hypersensitivity Allergy	Insect sting allergy, hay fever, asthma, food allergy Immediate hypersensitivity	Immediate, with 20 minutes
Type II	Cytotoxic hypersensitivity	Transfusion reactions, Rh disease of the newborn, Goodpasture's syndrome, Graves' disease	Variable
Type III	Immune-complex hypersensitivity	Systemic lupus erythematosus (SLE), rheumatoid arthritis, farmer's lung disease, serum sickness	Variable
Type IV	Cell-mediated hypersensitivity Delayed hypersensitivity	Organ transplantation rejection, poison ivy/oak/sumac, clothing and jewelry allergies, granulomatous reaction	12 hours to several weeks

Type I: Anaphylactic Hypersensitivity (Allergy)

Type I hypersensitivity—which caused Mr. C.'s reaction to the bee sting—is also called **immediate hypersensitivity** because its symptoms develop rapidly, usually within 10 to 20 minutes. Its more common name is **allergy.** The antigens that stimulate it are called **allergens.** Our reaction to the triggering of a type I hypersensitivity is called **anaphylaxis** (*ana*, "away from"; *phylaxis*, "protection").

Type I hypersensitivities, such as Mr. C.'s, begin when a person is exposed to an allergen. Rather than stimulating a protective immune response, the allergen sensitizes the person. Subsequent exposure to the same allergen triggers a harmful allergic reaction.

Allergic reactions can be local, as Mr. C.'s was initially, or they can be systemic and life threatening, as Mr. C.'s became. Systemic reactions affect the air passages that supply the lungs and blood vessels throughout the body. People who are allergic to a particular substance may have a localized reaction after one exposure to it and a systemic reaction after another.

Now let's examine how allergens cause type I hypersensitivity.

Immunological Basis of Type I Hypersensitivity. In Mr. C.'s case the allergen was **phospholipase A,** an enzyme in bee venom. Phospholipase A from a previous bee sting had stimulated his B cells to produce immunoglobulin E (IgE) antibodies against this allergen (Chapter 17). The constant region of these IgE antibodies bound to the surfaces of mast cells and basophils (which are powerful inflammatory mediators; Chapter 16).

Then, when Mr. C was stung again, phospholipase A in the bee's venom bound to the antigen-binding sites on two adjacent IgE molecules (**Figure 18.1**). That binding triggered the mast cell or basophil to release inflammatory mediators, including histamine, prostaglandins, and leukotrienes.

Histamine acted almost immediately. It caused blood vessels to dilate, which in turn constricted the passages that deliver air to the lungs. The leukotrienes acted more slowly. They caused Mr. C's prolonged respiratory difficulty.

Inflammatory mediators caused all of Mr. C.'s allergic reactions. For example, leukotrienes made Mr. C.'s bronchial smooth muscles contract, which caused his dry spasmodic cough. Histamine increased vascular permeability, which caused edema of nearby tissues and many symptoms: Edema of his skin caused the raised itchy rash (**hives**). Edema of his lips caused them to become large and pale. Edema of his intestinal tract produced nausea. Edema of his **larynx** narrowed the airway within his larynx (voice box), causing his hoarseness and the itching under his chin.

Anaphylactic Reactions. Mr. C.'s reaction to these inflammatory mediators was potentially life threatening. Laryngeal edema could have completely obstructed his airway, causing immediate death from suffocation. Increased vascular permeability could have caused a very rapid loss of fluid from his bloodstream, causing possible sudden death from **anaphylactic shock** (a precipitous fall in blood pres-

TABLE 18.1 Hypersensitivity (continued)

Antibodies Involved	Action	Damage Caused by Antibody Action	Treatment
IgE	IgE antibodies link mast cells and allergen	Mast cells and basophils release chemicals (e.g., histamine) during degranulation	Epinephrine, antihistamines, immunotherapy (desensitization)
IgG, IgM	IgG and IgM mark normal human cells as targets for immune effectors	Complement and phagocytosis	Varies; supportive treatment only for transfusion reactions, exchange transfusion for Rh disease; insulin for diabetes
IgG, IgM	Antigen-antibody complexes are distributed throughout the body	Complement and phagocytosis stimulated by antigen-antibody complexes	Varies; often steroids to suppress further antibody production
None	Lymphokines and macrophages interact, causing inflammation	T cells activated	Varies; often steroids to suppress T-cell function

sure). Each year about 40 people in the United States die from anaphylactic reactions caused by bee stings.

The epinephrine Mr. C. received in the emergency room reversed his anaphylactic reactions almost immediately by constricting his blood vessels and relaxing his bronchial muscle. The antihistamine medication he took later blocked the effects of histamine by competing for its binding sites on target cells. Mr. C.'s chances of suffering an anaphylactic reaction if he suffers another bee sting are about 60 percent. If used immediately, the preloaded syringe of epinephrine prescribed for him would counteract it.

Immunotherapy. The immunotherapy his physician suggested could desensitize him to bee venom. The procedure would involve injecting increasing amounts of bee venom until he could tolerate a substantial dose. Then he would be protected from bee sting anaphylaxis.

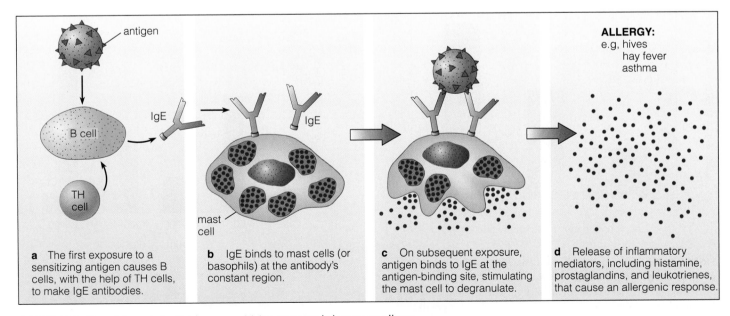

a The first exposure to a sensitizing antigen causes B cells, with the help of TH cells, to make IgE antibodies.

b IgE binds to mast cells (or basophils) at the antibody's constant region.

c On subsequent exposure, antigen binds to IgE at the antigen-binding site, stimulating the mast cell to degranulate.

d Release of inflammatory mediators, including histamine, prostaglandins, and leukotrienes, that cause an allergenic response.

FIGURE 18.1 Type I (anaphylactic) hypersensitivity, commonly known as allergy.

The immunological basis of this seemingly paradoxical form of treatment is thought to be the following: Small does of venom stimulate the production of IgG antibodies, which protect by binding to phospholipase A molecules, thereby preventing them from binding to IgE on mast cells. Immunotherapy is effective in preventing adverse reactions to bee stings in 97 to 100 percent of patients.

Allergy. Type I hypersensitivity reactions are very common. Up to 20 percent of the population of the United States is affected by some form of them. The tendency toward allergies is inherited. Perhaps it involves a genetic capacity to produce to excessive amounts of IgE antibodies (**Figure 18.2**).

Allergens can enter our bodies by several routes. Some, such as insect venoms and penicillin or other medications, are injected into tissues beneath the skin. These allergens enter the circulation quickly and are particularly likely to cause life-threatening anaphylaxis. Others, such as pollen, mold spores, animal dander, and feces of microscopic mites, which are inhaled (**Figure 18.3**), usually provoke localized respiratory allergies. Still other allergens, including medications and foods, are ingested. They can provoke either localized or generalized allergic reactions.

Inhaled allergens that primarily affect the upper respiratory tract cause **allergic rhinitis** (called **hay fever** when it occurs seasonally due to pollen). It's the most common allergic disorder. The allergen acts on mast cells in the mucous membranes of the nose. The inflammatory mediators they release cause nasal tissues to swell and produce excess mucus. The result is nasal congestion, itching, and mucus discharge. Because allergic rhinitis is mediated primarily by histamine, antihistamines are usually an effective treatment, but nasally administered steroids or sodium cromolyn are more effective. Steroids suppress the immune response. Sodium cromolyn prevents release of inflammatory mediators by mast cells.

Asthma. Inhaled allergens that affect the lower respiratory tract can cause **asthma,** a breathing disorder that results from constriction of the smooth muscle surrounding bronchial passages causing the airways themselves to swell. The consequence is difficult breathing. Asthma is a chronic disorder that affects most sufferers intermittently throughout their lives. In addition to allergens that cause hay fever, respiratory irritants, such as air pollutants or viral infections, can precipitate asthma attacks. Leukotrienes, not histamine, are primarily responsible for asthma, so antihistamines are not an effective treatment. Asthma is most commonly treated with inhaled epinephrine-like drugs, sodium cromolyn, and steroids. Asthma is a common and often relatively mild disease, but it can be life threatening. For unknown reasons, the incidence of fatal cases of asthma have increased significantly in recent years.

Allergy Testing. To avoid exposure or to plan treatments, it's sometimes important to determine which allergens are causing a patient's symptoms. Some patients with severe allergic rhinitis who don't respond adequately to

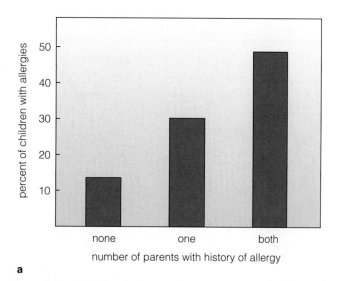

a

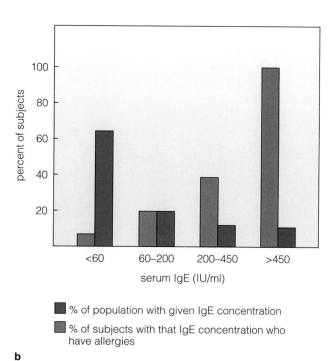

b

FIGURE 18.2 Predisposition to allergies. (a) A child's risk of becoming allergic rises if one parent is allergic and even more if both are allergic. (b) Risk of being allergic rises dramatically as serum IgE rises. Almost everyone with serum IgE levels greater than 450 IU/mL is allergic.

■ % of population with given IgE concentration
■ % of subjects with that IgE concentration who have allergies

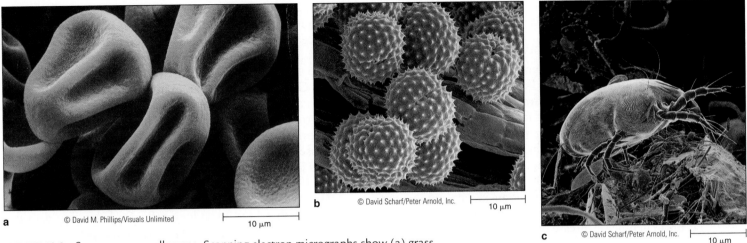

a   © David M. Phillips/Visuals Unlimited   ⊢————⊣ 10 μm

b   © David Scharf/Peter Arnold, Inc.   ⊢————⊣ 10 μm

c   © David Scharf/Peter Arnold, Inc.   ⊢————⊣ 10 μm

FIGURE 18.3   Some common allergens. Scanning electron micrographs show (a) grass pollen; (b) ambrosia (ragweed pollen), the most common cause of hay fever in the East and Midwest of the United States; and (c) the house dust mite. The mite's fecal pellets, which are about the same size as pollen grains, become airborne and cause allergies.

medication want to try immunotherapy, even though de-sensitization to pollens isn't as effective as desensitization to bee venom. The first step is skin testing.

Small amounts of different allergens are placed on the patient's skin (**Figure 18.4**). If a person is allergic to some of them, sensitized mast cells are stimulated to produce small hivelike swellings where the allergen in applied. Although skin testing is very accurate, it is uncomfortable and in rare case causes life-threatening anaphylaxis. To avoid these problems, it's now becoming more common to determine IgE levels in the patient's blood.

Food Allergies. Ingested allergens can also cause serious reactions. Most so-called food allergies are not allergies at all. They are a result of irritation of the intestinal tract or problems with digestion. However, a few adverse reactions to food are caused by type I hypersensitivity. Highly allergenic foods such as peanuts or shellfish can cause hives or systemic anaphylactic reactions identical to Mr. C.'s.

Drug Allergies. Type I hypersensitivity reactions to ingested medications are more common than to food. Penicillin, for example, can act as a hapten that becomes a potent allergen when it binds to human albumin (Chapter 17). Usually,

allergic drug reactions cause only hives, but they can cause systemic anaphylaxis, particularly if the drug is injected rather than ingested. Each exposure to penicillin carries a risk of provoking allergy, even in people who have taken the drug safely many times before.

Type II: Cytotoxic Hypersensitivity

Type II hypersensitivity occurs when the immune system mistakenly attacks some of our own cells as though they were invaders. It is responsible for three disorders: autoimmune diseases, transfusion reactions that follow transfusions of mismatched blood, and hemolytic disease of the

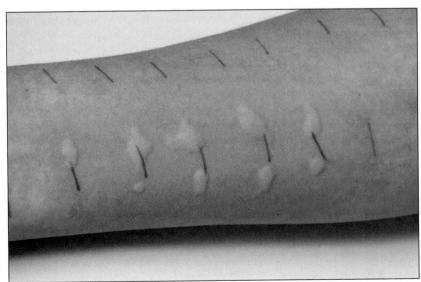

FIGURE 18.4   Allergy skin tests. Tests are usually done on the forearm (as in this photo) or the back. In the test shown here, hivelike reactions have developed on either end of some of the black marks where the allergens were placed.

© SIU Biomed Comm/Custom Medical Stock Photo

newborn. Type II hypersensitivity is also called **cytotoxic hypersensitivity** because cells affected by it usually die.

Immunological Basis of Type II Hypersensitivity.

Type II hypersensitivity occurs when IgM or IgG antibodies bind to some of our body's own cells, instead of to pathogenic microbes (**Figure 18.5**). The cells to which they bind could suffer any of four possible consequences: (1) altered function, (2) being attacked by phagocytes and killer cells (Chapter 16), (3) being clumped by the multivalent IgM, or (4) being lysed by membrane attack complex (through the terminal complement pathway).

The first two consequences occur in autoimmune disorders. The third occurs in transfusion reactions. The last occurs in hemolytic disease of the newborn. We'll consider all these briefly.

Autoimmune Diseases. Autoimmune diseases occur when immune tolerance fails and antibodies are produced against the body's own constituents. Some but not all autoimmune diseases are caused by type II hypersensitivity.

Graves' disease is an example of altered cell function. It's caused by antibodies binding to the cells that produce thyroid hormone and stimulating them to produce more hormone. The resulting high levels of thyroid hormone cause many physiological abnormalities, including weight loss and protruding eyes.

Goodpasture's syndrome is an example of cells being attacked and killed. In this disease the body produces antibodies that bind to kidney cells. Such binding activates complement, which attracts activated leukocytes that cause a destructive inflammatory reaction. The condition can

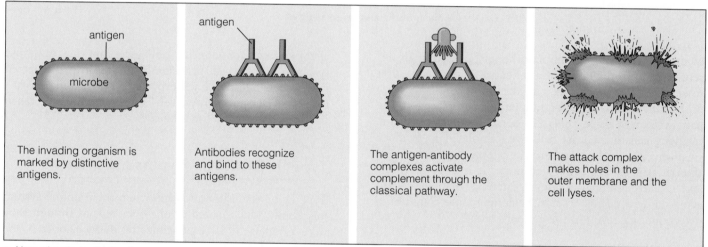

The invading organism is marked by distinctive antigens.

Antibodies recognize and bind to these antigens.

The antigen-antibody complexes activate complement through the classical pathway.

The attack complex makes holes in the outer membrane and the cell lyses.

a Normal antimicrobial action

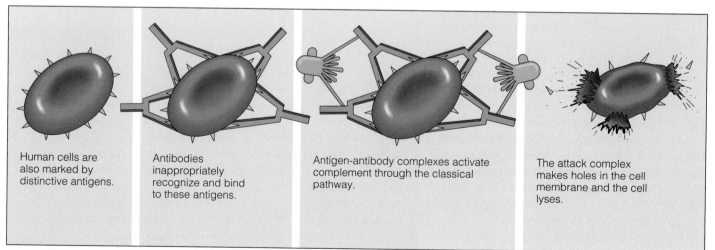

Human cells are also marked by distinctive antigens.

Antibodies inappropriately recognize and bind to these antigens.

Antigen-antibody complexes activate complement through the classical pathway.

The attack complex makes holes in the cell membrane and the cell lyses.

b Hypersensitivity

FIGURE 18.5 Type II (cytotoxic) hypersensitivity resembles the normal immunological events that destroy pathogenic microorganisms.

persist for months or years. Unless treated by dialysis or a kidney transplant, many patients die of renal failure.

Transfusion Reactions. Transfusion reactions are an example of cells being clumped by multivalent IgM. They are a consequence of the immune system's reacting to foreign antigens on transfused blood cells. **Erythrocytes** (red cells) in blood bear a variety of antigens that vary according to blood type. The best known of these are those that constitutes the ABO antigen system. Separate genes determine which of two antigens (A or B) are present on a person's erythrocytes (**Table 18.2**). People who produce only the A antigen are said to have type A blood. Those who produce only B antigens have type B blood. Those who produce both have type AB blood, and those who produce neither have type O blood.

Because of immunological tolerance, people don't have antibodies against the antigens they produce. But people who lack A or B antigens do produce antibodies (primarily of the IgM class) against them even if they have never been exposed to another person's blood. So almost everyone except those with type AB blood is likely to experience a transfusion reaction on receiving blood with A or B antigens that he or she lacks. If the antibodies in the transfusion recipient's serum react with antigens on the donor's red cells, they can provoke two disastrous complications (**Figure 18.6**).

The first complication occurs when the transfused erythrocytes agglutinate (clump together) because multivalent IgM antibody molecules form interconnecting networks of cells (Table 17.7). The second complication occurs

TABLE 18.2 ABO Antigen System

Blood Type	Antigen(s) Person Produces		Blood Type Person Can Safely Receive
	A	B	
A	+	−	A or O
B	−	+	B or O
A B	+	+	A, B, or O
O	−	−	O

when antigen-antibody complexes activate complement to form membrane attack complexes that lyse erythrocytes. The number of circulating red cells available to carry oxygen decreases, and large amounts of hemoglobin and other potentially damaging cell contents are released into the circulation. Together these complications cause a **transfusion reaction.** The symptoms are rapid onset of fever, vomiting, and shock. Thus people cannot safely receive blood containing an antigen that they lack (Table 18.2).

Hemolytic Disease of the Newborn. In addition to the A and B antigens, human erythrocyte may carry a third antigen, designated Rh. Blood is termed Rh positive if red blood cells carry this antigen and Rh negative if they don't. Rh antigens cause the most severe form of

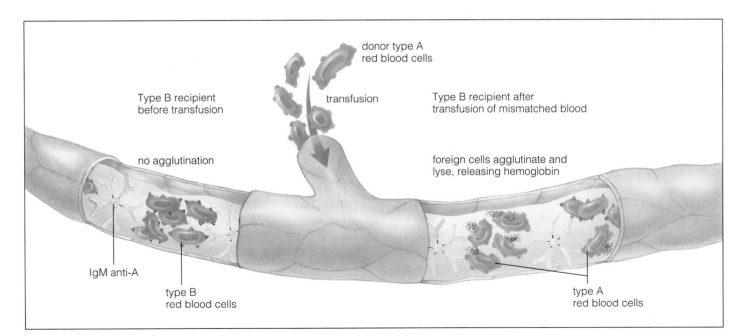

FIGURE 18.6   Consequences of a mismatched blood transfusion.

SHARPER FOCUS

BLOOD TRANSFUSION

The idea of replacing blood lost as the result of injury arose centuries ago. It seemed the obvious way to prevent death from massive bleeding. As early as the 1600s, physicians tried giving animal or human blood to bleeding humans. But their efforts at transfusion were disastrous failures. Occasionally there were lifesaving miracles, but the overall results were so bad that transfusions were banned in most European countries. Only the Incas were routinely able to perform successful blood transfusions. Now we know why. All native South Americans have type O blood, so they can give and receive blood without immunological complications.

The rest of the world had to wait until the early 1900s, when Karl Landsteiner discovered the ABO blood-antigen system. His work made blood transfusions possible. But

Landsteiner's discovery was not put to practical use until World War I, when physicians began to collect blood, type it according to the ABO system, and transfuse many of the more than 20 million soldiers who were seriously wounded. Quickly another important practical discovery was made: Sodium citrate would prevent blood from clotting, allowing blood to be stored for prolonged periods in blood banks in anticipation of massive needs for it.

Now more than 3 million blood transfusions are performed each year in the United States. Many of these are transfusions of whole blood. But other types of transfusions are also done routinely. Because blood can now be separated into its various lifesaving components—oxygen-carrying erythrocytes, blood-clotting platelets, and serum proteins such as immunoglob-

ulins or clotting factors—each of these blood products can be administered selectively. Patients who lose blood during surgery, for example, usually receive packed red blood cells to replenish their lost oxygen-carrying capacity. Patients who begin to bleed from treatments such as leukemia therapy receive a platelet transfusion. Patients with **hemophilia** (a genetically determined bleeding disorder) receive the clotting proteins that they are unable to produce for themselves.

But blood transfusion remains a risky business. To prevent the potentially fatal mistake of giving a patient incompatible blood, checks and double-checks are always done. Before any transfusion, two health-care professionals must check the identification code on the unit of blood and compare it with the patient's hospital identification number.

Then both must sign to verify their identification. The patient is also observed closely during a transfusion, so it can be stopped if signs of a transfusion reaction develop.

Another risk is transmission of infection. Many viral pathogens survive in banked blood and can be spread from donor to recipient. For years, hepatitis B was the most feared infectious complication of blood transfusion, but today HIV is the greater concern. To make transfusions safer, blood donors are carefully questioned for their risk of infection and donated blood is screened for infection-related antibodies. Still, no blood can ever be certified as being totally pathogen-free.

Blood transfusion remains today a potentially lethal but lifesaving form of therapy. Like all medical treatments, its likely benefits must be weighed against the known risks.

---

hemolytic (*hemo*, "blood"; *lytic*, "related to lysis") **disease** of the newborn, a potentially fatal disorder. Let's examine what causes it.

When anti-Rh antibodies bind to an Rh-positive red blood cell, they trigger the complement cascade that results in cell lysis. Anti-Rh antibodies are mainly IgG, which can readily cross the placenta. They are produced by Rh-negative people who have been exposed to Rh-positive blood. Such a person is said to become **Rh immunized.** Rh immunization usually occurs when blood from an Rh-positive baby enters the circulation of an Rh-negative mother during pregnancy or around the time of birth (**Figure 18.7**).

If such an Rh-immunized mother carries another Rh-positive baby, life-threatening hemolytic disease of the newborn may occur. Anti-Rh IgG antibodies cross the placenta and enter the fetal circulation. They bind to the baby's Rh-positive cells and trigger their lysis. The developing baby may die from a profound anemia, or it may survive to birth and then die of shock or suffer a type of permanent brain damage from lysis-produced toxins. Severely affected newborns are treated by **exchange transfusion.** The baby's Rh-positive red blood cells are removed and replaced by Rh-negative cells, which cannot be damaged by the maternal antibodies in the baby's circulation (Chapter 17).

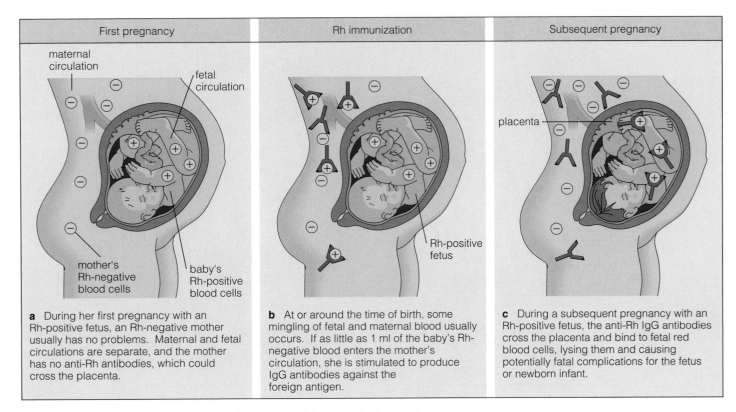

First pregnancy	Rh immunization	Subsequent pregnancy

maternal circulation

fetal circulation

mother's Rh-negative blood cells

baby's Rh-positive blood cells

placenta

Rh-positive fetus

a During her first pregnancy with an Rh-positive fetus, an Rh-negative mother usually has no problems. Maternal and fetal circulations are separate, and the mother has no anti-Rh antibodies, which could cross the placenta.

b At or around the time of birth, some mingling of fetal and maternal blood usually occurs. If as little as 1 ml of the baby's Rh-negative blood enters the mother's circulation, she is stimulated to produce IgG antibodies against the foreign antigen.

c During a subsequent pregnancy with an Rh-positive fetus, the anti-Rh IgG antibodies cross the placenta and bind to fetal red blood cells, lysing them and causing potentially fatal complications for the fetus or newborn infant.

FIGURE 18.7   Hemolytic disease of the newborn, a type II hypersensitivity reaction.

Type III: Immune-complex Hypersensitivity

Type III hypersensitivity, also called immune-complex hypersensitivity, resembles type II hypersensitivity in that antibodies react with components of a person's own body. It differs from type II hypersensitivity in that the reacting body components are soluble molecules in the circulation rather than cells.

Immunological Basis of Type III Hypersensitivity. Circulating, soluble antigen-antibody complexes initiate the type III hypersensitivity reaction. Many such complexes are eliminated by macrophages, but some escape. These surviving complexes then lodge in the capillaries of normal tissues, where they activate complement and provoke an inflammatory response that attracts activated leukocytes. The leukocytes release enzymes that damage nearby tissues (**Figure 18.8**).

Systemic Lupus Erythematosus. The autoimmune disorder **systemic lupus erythematosus (SLE)** is an example of type III hypersensitivity. Components of the cell nucleus enter the circulation and react with antibodies.

The symptoms of SLE are highly variable because the antigen-antibody complexes can be deposited anywhere in the body. Usually they deposit in the kidneys, the skin, and the joints, but any organ, including the brain, can be affected. Kidney inflammation is the most common cause of death from SLE, but skin inflammation often causes a disfiguring facial "butterfly rash" (**Figure 18.9**). Inflammation of the joints causes arthritis. Inflammation of brain tissue causes disorders in thinking. There is no satisfactory treatment for SLE.

Rheumatoid arthritis is another form of type III hypersensitivity. It's triggered by IgG antibodies reacting with rheumatoid factor, an antigen that persons with rheumatoid arthritis produce. Joints throughout the body may be damaged or completely destroyed.

Chronic infections can also cause type III hypersensitivity. For example, chronic viral hepatitis can cause a type III destructive inflammation of blood vessels. In other cases, offending antigens are introduced into the body in the course of medical treatments. **Serum sickness** can occur when proteins from animal serum (for example, horse serum used to treat bites of venomous snakes) are used in medical therapy. It produces a hivelike rash along with swollen and tender joints.

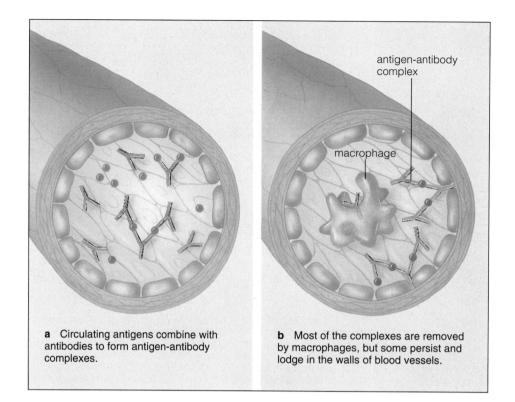

a Circulating antigens combine with antibodies to form antigen-antibody complexes.

b Most of the complexes are removed by macrophages, but some persist and lodge in the walls of blood vessels.

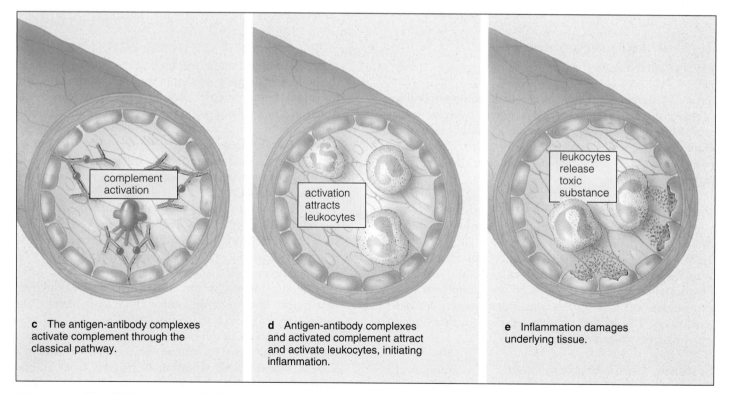

c The antigen-antibody complexes activate complement through the classical pathway.

d Antigen-antibody complexes and activated complement attract and activate leukocytes, initiating inflammation.

e Inflammation damages underlying tissue.

FIGURE 18.8 Type III (immune-complex) hypersensitivity reactions involve antigens that are free in the circulation.

(Art by Carlyn Iverson.)

FIGURE 18.9   A butterfly-shaped rash characteristic of systemic lupus erythematosus.

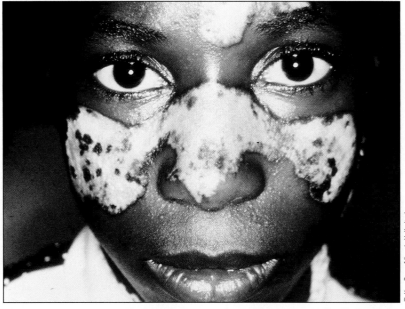

Type IV: Cell-mediated (Delayed) Hypersensitivity

Type IV hypersensitivities are initiated by certain TH lymphocytes called TD because type IV reactions are delayed. They occur from 12 hours to several weeks after exposure. The TD cells produce lymphokines that cause tissue damage.

Immunological Basis of Type IV Hypersensitivity. Type IV is the least understood of the hypersensitivities. The reaction begins when an antigen is presented to a TD cell that bears a matching antibody receptor. The TD cell proliferates into an activated clone, causing the person to become sensitized. On reexposure to that antigen, the activated TD cells release lymphokines that provoke inflammation (**Figure 18.10**).

The impacts of type IV hypersensitivities are quite variable. They range from the everyday annoyance of **contact hypersensitivities,** such as reactions to poison ivy and poison oak (**Figure 18.11**), to the life-threatening rejection of transplanted organs. The **tuberculin skin test,** used to diagnose tuberculosis, is also an example of type IV hypersensitivity. Probably the most significant form of type IV hypersensitivity is the **granulomatous reaction.** It's responsible for the tissue damage caused by certain chronic infections, including tuberculosis.

Now let's briefly consider organ transplantation, because organ rejection is a result of type II and type IV hypersensitivity.

ORGAN TRANSPLANTATION

Organ transplantation has become an important tool of modern medicine. There are four types of grafts (**Table 18.3**): **autografts** (a transplant from one part of a person's body to another), an **isograft** (a transplant

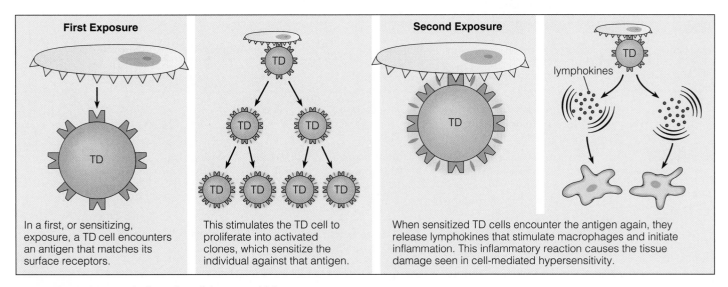

First Exposure		Second Exposure	

In a first, or sensitizing, exposure, a TD cell encounters an antigen that matches its surface receptors.

This stimulates the TD cell to proliferate into activated clones, which sensitize the individual against that antigen.

When sensitized TD cells encounter the antigen again, they release lymphokines that stimulate macrophages and initiate inflammation. This inflammatory reaction causes the tissue damage seen in cell-mediated hypersensitivity.

FIGURE 18.10   Type IV (cell-mediated) hypersensitivity.

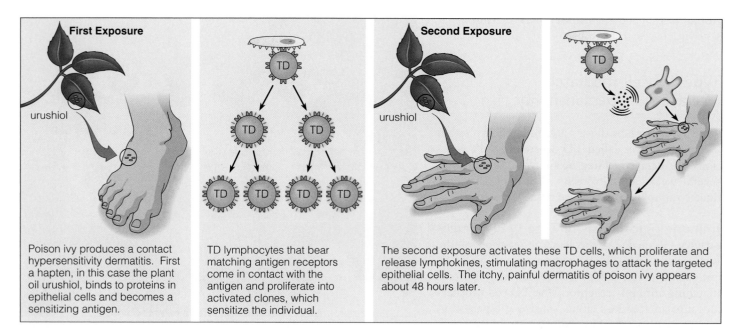

First Exposure

urushiol

Poison ivy produces a contact hypersensitivity dermatitis. First a hapten, in this case the plant oil urushiol, binds to proteins in epithelial cells and becomes a sensitizing antigen.

TD lymphocytes that bear matching antigen receptors come in contact with the antigen and proliferate into activated clones, which sensitize the individual.

Second Exposure

urushiol

The second exposure activates these TD cells, which proliferate and release lymphokines, stimulating macrophages to attack the targeted epithelial cells. The itchy, painful dermatitis of poison ivy appears about 48 hours later.

FIGURE 18.11   Contact hypersensitivity, a kind of type IV hypersensitivity that produces an itchy dermatitis.

from one identical twin to another), an **allograft** (a transplant between members of the same species), and a **xenograft** (a transplant from between two species).

Autografts and isografts are almost always successful, because no foreign antigens are introduced into the recipient. But allografts and xenografts, without the use of drugs, lead to rejection.

Rejection can occur quickly—sometimes within a few minutes to 48 hours after the transplant. Such rapid rejection is caused by antibody-mediated type II hypersensitivity. But usually rejection occurs weeks to months later as a result of cell-mediated type IV hypersensitivity. (The eye is not subject to rejection because it is isolated from the immune system.)

TABLE 18.3 Types of Tissue Grafts

Name	Source of Tissue	Example	Foreign Antigens	Outcome
Autograft (auto = self)	Transplant from one part of a person's body to another	Skin graft for burns	None	No rejection
Isograft (iso = same)	Transplant between genetically identical individuals such as identical twins	Usually bone marrow; rarely, kidney	None	No rejection
Allograft (allo = different)	Transplant from a genetically different member of the same species, such as a nonidentical relative or a recently dead person	Bone marrow, kidney, heart, liver, lung, cornea	Variable; depends upon the closeness of tissue match between donor and recipient	Rejection would normally occur but is often prevented with immunosuppressive medication
Xenograft (xeno = foreign)	Transplant from nonhuman primate to human	Baboon heart to infant with fatal heart defect (experimental)	A great many foreign antigens	Rejection appears inevitable

If the transplanted organ bears only a few antigens that the recipient recognizes as foreign, the graft stands a reasonable chance of survival. The antigens that differ most among humans are the major histocompatibility complex (MHC) antigens (Chapter 17), which are encoded by a group of genes called the **human leukocyte antigen (HLA)** complex. These must match as closely as possible to increase the probability of a successful transplant. These odds can be increased by using immunosuppressive drugs.

Immunosuppression

Immunosuppressive drugs inhibit the immune response and thus block organ rejection. In doing so, most of them leave the patient vulnerable to life-threatening infections. But a drug called **cyclosporine,** the natural product of a fungus, limits that vulnerability. It interferes selectively with T-cell function, leaving B-cell function nearly normal. In other words, cyclosporine blocks the cell-mediated immunity that leads to organ rejection while sparing the body's innate and antigen-mediated defenses against infection.

So far in this chapter we've discussed how our immune system can malfunction and harm us. Now we'll turn to the opposite hazard: how the immune system can become deficient and fail to protect us. Such defects are called **immunodeficiencies.**

IMMUNODEFICIENCIES

All aspects of the immune system can become deficient. In some cases the entire system is unable to function. In other cases, only a single component is defective. For example, a patient might be unable to produce a certain type of antibody or have an abnormality in one particular type of T cell. Immunodeficiencies can be categorized according to the component that is defective: T cells, B cells, both T and B cells, phagocytes, or complement (**Table 18.4**).

All immunodeficiencies result in recurrent infection. They can appear in anyone, from newborns to the elderly. Those that are **congenital** (inherited) usually develop in early life. **Acquired immunodeficiencies,** which develop later, are caused by infection, cancer, or the side effects of immunosuppressive medications.

Congenital immunodeficiencies often lead to death from infection early in life. The most profound ones,

TABLE 18.4 Immunodeficiency Disorders: Selected Examples

Disorder	Immune Defect	Congenital vs. Acquired	Clinical Manifestations
Severe combined immunodeficiency (SCID)	Deficiency of B and T cells	Congenital	Overwhelming infections that cause death in early infancy unless patient is maintained in germ-free environment or receives a bone marrow transplant
X-linked agammaglobulinemia	Severe deficiency to near absence of B cells	Congenital (inherited on the X chromosome, so it affects males only)	Frequent, often life-threatening bacterial infections that begin in infancy
IgA deficiency	Selective inability to produce IgA	Congenital	Repeated bacterial infections of the respiratory and gastrointestinal tracts
DiGeorge's syndrome	Abnormal development of T cells associated with abnormal thymus	Congenital	Severe immunodeficiency associated with other developmental abnormalities, often leading to death in infancy
AIDS	Severe deficiency of TH lymphocytes	Acquired through infection by HIV	Recurrent bacterial, viral, fungal, and protozoal infections beginning when TH cell count falls below 500 per microliter of blood
Chronic granulomatous disease	Deficiency in phagocyte function—neutrophils unable to kill ingested pathogens	Congenital; inherited on the X chromosome	Recurrent bacterial infections, both blood-borne and soft tissue, beginning during the first year of life
C5 dysfunction	Abnormal function of the C5 complement protein	Congenital	Recurrent bacterial infections

CASE HISTORY

David—Life in a Germ-free World

The best-known patient with SCID was a boy from Texas known to the public by his first name, David. At birth, David was diagnosed as having SCID. His **prognosis** (probable outcome) was dismal. SCID babies usually die from infection within weeks or months. David didn't die during infancy. Then he was placed in a germ-free environment and protected from all contact with infection-causing agents. Pictures of David show him sitting alone inside a plastic enclosure or wearing a protective "spacesuit."

© Corbis/Sygma

David in his germ-free environment.

When he was 12, David left his germ-free enclosure to receive a bone marrow transplant from his sister that might have restored his immune system. But a few months later David died. The transplant hadn't failed. Rather, David probably died from cancer caused by the Epstein-Barr virus (Chapter 27), an unusual complication. The virus, which usually does not cause cancer, was in the bone marrow donated by his sister. In David's immune-disordered system, however, the virus was fatal.

The technology of protective isolation allowed David to remain germ-free and therefore alive, but his life in the germ-free environment was tragically restricted.

called **severe combined immunodeficiencies** (**SCID**), disable both B-cell and T-cell immunity. Patients with SCID die from infection during early infancy unless extreme measures are taken to restore their immune function or to protect them from all contact with microorganisms.

Acquired immunodeficiencies are far more common. Patients who receive long-term therapy with systemic steroids and persons with **leukemias** (blood cancers) and **lymphomas** (lymph node cancers) are at particular risk for acquired immunodeficiencies. Although the reasons aren't clear, immune defenses seem to decline when a person becomes gravely ill for any reason. That's why many people with serious illnesses become infected and die. When the serious illness involves components of the immune system, the risk of life-threatening infection is greatly increased.

Acquired immunodeficiency syndrome (AIDS) is a special example of an acquired immunodeficiency. Because human immunodeficiency virus (HIV) infects and kills TH cells, it affects both humoral and cell-mediated immunity. Untreated, most patients die from the infection. HIV and AIDS are discussed in Chapter 27.

Finally, let's consider the question of whether cancer is a result of an inadequate immune response.

CANCER AND THE IMMUNE SYSTEM

Cells change fundamentally when they undergo **transformation** (become cancerous). They multiply rapidly to produce large numbers of undifferentiated cells that form **tumors.** These crowd and eventually kill normal neighbor cells. In a sense, transformed cells become foreign cells. Many even produce new antigens on their surfaces. Unsurprisingly, cytotoxic T cells, natural killer (NK) cells, and macrophages destroy antigen-marked transformed cells. Some scientists believe they play a dominant role in controlling cancer.

The **immune surveillance theory** proposes that transformation occurs frequently, but our immune system usually eliminates transformed cells before they cause cancer. The fact that people with defective immune systems are at high risk of developing cancer supports the theory

But the immune surveillance theory must be reconciled with the fact that tumors do develop and cause fatal disease in otherwise healthy people. Possibly, cancer cells fail to stimulate an effective immune response until the tumor is too large to be controlled, or the cancer antigens are someway hidden from the immune system.

Many cancer immunologists believe that enhancing the immune system may be the ideal tool to treat widespread cancers.

SUMMARY

Case History: It's Nothing—Just a Bee Sting (p. 433)

1. Bee stings can lead to potentially lethal anaphylactic reactions.

Immune System Malfunctions (p. 433)

2. Immunological disorders are caused by a malfunction of the immune system, leading to either hypersensitivity (an inappropriate immune response) or immunodeficiency (an inadequate immune response).

Hypersensitivity (pp. 433–445)

3. There are four types of hypersensitivity reactions.

Type I: Anaphylactic Hypersensitivity (Allergy) (pp. 434–437)

4. Type I is also called immediate hypersensitivity because the symptoms develop within 10 to 20 minutes. The common name is allergy. Antigens that stimulate type I reactions are called allergens.

5. The first exposure to a sensitizing antigen causes B cells to make IgE antibodies. IgE binds to mast cells; on subsequent exposure, antigen binds to bound IgE, stimulating the mast cell to release inflammatory

mediators, including histamine, and leukotrienes.

6. Signs and symptoms of type I anaphylactic reactions usually include local inflammation, pain, redness, warmth, and swelling. There may also be bronchial constriction, a dry spasmodic cough, and different types of edema, causing hives, nausea, and hoarseness. Severe laryngeal edema can cause suffocation. Anaphylactic shock is a precipitous fall in blood pressure that can cause sudden death.

7. Treatment consists of administering epinephrine and antihistamine. Individuals who are at frequent risk of bee sting may undergo immunotherapy.

8. Allergens can enter by injection (insect venom, medications), inhalation (pollen, animal dander, feces of microscopic dust mites), or ingestion (medications, foods, food additives).

9. Inhaled allergens that affect the upper respiratory tract cause allergic rhinitis (hay fever), the most common type I reaction. Treatment is usually with antihistamines.

10. Inhaled allergens that affect the lower respiratory tract can cause asthma, a breathing disorder resulting from constricted bronchial passages. Treatment is usually with inhaled epinephrine-like drugs, cromolyn, and occasionally steroids. Asthma is a chronic condition that can be life threatening.

11. True food allergies produce hives or systemic anaphylactic reactions. Shellfish and peanuts are highly allergenic foods.

12. Penicillin is a hapten that can combine with serum albumin to become a powerful allergen.

Type II: Cytotoxic Hypersensitivity (pp. 437–441)

13. Cytotoxic hypersensitivity occurs when IgM or IgG antibodies bind abnormally to antigens on the surface of human cells.

14. One outcome is cell-bound antibodies activating complement and lysing the target cell with membrane-attack complex made through the terminal complement pathway. This occurs in transfusion reactions and hemolytic disease of the newborn.

15. Graves' disease is an example of altered cell function. The thyroid gland becomes hyperactive.

16. Goodpasture's syndrome is an example of cells being attacked. Kidney cells are affected.

17. Transfusion reactions are examples of cells being clumped by IgM antibodies.

18. If antigens on the erythrocytes in a transfusion react with antibodies in the recipient, the transfused erythrocytes lyse or agglutinate.

19. People with type A blood can receive only type A; people with type B can receive only type B. People with type AB are universal recipients, but they can donate blood only to others with type AB blood. People with type O blood are universal donors; they can receive only O blood.

20. A person with the Rh antigen is said to be Rh positive; one who doesn't is Rh negative. Rh antigens cause a potentially fatal hemolytic disease of the newborn.

21. During a first pregnancy, an Rh-negative mother usually has no problems with an Rh-positive fetus. At or around the time of birth, some mingling of blood between mother and child typically occurs and the mother begins to produce anti-Rh antibodies. During a subsequent pregnancy with an Rh-positive fetus, the anti-Rh antibodies cross the placenta, bind to the fetal erythrocytes, and lyse them. Severely affected babies are given exchange transfusions.

Type III: Immune-Complex Hypersensitivity (pp. 441–443)

22. A type III hypersensitivity reaction is initiated when antibodies bind to circulating antigens and form antigen-antibody complexes that remain soluble. Macrophages normally remove these complexes, but if they persist they lodge in capillaries and activate complement. They initiate an inflammatory response attracting leukocytes, which degranulate, releasing enzymes that destroy surrounding tissue.

23. Type III hypersensitivity is the basis of systemic lupus erythematosus (SLE), an autoimmune disorder. Wherever the circulating antigen-antibody complexes lodge, they cause a destructive inflammatory response (typical sites are the skin, joints, and kidneys).

24. Many kinds of antigens provoke type III hypersensitivity reactions, including self antigens, microbial antigens, and antigens from plants and animals.

25. Chronic viral hepatitis yields antigens that causes type III hypersensitivity. Serum sickness is a type III response provoked by antigens in animal serum.

Type IV: Cell-Mediated (Delayed) Hypersensitivity (p. 443)

26. Type IV hypersensitivity is initiated by TD cells; it is called cell-mediated hypersensitivity. It is also called delayed hypersensitivity because the reaction occurs from 12 hours to several weeks after exposure.

27. Type IV reactions begin when an antigen is presented to a TD cell with a matching antibody receptor. The TD cell is stimulated to proliferate into an activated clone, which sensitizes the individual to that particular antigen. When the person encounters the antigen again, the TD cells release lymphokines that stimulate macrophages and provoke inflammation.

28. Contact hypersensitivity is a common form of type IV reaction. It produces an itchy dermatitis (skin rash) from exposure to certain compounds in plants (poison ivy, poison oak), clothing, and jewelry.

29. The irritating agents are haptens that attach to skin proteins. The hapten-protein complexes sensitize TD cells, and on subsequent exposure the TD cells initiate an inflammatory response.

30. The tuberculin skin test is a delayed hypersensitivity reaction.

31. The granulomatous reaction causes most of the tissue damage in tuberculosis, leprosy, and schistosomiasis.

Organ Transplantation (pp. 443–445)

32. Rejection of organ transplants (tissue grafts) is due to type II and type IV hypersensitivities. There are four types of grafts: autografts, isografts, allografts, and xenografts.

33. MHC antigens determine tissue compatibility (histocompatibility).

Immunosuppressive drugs, such as cyclosporine, and antibodies directed against T cells help block immunological rejection.

Immunodeficiencies (pp. 445–446)

34. Immunodeficiencies are disorders of immune function. They can be defects in T-cell function, B-cell function, both T- and B-cell function, phagocyte function, or complement. Immunodeficiencies are characterized by recurrent infection.

35. Congenital immunodeficiencies are inherited or develop before birth. Several profound congenital immunodeficiencies that disable both T and B cells are grouped under the name severe combined immunodeficiencies (SCID). These patients usually die early in infancy unless extreme measures are taken.

36. Acquired immunodeficiencies develop after birth as a result of infection, cancer, or the side effects of immunosuppressive medications. Patients receiving long-term therapy with systemic steroids and patients with leukemia or lymphoma are at particular risk. AIDS is an acquired immunodeficiency caused by a virus.

Cancer and the Immune System (p. 447)

37. Cells that undergo transformation form large growths called tumors. Tumor-specific antigens appear on the cancer cells, marking them for destruction by T cells, NK cells, and macrophages.

38. The immune surveillance theory proposes that transformation occurs frequently, but the immune system eliminates most malignant cells before they cause cancer. This theory must be reconciled with immunological escape, the fact that people who are otherwise healthy develop cancer.

REVIEW QUESTIONS

Immune System Malfunctions

1. Explain the difference between an immunological disorder and an immunodeficiency.

Hypersensitivity

2. What is hypersensitivity? Name the four types.

3. Why is anaphylactic hypersensitivity also called immediate hypersensitivity? What is the common name by which people know anaphylactic hypersensitivity?

4. Explain the immunological basis of anaphylactic hypersensitivity. Which antibodies are involved?

5. What are the signs and symptoms of anaphylactic hypersensitivity? What causes anaphylactic shock? Describe the treatment for anaphylactic hypersensitivity caused by insect venom.

6. Give some examples of injected, inhaled, and ingested allergens.

7. Explain the difference in cause, signs and symptoms, and treatment of hay fever and asthma.

8. Explain this sentence: Penicillin should be used judiciously to prevent drug sensitization.

9. Explain the immunological basis of cytotoxic hypersensitivity, including the three possible outcomes. Which antibodies are involved?

10. Describe the ABO antigen system and blood typing. Define universal recipient and universal donor.

11. What happens if there is immunological incompatibility in a blood transfusion? What is a transfusion reaction?

12. What is the Rh system of blood typing and why is it important? How does hemolytic disease of the newborn occur? Why is an exchange transfusion the best treatment for a severely affected newborn?

13. What is an autoimmune disorder? Describe Goodpasture's syndrome and the type II mechanism that causes it. Do the same for Graves' disease.

14. What is the basic difference between type II hypersensitivity and type III? Describe the immunological basis of type III hypersensitivity. Which antibodies are involved?

15. Why is systemic lupus erythematosus (SLE) classified as an autoimmune disorder? What happens at the cellular level to cause the signs and symptoms of SLE?

16. Explain this statement: The treatment of SLE presents the same issues raised by all autoimmune disorders.

17. What kinds of antigens besides self antigens cause type III reactions? Give some examples of type III hypersensitivity (other than SLE).

18. How is type IV hypersensitivity different from types I, II, and III? Why is type IV also called delayed hypersensitivity?

19. What is the immunological basis of type IV reactions?

20. Explain how organ transplants are rejected because of cell-mediated hypersensitivity.

21. Define these terms and give an example of each: autograft, isograft, allograft, xenograft.

22. Define histocompatibility. What is a person's tissue type? How is it determined?

23. Explain the two major ways of blocking organ rejection.

24. What is contact hypersensitivity? Give some examples.

25. Explain why someone may not be affected at first by exposure to poison ivy but inevitably will. Describe the events at a cellular level.

26. Explain this statement: The tuberculin skin test is a delayed hypersensitivity reaction.

27. What is a granulomatous reaction?

Immunodeficiencies

28. How are immunodeficiency disorders usually categorized?

29. For each of the following congenital immunodeficiency disorders, identify the breakdown of immune function:
    a. severe combined immunodeficiencies (SCID)
    b. X-linked agammaglobulinemia
    c. DiGeorge's syndrome
    d. chronic granulomatous disease
    e. C5 deficiency

30. Who is at most risk for acquired immunodeficiencies? What is the breakdown in immune function in AIDS?

Cancer and the Immune System

31. Define the following terms: transformation, tumor, tumor-specific antigens.

32. What is the immune surveillance theory? What evidence is in its favor? What is immunological escape? Give an example.

33. What is cancer immunotherapy? Give an example.

CORRELATION QUESTIONS

1. In what way are allergies and asthma similar? In what way do they differ?

2. Would you advise a woman with Graves' disease to breastfeed her baby? Explain.

3. Is it possible that a person with type O blood would not suffer a transfusion reaction on receiving type AB blood? Explain.

4. If a person developed a rash when exposed to a certain plant, how would you decide whether it was the result of an allergy or contact dermatitis?

5. Is it possible that a flu vaccine could decrease a person's chances of contracting bacterial pneumonia? Explain.

6. How might it be possible that an immunotoxin would be effective against a tumor, considering that immune surveillance must have failed?

ESSAY QUESTIONS

1. Write a case history of yourself or someone you know who suffers from a type I hypersensitivity reaction. Explain the immunological basis of the various signs and symptoms.

2. Discuss desensitization with respect to when it is worthwhile and how it works. Speculate about why it sometimes is ineffective.

SUGGESTED READINGS

Carpenter, C. B. 1989. Immunosuppression in organ transplantation. *New England Journal of Medicine* 322:1224–26.

Lichtestein, L. 1993. Allergy and the immune system. *Scientific American* 269 (September): 116–25.

Lockey, R. 1990. Immunotherapy for allergy to insect stings. *New England Journal of Medicine* 323:1627–29.

Parham, P. 2000. *The immune system.* New York: Garland Publishing.

For additional readings, go to InfoTrac College Edition, your online research library at: http://www.infotrac.thomsonlearning.com

NINETEEN

Diagnostic Immunology

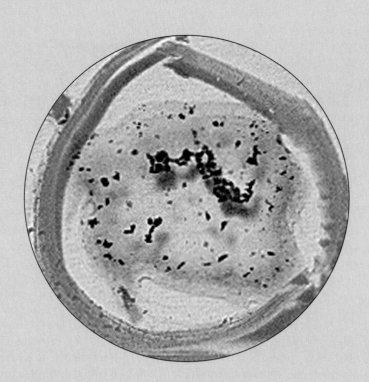

CHAPTER OUTLINE

LEARNING GOALS

To understand:

- *How antigen-antibody reactions can be used to diagnose diseases*
- *The methods for detecting antigen-antibody reactions—precipitation,* *agglutination, and complement fixation*
- *The procedures for performing various immunoassays, including radio-* *immunoassays, immunofluorescence assays, and enzyme-linked immunosorbent assays (ELISA)*
- *The uses of fluorescent antibodies*

Diagnostic Immunology at Home: Pregnancy Tests

C. G. was a 22-year-old woman who thought she might be pregnant. Although her menstrual periods were often irregular, this one was several weeks late. She called her doctor for an appointment, but none was available for 2 weeks. C. G. made the appointment but decided to try a home pregnancy test to find out sooner whether she was pregnant.

At the drugstore, C. G. found more than a dozen different brands of home pregnancy tests. The package information revealed that all the kits were basically alike. They contained a plastic dipstick to be immersed in urine. An indicator dot changed color if the woman was pregnant. C. G. bought the least expensive kit, took it home, and carefully followed the simple directions. There was no color change, indicating C. G. was not pregnant.

Although she was confident she had used the test exactly according to directions, C. G. decided to keep her doctor's appointment anyway. Perhaps a more sensitive test would show an early pregnancy that her home kit might have missed. The doctor sent C. G. to a small clinical laboratory where she was asked to provide another urine specimen. Curious about the whole process, C. G. asked the lab technologist how she planned to test the specimen. The technologist explained that pregnancy tests are based on the presence or absence of a hormone called human **chorionic gonadotropin** (HCG), which is produced by the placenta and therefore found only in pregnant women. The test, she explained, contains a monoclonal antibody linked to an enzyme and a dye system. If HCG is in the urine, it binds to the antibody and the enzyme causes the dye to change color. In fact, she told C. G., the clinical lab test is just like the home pregnancy tests sold in the drugstore. All can detect even early pregnancies. The lab test confirmed C. G.'s home result—no color change, no pregnancy.

Case Connections

- C. G.'s pregnancy test used a reagent (a monoclonal antibody) for detecting the presence of one particular protein and a procedure (ELISA) which caused the dipstick to change color. As we will see in this chapter, this reagent and procedure are also used to identify pathogens.
- You will learn about the reagent and the procedure and how they work.
- You will also learn how the ELISA procedure is able to detect minuscule amounts of protein such as the levels of chorionic gonadotropin in urine at a very early stage of pregnancy.

Courtesy Carter Products, division of Carter-Wallace, Inc.

Home pregnancy kit.

DIAGNOSTIC IMMUNOLOGY

In Chapters 16 and 17 we saw how immunological reactions defend us against microbial infections. And in Chapter 18 we saw how certain inappropriate immunological reactions can cause disease. Now we'll examine a third aspect of the immunological reactions: How antigen-antibody reactions can be used to diagnose disease. This field is called **diagnostic immunology.** Many diagnostic tests that employ antigen-antibody reactions—such as the pregnancy test, for example—use **monoclonal antibodies** (preparations of identical antibody molecules that bind to one specific epitope; see the boxes in this chapter). Because such preparations are pure, they give more reliable results.

Diagnostic immunology is traditionally called **serology** because its methods test for the presence of antibodies or antigens in **serum** (the cell-free liquid component of blood). Serological tests, or assays, involve taking a specimen, such as blood or tissue, from a patient and mixing it with a **reagent** (something that participates in a chemical reaction) that contains known concentrations of an antigen or an antibody. A reagent containing an antibody is used to assay for antigens. One containing an antigen is used to assay for antibodies. Both kinds of tests are useful clinically.

Being able to assay antigens or antibodies allows a clinician to diagnose a present or past infection. For example, hepatitis A (infectious hepatitis) is usually diagnosed by detecting antibodies—those that the body makes against the virus. Hepatitis B (serum hepatitis), on the other hand, is usually diagnosed by detecting antigens—those from the viral coat. Many noninfectious disorders, such as autoimmune disorders, are diagnosed by identifying abnormal antibodies.

Some serological tests are qualitative; they reveal only whether or not a particular antigen or antibody is present. But others are quantitative. They measure the concentration of a particular antigen or antibody. Both types have their uses. For example, qualitative serological tests are used to diagnose syphilis because they are highly specific and reliable. Quantitative serological tests, by measuring the rise or fall of the concentration of antigens or antibodies, are used to follow the course of an infection and document a cure (Chapter 24).

DETECTING ANTIGEN-ANTIBODY REACTIONS

All serological tests involve a reaction between an antigen and an antibody, but they differ in the way they detect that such a reaction has taken place. There are the three basic ways to do this—precipitation reactions, agglutination reactions, and complement fixation reactions. We'll consider them sequentially.

Precipitation Reactions

Some reactions between soluble antigen and antibody molecules can be detected because they form a visible insoluble precipitate. Such reactions are called **precipitation reactions** (or **precipitin reactions**). The precipitates are composed of huge macromolecular lattices that form when many antigen and antibody molecules become linked together. Usually the precipitate is detected as a cloudiness that forms in the reaction mixture.

For a precipitation reaction to occur, both antibody and antigen must have at least two binding sites because at least two connections to each reactant are needed to build a molecular lattice. In addition, both reactants must be present in optimal (approximately equal) proportions. If either reagent is present in excess, it will cover all the binding cites on the other reagent, making it impossible to build a lattice (**Figure 19.1**). The range of proportions at which a lattice can form is called the **equivalence zone.**

A simple qualitative precipitation reaction can be performed just by mixing a solution of antigen and antibody in a test tube. To ensure that the proportions are right for

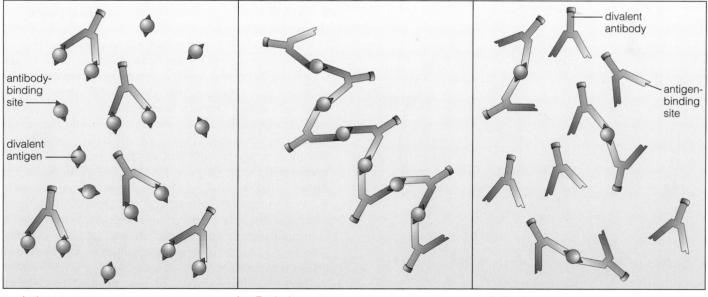

a Antigen excess   **b** Equivalence zone   **c** Antibody excess

FIGURE 19.1 Precipitation reactions. In this case both antibody and antigen are divalent (have two binding sites). (a) Excess antigen quickly binds to all antigen-binding sites on antibody molecules, stopping the formation of a lattice. (b) In the equivalence zone, antibody and antigen are present in optimal proportions to form the cross links necessary to build a large molecular lattice. (c) Excess antibody quickly binds to all antibody-binding sites on antigens stopping the formation of a lattice.

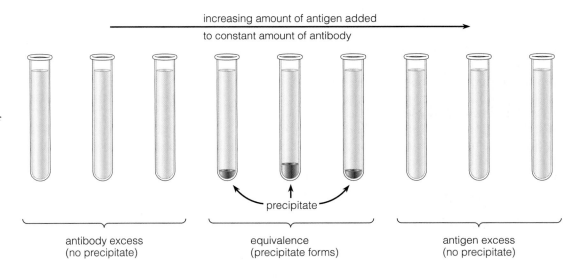

FIGURE 19.2 Liquid precipitation test. Increasing amounts of antigen are added to a constant amount of antibody in a test tube. If they can react with each other, a precipitate will form in the zone of equivalence.

increasing amount of antigen added
to constant amount of antibody

precipitate

antibody excess
(no precipitate)

equivalence
(precipitate forms)

antigen excess
(no precipitate)

forming a precipitate, different concentrations (usually obtained by dilutions) of one of the reactants are added to constant amounts of the other. For example, increasing amounts of an antigen may be added to constant amounts of a solution of antibody in test tubes (**Figure 19.2**). If antibody and antigen can react, a visible precipitate will form in the zone of equivalence. These tubes become cloudy; later the precipitate settles to the bottom of the tubes.

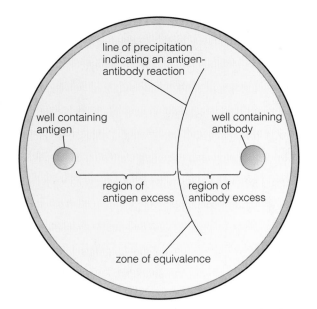

line of precipitation
indicating an antigen-
antibody reaction

well containing
antigen

well containing
antibody

region of
antigen excess

region of
antibody excess

zone of equivalence

FIGURE 19.3 Gel precipitation test. Here a source of antigen and antibody are added to separate wells cut in an agar-filled petri dish. As they diffuse toward each other, they form two overlapping gradients of concentration. A zone of equivalence forms somewhere within them. If antibody and antigen can react, a line of precipitation forms in the zone of equivalence.

Simple Immunodiffusion Tests. **Immunodiffusion tests** eliminate the necessity of making dilutions. These tests are done in a gel slab—agar in a petri plate, for example. Preparations of antibodies and antigens are placed in separate wells cut into the gel. Then both antibodies and antigens automatically dilute themselves as they diffuse out of the wells and into the gel, forming gradients of concentration. Somewhere in the region where the gradients overlap there is a zone of equivalence. If antigen and antibody can react, a precipitate forms in this zone of equivalence. It's visible as a sharp line of cloudiness in the otherwise clear gel (**Figure 19.3**).

Slightly more complicated immunodiffusion tests are used to answer more complicated questions than just whether a particular antigen or antibody is present in a solution. For example, the **double-diffusion** method (also called the **Ouchterlony** method for the person who developed it) can be used to determine if two antigens are identical, similar, or different. Antibody is placed in a central well and antigens are placed in two surrounding wells (**Figure 19.4**). If one continuous line forms (actually two lines that join precisely end to end), the two antigens are identical. If two separate lines form, making a crossing pattern, the antigens are different. A continuous line with a **"spur"** indicates partial identity: The two antigens have some epitopes in common and others that differ.

Immunodiffusion tests can also be used to estimate the relative concentration of an antigen (or antibody) in a series of samples. Such a test is called **radial diffusion** (**Figure 19.5**). To test relative concentrations of an antigen, antibody is incorporated uniformly throughout the gel and the samples of antigen are added to wells cut into the gel. Instead of a line, a ring of precipitate

DOUBLE IMMUNODIFFUSION PRECIPITATION ASSAY

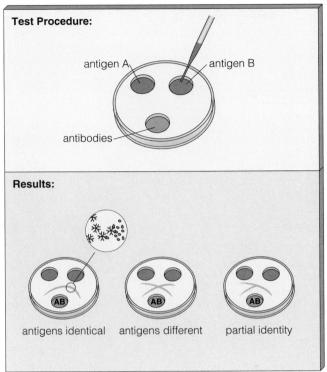

SINGLE IMMUNODIFFUSION PRECIPITATION ASSAY

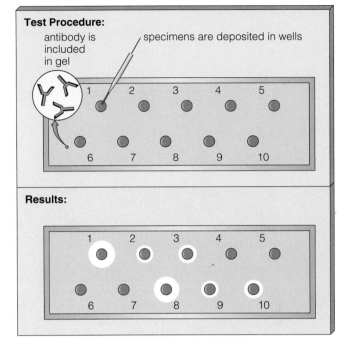

FIGURE 19.4 Double-diffusion method. Two sources of antigens and one of antibodies are placed in nearby wells (holes cut in agar). Antigens and antibodies diffuse toward each other, forming a line of precipitation where they meet in optimal proportions. Identical antigens form a single continuous line. Completely different antigens form two crossing lines. Partially identical antigens form a continuous line with a "spur" or branch.

FIGURE 19.5 Radial-diffusion method. A single type of antibody is incorporated uniformly in the gel, and wells are filled with clinical specimens that might contain matching antigen. A circular band of visible precipitate forms around each well that contains matching antigen. The more antigen present, the larger the band. In this assay, for example, samples 1 and 8 had relatively large concentrations of antigen; samples 2, 3, 9, and 10 had lower concentrations; and samples 4, 5, 6, and 7 had no detectable antigen.

forms around the wells that contain antigen. The diameter of this ring is roughly proportional to the logarithm of the concentration of antigen in the well.

Tests That First Separate Antigens. To detect individual antigens in a complex mixture, it's sometimes necessary to separate (fractionate) them before doing the test. There are many variations on this general theme of separate-then-test. We'll consider two of them: immunoelectrophoresis and Western blots.

Immunoelectrophoresis. In immunoelectrophoresis tests, antigens are first separated by **electrophoresis:** The mixture of antigens is added to a well in gel slab and an electric field is applied across the gel. Depending upon the charge they carry and their size, the antigens are pulled by the electric field to different locations on the gel. Then the electric field is turned off and an antibody mixture is added to a trough that runs parallel to the direction of the field. Antigens and antibodies diffuse toward

one another, forming a series of lines of precipitation. The location of the lines identifies the various antigens. The use of immunoelectrophoresis to identify various protein antigens in human serum is illustrated in **Figure 19.6.** The pattern and intensity of such lines is useful for certain diagnoses.

Western Blots. **Western blots** (immunoblots) are a more sensitive modification of immunoelectrophoresis. They are useful for certain diagnoses, including a confirmatory diagnosis of human immunodeficiency virus (HIV) infection. The mixture of antigens in a clinical sample (for example, serum from a patient who might have HIV) is separated on a gel by electrophoresis, much like the first step of immunoelectrophoresis. Then the antigens are transferred (blotted) onto a sheet of nitrocellulose. Finally, labeled (usually with a dye) antibody (anti-HIV antibody in this case) is added to the sheet. If the antigen is indeed present, antibody will bind to it, forming a colored spot.

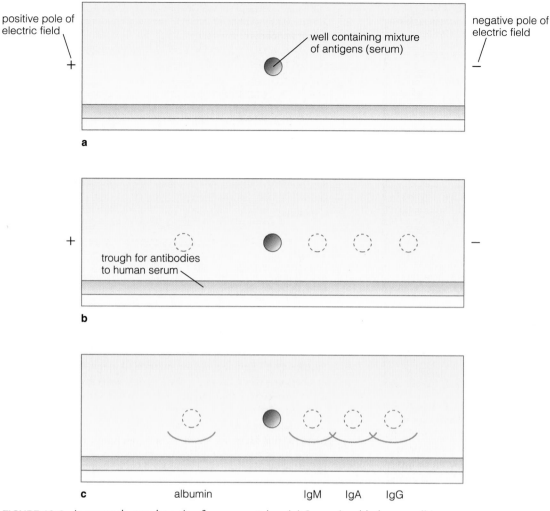

positive pole of electric field

well containing mixture of antigens (serum)

negative pole of electric field

+ −

a

+ −

trough for antibodies to human serum

b

c   albumin   IgM   IgA   IgG

FIGURE 19.6 Immunoelectrophoresis of serum proteins. (a) Serum is added to a well in an agar slab and an electric field is applied. (b) The electric field pulls the proteins to various positions on the slab (indicated by the dotted circles), but they are not visible. Antibodies to human serum are added to a trough that runs the length of the slab. (c) Antigens and antibodies diffuse into the agar, forming precipitation lines where they meet in zones of equivalence.

Agglutination Reactions

Agglutination (also called agglutinin) reactions are similar to precipitation reactions. They too combine antigen and antibody molecules into a visible, easily detectable network. Agglutination reactions differ, however, in that the antigen or antibody is attached to a large particle, such as a cell or a latex bead instead of a soluble molecule. Therefore agglutination reactions produce a visible clump of particles instead of a precipitate. Agglutination reactions are a classic serological technique; some of these assays have been used since the late 1800s.

There are as many variations of agglutination reactions as there are of precipitation reactions. We'll consider some of them here to illustrate the general principles.

Agglutination reactions can be either direct or indirect. A **direct agglutination reaction** involves antigens or antibodies that are naturally a part of a larger particle, such as a microorganism or an erythrocyte. If the particle is an erythrocyte, the reaction is called **hemagglutination.** In contrast, in an **indirect agglutination reaction** an antigen or antibody is first adsorbed onto a particle such as a latex bead. And sometimes, such as in the **Coombs test,** anti-

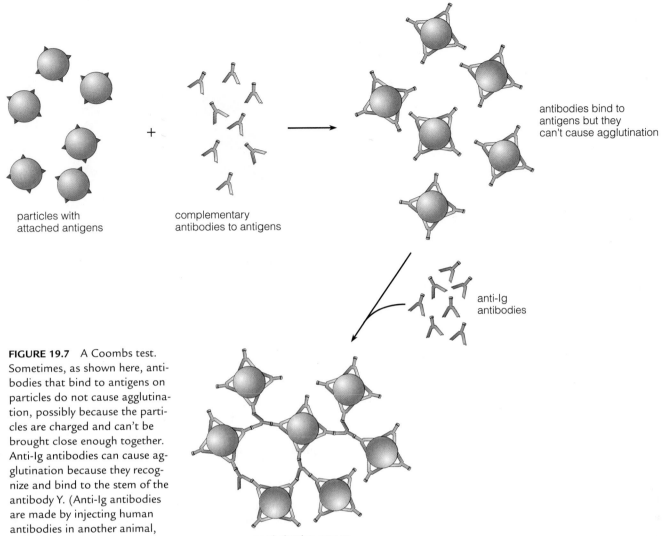

particles with
attached antigens

complementary
antibodies to antigens

antibodies bind to
antigens but they
can't cause agglutination

anti-Ig
antibodies

agglutination occurs

FIGURE 19.7 A Coombs test. Sometimes, as shown here, antibodies that bind to antigens on particles do not cause agglutination, possibly because the particles are charged and can't be brought close enough together. Anti-Ig antibodies can cause agglutination because they recognize and bind to the stem of the antibody Y. (Anti-Ig antibodies are made by injecting human antibodies in another animal, such as a rabbit.)

immunoglobulins (antibodies against antibodies) have to be added to cause agglutination (**Figure 19.7**).

Agglutination tests are extremely sensitive. They can detect antibody at concentrations as low as 1 μg/ml. Many such tests are designed to diagnose present or past microbial infection by identifying antibodies produced in response to that infection. In such a test, the test reagent contains known microbial antigens. If the antigens in the test reagent agglutinate with antibodies in the patient's serum, the test is positive. It shows that the person has probably been infected by the microorganism in question.

Qualitative agglutination reactions are usually performed on a microscope slide. They're easy to do. A drop of the clinical specimen to be tested for antibodies is combined with a drop of the test reagent antigens. If agglutination occurs (clumps form), the test is positive,

indicating antibodies are present. If no agglutination occurs, the test is negative (**Figure 19.8**).

Qualitative slide agglutination tests are widely used in clinical medicine. For example, an indirect latex agglutination slide test is still used to diagnose strep throat. Material swabbed from the patient's throat is mixed with a test reagent containing particle-bound antibodies against the strep bacterium. Before monoclonal antibody technology became cost-effective, the test for pregnancy was also based on agglutination, and blood type is still determined by a slide agglutination test. Test reagents containing antibodies against the A, B, and Rh red blood cell antigens are combined with a patient's blood cells. A positive result is indicated by hemagglutination (**Figure 19.9**).

Quantitative agglutination tests involve progressive dilutions of a clinical sample until agglutination with the test

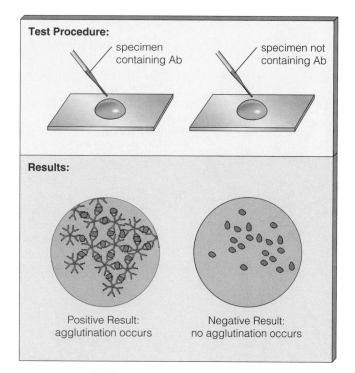

FIGURE 19.8 Qualitative agglutination assays. These tests determine the presence or absence of antibody or antigen in a test specimen. In this case the test is designed to detect the presence of antibodies (Ab) in the sample. If agglutination occurs, the mixture of specimen and test reagent forms clumps and the test is positive. If no agglutination occurs, the test is negative.

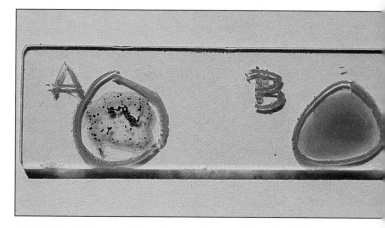

FIGURE 19.9 Hemagglutination reactions for blood typing. In this example the patient's antigen-bearing red blood cells were combined with two test reagents, one containing antibodies against antigen A and the other containing antibodies against antigen B. Agglutination occurred with the anti-A reagent but not anti-B. This shows that the patient has type A blood.

reagent no longer occurs. Results are usually expressed as a **titer**—the highest dilution of a test serum that causes agglutination (**Figure 19.10**).

Agglutination reactions to detect antibodies also give useful information. For example, if no antibodies against a microorganism are detected in a specimen, the person probably has never been infected by that microorganism and remains susceptible to infection. A series of tests over time that show a rising antibody titer or conversion from a negative to a positive response indicates a recent infection. However, it takes up to 6 weeks for antibody titers to rise, so these tests may not help in making decisions about treatment.

Complement Fixation Reactions

Antibodies bound to antigen can activate complement and initiate the formation of a membrane-attack complex that will lyse a target cell (Chapter 16). **Complement fixation** assays take advantage of this phenomenon to detect antibody present in a clinical specimen. Moreover, complement fixation assays are quantitative: The concentration of

antibody can be determined. Like agglutination assays, complement fixation assays are extremely sensitive.

Complement fixation assays entail three steps:

1. A clinical specimen suspected of containing a certain antibody is added to a test reagent that contains the corresponding antigen. If antibody is present, antigen-antibody complexes form.

2. A measured amount of complement (usually from a guinea pig) is added. If antigen-antibody complexes have formed, complement will bind to them, initiating the complement cascade and using up the added complement. The complement is said to be **fixed.** But if no complexes are formed, free complement remains in the mixture.

3. An **indicator system** is added to test for free complement. This system consists of erythrocytes (usually from sheep) that are coated with anti-erythrocyte antibody. If free complement is present, it binds to the antibody-coated erythrocytes, activating a complement cascade and lysing them. So lysis means that there wasn't any antibody in the clinical sample. No lysis means that antibody was present (**Figure 19.11**). Complement fixation tests can be quantitative because a measured amount of complement is added and the amount of lysis can be measured.

Because several steps are involved, complement fixation assays are tedious. They can take at least 48 hours to complete, so they are being replaced by simpler, less-expensive methods. However, complement fixation is still used to diagnose respiratory syncytial virus, which causes

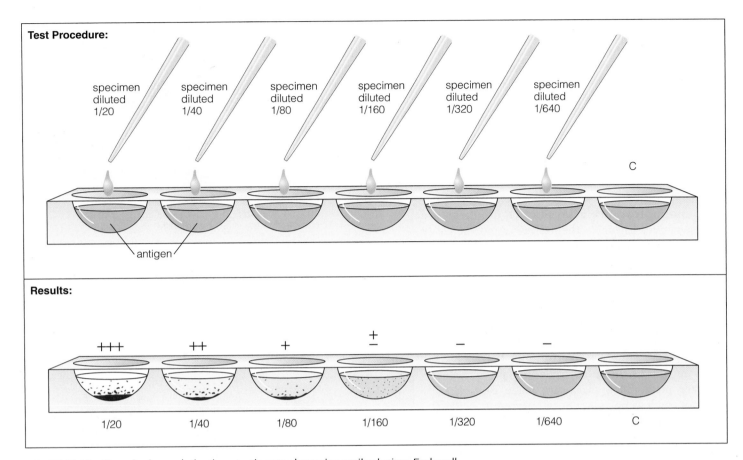

FIGURE 19.10 Quantitative agglutination reactions to determine antibody titer. Each well contains an equal amount of a certain type of microbial antigen bound to a carrier particle and suspended in a salt solution. A serum specimen taken from a patient suspected of being infected by that particular microorganism is prepared in progressively more dilute suspensions and added to each well. If antigen-antibody binding occurs, clumps form and the test is scored as positive. If no clumps appear, no agglutination has taken place and the test is scored as negative. (To aid scoring a control, test [C] is included to which no specimen is added.) Strengths of reactions are scored by pluses and minuses: +++, very strong; ++, strong; +, weak; +/−, probably positive; −, negative. The antibody titer in this test is 1:160.

pneumonia in infants (Chapter 22). Complement fixation tests are also used by epidemiology labs to chart the yearly progress of influenza (Chapter 22), and they're sometimes used to diagnose Q fever (Chapter 22) and certain fungal infections (Chapter 22).

IMMUNOASSAYS

In recent years a powerful set of immunoassays has been developed. Unlike the methods we have just discussed, these assays measure antigen-antibody reactions directly by employing tagged reactants. Because even minute amounts of these tags can be detected, the assays are extremely sensitive. They're many times more sensitive than

methods we have just discussed. These methods have many possible variations (**Table 19.1**). They can be used to detect antigens or antibodies, and either antigen or antibody can be tagged. The tagging can be done by binding a radioactive compound, an enzyme, or a fluorescent compound to one of the reactants. The assay can be either direct or indirect. In a direct assay, one of the reactants is tagged. In an indirect assay an immunoglobulin (Ig) that binds to a reacting antibody is tagged.

The sensitivity of these assays depends upon the sensitivity with which the tag—a radioactive compound, a fluorescent compound, or an enzyme—can be detected. Minuscule amounts of radioactive compounds can be detected, so **radioimmunoassays** (which use such tags) are extremely sensitive. Not quite such small amounts of fluorescent compounds can be detected, so

FIGURE 19.11 Complement fixation test. Here, patient A has a positive complement fixation test, indicating that her serum contains antibodies against the microorganism being tested for. Patient B, on the other hand, has a negative test, indicating the absence of antibodies against the test microorganism.

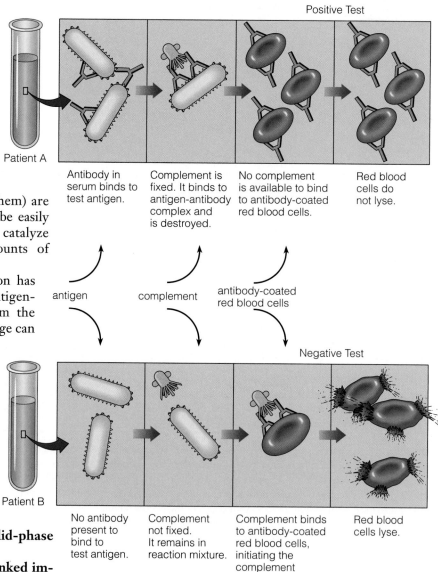

Positive Test

Patient A

| Antibody in serum binds to test antigen. | Complement is fixed. It binds to antigen-antibody complex and is destroyed. | No complement is available to bind to antibody-coated red blood cells. | Red blood cells do not lyse. |

antigen   complement   antibody-coated red blood cells

Negative Test

Patient B

| No antibody present to bind to test antigen. | Complement not fixed. It remains in reaction mixture. | Complement binds to antibody-coated red blood cells, initiating the complement cascade. | Red blood cells lyse. |

immunofluorescence assays (which use them) are less sensitive. Enzymes themselves cannot be easily detected, but small amounts of them can catalyze chemical reactions, producing large amounts of products that can be easily detected.

To know if an antigen-antibody reaction has occurred in an immunoassay, the tagged antigen-antibody product has to be separated from the tagged reactant. This seeming major challenge can be done very simply by performing the assay in one of the small wells in a microtiter plate. Such plates are made of plastics (such as polyvinyl or polystyrene) that bind a layer of protein to their surface. By so binding the untagged reactant to the surface (and blocking the rest of the surface with an inert protein), the tagged antigen-antibody product will also be bound and the unused reactant can be easily washed away. Such assays that depend upon the plastic surface for separating reactants and products are called **solid-phase immunoassays** or **immunosorbent assays.**

Of all possible immunoassays, **enzyme-linked immunosorbent assays** (**ELISA**) are becoming dominant. They exceed immunofluorescence assays in sensitivity, and they avoid the expensive measuring instruments and problems of disposal of radioactive waste associated with radioimmunoassays.

Figure 19.12 outlines the steps of an indirect ELISA for detecting in human serum antibodies to a particular antigen. It illustrates the principles of such assays.

ELISA is widely used in clinical diagnosis. The pregnancy test discussed in the beginning of this chapter is an ELISA. ELISA is also used to diagnose viral hepatitis (Chapter 23), herpes simplex infections (Chapters 24 and 26), rubella virus (Chapter 26), and rotavirus infection (Chapter 23). However, DNA probes (Chapter 10) are rapidly replacing immunoassays for clinical diagnosis. Soon immunoassays may become exclusively tools of research, not of clinical medicine.

FLUORESCENT ANTIBODIES

In the preceding section, we saw how antibodies tagged with a fluorescent compound are used in immunofluorescence assays. These so-called **fluorescent antibodies** have other uses as well. They are used to visualize specific antigens in tissues or on the surfaces of microorganisms. They are also used to sort out or separate from a mixture of different types of cells those that have a specific antigen on their surface. Let's examine these various uses of fluorescent antibodies.

Fluorescent antibodies are made by covalently binding a fluorescent compound such as **fluorescein iso-**

SHARPER FOCUS

MONOCLONAL ANTIBODIES

Each B lymphocyte produces only one kind of antibody—one that recognizes and binds to a single kind of epitope (Chapter 17). So a single clone of B cells produces usable amounts of one kind of antibody molecules, all with identical antigen-binding sites. Such a preparation is called a **monoclonal antibody.** The term monoclonal is used to distinguish these preparations from the product of our body's normal immune response—against an invading bacterium, for example. Such an infection presents several thousand different epitopes that activate several thousand different B-cell clones. They produce **polyclonal antibodies.**

The many components of polyclonal antibodies are useful for combating microbial infection. But single-component monoclonal antibodies with known specificities are immensely practical tools for clinicians and scientists. Making monoclonal antibodies is not as simple as it sounds. It's not possible to do it by cloning a B cell in the laboratory, because normal B cells don't multiply or produce antibodies under laboratory conditions of cell culture.

César Milstein and Georges Köhler of the Medical Research Council Laboratory in Cambridge, England, were the first to solve this problem. In 1984 they received the Nobel Prize for their accomplishment. They fused antibody-producing B cells from a mouse with a line of cancer cells, producing what they called **hybridoma cells.** Hybridoma cells can be grown in cell culture, and they do produce antibodies.

The hybridoma procedure did solve the problem of producing monoclonal antibodies, but it is complex and cumbersome. Hybridoma cells have to be grown for several months in culture and in mice. Then cells that produce the desired monoclonal antibody must be selected from a complex mixture of many other cells producing other antibodies. Moreover, mouse cells can't produce human antibodies that are essential for certain therapeutic applications.

These problems were overcome with the advent of recombinant DNA technology (Chapter 7). Antibody-encoding genes are cloned into mammalian cells in tissue culture or into *Escherichia coli,* which then produces monoclonal antibodies. Human, as well as animal, monoclonal antibodies can be produced this way. The recombinant DNA procedure has many advantages over the hybridoma procedure. It is simpler, faster, and less expensive.

But producing significant amounts of antibody by recombinant DNA technology remains a challenge. Except for those used therapeutically, most monoclonal antibodies are still made using the mouse system.

thiocyanate to an antibody molecule. When such a fluorescent antibody is irradiated with invisible ultraviolet light, it fluoresces (gives off) visible green light. Other fluorescent compounds can be used to produce fluorescent antibodies that fluoresce yellow, orange, or red light. Therefore fluorescent antibodies can be used as highly specific stains (Chapter 4) to visualize the location or presence of antigens. For example, they can be used to demonstrate the presence of a specific protein in the skin or kidney of patients with systemic lupus erythematosus (Chapter 18), and they can be used to detect and identify specific microbial cells in a mixture (**Figure 19.13**).

Fluorescent antibodies are also used in cell-sorting machines to identify and separate out cells with one particular antigen on their surface. For example, they can be used to separate CD4 cells from other lymphocytes.

TABLE 19.1 Possible Variations in Immunoassays

Property of Assay	Possible Variations
Component measured	Antigen or antibody
Component tagged	Antigen or antibody
Method of tagging	Radioactive compound (radioimmunoassay)
	Enzyme (enzyme immunoassay)
	Fluorescent compound (immunofluorescence assay)
Direct assay	Test antigen or antibody is tagged
Indirect assay	Anti-Ig is tagged

LARGER FIELD

THE IMMUNE SYSTEM IN A BOTTLE

Monoclonal antibodies are useful because of what antibodies can do—recognize and bind to one specific type of antigen molecule in a sea of closely related antigens. This remarkable specificity has many clinical applications. Monoclonal antibodies tagged with radioactive or fluorescent markers can detect extremely small quantities of a substance in the body, such as an antibody, a microorganism, or a drug. For example, **digitalis** (a powerful heart medication) is used in such extremely small doses that until recently it was impossible to monitor concentrations of it in blood.

Now, with the help of a monoclonal antibody-based drug assay, it possible to measure digitalis levels precisely and fine-tune dosage of this potentially toxic but lifesaving medication for heart patients.

Monoclonal antibodies are also used in medical research, which also often depends upon identifying and precisely measuring small quantities of a specific biological substance. For example, monoclonal antibodies were the key to unraveling the complexity of the major histocompatibility complex (MHC) antigen system (Chapter 17). Human populations carry many different MHC

genes, which code for similar but significantly different MHC antigens. Determining which of these antigens were identical and which were ever-so-slightly different became possible by using monoclonal antibodies.

Monoclonal antibodies may someday play a significant role in medical treatment, especially of cancer. The problem with most cancer treatments is lack of specificity—normal cells are destroyed along with the cancer cells. Monoclonal antibodies directed against tumor antigens could be coupled with lethal drugs or radioactive substances, possibly creating the magic bullet that cancer therapy

has sought for so long (Chapter 18). This ideal form of cancer treatment still remains a dream of the future, but the clinical trials are under way.

The ease with which monoclonal antibodies can be produced through genetic engineering may further extend their use in research, diagnosis, and treatment. According to Richard Lerner from the Scripps Clinic in La Jolla, California, one of the developers of the new recombinant DNA technique for manufacturing monoclonal antibodies, the new technology "opens the way to doing in a bottle exactly what the immune system would do."

1. A test antigen is added to the well. It adsorbs to plastic surface.

2. A protein such as gelatin is added to block the uncoated surface.

3. A patient's serum is added. The complementary antibody it contains binds to the antigen.

4. Anti-Ig to which an enzyme is bound is added. It binds to the antibody.

5. The enzyme's substrate is added. The enzyme converts it into a colored product.

6. The color that develops in the well means that the serum *did* contain antibody against the test antigen.

FIGURE 19.12   Indirect ELISA test for serum antibodies against a particular antigen. Between each step, the surface is washed to remove any unbound reagents.

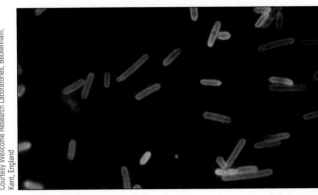

Courtesy Wellcome Research Laboratories, Beckenham, Kent, England

FIGURE 19.13 Fluorescent antibodies. Two species of rod-shaped bacteria are present in this sample. Antibodies against one species are tagged with a dye called isothiocyanate, which fluoresces green. Antibodies against the other species are tagged with a dye called rhodamine B, which fluoresces red. As a result, when irradiated with ultraviolet light, the cells of one species glow green and the others glow red.

SUMMARY

Diagnostic Immunology (pp. 452–453)

1. The immunological reactions that defend against infection can be used to diagnose many disorders.

2. Diagnostic immunology is also known as serology. The term serology comes from serum—the cell-free liquid component of blood—because many serological tests detect antibodies in serum.

3. Serological tests can test for antigens or antibodies. They can be qualitative or quantitative.

4. Serological tests allow a clinician to diagnose a present or past infection and autoimmune disorders.

Detecting Antigen-Antibody Reactions (pp. 453–459)

5. Antigen-antibody reactions can be detected by precipitation, agglutination, or complement fixation.

Precipitation Reactions (pp. 453–456)

6. Precipitation (also called precipitin) reactions depend upon the tendency of some antigens and antibodies to form lattices, huge interlocking macromolecular webs that become visible as they precipitate, forming cloudy regions in the test mixture. For a precipitate to form, antigen and antibody must be present in at optimal proportions called the equivalence zone.

7. When a precipitation reaction is done in liquid, several concentrations of one or the other reactant must be used to find the equivalence zone.

8. An immunodiffusion assay is performed in a gel. Antibodies and antigens are placed in wells. They form concentration gradients in the gel as the diffuse through it. A precipitation line forms in the zone equivalence between the two gradients.

9. A double diffusion test is used to determine whether two antigens are identical, different, or partially identical.

10. A radial-diffusion test determines the relative concentration of antigen (or antibody) in a set of samples.

11. In some diffusion tests, antigens are separated before the test is done. In immunoelectrophoresis tests, they are separated according to how they move in a gel to which an electric field has been applied.

12. In Western blots, antigen are separated by electrophoresis, blotted to a nitrocellulose sheet, and then detected with a colored antibody.

Agglutination Reactions (pp. 456–458)

13. Agglutination reactions produce a visible clump of antigen, antibody, and attached particles. Direct agglutination reactions involve antigens or antibodies that are part of a particle, such as a microorganism or erythrocyte. If the particle is an erythrocyte, the reaction is called hemagglutination.

14. Indirect agglutination reactions involve antigens or antibodies adsorbed onto a particle such as a latex bead.

15. Agglutination assays are very sensitive.

16. Qualitative agglutination tests are performed on a microscope slide. A drop of the test specimen is added to a drop of reagent containing microbial antigens. If clumping occurs, the test is positive.

17. The qualitative slide agglutination test is still used widely for diagnosing strep throat, determining pregnancy, and typing blood.

18. Quantitative agglutination tests involve progressive dilutions of a clinical sample until agglutination with the reagent no longer occurs. Results are expressed as a titer—the highest dilution of a test serum that gives a positive agglutination response.

19. Agglutination reactions to detect antibodies in response to bacterial infection have many applications.

Complement Fixation Reactions (pp. 458–459)

20. Complement fixation assays are based upon antibodies binding to antigens and activating complement to initiate the formation of a membrane attack complex that will lyse a target cell.

21. Complement fixation assays are quantitative: They can measure antibody concentration.

22. Complement fixation reactions involve several steps: first the antigen-antibody reaction; then the addition of complement, which will bind (be fixed) to the antigen-antibody complex if it formed; and finally the addition of a test system consisting of antibody-covered sheep red blood cells. If complement has been fixed, nothing happens. If it hasn't, the complement cascade is activated and the red blood cells lyse.

Immunoassays (pp. 459–460)

23. A powerful and sensitive set of immunoassays have been developed that use tagged reactants. Reactants can be tagged with radioactive compounds, fluorescent compounds, or enzymes.

24. The sensitivity of the assays depends upon the sensitivity with which the tag can be detected.

25. In such assays the tagged antigen-antibody complex that forms must be separated from the excess tagged reactant.

26. Solid-phase or immunosorbent assays take advantage of the ability of plastics to absorb proteins in order to separate antigen-antibody complexes from excess reactant.

27. Enzyme-linked immunosorbent assays (ELISA) have many advantages. They are highly sensitive; they do not require expensive equipment; and they avoid the problem of disposal of radioactive waste.

Fluorescent Antibodies (pp. 460–463)

28. Fluorescent antibodies are used to visualize antigens in tissues on the surface of microorganisms. They are also used to separate different cell types.

29. They can be used as highly specific stains for microscopy.

REVIEW QUESTIONS

Diagnostic Immunology

1. What is serology, and what is the basis of the term?

2. What are monoclonal antibodies? How are they made?

Detecting Antigen-Antibody Reactions

3. What is the basis of precipitation reactions?

4. What does zone of equivalence mean?

5. What is the advantage of performing a precipitation reaction in a gel?

6. What does a double-diffusion reaction determine?

7. Why does a spur form in a double-diffusion reaction between two partially identical antigens?

8. What can a radial-diffusion reaction determine that a double-diffusion reaction can't?

9. What properties of antigens permit their separation in an immunoelectrophoresis test?

10. What does blot mean as applied to a Western blot?

11. What's the difference between an agglutination reaction and a precipitation reaction?

12. When an agglutination reaction uses red blood cells, what is it called?

13. What does titer mean as applied to an agglutination reaction?

14. What does fix mean as applied to the complement fixation reaction?

Immunoassays

15. What kinds of tags can be used in immunoassays?

16. Explain the value of doing immunoassays in plastic containers?

17. What does ELISA stand for?

18. Why is an ELISA so sensitive?

19. What are some advantages of ELISA over radioimmunoassays?

Florescent Antibodies

20. What are some of the uses of fluorescent antibodies?

21. What source of irradiation is used to view a fluorescent antibody-stained specimen?

CORRELATION QUESTIONS

1. In an immunodiffusion test, what factor would cause the precipitation line to lie closer to the antigen well?

2. Why does a radial-diffusion test form a zone of precipitation instead of a line?

3. What would happen if too much complement were added in a complement fixation reaction?

4. What would happen if one neglected to add gelatin in the course of doing an ELISA?

5. Would it be possible to identify three different bacterial species in a mixture of cells using fluorescent antibodies? Explain.

6. In the immunoelectrophoresis assay shown in Figure 19.6, why did albumin move to one side of the well and IgM to the other?

ESSAY QUESTIONS

1. Discuss the relative advantages of using immunologically based tests and DNA probes for clinical diagnosis.

2. Discuss the various ways that immunologically based tests are used in clinical medicine.

SUGGESTED READINGS

Benjamini, E., G. Sunshine, and S. Leskowitz. 1996. *Immunology: A short course.* New York: Wiley-Liss.

Rose, N. R., R. G. Hamilton, and B. Detrick, eds. 2002. *Manual of clinical laboratory immunology.* 6th ed. Washington, D.C.: ASM Press.

For additional readings, go to InfoTrac College Edition, your online research library at: http://www.infotrac.thomsonlearning.com

TWENTY

Preventing Disease

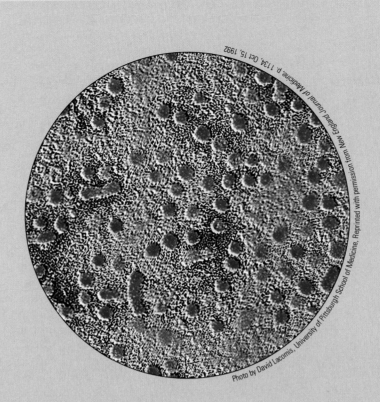

Photo by David Lacomis, University of Pittsburgh School of Medicine. Reprinted with permission from New England Journal of Medicine, p. 1134, Oct. 15, 1992.

CHAPTER OUTLINE

LEARNING GOALS

To understand:

- *How epidemiology contributes to our understanding of disease*
- *How epidemiologists collect information and why statistics are central to epidemiology*

- *The types and uses of epidemiology— descriptive epidemiology, surveillance epidemiology, field epidemiology, and hospital epidemiology*
- *How public health organizations help prevent disease*

- *How controlling reservoirs and disease transmission help prevent disease*
- *The types and uses of vaccines*

Epidemiology Stops an Epidemic

On January 13, 1993, a physician in Washington State reported to the state department of health that an unusually large number of patients with bloody diarrhea had come to the emergency room where he worked. By culture, stool samples from the victims were found to contain a newly recognized pathogen, *Escherichia coli* O157:H7, which was known to cause bloody diarrhea and a potentially lethal illness called hemolytic-uremic syndrome (HUS; Chapter 23, Infections of the Digestive System). A team of epidemiologists was assigned to investigate. The size and extent of the epidemic proved to be much greater than what the physician had seen and reported. There were at least 720 cases spread over four Western states.

It was critically important to identify quickly the source of the bacteria that were causing the epidemic in order to stop it. To do so, the epidemiologists, using a time-tested tool of their trade, initiated a case-control study. They interviewed **cases** (people who had the disease) and **con-trols** (similar people who were well). Soon they discovered that all the cases had one experience in common: They had eaten a hamburger at one of the restaurants belonging to a fast-food chain (designated "chain A" by the epidemiologists). Unused hamburger patties at the restaurants were found to be contaminated with *E. coli* O157:H7. The source had been located. By January 18 a recall of unused hamburger patties from "chain A" was underway; 171 victims of the outbreak had been hospitalized, and four of them had died. But the rapid identification of the source of infection using the methods of epidemiology had avoided an even more devastating epidemic.

The contaminated patties were eventually traced back to a food processing plant, which was found to be contaminated with *E. coli* O157:H7. The contamination must have come from infected cattle, the principal reservoir of this pathogen. Studies have shown that 1 to 2 percent of healthy cattle carry *E. coli* O157:H7 in their intestinal tracts. Organisms from the intestines can contaminate the carcass at slaughter, and those on the surface are spread throughout the meat when it is ground. Meat from only a few infected animals can contaminate large amounts of ground beef when it is mixed. Subsequent outbreaks of *E. coli* O157:H7–caused disease have been traced to a variety of foods that were probably contaminated by cow manure, including raw milk and apple cider. Presumably the cider was made from apples harvested over an orchard floor where contaminated cattle grazed.

Case Connections

- As we'll see, the methods used by epidemiologists to stop the *E. coli* O157:H7 outbreak in 1993 are markedly similar to those used by John Snow 150 years previously (see Case History: Cholera Epidemic on Broad Street) to stop a cholera epidemic in London: Identify the source of infection and eliminate it.
- The methods of epidemiology, as illustrated by this case history, are designed to prevent, not cure, disease.

PREVENTING DISEASE

The principal means we use to prevent disease are epidemiology and public health measures. Epidemiology is the study of when and where diseases occur and how they're transmitted. Public health programs develop and implement ways to prevent and control disease. In other words, epidemiology generates the information needed to carry out effective public health programs. The goal is preventing disease.

Prevention is the best way to fight disease. It eliminates the suffering of human illness. Treatment only decreases it. Prevention is also more economical. A single sewer system or immunization program can prevent thousands of illnesses. And prevention is effective. It's largely responsible for the good health that most people in industrialized countries enjoy today.

In this chapter we'll discuss epidemiology and public health.

EPIDEMIOLOGY

The distinctive feature of epidemiology is its focus on populations, rather than individuals.

Cholera Epidemic on Broad Street

For centuries cholera has been a devastating epidemic disease. The special horror of this deadly disease is its suddenness. A healthy person can go to work in the morning, fall victim to the profuse, watery diarrhea, and be dead from dehydration and shock by nightfall.

Cholera was widespread in Europe in the 1800s. It couldn't be controlled because no one knew what caused the outbreaks. But John Snow, an English physician, decided to take a direct and practical approach. Instead of looking for the cause of the disease, he looked for its source and eliminated it.

In 1854 cholera broke out in the London neighborhood where Broad Street joined Cambridge Street. It was devastating. Snow called it "the most terrible outbreak of cholera that ever occurred in this kingdom." During the first 10 days of September, more than 500 people died who lived in an area 250 yards by 250 yards. Had people not begun fleeing the neighborhood, the number of deaths would certainly have been even greater.

As the epidemic raged, Snow began a systematic investigation to find the source. He reviewed all the records of deaths registered to cholera. He collected information about each victim, and he interviewed all the survivors who remained in the neighborhood. Then he made a map showing where deaths had occurred. It revealed that nearly every cholera death occurred near the Broad Street water pump. Interviews with survivors and others in the neighborhood strengthened his suspicion that the Broad Street pump was the source of cholera. Some victims did live nearer other pumps, but he found that they had used the Broad Street pump because it was on their way to school or because they preferred its water.

Information about people who remained well also implicated the pump. A workhouse in the heart of the Broad Street neighborhood reported only 5 cholera deaths among its more than 500 inmates. The workhouse didn't take its water from the Broad Street pump. It had its own water supply. And none of the employees of a nearby brewery became ill. They never went to the pump for water because the proprietor allowed them to drink beer. Snow concluded that the water from the Broad Street pump had to be the source of the epidemic.

From previous investigations, Snow had suspected that water contaminated by fecal matter was the source of cholera. There were no obvious signs of fecal contamination of the water from the Broad Street pump, but Snow was convinced by his investigation. He went to the local parish church Board of Guardians and convinced them to remove the handle of the Broad Street pump.

The cholera epidemic at Broad Street ended, but that wasn't the end of cholera. London was visited by many similar outbreaks over the next decade. Even knowing that contaminated water was a source of cholera could not solve the problems caused by lack of public sanitation. But without doubt, Snow's work was a turning point. His method of drawing conclusion about a disease by collecting information about many individuals was a new approach to medical research. It was the beginning of the science of epidemiology.

Sketch from 1866 showing King Cholera pumping drinking water.

Several terms are used to describe the way infectious diseases affect populations. An **epidemic** is a disease outbreak that affects many members of a population within a short time. Cholera, diphtheria, and polio are diseases that occur as epidemics. A **pandemic** is an epidemic that spreads worldwide. Acquired immunodeficiency syndrome (AIDS) today is a pandemic. Other diseases are **endemic:** they're always present in a population at about the same level. Gonorrhea is an endemic disease. Other diseases, such as tetanus and trichinosis, are **sporadic.** They occur only occasionally in a population. Epidemics are much less common today than they were in the past, but they still occur. For example, South America experienced a cholera epidemic in 1992.

The word epidemiology was coined to describe the science of epidemic disease. But modern epidemiology also addresses a wide variety of other health-related questions. For example, epidemiological studies helped prove that **pellagra** (a fatal disease of neurological deterioration) is due to a vitamin deficiency. Epidemiology also provided the irrefutable evidence that smoking is harmful and that fluoride can prevent tooth decay. Epidemiological studies even help determine which tests should be included in routine medical checkups. Epidemiological research has led to many major medical advances (**Table 20.1**) and continues to do so.

Relatively recently, epidemiologists helped solve the mystery of three "new" diseases—legionellosis (Chapter 22), Lyme disease (Chapter 27), and AIDS (Chapter 27). In all three cases a disease was known but the cause wasn't. Epidemiological investigations identified the mode of transmission of all three—inhaling contaminated air for legionellosis, a tick bite for Lyme disease, and sexual transmission or exposure to infected blood for AIDS. Thus it became possible to prevent their transmission before their causes were known. Eventually, microbiological research identified the causes of the three diseases—bacteria for legionellosis and Lyme disease and a retrovirus for AIDS.

The Methods of Epidemiology

Epidemiology isn't a laboratory science. It's an information science.

Sources of Information. Much of the information epidemiologists use comes from the public record: vital statistics, census data, and disease reports to public health authorities. Epidemiologists also collect their own information from questionnaires, surveys, and hospital records.

Vital statistics—records of births, deaths, and other human events such as marriages and divorces—are kept by almost all governments. They are the starting points for most epidemiological studies. Snow started his with such statistics. He used death certificates that listed cholera as the cause of death.

A census also generates useful information for epidemiologists. Knowing the number of people living in an area and their distribution by age, race, sex, occupation, national origin, marital status, and income is invaluable because diseases are often unevenly distributed among

---

TABLE 20.1 Significant Medical Advances Made by Epidemiology

Year	Discovery
1700–1713	Bernardino Ramazzini (Italian physician) described the first occupational diseases by studying the diseases specific to different tradesmen—painters, gilders, and so on.
	Similar discoveries were made later that painters suffered illness from exposure to lead-based paint and surgeons and pharmacists experienced other illnesses from exposure to mercury.
1847	Oliver Wendell Holmes (U. S. physician) and Ignaz Semmelweiss (Austrian physician) discovered at the same time that puerperal fever occurs when birth attendants do not wash their hands after examining infected patients or doing autopsies.
1854	John Snow (English physician) showed that cholera epidemics are related to a contaminated water supply.
1900	Walter Reed (U. S. surgeon) discovered that yellow fever is transmitted by mosquitoes.
1926	Joseph Goldberger (U. S. physician and microbiologist) discovered that pellagra, a potentially fatal disorder manifested by rash, intestinal upset, and mental deterioration, is caused by a dietary deficiency of niacin.
1955	Evarts Ambrose Graham (U. S. surgeon) and E. Cuyler Hammon (U. S. statistician) proved that smoking is a significant risk factor for premature death from lung cancer.

subpopulations. Tuberculosis, for example, is most common among the elderly, the poor, and recent immigrants. Coccidioidomycosis and blastomycosis are limited to people who live in particular geographical areas (Chapter 22).

Government agencies also record all cases of certain diseases. Physicians who treat patients with these **notifiable diseases** must report them to their local department of public health when the diagnosis is made (**Table 20.2**).

TABLE 20.2 Reportable Diseases in California

Requirements

Report Immediately by Phone to the County Health Department

- Anthrax
- Botulism
- Cholera
- Dengue fever
- Diarrhea of the newborn, outbreaks
- Diptheria
- Rabies, human or animal
- Yellow fever

Report within 1 Working Day by Phone or Mail

- Amoebiasis
- Campylobacteriosis
- Conjunctivitis, acute, infectious of the newborn
- Encephalitis, viral, bacterial, fungal, parasitis (specify)
- *Haemophilus influenzae,* invasive disease
- Hepatitis A
- Listeriosis
- Malaria
- Measles
- Meningitis, viral, bacterial, fungal, parasitic (specify)
- Pertussis
- Q fever
- Relapsing fever
- Salmonellosis
- Shigellosis
- Streptococcal infections (outbreaks and food handlers only)
- Syphilis
- Trichinosis
- Typhoid fever, cases and carriers

Report within 7 Calendar Days

- AIDS
- Alzheimer's disease and related conditions
- Brucellosis
- Chancroid
- Chlamydial infection (*Chlamydia trachomatis*)
- Coccidioidomycosis
- Cryptosporidiosis
- Cysticercosis
- Disorders characterized by lapses of consciousness
- Food poisoning (report by telephone outbreaks of two or more cases)
- Giardiasis
- Gonococcal infections
- Granuloma inguinale
- Hepatitis, non-A, non-B (including hepatitis C)
- Hepatitis, unspecified
- Hepatitis B, cases and carriers
- Hepatitis delta (D)
- Kawasaki syndrome
- Legionellosis
- Liptospirosis
- Lyme disease
- Lymphogranuloma venereum
- Mumps
- Nongonococcal urethritis (not chlamydial)
- Pelvic inflammatory disease
- Reye's syndrome
- Rheumatic fever, acute
- Rocky Mountain spotted fever
- Rubella (German measles)
- Tetanus
- Toxic shock syndrome
- Tuberculosis
- Tularemia
- Typhus fever

Note: Data from Title 17, Section 2500, California Code of Regulations. Each state makes its own regulations.

Some of these diseases must be reported immediately by telephone, others within 1 working day, and a third group within 7 calendar days. Such reporting is essential for stopping potential epidemics.

Local public health statistics are forwarded to state agencies and to the Centers for Disease Control and Prevention (CDC) in Atlanta, Georgia. Epidemiologists at the CDC prepare the **Morbidity and Mortality Weekly Report** (**MMWR**). (Morbidity is illness and disability; mortality is death). MMWR is a compilation of weekly and cumulative annual statistics on reportable diseases from around the United States (**Figure 20.1**). In addition to disease statistics, each issue of MMWR contains three or four short articles on morbidity and mortality trends in the population of the United States. It's available in printed form or on the Internet (www.cdc.gov).

Epidemiologists also use surveys, questionnaires, interviews, and hospital records. In contrast to data from the public record, this information must be gathered specially for each project.

Uses of Statistics. Epidemiologists usually express health information as a **rate** (the ratio of the number of people in a particular category to the total number of people in the population being studied). For example, the death rate from cholera of residents of the Broad Street workhouse was

1 percent. That's more meaningful than saying five deaths occurred there. Rates are especially useful for making comparisons, for example, between incidence of cholera in the workhouse and in the rest of the Broad Street neighborhood. Two kinds of rates are commonly used in epidemiology: incidence rate and prevalence rate.

Incidence rate is the rate of acquiring a disease or condition during a certain period (for example, a week, a month, or a year). Changes in incidence rates reveal the growth or decline of an epidemic. For example, at one point in the AIDS epidemic, the incidence rate among the general population was estimated to be 0.003 percent (3 people in 100,000) per year. In contrast, the incidence rate among homosexual men was estimated to be 1 to 3 percent (3000 people in 100,000) per year.

Prevalence rate is the rate of having a certain disease at any particular time. Thus prevalence rate equals incidence rate times the average duration of illness. Prevalence rates are particularly high for long-lasting diseases such as human immunodeficiency virus (HIV) infection. For example, at the time in the United States that the incidence rate for AIDS was 0.003 percent, about 1.25 million Americans had the disease. Its prevalence rate was (1.25/250) approximately 0.5 percent. In contrast, prevalence rate for a disease such as cholera, which results in rapid recovery or death, is about the same incidence rate.

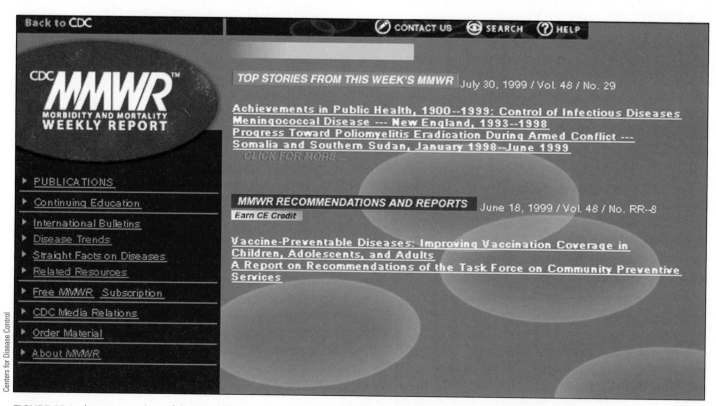

FIGURE 20.1   Internet version of the Morbidity and Mortality Weekly Report (MMWR).

Births and deaths are also expressed as rates. Death rates are usually subdivided by age group. Such age-adjusted breakdown provides useful information. We expect a high death rate among the elderly, but not among infants or young adults. An elevated death rate among the young calls for an epidemiological investigation.

Incidence and prevalence rates can also be age adjusted to show which age groups are particularly vulnerable to a disease. For example, age-adjusted incidence rates show that shigellosis is largely a disease of children (**Figure 20.2**).

Types of Epidemiological Studies

There are many kinds of epidemiological studies. In this section we'll look at four of them: descriptive epidemiology, surveillance epidemiology, field epidemiology, and hospital epidemiology.

Descriptive Epidemiology.
Descriptive epidemiology seeks to provide general information about a disease: What are its incidence and prevalence rates? In what parts of the country is it most common? What's the death rate from it in the general population and in selected subpopulations, such as different ethnic groups? What's the ratio of fatal to nonfatal cases? How do factors such as age, geographical distribution, race, economic status, and sex affect the likelihood that a person will contract the disease? A descriptive study might also consider where victims seek medical treatment and what type they receive. Descriptive studies often take years to complete, but they can be extremely valuable. They reveal

needs for better health care. Some descriptive epidemiological information about tuberculosis is shown in **Figure 20.3.**

Descriptive studies can be particularly useful in showing how nonmicrobial factors affect a disease. For example, Koch proved that *Mycobacterium tuberculosis* causes tuberculosis. But epidemiological research showed that people who live in overcrowded conditions, who have poor nutrition or general health, or who suffer from alcoholism or AIDS are particularly vulnerable. Thus poverty, malnutrition, ill health, alcoholism, and AIDS also "cause" tuberculosis.

Surveillance Epidemiology.
Surveillance epidemiology tracks epidemic diseases. It monitors situations that might lead to an epidemic and follow the progress of an epidemic once it begins.

Traditionally, epidemiological surveillance tracked individuals who were a risk to public health. For example, people who entered a smallpox-free country from one where smallpox occurred were monitored for several weeks to ensure they weren't infected.

Today, epidemiological surveillance includes more sophisticated techniques, some of which were crucial in eradicating smallpox worldwide. An effective vaccine for smallpox had been available for more than a hundred years. If everyone had been vaccinated, the disease would have been eradicated a century ago. But the financial and organizational problems of vaccinating the entire world population were overwhelming, particularly in developing countries, where smallpox was most prevalent.

In the 1970s the World Health Organization's smallpox eradication team developed reliable ways of reporting

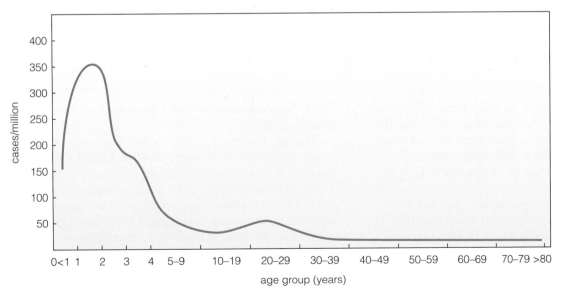

FIGURE 20.2   Age-adjusted incidence rates of shigellosis. Unadjusted rates suggest that the disease is relatively rare—fewer than 50 cases per million population. But age-adjusted rates reveal that it is a significant illness among children—more 350 cases per million at age 2.

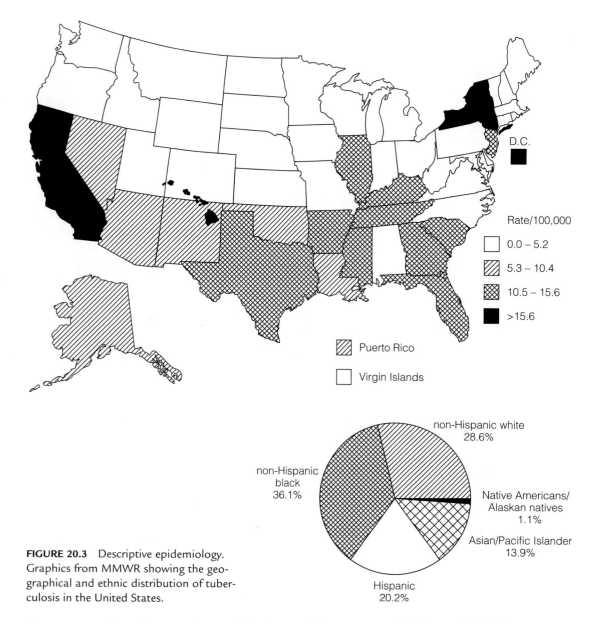

FIGURE 20.3 Descriptive epidemiology. Graphics from MMWR showing the geographical and ethnic distribution of tuberculosis in the United States.

all new cases of smallpox, so they could maintain effective epidemiological surveillance over potential outbreaks. If smallpox were found, affected persons were isolated and exposed persons were vaccinated. These procedures eliminated smallpox from many areas by vaccinating as little as 6 percent of the population.

The procedures were simple, but the logistics were immense. Finally, in 1977, smallpox was eradicated (**Figure 20.4**), although some fear it might be brought back by terrorism.

Surveillance epidemiology has also been used to control (but so far not eradicate) malaria, diphtheria, measles, and polio. Although polio was expected to be eradicated by 2000 and measles soon thereafter, both still exist.

Field Epidemiology. Field epidemiology investigates disease outbreaks.

Disease outbreaks come to the attention of public health authorities in a variety of ways. Occasionally there are reports of an unusual number of cases of a particular disease in a certain area. AIDS was first noticed when many cases of a rare disease, *Pneumocystis carinii* pneumonia, were reported among young men (Chapter 27). Investigation revealed that all these patients had other symptoms of immunodeficiency and they all were homosexual. With further investigation, epidemiologists concluded that some infectious agent, transmitted by sexual contact, was destroying the patients' immune system. Their conclusion was later shown to be correct when virologists identified the HIV retrovirus that causes AIDS.

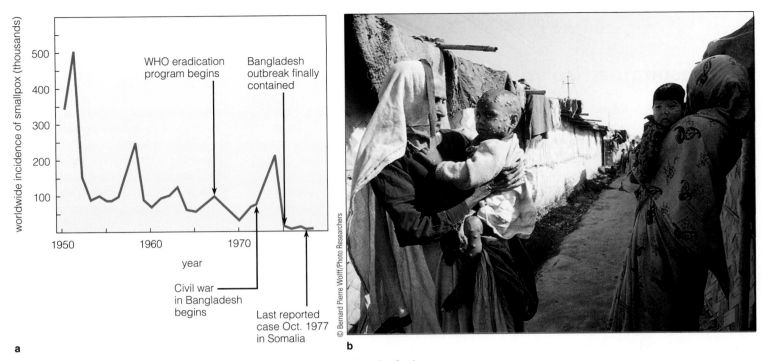

FIGURE 20.4 Eradicating smallpox. (a) Worldwide incidence rate during the final years that smallpox existed. (b) Bangladeshi refugees with smallpox in a Bihari relief camp during the last outbreak of the disease. Eradicating smallpox from Bangladesh, the last place it existed, required the cooperation of 100 Bangladeshi epidemiologists, 12,000 local workers, 13,000 village recruits, and 75 international health workers.

Other times, large numbers of people who have been together at some gathering simultaneously become ill. These outbreaks are known as **common source epidemics.** Usually these epidemics turn out to be food poisonings. Typically they occur when *Staphylococcus aureus*–contaminated food has been served at a large gathering. Those who eat the tainted food suffer from diarrhea and vomiting within hours (Chapter 23). The outbreak of legionellosis in a Philadelphia hotel was a case of an airborne common source epidemic (Chapter 22).

Public complaints also alert public health authorities of developing epidemics. Residents of Old Lyme, Connecticut, complained to the local health department and to physicians at the Yale University Medical Center of an extraordinary number of cases of arthritis in their community. At first, medical experts were skeptical. But in 1977 field epidemiologists confirmed a statistically significant increase in serious cases of arthritis. The source was traced to a tick bite (Chapter 27). Later the causative microorganism, *Borrelia burgdorferi*, was identified.

Sometimes hospitals report unusual numbers of cases of a particular illness. This happened in Los Angeles County during the 1980s. An unusually high number of cases of listeriosis occurred among new mothers and their infants (see Larger Field: Listeriosis Outbreak in

Chapter 24). Epidemiological investigation revealed that all these women had recently eaten a particular brand of unpasteurized soft cheese. Cultures from containers of cheese still on grocery store shelves were positive for *Listeria monocytogenes.* The cheese was removed, and the epidemic ended.

Field epidemiology is detective work. Epidemiologists collect as much information as they can about the victims, their families, and their neighbors (**Table 20.3**). Then they try to deduce the cause of the epidemic and how it can be controlled (**Figure 20.5**). Much new information about disease transmission comes from field studies. For example, while investigating an outbreak of infectious hepatitis, field epidemiologists found that all victims had recently eaten shellfish. Thus they discovered that shellfish harvested from contaminated waters could spread viral hepatitis.

Hospital Epidemiology. Hospital settings are a particular focus of epidemiologists because concentrations of sick people greatly facilitate disease transmission. Approximately 8 percent of all inpatients acquire a new infection while in the hospital. And at least 20,000 people in the United States die of **nosocomial** (hospital-acquired) infections annually. They cost billions of dollars each year.

CASE HISTORY

Respiratory Isolation

It was a typical busy January evening in the pediatric emergency department, and T. D., a young doctor, picked up the chart of yet another baby with a cough. As soon as he saw this baby, T. D. knew she was sicker than most. The baby was struggling to breathe. She had to be admitted to the hospital immediately.

T. D. called the pediatric ward for a bed—often a problem during the winter when the hospital is full to overflowing with children suffering from various infections. The nurse's first question was predictable: Would the baby need isolation? Yes, he told her, the baby probably had respiratory syncytial virus (RSV). She would require respiratory isolation to protect other patients from this highly contagious infection. This need made the bed situation even worse because it meant the baby would have to be housed alone in a room designed to hold four cribs. The nurse told T. D. she would see if she could move children around the ward, putting patients who did not require isolation together to make room for his new patient. If there was no way to clear a room, the infant would have to be transferred to another hospital.

Twenty minutes later the nurse phoned T. D. to say she had managed to clear space for his baby. Two days later, when the baby's nasal smear came back positive, confirming the RSV diagnosis, the infant was moved into a room with other known RSV patients. Because they all had the same infection, they did not need to be isolated from one another. This opened space for more patients who might require respiratory isolation.

TABLE 20.3 Ten Basic Questions Asked in an Epidemiological Interview

Question	How It Might Help the Investigator
1. Have you have recent contact with a person who had a similar illness?	"Yes" would suggest an infectious disease.
2. Where do you live?	Some diseases are found only in certain geographical areas, such as coccidioidomycosis (San Joaquin valley fever in the San Joaquin Valley of California).
3. Who lives at home with you?	An adult with young children might have contracted one of the diseases commonly spread in day-care centers, such as hepatitis or giardiasis.
4. Do you have any pets?	"Yes" might suggest a zoonotic disease, such as toxoplasmosis among cat owners.
5. What is your work or daily routine?	This might provide clues to occupational illnesses, such as cancer among workers exposed to asbestos.
6. Where do you work or study?	This might provide a connection to others with the same disease, such as children attending a school where a measles outbreak is occurring.
7. What are your hobbies?	Leisure activities might involve disease exposure; hunters can contract tularemia through contact with wild animals.
8. Have you traveled away from home recently?	International travelers may contract a variety of diarrheal illnesses, such as traveler's diarrhea or shigellosis.
9. Are your immunizations up-to-date?	Unimmunized people are at risk for diseases such as measles that immunized people are unlikely to contract.
10. Do you use illegal drugs?	People who inject drugs illegally often use contaminated needles, putting them at risk for many bloodborne infections, including HIV and hepatitis B.

a

b

c

FIGURE 20.5   Some of the varied activities of field epidemiologists. (a) Searching for infected insect vectors. (b) Collecting information. (c) Interpreting the data.

Nosocomial Infections. Even in the best hospitals, nosocomial infections are almost inevitable and they're usually hard to treat. There are three reasons for this: immunocompromised patients, invasive medical procedures, and antibiotic resistance. We'll consider each of these briefly.

Immunocompromised Patients. Many patients in a modern hospital are immunocompromised. They include victims of advanced cancer, recipients of organ transplants, patients in kidney failure, and the frail elderly. In fact, all gravely ill people have weakened immune defenses (Chapter 18). Moreover, some treatments, including anticancer medications and steroids, further depress immune function. For these patients, even microorganisms from their own normal flora may cause life-threatening infections (Chapter 15).

Invasive Medical Procedures. Invasive medical procedures foster infection because they bypass the body's surface defenses (Chapter 14). Such procedures, including blood drawing, intravenous lines, and urinary catheterization, are routine in hospitals. Then there are more specialized invasive procedures, such as **endoscopy, bronchoscopy,** and **laparoscopy,** in which instruments are introduced into the gastrointestinal tract, the respiratory tract, and the abdominal cavity, respectively. All surgical procedures carry some risk of infection.

Antibiotic Resistance. Antibiotic-resistant strains of bacteria tend to accumulate in hospitals. If they are transmitted to an immunocompromised patient, antibiotics may not be effective and death can result.

LARGER FIELD

MMWR ALERT

The Morbidity and Mortality Weekly Report contained the following story of a hospital epidemic in Wisconsin. From October 1986 through June 1988, nearly 9 percent of patients who underwent endoscopy of the upper gastrointestinal tract became infected with a particular strain of *Pseudomonas aeruginosa*. This is a much higher incidence rate than normal.

P. aeruginosa is a notorious nosocomial pathogen, capable of surviving for extended periods in almost any liquid environment, even some disinfectant solutions. Although not highly virulent, *P. aeruginosa* can cause serious, even fatal, infections in immunocompromised patients. An epidemiological investigation revealed that the apparatus for sterilizing the endoscopes was faulty. (Endoscopes are delicate tubular instruments made of flexible light-emitting fibers that enable a physician to look inside the body—at an ulcer in the stomach or a tumor in the intestines, for example.) After use, endoscopes were placed in an apparatus where they were flushed with a detergent solution, disinfected with a chemical germicide, and rinsed with tap water. But in the Wisconsin hospital, the equipment had become contaminated. A thick film of *P. aeruginosa* was growing in the detergent holding tank, inlet water hose, and air vents. Only when hospital personnel began to rinse and dry the endoscopes by hand was the outbreak brought under control.

Types of Nosocomial Infections. A patient can acquire a nosocomial infection in almost any part of the body, but certain sites are particularly vulnerable. Urinary tract infections, which usually result from **urinary catheterization** (passing a catheter or plastic tube through the urethra to drain urine from the bladder), account for nearly half of nosocomial infections.

Surgical wound infections are the second most common type. All operating room personnel practice strict aseptic techniques, including **scrubbing** (meticulous hand washing) and wearing sterile masks and gowns. All instruments are sterilized, and the air is filtered to remove microorganisms. Still, at least 10 percent of patients who undergo surgery develop an infection. The infection rate is highest in cases where bacterial contamination is most likely, such as operations that open the microbe-laden intestinal tract or those that repair dirty wounds.

Pneumonias rank closely behind wound infections as the third most common type of nosocomial infection. Some pneumonias occur as the direct result of medical intervention. For example, mechanical respirators and **intubation** (passing a plastic tube down the patients airway) to help a patient breathe introduce microorganisms just as catheters do. Pneumonias also occur in severely ill patients who inhale microbe-containing secretions from their upper respiratory tract into their lungs.

Skin infections also occur in hospitals. Babies in newborn nurseries and burn victims are particularly vulnerable. Nursery epidemics due to *Staphylococcus aureus* can be troublesome and unsightly. They cause blistering and peeling, but they are seldom fatal. Nosocomial skin infections of burn victims, however, can be life threatening because they often progress to become systemic. The loss of skin, a crucial part of our first line of defense, makes infection almost inevitable. Despite specialized care and scrupulous aseptic technique, infection is the most common cause of death among burn victims who survive the initial injury.

Hospital Epidemiologists and Infection Control. Because nosocomial infections are a constant threat, most hospitals have epidemiologists, often nurses, who work with doctors on an infection-control team. Hospital epidemiologists are specially trained to recognize and interrupt a potential hospital epidemic.

Once an epidemic is discovered, the team informs the staff and enforces infection-control measures. Measures to control a *Staphylococcus aureus* outbreak in a nursery, for example, might involve calls for the staff to take greater care with hand washing, monitoring their success by taking cultures from workers' hands, and culturing nasal sample to determine which staff members are *S. aureus* carriers.

Other infection-control measures might include isolating infected patients, restricting the transfer of patients from one part of the hospital to another, and even closing certain areas until the outbreak is contained.

In addition to monitoring potential epidemics, hospital epidemiologists enforce the CDC program of **Universal Blood and Body Fluid Precautions.** This program was initiated in 1983 in response to the threat the AIDS epidemic posed to health-care workers. It directs workers to treat all patients as though they were infected and to wear gloves, gown, and goggles whenever they come in direct physical contact with blood or other secretions that might contain the AIDS virus. Despite these precautions, a few cases of occupationally acquired AIDS are inevitable. Accidents such as needle sticks or inadvertent cuts do occur, especially during surgery. The infection-control team plays an indispensable role in hospital safety. It's a challenging and often unpopular job.

Now let's turn to the other major way to control disease: public health.

PUBLIC HEALTH

As we've noted, public health implements disease **prophylaxis** (prevention) suggested by epidemiological studies. There are two principal methods: (1) limiting exposure to pathogenic microorganisms by decreasing or eliminating the pathogen's reservoir (Chapter 15) or by interrupting disease transmission, and (2) stimulating the body's natural immune defenses by immunization.

We'll discuss each of these after we say a few words about public health organizations.

Public Health Organizations

Many agencies at local, state, national, and global levels safeguard the public health. City and county departments of public health help citizens and practicing physicians in numerous ways—from inspecting food distributors to providing free immunizations. They track reportable diseases and make information about current disease trends available to health-care providers. For example, if a doctor has a patient with an animal bite, he or she can learn about recent rabies cases in the area by telephoning the local health department.

Local health departments report to state public health agencies. Because state laws govern disease reporting, state health departments play a critical role in tracking reportable diseases. Most state health departments publish a bulletin similar to MMWR and do diagnostic testing for certain diseases. In California, for example, Lyme disease serology (Chapter 27) and rabies pathology (Chapter 25) tests are among those performed by state labs.

The United States Public Health Service (USPHS) is our national public health agency. The CDC is one of its agencies. They monitor infectious and other preventable diseases and disseminate the information through the MMWR. The CDC also does research. It was the CDC, for example, that isolated the bacterium that causes legionellosis (Chapter 22). In addition, the CDC issues recommendations on the use of antibiotics and vaccines. It plans and executes national public health education campaigns, and it trains epidemiologists.

The international organization most concerned with improving public health is the United Nations–sponsored World Health Organization (WHO). WHO has undertaken ambitious programs to improve the health of people in the world's poorest nations. Its smallpox eradication program was history-making. Some of its current programs include campaigns to increase childhood immunization and breastfeeding and educational programs to improve infant nutrition.

Limiting Exposure to Pathogenic Microorganisms

Infections spread in an astonishing number of ways (Chapter 15). A drink of water, a mosquito bite, or even a conversation with a friend can transmit disease. Ways of preventing exposure to pathogenic microorganisms are equally varied. They range from closing the screen door to filtering the water supply of a city. Public health practices that have most dramatically decreased the incidence of disease include providing clean water and clean food, promoting personal cleanliness, insect control, and prevention of sexually transmissible and respiratory diseases.

Clean Water. Many diseases, including cholera, typhoid fever, and viral and bacterial diarrheas, are spread when human sewage contaminates the water supply. Such diseases can be prevented by treating sewage and supplying clean public water.

The history of municipal sewage and water supplies makes interesting—but at times horrifying—reading. Until the late nineteenth century, cities were dangerous places to live. Raw sewage collected in open cesspools. Clean drinking water wasn't available. Disease was so rampant that more people died in cities than were born in them. Urban populations were maintained only by constant immigration from the countryside. The invention of the toilet made the situation worse because sewage disposal systems did not exist (see Larger Field: One Step Forward, Two Steps Back). Many people opposed public programs to dispose of sewage and supply clean water because they were expensive and considered to be governmental intrusion into people's private lives. Only after they had been instituted, however, did self-sustaining urban life become possible.

Still, worldwide, contaminated water and waterborne infections are quite common. Cholera and typhoid fever still exist. Diarrhea remains a leading cause of death among infants and young children. But even if clean water

ONE STEP FORWARD, TWO STEPS BACK

We tend to think of toilets as an improvement over chamber pots and outhouses. In many respects they certainly are. But the early ones caused the death of thousands of people. The toilet, or water closet, was patented and first manufactured in England during the 1770s. Before this time, human excrement was collected and disposed of as night soil, so the waste and bacteria remained in relatively contained areas. Toilets, on the other hand, drained into sewers, which emptied into rivers. Drinking water was supplied by private companies that collected river water—now contaminated—and transported it back to cities. When the toilet came into European homes, so did higher risks of cholera and typhoid fever.

As early as the 1820s, some physicians and sanitary engineers suspected that drinking water contaminated by human waste was a threat to health. As one London engineer put it, "A stream that receives daily the evacuations of a million human beings with all the filth and refuse of various offensive manufacturers cannot require to be analyzed, except by a lunatic, to determine whether it ought to be pumped as a beverage for the inhabitants of the Metropolis of the British empire." Still, many were unconvinced. Even when microscopic examination revealed that London's water supply was teeming with "animalcules," some experts declared this was "no more harmful than eating fish." Sanitary reform was slow, advancing only as the germ theory of disease gained acceptance in the late 1800s. At that time, European cities began to develop sewage-treatment and water-filtering systems to furnish safe water.

Early water closets.

---

isn't available, individuals can protect themselves against most infections (except for some viral diseases such as hepatitis A) by boiling their drinking water.

Clean Food. Microorganisms in food make it spoil, and they can also cause disease. Many early food-preservation techniques, including canning, were developed before people understood why they worked. But methods of controlling specific foodborne diseases were devised only after the germ theory of disease was accepted.

Milk is particularly likely to transmit such infections as tuberculosis, typhoid fever, scarlet fever, brucellosis, and diphtheria. These illnesses were rampant before pasteurization (Chapter 9). Pasteur developed the process in 1864 to kill the wine-spoilage microorganisms. Only years later was it applied to milk. Chicago passed the first law requiring pasteurization of milk in 1908. Today in the United States pasteurization is almost universal. And milkborne diseases have become rare.

If food is adequately cooked, diseases such as trichinosis, salmonellosis, and tapeworm infection can be prevented (Chapter 23). Refrigeration can also prevent foodborne diseases, even though it only slows the growth of microorganisms. For example, refrigeration prevents staphylococcal food poisoning because the bacteria can't grow and produce their toxins (Chapter 23).

Personal Cleanliness.

Many disease-causing microorganisms are transmitted from person to person by dirty hands or contaminated fomites (Chapter 15). Probably the single most important habit of personal cleanliness for preventing the spread of disease is the simple act of hand washing. Even though we all know pathogens aren't visible to the naked eye, studies show that even nurses and doctors in hospitals often fail to wash their hands unless they can see that they're dirty. The importance of hand washing in preventing illness in the home, in day-care centers, and in hospitals can't be overemphasized.

Insect Control.

Insect control is an effective way to control diseases such as malaria and yellow fever that are transmitted by biting insects (Chapter 15). Early public health programs concentrated on draining swamps (to destroy breeding grounds), screening living areas, and using mosquito netting.

When powerful insecticides such as dichlorodiphenyltrichloroethane (DDT) became available in the 1940s, insect control became much more effective. At first DDT seemed miraculous. It was inexpensive, and it could virtually eliminate huge insect populations. But unfortunately, pesticide-resistant insects soon emerged, as did serious environmental problems. It's now generally accepted that insecticides can't provide the final answer to insectborne diseases. In 1987, after having spent more than $3 billion on insecticides over 25 years, public health planners in India announced that they would abandon the use of pesticides in their antimalaria campaign. Instead they would concentrate on old environmental measures, such as draining stagnant waters, and they would work to develop new ones.

Prevention of Sexually Transmissible Disease.

Sexually transmissible diseases, including syphilis and gonorrhea, have caused suffering and death for centuries, and now AIDS has joined the list. All sexually transmissible diseases are completely preventable by interrupting the chain of transmission. But interruption must be done on an individual basis by limiting sexual exposure or using condoms. Unfortunately, individual efforts to control disease are always less effective than public health projects. Therefore sexually transmissible infections continue to be a major problem in developed, as well as developing, countries.

Syphilis and gonorrhea continue to be controlled by early diagnosis and treatment. For AIDS, however, prevention is crucial because no cure exists. So far, AIDS prevention programs have relied on public education—encouraging people to limit their sexual exposure and to use condoms.

Prevention of Respiratory Diseases.

Pathogens can be excluded from the water we drink and the food we eat. But it's almost impossible to exclude them from the air we breathe (**Figure 20.6**).

Spread of some respiratory diseases can be minimized by isolating infected individuals. Before childhood immunization for diphtheria became common, isolation was the major way the disease was controlled. Children with diphtheria were taken out of their homes and isolated in a hospital diphtheria ward to decrease the likelihood that other family members would be infected. But isolation can't stop transmission of diphtheria because patients often begin to exhale pathogens before they show signs of illness. Moreover, healthy carriers transmit the pathogen.

Facemasks slow the spread of some respiratory diseases. But immunization is the only effective way to stop transmission of respiratory pathogens.

IMMUNIZATION

Adaptive immune defenses offer effective protection against pathogens once we've been exposed to them (Chapter 17). Immunization can do the same without the hazard of illness. There are two approaches to immunization: active and passive. Active immunization stimulates a person's own immune system. Passive immunization supplies preformed antibodies from an immune person or animal. We'll consider both approaches here.

Active Immunization

Certain infectious diseases afflict a person only once during a lifetime. That's because memory B lymphocytes produced during the first infection can mount an immune response so quickly that a second infection can't get started (Chapter 17). Active immunization mimics a once-in-a-lifetime infection without causing disease. It's also called vaccination. It's done more than anything else, except public sanitation, to control infectious disease.

A good vaccine must be safe and relatively free of side effects. However, no vaccine is completely safe. The vaccine's safety must be weighed against the likelihood and hazard of contracting the natural disease. For example, variolation, the first vaccine against smallpox, had a 1 to 2 percent fatality rate. But for many people the risk was acceptable considering the horror and prevalence of smallpox. Even Jenner's smallpox vaccination led to fatal complications in approximately 1 out of every 1 million people immunized. Vaccines in use today must meet high safety standards, but risk can never be completely

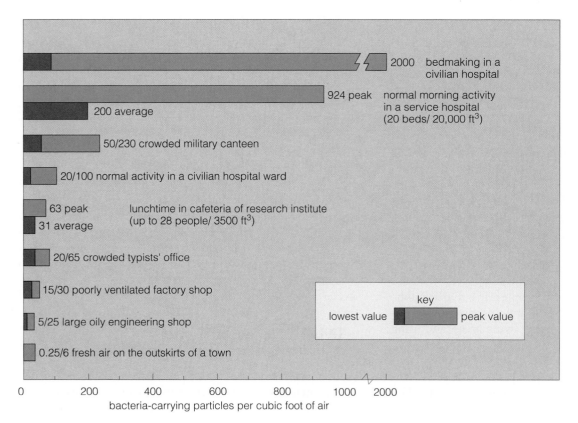

FIGURE 20.6   Bacteria in the air. (Results of a study by the U.S. armed forces.) The highest bacterial concentrations are the result of human activities.

eliminated. Oral polio vaccine, for example, caused paralytic polio in approximately 1 out of 7 million people. That's why it's been replaced in the United States by the killed vaccine for infants.

Vaccines have also been accused of causing a variety of unrelated illnesses. In response, Gordon Ada of The John Curtin School of Medical Research recently stated, "Overall, there is no firm scientific or clinical evidence that administration of any vaccine causes a specific allergy, asthma, autism, multiple sclerosis, or sudden infant death syndrome."

Besides being safe, a vaccine must be highly immunogenic. That is, it must stimulate an immune response strong enough to confer assured protection against the natural infection. No vaccine, however, provides complete protection to every person immunized. The measles vaccine, for example, protects 90 to 95 percent of those immunized. Nevertheless, this level of protection was considered sufficient to control measles by means of **herd immunity** (prevention of epidemics due to the scarcity of susceptible hosts). This proved not to be the case in the late 1980s, when numerous measles outbreaks occurred in schools. Epidemiologists concluded that the 5 to 10 percent of immunized students who were unprotected was

sufficient to fuel an epidemic, so a second dose of measles vaccine was recommended. The second round of immunization can be expected to protect 90 to 95 percent of those who remained unprotected after their first immunization, leaving a negligible number of people unprotected and thus ensuring herd immunity.

To be effective, vaccines must be appropriately administered. And every immunizing agent has its own preferred route—either **orally** (by mouth), **subcutaneously** (below the skin), or **intramuscularly** (into muscle). The Sabin polio vaccine, for example, is administered orally, whereas the Salk polio vaccine is administered subcutaneously. To maintain their effectiveness, vaccines must also be properly stored and administered while still active.

Types of Vaccines. Some vaccines, such as the one for measles, stimulate both antibody and cell-mediated immune defenses. Others, such those for tetanus and diphtheria, stimulate primarily antibody-mediated immunity.

All vaccines contain one or more microbial antigens that stimulate an immune response, but the nature and source of the antigens differ. In the following sections, we'll consider the various kinds of vaccines: attenuated, inactivated, acellular, and DNA based.

LARGER FIELD

THE FIRST VACCINATION

Smallpox, a disfiguring and often fatal disease, was the first disease against which a safe and effective immunizing agent was developed. It is also the first and, so far, only disease to be eliminated from Earth as the result of public health efforts.

In Asia and Africa, active immunization against smallpox was practiced for centuries as folk medicine. In a practice called variolation, crusts from a smallpox blister were inoculated into a healthy person through a tiny cut in the skin or into the lining of the nose, to induce smallpox. Usually smallpox acquired from variolation was mild—only a few blisters appeared—but 1 to 2 cases in every 100 died. On the other hand, half the people who were naturally infected died, and variolation conferred lifelong immunity. When variolation was introduced in Europe in the late 1700s, Catherine the Great of Russia and Louis XVI of France thought the risk worthwhile for themselves.

About this time, Edward Jenner was a boy in England. He underwent variolation and suffered an unusu-

ally severe case of smallpox. He never forgot the experience. Later he became a country physician, and, like other country people, he was familiar with the folk wisdom that dairymaids who had recovered from a mild disease called cowpox were protected for life against smallpox. Most

physicians considered the cowpox-smallpox connection an old wives' tale. But Jenner was intrigued, and in 1796 he performed an experiment that showed cowpox infection could indeed prevent smallpox.

Jenner took crusts from the hand of a dairymaid, Sarah Nelms, who was suf-

fering from cowpox and inoculated them into the skin of James Phipps, a healthy 8-year-old who had never had smallpox. James became mildly ill for 1 day. Then James underwent variolation. He failed to contract even a mild case of smallpox. Jenner inoculated James again, and

again James suffered no effect. Jenner concluded, correctly, that cowpox had made young James immune to smallpox.

The cowpox virus became the first safe vaccine, and vaccination replaced variolation as a much safer way to prevent smallpox. We know now that cowpox

infection provides immunity against smallpox because cowpox and smallpox are closely related viruses that share certain antigens.

We no longer vaccinate against smallpox because the disease no longer exists, but with the threat of bioterrorism it might begin again, either routinely or to protect groups at special risk. Certainly it would be used to control outbreaks.

Another interest in the vaccinia virus that was long used for smallpox vaccination has developed. Using genetic engineering, researchers have identified genes in many pathogens that encode antigens that might be used for vaccines. Some of these genes have been transferred into the vaccinia virus and expressed there. Because we have a great deal of practical information about immunization with vaccinia—accumulated during the worldwide smallpox eradication program—vaccinia might become an ideal vehicle for immunizing against diseases other than smallpox. This is only one of the ways recombinant DNA technology may soon be used to develop new vaccines (Chapter 7).

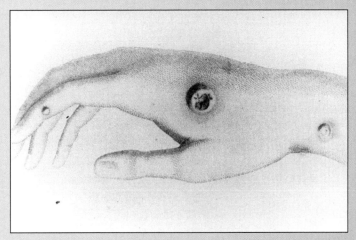

Cowpox crusts on Sarah Nelms's hand.

Attenuated Vaccines. **Attenuated vaccines** contain live cells of a pathogen that have been genetically altered to cause a limited infection without serious illness. When they were first developed during the 1800s, the basis of attenuation was unknown. The accepted approach was to cultivate the pathogen under abnormal conditions—at an unfavorable temperature or in an unusual host—until it lost virulence.

For example, Pasteur attenuated his anthrax vaccine by growing *Bacillus anthracis* at the highest temperature it could tolerate. Now we know that high temperature causes the anthrax bacterium to lose the plasmid that encodes anthrax toxin.

Attenuated vaccines are usually highly effective because the organism multiplies within the host, producing antigens and thereby stimulating a strong immune response. They also tend to produce long-lasting immunity. Most attenuated vaccines now used are for viral illnesses. They include polio, mumps, measles, and rubella vaccines.

Inactivated Vaccines. **Inactivated vaccines** contain cells of pathogenic microorganisms that have been killed, either by heat or by a chemical such as formalin, phenol, or acetone. Inactivated vaccines are quite safe, but they tend to be less effective than attenuated vaccines. The process of inactivation can partially destroy the antigens that stimulate immunity. Also, because the microorganisms are dead, they can't multiply in the host; the vaccine itself must contain enough antigen to produce a protective immunological reaction. Most inactivated vaccines must be administered more than once, in **booster** doses. Inactivated vaccines are used against many bacterial illnesses, including whooping cough and typhoid fever, as well as a few viral illnesses, including rabies and hepatitis A.

Acellular Vaccines. Both attenuated and inactivated vaccines are whole-cell vaccines. They contain a complete microorganism. But not all of the microorganism's components contribute to stimulating an immune response, and some of them cause side effects. Acellular vaccines contain only certain components of the microorganism, ideally those that stimulate maximum immune response without causing side effects. The biggest recent advance has come in this area with the introduction of the acellular whooping cough vaccine, which is virtually free of side effects.

Toxoids were the first acellular vaccines. They are purified exotoxins that have been inactivated by heat or chemicals (**Figure 20.7**). Toxoids are effective only against diseases such as tetanus and diphtheria, which are caused by exotoxins rather than by the microorganisms themselves (Chapter 15). Toxoids stimulate the production of antibodies called **antitoxins,** which combine with toxin molecules and inactivate them.

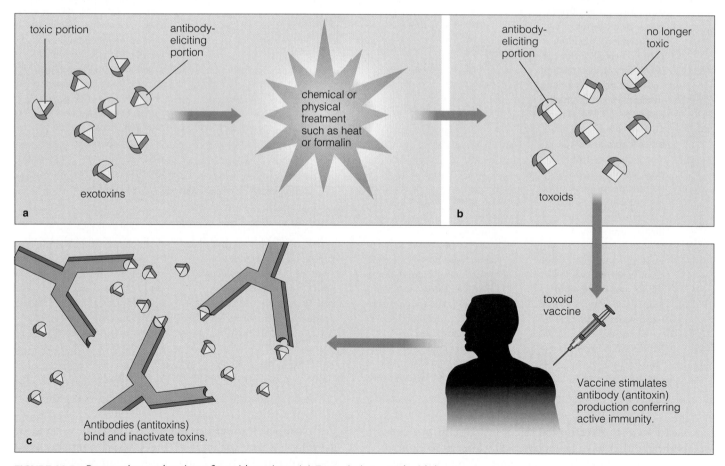

FIGURE 20.7   Preparation and action of toxoid vaccines. (a) Exotoxin is treated with heat or a chemical compound such as formalin. (b) Treatment produces a toxoid, which is harmless but still immunogenic. (c) When injected as a vaccine, the toxoid stimulates the formation of antibodies called antitoxins. The antitoxins react with natural exotoxins to inactivate them and thus confer active immunity.

Other acellular vaccines, called **subunit vaccines,** contain purified protein components, which are antigenic and thereby confer immunity. The vaccine used against pneumococcal pneumonia contains purified capsules from various strains of the causative bacterium, *Streptococcus pneumoniae.*

Sometimes a purified cellular component is too weakly antigenic to be an effective vaccine by itself. This is the case with the polysaccharide capsule from *Haemophilus influenzae* type b (Chapter 22). This problem can often be solved by combining the weak antigen with a protein, such as diphtheria toxoid, thereby making it a much more powerful vaccine, termed a **conjugate vaccine.**

An alternative to purifying a cellular component from an intact organism is to produce it by recombinant DNA technology. Such a genetic engineering approach has distinct advantages. Production is less hazardous and less expensive, and the component can be produced in inactive form so no further inactivation step is required. This approach has already been successful. The vaccines now used against pertussis (Chapter 22) and hepatitis B (Chapter 23) are made in this simple, effective way.

Now it appears that an even simpler approach might be using antigen-encoding DNA.

DNA-based vaccines. In 1990 researchers at Vical, a startup biotechnology company in San Diego, made a remarkable chance discovery. They found that mice would make microbial proteins if cloned DNA was injected into their muscles. This observation opened the possibility that mice, and maybe humans as well, could make their own vaccines if injected with appropriate microbial DNA. Confirmation came rather quickly. In 1993 naked DNA was shown to work as a vaccine that protected mice against influenza virus. Now DNA vaccines for humans has become a very active field of research. It even has its own Web site (www.genweb.com/Dnavax/dnavax.html). DNA vaccines against several diseases are now being evaluated in human trials.

Table 20.4 summarizes the current status of vaccines against human pathogens.

Active Immunization and Public Health Policy.

Global immunization programs sponsored by the World Health Organization (WHO) have made spectacular progress in recent decades. Eradicating smallpox is its most stunning success, but the Expanded Program on Immunization has made significant progress against the six preventable diseases that cause the greatest worldwide childhood death and disability—measles, diphtheria, pertussis, tetanus, polio, and tuberculosis. In 1974 only 5 percent of the children in the developing world were immunized against all six of these diseases. By 1988 almost 50 percent were, saving 1.4 million lives per year, at a cost of less than 50 cents per child. Still, about 2 million children in developing countries die each year from measles, for example.

In the United States the U.S. Public Health Service has established immunization goals for all Americans, because younger children in the U.S. are the most adequately immunized and U.S. adults the least. For example, 77 percent of 2-year-olds are completely immunized; 97 percent of kindergartners have received basic immunizations, but only 20 percent of adults who need protection against influenza and pneumonia have been immunized. In addition, there is still a long deadly list of pathogens for which no effective vaccine exists (**Table 20.5**).

Passive Immunization

Active immunization stimulates the recipient's own immune system to mount a protective response that prevents later infection. Passive immunization is a kind of antibody transfusion. Antibodies produced by a human or animal donor are collected, purified, and administered to the recipient. This type of immunization is called passive because protection does not require participation of the recipient's immune system. However, the protection lasts only as long as the antibody molecules survive in the recipient—months with antibodies from another human, but only weeks with antibodies from animals.

The antibody preparations used may be gamma globulin—a collection of antibodies from the pooled serum of many different donors—or special preparations that contain high titers of specific antibodies. For example, tetanus immune globulin contains high concentrations of antibodies against tetanus toxin, and varicella zoster immune globulin contains high concentrations of antibodies against the virus that causes chickenpox and shingles.

Passive immunization has certain advantages. Even severely immunosuppressed patients can be protected, and protection is immediate. But passive immunization also has disadvantages. Protection lasts only a short time, and if animal antibodies are used, there is a risk of serum sickness (Chapter 18). In general, other forms of prevention or treatment usually work better than passive immunization.

Passive immunization can occasionally be useful in preventing tetanus (Chapter 25). Most of us who have received the tetanus toxoid vaccine only need a booster dose of the toxoid after receiving a tetanus-prone wound (such as stepping on a dirty nail). Because the immune system has responded to the vaccine once, the booster dose stimulates a memory response immediately—in plenty of time

TABLE 20.4 Vaccines Against Human Pathogens

Pathogen	Disease	Type of Vaccine	Status	Chapter
Bacteria				
Bacillus anthracis	Anthrax	Inactivated	Available	27
Bordetella pertussis	Whooping cough	Inactivated, subunit	Available	22
Borrelia burgdorferi	Lyme disease	Subunit	Available	27
Clostridium tetani	Tetanus	Toxoid	Available	25
Corynebacterium diphtheria	Diphtheria	Toxoid	Available	22
Coxiella burnetii	Q fever	Inactivated	Available	22
Haemophilus influenzae	Meningitis, epiglottitis, pneumonia	Conjugated	Available	25 22 22
Mycobacterium leprae	Leprosy	Inactivated	Phase 3[a]	26
Mycobacterium tuberculosis	Tuberculosis	Attenuated	Available	22
Neisseria meningitidis	Meningitis			25
Serogroup B		Subunit, conjugated	Available	
Serogroup C		Conjugated	Phase 3	
Salmonella typhi	Typhoid fever	Attenuated, conjugated	Available	23
Staphylococcus aureus	Impetigo, toxic shock	Conjugated	Phase 3	26
Streptococcus pneumoniae	Pneumonia, otitis media, meningitis	Conjugated	Available	22
Vibrio cholerae	Cholera	Attenuated, inactivated	Available	23
Viruses				
Adenovirus	Respiratory disease	Attenuated	Available	22
Hepatitis A	Liver disease, cancer	Attenuated, inactivated	Available	23
Hepatitis B	Liver disease, cancer	Subunit	Available	23
Influenzavirus A	Respiratory disease	Attenuated, inactivated, subunit	Available	22
Influenzavirus B	Respiratory disease	Subunit	Phase 3	22
Japanese encephalitis virus	Encephalitis	Inactivate	Available	25
Measles virus	Measles	Attenuated	Available	26
Mumps virus	Mumps	Attenuated	Available	23
Poliovirus	Polio	Attenuated, inactivate	Available	25
Rabies virus	Rabies	Inactivated	Available	25
Rubella virus	German measles	Attenuated	Available	26
Vaccinia virus	Smallpox	Attenuated	Available	26
Varicella zoster virus	Chicken pox	Attenuated	Available	26
Yellow fever virus	Yellow fever	Attenuated	Available	27
Protozoan				
Leishmania	Leishmaniasis	Attenuated, inactivate	Phase 3	12
Fungus				
Coccidioides immitis	San Joaquin Valley fever	Inactivated	Phase 3	22

[a]Phase 3 indicates the vaccine are being evaluated in the final stages of clinical trials and should be available in 5 to 10 years.

TABLE 20.5 Some Pathogens for Which No Effective Vaccine Exists

Borrelia burgdorferi

Chlamydia spp.

Coccidioides immitis

Cytomegalovirus

Dengue fever virus

Epstein-Barr virus

Group A streptococcus

Hepatitis C, D, and E viruses

Herpes simplex virus types 1 and 2

HIV (AIDS)

Human papillomavirus

Legionella pneumophila

Leishmania spp.

Mycoplasma pneumoniae

Neisseria gonorrhoeae

Parainfluenza virus

Plasmodium spp.

Pseudomonas aeruginosa

Respiratory syncytial virus

Rickettsia rickettsii

Rotavirus

Shigella spp.

Toxoplasma gondii

Treponema pallidum

Data from Young, P. 1997. Bright Outlook on Direct DNA Immunizations. *ASM News* 63:659–63.

to protect against tetanus. A never-immunized person, however, who will mount a primary rather than a memory immune response when receiving the vaccine, needs immediate protection against the potentially fatal tetanus toxin. Thus unimmunized people who suffer a dirty wound should receive tetanus immune globulin (passive immunization) at the same time they receive their first dose of tetanus toxoid vaccine (active immunization). They should then go on to receive the full series of tetanus boosters, so they will be adequately protected in the future.

SUMMARY

Preventing Disease (p. 466)

1. Epidemiology is the study of when and where diseases occur and how they are transmitted. Public health organizations develop and implement plans to prevent and control disease.

Epidemiology (p. 466–477)

2. Epidemiology focuses on populations rather than individuals. Disease can be transmitted in an epidemic, affecting many members of a population in a short time. If spread worldwide, it becomes a pandemic. Some diseases are endemic, always present in a population; others are sporadic, occurring only occasionally in a population.

The Methods of Epidemiology (pp. 468–471)

3. Many sources of epidemiological information are part of the public record. These include vital statistics and census data. Government agencies track notifiable, or reportable, diseases.

4. The Centers for Disease Control and Prevention (CDC) is the national agency that maintains and reports epidemiological trends in disease. It publishes the *Morbidity and Mortality Weekly Report* (MMWR).

5. Epidemiologists use surveys, questionnaires, interviews, and hospital records to gather specific information.

6. Epidemiologists usually express health information about a population as a rate—the ratio of the number of people in a particular category to the total number of people in the population being studied. Rates allow comparisons to be made.

7. Incidence rate is the rate of developing a disease or condition during a certain period.

8. Prevalence rate is the rate of people who have a certain disease at any particular time.

9. Age-adjusted rates are usually more informative than raw data.

Types of Epidemiological Studies (p. 471)

10. Retrospective studies analyze events that have already occurred. Prospective studies record events as they are happening and analyze them at the end of a certain period. Experimental studies deliberately influence events to see what the outcome will be.

11. Descriptive epidemiology provides general information about a disease, including how nonmicrobial factors contribute to infectious disease.

12. Surveillance epidemiologists track epidemic diseases by monitoring situations that might lead to epidemics and following the progress of an epidemic if one begins. Surveillance epidemiology played an important role in global eradication of smallpox.

13. Field epidemiologists investigate outbreaks of disease to determine the source of the epidemic and how it can be controlled. Field epidemiologists discovered the sources of AIDS, Lyme disease, and legionellosis.

14. Nosocomial infections are hospital-acquired infections.

15. Urinary tract infections, usually caused by catheterization, account for nearly half of all nosocomial infections.

16. Surgical wound infections are the second most frequent type of nosocomial infection.

17. Skin infections, especially among newborns and burn victims, are a close fourth among nosocomial infections.

18. Hospital staff epidemiologists are specially trained to recognize and interrupt a potential hospital epidemic. Staff epidemiologists communicate the problem and enforce routine infection-control policies such as hand washing.

19. The Universal Blood and Body Fluid Precautions, issued by the CDC to protect hospital workers, apply to all patients because it is impossible to tell which patients are HIV infected.

Public Health (pp. 477–479)

20. Public health deals with prophylaxis. The two major methods are limiting exposure to a pathogen and immunization.

Public Health Organizations (p. 477)

21. Agencies to protect the public health exist at the local, state, national, and global levels. The United States Public Health Service (USPHS) is our national public health agency. The Centers for Disease Control and Prevention (CDC) is one of its arms. The World Health Organization (WHO) is the international organization concerned with improving health.

Limiting Exposure to Pathogenic Microorganisms (pp. 477–479)

22. Waterborne diseases can be prevented by clean water supplies and adequate sewage treatment.

23. Heat kills most foodborne diseases; refrigeration slows the growth of microorganisms. Pasteurization makes milk—which is particularly likely to carry infection—safe to drink.

24. Probably the single most important habit of personal cleanliness that can prevent the spread of disease is hand washing.

25. Diseases transmitted by insect bites can be controlled by decreasing or eliminating the reservoirs.

26. All sexually transmissible diseases are preventable by interrupting the chain of transmission. This depends upon ongoing individual effort. AIDS prevention is imperative because there is no cure.

27. Immunization is the best way to prevent transmission of respiratory diseases because it is virtually impossible to eliminate pathogens from the air we breathe.

Immunization (pp. 479–485)

28. Immune defenses can be augmented by immunization.

Active Immunization (pp. 479–483)

29. In active immunization, or vaccination, a person's own immune system is artificially stimulated to ward off disease. A vaccine is an agent capable of inducing active immunity without causing disease.

30. Herd immunity refers to prevention of epidemics because of scarcity of susceptible hosts.

31. Every vaccine has a preferred route of administration, either orally, subcutaneously, or intramuscularly.

32. Attenuated vaccines contain live microorganisms that are weakened. They cause a limited infection, usually without serious illness, and stimulate a powerful and long-lasting immune response because the live organisms multiply in the body, producing more antigens. Most attenuated vaccines are used for viral illnesses.

33. Inactivated vaccines contain killed microorganisms. The dose must contain enough antigen to produce a protective immunological response because the cells cannot multiply to produce more. Most inactivated vaccines require a booster shot.

34. Genetic engineering is changing the way vaccines are made. For example, a genetically engineered vaccine for hepatitis B contains no contaminated blood products, making hepatitis B immunization more acceptable to many patients.

35. The new vaccine for *Haemophilus influenzae* type b is a protein conjugate vaccine. The polysaccharide antigen is conjugated (combined) with a protein, such as diphtheria toxoid, which transforms it into a much more powerful antigen.

36. Toxoid vaccines are made of toxins modified by heat or chemicals so they remain immunogenic but are harmless. Toxoids stimulate the production of antibodies called antitoxins. These vaccines are used against tetanus and diphtheria.

37. The possibility of using naked DNA as vaccine holds great promise.

38. The USPHS has established immunization goals for all persons in the U.S. Even in the United States, more widespread active immunization would improve public health.

Passive Immunization (pp. 483–485)

39. In passive immunization, antibodies produced in a donor human or animal are collected, purified, and administered to the recipient.

40. Passive protection lasts only as long as the antibody molecules survive—weeks to months. Gamma globulin, a collection of various antibodies from the pooled serum of many donors, or special preparations with high titers of specific antibodies are used.

41. Passive immunization can be used with immunosuppressed patients. Protection is immediate. Because protection lasts only a short time, antibodies must be administered after every known exposure. With animal antibodies, there is a risk of serum sickness.

42. For some diseases, passive immunization is best. An injection of gamma globulin for someone exposed to hepatitis A is almost certain to provide needed protection. Someone who has never been immunized against tetanus and receives a tetanus-prone wound should receive a shot of tetanus immune globulin and be actively immunized.

REVIEW QUESTIONS

Preventing Disease

1. What is epidemiology? How does it differ from public health?

2. Distinguish among these patterns of disease transmission in a population—epidemic, pandemic, endemic, sporadic.

Epidemiology

3. Why are reliable sources of information important in epidemiology? Why are vital statistics and the census useful? Name some other sources of information epidemiologists use.

4. What are notifiable diseases? What is the CDC? The MMWR?

5. Explain this statement: Statistics is the language of epidemiology.

6. Define these terms: rate, incidence rate, prevalence rate, age-adjusted rate.

7. What is the difference between a retrospective study and a prospective study? What is an experimental study?

8. What is descriptive epidemiology?

9. What is surveillance epidemiology? Why was effective surveillance epidemiology critical in eradicating smallpox worldwide?

10. Explain this statement: Field epidemiologists are like detectives.

11. Describe the most common kinds of nosocomial infections and, in each case, the microorganisms involved.

12. Describe the work of a hospital epidemiologist. What are Universal Precautions?

Public Health

13. Give some examples of how local, state, national, and global agencies help scientists and clinicians improve public health.

14. Explain this statement: Public health deals with prophylaxis.

15. What are some the ways that public health has diminished reservoirs and interrupted transmission in each of these areas: clean water, clean food, personal cleanliness, insect control, prevention of sexually transmissible diseases, and prevention of respiratory diseases.

Immunization

16. Define: immunization, vaccine, immunogenic, herd immunity.

17. What is the difference between active immunization and passive immunization?

18. Explain this statement: All good vaccines are relatively safe and relatively effective.

19. What is an attenuated vaccine? What is an inactivated vaccine? What are the advantages and disadvantages of each?

20. Give an example of how genetic engineering has made a vaccine safer. Give an example of how genetic engineering has made a vaccine more effective.

21. How can naked DNA act as a vaccine?

22. What are toxoid vaccines, and how are they produced? Give some examples of toxoid vaccines.

23. Explain this statement: Even in the United States, more widespread immunization would improve public health.

24. What is passive immunization? What are its advantages and disadvantages compared with active immunization? Give an example of when passive immunization is best.

25. What is gamma globulin? What is tetanus immune globulin?

CORRELATION QUESTIONS

1. Why is the prevalence rate for AIDS higher than the prevalence rate for cholera?

2. What is the significance for an increasing incidence rate for a particular disease? Explain.

3. Does the incidence rate give any information about the size of the population for which it was determined? Explain.

4. What data would you need to determine whether a disease was endemic or epidemic?

5. Knowing the prevalence rate and incidence rate of a disease, what other important information could you calculate about the disease? Explain.

6. Would it be possible to eradicate a disease by passive immunization? Explain.

ESSAY QUESTIONS

1. Discuss John Snow's study of the Broad Street cholera epidemic from the standpoint of how it defines the field of epidemiology.

2. Sketch some scenarios illustrating how age-adjusted death rates might be essential to reaching a valid conclusion in an epidemiological study.

SUGGESTED READINGS

Gordis, L. 2000. *Epidemiology.* 2nd ed. Philadelphia: W. B Saunders.

Snow, J. 1965. *Snow on cholera.* New York: Hafner.

Taubes, G. 1997. Salvation in a snippet of DNA. *Science* 278:1711–14.

For additional readings, go to InfoTrac College Edition, your online research library at:
http://www.infotrac.thomsonlearning.com

TWENTY-ONE

Pharmacology

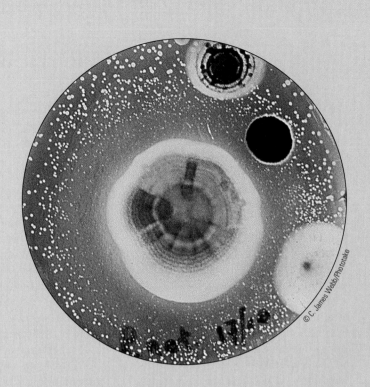

CHAPTER OUTLINE

*Case History: Scarlet Fever—
Then and Now*

**PRINCIPLES OF
PHARMACOLOGY**
Drug Administration
Drug Distribution
Eliminating Drugs from the Body
Side Effects and Allergies
Drug Resistance
The Original Wonder Drug
Drug Dosage
The Serum Killing-Power Test

**TARGETS OF ANTIMICROBIAL
DRUGS**
The Cell Wall
Cell Membranes
Protein Synthesis
The Medical Letter
Nucleic Acids
Folic Acid Synthesis

ANTIMICROBIAL DRUGS
Antibacterial Drugs
The Discovery of Streptomycin
Antimicrobial Drugs in U.S. History
Antimycobacterial Drugs

Antifungal Drugs
Antiparasitic Drugs
Antimalarial Drugs
Antiviral Drugs

SUMMARY

REVIEW QUESTIONS

CORRELATION QUESTIONS

ESSAY QUESTIONS

SUGGESTED READINGS

LEARNING GOALS

To understand:

- *How antimicrobial drugs are selected and administered*
- *How drugs become distributed throughout the body and how they are eliminated from it*

- *How antimicrobial drugs act and how microorganisms become resistant to them*
- *How the sensitivity of microbes to drugs is tested*

- *The properties of the drugs commonly used to treat bacterial, mycobacterial, fungal, parasitic, and viral infections*

Scarlet Fever—Then and Now

T. M., a 6-year-old kindergartner, came home from school tired and listless, with a fever of 102°F. His mother put him to bed and gave him the prescribed dose of children's acetaminophen to relieve the fever. That night T. M. refused dinner, complaining that he wasn't hungry and it hurt to swallow. In the morning his father noticed he had a rash over most of his body. T. M's tongue was bright red and swollen. His father called the pediatrician for an appointment that day.

When T. M.'s pediatrician examined him, he still had a temperature and appeared tired and uncomfort-

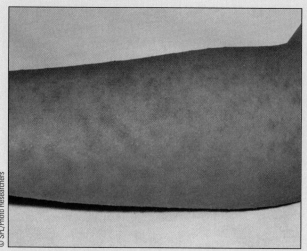

The rash typical of scarlet fever.

able. His throat was bright red, with white patches on his tonsils and tiny red spots on the soft palate. His tongue was inflamed, swollen, and bumpy. Lymph nodes in his neck were enlarged and tender. The bright red rash covering most of his body felt like coarse sandpaper. From these signs and symptoms, T. M.'s pediatrician made a probable diagnosis of **scarlet fever**—a clinical syndrome caused by the highly pathogenic bacterium *Streptococcus pyogenes.*

Her diagnosis was confirmed by a GenProbe test, which depends upon information gained from genomics (Chapter 7) done on cells swabbed from T. M.'s throat. Since T. M. had no history of drug allergies, the physician prescribed penicillin, the preferred drug for *S. pyogenes* infections. T. M. received a liquid preparation of penicillin VK, a potassium salt of the orally absorbed penicillin V, twice a day. Twelve hours after the penicillin therapy began, T. M. felt much better. His recovery was complete and uneventful. Soon he was back at school.

Fifty years ago, a child who contracted scarlet fever became extremely ill and often died. Today most children become only mildly ill and recover quickly with antibiotic therapy. To avoid the frightening associations that a diagnosis of scarlet fever can still hold for a parent, physicians sometimes use the term **scarletina.** Antibiotics have played a major role in controlling scarlet fever, but the principal difference between scarlet fever today and yesterday is the diminished virulence of *Streptococcus pyogenes.* However, its virulence is rising again, and there is concern that antibiotic-resistant strains of streptococci may become prevalent. Strains resistant to macrolide antibiotics first appeared in 2001; now almost half of clinical isolates are resistant to these antibiotics. Fortunately, *S. pyogenes* remains sensitive to penicillins.

Case Connections
- Although scarlet fever itself has changed, availability of effective drugs to fight this and other infectious diseases exerted the major impact.
- This chapter considers these drugs that have profoundly changed our lives—how they're used, how they act, and why some are losing their effectiveness.

PRINCIPLES OF PHARMACOLOGY

Antimicrobial drugs, including **antibiotics** (naturally occurring antimicrobial agents; **Figure 21.1**) and those made by chemical synthesis, have changed the face of modern medicine within the past 50 years. Before these drugs were available, there was no way, other than by supportive care, to treat an infectious disease once it began. But now most infectious diseases can be treated and cured—many quickly and painlessly. The impact of these drugs has been profound. For example, antimicrobial drugs have added 10 years to the life expectancy of persons in the United States. (Finding a cure for all cancers would probably add only about 2 years.) You might be among the many readers of this

FIGURE 21.1 Antibiotic-producing microorganisms. A plate inoculated with soil often has colonies, like the large one in the center of the plate (with brown spore-covered center and colorless spore-free edge). Note how the small white colonies become smaller the closer they are to the antibiotic-producing colony.

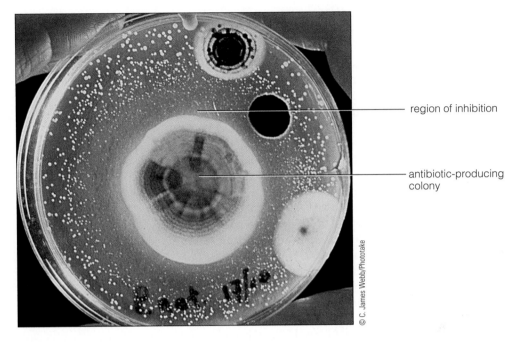

region of inhibition

antibiotic-producing colony

© C. James Webb/Phototake

textbook who contributed to this statistic by surviving a potentially lethal childhood infection thanks to treatment with one of the drugs we'll discuss in this chapter.

Pharmacology is the study of **drugs** (any chemicals that have a physiological effect on living things). This broad class of chemicals, which includes alcohol, caffeine, and penicillin, is subdivided by use. For example, drugs, often called **agents,** used to treat any disease are called **chemotherapeutic agents.** Those used to treat infectious diseases are called **antimicrobial agents.** Antimicrobial agents are further subdivided according to their source and antimicrobial activity (**Table 21.1**).

It's essential to know something about pharmacology in order to treat microbial infections. For example, T. M.'s pediatrician had to have the knowledge to choose and administer an effective drug. In this chapter we'll consider the basics of the pharmacology and how antimicrobial agents act. Then we'll survey the antimicrobial agents now being used. We'll start by considering how drugs are administered.

Drug Administration

The goal of drug administration is to get an effective concentration of the drug quickly to the site of the infection. There are many possible routes. We'll discuss the major ones in the following paragraphs.

Minor infections on the body surface, such as impetigo or athlete's foot (Chapter 26), can be treated by **external therapy,** also called **local** or **topical therapy.** In such cases the drug is applied directly to the infected area. But most infections require **systemic therapy** (the drug enters the patient's bloodstream). Systemic therapy can be achieved by intravenous, intramuscular, oral, or inhalation administration.

Intravenous (IV) administration introduces the drug directly into a vein. This is the fastest way to get high levels of a drug into the bloodstream, but it's technically difficult and painful. A needle or plastic catheter must be introduced into the patient's vein and remain there throughout therapy, so there is a risk of a **superinfection** (an added infection). Intravenous drug therapy is therefore used only for serious infections of hospitalized patients.

TABLE 21.1 Some Kinds of Antimicrobial Drugs

Type of Antimicrobial Drug	Defining Properties
Distinguished by Source	
Antibiotic	Metabolic product of a microorganism
Synthetic drug	Drug produced by chemical synthesis
Semisynthetic drug	Antibiotic that has been modified chemically
Distinguished by Activity	
Antibacterial agents	Used against bacterial infections
Antimycobacterial agents	Used against tuberculosis and related infections
Antifungal agents	Used against fungal infections
Antiviral agents	Used against viral infections
Antiparasitic agents	Used against protozoal and helminthic infections

Intramuscular (IM) administration introduces the drug into muscle, usually by injection into the arm or buttock. The drug reaches peak levels in the bloodstream within 15 minutes because muscle is richly supplied by blood vessels. But IM administration can be painful, and the injection must be done by a trained professional.

In **oral administration** (or **PO,** from the Latin *per os,* "by mouth"), the drug is swallowed. When it enters the gastrointestinal tract, it is absorbed into the bloodstream. This is the simplest and most common way to administer a drug, and it's painless. But oral administration is slow and inefficient. Only a fraction of the dose reaches the bloodstream (**Figure 21.2**). Moreover, orally administered drugs usually must be taken several times daily on consecutive days. Often the result is dosage errors and failure to complete the full course of treatment.

T. M.'s pediatrician decided on oral administration of penicillin for several reasons. T. M.'s illness was not serious enough to require hospitalization. That eliminated intravenous administration. Intramuscular injection of a benzathine salt of penicillin G would have been a good choice because it dissolves slowly and would provide a continuous source of penicillin for about 10 days—the necessary time of treatment. This route of administration was rejected because T. M.'s parents didn't want him to receive the painful shot. Oral therapy with penicillin VK is also effective. When given orally, penicillin VK is the preferred form because it's more resistant than penicillin G to destruction by stomach acid. Still, only about a third of the ingested dose actually enters the bloodstream. In contrast, about two-thirds of an IM-administered dose of penicillin VK enters the bloodstream. Oral administration was effective for T. M. because his parents remembered to give it to him twice a day during the prescribed 10-day course.

Even after a drug enters the bloodstream, there are barriers that might prevent it from reaching the site of infection. Now let's consider these barriers.

Drug Distribution

The two major barriers to drug distribution are cell membranes and proteins that bind to drugs (**Figure 21.3**).

Cell membranes can prevent the free passage of drugs into and out of cells. Therefore they're barriers to the entrance of drugs into tissues, such as those that surround blood vessels. These cell membranes, like those that surround microorganisms, are composed of a phospholipid bilayer with embedded proteins (Chapter 4). Drugs cross human membranes by passing through one or the other of these two components. Those drugs that dissolve in the phospholipid can pass directly through it. But most drugs are not soluble in lipids. Such drugs can pass only through

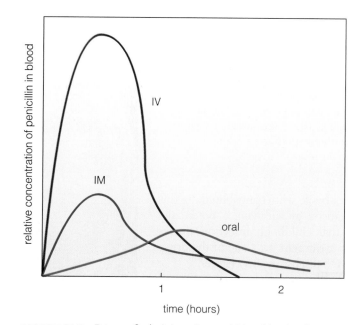

FIGURE 21.2   Route of administration and blood levels of penicillin. Intravenous administration produces the highest blood levels within the shortest period. Intramuscular administration produces lower levels in less than an hour. Orally administered penicillin takes more than an hour to reach its maximum blood level, which is somewhat lower than that reached by intramuscular injection.

the protein part of a membrane. Some of them pass through protein-lined pores. Others pass through enzyme-catalyzed transport systems. Drug delivery to the central nervous system and eye is particularly difficult because these organs are surrounded by nearly impermeable membranes. As a result, nervous system and eye infections are difficult to treat with antimicrobial agents. Especially high dosages must be used.

The second barrier, drug-binding proteins, form protein-drug complexes in the blood. In this bound form the drug cannot cross either component of the membranes. But such binding is only partial, and it's reversible. As a result, some unbound drug is always present. And as the unbound drug molecules leave the immediate vicinity, some previously bound drug molecules are released from the complex. Thus protein-binding slows drug distribution, but it doesn't stop it.

T. M.'s physician was confident that adequate amounts of penicillin would be distributed to his throat tissues. In fact, it reaches effective antimicrobial concentrations in almost all tissues except those in the brain and eye. About 65 percent of penicillin is reversibly bound to the plasma protein albumin.

There's a constant race between a drug's being distributed to where it's needed in the body and its being eliminated from the body. Now we'll consider the issue of elimination.

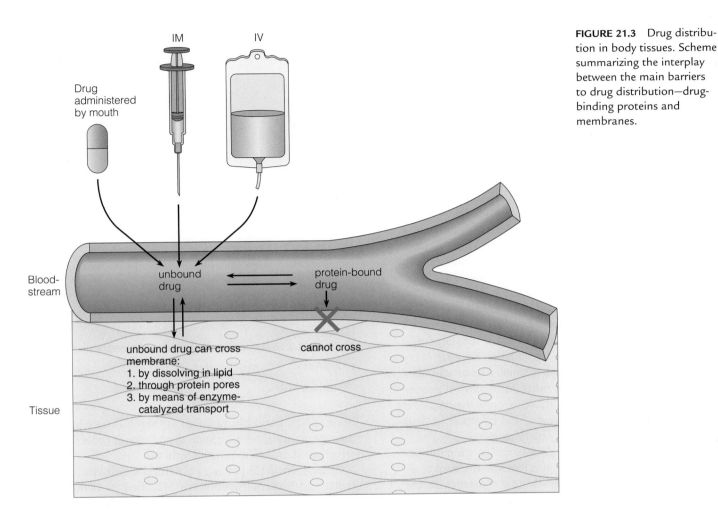

FIGURE 21.3 Drug distribution in body tissues. Scheme summarizing the interplay between the main barriers to drug distribution—drug-binding proteins and membranes.

Eliminating Drugs from the Body

Drugs can be eliminated in one of two ways. They can be converted metabolically to a different compound or they can leave the body. Drug metabolism occurs principally in the liver. Usually the metabolic product is inactive as a drug. But sometimes, metabolic products are as active or even more active than the drug itself. For example, the drug sulfamidochrysoidine (**Prontosil**), one of the first successful antimicrobial agents, has no antimicrobial activity. But its metabolic product, sulfa, is a potent antimicrobial agent.

Most drugs leave the body by excretion, usually by the kidneys into the urine. A few drugs are excreted by the liver into the bile and eliminated in the feces. It's important to know how a drug is excreted, because a drug administered to someone whose liver or kidneys are not functioning normally might rise to toxic levels.

Before prescribing penicillin, T. M.'s physician considered the fact that penicillin is excreted rapidly through the kidneys. This means levels of penicillin in the bloodstream decline rapidly (Figure 21.2). Within an hour and a half after a person receives a soluble form of penicillin intra-

muscularly, most of it is in the urine. T. M., a healthy child with normally functioning kidneys, could be expected to excrete his oral dose of penicillin almost as rapidly. Thus to maintain adequate levels of penicillin to combat the infection, he would have to take his medication regularly and often.

In the early days of penicillin, when the drug was scarce and expensive, it was routinely recovered from patients' urine so it could be reused. Today **probenecid**, a drug that slows the rate of excretion, is sometimes administered along with penicillin to prolong its action.

Now let's turn our attention to the hazards of antimicrobial drugs. They can hurt, as well as help.

Side Effects and Allergies

An effective antimicrobial drug must exhibit selective toxicity. That means it must inhibit or kill microorganisms while leaving human cells relatively unharmed. Many antimicrobial drugs do this. They exert selective toxicity because they attack microbial targets, such as the bacterial peptidoglycan cell wall, that do not exist, or bacterial ribosomes, that are substantially different in human cells (Chapter 4).

Most antimicrobial agents do have some **side effects** (secondary usually adverse actions). Some of these are explainable by similarities between the drug's target in microorganisms and in human cells. But the exact basis of most side effects is poorly understood. Although side effects exist, their danger must be weighed against the potential benefit of the drug. For example, when streptomycin was the only agent available to treat tuberculosis, some people willingly accepted the possible side effect of deafness in order to survive of a deadly infection.

T. M.'s physician had to weigh the possible side effects of penicillin against its ability to combat T. M.'s infection. Because penicillin's target, peptidoglycan, does not exist in humans, the drug is almost completely nontoxic. Only in exceptional cases does it cause side effects, but it can cause life-threatening allergic reactions (type I hypersensitivity reactions; Chapter 18, **Figure 21.4**). For this reason T. M.'s physician inquired carefully about any evidence suggesting that T. M. might be allergic to penicillin. Because every exposure to an antibiotic carries a small risk of stimulating drug allergy, penicillin, like other antimicrobial agents, should never be administered needlessly. For example, it should never be prescribed to treat a viral infection, such as a cold, for which it is completely ineffective.

Even wonderfully effective drugs such as penicillin become useless when bacterial strains become resistant to it. We'll discuss the phenomenon of drug resistance in the following section.

Drug Resistance

Drug-resistant microbes can grow and reproduce in the presence of a particular drug that would inhibit or kill sensitive strains. Resistance can be either natural or acquired. Some species of microorganisms are innately resistant to particular drugs. But occasionally, certain cells of a drug-sensitive species may undergo a genetic change and become resistant. Such acquired drug resistance is passed on to the cell's progeny, producing a drug-resistant strain.

Natural Drug Resistance. Natural resistance to antimicrobial agents occurs for various reasons. Some species lack the target that the antimicrobial drug attacks. For example, penicillin interferes with peptidoglycan cell wall synthesis, so any organism that lacks a peptidoglycan cell wall is naturally resistant. Therefore viruses, fungi, protozoa, and wall-less bacteria, such as the mycoplasmas, are resistant to penicillin. Other species are naturally resistant because the antimicrobial drug is not able to enter the cell and exert its damaging effect. For example, Gram-negative

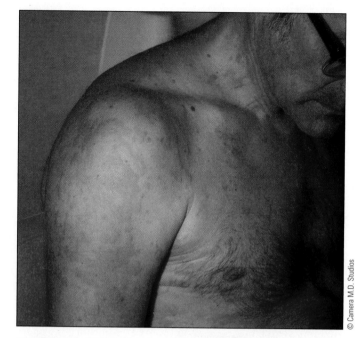

FIGURE 21.4   This patient's rash is a type I hypersensitivity reaction to penicillin.

bacteria are naturally resistant to many drugs, including most penicillins, because such drugs cannot cross their relatively impermeable outer membrane.

Based on their pattern of natural sensitivity or resistance, antibacterial agents are classified as being either **narrow spectrum** or **broad spectrum** (**Figure 21.5**). Narrow-spectrum antibacterial drugs affect only a single microbial group. They are the best choice when the infecting microorganism is known to belong to that sensitive group. Because they target the pathogen fairly specifically, they leave most of the body's normal protective microbial biota unharmed. Most penicillins are narrow-spectrum drugs; they kill only Gram-positive bacteria. *Streptococcus pyogenes*, which causes scarlet fever, belongs to that group. Other antibiotics, such as polymyxin B, are effective only against Gram-negatives. Isoniazid is a narrow-spectrum drug that's effective only against the mycobacteria.

Broad-spectrum antibacterial drugs are effective against more than a single microbial group. Ampicillin, a semisynthetic derivative of penicillin, affects both Gram-positive and Gram-negative bacteria. Chloramphenicol, tetracycline, and streptomycin have even broader spectra.

When treating an infection with an unknown cause, broad-spectrum agents are a good choice. Ear infections, for example, can be caused by either Gram-positive or Gram-negative species, so the broad-spectrum drug ampicillin is a good choice; the narrow-spectrum penicillin VK is not. However, there are drawbacks to broad-spectrum drugs. Because they are effective against so many kinds of

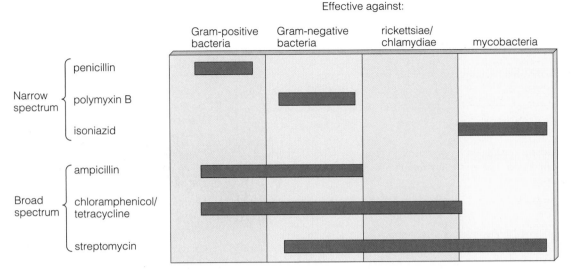

Effective against:

FIGURE 21.5 Antibacterial
drug spectrum. Narrow-
spectrum antibacterials
act on only one of these
bacterial groups. Broad-
spectrum drugs act on
more than one.

microorganisms, they significantly alter the body's normal microbial biota (Chapter 14). Such changes can cause unpleasant symptoms and allow another microorganism to initiate an infection. For example, children treated with ampicillin for ear infections often suffer from diarrhea caused by disruption of the normal intestinal biota or from rashes caused by a yeast infection.

Acquired drug resistance of pathogens is a major concern of modern medicine. Now we'll consider how it develops.

Acquired Drug Resistance. Bacterial strains become drug-resistant by mutations and genetic exchange (Chapter 6). Antibiotics don't cause these genetic changes, but they favor the resistant strains once the change has occurred. In the presence of an antibiotic, resistant strains survive, multiply, and become dominant. This process has had a huge impact on the usefulness of antibiotics. For example, during the 1940s, nearly all strains of the virulent pathogen *Staphylococcus aureus* were sensitive to penicillin. Today more than 90 percent of them are penicillin-resistant.

Mechanisms. There are many different ways that microorganisms become drug-resistant, but they fall into three broad classes: (1) acquiring enzymes that can inactivate or destroy the drug, (2) changing their cellular target upon which the drug acts, and (3) excluding the drug from the cell or removing it once it has entered (**Table 21.2**). We'll consider each of these briefly.

Acquiring an enzyme that destroys a drug is probably the most common mechanism of drug resistance. Some of these enzymes add a functional group, such as acetate or phosphate group, to the drug. Others destroy a chemical bond within the drug. For example, the bacterial enzyme beta-lactamase breaks a bond in the beta-lactam ring of the penicillin molecule, thereby inactivating it.

Changing the target of drug action is another route to drug resistance. Penicillin resistance of the respiratory pathogen *Streptococcus pneumoniae* is an example. Such resistance is acquired when the bacterium undergoes a series of mutations in their **penicillin-binding proteins** (the enzymes that catalyze formation of the peptidoglycan cell wall) that make them less able to bind penicillin.

Excluding or removing a drug is an effective route to resistance. Penicillin-resistant strains of *Neisseria gonorrhoeae* exclude the antibiotic: Changes in their outer membrane porin proteins prevent penicillin from entering the cell. Many Gram-negative bacteria remove tetracycline by acquiring a membrane transport system that pumps molecules of the antibiotic out of the cell.

Genetics. There are two genetic routes to acquiring drug resistance. Either mutations occur on the microorganism's chromosome or the microorganism acquires a resistance-conferring plasmid.

Some chromosomally drug-resistant strains have emerged gradually, over decades of antibiotic use. A series of mutations renders a bacterial strain gradually more and more drug-resistant. Higher and higher doses of antibiotics are required to cure infection, until resistance becomes so great that the drug is no longer effective. Penicillin-resistant *Streptococcus pneumoniae* with altered penicillin-binding proteins and penicillin-resistant *Neisseria gonorrhoeae* with altered outer-membrane porins are both examples of gradual chromosomally mediated resistance.

Drug resistance due to plasmid-borne genes occurs almost immediately after a microorganism acquires a plasmid. Plasmids can be passed rapidly from one strain of microorganism to another, and even from one species to

TABLE 21.2 Mechanisms of Acquired Drug Resistance

Mechanism	Antimicrobial Agent	Drug Action	Mechanism of Resistance
Destroys drug	Aminoglycosides	Binds to 30S ribosome subunit, inhibiting protein synthesis	Plasmids encode enzymes that chemically alter the drug (e.g., by acetylation or phosphorylation), thereby inactivating it.
	Beta-lactam antibiotics (penicillins and cephalosporins)	Binds to penicillin-binding proteins, inhibiting peptidoglycan synthesis	Plasmids encode beta-lactamase, which opens the beta-lactam ring, inactivating drug.
	Chloramphenicol	Binds to 50S ribosome subunit, inhibiting formation of peptide bonds	Plasmids encode an enzyme that acetylates the drug, thereby inactivating it.
Alters drug target	Aminoglycosides	Binds to 30S ribosome subunit, inhibiting protein synthesis	Bacteria make an altered 30S ribosome that does not bind to the drug.
	Beta-lactam antibiotics (penicillins and cephalosporins)	Binds to penicillin-binding proteins, inhibiting peptidoglycan synthesis	Bacteria make altered penicillin-binding proteins that do not bind to the drug.
	Erythromycin	Binds to 50S subunit, inhibiting protein synthesis	Bacteria make a form of 50S ribosome that does not bind to the drug.
	Quinolones	Binds to DNA topoisomerase, an enzyme essential for DNA synthesis	Bacteria make an altered DNA topoisomerase that does not bind to drug.
	Rifampin	Binds to RNA polymerase, inhibiting initiation of RNA synthesis	Bacteria make an altered polymerase that does not bind to drug.
	Trimethroprim	Inhibits the enzyme dihydrofolate reductase, blocking the folic acid pathway	Bacteria make an altered enzyme that does not bind to the drug.
Inhibits drug entry or removes drug	Penicillin	Binds to penicillin-binding proteins, inhibiting peptidoglycan synthesis	Bacteria change shape of outer-membrane porin proteins, preventing drug from entering cell.
	Erythromycin	Binds to 50S ribosome subunit, inhibiting protein synthesis	New membrane transport system prevents drug from entering cell.
	Tetracyclines	Binds to 30S ribosome subunit, inhibiting protein synthesis by blocking tRNA	New membrane transport system pumps drug out of cell.

another, by genetic conjugation (Chapter 6). Plasmids that carry genes encoding drug resistance are called R factors (R for resistance). A single R factor may carry genes that encode resistance to several different antibiotics. On acquiring such a plasmid, a microorganism becomes a **multiple-drug–resistant strain.** Penicillin resistance from beta-lactamase is an example of plasmid-borne resistance. Plasmid-borne drug resistance was recognized soon after the antibiotic era began in the 1940s. In some cases the spread of drug-resistant strains of pathogens has been dramatic and sudden. (See Sharper Focus: A Game We May Never Win in Chapter 24.)

Slowing Resistance. Each year, huge quantities of penicillin and other antibiotics are prescribed. At least a third of all hospital patients receive one or more courses of antibiotic therapy during their stay. In addition, antimicrobials are widely used as additives to animal feed. The result of such wide distribution of drugs is a rising tide of drug resistance that makes it difficult to treat certain infections (**Figure 21.6**). Worse, the more antibiotics we use, the more rapidly drug-resistant strains will come to predominate over sensitive strains. Thus our best opportunity to slow the emergence of drug resistance and preserve the efficacy of antimicrobial drugs—a precious resource—is to limit their use.

SHARPER FOCUS

THE ORIGINAL WONDER DRUG

When penicillin became available for widespread clinical use in the 1940s, the treatment of infectious diseases was transformed. People with potentially fatal infections were cured within hours. Penicillin was called a "wonder drug," and justly so. Penicillin remains the most widely used of all antibiotics. It's the drug of first choice for treatment of infections caused by *Streptococcus pyogenes,* including streptococcal pharyngitis (strep throat), streptococcal wound infections, and scarlet fever.

Penicillin, which is produced by fungi belonging to the genus *Penicillium,* kills bacterial cells by interfering with the incorporation of new peptidoglycan units into their cell wall. Normally, as bacterial cells grow, chemical bonds in the peptidoglycan wall are broken to accommodate new peptidoglycan units. But penicillin binds and inactivates the enzymes, called penicillin-binding proteins, that catalyze the resealing process. Then the breaks cannot be resealed. The wall becomes weakened, and the cell lyses.

The penicillin molecule consists of two fused rings and a changeable R group. One of these rings, the beta-lactam ring, can be destroyed by beta-lactamase (also called penicillinase). This enzyme is produced by certain strains of penicillin-resistant bacteria. Different R groups on the beta-lactam ring modify penicillin's antimicrobial action and its sensitivity to beta-lactamases. The first penicillin to be isolated (Chapter 1) was penicillin G. Its R group is a benzyl ring. But *Penicillium* itself can add different R groups, and R groups can be changed chemically, producing so-called semisynthetic penicillins. The result is a set of "penicillins" with somewhat different properties (Figure 21.11). The R group in penicillin V makes the drug more resistant to stomach acid and therefore more suitable to oral administration. The R group in amoxicillin makes it effective against Gram-negative bacteria, which are naturally resistant to penicillin G. Still other side chains—such as the one methicillin has—make the beta-lactam ring more resistant to beta-lactamases.

The second ring of penicillin is the thiazolidine ring that carries a carboxyl group. This acid group can form salts with bases such as procaine, benzathine, or potassium. Different salts have special therapeutic purposes. For example, the procaine salt of penicillin G dissolves rapidly and therefore acts quickly. In contrast, the benzathine salt dissolves slowly and therefore continues to release low levels of penicillin for several weeks. The potassium salt of penicillin V, called penicillin VK (K is the chemical notation for potassium), is one of the most common forms of penicillin administered orally.

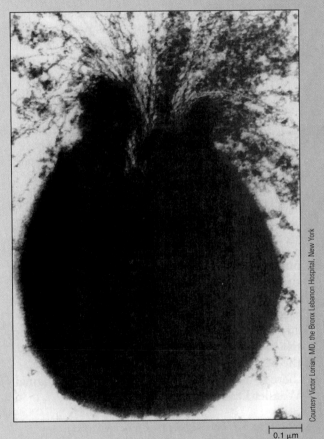

a Rupture of the cell wall of a penicillin-treated bacterium.

Courtesy Victor Lorian, MD, the Bronx Lebanon Hospital, New York

0.1 μm

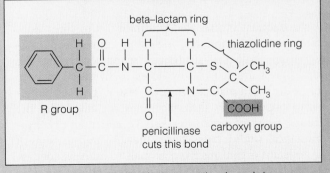

b Structure of penicillin G. The R group is a benzyl ring.

CASE HISTORY

The Serum Killing-Power Test

A simple determination of a microorganism's susceptibility to an antibiotic is usually all a clinician needs to guide therapy. But sometimes more information is called for. Such was the case of a young woman with endocarditis, an infection of the lining of the heart (Chapter 27). In many ways she was a typical endocarditis patient. She had a mild heart defect that made her prone to infection. And she was infected by *Streptococcus viridans*, one of the most common microbial causes of endocarditis. But her case was unusual because her strain of *S. viridans* was relatively resistant to penicillin—the first-line drug for treating endocarditis. Because endocarditis infections are difficult to eradicate under the best of circumstances, her doctors were concerned. They therefore requested a serum killing-power test. A sample of the young woman's serum was obtained shortly after she received a dose of penicillin and was added to cells of *S. viridans* that had been isolated from her blood. The results indicated that her undiluted serum was capable of killing the bacteria, but even a twofold dilution of the serum caused the drug concentration to fall to an ineffective level. Because the margin of safety was so narrow, treatment was continued for longer than usual. The young woman recovered completely.

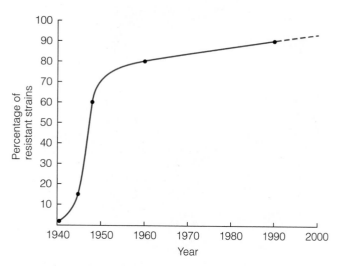

FIGURE 21.6 Rise in the proportion of penicillin-resistant strains of *Staphylococcus aureus* since it was introduced for clinical use in 1940.

One frequently discussed possibility is to limit nonmedical uses of antibiotics, including animal feed additives. About half the antibiotics manufactured in the United States are fed in low doses to livestock to keep them healthy and make them grow faster. The antibiotic-sensitive organisms in their intestines are killed. But the antibiotic-resistant organisms survive and multiply. Epidemiological studies have shown that the transmission of drug-resistant organisms from animals to humans does indeed occur. An outbreak of drug-resistant *Salmonella newport* infection that caused 17 severe illnesses and 1 death was traced to contaminated hamburger meat from cattle that received antibiotics in their feed.

It is widely acknowledged that the medical use of antibiotics ought to be more selective and thoughtful. Probably the greatest misuse is administering antibacterial agents, such as penicillin, for untreatable viral infections, such as colds. In many countries these agents can be bought without prescription and taken by anyone who feels ill. Even in countries such as the United States, where antibiotic use is regulated, physicians too often prescribe indiscriminately or give in to the demands of patients.

In addition, a course of antibiotics ought to be prescribed at a high enough dosage and taken long enough to eradicate the infection. This prevents the spread of the drug-resistant organisms that survive partial treatment. Another approach to slowing drug resistance in the clinical setting is **combined therapy** (administering two different drugs at the same time). It is extraordinarily improbable that a pathogen would simultaneously gain resistant to two different drugs. But, in spite of combined therapy, highly drug-resistant strains continue to appear. Usually they develop when patients do not follow instructions to take both drugs.

In response to the crisis of increased antibiotic resistance, the Centers for Disease Control and Prevention (CDC) has initiated an educational program, Campaign to Prevent Antimicrobial Resistance. Detailed information is available at: http://www.cdc.gov/drugresistance/health care.

Drug Dosage

After deciding which drug to use and how to administer it, the clinician must decide on **dosage** (the quantity of a drug to be administered). The aim is to deliver an effective concentration of the drug to the site of the infection. The concentration that gets to the site depends upon the route of administration, drug distribution, and drug elimination. What constitutes an effective concentration depends upon the susceptibility of the infecting organism to the drug. So, in order to decide on a proper drug dosage, the pathogen's susceptibility must be considered.

Often clinicians can estimate the pathogen's antibiotic susceptibility from recent experience with similar infections. But sometimes it must be determined using one of three kinds of laboratory tests—disc diffusion, broth dilution, and serum killing power—designed primary for bacterial pathogens. We'll consider them here.

Disc-Diffusion Method.
The **disc-diffusion method,** sometimes called the **Kirby-Bauer method,** uses filter-paper discs impregnated with antimicrobial agents. A petri plate is inoculated with the pathogen, and discs, each impregnated with known quantities of antimicrobial agents,

are placed on its surface (**Figure 21.7**), and drug from the disc diffuses through the agar medium.

When the plate has been incubated long enough for the pathogen to produce a confluent lawn of growth, results can be interpreted. Some of the drug discs will be surrounded by a clear halo. That's where bacterial growth has been inhibited by the antimicrobial agent. The more sensitive the microorganism is, the larger the halo will be. So the size of each halo is measured and compared with susceptibility standards for each drug. Based on these comparisons, the microorganism is judged to be sensitive, intermediate, or resistant to each drug.

These results can be extremely useful. Knowing whether an organism is sensitive, intermediate, or resistant to various antimicrobial agents is almost always sufficient to choose the proper drug and its dosage. Disc-diffusion tests have the added advantage of being technically easy and relatively inexpensive to perform. Disc diffusion is by far the most common type of susceptibility test used in clinical medicine.

Broth-Dilution Method.
The **broth-dilution method** is more complicated and expensive than the disc-diffusion method, but it yields the more precise results needed to treat critically ill patients. The method tests a drug's

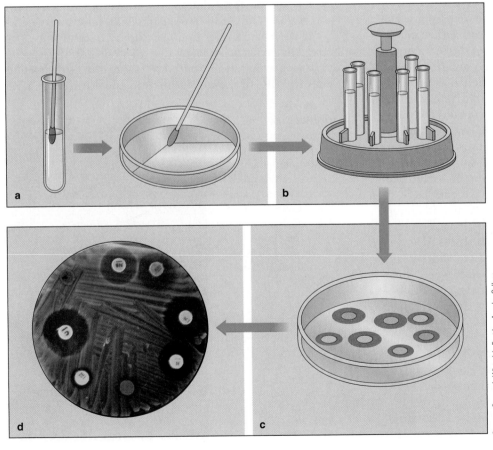

FIGURE 21.7 Disc-diffusion susceptibility test. (a) A culture of the microorganism to be tested is spread with a cotton swab over the surface of an agar-solidified medium in a petri dish. (b) Discs impregnated with antimicrobial agents are placed on the agar surface, sometimes with a device such as the one shown here. The petri dish is incubated and the drugs diffuse out from the disc. (c, d) When the microorganism forms a confluent lawn of growth, clear zones in which growth did not occur are visible around the discs impregnated with agents to which the test organism is susceptible; larger zones reflect greater susceptibility.

Courtesy George A. Wistreich, East Los Angeles College

FIGURE 21.8 MIC and MBC testing. (a) The minimum inhibitory concentration (MIC) is 2 μg/ml, the concentration of a drug that stops microbial growth in the tube. (b) The minimum bactericidal concentration (MBC) is 8 μg/ml, the concentration that kills cells. They don't grow on drug-free plates after having been exposed to this concentration of drug.

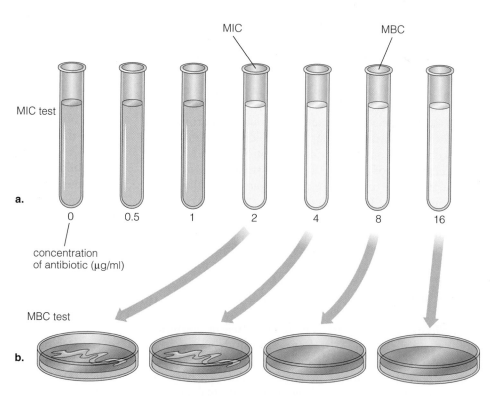

ability to prevent growth of bacteria in liquid cultures. A series of tubes containing decreasing concentrations of a drug is prepared and inoculated with the microorganism to be tested (**Figure 21.8**).

After growth has occurred, making some tubes turbid, results can be interpreted (**Figure 21.8a**). The clear test tube with the lowest drug concentration contains the **minimum inhibitory concentration** (**MIC**) of the drug. More information can be obtained by taking microbial cells from the clear tubes and attempting to grow them on a drug-free petri plate (Figure 21.8b). If organisms do grow on the petri plate, the tube from which the bacteria were taken contained a **bacteriostatic concentration** (capable of stopping the growth of the cells but not killing them) of the drug. If organisms do not grow on the petri plate, the tube contained a **bactericidal concentration** (lethal concentration) of the drug. The tube with the lowest bacteriocidal drug concentration contains the **minimum bactericidal concentration** (**MBC**) of the drug.

Knowing the MIC and MBC gives the clinician valuable information. The MIC and MBC can be compared with serum drug levels to see if inhibitory or bactericidal concentrations are being reached in the patient's body. In addition, relative values of MIC and MBC provide information about the drug itself. If the MBC is much higher than the MIC, the drug is **bacteriostatic.** If the MIC and MBC are about the same, the drug is bactericidal. Many infections can be treated effectively by bacteriostatic drugs because they tip the balance in favor of the patient's immune system. But, a person with a weak immune system may need a bactericidal drug.

Serum Killing Power. In particularly critical cases, a **serum killing-power test** may be done. Some of the patient's own drug-containing serum is withdrawn and tested to see if it kills the infecting microorganism.

The trend today is toward more automated versions of broth-dilution tests (**Figure 21.9**). In some systems, only the reading of final test results is automated. In others all aspects of the test—dilutions, inoculations, incubations, and reading and printing of results—are fully automated.

It wasn't necessary to do sensitivity testing in T. M.'s case. His pediatrician knew that all strains of *Streptococcus pyogenes* are uniformly and highly susceptible to penicillin with an MBC of about 1 ng/ml. And safe drug levels in the

FIGURE 21.9 The broth-dilution test can be automated by using multitipped pipettes to fill depressions in plastic plates (microtiter plates). After incubation, MIC is determined on the same basis as the manual method (Figure 21.8).

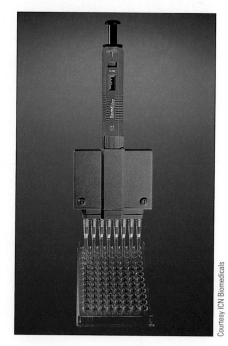

bloodstream reach 100 μg/ml—100,000 times greater than the MBC. Thus it was only necessary to swab bacteria from T. M.'s throat to confirm that *S. pyogenes* was the cause. It was not necessary to determine the strain's sensitivity to penicillin.

Earlier in this chapter, we discussed the side effects of drugs. We emphasized that a good antimicrobial drug has selective toxicity. And selective toxicity depends upon the drug's acting on a target within the microbial cell that is different or lacking within our own cells. Now let's consider some of these special drug targets.

TARGETS OF ANTIMICROBIAL DRUGS

There are five principal targets of antimicrobial drugs: the cell wall, cell membranes, protein synthesis, nucleic acid synthesis, and synthesis of **folic acid** (an essential cofactor). We'll consider each of these targets briefly (**Figure 21.10**).

The Cell Wall

Because the peptidoglycan cell wall is unique to bacteria, it's a perfect target for antimicrobial agents. Agents that destroy peptidoglycan or block its synthesis kill bacteria by causing them to lyse. These agents have no effect on eukaryotic cells, including human cells, because all eukaryotic cells lack peptidoglycan. These agents are also without effect on wall-less bacteria, such as the mycoplasmas.

The penicillins and many other drugs fall into this category. They, along with the cephalosporins, block peptidoglycan synthesis at the final step of cross-linking the mesh. Other drugs in this category, such as vancomycin and bacitracin, interfere with earlier steps of synthesizing linear peptidoglycan strands.

Cell Membranes

Cell membranes are not ideal targets of drug action because all cells have them. But there are some differences between microbial and mammalian cell membranes that provide the basis for some selective toxicity. Most drugs that target cell membranes are highly toxic when administered systemically.

Some these agents are detergents that can destroy the outer membrane of Gram-negative bacteria. Polymyxin B is an example. It's useful when administered topically to control infections of the body surfaces. Other agents se-

lectively damage sterols in the plasma membrane of eukaryotic microorganisms. These drugs are selectively toxic to fungi because they do greater damage to ergosterol, which is found in fungal cell membranes, than to cholesterol, which is found in human cell membranes. Even so, their selective toxicity is relatively poor. Some of these agents are used topically, but others are administered systemically to treat life-threatening fungal infections. Amphotericin B is an example. It has many serious side effects, including fever, chills, vomiting, and kidney failure. But it does save lives.

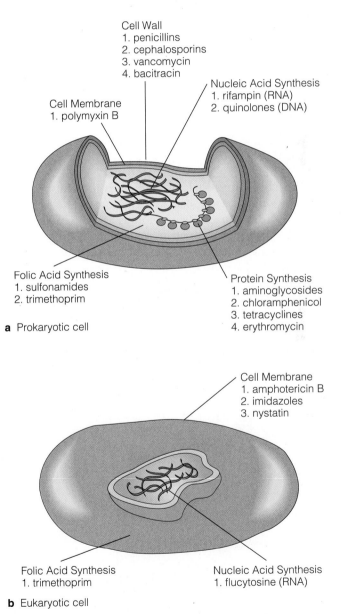

FIGURE 21.10 The targets of antimicrobial drugs. The targets in prokaryotic cells (a) and eukaryotic cells (b) and the drugs that act on them.

Curiously, humans, as well as other animals, make a vast array of proteins that can kill bacteria by attacking their membranes.

Protein Synthesis

Antimicrobial drugs that interfere with protein synthesis do so by acting on the bacterial ribosome. The selective toxicity of these drugs depends upon the differences between the 70S prokaryotic ribosome and the 80S eukaryotic ribosome (Chapter 4). But the mitochondria of eukaryotic cells have bacteria-like 70S ribosomes, so these drugs are somewhat toxic to humans.

Many widely used antimicrobial drugs belong to this group. Various members affect different aspects of ribosomal function (**Figure 21.11**). Members of the aminoglycoside group, including streptomycin, bind to the 30S subunit of the 70S ribosome, slowing its rate of making protein and causing it to make faulty proteins. As a result, aminoglycosides are bactericidal. The tetracyclines also bind to the 30S ribosome subunit. They interfere with the attachment of the amino acid–carrying transfer RNA (tRNA) molecule to the growing chain. The effect of tetracycline is reversible; protein synthesis begins again when tetracycline levels fall. Erythromycin and chloramphenicol both bind to the 50S ribosome subunit of 70S ribosomes. Chloramphenicol specifically inhibits formation of peptide bonds.

Nucleic Acids

All cells, both prokaryotic and eukaryotic, make nucleic acids. But some essential enzymes—the **topoisomerases** that unwind existing DNA chains and the **polymerases** that extend the new chains—are significantly different in microorganisms and human beings. The antibiotic rifampin selectively inhibits bacterial RNA polymerase. The quinolones selectively inhibit a microbial topoisomerase.

Differences also exist in the pathways through which nucleotides, the building blocks of nucleic acids, are made. Certain drugs, such as flucytosine, exert their effect here. For example, fungi, which are sensitive to **flucytosine** (5-fluorocytosine), have an enzyme (cytosine deaminase) that converts the drug to 5-fluorouracil. And 5-fluorouracil becomes incorporated into RNA, making it nonfunctional. Humans lack the enzyme that converts flucytosine into 5-fluorouracil, so they are unaffected by flucytosine. (See Case History: Death by Mutation in Chapter 6.)

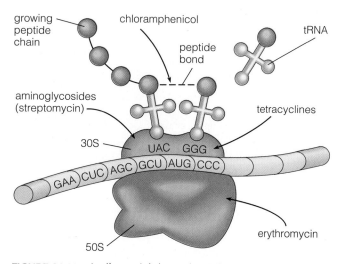

FIGURE 21.11   Antibacterial drugs that inhibit protein synthesis. Site of action of some of these drugs on the 70S prokaryotic ribosome.

Folic Acid Synthesis

A major group of antimicrobials acts by interfering with the production of the essential vitamin **folic acid.** These drugs are selectively toxic to microorganisms because humans do not synthesize folic acid. Instead, we obtain it in our diet. Because drugs in this category resemble natural metabolites but block steps in their metabolism, they are called **antimetabolites.**

The sulfonamides inhibit a key enzyme in the folic acid pathway because they closely resemble the natural metabolite para-aminobenzoic acid (PABA), an intermediate in the pathway that produces folic acid. By competing with PABA, sulfas stop production of folic acid and thereby stop microbial growth. Another drug, trimethoprim, blocks the same pathway at a different point in the pathway.

ANTIMICROBIAL DRUGS

New antimicrobial agents are developed each year. Others pass out of common use because of emerging microbial resistance or the availability of better agents. In

this section we'll discuss the antimicrobial drugs most commonly used in clinical medicine today. We'll group them according to the type of infection they're usually used to treat—bacterial, mycobacterial, fungal, parasitic, malarial, and viral.

Antibacterial Drugs

The most effective antimicrobial chemotherapeutic agents in use today are antibacterial drugs (**Table 21.3**).

Penicillins. All the penicillins share the same basic structure, which consists of a system of welded five- and six-membered rings. The various individual penicillins differ with respect to the kind of chemical group, termed an **R group,** that's attached to the ringed structure (**Figure 21.12**). Penicillins with a vast variety of R groups have been made—some, termed **natural penicillins,** by altering

TABLE 21.3 Antimicrobial Drugs

Drug	Antibiotic/Synthetic	Mode of Action	Spectrum
Antibacterial Drugs			
Penicillins	Antibiotics and semisynthetics	Inhibit synthesis of peptidoglycan	Antibiotics control mostly Gram-positives; semisynthetics inhibit some Gram-negatives as well
Cephalosporins	Antibiotics and semisynthetics	Inhibit synthesis of peptidoglycan	Antibiotics control mostly Gram-positives; semisynthetics inhibit many Gram-negatives as well
Sulfonamides	Synthetic	Inhibits synthesis of folic acid	Gram-positives, Gram-negatives, chlamydiae, but many now resistant
Aminoglycosides	Antibiotics and a few semisynthetics	Bind to 30S ribosomal subunit and inhibit protein synthesis	Gram-negatives
Chloramphenicol	Antibiotic now made by chemical synthesis	Inhibits formation of peptide bonds and thereby protein synthesis	Gram-positives, Gram-negatives, chlamydiae, rickettsiae, mycoplasmas
Tetracyclines	Antibiotics and semisynthetics	Binds to the 30S ribosome inhibiting recognition of aminoacyl-tRNA and thereby protein synthesis	Gram-positives, Gram-negatives, chlamydiae, rickettsiae, mycoplasmas
Erythromycin	Antibiotic	Binds to the 50S ribosome and inhibits protein synthesis	Gram-positives
Quinolones	Synthetics	Binds to DNA topoisomerase and inhibits DNA synthesis	Gram-positives, Gram-negatives, mycoplasmas, mycobacteria
Vancomycin	Antibiotic	Inhibits synthesis of peptidoglycan	Gram-positives
Antimycobacterial Drugs			
Isoniazid	Synthetic	Inhibits synthesis of mycolic acid	Mycobacteria
Rifampin	Semisynthetic	Binds to RNA polymerase and inhibits synthesis of RNA	Mycobacteria
Ethambutol	Synthetic	Inhibits incorporation of mycolic acids into cell wall	Mycobacteria

(continued)

TABLE 21.3 Antimicrobial Drugs (continued)

Drug	Antibiotic /Synthetic	Mode of Action	Spectrum
Antifungal Agents			
Nystatin	Antibiotic	Interacts with ergosterol, disrupting fungal membranes	Primarily *Candida albicans*
Amphotericin B	Antibiotic	Disrupts fungal membranes	Many fungi, including those that cause systemic infections
Imidazoles and triazoles	Synthetic	Inhibit synthesis of sterols in fungal membranes	Many fungi
Flucytosine	Synthetic	Converted to 5-fluorouracil, which becomes incorporated into RNA, making a faulty product	Many fungi, including those that cause systemic infections
Griseofulvin	Antibiotic	Prevents cell division	Fungi that cause ringworm
Antiparasitic Drugs			
Mebendazole	Synthetic		Roundworms
Metronidazole	Synthetic	Robs cell of reducing power	Anaerobic bacteria and many protozoa
Chloroquine	Semisynthetic		Malaria-causing protozoa
Antiviral Drugs			
Amantadine	Synthetic	Inhibits uncoating of virus	Prevents development of influenza A
Acyclovir	Synthetic	Inhibits synthesis of viral DNA	Herpesviruses
Ribavirin	Synthetic	Inhibits synthesis of viral RNA	Respiratory syncytial virus
Zidovudine, didanosine, zalcitabine	Synthetic	Nucleoside analogue inhibitors of reverse transcriptase	Human immunodeficiency virus (HIV)
Delavirdine, nevirapine	Synthetic	Nonnucleoside analogue inhibitors of reverse transcriptase	HIV
Indinavir, nelfinavir, ritonavir	Synthetic	Inhibitors of HIV protease	HIV
Interferons	Cell products	Natural antiviral agents	Used to treat chronic viral hepatitis, genital warts, Kaposi's sarcoma

the conditions of fermentation, and others, termed **semisynthetic penicillins,** by chemically switching R groups of the fermentation product.

All these penicillins act the same way and are bactericidal (see Sharper Focus: The Original Wonder Drug). But the different penicillins have varying properties. One group is resistant to the action of **beta-lactamase,** the enzyme produced by some penicillin-resistant pathogens that destroys most penicillins. (Infections by beta-lactamase–producing bacterial strains are also treated by coadministering clavulanic acid, a drug with no antimicrobial activity that inactivates beta-lactamase.) Another group is broad spectrum, acting against Gram-negative, as well as Gram-positive, microbes. The properties of some of the most commonly used penicillins are summarized in **Table 21.4.**

Cephalosporins. Cephalosporins are closely related to the penicillins. They have the same beta-lactam ring, but it's welded to a six-membered ring in place of penicillin's five-membered thiazolidine ring (**Figure 21.13**). Their mode of action is identical to that of the penicillins.

Cephalosporins are broad-spectrum drugs, but they are particularly effective against the Gram-positives. As with the penicillins, chemical modification of natural cephalosporins has produced numerous semisynthetic antibiotics. Each generation of them (there are now four) has been more active against Gram-negative organisms than its predecessor. Third- and fourth-generation cephalosporins are widely used to treat many life-threatening infections, including meningitis, pneumonia, and sepsis. Cephalosporins are more resistant than penicillins to beta-lactamases, making them much more effective against most strains of *Staphylococcus aureus*. Some of the newer cephalosporins have additional advantages, such as good penetration into the central nervous system or unusually long persistence in the bloodstream. As a result, they need not be administered as frequently.

Side effects of the cephalosporins, like the penicillins, are minimal, but they are highly allergenic. And because of the chemical similarity between cephalosporins and penicillins, some patients who are allergic to penicillin are allergic to cephalosporins on first exposure.

Sulfonamides. The sulfonamides, or sulfa drugs, were the first antimicrobial agents (after salvarsan) used to cure infections in humans. These drugs are synthetic organic chemicals first produced by the German chemical company I. G. Farben as dyes.

Sulfonamides (Figure 21.13) are a large family of drugs that act by interfering with the bacterial cell's ability to synthesize folic acid. They are broad spectrum and bacteriostatic.

The history of sulfonamides illustrates how some antimicrobial drugs become less useful with time. When sulfonamides first became available in the 1930s, they

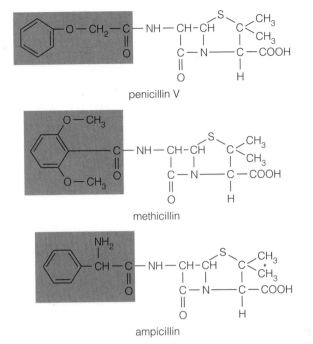

penicillin V

methicillin

ampicillin

FIGURE 21.12   Chemical structures of some of penicillins. They differ only in their R groups (colored).

TABLE 21.4 The Penicillins

Class	Name[a]	Comments
Natural		Effective against strains of sensitive of Gram-positive bacteria.
	Penicillin G	Used against systemic infections of Gram-positive bacteria, including anthrax.
	Penicillin VK	Can be administered by mouth. Used against *Streptococcus pyogenes* and pneumococcus.
Beta-lactamase-resistant		Effective against many Gram-positive bacteria, including those that produce beta-lactamase.
	Nafcillin	Used against *Staphylococcus* infections.
	Oxacillin	Used against various bacterial infections.
	Cloxacillin	Used against various bacterial infections.
	Dicloxacillin	Used against *Staphylococcus* skin infections and osteomyelitis.
Broad spectrum		Effective against many Gram-positive and Gram-negative bacteria.
	Ampicillin	Used against many bacterial infections and as endocarditis prophylaxis.
	Amoxicillin	Uses similar to those of ampicillin; in addition, used to treat gonorrhea.
	Ticarcillin	Used against various bacterial infections, including *Pseudomonas aeruginosa*.
	Piperacillin	Uses similar to ticarcillin.
	Mezlocillin	Uses similar to ticarcillin.

[a]These are generic names. The same products are sold under a variety of trade names.

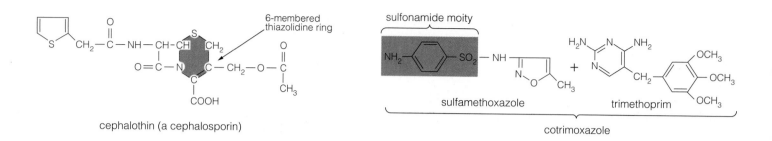

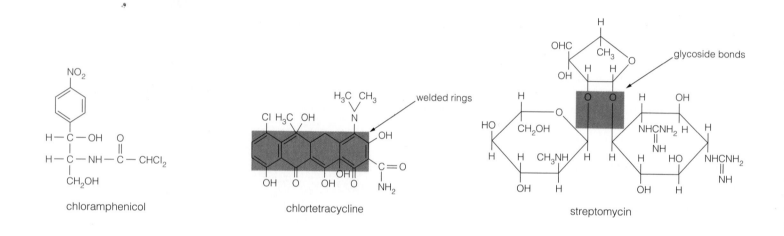

chloramphenicol

chlortetracycline

streptomycin

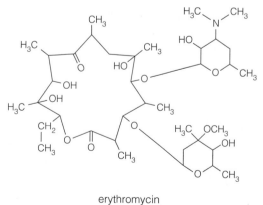

erythromycin

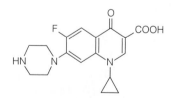

ciprofloxacin
(a quinolone)

FIGURE 21.13   Chemical structures of some major antibacterial agents.

LARGER FIELD

THE DISCOVERY OF STREPTOMYCIN

Albert Schatz received his B.Sc. and Ph.D. degrees from Rutgers University. His undergraduate major was soil science, and his graduate work was in soil microbiology. The search for a new antibiotic challenged him because as a young boy he knew people who died from what was then called blood poisoning, as well as ear infections, pneumonia, diphtheria, whooping cough, and tuberculosis. When he was in grade school, a classmate died from one disease or another almost every year. For his research that led to the discovery of streptomycin, he has received honorary degrees and medals and has been named an honorary member of medical and scientific societies. In 1993 Rutgers University awarded him the Rutgers Medal at the fiftieth anniversary celebration of the discovery of streptomycin. He wrote the following description of the events leading to his momentous discovery for this book.

I began the research that led to the discovery of streptomycin in 1943 when I was working for my Ph.D. degree at Rutgers University. At that time, the discovery of penicillin had motivated a search for other antibiotics. Several had been discovered but were too toxic to be used. I was aware of that when I spent 6 months as a bacteriologist in army hospitals, where I saw deaths caused by Gram-negative bacteria against which there were no effective drugs. Therefore, during off-duty hours,

I isolated and tested soil fungi and actinomycetes to see which ones inhibited the growth of Gram-negative bacteria on agar plates. When I was discharged and resumed graduate work, I continued the search that I had begun in army hospital laboratories.

Shortly after I returned to Rutgers, Drs. William Feldman and H. Corwin Hinshaw at the Mayo Clinic asked Dr. S. A. Waksman, chairman of the Department of Soil Microbiology, to look for an antibiotic to control tuberculosis. Dr. Waksman was reluctant to take on that project because he was afraid of tuberculosis, which was also known as The Great White Plague. He did not want the tubercle bacillus, *Mycobacterium tuberculosis hominis,* which had killed more than a billion people, in his third-floor laboratory, right next to his office. Like Gram-negative bacteria, the tubercle bacillus was resistant to all chemotherapy. But I had handled pathogenic bacteria in the army and was confident I could safely work with the tubercle bacillus. When I informed Dr. Waksman that I wanted to take on the TB problem as part of my Ph.D. research, he let me do that. But he transferred me to a basement laboratory and insisted that I never bring any TB cultures up to the third floor.

Dr. Waksman and others warned me that there was little likelihood of finding a cure for tuberculosis. The tubercle bacillus has a waxy coating, which is why it grows slowly and requires a special stain to make it visible through a microscope. It was assumed that no drug could get through that protective waxy coating. However, I reasoned that food had to get into the cells and waste products had to get out— otherwise neither the tubercle bacillus nor tuberculosis would exist.

Dr. Feldman provided me with a virulent strain of human tuberculosis, with which he was working. He subsequently contracted tuberculosis but recovered after 2 years of treatment. I feel good that no one in the building where I worked contracted tuberculosis. The laboratory I worked in was not equipped with safety equipment that is now used. It had no ultraviolet light, no special incubator, and no positive air pressure to continuously blow the air through a filter. Eventually I isolated two strains of the actinomycete *Streptomyces griseus.* Both produced a new antibiotic that I called streptomycin. One strain came from a heavily manured field soil. The other came from an agar plate that my fellow graduate student Doris Jones had streaked with a swab of a healthy

Albert Schatz.

chicken's throat. At about 2:00 P.M. on October 19, 1943, I knew that I had found a new antibiotic.

But would it control tuberculosis in vivo? Drs. Feldman and Hinshaw would find that out with guinea pigs. My job was to produce enough streptomycin for their initial tests. To do that, I had to run three stills around the clock. At night, I drew lines, with a red glass-marking pencil, on the flasks from which I was distilling and went to sleep on a wooden bench in the laboratory. The night watchman checked the flasks periodically and woke me up when the liquid in the flasks went down to the lines I had drawn. I then added more liquid and went back to sleep. I also had to save, purify, and reuse the solvents I worked with because during World War II they were rationed. When the Mayo Clinic tests were over, I was exhausted. But I knew I would have an acceptable Ph.D. dissertation.

SHARPER FOCUS

ANTIMICROBIAL DRUGS IN U.S. HISTORY

In the 1920s, when Calvin Coolidge was president of the United States, 16-year-old Calvin Jr. became ill. He had been playing tennis and developed a blister on his toe. The blister became infected and the infection spread. In spite of the best medical attention, the boy died. The diagnosis was streptococcal septicemia, an overwhelming infection caused by the bacterium *Streptococcus pyogenes*. Ten years later, in 1936, another president's son suffered from an infection caused by the same bacterium. Franklin Roosevelt, Jr., then a student at Harvard, was admitted to the hospital with a streptococcal throat infection. Newspapers around the country carried the story and public concern grew—the memory of young Coolidge's death was still fresh. Soon, however, the press reported that FDR Jr. was being treated with a new drug from Germany that could kill streptococcal bacteria. The drug was Prontosil, an antimicrobial of the sulfonamide family. Its use to cure FDR Jr.'s strep throat marked the beginning of the antimicrobial era in the United States.

It's hard to imagine what life was like before the development of antimicrobial drugs, only 60 years ago. Anyone—young or old—could suddenly be taken ill and die from infection. Medical science could do little, if anything, to save a person. Today when people become ill they believe that if they consult a doctor they can be cured. Often they're right. This confidence is largely the result of antimicrobial drugs. Walsh McDermott, a doctor who lived through the transformation of medical practice that resulted from antimicrobial therapy, described it as a "historic watershed": "One day we could not save lives, or hardly any lives; on the very next day we could do so across a wide spectrum of diseases. This was an awesome acquisition of power. . ."

Obviously, antimicrobial drugs help people hospitalized with pneumonia or meningitis. But patients suffering from diseases as different as cancer, diabetes, and stroke often depend just as much upon antimicrobial drugs. People who are seriously ill for any reason are prone to develop infections. Without antimicrobial drugs, modern surgery would be severely limited and organ transplantation could never have been attempted.

Antibiotics have also played an important role in other medical breakthroughs. In developing a new live viral vaccine, for example, the first step is to grow a tissue culture of individual animal cells in the laboratory. But until antibiotics were added to the culture medium, contaminating microorganisms almost always overwhelmed the animal cells. Thus antibiotics made the perfection of tissue-culture techniques possible, leading to the development and mass production of live viral vaccines. Indirectly, then, antibiotics are responsible for the virtual disappearance of polio, measles, and mumps.

Walsh McDermott and physicians of his generation were awestruck by the power of antibiotics, but we now know that there are serious limitations to the use of this power. Antimicrobial chemotherapy is highly effective only against infections caused by bacteria. In spite of recent advances, most patients with life-threatening fungal and viral infections still die from them, just as young Coolidge died from his streptococcal infection in the 1920s. Even more distressing, physicians today can see the power to cure bacterial infections beginning to ebb against mounting antibiotic resistance. Antibiotics that were once effective in treating potentially fatal infections have now become almost useless. The sulfa drugs that miraculously saved FDR Jr.'s life in 1936, for example, are seldom used today because almost all the bacteria against which they were once effective are now resistant.

New drugs have been developed, however, and most bacterial infections can still be treated successfully by one antimicrobial agent or another. But drug selection must be made carefully, and practitioners must stay up-to-date on constantly changing patterns of antibiotic resistance. It is inevitable that microbes will eventually develop resistance to today's drugs.

dramatically reduced death rates from infections caused by staphylococci and streptococci. But with extensive use, microbial resistance emerged and spread rapidly. Today none of these infections is routinely treated with sulfonamides. Moreover, sulfonamides cause more allergies and side effects than newer drugs. Today the use of sulfonamides by themselves is generally limited to treatment of urinary tract infections. But in combination with another antimicrobial drug, trimethoprim (a mixture called **Cotrimoxazole;** Figure 21.13), they are also used to treat minor respiratory infections and infections caused by the opportunistic pathogen *Pneumocystis carinii*. Cotrimoxazole is many times more active than either of its constituents alone. That's because each of them inhibits the synthesis of folic acid, but at different steps in the pathway.

Aminoglycosides.
The aminoglycosides were discovered in a deliberate, intensive search for antimicrobials effective against Gram-negative organisms and mycobacteria (see Larger Field: The Discovery of Streptomycin). In 1943 a strain of *Streptomyces griseus* was isolated that produced streptomycin. Related drugs—including amikacin, gentamicin, kanamycin, spectinomycin, and tobramycin—were later additions to this drug family.

The aminoglycosides are composed of amino sugar molecules connected to one another by glycoside bonds (Figure 21.13). All aminoglycosides must be injected. They are not well absorbed when taken orally, and they do not penetrate the central nervous system. So these antibiotics are used only to treat serious infections in hospitalized patients. But they are extremely effective against certain infections, such as kidney infections caused by Gram-negative bacteria. Their use is limited primarily by their toxicity. All aminoglycoside antibiotics can irreversibly damage the inner ear. Another drawback is the rapid emergence of drug-resistant strains. Resistance to streptomycin develops so readily that it must be used in combination with another antimicrobial agent.

Chloramphenicol.
Chloramphenicol is an antibiotic produced by *Streptomyces venezuelae*. It's unique among antibiotics because its chemical structure is simple enough for drug companies to produce it synthetically (Figure 21.13).

When chloramphenicol was discovered, it seemed to be the ideal drug. It's effective when taken orally. It penetrates well into protected areas such as the brain and the interior of the eye. It's stable enough to be stored for long periods without refrigeration. It has an extremely broad spectrum that includes Gram-positive and Gram-negative bacteria, rickettsiae, chlamydiae, and mycoplasmas. Drug allergy is rare, and in the vast majority of patients it has few side effects. But then chloramphenicol's reputation crashed.

In rare cases chloramphenicol is fatal because it can cause **aplastic anemia** (the bone marrow completely stops producing blood cells). Although this reaction occurs in fewer than 1 out of 30,000 drug recipients, it is inevitably fatal. Another life-threatening form of chloramphenicol toxicity is gray baby syndrome. Newborns who fail to metabolize the drug at a normal rate die from multiple toxic complications. Because of its potentially lethal side effects, chloramphenicol is used in the United States only to treat seriously ill hospitalized patients. It's not given to infants. Still, chloramphenicol is the preferred drug for treating typhoid fever. It's also used for life-threatening infections caused by certain anaerobic species, especially the *Bacteroides* spp.

Tetracyclines.
The tetracycline antibiotics, with four welded rings (Figure 21.13), were discovered in the 1950s. They were the first to be designated broad spectrum because they were effective against so many different types of bacteria: Gram-positive and Gram-negative bacteria, chlamydiae, rickettsiae, and mycoplasmas. The natural tetracyclines, including chlortetracycline and oxytetracycline, are produced by *Streptomyces* spp. Other drugs of this family are produced semisynthetically. Tetracycline itself can be produced by either fermentation or semisynthetically.

The tetracyclines interfere with bacterial protein synthesis by blocking the ribosomes' acceptance of aminoacyl-tRNA (Chapter 5). Tetracyclines are well absorbed orally and widely distributed throughout the body, but they enter the brain poorly. Side effects include gastrointestinal pain and diarrhea. The drugs themselves irritate the intestine, and they cause further irritation by dramatically changing its normal biota. Tetracyclines also cause increased sensitivity to sunlight. Because they permanently stain developing teeth brown, they cannot be given to children or pregnant women (**Figure 21.14**). Tetracycline allergies, however, are rare.

Because they are easy to administer and have an extremely broad spectrum, tetracyclines were widely used clinically. They were also added to animal feed to promote the growth of livestock. Consequently, tetracycline-resistant strains of pathogens emerged. Today the tetracyclines are used much less often than they were 25 years ago. They're used primarily to treat acne and infections caused by chlamydiae, rickettsiae, and mycoplasmas.

Macrolides.
Erythromycin (Figure 21.13), the first macrolide antibiotic to be discovered, has been a mainstay of antimicrobial therapy since its discovery in 1952. It is produced by *Streptomyces erythreus*.

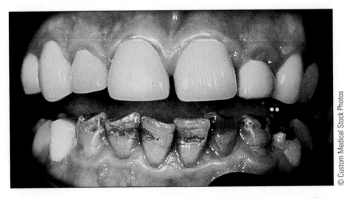

FIGURE 21.14 A child's teeth stained brown because tetracycline was administered when they were developing.

Erythromycin is effective only against Gram-positive bacteria. It is widely used to treat strep throat and other common respiratory infections, particularly in patients who are allergic to penicillin. If T. M. had been allergic to penicillin, he would probably have been given erythromycin for his scarlet fever. Recently, however, erythromycin-resistant strains of *Streptococcus pyogenes* have become prevalent. Erythromycin remains the preferred drug for treating mycoplasma pneumonia, legionellosis, and carriers of diphtheria and whooping cough.

Erythromycin is absorbed well when administered orally but often causes stomach pain, nausea, and vomiting. These side effects are minor but they limit the drug's use. Serious side effects are rare. Allergic reactions are uncommon.

Other macrolide antibiotics include dirithromycin, azithromycin, and clarithromycin, a semisynthetic macrolide used increasingly for minor respiratory infections such as ear infections. It causes less nausea and vomiting than does erythromycin.

Quinolones. The quinolones are a relatively new and useful family of synthetic antimicrobial agents. These broad-spectrum agents block DNA replication in bacteria by inhibiting the unwinding enzyme DNA topoisomerase. They were derived from nalidixic acid, a narrow-spectrum antibiotic (see Other Antibacterial Drugs). The quinolones are extremely versatile, acting against Gram-positives, many Gram-negatives, and intracellular pathogens such as species of *Mycoplasma*, *Legionella*, *Brucella*, and *Mycobacterium*. They are easily administered orally and have relatively few side effects. Moreover, drug resistance does not emerge readily. Drugs from this family include the widely used ciprofloxacin, which is the drug of choice against anthrax (Figure 21.13).

Other Antibacterial Drugs. Some antibacterial agents that are used rarely or under limited circumstances remain clinically important. They include nalidixic acid and nitrofurantoin, vancomycin, bacitracin, and polymyxin B.

Nalidixic acid and nitrofurantoin are synthetic antimicrobials that inhibit DNA synthesis. Because they are concentrated by the kidneys, they are particularly effective against urinary tract infections. That's their only use.

Vancomycin is an antibiotic produced by *Streptomyces orientalis*. It acts on Gram-positive bacteria by interfering with peptidoglycan synthesis. Vancomycin is not related chemically to any other antimicrobial agent. Because it is not well absorbed orally, it must be administered intravenously. It also has a fairly high incidence of toxic side effects, including damage to the ears and kidneys. Nevertheless, vancomycin is extremely useful for treating life-threatening infections caused by multiple-drug–resistant strains of *Staphylococcus aureus*. In many cases it is the only effective drug available.

Bacitracin and polymyxin B are polypeptide antibiotics that destroy lipid membranes. Both are highly toxic if used systemically, but they're safe and effective when used topically. Ointments containing one or a mixture of these two drugs are available without prescription.

Antimycobacterial Drugs

Mycobacterium spp. cause tuberculosis and leprosy (Chapters 22 and 26). Treating these infections is difficult for several reasons:

1. Mycobacteria are resistant to most antimicrobial drugs, in part because the mycolic acid layer that surrounds the cell wall is nearly impermeable.

2. Mycobacteria grow very slowly, so antimicrobial therapy must continue for months or years.

3. Antibiotic-resistant mycobacteria develop readily, so at least two drugs must usually be used simultaneously to prevent the emergence of drug-resistant strains.

4. *Mycobacterium tuberculosis* is an intracellular pathogen, so only drugs that can enter human cells are effective.

Despite these difficulties, many antituberculosis drugs have been discovered. Most cases of tuberculosis can still be treated effectively, but strains resistant to known drugs are becoming increasingly prevalent. Research to discover new antimycobacterial drugs is now intense.

Isoniazid. Isoniazid is a synthetic drug that resembles nicotinamide (**Figure 21.15**). It's the most useful drug currently available for treating tuberculosis. Curiously, isoniazid is not active itself. It becomes activated by a mycobacterial enzyme, a peroxidase. Then it becomes bacteriostatic for resting cells and bactericidal for growing cells. About 95 percent of *M. tuberculosis* strains are still isoniazid-sensitive.

Isoniazid is well absorbed orally, and it readily penetrates into cells. It's inactivated in the liver at a rate that varies somewhat from person to person. Then it's excreted in the urine. Occasionally isoniazid causes liver damage, which can be fatal in extreme cases. So people taking isoniazid are checked regularly for early signs of liver toxicity.

Isoniazid is also effective for preventing tuberculosis in people who have been infected but have not yet suffered any illness. Because the number of bacteria in these patients is relatively small, isoniazid is the only drug administered. But after disease has developed, isoniazid is always administered in combination with another drug.

Rifampin. Rifampin is a semisynthetic antibiotic (Figure 21.15). It's the drug that's most often administered in combination with isoniazid. Rifampin is made from rifamycin B, an antibiotic produced by *Streptomyces mediterranei*. Rifampin acts against many Gram-positive and Gram-negative bacteria, so it's sometimes used to treat infections other than tuberculosis. It is well absorbed orally, and it penetrates almost all parts of the body and individual human cells. Unlike isoniazid, rifampin has relatively few toxic side effects. But like isoniazid, it's used as part of a multiple-drug therapy.

Ethambutol. Ethambutol is a relatively simple synthetic drug (Figure 21.15) that's effective only against mycobacteria. But almost all strains of *Mycobacterium tuberculosis* are sensitive to it. Its precise mechanism of action is not understood, but it does inhibit the incorporation of mycolic acid units into the mycobacterial cell wall. It's well absorbed orally, and toxic side effects are rare. Infected persons who enter the United States from a country where drug-resistant strains of *M. tuberculosis* are common are usually treated with a combination of isoniazid, rifampin, and ethambutol.

Antifungal Drugs

The similarities between fungi and human cells make fungal infections difficult to treat. Systemic fungal infections are a particular problem. Such therapy is further complicated by the fact that there are no good laboratory tests for the susceptibility of pathogenic fungi to antimicrobial drugs. However, fungal infections of the skin and mucous membranes can be treated safely and effectively with a variety of drugs.

Nystatin. Nystatin is a member of the **polyene** (many double bonds; **Figure 21.16**) family of antibiotics. It's produced by *Streptomyces noursei* and named after its place of discovery—the New York State Health Laboratory. Nystatin damages the cytoplasmic membrane, allowing cell contents to leak out. It is selectively toxic for fungi because it interacts primarily with ergosterol, a membrane sterol produced by fungi but not humans. Nystatin is used mainly against *Candida albicans*, to treat vaginal and skin infections, for example, and intestinal infections in very ill, immunocompromised patients. For gastrointestinal infections, it is administered orally. Because it can't cross the intestinal wall, it acts locally on the intestinal lining.

Amphotericin B. Amphotericin B is also a polyene antibiotic (Figure 21.16). It, too, disrupts fungal cell membranes. But unlike nystatin, it's used to treat systemic infections. Fungal cells are more susceptible than human cells to its destructive action, but human cells do suffer substantial damage. About half the patients who receive an intravenous dose of amphotericin B experience chills,

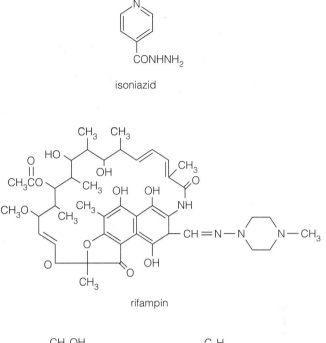

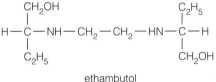

FIGURE 21.15 Chemical structures of major antimycobacterial agents.

vomiting, and fever. Many continue to suffer for weeks or months. More than 80 percent also sustain some permanent kidney damage. Despite what would seem to be an unacceptable level of toxicity, amphotericin B is used to treat life-threatening systemic fungal infections such as cryptococcosis and mucormycosis. It's also used to treat coccidioidomycosis, histoplasmosis, and blastomycosis if the patient is seriously ill or immunocompromised (Chapter 22). There are no better drugs available.

Imidazoles and Triazoles. In recent years the use of imidazoles (Figure 21.16) and triazoles has been increasing. These agents—which include ketoconazole, miconazole, and terconazole—inhibit the synthesis of sterols that occur in the cytoplasmic membranes of fungi. Some of these agents are used topically to treat local infections. Others are used systemically. Ketoconazole, in particular, is used instead of the far more toxic amphotericin B for many systemic mycoses. Miconazole, on the other hand, is used only for fungal infections of the skin, such as athlete's foot and ringworm. Fluconazole has come into common use for systemic treatment of candidiasis, for example vaginal yeast infections.

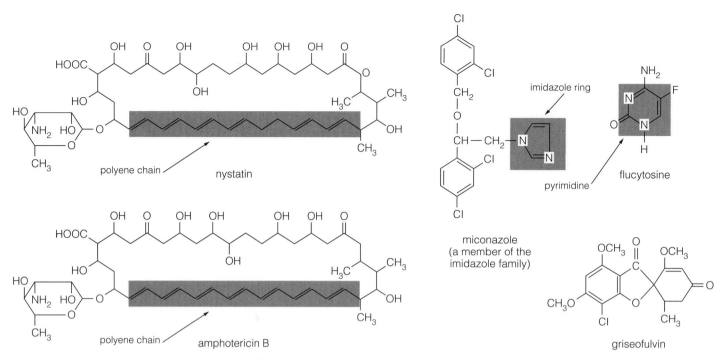

FIGURE 21.16   Chemical structures of major antifungal agents.

Flucytosine. Flucytosine is synthetic pyrimidine analogue (Figure 21.16) used to treat systemic fungal infections. Flucytosine is much less toxic than amphotericin B, but it's also much less effective. Many fungal species are naturally resistant to its action. Others rapidly become resistant. Flucytosine is usually administered in combination with amphotericin B.

Griseofulvin. Griseofulvin is an antibiotic isolated from *Penicillium griseofulvum* in the 1930s. It was generally ignored because it didn't kill bacteria. But later it was found to be effective against fungi that cause infections of the skin, hair, or nails, known as ringworm. It acts only on these fungi, and does so by interfering with cell division.

Most ringworm infections are easily cured by topical creams (such as miconazole from the imidazole group). But others can be eradicated only by prolonged oral treatment with griseofulvin. Griseofulvin does not produce serious side effects or stimulate drug resistance.

Antiparasitic Drugs

Finding drugs that are selectively toxic to protozoa and helminths is difficult. However, some effective chemotherapeutic agents are available.

Mebendazole. Mebendazole (**Figure 21.17**) is an anti-helminthic agent that was developed during the 1970s. It is effective against many types of roundworms, including whipworm, hookworm, pinworm, and Ascaris (Chapter 23). Mebendazole is used for veterinary, as well as human, medicine. It interferes with the helminth's—but not a mammalian cell's—ability to take up glucose. Less than 10 percent of the drug is absorbed from the gastrointestinal tract after oral administration, so most remains in the intestines where these parasitic worms live. In the United States, mebendazole is often used to treat pinworms, a minor infection seen frequently in children. Because so many different worm species are sensitive to mebendazole, it is a good drug for patients simultaneously infected by several types of roundworm. Such conditions are commonly encountered in developing countries.

Metronidazole. Metronidazole (Figure 21.17) has an unusual spectrum. It is effective against obligate anaerobic bacteria, as well as many protozoal parasites. Its mode of action is not completely understood, but it is known that microorganisms expend electrons in reducing this drug, thus robbing the cell of reducing power and ability to generate adenosine triphosphate (ATP; Chapter 5). Moreover, reduced forms of the drug are cytotoxic.

Metronidazole is used to treat infection by several species of protozoa, including *Trichomonas vaginalis*, a protozoan transmitted by sexual contact; *Entamoeba histolytica*, the agent of amoebic dysentery; and *Giardia lamblia*, a common cause of diarrhea. It is also used to treat certain bacterial infections. Metronidazole is well absorbed orally, and it has few side effects. However, it does cause cancer in animals and at an unusually high rate. These laboratory results dictate caution. Metronidazole should not be used during pregnancy.

Antimalarial Drugs

Malaria (Chapter 27) is one of the world's most devastating diseases. It infects 500 million people yearly and kills an estimated 2200 children daily. It kills more people now than it did 30 years ago, in part because malarial parasites are becoming increasingly resistant to once-effective drugs.

Chloroquine. Chloroquine (Figure 21.17) is still the most effective drug against malaria in regions of the world where malarial parasites remain sensitive to it. These areas now include Central America, the Caribbean, North Africa, the Middle East, and parts of India. In other areas, strains are completely resistant. Chloroquine is a synthetic compound derived from quinine, a toxic compound extracted from the bark of the South American cinchona tree that has been used for at least 350 years to treat malaria. Chloroquine's mode of action is unknown, but it is highly concentrated within red blood cells, the site of malarial infection.

Other Antimalarial Drugs. A variety of drugs used singly and in combinations are being used to replace chloroquine. Their properties are summarized in **Table 21.5.**

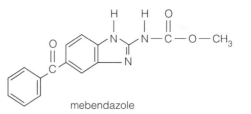
mebendazole

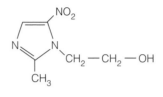
metronidazole

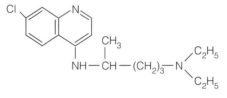
chloroquine

FIGURE 21.17 Chemical structures of antihelminthic and antiprotozoal agents.

TABLE 21.5 Antimalarial Drugs

Drug	Comments
Artisinin	A naturally occurring compound from the Quinghaosu plant used in China for at least 2000 years. Effective but toxic. Resistance develops easily. Not available in North America and Europe.
Atovaquone	A member of a class of ubiquinone analogues first described in the 1920. Effective, but resistance develops quickly.
Chloroquine	The mainstay of antimalarials until resistant strains developed.
Doxycycline	An alternative to mefloquine.
Malarone	A combination of atovaquone and proguanil.
Mefloquine	Related to quinine. The drug of choice for malaria prevention in chloroquine-resistant areas.
Primaquine	An effective prophylactic.
Proguanil	Some advise as a replacement for chloroquine.
Quinine	A toxic alkaloid from the bark of the cinchona tree used to treat malaria for over 350 years.
Tafenoquine	Similar to primaquine but less effective.

Antiviral Drugs

A virus and the cell it infects are so intimately associated that it's extremely difficult to achieve selectively toxicity with chemotherapeutic agents, but a few antiviral drugs are now available. Their success raises hopes that we may soon enter an era of antiviral chemotherapy.

Amantadine. Amantadine (**Figure 21.18**) is an orally administered synthetic drug long known to be clinically effective against influenza A virus (Chapter 22). It interferes with viral replication at an early stage, probably at uncoating. It prevents the development of influenza A in 80 percent of people who have recently been exposed to the virus. But it is much less effective once symptoms appear.

Because amantadine is only truly effective before people become ill, it's difficult to know when to use it. It's obviously impractical for everyone to take amantadine daily during the winter flu season. However, it can be used effectively under special circumstances. For example, highly susceptible individuals, such as nursing home residents, might be given amantadine after one case of influenza has been diagnosed.

Some people feel amantadine is beneficial if started during the first 2 days of influenza symptoms. Fever disappears sooner, and the illness may be shortened by as much as 50 percent. The problem is that it is very difficult to know whether a patient is infected with influenza A, and it's useless against other viruses that cause similar illnesses. Moreover, amantadine has some unpleasant side effects, including anxiety, headache, and insomnia.

Acyclovir. Acyclovir is an analogue of the nucleoside guanosine (a purine; Figure 21.17, Chapter 6). It inhibits the synthesis of viral DNA but not human DNA. Acyclovir's selective toxicity is due to an enzyme found only in virus-infected human cells. The enzyme transforms acyclovir into the actual DNA inhibitor.

Acyclovir is effective against DNA viruses of the herpes family. It's widely used to treat genital herpes infections (Chapter 24). Administered orally and applied directly to infected skin, it shortens the period of painful genital blisters. Unfortunately, however, it does not prevent recurrences. Administered intravenously, it's also used to treat life-threatening herpes simplex encephalitis. Unfortunately, herpesvirus strains that are resistant to acyclovir have already appeared. They cause severe disease in immunocompromised people such as AIDS patients.

Acyclovir is somewhat effective against the varicella zoster herpesvirus that causes chickenpox and shingles (Chapter 26). It doesn't cure chickenpox, but it seems

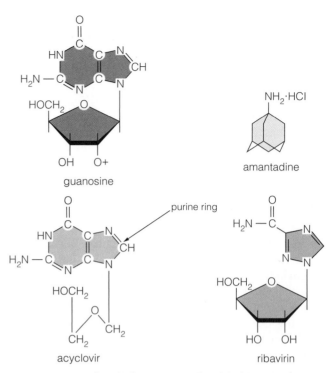

FIGURE 21.18   Chemical structures of antiviral agents.

to shorten the period of fever and decrease the number of blisters. Some physicians recommend acyclovir for high-risk patients, although this is far less of a problem now that the varicella vaccine is available to prevent infection.

Ribavirin. Like acyclovir, ribavirin closely resembles the nucleoside guanine (Figure 21.17). It also interferes with synthesis of viral nucleic acid, but with RNA not DNA. The exact mechanism is unknown.

Anti-HIV Drugs. Recently a set of antiviral agents have been developed that are remarkably effective against human immunodeficiency virus (HIV) infections. They seem almost to cure them. These agents act against either of two enzymes that are found in the HIV virion and that are essential for its replication—reverse transcriptase and HIV protease. The agents that act against them are called reverse transcriptase inhibitors and protease inhibitors. Most treatment regimens administer two reverse transcriptase inhibitors and one or two protease inhibitors. The biochemical basis of their action is discussed in some detail in Chapter 27. We'll say something about the drugs themselves here.

Reverse Transcriptase Inhibitors. Most reverse transcriptase inhibitors are nucleoside analogues. They become incorporated into a growing strand of viral DNA, and, in so doing, they block further elongation of it. The first such inhibitor, zidovudine (Azidothymidine [AZT]), is an analogue of thymine nucleoside, thymidine (**Figure 21.19**). Other drugs in this class are analogues of other nucleosides. For example, didanosine is an analogue of inosine and zalcitabine is an analogue of cytosine.

Two other drugs, delavirdine and nevirapine, inhibit reverse transcriptase by different mechanisms. They're called nonnucleoside analogue inhibitors.

Protease Inhibitors. Three protease inhibitors, indinavir, nelfinavir, and ritonavir, are now available to treat HIV infections. These drugs inhibit the activity of HIV protease by binding to the enzyme's active site. Like the reverse transcriptase inhibitors, they inhibit viral replication.

Interferons. Human cells produce **interferons** (small glycoproteins that stimulate other cells to make antiviral agents; Chapter 16). Large amounts of these compounds can be made by recombinant DNA technology. Unfortunately, interferons have been ineffective in treating many viral infections, and they have significant toxicity. They cause symptoms that mimic viral infection. However, there are some clinical uses. They're used to treat chronic viral hepatitis (Chapter 23) and genital warts (Chapter 24), and they're helpful for patients suffering from the AIDS-related cancer Kaposi's sarcoma (Chapter 18).

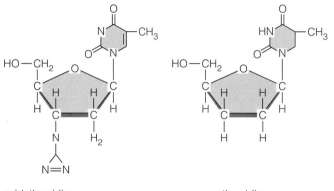

FIGURE 21.19 Structures of AZT and thymidine, its natural analogue.

SUMMARY

Principles of Pharmacology (pp. 490–501)

1. Pharmacology is the study of drugs, any chemicals that have a physiological effect. Drugs used to treat disease are called chemotherapeutic agents. Chemotherapeutic agents used to treat infections are called antimicrobial agents.

2. Antibiotics are metabolic products of one microorganism that kill or inhibit the growth of other microbes. Synthetic drugs are produced chemically. Semisynthetic antibiotics have been chemically modified.

3. Antimicrobial agents can be classified according to their targeted microorganism. Thus there are antibacterials, antimycobacterials, antivirals, antifungals, and antiparasitics.

Drug Administration (pp. 491–492)

4. External (local, topical) therapy is the application of a drug directly to the infected area. Most infections require systemic therapy; administration can be intravenous (IV), intramuscular (IM), or oral (PO).

Drug Distribution (pp. 492–493)

5. There are two major barriers to drug distribution—membranes and drug-binding proteins.

Eliminating Drugs from the Body (p. 493)

6. Eventually all drugs are eliminated from the body, usually in the urine or bile.

Side Effects and Allergies (pp. 493–494)

7. An effective drug is selectively toxic. Most antimicrobial drugs have side effects.

Drug Resistance (pp. 494–498)

8. A microorganism is drug-resistant if it can grow in the presence of a drug. Natural drug resistance is an intrinsic property of a microbial species. Acquired drug resistance is a property gained by individual strains.

9. Narrow-spectrum antimicrobial drugs affect only a single microbial group. Broad-spectrum antimicrobials affect more than one microbial group. Broad-spectrum drugs may alter the normal biota, causing a superinfection.

10. Microbial strains acquire drug resistance through genetic change.

11. There are three mechanisms of acquired drug resistance: the microorganism produces an enzyme that destroys the drug, the target of drug action changes, or the drug is kept out of the cell.

12. Drug resistance can be encoded by genes on the microbial chromosomes or on plasmids. Chromosomally encoded resistance occurs by mutation.

13. Plasmid-encoded resistance can be spread rapidly from one strain to another strain or species by conjugation. Plasmids that carry genes encoding drug resistance are called R factors.

14. Acquired drug resistance can be slowed by limiting nonmedical uses of antibiotics, being more selective in the medical use of antibiotics, and, under some conditions, administering two drugs at the same time (combination therapy).

Drug Dosage (pp. 499–501)

15. Drug dosage depends upon route of administration, drug distribution, drug elimination, and antimicrobial susceptibility.

16. The simplest and most commonly used antimicrobial susceptibility test is the disc-diffusion susceptibility test (or Kirby-Bauer test). In the broth-dilution susceptibility test, the clear test tube with the lowest drug concentration contains the minimum inhibitory concentration (MIC) of that drug. The tube with the lowest drug concentration that is bactericidal contains the minimum bactericidal concentration (MBC) of that drug. In the serum killing-power test, some of the patient's serum is tested in the laboratory to see if it kills the infecting microorganism.

Targets of Antimicrobial Drugs (pp. 501–503)

The Cell Wall (p. 501)

17. Drugs that inhibit peptidoglycan synthesis have high selective toxicity and are bactericidal; penicillins, cephalosporins, vancomycin, and bacitracin are examples.

Cell Membranes (pp. 501–502)

18. Drugs that damage the cell membrane cause increased permeability, allowing vital molecules to leak out. They have limited selective toxicity and are microbiocidal; polymyxin B and amphotericin B are examples.

Protein Synthesis (p. 502)

19. Drugs that inhibit protein synthesis by affecting the 70S bacterial ribosome are usually bacteriostatic. The

aminoglycosides bind to the 30S subunit and are bactericidal. Tetracyclines bind to the 30S subunit and are bacteriostatic. Erythromycin and chloramphenicol bind to the 50S subunit and are bacteriostatic.

Nucleic Acids (p. 502)

20. Drugs that inhibit nucleic acid synthesis include the quinolones, which inhibit topoisomerase; rifampin, which inhibits bacterial RNA polymerase; and flucytosine and other antimetabolites, which interfere with the synthesis of functional nucleic acid.

Folic Acid Synthesis (pp. 502–503)

21. Drugs that inhibit folic acid synthesis are antimetabolites. The sulfonamides are competitive inhibitors of the enzyme substrate PABA; trimethoprim is a competitive inhibitor of another intermediate of the folic acid pathway.

Antimicrobial Drugs (pp. 503–515)

22. Antimicrobial drugs can be classified by the infection they are usually used to treat.

Antibacterial Drugs (pp. 503–510)

23. Penicillins interfere with cross-linking peptidoglycan. Natural penicillins (G and V) are most effective against Gram-positives. Ampicillin and amoxicillin are broad-spectrum semisynthetics that act against both Gram-positives and many Gram-negatives. Side effects are negligible, but allergies are common.

24. Cephalosporins also interfere with cross-linking peptidoglycan. They are broad spectrum and can be natural or semisynthetic. The third generation is particularly effective against meningitis, pneumonia, and sepsis; some penetrate the central nervous system well, and others persist in the bloodstream. Side effects are negligible, but people allergic to penicillin can be allergic on first exposure.

25. Sulfonamides are broad-spectrum synthetics that act on Gram-positives, some Gram-negatives, and chlamydiae. They interfere with folic acid synthesis. Today they are generally used by themselves or in combination with trimethoprim (cotrimoxazole) to treat urinary tract infections. Side effects include crystals in urine and depressed stem cell production; allergy is common.

26. Aminoglycosides, such as streptomycin and neomycin, developed to control Gram-negatives, must be administered intravenously or intramuscularly. Major side effects include loss of hearing and kidney function. Drug resistance emerges rapidly.

27. Chloramphenicol is an antibiotic that can be manufactured synthetically. It can be taken orally, penetrates the brain and eye, and can be stored for long periods with refrigeration. It is extremely broad spectrum. It has few side effects, and rarely causes allergy. But rarely it can be fatally toxic. In the United States, it is used only with hospitalized patients and never with infants.

28. Tetracyclines, the first antibiotics designated broad spectrum, are used to treat acne and infections caused by chlamydiae, rickettsiae, and mycoplasmas. They can be natural or semisynthetic. They cause increased sensitivity to sunlight. Because they stain developing teeth, they cannot be used with children or pregnant women. Resistant strains have emerged.

29. Erythromycin is a narrow-spectrum macrolide antibiotic effective against only Gram-positives. It is often used for patients allergic to penicillin and is prescribed for strep throat and other common respiratory infections. Side effects include stomach upset and, rarely, liver damage. Allergies are rare.

30. The quinolones are broad-spectrum drugs derived from a narrow-spectrum antibiotic, nalidixic acid. They act against Gram-positives, many Gram-negatives, and intracellular pathogens. They are adminis-

tered orally and have relatively few side effects; allergy is rare, and drug resistance is slow to emerge.

31. Nalidixic acid and the synthetic drug nitrofurantoin are narrow-spectrum drugs that tend to concentrate in the kidneys, which makes them most suitable for treating urinary tract infections. They inhibit DNA synthesis.

32. Vancomycin acts on Gram-positives. It must be administered intravenously and has serious side effects, including damage to the ears and kidneys. But it can be the only treatment for life-threatening infections caused by multiple-drug–resistant strains of *Staphylococcus aureus.*

33. Bacitracin and polymyxin B, which are highly toxic if used systemically, are effective topically for minor skin infections. Ointments containing one or both of these are available without prescription.

Antimycobacterial Drugs (pp. 510–511)

34. Isoniazid, a synthetic, is the main drug used in treating tuberculosis. It is bacteriostatic for resting cells and bactericidal for growing cells. It can cause liver damage, which is fatal in extreme cases.

35. Rifampin, usually used in combination with isoniazid, is semisynthetic and acts by inhibiting the synthesis of RNA. It is also used to treat many Gram-positive and Gram-negative bacteria. Side effects are negligible.

36. Ethambutol, a synthetic, is effective only against mycobacteria. It is used in combination with isoniazid and rifampin. Side effects are negligible.

Antifungal Drugs (pp. 511–512)

37. Nystatin belongs to the polyene family of antibiotics. It disrupts the cytoplasmic membrane and is selectively toxic. It is used locally, primarily to treat *Candida albicans* infections.

38. Amphotericin B is a polyene that also disrupts cell membranes. Side effects include chills, vomiting, fever, and, in 80 percent of patients, kidney damage. Nevertheless, it is the preferred treatment for certain systemic infections, such as cryptococcosis and mucormycosis.

39. The imidazoles and triazoles both inhibit the synthesis of plasma membrane sterols. Some agents are used topically (miconazole), whereas others are used systemically (ketoconazole).

40. Flucytosine is a synthetic drug used to treat systemic infections. Because drug resistance emerges rapidly, it is usually used in combination with amphotericin B.

41. Griseofulvin acts only on fungi that cause ringworm infections. It is administered orally when topical preparations do not work; it interferes with cell division. Side effects are negligible, and there is no serious drug resistance.

Antiparasitic Drugs (pp. 512–513)

42. Mebendazole is used for many roundworm infections. In the United States it is used mostly to treat pinworms. It interferes with the helminth's ability to take up glucose.

43. Metronidazole is effective against protozoal parasites and is used to treat trichomoniasis, amoebic dysentery, and *Giardia lamblia* diarrhea. It is also effective against obligate anaerobic bacteria. Side effects are few, though it can cause cancer in laboratory animals and possibly birth defects in humans.

Antimalarial Drugs (p. 513)

44. Chloroquine is the most effective drug for treating malaria, especially the most virulent form. It is synthetic and selectively toxic because it concentrates in red blood cells. Side effects include mild headache and abdominal discomfort. Resistant strains have emerged.

Antiviral Drugs (pp. 514–515)

45. Amantadine, a synthetic drug, interferes with viral replication. It is most useful in preventing influenza A, especially if administered soon after exposure. Side effects include anxiety, headache, and insomnia.

46. Acyclovir interferes with the synthesis of viral DNA. It is used to treat herpes infections orally and topically and, in serious systemic cases, intravenously. It is somewhat effective against the varicella zoster herpesvirus.

47. Ribavirin also interferes with viral nucleic acid replication. It is inhaled as a mist for respiratory syncytial virus infections. Immediate side effects are minimal, but long-term effects are not known.

48. There are two kinds of anti-HIV agents: reverse transcriptase inhibitors and HIV protease inhibitors. Usually 3 or 4 drugs are given simultaneously.

49. Interferons can be made by recombinant DNA technology. In the United States, interferons are used to treat chronic viral hepatitis and genital warts. They have significant side effects.

REVIEW QUESTIONS

Principles of Pharmacology

1. What is pharmacology? What is the difference between chemotherapeutic agents and antimicrobial agents? Between antibiotics, synthetic drugs, and semisynthetic drugs?

2. What is the difference between local, or topical, therapy and systemic therapy? Describe the three ways of administering drugs systemically and the advantages and disadvantages of each.

3. What are the barriers to drug distribution in the body?

4. Why is it important to know how and in what period of time drugs are eliminated from the body?

5. What are drug side effects? Give an example of how side effects must be weighed against drug benefits.

6. What determines if a microorganism is drug resistant? What is the difference between natural and acquired resistance? Why are some species naturally resistant? How do species acquire resistance?

7. What is the difference between narrow- and broad-spectrum drugs?

8. Compare and contrast chromosomally mediated resistance and resistance due to plasmid-borne genes.

9. How can drug resistance be slowed, and why it is important to do so?

10. Define drug dosage and antimicrobial susceptibility.

11. How would you perform a disc-diffusion susceptibility test? A broth-dilution susceptibility test? What is the difference between a bacteriostatic and a bactericidal drug?

12. Discuss the significance of MIC and MBC. What information does a serum killing-power test give a clinician?

Targets of Antimicrobial Drugs

13. Give some examples of drugs that act this way:
    a. inhibit cell-wall synthesis
    b. disrupt the cell membrane
    c. inhibit protein synthesis
    d. inhibit nucleic acid synthesis
    e. inhibit folic acid synthesis

Antimicrobial Drugs

14. Tell whether each of the following antibiotics is narrow or broad spectrum, what the mode of action is, how the drug is used, and whether allergy or side effects are a concern:
    a. penicillins
    b. cephalosporins
    c. sulfonamides
    d. aminoglycosides
    e. chloramphenicol
    f. tetracyclines
    g. erythromycin
    h. quinolones
    i. nalidixic acid and nitrofurantoin
    j. vancomycin
    k. bacitracin and polymyxin B

15. Why is tuberculosis difficult to treat? How does each of these antimycobacterials act: isoniazid, rifampin, ethambutol? Why are these drugs usually used in combination?

16. Why is it difficult to develop effective antifungal agents? Discuss the modes of action, use, and side effects for each of the following antifungals:
    a. nystatin   d. flucytosine
    b. amphotericin B   e. griseofulvin
    c. imidazoles and
       triazoles

17. What are the modes of action for these antiparasitics: mebendazole, metronidazole, chloroquine? How is each used? What are the side effects?

18. Why is it difficult to develop selectively toxic antiviral agents? What is the mode of action of these drugs:
    a. amantadine
    b. acyclovir
    c. ribavirin
    d. interferons
    e. azidothymidine (AZT)
    f. dideoxyinosine (ddI)
    g. indinavir

CORRELATION QUESTIONS

1. Is their any similarity in the basis for selective toxicity of penicillin and AZT? Explain.

2. Does the following statement reasonable to you? One consideration for deciding whether penicillin VK should be administer PO or IM is cost of the drug. Explain.

3. If you knew the cause of bacterial infection, would you prefer to treat it with a broad-spectrum or narrow-spectrum antibiotic? Why?

4. When chlortetracycline was first introduced, it was thought resistant strains would never develop, because even if a huge population of a particular bacterial culture were exposed to it, no resistant strain appeared. However, rather quickly after it was used therapeutically, resistant strains did appear. How would you explain this seeming contradiction?

5. Would you expect resistance to flucytosine to occur as a result of a chromosomal mutation or of acquiring a plasmid? Explain.

6. What property would you expect a broad-spectrum penicillin to have that a narrow-spectrum penicillin lacks?

ESSAY QUESTIONS

1. "It is inevitable that microbial resistance will eventually develop to today's drugs." Discuss this statement from a historical point of view and from an epidemiologist's point of view. Why is drug resistance a major and urgent problem?

2. Do you support or oppose the use of antibiotics in livestock? Explain your position.

SUGGESTED READINGS

Barrlett, J. G., and R. D. Moore. 1998. Improving HIV therapy. *Scientific American* July: 84–87.

Chopra, I., and M. Roberts. 2001. Tetracycline antibiotics: Mode of action, applications, molecular biology, and epidemiology of bacterial resistance. *Microbiology and Molecular Biology Review* 65:232–60.

Levy, S. B. 1998. The challenge of antibiotic resistance. *Scientific American* March:46–53.

Williams, R. J., and D. L. Heymann. 1998. *Containment of antibiotic resistance. Science* 279:1153–54.

Witte, W. 1998. Medical consequences of antibiotic use in agriculture. *Science* 13:996–97.

For additional readings, go to InfoTrac College Edition, your online research library at: http://www.infotrac.thomsonlearning.com

TWENTY-TWO

Infections of the
Respiratory System

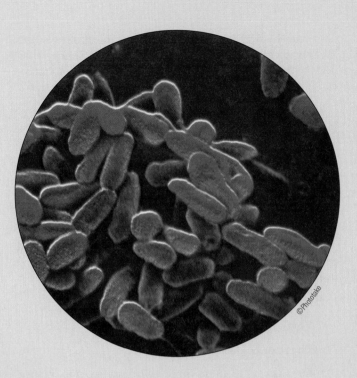

CHAPTER OUTLINE

CLINICAL SCIENCE
   Diagnosis
   Prognosis and Treatment
   Types of Infection
   Progress of an Infection

ORGANIZATION OF
CHAPTERS 22 THROUGH 27

THE RESPIRATORY SYSTEM
   Structure and Function
   Fighting for Every Breath
   Defenses and Normal Biota:
      A Brief Review
   Clinical Syndromes

UPPER RESPIRATORY
INFECTIONS
   Bacterial Infections
   Viral Infections

LOWER RESPIRATORY
INFECTIONS
   Bacterial Infections
   Deal Me Out
   Courage—and Lead Weights
   Viral Infections
   *Influenza and Asthma: A Dangerous
      Combination*
   Fungal Infections

SUMMARY

REVIEW QUESTIONS

CORRELATION QUESTIONS

ESSAY QUESTIONS

LEARNING GOALS

To understand:

- The anatomy and function of the respiratory system and its defenses against microorganisms

- The clinical syndromes that characterize respiratory infections

- The bacterial and viral causes of upper respiratory infections; their diagnosis, prevention, and treatment

- The bacterial, viral, and fungal causes of lower respiratory infections; their diagnosis, prevention, and treatment

This and the next five chapters consider infectious diseases. We'll begin with a brief discussion of **clinical science,** which deals with the way that diseases are diagnosed and treated.

CLINICAL SCIENCE

A **clinician** (a practitioner of clinical science) relies on careful observation and his or her past experience when treating patients. The practice of clinical science is imprecise and can be frustrating because the course of an illness proceeds somewhat differently in each patient.

Nevertheless, clinical science has developed logical, time-proven ways to help patients. The clinician begins treatment by asking three questions:

1. What's wrong with this person?

2. What's likely to happen next?

3. What can be done to help?

The answers to these three questions provide, respectively, a **diagnosis,** a **prognosis,** and a **treatment plan.**

Diagnosis

Diagnosis is fundamental because both prognosis and a treatment plan depend upon it. In making a diagnosis, a clinician focuses first on determining **etiology** (the cause of the patient's illness). Presented with the effect of an illness the clinician relies on knowledge and previous experience to infer its probable cause. That's the essence of clinical diagnosis: Working backward from what one sees to what one supposes to be its cause. The clinician relies on the **clinical presentation** (the available, relevant information about the ill person) obtained by taking a **medical history,** performing a **physical examination,** and analyzing diagnostic studies such as laboratory tests or x-rays.

In taking a medical history, the clinician asks such questions as: "Do you have any pain?" "Where is that pain?" and "When did you become ill?" The history reveals the person's **symptoms** (what the patient has experienced).

The physical examination includes measuring vital signs (body temperature, pulse, respiration, blood pressure), height and weight, as well as making clinician's **observations** (what he or she can see, such as color and effort of breathing), **auscultations** (what he or she can hear, such as the heartbeat or breathing sounds), and **palpations** (what he or she can feel, such as the size of the liver or masses in the abdomen). All these are **clinical signs.**

A clinical presentation may seem inconsistent or confusing. To make sense of it, the clinician makes a **differential diagnosis** (a list of possible diagnoses ranked in order of their probability). As additional information becomes available, the clinician refines the list. While working toward a **definitive diagnosis,** the clinician must make decisions about prognosis and treatment based upon his or her **working diagnosis** (his or her best guess at the time).

A definitive diagnosis usually has two components— and **anatomical diagnosis** (identifying the part of the body that is affected) and an **etiological diagnosis** (identifying the cause of the illness). Let's consider each of them.

Anatomical Diagnosis. Often the clinical presentation immediately suggests an anatomical diagnosis. A patient's **chief complaint** (the first statement to the clinician) may lead directly to an anatomical diagnosis. For example, if the chief complaint is "I'm having trouble breathing," the clinician immediately focuses on the respiratory system. The anatomical diagnosis is refined as the clinician determines which part of the respiratory system is affected. Such refinement usually depends upon clinical signs, as well as symptoms. For example, difficult breathing along with abnormal breath sounds heard through a stethoscope implicate lung involvement. Such combinations of signs and symptoms are called **clinical syndromes.** If they implicate one particular structure, they are called **anatomical syndromes.**

Etiological Diagnosis. The first step in making an etiological diagnosis is to assign the cause of illness to some broad category, such as **infection** (illness caused by microorganisms), **trauma** (injury), **malignancy** (cancer), **malnutrition** (poor nutrition), or **toxicity** (poisoning). Sometimes the etiology is all too obvious, for example trauma from an accident. Other times the cause may be obscure, for example, toxicity from exposure to environmental poisons. In this and the following five chapters we'll consider only those human illnesses caused by infection. The principal clues of infectious disease are fever, inflammation, and recent exposure to a communicable disease.

An anatomical diagnosis of infection is inadequate for making a prognosis and a treatment plan. The particular **etiological agent** (the causative microorganism) must be identified, usually by tests done in a clinical laboratory. But such identification may not always be practical. For example, the causative microorganisms might be multiplying deep within human tissues, where they can't be readily recovered; such is often the case with pneumonia. Or the etiological agent might be one of the many viruses for which there is no readily available laboratory test; such is often the case with the common cold. In such situations, knowledge of which microorganisms usually infect various parts of the body, along with information about the patient's age and previous state of health, usually allows the clinician to formulate a fairly reliable working diagnosis. For example, if the signs and symptoms suggest pneumonia, the clinician knows that *Streptococcus pneumoniae* must be considered. If the patient is an unimmunized child, *Haemophilus influenzae* is also a candidate. If the patient has acquired immunodeficiency syndrome (AIDS) and poor immune defenses, *Pneumocystis carinii* is a strong possibility.

Correlating Anatomical and Etiological Diagnoses.

A few clinical syndromes (for example, chickenpox and measles) have only a single etiological agent. For this class of infections, a conclusive diagnosis can be made without laboratory tests. Usually, however, it's not so easy to make an etiological diagnosis based solely on a history and physical examination. Many anatomical syndromes, including **pharyngitis** (sore throat), can be caused by many different etiological agents. Conversely, many etiological agents, including group A streptococci, can cause many different anatomical syndromes.

Prognosis and Treatment

Accurate diagnosis is often essential to determining prognosis and treatment. Consider, for example, strep throat and infectious mononucleosis, two common illnesses that can be difficult to distinguish clinically. A patient with either of these illnesses, particularly a teenager or a young adult, is likely to come to a clinician with the symptom of sore throat and clinical signs of a bright red throat with white patches of pus on the tonsils and large, tender, swollen lymph nodes in the neck. But these two ailments differ in their prognosis and proper treatment. For strep throat, prognosis is guarded but treatment is excellent. For infectious mononucleosis (mono), the concerns are reversed: Prognosis is good, but treatment is extremely limited. To determine the disease's prognosis and proper treatment, the etiological agent must be determined by laboratory tests.

Types of Infections

In addition to anatomical and etiological diagnoses, clinicians describe infections in other ways (**Table 22.1**). They use terms such as **acute, chronic,** and **persistent** to indicate the suddenness with which symptoms appear and how long they last. They use many other terms as well: **symptomatic** and **asymptomatic** to describe if the infection causes disease; **local** and **systemic** to tell whether infection occurs in one part of the body or everywhere; and **primary, secondary,** and **mixed** to describe relationship between one infection and another.

TABLE 22.1   Categories of Human Disease

Category	Definition	Example
Infection	Illness caused by multiplication of microorganisms in the body	Streptococcal pharyngitis
Trauma	Injury inflicted by a physical agent	Broken bone
Malignancy	Uncontrolled multiplication of the body's own cells	Brain tumor
Metabolic defect	Illness caused by inability to perform normal biochemical transformations	Diabetes
Inherited or congenital defect	Abnormality caused by abnormal genes (inherited) or problem in development before birth (congenital)	Cystic fibrosis spinabifida
Immunological disorder	Illness caused by weakness or malfunction of the immune system	Allergy
Nutritional disorder	Illness caused by poor nutrition	Obesity, vitamin deficiency
Degenerative disorder	Decrease in function associated with advancing age	Osteoarthritis
Toxicity	Debility caused by poisoning	Lead poisoning
Iatrogenic disease	Disease that results from medical intervention	Nasocomial infection
Idiopathic	Unknown cause	Nephrotic syndrome (childhood kidney disease)

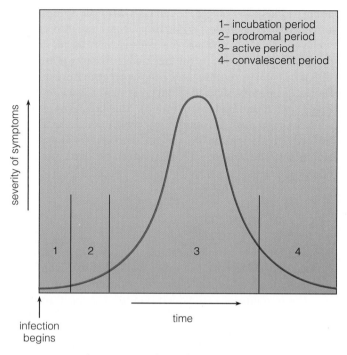

FIGURE 22.1   The progress of an infection.

Progress of an Infection

A set of terms is used to describe the progress of an infection as symptoms come and go (**Figure 22.1**). The **incubation period** is the interval between infection and onset of symptoms. It's usually followed by the **prodromal period** (the time of early, mild symptoms), the **active period** (when symptoms reach their peak), and the **convalescent period** (when symptoms diminish and cease). Some infections, such as pertussis (see the beginning of Chapter 15), go through all these stages; others go through only some of them.

ORGANIZATION OF CHAPTERS 22 THROUGH 27

Perhaps the best way to learn about the infectious diseases is to adopt the perspective of a clinician. That's why this and the following five chapters are organized by anatomical syndrome. One chapter is devoted to each of the major human organ systems: respiratory, digestive, genitourinary, nervous, skin (and conjunctivae), and cardiovascular and lymphatic. In each chapter we'll begin by briefly reviewing the organ system's anatomy, physiology, and natural defenses. Then we'll consider the common clinical syndromes associated with infections of that organ system. Finally we'll discuss the microorganisms and diseases that are most

likely to cause these syndromes. Throughout we'll examine case histories that illustrate important points.

Organizing our discussion of infectious diseases by organ systems presents a few problems because some affect more than one system. For example, the etiological agent of typhoid fever (Chapter 23) passes through two organ systems: It enters through the mouth (digestive system), then enters the bloodstream (the cardiovascular system), and finally reenters the digestive system to exit through the anus. Other infections affect many organ systems. We'll discuss most of them in Chapter 27 (Cardiovascular and Lymphatic Systems).

THE RESPIRATORY SYSTEM

The respiratory system extends from the nose to the millions of alveolar air sacs that lie deep within the lungs. This organ system exchanges oxygen from the air for carbon dioxide produced in the tissues. If it stops functioning for more than just a few minutes, permanent brain damage or death is inevitable. The entire respiratory tract is constantly exposed to air and the microorganisms it contains. Unsurprisingly, this organ system is infected more frequently than any of the others.

We'll begin by considering the structure of the respiratory system and the functions of its various parts.

Structure and Function

We'll consider the respiratory system to consist of two parts: **the upper respiratory system** and **the lower respiratory system** (**Figure 22.2**). The upper respiratory system lies within our head and neck. Air entering it first passes through the **nasal cavity,** where it is warmed and filtered. Then it enters the **nasopharynx,** a chamber that includes the uppermost part of the throat, called the **pharynx.** The nasopharynx connects with other air-containing spaces in the head: the **sinuses** (cavities within the bones of the skull) and the **middle ear** (the small compartment between the eardrum and the inner ear). Air doesn't flow through these spaces, but it can enter them and bring infection. The **adenoids** (Chapter 17) are located in the nasopharynx. Just below the nasopharynx, the back of the mouth connects to the respiratory system. Then for a short stretch, called the **oropharynx,** the respiratory and digestive systems are joined: Both air and food pass through the oropharynx. The **tonsils** (Chapter 17) are located in the oropharynx. The respiratory and digestive systems separate again at the **epiglottis.** This flaplike structure diverts food into the digestive tact by closing off the respiratory tract

CASE HISTORY

Fighting for Every Breath

J G., a 10-year-old boy, had had a mild cold for several days when his temperature suddenly rose to 103°F. He developed a shaking chill. The next day he said the right side of his chest hurt. Coughing or taking a deep breath caused sharp pain. His breathing had become rapid and shallow. He looked anxious and much sicker than he did the day before.

J. G.'s mother realized that he needed immediate attention. She took him to a hospital emergency room. J. G. was so weak and uncomfortable that he had to be taken to the examining room in a wheelchair. His temperature was 104°F. Even though his respirations were rapid and shallow, he was working hard to breathe.

The examining physician tapped on his chest. Instead of the normal hollow thump, she heard a dull note over his lower right chest. The sound suggested that this area was not filled with air, so she used a stethoscope to examine it further. Instead of hearing the normal soft flow of air entering and leaving the lungs, she heard no breath sounds at all in some places and wet crackling sounds called **rales** in others. An x-ray of J. G.'s chest showed a black left lung. That's the normal appearance indicating that the x-rays are traveling mainly through air. But the lower lobe of his right lung was white, the characteristic appearance of x-rays passing through

fluid. J. G. continued to complain of severe chest pain. He was unable to cough up sputum for microbiological examination. A sample of blood was drawn, under sterile conditions to prevent contamination by skin bacteria, and sent to the microbiology laboratory for culture. A second sample was sent to the hematology laboratory. The hematology laboratory immediately reported an unusually high number of neutrophils, many of which looked immature, indicating the adaptive immune system had been stimulated to produce new cells.

Based on J. G.'s history and physical examination, his attending physician felt confident in diagnosing pneumonia. Because his illness was severe, he was admitted to the hospital. He was given a large dose of cephalosporin intravenously so it would enter his bloodstream immediately. The medication was repeated every 6 hours, and by the next day he felt much better. His temperature had returned to normal. Later the microbiology laboratory reported that his blood culture grew a Gram-positive coccus, *Streptococcus pneumoniae*, that was sensitive to penicillin and cephalosporin.

Within 2 days, J. G. was breathing comfortably. He had no more episodes of fever, so he was released to continue his recuperation at home. He was given a prescription for oral penicillin. A respiratory therapist in-

structed his mother in **percussion** (gently hitting the chest with cupped hands) and **postural drainage** (positioning the patient so fluid drains more readily). These procedures helped J. G. loosen and cough up the fluid from his infected lung. J. G. returned to school the following week. He had no further problems from his brief but serious illness.

Case Connections

■ As you read this chapter, note particularly the information that allowed J. B.'s physician to make his anatomical and etiological diagnoses as well as his prognosis and treatment plan.

■ You will learn how the infecting microorganism probably evaded J. B.'s defenses, establishing an infection.

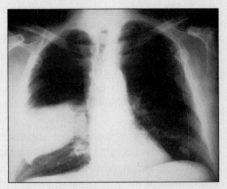

Chest x-ray of a patient with pneumococcal pneumonia. The white area in the lower part of the right lung is filled with fluid as a result of the infection.

during swallowing. Except for this brief period, the epiglottis is open, allowing air to enter the lower respiratory tract.

The lower respiratory system begins at the **larynx** (voice box), which connects to the **trachea** (windpipe). Near the lungs, the trachea branches into two smaller airways, the right and left **primary bronchi.** Within the lungs, the pri-

mary bronchi branch progressively into smaller **secondary bronchi** and then even smaller **tertiary bronchi.** Finally the tertiary bronchi branch repeatedly, becoming tiny tubes called **bronchioles,** each of which terminates in minute air sacs called **alveoli.** A lung contains millions of them. Collectively, alveoli make up the tissue of the lungs and fulfill its

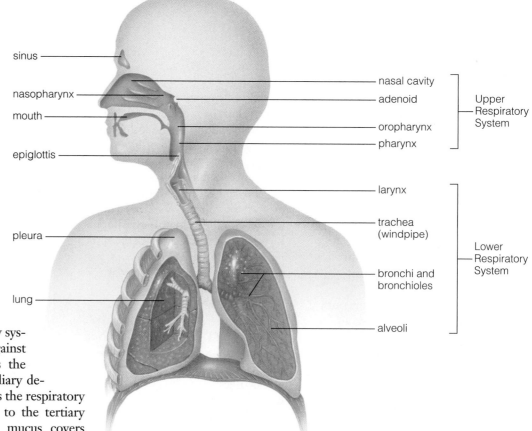

FIGURE 22.2 Anatomy of the respiratory system.

(Art by Carlyn Iverson.)

function, exchanging oxygen and carbon dioxide across their thin walls. Each lung is surrounded by a smooth membrane called the **pleura.**

Now let's examine how the respiratory system protects itself from the bacteria that enter it with each breath we take.

Defenses and Normal Biota: A Brief Review

Unsurprisingly, the respiratory system has effective defenses against the microorganisms. Perhaps the most important is the mucociliary defense (Chapter 16), which lines the respiratory system from the nasal cavity to the tertiary bronchi. Microbe-entrapping mucus covers the entire region, and ciliated epithelial cells below the mucus constantly move it, like a moving belt of flypaper, toward the nose and mouth, where it and the particles it contains are eliminated from the body. The mucociliary system keeps the airways clean. The bronchioles and alveoli don't have a mucociliary defense, but they do contain phagocytic macrophages and secretory IgA (Chapter 17).

In spite of its impressive antimicrobial defenses, microorganisms do inhabit the upper respiratory system. Being warm, moist, and nutrient-rich, it offers a favorable environment for microbial growth. Indeed, the upper respiratory system is densely colonized by commensal microorganisms, including streptococci, lactobacilli, and some Gram-negatives such as *Moraxella catarrhalis* (Chapter 14). But that's as far as they normally get. In healthy people the lower respiratory system (as well as the sinuses and middle ear) is sterile.

Occasionally, however, the defenses break down and infection occurs. Now we'll consider the consequences.

Clinical Syndromes

Respiratory infections express themselves in remarkably diverse ways (**Table 22.2**).

Upper Respiratory System. If the membranes that line the nose are infected, the condition is called **rhinitis** (nasal inflammation). The infected membranes swell, stimulating excess production of mucus. Rhinitis, also called a **cold,** is the most common of all respiratory syndromes. Most of us experience it at least once a year.

Infection of the adenoids, called **adenoiditis,** occurs frequently in certain children. Infection of the throat, called **pharyngitis,** or **tonsillitis** if the tonsils are primarily infected, causes sore throat and sometimes fever. An infected throat is usually red; it may be covered by a milky white exudate, ulcers, blisters, or even a grayish membrane. A skilled clinician can often identify the infecting microorganism by examining the appearance of the throat.

Sinusitis and **otitis media** are infections of the sinus or middle ear, respectively. They occur when these spaces fill with fluid, which often happens when rhinitis closes their openings to the nasopharynx. **Purulent** (pus-producing) **sinusitis** typically causes fever and headache. Otitis media typically causes pain, temporary hearing loss, and fever.

Lower Respiratory System. Infection of the epiglottis, called **epiglottis,** can cause the structure to swell suddenly to many times its normal size, stopping airflow

Labels in figure: sinus, nasopharynx, mouth, epiglottis, pleura, lung, nasal cavity, adenoid, oropharynx, pharynx, larynx, trachea (windpipe), bronchi and bronchioles, alveoli, Upper Respiratory System, Lower Respiratory System

TABLE 22.2 Clinical Syndromes of Respiratory Infections

Syndrome	Region Affected	Signs and Symptoms	Causative Agents
Rhinitis	Nasal cavity	Nasal discharge, sneezing	Rhinovirus, coronavirus, and others
Adenoiditis	Adenoids	Nasal discharge, obstruction of nasal passages	Bacterial pathogens, including *Haemophilus influenzae* and *Streptococcus pneumoniae*
Pharyngitis	Pharynx	Sore throat, fever	*Streptococcus pyogenes* and many viruses
Tonsilitis	Tonsils	Sore throat, fever	*S. pyogenes* and many viruses
Sinusitis	Sinuses	Fever, headache	Bacterial pathogens, including *H. influenzae*, *S. pneumoniae*, *Moraxella catarrhalis*
Otitis media	Middle ear	Ear pain, fever, temporary hearing loss	Bacterial pathogens, including *H. influenzae*, *S. pneumoniae*, *M. catarrhalis*, and many viruses
Epiglottitis	Epiglottis	Fever, sudden obstruction of upper airway	*H. influenzae*
Laryngitis	Larynx	Hoarse voice	Many viruses
Laryngotracheo-bronchitis (croup)	Larynx, trachea, and primary bronchi	Hoarse, barking cough	Parainfluenza virus
Bronchitis	Bronchi	Cough that produces infected mucus, fever	Many bacterial pathogens
Bronchiolitis	Bronchioles	Wheezing, rapid breathing	Respiratory synctyial virus (RSV)
Pneumonia	Lungs	Cough, fever, difficult breathing	Many bacterial, viral, and fungal pathogens
Pleurisy	Lungs and pleura	Cough, fever, labored and painful breathing	Many bacterial, viral, and fungal pathogens

completely and causing sudden death. Infection of the larynx, called **laryngitis,** or **laryngotracheobronchitis (croup)** if the tissues below it are also involved, can also cause swelling and stop breathing. Typically, however, laryngitis and croup are milder infections that only produce hoarseness or a characteristic barking cough. Any severe narrowing of the epiglottis or the larynx causes a clinical sign called **stridor,** a whistling sound when the person breathes in.

Infection within the lungs of bronchi, called **bronchitis,** causes their membranes to swell and produce **phlegm** (thick mucus). A cough that brings it up is typical of bronchitis, as is fever. Infection of the bronchioles, called **bronchiolitis,** causes inflammation that narrows these tiny collapsible airways. Air continues to enter the lungs freely but has difficulty getting out. Thus clinical signs of bronchiolitis include **wheezing** (a musical noise heard during expiration) and **tachypnea** (*tachy,* meaning "fast"; *pnea,* meaning "breathing").

Infection of lung tissue, called **pneumonia,** causes fluid and microorganisms to fill the alveoli so that normal gas exchange cannot take place. Typically the patient's symptoms are fever, tachypnea, and a cough. Infection of the pleura, called **pleurisy,** is associated with painful breathing.

Now we'll examine the various infections of the respiratory system, beginning first with the upper respiratory system.

UPPER RESPIRATORY INFECTIONS

Bacteria and viruses cause most upper respiratory infections. Some of these are minor, but others can cause sudden death by completely blocking vital airways.

Bacterial Infections

Bacteria cause the most serious respiratory infections, but now most of them can be prevented or effectively treated.

Epiglottitis. The epiglottis is prone to infection by *Haemophilus influenzae* (Bergey's Gammaproteobacteria, Chapter 11). This Gram-negative, rod-shaped bacterium is nonmotile and does not form endospores. It requires two unusual growth factors: **hemin** (an iron-containing compound from cytochromes and hemoglobin), and nicotinamide adenine dinucleotide (NAD) (or nicotinamide adenine dinucleotide phosphate [NADP]; Chapter 5). Both are present in red blood cells, leading to the name *Haemophilus* ("blood-loving"). The species name *influenzae* reflects the fact that this bacterium was once thought to cause influenza. But it does not. Influenza is caused by a virus. In nature, *H. influenzae* exists only in humans. It can be cultured in the laboratory but with difficulty. It grows best on a chocolate agar medium containing red blood cells broken down by heat to make the growth factors readily available.

Some strains of *Haemophilus influenzae* produce a polysaccharide capsule, making them resistant to phagocytosis and lysis by complement. Thus they are able to invade human tissues. According to their capsular antigen, strains of *H. influenzae* are designated as types a through f. Type b is by far the most important. It causes more than 95 percent of the serious infections, including epiglottitis. Capsuleless strains cannot invade tissues, but they can infect mucous membrane–covered surfaces.

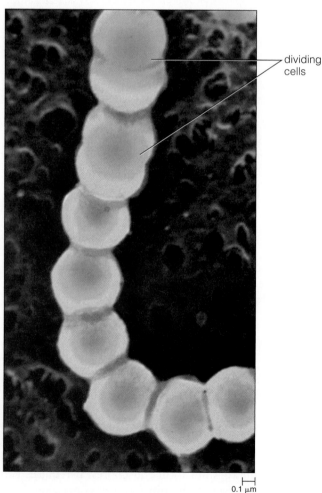

FIGURE 22.3   *Streptococcus pyogenes*. Note the two dividing cells.

Epiglottitis is probably the most dramatic clinical syndrome caused by *H. influenzae* type b. Children between 4 and 6 years are the most vulnerable. Most adults have protective antibodies against the bacterium, and infants are protected by transplacental maternal antibodies. Older drugs, such as ampicillin, are almost useless because of plasmid-borne antibiotic resistance. But chloramphenicol and the new cephalosporins, such as the ceftriaxone, remain effective.

Fortunately, epiglottitis is now nearly eliminated because an effective vaccine is now available. Today babies are routinely immunized with a conjugate vaccine called Hib (for *Haemophilus influenzae* type b), which they receive in a series of injections beginning at age 2 months. Side effects are minimal. Pediatric wards are now empty of patients with epiglottitis (and other *Haemophilus meningitis*–caused diseases as well; Chapter 25).

Streptococcal Pharyngitis. *Streptococcus pyogenes* (*pyogenes* means "pus forming") is a common cause of pharyngitis (**strep throat**), as well as other clinical syndromes (Chapter 24 and 26). This chain-forming, lactic acid bacterium (Bergey's Firmicutes, Chapter 11) is also known as group A beta-hemolytic streptococcus (**Figure 22.3**). Group A refers to the type of polysaccharide antigen that is present on their cell surface (types A through S exist), and beta-hemolytic refers to its effect on red blood cells (see Figure 3.21). Beta-hemolytic stains lyse red blood cells, so on blood agar medium, their colonies are surrounded by a clear halo. Bacteria that are beta-hemolytic and are also sensitive to the antibiotic bacitracin are likely to be *Streptococcus pyogenes*. Such a tentative identification can be confirmed serologically by demonstrating the presence of group A polysaccharide antigens.

Humans are the only natural reservoir for group A streptococci. Although some healthy people are asymptomatic carriers of the pathogen, most people with *S. pyogenes* suffer significant symptoms.

The Clinical Syndrome. Strep throat is usually transmitted from person to person by respiratory droplets. But contaminated food, particularly unpasteurized milk, can also spread the disease. Strep throat is common in school-age children. Typically it causes severe sore throat, fever, chills, and a headache. Usually the pharynx is severely inflamed. A whitish exudate covers the tonsils, and the lymph nodes in the neck are swollen and tender. The symptoms of strep throat resemble those of infectious mononucleosis (caused by Epstein-Barr virus), as well as infections caused by many other microorganisms. So streptococcal pharyngitis can be definitively diagnosed only by identifying *S. pyogenes* by means of laboratory tests.

It may take up to two days for a throat culture to reveal bacterial growth, but latex agglutination kits (Chapter 19) that detect streptococcal antigens in a throat swab can provide a preliminary diagnosis within minutes, enabling the clinician to make an immediate decision about antibiotic treatment.

Complications and Changing Virulence. Streptococcal pharyngitis is usually a self-limited illness. Most people recover within a few days without medical treatment, but serious complications can occur. For example, certain strains of *S. pyogenes* produce an erythrogenic (*erythro*, meaning "red"; *genic*, meaning "producing") toxin that kills cells and causes intense inflammation. Infection by such strains develops into **scarlet fever,** a clinically identifiable syndrome of rash and fever (Chapter 21) that at the turn of the century was a life-threatening illness. Fortunately, today's cases are relatively mild (Chapter 26).

Other complications occur when *S. pyogenes* spreads from the throat into the sinuses, the middle ear, the lungs, or even into the bloodstream, where it can cause a life-threatening septicemia. Like scarlet fever, fatal systemic infections are far less common than they were 50 years ago. But some recent deaths, including that of Jim Henson, creator of the Muppets, may be a sign that killer strains of streptococcus are returning. (Henson's illness is called toxic shock–like syndrome [TSLS].)

Researchers theorize that changing streptococcal virulence is a consequence of which of three types of erythrogenic toxins (designated A, B, and C) a particular strain produces. Some strains don't produce any toxin; they do not produce scarlet fever or other toxin-associated complications. Strains that produce type A toxin are the most virulent; those that produce types B and C are less likely to cause lethal complications. Apparently, strains that produce type A toxin were common early in the twentieth century and became relatively uncommon by mid-century. It's possible that these strains are on the rise again.

Life-threatening complications can occur after a streptococcal infection is over. These postinfectious complications are immunologically mediated (Chapter 18). They include rheumatic fever and acute poststreptococcal glomerulonephritis.

Rheumatic fever causes inflammation in many organs of the body, including the joints, skin, and brain, as well as life-threatening damage to heart valves. Rheumatic fever is the leading cause of heart disease among children in developing countries. More than 6 million children are affected in India alone. But rheumatic fever can be prevented if strep throat is treated with penicillin within the first 10 days of onset of the illness. Patients who do develop rheumatic fever are at particular risk for a second streptococcal infection, which can reactivate the disease and cause fatal heart damage. To prevent this potential disaster, rheumatic fever patients should receive monthly injections of benzathine penicillin.

Acute poststreptococcal glomerulonephritis causes sudden kidney failure. The urine becomes scant and dark colored. Body parts swell from retained fluid, and blood pressure rises alarmingly. This life-threatening condition often requires hospitalization and intensive care. Fortunately, effects of poststreptococcal glomerulonephritis are temporary. Children who survive the initial kidney shutdown do not suffer lasting damage. Poststreptococcal glomerulonephritis is a misdirected immune response triggered by the infection. In fact, other bacteria and viruses trigger similar responses, sometimes even during the period of acute infection. For this reason the disease is more appropriately called **infectious glomerulonephritis.**

Because strep throat can lead to such serious complications, any child with fever and sore throat should be examined by means of a throat swab. If the resulting culture reveals the presence of *S. pyogenes*, an antibiotic should be given. Such treatment—either with penicillin or erythromycin for penicillin-allergic patients—can prevent rheumatic fever but not infectious glomerulonephritis.

Diphtheria. Diphtheria is usually a potentially lethal pharyngitis, caused by the Gram-positive bacterium *Corynebacterium diphtheriae* (Bergey's Actinobacteria, Chapter 11). Groups of its irregularly shaped rod-shaped cells sometimes look like small stacks of coins or clusters of Chinese characters (**Figure 22.4a**). *C. diphtheriae* can be cultivated aerobically in the laboratory on a special medium called Loeffler's medium (named for Frederick Loeffler, who proved in 1884 that *C. diphtheriae* causes the diphtheria). This medium does not support the growth of streptococci, which, on most media, overwhelm the slower-growing *C. diphtheriae*. It can also be grown on media containing cysteine and potassium tellurite. On these media, *C. diphtheriae* colonies are dark gray to black.

Virulent strains of *C. diphtheriae* produce **diphtheria toxin,** which causes almost all the symptoms of diphtheria. Curiously, the gene that encodes diphtheria toxin (called *tox*) isn't carried by *C. diphtheriae*. Instead it is carried by a temperate bacteriophage. Only strains of *C. diphtheriae* infected by this phage produce toxin and cause diphtheria. Presence of the toxin can be identified by an immunodiffusion assay (Chapter 19).

Diphtheria toxin is a typical AB toxin (Chapter 15). Fragment B binds to receptors on human cells and allows the toxin to enter. Inside the cell, fragment A interferes with protein synthesis by inactivating one specific protein necessary for polypeptide elongation. This target protein is present only in eukaryotic cells.

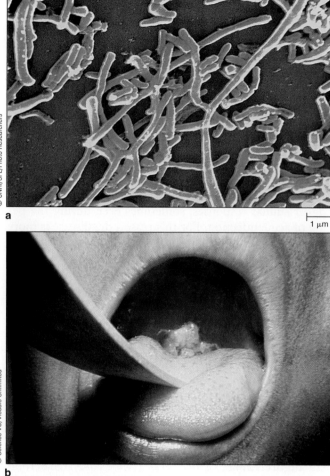

© CNRI/SPL/Photo Researchers

a

⊢ 1 µm ⊣

© Science VU/Visuals Unlimited

b

FIGURE 22.4 Diphtheria. (a) Cells of the causative bacterium, *Corynebacterium diphtheria*. (b) A patient with a typical diphtherial pseudomembrane on the tonsils.

Diphtheria antitoxin (antibodies against diphtheria toxin) can prevent diphtheria. They bind to the toxin and thereby prevent it from entering susceptible cells. Antitoxin is produced during an infection. Therefore some people recover from diphtheria without treatment, but the risk is great. Patients should be treated with antitoxin produced artificially in horses. Serum sickness is a potential but acceptable risk because the disease is life threatening (Chapter 18). If antitoxin is administered before a significant amount of toxin has entered host cells, recovery is almost certain. But with every day that antitoxin therapy is delayed, risk of death increases. Erythromycin will stop further production of diphtheria toxin and prevent transmission of the disease to others. But it does not neutralize the toxin already formed.

The Clinical Syndrome. Human beings are the only natural hosts and reservoir for *Corynebacterium diphtheriae*. It's present in the throats of healthy carriers and can be trans-

mitted to others by respiratory droplets. It's noninvasive. It infects only the mucous membranes of the pharynx. The toxin it produces there does all the damage. The throat swells and becomes covered by a tough, grayish **pseudomembrane** composed of dead human cells and microorganisms, a clinical finding called **membranous pharyngitis (Figure 22.4b)**. Lymph nodes in the neck swell markedly.

Together the swollen throat tissue and pseudomembrane can obstruct the airway, leading to sudden death by suffocation. Diphtheria toxin can kill in other ways as well. It can enter the bloodstream and damages distant organs such as the heart, kidneys, brain, or nerves. These organs may not be affected until several weeks after the original throat infection. At this advanced stage of the disease little can be done to help the patient.

C. diphtheriae can also infect the skin, causing **cutaneous diphtheria**. This disease usually occurs in the tropics, but it's sometimes seen in cooler climates among people with generally poor health. Cutaneous diphtheria usually begins in a wound already infected by other bacteria. Then the wound develops a gray membrane which fails to heal unless treated with antibiotics. Fortunately, diphtherial skin infections do not usually cause toxin-mediated damage to distant organs.

Epidemiology and Prevention. Diphtheria was once a dreaded disease of children, but today few doctors in the developed world have ever seen a case. Treatment with horse antitoxin became possible in the 1890s, but the disease was not conquered until **diphtheria toxoid** made wide-scale immunization possible in 1923. Before 1923, more than 100,000 children in the United States contracted diphtheria every year, and more than 10,000 of them died. Now the disease is almost unknown in the United States. However, it remains a major public health problem in the parts of the world where immunization is not available. Tens of thousands of children still die from it annually.

Lifesaving diphtheria toxoid is made by treating diphtheria toxin with formaldehyde. This treatment destroys the toxin's disease-causing properties but not its ability to stimulate production of diphtheria antitoxin. In the United States, most infants are immunized with diphtheria toxoid during their first 6 months as part of the diphtheria-tetanus-pertussis (DTP) vaccine. Such widespread immunization has made toxin-producing strains of *C. diphtheriae* so rare that the entire population benefits from herd immunity. Herd immunity against diphtheria probably sets in when more than 75 percent of children are immunized. But immunization with diphtheria toxoid does not confer lifelong immunity. Adults should have a booster dose at least every 10 years, especially if they are traveling to developing countries where diphtheria is still common.

Viral Infections

It is safe to say that every reader of this textbook has had a viral upper respiratory infection at some time—and probably many more than one. Such infections are extraordinarily common. Fortunately, they are much less serious than upper respiratory bacterial infections. Nevertheless, prevention and treatment are major challenges.

The Common Cold. The clinical syndrome of a cold, or upper respiratory infection (URI), is all too familiar to all of us. They include sneezing, **rhinorrhea** (the production of excess nasal mucus), and nasal congestion, sometimes accompanied by a sore throat, fever, and headache. The feeling of general discomfort that accompanies most colds is partly due to interferons we produce to combat the infection (Chapter 16). Typically a cold lasts about a week. Colds are usually considered to be trivial illnesses, but they cause the loss of more school and work days than any other illness. Economically, cold are serious illnesses.

Colds are caused by more than 200 types of viruses. We'll examine a few of them here.

Rhinoviruses. Rhinoviruses are named for their portal of entry, the nose (*rhino* means "nose"). They are the major cause of colds, but they only cause 25 to 50 percent of them. Rhinoviruses are single-stranded RNA viruses belonging to the picornavirus group. They are extremely small—only 25 to 30 nm in diameter—with icosahedral symmetry (Figure 1.7). There are about 100 different serotypes of rhinoviruses, each with a different antigen in its protein capsid, and new types continue to be identified.

Rhinoviruses occur in all populations, especially between October through April in northern temperate climates. Their only reservoir is human beings, and they are usually transmitted by direct hand-to-hand contact or by fomites. Transmission by respiratory droplets (sneezing or coughing) does occur, but it is much less common. As a result, the best way to avoid colds is frequent hand washing and disinfecting fomites.

Type-specific IgA antibodies in the nasal mucosa are produced in response to rhinovirus infections. These antibodies inactivate virus particles before they initiate infection. Therefore people who recover from a cold are protected against that particular type of rhinovirus for about 18 months. But complete immunity against rhinoviral infection would require antibodies against every rhinovirus serotype. Because children have had the least opportunity to develop protective antibodies, they are the most susceptible to rhinovirus infection. They typically get several colds a year.

As we've all heard, there's no cure for the common cold. Antibiotics are useless. Nonprescription remedies alleviate some of the symptoms, but they don't shorten the underlying illness. And no prevention or cure for the common cold is on the horizon. The large number of rhinoviral serotypes and the relatively short duration of natural immunity make vaccination impractical. Hand washing and using antiviral chemicals to decontaminate fomites help, but they don't have a major effect on frequency of colds. Interferons, which are now available through recombinant DNA technology, seemed promising because they are natural antiviral agents. Interferon-alpha administered intranasally does prevent rhinoviral infections in nearly 80 percent of household contacts after one family member comes down with a cold. Unfortunately, however, spread of colds caused by other viruses is not stopped by interferon. At best, this expensive treatment prevents only about 40 percent of colds.

Other Causes of the Common Cold. Coronaviruses also cause the common cold. These medium-size RNA viruses (about 100 nm in diameter) are named for the prominent glycoprotein spikes on their outer surface, which resemble the solar corona (**Figure 22.5**). Coronaviruses are extremely difficult

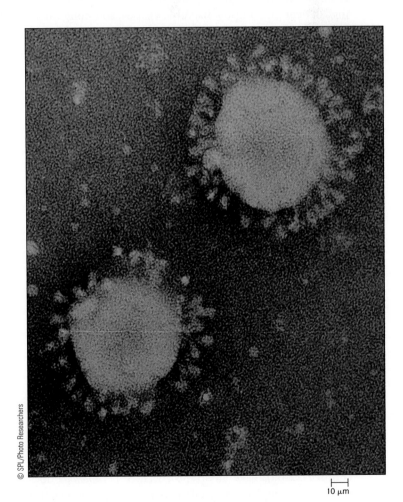

© SPL/Photo Researchers

10 μm

FIGURE 22.5 Coronaviruses. Note the glycoprotein spikes on their outer surfaces, which resemble the sun's corona.

to isolate in cell culture, but serological studies indicate that they probably cause 10 to 15 percent of colds in adults. They also cause pneumonia and intestinal infections.

The list of other microorganisms that can cause the common cold is quite long. It includes coxsackieviruses, echoviruses, adenoviruses, myxoviruses, and even some bacteria such as *Mycoplasma pneumoniae* and *Coxiella burnetii*. These microorganisms, which usually cause other clinical syndromes, are discussed later in this chapter. No microorganism can be isolated from almost half of cold-sufferers, so it's likely that more cold-causing microorganisms remain to be identified.

LOWER RESPIRATORY INFECTIONS

Unlike the upper respiratory tract, the lower respiratory tract has numerous highly branched airways. Infections can cause swelling that obstructs these airways, but breathing isn't cut off abruptly. Bacteria, viruses, and fungi can all infect the lower respiratory tract. We'll discuss, in turn, the representatives of these three groups that infect the lower respiratory tract and the diseases they cause.

Bacterial Infections

When bacteria infect the lower respiratory tract, they usually cause pneumonias. Their symptoms and severity vary widely.

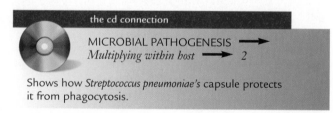

the cd connection

MICROBIAL PATHOGENESIS ➡
Multiplying within host ➡ 2

Shows how *Streptococcus pneumoniae's* capsule protects it from phagocytosis.

Pneumococcal Pneumonia. Pneumococcal pneumonia is a serious, life-threatening disease caused by *Streptococcus pneumoniae*, commonly called **the pneumococcus.** Untreated, about 30 percent of patients are killed by this infection

The Organism. Streptococcus pneumoniae is Gram-positive (Bergey's Firmicutes, Chapter 11). Its cells are nearly round (with slightly pointed ends), and they usually occur in pairs. They are described as being **lancet-shaped diplococci** (**Figure 22.6a**). Like all streptococci, *S. pneumoniae* does not produce the enzyme catalase, but unlike the others, it is very sensitive to a chemical called optochin. Its cells lyse

when exposed to a disc impregnated with this chemical (**Figure 22.6b**). Colonies of *S. pneumoniae* grown on a blood agar medium are surrounded the green halo characteristic of alpha-hemolysis.

Humans are the sole reservoir for *S. pneumoniae*. Still, it's rather common. About 10 percent of healthy adults harbor it in their throats. Not all these strains are virulent. Only those strains with a polysaccharide capsule can cause disease because the capsule protects these strains from phagocytosis. Their thick capsule, which is revealed by negative staining (Chapter 3) as a bright halo (**Figure 22.6c**), gives their colonies a characteristic shiny, mucoid appearance.

The polysaccharides that compose these capsules vary from one strain of bacteria to another in chemical composition and antigenicity. So far, more than 80 pneumococcal serotypes have been isolated and identified by number. Serotypes are easy to identify because type-specific antibody causes their capsule to swell, a response called the **quellung reaction** (**Figure 22.6d**).

The Clinical Syndrome. Now let's refer back to the case history at the beginning of this chapter. Recall that the emergency room physician immediately suspected pneumococcal pneumonia, even though viruses are the most common cause of pneumonia. This is because she noted that J. G.'s signs and symptoms were typical of pneumococcal pneumonia: He had a severe illness with high fever and chest pain; his x-ray showed that an entire lobe of his lung was involved. Also, pneumococcal pneumonia is by far the most common pneumonia serious enough to require hospitalization.

The diagnosis was confirmed when *S. pneumoniae* was isolated from a sample of J. G.'s blood, but this is usually not possible. In most patients, even those who are critically ill, pneumococci are not present in blood; they remain exclusively in the lungs, where they are difficult to recover. Coughed up sputum may contain some pneumococci, but they are usually heavily contaminated with normal mouth microorganisms. Pneumococci can be recovered by **aspirating** (inserting a needle to draw out fluid) the lungs or trachea, but such procedures are painful and potentially dangerous, so they are seldom attempted. As a result, microbiological diagnosis of pneumococcal pneumonia is rarely possible.

J. G. probably was infected when he inhaled respiratory droplets that had been exhaled by someone nearby. That's the way the disease is usually transmitted. These infected droplets probably came to rest on the respiratory epithelial surfaces that are protected by mucociliary defenses. If so, in J. G's case they failed. They may have been weakened by his recent cold. Less likely, the pneumococci drifted all the way to the lower respiratory tract. There they would have to avoid phagocytosis by alveolar macrophages and inactivation by secretory IgA to cause pneumonia.

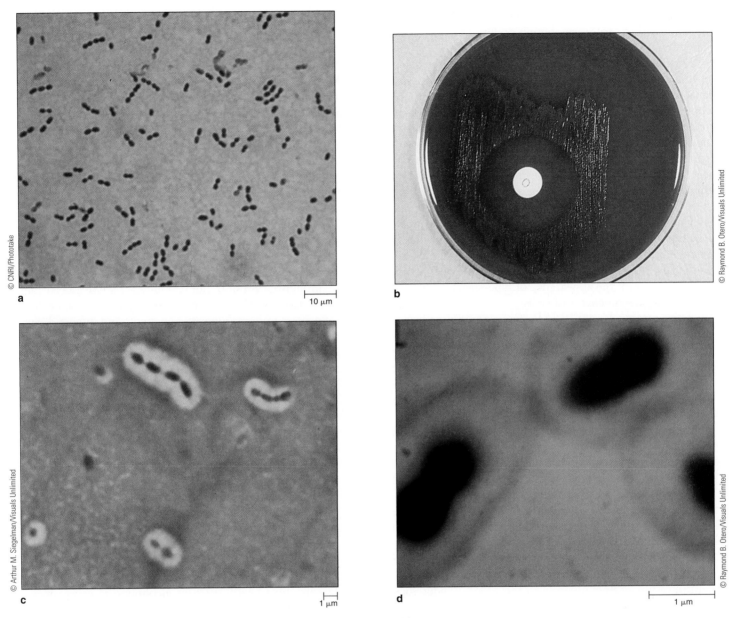

FIGURE 22.6 *Streptococcus pneumoniae.* (a) Stained micrograph showing typical lancet-shaped diplococci. (b) Sensitivity to optochin. Note how colonies fail to grow on an agar plate near the optochin-impregnated disc. (c) Negative staining revealing the capsule as a clear halo around cells. (d) Pneumococcal capsules swollen by exposure to type-specific antibodies (the quellung reaction).

Having evaded J. G.'s defenses, the pneumococci multiplied in his lung and triggered their typical intense inflammatory response (Chapter 16): Leukocyte production was stimulated (as reflected by J. G.'s blood count); small blood vessels in the lungs leaked, filling nearby alveoli with fluid, blood cells, and serum proteins.

This inflammatory response accounts for almost all the clinical findings of pneumococcal pneumonia. The af-fected region of the lung becomes **consolidated** (fluid-filled), giving the impression of a solid organ when tapped, listened to, or penetrated by x-rays. Such consolidation halts normal gas exchange, causing the characteristic labored breathing. Ironically, the intense inflammatory response rarely stops the infection. The huge numbers of phagocytes it activates are relatively ineffective against pneumococcus because of its thick polysaccharide capsule.

The pneumococci continue to multiply. They spread through the lung to nearby lymph nodes. They may even enter the bloodstream, as they did in J. G.'s case.

Fortunately, J. G.'s pneumococcal pneumonia was successfully treated. If untreated (the rule before antibiotics), it can become a dreadful disease. Those who succumb suffer from unrelenting fever and progressively worsening respiratory problems. Breathing becomes so labored they struggle for each breath. Eventually, for reasons not fully understood, they suffer a sudden decrease in blood pressure, sometimes accompanied by heart failure. Death can occur within days, but patients are often sick for a week or more before they die.

The majority of untreated patients do survive because their adaptive immune defenses save them. Such patients usually experience a **"crisis"** 6 to 10 days into their illness. They develop a drenching sweat and suddenly their fever disappears, indicating the beginning of recovery. The crisis coincides with the appearance of type-specific antibody against the pneumococcal capsule.

This anticapsular antibody forms an opsonizing bridge between bacterium and phagocyte that greatly facilitates phagocytosis (Chapter 16). It also activates complement through the classical pathway, creating more opsonizing bridges. Then the phagocytes can eliminate the pneumococci. At the same time, macrophages migrate to the consolidated lung and clear away the debris. Eventually, healing is complete. But that's not the case with other types of pneumonia: They can cause permanent lung damage.

The spleen also plays a vital role in recovery from pneumococcal pneumonia. It filters the pneumococci out of the blood and presents them to splenic macrophages, which process pneumococcal antigen and present it to the millions of lymphocytes that reside in the spleen. Then they start to make type-specific antibody. So people without a normal spleen are highly vulnerable to **pneumococcal sepsis,** a life-threatening condition caused by multiplication of pneumococci in the blood. Under such conditions, pneumococci can attain concentrations as high as a million cells per milliliter of blood.

Patients who survive a pneumococcal infection are permanently immune to that particular pneumococcal serotype. But, because there are 80 pneumococcal serotypes, repeated pneumococcal infections are common.

Treatment and Prevention. Sixty years ago, before antibiotics, a severely ill patient such as J. G. with pneumococci in his or her blood almost surely would have died. Today less than 5 percent of patients who receive timely and appropriate antibiotic treatment die. Most patients who succumb are weakened by advanced age, serious illness, or immune deficiency.

Penicillin has been our reliable protection against pneumococci since the 1940s. In the 1970s, however, a few penicillin-resistant strains of pneumococci appeared in patients who had recently been treated with penicillin or had become infected in a hospital. Since then, reports of penicillin-resistant pneumococci have become more frequent. Many of these strains are highly virulent and resistant to multiple antibiotics, including the usually reliable fallback antibiotic vancomycin. Most infections by these organisms have been fatal. Such cases remain rare, but the trend is alarming.

The other protection against pneumococcal pneumonia is immunization. It's already recommended that highly susceptible people (those over age 65 or of those with a chronic illness or immune deficiency) be immunized. The vaccine in use for this purpose (Pneumovax) contains polysaccharide antigens from 23 different pneumococcal stereotypes. Most adults who receive the vaccine are protected, and side effects are negligible (redness and pain at the immunization site). But pneumococcal polysaccharides are relatively weak immunizing agents, so this vaccine does not protect immunosuppressed patients and infants. However, a new highly effective conjugate vaccine (Prevnar) is now being used for infants. It might soon be routinely administered to all children, as DTP now is (Chapter 17).

Other Acute Bacterial Pneumonias. Approximately 90 percent of acute bacterial pneumonias are caused by the pneumococcus. The other 10 percent, which present an identical clinical syndrome, are caused by *Haemophilus influenzae, Klebsiella* spp., *Staphylococcus aureus, Streptococcus pyogenes, Escherichia coli* (Chapter 23), or *Proteus* spp. (Chapter 24). The prognosis for these pneumonias, even with antibiotic treatment, is usually worse than for pneumococcal pneumonia: Permanent lung damage often results.

Klebsiella spp. account for less than 5 percent of bacterial pneumonias, but many strains are resistant to multiple antibiotics. *Klebsiella* pneumonia is most common in people with poor health, such as alcoholics. It causes widespread lung destruction. Even with optimal medical treatment, most patients die.

Staphylococcus aureus causes a small percentage of bacterial pneumonias. This infection is highly damaging to the lungs, often forming abscesses or holes where lung tissue is completely destroyed. It is most common in weakened hosts—the very young, the very old, those recovering from acute infections such as influenza, and those with chronic illnesses. Mortality is high. *S. aureus* is a much more common cause of other clinical syndromes (Chapters 23 and 26).

Mycoplasmal Pneumonia. Mycoplasmal pneumonia usually comes on gradually. It's a mild disease, with few definite signs and symptoms. Chest x-rays reveal a diffuse patchy pattern rather than consolidation of a lobe. It's rarely fatal, though unusually severe cases do occur.

Because its clinical pattern is so different from pneumococcal pneumonia, it is called **atypical** or **primary atypical pneumonia** (PAP) (or sometimes walking pneumonia).

Mycoplasma pneumoniae is the causative agent (**Figure 22.7**). It shares the properties of other mycoplasmas, including lack of a cell wall and being difficult to culture (Bergey's Firmicutes, Chapter 11). It usually grows in the trachea and is transmitted in respiratory droplets. More than 90 percent of the people it infects fail to develop pneumonia. It can cause as many as 60 percent or as few as 10 percent of community-acquired pneumonias, most of which occur in school-age children and teenagers. They suffer a headache, low-grade fever, and persistent dry cough for as long as 3 weeks. The extent of the illness can be shortened by antibiotic therapy. Because they are wall-less organisms, mycoplasmas are not sensitive to the penicillins or cephalosporins. Tetracycline is the treatment of choice, except for pregnant women and young children. Erythromycin is best for them.

Other infectious agents, including viruses, chlamydiae, and rickettsiae, cause similar atypical pneumonias.

Chlamydial Pneumonia (Ornithosis or Psittacosis).

People with chlamydial pneumonia develop fever, headache, chills, and cough. If the illness progresses, they develop a persistent high fever, mental confusion, and marked shortness of breath. The disease can be life threatening.

Chlamydiae are extremely simple bacteria with a complex life cycle (Bergey's Chlamydiae, Chapter 11). Because they can't generate their own adenosine triphosphate (ATP), they are obligate intracellular parasites. They can be propagated in the yolk sacs of chick embryos or in cultures of human cells, but culturing is slow and expensive. Therefore *Chlamydia psittaci* infections are usually diag-

nosed serologically. The patient's blood is tested to see if antibody titer has risen during the illness.

C. psittaci commonly infects all types of birds—wild birds, farm birds, and pets. Infected birds usually do not appear to be ill. But when stressed, for example by shipment, the disease spreads. Then the chlamydiae invade various organs and are excreted in the bird's droppings. If a human inhales dust from these droppings, he or she can develop pneumonia. Occasionally the disease is transmitted to other people in respiratory droplets. The illness is

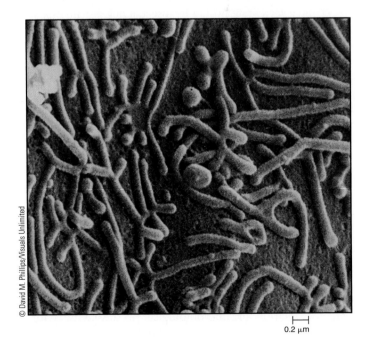

© David M. Phillips/Visuals Unlimited

0.2 μm

FIGURE 22.7 *Mycoplasma pneumoniae.* Scanning electron micrograph showing the irregular shapes of these wall-less bacteria.

also called **ornithosis** (*ornis* means "bird" in Greek) because of its association with birds, or **psittacosis** (*psittakos* means "parrot" in Greek) because it was first described in parrots. Sometimes it is called **parrot fever.**

A history of exposure to sick birds or employment in the poultry industry can be critical clues in diagnosing patients with pneumonia. In fact, the disease is an occupational hazard for bird handlers and workers in poultry slaughterhouses. There is no vaccine, but tetracycline and erythromycin are effective treatments. The disease is uncommon. Fewer than 100 cases a year are reported in the United States. In a bad epidemic, up to 20 percent of patients die.

Q Fever.
Q fever is an atypical pneumonia. Clinically you can't tell it from mycoplasmal pneumonia or ornithosis. Definitive diagnosis depends upon doing specific serological tests, but the reagents for these tests are not readily available. Q fever is rare. It almost never causes death. Occasional epidemics do occur among people who raise cattle and sheep, so a history of exposure to livestock is critical to diagnosing Q fever. The disease was first described in the late 1930s as an ailment of meat-packinghouse workers. The mysterious illness was named Q (for query) fever.

The causative bacterium is *Coxiella burnetii*, a member of the rickettsia family of small intracellular parasites (Bergey's Gammaproteobacteria, Chapter 11). (Other rickettsial diseases are discussed in Chapter 27.) *C. burnetii* was named after an American named Cox and an Australian named Burnet who made significant contributions to the study of this microorganism. Like most rickettsiae, it requires both an insect and a vertebrate host. The insect, a tick, usually transmits it to an animal by biting. Humans can also acquire the disease directly from a tick bite. But humans usually become infected by inhaling the microorganism from dried infected animal placentas, feces, or amniotic fluid. We can also be infected by drinking milk from infected animals. *C. burnetii* can survive the pasteurization that kills tuberculosis organisms.

Human infection with *C. burnetii* is usually asymptomatic, but it may cause Q fever. Tetracycline is used to treat serious infections, but antibiotic treatment is not as effective against Q fever as it is against most rickettsial infections.

Legionellosis.
Legionellosis was first discovered in July 1976. At an American Legion convention in Philadelphia, 182 people became ill with a mysterious form of pneumonia. When the outbreak was over, 29 were dead. No pathogen could be identified. Fearing further fatal epidemics, the public was alarmed. It took 6 months of intensive investigation by microbiologists at the Centers for Disease Control and Prevention (CDC) to isolate a previously unknown bacterium that they named *Legionella pneumophila* (Bergey's Gammaproteobacteria, Chapter 11).

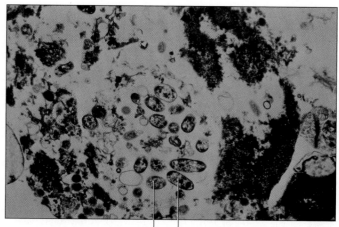

Legionella pneumophila cells

FIGURE 22.8 *Legionella pneumophila.* This section of human lung shows an alveolar macrophage filled with cells of this bacterium.

Why was *L. pneumophila* not discovered until almost 100 years after most other bacterial pathogens? First, the microbiological cause of most cases of pneumonia is not determined. Earlier *Legionella* infections were probably misdiagnosed clinically as other types of pneumonia. If the legionellosis (or legionnaires' disease) epidemic in Philadelphia had not been so spectacular, we might still not have identified the pathogen. Second, *L. pneumophila* is unusually difficult to identify in the laboratory. It stains poorly or not at all by the Gram technique. So special stains, such as silver-impregnation stains or direct immunofluorescence (Chapters 3 and 19), are required to detect it in clinical samples. Also, the organism is very fastidious in its growth requirements. It was finally isolated by infecting embryonated hens' eggs and guinea pigs.

L. pneumophila is a small, aerobic, motile, rod-shaped bacterium. Macrophages phagocytize it, but they can't kill it, so *L. pneumophila* multiplies intracellularly (**Figure 22.8**). Cell-mediated immunity, as well as antibodies, contribute to recovery from legionellosis. Between 1 and 25 percent of community-acquired pneumonias are caused by *L. pneumophila*.

L. pneumophila lives in natural and artificial water supplies. It's tough. It can survive heat, chlorination, and being aerosolized. The probable source of the Philadelphia epidemic, and several others, was *L. pneumophila* growing in water-cooled air-conditioning systems.

Most people infected by *L. pneumophila* experience only minor symptoms. But some develop a life-threatening pneumonia, such as the one that hit the legionnaires. In such cases, onset is sudden. Patients experience weakness,

LARGER FIELD

COURAGE—
AND LEAD WEIGHTS

Researchers working on tuberculosis have traditionally used the drug-sensitive Erdman strain of *Mycobacterium tuberculosis*. But with the possibility of a major outbreak of untreatable TB, the fastest progress is likely to come from studying the resistant strains directly. Some microbiology laboratories, called BL-3 labs, are specially equipped for the study of dangerous pathogens. But even with safeguards, researchers are uneasy dealing with bacteria that may cause untreatable tuberculosis. Ian Orme, a research microbiologist at Colorado State University's BL-3 tuberculosis lab, said in an interview with *Science* magazine, "To tell you the truth, we're fairly nervous about doing experiments on aerosolized multiple-drug-resistant strains." His lab uses an aerosol machine sealed with a 10-inch rubber gasket to prevent leaks. But, said Orme, "I think we're going to put lead weights on top of the gasket."

---

headache, high fever, cough, and shaking chills. X-rays show consolidation of an entire lobe. The elderly, and especially those with another serious illness or a weakened immune system, are most often affected. Smokers and alcoholics are particularly susceptible. *L. pneumophila* is resistant to penicillin and cephalosporin, the drugs usually used to treat severe pneumonia when the causative organism is unknown. Thus the clinician must suspect legionnaires' disease in order to begin effective antimicrobial therapy with erythromycin. Legionellosis can be diagnosed in a number of ways: by antigen detection tests, such as radioimmunoassay or enzyme-linked immunoassays; by direct fluorescent antibody stains of clinical specimens; or by culture in a special medium (buffered charcoal-yeast-extract agar).

Pertussis (Whooping Cough). Pertussis (which means "intensive cough") is a tracheobronchitis of infants and young children. It produces prolonged and uncontrollable fits of coughing (see Chapter 15). It's highly contagious and it can be fatal, especially to infants. By the time pertussis is diagnosed, pertussis toxin has already caused considerable damage, so antibiotic treatment does little to shorten the illness. The only treatment is supportive nursing care. Erythromycin can be administered, but only to prevent transmission to others.

Pertussis is caused *Bordetella pertussis*, a small Gram-negative coccobacillus (Bergey's Betaproteobacteria, Chapter 11). It's difficult to culture in the laboratory because of its highly fastidious growth requirements. It will grow on the complex Bordet-Gengou medium, a potato-glycerol-blood agar named after the microbiologists who in 1906 first identified the microorganism. *B. pertussis* can usually be identified by fluorescent antibody tests.

B. pertussis is a highly evolved pathogen. It produces a protein called **filamentous hemagglutinin** with which it attaches to susceptible cells; an exotoxin called **pertussis toxin,** which causes most of the disease symptoms; and a cytotoxin that kills tracheal cells. Its pathogenesis is discussed in Chapter 15. Humans are its sole reservoir.

Pertussis was once a common childhood illness, but now only a few thousand cases are reported annually in the United States. This dramatic change is due to the DTP vaccine. Most infants were immunized during their first 6 months. They received booster doses in their second year and when they entered school. In spite of its effectiveness, the vaccine was criticized for its side effects. Most children develop a slight fever, as well as pain and redness at the injection site, but a small number of infants suffer more serious complications, including convulsions. Extremely rare (and unconfirmed) cases of brain damage were reported. Because of these side effects, pertussis immunization was suspended or drastically curtailed in Sweden and the United Kingdom, but soon both countries witnessed an alarming increase in pertussis cases.

Now this problem seems to have been solved by a recently developed acellular vaccine. It's combination with diphtheria and tetanus vaccine, called DTaP, is given to infants at 2, 4, 6, and 15 months. Protection is excellent and side effects have been reduced dramatically. But in spite of widespread immunization, scattered cases of pertussis continue to occur each year in the United States.

Tuberculosis. Tuberculosis (TB) is—and has been throughout history—one of the most widely distributed diseases afflicting humans. It remains the leading killer among all infectious diseases, with an estimated 10 million new cases and 3 million deaths annually. Worldwide, TB accounts for about a quarter of avoidable adult deaths. Throughout the twentieth century, the incidence of tuberculosis declined continuously in the United States. But in the last few years the decline has stopped, and its incidence is now beginning to rise.

Robert Koch identified *Mycobacterium tuberculosis*, sometimes called the tubercle bacillus, as the cause tuberculosis (see Koch's Postulates, Chapter 15). This a rod-shaped obligate aerobe has an unusual waxy cell wall (Bergey's Actinobacteria, Chapter 11) that affects many of its properties, including its slow growth and unusual staining properties. Special staining techniques, called acid-fast stains, are used to detect its cells in clinical specimens (Chapter 3). Its waxy wall also allows it to survive prolonged drying: *M. tuberculosis* cells that enter the air when a tuberculosis patient coughs can remain infectious for up to 8 months. The mycolic acid components of the waxy wall also add to the microorganism's disease-causing potential. They protect *M. tuberculosis* cells against the potentially lytic enzymes and oxidants within the phagocytes that engulf them. One of the mycolic acids, called **cord factor,** that causes *M. tuberculosis* cells to clump in cordlike masses (see Figure 11.14) is essential for virulence. Strains the lack it do not cause disease. *M. tuberculosis* grows best where oxygen concentrations are high, so it tends to infect the upper lobes of the lungs. But even under the most favorable conditions, it grows very slowly. A typical doubling time is approximately 20 hours (compared with 20 minutes for *Escherichia coli*). In practical terms, the slow growth means laboratory identification can take up to 6 weeks. Disease usually develops slowly as well, but occasionally it can progress extremely rapidly. It does in persons with AIDS.

The Clinical Syndrome. We're discussing tuberculosis in this chapter because it is usually transmitted by respiratory droplets and the lungs are most often affected. But *M. tuberculosis* can infect almost every part of the body: the lymphatic system, the genitourinary system, the skeletal system, and the nervous system. So it produces many different clinical syndromes.

Primary infection usually begins when someone inhales *M. tuberculosis* from an infected person's respiratory secretions (**Figure 22.9**). The microorganisms that reach the lungs are phagocytized by alveolar macrophages, but they are not killed. They continue to multiply slowly inside and outside host cells. Some of them enter the bloodstream and lymphatic circulation and are spread throughout the body.

Then, several weeks later, a critical change occurs. T lymphocytes, stimulated by mycobacterial antigens, begin to proliferate and secrete lymphokines. These activate macrophages that trigger cellular immunity against tuberculosis and help control the infection. The immune re-

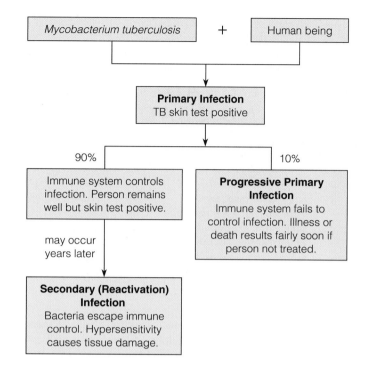

FIGURE 22.9   Possible outcomes of tuberculosis infection.

sponse also stimulates tuberculin hypersensitivity, a form of type IV delayed hypersensitivity (Chapter 18), which is the basis for the tuberculin skin test and also the cause of most of the tissue damage associated with severe cases.

More than 90 percent of the people who contract a primary tuberculosis infection do not become ill. The activated macrophages engulf the bacteria and isolate them within nodules called **tubercles** or **granulomas.** A tubercle is an inflammatory lesion. It contains a few phagocytized mycobacteria, but it is composed mostly of activated macrophages and lymphocytes. Often host cells in the center of a tubercle die. The dead tissue looks dry and crumbly, like cheese, a phenomenon called **caseation necrosis** (cheeselike death). A few bacteria survive in the caseous center and persist for years.

If immune defenses control the primary infection, the victim does not become ill, but the walled-off tubercle bacilli persist indefinitely. In time, tubercles become calcified. These structures are called **Ghon complexes (Figure 22.10a).** Months or years later, *M. tuberculosis* cells can escape from old tubercles. Macrophages, activated during the primary infection, prevent the escaped bacteria from entering the blood and lymphatic circulation, but the hypersensitivity they trigger can destroy the lungs (**Figure 22.10b**). These patients are said to have **reactivation tuberculosis.** They suffer fever, fatigue, weight loss, and a chronic cough—a clinical syndrome once known as **consumption.** Symptoms may develop over many years. But without treatment, the patient becomes debilitated and dies.

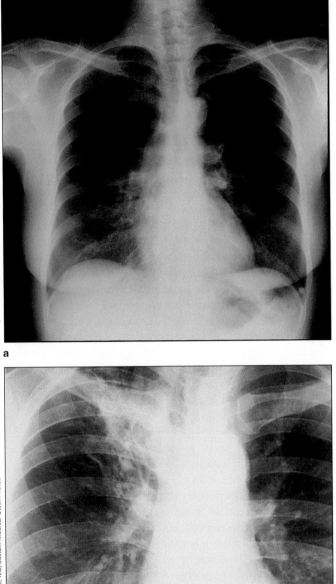

a

b

FIGURE 22.10   X-rays of patients with tuberculosis. (a) This patient with a controlled infection has calcified tubercles (Ghon complexes). (b) This patient with reactivation tuberculosis has areas of tissue destruction, in the upper lobes of the lungs.

Malnutrition, old age, and infections such as AIDS make a person particularly susceptible to reactivation tuberculosis.

In about 10 percent of cases—usually young children or adults with a weakened immune system—cellular immunity fails to control primary tuberculosis. Then the in-fection turns into **progressive primary tuberculosis.** It often spreads through the lungs or into the nearby pleura, destroying lung tissue and sometimes creating large bacteria-containing cavities. Occasionally, primary tuberculosis spreads beyond the respiratory system, to the bones and kidneys, for example. Or it can spread to the membranes in the central nervous system, causing **tuberculous meningitis** (Chapter 25). Sometimes, particularly in children, it spreads throughout the blood and lymph and establishes many tiny infected areas scattered through the body. This is a life-threatening form of the disease called **miliary tuberculosis** because the infection sites look like seeds of millet (grain).

Without treatment, patients with progressive primary tuberculosis often die. Death from meningitis or miliary tuberculosis occurs within weeks to months. But if the disease stays within the lungs, patients usually become ill more gradually. Eventually they suffer from fever and increasing fatigue. Those with large pulmonary cavities cough, often producing sputum tinged with blood. They spew enormous numbers of infectious *M. tuberculosis* cells into the air.

Prevention and Treatment. A tuberculosis vaccine is available, but it is not very reliable. This live vaccine, called bacille Calmette-Guerin (BCG), is made from an attenuated strain of the closely related bacterium ***Mycobacterium bovis.*** BCG is used widely around the world, and some clinical studies show that it stimulates immunity in 80 percent of recipients. But others show no benefit at all. There's fairly good evidence, though, that it reliably protects children from miliary TB and TB meningitis. Immunization with BCG makes sense in places where enormous numbers of people die from tuberculosis each year. The World Health Organization recommends that it be given to all children in developing countries.

BCG vaccination is not routine in the United States, largely because it interferes with **tuberculin skin testing** that public health authorities rely on. The test involves injecting a purified mycobacterial antigen into the skin. Anyone previously infected with tuberculosis mounts a delayed hypersensitivity reaction. Within 48 hours, a firm, raised area appears at the injection site. About 5 to 10 percent of people in the United States show such a positive test. Anyone who does should be followed up with a chest x-ray or culture to see if the disease is active.

Skin-test screening is an effective program of tuberculosis control because the disease can be cured with antimycobacterial drugs. Individuals who test positive but do not have active tuberculosis are treated with a single drug, usually isoniazid, for 1 year (Chapter 21). This is prophylactic treatment because it prevents reactivation infection. People with active infections receive multiple drug therapy for as long as 2 years. This treatment cures the infection and protects the community because it keeps people with active disease from spreading it to others.

TB in the Twenty-First Century. Until recently, tuberculosis seemed to be a vanishing problem in the United States. Effective screening and treatment were readily available, and every year the number of new cases dropped. Public health officials projected that tuberculosis would be eliminated from the United States by 2010. So research funding was cut back, the government stopped monitoring the appearance of drug-resistant strains, and pharmaceutical companies curtailed research into antimycobacterial drugs.

But tuberculosis case rates began to increase by 3 to 6 percent each year in the late 1980s. This resurgence reached its peak in 1992, when 23,636 case were reported. Since that time the incidence has continued to decline. By 1997 it had decline by 26 percent.

A continuing concern, however, is the appearance of new strains of *M. tuberculosis* that are resistant to most currently available drugs. New effective anti-TB drugs are needed.

The final conquest of tuberculosis may depend more than anything else upon microbiological research, which is proceeding at a rapid rate. Now the complete genome sequence of *M. tuberculosis* has been determined. The structure of this organism's unique cell wall is becoming understood (see Sharper Focus: A Matched Team, in Chapter 4), and the mechanism by which some strains of *M. tuberculosis* become resistant to isoniazid has been discovered (see Sharper Focus: Fighting TB—New Hope, in Chapter 7). There is reason to hope that incidence of this dread disease will continue to decline, at least in the developed countries.

Bovine Tuberculosis.

Mycobacterium tuberculosis is the major cause of tuberculous infection, but it's not the only one. A closely related bacterium, *Mycobacterium bovis*, which usually infects cows, causes a disease called **bovine tuberculosis** in people. Humans usually get it by drinking unpasteurized milk from infected cows. So, *M. bovis* usually enters the human body through the gastrointestinal system. Then it infects lymph nodes and bone. Bovine tuberculosis is less of a problem now than it used to be because it's been controlled by pasteurizing milk and monitoring dairy herds.

A group of bacteria called atypical mycobacteria also cause a form of tuberculosis. These include *M. avium*, *M. intracellulare*, and *M. kansasii*. They inhabit soil, water, or other animals, and they're not transmitted from person to person. They cause several clinical syndromes, including lymph node infections in children, chronic ulcers of the skin, and a disease indistinguishable from pulmonary tuberculosis. These infections are rare and less severe than tuberculosis, but they can progress rapidly and be life threatening in AIDS patients.

Viral Infections

Viral infections of the lower respiratory tract usually cause relatively mild disease. Some, however, some can be life threatening.

Influenza (Flu).

Almost everyone on Earth has been infected by the influenza virus, and closely related viruses infect many animal species. Influenza got its name in the 1400s when people in Italy attributed an outbreak to the "influenza" (influence) of the stars. Influenza can kill. Thousands of people die of it each year. Influenza is caused by influenza virus. (Properties of the virus are discussed in Chapter 13; see Figure 13.17.)

Epidemiology. Infection with the influenza virus stimulates the production of antibodies that prevent reinfection, yet people suffer from influenza again and again. That's because the HA and NA antigens on the virus surface change frequently, creating new strains that are not inactivated by older antibodies. Strains of influenza are identified by the HA and NA antigens using abbreviations and numbers. For example, the strain that caused most cases of influenza A in the winter of 1989–90 was H_3N_2. New strains are also named for the place they are first identified, for example, Hong Kong or Beijing.

When influenza antigens change slightly by antigenic drift (see Influenza, Chapter 13), previously infected people have limited protection against the new strain. But radical changes that occur by antigenic shift make everyone susceptible. Such changes occur about every 10 years. They set the stage for an influenza pandemic.

By far the most disastrous pandemic occurred in 1918, just before the end of World War I. It was probably the most devastating epidemic ever to afflict the human race. At least 20 million people died in only 120 days, including 12.5 million in India and half a million in the United States. In Alaska, more than half the population died. In major U.S. cities, morgues overflowed and coffins became scarce. The 21,000 deaths reported during the week of October 23, 1918, are the highest weekly mortality rate ever recorded in the United States (**Figure 22.11**).

Animal reservoirs are critical to the epidemiology of human influenza. Influenza A is widespread in birds. Although birds can't transmit the virus directly to people, they can exchange it with pigs. So can people. Recombination between human and avian strains can occur in pigs, creating the major antigenic shifts that lead to pandemics. The exceptionally virulent 1918 strain probably arose this way. That's why it's sometimes called swine flu.

Even during nonepidemic years, influenza outbreaks are major economic and public health problems. Approximately 20,000 people in the United States, most of them

© Jon Levy/Gamma Press USA

FIGURE 22.11 Hoping to escape death from influenza during the 1918 pandemic, this telephone operator wears a gauze mask.

elderly or in poor health, die of influenza every year. The cost of influenza in absence from work, visits to the doctor, and admissions to the hospital exceeds $5 billion annually in the United States alone.

Clinical Syndromes. Illness from influenza is severe and disabling but usually brief. Symptoms (fever, headache, muscle aches, and cough) appear abruptly. The most common clinical syndrome is a tracheobronchitis in which the ciliated epithelial cells lining both airways are infected and killed by the virus. This weakens the mucociliary defenses of the respiratory tract, making the person vulnerable to secondary bacterial infection. Influenza is hard to diagnose clinically, because other viruses cause identical symptoms. Definitive diagnosis requires culturing the virus or demonstrating viral antigens by immunofluorescent stains. Such diagnoses are not particularly useful clinically because influenza infections are self-limiting, and specific treatment is seldom prescribed. But diagnoses are done for epidemiological purposes to follow the progress of influenza outbreaks that occur each winter.

Influenza can progress to pneumonia. Such a consequence is uncommon but life threatening. Death can occur within hours. During the 1918 pandemic, about 20 percent of the victims developed viral pneumonia. That's why so many young healthy adults died. Today most influenza-associated deaths are actually caused by secondary bacterial pneumonias. These can be prevented by antibiotics. But if a 1918-like, pneumonia-causing strain of influenza were to reappear again, antibiotics would be of little value.

Prevention and Treatment. Vaccination protects about 70 percent of recipients from influenza. But the changing HA and NA antigens of the influenza virus require that a new vaccine be produced each year before the flu season begins. Manufacturers collect information from around the world to design the next year's vaccine. Because the capacity for producing vaccine is limited, influenza immunization is recommended only for high-risk individuals—the elderly, the chronically ill, the institutionalized, and health-care professionals. Usually vaccine side effects are minor. But the 1976 swine flu immunization program (which was recommended for everyone) was associated with a significant number of cases of **Guillain-Barré syndrome** (temporary paralysis). **Amantadine** (an antiviral drug) is about as effective as immunization in preventing influenza A, although it too has side effects and must be taken throughout the flu season to be effective. It can speed recovery if administered during the first 2 days of illness (Chapter 21). Two new antiviral agents, **zanamivir** and **oseltamivir,** which act by inhibiting neuraminidase, are now available to treat influenza. They must be administered early in the infection, and their benefit is limited.

Croup. Croup is a childhood laryngotracheobronchitis. It narrows the airway at and below the vocal cords. Severe cases resemble epiglottitis, but cases are milder and the onset is more gradual. The typical croup patient is a toddler who becomes hoarse and develops a loud, barking cough. Symptoms are worst at night, when mucus accumulates in the swollen airways.

Croup is caused by parainfluenza viruses, which, along with the viruses that cause measles and mumps, belong to the paramyxovirus family. These viruses contain a single piece of single-stranded RNA surrounded by a helical capsid and a lipid envelope. Spike proteins protruding from the lipid layer allow the virus to adhere to human cells. Parainfluenza viruses are transmitted by respiratory droplets. Some of these viruses cause the common cold.

Croup epidemics occur during the fall. There are no effective antiviral agents or vaccines against parainfluenza virus, so treatment is limited to supportive nursing care and agents that minimize the swelling of affected tissues. Moisture in the air thins the secretions, making them easier to clear with a cough. Thus home treatment consists of a humidifier in the child's room or a brief visit to a steamy bathroom.

Natural infection does not produce immunity, so children may suffer recurrent episodes of croup.

CASE HISTORY

Influenza and Asthma: A Dangerous Combination

E. E. was a 23-year-old man who came to his doctor's office in mid-January because he had problems breathing. He had felt well until 3 days earlier, when suddenly at 1 o'clock in the afternoon he developed a high fever, a shaking chill, and a headache. All the muscles of his body ached. Although he seldom missed work due to illness, he felt unable to continue at his job and went home early. By the time he reached his apartment he had developed a dry cough.

E.E. took medication for his fever and went to bed. The fever and body aches came and went throughout the night. By morning his cough was clearly worse. He called in sick to work. He took some cough medicine he had purchased at the grocery store and continued to rest in bed. At times his fever climbed to 103°F. He didn't feel well enough to eat or drink much. He stayed in bed all day.

That night he awoke about midnight unable to breathe. He sat up in bed struggling to breathe more deeply.

He realized he was having the worst asthma attack he had experienced in many years. He searched his medicine cabinet for leftover asthma medication but found only an expired, almost-empty inhaler. He considered going to an emergency room but decided to wait to see his doctor in the morning. He spent the rest of the night sitting up in bed, using every muscle in his chest as he tried to get a deep breath.

Early the next morning when E. E. called his doctor for an appointment he couldn't finish a sentence without stopping to catch his breath. The nurse told him to come in immediately. His neighbor gave him a ride. On arrival at the doctor's office the nurse attached a pulse oximeter to E. E.'s finger to determine how well his lungs were delivering oxygen to his bloodstream. She found that his oxygen saturation was only 86 percent. He was in a dangerous state of respiratory distress. E. E. still had a fever of 103°F, and he was slightly dehydrated. He looked frightened and quite ill.

The doctor treated E. E. immediately with an aerosolized medication to open the airways that were narrowed by his asthma. He improved within minutes. But even repeated aerosol treatments failed to stabilize his breathing and improve the oxygen saturation of his blood enough for him to return home. As E. E.'s doctor prepared to admit him to the hospital, he asked if E. E. had come in the previous fall to get a flu shot, a precaution recommended for all patients with a history of asthma. E. E. admitted that he hadn't gotten around to it this year.

E. E. remained in the hospital for 2 days, improving gradually as he received round-the-clock aerosol treatment for his asthma. A nasal swab performed in the doctor's office confirmed the suspected diagnosis of influenza A. He was discharged with medication to use at home in case of another asthma attack and a reminder to come in for a flu shot the following October.

Bronchiolitis. Bronchiolitis is a common syndrome of respiratory infection in infants caused by respiratory syncytial virus (RSV). This paramyxovirus-like parainfluenza gets its name from the **syncytia** (large, abnormal cells with multiple nuclei) that develop in infected respiratory tissue.

Outbreaks of RSV infection occur during the late winter and early spring, just as the flu season is ending. RSV infects all ages (almost everyone has been infected by the age of 4), but most older children and adults suffer only coldlike symptoms. Infants under 12 months suffer from the typical clinical syndrome of bronchiolitis. The typical patient is a wheezy infant who begins to breathe faster and faster over a period of several days. These babies usually can be cared for at home. But they require hospitalization if their lungs are so severely affected that they need extra oxygen or if they are breathing too fast to eat. Young infants may stop breathing. A few older infants also die, but this is rare.

The virus is transmitted principally by hand contact and to a lesser extent by respiratory droplets. It spreads quickly within a family and also in pediatric wards and nurseries. Epidemics in intensive care nurseries, where babies are small and sick, are devastating. Almost every infant will become infected. A few are sure to die. Transmission can be decreased by careful hand washing.

RSV infection is a significant public health problem. Although only 1 percent of infected babies become ill enough to need hospitalization, infections are so common that these patients account for nearly half of all admissions to pediatric units during the peak season. Natural infection does not produce lasting immunity, and attempts to develop a vaccine have been unsuccessful. Treatment consists mainly of supportive care and administering oxygen.

Hantavirus Pulmonary Syndrome. In May 1993 a mysterious respiratory illness appeared among otherwise healthy young people in the Four Corners area of the U.S. Southwest. The initial symptoms—fever, muscle aches, and respiratory distress—were not alarming. But about 70 percent of the victims died within 5 to 6 days, often within hours of taking a dramatic turn for the worse. Death was caused by catastrophic lung failure. Capillaries leaked profusely, and fluid filled the air spaces. By November 1993, 45 cases had been reported in 12 states; 27 people died.

Researchers identified the cause as a new hantavirus strain. They named it the Sin Nombre (without name) virus (SNV) (**Figure 22.12**). It occurs principally in long-tailed deer mice, which shed the virus in their urine, feces, and saliva. People contract the disease by inhaling aerosolized particles of the virus. Finding that a hantavirus caused this respiratory disease was a surprise because hantaviruses had previously been associated with kidney disease. These viruses are a group of arboviruses named for the Hantaan River in North Korea. Westerners first encountered them during the Korean War, when about 3000 soldiers were infected and almost 300 died of a hemorrhagic kidney disease. Other strains, discovered later, also attacked the kidneys.

Virologists do not yet know whether the hantavirus pulmonary syndrome is an emerging disease or one that has gone unrecognized. Every year, thousands of people in the United States die of acute respiratory distress syndrome (ARDS). Maybe some of these cases are caused by hantavirus. The virus has been isolated, and epidemiologists are evaluating how great a public health risk it is.

Fungal Infections

Fungi commonly infect the lower respiratory tract, but they rarely cause serious illness. When they do, infections that begin in the respiratory system spread throughout the body. These infections are called systemic mycoses (mycosis means fungal infection). We'll discuss these serious fungal infections here.

Histoplasmosis. Histoplasmosis is the most common respiratory disease caused by a fungus. It's caused by *Histoplasma capsulatum*, a dimorphic fungus that grows as a yeast above 35°C and produces mycelia below. When inhaled, conidia initiate infection. *H. capsulatum* is found in the soil, particularly where bat or bird droppings are plentiful. Bats are hosts for *H. capsulatum*, but birds are not. Bird droppings, however, stimulate its growth in the soil. People are likely to become infected when they explore caves, clean chicken coops, or enter attics, barns, or other bird-roosting areas.

Histoplasmosis occurs worldwide, but it's much more common in certain areas. In the United States it occurs in the Ohio–Mississippi River valley and is relatively uncommon elsewhere (**Figure 22.13**). Skin tests show that 80 percent

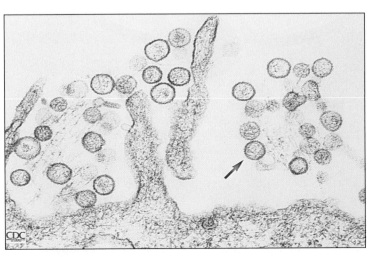

FIGURE 22.12 The Sin Nombre hantavirus (arrow) in a section of a cultured cell.

histoplasmosis

coccidioidomycosis

FIGURE 22.13 Geographical distribution of histoplasmosis and coccidioidomycosis in North America.

of long-term residents of the Ohio–Mississippi River valley have been infected. Elsewhere in the United States the rate is only about 20 percent.

Only about 1 percent of those infected become ill. They develop a flulike illness, and most recover without treatment. However, in people with a weakened immune system (AIDS or cancer patients and people taking immunosuppressive medication) the disease progresses. Some develop a chronic respiratory illness similar to pulmonary tuberculosis. In others, infection spreads throughout the body. Diagnosis rests on serology, or finding the organism in infected tissue. Histoplasmosis is treated with amphotericin B or ketoconazole, but curing immunosuppressed patients with widespread disease is difficult.

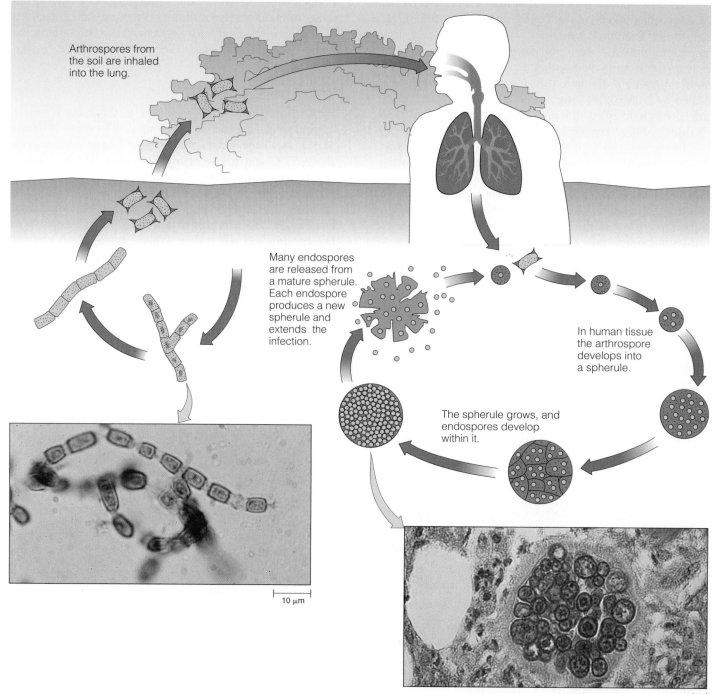

Arthrospores from the soil are inhaled into the lung.

Many endospores are released from a mature spherule. Each endospore produces a new spherule and extends the infection.

In human tissue the arthrospore develops into a spherule.

The spherule grows, and endospores develop within it.

10 μm

10 μm

FIGURE 22.14   Life cycle of *Coccidioides immitis*.

Coccidioidomycosis. Coccidioidomycosis (also called San Joaquin Valley fever) occurs in the San Joaquin Valley of California, the U.S. Southwest, and parts of Mexico and South America that have a semiarid climate and alkaline soils (Figure 22.12). Most people in these areas have been infected at some time during their lives. Nearly all remain well or suffer only a mild respiratory illness, but a small number develop a chronic respiratory illness similar to tuberculosis. A smaller number develop a disseminated disease, which can lead to chronic meningitis (Chapter 25).

Coccidioidomycosis is caused by *Coccidioides immitis*, a dimorphic fungus. It usually grows as a mycelium in the soil (**Figure 22.14a**) and produces arthrospores that remain viable for hundreds of years. These spores are small enough to be inhaled into the alveoli. There they develop into thick-walled spherules, which contain endospores that, when released, develop into new spherules (**Figure 22.14b**). Disseminated coccidioidomycosis is treated with amphotericin B, but it's usually fatal. No vaccine exists yet.

Blastomycosis. Blastomycosis is a disease of humans and other animals, especially dogs and cats. In humans, infection begins in the lungs, causing a clinical syndrome that can resemble pulmonary tuberculosis. More often, it affects distant organs—especially skin, bone, and testes. Cutaneous blastomycosis causes raised, wartlike lesions, usually on exposed areas such as the face or hands. Bone and testicular lesions cause pain, swelling, and sometimes drainage.

Blastomycosis is caused by *Blastomyces dermatitidis*, another dimorphic fungus. In the soil it grows as a mycelium, producing conidia that infect humans when inhaled. In tissues it grows as a yeast. *B. dermatitidis* appears to be limited to North America and Africa, but its exact geographical distribution is unknown. It's very difficult to isolate from nature.

Blastomycosis is diagnosed by identifying the fungus in affected tissue. Amphotericin B is the treatment of choice. Without treatment, systemic blastomycosis is rapidly fatal.

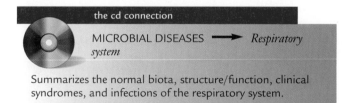

the cd connection

MICROBIAL DISEASES ➞ *Respiratory system*

Summarizes the normal biota, structure/function, clinical syndromes, and infections of the respiratory system.

***Pneumocystis* Pneumonia.** *Pneumocystis* pneumonia a life-threatening pneumonia in immunocompromised patients. It's caused by *Pneumocystis carinii*. Because it was traditionally classified as a protozoan, its cellular forms have been given protozoan names. But its ribosomal RNA sequence reveals that it's a fungus. *P. carinii* is an opportunistic pathogen. Very little is known about its reservoir, its life cycle in nature, or its mode of transmission. It is, however,

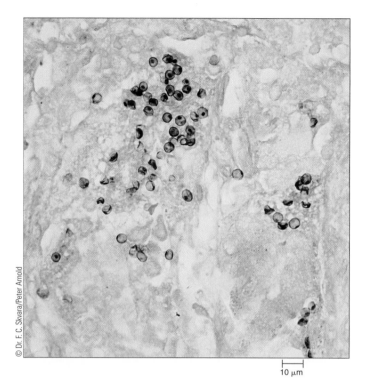

FIGURE 22.15 Tissue biopsy of a patient with *Pneumocystis* pneumonia. The purple structures are stained *Pneumocystis carinii* cysts.

found in the lungs of many human beings. The vast majority of healthy children under the age of 4 have evidence of past infection by *P. carinii*. This suggests it's inhaled during infancy or early childhood without causing disease. But if the immune system is significantly weakened, the infection may be reactivated, leading to a severe lung infection (**Figure 22.15**) manifested by fever, cough, tachypnea (rapid breathing), and death.

Pneumocystis pneumonia used to be a rare disorder seen on cancer wards, in patients receiving immunosuppressive medications, and occasionally in severely malnourished infants. Since the beginning of the HIV/AIDS epidemic, however, it has become a major killer, causing more deaths in the United States than do "traditional" infections such as tuberculosis. Moreover, AIDS patients present a slightly different clinical syndrome than do others with *Pneumocystis* pneumonia. They seem to get sick more slowly and survive longer without treatment. One thing is clear, however: People who develop *Pneumocystis* pneumonia often die from it. It is the most common cause of death among people with AIDS.

Trimethoprim-sulfamethoxazole and pentamidine isethionate, another antimicrobial agent, are effective treatments. Unfortunately, many AIDS patients cannot tolerate these medications.

Table 22.3 summarizes what we've discussed about the infections of the upper and lower respiratory tracts.

TABLE 22.3 Infections of the Respiratory System

Infection	Causative Agent	Mode of Transmission	Symptoms	Prevention and Treatment
Upper Respiratory Infections				
Bacterial				
Epiglottis	*Haemophilus influenzae* type b	Respiratory droplets	Fever, sudden and severe respiratory distress	Vaccination; treatment with antibiotics and surgical intervention to secure airway
Streptococcal pharyngitis	*Streptococcus pyogenes*	Respiratory droplets	Sore throat, fever, chills, headache; may develop into scarlet fever, rheumatic fever, or acute poststreptococcal glomerulonephritis	Treatment with penicillin or erythromycin
Diptheria	*Corynebacterium diptheriae*	Respiratory droplets	Swollen throat covered with pseudomembrane that obstructs airway; toxin may damage heart, kidneys, brain, or nerves	Toxoid vaccination; treatment with antitoxin effective if administered early enough
Viral				
Common cold	Rhinovirus, coronavirus, others	Primarily hand-to-nose	Sneezing, nasal discharge	Prevention by hand washing
Lower Respiratory Infections				
Bacterial				
Pneumonococcal pneumonia	*Streptococcus pneumoniae*	Respiratory droplets	Fever, painful breathing, often life threatening	Pneumovax; treatment with penicillin or other antibiotics
Other typical bacterial pneumonias	Many, including *Haemophilus influenzae, Staphylococcus aureus, Klebsiella* spp., *Streptococcus pyogenes*	Respiratory droplets	Fever, difficulty breathing, often life threatening	Treatment with antibiotics appropriate to the infecting organism
Mycoplasma pneumonia (primary atypical pneumonia)	*Mycoplasma pneumoniae*	Respiratory droplets	Headache, low fever, persistent dry cough, not life threatening	Treatment with erythromycin or tetracycline
Ornithosis (psittacosis)	*Chlamydia psittaci*	Inhaling organisms shed by birds; respiratory droplets	Headache, fever, chills, cough, occasionally life threatening	Prevention by controlling infection in birds; treatment with erythromycin or tetracycline
Q fever	*Coxiella burnetii*	Inhalation	Headache, fever, cough, chills, rarely life threatening	Treatment with tetracycline
Legionellosis	*Legionella pneumophilia*	Inhaling water droplets contaminated with bacteria	Weakness, headache, cough, high fever, chills; life threatening in elderly and weakened individuals	Treatment with erythromycin

TABLE 22.3 Infections of the Respiratory System (continued)

Infection	Causative Agent	Mode of Transmission	Symptoms	Prevention and Treatment
Pertussis (whooping cough)	*Bordetella pertussis*	Respiratory droplets	Coughing paroxysms; life threatening in infants	DTP vaccination; treatment with supportive care
Tuberculosis	*Mycobacterium tuberculosis*	Inhalation	Pulmonary tuberculosis characterized by fatigue, cough, fever, weight loss; many other syndromes with dissemination to other organ systems	BCG vaccination or tuberculin skin test; asymptomatic individuals treated with isoniazid; actively infected individuals treated with multiple antibiotics
Viral				
Influenza	Influenze virus	Respiratory droplets	Fever, headache, muscle aches, cough; pneumonia uncommon but life threatening	Vaccination; treatment with amantadine
Croup (laryngotracheobronchitis)	Usually parainfluenza virus	Respiratory droplets	Hoarseness, barking cough, stridor in severe cases	Treatment with supportive care
Bronchiolitis	Usually RSV	Close personal contact	Wheezing, rapid breathing	Treatment with supportive care; ribavirin in life-threatening cases
Fungal				
Histoplasmosis	*Histoplasma capsulatum*	Inhalation	Tuberculosis-like	Treatment with amphotericin B, ketoconazole
Coccidioidomycosis (San Joaquin Valley fever)	*Coccidioides immitis*	Inhalation	Tuberculosislike; occasionally a fatal chronic meningitis	Treatment with amphotericin B
Blastomycosis	*Blastomyces dermatitidis*	Inhalation	Tuberculosislike; often disseminates to skin, bone, testes, causing painful lesions and, without treatment, death	Treatment with amphotericin B
Pneumocystis pneumonia	*Pheumocystis carinii*	Unknown; perhaps inhaled in infancy and reactivated with weakened immune system	Fever, cough, difficulty breathing; frequently life-threatening	Treatment with trimethoprim-sulfamethoxazole, pentamidine isethionate

SUMMARY

The Respiratory System (pp. 522–525)

Structure and Function (pp. 522–524)

1. The upper respiratory system consists of the nasal cavity, the nasopharynx, the pharynx, and the epiglottis.

2. The lower respiratory system begins at the larynx. The trachea branches into two smaller airways, the bronchi, which branch into secondary and tertiary bronchi and then bronchioles. The bronchioles terminate in the alveoli, tiny air sacs that make up the lungs. The lungs are surrounded by a membrane called the pleura.

Clinical Syndromes (pp. 524–525)

3. Clinical syndromes of the respiratory system include rhinitis, pharyngitis, epiglottitis, bronchitis, and pneumonia (infection of the lungs with fluid and microorganisms replacing the air that normally fills the alveoli).

4. Clinical syndromes can be caused by many different microorganisms. Pneumonia, for example, can be bacterial, viral, or fungal.

Upper Respiratory Infections (pp. 525–530)

Bacterial Infections (pp. 525–528)

5. Encapsulated strains of *Haemophilus influenzae* cause epiglottitis, a life-threatening infection. Nonencapsulated strains cause mild infections such as sinusitis. *H. influenzae* was the scourge of pediatric wards until 1990 when the *Haemophilus influenzae* type b (Hib) vaccine was developed.

6. *Streptococcus pyogenes* (also called group A beta-hemolytic streptococcus) is highly virulent. It causes many clinical syndromes, including streptococcal pharyngitis (strep throat). Complications include scarlet fever and toxic shock–like syndrome (TSLS).

Rheumatic fever can develop after the streptococcal infection is over.

7. *Corynebacterium diphtheriae* is noninvasive, but it infects the mucous membranes of the pharynx and produces diphtheria toxin, which causes all the symptoms of diphtheria. The throat swells and becomes covered with a grayish pseudomembrane of dead cells. Treatment is with diphtheria antitoxin, produced artificially in horses. Infants can be immunized as part of the diphtheria, tetanus, pertussis (DTP) vaccine.

Viral Infections (pp. 529–530)

8. Rhinoviruses cause the clinical syndrome of a cold or upper respiratory infection (URI). Their only reservoir is human beings. Transmission is usually by direct contact or fomites. Transmission can be interrupted by frequent hand washing. Recovery depends upon the immune system; antibiotics are useless.

9. The common cold is also caused by other viruses, primarily coronaviruses.

Lower Respiratory Infections (pp. 530–545)

Bacterial Infections (pp. 530–538)

10. *Streptococcus pneumoniae* (the pneumococcus) causes pneumococcal pneumonia. Penicillin is the drug of choice, and recovery is complete; however, penicillin-resistant strains are increasing. A conjugated vaccine is available.

11. Acute bacterial pneumonia caused by other bacteria, such as *Klebsiella* spp. and *Haemophilus influenzae*, often leaves permanent lung damage.

12. *Mycoplasma pneumoniae* causes a mild pneumonia: gradual onset, few definite signs and symptoms, a patchy x-ray pattern. It is called atypical pneumonia. Treatment is with tetracycline.

13. Humans can contract a pneumonia by inhaling *Chlamydia psittaci* from bird droppings. The illness is also called ornithosis and psittacosis.

14. *Coxiella burnetii* causes Q fever, which is clinically indistinguishable from mycoplasmal pneumonia or ornithosis. Humans become infected by inhaling microorganisms from infected animal placentas or feces. *C. burnetii* requires both an insect and vertebrate host.

15. *Legionella pneumophila* causes legionellosis, a pneumonia named for a virulent outbreak at an American Legion convention. *L. pneumophila* lives in water supplies and can be inhaled when aerosolized. The elderly, individuals with weakened immune systems, smokers, and alcoholics are most susceptible.

16. *Bordetella pertussis* causes pertussis (whooping cough), a highly contagious and often fatal childhood infection. The pertussis vaccine (whole cell) is criticized for its side effects and has been replaced by an acellular vaccine. The combination vaccine is DTaP.

17. *Mycobacterium tuberculosis* causes tuberculosis. *M. tuberculosis* has unusual waxy cell walls that resist drying. Cells remain viable for up to 8 months. Primary infection produces tubercles or granulomas that become calcified (Ghon complexes) in patients who overcome infection. Bacilli in old tubercles may escape to cause a secondary, or reactivation, infection. The clinical syndrome includes fever, fatigue, weight loss, and a chronic cough. There is a vaccine (BCG), but it is not reliable. The United States relies on a tuberculin skin-testing program. New drug-resistant strains of *M. tuberculosis* are emerging.

18. *Mycobacterium bovis* causes the human disease bovine tuberculosis in people who drink unpasteurized milk from infected cows. The symptoms resemble pulmonary tuberculosis.

Viral Infections (pp. 538–541)

19. Almost everyone has been infected by the influenza virus. Illness may be severe (fever, headache, muscle aches, cough), but it is usually brief.

The virus's structure makes it prone to minor genetic changes (antigenic drift) and radical changes (antigenic shift). These generate new strains. An antigenic shift that occurred in 1918 produced a strain that caused a devastating pandemic. A vaccine is available for high-risk individuals.

20. Some parainfluenza viruses cause the common cold. Parainfluenza type 1 is the most common cause of childhood croup, which causes upper airways to narrow and produces a hoarse cough. Severe cases resemble epiglottitis. Treatment is limited to supportive care.

21. Respiratory syncytial virus (RSV) is the most common cause of fatal lower respiratory infection in infants and young children (though all ages can be infected). Outbreaks occur yearly, as the flu season is ending. It is transmitted primarily by hand contact. It spreads quickly within a family, pediatric wards, and intensive care nurseries. Treatment consists of supportive care, including administering oxygen.

22. A new hantavirus strain was identified as the cause of an outbreak of respiratory illness in the U.S. Southwest. Previously known strains attack the kidneys. Symptoms include fever, muscle aches, and respiratory distress, and a death rate of 70 percent due to lung failure. People become infected by inhaling particles of the virus, which is shed by certain rodents.

Fungal Infections (pp. 541–545)

23. Histoplasmosis, caused by *Histoplasma capsulatum*, is the most common fungal respiratory disease. People become infected when they inhale spores where bat and bird droppings are plentiful. Only about 1 percent of infected people become ill (with flulike symptoms), and in the United States it occurs largely in the Ohio–Mississippi River valley. Immunocompromised patients develop chronic or disseminated infection.

24. *Coccidioides immitis* causes coccidioidomycosis (also called San Joaquin Valley fever). Humans become infected when they inhale airborne arthrospores. Most people in endemic areas get a mild case at some time, but a few develop a chronic infection similar to tuberculosis or a disseminated infection that can be fatal.

25. *Blastomyces dermatitidis* causes blastomycosis. Humans and other animals become infected when they inhale conidia from infected soil. The clinical syndrome resembles tuberculosis, though the skin, bone, and testes may also be affected. Without treatment (usually with amphotericin B), systemic blastomycosis is fatal.

26. *Pneumocystis carinii*, traditionally classified as a protozoan, has been shown by its ribosomal RNA sequence to be a fungus. It is an opportunistic pathogen that causes *Pneumocystis* pneumonia, a life-threatening infection in immunocompromised patients. Once a rare disease, it is now the most common cause of death among people with AIDS.

REVIEW QUESTIONS

The Respiratory System

1. Sketch the respiratory system, and label these parts:
   a. nasal cavity
   b. nasopharynx
   c. pharynx
   d. epiglottis
   e. larynx
   f. trachea
   g. primary bronchi
   h. secondary bronchi
   i. tertiary bronchi
   j. bronchioles
   k. alveoli
   l. lungs
   m. pleura

2. Why is the respiratory system likely to be infected?

3. What is the function of the respiratory system?

4. Briefly review the normal biota and defenses of the respiratory system.

5. What is a clinical syndrome? Describe these clinical syndromes of the respiratory system:
   a. pharyngitis
   b. bronchitis
   c. epiglottitis
   d. bronchiolitis
   e. laryngitis
   f. pneumonia

6. Explain this statement and give an example: Any of the respiratory clinical syndromes can be caused by many different microorganisms.

Upper Respiratory Infections

7. Why is the *Haemophilus influenzae* type b (Hib) vaccine such a breakthrough? What does the "type B" designation mean?

8. Why should strep throat never be ignored? How is it diagnosed? What is the causative agent? Why do researchers think streptococcal virulence may be changing in the United States?

9. Describe how diphtheria toxin causes the various clinical manifestations of diphtheria. What is the treatment? What is cutaneous diphtheria?

10. Why is penicillin useless in treating the common cold? What interrupts the transmission of a cold? Name the most common causative agents.

11. Describe the primary causative agent, the clinical manifestations, and the treatment for acute bacterial pneumonia.

12. Why is mycoplasmal pneumonia also called atypical pneumonia? Why is penicillin not the drug of choice?

13. Name the microorganism that causes ornithosis and describe its life cycle. Name the microorganism that causes Q fever and describe its life cycle. In what ways are both these infections occupational hazards?

14. Why was *Legionella pneumophila* only discovered about 100 years after most other bacterial pathogens?

15. Describe the clinical syndrome of pertussis and its treatment. Why is the pertussis vaccine controversial, and what is being done to resolve the controversy?

16. List reasons why tuberculosis is the leading killer among all infections today. Draw on everything you know about the causative microorganism, the clinical syndrome, and public health issues.

17. Why is bovine tuberculosis an atypical tuberculosis?

Lower Respiratory Infections

18. Why is the influenza virus prone to antigenic drift and antigenic shift? Could the flu pandemic of 1918 happen again? Explain.

19. Name the causative agent of croup and describe the symptoms and its treatment.

20. Why is prevention so important with respiratory syncytial virus (RSV)?

21. When a mysterious respiratory illness with a 70 percent fatality rate broke out in the U.S. Southwest in the early 1990s, epidemiologists were called in. What have they—and virologists—since learned?

22. How are histoplasmosis, coccidioidomycosis (San Joaquin Valley fever), and blastomycosis transmitted? Describe these infections and their treatment.

23. Why has *Pneumocystis carinii* been reclassified as a fungus? Why is it not surprising that *Pneumocystis* pneumonia is the most common cause of death among people with AIDS?

CORRELATION QUESTIONS

24. Why don't histoplasmosis and coccidioidomycosis usually occur in the same place?

25. Do you think smokers would be more susceptible than nonsmokers to pneumococcal pneumonia? Explain.

26. What are the similarities and differences between the ways the *Streptococcus pneumoniae* and *Corynebacterium diphtheria* cause disease?

27. Would a new, powerful cephalosporin be a good candidate to treat ornithosis? Why?

28. The vaccine against diphtheria has been wonderfully effective. Would it be a good idea to take the same approach to develop a vaccine against pneumococcal pneumonia? Explain.

29. Why does *Haemophilus influenzae* have a greater potential than *Streptococcus pneumoniae* for causing sudden death from respiratory failure?

ESSAY QUESTIONS

1. Discuss methods that could be used to control TB today.

2. Discuss the possibility of the world experiencing a pandemic of the severity of the one that occurred in 1918.

SUGGESTED READINGS

Austrian, R. 1985. *Life with the pneumococcus—notes from the bed side.* Philadelphia: University of Pennsylvania Press.

Edman, J. C., J. A. Kovacs, H. Masur, D. V. Santi, H. J. Elwood, and M. L. Sogin. 1988. Ribosomal RNA sequence shows *Pneumocystis carinii* to be a member of the fungi. *Nature* 334:519–22.

Finegold, S. M. 1988. Legionnaire's disease—still with us. *New England Journal of Medicine* 318:571–73.

Gordan, R. D. 1990. Prophylaxis and treatment of influenza. *New England Journal of Medicine* 322:443–50.

Karzin, D. T., and K. M. Edwards. 1988. Diphtheria outbreaks in immunized populations. *New England Journal of Medicine* 318:41–43.

Musher, D. M. 1983. Haemophilus influenza infections. *Hospital Practice* August:158–73.

Ryan, F. 1993. *The forgotten plague.* New York: Little, Brown and Co.

For additional readings, go to InfoTrac College Edition, your online research library at: http://www.infotrac.thomsonlearning.com

TWENTY-THREE

Infections of the Digestive System

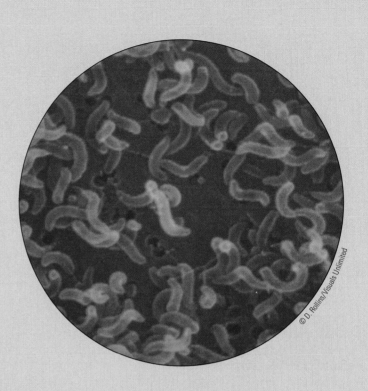

© D. Rollins/Visuals Unlimited

CHAPTER OUTLINE

LEARNING GOALS

To understand:

- The anatomy and function of the digestive system and its defenses against microorganisms
- The clinical syndromes caused by infections of the digestive system
- The bacterial and viral causes of oral cavity and salivary gland infections; their diagnosis, prevention, and treatment
- The bacterial, viral, protozoal, and helminthic causes of intestinal infections; their diagnosis, prevention, and treatment
- The viral and helminthic causes of liver infections and their clinical syndromes; their diagnosis, prevention, and treatment

Good Reason To Be Worried

A R., a 3-year-old boy, was brought to a pediatric emergency room in late September with a high fever and diarrhea. His mother gave a brief history of his illness. He became ill suddenly the day before. His temperature rose to 105°F. He was listless. He refused to eat. He vomited twice. In the morning he seemed lethargic. Once he trembled uncontrollably and lost consciousness. Shortly after this convulsion, A. R. passed a large watery stool. During the next few hours, he passed many smaller stools streaked with blood and mucus. A. R. lived in a small apartment with 10 others, several of whom had recently had diarrhea.

A. R. moaned and cried quietly as the emergency room physician examined him. But he did not speak and barely struggled. His temperature was 104.5°F. He seemed to be mildly de-

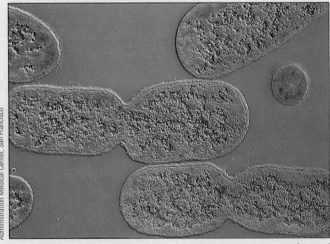

From *Transactions of the Royal Society of Tropical Medicine and Hygiene*, V. 74, No. 4, 1980, 429–433. Photo courtesy of Robert L. Owen, MD, Veterans Administration Medical Center, San Francisco

Shigella, the most common cause of dysentery in the United States.

hydrated. His mouth was dry. When he cried, there were no tears. Laboratory analysis showed only a slight elevation in the number of A. R.'s circulating white blood cells, but there was an exceptionally high number of immature neutrophils—35 percent compared with a normal 1 percent. A mucus-containing sample of his stool was stained with methylene blue and examined under the microscope. It revealed sheets of polymorphonuclear leukocytes (evidence of colonic inflammation). Samples of A. R.'s blood and stool were sent to the microbiology laboratory for culturing.

A. R. was admitted to the hospital. His high fever, appearance, and especially the blood and mucus in his stool suggested a working diagnosis of dysentery. Such a colitis syndrome is usually caused by a bacterial pathogen. In the United States it's most commonly caused by species of *Shigella*. A. R. was given trimethoprim-sulfamethoxazole (trimethoprim-sulfa), and an intravenous line was inserted to rehydrate him.

A. R.'s worried parents were told that their son was expected to recover. But they stayed at his bedside during most of his weeklong hospitalization. His recovery was

gradual. The day after he was admitted, A. R. was more alert and responsive, but he still had a temperature of 103°F and severe diarrhea. He had more than 20 small mucus-containing stools that day alone. Twenty-four hours after admission, the clinical microbiology laboratory returned its preliminary report: a Gram-negative, non–lactose-fermenting rod in A. R.'s stool sample was identified as the probable pathogen.

Over the next few days, A. R.'s temperature gradually returned to normal. The number of stools decreased. At the end of a week, when he was discharged from the hospital, he was thinner but energetic and otherwise in good health. The final report from the microbiology laboratory identified the stool culture as *Shigella sonnei*, the most common species of *Shigella* in the United States. Sensitivity studies showed that it was sensitive to trimethoprim-sulfa. The blood culture grew no microorganisms, indicating that A. R.'s infection was limited to his intestines.

Case Connections

- In this chapter we'll learn more about shigellosis, how it causes disease, and why it is so highly infectious.
- We'll also learn that *Shigella*'s ability to make a toxin and cause disease was probably transferred to a harmless strain of *Escherichia coli*, converting it into the deadly *E. coli* O157:H7.

THE DIGESTIVE SYSTEM

The digestive system—also called the **gastrointestinal system** (*gaster* is Greek for "stomach")—supplies the body with essential nutrients. Much of this system is continually exposed to huge numbers of microorganisms.

First we'll consider the various parts of the digestive system and what they do. Then we'll consider the infectious diseases that afflict them.

Structure and Function

The digestive system is composed of the alimentary tract and associated organs (**Figure 23.1**). The **alimentary tract** is a tubelike structure that's about six times longer than a human being. It runs from the mouth to the anus and consists of the oral cavity (mouth); pharynx (throat), which is also part of the respiratory system; esophagus; stomach; small intestine; and large intestine. The associated organs include the teeth, tongue, salivary glands, liver, gallbladder, and pancreas. Together these organs consume food, digest it, absorb nutrients, and eliminate unabsorbed waste.

Digestion is partly mechanical: Chewing breaks up pieces of food, and **peristalsis** (intestinal movement that propels food through the tract) breaks the pieces into smaller particles. The digestive process is also chemical: Food, which is composed largely of macromolecules, is hydrolyzed into smaller molecules by enzymes and stomach acids. **Bile** (a biological detergent made by the liver) breaks up lumps of fat. The products of digestion are small-molecule nutrients. These are absorbed by the cells that line the intestines.

Let's follow a mouthful of food though the digestive system. In the mouth it's chewed and exposed to digestive enzymes in saliva (the secretion of the **salivary glands**). Then it's swallowed into the **pharynx,** passes through the **esophagus,** and empties into the highly muscular stomach,

where enzymes, acid, and strong mixing action reduce the food to a liquid called **chyme.** Chyme leaves the stomach through a narrow muscular neck to enter the **duodenum** (the first part of the small intestine), where enzymes from the pancreas and bile (which is stored in the **gallbladder**) continue the digestive process. Finally, in regions of the small intestine called the **jejunum** and the **ileum,** additional enzymes produced within the intestinal lining complete the process of digestion. Most products of digestion, such as simple sugars and amino acids, are absorbed there in the small intestine. The indigestible components then pass into the **large intestine,** or **colon,** where they move slowly as water and certain vitamins are absorbed. The nutrient-depleted intestinal contents are then stored in the rectum until they are eliminated through the anus as feces.

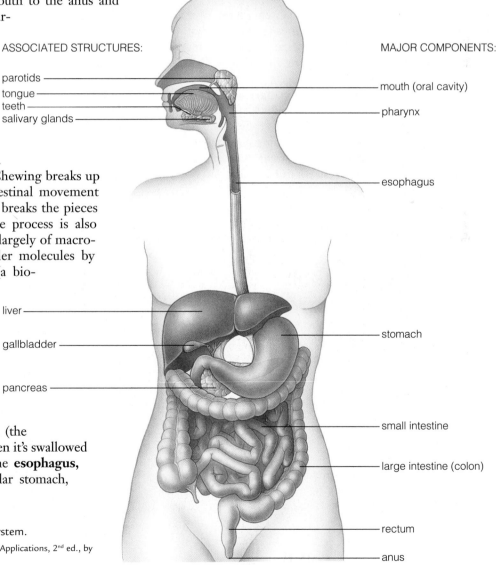

ASSOCIATED STRUCTURES:

parotids
tongue
teeth
salivary glands

liver

gallbladder

pancreas

MAJOR COMPONENTS:

mouth (oral cavity)

pharynx

esophagus

stomach

small intestine

large intestine (colon)

rectum

anus

FIGURE 23.1   Anatomy of the digestive system.

Defenses and Normal Biota: A Brief Review. Like other body surfaces, the lining of the intestinal tract has multiple defenses (Chapter 14). These include a seamless surface of epithelial cells, the muscular motions of chewing and peristalsis, and bactericidal chemicals such as hydrochloric acid, bile, and secretory IgA (Chapter 17).

Millions of microorganisms constantly enter the digestive system, and huge populations of commensal microorganisms become established there. Some of these, such as streptococci, anaerobes, and enterobacteria, are important to good health (Chapter 14). But certain associated organs—the liver, gallbladder, and pancreas—are normally free of microorganisms. These organs are protected by both the innate and adaptive immune defenses (Chapters 16 and 17).

Clinical Syndromes

Pathogens also enter the digestive system. Some of these can cross the epithelial layer that lines the gastrointestinal system and spread to other organs, causing enteric fevers, such as typhoid. But most pathogens, including A. R.'s, remain within the digestive tract.

The pathogens that remain in the digestive tract cause a variety of clinical syndromes (**Table 23.1**). **Gastritis** (infection or inflammation of the stomach) causes pain in the upper abdomen and occasionally bleeding. **Gastroenteritis** (*enteron* is Greek for "intestine") is characterized by diarrhea and sometimes nausea, vomiting, and crampy ab-

dominal pain. It can be caused by viral or bacterial pathogens (discussed in this chapter) or by bacterial toxins produced outside the body and ingested in food (Chapter 15). Although gastroenteritis can be intensely uncomfortable, it is seldom life threatening, at least to adults.

Colitis primarily involves the colon. It is also called **enterocolitis** because the lower small intestine may be involved as well. Sometimes it's called **dysentery.** Unlike gastroenteritis, colitis typically damages the intestinal wall, so stools of these patients often contain blood and mucus.

The associated organs also become infected by pathogens that produce characteristic clinical syndromes. Teeth can be attacked by certain lactic acid bacteria in the mouth, which produce acidic metabolic end-products that cause **dental caries** (cavities). The crevices around the teeth can be attacked by anaerobes that proliferate and cause **periodontal disease** (*peri*, meaning "around"; *dontal*, meaning "teeth"), or **periodontitis** (destruction of gum and bone tissue). The **parotids** (a pair of salivary glands located over the jaw just below the ear) can be infected, producing **parotitis** (a distinctive syndrome of facial swelling). In addition, the inside of the mouth can become infected, causing **stomatitis** (*stoma*, meaning "mouth"; *itis*, meaning "inflammation"), a painful inflammation usually characterized by blisters or ulcers

Liver damage produces **hepatitis** (*hepar*, Greek for "liver"). It can be caused by infection, toxic chemicals, blockage of the bile drainage system, and hereditary disease. Patients with hepatitis become **jaundiced** (yellow)

TABLE 23.1 Clinical Syndromes of Digestive System Infections

Syndrome	Region Affected	Signs and Symptoms	Example
Enteric fever	Digestive system is the portal of entry, but entire body becomes infected	Fever, other systemic symptoms	Typhoid fever
Gastritis	Stomach	Upper abdominal pain, occasional bleeding	Associated with *Helicobacter pylori* colonization
Gastroenteritis	Stomach and intestine	Diarrhea, nausea, vomting, abdominal pain	Salmonellosis, rotavirus diarrhea
Colitis (dysentery, enterocolitis)	Colon	Diarrhea with blood and/or mucus	Shigellosis, amebiasis
Dental caries	Teeth	Enamel destruction leading to pain and tooth loss	Associated with *Streptococcus mutans* colonization
Periodontal disease	Gums and bone supporting teeth	Loosening and eventual loss of teeth	Associated with *Bacteriodes gingivalis* colonization
Parotitis	Parotid (salivary) glands	Swelling and pain below ear and over angle of the jaw	Mumps
Hepatitis	Liver	Jaundice, loss of appetite, dark urine	Hepatitis A

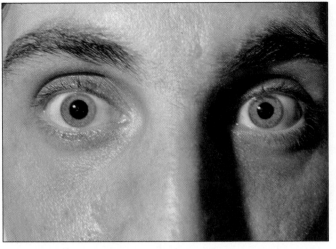

FIGURE 23.2 Jaundice. Note the yellowed skin and eyes.

because bilirubin builds up in their bodies (**Figure 23.2**) and colors their urine. Such patients typically lose their appetite, experience changes in their sense of taste and smell, and, if damage is severe, have uncontrollable bleeding.

Now we'll examine the various infections that cause these clinical syndromes.

INFECTIONS OF THE ORAL CAVITY AND SALIVARY GLANDS

Our oral cavity sustains a dense microbiota. Over our lives we suffer considerable damage from bacterial infections of our mouths. They can destroy our teeth, gums, and part of our jaw bone. In contrast, the salivary glands are normally sterile. But they too can become infected, most commonly by mumps virus.

Bacterial Infections

Many microbial species in the mouth have their own ecological niche. *Streptococcus salivarius* lives on the tongue. *Streptococcus mutans* and *Streptococcus sanguis* attach to the teeth. *Streptococcus mitior* lives on the inner surface of the cheek. *Bacillus melaninogenicus* and other anaerobic bacteria grow in the crevices between teeth and gums. Most bacteria in our mouths are harmless commensals, but some are links in the chain of events that leads to dental caries or periodontal disease.

Dental Caries.
Dental caries are a problem today. More than 6 percent of health expenditures in the United States are to restore or replace teeth damaged by caries. In contrast, caries were almost unknown

among early humans, and until recently they were uncommon among people in developing countries.

Dental caries are caused by *Streptococcus mutans* and eating sucrose (table sugar). *S. mutans* has adhesins on its pili that allow it to attach firmly to **tooth enamel** (the hard material that covers exposed surfaces of the teeth; Figure 14.10). Once attached to a tooth surface, it is difficult to dislodge.

Two properties of *S. mutans* make it **cariogenic** (caries-causing) when sucrose is present. As it metabolizes sucrose, it polymerizes the glucose moieties into glucan and ferments the fructose moieties to lactic acid. The glucan forms the mesh that, together with *S. mutans*, other bacteria, and debris, makes up **dental plaque.** The lactic acid accumulates in the plaque and destroys **dental enamel** (the tooth surface). When the supply of sucrose has been exhausted, acidity diminishes because it is neutralized by saliva. The minerals in saliva can repair minor damage to enamel done by lactic acid, but if acidic conditions persist for too long or occur too frequently, damage cannot be reversed. When decay penetrates the dental enamel, the tooth can be killed if it isn't protected by a filling.

A person's susceptibility to tooth decay is affected by when he or she acquires *S. mutans*. Almost all adults carry this bacterium, but studies show that children who acquire it before the age of 2 have eight times as many caries as children who acquire it after age 4. Children usually acquire *S. mutans* from their intimate caretakers, so repair of active caries in new mothers protects their infants.

Eating sucrose is critically important to tooth decay. Except for sugar cane, sugar beets, and a few other plants, most naturally occurring foods owe their sweetness to fructose. That's why early humans didn't suffer from dental caries. They didn't eat sucrose-containing foods.

Yet a third factor determining susceptibility to caries is the teeth themselves. Some people are more caries-prone than others because of genetic factors. Fluoride ion is also important. People who take in adequate amounts of fluoride, either in drinking water or as part of dental care, develop fluoride-containing dental enamel that is stronger and more acid resistant. Fluoridation of public water supplies decreases the incidence of caries by 60 percent. It's extremely cost effective.

Research on *S. mutans* may soon lead to the development of a vaccine. Oral administration of immunogenic proteins from *S. mutans* could stimulate the production of IgA salivary antibodies against the bacterium and possibly eliminate it from the mouth. Alternatively, antibodies against the enzyme that makes glucan could eliminate *S. mutans*' ability to form caries.

Periodontal Disease.
Periodontal disease is the leading cause of tooth loss. It has been for centuries. Ancient Egyptian and Chinese scholars describe it in detail. Until

FIGURE 23.3   Periodontal disease. (a) Teeth of a person with advanced periodontal disease. (b) Sketch showing loss of bone and receding gum tissue around a diseased tooth.

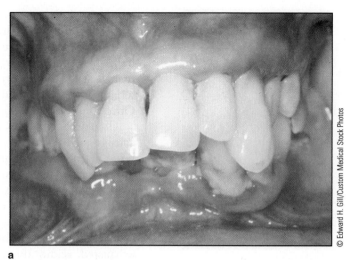

a

b

© Edward H. Gill/Custom Medical Stock Photos

recently tooth loss was considered an unavoidable problem of aging, but microbiological research shows that it is a bacterial infection that can be prevented and treated.

Periodontal disease damages the **periodontium** (tissues that support the teeth). These include the tooth root, the **periodontal ligament** (collagen fibers that connect tooth to bone), the **bone socket** (where the tooth is embedded), and the **gingiva** (gums, which cover the tooth-supporting bone).

Periodontal disease begins when plaque accumulates in the **subgingival crevice** (channel between tooth and gingiva). The gums become inflamed (a condition called **gingivitis**), causing redness, swelling, and easy bleeding. If the infection proceeds, **periodontitis** results. Deep pockets form in the subgingival crevice. The periodontal ligament and surrounding bone are damaged. Eventually, otherwise healthy teeth loosen and fall out (**Figure 23.3**).

The microbiota in areas of periodontal damage differs significantly from the microbiota surrounding healthy teeth (mostly streptococci and actinomycetes). Approximately 300 species, including gram-negative anaerobic bacteria, are present. One of these, *Porphyromonas* (formerly *Bacteroides*) *gingivalis*, appears to be essential for periodontal disease to occur. *P. gingivalis* has a number of disease-causing capabilities, including the ability to invade gingival cells and to cause certain host cells to release cytokines and other compounds that lead to the destruction of connective tissue and bone. *P. gingivalis* is thought to enter the mouth by transmission from infected individuals.

The best way to prevent or slow the progress of periodontal disease is to remove plaque from gum margins with thorough brushing and flossing. But the growing understanding of the microbiology of periodontal disease may soon provide new approaches to diagnosis and treatment. Antibiotics might be able to control periodontal infections. Placing synthetic fibers filled with tetracycline into the periodontal pockets greatly reduces bacterial populations and improves gingivitis.

Viral Infections

Many viruses that cause **exanthems** (skin rashes) also cause **enanthems** (redness or blisters inside the mouth). These include measles virus and herpes simplex virus. But because both of them primarily cause skin infections, we'll discuss them in Chapter 26. Here we'll discuss mumps, the most common viral infection of the salivary glands.

Mumps. Mumps is a systemic viral infection that usually causes parotitis. One or both of the parotid (salivary) glands below the ear swell (**Figure 23.4**), accompanied by mild pain and sometimes fever. Salivary glands beneath the tongue or the jaw may become inflamed, even if the parotids are not affected. Mumps is caused by mumps virus, which infects only humans and belongs to the paramyxovirus family. Thirty-five years ago, nearly every child in the United States got mumps before age 15. Today, with routine pediatric immunization, mumps is unusual.

The mumps virus, transmitted in the saliva or respiratory secretions of an infected person, enters the respiratory system. It replicates in the upper respiratory epithelium and then spreads directly to the salivary glands. Twenty-five percent of such infections cause no symptoms at all. After a few days, the virus enters the bloodstream. Occasionally it spreads to tissues throughout the body, causing serious complications. In about 10 percent of mumps patients, the virus affects the central nervous system. Most of these patients develop only a severe headache, but a few experience a temporary loss of balance, and a very few suffer permanent brain damage. In adult males the testes may be infected, producing **orchitis** (a painful testicular inflammation). Contrary to popular belief, however, it almost never results in sterility. The pancreas and thyroid can also be affected. In rare cases, mumps destroys the inner ear, causing deafness.

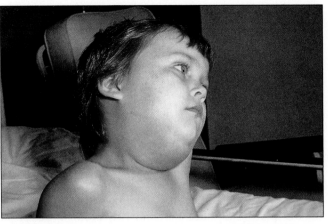

FIGURE 23.4   Mumps. Typically swelling occurs below ear and over the margin of the jaw where the parotid salivary gland lies.

Mumps vaccine, a live attenuated strain of mumps virus, is part of the measles, mumps, rubella (MMR) vaccine given at age 15 months and repeated at age 4½ years. Since the MMR vaccine was released in 1968, the incidence of mumps in the United States has decreased by 97 percent.

INFECTIONS OF THE INTESTINAL TRACT

Infections of the intestinal tract range from mild 1-day food poisoning to deadly cholera and typhoid fever. Intestinal infections can be bacterial, viral, protozoal, or helminthic. Nearly all are transmitted by the fecal-oral route.

Bacterial Infections

Bacteria cause most lethal infections of the intestinal tract. With improved sanitation and clean drinking water, many have become uncommon in developed countries. Some can be treated with antimicrobial agents.

Shigellosis. A. R.'s illness, discussed in the opening of this chapter, was a typical case of shigellosis. The causative bacteria, *Shigella* spp., belong to the Enterobacteriaceae (Bergey's Gammaproteobacteria, Chapter 11). Because these bacteria are Gram-negative rods, or bacilli, another name for A. R.'s illness is bacillary (rod-shaped) dysentery.

Intestinal bacterial infections are diagnosed by identifying pathogens present in stool specimens. Such identification presents a special challenge because, unlike blood and urine, which are normally sterile, intestinal samples always contain large numbers of bacteria. Identifying pathogens is usually

based on their ability to grow on differential media or on their distinctive biochemical characteristics. For example, a bacterium from A. R.'s stool grew on differential medium, but it did not ferment lactose, did not produce gas from carbohydrate, did not produce hydrogen sulfide from thiosulfate, and was motile. Those properties identified it as belonging to the genus *Shigella*. Immunological testing of O antigens on its cell surface showed it to be *Shigella sonnei*, not *Shigella dysenteriae*, *Shigella flexneri*, or *Shigella boydii*.

Shigella spp. can be highly virulent, as was the causative agent of A. R.'s case. Such virulent strain adhere by adhesins on their pili to carbohydrate receptors on cells of the colon. Then they invade these host cells by stimulating phagocytosis (even though colon epithelial cells are not normally capable of it). Once phagocytized, the bacterium lyses the phagosome to enter the cytoplasm (**Figure 23.5**). There it multiplies rapidly and produces toxin. About 6 hours later, the host cell lyses, releasing many bacterial cells that infect neighboring epithelial cells. This pattern of invasion, multiplication, and toxin production makes characteristic patchy areas, called **microabscesses,** in the colon. *Shigella* cells stay in the intestinal wall. That's why blood samples of shigellosis patients such as A. R. are usually free of bacteria.

Most *Shigella* spp. produce potent toxins. The one produced by *S. dysenteriae* is **Shiga toxin.** Those produced by other species are called **Shigalike toxins.** All of them are two-subunit protein toxins (Chapter 15). They damage blood vessels in the intestinal wall and cause intense inflammation of the intestine. Bleeding and inflammation cause the stools to be streaked with blood and to contain the strings of mucus that were present in A. R.'s stool. The toxin can cause the watery diarrhea that A. R. had. The toxin can also spread out of the intestine. Its effect on the nervous system contributes to the convulsions, such as A. R.'s, that are common in children with shigellosis.

Transmission and Epidemiology. Shigella grows only in the intestinal tract of humans. It's transmitted by the fecal-oral route. Flies, fingers, and food are the usual vehicles. But because *Shigella* cells survive for a long time in contaminated water or on fomites, they too transmit it. People who live in crowded conditions where cleanliness is difficult are particularly likely to contract shigellosis. Most of their infections, like A. R.'s, could be prevented by the most basic form of personal hygiene: good hand washing.

Children are far more likely than adults to get shigellosis. Those under 5 years old account for about half the reported cases, because they are too young to follow good hygiene habits and they are more susceptible to *Shigella* infection.

Many fecal-oral infections, including cholera and typhoid fever, have been nearly eradicated from industrialized

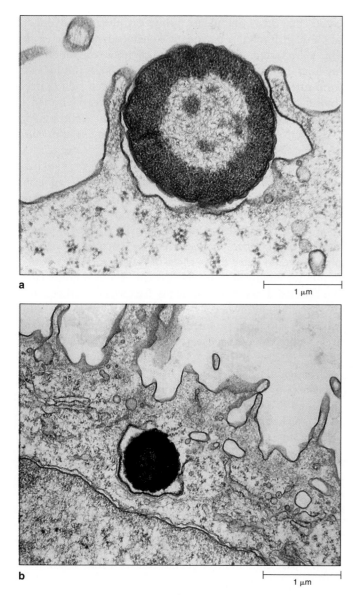

FIGURE 23.5 *Shigella*. (a) *Shigella flexneri* cell being engulfed into an epithelial cell. (b) *S. flexneri* escaping into the cell's cytoplasm by inducing lysis of the host-cell phagocytic vacuole.

Clinical Syndrome and Treatment. A. R. manifested all the classical symptoms of shigellosis, but many victims, particularly adults in the United States, manifest only a watery diarrhea. Because the symptoms are so variable, shigellosis can be diagnosed with certainty only by culturing *Shigella* from the patient's stool or from a rectal swab.

Shigellosis is usually a self-limited disease, even in children. But as A. R.'s case showed, it can be life threatening. Probably the greatest threat to A. R.'s survival was **dehydration** (depletion of the body's water reserves). More than a 15 percent loss of body weight through dehydration is usually fatal. The associated decreased volume in the circulatory system lowers blood pressure, limits blood supply to vital organs, and causes shock and ultimately death. When A. R. was admitted to the hospital, the physician estimated that dehydration had decreased his weight by 5 to 10 percent.

Dehydration can be treated effectively by fluid therapy, but severe *Shigella* infections require antimicrobial treatment as well. Unfortunately, multiply antibiotic-resistant strains are now common. Trimethoprim-sulfamethoxazole, as was given to A. R., is the drug of choice. But for strains resistant to this combination, a cephalosporin should be given. Drug-resistant *Shigella* spp. have become so common that all clinical isolates should be tested for their susceptibility to antimicrobial agents.

Typhoid Fever. Typhoid fever is a potentially fatal enteric fever caused by *Salmonella typhi*. Like most other intestinal infections, typhoid fever is transmitted by the fecal-oral route, usually by consuming contaminated food or water. But direct person-to-person transmission does occur, although rarely. Symptoms typically begin a week or two after infection, when bacterial cells enter the bloodstream. The most dramatic symptom is a high fever—usually over 104°F. It continues for days or weeks. Patients become tired and confused and lose their appetite. They suffer from headache and other aches and pains. Some develop **rose spots** (a characteristic faint rash). Late in the illness, bacteria that have infected the liver are excreted in the bile. They reenter the gastrointestinal tract and are shed in the feces.

Most people with typhoid fever are seriously ill for about a month. The majority recover, but up to 10 percent die from complications such as intestinal bleeding or perforation (rupture) of the intestinal tract. Antibiotic treatment shortens the illness and improves the chances of survival. However, many *S. typhi* strains have become resistant to several antimicrobial drugs. Chloramphenicol, in spite of its toxicity, remains the drug of choice (Chapter 21).

Most typhoid fever patients stop excreting *S. typhi* in their feces when they recover, but about 3 percent develop a chronic gallbladder infection and become persistent carriers. Some excrete huge numbers of *S. typhi* cells for years. Typhoid carriers must register with the local public health de-

countries, but not shigellosis. In the United States, thousands of cases are reported annually. Shigellosis is difficult to eradicate partly because it is so infectious. A person must ingest thousands to millions of bacterial cells to contract typhoid fever or cholera, but only 200 cells are sufficient to cause shigellosis. In homes with an infected family member, large numbers of live bacteria can be recovered from the floor and around the toilet. Shigellosis was once known as **asylum dysentery** because of mass outbreaks that occurred in mental institutions. Today shigellosis spreads rapidly through day-care centers.

LARGER FIELD

TYPHOID MARY

In 1900 Mary Mallon, an Irish immigrant, was working as a cook for a wealthy Mamaroneck, New York, family. A young visitor to the home contracted typhoid fever, and shortly after, Mary left to take a job with a family in New York City. Just before Christmas 1901, the New York City family's laundress died of typhoid. After her death, Mary took another position. In Dark Harbor, Maine, Mary was the cook for a lawyer's family. Within 2 weeks of her arrival, seven of the eight people in the household became ill with typhoid fever. Mary stayed to nurse them, and the lawyer rewarded her with an extra $50 before she left. There was also a waterborne outbreak of typhoid in Ithaca, New York, and another cluster of cases among servants in a house in Sands Point, Long Island, where

Mary was working in 1904. Finally, in 1906, an outbreak of six cases of typhoid in Oyster Bay, New York, prompted an epidemiological investigation.

When the health authorities in Oyster Bay could not find the source of typhoid fever, they contacted George Soper, a sanitary engineer from the New York City Health Department who was known for his talents as an investigative epidemiologist. Dr. Soper checked all the likely sources of typhoid—from the family's milk, water, well, and cesspool to clams from the local bay—with no better success than the local authorities. Eventually his investigation led him to Mary Mallon. Six months later he traced her back to New York City, where she was cooking for a family on Park Avenue under an assumed name. Her employer's daughter had recently died of ty-

phoid. When Soper presented Mary with evidence of her past association with typhoid, she attacked him with a cleaver and ran to a nearby outhouse to hide. Police finally took her—kicking, screaming, and biting—to Riverside Hospital for Communicable Disease on North Brother Island in the East River. Microbiological examination of her stools confirmed that Mary was indeed a typhoid carrier. Her story created a sensation in the local press and among wealthy New Yorkers, who worried that their Irish help might be exposing them to disease and death.

At Riverside Hospital, Dr. Soper explained to Mary that she could probably be cured by having her gallbladder removed, but she refused surgery. She also refused to stop working as a cook, which was her sole means of support and a great source of

enjoyment. As a result, Mary was kept at Riverside as a prisoner-patient for 3 years. She was trained as a hospital laundress and released in 1910 on the promise that she would not return to cooking and would report regularly to the health department. But Mary disappeared and returned to working as a cook under the name Mrs. Brown. A typhoid epidemic—25 infections and 2 deaths—at the Sloane Hospital for Women led authorities to Mary 3 years later.

Mary was returned to Riverside, where she was quarantined for the remainder of her life—25 years. She worked as a laboratory technician and lived in her own cottage near the hospital until paralyzed by a stroke in 1932. Mary Mallon died in November 1938. In all, 1300 cases of typhoid were linked to her.

partment and report at intervals to have their stool tested for the continued presence of *S. typhi*. As long as they remain carriers, they may not work at certain occupations, including those involving food or children. This infringement of personal liberty is considered justified to prevent epidemics of typhoid fever (see Larger Field: Typhoid Mary). Usually typhoid carriers stop excreting bacteria if they are treated with antibiotics and have their gallbladder removed. *S. typhi* is unusual among *Salmonella* spp. because it infects only humans.

Typhoid fever has become a rare disease in developed countries. In the United States there are only a few hundred cases annually. Most of these are acquired by people traveling overseas. Our freedom from typhoid fever is a direct consequence of safe drinking water (**Figure 23.6**). Vaccines are available and recommended for people at high

risk of exposure, such as travelers to developing countries, but they are only between 17 and 66 percent effective.

Salmonella spp., like *Shigella* spp., are Gram-negative rods and belong to the Enterobacteriaceae (Bergey's Gammaproteobacteria, Chapter 11). Unlike *Shigella* spp., *Salmonella* spp. infect many animals, as well as humans. Species of *Salmonella* are identified by biochemical properties: They do not ferment lactose; they do produce hydrogen sulfide from thiosulfate and visible gas from fermentable sugars, and most are motile (**Figure 23.7**).

Salmonella has more than 2000 distinct strains that can be distinguished serologically on the basis of two major surface antigens, the O (lipopolysaccharide) and H (flagellar) antigens. Many of these **serotypes,** or **serovars,** were given species names when isolated, but taxonomists have

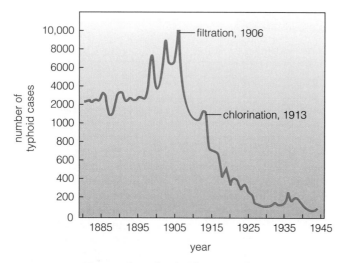

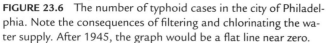

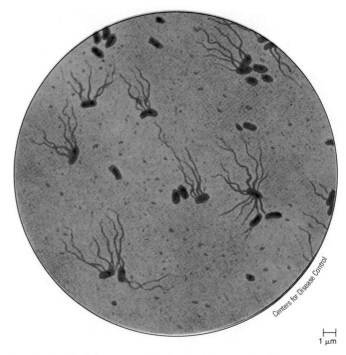

FIGURE 23.6   The number of typhoid cases in the city of Philadelphia. Note the consequences of filtering and chlorinating the water supply. After 1945, the graph would be a flat line near zero.

FIGURE 23.7   *Salmonella typhi.* These cells are stained to show their flagella.

now grouped them into just a few species: *Salmonella typhi*, *Salmonella choleraesuis*, and *Salmonella enteritidis*. However, species names for serotypes continue to be used in the medical literature. Identification of serotypes is not essential clinically, but it is useful epidemiologically. Infections with the same serotype are likely to come from the same source.

Salmonellosis.

Salmonellosis is often called a food poisoning because people usually contract it by consuming contaminated food. In fact, it is a true infection caused by bacteria multiplying in the intestines (**Table 23.2**). It's caused by serovars traditionally designated *Salmonella enteritidis* or *Salmonella choleraesuis*.

Infection begins when a person ingests large numbers—usually millions—of these cells. They invade the epithelium of the small intestine and multiply. Unlike *Salmonella typhi*, they usually stay there. About a day after eating contaminated food, gastrointestinal symptoms begin. They almost always include diarrhea, but abdominal cramps, fever, nausea, and vomiting are common. Most healthy adults recover without treatment within a few days, but the very young,

TABLE 23.2 Food Poisoning by Bacteria

Organism	Infection or Intoxication	Foods
Salmonella spp. (*Salmonella enteritidis*)	Infection—bacteria multiply in the small intestine	Meat, poultry, eggs
Staphylococcus aureus	Intoxication—preformed toxin ingested in food	Whipped cream, eggs, mayonnaise, meat (especially ham)
Bacillus cereus	Intoxication—preformed toxin ingested in food	Rice, meat, vegetables
Clostridium perfringens	Intoxication—bacteria are ingested in food and toxin released when spores form in intestine	Meat
Clostridium botulinum (Chapter 25)	Intoxication—preformed toxin ingested in food	Nonacidic canned foods, especially green beans
Vibrio parahaemolyticus	Infection—bacteria multiply in stomach	Fish

LARGER FIELD

NO ROOM FOR SALMONELLA

Microbiologists have long known that the natural microbial biota of the intestines protects us and other animals against enteric infections. Overgrowth of *Clostridium difficile* following antibiotic therapy is just one bit of supporting evidence. But like the weather, this knowledge has stimulated more talk than action. Recently this all changed. Microbiologists at the U.S. Department of Agriculture devised a mixture of 29 different strains of bacteria that protect chickens from being infected by *Salmonella*. And when chickens are protected, so are we. Between 2 and 4 million people in the United States are infected by *Salmonella* annually—a large portion of them by *salmonella*-contaminated poultry and eggs.

The 29-strain bacterial cocktail, called PREEMPT, mimics the normal microbial biota of a healthy chicken. Among other microbes, it contains three strains of *Enterococcus faecalis*, three of *Enterococcus faecium*, two of *Lactobacillus*, and two of *Escherichia coli*. The cocktail works, and it's easy to administer. In trials on 80,000 chickens, subsequent infection rates with *Salmonella* were reduced almost to zero. It also protects against *Campylobacter* and *Listeria*. It's administered to newly hatched chicks by spraying them. Then the chicks inoculate themselves by preening. To quote Caroline Dewaal, food safety director for the Center for Science in the Public Interest, "It's a milestone."

very old, or immunosuppressed hosts may die. In these few cases, salmonellosis causes severe dehydration or becomes a systemic bloodborne infection.

Salmonellosis is a major public health concern. About 40,000 cases are reported yearly to public health departments. Undoubtedly millions of cases go unreported. Most of us have had this disease at one time or another. Few people die, but many workdays are lost and the cumulative burden of human suffering is immense.

Salmonellosis is widespread because *Salmonella* is widespread. Many serotypes infect domestic livestock. Meat sold in supermarkets, unpasteurized milk, and even Grade A shell eggs often harbor the pathogen. Of the 4 billion chickens consumed in the United States each year, at least 1.4 billion are contaminated with *Salmonella* (**Figure 23.8**). Cooking to 145°F does kill these bacteria, but often food, poultry in particular, is not cooked long enough to kill bacteria inside the body cavity. In addition, cutting boards or utensils can spread bacteria from uncooked meat to other foods. Salmonellosis can be prevented by using a nonporous cutting board that can be disinfected, by thoroughly cooking all foods containing meat or eggs, and by washing hands frequently. Pasteurized, and therefore *Salmonella*-free, shell eggs are becoming available in certain stores.

Salmonellosis is usually self-limiting. Most victims don't consult a doctor and don't need to. In fact, treating uncomplicated cases with antibiotics is **contraindicated** (medically inadvisable). Such treatment could cause pa-

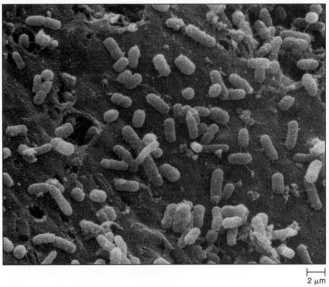

© David Scharf/Peter Arnold

2 μm

FIGURE 23.8 A scanning electron micrograph showing chicken meat covered with cells of *Salmonella*.

tients to become chronic carriers of *Salmonella*. But in serious cases, particularly when bacteria enter the bloodstream, antimicrobial treatment may be essential for recovery. Unfortunately, there is growing drug resistance among *Salmonella* strains. More than one-fourth of those isolated from hospitalized patients are resistant to two or more antimicrobial agents. This increasingly serious drug resistance in enteric pathogens may be partly due to the widespread use of antibiotics in animal feed (Chapter 21).

Escherichia coli Diarrhea and Hemolytic-Uremic Syndrome.

Escherichia coli is abundant in the large intestine of humans. Most strains of *E. coli* are harmless commensals. They help protect us from infection by enteric pathogens such as *Shigella* and *Salmonella* (Chapter 14). But some *E. coli* strains are virulent pathogens. They are a major cause of infant diarrhea, traveler's diarrhea, and hemolytic-uremic syndrome (HUS). *E. coli* is also an important pathogen of the urinary tract (Chapter 24). Pathogenic strains of *E. coli* are distinguished and designated by the types of two antigens they produce. One, designated O, is part of the outer membrane, and the other, designated H, is part of the flagellum. Particular forms of these antigens are assigned numbers. For example, one highly pathogenic strain that we'll return to later is designated O157:H7.

Pathogenic strains of *E. coli* have virulence factors that nonpathogenic strains lack: They can adhere to human cells, invade human tissue, and produce toxins. Depending upon the particular set of virulence factors a pathogenic strain has, it is classified as being enterohemorrhagic, enteropathogenic, enterotoxigenic, enteroinvasive, or enteroaggregative (**Table 23.3**).

Enterotoxigenic strains of *E. coli* are the primary cause of traveler's diarrhea and infant diarrhea in developing countries. They contain plasmids that encode enterotoxins that cause diarrhea (Chapter 15). There are two of them. One, termed **heat labile toxin,** is destroyed by heat. The other, termed **heat stable toxin,** is not. Heat labile toxin is similar to the toxin that causes cholera. But it is not as powerful. Infants in countries with poor sanitation are susceptible to these strains because they have not yet developed immunity. Travelers to these countries are infected for the same reason. During the Gulf War in 1990, some U.S. units had rates of this diarrhea as high as 10 percent per week. As with all types of diarrhea, patients who are significantly dehydrated should receive fluid replacement. But ways to prevent and treat traveler's diarrhea are controversial. Most experts recommend against preventative medication. However, treatment with bismuth salicylate or an antimicrobial agent such as trimethoprim-sulfamethoxazole or ciprofloxacin should be started if diarrhea starts. With early treatment, illness that would otherwise last 3 to 5 days may be over within hours.

Enteroinvasive strains of *E. coli* cause a dysentery almost identical to shigellosis. They produce plasmid-encoded proteins that allow the bacteria to invade human cells. But these strains of *E. coli* are not as virulent as *Shigella*, and many more bacterial cells are required to initiate an infection.

Enteropathogenic strains of *E. coli* cause diarrhea in newborn infants, sometimes starting epidemics in hospital nurseries. This type of pathogenic *E. coli* is not well understood. Somehow it destroys the **microvilli** (tiny projections that increase the surface available to absorb nutrients) in the intestinal epithelium. Enteropathogenic strains produce adhesins that are part of the outer membrane, not pilus proteins like most adhesins.

Enteroaggregative strains of *E. coli* cause a chronic watery diarrhea in infants.

One enterohemorrhagic strain of *E. coli*, O157:H7, has been much in the news recently. It had been recognized as a human pathogen since the mid-1980s, but its status changed abruptly in January 1993, when it was identified as the cause of an epidemic of serious disease in Washington State. The symptoms were bloody diarrhea and a relatively unusual illness called **hemolytic-uremic syndrome (HUS)**. HUS is a life-threatening condition that includes severe anemia and kidney failure in children. Eventually 720 cases were identified in four Western states. Of these, 520 were shown to have *E. coli* O157:H7 in their stools, 171 were hospitalized, and 4 died. The source of *E. coli* O157:H7 was a hamburger patty–making facility in Los

TABLE 23.3 Types of *Escherichia coli* That Cause Diarrhea

Types of *E. coli*	Diseases	Type of Diarrhea	Virulence Factors
Enterohemorrhagic	Hemorrhagic colitis and hemolytic uremic syndrome	Bloody or nonbloody	Adherence, toxin production
Enteropathogenic	Infant diarrhea	Watery	Adherence
Enterotoxigenic	Traveler's diarrhea, infant diarrhea in developing countries	Watery	Adherence, toxin production
Enteroinvasive	Diarrhea with fever in all ages	Bloody or nonbloody	Adherence, invasion of mucosa
Enteroaggregative	Chronic diarrhea in infants	Watery	Adherence

Angeles that had distributed patties to a fast-food restaurant chain in the Northwest. Since then, there have been outbreaks of *E. coli* O157:H7 traced to contaminated unpasteurized milk, apple juice, salad mixes, salami, yogurt, and water. The main reservoir of *E. coli* O157:H7 is healthy cattle. A small percentage of them carry the bacterium and pass live cells in their feces. Meat becomes contaminated by contact with fecally soiled hides during slaughter, fruit and vegetables by contact with contaminated pastures.

The deadly consequences of *E. coli* O157:H7 come from the Shigalike toxin (called Vero toxin) it produces. An innocuous strain of *E. coli* might have acquired the ability to produce it by plasmid transfer from a strain of *Shigella*. Once HUS symptoms appear, antibiotic treatment is useless. Patients' vital functions are maintained in an intensive care unit until they recover. Still, some children die. The best defense against *E. coli* O157:H7 is prevention—avoiding contaminated food.

Cholera.

Cholera is one of the world's most dreaded diseases. Victims can show the first symptoms in the morning and be a shriveled, dehydrated corpse that evening. During the 1800s a series of cholera pandemics started in India and swept across the globe. It was stopped in developed countries when modern systems of sewage disposal and public sanitation were adopted (Chapter 20). But the disease has continued elsewhere. In 1947 an epidemic in Egypt left 20,000 people dead. In endemic regions of India and Bangladesh, cholera annually claims many victims, most of them children. Recently cholera has become established in the Western Hemisphere.

Cholera is caused by *Vibrio cholerae* (Bergey's Gammaproteobacteria, Chapter 11; Figure 11.6a). Humans are its major reservoir, but it persists and multiplies in freshwater supplies and in seawater. As a result, most cases of cholera occur where public sanitation is inadequate, but infected shellfish are also a significant source of infection. The few cases of cholera that have been reported in the United States in recent years were caused by inadequately cooked shellfish caught in estuaries that contained *V. cholerae*.

The most significant characteristic of virulent strains is their ability to produce the cholera exotoxin, which causes all the symptoms of cholera. By raising the level of cyclic adenosine monophosphate (AMP; Chapter 5), it stimulates epithelial cells to secrete large quantities of chloride ion into the intestine. This causes water, sodium ion, and other electrolytes to follow and leave the body as diarrhea. The loss of water can reach astounding proportions—as much as a liter an hour. Diarrhea can be so profuse that the stools lose the appearance of feces and resemble water flecked with small particles of mucus. These are the characteristic **rice water stools** of cholera.

Symptoms and Treatment. Cholera exerts its lethal effect through dehydration, which is sudden and dramatic. Because they have a smaller reserve of body water, children are the most vulnerable to death from dehydration. For them the death rate from cholera is 60 percent without treatment.

Once the infection begins, the only effective treatment is replacing lost fluid. Traditionally, water and electrolytes were administered intravenously, but in most developing countries this kind of treatment, which requires sterile supplies and trained personnel, is impractical. Oral rehydration therapy (ORT), which was developed in the 1960s, is much cheaper, easier, and more effective.

The fluids used for ORT contain glucose and sodium, which allow fluids to be absorbed in the large intestine more rapidly than enterotoxin causes them to be secreted by the small intestine. ORT can drop the death rate from cholera to less than 1 percent. Cholera patients should also receive an antibiotic, preferably tetracycline, to stop *Vibrio cholerae* from multiplying in the intestine. Such treatment shortens the illness and decreases the likelihood of infecting others. But prevention is better than the best treatment. It depends upon adequate sanitation and clean drinking water. Until that's possible, preventing cholera depends upon developing a cheap and effective vaccine. The vaccine currently available has to be administered by injection, and it's not very effective.

Recent Epidemiology. Cholera has always been endemic in many parts of India and Asia. Over centuries, pandemics began in these regions and spread as far as the Americas. They would wane within a few years, leaving areas outside Asia completely cholera-free. But in 1961, the pattern changed. A new strain, designated *V. cholerae* 01 biotype El Tor, appeared. This strain spread through the Middle East. It reached Africa in 1970, where it invaded 29 countries and did not go away. Cases are still occurring in these countries.

In January 1991 the El Tor biotype reached Peru. By mid-February, more than 10,000 cases were reported each week. And El Tor continued to spread. By the end of 1992, more than 700,000 cholera cases had been reported in 21 countries in the Western Hemisphere and more than 6000 people had died (**Figure 23.9**). The epidemic spread through contaminated water, raw vegetables that had been irrigated with sewage, and shellfish from contaminated waters. Now cholera is probably in the Americas to stay. Reservoirs have been established in plankton, shellfish, water, and humans.

Health officials do not even hope to stem the El Tor epidemic by ensuring clean water for everyone. It's far too expensive. Costs are estimated at $200 billion. But short-term, cost-effective solutions, including repairing existing water systems, chlorinating water, and educating people to

SHARPER FOCUS

THE MEDICAL ADVANCE OF THE CENTURY

Human beings are approximately 60 percent water, so the body of a person weighing 70 kilograms (154 pounds) contains about 42 liters, or 93 pounds, of water. Most of this water is inside cells and the spaces between them. Only about 3.5 liters circulate in the bloodstream. In a day the average person takes in about 3 liters of fluid to replace the amount lost in urine, sweat, feces, and respiration. If more water is lost from the body than can be replaced, problems soon arise. The 70-kilogram person can lose 3 or 4 liters of water and suffer only from thirst and symptoms of mild dehydration—decrease in urine volume, dry mouth, and perhaps some light-headedness on standing. But when more than 8 or 9 liters of fluid are lost, life-threatening symptoms develop. Not enough fluid remains to maintain an adequate volume of circulating blood. Blood pressure falls. As a result, vital organs do not receive the oxygen and other nutrients they need. The heart beats faster in an attempt to compensate. If fluid losses exceed about 12 liters, the person goes into irreversible shock (the circulatory system shuts down). Death is the inevitable result.

Dehydration is a risk in all diarrheal illnesses, and the essential treatment is rehydration. It was therefore a medical breakthrough to find that oral rehydration therapy (ORT) could effectively replace traditional intravenous treatment for even the most serious cholera cases. The British medical journal *Lancet* announced that ORT was "potentially the most important medical advance in this century." Oral rehydration has saved not only the lives of many cholera victims, but also the lives of children dehydrated by other diarrheal illnesses. Making oral rehydration solutions available is one of the highest public health priorities of the World Health Organization (WHO); in spite of all efforts, however, diarrheal dehydration still kills about 3 million children each year.

boil household water and cook vegetables, are effective. Just as important is making oral rehydration therapy available to victims. That's the main reason that relatively few people have died.

Vibrio parahaemolyticus Gastroenteritis.

Vibrio parahaemolyticus gastroenteritis can be a mild, self-limited diarrhea or an explosive cholera-like illness. Usually the watery diarrhea is mild and accompanied by low-grade fever, abdominal cramps, nausea, and vomiting, and recovery without treatment occurs within a few days. Occasionally, however, dehydration occurs, requiring fluid replacement and antibiotic therapy.

The causative agent, *Vibrio parahaemolyticus*, is closely related to *Vibrio cholerae* (Bergey's Gammaproteobacteria, Chapter 11). This salt-requiring Gram-negative rod lives in coastal waters and estuaries around the world. In 1951 Japanese microbiologists discovered that it is a common cause of illness among people who eat fish from contaminated waters. Since then it has been found to cause about a quarter of all reported cases of diarrhea in Japan, where fish is often eaten raw. Outbreaks have also been reported around the world, including almost all coastal states of the United States. *V. parahaemolyticus* gastroenteritis is usually called a food poisoning, but it is a true infection.

V. parahaemolyticus multiplies extremely rapidly. Its doubling time under ideal conditions can be as short as 9 minutes, so contaminated fish can come to contain huge numbers of microorganisms within a short time.

Yersinia enterocolitica Enterocolitis.

Yersinia enterocolitica is a Gram-negative bacterium (Bergey's Gammaproteobacteria, Chapter 11). It belongs to the same genus as the bacterium that causes plague (Chapter 27). It is usually found in animals, but it can infect humans. Infection can come from wild or domestic animals, raw milk, oysters, and water.

Y. enterocolitica has a number of virulence factors: It can adhere to the intestinal epithelium, invade cells, and produce an enterotoxin. It's an unusual pathogen because it grows best at room temperature and continues to multiply at refrigerator temperatures. This may be why *Y. enterocolitica* infections are more common in cold climates, such as Northern Europe and the colder parts of North America.

Y. enterocolitica infects the gastrointestinal tract, usually causing an enterocolitis. Abdominal pain is common and can become so intense that appendicitis is suspected, causing patients to undergo surgery. In some cases, bacteria may enter the bloodstream and produce a systemic infection. Y. *enterocolitica* enterocolitis is most common in young children.

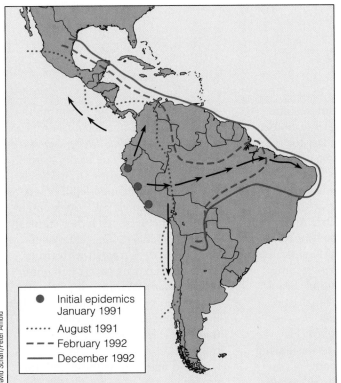

FIGURE 23.9 Cholera epidemic in Latin America, 1991–1992. The epidemic began in Peru in 1991. Outbreaks spread up and down the coast of South America, reaching Mexico by December 1992.

Legend:
- ● Initial epidemics January 1991
- ······ August 1991
- – – – February 1992
- —— December 1992

David Scharf/Peter Arnold

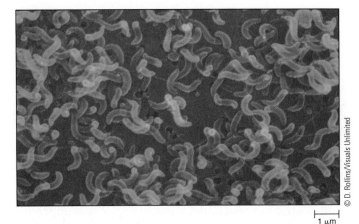

1 μm

© D. Rollins/Visuals Unlimited

FIGURE 23.10 *Campylobacter.* Scanning electron micrograph of cells of *C. jejuni.*

Campylobacteriosis. Campylobacteriosis is characterized by frequent episodes of bloody diarrhea, abdominal pain, and sometimes fever. The epithelial lining of the intestinal tract is usually damaged, probably by the invasion of these bacterial cells and their production of a cytotoxin. Campylobacteriosis is usually a non–life-threatening, self-limited illness that lasts less than a week. Treatment with erythromycin or tetracycline may speed recovery.

Campylobacter spp. (Bergey's Epsilonproteobacteria, Chapter 11) are slightly curved, Gram-negative, rod-shaped bacteria (**Figure 23.10**). They weren't recognized as being human pathogens until the 1970s, probably because they are so difficult to cultivate in the laboratory. They require microaerophilic conditions, unusually high incubation temperatures (42°C), and selective media. They cause more than 2 million illnesses each year in the United States—more than either *Salmonella* or *Shigella.*

Campylobacter jejuni is the species that causes most *Campylobacter* diarrheas. *Campylobacter fetus* more often causes a bloodborne infection in elderly or immunocompromised patients. *C. jejuni* grows in the intestinal tract of healthy cattle, sheep, dogs, cats, and poultry. Humans usually become infected by ingesting contaminated meat or milk. Person-to-person transmission can also occur.

Helicobacter pylori **Gastritis (Peptic Ulcer Disease).** Peptic ulcer disease was long thought to be the product of stressful living. But just within the past decade, medical science has come to realize that infection by the Gram-negative bacterium *Helicobacter pylori* (Bergey's Epsilonproteobacteria, Chapter 11) is the major factor. In most patients, *H. pylori* is necessary for development of peptic ulcers, but it's not sufficient. Most people infected by *H. pylori* develop chronic gastritis, and about 15 to 20 percent of them go on to develop ulcers. Ulcers can lead to gastric carcinoma, an aggressive and usually fatal form of cancer. *H. pylori* can be eliminated by antibacterial agents. When this is done, ulcers heal and symptoms disappear. Ulcers recur in less than 15 percent of these patients.

H. pylori is a curious organism. It lives only in the stomach and duodenum of humans, an extremely acidic and hostile environment. It has a small genome—only about a third as large as *Escherichia coli*'s. Apparently, *H. pylori* is a specialist. It has only those genes it needs to survive in this single inhospitable environment. One of these genes encodes an adhesin that allow it to attach to cells on the stomach epithelium (**Figure 23.11**). Another encodes the enzyme urease. Urease splits urea, forming carbon dioxide and ammonia, a base that neutralizes stomach acid in the bacterium's immediate vicinity. The presence of urease in *H. pylori* provides one means of detecting the infection. The patient is given labeled urea and his or her breath is tested for labeled carbon dioxide. It's also possible to culture the organism from a sample obtained by stomach biopsy, but most diagnoses are based on a blood test for antibodies against *H. pylori.*

All patients with peptic ulcer disease should be given antimicrobial therapy. The most effective is triple therapy

CASE HISTORY

Just Stress?

M.H. was a high-school student who found herself in an extraordinarily stressful situation. An excellent student, she was in the midst of applying to academically rigorous colleges while her family life continued to deteriorate. Her parents had been having marital problems for years. The threat of physical violence at home had dramatically increased in recent months. Her father, who had a frightening temper, threatened to kill her if he found her with a boyfriend he didn't approve of.

For 6 months, M. H. had suffered from increasingly intense abdominal pain. It was high in her abdomen and came at unpredictable times: sometimes at school, sometimes at home. A typical episode would last for several hours. Sometimes she could continue her usual activities, but during other episodes she was unable to remain in school or continue to study. Her physician learned little more from her physical examination. The part of her abdomen was slightly tender to pressure. Routine screening test were similarly unhelpful. Stool samples tested negative for the presence of blood and *Giardia*.

Only a few years ago, M. H. would have been diagnosed as having "functional abdominal pain," a condition thought to be caused by psychological factors. But recent publications lead her physician to order a blood test for *Helicobacter pylori* antibodies. The test came back positive. Subsequent endoscopy showed that she did not have a peptic ulcer but her stomach lining was rough and irritated. The presence in her stomach of *H. pylori* was established by culture. M. H. was treated with triple antimicrobial drug therapy to eliminate *H. pylori*. Within a few weeks her abdominal pain disappeared. She finished her senior year without further absences. She was admitted to the college of her choice and 2 years later is doing well, without abdominal pain.

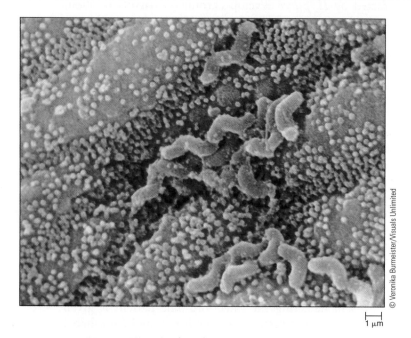

© Veronika Burmeister/Visuals Unlimited

1 µm

FIGURE 23.11   Peptic ulcer disease. A scanning electron micrograph of *Helicobacter pylori* cells on stomach epithelium.

with metronidazole, tetracycline, and bismuth subsalicylate (Pepto-Bismol). This treatment eradicates *H. pylori* in 80 to 95 percent of cases. Whether its better to treat patients with gastritis only remains controversial.

***Clostridium difficile* Diarrhea.** *Clostridium difficile* (Bergey's Firmicutes, Chapter 11) is a toxin-producing bacterium that causes one type of **iatrogenic** (medically induced) diarrhea—the unintended result of antibiotic therapy. Small populations of this organism normally inhabit human intestines (Chapter 14). During antibiotic therapy, many other bacteria are eliminated from the intestinal tract, but the relatively drug-resistant *C. difficile* survives and flourishes with the resulting diminished competition. The result is usually a relatively mild diarrhea that resolves when antibiotic therapy is discontinued. But sometimes *C. difficile* causes a life-threatening enterocolitis, or a severe, persistent diarrhea. In these cases the usual treatment is vancomycin. It eradicates the *C. difficile* and controls the diarrhea. *C. difficile* is a significant cause of nosocomial infection (Chapter 20).

Staphylococcal Food Poisoning. The gastroenteritis caused by staphylococcal food poisoning is usually accompanied by vomiting, diarrhea, and crampy abdominal pain. Because the toxin is already in the food when it is eaten, the onset of illness is quite rapid—within 2 to 6 hours. That compares with 1 to 2 days for a foodborne infection such as salmonellosis. Staphylococcal food poisoning is usually brief and self-limited. It is the most commonly reported food poisoning in the United States. That's probably because it often occurs in large outbreaks—at picnics or other social gatherings—making it easy to identify.

The causative bacterium is *Staphylococcus aureus* (Bergey's Firmicutes, Chapter 11; Figure 11.10a). It causes various clinical syndromes (Chapter 26). Only strains that produce a heat-stable protein enterotoxin cause food poisoning.

Most staphylococcal food poisoning results from improper preparation or storage of food. Staphylococci are usually introduced into food (often whipped cream, eggs, mayonnaise, or meat, especially ham) from the body of the person preparing it. Enterotoxin-producing staphylococci are on many healthy people, in the mucous membranes of the nose or in tiny cuts in the skin (Chapter 14). They multiply rapidly in warm food, producing enterotoxin as they do. There is no obvious evidence of staphylococcal growth. The food smells and tastes normal. But once formed, the toxin cannot be destroyed by cooking, except by the high temperature of a pressure cooker.

Prevention is the key to controlling staphylococcal food poisoning. Antibiotic treatment is useless because illness is caused by a toxin, not bacterial growth. Fluid replacement is necessary only if severe diarrhea causes significant dehydration. But prevention is relatively easy. Careful hand washing can prevent staphylococci from getting into the food. And even if they do, refrigeration at 5°C will prevent their growth and toxin production.

Other Food Poisonings.
Several spore-forming bacteria (Bergey's Firmicutes, Chapter 11) cause food poisonings of varying degrees of severity. The heat resistance of the spores they produce explains much of the pattern of these diseases.

Bacillus cereus. *Bacillus cereus* lives in soil, water, and the gastrointestinal tract of humans and animals, so it often finds its way into food. If the food is left unrefrigerated after cooking, the spores germinate and the vegetative cells produce an enterotoxin. This food poisoning is relatively uncommon, and the gastroenteritis it produces is usually mild and brief.

There are two forms of this illness, associated with two different enterotoxins. The heat-stable enterotoxin causes vomiting. This form of the disease is usually transmitted in contaminated rice. The heat-labile enterotoxin primarily causes diarrhea and is usually transmitted by contaminated meat and vegetables.

Clostridium perfringens. Gastroenteritis is only one of many clinical syndromes caused by *Clostridium perfringens* (Chapter 26). It lives in the gastrointestinal tract of animals and humans and is common in feces-rich soils. It usually contaminates meat. Vegetative cells are killed by cooking, but if meat or gravy is kept warm, spores can germinate and produce new vegetative cells that are ingested along with the food. When these vegetative cells once again form spores, they release an enterotoxin in the intestinal tract. The gastroenteritis it causes is characterized principally by diarrhea. Illness lasts less than a day and seldom causes clinically significant dehydration. Serological studies show that many people in the United States are exposed to the *C. perfringens* enterotoxin. But this type of food poisoning is seldom reported, perhaps because noticeable illness occurs only if enormous numbers of microorganisms are ingested.

Clostridium botulinum. *Clostridium botulinum* causes a foodborne intoxication, the result of a neurotoxin rather than an enterotoxin. **Botulism,** the disease it causes, primarily affects the nervous system (Chapter 25).

Viral Infections

Viral gastroenteritis is familiar to all of us as "stomach flu." Only the common cold occurs more often. The two major causes of viral gastroenteritis are rotavirus and Norwalk agents, but definitive microbiological diagnoses are seldom made. Other viral pathogens may also cause this disease.

Rotavirus Gastroenteritis.
Watery diarrhea, often accompanied by fever, nausea, and vomiting, begins about 48 hours after rotavirus infection. It's not clear how the virus acts, but it damages the epithelial cells of the intestinal wall. The seriousness of the disease depends upon how much fluid is lost from the body. The vast majority of children in the U.S. recover uneventfully at home. But because the disease is so common, about half of all children hospitalized for dehydration are infected by rotavirus. Children in developing countries who may be malnourished or infected by parasites often suffer severe dehydration and die from rotavirus infections.

In spite of its clinical importance, rotavirus was not identified until 1973. It is a reovirus. Electron micrographs (**Figure 23.12**) show a wheel-shaped double capsid surrounding the virion (*rota* means "wheel" in Latin). It is relatively resistant to environmental damage, including changes in temperature and exposure to solvents such as ether.

Rotavirus is transmitted by the fecal-oral route. It is highly infectious, and diarrhea often spreads rapidly through a family. Epidemics also occur in day-care centers and hospitals. Infants and young children are most commonly affected, partly because they are the most susceptible (previous infection provides limited immunity) and partly because of their inability to maintain good hygiene. Outbreaks of rotavirus infection are most common in the winter.

Rotavirus can now be diagnosed fairly readily and inexpensively by means of an enzyme-linked immunosorbent assay (ELISA; Chapter 19). But it's seldom used clinically because definitive diagnosis does not help with prognosis or treatment. All acute gastroenteritis syndromes improve within a few days, and all must be treated with rehydration therapy if dehydration is severe. There's no specific antiviral therapy for rotavirus infection.

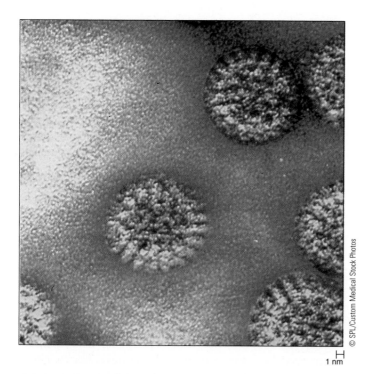

1 nm

FIGURE 23.12 Rotavirus. An electron micrograph showing its wheel-like appearance.

Prevention is the only way to control rotavirus infections. Breastfeeding probably helps prevent disease in nursing infants, because both breast milk and colostrum contain IgA antibodies that neutralize rotavirus virions (Chapter 17). Standard habits of good hygiene help. A promising and effective oral vaccine was licensed but was withdrawn from the market because it was associated with increased cases of a rare bowel obstruction.

Norwalk Agents Gastroenteritis. Norwalk agents are named for an epidemic of gastroenteritis that occurred in an elementary school in Norwalk, Ohio, in 1969. They are particularly small viruses classified as calcilike agents because they resemble the calcivirus family of animal pathogens.

Probably a third of all gastroenteritis outbreaks are caused by Norwalk agents. The virus is transmitted by the fecal-oral route. It causes the usual clinical picture of gastroenteritis—nausea, vomiting, crampy abdominal pain, and diarrhea. Unlike rotavirus, however, Norwalk agents mostly infect older children and adults, rather than infants. The incubation period is about 48 hours. The illness is self-limited and usually brief. Fluid-replacement therapy is rarely needed. Immunity after infection is not long lasting. Outbreaks can be prevented only by careful attention to hand washing and general hygiene.

In late 2002 and early 2003 Norwalk agents gained national attention as the cause of epidemics of gastrointestinal illess affecting hundreds of passengers on cruise ships.

Protozoal Infections

Protozoal infections of the intestinal tract (also transmitted by the fecal-oral route) cause a range of clinical syndromes. Some cause mild diarrheas. Others cause life-threatening systemic infections. Some occur primarily in developing countries. Others are common in the United States.

Amoebic Dysentery. *Entamoeba histolytica* is an amoeba that lives in the intestine of humans. Most people who harbor it suffer no or only mild and occasional problems. These include abdominal distention, loose stools, or constipation. But *E. histolytica* can kill human cells, causing either **amoebic dysentery** (an invasive colitis) or **extra-intestinal amebiasis** (a widespread infection). Amoebic dysentery is characterized by bloody mucoid stools, fever, and abdominal pain. The most common life-threatening complication of extraintestinal amebiasis is liver abscess.

E. histolytica exists as a **trophozoite** and as a **cyst** (a hardier resting form) (**Figure 23.13**). Cysts are transmitted to a new host by the fecal-oral route and become trophozoites in the intestines.

Like most fecal-oral infections, *E. histolytica* occurs where sanitation is poor. On average, 10 to 15 percent of the population in developing countries is infected by *E. histolytica*. Probably less than 2 percent are infected in the United States. Because infection is relatively rare here, it often goes undiagnosed. Serological tests are useful, but definitive diagnosis requires identification of the microorganism in stool samples. And identification is difficult. Trophozoites disintegrate within an hour after the stool is passed, and many nonprescription medications used to treat diarrhea destroy them. Moreover, *E. histolytica* resembles other harmless amoebae that commonly inhabit the intestinal tract. Nevertheless, prompt diagnosis is important because metronidazole in combination with iodoquinol can cure even the most extensive infections.

Giardiasis. Antony van Leeuwenhoek suffered an acute case of diarrhea in 1681. He examined a drop of his stool with his simple microscope and found it swarming with *Giardia lamblia* (Chapter 1). He assumed they caused his illness. But until the last few decades, modern medicine considered *G. lamblia* to be a harmless commensal. It is, in fact, a significant and common intestinal pathogen. Today when doctors in the United States encounter a case of diarrhea that lasts longer than a week, they, like Leeuwenhoek, look for *G. lamblia*.

G. lamblia occurs as a trophozoite that causes infection and a cyst that is passed from host to host by the fecal-oral route. The trophozoite has two nuclei, multiple flagella, and a disc-shaped ventral sucker on its underside that binds the parasite firmly to the intestinal epithelium (**Fig-**

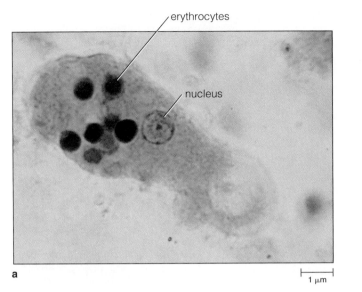

erythrocytes

nucleus

a   1 μm

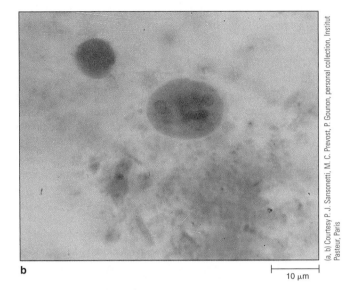

b   10 μm

(a, b) Courtesy P. J. Sansonetti, M. C. Prevost, P. Gounon, personal collection, Institut Pasteur, Paris

FIGURE 23.13   Stained preparations of *Entamoeba histolytica* seen under the light microscope. (a) A motile amoeba-like trophozoite that has ingested erythrocytes. (b) A cyst.

ure 23.14). Trophozoites multiply by binary fission in the intestine. Cysts are oval-shaped. They form in the large intestine and are shed in feces.

Many people infected by *G. lamblia* remain symptom-free. Others, however, suffer a sudden, unpleasant illness that begins about 2 weeks after infection. The first sign is usually an explosive, foul-smelling, watery diarrhea followed by a bloated abdomen and copious amounts of foul-smelling intestinal gas. These symptoms usually last only a few days. Then the long-term symptoms begin. They include abdominal pain, nausea, and occasional episodes of diarrhea.

There are several patterns of transmission, which put certain groups at high risk. Campers are a high-risk group because mountain streams can be contaminated with hu-

man feces or with the feces of *Giardia*-carrying animals, especially beavers. Entire cities can be at risk of waterborne epidemics because *Giardia* cysts are resistant to chlorine, which is often used to purify municipal water. In the United States, children are the most likely to get *Giardia* infections by person-to-person contact at day-care centers or at home. Individuals who practice oral-anal sex are at particular risk. Giardiasis used to be diagnosed by examining stool samples for cysts, but now most labs use an ELISA test that detects *Giardia* antigens.

There are several approaches to preventing giardiasis. Campers should boil, filter, or chemically treat their drinking water. Municipal water districts must filter water supplies. Good hygiene, especially hand washing, helps reduce the spread of *Giardia* at home and in day-

From *Transactions of the Royal Society of Tropical Medicine and Hygiene*, Vol. 74, No. 4, 1980 429–433. Photo courtesy Robert L. Owen, MD, Veterans Administration Medical Center, San Francisco

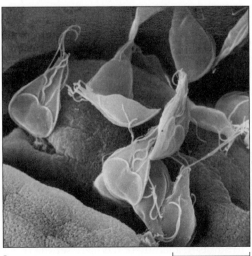

a   10 μm

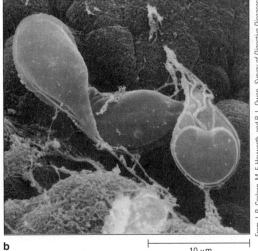

b   10 μm

From J. R. Carlson, M. F. Heyworth, and R. L. Owen, *Survey of Digestive Diseases* 2:201–213, 1984. Used by permission of S. Karger AG, Basel, Switzerland. Photo courtesy Robert L. Owen, MD.

FIGURE 23.14   *Giardia lamblia.* (a) Scanning electron microscopy of a mass of trophozoites on the intestinal epithelium. (b) The disc-shaped ventral sucker (cell on the right).

care centers. High-risk sexual behavior should be avoided. Giardiasis can be cured by the antimicrobial drugs metronidazole and quinacrine.

Balantidiasis. *Balantidium coli* is the only ciliate protozoan known to infect humans. Like *Entamoeba histolytica* and *Giardia lamblia*, it occurs as a metabolically active trophozoite and a resting cyst. The trophozoite causes the disease. The cyst spreads the infection by the fecal-oral route.

B. coli invades the epithelium of the colon and the lower small intestine, causing the colitis called **balantidiasis.** Diarrheal stools typically contain blood and pus. Patients often suffer from abdominal pain, nausea, and loss of appetite. Patients may have long symptom-free periods. Some people are healthy carriers of *B. coli*.

B. coli is found worldwide, but symptomatic infections are usually restricted to tropical countries. Unlike giardiasis, balantidiasis is rarely seen in North America. Animal reservoirs, especially pigs and monkeys, perpetuate the disease. Humans are usually infected when they ingest food or water contaminated by pig feces, but human-to-human transmission also occurs. Good hygiene is generally sufficient to prevent infection. Tetracycline, iodoquinol, or metronidazole is an effective cure.

Cryptosporidiosis. Cryptosporidiosis is receiving increased attention. It has recently caused epidemics in metropolitan areas (Chapter 9), and it causes severe problems for immunocompromised people, especially those with acquired immunodeficiency syndrome (AIDS). *Cryptosporidium*, the protozoan that causes the infection, inhabits the intestinal tract of many kinds of animals, including fish, reptiles, and mammals. People become infected by fecal-oral contamination from animal reservoirs or other humans.

Cryptosporidium invades the intestinal epithelium and multiplies there, causing enterocolitis, usually accompanied by abdominal pain and watery, bloodless diarrhea. People with a normal immune system recover in 10 days or less, but AIDS patients can develop severe life-threatening diarrhea that can last for months. There are no known antimicrobial cures. Treatment is limited to rehydration therapy if fluid loss becomes dangerous. Prevention—including good hygiene and avoiding high-risk sexual behavior—is best.

Helminthic Infections

When people talk about "having parasites," they usually mean parasitic worms—helminths (Chapter 12). Most of these infections begin when people consume feces-contaminated material or infected meat or fish. They

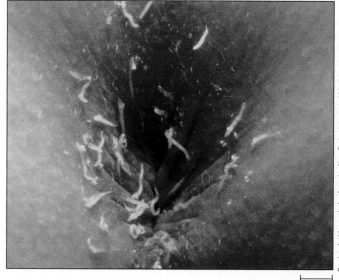

Reprinted with permission from the *New England Journal of Medicine*, Vol. 328, No. 13, p. 927. Photo by Martin Weber, MD, mrc-Labs, Fajara, Banjul, the Gambia, West Africa.

⊢——⊣
10 mm

FIGURE 23.15   Female pinworms leaving the anus of a 5-year-old child to deposit their eggs.

might remain in the intestinal tract or travel to the lungs, liver, muscles, or brain.

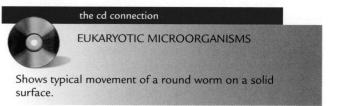

the cd connection

EUKARYOTIC MICROORGANISMS

Shows typical movement of a round worm on a solid surface.

Helminthic infections are most common in countries with poor sanitation, but they also occur in the United States. The worldwide burden of these infections is enormous. In some parts of the world, up to 90 percent of the population suffers from debilitating or life-threatening worm infestations. Most helminthic infections can be cured by one or another of the many new antiparasitic drugs.

In the following sections we'll briefly consider some of these worms and the diseases they cause.

Pinworm. Pinworms cause the most common helminthic infection in the United States. These small white roundworms (**Figure 23.15**) mate in the intestinal tract. Then the females crawl out and lay their eggs on skin near the anus. Parents who look for them may see these worms crawling across the skin of their infected child. Eggs perpetuate infection by fecal-oral transmission. They remain alive for more than a week.

Pinworms usually infect young children, but the infection can spread to adults in a family or day-care setting. For-

LARGER FIELD

WORM WORRIES

The idea of worms burrowing through our internal organs is unpleasant. Yet this is what happens in visceral larva migrans, a clinical syndrome most commonly caused by *Toxocara canis,* the canine roundworm. *T. canis*–infected dogs excrete the eggs in their feces. After maturing in soil for about 3 weeks, the eggs can infect humans or any other animals that happen to ingest them. In humans, a dead end for these parasites, the eggs hatch in the small intestine, burrow their way through the intestinal wall, and then travel through tissues, including the lung and liver.

Most clinically diagnosed cases of visceral larva migrans occur in young children, particularly those who eat dirt. Rarely do children become seriously ill with visceral larva migrans, but a milder, symptomless infection may be fairly common. Although we carefully dispose of human excrement—dramatically decreasing infestations with human worms such as *Ascaris*—dog feces are everywhere. Studies show that up to 30 percent of soil samples in public playgrounds and parks are contaminated with eggs of *T. canis.* This is not surprising, because *T. canis* infects up to 80 percent of puppies and at least 20 percent of adult dogs kept as pets in the United States. Studies show that many people who never noticed symptoms have detectable antibodies against *T. canis*—evidence that they have had visceral larva migrans. The extent of the problem is unknown.

tunately, the infection is not serious. Many people suffer no symptoms at all. Others complain only of perianal itching. The infection is easily treated with pyrantel pamoate or mebendazole, but eliminating eggs from the home and thus preventing reinfection is much more difficult.

Ascaris lumbricoides. *Ascaris lumbricoides,* another roundworm, has a rather complex life cycle. When eggs are ingested, they hatch in the intestine, producing larvae that burrow through the intestinal wall and enter the bloodstream. When they reach the lungs, they break into the alveoli and mature there in about 3 weeks. Then they are coughed out of the lungs and swallowed. When they reenter the digestive tract, they grow into worms that can be 12 inches long (see opening of Chapter 12). These produce the infective eggs that are shed in feces.

Although rare in the United States, *Ascaris* is extremely common in parts of the world with poor sanitation. Probably more than a billion humans suffer from ascariasis. Surprisingly, they can harbor large populations of these worms without suffering any symptoms. But the infection can be uncomfortable or dangerous. When large numbers of larvae migrate through the lungs, they cause an asthmalike cough. When mature worms crawl into the bile duct, they can block the normal flow of bile. Occasionally they form an enormous tangled mass in the intestine, creating an obstruction.

Ascaris infections can be effectively treated with antiparasitic drugs, including mebendazole, pyrantel pamoate, and piperazine. When heavily infested patients are treated, they pass masses of dead worms.

Hookworm. There two quite similar species of hookworm, *Ancylostoma duodenale* and *Necator americanus.* These roundworms are found in different parts of the world: *A. duodenale* in the Old World, and *N. americanus* in the Americas. Their eggs hatch in the soil and develop into larvae that can penetrate human skin. People usually become infected by walking barefoot in dirt containing these larvae. They make their way through the bloodstream to the alveoli, where they are coughed up and swallowed into the intestinal tract. They live and mate there, producing eggs that are passed in the feces. Diagnosis depends upon finding hookworm eggs in the feces.

Hookworms cause more serious disease than pinworms or *Ascaris* because adult worms have mouthparts called biting plates (**Figure 23.16**). The worm uses them to suck blood from the intestines of their host. Heavy infestations cause profound anemia. Victims suffer weakness, fatigue, and sometimes malnutrition, nausea, vomiting, and diarrhea.

Hookworms inhabit warm climates. *N. americanus* was once common in the southern United States, but improved sanitation has almost eliminated it. Infections can be treated by mebendazole or pyrantel pamoate.

Strongyloides stercoralis. *Strongyloides stercoralis,* like hookworm, is a roundworm with a larvae that can penetrate human skin (Figure 1.8). Also like hookworm, it enters the blood, spreads to the lungs, and produces adult worms in the intestinal tract. But it differs from hookworm in two important respects. It can reproduce without a human host, and its eggs hatch into larvae within the intestine. As a result, a

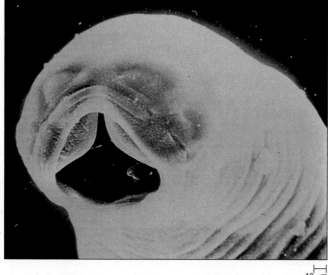

© Marsik/Visuals Unlimited

10 µm

FIGURE 23.16 Scanning electron microscopy of the hookworm *Necator americanus* showing its biting plates.

person can suffer continuous reinfection as larvae are passed and immediately reenter the body by penetrating the skin. This pattern leads to extremely heavy infestations.

The disease, **strongyloidiasis,** usually occurs in warm, moist climates, but it does occur sometimes in the United States and Canada. Patients may have respiratory complaints when the larvae migrate through the lungs and bloody diarrhea accompanied by malnutrition when they invade the intestinal wall. People with compromised immune systems suffer extremely debilitating infections that can be fatal. Treatment is with mebendazole or pyrantel pamoate.

Whipworm. Whipworm (*Trichuris trichiura*) is a whiplike roundworm. It's similar to *Ascaris* but has a simpler life cycle. Ingested eggs hatch in the small intestine and penetrate the intestinal wall. They mature, mate, and lay eggs there.

Infections with a small number of whipworms are usually asymptomatic, but heavier infestations can cause bloody diarrhea, abdominal pain, and weight loss. Extremely large numbers can obstruct the intestines. Treatment with mebendazole is effective.

Trichinella spiralis. *Trichinella spiralis* is the roundworm that causes trichinosis. Its life cycle is described in Chapter 12 (Figure 12.21). The symptoms of the disease—fever, muscle pain, and malaise—are caused by larvae migrating through the tissues. Symptoms are mild if only a few larvae are ingested, but they can become severe if large numbers are. In severe cases the heart, lungs, diaphragm, and brain are damaged. Trichinosis can be fatal within a month or two after infection.

Treatment with thiabendazole, mebendazole, and possibly steroids to decrease inflammation is only moderately effective.

Tapeworms. *Taenia* spp. are the most common tapeworms. The life cycles of these flatworms (Platyhelminthes) is described in Chapter 12. The two most important species clinically are *Taenia solium*, the pork tapeworm, and *Taenia saginata*, the beef tapeworm.

These tapeworm infections cause surprisingly few symptoms. Even patients who harbor enormous tapeworms are seldom aware of their infection, but some abdominal pain, diarrhea, and indigestion may occur. A much more serious condition occurs, however, when people ingest the pork tapeworm eggs as the result of human fecal contamination. In this infection, called **cysticercosis,** larvae become encysted in human tissues. If that happens in vital tissues such as the brain or the eye, the infection can be disabling or life threatening. *Taenia* infections are treated with niclosamide or praziquantel. Cysticercosis may require surgery.

Echinococcus granulosus is a tapeworm that ordinarily infects carnivores and herbivores (Chapter 12). The cysts that develop when humans are infected can become as large as 12 inches in diameter. They cause significant damage to vital organs. Sheepherders are the usual victims. They become infected when they ingest food or water contaminated by the feces of their sheepdogs. The disease is limited mainly to sheep-raising areas. The only treatment is surgical removal of the cyst.

INFECTIONS OF THE LIVER

Now we'll consider some viruses and helminths that infect the liver. Sometimes the damage is minor; in other cases it can be life threatening.

Viral Infections

Many different viruses—some identified only recently—infect the liver and cause hepatitis. The viruses all cause symptoms of liver damage, but their severity differs.

Hepatitis A Virus. Hepatitis A virus (HAV) is a small nonenveloped virus that contains single-stranded RNA. It belongs to the picornavirus family and the genus Enterovirus, which includes other viruses that enter the body through the gastrointestinal tract—such as poliovirus, the coxsackieviruses, and the echoviruses.

HAV spreads readily from person to person by the fecal-oral route. Huge numbers of virions are excreted in the feces of infected persons, even before they become ill. It causes hepatitis A (also called **infectious hepatitis**), the most common form of hepatitis, which tends to occur sporadically rather than as part of a large-scale outbreak. Transmission is primarily through close personal contact, such as from an infected family member. Families with young children in day care are at particularly high risk. Hepatitis A can also be contracted by oral-anal sexual contact. A study in Seattle, before the safe-sex campaign of the AIDS era, showed that 22 percent of susceptible gay men contracted hepatitis A each year.

After ingesting the hepatitis A virus, people feel well for about 25 days. During this period, virus from the gastrointestinal tract enters the bloodstream and becomes concentrated in the liver. It multiplies there, killing many **hepatocytes** (liver cells) as it does. This liver damage causes the typical symptoms of hepatitis: nausea, vomiting, loss of appetite, fatigue, disorders of taste and smell, dark urine, and jaundice, a yellow discoloration of the skin and whites of the eyes. The liver also becomes enlarged and tender. Other than the mysterious loss of taste and smell, the symptoms are all the consequence of diminished liver function. The damaged liver excretes bilirubin more slowly, so it accumulates in body tissues and urine, darkening them or turning them yellow. If the liver becomes severely damaged, it can't produce normal quantities of blood clotting factors. Then life-threatening bleeding can occur. Hepatitis A is usually a benign and self-limiting illness, but fatal complications sometimes occur.

Many people infected by HAV remain asymptomatic. Asymptomatic hepatitis is rare in adults but common in young children who can then transmit the virus to adults. Therefore hepatitis A epidemics are quite common in infant day-care centers. They are usually recognized only when the adult day-care workers and parents become jaundiced.

Hepatitis A is distinguished from other types of hepatitis by the presence of antiviral antibodies. Usually diagnosis rests on detecting **anti-HA** (the antibody against the major HAV antigen). By the time a patient becomes ill enough to consult a doctor, extremely high levels of anti-HA are usually present. Low titers of antibodies persist indefinitely and provide lifelong immunity.

This means that immunization is possible. But until recently, the only method available was passive immunization (administration of anti-HA antibodies). In 1995 the situation changed when a vaccine was approved by the U. S. Food and Drug Administration. Now two vaccines are available. They are now recommend for people over 2 years of age who are residents of or travelers to regions with high endemic rates of hepatitis A.

Hepatitis B Virus. Hepatitis B virus (HBV) causes a clinical syndrome that's difficult to distinguish from hepatitis A. But the viruses themselves, the circumstances, and the consequences of infection are quite different. HAV causes a relatively harmless infection. HBV is second only to tobacco as a known cause of human cancer.

Hepatitis B virus is a DNA-containing virus belonging to the hepadnavirus family (Chapter 13). Its virions are sometimes called Dane particles. It has three antigens, HBsAg, HBcAg, and HBeAg, which sometimes assemble to form empty shells that resemble spheres or tubules. These, along with the intact virion, are seen in the serum of HBV patients (**Figure 23.17**).

HBV has a very long incubation period, an average of 75 days. Because it is bloodborne, it's is sometimes called **serum hepatitis.** Transmission often occurs through exchange of contaminated blood or blood products: via transfusions, shared needles among drug users, and accidental needle sticks in hospitals. It can also be transmitted by intimate contact, including sexual contact, birth, and

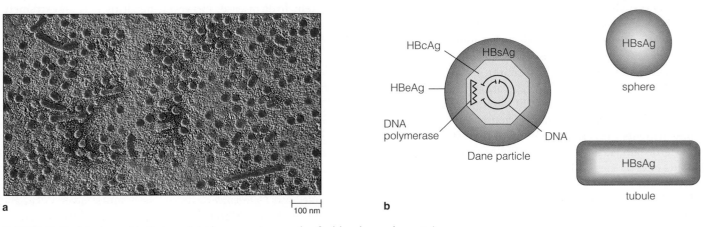

Photo by David Lacomis, University of Pittsburgh School of Medicine. Reprinted with permission from the *New England Journal of Medicine*, p. 1134, October 15, 1992.

FIGURE 23.17 The hepatitis B virus. (a) Electron micrograph of a blood sample containing Dane particle and empty spheres and tubules.(b) Sketch of these particles.

mother-infant contact. The virus is more highly infectious that HIV. In other respects, however, their transmission is so similar that HBV has been called AIDS's twin.

About half the people infected with HBV never show symptoms, and 99 percent of those who do survive the initial illness. But a few patients develop **fulminant hepatitis.** This disastrous condition causes total liver failure and death within a few days. HBV may also become a lifelong, life-threatening problem if chronic infection occurs, and that's relatively common. About 10 percent of infected adults and as many as 80 percent of infected infants develop it. The virus persists and may initiate a slow but ultimately fatal **cirrhosis** (liver destruction). It also increases a person's risk of developing liver cancer by as much as 300 times. Where HBV is widespread, liver cancer is common.

HBV is a major public health concern. Worldwide, there are more than 300 million HBV carriers. In the United States, infection rates soared during the 1980s, and by the early 1990s hepatitis B became the second most common reportable infection (after gonorrhea). Steroids and alpha interferon may have some value in treating hepatitis B, but these treatments are still experimental and the focus remains on prevention.

Progress has been made. Sensitive serological tests have been developed to detect HBV-contaminated blood so it can be discarded. Passive immunization is available to protect an at-risk person such as a hospital worker who receives an accidental needle stick with infected blood or infants born to chronically infected mothers. But the biggest step in preventing hepatitis B was the release of an effective hepatitis B vaccine in the early 1980s. Then an even newer vaccine, which is currently in use, was the first to be produced by recombinant DNA technology. This vaccination is safe and effective, but it still remains expensive. Nevertheless, in 1992 it was recommended for immunization of all newborns. Physicians are also vaccinating older children, particularly adolescents who are at increased risk by being sexually active.

The new HBV vaccine could virtually eradicate hepatitis B within the next generation. But immunization is lagging because the vaccine is expensive and parents feel babies get "so many shots." Moreover, three shots are required for full protection. But regulations are just now being implemented to require HBC immunization to enter school. That should be a major step toward eradication. Some patients with chronic HBV infection are being given interferon-alpha, but the treatment has many side effects and its effectiveness is limited.

Hepatitis C Virus. Even after most blood contaminated with the hepatitis B virus could be identified and discarded, 5 to 10 percent of patients receiving blood transfusions still developed hepatitis. In 1989 a new hepatitis virus was identified and named hepatitis C virus (HCV).

Like HBV, HCV is transmitted by contaminated blood or sexual contact. But epidemiology is complicated by the unusually long interval—up to 6 months—between exposure to HCV and the onset of illness. Only about half the patients infected by HCV develop hepatitis, but about half of these develop a chronic infection and liver disease. Although much less common than hepatitis B, hepatitis C is a serious disease that merits careful public health attention. Eliminating infected blood from the transfusion pool should prevent transfusion-acquired hepatitis C, but most cases are acquired by other means. New strategies must be developed to control this disease. Evidence suggests that alpha interferon may be useful in treating chronic liver disease caused by HCV.

Hepatitis Delta Agent. In 1977 researchers found a new antigen—delta antigen—and antibodies against it in patients with HBV. This antigen is part of another RNA-containing virus called hepatitis delta. This virus infects only people who are already infected by HBV. The disease it causes, type D hepatitis, is a severe, acute, and chronic liver disease. The acute infection is often fulminant. Chronic hepatitis delta in most cases progresses to cirrhosis and fatal liver disease. No treatment or prevention currently exists. Any chronic carrier of HBV is at risk, but uninfected people can be protected by vaccination against HBV.

Hepatitis E Virus. Tens of thousands of people in Asia, Africa, and India contract infectious hepatitis but have no detectable antibodies against the usual agent, HAV. These cases are all transmitted by the fecal-oral route. They often occur in epidemics that can be traced to contaminated water. The only known cases in Western countries were acquired during travel to the affected areas. The RNA-containing virus that causes this infection has recently been identified and named hepatitis E virus (HEV).

Like hepatitis A, hepatitis E is usually benign and self-limited. It is not known to cause chronic infection or persistent liver disease. However, hepatitis E is an extremely serious disease for pregnant women. About 20 percent of these victims of HEV victims die from acute hepatitis.

Helminthic Infections

Some species of trematodes, or flukes, of the flatworm (Platyhelminthes) family cause infections that do their primary damage to the liver (Chapter 12).

Liver Fluke Infections. *Fasciola hepatica,* the sheep liver fluke, and *Opisthorchis sinensis,* the Chinese liver fluke, are two of the most common causes of liver fluke infection. These worms enter a new human host when an infective larval form, called a metacercaria, is ingested with water plants or freshwater fish. Larvae penetrate the intestinal wall

FIGURE 23.18 Life cycle of *Fasciola hepatica,* the sheep liver fluke.

(Art by Carlyn Iverson.)

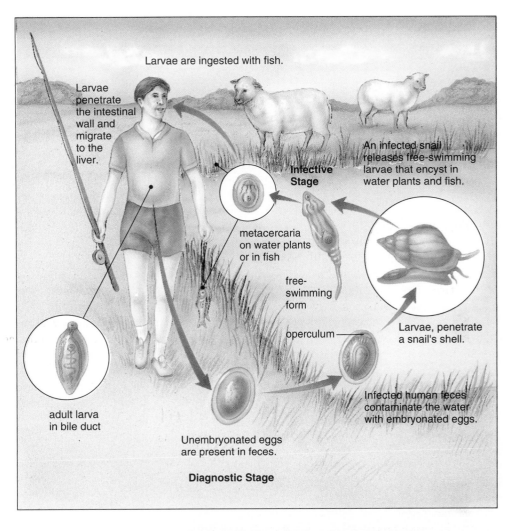

and migrate to the liver, where they enter the bile ducts and develop into egg-producing adult flukes. Eggs are passed in the feces. If egg-containing feces contaminate water, the larvae penetrate the shell of an aquatic snail, where the worm completes its life cycle. Infected snails release free-swimming larvae that encyst in water plants or fish, completing the cycle of infection when people eat them (**Figure 23.18**).

Because these flukes invade the liver, they cause signs and symptoms of liver damage: liver enlargement, tenderness, and jaundice. Liver fluke infection may cause extensive cell death, called liver rot, or predispose the infected person to the development of bile duct cancer. *F. hepatica,* which infects sheep, as well as humans, causes infection in sheep-raising areas of Egypt, Japan, Latin America, and countries of the former Soviet Union. *O. sinensis* occurs in China, Japan, Korea, and Vietnam. Both infections occur in the United States only among people who have traveled to the affected areas. The drug of choice for both infections is praziquantel. Prevention depends upon sanitation to prevent fecal contamination of water.

Table 23.4 summarizes the infections of the digestive system.

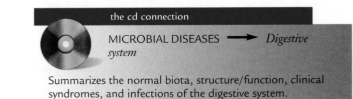

the cd connection

MICROBIAL DISEASES ⟶ *Digestive system*

Summarizes the normal biota, structure/function, clinical syndromes, and infections of the digestive system.

TABLE 23.4 Infections of the Digestive System

Infection	Causative Agent	Mode of Transmission	Symptoms	Prevention and Treatment
Oral Cavity and Salivary Gland Infections Bacterial				
Dental caries	*Streptococcus mutans* and others	Early colonization	Tooth enamel decay, eventual tooth loss	Prevent by limiting sucrose, cleaning frequently, adding fluoride to drinking water; treat by repairing cavity with filling

continued

TABLE 23.4 Infections of the Digestive System *(continued)*

Infection	Causative Agent	Mode of Transmission	Symptoms	Prevention and Treatment
Periodontal disease	*Bacteriodes gingivalis* and others	Early colonization	Destruction of tissues that support teeth	Prevent by cleaning carefully around gum margin
Viral				
Mumps	Mumps virus	Respiratory droplets	Swelling of parotids and other salivary glands; occasionally other organs involved, such as brain or testes	Prevent with vaccination
Intestinal Tract Infections **Bacterial**				
Shigellosis	*Shigella* spp.	Fecal-oral route	Dysentery; fever, abdominal pain, possibly seizures in children	Prevent with good hygiene; treat with rehydration and antibiotics
Typhoid fever	*Salmonella typhi*	Fecal-oral route	Enteric fever with headache, nausea, fatigue, confusion; complications include intestinal bleeding or perforation	Prevent with good hygiene and clean water supplies; treat with antibiotics
Salmonellosis	*Salmonella enteritidis*	Contaminated foods	Watery diarrhea, vomiting, abdominal pain; complications include dehydration and bloodborne infection	Treat with rehydration and antibiotics
Traveler's diarrhea	*Escherichia coli* and other Gram-negative rods	Fecal-oral route	Watery diarrhea, nausea, vomiting; complications include dehydration	Prevent with good hygiene, boiling water, and cleaning or cooking food; treat with rehydration and antibiotics for severe cases
Cholera	*Vibrio cholerae*	Fecal-oral route	Profuse watery diarrhea, often life-threatening dehydration	Prevent with good hygiene and clean water supply; treat with rehydration
Vibrio parahaemolyticus gastroenteritis	*V. parahaemolyticus*	Eating undercooked, contaminated fish	Watery diarrhea, nausea, vomiting; complications include dehydration	Prevent by refrigerating and cooking fish
Yersinia enterocolitica diarrhea	*Y. enterocolitica*	Fecal-oral route	Bloody diarrhea, fever, abdominal pain	Prevent with good hygiene; treat with antibiotics
Campylobacteriosis	*Campylobacter jejuni*	Fecal-oral route	Bloody diarrhea, fever, abdominal pain	Prevent with good hygiene; treat with antibiotics in severe cases
Gastritis, peptic ulcer	*Helicobacter pylori*	Unknown	Colonization of the gastric lining by *H. pylori* causes ulcer disease	Treatment remains controversial
Clostridium difficile diarrhea	*C. difficile*	Opportunistic pathogen—iatrogenic after antibiotic treatment; also nosocomial transmission	Colitis with possibly severe diarrhea	Treat with the antibiotic vancomycin in severe cases

TABLE 23.4 Infections of the Digestive System *(continued)*

Infection	Causative Agent	Mode of Transmission	Symptoms	Prevention and Treatment
Staphylococcal food poisoning	*Staphylococcus aureus*	Eating food that contains preformed toxin	Intoxication that causes early onset of watery diarrhea and vomting	Prevent by properly preparing and refrigerating food; treat by rehydration in severe cases
Bacillus cereus food poisoning	*B. cereus*	Eating contaminated food	Intoxication, usually mild; one form causes principally diarrhea, and another principally vomiting	Prevent by properly refrigerating food
Clostridium perfringens food poisoning	*C. perfringens*	Eating contaminated food	Usually a mild diarrhea	Prevent by properly refrigerating food
Botulism	*Clostridium botulinum*	Eating contaminated food	Food poisoning with neurological manifesta-tions (Chapter 25)	Prevent by properly canning and cooking food
Viral				
Rotavirus diarrhea	Rotavirus	Fecal-oral route	Profuse watery diarrhea and vomiting; often severe dehydration; usually affects children	Prevent with good hygiene
Norwalk diarrhea	Norwalk agents	Fecal-oral route	Watery diarrhea and vomiting; often occurs in epidemics	Prevent with good hygiene
Protozoal				
Amebiasis	*Entamoeba histolytica*	Fecal-oral route	Colitis, with bloody diarrhea; may spread to distant organs, may cause liver abscesses	Prevent with good hygiene; treat with metronidazole and iodoquinal
Giardiasis	*Giardia lamblia*	Fecal-oral route	Diarrhea, abdominal pain, foul-smelling intestinal gas	Prevent with good hygiene; treat with metronidazole
Balantidiasis	*Balantidium coli*	Fecal-oral route	Diarrhea with blood and pus, abdominal pain, loss of appetite	Prevent with good hygiene; treat with tetracycline or metronidazole
Cryptosporidiosis	*Cryptosporidium* spp.	Fecal-oral from other humans or animal reservoirs	Diarrhea, often profuse and chronic in AIDS patients	Prevent with good hygiene; treatment limited to rehydration
Helminthic				
Pinworms	*Enterobius vermicularis*	Fecal-oral route	Asymptomatic or anal itching; common in U.S. children	Prevent with good hygiene; treat with pyrantel pamoate or mebendazole
Ascariasis	*Ascaris lumbricoides*	Fecal-oral route	Often asymptomatic but can cause cough during lung migration, bile duct or intestinal obstruction	Prevent with good hygiene; treat with mebendazole, pyrantel pamoate, or piperazine

continued

TABLE 23.4 Infections of the Digestive System *(continued)*

Infection	Causative Agent	Mode of Transmission	Symptoms	Prevention and Treatment
Hookworm	*Ancyclostoma duodenale* or *Necator americanus*	Worm larvae penetrate skin	Nausea, vomiting, abdominal pain; significant anemia can be a complication	Prevent by not walking barefoot on contaminated ground and not contaminating soil with infected feces; treat with mebendazole or pyrantel pamoate
Strongyloidiasis	*Strongyloides stercoralis*	Worm larvae penetrate skin	Abdominal pain, diarrhea, malnutrition, coughing, or wheezing due to lung migration	Prevention same as for hookworm
Whipworm	*Trichuris trichiura*	Fecal-oral route	Usually asymptomatic, but may cause bloody diarrhea, weakness, weight loss	Prevent with good hygiene; treat with mebendazole
Trichinosis	*Trichinella spiralis*	Eating infected meat that is incompletely cooked	Fever, muscle pain	Prevent by cooking meat well and not feeding pigs garbage; treatment with thiabendazole mebendazole and steroids only moderately effective
Tapeworm infections	*Taenia saginata* (beef tapeworm) and *T. solium* (pork tapeworm)	Eating incompletely cooked, infected meat	Often asymptomatic except for passing worm proglottids; may cause malnutrition and weight loss	Prevent by eating well-cooked meat; treat with niclosamide or praziquantel
Echinococcosis (hydatid disease)	*Echinococcus granulosus*	Fecal-oral route, from dogs	Tissue destruction due to growth of cysts	Prevent by good hygiene; treat by surgically removing cysts

Infections of the Liver
Viral

Infection	Causative Agent	Mode of Transmission	Symptoms	Prevention and Treatment
Hepatitis A	Hepatitis A virus (HAV)	Fecal-oral route	Jaundice, nausea, vomiting, loss of appetite; usually not life threatening, no chronic infection or late complications	Prevent by good hygiene and new vaccine; only treatment is supportive care
Hepatitis B	Hepatitis B virus (HBV)	Contaminated blood or blood products, sexual contact, mother-infant contact during pregnancy or at time of birth	Jaundice, fulminant infection with liver failure may be fatal; late complications include chronic infection that may lead to cirrhosis or liver cancer	Prevent by vaccine and avoiding high-risk behaviors such as illicit intravenous drugs and unprotected sex; treatment is supportive only
Hepatitis delta	Hepatitis delta agent	Unknown	Coinfection with HBV required; may cause fulminant acute hepatitis or chronic infection	No prevention or treatment aside from avoiding HBV infection
Hepatitis C	Hepatitis C virus (HCV)	Contaminated blood or blood products, sexual contact	Acute hepatitis like type A or B and a chronic infection that may lead to liver damage like type B	Avoid high-risk behaviors (see hepatitis B)

TABLE 23.4 Infections of the Digestive System *(continued)*

Infection	Causative Agent	Mode of Transmission	Symptoms	Prevention and Treatment
Hepatitis E	Hepatitis E virus (HEV)	Fecal-oral route	Usually asymptomatic or mild infection like type A, but often fatal in pregnant women	Prevent by good hygiene (see hepatitis A)
Helminthic				
Liver fluke infections	*Fasciola hepatica* and *Opisthorchis sinensis*	Eating larval worms in fish or water plants	Enlarged, tender liver; jaundice	Prevent by avoiding fecal contamination of water; treat with praziquantel

SUMMARY

The Digestive System (pp. 551–553)

Structure and Function (pp. 551–552)

1. The digestive system consists of the tubelike alimentary tract (consisting of the oral cavity, pharynx, esophagus, stomach, small intestine, and large intestine) and associated organs (the teeth, tongue, salivary glands, liver, gallbladder, and pancreas).

2. Together these organs consume food, digest it, absorb nutrients, and eliminate unabsorbed waste.

Clinical Syndromes (pp. 552–553)

3. Clinical syndromes of the intestinal tract include gastritis (upper abdominal pain), gastroenteritis (diarrhea and sometimes vomiting and cramping pain), and colitis (also called dysentery, which is characterized by diarrhea containing blood and mucus).

4. Gastroenteritis can also be caused by bacteria that produce toxins outside the intestinal tract, which are then ingested in food. These are intoxications (food poisoning), not infections.

5. Clinical syndromes of the associated structures include dental caries, periodontal disease, parotitis, and hepatitis.

Infections of the Oral Cavity and Salivary Glands (pp. 553–555)

Bacterial Infections (pp. 553–554)

6. *Streptococcus mutans*, the primary cause of dental caries, produces glucans (the basis of dental plaque) and lactic acid (which can penetrate tooth enamel).

7. *Porphyromonas gingivalis* is the primary cause of periodontal disease. Infection begins when plaque accumulates at the gum margin, causing gingivitis. Eventually, deep pockets form in the infected tissue where anaerobic bacteria multiply (periodontitis). Teeth eventually become loose and fall out.

Viral Infections (pp. 554–555)

8. The mumps virus is transmitted through saliva or respiratory secretions of an infected person. It causes parotitis, usually with mild pain and fever. After local multiplication, the virus enters the bloodstream, occasionally causing complications. In adult males the testes may be painfully infected. Immunization for mumps is part of the MMR (measles, mumps, rubella) vaccine.

Infections of the Intestinal Tract (pp. 555–570)

9. Transmission of nearly all intestinal tract infections is by the fecal-oral route. The infections range from mild to deadly.

Bacterial Infections (pp. 555–565)

10. *Shigella* spp. cause shigellosis, or bacillary dysentery. More than half of the reported cases are in children under age 5. Transmission is by the oral-fecal route, and it is highly communicable, spreading rapidly through a family or day-care center. The bacteria produce a potent Shiga (or Shigalike) toxin. More drug-resistant strains are emerging.

11. *Salmonella typhi* causes typhoid fever, which begins with a high fever that lasts for days or weeks and develops into headache, other aches and pains, and sometimes a rash called rose spots. Patients shed the bacteria in feces and thereby infect others through the fecal-oral route. A small number become carriers. Many *S. typhi* strains are drug-resistant.

12. Salmonellosis is most often caused by *Salmonella enteritidis* or *Salmonella choleraesuis*. It is spread mainly by contaminated food. Symptoms include diarrhea, cramps, fever, and vomiting. The infection is usually self-limiting. Antibiotic treatment is contraindicated unless there are complications.

13. Most strains of *Escherichia coli* are harmless, but some are a major cause of traveler's diarrhea, infant diarrhea in developing countries, and hemolytic-uremic syndrome. One strain, designated O157:H7, has caused recent outbreaks and deaths in the United States.

14. *Vibrio cholerae* causes cholera, with its characteristic rice water stools. Diarrhea is so profuse that dehydration is sudden, and death often follows. Treatment consists of oral rehydration and tetracycline. In January 1991 the El Tor strain was found in Peru. Now cholera is probably in the Americas to stay.

15. *Vibrio parahaemolyticus* causes either a mild, self-limiting diarrhea or a cholera-like illness. About a quarter of all reported diarrhea cases in Japan are caused by *V. parahaemolyticus.*

16. *Yersinia enterocolitica* causes an enterocolitis. Abdominal pain can be so severe that appendicitis is suspected. Infection can come from wild or domestic animals, raw milk, oysters, or water.

17. *Campylobacter jejuni* probably causes more diarrheal illness in the United States than either *Salmonella* or *Shigella*. Campylobacteriosis is characterized by bloody diarrhea, abdominal pain, and sometimes fever. It is usually self-limiting.

18. *Helicobacter pylori* has been established as a cause of gastritis, which can lead to peptic ulcers and gastric carcinoma.

19. *Clostridium difficile* causes diarrhea as a result of antibiotic therapy. When therapy stops, the problem is usually resolved. For severe or chronic diarrhea, treatment is with vancomycin.

20. *Staphylococcus aureus* causes the most commonly reported food poisoning in the United States. Onset is sudden, with vomiting, diarrhea, and crampy abdominal pain. Staphylococci are accidentally introduced into food (most often whipped cream, eggs, mayonnaise, or ham) by someone carrying an enterotoxin-producing strain of *S. aureus*. The bacteria multiply in warm food. Prevention is possible through careful hand washing during food preparation and through refrigeration.

21. *Bacillus cereus* causes a mild, brief food poisoning. Spores of *B. cereus* not killed during cooking germinate to produce toxin if the food (usually meat, vegetables, or rice) is not refrigerated.

22. *Clostridium perfringens* can cause a food poisoning so brief and mild that it is seldom reported.

Viral Infections (pp. 565–566)

23. Rotavirus causes a gastroenteritis characterized by watery diarrhea, fever, and vomiting—commonly called "the stomach flu." It is highly infective by the fecal-oral route. The best prevention is hand washing. Malnourished children in developing countries become dehydrated and die from rotavirus infection.

24. Norwalk agents are extremely small viruses named for an epidemic of gastroenteritis that occurred in Norwalk, Ohio. Norwalk agents affect adults more often than children. Usually the illness is brief and self-limiting.

Protozoal Infections (pp. 566–568)

25. *Entamoeba histolytica* is an amoeba normally found in the intestines. Where sanitation is poor, it can cause amoebic dysentery (bloody mucoid stools, fever, and abdominal pain) and amebiasis (the parasites enter the bloodstream and can cause fatal liver abscesses). *E. histolytica* exists as a motile trophozoite and as a cyst, which is passed in the feces. Treatment is with metronidazole in combination with iodoquinol.

26. *Giardia lamblia*, a flagellated protozoan, causes giardiasis, which begins with an explosive diarrhea and copious amounts of foul-smelling intestinal gas. The first few such days are followed by a week or more of abdominal pain, nausea, and occasional diarrhea. *G. lamblia* has a sylvatic cycle, putting campers at special risk. When it occurs in the United States, it is most often in children. Treatment is with metronidazole or quinacrine.

27. *Balantidium coli* is a ciliate protozoan that causes balantidiasis, most often seen in tropical countries. Infected individuals may have long symptom-free periods. Other carriers may be entirely healthy. Metronidazole or tetracycline is effective.

28. *Cryptosporidium* is a protozoan that causes cryptosporidiosis, normally a mild enterocolitis. It has caused recent epidemics in the United States. AIDS patients and other immuno-compromised individuals can develop life-threatening diarrhea that lasts for months. Treatment is limited to rehydration.

Helminthic Infections (pp. 568–570)

29. Pinworm (*Enterobius vermicularis*) is a nematode that causes the most common helminthic infection in the United States. Infection comes from ingesting pinworm eggs, which can survive for up to a week on fomites. Children are most commonly infected, though adults who handle children are at risk. Perianal itching is the primary symptom. Treatment is with pyrantel pamoate or mebendazole.

30. *Ascaris lumbricoides* is a nematode. Infection comes from ingesting eggs, which hatch and burrow through the wall of the intestine to enter the bloodstream. They travel to the lungs, where they mature in 3 weeks. Larvae are coughed out of the lungs, swallowed, and reenter the digestive tract. Worms that mature in the intestines grow to 12 inches or more and produce eggs that are passed in the feces. Treatment is with various antiparasitic drugs.

31. Hookworm (*Ancylostoma duodenale* and *Necator americanus*) are nematodes. Their eggs hatch in soil, and larvae penetrate human skin. Larvae travel to the lungs, where they get coughed up and swallowed into the intestinal tract. Mature worms in the intestines produce eggs that are passed in the feces. Hookworms attach to the intestines and suck blood. Heavy infestations can cause diarrhea, vomiting, and even malnu-

trition and anemia. Mebendazole and pyrantel pamoate are effective.

32. *Strongyloides stercoralis*, a nematode, penetrates the skin, travels to the lungs, and produces adult worms in the intestinal tract. Eggs can hatch into infective larvae in the intestine, so the person can suffer continuous reinfection. Immunocompromised individuals can suffer a fatal diarrhea. Treatment is with mebendazole or pyrantel pamoate.

33. Whipworm (*Trichuris trichiura*), a nematode, hatches in the small intestine and burrows into the wall. It never leaves the intestine. Heavy infestations can cause bloody diarrhea, abdominal pain, and weight loss. Treatment is with mebendazole.

34. *Trichinella spiralis*, a nematode, causes trichinosis. Infection is by ingesting encysted larvae in meat, usually pork. The larvae develop into worms in the small intestine and produce more larvae that enter the bloodstream and travel to muscle, where they form cysts. Symptoms are fever, muscle pain, and malaise. In severe cases, larvae enter the heart, lungs, and brain. Trichinosis can be prevented with thorough cooking or freezing.

35. *Taenia* spp. are tapeworms (flatworms). People are infected by eating encysted larvae in pork (*T. solium*) or beef (*T. saginata*). The larvae mature in the intestines, where the adult attaches itself by suckers on its scolex and grows by adding proglottids. Mature proglottids, which contain infective eggs, are passed in the feces. Symptoms may include abdominal pain, diarrhea, and indigestion. A life-threatening infection called cysticercosis may develop if ingestion of eggs occurs as a result of human fecal contamination. Thoroughly cooking meat is the best prevention.

36. *Echinococcus granulosus* is a fluke that causes hydatid disease, which is endemic to sheep-raising areas. A person ingests eggs from dog feces. Hydatid cysts develop in their tissues, sometimes to 12 inches in diameter.

Symptoms depend upon the location. Surgical removal of the cyst is the only treatment.

Infections of the Liver (pp. 570–573)

37. Hepatitis A virus (HAV) causes type A hepatitis (infectious hepatitis). Transmission is usually by the fecal-oral route. Symptoms include jaundice, fatigue, loss of appetite, vomiting, and disorders of taste and smell. Illness is usually self-limiting, and the virus is eliminated from the body. Two vaccines are now available.

38. Hepatitis B virus (HBV) causes hepatitis B (serum hepatitis). Transmission is through contaminated blood or sexual contact. The clinical syndrome is indistinguishable from that of hepatitis A, but HBV causes cancer, fulminant hepatitis (total liver failure), and chronic infection, leading to death. HBV vaccination is part of the routine infant inoculation schedule.

39. Hepatitis D affects only those already infected with hepatitis B; it is often fatal.

40. Hepatitis C virus (HCV) causes hepatitis C. Transmission in 70 percent of cases is by blood or sexual contact. About one-fourth of infected individuals develop chronic liver disease, but symptoms do not appear for 6 months after infection. Interferon-alpha may be useful in treatment.

41. Hepatitis E (HEV) virus was only recently identified. It causes hepatitis E. Transmission is by the fecal-oral route. Hepatitis E is self-limiting and benign except in pregnant women, about 20 percent of whom die from acute hepatitis.

42. *Fasciola hepatica* (sheep liver fluke) and *Opisthorchis sinensis* (Chinese liver fluke) invade the liver and cause jaundice and liver rot or predispose the person to bile duct cancer. Infection comes from eating the larval form in water plants or fish. The worm requires a particular aquatic snail to complete its life cycle.

REVIEW QUESTIONS

The Digestive System

1. Describe the digestive system, naming the six associated structures.

2. What is the function of the digestive system? Explain this statement: Part of the digestive process is mechanical and part is chemical.

3. Describe the following clinical syndromes:
   a. gastritis
   b. gastroenteritis
   c. colitis (dysentery)
   d. dental caries
   e. periodontal disease
   f. parotitis
   g. hepatitis

4. What is the difference between an intoxication and food poisoning?

Infections of the Oral Cavity and Salivary Glands

5. Explain how *Streptococcus mutans* causes dental caries. What new treatments are being researched?

6. What is periodontal disease, and how does it develop? How is it treated?

7. What causes the mumps? Is this infection serious? Explain.

Infections of the Intestinal Tract

Bacterial Infections

8. Describe the cause, transmission, pathogenic mechanisms, and symptoms of shigellosis. How and why has treatment changed, and what are the prospects for the future?

9. Discuss the prevention of typhoid fever, including problems with carriers. What problems are arising with treatment today?

10. Is salmonellosis a food poisoning or an infection? Why might there be any confusion? Why is antibiotic treatment normally contraindicated?

11. How are intestinal diseases are caused by *Escherichia coli*?

12. How can cholera be prevented? Describe the recent epidemiology of cholera.

13. Describe the clinical syndrome caused by *Vibrio parahaemolyticus* infection. Who is at risk?

14. What is distinctive about the enterocolitis caused by *Yersinia enterocolitica*? What are the sources of infection?

15. Is infection by *Campylobacter jejuni* a major public health concern? Explain.

16. Does *Helicobacter pylori* cause gastritis, peptic ulcers, and gastric carcinoma? Explain your answer.

17. What is an iatrogenic illness? How does *Clostridium difficile* cause an iatrogenic diarrhea? How is the illness usually resolved?

18. Describe how *Staphylococcus aureus* can come to contaminate food and what circumstances would then lead to food poisoning. What are the symptoms?

19. How does *Bacillus cereus* cause food poisoning? How is this mechanism the same as or different from the one that causes *Clostridium perfringens* food poisoning?

Viral Infections

20. Describe the cause, symptoms, prevention, and treatment of the gastroenteritis we commonly call the stomach flu.

21. What are Norwalk agents? How is the illness they cause like and unlike rotavirus infection?

Protozoal Infections

22. What are the symptoms of amoebic dysentery? Why is diagnosis difficult? What is the treatment?

23. Explain this statement: *Giardia lamblia* has a sylvatic cycle. Describe symptoms, prevention, and treatment of giardiasis.

24. How does *Balantidium coli* cause infection? Where is balantidiasis most often seen?

25. Why are there more cases of cryptosporidiosis today than in the past? What preventive measures can high-risk individuals take?

Helminthic Infections

26. What is the most common helminthic infection in the United States? How serious is it? Explain. Why is it highly infectious? Name the microorganism that causes it.

27. Describe the life cycle of *Ascaris lumbricoides*. How does the life cycle cause symptoms in an infected individual? What kind of helminth is *A. lumbricoides*?

28. How are the two species of hookworm different? How are they alike? Trace the life cycle, from larvae to egg-producing adult. How is hookworm diagnosed and treated? Why might iron supplements be required?

29. How is *Strongyloides stercoralis* like hookworms? How is it different? What is autoinfection?

30. Describe a whipworm infection—from transmission through treatment.

31. How does *Trichinella spiralis* cause illness? What are the symptoms? Can trichinosis be fatal? Explain. What kind of helminth is *T. spiralis*? Why is trichinosis relatively rare in the United States today?

32. What kind of helminths belong to the *Taenia* spp.? How do *T. solium* and *T. saginata* cause illness? How is infection diagnosed? What is cysticercosis?

33. How does *Echinococcus granulosus* cause disease? Who is most at risk? What is the treatment? What kind of helminth is *E. granulosus*?

Infections of the Liver

34. For each type of hepatitis (A, B, C, D, and E), give the following:
   a. the causative agent
   b. mode of transmission
   c. individuals most at risk
   d. clinical syndrome and complications (if any)
   e. treatment
   f. prevention

35. What kind of helminths are *Fasciola hepatica* and *Opisthorchis sinensis*? Describe their life cycle. What are the symptoms and treatment? Why do infections in the United States occur only among people who have traveled to the infected areas?

CORRELATION QUESTIONS

1. If a human skeleton was found with a few missing teeth but caries-free remaining teeth, what would you guess about when the person lived and how old he or she was at death? Explain.

2. A group of people who shared the same meal later experienced gastroenteritis. What question would you ask them to decide whether it was caused by *Staphylococcus aureus* or *Salmonella enteritidis*?

3. If a patient being treated with tetracycline for periodontal disease developed a mild diarrhea, what do you think might be the cause? Explain.

4. Why do think most bacteria that cause intestinal infections are Gram-negative? Explain.

5. Would you recommend that a child be vaccinated against HBV? Why?

6. Do you think it will be possible to eradicate cholera? Explain.

ESSAY QUESTIONS

1. Compare the respiratory and digestive systems with respect to their vulnerability to infection and its consequences.

2. Develop a convincing scenario for how *Escherichia coli* O157:H7 might have arisen from a nonpathogenic strain.

SUGGESTED READINGS

Alter, M. J. 1989. Hepatitis C: And miles to go before we sleep. *New England Journal of Medicine* 321:1538–39.

Cover, T. L., and R. C. Aber. 1989. Medical progress: *Yersinia enterocolitica. New England Journal of Medicine* 321:16–24.

Gordon, R. 1989. Tales of Typhoid Mary. *Hippocrates*, March/April, 102–3.

Grady, G. F. 1986. The here and now of hepatitis B immunization. *New England Journal of Medicine* 315:250–51.

Lamont, J. and H. F. Jenkinson. 1998. Life below the gum line: pathogenic mechanisms of *Porphyromonas gingivalis. Microbiology and Molecular Biology Review* 62:1244–63.

Mahmoud, A. A. F. 1989. Parasitic protozoa and helminths: Biological and immunological challenges. *Science* 246:1015–22.

Shaw, J. H. 1987. Medical progress: Causes and control of dental caries. *New England Journal of Medicine* 317:996–1004.

Williams, R. C. 1990. Medical progress: Periodontal disease. *New England Journal of Medicine* 322:373–82.

For additional readings, go to InfoTrac College Edition, your online research library at: http://www.infotrac.thomsonlearning.com

TWENTY-FOUR

Infections of the Genitourinary System

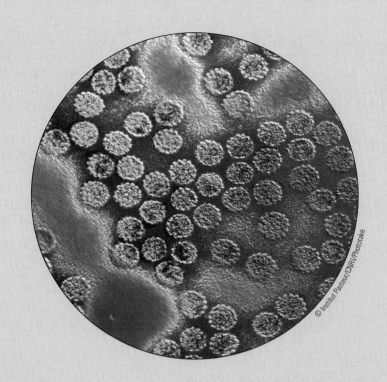

© Institut Pasteur/CNRI/Phototake

CHAPTER OUTLINE

LEARNING GOALS

To understand:

- The anatomy and function of the urinary and reproductive systems and their defenses against microorganisms
- The clinical syndromes that characterize genitourinary infections
- The bacterial causes of urinary tract infections and their clinical syndromes, prevention, and treatment

- The bacterial and viral causes of sexually transmissible diseases (STDs) and their clinical syndromes, prevention, and treatment
- The bacterial, fungal, and protozoal causes of female reproductive tract infections and their clinical syndromes, prevention, and treatment

- The causes of male reproductive tract infections and their clinical syndromes, prevention, and treatment
- The bacterial and viral causes of infections transmitted from mother to infant and their clinical syndromes, prevention, and treatment

You Can't Be Too Careful

One morning, A. V., a 25-year-old married woman, noticed some discomfort on urination. Over the next few hours, her discomfort became a burning sensation. She had to urinate often and suddenly, but each time she passed very little urine. Her symptoms worsened during the day. The next morning, she had a fever of 101°F. She felt nauseated, vomited after breakfast, and noticed pain on her right side. Then A. V. called her doctor.

A. V.'s physician recorded her history: fever, **dysuria** (painful urination), urinary frequency, and urinary urgency. When he examined her, she had a fever of 102°F and appeared quite ill. He noted tenderness when he pressed on the part of the abdomen overlying the bladder. When he tapped on the right side of her back near her kidney—where the ribs meet the spine—A. V. winced in pain. Her physician asked A. V. to collect a urine specimen for urinalysis. Under the microscope the specimen showed large numbers of bacteria and leukocytes. The leukocytes were clumped together in long masses called **casts.** The rest of the urine sample was sent to a microbiology laboratory for culture.

A. V.'s physician made a preliminary diagnosis of urinary tract infection (UTI). The fever and tenderness around her right kidney further suggested that it was probably **pyelonephritis** (kidney infection). He prescribed cephalexin, an oral antibiotic. He wanted to see her again the next morning. Twenty-four hours later A. V. felt and looked much better. Her temperature had returned to normal. Her urinary symptoms were better but not gone. Her physician asked for another urine specimen. It contained a few white blood cells but no bacteria. A. V. was told to continue taking the medicine for 10 days and return once more. Urine samples from her second and third visits were also sent to the laboratory for culture.

The laboratory reported that A. V.'s first urine sample contained a pure culture of *Escherichia coli* with more than 100,000 bacterial cells per milliliter of urine. Fortunately, the strain of *E. coli* was sensitive to cephalexin. Her second and third samples were sterile.

Case Connection

- In this chapter you will learn why women are more susceptible than men to pyelonephritis.
- You will also learn how pyelonephritis is diagnosed and what complications it can lead to.
- You will also learn why A. V.'s physician prescribed a relatively long (10-day) treatment.

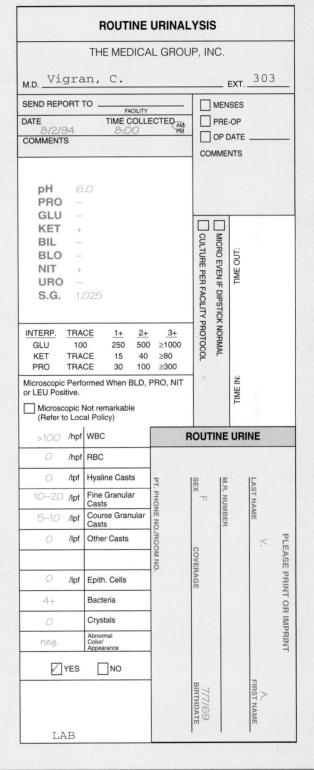

Laboratory slip for urinalysis.

FIGURE 24.1 Organs of the genitourinary system. (a) The urinary system. (b) The male reproductive system. (c) The female reproductive system.

(Art by Kevin Somerville from Biology: Concepts and Applications, 2nd ed., by C. Starr, Brooks/Cole, 1994. All rights reserved.)

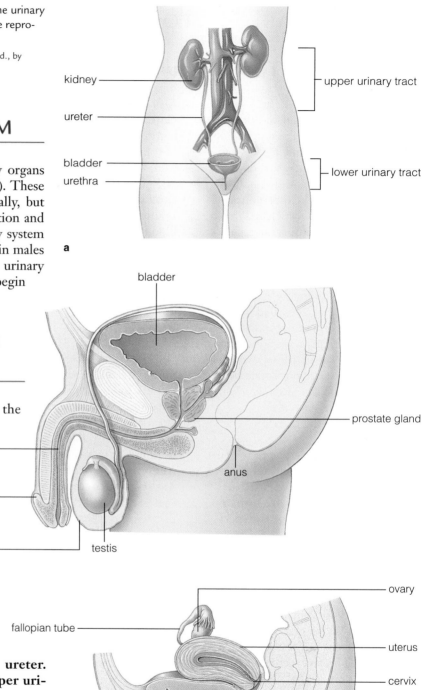

THE GENITOURINARY SYSTEM

Together the **genital** (reproductive) and urinary organs make up the **genitourinary** system (**Figure 24.1**). These two sets of organs are closely related anatomically, but they fulfill entirely different functions: reproduction and excretion of metabolic wastes. The genitourinary system is the only organ system that differs significantly in males and females. In females the reproductive and urinary tracts are completely separate. In males they begin separate and then join.

STRUCTURE AND FUNCTION OF THE URINARY SYSTEM

Let's begin by considering the organs that are the same in males and females. The two kidneys clear metabolic waste from the body by removing impurities from the blood and concentrating them in the urine. Blood flowing into a kidney enters a cluster of blood vessels called the **glomerulus.** There the blood is filtered through a porous membrane under high pressure. The filtrate then passes through a series of microscopic tubules, where it is processed: Certain molecules are added, and others, including water, are reabsorbed into the bloodstream. The processed product of the tubules is urine. It leaves the kidney through a tube called a **ureter.** The kidneys and their ureters constitute the **upper urinary tract.**

The ureters transport the urine to the **bladder,** the saclike container where urine is stored. It leaves the body through a tube called the **urethra.** Here males and females differ. The urethra of a female is only a few centimeters long. And it transports only urine. The male urethra is longer. It leads to the head of the penis and transports **semen** (sperm-containing fluid) in addition to urine. The bladder and urethra constitute the **lower urinary tract.**

STRUCTURE AND FUNCTION OF THE REPRODUCTIVE SYSTEM

The reproductive systems of males and females produce, store, and transport the **gametes** (**ova** or **sperm**). The female reproductive tract also provides a place for the fetus to develop.

In males the two **testes,** contained within the saclike **scrotum,** produce sperm. They are carried in liquid through a series of ducts to the urethra. Along the way, several glands, including the **prostate gland,** add other substances to the liquid. The sperm-containing product is **semen.** It leaves the body through the urethra by ejaculation. Because it transports both urine and semen, the male urethra belongs to both the urinary and genital systems.

In females, two **ovaries** (organs within the abdominal cavity) produce **ova** (eggs). They pass from the ovaries through the **fallopian tubes** to the **uterus.** The uterus ends at the **cervix,** a necklike extension that projects into the **vagina.** If the ovum unites with a sperm cell, the resulting **embryo** develops within the uterus. Contents of the uterus, including menstrual secretions, as well as products of conception, leave the body through the vagina. Sperm enter the female body through the vagina on their way to fertilize the ovum.

Defenses and Normal Biota: A Brief Review

The genitourinary system transports materials from inside the body to the outside. In this respect it differs fundamentally from the respiratory and gastrointestinal systems. Because they transport materials from outside the body to the inside, they are constantly exposed to the microorganisms and are heavily colonized by commensal species (Chapter 14). In contrast, the genitourinary system comes in contact with microorganisms only where its outflow meets the skin. So the genitourinary system is largely free of microorganisms. The vagina is a notable exception. It is heavily colonized by microorganisms (Chapter 14).

When pathogens enter the normally sterile genitourinary system, they cause disease. Some enter where the system opens at the skin and make their way against the flow of urine. Others readily enter the female genital tract during childbirth or nonsterile abortion. Many are adapted to enter during sexual intercourse. These cause the **sexually transmissible diseases** (**STDs**). Pathogens from the female reproductive tract can also be transmitted to a baby during birth.

The urinary tract is defended by an outer layer of tightly joined epithelial cells that line the urethra and also by the flow of urine that flushes out microorganisms. The vagina is further defended by its low pH.

Clinical Syndromes

The urinary tract is infected often, usually by bacterial pathogens (**Table 24.1**). **Cystitis** (infection of the bladder) is associated with **urethritis** (inflammation of the urethra), which causes frequent and painful urination. **Pyelonephritis** (infection of the kidneys) is associated with fever and flank pain, as well as urethral symptoms. But often it's difficult to determine which organs are involved. Then it's simply called a **urinary tract infection** (**UTI**).

All organs of the female genital tract are vulnerable to infection. When the vagina is infected, the syndrome is called **vaginitis,** which is characterized by vaginal irritation and discharge. It's uncomfortable but not life threatening. Infection of the lining of the uterus is called **endometritis;** infection of the fallopian tubes is called **salpingitis;** and infection of the ovaries is called **oophoritis.** When the uterus, fallopian tubes, ovaries, and abdominal cavity itself and are involved, it's commonly called **pelvic inflammatory disease** (**PID**). PID is often associated with fever and abdominal pain. It can be life threatening.

In this chapter we'll discuss the microbial diseases that cause these various clinical syndromes. We'll also discuss the STDs. Because they cause a variety of clinical syndromes, their only common property is their mode of transmission. Then we'll discuss **perinatal** (around birth) infections, which are usually transmitted from mother to infant. We'll start by considering infections of the urinary tract.

URINARY TRACT INFECTIONS

All the major pathogens that infect the urinary tract are bacteria. They enter through the urinary tract itself or from the bloodstream. Some can damage the kidneys permanently or multiply in the bloodstream, causing a life-threatening septicemia.

Urinary tract infections are extremely common. A. V.'s case was typical in many respects: Most victims are females, and they usually become infected by bacteria, most commonly *Escherichia coli*, from their own gastrointestinal tract. Bacteria enter the urethra from fecal contamination of the surrounding skin. Sometimes sexual intercourse forces bacteria into the female urinary tract. Infection is usually confined to the bladder, but occasionally bacteria enter the ureters and pass into the kidneys, as they did in A. V.'s case.

Females are more vulnerable than males to UTIs because their urethras are shorter and relatively close to the

TABLE 24.1 Clinical Syndromes of Infection of the Genitourinary System

Syndrome	Region Affected	Signs and Symptoms	Causative Agents
Cystitis	Bladder	Painful and frequent urination	Many organisms, including *Escherichia coli*
Urethritis	Urethra	Painful and frequent urination; often a discharge in males	Many organisms, including *Neisseria gonorrhoeae* in males
Pyelonephritis	Kidneys	Fever, flank pain	Many organisms, including *E. coli*
Urinary tract infection (UTI)	Any or all parts of the urinary system (bladder, urethra, kidneys)	Any or all of the symptoms of cystitis, urethritis, pyelonephritis	Many organisms, including *E. coli*, *Pseudomonas* spp., *Proteus* spp., *Klebsiella* spp.
Vaginitis	Vagina	Vaginal irritation, often a discharge	Many organisms, including *Candida albicans*, *Trichomonas vaginalis*, *Gardnerella vaginalis*
Endometritis	Uterus	Fever, tender and enlarged uterus, abdominal pain	Many organisms, including *E. coli*, *N. gonorrhoeae*, group B streptococci
Salpingitis/oophoritis	Fallopian tubes/ovaries	Fever, abdominal pain; complications include scarring and infertility	Many organisms, including *N. gonorrhoeae*, *Chlamydia trachomatis*
Pelvic inflammatory disease (PID)	Vagina, uterus, fallopian tubes, ovaries, abdominal cavity itself	Fever, abdominal pain; can be life threatening	Many organisms, including *N. gonorrhoeae*, *C. trachomatis*

anus. In both males and females, any interference with the normal flow of urine adds to vulnerability. A partial obstruction of the urinary tract (for example, by an enlarged prostate gland) or an inability to completely empty the bladder (for example, due to a spinal cord injury) often results in recurrent UTIs. Hospitalized patients who have urinary catheterization are particularly vulnerable to UTIs. UTIs are the most common type of nosocomial infection (Chapter 20).

Now let's consider how a UTI is diagnosed.

Diagnosis

A preliminary diagnosis of UTI is often based upon a clinical history, physical examination, and urinalysis. That's what A. V.'s physician did. A normal urine sample contains very few bacteria or blood cells. But urine from a person with a UTI typically contains large numbers of bacteria, as well as white and red blood cells. The white blood cell casts seen in A. V.'s urine strongly suggested pyelonephritis because such casts form in the tubules of the kidney. An even quicker office test for UTI is an inexpensive dipstick test. If it detects **leukocyte esterase** (an enzyme from white blood cells), nitrites (produced by bacterial metabolism), and blood in a sample of urine, the patient probably has a UTI.

A definitive diagnosis of UTI requires the culturing of bacteria from the urine sample. The sample has to be handled carefully because urine is an excellent culture medium for bacteria. If a sample is allowed to stand at room temperature for even a brief time, the small number of skin contaminants in any urine sample begin multiplying rapidly. So a urine sample must be collected to minimize contamination and refrigerated until it is cultured. Even then only finding a pure culture with more than 100,000 bacterial cells per milliliter is meaningful. Fewer cells or a mixed culture generally indicates that the sample was contaminated.

Lower versus Upper Urinary Tract Infections

The seriousness of a UTI depends upon the extent of the infection. Infections that involve only the lower urinary tract are common. They can be uncomfortable, but they don't have serious long-term consequences. Their typical symptoms (urgent and frequent urination) are a natural defense mechanism. Each time the bladder is emptied, bacteria are eliminated. Normally less than 1 ml of infected urine is left behind to inoculate the sterile urine coming from the ureters, so frequent urination can control the infection. Thus patients with UTIs should drink lots of fluid.

Infections such as A. V.'s that also involve the kidneys and ureters are less common but more worrisome than those restricted to the lower urinary tract. Patients with upper urinary tract infections tend to be sicker with fever, vomiting, and flank pain. There is also a hazard that bacteria from the kidneys may enter the bloodstream and cause a life-threatening infection. That's why A. V.'s physician followed her progress closely after diagnosing pyelonephritis. That's also why he treated her with antibiotics for 10 days instead of the usual 3-day treatment for uncomplicated upper UTIs. Repeated episodes of pyelonephritis may cause lasting kidney damage, which could cause kidney failure, necessitating dialysis or a kidney transplant.

Treatment

UTIs are usually diagnosed before the results of the urine culture are available, so treatment must be started before the antibiotic sensitivity of the pathogen is known. Moreover, many strains of the intestinal bacteria that cause UTIs carry R factors, making them resistant to several antibiotics (Chapter 21). So a physician has to make an educated guess about which antibiotic to prescribe. Some studies show that about 90 percent of *Escherichia coli* strains are sensitive to cephalexin, so prescribing this drug for A. V.'s infection was a good bet but far from a sure one. Moreover, patterns of antibiotic resistance change constantly. Because no antibiotic is effective against all strains of common UTI pathogens, a physician must prescribe on the basis of known patterns of antibiotic resistance in the community.

Bacteria That Cause UTIs

Escherichia coli is by far the most common pathogen of the urinary tract. It causes 90 percent of first UTIs in healthy people and most nosocomial UTIs. Its critical virulence factor is adherence. Each cell of a **uropathogenic** strain (one capable of causing a UTI) has from 10 to 200 adhesin-bearing pili that can bind firmly to matching cell receptors on the urinary epithelium. Women who have greater numbers of epithelial receptors in their urinary tract are vulnerable to repeated urinary tract infections.

Other bacteria cause UTIs that are clinically indistinguishable from A. V.'s *E. coli* infection. These include species belonging to the enteric bacterial genera *Proteus* and *Klebsiella*, as well as to *Pseudomonas*. Because *Proteus* and *Klebsiella* form ammonia from urea, they make the urine alkaline. This fosters the formation of **kidney stones** (solid material that can block urinary flow through the ureters). Some Gram-positive bacteria, particularly enteric streptococci, also cause UTIs.

Now we'll consider an infection of the urinary system that doesn't enter the body through an opening of the genitourinary tract.

Leptospirosis

Leptospirosis is caused by *Leptospira interrogans* (**Figure 24.2**), a member of the spirochete family (Bergey's Spirochaetes, Chapter 11). It usually enters humans through mucous membranes or a break in the skin, usually from animal urine, urine-contaminated water, or soil. Many animals, including dogs, cats, cattle, rodents, and various wild animals, are reservoirs of *Leptospira*, but rats are the principal one. Once this highly motile organism penetrates mucous membranes or damaged skin, it enters the bloodstream. Then it's distributed throughout the body. Those cells that reach the kidneys multiply there and are excreted in the urine.

Probably most people infected by *L. interrogans* are asymptomatic, and even those who become ill suffer only flulike symptoms: fever, headache, muscle pain, and reddened eyes. They usually recover within a week of their first symptoms. Some leptospirosis patients, however, progress to life-threatening liver and kidney damage, producing a rash, bleeding, and shock. Severe cases are usually marked by jaundice. About 10 percent of these patients die. But if the patient recovers, liver and kidney function returns to normal.

Diagnosis of leptospirosis is difficult. *L. interrogans* cells are so small that they are barely visible by light microscopy,

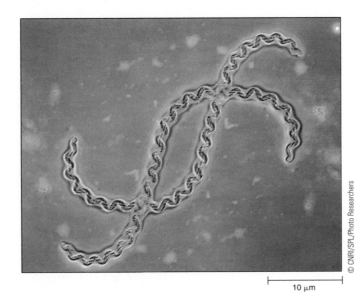

FIGURE 24.2 *Leptospira interrogans*. It gets its species name from the hooked end that resembles a question mark.

and culturing it requires special media not available in most clinical laboratories. As a result, diagnosis usually depends upon detecting an increasing titer of antibodies against *L. interrogans*. Treatment with penicillin or tetracycline, if started early, can shorten the illness and prevent serious complications.

Less than 100 leptospirosis cases are reported annually in the United States. But these are probably only the most severe ones. At one time leptospirosis occurred mainly among packinghouse workers, sewer workers, and people who lived in rat-infested housing. Today it's most common in young people who swim or wade in infected water. Leptospirosis is a common disease in tropical countries. In Thailand and Vietnam, for example, almost a quarter of the population shows serological evidence of having been infected by *Leptospira*. Large doses of penicillin G are the preferred treatment.

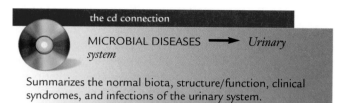

the cd connection

MICROBIAL DISEASES ⟶ *Urinary system*

Summarizes the normal biota, structure/function, clinical syndromes, and infections of the urinary system.

SEXUALLY TRANSMISSIBLE DISEASES (STDs)

Until the 1970s many clinicians believed that STDs would soon be a eliminated by the use of antibiotics. But sadly, the problem didn't go away. It worsened. New sexually transmissible pathogens have been recognized, and many of these are viral, so they aren't treatable by antibiotics (**Table 24.2**).

TABLE 24.2 Some Microorganisms That Cause Sexually Transmissible Infection

Causative Agent	Infection	Comments
Bacteria		
Neisseria gonorrhoeae	Gonorrhea	Most common reportable STD in the United States; usually symptomatic in men and asymptomatic in women; new antibiotic-resistant strains appearing.
Treponema pallidum	Syphilis	Manifests many clinical syndromes; disease progresses through stages over several decades; treatable with penicillin.
Chlamydia trachomatis	PID in women and nongonococcal urethritis (NGU) in men; lymphogranuloma venereum (LGV)	Serovars D–K cause the most commonly transmitted STD in the United States; mild illness and difficult to diagnose, though treatable. LGV is rare, painful, and treatable.
Ureaplasma urealyticum, Mycoplasma hominis	Urethritis, vaginitis	Widespread, often asymptomatic, but can cause PID in women and NGU in men.
Haemophilus ducreyi	Chancroid	Open sores on genitals can lead to scarring without treatment; on the rise in U.S. inner cities; treatable.
Calymmatobacterium granulomatis	Granuloma inguinale	Draining ulcers that can persist for years; treatable; rare in U.S.
Viruses		
Herpes simplex virus	Genital herpes simplex	Painful blisters; enters latent stage, with reactivation due to stress; also oral, pharyngeal, and rectal herpes from sexual contact. No cure. Extremely prevalent in U.S.
Human papillomarivus	Condyloma acuminata; probably predisposes to cervical cancer	Genital warts, though often asymptomatic. No cure. Very common in U.S.

Moreover, the "traditional" bacterial STDs have made a stunning comeback. In the United States today, more than 12 million STDs are diagnosed each year, at an estimated cost of $3.5 billion. Most people in the U.S. contract at least one STD before the age of 35. Those at highest risk are people with multiple sexual partners who do not use condoms.

The consequences of STDs are most severe for women and children. STDs can block fallopian tubes (causing infertility and life-threatening tubal pregnancies) and cause cancer of the female genital tract, fetal death, birth defects, and newborn blindness. Women are particularly vulnerable because transmission of STDs is more efficient from men to women than vice versa. Also, early diagnosis of most STDs is more difficult in women.

Acquired immunodeficiency syndrome (AIDS)—the fatal and incurable STD that erupted in the 1980s (Chapter 27)—has had a profound effect on the epidemiology of all STDs. Because of the common mode of transmission, people with one STD are likely to have another, sometimes with disastrous consequences. For example, AIDS patients with syphilis suffer unusually severe symptoms. And STDs—such as syphilis, chancroid, and herpes—that cause genital ulcers facilitate the sexual transmission of AIDS. On the other hand, the AIDS epidemic focused long-overdue attention on safer sex practices that could limit the transmission of all STDs. Safer sex means reducing the number of sexual contacts (**Figure 24.3**) and using condoms to decrease the chance of infection with each contact.

Bacterial Infections

The most common STDs are caused by bacteria. They are all curable by antibiotics.

Gonorrhea. Gonorrhea is caused by *Neisseria gonorrhoeae*, which is often called the **gonococcus** (Bergey's Betaproteobacteria, Chapter 11). It remains one of the most common reportable diseases in the United States. In the 1970s it reached almost epidemic proportions. With increased sexual freedom along with decreased use of condoms (because of the availability of birth control pills), it reached a maximum of more than 450 reported cases for every 100,000 people in the United States. Then the rate underwent a steady decline, reaching 122 cases per 100,000 in 1997. In 1998 the rate abruptly increased by 7.9 percent and since that time has

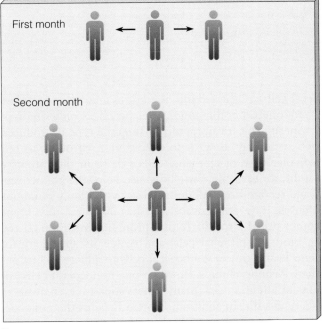

a

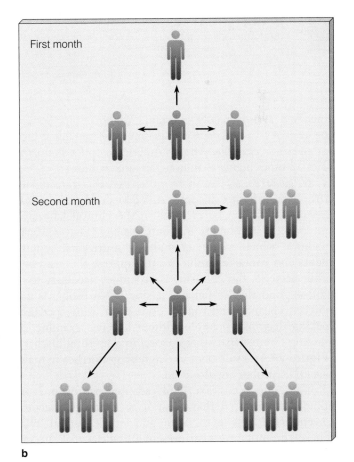

b

FIGURE 24.3 How the number of sexual contacts affects the spread of STDs. (a) If people have 2 sexual contacts a month, 1 infected person can infect 8 others in 2 months. (b) If people have 3 sexual contacts a month, 1 infected person can infect 15 others in 2 months.

SHARPER FOCUS

A GAME WE MAY NEVER WIN

Overall, gonorrhea rates in the 1990s fell, probably in part because of increased use of condoms and more cautious sexual behavior fostered by the AIDS epidemic. But at the same time, antibiotic resistance in *Neisseria gonorrhoeae* became a serious problem. And it's growing worse in the twenty-first century.

Less than 15 years ago, when almost all strains of *N. gonorrhoeae* isolated in the United States were drug-sensitive, gonorrhea could be treated by a single intramuscular injection of penicillin or, for penicillin-allergic patients,

a brief course of oral tetracycline. Now neither of these drugs can be counted on to cure a high enough percentage of patients to justify their use as a first-line agent. This might seem like an overreaction because only a little more than 8 percent of gonorrhea cases diagnosed in 1990 were caused by drug-resistant strains. But research shows that 2 percent treatment failures are enough to spread a resistant strain throughout a community. The treated patient who has not been cured will pass the resistant strain to his or her next sexual contact, who

will in turn spread the drug-resistant infection to someone else. Thus even a low failure rate in treatment is unacceptable from a public health standpoint.

There is a little more good news, though. We have new drugs for treating penicillin- and tetracycline-resistant strains of *N. gonorrhoeae*. Cure is virtually certain after a single—if extremely painful—injection of ceftriaxone, from the cephalosporin family, or oral administration of the closely related cefixime. Newer drugs of the fluoroquinolone family are also effective but are considered second-line agents because

they do not cure syphilis, which also infects many gonorrhea patients.

These new drugs, however, are much more expensive than penicillin. Curing a single patient with ceftriaxone may cost as much as $80, compared with less than $1 for penicillin. This is a major problem in publicly funded clinics. Moreover, no one knows how long these new drugs will remain effective as drug-resistant strains continue to emerge. In trying to control gonorrhea with antimicrobial treatment, we may be playing a game of catch-up that we can never win.

---

remained approximately constant. Young adults between the ages of 16 and 25 and ethnic minorities remain at highest risk. For example, African Americans suffer gonorrhea rates 32 times higher than non-Hispanic whites.

N. gonorrhoeae, like many other sexually transmissible pathogens, is fragile. It's easily killed by drying or exposure to sunlight. It can be cultured in the clinical laboratory but only in a complex medium, usually Thayer-Martin medium, which meets its multiple nutritional requirements and contains antibiotics to suppress the growth of other bacteria. But culturing is no longer necessary to diagnose gonorrhea in females. Using DNA probes is now quicker, cheaper, and can be done on a urine specimen, making diagnosis possible without a pelvic examination. Finding Gram-negative diplococci in a urethral discharge (**Figure 24.4**) is sufficient to diagnose gonorrhea in males, but DNA probes are also used.

The gonococcus is a highly adapted pathogen. It has adhesin-bearing pili that allow it to attach to receptors on epithelial cells in the male and female genital tracts. These receptors are also present on surface of the eye, throat, rectum, and sperm. By attaching to sperm, gono-

coccal cells are carried into the uterus and fallopian tubes. At any one time, the gonococcus produces only one type of adhesin. But it can rapidly change the type it produces. Such switching allows it to adhere more strongly to one particular type of epithelial cells—those in the throat or rectum, for example. Switching also helps the gonococcus escape recognition and destruction by the body's immune defenses. The gonococcus also has an outer membrane protein called **Protein II** that promotes infection. It too can bind to epithelial cells and switch its specificity for them. It also causes gonococcal cells to adhere to one another, forming clumps. So when one gonococcal cells adheres to a human cell, many others stick to it, creating an infectious unit. In addition, Protein II helps the gonococcus avoid being phagocytized.

Gonorrheal endotoxin (the toxic component of the outer membrane) does most of the gonococcus' damage to host cells. Endotoxin can kill cells, cause inflammation, and paralyze the cilia that protect the fallopian tubes. Then the gonococcus and other bacteria can multiply there. As a result, mixed infections by *N. gonorrhoeae* and other bacteria are common in the fallopian tubes.

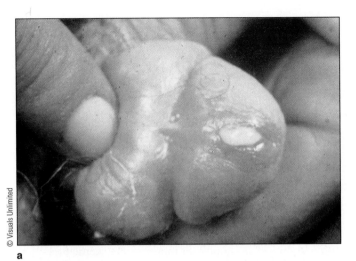

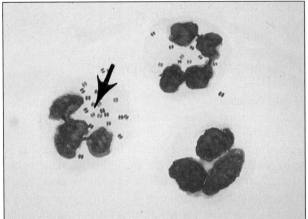

FIGURE 24.4   Gonorrhea. (a) Pus-laden urethral discharge from the penis. (b) A Gram stain of the discharge. Arrow points to *Neisseria gonorrhoeae* diplococci. The larger cells are polymorphonuclear leukocytes.

Clinical Syndromes. N. gonorrhoeae infects only humans, and it is transmitted only by direct body contact, usually sexual intercourse or birth. It is highly contagious. About 20 percent of men and 50 percent of women contract gonorrhea after a single exposure. Infection usually begins in the lower genital tract—the urethra in males and the cervix in females, but it can begin in the throat or the rectum if they are points of sexual contact.

Two to seven days after an infecting sexual contact, most men experience painful urination and a discharge of pus from the urethra. If antibiotics are administered, these symptoms resolve rapidly. Without treatment, they can continue for months. Infected women may notice an abnormal vaginal discharge, or they may notice no symptoms at all. Without symptoms, people are not likely to seek medical treatment. But they can spread the pathogen to new sexual partners. In addition, they are at risk for serious complications.

Sometimes such complications are restricted to the reproductive organs. Women can develop pelvic inflammatory disease with fever, dysuria, and severe abdominal pain. The fallopian tubes can become inflamed, causing scarring and infertility. If the gonococcus spreads into the abdominal cavity, death can result. Men can suffer infertility if the infection spreads to the tubes that transport sperm from the testes to the urethra.

A small percentage of infections lead to systemic illness. The gonococcus enters the bloodstream, causing **purulent** (pus-producing) arthritis and a rash. Rarely the heart, liver, or meninges become infected. Such disseminated disease is difficult to diagnose because there is often no sign of genital tract infection.

Infants can also become victims. They become infected while passing through an infected birth canal. They can suffer blindness or a life-threatening systemic infection (Chapter 26).

Prevention and Treatment. Gonorrhea has no reservoir outside infected human beings, and reliable, inexpensive diagnosis and treatment have been available for years. So it should be easy to control or even eliminate the disease. But many factors contribute to its perpetuation.

First, N. gonorrhoeae's remarkable ability to change its pili and outer membrane proteins makes natural immunity poor. Many people suffer repeated gonorrheal infections. So the prospects for an effective vaccine are also poor.

The discovery of penicillin was a major step forward in controlling gonorrhea. At that time, a previously untreatable disease became curable with a single injection. Recently, though, antibiotic-resistant strains of N. gonorrhoeae have become more common (see Sharper Focus: A Game We May Never Win). Gonococci can become penicillin-resistant in either of two ways: by chromosomal mutations or by acquiring a plasmid (Chapter 21). Chromosomal mutations confer incremental steps toward resistance. Strains with only a few mutations remain sensitive to high concentrations of penicillin, but those that have accumulated many mutations may be completely resistant. Plasmids that confer resistance encode a beta-lactamase, an enzyme that destroys penicillin. Strains that acquire them immediately become highly penicillin-resistant. As a result of resistance, penicillin is no longer used routinely to treat gonorrhea.

Controlling gonorrhea is also complicated by the simple fact that it is sexually transmitted. People find it difficult to

CASE HISTORY

The Stages of Syphilis

Patient Number 1

A 22-year-old student was examined at a county-sponsored health clinic. He had noticed an open sore on his penis several days earlier, but it caused him no discomfort. The examining physician noted that the sore was about 1 cm in diameter with a weepy bright red base and a raised border (Figure 24.6). The doctor also found one enlarged lymph node in the patient's groin. The young man reported sexual contact with a new partner about 3 weeks earlier.

Patient Number 2

A 53-year-old businessman made an appointment to see a private dermatologist. He was developing a red bumpy rash all over his body, even on the palms of his hands and the soles of his feet (Figure 24.7). He had also noticed a few irregular bald patches on his scalp, and recently he had been feeling unusually tired and occasionally feverish. The physician who examined him found enlarged lymph nodes throughout his body and several white patches inside his mouth. After being pressed by the physician, the man admitted to sexual contact with a prostitute about 3 months earlier.

Patient Number 3

A 60-year-old housewife was taken by her sister to see a psychiatrist. Over the last several years, the sister explained, the patient had become increasingly irritable and her personality had changed. She made inappropriate comments at social gatherings and was becoming careless about her dress and grooming. She was forgetful and could no longer do simple tasks such as balancing her checkbook. Sometimes she seemed depressed, but at other times she was extremely expansive, boasting of grandiose plans. The sister stated that the patient's health had always been quite good and she seldom consulted a doctor. When asked specifically, she did remember some sort of treatment for venereal disease about 40 years before.

deal objectively and openly with diagnosis and treatment. Until recently, for example, college textbooks stated that children could acquire gonorrhea by nonsexual routes. Today we acknowledge that these cases are the result of sexual abuse. Similar emotional reactions led to the myth that gonorrhea could be acquired from fomites such as toilet seats.

When a case of gonorrhea is identified, it must be reported to the public health department. Until recently, the infected person was asked to name his or her recent sexual contacts so these people can be tested and, if necessary, treated. However, out of embarrassment, neither patients nor physicians always cooperated. About 300,000 cases of gonorrhea are reported to public health departments each year, but the total number of cases in the United States is estimated to be three or four times that. Gonorrhea and most other STDs can usually be prevented with condoms (**Figure 24.5**).

Syphilis. The three people in Case History: The Stages of Syphilis all suffered from syphilis. This disease appeared suddenly in Europe during the 1490s. No one knows why. *Treponema pallidum* (Bergey's Spirochaetes, Chapter 11), the spirochete that causes it, is almost identical to *Treponema pertenue*, which causes the much milder disease called yaws. Possibly, *T. pertenue* quickly acquired increased virulence, becoming *T. pallidum.* Some medical historians theorize that the infection was imported to Europe from the Americas. Within a decade after its arrival in Europe, a devastating epidemic spread as far as China. Because syphilis caused disfiguring sores, it was called **great pox** to distinguish it from smallpox.

In the 1940s, when penicillin became available, syphilis could be cured. Even so, syphilis continued to be a major public health problem in the United States. During the 1970s and 1980s, syphilis rates rose—mainly among homosexual men. It reached a peak of more than 20 cases per 100,000 in 1990. Since then the rate has declined rapidly. By 1996 it reached the national health objective of fewer than 4 cases per 100,000, which was set for the year 2000. By the year 2000 it had fallen to 2.2 per 100,000. Still, it remains more prevalent in certain ethnic groups. The rate for non-Hispanic African Americans is 50 times greater than the rate for non-Hispanic whites. In addition, there have been recent outbreaks among men who have sex with men.

Diagnosis. Treponema pallidum is a highly motile spirochete. Darkfield microscopy of a smear from Patient Number 1's open lesion would have surly revealed motile spirochetes (**Figure 24.6**) and provided a diagnosis.

LARGER FIELD

THE NEAPOLITAN-FRENCH-WEST INDIAN-SPANISH-CANTON-CHINESE DISEASE

Europeans did not know the cause of the devastating disease that swept across the continent in the 1490s or even what to call it. When Charles VIII of France attacked the city of Naples in 1495, the Neapolitans sent infected prostitutes out to entertain the besieging troops.

Naples nevertheless fell quickly—as did Charles's victorious troops on their return home. They became acutely ill with disfiguring skin lesions, and many died of what Charles called the Neapolitan disease. When Charles's mercenaries returned to their homes in England, Poland,

Russia, and Scandinavia, the infection they brought with them came to be called the French disease. No country wanted to claim this disaster as its own. Spaniards called it the West Indian disease, whereas the French called it the Spanish disease. The Chinese called it the

Canton disease, and the Japanese, the Chinese disease. In 1530 an Italian physician wrote a poem about this infection as a curse from the Sun God. The literary victim—a shepherd named Syphilis—gave the disease its modern name.

T. pallidum is difficult to culture. It infects only human beings and is extremely fragile. Until recently it couldn't be grown in the laboratory, and it's still not cultivated routinely for diagnostic purposes. If darkfield microscopic observation fails, serological testing is required.

There are several serological tests, including an enzyme-linked immunosorbent assay (ELISA) test (Chapter 19), available to diagnose syphilis.

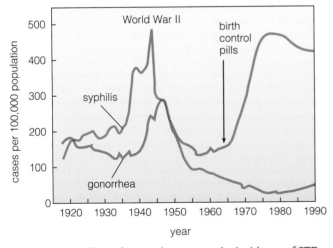

FIGURE 24.5 Effect of external events on the incidence of STDs. Both gonorrhea and syphilis rose dramatically during World War II, then fell with the availability of antibiotics. During the late 1960s and 1970s, after the introduction of birth control pills, the incidence of gonorrhea rose again. In contrast, syphilis remained controlled by penicillin.

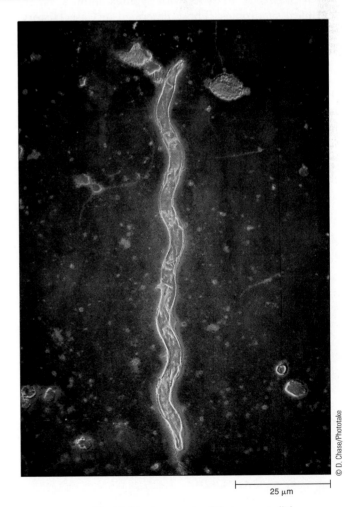

FIGURE 24.6   Darkfield micrograph of *Treponema pallidum*.

TABLE 24.3 Stages of Syphilis

Stage	Time after Infection	Manifestation	Diagnosis	Treatment	Infectious
Primary	2 to 3 weeks	Chancre at site of infection	Observation of spirochetes from a chancre under dark-field microscopy or presence of a chancre with positive syphilis serology	Penicillin	Yes
Secondary	8 to 11 weeks	Skin rash, fatigue, fever, enlarged lymph nodes	Fluorescent antibody assays, serology	Penicillin	Yes
Latent	After secondary	None	Serology	Penicillin	For first 4 years; not thereafter
Tertiary	5 to 40 years	Neurological symptoms, blood vessel damage, gummas	Serology, but may not be reliable	Penicillin, but only gummas respond to treatment	No

Clinical Syndromes. Patients with syphilis show a variety of clinical syndromes. Those described in Case History: The Stages of Syphilis are only three examples. Syphilis is characterized by a progression of changing signs and symptoms that occur over many years. These are divided into four stages—primary, secondary, latent, and tertiary (**Table 24.3**). Not all patients, however, pass through these stages. The disease can be stopped at any of the first three stages by an appropriate antibiotic.

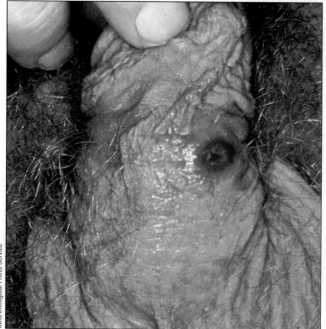

CDC/Biological Photo Service

FIGURE 24.7 Typical chancre of primary syphilis.

Because it is so fragile, *T. pallidum* is transmitted only by direct body contact, usually sexual. Infection occurs when it crosses a mucous membrane or a break in the skin.

Primary syphilis begins a few weeks to a month later, when a **chancre** (a weepy ulcer with raised borders) appears at the entry site. This was the painless sore that led Patient Number 1 to visit the health clinic. Fluid from the center of the chancre contains enormous numbers of spirochetes. With or without treatment, the chancre heals in a few weeks. A chancre on the penis is so noticeable that men usually seek medical attention (**Figure 24.7**), but women are often unaware of the chancre because it typically occurs on the cervix or within the vagina. So they are less likely to seek treatment.

Secondary syphilis usually begins 6 to 8 weeks after the chancre appears. Its signs and symptoms affect the skin and mucous membranes, so people sometimes consult a dermatologist, as Patient Number 2 did. The skin lesions are extremely variable, making diagnosis difficult. These lesions and those on mucous membrane lesions, such as the ones in Patient Number 2's mouth, contain many spirochetes. Contact with them can spread syphilis. The patient also has systemic symptoms, which include fatigue, fever, and enlarged lymph nodes. As bacteria spread throughout the body, they damage blood vessels, causing a characteristic rash (**Figure 24.8**). At this stage, diagnosis can be made by identifying motile spirochetes from mucous membrane lesions in the mouth if the patient has such lesions. But even then, special fluorescent antibody markers must be used to distinguish *T. pallidum* from other spirochetes that normally inhabit the mouth.

Latent syphilis ensues if secondary syphilis is not treated. This stage can last for years. Initially the only ev-

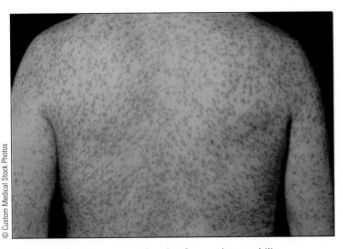

FIGURE 24.8   Disseminated rash of secondary syphilis.

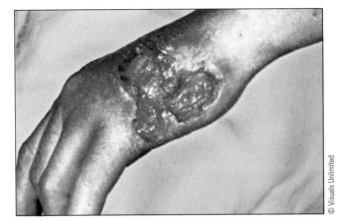

FIGURE 24.9   Gummas typical of late syphilis (tertiary stage).

idence of infection is a positive serological test and the ability to infect others. Then after about 4 years, most patients are no longer even infectious.

Tertiary, or **late, syphilis** develops in a minority of patients after 5 to 40 years of latent syphilis. It can progress to severe illness or death. Like secondary syphilis, it can be expressed in an astonishing number of ways. About 8 percent of people show some form of neurosyphilis, which may resemble almost any neurological disease. Patient Number 3, for example, had a form called **general paresis.** (In spite of what her sister remembered, she obviously had not received adequate treatment for her syphilis.) Another 10 percent of patients suffer from **cardiovascular syphilis,** in which large blood vessels are damaged. This form is fatal. The most common form of late syphilis is characterized by **gummas,** soft granulomas that usually replace skin or bone (**Figure 24.9**) but can occur in any organ. They are life threatening only if they destroy a vital organ. Gummas disappear with adequate antibiotic therapy, but damage they cause cannot be reversed. Late syphilis is usually not infectious, and sometimes serological tests are no longer positive.

Prenatal, or **congenital, syphilis** occurs when mothers with *T. pallidum* in their bloodstream infect their babies before birth. Bacteria cross the placenta during the middle months of pregnancy. About half of these infected fetuses die before or at birth. Some that survive have birth defects, such as tooth and bone deformities or deafness. Some develop mucous membrane lesions such as **snuffles** (a thick nasal discharge containing many *T. pallidum* cells). Some infants appear normal initially but at age 1 or 2 develop painful bone lesions and many other problems.

Prevention and Treatment. Preventing syphilis, like all other STDs, requires limiting sexual exposure and using condoms. Early diagnosis and treatment also decreases transmission. Congenital syphilis is preventable by prenatal testing of mothers and early treatment (before the second trimester, when syphilis crosses the placenta).

Natural immunity to *T. pallidum* is poor, So a vaccine seems unlikely. But *T. pallidum* is one of the few microorganisms that has remained completely sensitive to penicillin. Most infections can be cured with a single injection of benzathine penicillin. Patients with neurosyphilis, however, particularly those who also have AIDS, require hospitalization for a 10-day course of intravenous penicillin.

Chlamydia. Infection by *Chlamydia trachomatis* (Bergey's Chlamydiae, Chapter 11), called simply **chlamydia,** is the most prevalent STD worldwide. The actual number of cases is not known with certainty because compliance with local health laws requiring that health-care providers and laboratories report cases are quite variable. In addition, many cases are asymptomatic. But symptoms that do occur appear gradually in men and women 1 to 3 weeks after unprotected sexual contact with an infected partner. The illness is mild, consisting of dysuria and a small amount of mucoid discharge from the urethra or the vagina. But many women suffer infections of the uterus and fallopian tubes. These can have serious consequences. Chlamydia is the leading cause of female infertility and **ectopic pregnancy** (a life-threatening condition in which pregnancy develops in a fallopian tube, eventually rupturing it and causing massive abdominal bleeding) in this country. Moreover, infected women transmit chlamydia to their infants at the time of birth, causing neonatal eye infections and pneumonia.

Because it has been so difficult to diagnose, chlamydia was not recognized as an important STD until recently. It cannot be identified by routine bacterial culture. It has to be propagated in tissue culture. But with the advent of rapid antigen detection tests in the mid-1980s (Chapter 19), the prevalence and importance of chlamydia became apparent. Because symptoms are so subtle (only half of women with chlamydia-caused infertility remember any illness), everyone who is sexually active should be regularly screened. Screening females, however, is not routine because sample collection for the antigen test requires a pelvic examination. So screening is limited to those at highest risk. Now that might change completely because a polymerase chain reaction (PCR) test (Chapter 7) has been developed that can detect chlamydia in a urine sample. Plans are now in place for mass screening of at-risk populations. The impact on prevalence of chlamydia could be profound.

Chlamydia can be cured with doxycycline, erythromycin, or azithromycin.

Lymphogranuloma Venereum.

Certain strains of *Chlamydia trachomatis* cause a different STD called **lymphogranuloma venereum.** This rare disease (fewer than 500 cases are reported in the United States each year) produces a distinctive clinical syndrome. One to four weeks after sexually transmitted infection, a small sore usually appears on the genitals. The person experiences fever and headache. The lesion heals without treatment, but then lymph nodes in the groin enlarge and may become painful. Left untreated, the nodes may become draining ulcers. Scarring and lymphatic obstruction damage the external genitalia. Diagnosis is made by culturing organisms from the infected nodes. Treatment with doxycycline or erythromycin is effective.

Nongonococcal Urethritis.

Nongonococcal urethritis is caused by *Ureaplasma urealyticum* and *Mycoplasma hominis.* These mycoplasmas (Bergey's Firmicutes, Chapter 11; the terms mycoplasmas and Mollicutes are synonyms) are common residents of the urethra or vagina of sexually active men and women. More than half of all people with five or more lifetime sexual partners are colonized by these bacteria. Most have no symptoms, but some men suffer from a urethral discharge. These mycoplasmas may cause symptoms in women as well. Mycoplasmas are so widespread it's not practical to screen for them or treat them routinely. When men seek care for urethral discharge, they are usually given doxycycline. It's effective against the mycoplasmas, as well as many other sexually transmissible pathogens.

Chancroid.

Chancroid is an STD caused by *Haemophilus ducreyi* (Bergey's Gammaproteobacteria, Chapter 11). This bacterium can enter the body only through mucous membranes or breaks in the skin, so chancroid is less communicable than gonorrhea or syphilis. The first sign of chancroid is a soft chancre on the genitals that resembles the chancre of syphilis but is softer and can be quite painful. The chancre eventually heals spontaneously, but antibiotic treatment speeds recovery and prevents scarring and damage to the genitals.

Chancroid is fairly common in tropical countries. It was rare in the United States until the mid-1980s. By 1990 more than 5000 cases of chancroid were being reported, compared with fewer than 1000 per year in the early 1980s. Chancroid appears to be closely linked to the crack cocaine epidemic. Most patients are prostitutes and cocaine addicts. Because symptom-free infections are rare and ulcers are painful, the spread of chancroid means people continue sexual activity despite considerable pain. A single dose of ceftriaxone cures chancroid in otherwise healthy patients.

Granuloma Inguinale.

Granuloma inguinale is caused by *Calymmatobacterium granulomatis,* a Gram-negative rod. It infects the skin and mucous membranes of the genital organs. It's not highly communicable because the bacteria can enter only through broken skin or mucous membrane. It causes a raised lesion at the site of entry. These eventually develop into open, draining ulcers that may persist for years. Sometimes they spread to the legs or abdomen. Diagnosis is made on the basis of characteristic intracellular inclusions called Donovan bodies that are visible under a light microscope in scrapings from the ulcers. Granuloma inguinale can be cured by several antimicrobial agents, including ampicillin, tetracycline, and trimethoprim-sulfamethoxazole. The infection is extremely rare in the United States. It occurs in Asia, Africa, and South America.

Viral Infections

Viral STDs affect sexually active adults and newborns. They can predispose a person to cancer. Unlike bacterial STDs, viral STDs are not curable.

Herpes.

Herpes simplex virus (HSV) is by far the most prevalent sexually transmitted viral pathogen in industrialized countries. At least 20 percent of people in the United States between ages 16 and 40 have serological evidence of past infection. Before 1970 little attention was paid to this pathogen, but during the 1970s the incidence of genital herpes simplex in the United States increased dramatically. Today the 20 million symptomatic episodes of genital herpes that occur annually in adults are a serious public health concern, causing considerable pain and disability. But the real horror of herpes is infection of newborns.

HSV is a DNA virus that belongs to the herpes family. In every person it infects it has an active and latent phase (Chapter 15). During its active phase, it multiplies explosively. Between 50,000 and 200,000 new virions are produced from each infected cell. Because HSV inhibits its host cell's metabolism and degrades its DNA, the cell dies. Such active infections may be symptom-free, or they may form painful blisters in the infected tissue. During latent infections, the viral genome becomes incorporated into the nucleus of the host cell; viral genes are not expressed and the patient is symptom-free. But latent infections may become activated later, causing a new active infection.

HSV affects epithelial cells and neurons. Infection begins when virus is introduced directly into a break in the skin or mucous membrane. The viruses enter nearby nerve endings. They spread infection along the nerve fiber to the body of the neuron, where a latent infection is established. The infected nerve cell does not die. When the virus is reactivated, it attacks the skin or mucous membrane that the nerve supplies. This phenomenon accounts for one of the most distinctive features of HSV infection: Its tendency to recur at the same site again and again.

There are two types of herpes simplex virus: HSV-1 and HSV-2. Their DNA sequences are about 50 percent identical, but they bind to different host-cell receptors. HSV-1 typically infects the mouth and face (Chapter 26). HSV-2 typically infects the genital tract. But the reverse can occur. Each type can be transferred from mouth to genitals and vice versa. The lesions they cause are clinically indistinguishable, but genital recurrences are much more frequent when the infection is caused by HSV-2.

In some immunocompromised patients, these viruses can affect almost every organ of the body, with fatal consequences.

Clinical Syndromes. HSV-2 is transmitted to a new host during sexual contact. It infects the external genitalia, the urethra, and the cervix. Rectal and pharyngeal herpes are also caused by sexual contact.

Genital herpes causes a painful rash of tiny fluid-filled vesicles. The fluid contains infectious viruses. The initial infection by HSV (called a **primary infection**) usually causes fever, headache, and muscle aches, in addition to the rash. A reactivated latent infection usually produces rash alone. Stress, fever, or trauma (all of which can temporarily depress the immune response) stimulate reactivation, which is common. Typically a person with a genital infection by HSV-2 will suffer four recurrences during the next year. Recurrences are usually less severe and briefer than the primary infection, but they can be extremely painful. Both a primary infection and reactivation can occur without symptoms. Therefore apparently healthy people can transmit herpes to their sexual partners or their newborns.

Neonatal herpes is a devastating disease (**Figure 24.10**). It spreads to the brain and other internal organs. Most babies that don't die suffer serious lifelong disabilities, including blindness, deafness, and profound retardation. Babies can be infected after birth by close contact with infected adults, but most are infected during passage through the birth canal. Delivery by cesarean section prevents these tragic infections, but most affected babies are born to asymptomatic women who are unaware of their infection.

Prevention and Treatment. Genital herpes, like other STDs, can be prevented by avoiding sexual contact with an infected person. However, many people who have the infection are not aware of it. Once infected, they have the potential to transmit infection for the rest their lives. Moreover, condoms are not always protective because a woman's external genitalia and the parts of the penis not covered by a condom can spread the virus. Still, a condom provides some protection.

Genital herpes can't be cured. Acyclovir does shorten the duration of a primary infection, but it doesn't prevent a latent infection from being established. It has little effect on recurrent infections. However, given on a daily basis, it does help those few people who experience such frequent recurrences that they are in almost constant pain. To be effective, therapy must be continued indefinitely.

Acyclovir and another antiviral agent, vidarabine, are also used to treat life-threatening HSV infections that involve the brain or other internal organs or that occur in immunocompromised patients. Acyclovir has decreased the mortality rate of neonatal herpes from 65 percent to 25 percent, but, tragically, most of the survivors suffer brain damage. Unfortunately, acyclovir-resistant strains of HSV have appeared in AIDS patients.

Genital Warts. Genital warts (**condyloma acuminata**) are transmitted during sexual contact or during passage through the birth canal. They appear on the skin or mucous membranes of the external genitalia, including the penis, vulva, cervix, and perianal skin. They

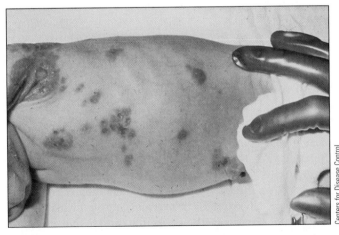

FIGURE 24.10 Neonatal herpes. This 10-day-old baby covered with vesicles is likely to die or be permanently handicapped from the disease.

Centers for Disease Control

SHARPER FOCUS

SAFER SEX—WHAT DOES IT REALLY MEAN?

Safer sex has become a familiar phrase during the AIDS epidemic, but what does it mean? In fact, safer sex refers to behaviors that keep people safer from all STDs, not just AIDS. There are only two sure ways to avoid sexually transmitted infection—either abstain from sex altogether or establish a permanent relationship in which both partners have sex only with each other. The key to this second option is permanence. Many people have relationships with a single partner, but the partner changes over weeks, months, or years.

This serial monogamy puts people at extremely high STD risk—maybe even at higher risk than those who practice promiscuous, multiple-partner sex if the total number of contacts over time is high. Even in committed relationships, therefore, condoms must be used for all sexual contact that involves penile penetration of any body orifice. Latex condoms with spermicides are the best, but the most important message of the safer sex campaign is to use some type of condom, because any condom is far better than none.

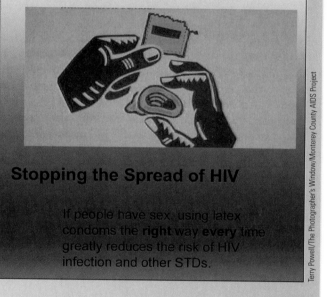

Para detener el VIH

Si se tienen relaciones sexuales, los condones de latex, usados en **cada acto sexual** y de la **manera correcta**, reducen mucho el riesgo de contraer el VIH y otras enfermedades de transmision sexual.

Stopping the Spread of HIV

If people have sex, using latex condoms the **right** way **every** time greatly reduces the risk of HIV infection and other STDs.

Terry Powell/The Photographer's Window/Monterey County AIDS Project

---

are caused by human papillomavirus (HPV), a DNA virus of the papovavirus family (**Figure 24.11**).

HPV can't be cultivated in the laboratory, but nearly 50 different strains have been distinguished by DNA sequence. Specific strains of HPV cause warts on specific parts of the body, for example, the external genitalia, the upper airways, the soles of the feet, or other areas of skin. Some strains cause wartlike growths, called laryngeal papillomatosis, on the vocal cords. They usually infect children and can cause hoarseness or even respiratory obstruction.

Genital warts is not a reportable disease, but it's quite common. Up to 70 percent of female patients at STD clinics and 20 percent at family planning clinics carry HPV in their genital tract. Many of these infections are asymptomatic.

HPV infection used to be considered only a cosmetic problem, but it can lead to cancer. Cervical and vulvar carcinoma in women and squamous cell carcinoma of the penis and rectum in men have been clearly associated with HPV. The association between HPV and cervical cancer is especially significant because this malignancy is so common among sexually active women between the ages of 18 and 30.

Like all STDs, genital warts can be largely prevented by safer sex practices. Condoms are effective. Genital warts can usually be controlled or eliminated by freezing them with liquid nitrogen or painting them with liquid podophyllin or trichloroacetic acid. These treatments remove the warts, but they don't eliminate the infection or, probably, the risk of cancer. Women with a history of genital warts should have annual Papanicolaou (pap) smears to detect early cervical cancer.

Other Viral STDs. Two other viruses are often sexually transmitted—hepatitis B and human immunodeficiency virus (HIV). Both are considered in other chapters (hepatitis B in Chapter 23 and AIDS in Chapter 27) because their principal targets are not the urogenital system.

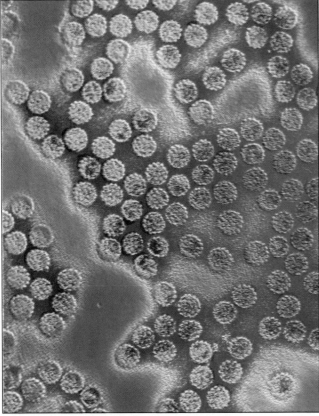

© Institut Pasteur/CNRI/Phototake

a

10 μm

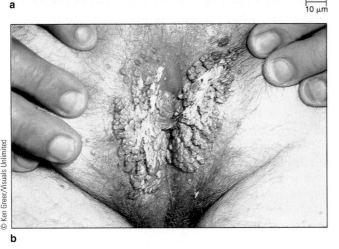

© Ken Greer/Visuals Unlimited

b

FIGURE 24.11 Human papillomavirus. (a) Electron micrograph of virions. (b) Genital warts.

INFECTIONS OF THE FEMALE REPRODUCTIVE TRACT

Infections of the female genital tract can be caused by bacterial, fungal, or protozoal pathogens. Some are sexually transmissible. Others are not.

Bacterial Infections

Bacteria that infect the lower or upper female reproductive tracts produce clinical syndromes that range from a relatively harmless vaginitis to serious illnesses such as pelvic inflammatory disease and toxic shock syndrome.

Gardnerella Vaginitis. Gardnerella vaginitis is caused by *Gardnerella vaginalis*, a motile, pleomorphic, Gram-positive bacterium that was once considered to be a species of *Haemophilus*.

G. vaginalis is present in vaginal secretions of up to 40 percent of sexually active, asymptomatic women (**Figure 24.12**). In women with vaginitis, it's incidence is higher. As *G. vaginalis* proliferates, so do large numbers of anaerobic bacteria. The population of lactobacilli decreases, and the pH of the vagina becomes abnormally high. These abnormalities occur together, but cause and effect are not clear. Therefore the clinical syndrome is also called **bacterial vaginitis.** Gardnerella vaginitis causes a fishy-smelling vaginal discharge because it contains putrescine and cadaverine—amines that are also found in rotting fish.

Many women visit the doctor complaining of vaginal irritation and discharge. This clinical syndrome can be caused by many different pathogens. Diagnosis depends upon examining the vaginal discharge under a microscope, a procedure called a vaginal wet mount. It can be done in a clinical laboratory or in the clinician's office. Gardnerella

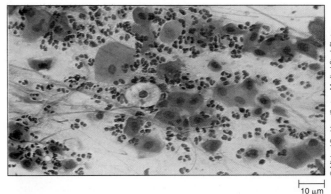

© Michael Davidson/Custom Medical Stock Photos

10 μm

FIGURE 24.12 *Gardnerella vaginalis.* Gram stain of vaginal discharge. The large red cells a from the vaginal epithelium. Small purple cells are *G. vaginalis.*

CASE HISTORY

Toxic Shock Syndrome

L. was a 19-year-old woman who came to an emergency clinic complaining of a high fever. She had been perfectly well until a few hours earlier, when she started to think she might be coming down with the flu because she ached all over and felt feverish. Her mother, concerned about the sudden change, insisted she see a doctor. As part of the history, the admitting physician noted that L. L. had begun her menstrual period 3 days before and was using vaginal tampons.

The physician noted that L. L. had a fever of 105°F, but he was more concerned by the bright red rash that looked like sunburn over most of her body (Figure 24.13). Her blood pressure was slightly low. Because of her history of tampon use, the physician considered the possibility that L. L. was suffering from toxic shock syndrome. He admitted her to the hospital for observation—a decision that saved L. L.'s life. Within 2 hours she became unresponsive and her blood pressure was too low to be measured with a standard blood pressure cuff. L. L. was in shock.

She was transferred to the intensive care unit, where heroic efforts maintained her vital functions. Intravenous fluids and medications sustained her blood pressure. A mechanical respirator supported her breathing. She was treated with nafcillin, an antimicrobial effective against *Staphylococcus aureus*. After about a week, L. L.'s condition stabilized. At this time her skin began to peel, particularly on her palms and soles. Eventually she was discharged in good condition. The only permanent reminder of her near-fatal illness was the loss of two toes because of poor blood supply while she was in shock.

---

vaginitis is diagnosed if (1) more easily recognizable vaginal pathogens, such as *Trichomonas vaginalis* and *Candida albicans*, are absent, and (2) **clue cells** (sloughed vaginal epithelial cells surrounded by many *G. vaginalis* cells) are present. Gardnerella vaginitis is treated with metronidazole. It decreases the population of vaginal anaerobes and helps reestablish a normal vaginal biota.

Toxic Shock Syndrome. Toxic shock syndrome is caused by strains of *Staphylococcus aureus* that produce a potent toxin called **toxic shock syndrome–associated toxin**

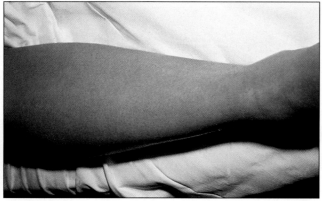

Centers for Disease Control

FIGURE 24.13 A bright red rash of a patient with toxic shock syndrome.

(**TSST**). When these bacteria multiply in the body, they produce TSST, which is absorbed into the bloodstream. It causes an illness like L. L.'s. Toxin-producing staphylococci can multiply in many parts of the body and cause toxic shock syndrome in men, women, or children (**Figure 24.13**). The typical toxic shock patient, however, is a menstruating woman like L. L. who is using vaginal tampons.

Toxin-producing strains of *S. aureus* are part of the normal vaginal biota of some women, but their numbers are too small to cause disease. Tampons, however, can set the stage for trouble. When inserted, they can abrade the vaginal wall, which speeds the uptake of toxin. And a blood-soaked tampon is an ideal environment for rapid multiplication of *S. aureus*. In general the longer a tampon is in the vagina, the more likely it is to foster staphylococcal growth.

Illnesses like L. L.'s were first described in the medical literature in 1978. Epidemiologists found that toxic shock was most likely to strike women who used a particular brand of super-absorbent tampon, probably because they were kept longer in the vagina. Taking these tampons off the market and publicizing the importance of changing tampons regularly brought this life-threatening syndrome under control. Today most cases are still related to tampons, but the disease has become rare. Fewer than 3 in 100,000 menstruating women develop toxic shock syndrome each year.

Pelvic Inflammatory Disease (PID). Sometimes bacteria from the vagina travel upward and infect the uterus, the fallopian tubes, and the ovaries, causing pelvic inflammatory disease. Infection may be asymptomatic, but often fever, abnormal vaginal bleeding, and severe lower abdominal pain occur. These infections occur commonly in young sexually active women. Each year, more than 1 million women are affected in the United States. If untreated, PID may be fatal. Even with treatment, serious complications occur. Each episode of PID carries a 15 percent risk of infertility from scarred fallopian tubes. It can also lead to increased risk of ectopic pregnancy.

Several pathogens, most sexually transmissible, can cause PID. *Chlamydia trachomatis,* followed by *Neisseria gonorrhoeae* and *Mycoplasma hominis,* is the most common cause in the United States. Women who have multiple sexual partners are at highest risk. The contraceptive intrauterine device (IUD) was withdrawn from the market largely because of its association with PID. Cervical cultures are taken at the time of diagnosis, but results are not available for several days. Treatment—which must be begun immediately—is therefore designed to eliminate all the most likely infecting microorganisms. Patients who are seriously ill are hospitalized to be treated with intravenously (IV) administered antibiotics. Patients with milder disease can be treated with a single dose of intramuscularly or orally administered antibiotic.

Infections after Childbirth or Abortion. Endometritis is particularly likely to occur after childbirth or abortion because trauma to the epithelial lining of the uterus makes it particularly vulnerable to infections. It is characterized by fever and a tender, enlarged uterus. If a bloodborne infection develops, it is called **puerperal sepsis.** *Streptococcus pyogenes* was once the most common nosocomial cause of endometritis (see Larger Field: Childbed Fever, in Chapter 1), but today *Escherichia coli, Neisseria gonorrhoeae, Chlamydia trachomatis, Clostridium* spp., or anaerobic streptococci are usually responsible. Thus antimicrobial treatment must be effective against Gram-positives, Gram-negatives, and anaerobes. Cephalosporins are the most commonly prescribed treatment.

The vulnerable uterus becomes infected during the pelvic examinations performed during labor or when instruments such as forceps are inserted to help deliver the baby. Risk increases greatly if part of the placenta remains in the uterus after delivery. These medical disasters are much less common today than they were in Semmelweis's time (Chapter 1, "Childbed Fever"), but they still occur. Young, healthy mothers still die despite antimicrobial treatment.

Abortions, either spontaneous or induced, also provide an opportunity for microorganisms to enter the uterus. Infections were common when abortion was illegal because many illegal abortions were not performed under sterile conditions and did not completely empty the uterus. Also, women were afraid to seek medical treatment if infection occurred.

Fungal and Protozoal Infections

A fungus causes many cases of vaginitis. A protozoan can cause the same syndrome.

Candidiasis. The yeastlike fungus *Candida albicans* in small numbers is a normal, harmless inhabitant of the vagina, but it can proliferate and cause a vaginitis called candidiasis. The symptoms are a thick white vaginal discharge and severe vaginal itching. *Candida* cells can be seen under the microscope if the vaginal secretions are first mixed with potassium hydroxide (KOH) to destroy obscuring epithelial cells (**Figure 24.14**). Local treatment with antifungal medications such as nystatin or the newer terconazole is usually effective.

Certain factors predispose a woman to vaginal candidiasis. These include the change in hormone levels that accompany pregnancy and the use of oral contraceptives. Broad-spectrum antibiotics can also causes candidiasis by altering the vagina's normal microbial biota, thereby allowing overgrowth of *C. albicans* (Chapter 14).

Trichomoniasis. *Trichomonas vaginalis,* a flagellate protozoan, causes trichomoniasis, which is sexually transmissible. We discuss it here instead of with other STDs

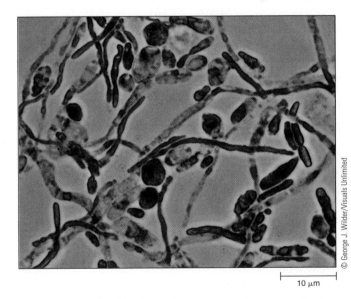

10 μm

© George J. Wilder/Visuals Unlimited

FIGURE 24.14 *Candida albicans.* A wet mount from a sample obtained during a pelvic examination of a woman complaining of vaginal itching. Note the hyphae, as well as budding yeasts.

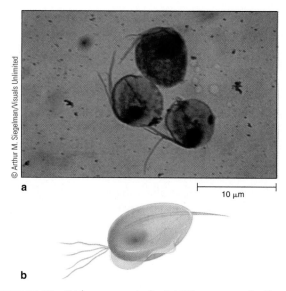

a

10 μm

b

FIGURE 24.15 *Trichomonas vaginalis.* (a) Wet mount of cells. (b) Sketch showing surface properties.

because its primary clinical manifestation is a painful vaginitis characterized by a copious, frothy vaginal discharge. Some estimates put infection by *T. vaginalis* at 25 percent of women in the United States. Because infection is so widespread and often asymptomatic, it was once thought to be part of the normal vaginal biota. *T. vaginalis* usually infects the urethra of males, but it causes no symptoms unless it is accompanied by a bacterial infection.

Trichomoniasis is usually diagnosed by examining infected vaginal secretions under a microscope. The protozoa can be readily spotted because they move in a characteristic jerky way (**Figure 24.15**). *T. vaginalis* is easy to eradicate with oral metronidazole. To prevent reinfection, a woman's asymptomatic male sexual partner must also be treated.

INFECTIONS OF THE MALE REPRODUCTIVE TRACT

Men who seek medical attention for a reproductive tract infection usually complain of urethritis. They describe dysuria, urethral itching, and/or urethral discharge. Many pathogens can cause urethritis, and most are sexually transmissible. Urethritis is usually not a serious condition.

The classical cause of male urethritis is *Neisseria gonorrhoeae*. Gonococcal urethritis typically causes a copious, purulent urethral discharge loaded with intracellular Gramnegative diplococci. Most cases of urethritis, however, are caused by other pathogens. Until fairly recently, these others were lumped together under the heading nongonococcal

urethritis (NGU). If the infection wasn't gonorrhea, it usually wasn't possible to determine exactly what it might be.

Today most pathogens that cause nongonococcal urethritis are identified. By far the most common is *Chlamydia trachomatis*. It causes 30 to 50 percent of NGU. *Ureaplasma urealyticum* is a close second, causing 20 to 30 percent of cases. Other mycoplasmas may also cause NGU. A small number of cases, probably less than 5 percent, are attributable to nonbacterial pathogens—herpes simplex virus and *Trichomonas vaginalis*.

INFECTIONS TRANSMITTED FROM MOTHER TO INFANT

Several microorganisms that affect the mother cause much more serious infection in the fetus or newborn. Some but not all of these pathogens are sexually transmissible. They can be bacterial or viral.

Bacterial Infections

Some of the most devastating bacterial infections of newborns are acquired from the mother during pregnancy or at the time of birth.

Listeriosis. Listeriosis is caused by *Listeria monocytogenes* (Bergey's Section 14, Chapter 11). It takes many forms, including meningitis (Chapter 25), septicemia (Chapter 27), and endocarditis (Chapter 27). *L. monocytogenes* is an opportunistic pathogen. It rarely causes serious disease in healthy, nonpregnant adults, but people with compromised immune function and pregnant women are at increased risk. The illness is mild. But bacteria that enter the bloodstream can infect the placenta. Then they can cross the placenta to infect the fetus, or they can infect the newborn when membranes rupture during delivery. Fetal infections often cause abortion or stillbirth. Babies with listeriosis can suffer from widespread abscesses, meningitis, or septicemia, and many die, even with early treatment.

Group B Streptococcal Infection. Group B streptococci (also known as *Streptococcus agalactiae*) are significant pathogens of the female reproductive tract and newborn infants. They were first recognized for causing puerperal sepsis. Now they are more important as causes of life-threatening neonatal infections.

Group B streptococci are commonly found in the throat, gastrointestinal tract, and vagina of adults, and 15 to 20 percent of pregnant women probably carry them in their reproductive tract. Babies are infected when fetal membranes rupture during delivery. Pneumonia, meningitis, and sepsis are the most common consequences. Two of every 1000 babies are born infected by this organism, and nearly a quarter of these die in spite of antibiotic treatment. Many others are left permanently handicapped after surviving neonatal meningitis. Group B strep also infects new mothers. It causes one out of five cases of postpartum endometritis and is the second most common cause of systemic infection following cesarean section. Recent studies suggest that high-risk mothers and their infants may be protected by receiving ampicillin intravenously during labor and delivery.

Viral Infections

Many viruses—including herpes simplex (discussed earlier), rubella (Chapter 26), and cytomegalovirus—cause devastating infections and birth defects when transmitted from mother to fetus or newborn.

Cytomegalic Inclusion Disease. Cytomegalovirus (CMV), a member of the herpesvirus group, is an extremely common cause of human infection around the world. Nearly 100 percent of people in developing countries and 75 percent of people in industrialized countries show evidence of CMV infection. Like other herpesviruses, cytomegalovirus establishes a latent infection that can be reactivated. Cells infected with the virus swell and develop inclusion bodies in their nuclei and cytoplasm, which accounts for the virus's name, cytomegalovirus (*cyto*, "cell"; *megalo*, "large": large cell virus), and the disease it causes, cytomegalic inclusion disease.

CMV is transmitted sexually, by close contact such as the exchange of saliva, and in infected blood. Infection in healthy children and adults is usually asymptomatic, but it may cause a brief, mononucleosis-like illness. If a first infection occurs during pregnancy, however, and the virus infects the fetus, it can cause spontaneous abortion, stillbirth, or birth defects such as blindness, deafness, and mental retardation. Approximately 10 percent of infected infants show clinical evidence of disease (**Figure 24.16**). Infected infants may excrete the virus in their urine and saliva for years, and toddlers often transmit the virus to their caretakers or to one another. Day-care centers are hotbeds of CMV transmission. Pregnant women without serological evidence of previous infection probably shouldn't work there. The only antiviral agent that appears to be at all useful in treatment is ganciclovir.

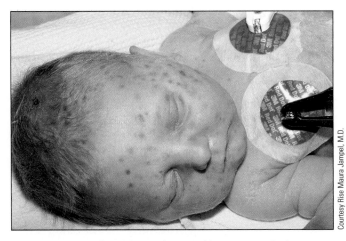

FIGURE 24.16 Baby with a rash caused by cytomegalovirus.

Courtesy Rise Maura Jampel, M.D.

CASE HISTORY

A Good Outcome to a Bad Scare

Baby Boy W. was born to a healthy 25-year-old woman who already had a 2-year-old child. She had seen the doctor frequently during her pregnancy because of concerns about her uterus and placenta, but she and her obstetrician were pleased when she went into labor normally just a few days after her due date. The mother seemed well and did not have a fever. There was no reason to suspect that Baby Boy W. would be at especially high risk for illness.

After his birth, Baby Boy W. nursed a little and then seemed sleepy, as newborn babies usually do. Later in the evening his mother called for help from one of the nurses because she was having trouble waking him up to feed him. This still didn't seem very worrisome in such a young baby. By the next morning, though, there was no doubt that there was something very wrong with the baby. He had a fever. His mother was barely able to wake him at all, and he was pale and clammy. When a doctor came to see him, he noticed that the baby was having small jerky movements that looked like seizures.

The doctor immediately transferred the baby to an intensive care unit and began diagnostic tests. Blood was taken for a complete blood cell count and blood culture, as well as for measurement of glucose and electrolytes. A lumbar puncture was also performed. The cerebrospinal fluid was cloudy. It was obvious that the baby had neonatal meningitis and might not survive. Antibiotic treatment was started as soon as an intravenous line could be placed. When lab results became available, they confirmed the grim situation. The white blood cell count in the cerebrospinal fluid was very high, indicating a rapidly progressing meningitis. And the white blood cell count in the baby's blood was quite low, indicating that his body was not responding well to fight the infection.

Baby Boy W. had a long and rocky course in the hospital. During the first few days, his seizures continued and had to be controlled with medication. When the results of the cerebrospinal culture became available a few days later, they showed that his meningitis was due to a group B streptococcus—one of the most dangerous infections that infants can acquire from their mother's during birth. His fever continued for more than a week. He required 2 weeks of intravenous antibiotic treatment before he was well enough to return home.

At that time doctors were still concerned that he might not develop normally, but by 4 months of age the baby was doing extremely well. He had had no more seizures and was off all medication. Even more important, he was smiling, babbling, and responding happily to both his parents. But a year later, when Baby Boy W began to walk, it became evident that he would have mild cerebral palsy, a lifelong but relatively minor complication of the infection that so easily could have killed him.

Nosocomial transmission of CMV in blood transfusions presents special ethical problems. Like other viral pathogens, CMV can be detected in banked blood. But because the pathogen is so widespread, it is not practical to discard all contaminated blood. It's been suggested that particularly susceptible patients—for example, severely ill newborns—should receive only CMV-negative blood. But this raises a difficult question. Should health-care providers knowingly administer infected blood to some people, reserving noninfected blood for others?

the cd connection

MICROBIAL DISEASES ⟶
Reproductive system

Summarizes the normal biota, structure/function, clinical syndromes, and infections of the reproductive system.

Table 24.4 summarizes infections of the genitourinary system.

TABLE 24.4 Infections of the Genitourinary System

Infection	Causative Agent	Mode of Transmission	Symptoms	Prevention and Treatment
Urinary System				
Urinary tract infection	Escherichia coli, Proteus spp., Klebsiella spp., Pseudomonas spp., enterococci, Staphylococcus saprophyticus	Fecal skin contaminants travel upward	Urinary frequency, urgency, dysuria; fever and flank pain in cases of pyelonephritis	Antibiotic choice depends upon strain sensitivity; common drugs include ampicillin, trimethoprim-sulfamethoxazole, nalidixic acid
Leptospirosis	Leptospira interrogans	Exposure to animal urine	Flulike illness; jaundice and kidney failure in severe cases	Penicillin, tetracycline
Sexually Transmissible Infections Bacterial				
Gonorrhea	Neisseria gonorrhoeae	Sexual or perinatal	Urethral discharge; vaginal discharge and pelvic pain; rare systemic infections produce arthritis or rash	Ceftriaxone, cefixime
Syphilis	Treponema pallidum	Sexual or perinatal	Enormous variety of clinical manifestations depending upon disease stage; no symptoms during latent stage; prenatal infection causes birth defects, snuffles, bone pain	Penicillin
Chlamydia	Chlamydia trachomatis serovars D-K	Sexual or perinatal	Mild dysuria, mucoid discharge leading to PID, NGU; high likelihood of female infertility and infection of newborns	Doxycycline, erythromycin, azithromycin
Lymphogranuloma venereum (LVG)	Chlamydia trachomatis serovars L_1-L_3	Sexual	Genital sore, enlarged groin lymph nodes	Doxycycline, erythromycin
Genital mycoplasma	Ureaplasma urealyticum, Mycoplasma hominis	Sexual	Usually asymptomatic; may lead to PID, NGU	Doxycycline
Chancroid	Haemophilus ducreyi	Sexual	Painful genital ulcer	Ceftriaxone
Granuloma inguinale	Calymmatobacterium granulomatis	Sexual	Persistent draining ulcers	Ampicillin, tetracycline, trimethoprim-sulfamethoxazole
Viral				
Genital herpes	Herpes simplex virus (usually HSV-2)	Sexual or perinatal	Painful genital blisters; systemic infection in newborns and immuno-compromised patients	Acyclovir lessens symptoms, but does not cure infection or prevent recurrence
Genital warts condyloma acuminata)	Human papillomavirus (HPV), particularly types 6, 11, 16	Sexual	Warts on external genitalia; predisposition to cervical and other cancers	Liquid nitrogen or podophyllin decreases warts, but no cure for infection and doesn't reduce risk of cancer

continued

TABLE 24.4 Infections of the Genitourinary System (continued)

Infection	Causative Agent	Mode of Transmission	Symptoms	Prevention and Treatment
Infections of the Female Reproductive System				
Bacterial				
Bacterial vaginosis	Overgrowth of *Gardnerella vaginalis*, associated with increase in vaginal anaerobes and decrease in lactobacilli	Sexual	Foul-smelling vaginal discharge; vaginal irritation	Metronidazole
Toxic shock syndrome	Toxin-producing *Staphylococcus aureus*, usually growing in vagina	Overgrowth of normal flora, usually associated with tampon use	Flulike illness, rash, shock	Nafcillin or other penicillinase-resistant penicillin; aggressive supportive care
Endometritis	*Escherichia coli*, *Neisseria gonorrhoeae*, *Chlamydia trachomatis*, *Clostridium* spp., anaerobic streptococci, group B streptococci	Bacterial contamination of uterus after childbirth or abortion	Uterine tenderness, fever, life-threatening bloodborne infection	Cephalosporin
Fungal				
Vaginal candidiasis	*Candida albicans*	Overgrowth of normal vaginal flora	Vaginal itching, thick white discharge	Nystatin, terconazole
Protozoal				
Trichomoniasis	*Trichomonas vaginalis*	Sexual	Vaginal itching, copious vaginal discharge	Metronidazole
Perinatally Transmitted Infections				
Bacterial				
Listeriosis	*Listeria monocytogenes*	Foodborne, and possibly sexual; also perinatal	Meningitis, endocarditis, septicemia in newborns and innumosuppressed adults	Penicillin, gentamicin
Group B streptococcal disease	Group B streptococcus	Perinatal	Septicemia, meningitis, pneumonia of newborns	Ampicillin
Viral				
Cytomegalic inclusion disease	Cytomegalovirus (CMV)	Close contact (exchange of saliva); blood transfusion; perinatal	Mononucleosis-like illness; stillbirth, birth defects	Ganciclovir

SUMMARY

The Genitourinary System (p. 584)

1. The genital and urinary tracts are closely related anatomically, but they fulfill different functions—reproduction and excretion of metabolic wastes.

2. The upper urinary tract includes the kidneys and ureters. The lower urinary tract includes the bladder and urethra.

3. The male reproductive system includes the testes (within the scrotum), the urethra (part of both the urinary and reproductive systems because it transports urine and semen), and a series of ducts and glands, including the prostate gland.

4. The female reproductive system includes the ovaries, the fallopian tubes, the uterus, and the vagina.

5. Clinical syndromes of the urinary tract include cystitis, urethritis, pyelonephritis, and urinary tract infection (UTI). Clinical syndromes of the female genital tract include vaginitis, endometritis, salpingitis, oophoritis, and pelvic inflammatory disease (PID).

Urinary Tract Infections (pp. 585–588)

6. All major pathogens of the urinary tract are bacteria. Diagnosis is based upon a clinical history, physical examination, and urinalysis. Upper urinary tract infections are much more serious than lower urinary tract infections.

7. It's sometimes difficult to tell how much of the urinary tract is affected, but treatment of both lower and upper urinary tract infections is the same: broad-spectrum penicillins, sulfa drugs, nalidixic acid, and nitrofurantoin.

8. *Escherichia coli* is by far the most common cause of UTIs, including nosocomial UTIs. Virulence factors of uropathogenic strains include adherence, hemolysins and colicins, serum resistance, and R factors.

9. Other bacterial species cause infections that are indistinguishable from *E. coli* infection. They include *Proteus*, *Pseudomonas*, and *Klebsiella* spp. and certain enteric streptococci.

10. *Leptospira interrogans* causes a systemic infection called leptospirosis. It is transmitted when urine from an infected animal enters the body directly through a break in the skin or indirectly through soil or water. Most infections are asymptomatic and self-limiting, but they can be fatal when liver or kidney damage occurs. Diagnosis depends upon serological tests; treatment is with penicillin or tetracycline.

Sexually Transmissible Diseases (pp. 586–599)

11. The epidemiology of STDs in the United States has changed recently with the emergence of new viral infections and increased incidence of others. People with one STD are likely to become coinfected.

12. Bacterial STDs are curable; viral STDs are not.

13. Women are more vulnerable to infection by STDs, and the consequences are greater (infertility, life-threatening pregnancies, fetal death, cancer). Children can suffer birth defects and newborn blindness.

14. Safer sex means reducing the number of sexual contacts and using condoms to help prevent infection with each contact.

Bacterial Infections (pp. 589–596)

15. *Neisseria gonorrhoeae* causes gonorrhea. It is transmitted only by direct contact. Virulence factors include adhesin-bearing pili, Protein II, and endotoxin. Symptoms in men—painful urination and a urethral discharge—appear 2 to 7 days after exposure. Infections in women are usually asymptomatic. *N. gonorrhoeae* causes blindness in newborns. Treatment is with penicillin or tetracycline; drug-resistant strains are becoming more common.

16. *Treponema pallidum* causes syphilis. Diagnosis is by darkfield microscopy or serological testing. Syphilis has four stages: primary, secondary, latent, and tertiary. Treatment with penicillin is effective during the first three stages. Tertiary syphilis may be fatal or benign, characterized by gummas, soft granulomas in nonvital organs. Prenatal (congenital) syphilis results in fetal death or birth deformities, such as snuffles. Drug-resistant strains have not emerged.

17. *Chlamydia trachomatis* causes chlamydia, the leading cause of infertility in women. Symptoms are subtle in both males and females (mild dysuria and a small discharge), which makes screening critical. Women under 25 and pregnant women are most at risk. *C. trachomatis* causes neonatal blindness and pneumonia. Treatment is with doxycycline, erythromycin, or azithromycin. Certain serotypes cause lymphogranuloma venereum (LGV), which can lead to painful ulcers.

18. *Ureaplasma urealyticum* and *Mycoplasma hominis* are part of the normal urethral and vaginal flora of most sexually active men and women. They sometimes cause a nongonococcal urethritis. Treatment for men with a discharge is doxycycline.

19. *Haemophilus ducreyi* causes chancroid. It is characterized by a soft chancre, which eventually heals; but without treatment (with ceftriaxone), it can leave scarring. In the United States chancroid is closely linked to the crack cocaine epidemic.

20. *Calymmatobacterium granulomatis* causes granuloma inguinale. Infection begins with lesions that develop into spreading ulcers. Diagnosis is based upon the presence of Donovan bodies. Treatment is with antibiotics, including ampicillin. Granuloma inguinale is rare in the United States.

Viral Infections (pp. 596–599)

21. Herpes simplex virus (HSV) causes herpes, a highly communicable STD. Both epithelial and nerve cells are affected. The infection is both active (characterized by painful blisters) and latent. Acyclovir and vidarabine may be prescribed in severe cases. Neonatal herpes causes death, blindness, deafness, and profound retardation.

22. Human papillomavirus (HPV) causes condyloma acuminata (genital warts). The cosmetic problem is minor compared with the association that apparently exists between HPV and cervical cancer. Because there is no cure, lifelong monitoring through pap smears is critical.

Infections of the Female Reproductive Tract (pp. 599–601)

23. *Gardnerella vaginalis* causes a common vaginitis. The discharge is odorous. An overgrowth of *G. vaginalis* is accompanied by an unusually large anaerobic bacterial population. Diagnosis is by the presence of clue cells in a vaginal wet mount. Treatment is with metronidazole to reestablish the normal vaginal flora.

24. Toxic shock syndrome—a life-threatening infection—is caused by strains of *Staphylococcus aureus* that produce toxic shock syndrome–associated toxin (TSST). Most at risk are menstruating women using vaginal tampons; there are few cases today.

25. Pelvic inflammatory disease (PID) occurs when bacteria from the vagina travel up to infect the uterus, fallopian tubes, and ovaries. Pathogens include *Chlamydia trachomatis*, *Neisseria gonorrhoeae*, and *Mycoplasma hominis*. Diagnosis is by cervical culture, but treatment must begin immediately with ceftriaxone and oral tetracycline.

26. Endometritis is an infection of the uterus caused most often today by *Escherichia coli*, *Neisseria gonorrhoeae*, *Chlamydia trachomatis*, *Clostridium*

spp., and anaerobic streptococci. Treatment is with cephalosporins. Uterine infections occur after birth (when the uterine lining is disturbed) or after abortions not performed under sterile conditions.

27. The major fungal cause of vaginitis is *Candida albicans*, an opportunistic pathogen. Candidiasis is characterized by a thick white vaginal discharge and severe vaginal itching. A hormonal imbalance and antibiotic therapy put a woman at risk.

28. *Trichomonas vaginalis* is a protozoan that causes trichomoniasis. It may be asymptomatic or characterized by a painful vaginitis and copious, frothy discharge. In males, infection is asymptomatic. Treatment is with metronidazole.

Infections of the Male Reproductive Tract (p. 602)

29. The classical male reproductive tract infection is caused by *Neisseria gonorrhoeae* and characterized by a copious, purulent urethral discharge.

30. Most infections, however, are nongonococcal urethritis (NGU). Most cases are caused by *Chlamydia trachomatis* and *Ureaplasma urealyticum*. Occasionally the cause is nonbacterial—*Trichomonas vaginalis* or herpes simplex virus.

Infections Transmitted from Mother to Infant (pp. 602–606)

31. Pathogens are transmitted across the placenta, during passage through the birth canal, or after birth through breast milk or intimate contact.

32. *Listeria monocytogenes* causes listeriosis. Pregnant women are most at risk. The illness is mild, but it causes abortion, stillbirth, or—in the newborn—septicemia, meningitis, and often death, despite treatment. Immunocompromised individuals are also at high risk.

33. Group B streptococci cause neonatal group B strep infections when

fetal membranes rupture during delivery. The most common forms are pneumonia, meningitis, and sepsis. Sepsis causes death despite treatment in one-fourth of the cases and permanent handicaps in survivors. Ampicillin administered intravenously to high-risk mothers and infants during delivery appears to be helpful.

34. Cytomegalovirus (CMV), a member of the herpes group, causes cytomegalic inclusion disease. CMV is transmitted sexually, by close contact (exchanging saliva), and in blood. Infection is usually asymptomatic, but if first infection occurs during pregnancy, it can cause abortion, stillbirth, or birth defects. Handicapped and asymptomatic children shed the virus for years, infecting adult caretakers and other children. Blood-bank blood can contain CMV.

REVIEW QUESTIONS

The Genitourinary System

1. Name the organs in the lower urinary tract, the upper urinary tract, the male reproductive system, and the female reproductive system. What are the functions of the urinary and reproductive systems?

2. Describe the following clinical syndromes:
   a. cystitis
   b. urethritis
   c. pyelonephritis
   d. urinary tract infection (UTI)
   e. pelvic inflammatory disease (PID)
   f. vaginitis

3. What are perinatal infections?

Urinary Tract Infections

4. Why are women anatomically more prone to urinary tract infections than men? Name some causes of recurrent UTIs.

5. Why should patients with a UTI drink large quantities of water?

What are the general symptoms of lower urinary tract infections? Of upper urinary tract infections?

6. Why are sulfa drugs particularly effective against urinary tract pathogens?

7. Describe the virulence factors in uropathogenic strains of *Escherichia coli*.

8. Why do so many cases of leptospirosis go undiagnosed?

9. Explain this statement: Until the 1970s, many clinicians believed STDs would be a problem of the past.

10. What is safer sex?

11. Why are women more vulnerable to infection by an STD than men? Why are the long-term consequences of infection greater for a woman? Explain this statement: Children are victimized by STDs.

Sexually Transmissible Diseases (STDs)

12. Describe the virulence factors of the gonococcus. How is gonorrhea transmitted? Describe the symptoms in men and in women. Is gonorrhea treatable? Explain.

13. How is syphilis diagnosed? Describe primary, secondary, latent, and tertiary syphilis. What is the drug of choice for treating syphilis? Describe congenital syphilis.

14. What group is at highest risk for chlamydia? What are the symptoms and possible long-range complications for women? For men? Describe prevention and treatment.

15. Describe the clinical syndromes caused by *Ureaplasma urealyticum* and *Mycoplasma hominis*.

16. Why is chancroid less communicable than gonorrhea and syphilis?

17. How is granuloma inguinale diagnosed? What are the symptoms and treatment?

18. How prevalent are genital herpes infections in the United States? Explain how the herpes simplex virus causes both active and latent infections. Describe the clinical syndrome of primary infection and recurrent infections. Why are condoms not as helpful in preventing herpes infections as they are in preventing some other STDs? How can neonatal herpes be prevented? What is the treatment and how effective is it for infants? For adults?

19. What is the treatment for genital warts, and how effective is it?

Infections of the Female Reproductive Tract

20. How does *Gardnerella vaginalis* cause vaginitis? How is it diagnosed? What is the treatment?

21. Describe the pathogenesis of toxic shock syndrome. Who is most at risk? What are the symptoms and treatment?

22. What significant nonfatal complications can result from pelvic inflammatory disease? How is PID transmitted? Discuss the prevention and treatment.

23. Name some situations that often lead to uterine infections. What is the treatment?

24. In what way is *Candida albicans* an opportunistic pathogen? What are the symptoms of candidiasis? How is it treated?

25. Describe the microorganism *Trichomonas vaginalis*. How is trichomoniasis transmitted? What are the symptoms and treatment?

Infections of the Male Reproductive Tract

26. Discuss the symptoms of male urethritis. What is the most common cause of male urethritis? Name the pathogens that most commonly cause nongonococcal urethritis (NGU).

Infections Transmitted from Mother to Infant

27. How is *Listeria monocytogenes* an opportunistic pathogen? Who is at highest risk? What forms does listeriosis take? How is a fetus or newborn affected?

28. How do infants acquire group B streptococcus infection? What are the risks to new mothers? How can infection be prevented?

29. How widespread is cytomegalic inclusion disease? Discuss the transmission, symptoms, prevention, and treatment.

CORRELATION QUESTIONS

1. What questions would you ask and what tests would you request to determine the cause of a woman's abnormal vaginal discharge?

2. What questions would you ask and what tests would you request to determine the cause of a chancre?

3. Some say that Henry VIII of England died of syphilis. What historical facts would you seek to prove or disprove this contention?

4. In certain ways, disease caused by *Gardnerella vaginalis* resembles disease caused by *Clostridium difficile*. Explain.

5. What infections of the genitourinary tract can be diagnosed by using a light microscope?

6. In what way is the transmission of gonorrhea and syphilis identical?

ESSAY QUESTIONS

1. Discuss the factors that cause new STDs to arise and existing STDs to become epidemic.

2. Discuss the ethical questions related to being able to test blood for the presence of cytomegalovirus. Include the advice you would offer to address the problem.

SUGGESTED READINGS

Cates, W., Jr., and A. R. Hinman. 1991. Sexually transmitted diseases in the 1990s. *New England Journal of Medicine* 325:1368–70.

Handsfield, H. H. 1991. Recent developments in STDs: I. Bacterial diseases. *Hospital Practice* 47–56.

Handsfield, H. H. 1991. Recent developments in STDs: II. Viral and other syndromes. *Hospital Practice* 175–200.

Luby, J. 1982. Therapy in genital herpes. *New England Journal of Medicine* 306:1356–57.

Morbidity and Mortality Weekly Report. 1990. Progress toward achieving the 1990 objectives for the nation for sexually transmitted diseases. 39:53–57.

Schaechter, J. 1989. Why we need a program for the control of *Chlamydia trachomatis. New England Journal of Medicine* 320:802–4.

Witook, E., III, and C. M. Marra. 1992. Acquired syphilis in adults [review article]. *New England Journal of Medicine* 327:1060–69.

For additional readings, go to InfoTrac College Edition, your online research library at: http://www.infotrac.thomsonlearning.com

TWENTY-FIVE

Infections of the Nervous System

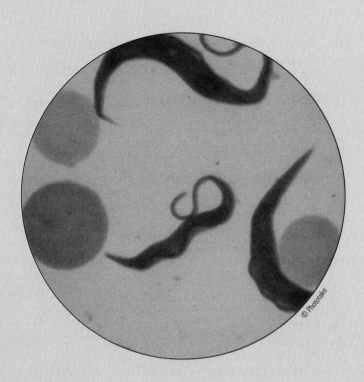

© Phototake

CHAPTER OUTLINE

LEARNING GOALS

To understand:

- *The anatomy and function of the nervous system and its defenses against microorganisms*
- *The clinical syndromes caused by nervous system infections*

- *The bacterial, viral, and fungal infections of the meninges—their diagnosis, prevention, and treatment*

- *The bacterial, viral, prion-associated, and protozoal infections of neural tissue—their diagnosis, prevention, and treatment*

Twenty-Four Hours

C. H. was a healthy 19-year-old male college student. One evening he suddenly developed a severe pounding headache. Within an hour he had chills and a fever and felt sicker than he ever had before in his life. He asked his roommate to drive him to an emergency clinic. There the examining physician found no abnormality except for his temperature of 105°F. C. H. was told he only had a flulike illness but that he should return if he didn't feel better in a day or so.

The next day C. H. was much worse. In the morning, red spots appeared on his arms and legs. During the day, they grew and changed to a purplish color. He felt nauseated and gagged for no apparent reason. He could not eat or drink. His headache grew even more painful. By evening his back hurt and his neck felt stiff.

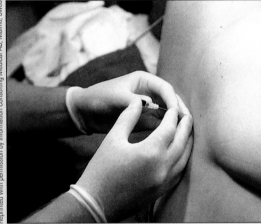

Lumbar puncture. The patient's back is bent to widen the spaces between the vertebrae and allow the needle to enter the spinal canal. Because the needle is inserted below the end of the spinal cord, the procedure is safe.

Reprinted with permission by Information Consulting Medical AB, Malmo, Sweden

He became extremely weak. In just 24 hours, C. H. had gone from being completely well to becoming critically ill. Again his roommate drove him to the emergency clinic.

C. H. could barely sit up during the examination. His temperature had fallen to 101°F, but he looked extremely ill. His skin was pale with slightly raised purple spots, some of which were nearly 1 cm in diameter. They covered his trunk and extremities. His mouth was dry. He appeared to be slightly dehydrated. He could move his head easily from side to side but not forward. It was too painful. The stiff neck suggested meningismus (irritation of the meninges).

C. H.'s physician recommended lumbar puncture because that was the only way to diagnose what she thought might be a life-threatening infection. Her suspicions were confirmed. Cerebrospinal fluid (CSF) is normally clear, but C. H.'s was cloudy—a sure sign of meningitis. In addition, the pressure of his CSF was much higher than normal. That caused his headache. Samples of C. H.'s CSF, blood, and urine were sent to a clinical laboratory for analysis.

C. H.'s blood contained more leukocytes than normal. That suggested an infection. Normally CSF doesn't contain any leukocytes. C. H.'s CSF was packed with them (more than 15,000 leukocytes per mm³). Because the number of leukocytes exceeded 1000, C. H.'s physician diagnosed **purulent** meningitis. Its usual cause is a bacterial infection. (Counts less than 1000 indicate nonpurulent meningitis, which is usually nonbacterial.) A Gram stain of the CSF revealed many Gram-negative diplococci. A diagnosis of meningitis caused by *Neisseria meningitidis* was almost certain. Confirmation came from culturing the bacteria in his CSF.

As soon as the CSF cell count and Gram stain was completed, C. H.'s physician gave intravenous penicillin and a cephalosporin antibiotic. She admitted C. H. to the hospital and placed him in respiratory isolation. Four hours after his first dose of antibiotics, C. H. was feeling well enough to watch television, but he still complained of a headache and stiff neck.

A few days later the cephalosporin was discontinued, but C. H. continued on intravenous penicillin therapy for 10 days. Then all the signs and symptoms of his illness disappeared. He was discharged from the hospital in his normal state of good health.

Case Connections

- As you read this chapter, ask yourself which of C. H.'s signs and symptoms would lead to a diagnosis of meningitis instead of encephalitis or myelitis.
- Also ask which of C. H.'s signs and symptoms would suggest a bacterial infection instead of a viral infection or the consequences of a neurotoxin.

THE NERVOUS SYSTEM

The nervous system integrates and coordinates all sensory and motor activities. Its largest organ, the brain, performs higher functions that allow us to appreciate life through consciousness, thought, and memory. Brain function is central to our concept of humanity. A person may be considered dead when the brain stops functioning, even if "vegetative functions" such as respiration and heartbeat continue.

Structure and Function

The nervous system is composed of neurons and neuroglia. **Neurons** are the nerve cells. **Neuroglia** are the cells that support neurons and bind them together. Neurons are unlike most other cells in the body because they don't undergo a cycle of death and replacement. Throughout our lives we develop only a few more neurons than those present at birth. Damage to the nervous system is therefore likely to cause permanent disability.

The nervous system has two parts: the **central nervous system** (**CNS**) and the **peripheral nervous system** (**PNS**). The CNS includes the **brain** and the **spinal cord.** The PNS is composed of the **nerves** that run throughout the body. Nerves are bundles of neurons and supporting tissue that transmit messages of sensation from sense organs to the brain and commands for movement and other functions from the CNS (**Figure 25.1**).

The CNS is well protected against infection and injury. The brain is protected by the skull, and the spinal cord by the many small interconnected bones called **vertebrae.** Beneath these bones, both the brain and the spinal cord are further protected by three membranes called the **meninges.** The outermost and thickest membrane is called the **dura mater.** The middle layer is called the **arachnoid.** And the innermost and thinnest membrane is called the **pia mater.**

A clear liquid called **cerebrospinal fluid** (**CSF**) circulates around the brain and spinal cord between the pia mater and the arachnoid. CSF adds to the protection of the CNS because it absorbs some of the shock of an injury. Samples of CSF can be safely removed from the spinal column below the point at which the spinal cord ends by a procedure called a **lumbar puncture,** or **spinal tap** (see photo accompanying Case History: Twenty-four Hours).

In addition to the protection against mechanical injury we have just discussed, the CNS is protected against chemical injury by the **blood-brain barrier.** This barrier is provided by the thickened outer layer of the capillaries (the smallest blood vessels) that supply blood to the CNS. The layer blocks the passage of many toxins into the brain but not the entry of essential molecules, such as oxygen, carbon dioxide, glucose, and amino acids. Unfortunately, the layer also blocks the passage of many antibiotics. As a result, CNS infections are difficult to treat.

Unless infected, the CNS is free of microorganisms. Infections can occur when massive trauma opens the skull. They can also be carried to the CNS in the bloodstream or the cerebrospinal fluid, and they can spread to the CNS from nearby sites, such as the sinuses or middle ear.

In summary, the CNS is the body's command center. Orders are sent out from it and sensations are received by it through nerves. It's well protected by bones, the meninges, the CSF, and the blood-brain barrier. In spite of this elaborate protection, infections do occur.

Clinical Syndromes

Most infections of the nervous system affect the CNS. The most common clinical syndromes they cause are meningitis, encephalitis, and myelitis (**Table 25.1**).

Meningitis, which C. H. suffered, is inflammation of the meninges. His normally clear CSF became cloudy with white blood cells, which were attracted to the site of inflammation. Swelling around the brain increased pressure inside the skull, causing his intense headaches. Inflammation and irritation of the spinal meninges affected nearby muscles, causing his stiff neck. His fever was typical of most cases of meningitis. C. H. suffered no loss of brain function, but many patients with meningitis do because inflammation of the meninges can diminish the flow of blood that the brain needs to function properly.

Encephalitis is inflammation of brain tissue. It does affect brain function, changing a patient's state of consciousness or behavior. Many other disorders, even high fever, cause clinical syndromes that resembles infectious encephalitis. But when no other cause can be found, infection is a likely possibility.

Myelitis is inflammation of the spinal cord. It has many causes, one of which is infection. Because the nerves in the spinal cord carry all messages for sensory and motor function, myelitis causes different symptoms depending upon which messages are interrupted.

Some bacterial pathogens that do not enter the nervous system produce **neurotoxins** (highly destructive proteins; Chapter 15) that do enter and cause symptoms. Such **intoxications** (toxin-caused disorders) do not produce the typical syndromes of CNS infections described earlier. They cause a variety of symptoms that depend upon the action of the particular neurotoxin.

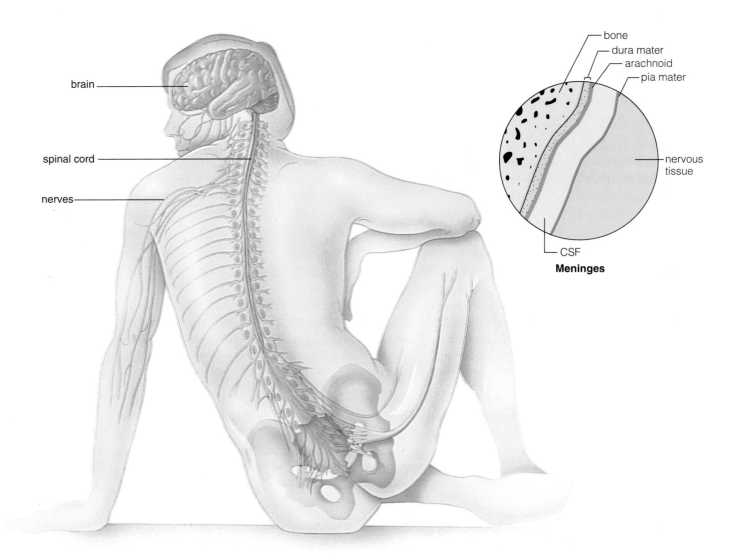

FIGURE 25.1   Organs of the nervous system.

(Art by Kevin Somerville from Biology: The Unity and Diversity of Life, 6ᵗʰ ed., by C. Starr and R. Taggart, Brooks/Cole, 1992. All rights reserved.)

TABLE 25.1   Clinical Syndromes of Nervous System Infections

Syndrome	Region Affected	Signs and Symptoms	Causative Agents
Meningitis	Meninges	Fever, headache, stiff neck, usually some disturbance of brain function	*Neisseria meningitidis, Haemophilus influenzae, Streptococcus pneumoniae, various viruses, Cryptococcus neoformans*
Encephalitis	Brain	Disturbance of brain function, usually fever	Rabies virus, arboviruses, *Trypanosoma brucei gambiense, Trypanosoma brucei rhodesiense*, prions
Myelitis	Spinal cord	Disturbance of nerve transmission	Poliovirus
Neurotoxin-caused damage to nervous tissue	Central and peripheral nervous system	Paralysis, rigid (tetanus) or flaccid (botulism)	*Clostridium tetani, Clostridium botulinum*

INFECTIONS OF THE MENINGES

Bacterial, viral, and fungal pathogens all infect the meninges and cause meningitis.

Bacterial Causes

Bacteria that infect the meninges all cause a similar set of symptoms.

Meningococcal Meningitis. The Gram-negative pathogen *Neisseria meningitidis* (commonly called the **meningococcus**) caused C. H.'s illness. Its only natural reservoir is the human body. C. H.'s illness was relatively mild, but often the meningococcus causes fatal disease, even with antibiotic treatment.

N. meningitidis can be cultured in the laboratory, usually by growing it on very rich medium called a **chocolate agar** (named for brown color of the heated blood these agars contain). The colonies that develop are nonhemolytic, and they grow best in an atmosphere enriched with 5 to 10 percent carbon dioxide. *N. meningitidis* can be distinguished from its close relative *Neisseria gonorrhoeae* (Chapter 24) by its ability to ferment maltose.

The meningococcus is a superbly adapted pathogen. It's readily transmitted from person to person in respiratory droplets (Chapter 15), and up to a third of humans are asymptomatic carriers of meningococcus. C. H. was probably infected by such a healthy person.

The meningococcus is also well equipped to avoid the body's antibacterial defenses. Not all the meningococcal cells that came to rest on C. H.'s respiratory epithelium were killed by IgA antibodies that reside there (Chapter 17). Some cells survived because *N. meningitidis* produces **IgA protease,** an enzyme that destroys IgA. Meningococcal cells also bear adhesins (Chapter 15) on their outer surface that allow them to stick so tightly to the epithelium that the mucociliary defenses can't sweep them away.

The meningococcal cells that stuck to C. H.'s nose and throat might have lived there harmlessly, making him just another asymptomatic carrier. Instead they became **invasive.** They entered the deeper tissues, where nutrients are abundant. But iron is an exception. It's not available to invading microorganisms because it's tightly bound to **transferrin** and **lactoferrin** (iron-carrier proteins). *N. meningitidis* is able to acquire this essential nutrient by binding to these proteins and stealing their iron (Chapter 15).

The meningococcal cells were also able to evade phagocytosis (Chapter 15) because their polysaccharide capsule protected them from otherwise certain destruction. Certain antibodies can overcome this protection (Chapter 16), but

C. H.'s adaptive immune system didn't have time to produce them. His infection was established before his immune system could mount a primary immune response, and it couldn't mount a rapid secondary immune response because C. H. had never before been exposed to this particular strain of *N. meningitidis*. It is now becoming standard practice to give meningococcal vaccine to all students entering college dormitories. Although it is effective, it does not cover all serotypes, so C. H.'s type of illness will not be completely eliminated by it

Meningococcal cells then entered C. H.'s bloodstream, which carried them to all parts of his body. Some meningococci in the capillaries of his skin died, releasing endotoxins, which stimulated host cells to release more toxic substances, such as tumor necrotizing factor (Chapter 15). These damaged blood vessels, allowing blood to seep into his tissues, producing his purplish spots (**petechiae**).

Although endotoxin caused damage, the intact meningococcal cells that entered C. H.'s central nervous system caused his worst symptoms. They multiplied rapidly in his meninges, causing inflammation, a life-threatening event. If his infection had not been stopped with antimicrobial drugs, C. H. most likely would have died. Seventy to ninety percent of patients do die from untreated meningococcal meningitis.

Acute Purulent Meningitis. C. H.'s illness showed most of the symptoms of acute purulent meningitis: sudden onset, high fever, stiff neck, headache, and worsening within 24 hours. But he didn't suffer an altered state of consciousness. That usually occurs because inflamed meninges disrupt the flow of blood to the brain, causing meningitis patients to suffer convulsions, extreme drowsiness (which may progress to coma), or agitated behavior similar to that caused by drug overdose. C. H.'s age was also atypical. Meningitis occurs most often in children or the elderly.

C. H.'s purplish skin rash alone allows a working diagnosis of meningococcal disease, but not necessarily meningitis. The rash is also seen in **meningococcemia,** an illness in which *N. meningitidis* invades the bloodstream but not the central nervous system.

Epidemiology. C. H.'s infection was an isolated case, but meningococcal infection can occur in devastating epidemics. Outbreaks are particularly likely where large numbers of people are crowded together under stressful conditions, such as military barracks. Entire cities may be affected by epidemics, as was the case in São Paulo, Brazil, during the early 1970s. The last nationwide epidemic in the United States occurred in the mid-1940s.

Epidemic disease may be unusually severe. Death can occur within hours even with antibiotic therapy, and many survivors suffer permanent brain damage. After epidemics, up to 90 percent of healthy people in the community become asymptomatic carriers. Family members of meningitis patients are almost certain to become carriers.

Carriers spread the infection. By treating them with the antibiotic rifampin, epidemics can be stopped. Rifampin, however, is not adequate for treating systemic meningococcal infection. Penicillin G is the drug of choice. Treating carriers is called **meningococcal prophylaxis.** Even if an epidemic is not in progress, close contacts of patients with meningococcal infection are sometimes given rifampin prophylaxis. Because C. H. lived with only two roommates, physicians decided to treat them with rifampin and observe them for early signs of illness. If he had lived in a large college dormitory, all his dormitory contacts might have been given rifampin.

A vaccine is now available to control meningococcal meningitis. Various strains of *N. meningitidis* produce distinct capsular antigens, the major types being A, B, C, Y, and W. The vaccine is effective against all of them except B, so vaccination can now be used along with rifampin prophylaxis to control most epidemics. It's not recommended for routine immunization of children, but it is given to all military recruits in the United States, as well as college freshmen who will be living in dormitories or other high-density housing.

Haemophilus influenzae **Meningitis.** *Haemophilus influenzae*, like *Neisseria meningitidis*, is an invasive bacterial pathogen with a polysaccharide capsule (**Figure 25.2**). It infects the respiratory tract (Chapter 22), as well as the meninges. Like *N. meningitidis* meningitis, *H. influenzae* meningitis is often transmitted by asymptomatic carriers. Rifampin is commonly prescribed to treat them.

Until about 10 years ago, *H. influenzae* caused 75 percent of all meningitis in infants and young children. That situation changed abruptly in late 1990 when an *H. influenzae* type b (Hib) polysaccharide-protein conjugate vaccine became available. Within a year, *H. influenzae* type b diseases in children under 5 fell about fourfold. Rates continue to fall as more children are immunized and herd immunity protects unvaccinated children. (Also see the discussion of Hib and epiglottitis in Chapter 22.)

Pneumococcal Meningitis. *Streptococcus pneumoniae* is yet another encapsulated bacterial pathogen that can cause meningitis. *S. pneumoniae* most commonly causes pneumonia (Chapter 22), but it also causes about half the cases of meningitis in adults over the age of 40. Pneumococcal meningitis often strikes people who are in a weakened state of health. Newborns, people with sickle cell disease, and alcoholics are particularly vulnerable. With the advent of a vaccine for *H. influenzae*, *S. pneumoniae* became one of the two most common causes of meningitis in children. But now

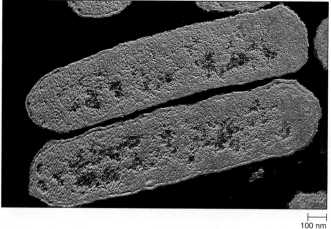

FIGURE 25.2   *Haemophilus influenzae* is an invasive bacterial pathogen that causes meningitis.

that infants are routinely vaccinated against pneumococcus, pneumococcal meningitis will undoubtedly decline as well.

Bacteria usually enter the central nervous system from the bloodstream, but they can also spread directly from the middle ear or sinuses. In spite of antibiotic treatment, 5 to 10 percent of patients die. Permanent brain damage occurs more commonly after *S. pneumoniae* than after either *Neisseria meningitidis* or *Haemophilus influenzae* meningitis.

Escherichia coli **Meningitis.** *Escherichia coli* causes less than 5 percent of bacterial meningitis cases, but newborn infants and patients who have undergone neurosurgery are particularly prone to this deadly infection. Even with appropriate treatment using ampicillin and gentamicin, half the patients die.

Viral Causes: Aseptic Meningitis

Sometimes patients with typical symptoms of meningitis symptoms are found to have no bacteria and only a few leukocytes (typically less than 500 per mm³) in their cerebrospinal fluid. The diagnosis in these cases is nonpurulent, or **aseptic, meningitis,** and viral infection is the presumed cause. Aseptic meningitis is fairly common, especially among children and young adults. Overall it accounts for about 40 percent of cases of meningitis.

Many different viruses can cause meningitis. The most common are the mumps virus (Chapter 23), members of the enterovirus group (including echoviruses, coxsackieviruses, and polioviruses), and arboviruses. The meningitis syndromes produced by different viruses are virtually indistinguishable, and usually the causative virus is not identified.

Fortunately it is not important to know which virus is the cause. Antimicrobial treatment is not effective, and un-

CASE HISTORY

Aseptic Meningitis

One Saturday afternoon in September, E. V., an eighteen-year-old girl came to an acute care clinic complaining of "the worst headache of my life." She had been ill for 2 days. She had a low-grade fever and some diarrhea. She had vomited once or twice and had no appetite. But her most painful symptom was her headache. It was severe when she first became ill. Now it was almost unbearable.

When she arrived at the clinic, E. V.'s temperature was 101°F. She looked quite uncomfortable, but she walked to the examining room without assistance and she told her story clearly. She didn't appear to be extremely ill, but her physical examination revealed one very worrisome finding: Her neck was stiff. She could move her head from side to side without difficulty, but bringing her chin down toward her chest was almost impossible. This finding meant that a spinal tap should be done to determine whether or not she had meningitis.

E. V. signed a consent for her spinal tap and underwent the procedure without complications. The spinal fluid looked clear. E. V. also had her blood drawn for a complete blood cell count. She waited for an hour for the lab results to come back. The blood count was almost normal, but her white blood cell count, at 17,000, was slightly elevated and she had relatively more lymphocytes than granulocytes. This all suggested a viral infection. The laboratory examination of her spinal fluid gave results that were decidedly abnormal. Her CSF contained 500 white blood cells per mm^3 (normally there are none). Because there were no red blood cells in the sample of her CSF, the laboratory technician knew it hadn't been accidentally contaminated with blood. A Gram stain failed to revealed the presence of bacterial cells.

These laboratory results confirmed a diagnosis of meningitis. But they didn't tell what type of microorganism was causing her infection. Was E. V. at the early stage of a life-threatening bacterial meningitis? Or was she suffering from a viral (aseptic) meningitis that would improve without treatment? A number of facts suggested that viral meningitis was the more likely cause: She didn't appear to be severely ill; the number of white cells in her blood was almost normal; and she had only a small number of white blood cells and no visible bacteria in her CSF. But even a small possibility of a life-threatening bacterial meningitis calls for precautionary treatment. E. V. was admitted to the hospital for intravenous treatment with a broad-spectrum antibiotic. She rested and received medication for her slowly improving headache. Three days later the results of her CSF culture were completed. No bacterial growth had occurred. This meant that antibiotic treatment could be stopped and E. V. could be sent home to rest while she recovered fully. A week later she was back attending her college classes and managing her regular busy schedule.

treated patients usually recover within a few days or weeks. It's only important to establish with certainty that the cause is not bacterial. As we've seen, bacterial meningitis is deadly and requires immediate treatment. Viral meningitis does not. If there is any doubt about the cause, the patient should be treated with antibiotics until the possibility of a bacterial infection is ruled out. Many of the viruses that cause meningitis also cause other infections of the nervous system.

Fungal Causes

Fungi cause meningitis that's very different from the disease caused by bacteria or viruses. Fungi cause **chronic meningitis,** a disease that comes on gradually and progresses slowly. Most cases are caused by either of two fungi—*Coccidioides immitis* and *Cryptococcus neoformans. C. immitis* causes coccidioidomycosis (Chapter 22). A very few of these patients develop chronic meningitis, which closely resembles cryptococcal meningitis, described next.

Cryptococcal Meningitis. Over days or weeks, patients with *C. neoformans* meningitis experience headaches and sometimes confusion, weakness, or loss of coordination. Sometimes they have a stiff neck. Their CSF contains only a few hundred leukocytes per cubic millimeter, most of which are mononuclear. The presence of *C. neoformans* cells in the CSF confirms the diagnosis of cryptococcal meningitis. Without treatment, 90 percent of patients die within a year from brain swelling. Treatment with amphotericin B,

CASE HISTORY

A Stiff Baby and a Limp Baby

The Stiff Baby

Baby Boy T was a full-term healthy baby born in a rural region of a developing country. He was delivered at home after a short and uneventful labor. The woman assisting the delivery cut the baby's umbilical cord with an unsterilized instrument and, as was customary in the area, dressed the cord with a paste made of clay. Baby Boy T was vigorous and nursed well. On the sixth day, however, he became unusually irritable. He cried a lot. He had trouble sucking from the breast and swallowing milk. Over the next few days, he became stiff. He suffered spasms that contracted muscles all over his body. His alarmed parents took Baby Boy T to the nearest health clinic. By then his body was completely rigid. His back was arched. He was so stiff he could be supported by one hand under his head and another under his feet (**Figure 25.3**). His grieving parents were told that nothing could be done. Soon Baby Boy T stopped breathing and died.

The Limp Baby

Baby Boy B was a normal newborn delivered in a large hospital in the United States. A day later he went home with his parents. He nursed well and grew rapidly. At 4 months, however, his mother began to notice subtle changes. He became listless and less active. He lost his developing ability to hold his head up. He became constipated. He breastfed poorly. He became very weak. Even his cry was weak. His parents took Baby Boy B to an emergency clinic, where the examining physician noted that his muscles lacked normal tone. When the doctor raised Baby Boy B's arm or leg and dropped it, it flopped back down on the examining table as if he were asleep. His breathing was weak. His gag reflex was absent. The doctor immediately admitted him to the hospital.

Baby Boy B's weakness worsened. He became unable to breathe, so he was transferred to the intensive care unit, put on a mechanical respirator, and fed intravenously (**Figure 25.4**). After several weeks of supportive care, his muscle strength slowly began to return. Eventually he was discharged from the hospital. Baby Boy B developed normally with no lasting effects from his illness.

FIGURE 25.3   A baby with tetanus is rigid from uncontrollable contractions of the back muscles.

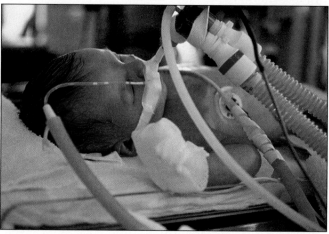

FIGURE 25.4   A baby with botulism is on a mechanical respirator because of flaccid paralysis of respiratory muscles.

sometimes in combination with 5-flucytosine, decreases the death rate but only to about 40 percent.

Cryptococcus neoformans is a yeast with a thick polysaccharide capsule (see Case History: The Capsule Is the Clue, in Chapter 4). Unlike most other fungal pathogens, it is not dimorphic. *C. neoformans* is plentiful in the soil and can reach astounding concentrations in bird droppings, especially those from pigeons.

If *C. neoformans* is inhaled into the lungs, it may only establish a respiratory infection with few if any symptoms.

But it can enter the bloodstream and spread to the central nervous system, causing deadly chronic meningitis. Fortunately, cryptococcal meningitis is rare. Only a few hundred people in the United States acquire this deadly disease each year. Most victims have a weakened immune systems. People with acquired immunodeficiency syndrome (AIDS) are particularly susceptible.

DISEASES OF NEURAL TISSUE

Bacterial, viral, and protozoal pathogens can all damage neural tissue. Some do so directly by infecting it. Others do so indirectly by producing neurotoxins. Prions also cause neurological disease.

Bacterial Causes

Some bacteria damage nervous tissue by producing neurotoxins.

Clostridium tetani: **Tetanus.** *Clostridium tetani* caused Baby Boy T's horrible illness and death. It's a Gram-positive, anaerobic, rod-shaped bacterium that moves by means of peritrichous flagella. It grows best at 37°C and a pH of 7.4, conditions that exist in the human body.

C. tetani is part of the normal intestinal biota of horses, cattle, and some humans. It produces endospores that cause a spore-containing *C. tetani* cell to look like a drumstick (**Figure 25.5**). These endospores are found almost everywhere. They're especially abundant in cultivated fields that have been fertilized with animal manure.

Tetanus, a *C. tetani* infection, develops when endospore-containing dirt enters a wound. Many such wounds become anaerobic because oxygen-carrying blood has been cut off and aerobic bacteria use up any residual oxygen. Decaying tissue is also anaerobic. Endospores germinated in the decaying tissue of Baby Boy T's umbilical cord. This is how newborns typically get tetanus. Adults usually become infected through wounds or by nonsterile surgical procedures. In the past, tetanus was a frequent, fatal complication of illegal abortions. It killed more soldiers than battle-inflicted injuries did. But immunization changed the picture completely. In World War II, only 12 U.S. soldiers contracted tetanus.

Tetanospasmin: The Tetanus Toxin. **C. tetani** has almost no invasive ability. It stays within the infected wound. There it produces a deadly neurotoxin, **tetanospasmin,** which enters the bloodstream and is carried to the CNS. It caused all Baby Boy T's signs and symptoms of tetanus

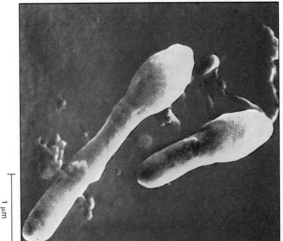

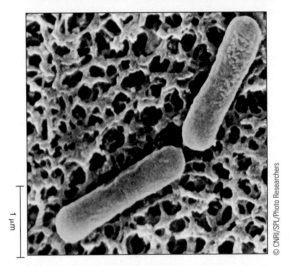

FIGURE 25.5 *Clostridium tetani* and *Clostridium botulinum*. The round spores of *C. tetani* occur at the end of the cell, causing it to swell. *C. botulinum* has oval spores near the end of the cell.

and his death. Tetanospasmin is one of the most potent toxins known. As little as 130 micrograms (a speck) can kill an adult.

Normally the CNS sends out two kinds of messages to muscles. One stimulates contraction. The other inhibits it. Tetanospasmin inhibits the inhibitory messages. As a result, muscles throughout the body receive only stimulatory messages and contract uncontrollably.

When Baby Boy T's muscles began to stiffen, the strong ones overpowered the weak ones. This caused his back to arch (medically called **opisthotonos**). The jaw muscles of adult tetanus victims contract, causing lockjaw (medically called **trismus**). Survivors of tetanus say these muscle spasms are excruciatingly painful. Some spasms are strong enough to break bones. Some prevent breathing, causing death, as they did for Baby Boy T.

Prevention and Treatment. By the time clinical tetanus is recognized, it is difficult to treat. After tetanospasmin has bound to nerve cells, there is no way to interfere with its action. But it does begin to wear off after about a week. If the victim survives that long, recovery is complete. Young adults are the most likely to recover, but their survival rarely exceeds 50 percent. Newborns are doomed. Even with the best treatment, Baby Boy T almost certainly would have died. Cases such as Baby Boy T's are tragically common. In some parts of the developing world, tetanus causes up to 10 percent of deaths in young babies.

Before clinical signs appear, tetanus can be effectively treated with tetanus immune globulin. Immune globulin contains high concentrations of **tetanus antitoxin** (antibodies against tetanus toxin). These antibodies bind to tetanospasmin and inactivate it before it attaches to nerve cells. It's better, of course, to prevent tetanus in the first place by immunizing with tetanus toxoid (Chapter 20).

Universal tetanus immunization as part of the diphtheria-tetanus-pertussis (DTP) inoculation has made tetanus exceedingly rare in the United States. The incidence rate is about 0.05 per 100,000 people, less than one in a million, and these cases are limited to people who lack immunity. It's a good idea to maintain immunity with regular tetanus immunization and get a booster after suffering a tetanus-prone wound.

Could Baby Boy T's death have been avoided? Yes, quite easily. If his mother had been adequately immunized, he would have been protected by her antibody passing through the placenta. Immunizing pregnant women could eliminate neonatal tetanus. If his umbilical cord had been cut with a sterile instrument, if the wound had been kept clean, or if tetanus antitoxin had been administered before he became ill, he would have lived.

Botulism.
The bacterium that made Baby Boy B ill was quite like the one that killed Baby Boy T. It too is a species of the genus *Clostridium*. It too is motile with peritrichous flagella. It too is rod-shaped, endospore-forming anaerobes, but their spores are slightly different (Figure 25.5).

The endospores of *C. botulinum* are found virtually everywhere. Endospores in soil can contaminate food such as vegetables and honey. Spores in ocean sediments can contaminate fish and seafood.

Botulinum Toxin. *Clostridium botulinum* produces **botulinum toxin** (a neurotoxin). It's the most poisonous natural substance known, even more toxic than tetanospasmin (although tiny amounts are injected subcutaneously as a beauty aid to remove wrinkles). A lethal dose of toxin for a human is measured in nanograms, an amount you can't even see. The genes encoding toxin production are carried on a prophage, so only phage-carrying strains of *C. botulinum* produce toxin.

Various strains of *C. botulinum* produce antigenically distinct forms of botulinum toxin, designated A through G. Only types A, B, and E cause human disease. Type A toxin is the most potent of all. Strains producing it are most common west of the Mississippi River. Other strains are more common in other places and in the oceans.

As the illnesses of Baby Boy T and Baby Boy B show, tetanospasmin and botulinum toxin have vastly different actions. Botulinum toxin affects the peripheral, not the central, nervous system. It enters the circulatory system and binds to neurons where they join to muscle. Botulinum toxin stops the release of acetylcholine—the molecular messenger that delivers the signal for a muscle to contract. The result is a limp, or flaccid, paralysis, the opposite of rigid paralysis caused by tetanospasmin. In adults, botulinum toxin first affects nerves that control the head. The result is double vision and also difficulty with speaking and swallowing. Later other muscles, including those that control breathing, may become paralyzed.

Clinical Syndromes. People can be exposed to the toxin in different ways. Baby Boy B swallowed *C. botulinum* cells that colonized his intestinal tract. The bacterial cells produced neurotoxin inside his body, which was absorbed into his bloodstream, causing his illness. This syndrome, which occurs in babies, is called **infant botulism.**

Infant botulism wasn't even known before 1976. Now it is recognized as the most common form of botulism in the United States. Several hundred cases are diagnosed each year. It's a mystery why this type of botulism affects only infants under 6 months of age. Differences in normal bacterial biota of the intestines probably account for the sensitivity of infants and the resistance of adults.

In adults, botulism is a form of food poisoning. In other words, **adult botulism** is like staphylococcal food poisoning (Chapter 23). It is an intoxication, not an infection. Bacterial cells do not multiply inside the body. The word botulism comes from the Latin word for "sausage," because in the 1700s botulism was associated with eating blood sausage.

Today botulism is usually acquired from eating home-canned food that was inadequately heated. Spores of *C. botulinum* that survive the process germinate in the food during storage, and vegetative cells produce toxin. Reheating destroys the toxin. But if reheating is insufficient, potent toxin remains in the food. Eating even a small amount of such contaminated food is enough to cause death. *C. botulinum* cannot grow in an acidic environment, so home-canned neutral vegetables such as green beans, peppers, and mushrooms are more dangerous than acidic foods, such as tomatoes and pickles.

Rarely, botulism is acquired through an infected wound. This occurs as it does with tetanus. Spores germinate in the anaerobic wound and produce toxin that enters the bloodstream.

LARGER FIELD

A BIZARRE AND DEADLY ENCOUNTER

People in the United States who stay close to home hardly ever die of rabies. Only four domestically acquired cases were reported to the Centers for Disease Control and Prevention (CDC) between 1980 and 1990, so it was a surprise when three such cases were reported within a 3-month period in 1991.

One case in Arkansas was typical. On August 17 a lifelong male resident of the state came down with a sore throat and had trouble swallowing. Two days later the man was treated with antibiotics. That night he began to pace and spit frequently. His facial muscles twitched uncontrollably. He was anxious and fearful. He was hospitalized. Physicians thought he might have a drug overdose or viral encephalitis. Rabies was considered unlikely because there was no history of animal bite. Although he remained alert for several days, he became increasingly agitated. He complained of itching all over his body, and he vomited repeatedly. At times his temperature rose at times to 106°F. He died on August 25.

Brain samples taken at autopsy were examined using monoclonal antibodies. The results showed that the man died of rabies—a strain commonly found in silver-haired bats. The man had lived in a previously abandoned rural house, so he certainly had had close contact with bats. A friend reported that a bat had landed on the man's mouth one night in early July. The man killed the bat, but he may have been scratched during the encounter.

After reviewing this case and the two others, CDC epidemiologists reaffirmed their recommendation that rabies prophylaxis is essential for anyone who is bitten, scratched, or has mucous membrane contact with any wild animal of a species known to carry rabies.

Baby Boy B survived because his vital functions were maintained until the effects of the toxin wore off. With good supportive care, such as Baby Boy B received, only 3 percent of victims of infant botulism die. The fatality rate of botulism for adults is a little higher. About 10 percent die. For adult botulism, a polyvalent botulism antitoxin from horse serum is recommended, but allergic reactions are common (Chapter 18).

Adult botulism is easily preventable by carefully following canning procedures and boiling canned food before eating. Infant botulism is more difficult to prevent because *C. botulinum* spores are everywhere. Most of us consume them regularly. Consuming honey is a major risk factor for babies because it often contains large numbers of *C. botulinum* endospores. Honey is not recommended for children under the age of 1.

Viral Causes

Many viruses infect neurons. Some, such as herpes simplex (Chapters 24 and 26) and varicella zoster (Chapter 26), infect other cells as well. Those that usually infect the nervous system cause devastating diseases.

Rabies Virus: Rabies. Rabies virus belongs to the rhabdovirus family of bullet-shaped viruses (**Figure 25.6**). It contains a single minus strand of RNA and has an envelope with projecting spikes, which bear the antigens that stimulate production of protective antibodies.

Rabies virus infects many different mammals, including humans. Rabies is usually transmitted by a bite when

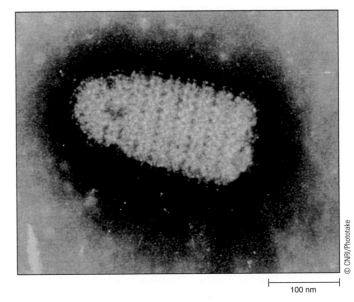

100 nm

© CNRI/Phototake

FIGURE 25.6 Electron micrograph of a single bullet-shaped rabies virus.

virus in the animal's saliva enters the wound. Other modes of transmission occur rarely, such as when an infected mother animal licks her young or when an animal breathes the air in a cave with a dense population of infected bats. One case of human rabies was caused by transplanting an infected cornea.

Clinical Syndrome. After rabies virus enters a wound, it multiplies in nearby muscle tissue. Eventually it enters a nerve and travels along it to the brain, where it exerts its deadly effect. The time between the bite and appearance of symptoms is quite variable, the average time being 2 months. But it can be as short as 9 days or as long as several years. In general the closer the bite is to the brain, the shorter the incubation time.

The symptoms of rabies are horrifying. The first are flulike. But even this **prodromal phase** (earliest symptoms) can produce strange sensations. The region around the bite may tingle or burn. The victim may feel nauseated or depressed or sense impending death. During the **excitation phase** (the next phase), severe neurological symptoms appear. The person loses muscle control. Speech and vision are impaired. Anxiety and apprehension may become unbearable. The classic symptom **hydrophobia** (fear of water) occurs when muscles that control swallowing contract abnormally. When a mouthful of water is taken, the patient either chokes violently or spits it out in an uncontrollable spasm. After this painful experience, the mere thought of drinking may trigger a similar attack. Throughout the excitation phase, the victim remains awake and alert. Finally the **paralytic phase** begins. Muscles weaken. Consciousness fades. The patient dies. Even with the best available supportive treatment, the fatality rate approaches 100 percent.

Rabies is diagnosed clinically and then confirmed by an immunofluorescent antibody test that detects rabies virus in infected nerve cells. Before this test became available in the late 1950s, diagnosis was based on finding characteristic structures, called **Negri bodies,** in infected nerve cells. Negri bodies develop where new virus particles are being assembled.

Epidemiology and Prevention. Humans are rarely infected by the rabies virus, which survives in infected animals. In rural areas, skunks, foxes, raccoons, coyotes, and bats are commonly infected. In urban areas, dogs and cats get rabies. Infected animals can usually be recognized by their abnormal behavior. Nocturnal animals appear in the daytime. Wild animals fearlessly approach humans. Gentle pets become aggressive.

In the United States, less than a dozen cases of human rabies are reported annually, and some of these were not acquired in this country. Mass immunization of pets is one reason for the low incidence. Still, an enormous reservoir exists among wild animals.

Because rabies has such a long incubation period, it can be prevented by active immunization even after infection. The first rabies vaccine was developed by Louis Pasteur in 1885. His vaccine was a weakened rabies virus from the brain and spinal cord of infected rabbits. His treatment consisted of 14 to 21 injections into the abdominal wall. The injections were painful, and hypersensitivity reactions were common.

Pasteur's treatment continued to be used into the mid-twentieth century. The currently used vaccine (HDCV) is prepared from virus grown in *h*uman *d*iploid *c*ell *c*ulture. This vaccine, which is administered as six injections, is reasonably free from side effects. If there's a real risk of rabies, a combination of passive immunization with human rabies immune globulin (a serum rich in antibodies against rabies) and active immunization with HDCV is administered.

Poliovirus: Poliomyelitis.

Polioviruses are plus-strand RNA viruses belonging to the picornavirus (meaning tiny RNA) family. They're closely related to the hepatitis A virus and coxsackie enteroviruses (Chapter 23).

Poliovirus has three different serotypes—1, 2, and 3. Because an antibody can neutralize only one type of virus, protection depends upon a person's having antibodies against all three viral types. Because only humans are naturally infected by poliovirus, polio vaccines are prepared from virus grown in cultured human cells.

Clinical Syndrome. Control of poliomyelitis, or polio as it is generally called, is one of the great success stories of microbiology and medical science. In the early 1950s, 39,000 cases of paralytic polio and 1900 deaths were reported in a single year in the United States. Paralyzing or killing young people, polio became the nation's most dreaded infectious disease (**Figure 25.7**). Then, with widespread use of vaccines in the mid-1950s, incidence of the disease declined rapidly. Now there are no new cases of polio in the United States caused by wild virus. The last case of polio (caused by wild virus) in the Western Hemisphere occurred in Peru in 1991. In 1994 an international commission declared the Western Hemisphere to be free of wild poliovirus. A few cases of polio do occur annually as a side effect of vaccination, but they too will end when polio is eradicated worldwide. The World Health Organization set a goal for eradicating polio by the end of the year 2000, but the goal of certifying the world as polio-free by the end of 2005 is still within reach.

During the polio epidemics of the 1950s, patients were kept alive in a mechanical respirator called an iron lung (Figure 25.7).

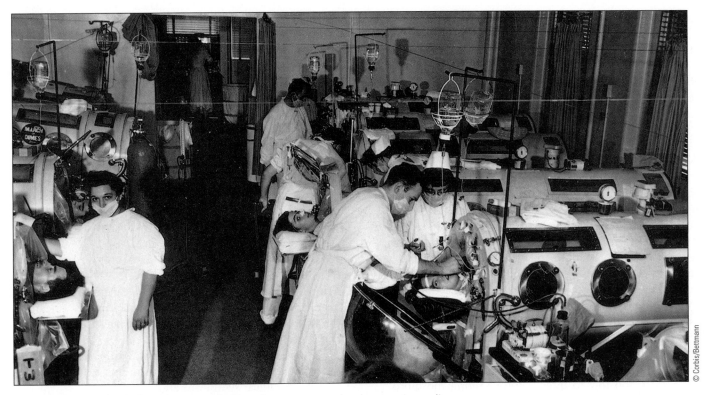

FIGURE 25.7 Mechanical respirators, called iron lungs, were used to keep patients alive during the polio epidemics of the 1940s and 1950s.

The Polio Vaccine. In 1955 Jonas Salk, a U.S. microbiologist, developed a vaccine against polio. To prepare this inactivated poliovirus vaccine (IPV), viruses are grown in human cell culture and treated with formalin. IPV is a trivalent vaccine. It contains all three serotypes of poliovirus.

In 1961 an oral polio vaccine (OPV), also called the Sabin vaccine after its discoverer, was adopted for mass immunization in the United States because it confers long-lasting immunity. OPV contains live attenuated viruses grown in cultured monkey cells. It is also a trivalent vaccine composed of live attenuated viruses. The only drawback to OPV is safety. One out of about 8 million recipients of OPV develops paralytic poliomyelitis. Prior administration of IPV decreases this incidence even further. Now all infants in the United States are immunized with IPV. OPV is used only in those parts of the world where polio remains active.

Although polio has been eradicated in the United States, vaccination must be continued until it is eradicated worldwide. Until that time there's danger of polio being imported from areas such as Africa, where it still exists.

Arboviruses: Encephalitis. Arboviruses are a diverse group of RNA viruses that share a common mode of transmission: They are all *ar*thropod-*bo*rne. Arboviruses include some togaviruses, flaviviruses, and bunyaviruses. More than 400 different arboviruses have been identified. They infect many vertebrate species and cause devastating diseases in humans, including yellow fever and dengue fever (Chapter 27).

Arboviruses also cause encephalitis. In the United States the most common ones are **eastern equine encephalitis (EEE)**, **western equine encephalitis (WEE)**, **California encephalitis (CE)**, and **St. Louis encephalitis (SLE)** viruses. Other encephalitic arboviruses are more common in other parts of the world, such as **Japanese B encephalitis** in Asia. Animals, especially horses and birds, are the main reservoir of these viruses. All are transmitted by mosquitoes, and some are also transmitted by ticks. Therefore epidemics of encephalitis occur during the summer months, when mosquitoes are plentiful.

Different viruses cause each type of encephalitis. Their severity varies. Some, such as EEE, are quite severe, resulting almost always in death or permanent brain damage. Others are much milder. Most are **subclinical** (without symptoms). When symptoms occur, they begin with fever, headache, and usually a stiff neck (indicating irritation of the meninges). These can progress to disturbed brain function, expressed as confusion, paralysis, convulsions, sleepiness, and sometimes coma. If illness is severe, brain function usually does not return to normal.

Diagnosis is difficult. The CSF contains only a few hundred leukocytes per cubic millimeter and no visible microorganisms. During epidemics, diagnosis is made from case history and physical examination. Only occasionally can antibodies against arbovirus be detected during or after the illness. Viral inclusion bodies in autopsy specimens can sometimes be seen by electron microscopy. Only one arbovirus encephalitis, Japanese B, can be prevented by vaccination. Once a person is infected, only supportive care can be offered.

West Nile Virus. In 1999 cases of encephalitis, caused by West Nile virus, occurred for the first time in the Western Hemisphere. New York City was the focus of the outbreak. Prior to that time West Nile virus was found only in Africa, western Asia, Europe, and the Middle East. Between 1999 and 2001, West Nile virus caused 18 deaths and 131 other illnesses in the northeastern United States. Then, the virus spread south and west. By the end of 2002, it had caused 241 deaths, 3,852 illnesses had been reported, and only 4 Western states were still virus-free.

West Nile virus is a flavivirus, closely related to Japanese encephalitis and St. Louis encephalitis virus. It, too, is spread by mosquitoes. Birds are its principal host, particularly crows, but at least 75 other species are known to be susceptible. Humans are an incidental host in which virus levels remain too low to infect another biting mosquito. Thus the virus cannot be spread from one human to another, directly or by means of a mosquito vector. Most humans who are infected by West Nile virus develop no symptoms; some develop a mild illness characterized by headache, body aches, skin rash, and swollen glands; a few develop encephalitis, some of whom die. At present there is no effective drug or vaccine available to control West Nile virus. But intensive efforts are under way to develop a vaccine; one has shown promise in animals and will be tested in humans in 2003.

Other Viruses That Cause Encephalitis. Many viruses produce a clinical syndrome that's indistinguishable from arbovirus encephalitis. Very rarely, each of the common viral pathogens of childhood—measles, mumps, chickenpox, and rubella—causes encephalitis. But these viruses infect so many people they account for a significant number of cases of encephalitis. Enteroviruses, such as coxsackieviruses and echoviruses, also cause encephalitis. Which virus causes the most cases is not known, but because most cases occur in the summer, arboviruses or enteroviruses are the most likely candidates.

Herpes simplex virus types 1 and 2 (Chapters 24 and 26) can also cause devastating encephalitis that usually results in death or profound brain damage. Diagnosing herpes encephalitis is critically important because it can be effectively treated with acyclovir or ganciclovir. If herpes encephalitis is suspected, **brain biopsy** (surgically removing a small piece of tissue for study) may be considered.

Prion Causes

Prions (Chapter 15) cause forms of encephalitis that are characterized by **spongiform encephalopathy** (meaning the damaged brain resembles a sponge because it is riddled with holes; **Figure 25.8**).

The various prion-caused diseases differ in their mode of transmission but not in their symptoms. All produce severe neurological symptoms and death.

Kuru, often called laughing death, occurs only among the Fore people of Papua New Guinea. The illness begins with clumsiness and progresses to total incapacitation and death within a year or two. Until 1957 the disease killed an incredible 2 to 3 percent of the population yearly. Then it was discovered that Kuru was transmitted by ritualistic cannibalism in which the Fore tribe honored their dead by eating parts of the body, including the brain. The practice has stopped, and Kuru has virtually disappeared.

Creutzfeldt-Jakob disease (CJD) (Chapter 25) can be transmitted by transplants, contaminated surgical instruments, and injection of growth hormone from human pituitaries. CJD can also be hereditary. Members of certain families have an increased likelihood of acquiring CJD, usually later in life. How can a prions cause hereditary, as well as infectious, disease? Evidence suggests that mem-

FIGURE 25.8 Comparison of (a) normal brain tissue and (b) the spongiform brain tissue taken from a patient with Creutzfeldt-Jakob disease.

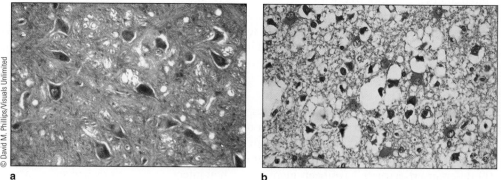

a

b

© David M. Phillips/Visuals Unlimited

© Frederick C. Skvara/Peter Arnold

bers of CJD-prone families carry mutant forms of a protein designated PrP. These forms are more likely to undergo a spontaneous conformational change (alpha helices to beta sheets) and become prions.

At least six prion diseases afflict animals. Scrapie, which afflicts sheep and goats, is the most thoroughly studied. It was named for the bizarre behavior of infected animals: They develop an intense itch, which drives them to scrape off their fleece. Other prion diseases afflict mink, deer, and elk, as well as domestic and great cats. Chronic wasting disease occurs in the deer and elk of certain portions of Colorado, Wyoming, and Nebraska. In some regions almost 15 percent of the deer are infected.

The most celebrated prion disease of animals is **mad cow disease** (bovine spongiform encephalopathy [BSE]). This deadly neurological disease suddenly appeared in dairy cattle in Britain in 1980. Since then, there have been more than 160,000 cases. BSE is thought to be a form of scrapie that was transmitted to cattle through processed sheep tissue in cattle feed. To stop the spread of BSE, cattle were slaughtered and adding animal products to cattle feed was prohibited. As a consequence, the incidence of BSE in cattle has almost disappeared. Still there is evidence that BSE has been transmitted to humans who ate contaminated beef and is expressed as **new variant Creutzfeldt-Jacob disease.** Because BSE might be expected to have a long incubation period in humans, we still don't know how many humans might be infected.

Protozoal Causes

Protozoal pathogens also cause devastating neurological diseases.

African Trypanosomiasis (Sleeping Sickness). Two subspecies of trypanosomes (Chapter 12; **Figure 25.9**)— *Trypanosoma brucei gambiense* and *Trypanosoma brucei rhodesiense*—cause **African trypanosomiasis** or **sleeping sickness.** This severe encephalitis is transmitted from one human to another by the large and aggressive tsetse fly (**Figure 25.10**). This fly is found only in equatorial Africa, and so is sleeping sickness. A huge reservoir of *T. b. rhodesiense* exists there in game animals, cattle, and sheep.

Trypanosomes multiply in the tsetse fly and become concentrated in its saliva. When the fly bites a human, a hard, red nodule forms at the site. Trypanosomes enter the bloodstream, and within a few days to weeks they spread to the spleen, lymph nodes, and liver. There they multiply, causing fever, malaise, and enlarged lymph nodes.

Eventually the trypanosomes invade the CNS, where they damage blood vessels, as well as brain tissue. The first signs are changed personality or a decreased ability to concentrate. Soon the infected person cannot speak or walk. Eventually the victim becomes extremely tired and (as the name of the disease implies) sleeps almost continuously. Coma and death follow. The disease caused by *T. b. gambiense* can last for several years, but the disease caused by

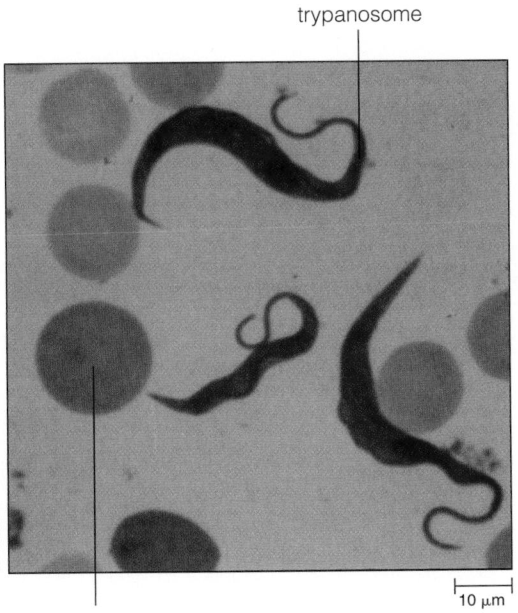

FIGURE 25.9 Trypanosomes in blood.

FIGURE 25.10 Tsetse fly (*Glossina morsitans*), the vector of sleeping sickness.

LARGER FIELD

LOUISE PEARCE'S MISSION

Louise Pearce, a young scientist from the Rockefeller Institute, left New York in 1920 with a small supply of a new drug—tryparsamide—that she hoped would cure African trypanosomiasis. Her destination was the Belgian Congo (now Congo), where sleeping sickness infected about 10 percent of the population. Leopoldville, the colonial capital, was a frontier town that buzzed with malaria-carrying mosquitoes and tsetse flies. Undeterred, Pearce began work at the local hospital in a crowded ward reserved for Africans. At bedside she wrote her clinical notes, recording the relentless progression of sleeping sickness. The suffering surrounding her affected her deeply. She had only enough tryparsamide to treat 77 patients, but she went ahead anyway. Amazingly, 80 percent of her patients recovered, including many with advanced brain involvement.

Tryparsamide was a significant advance in the treatment of trypanosomiasis, but it was not the final answer. Some patients became blind, and tryparsamide was useless against *Trypanosoma brucei rhodesiense*. It was, however, highly effective against *Trypanosoma brucei gambiense*—the species found in most of Africa. Tryparsamide is still used, but in 1991 a new effective drug called O-11 was announced. It may turn out to be the perfect cure Louise Pearce set out to find 82 years ago.

Louise Pearce.

Courtesy Rockefeller Archive Center, North Tarrytown, NY

T. b. rhodesiense usually ends in death within a few months. African trypanosomiasis is one of the few human diseases that is 100 percent fatal if untreated.

African trypanosomiasis is diagnosed by identifying parasites in the blood. Treatment is possible before the CNS is involved, but afterward only compounds that cross the blood-brain barrier can stop the disease. And these agents are highly toxic. Europeans were eager to find a cure for sleeping sickness because it was an obstacle to their colonization of Africa. Paul Ehrlich was looking for such a cure when he discovered salvarsan, which was effective against syphilis (Chapter 1). Several effective drugs—including melarsoprol, suramin, and tryparsamide—have been developed in this century. Still the search for better drugs continues (see Larger Field: Louise Pearce's Mission).

Sleeping sickness is a major cause of suffering and death in parts of Africa. At least a million people are infected at any one time. Related diseases in domestic animals cause major economic losses. Chances of preventing sleeping sickness appear to be bleak. Eliminating the tsetse fly is difficult because it inhabits millions of square miles. Control through vaccination seems unlikely because humans do not develop long-lasting natural immunity to the disease (see Sharper Focus: Anti-immune Trickery, in Chapter 17).

Table 25.2 summarizes the infections of the nervous system.

---

the cd connection

MICROBIAL DISEASES ⟶ *Nervous system*

Summarizes the normal biota, structure/function, clinical syndromes, and infections of the nervous system.

TABLE 25.2 Infections of the Nervous System

Infection	Causative Agent	Mode of Transmission	Symptoms	Prevention and Treatment
Infections of the Meninges				
Bacterial				
Bacterial meningitis	*Neisseria meningitidis, Haemophilus influenzae, Streptococcus pneumoniae, Escherichia coli*	Usually respiratory	Acute onset of fever, headache, stiff neck, disturbed brain function (usually)	Vaccines for *N. meningitidis* and *H. influenzae*; all treatable with antibiotics
Viral				
Viral (aseptic) meningitis	Mumps virus, arboviruses, enteroviruses, and others	Respiratory or fecal-oral	Acute onset of fever, headache, stiff neck	Prevention through good hygiene; no treatment
Fungal				
Cryptococcal meningitis	*Cryptococcus neoformans*	Inhaling fungus, particularly from bird droppings	Gradual onset of headache, confusion, weakness	Treatment with amphotericin B and 5-flucytosine
Neural Infections				
Bacterial				
Tetanus	Toxin of *Clostridium tetani*	Wound infection	Rigid paralysis	Prevention by toxoid immunization; supportive treatment
Botulism	Toxin of *Clostridium botulinum*	Usually food poisoning or intestinal infection; rarely, wound infection	Flaccid paralysis	Prevention by proper food preparation and storage; treatment by antitoxin and supportive care
Viral				
Rabies	Rabies virus	Usually animal bite	Anxiety, impaired swallowing, coma, death	Prevention by immunizing pets and postexposure immunization of humans; no effective treatment
Polio	Poliovirus	Fecal-oral	Most infections asymptomatic; flaccid paralysis	Prevention by immunization; supportive treatment only
Arbovirus encephalitis	Various alphaviruses, flaviviruses, bunyaviruses	Insect vector, usually mosquito	Fever, headache, stiff neck progressing to coma and death	Avoiding mosquito bites is best prevention; supportive treatment only
Prion				
Kuru	Prion-associated agent	Ritual cannibalism of infected brains	Tremor and weakness progressing to coma and death	Stopping cannibalism; supportive treatment only
Creutzfeldt-Jakob disease	Prion-associated agent	Unknown; some cases documented through exposure to extracts of infected brains	Dementia leading to coma and death	No systematic program for prevention; supportive treatment only
Protozoal				
African trypanosomiasis (sleeping sickness)	*Trypanosoma brucei gambiense* and *Trypanosoma brucci rhodesiense*	Bite of tsetse fly	Fever, enlarged lymph nodes progressing to somnolence, coma, and death	Prevention by avoiding fly bites; treatment with melarsoprol, suramin, tryparsamide

SUMMARY

Case History: Twenty-four Hours (p. 612)

1. C. H. showed all the signs and symptoms of meningitis. His high leukocyte count indicated it was purulent meningitis. Laboratory culture of his spinal fluid showed it was caused by *Neisseria meningitidis*.

The Nervous System (pp. 613–614)

2. The nervous system consists of the central nervous system (CNS) and the peripheral nervous system (PNS). The CNS includes the brain and spinal cord. They are covered by continuous membranes called the meninges. Cerebrospinal fluid (CSF) circulates around the brain and spinal cord. The PNS is made up of nerves (bundles of neurons and supporting tissue).

3. The outer layer of capillaries in the CNS is thickened, which limits permeability. This extra protection is referred to as the blood-brain barrier. Many toxins cannot enter the brain. Neither can many antibiotics, making CNS infections difficult to treat.

4. Clinical syndromes of the CNS include meningitis, encephalitis, and myelitis.

5. Diagnosis of nervous system infections often requires a sample of CSF. It is obtained from the lower spinal cord. The procedure is called the lumbar puncture, or spinal tap.

Infections of the Meninges (pp. 615–619)

6. *Neisseria meningitidis* causes meningococcal meningitis. Transmission is by respiratory droplets. Many people are asymptomatic carriers. Onset of acute purulent meningitis is sudden. Its symptoms are a high fever, stiff neck, headache, purplish skin rash, and often disturbed brain function. Children between 6 months and 2 years are at highest risk. Epidemic disease is unusually severe, causing death or permanent brain damage. Epidemics can be interrupted by meningococcal prophylaxis with rifampin. Vaccines exist for some strains, but not yet for type B. Type B causes most cases.

7. *Haemophilus influenzae* causes *H. influenzae* meningitis. Like the meningococcus, *H. influenzae* is invasive and has a polysaccharide capsule. *H. influenzae* once caused most meningitis in infants and young children. But in 1990, *H. influenzae* type b vaccine (Hib) was released and infection rates dropped dramatically.

8. *Streptococcus pneumoniae* causes pneumococcal meningitis, mainly in adults. This is the type of meningitis that most frequently causes brain damage.

9. *Escherichia coli* causes few cases of meningitis. It is a deadly disease for newborns and patients who have undergone neurosurgery.

10. Patients with meningitis symptoms that do not have bacteria in their CSF are diagnosed with nonpurulent, or aseptic, meningitis. Viral infection is presumed to be the cause. Mumps virus, members of the enterovirus group, and arboviruses cause most cases. Children and young adults are most at risk. Recovery occurs within a few weeks without treatment. It is vitally important to establish that bacteria are not the cause, because bacterial meningitis can be fatal.

11. Two fungi—*Coccidioides immitis* and *Cryptococcus neoformans*—cause chronic meningitis. A few coccidioidomycosis patients also develop chronic meningitis. *C. neoformans* is inhaled. If it spreads through the bloodstream to the CNS, the consequences are serious. These two fungi cause similar diseases. Onset is gradual. There may or may not be a stiff neck; headaches and other neurological symptoms such as confusion and loss of coordination appear. Treatment is with amphotericin B. The AIDS epidemic has led to an increase in the incidence of cryptococcal meningitis.

Diseases of Neural Tissue (pp. 619–626)

12. *Clostridium tetani* causes tetanus when endospore-containing dirt enters a deep wound. *C. tetani* produces the neurotoxin tetanospasmin. It causes rigid paralysis and excruciating muscle spasms. Before clinical signs appear, treatment with tetanus immune globulin is effective. Immunization is part of the diphtheria-tetanus-pertussis (DTP) inoculation. Immunity can be maintained with regular tetanus immunization.

13. *Clostridium botulinum* produces botulinum toxin, the most poisonous natural substance known. The toxin causes a flaccid (limp) paralysis that can be fatal if breathing muscles are affected. Infant botulism is caused by ingesting spore-laden food. Adults usually acquire botulism as a food poisoning. Spores can also produce toxin in a deep wound. Treatment may require a mechanical respirator. A botulism antitoxin is available. Preventive measures include not giving honey to infants under age 1, following proper home-canning procedures, and boiling home-canned food.

14. The rabies virus causes rabies. Infection is usually from the bite of an infected animal. The incubation period can be brief or up to several years. The prodromal phase is characterized by flulike symptoms and unusual sensations. During the excitation phase, severe neurological abnormalities, including hydrophobia, appear. In the paralytic phase, coma and death ensue. Rabies can be treated after exposure with human rabies immune globulin and six shots of

human diploid cell vaccine (HDCV).

15. Poliovirus causes poliomyelitis, a flaccid paralysis. Peak polio epidemics occurred in the early 1950s. An inactivated poliovirus vaccine (IPV) and later an oral polio vaccine (OPV) were developed. Now polio has been eradicated from the Western Hemisphere.

16. Arboviruses cause various types of encephalitis, including eastern equine encephalitis (EEE), western equine encephalitis (WEE), Japanese B encephalitis, and others. Each type has its own characteristics. Transmission is usually by a mosquito bite. Epidemics occur during the summer months. A vaccine is available only for Japanese B type. Treatment is supportive care.

17. Enteroviruses and some common viral pathogens of childhood—measles, mumps, chickenpox, and rubella—can cause encephalitis. Herpes simplex virus can cause an encephalitis that usually results in death or profound brain damage. Prompt diagnosis by brain biopsy is important. Acyclovir or ganciclovir are effective.

18. Prions cause spongiform encephalopathies. Kuru occurs in Papua New Guinea. Symptoms progress from clumsiness to total incapacitation and death. The disease was virtually eradicated when ritual cannibalism stopped. Creutzfeldt-Jakob disease follows a similar course. Animal prion diseases, such as mad cow disease, may be transmissible to humans.

19. Two protozoa, *Trypanosoma brucei gambiense* and *Trypanosoma brucei rhodesiense*, cause African trypanosomiasis or sleeping sickness. Transmission is by the tsetse fly. Early symptoms include fever, malaise, and enlarged lymph nodes. When pathogens reach the brain, the victim cannot concentrate, becomes extremely tired, and sleeps almost continuously. Coma and death follow.

Treatment before the brain is involved is possible with tryparsamide or O-11.

REVIEW QUESTIONS

Case History: Twenty-four Hours

1. Why did C. H. have a stiff neck?
2. Could C. H.'s disease have been prevented? How?

The Nervous System

3. Name the two parts of the nervous system, their main components, and their functions. What is the blood-brain barrier? How does it function to our advantage and disadvantage?
4. Describe these clinical syndromes: meningitis, encephalitis, myelitis. What are neurotoxins? How and why is a lumbar puncture performed?

Infections of the Meninges

5. What types of meningitis are caused by each of the bacteria listed below. How are they transmitted? What are their symptoms? How are they diagnosed? Which groups are at highest risk? How can they be prevented and treated?
   a. *Neisseria meningitidis*
   b. *Haemophilus influenzae*
   c. *Streptococcus pneumoniae*
   d. *Escherichia coli*
6. What properties make *Neisseria meningitidis* pathogenic?
7. How does aseptic meningitis differ from bacterial meningitis? Why is it crucial to distinguish between the two?
8. How is cryptococcal meningitis different from bacterial and viral meningitis?

Diseases of Neural Tissue

9. How do the illnesses caused by *Clostridium tetani* and *Clostridium*

botulinum differ? How does each cause illness? How are they transmitted? What are their symptoms? How can they be prevented and treated?
10. What is the clinical syndrome of rabies? How is rabies diagnosed and treated? How can it be prevented?
11. What is the clinical syndrome of poliomyelitis? Why was the Salk vaccine replaced with the Sabin vaccine?
12. What are some of the encephalitic arboviruses in the United States? Which is the most severe? What symptoms does it cause? How is it transmitted and treated?
13. What other viral pathogens can cause encephalitis? Why is prompt diagnosis of herpes simplex encephalitis crucial?
14. What is the epidemiology of Kuru? What do we know about Creutzfeldt-Jakob disease?
15. How is the trypanosomiasis caused by *Trypanosoma brucei rhodesiense* like that caused by *Trypanosoma brucei gambiense*? How is it different?

CORRELATION QUESTIONS

1. Can meningitis and encephalitis be distinguished on the basis of brain function being affected? Why?
2. Why do you think bacterial pathogens that cause damage to brain tissue do so by producing toxins?
3. If a patient has a stiff neck, headache, and a purplish rash, what illness do you think he or she might be suffering from? Why?
4. What are the similarities between the transmission of Kuru and mad cow disease?

ESSAY QUESTIONS

1. Discuss the possibility of eradicating rabies.
2. Discuss the bases for choosing between recommending that a child be given inactivated polio vaccine (IPV) or oral polio vaccine (OPV) now and 10 years ago.

SUGGESTED READINGS

Dowdle, W. R. 2001. Polio eradication: Turning the dream into reality. *ASM News* 67:397–402.

Monath, T. P. 1988. Japanese encephalitis: A plague of the orient. *New England Journal of Medicine* 319: 641–43.

Prusiner, S. B. 2001. Neurodegenerative diseases and prions. *New England Journal of Medicine* 344:1516–26.

Quegiliarillo, W., and W. M. Schield. 1992. Bacterial meningitis: Pathogenesis, pathophysiology, and progress. *New England Journal of Medicine* 326:864–72.

Seed, J. R. 2000. Current status of African trypanosomiasis. *ASM News* 66:395–402.

For additional readings, go to InfoTrac College Edition, your online research library at: http://www.infotrac.thomsonlearning.com

TWENTY-SIX

Infections of the Body's Surfaces

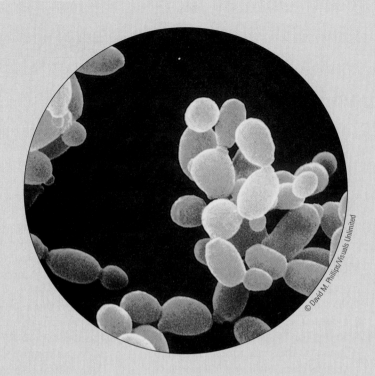

© David M. Phillips/Visuals Unlimited

CHAPTER OUTLINE

LEARNING GOALS

To understand:

- *The anatomy and function of the skin and eye and their defenses against microorganisms*

- *The clinical syndromes that characterize infections of the skin and eye*

- *The bacterial, viral, fungal, and arthropod causes of skin infections and their diagnosis, prevention, and treatment*

- *The bacterial, viral, and helminthic causes of eye infections and their diagnosis, prevention, and treatment*

A Lifelong Visitor

In 1994 R. C. was a 7-year-old boy whose parents took him to his pediatrician because he had a fever and an itchy, blistering rash. His fever was slight, and he appeared well. The physician examined the few dozen thin-walled vesicles on his trunk and immediately diagnosed **varicella** (usually called chickenpox). She told R. C.'s parents what to expect. His fever would continue, and he would develop more blisters over the next 5 days. The blisters would come in "crops" or "showers." Many would appear at the same time. Then, hours or days later, another set would appear, so R. C. would soon have many blisters at different stages of healing

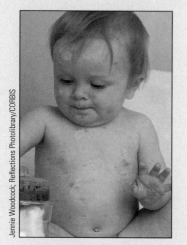

Jennie Woodcock: Reflections Photolibrary/CORBIS

Typical chickenpox rash.

all over his body, perhaps even in his mouth and throat. Those that were a few days old would begin to crust while new wet ones were still appearing. The wet ones would contain live virus (varicella zoster virus [VZV]). During the week that he had them, he could transmit the chickenpox to others. So he had to stay home from school. If any sores failed to heal or began to drain pus, he would have to return to the doctor. No treatment was prescribed, but the physician gave him medication for the itching. The physician cautioned R. C.'s parents not to administer aspirin for his fever, but they could give non–aspirin-containing medication. They were also told that R. C.'s younger sister might also come down with chickenpox in 2 to 3 weeks. By the end of the week, R. C. was indeed covered with blisters and scabs. He looked sicker than he really was. He recovered uneventfully and returned to school.

K. V., a healthy 72-year-old woman, experienced tingling in her skin along a narrow strip that extended from just below her left shoulder blade around her left side to her abdomen. By the next day the tingling became a searing pain. Then, a day later, tiny blisters appeared in the painful area. K. V.'s physician diagnosed **zoster** (commonly called shin-

gles), which is a reactivation of the VZV that earlier in her life caused her case of chickenpox. K. V.'s physician prescribed medication for the pain, which had become almost unbearable. The blisters healed in about a week, but the pain persisted. Six months later K. V. continued to suffer pain in that same strip of skin, but it had begun to diminish. Laboratory tests done at the time of K. V.'s illness were all normal. Her health otherwise remained good.

Case Connections

- Why did the physician caution R. C.'s parents not to give her aspirin? If you don't know, you'll learn later in this chapter.
- Chickenpox and zoster are caused by VZV. We'll discuss how one virus can cause two such different diseases later in this chapter.
- Might R. C. develop shingles later in life? Must K. V. have had chickenpox earlier in life? The connections between chickenpox and shingles are discussed in this chapter.
- In 1995, just a year after R. C.'s illness, a varicella vaccine became available and cases such as his have become rare in the United States. The vaccine can cause zoster, but less commonly than the natural infection.

THE BODY'S SURFACES

The body's surfaces—the skin and exposed surfaces of the eye—defend internal tissues against microorganisms. Without these tough, intact external coverings, microbial infections would overwhelm us.

Structure and Function of the Skin

First let's consider our largest organ, the skin (**Figure 26.1**). We'll start at its outer surface and then go deeper. The first layer, called the **epidermis,** is made up of several layers of epithelial cells that are tightly cemented together. The

FIGURE 26.1   Anatomy of the skin.

(Art by Robert Demarest from Biology: The Unity and Diversity of Life, 6th ed., by C. Starr and R. Taggart, Brooks/Cole, 1992. All rights reserved.)

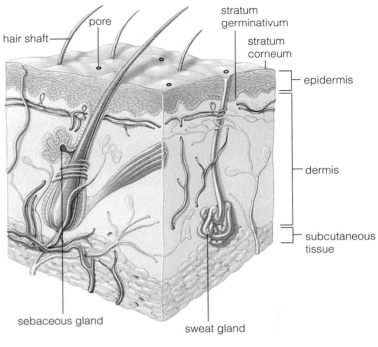

number of layers varies from place to place on our bodies. There are only a few layers in areas, such as the face, where the skin is thin. But there are many layers in areas, such as the palms of the hands and the soles of the feet, where it's thick.

The epidermis constantly replaces itself with cells produced by dividing cells in its innermost layer, called the **stratum germinativum.** As these new cells are pushed out toward the surface, they die and become packed with **keratin** (a waterproof protein). At the surface they form the tough layer called the **stratum corneum.** It's a formidable barrier to microorganisms.

Beneath the epidermis is a thicker layer of connective tissue called the **dermis.** It contains blood vessels, lymphatic vessels, nerve endings, sweat glands, sebaceous (oil) glands, and hair follicles. Sebaceous glands produce **sebum,** a fatty substance that lubricates the hair and skin. Both the sweat glands and hair follicles penetrate the epidermis. The tiny openings they form at the skin surface are entry sites for microorganisms. Moreover, the sebum and perspiration these two glands secrete foster the growth of microorganisms by providing nutrients and moisture. Most skin infections begin here. But these secretions also contain antimicrobial compounds. Sweat, for example, contains the enzyme **lysozyme,** which lyses Gram-positive bacteria. Sebum contains fatty acids, which are toxic to many microorganisms.

A layer of subcutaneous tissue lies beneath the dermis. Microorganisms can spread readily through this loosely organized layer of connective tissue.

Structure and Function of the Eye's Surface

Like the skin, the eye comes into direct contact with the environment (**Figure 26.2**). Most of its surface is protected by a loose layer of connective tissue covered with an epithelial membrane. Together these two layers are called the **conjunctiva** (*pl.*, conjunctivae). The conjunctiva is a barrier to microorganisms, but it doesn't cover the **cornea.** This transparent window through which we see is covered with its own dense layer of protective epithelium. Infections of the cornea are much

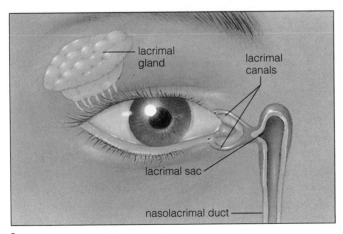

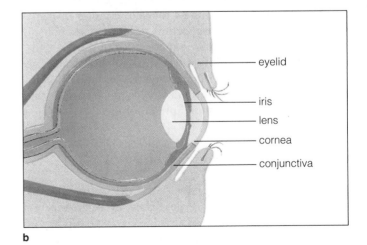

FIGURE 26.2   Anatomy of the eye.

(Art by Carlyn Iverson.)

TABLE 26.1 Classification of Skin Lesions

Basis of Classification	Name of Lesion	Characteristics
Color of lesion	Erythema	Red
	Petechiae	Purple (tiny)
	Purpura	Purple (large)
Texture of unbroken skin	Macules	Flat spots
	Papules	Small raised spots
	Maculopapular	Flat and bumpy regions
	Nodules	Larger, firm, round elevations
Contents of lesion	Vesicle	Small, water-filled
	Bulla	Large, water-filled
	Pustule	Pus-filled
Properties of broken skin	Erosion	Superficial skin loss
	Ulcer	Deeper skin loss
	Crusts	Lesions covered with dried blood or serum
Site of lesion	Exanthem	Skin rash caused by systemic infection
	Enanthem	Rash on mucous membrane caused by systemic infection

more serious than infections of the conjunctivae because they can cloud or scar this normally clear structure.

Defenses and Normal Microbiota: A Brief Review

Natural defenses—primarily the dryness of the skin and tears in the eye—make these surfaces inhospitable to microorganisms. Nevertheless, both sustain a characteristic microbiota. The microbiota of the skin consists of many commensals and a few opportunistic pathogens (Chapter 14). The greatest number and variety of microorganisms are found in relatively moist areas, such as under the arms, in the groin, and around the nose and mouth. The normal microbiota of the conjunctiva is much sparser, but it's otherwise similar to the skin microbiota.

Clinical Syndromes

Skin infections aren't usually described by anatomical syndromes. Instead, they're described by their appearance. That's logical because the infected tissue can be examined directly. Clinicians use the same set of terms to describe skin lesions (**Table 26.1**) regardless of whether or not they're caused by an infection. Appearance of skin

lesions is the key to diagnosing certain infections, such as measles, chickenpox, and scarlet fever.

In contrast, infections of the conjunctivae and cornea are usually described as clinical syndromes (**Table 26.2**). Infection of the conjunctivae causes **conjunctivitis,** or **pinkeye,** because superficial blood vessels are usually dilated. Pinkeye is extremely common, particularly in children. Infections of the cornea are called **keratitis.** It's much less common than conjunctivitis but more serious. If both surfaces are affected, the infection is called **keratoconjunctivitis.**

SKIN INFECTIONS

Bacteria, viruses, fungi, and arthropods all infect the skin.

Bacterial Causes

Most bacterial skin infections are localized, but some spread to become life-threatening systemic illnesses. Most can be treated effectively with antibiotics.

***Streptococcus pyogenes*: Impetigo, Erysipelas.** *Streptococcus pyogenes* (or group A streptococcus) is a versatile pathogen. It causes many different clinical syndromes.

TABLE 26.2 Clinical Syndromes of Eye Infections

Syndrome	Region Affected	Signs and Symptoms	Causative Agents
Conjunctivitis (pinkeye)	Conjunctivae	Inflammation of conjunctivae only; often reddening, discharge, discomfort but no threat to vision	Many agents, including *Neisseria gonorrhoeae*, strains of *Chlamydia trachomatis* that cause trachoma, herpes simplex virus, adenoviruses, enteroviruses, and other viruses
Keratitis	Cornea	Inflammation; may or may not be painful; deep lesions cause scarring and threaten vision	Same as for conjunctivitis
Keratoconjunctivitis	Conjunctivae and the cornea	Characteristics of both conjunctivitis and keratitis	Same as for conjunctivitis

We've already discussed strep throat (Chapter 22), scarlet fever (Chapter 21), and puerperal sepsis (Chapters 1 and 24). Later we'll discuss streptococcal septicemia (Chapter 27). But now we'll consider its many infections of the skin and underlying soft tissues.

The type of skin infection *S. pyogenes* causes depends upon the depth of the infection and the virulence of the infecting strain. The most common is a superficial, non–life-threatening infection called **impetigo,** which is usually seen in children (**Figure 26.3**). Typical impetigo is a small break in the skin that fails to heal. It grows larger and oozes a clear fluid. The fluid dries into a honey-colored crust that breaks when the lesion begins to ooze again. Usually these lesions contain *Staphylococcus aureus,* as well as *S. pyogenes. S. pyogenes* is usually the primary pathogen, but sometimes *Staphylococcus aureus* is, particularly in cases of blistered or **bullous impetigo.**

Impetigo can become extensive and troublesome. Occasionally it leads to poststreptococcal glomerulonephritis, a serious complication (Chapter 22). Because the open lesions of impetigo contain huge numbers of bacteria, it is readily transmitted by close contact or by fomites such as toys or clothing. It is most prevalent in warm, damp climates and among toddlers and young children. It can be controlled with oral or topical antibiotics.

Erysipelas is a deeper and more dangerous streptococcal skin infection. It affects the underlying dermis, as well as the epidermis (**Figure 26.4**), so it spreads quickly because bacteria move freely through lymphatic vessels in the dermis. Bacteria in the dermis may also enter the bloodstream, as they did in G. T.'s case. Then they initiate a systemic infection characterized by fever and malaise. If untreated, such infections are often fatal.

Streptococcal skin infections can extend into the subcutaneous layer. Then they can spread through the spaces between tissue planes at an astonishing rate, causing **streptococcal gangrene.** Inflammation can become so intense that blood vessels are destroyed and the overlying skin dies.

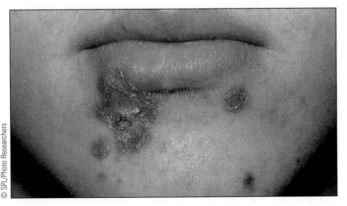

FIGURE 26.3 Impetigo. These lesion are covered with a honey-colored crust.

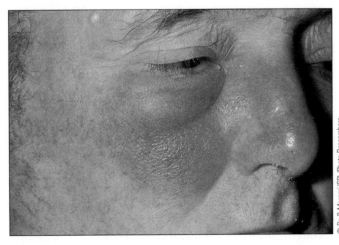

FIGURE 26.4 Erysipelas. Note the characteristic rash on this man's cheek.

CASE HISTORY

A Little Red Spot

G. T., a 45-year-old businessman, noticed while shaving one morning a small red bump near the right side of his nose. It looked like an ordinary pimple, so he wasn't concerned. He went to work as usual. During the day, the redness began to spread. By the time G. T. returned from work, he had a bright red patch the size of a dime on his cheek. It was hard, warm, slightly raised, and tender. The area had a distinct margin, almost as though someone had drawn a line around the affected area. On one side the skin was uniformly red. On the other it was entirely normal (Figure 26.3). G. T. went to bed hoping the annoying spot would be gone in the morning. But it wasn't. When he awoke, it had grown alarmingly. Now his entire cheek and the right side of his face were red and swollen. He felt sick, too. He took his temperature. It was 102°F. He tried to eat breakfast but felt nauseated and vomited. Instead of going to work, G. T. visited his doctor.

After taking a history and performing a physical examination, G. T.'s physician had him admitted to the hospital. He asked G. T. if he had a history of allergy to antibiotics. Then he wrote orders for G. T. to receive a cephalosporin antibiotic intravenously. At the hospital, a sample of G. T.'s blood was taken and sent to the microbiology to be cultured. After the first dose of antibiotic, the margin of the affected area stopped advancing. Within 24 hours his fever was gone and the redness slowly began to fade. Three days later G. T. was discharged with a prescription for oral penicillin. At the time of his discharge, his blood culture was reported to be growing *Streptococcus pyogenes*. He was diagnosed with having had erysipelas.

Case Connections
- Why did G. T.'s physician prescribe a cephalosporin antibiotic?
- How did the red area on G. T.'s face spread so rapidly?

Bacteria from these infections may also enter the bloodstream and cause a life-threatening systemic infection.

Virulence and Pathogenesis. Various strains of *Streptococcus pyogenes* have a vast repertoire of mechanisms for causing human disease. These include a surface M protein that protects the bacterium from phagocytosis (see Figure 15.9), numerous toxins, destructive enzymes, and the ability to provoke immunologically mediated tissue destruction (**Table 26.3**).

The M protein can be neutralized by antibody, so a person who has been infected by a given streptococcal serotype is immune from reinfection with the same type. But there are more than 80 different serotypes. As a result, few people are completely immune to streptococcal infection. M protein played a key role in G. T.'s infection. When the group A streptococci entered the dermal layer of his skin, they initiated an intense inflammatory response. But phagocytes couldn't destroy the bacteria because they were protected by an M protein. Phagocytosis would have been successful if G. T. had antibodies against that M protein. But he didn't. He probably would have died before the 7 to 10 days required to produce them through a primary immune response.

S. pyogenes produces many toxic substances that damage human tissues. We talked about the one of these, called erythrogenic toxin, that causes scarlet fever in Chapter 21. Other toxic products that these bacteria produce include **leukocidins** (which destroy leukocytes), **streptolysins** (which lyse red blood cells), **streptokinase** (which dissolves blood clots), and **hyaluronidase** (which dissolves hyaluronic acid, a polysaccharide that cements cells together). No one of these is essential for streptococci to cause disease, but collectively they contribute to the tissue damage that occurs.

Other possible consequences of *S. pyogenes* infection include rheumatic fever and acute poststreptococcal glomerulonephritis. These autoimmune reactions were discussed in Chapters 18 and 22.

Treatment and Prevention. Thanks to antibiotic treatment, G. T. missed less than a week of work. Without treatment he probably would have died, as many otherwise healthy young adults did in the era before antibiotics (see Sharper Focus: Antimicrobial Drugs in U.S. History, in Chapter 21).

Streptococcus pyogenes is one of the few Gram-positive cocci that has remained uniformly susceptible to small doses of penicillin. G. T. was initially treated with a cephalosporin antibiotic only because his physician couldn't definitely rule out infection by a penicillin-resistant strain of *Staphylococcus aureus*. When his blood culture confirmed the streptococcal infection, his treatment was switched to penicillin.

TABLE 26.3 Some Properties of Skin-Infecting Bacteria That Cause Disease

Organism	Property or Complication	Action
Group A streptococci	M protein	Allows streptococci to evade phagocytosis
	Erythrogenic toxin	Causes signs and symptoms of scarlet fever
	Leukocidin	Destroys leukocytes
	Streptolysins	Destroy red blood cells
	Streptokinase	Dissolves blood clots
	Hyaluronidase	Dissolves hyaluronic acid, which cements cells together
	Rheumatic fever	Inflames many organs, including heart valves
	Poststreptococcal glomerulonephritis	Inflames kidney tissue
Staphylococcus aureus	Alpha toxin	Damages cell membranes
	Delta toxin	Damages cell membranes
	Luekocidin	Destroys leukocytes
	Exfoliative toxin	Causes outer layer of skin to peel
	Coagulase	Causes plasma to clot
	Toxic shock syndrome–associated toxin (TSST)	Causes rash, falling blood pressure, and other manifestations of toxic shock syndrome
	Enterotoxin	Causes diarrhea and vomiting
Pseudomonas aeruginosa	Thick capsule	Inhibits phagocytosis; occludes respiratory passages in cystic fibrosis disease
Clostridium perfringens	Alpha toxin	Kills human cells
	Enterotoxin	Causes diarrhea and vomiting
Propionibacterium acnes	Fatty acids	Sebum trapped in pores is metabolized to irritating fatty acids, which initiate acne inflammation

***Staphylococcus aureus:* Impetigo, Boils, Abscesses.**
The skin and its underlying tissues are *Staphylococcus aureus*'s most common target. But as we've seen, it also causes food poisoning (Chapter 23) and toxic shock syndrome (Chapter 24). In fact, *S. aureus* can infect every organ in the body. The bones, joints, lungs, and heart are affected fairly often. In the United States, *S. aureus* causes more disease than any other bacterium.

Some stains of *S. aureus* are highly virulent. They ferment mannitol, produce coagulase and hemolysins, and grow as yellow colonies. Strains of *Staphylococcus epidermidis*, the other member of the genus, are avirulent or only weakly virulent. They cannot ferment mannitol or produce coagulase or hemolysins, and they grow as white colonies. *S. epidermidis* is present on almost everyone's skin and in their nasal passages. *S. aureus* is also present in the nasal passages of many healthy people.

Like streptococci, staphylococci cause an intense inflammatory reaction, and they produce a battery of toxins and enzymes that cause further damage. For example, **alpha** and **delta toxins** lyse human cells by attacking their membranes. **Leukocidins** destroy leukocytes, including phagocytes. **Exfoliative toxin** causes the outer layers of human skin to separate and peel away. Large quantities of it produce a horrifying clinical syndrome called staphylococcal **scalded skin syndrome** (**Figure 26.5**) in infants and the elderly. Fortunately, it is rare. Other toxins cause toxic shock syndrome (Chapter 24) and food poisoning (Chapter 23).

Pathogenesis. Staphylococcal skin infections differ depending upon which layer they attack. Superficial infections cause impetigo, which produces a clear exudate and no pus. Deeper staphylococcal infections typically do produce pus. There are several kinds of such pus-forming infections.

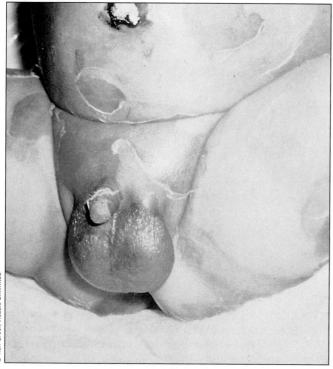

FIGURE 26.5 Staphylococcal scalded skin syndrome.

Folliculitis is a rash composed of tiny, pus-containing lesions within hair follicles. **Boils,** or **carbuncles,** are somewhat larger pus-containing lesions within the dermis. **Abscesses** are accumulations of pus that may penetrate into deeper tissues. In contrast, **cellulitis** is diffuse, extensive infection that spreads through the skin and underlying soft tissues; it is not pus-forming. Bacteria from a cellulitis infection can enter the bloodstream, leading to a life-threatening systemic infection.

Staphylococcal infections often occur when a foreign body such as a surgical suture or a splinter penetrates the skin. Such infections are more common in infants and the elderly because of their weak immune systems. If the immune system is vigorous, opsonization and phagocytosis quickly control staphylococcal infections.

Treatment. Staphylococcal infections are treatable with antibiotics, but large pus-containing lesions must be **lanced** (opened surgically) for the treatment to be effective. The greatest concern about staphylococcal infections is the emergence of antibiotic-resistant strains. Once all strains of *S. aureus* were sensitive to small doses of penicillin G. But today penicillinase-producing, and therefore penicillin-resistant, strains are common. Now staphylococcal infections are treated with semisynthetic penicillins, such as methicillin, or with cephalosporins. Recently, however, strains of methicillin-resistant staph-

ylococci have emerged, causing life-threatening infections and occasional hospital epidemics.

***Pseudomonas aeruginosa:* Folliculitis, *Pseudomonas* Infection.** *Pseudomonas aeruginosa* infects people who are already ill or injured. It's not highly virulent, but it's a successful opportunistic pathogen, because it has pili for adherence and an extracellular slime layer that interferes with phagocytosis. Burn patients are at particularly high risk because they have lost their skin's protection. Eighty percent of all burn fatalities are due to infection, and *P. aeruginosa* causes most of them. It is extremely difficult to keep a burn unit free of *P. aeruginosa.* It can survive in bedpans, respiratory equipment, whirlpool baths, and even disinfectant solutions. The pus exuding from the wounds of burn patients is often colored blue-green by a pigment that *P. aeruginosa* produces. If the infection becomes bloodborne, patients can die from Gram-negative shock. Preventing burn infections depends upon scrupulous cleanliness and burn **debridement** (removing dead or dying tissue).

Patients with cancer or diabetes are also vulnerable to *P. aeruginosa* infections, as are the airways of children with **cystic fibrosis** (a genetic disease of the lungs). Strains that infect these children produce an especially thick capsule that blocks the respiratory passages.

P. aeruginosa also causes less severe infections of otherwise healthy people. For example, it causes **hot-tub folliculitis,** a fairly common infection of hair follicles among people who bathe in hot tubs or spas contaminated by *P. aeruginosa.* The rash that results can be unsightly, but patients usually recover without antibiotic treatment. *P. aeruginosa* also contributes to **otitis externa** (swimmer's ear). It causes 10 to 20 percent of nosocomial infections.

Because it has minimal nutritional requirements and can tolerate a wide range of temperature, *P. aeruginosa* is found almost everywhere—in soil, water, and the gastrointestinal tract of humans and animals.

P. aeruginosa infections are extremely difficult to treat with antibiotics. A combination of antimicrobial compounds, including an aminoglycoside and a beta-lactam agent, are usually used.

***Clostridium perfringens:* Gas Gangrene.** *Clostridium perfringens* is an anaerobic, Gram-positive bacterium that infects deep wounds, usually dirt-contaminated wounds because *C. perfringens* is abundant in soil. *C. perfringens* grows only anaerobically, conditions that occur commonly in **necrotic** (dead and dying) tissue in deep wounds. The resulting infection initiates a chain reaction of tissue destruction and spreading infection. It produces a toxin in the wound, thereby killing nearby cells and enlarging the region where it can grow and produce even more toxin. Preventing infection depends upon timely debridement of deep wounds.

Clostridium perfringens infections can be completely asymptomatic, or they may cause **gas gangrene** (**Figure 26.6**). Gas gangrene often destroys large amounts of muscle, as well as skin. It's distinguished by bubbles of gas (hydrogen and carbon dioxide) produced by *C. perfringens*. Gas bubbles can be felt through the skin as **crepitance** (a crackling sensation) or seen in the fluid oozing from the wound. The overlying skin turns black and dies. Gas gangrene progresses rapidly. About a week after infection, symptoms (extensive tissue death, renal failure, and shock) begin abruptly. Death can occur within hours.

Patients can be treated with high-doses of penicillin, given clostridial antitoxin, and placed in an oxygen-enriched chamber (to stop growth of the anaerobe), but these measures have questionable value. The only truly effective way to control a gas gangrene infection is to remove all dead tissue, which usually means amputation. If amputation is impossible because of the wound's location (for example, the trunk), infection is often lethal in spite of medical therapy. Gas gangrene is occasionally caused by other toxin-producing species of *Clostridium*, such as *Clostridium septicum* and *Clostridium novyi*.

C. perfringens owes its pathogenicity to the toxins it produces. An enterotoxin causes clostridial food poisoning (Chapter 23). Alpha toxin is primarily responsible for the tissue damage caused by gas gangrene and infections that follow childbirth or abortion (Chapter 24).

Acne. At some time in our lives, most of us suffer from acne. Usually it is a minor annoyance. Small inflamed papules and pustules appear on the face and upper part of the body. They heal spontaneously within a few days. But acne can be painful and disfiguring. **Cystic acne** causes large tender nodules deep in the skin that leave lifelong scars (**Figure 26.7**).

Acne is not an infection. It's an inflammatory disorder, usually associated with an increased output of sebum stimulated by steroid hormones produced during adolescence. Acne develops when sebum becomes trapped in pores and is forced out into subsurface tissue. Then the commensals in the pores break down the sebum, producing fatty acids. They initiate the inflammatory response. The most important fatty-acid–producing commensal is the anaerobic Gram-positive rod *Propionibacterium acnes*.

Most cases of acne respond readily to treatment. Creams that peel away the outermost epidermal layers decrease the likelihood of pores becoming clogged. Antibiotics applied directly to the skin or taken orally decrease the population of *P. acnes*. The tetracyclines are especially effective because they penetrate sebum and are secreted in sweat. Thus they reach high concentrations in the skin.

The severest cases of cystic acne that do not respond to standard therapy are treated with a new drug called **isotretinoin** (Accutane). This compound, which is closely related to vitamin A, produces dramatic improvement. However, isotretinoin is a potent **teratogen** (producer of birth defects). Women are advised not to become pregnant while taking the drug. Still, a number of birth defects have occurred. The drug manufacturer has established extremely stringent guidelines for selecting and educating patients.

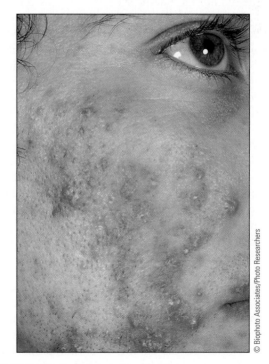

FIGURE 26.7 Cystic lesions caused by acne.

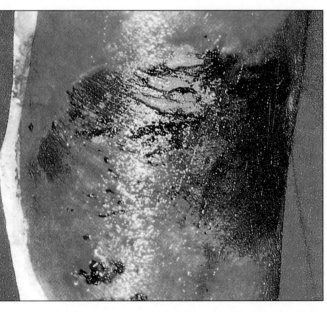

From *Infectious Diseases Illustrated: An Integrated Text and Color Atlas* by Harold P. Lambert and W. Edmund Farrar. Gower Medical Publishing, London, United Kingdom, 1982. Reprinted by permission.

FIGURE 26.6 Gas gangrene resulting from a bone-exposing fracture of the leg.

CASE HISTORY

Acne Can Be Controlled

H. J. was a 15-year-old girl who visited her doctor for treatment of her acne. She had worried about her acne for more than a year and had tried many cosmetics and over-the-counter products without improvement. She told her doctor that a friend advised her to go to the doctor and ask for a prescription for Accutane, a new medicine to cure acne.

Her doctor examined H. J.'s skin and found many acne lesions on her face, chest, and back. Some were comedones (blackheads or whiteheads); some were raised red pimples; some were pustules. Her doctor found no deep cystic lesions, nor scars from healed lesions. H. J.'s acne was most noticeable on her forehead and cheeks. She had at least two dozen acne lesions on her face; some contained pus. Her doctor concluded that H. J. had moderately severe pustular acne. She did not have cystic acne.

First her doctor talked to H. J. about acne. She explained that acne is very common among teenagers because the hormones that cause the physical development of puberty stimulate the skin to produce excess oil. Acne isn't caused by anything that a person does or fails to do, such as not eating a healthy diet or not washing well enough. Acne develops in the layers of the skin that are too deep to be affected by washing. Acne is a problem that some people are simply more prone to than others. It often runs in families. Most people have acne for several years during their adolescence. After that it usually goes away. Sometimes acne causes lifelong scarring, but even when it doesn't, it deserves medical treatment.

H. J.'s doctor explained that there are many effective medications for treating acne. Some of them cause the skin to peel slightly, so that pores don't become blocked with oil and cause acne lesions. Others help control the bacteria that cause the inflammation and the pus-containing lesions. Some of these medicines are applied directly to the skin, and others are taken by mouth. All these medicines take 6 to 8 weeks to work, so a person who starts treatment for acne has to be patient and continue to use the medicine every day. Remembering to continue the medicine is often more difficult after the acne begins to improve. Her doctor told H. J. that one oral medicine, called Accutane, is sometimes used for patients with the deep, scarring acne, called cystic acne, or for those who fail to improve with other medicines. Accutane is more dangerous than other acne medicines, so people who use it must have their blood drawn regularly to ensure that they are not suffering liver damage and that their blood lipids are not rising too high. A woman taking Accutane is required to start birth control pills or shots, because if she becomes pregnant while taking the drug, the fetus is likely to be severely malformed. In addition, people taking Accutane almost always develop very dry skin, dryness of their eyes, and cracking of their lips.

Her doctor told H. J. that she thought she would do very well on a combination of an oral antibiotic and a cream that would cause mild peeling of the skin. She wrote H. J. a prescription for doxycycline pills (an antibiotic that is highly concentrated in skin secretions) and one for a cream called tretinoin that causes the skin to peel. Her doctor warned H. J. that she ought to begin to use the tretinoin gradually, so that her skin would not become too irritated. She also advised her that tretinoin can sometimes make acne worse for the first week or two before it starts to improve. Then the kind of acne she had ought to respond very well to these safe medications. She encouraged H. J. to try them and come back to see her in 2 months.

H. J. kept her return appointment 2 months later. Her acne was much improved. She still had a few red spots on her forehead and cheeks, but she had no pustules. Her acne was barely noticeable. H. J. was delighted, and her doctor was pleased as well. She gave H. J. refills on both the medications and told her to come back for another visit in 4 months.

Case Connections
- Although acne is not an infection, *Propionibacterium acnes* plays an important role. What is that role?
- What's the rationale for treating acne with creams? Why are tetracycline antibiotics helpful in controlling acne?

Leprosy (Hansen's Disease). Leprosy is the term used in scientific literature for the disease caused by *Mycobacterium leprae*, but clinicians use the less distressing term **Hansen's disease** (for Norwegian scientist Gerhard Hansen, who identified the causative microorganism in 1878). *M. leprae* infects both the skin and peripheral nerves. The combined effect is a chronic skin rash with loss of sensation, such as a decreased ability to perceive touch or temperature.

M. leprae was the first microorganism to be clearly identified as a cause of human disease. Still, surprisingly little is known about it after more than a hundred years, although its genome has been sequenced recently. *M. leprae* is closely related to *Mycobacterium tuberculosis*, the organism that causes tuberculosis (Chapter 22). Both have a waxy cell envelope that can be stained only by an acid-fast technique. Both grow slowly. *M. leprae* grows best slightly below body temperature, which accounts for its tendency to infect cooler body parts, such as the nose, ears, toes, and fingers.

Research on *M. leprae* was difficult because it could not be grown in the laboratory, not even in human tissue culture. In 1971, however, researchers discovered that it would grow in artificially infected armadillos (not a result of an alphabetically directed search, armadillos are prone to a natural infection very similar to leprosy, probably because their body temperature is low). This led to the development of an experimental vaccine. Recombinant DNA technology may eventually provide a cheap and plentiful supply of vaccine.

Clinical Syndrome. Leprosy is an ancient disease described in the Bible and feared throughout history as contagious and disfiguring. Infected people were required to wear a bell to warn others of their approach. Although not as contagious as once believed, leprosy does spread. Family members of a leprosy patient have about a 10 percent chance of contracting the disease. It's about as communicable as active tuberculosis. The disease is probably transmitted from person to person by infected nasal secretions and by direct skin-to-skin contact.

About 3 to 5 years after infection, patients develop **indeterminate leprosy.** Its symptoms are a few innocent-looking skin lesions. Nerve function remains normal. A skin biopsy, however, will show early evidence of nerve damage, confirming a diagnosis of leprosy. Untreated, the infection usually progresses to either **tuberculoid** or **lepromatous leprosy.**

Patients with tuberculoid leprosy mount a vigorous cell-mediated immune response that holds their infection in check. Skin biopsies reveal tuberculoid granulomas similar to those seen in tuberculosis (Chapter 22) but no actively multiplying bacteria. A **lepromin test** (leprosy skin test) is positive. Skin damage is usually limited, consisting of a few erythematous or depigmented lesions. Sensation can be completely lost in these area.

Patients whose immune systems fail to mount an effective cell-mediated defense may develop lepromatous leprosy and become grossly disfigured (**Figure 26.8**). They have many *M. leprae* cells in their skin and nasal secretions, so these patients can infect others. Their lepromin test is negative because they lack a cell-mediated immune response against *M. leprae*. Interestingly, they do have high antibody titers against *M. leprae*, but these have no protec-

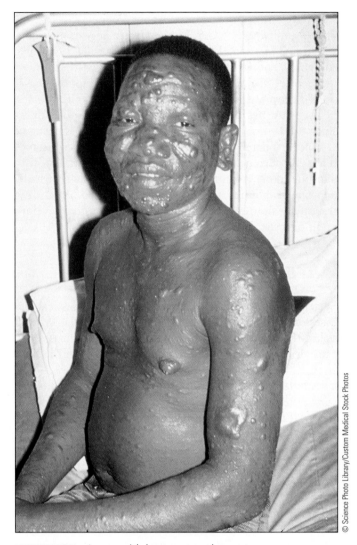

FIGURE 26.8 A man with lepromatous leprosy.

tive value. Skin damage is extensive. There are many flat or raised lesions, and underlying bone or cartilage is often destroyed. Facial features become thickened, taking on the classical leonine (lionlike) appearance of leprosy. The nose may collapse because of extensive tissue destruction. Fingers or toes may be lost. Loss of sensation is usually patchy.

No one knows why some people develop tuberculoid leprosy and others lepromatous leprosy. Some specialists believe that tuberculoid disease inevitably drifts toward lepromatous disease as cell-mediated defenses are exhausted. Others believe the patient's genetic makeup determines what type he or she develops.

Epidemiology and Treatment. Leprosy is a major public health problem in tropical countries and parts of Asia. About 12 million people are infected worldwide. In the United States it is rare but on the rise because of infected people who have entered the country. About 250 new cases are reported annually. The U.S. Public Health Service maintains a National Hansen's Disease Center at Carville, Louisiana. Anyone is entitled to free treatment there or at six outlying centers around the country. Treating imported cases should prevent spread of leprosy within the United States.

Leprosy has been treatable with antimicrobial drugs since the 1940s. Dapsone, the original antileprosy drug, is a sulfone related to the sulfonamides. Drug resistance tends to develop during the prolonged treatment with dapsone alone, so now a combination of drugs—usually dapsone, rifampin, and clofazimine, is used. Multidrug treatment shortens treatment and prevents the development of resistant strains.

Viral Causes

Systemic viral infections include many of the classical diseases of childhood—chickenpox, measles, rubella—as well as life-threatening infections such as smallpox. Viruses also cause warts, the most common benign tumors of the skin.

Varicella Zoster: Chickenpox, Shingles.
The two very different clinical syndromes described in Case History: A Lifelong Visitor are caused by the same infectious agent—the varicella zoster virus (VZV). It's a member of the herpesvirus family that infects only humans. Most of us have been infected by this virus by the time we reach adulthood, but this pattern will change with widespread use of the varicella vaccine licensed in 1995 and recommended for all children between 1 and 2 years of age.

Clinical Syndromes. Like the herpes simplex virus (Chapter 24), varicella zoster establishes a latent infection in nerve cells that can be reactivated later. People infected for the first time develop a generalized infection called varicella, or chickenpox, that produces blisters all over the body. Recovery is complete, but the virus remains latent in **spinal ganglia** (large masses of neurons lying near the spinal cord). If infection is reactivated in one of these ganglia, it travels down the associated neurons to produce new chickenpox-like blisters on the skin that the nerves supply. Because a **dermatome** (the area of skin supplied by a single sensory nerve) covers a stripelike area, so does shingles. Usually only one or two dermatomes are involved.

Varicella is usually a mild disease in children. R. C.'s case was typical. Because blisters occur in the outer skin layer, significant scarring is uncommon. But because the skin is broken, more serious streptococcal or staphylococcal infections can occur. Shortly after recovery, a few children develop a life-threatening illness called **Reye's syndrome,** in which liver and brain functions rapidly deteriorate. Most children who develop Reye's syndrome have taken aspirin during a preceding viral illness, usually either chickenpox or influenza. That's why R. C.'s physician warned so firmly against giving him aspirin.

Chickenpox is more serious for adults. They are much more uncomfortable and may contract a life-threatening viral pneumonia. Adults with chickenpox often require hospitalization. During pregnancy, chickenpox is especially serious. Women infected during the first 3 months of pregnancy may suffer abortions or deliver malformed babies. Women infected at term may deliver babies with active chickenpox who have a one in three chance of dying.

For all age groups there are only about 100 deaths per year from chickenpox in the United States, but this number seems more significant now that the disease is preventable by vaccination.

Reactivation of a latent varicella zoster infection occurs most often in patients like K. V., who are over the age of 40. But anyone who has had chickenpox may develop shingles. For people who are otherwise well, shingles usually occurs once, lasts only briefly, and is not life threatening. However, it can be very painful. The most significant complication is persistent pain in the affected nerves, a syndrome called **postherpetic neuralgia.** It can disable elderly victims.

Cell-mediated immunity is crucial for normal recovery from varicella zoster infections. For people with impaired T-lymphocyte function—cancer patients receiving immunosuppressive chemotherapy or people with acquired immunodeficiency syndrome (AIDS)—such infections can be fatal.

Prevention and Treatment. A live attenuated varicella zoster vaccine called the **Oka strain** has been developed in Japan. In 1995 it was approved for use in the United States.

Now it's recommended for all children over age 1. Preventing chickenpox is financially important. Parental time off work while children must stay home from school or day care costs more than $380 million annually. It's also important for reducing childhood hospitalization and mortality. Chickenpox resulted in approximately 100 deaths annually before the vaccine became available.

Preventing chickenpox in infants less than 1 year old (who are too young to receive vaccine) can be critical. Passive immunization with varicella zoster immune globulin (VZIG) can help them significantly if it's administered within 3 days of exposure. VZIG is obtained from recovering zoster patients who have extremely high titers of antiviral antibodies. Acyclovir is somewhat effective.

Gingivostomatitis, Fever Blisters.
Gingivostomatitis and fever blisters are generally caused by herpes simplex virus type 1 (HSV-1). Like herpes simplex virus type 2 (HSV-2), HSV-1 is a large, double-stranded DNA virus that infects the skin and mucous membranes and estab-

lishes a latent infection in nerves. The two viral types are distinguished by specific glycoproteins. HSV-1 usually causes infections above the waist, whereas HSV-2 causes most infections below the waist (Chapter 24).

HSV causes vesicles on infected skin. When the vesicles occur on mucous membrane surfaces such as the inside of the mouth, they rapidly evolve into ulcers. Recurrent lesions near the lips are sometimes called cold sores or fever blisters (**Figure 26.9**). Vesicles are painful when they first appear, but they crust and heal within about a week. Fluid in the vesicles contains live virus, so herpes spreads readily when this infectious material comes in contact with the mucous membranes or broken skin of a susceptible host.

The most common clinical syndrome of primary HSV-1 infection is gingivostomatitis (gingiva, gums; stoma, mouth). It is a moderately severe illness characterized by fever and painful blisters on the mouth and gums. Toddlers, who exchange saliva as they play closely together or mouth each other's toys or bottles, are commonly affected, especially if they live in crowded conditions. Epidemiologists estimate that 80 percent of low-income children are infected, compared with only 40 percent of children from more affluent families. Adults who have never been infected remain susceptible to herpes gingivostomatitis. Some studies show that half of college students are susceptible.

Acyclovir, which is used routinely for primary genital herpes, is probably effective in shortening oral herpes infections as well. Extensive clinical studies, however, have not been undertaken and the drug is not used routinely in children with gingivostomatitis.

After a primary oral HSV infection, the virus remains latent in nearby nerve cells. Fever, sunlight, emotional stress, or other stimuli can trigger reactivation at the same site any number of times. Then a cluster of tiny thin-walled blisters on a bright red base appears on the lips (Figure 26.9). Like primary lesions, reactivation vesicles contain live virus, and patients with cold sores can spread infection. Viral reactivation can also be asymptomatic, so apparently healthy people shed contagious viral particles.

HSV-1 can infect any part of the body if virus-containing secretions come in contact with a break in the skin. For example, **herpetic whitlow** is a finger infection seen in medical personnel caring for herpes patients and in children with oral herpes who suck their fingers. **Herpes gladiatorum** is seen in wrestlers who sustain prolonged skin-to-skin contact with one another. The most serious HSV-1 infection is **herpes keratitis,** which affects the surface of the eye.

Measles (Rubeola).

Measles (rubeola) is caused by measles virus, which belongs to the paramyxovirus family. Its virions consist of a coiled RNA nucleocapsid surrounded by a lipid envelope bearing many short protein

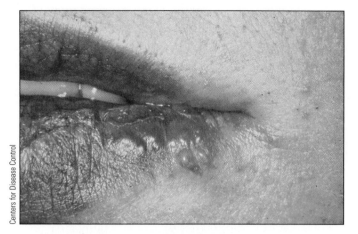

Centers for Disease Control

FIGURE 26.9   Herpetic vesicles on the lips are commonly called cold sores or fever blisters.

projections. By forming bridges between cells, these proteins enable the measles virus to agglutinate red blood cells and to fuse human cells to one another. Such cell fusion creates the giant multinucleated cells seen in measles-infected tissue. Measles virus is extremely fragile. It can survive for only a short time outside the human body, which is its only reservoir.

Measles follows a typical pattern. A susceptible person (usually a child) inhales the virus, which multiplies in the respiratory tract for several days. Then it spreads throughout the body and multiplies in lymphoid tissue. The first symptoms—fever, cough, **coryza** (runny nose), and conjunctivitis—usually appear on day 11 of the infection. That's when **Koplik spots** appear inside the mouth. They look like tiny white flecks of sand on a bright red base. This enanthem provides a sure diagnosis of measles even before the rash appears on day 14 on the face. Then it gradually spreads downward to cover the entire body (**Figure 26.10**). As it progresses, the individual spots become so numerous they merge. At its peak, the entire surface of the skin is deep reddish-purple and puffy. By the time the rash reaches the legs, the child is usually beginning to feel better. The rash fades in the same order as it appeared, from head to foot.

Measles is often a serious illness. The patient feels miserable. Complications are common, especially ear infections and pneumonia caused by secondary bacterial invaders. Bacterial pneumonia is the leading cause of death among children weakened by measles. Permanently handicapping complications also occur. During recent epidemics in the United States, 2 to 3 children per 1000 suffered brain damage from measles encephalitis.

Cell-mediated immunity is essential for normal recovery. Because cell-mediated immunity is depressed by malnutrition, measles is a terrible killer of children in developing countries. The World Health Organization

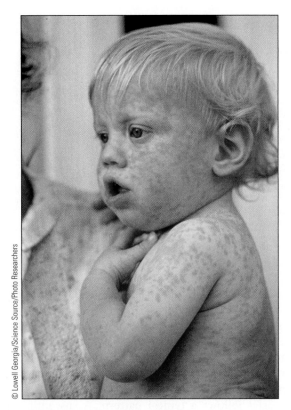

FIGURE 26.10   A toddler with a typical measles rash.

estimates that every minute three children die of measles and three more are permanently disabled.

Epidemiology. To survive, measles virus must sustain an unbroken chain of human infection. At first it might appear that the odds are against the virus. It can live within a host for only a few weeks, and an infected person can spread the virus for only about 1 week (from first symptoms develop until about 4 days after rash appears). During this brief period, the virus must be transmitted directly to a new host—it cannot survive on fomites or be transmitted by an uninfected carrier. Moreover, the pool of susceptible people is limited. Anyone previously infected is immune for life. If the chain of infection were broken, measles would be eradicated.

But measles virus is successful because measles is so highly communicable. When exposed, more than 90 percent of susceptible people become infected. And contact need not be intimate. For example, a person can become infected by being on the same airplane as someone who has measles. Because it is so communicable, measles is almost exclusively a disease of children. In unimmunized populations, virtually everyone has had measles by adulthood, so adults are immune.

Availability of new hosts does limit the spread of measles. Epidemiologists calculate that a population of 300,000 to 500,000 is probably necessary to sustain the

measles virus. In smaller isolated communities the disease disappears. But when the measles virus is reintroduced after many new births, an epidemic will spread with unbelievable rapidity. Such an epidemic occurred in Greenland in 1951. Of 4600 susceptible people, all but 5 contracted measles within 6 weeks. Similar epidemics devastated native populations in the Americas when measles was first introduced from Europe.

The epidemiology of measles changed in 1963 when a vaccine with live attenuated measles virus was developed. Widespread immunization of 15-month-old children as part of the measles-mumps-rubella (**MMR**) vaccine has reduced the incidence of measles in the United States by more than 99 percent. As a result, many young parents and health professionals are unfamiliar with measles. Some mistakenly think of it as a mild illness like chickenpox.

In inner cities, many preschool children are still not immunized against measles. Because the law requires proof of measles immunization for school, children over 5 are immunized, but their younger brothers and sisters often remain unvaccinated. Until all children receive both immunizations, measles cannot be eradicated from the United States.

The struggle to control measles in the developing world is more desperate. Malnourished children are often infected before their first birthday, when they are too young to be adequately protected by the present vaccine. A new vaccine is needed to protect infants as young as 4 to 6 months. In theory, measles, like smallpox, could be eradicated by a worldwide vaccination program. In reality, public health experts believe that until people around the world are provided with basic necessities—clean water, adequate food, and basic health care—measles will continue to kill children.

Subacute Sclerosing Panencephalitis. Subacute sclerosing panencephalitis (SSPE) is an extremely rare but fatal form of measles that affects the nervous system. It may be caused by mutant strains of the measles virus.

SSPE typically occurs in a child who contracts measles at an unusually young age. The patient appears to recover completely, but the virus replicates in brain cells. Five or more years later, symptoms of severe brain damage begin to appear. The disease progresses to seizures, paralysis, and eventually coma. Diagnosis is confirmed by cultivating measles virus from the victim's brain at autopsy. There is no treatment for SSPE. It is almost always fatal. Incidence of SSPE decreases markedly where measles immunization is widespread. In the United States it has virtually disappeared.

Rubella (German Measles).
Rubella (also called German measles and 3-day measles) is not related to rubeola (also called true measles). Rubella virus is a togavirus, as are arboviruses (Chapter 25).

Rubella is one of many mild, rash-producing illnesses of childhood. The virus is inhaled, and after about 2 weeks a mild fever and maculopapular rash appear. Lymph nodes are often enlarged and tender. The rash consists of faint spots that appear first on the face and gradually spread downward over the rest of the body (**Figure 26.11**). It lasts less than 3 days. Most people with rubella are not very ill, although adult women may complain of stiffness and swelling in their joints. Complications are rare. Because rubella is not as highly communicable as measles or chickenpox, a significant number of people reach adulthood without contracting the disease. People who have recovered from rubella are immune for life.

The rubella virus is dangerous because it is a powerful teratogen. Infection during the first 3 months of pregnancy is highly likely to cause abortion or birth defects of the eye, heart, and brain. A major rubella epidemic swept the United States in 1964. More than 20,000 babies were born with defects. Thousands more died before birth. The disaster spurred development of an attenuated vaccine.

The goal of preventing rubella-caused birth defects presented special problems. Immunizing women of childbearing age seemed dangerous. The live attenuated virus itself might cause birth defects. Immunizing children might produce a weak immunity that could wear off by adulthood, thereby increasing the frequency of the disease in pregnant women. In spite of these fears, a rubella vaccine (part of MMR) was introduced in 1969, and it's worked well. Childhood rubella and congenital rubella have declined substantially. Still, women of childbearing age are screened serologically to see if they are protected. If they aren't, they are immunized before becoming pregnant.

Smallpox. In the second edition of this book we discussed smallpox in one brief paragraph as a historical footnote. The disease, which throughout history had claimed hundreds of millions of lives, had been eradicated worldwide in 1977. It no longer seemed relevant. Smallpox was the first and so far the only infectious disease to be eliminated by human effort (Chapter 20). Unfortunately, this monumental human achievement could now be in jeopardy. Some feel that smallpox could be used as a weapon of bioterrorism.

One might ask where the virus that causes a nonexistent disease could be obtained. Although the disease has been eradicated, the virus still exists. Officially there are two sources of the virus—at the Centers for Disease Control and Prevention in Atlanta, Georgia, and at a Russian laboratory in Siberia. There are also the human corpses buried in permafrost or in very dry climates that could contain live virus. It's possible that the virus could be synthesized in a laboratory because its DNA sequence is available in the open literature. And, of course, stocks of the virus could have been stolen. Because smallpox vaccination only provides 7 to

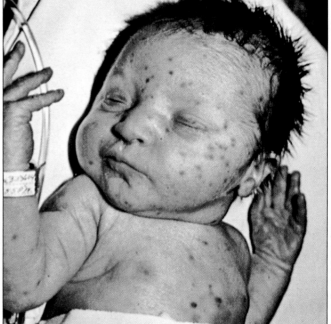

FIGURE 26.11   German measles. This congenitally infected infant has a typical "blueberry muffin" rash and probably other less obvious but more serious birth defects.

10 years' protection and routine vaccination stopped in 1980, almost the entire world is now susceptible to smallpox.

Smallpox was caused by **variola** (smallpox virus), a large double-stranded DNA virus that belongs to the poxvirus family. Other poxviruses cause animal diseases, such as monkeypox, cowpox, and even camelpox.

One of these, **vaccinia** (cowpox), is similar enough to confer effective immunity against smallpox and causes few symptoms. It can be used for immunization to control smallpox and was used to eradicate this deadly and disfiguring disease.

The severity of smallpox depended upon the viral strain causing the infection. The most virulent strain, **variola major,** produced a high fever and severe blistering rash, killing about half its victims. The pustules of smallpox originated deep in the skin, and most survivors were permanently scarred. In fatal cases, the blisters were often so numerous they touched one another, covering the entire body (see Figure 20.4). A much milder form, variola minor, had a mortality rate of less than 1 percent. Survival seems to depend entirely upon the virulence of the infecting strain. It has never been established that any medical treatment helped people recover.

Smallpox had nearly the same epidemiology as measles. The virus was inhaled, and an unbroken chain of human infection had to be maintained to perpetuate the virus. A significant difference, however, is that smallpox was much less communicable. Close personal contact was necessary for transmission. On the other hand, relatively low communicability was offset by increased hardiness. The virus could survive on fomites and in the environment for days to weeks. Lower communicability also meant that epidemics progressed less rapidly but lasted longer than measles epidemics. Because smallpox depended on a continuing chain of human transmission, eradication was possible and successful.

Warts. Warts are caused by human papillomaviruses (HPV). These DNA viruses belong to the papovavirus family. Some strains cause genital warts (condyloma acuminata) and laryngeal papillomatosis (Chapter 24). Other strains cause the common warts that occur on most skin surfaces and plantar warts on the soles of the feet. Genital warts are sexually transmitted, but no one knows how other forms are transmitted. Warts are not highly contagious. They spread if infected skin cells are directly inoculated into a break in the skin of another person. This association with trauma may explain why warts are most common on the hands and feet.

Warts appear on the skin without warning, usually before puberty. They disappear without treatment, usually after a few years. Sometimes a single wart will develop into clusters of warts, probably by viral autoinoculation. Skin warts are harmless and not linked to increased risk of cancer, but people often want them removed for cosmetic reasons. They can be eliminated by liquid nitrogen (which kills infected cells), keratolytic chemicals (which destroy keratin and thereby dissolve the infected cells), or surgery. Removal leaves no scar because warts occur in the epidermis, which is regularly replaced.

A possible link between HPV and cancer has stimulated interest in these organisms. Thirty percent of people with a rare syndrome of persistent warts (not common warts, but a particular type of warty growth) eventually develop skin cancers, and viral DNA is found in the malignant cells.

Fungal Causes

A variety of fungi survive on our body. They're annoying, but most are not serious for otherwise healthy people.

The Dermatophytes: Ringworm. The dermatophytes are a group of fungi that infect the body's outermost surfaces, causing athlete's foot, jock itch, and ringworm. The medical names for these infections use two words: **tinea** (meaning ringworm) followed by the affected region (**Table 26.4**). Despite its name, ringworm has nothing to do with worms. By middle age about 90 percent of men have been infected at least once. For some reason, women are not affected as often. Ringworm can be annoying and uncomfortable, but it never penetrates into deeper tissues or becomes life threatening.

Dermatophytes belong to three genera: *Trichophyton*, *Microsporum*, and *Epidermophyton* (**Figure 26.12**). All produce enzymes that digest keratin, and all are restricted to keratin-containing surfaces—the stratum corneum, hair, and nails. There are more than 40 species of dermatophytes, but fewer than 10 are found in the United States. Some live only on humans. Others live on animals or in the soil. Infection is most common in tropical climates and in crowded living conditions.

The body's immune response influences both the symptoms of dermatophyte infections and the chances for com-

TABLE 26.4 Ringworm Infections

Infection	Affected Region
Tinea pedis	Feet
Tinea cruris	Groin
Tinea capitis	Scalp
Tinea unguinum	Nails
Tinea corporis	Other parts of the body

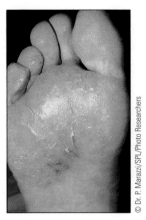

FIGURE 26.12   Athlete's foot. (See Figure 1.5 for a photo of *Epidermophyton floccosum.*)

plete recovery. Cell-mediated immunity is activated, but it's a mystery how the long hyphae of these fungi (too large to be phagocytized) are destroyed. Some ringworm infections, such as those contracted from animals, stimulate a powerful immune reaction. The infected areas are highly inflamed. They become bright red, weepy, and unsightly, but they usually heal completely within a week or two. Other infections, such as athlete's foot, stimulate a comparatively weak immune reaction with little inflammation. They cause only mild cracking and peeling, but they can persist for years. These long-lasting but relatively mild infections are a well-adapted symbiosis between humans and microorganisms. Curiously, acute infections usually occur in children. Athlete's foot usually doesn't develop until after puberty.

Because dermatophyte infections are superficial, most can be cured with topical antifungal agents such as miconazole or clotrimazole. Infections that affect thickened areas such as the scalp or nails often fail to respond to topical therapy. They are usually treated orally with griseofulvin. Griseofulvin is especially effective against dermatophytes because it concentrates in the stratum corneum. Although it has some toxic side effects, including headache, it can usually be used quite safely for the weeks to months required to eliminate dermatophytes.

Candida albicans: **Candidiasis.** *Candida albicans* and other closely related *Candida* species are commensal fungi that colonize the skin, mucous membranes, and gastrointestinal tract of almost every human being (Chapter 14).

Candida is also an opportunistic pathogen. Many minor disruptions of the body's equilibrium (such as pregnancy, antibiotic therapy, and oral contraceptives) predispose a person to **candidiasis** (symptomatic infections of body surfaces). Infants and the elderly are especially susceptible, as are people with immunodeficiency disorders. The most common sites of infection are the vagina (*Candida* vaginitis; Chapter 24), the mouth (thrush; **Figure 26.13**), and the diaper area of infants. Thrush, which often occurs in healthy infants, causes white plaques on the mucous membrane.

The lesions look like milk that can't be wiped off. *Candida* diaper dermatitis is a fiery red, raised rash, often with small pimples, or satellite lesions, just beyond its border. Because *Candida* thrives in a moist, warm environment, any part of the skin that remains wet may be infected. Candida of the hands, for example, is often seen in people who have jobs washing dishes. Superficial candidiasis is treated with antifungal creams containing nystatin or clotrimazole. Thrush infections are treated with a topically applied liquid.

Another type of local infection, called chronic mucocutaneous candidiasis, occurs only in people with defective

FIGURE 26.13   *Candida albicans* infection of the tongue.

cell-mediated immunity, and even then, rarely. In this disorder the *Candida* grows inside cells rather than in the spaces between them. The disease affects only the skin and mucous membranes, so it is not life threatening. But it can be long-lasting and cause grossly disfiguring warty lesions.

Occasionally, *Candida* invades deeper tissues of patients with serious immunodeficiency diseases. Then it does cause a life-threatening systemic infection that can affect any organ. Systemic candidiasis can be the fatal event for people with terminal illnesses such as AIDS or disseminated cancer. Systemic antifungal agents, such as amphotericin B, may be prescribed, but treatment is seldom effective.

Arthropod Causes

We've discussed biting arthropods in other chapters because many are vectors of infection. Here we'll discuss those few species that cause true skin infections. They live on body surfaces, cause symptoms, and are transmitted from person to person.

Scabies. Scabies is a skin infection caused by the mite *Sarcoptes scabiei*. It's transferred from person to person by close personal contact or by fomites such as clothing or bedding. Scabies is common worldwide. Its incidence in United States rises and falls. Epidemics last about 15 years. Then another 15 years pass before a new outbreak begins. A major U.S. epidemic began in the mid-1970s ended in the early 1990s.

S. scabiei lives in the epidermis of human skin. The female burrows into the stratum corneum, where it lays 15 to 20 eggs. These hatch into larvae a few days later, and a few days after that the larvae become mature mites. They dig their own skin burrows and continue the infectious cycle. These creatures travel across the human body at the rapid rate of about an inch per minute. They prefer the wrists and the spaces between the fingers, but they may be found anywhere. Surprisingly, most people with scabies harbor only about 11 adult female mites. Scabies resembles many other itchy skin diseases, so definitive diagnosis depends upon recovering mites or their eggs and identifying them under the microscope.

Scabies is a harmless infection that's limited to the very surface of the skin, but the itching it causes (which is worst at night) can be almost intolerable. Untreated, the infestation may continue for years. That's why it's sometimes called the **"seven-year itch."** Itching leads to severe scratching that can break the skin, allowing secondary bacterial infections to develop.

The mites themselves do no harm, but they trigger a destructive hypersensitivity reaction that does. When a person is first infected, the mites multiply without causing symptoms. Once the immune system is stimulated, however, the mites and their feces become a powerful irritant. Treatment with the arachnicide gamma benzene (Crotamiton or Kwell) kills the mites. But dead mites and their fecal pellets continue to cause irritation until the infected layer of skin is completely replaced several weeks later.

Pediculosis (Lice). **Pediculosis** is an infestation with lice. These are blood-sucking insects that live on human skin. The three main types are *Pediculus humanus capitis* (the head louse), *Pediculus humanus corporis* (the body louse), and *Phthirus pubis* (the pubic or crab louse; **Figure 26.14**). All types cause itching that can lead to skin breakdown and bacterial superinfection. Only body lice transmit microorganisms that cause epidemic diseases, such as typhus (Chapter 27).

Lice are usually transferred by direct body contact or by fomites such as clothes, hats, and bedding. They are most common among people living in unsanitary and crowded conditions. Pubic lice are usually transmitted by sexual contact. They can also be transmitted by fomites such as bedding and even toilet seats.

Recently, head lice reached epidemic proportions among children in the United States. Once an infestation develops at a school or another location where children play together, control can be difficult. Children must be examined frequently and treated. Lice are large enough to see with the unaided eye, but they are usually missed because they move so rapidly. So diagnosis is made by finding their eggs, called **nits,** which are glued to hair shafts. Effective treatment has become more difficult. All lice were once killed by insecticides, such as pyrethrins, but now many head lice are resistant to these agents.

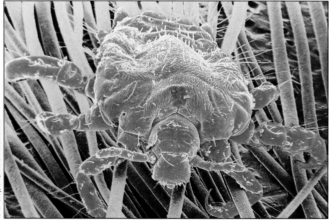

FIGURE 26.14 Scanning electron micrograph of a pubic louse clinging to pubic hair.

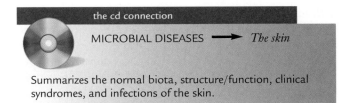

the cd connection

MICROBIAL DISEASES ⟶ *The skin*

Summarizes the normal biota, structure/function, clinical syndromes, and infections of the skin.

EYE INFECTIONS

We'll discuss here the bacteria and viruses that infect the eye's surfaces, the conjunctivae and cornea, along with helminthic infections that affect the skin and the eyes.

Bacterial Causes

Bacteria regularly come in contact with the surface of the eye. The cleansing antibacterial action of tears usually prevents infection, so any blockage of the tear ducts increases the likelihood of bacterial conjunctivitis. In early infancy the problem may be congenitally blocked ducts (Chapter 14). Also colds or allergies that cause nasal congestion may block the duct where it opens into the nose. Common causes of bacterial conjunctivitis include *Streptococcus* spp., *Staphylococcus aureus*, and *Haemophilus* spp. These infections cause conjunctival redness and a pus-containing discharge. But they usually don't affect the cornea, so they don't threaten vision. They're all easily cured with antibacterial eye drops, especially those that contain sulfacetamide, neosporin, or gentamicin.

Contact lenses predispose a person to bacterial conjunctivitis, especially if they aren't properly cleaned or if they're kept in the eyes too long. *Pseudomonas aeruginosa* is the usual cause of such **contact lens conjunctivitis.** It can spread to the cornea and threaten vision. Antibacterial eye drops usually cure it, and disinfecting contact lenses prevents it.

Inclusion Conjunctivitis, Trachoma.
We've already discussed how *Chlamydia trachomatis* infects the genital tract and is then transmitted to infants at birth (Chapter 24). But more commonly, it infects the eye. This infection is called **inclusion conjunctivitis** because of the intracellular inclusion bodies that it forms. The eyelid becomes swollen, and there may be a thick purulent discharge. Infection can be prevented by treating the mother for chlamydia during pregnancy, but antibiotic eye drops administered at birth do not prevent inclusion conjunctivitis. Infants may be given oral erythromycin for 2 weeks, but the infection usually resolves spontaneously. The cornea is not damaged. Rarely, sexually active adults may contract inclusion conjunctivitis from infected genital secretions.

Other strains of *C. trachomatis* cause a keratoconjunctivitis called **trachoma** (**Figure 26.15**). It's the most common cause of preventable blindness in the world. About 2 million people, of all ages and mostly living in developing countries, are blind from trachoma. It begins when *C. trachomatis* comes in contact with the surface of the eye. As it multiplies on the conjunctivae, the eye becomes inflamed and bumps form on the normally smooth conjunctivae (*trachoma* means "rough" in Greek.) Inflammation progresses gradually over many years. Eventually the conjunctivae become scarred, and if inflammation damages the cornea or if the eyelids become too distorted to close, blindness results.

Infectious *C. trachomatis* cells are shed from the eye. Thus transmission occurs by direct contact with these ocular secretions, by fomites such as towels, water, or flies. Trachoma is common where living conditions are crowded and unsanitary. It's found most widely in the Middle East, Africa, and Asia. It's also fairly common among Native Americans in the southwestern United States.

Chlamydiae are sensitive to many different antibiotics, including erythromycin and tetracycline. But unfortunately these agents do not eradicate *C. trachomatis* from the eye. Antibiotics do inhibit bacterial multiplication and moderate the resultant inflammation, so if antibiotic ointments are used early, they can preserve sight. Sometimes treatment is administered to entire populations that are at high risk. Once the cornea has been scarred, however, only a corneal transplant can restore vision.

Neonatal Gonorrheal Ophthalmia.
Neisseria gonorrhoeae can also be transmitted from an infected mother to the eyes of her newborn as it passes through the birth canal (Chapter 24). This infection, called **neonatal gonorrheal ophthalmia,** was once a common cause of blindness. Typically a thick, purulent discharge appears and eyelids swell

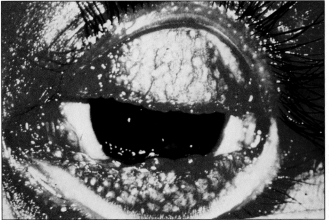

© AFIP/Visuals Unlimited

FIGURE 26.15 Eye of a person with trachoma.

2 to 5 days after birth. The cornea becomes scarred and occasionally perforated. Blindness results.

Today routine prenatal care includes a cervical culture for *Neisseria gonorrhoeae*, allowing diagnosis and cure before delivery. Still, the law in most states requires that every newborn be treated with antibacterial eye drops (formerly silver nitrate, but now usually erythromycin) to prevent gonococcal blindness. This prophylaxis prevents infection in industrialized countries. Infants who do become infected must be treated with systemic antibiotics such as ceftriaxone.

Viral Causes

Viral infections of the conjunctiva are extremely common, but usually they're not serious. Infecting viruses include the adenoviruses and enteroviruses. They cause the common and highly contagious **pinkeye** epidemics that race through schools and day-care centers (**Figure 26.16**). Clinically, viral conjunctivitis cannot be reliably distinguished from bacterial conjunctivitis. Viral conjunctivitis, however, tends to cause more intense redness and less purulent discharge than bacterial conjunctivitis. No treatment is effective, but the infection resolves spontaneously within a few days. Because the cornea is not involved, it does not threaten sight.

In contrast, herpes simplex virus (HSV) does cause a sight-threatening infection. It causes **herpetic keratitis,** an ulceration of the cornea. It usually begins after some sort of minor trauma to the eye, such as a scratch or sunburn. The infection causes irritation and excess tear production but surprisingly little pain. Usually only one eye is affected and the condition resolves spontaneously

without permanent damage. But if deeper layers of the cornea are infected, scarring and loss of vision may occur. Eye drops containing the antiviral agent iododeoxyuridine shorten the infection, but they do not prevent the recurrences that are typical of HSV infections. If an incorrect diagnosis leads to prescription of steroid-containing eye drops, which suppress immune function, the infection may progress rapidly, leading to corneal destruction and blindness. As in all cases of permanent damage to the cornea, sight can be restored only by a transplant.

Helminthic Causes

Some nematodes (roundworms) also infect the skin and eyes. Unlike conjunctivitis and keratitis, which begin when pathogens come into direct contact with exposed eye surfaces, these are systemic diseases. The pathogens reach the skin and eye from the blood and lymph.

Onchocerciasis (River Blindness).

Onchocerciasis (river blindness) is a serious disease of the skin and eyes that affects more than 18 million people in Africa and Latin America. It often causes blindness, and in some areas of Africa more than half the male population becomes totally blind by age 50 (**Figure 26.17**).

The pathogen, *Onchocerca volvulus*, is transmitted by black flies and buffalo gnats of the genus *Simulium*. When they bite, these insects leave larval worms under the skin. The larvae grow into mature nematodes. About a year later they produce millions of motile larvae called **microfilariae.** These larvae migrate through subcutaneous tissues, stimulating an inflammatory reaction that

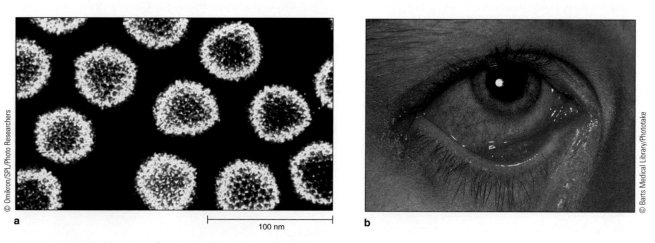

a

100 nm

b

FIGURE 26.16   Epidemic viral conjunctivitis (pinkeye). (a) Electron micrograph of causative agent, adenovirus. (b) Eye of an infected person.

LARGER FIELD

SOME HOPE

If river blindness (onchocerciasis) is diagnosed early, it often can be cured. But usually river blindness is diagnosed only by finding microfilariae in the skin or eyes or by identifying adult worms in subcutaneous nodules. These symptoms appear only after several months. Surgery is required to correct them. A blood test would detect early cases, but traditional serological methods have not worked. The answer seems to lie in genetic engineering.

Researchers have identified a critical onchocercal antigen, OV-16, and cloned its DNA. This allows them to mass-produce the antigen and design an immunoassay to detect the low levels of antibody produced during early infection. In other words, the assay can identify people who are infected by the river blindness pathogen but do not yet have visible manifestations of disease. Field trials in Mali, where onchocerciasis is common, show that the test is very specific for *Onchocerca volvulus* and can diagnose infection up to a year earlier than any other method. This raises hopes that screening programs, along with drug treatment, may at last bring this infection under control.

causes itching, rash, and marked thickening of the inflamed skin. Those that migrate to the eyes and invade the cornea can cause blindness. The infection is perpetuated when a biting black fly ingests larvae and then bites someone else.

Diagnosis is difficult; but, if recognized in time, onchocerciasis is treatable. The adult worms create clearly visible subcutaneous nodules that can be removed surgically, thus preventing production of more microfilariae. Antihelminthic drugs—including suramin, diethylcarbamazine, and mebendazole—kill the adult worms. The drug ivermectin, originally used for veterinary infections, also shows great promise. It kills migrating microfilariae.

Loa loa: **Loaiasis.** *Loa loa* is another nematode that infects the skin and eyes. The infection—called **loaiasis**—occurs only in Africa. Transmission is by an insect vector, the mango fly. Larvae are deposited under the skin and grow into adult worms that migrate through subcutaneous tissues and begin to produce microfilariae. The adult worms can produce microfilariae for as long as 17 years. Skin and eyes may be irritated or itchy. Sometimes migrating adult worms create painful lumps under the skin called **Calabar swellings,** or worms migrate through the conjunctivae, where they are easily visible. But many infections are asymptomatic, and blindness or other serious tissue damage does not occur. The infection can be treated with diethylcarbamazine. Worms that are visible in the eye can be removed surgically.

Table 26.5 summarizes skin and eye infections.

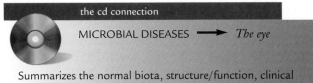

the cd connection

MICROBIAL DISEASES ⟶ *The eye*

Summarizes the normal biota, structure/function, clinical syndromes, and infections of the eye.

FIGURE 26.17   River blindness. Statue at the Carter Center in Atlanta, Georgia, of a boy leading a man blinded by the filarial parasite *Onchocerca volvulus*.

Courtesy John Ingraham

TABLE 26.5 Infections of the Body's Surfaces: Skin and Eye

Infection	Causative Agent	Mode of Transmission	Symptoms	Prevention and Treatment
Skin Infections				
Bacterial				
Impetigo, erysipelas, streptococcal gangrene	Group A streptococci	Close contact, fomites	Range from annoying superficial skin lesions to life-threatening systemic infections	Some strains preventable with good hygiene; all strains treatable with penicillin
Impetigo, boils, abscesses	*Staphylococcus aureus*	Close contact, fomites, foreign body in a wound	Range from superficial lesions to systemic infections	Some preventable with good hygiene; treatable with cephalosporins or penicillinase-resistant penicillins
Folliculitis, otitis externa	*Pseudomonas aeruginosa*	Fomites	Range from superficial rash to fatal systemic infections in compromised hosts, especially burn patients	Scrupulous hygiene to prevent nosocomial infections; treatment with combined drugs often ineffective in serious cases
Gas gangrene	*Clostridium perfringens*	Wound contamination	Tissue crepitance and death	Prevent with prompt debridement; treat with penicillin, amputation
Acne	*Propionibacterium acnes*	Acquired as a component of normal flora	Inflamed papules and pustules; painful cysts in severe cases	No effective prevention; treat with topical or oral antibiotics, agents to peel the skin surface, and isotretinoin in severe cases
Hansen's disease (Leprosy)	*Mycobacterium leprae*	Direct skin-to-skin contact and nasal secretions	Chronic skin rash and loss of sensation, leading to tuberculoid or lepromatous leprosy	Prevent by avoiding contact; treat with combined dapsone, rifampin, and clofazimine
Viral				
Varicella (chickenpox)	Varicella zoster virus	Respiratory	Crops of blisters over entire body; mild in children but risk of Reye's syndrome if child takes aspirin; can be life threatening in adults	Prevent in high-risk hosts by immunizing with varicella zoster immune globulin (VZIG); acyclovir marginally effective
Zoster (shingles)	Varicella zoster virus	Reactivation of earlier infection	Blisters appear in striped patterns; can be extremely painful but usually not life threatening	No effective prevention; treat for relief of symptoms only
Fever blisters, gingivostomatitis, whitlow	Herpes simplex type 1 virus (HSV-1)	Direct contact of infectious materials across mucous membranes	Fluid-filled vesicles; recurrent infection at same site	Resolve spontaneously; rarely, treat with acyclovir
Measles	Measles virus	Respiratory	Cough, coryza, conjunctivitis, Koplik spots followed by erythematous maculopapular rash; life threatening in malnourished children but rarely disabling or fatal in normal children	Effective vaccine since 1963 (MMR)
Rubella (German measles)	Rubella virus	Respiratory	3-day rash with mild illness, but a powerful teratogen	Effective vaccine since 1963 (MMR—measles, mumps, rubella)
Smallpox	Variola (smallpox virus)	Respiratory	Variola major produces high fever, severe blistering, scars, and often death; variola minor is mild form	Effective vaccine using cowpox virus discovered by Jenner in 1796
Warts	Human papillomavirus (HPV)	Unknown	Usually a cosmetic problem	Disappear without treatment; can be removed with liquid nitrogen

TABLE 26.5 Infections of the Body's Surfaces: Skin and Eye (continued)

Infection	Causative Agent	Mode of Transmission	Symptoms	Prevention and Treatment
Fungal				
Ringworm (athlete's foot)	Dermatophytes	Person-to-person, animal-to-person, fomites	Can be bright red weepy lesions or mild cracking and peeling of skin	Prevent by avoiding contact; treat with antifungals such as miconazole or, in resistant cases, oral griseofulvin
Candidiasis (thrush, vaginitis, chronic mucocutaneous candidiasis)	*Candida albicans*	Opportunistic	White plaques on mucous membranes; rash and warty skin growths in chronic mucocutaneous type; can be life threatening in immunocompromised hosts	Treat with topical antifungals containing nystatin or clotrimazole; amphotericin B used for systemic infections but seldom effective
Arthropod-caused				
Scabies	*Sarcoptes scabiei*	Close contact, fomites	Intense itching and rash	Treatment with the arachnicide gamma benzene
Pediculosis (lice)	*Pediculus* spp., *Phthirus pubis*	Close contact, fomites	Itching and skin irritation; body lice can be vectors for typhus	Effective treatment difficult because most lice are now resistant to gamma benzene
Eye Infections *Bacterial*				
Inclusion conjunctivitis	*Chlamydia trachomatis*	Usually occurs in infants who acquire infection at birth	Eyelid becomes swollen; purulent discharge	Prevent by treating mother for chlamydia, giving oral erythromycin; usually resolves spontaneously
Trachoma	*Chlamydia trachomatis*	Fomites including towels, water, or flies	Most common preventable blindness	Many antibiotics
Neonatal gonorrheal ophthalmia	*Neisseria gonorrhoeae*	Usually occurs in infants who acquire infection at birth	Swelling and purulent discharge	Prevent by treating mother; administer silver nitrate or antibiotic eye drop prophylaxis; treat infected newborns with ceftriaxone
Viral				
Viral conjunctivitis (pinkeye)	Many, including adenoviruses and enteroviruses	Direct contact, highly contagious; often occurs in school epidemics	Redness, discomfort; not serious	Resolves spontaneously without treatment
Herpes keratitis	Herpes simplex virus	Begins with introduction of HSV into the eye; associated with minor trauma	Irritation, tearing, little pain; can threaten sight	Usually self-limited; treat with iododeoxyuridine
Helminthic				
Onchocerciasis (River blindness)	*Onchocerca volvulus*	Black flies and buffalo gnats	Migrating microfilariae cause skin irritation, itching, and thickening; microfilariae that reach cornea cause blindness	If diagnosed early, treat with various antihelminthics, such as suramin and ivermectin
Loaiasis	*Loa loa*	Mango fly	Painful subcutaneous lumps; worms may be visible in conjunctivae; does not threaten vision	Treat with diethylcarbamazine; remove visible worms surgically

SUMMARY

Case History: A Lifelong Visitor (p. 632)

1. Chickenpox virus usually infects humans early in life, causing varicella (chickenpox) from which most people recover completely. But the virus remains in the body and can reactivate years later, causing zoster (shingles).

The Body's Surfaces (pp. 632–634)

2. The body's surfaces are the skin, the surface of the eye (the conjunctiva), and the cornea. The skin consists of an upper layer (epidermis) and a lower layer (dermis). Beneath is the subcutaneous tissue, composed of connective tissue and lymphatic vessels.

3. Skin infections are described by appearance. A lesion (any abnormality) can be described by color (erythema, petechiae, purpura) or texture and content (macules, papules, nodules, vesicles, bullas, pustules). Breaks in the skin include an erosion, ulcer, and crust.

4. Clinical syndromes of the eye include conjunctivitis (pinkeye), keratitis, and keratoconjunctivitis.

Skin Infections (pp. 634–649)

Bacterial Causes (pp. 634–642)

5. *Streptococcus pyogenes* (group A streptococcus) causes impetigo, erysipelas, and streptococcal gangrene. Transmission is by close contact or fomites. Virulent group A streptococci have M protein and produce toxins and destructive enzymes. Treatment is with penicillin or erythromycin, but drug-resistant strains are emerging.

6. *Staphylococcus aureus* causes folliculitis, impetigo, boils (carbuncles), and abscesses. Strains that produce exfoliative toxin cause staphylococcal scalded skin syndrome. Staphylococcal cellulitis is a diffuse infection that can be life threatening. Infec-

tion usually results from the presence of a foreign body in a wound or a person's being in a weakened condition. Treatment is with methicillin or cephalosporins; however, methicillin-resistant staphylococci have emerged. Pus-containing lesions must be lanced.

7. *Pseudomonas aeruginosa* is an opportunistic pathogen that causes most of the infections fatal to burn victims and 10 to 20 percent of nosocomial infections. Cancer and diabetes patients are at special risk of *Pseudomonas* infections, and children with cystic fibrosis carry a strain that clogs airways. Treatment is with aminoglycoside and a beta-lactam antibiotic, but it is rarely effective because patients are usually already seriously ill.

8. *Clostridium perfringens* causes gas gangrene, characterized by gas bubbles, crepitance, and dead, black skin. Death can occur within days from shock and renal failure. The only effective treatment is amputation. If location makes surgical removal impossible, treatment is with high-dose penicillin, clostridial antitoxin, and a hyperbaric oxygen chamber, but usually the infection is fatal.

9. *Propionibacterium acnes* is largely responsible for minor acne or cystic acne (large tender nodules deep in the skin that leave scars). Acne is an inflammatory disorder brought on by increased production of sebum in adolescence. Commensals break down the sebum, producing fatty acids, which initiate inflammation. Treatment is with tetracycline and a new drug called isotretinoin.

10. *Mycobacterium leprae* causes leprosy (also called Hansen's disease), an infection of the skin and peripheral nerves. Transmission is probably by infected nasal secretions and direct skin contact. Incubation is 3 to 5 years. The first stage is indeterminate leprosy, characterized by a few depigmented macules. If the immune system mounts an effective response, the person develops tuberculoid lep-

rosy, with minor lesions and often complete loss of sensation in the affected area. If the immune system is overwhelmed, the patient develops lepromatous leprosy, with extensive disfigurement and patchy loss of sensation. Treatment is with dapsone.

Viral Causes (pp. 642–646)

11. Varicella zoster virus causes an active infection—varicella (chickenpox)—and a latent infection that can reactivate later to become zoster (shingles). Varicella is a mild disease in children, characterized by epidermal blisters. Reye's syndrome, a life-threatening complication, is preventable by not giving aspirin. Varicella in adults can cause viral pneumonia. In pregnant women it can cause abortion, birth defects, or a fatally infected infant. Zoster usually occurs in only one or two dermatomes, but it can produce a painful postherpetic neuralgia. Immunization with varicella zoster immune globulin within 3 days of exposure is effective. Treatment with acyclovir on the first day is minimally effective. A varicella zoster vaccine called the Oka strain is now available.

12. Herpes simplex type 1 virus causes fever blisters and gingivostomatitis, which is characterized by fever and painful blisters on the mouth and gums. Gingivostomatitis spreads rapidly among young children. When HSV-1 infects the fingers, the infection is called herpetic whitlow. The active infection can be asymptomatic, so apparently healthy people shed the virus. Acyclovir can probably shorten the duration in adults, but it is not routinely used for children.

13. The measles (rubeola) virus causes measles, a serious and highly communicable disease. The virus is inhaled. Symptoms, including Koplik spots, usually appear on day 11. By day 14 the rash appears and begins to move from the face down the body. Complications include bacterial pneumonia and brain damage. Immunization in the United States

is with the measles-mumps-rubella (MMR) vaccine at 15 months and at age 5 upon entering school. A new vaccine is needed for children under 1 year who are not protected by the available vaccine. Theoretically, because humans are the only reservoir, it should be possible to eradicate measles.

14. Subacute sclerosing panencephalitis (SSPE) is a rare but usually fatal form of measles. The child appears to recover completely from a normal case of the measles, but measles virus replicates in brain cells. Five or more years later, symptoms of severe brain damage appear. There is no treatment.

15. Rubella virus causes rubella (German measles), which is not related to rubeola. In children, rubella is a mild illness; the rash lasts 3 days. Infection during the first 3 months of pregnancy, however, can cause abortion or birth defects. Vaccination is with MMR at 15 months. Rubella serology is routinely performed at the first prenatal exam.

16. Smallpox caused by variola virus, once a disfiguring and deadly disease, was eradicated worldwide in 1979. Now there's a possibility it could return if used by bioterrorists.

17. Certain strains of human papillomavirus (HPV) cause common warts and plantar warts. Mode of transmission is unknown. Common warts can be removed by surgery or keratolytic chemicals, though they usually disappear without treatment after a few years. A rare syndrome of persistent warts is associated with skin cancer.

Fungal Causes (pp. 646–648)

18. Dermatophytes are fungi that cause tinea (ringworm), an annoying but superficial infection. Dermatophytes digest keratin. Men are more commonly affected than women. Treatment is with topical antifungal agents such as miconazole or clotrimazole or orally with griseofulvin.

19. *Candida albicans* is an opportunistic pathogen that causes candidiasis. The most common sites of infection are the vagina (*Candida* vaginitis), the mouth (thrush), and the diaper area (*Candida* diaper dermatitis). Treatment is with antifungal creams; thrush is treated with a topically applied liquid. In chronic mucocutaneous candidiasis, the yeast form grows as an intracellular parasite. Systemic candidiasis strikes immunodeficient patients; treatment with amphotericin B is rarely effective.

Arthropod Causes (pp. 648–649)

20. *Sarcoptes scabiei* is a mite that causes scabies, characterized by severe itching, especially at night. The disease is a hypersensitivity reaction to the mite and its feces. Secondary infections develop. Epidemics occur at 15-year intervals and last about 15 years. Treatment is with gamma benzene (Crotamiton or Kwell).

21. Three types of human lice cause pediculosis: *Pediculus humanus capitis* (the head louse), *Pediculus humanus corporis* (the body louse), and *Phthirus pubis* (the pubic or crab louse). All types cause itching and can lead to bacterial superinfections. Head lice among children has reached epidemic proportions in the United States. Body lice are the only ones that can transmit microorganisms that cause epidemic disease. Pubic lice are transmitted by sexual contact or sometimes fomites. Lice can be killed by gamma benzene or pyrethrins.

Eye Infections (pp. 649–651)

22. *Streptococcus* spp., *Staphylococcus aureus*, and *Haemophilus* spp. commonly cause bacterial conjunctivitis. Symptoms include conjunctival redness and a pus-containing discharge; vision is not threatened. Treatment is with sulfacetamide, neosporin, or gentamicin. Contact lens conjunctivitis is usually caused by *Pseudomonas aeruginosa*. Vision can be threatened.

23. Strains of *Chlamydia trachomatis* that infect the genital tract cause inclusion conjunctivitis in newborns. Infection can be prevented by treating the mother during pregnancy. Other strains cause trachoma, a preventable blindness found in Africa, Asia, the Middle East, and among Native Americans in the United States. Chlamydiae multiply in the conjunctivae, causing inflammation, bumps on the surface, scarring, and blindness if the cornea is affected or the eyelids cannot close. Organisms are shed from the eye. Transmission is by direct contact or fomites. Erythromycin and tetracycline can control infection but cannot eradicate the organism. High-risk populations can use antibiotic ointments prophylactically.

24. *Neisseria gonorrhoeae* causes neonatal gonorrheal ophthalmia, once a common cause of blindness. The infection is prophylactically prevented in industrialized countries by routinely treating newborns with erythromycin or silver nitrate.

25. Viral infections of the conjunctivae are common and rarely serious. Viral conjunctivitis tends to cause more redness and a less purulent discharge than bacterial conjunctivitis. The infection resolves by itself in a few days; eyesight is not threatened.

26. Herpes simplex virus type 1 causes herpetic keratitis, which can be serious because the cornea is involved. Usually a minor trauma produces an infection that resolves spontaneously but recurs; eye drops with iododeoxyuridine can shorten the time of infection. If deeper layers of the cornea are involved, scarring and blindness can occur. If a misdiagnosis leads to prescribing steroid eye drops, the infection can spread rapidly and destroy the cornea. Sight can be restored only by a transplant.

27. *Onchocerca volvulus* is a nematode that causes onchocerciasis (river blindness) in Africa and Latin America. Infection is transmitted by black flies and buffalo gnats that deposit larval worms when they bite. The

larvae mature into nematodes, which produce microfilariae that migrate through subcutaneous tissue, causing itching, rash, or thickening of the skin. Microfilariae that reach the cornea cause blindness. If diagnosed in time, treatment with suramin, diethylcarbamazine, or mebendazole will kill the adult worms; ivermectin kills the microfilariae.

28. *Loa loa* is a nematode that causes loaiasis in Africa. Transmission is by the mango fly. Infection can be asymptomatic, or migrating adult worms can cause painful lumps in the skin (Calabar swellings); worms that migrate through the conjunctivae are visible. Treatment is with diethylcarbamazine; worms in the eye can be removed surgically.

REVIEW QUESTIONS

Case History: A Lifelong Visitor

1. Describe the connection between chickenpox and zoster.

The Body's Surfaces

2. Name and describe the layers and sublayers of the skin. What parts of the eye does the conjunctiva cover?

3. How are infections of the skin described? Define these terms:
   a. lesion   i. vesicle
   b. erythema   j. bulla
   c. petechiae   k. pustule
   d. purpura   l. erosion
   e. macules   m. ulcer
   f. papules   n. crust
   g. nodules   o. exanthem
   h. maculopapular   p. enanthem

4. Describe these clinical syndromes of the eye: conjunctivitis, keratitis, keratoconjunctivitis.

Skin Infections

Bacterial Causes

5. Describe the virulence factors of group A streptococcus. Name and describe streptococcal skin infec-

tions and their transmission, prevention, and treatment.

6. What are the virulence factors of *Staphylococcus aureus*? Name and describe staphylococcal skin infections and their transmission, prevention, and treatment.

7. Why are *Pseudomonas* infections difficult to control, especially in a hospital? Which groups are at highest risk of *Pseudomonas* infection? Describe the treatment.

8. How does *Clostridium perfringens* cause infection? Describe the symptoms and treatment of gas gangrene.

9. Explain this statement: Acne is not an infection. What dangers are involved with using the new drug isotretinoin?

10. Describe *Mycobacterium leprae*. Describe the development of leprosy, its treatment, and its epidemiology in the United States.

Viral Causes

11. Compare and contrast the seriousness of varicella infection in children, adults, and pregnant women. Describe the pathogenesis of shingles. Discuss the prevention and treatment of chickenpox, including the risks and benefits of the new Oka strain vaccine.

12. Describe the infections caused by herpes simplex type 1. How is HSV-1 transmitted?

13. Describe the transmission, development, basis for recovery, and potential complications of rubeola (measles). Why is measles a childhood killer in developing countries, and what are some solutions? How is measles controlled in the United States? Why is measles eradicable?

14. Describe the pathogenesis of subacute sclerosing panencephalitis (SSPE).

15. Compare and contrast rubella (German measles) and rubeola (true measles). What can be done to prevent congenital rubella?

16. Describe the infections caused by human papillomavirus and their

treatment. Why is there new research interest in HPV?

Fungal Causes

17. What are dermatophytes, and what infections do they cause? Explain this statement and give some examples: The body's immune response influences both the symptoms of dermatophyte infections and the chances for complete recovery.

18. Name and describe the major skin infections caused by *Candida albicans*. What is the treatment? Who is most at risk for chronic mucocutaneous candidiasis? For systemic candidiasis?

Arthropod Causes

19. Name the microorganism that causes scabies and describe the pathogenesis. Explain how it damages the skin.

20. What are the most common forms of pediculosis, and how is each transmitted and treated?

Eye Infections

21. What are the most common causes of bacterial conjunctivitis? What is the treatment? How does *Pseudomonas aeruginosa* cause eye infections, and how serious are they?

22. Describe the cause, prevention, and treatment of inclusion conjunctivitis. Discuss the transmission, pathogenesis, prevention, and treatment of trachoma. In what parts of the world is trachoma most prevalent?

23. Describe the cause, prevention, and treatment of neonatal gonorrheal ophthalmia.

24. Discuss the transmission, pathogenesis, clinical syndrome, and treatment of onchocerciasis.

25. What are the symptoms of loaiasis? What kind of organism causes this infection? What is the treatment?

CORRELATION QUESTIONS

1. Compare *Streptococcus pyogenes* and *Staphylococcus aureus* with respect to appearance of cells, susceptibility to phagocytosis, and susceptibility to antibiotic therapy.

2. Just by looking at it, how could you tell whether a pus-exuding wound was infected with *Pseudomonas aeruginosa* or *Staphylococcus aureus*?

3. Name a skin condition for which the fatty acid in sebum is beneficial and one for which it potentiates disease.

4. How could you distinguish between gangrene caused by *Streptococcus pyogenes* and that caused by *Clostridium perfringens*?

5. If you were a public health officer in charge of an area in which a war had just ended, would you be more concerned about infestations of *Pediculus humanus corpus* or of *Phthirus pubis*? Why?

6. What is the common mechanism by which microbial infections cause blindness?

ESSAY QUESTIONS

1. Discuss the various factors that should be considered before recommending a vaccination program for chickenpox.

2. Discuss what would have to be done to eradicate measles worldwide.

SUGGESTED READINGS

Advisory Committee. 1989. Measles prevention: Recommendations of the immunization practices. *Morbidity and Mortality Weekly Report* 38:S–9.

Brumfitt, W., and J. Hamilton-Miller. 1989. Methicillin-resistant *Staphylococcus aureus*. *New England Journal of Medicine* 320:1188–96.

Chickenpox: Reexamining our options. 1991. *New England Journal of Medicine* 325:1577–79.

Howley, P. M. 1986. On human papillomaviruses. *New England Journal of Medicine* 315:1089–90.

Lobess, E., N. Weiss, M. Karanis, H. Taylor, E. Ottesen, and T. Nutman. 1991. The immunogenic *Onchocerca volvulus* antigen: A specific and early marker of infection. *Science* 251:1603–5.

Shepard, C. C. 1982. Leprosy today. *New England Journal of Medicine* 397:1640–41.

Tucker, J. B. 2001. *Scourge. The once and future threat of smallpox*. New York: Atlantic Monthly Press.

Weller, T. H. 1983. Varicella and herpes zoster: Changing concepts of the natural history, control and importance of a not-so-benign virus. *New England Journal of Medicine* 309:1362–68; 1434–40.

For additional readings, go to InfoTrac College Edition, your online research library at:
http://www.infotrac.thomsonlearning.com

TWENTY-SEVEN

Systemic Infections

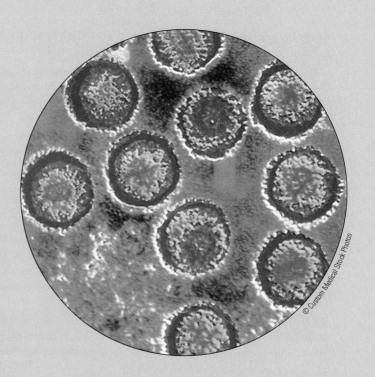

© Custom Medical Stock Photos

CHAPTER OUTLINE

LEARNING GOALS

To *understand*:

- *The anatomy and function of the cardiovascular and lymphatic systems and their defenses against microorganisms*

- *The clinical syndromes that characterize cardiovascular and lymphatic infections*

- *The principal microbial infections of the heart; their diagnosis, prevention, and treatment*

- *The bacterial, viral, protozoal, and helminthic causes of systemic cardiovascular and lymphatic infections; their diagnosis, prevention, and treatment*

- *What is known about the human immunodeficiency virus (HIV), the clinical syndromes of AIDS, and its transmission, prevention, and treatment; how the epidemiology of AIDS is changing and what future prospects might be*

Risky Business

C. D. was a 24-year-old woman who had been using heroin intravenously for several years. She felt well until about 3 weeks before her admission to the hospital. First she noticed she was running a fever, but she decided to ignore it. She became increasingly weak, tired, and pale. Then she noticed red spots, some slightly raised, on her palms and soles. Numerous splinterlike discolorations appeared under her fingernails. She began to cough. Breathing became difficult. When she became so weak she could hardly walk, her friends insisted she go to the emergency room.

The attending physician was concerned when he learned of C. D.'s intravenous drug abuse and continuing fever. He questioned her carefully about a history of heart disease but discovered nothing. She had a temperature of 102°F. She was very pale (laboratory tests later confirmed anemia). The discolorations on her palms, on her soles, and under her fingernails suggested that **emboli** (tiny clots) had entered her bloodstream and were clogging small vessels. Most worrisome was her significant heart murmur, suggesting damaged heart valves. C. D. showed signs of heart failure. She was admitted to the intensive care unit with

a diagnosis of probable bacterial endocarditis.

Her physician drew several blood samples for bacterial culture and ordered an **echocardiogram** (a sound-wave image of the heart). It revealed **vegetations** (abnormal growths) on several severely damaged heart valves. The culture results would not be available for several days, but the clinical presentation strongly suggested endocarditis. So C. D. was immediately treated with a combination of drugs, including methicillin. The tentative diagnosis of bacterial endocarditis was confirmed later when *Staphylococcus aureus* was found to be present in C. D.'s blood sample.

C. D.'s physician hoped to control her infection with antibiotics and stabilize her condition so that she would be strong enough to undergo surgery

necessary to replace her damaged heart valves. But C. D. died 3 days later.

Case Connections

- Which of C. D.'s symptoms led her physician to suspect endocarditis?
- Why was C. D.'s infection seemingly resistant to antibiotic therapy?
- Why did C. D. have a cough when her heart was infected?
- What was the immediate case of C. D.'s death?
- We'll consider all these questions in this chapter.

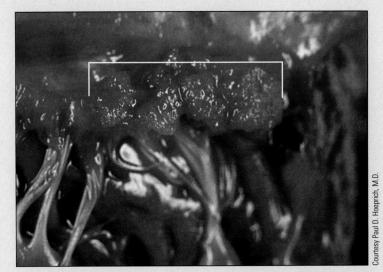

A postmortem photo of heart valves from a patient with endocarditis shows a large vegetation adhering to a valve (white bracket).

Courtesy Paul D. Hoeprich, M.D.

THE CARDIOVASCULAR AND LYMPHATIC SYSTEMS

The cardiovascular and lymphatic systems transport fluids through the body. Because these two systems interconnect all the organs and tissues of the body, they can become a deadly distribution network for pathogens.

Structure and Function

The cardiovascular system carries blood from the heart to the tissues and back again. In the process, some fluid is lost to the tissues. The lymphatic system recovers this lost fluid and returns it to the bloodstream. It also helps fight microbial infection (**Figure 27.1**).

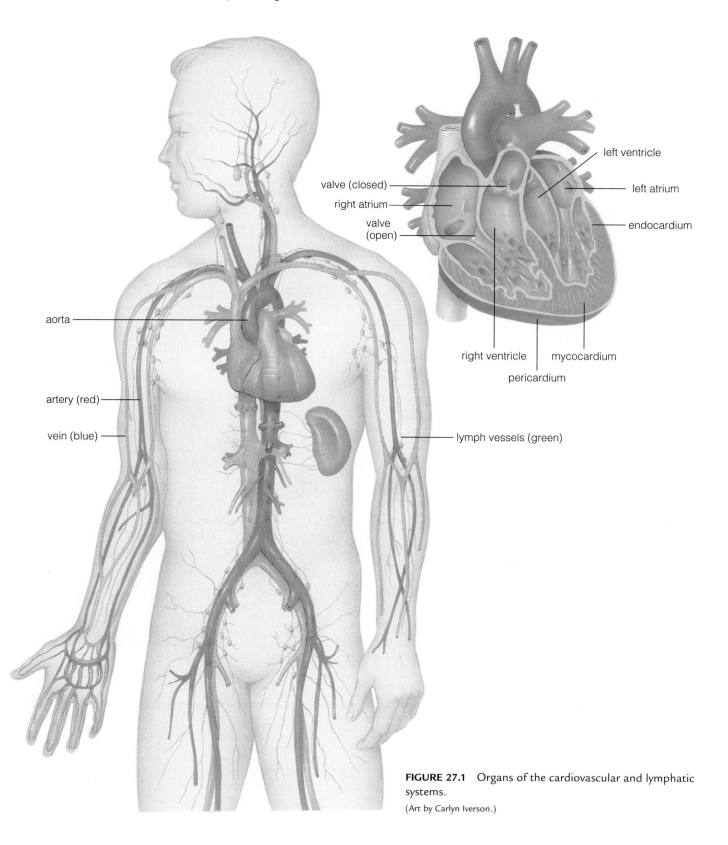

valve (closed)
right atrium
valve (open)

left ventricle
left atrium
endocardium

right ventricle
pericardium
mycocardium

aorta

artery (red)

vein (blue)

lymph vessels (green)

FIGURE 27.1 Organs of the cardiovascular and lymphatic systems.

(Art by Carlyn Iverson.)

The Cardiovascular System. The cardiovascular system consists of the heart, blood vessels, and blood. The heart is a four-chambered muscle that pumps blood through the blood vessels. The upper two chambers are thin-walled **atria** (*sing.*, atrium). They receive blood returning to the heart. The lower chambers are the thick-walled **ventricles,** which, when the heart contracts, pump blood out of the heart. Each of the four chambers has a **valve** at its outlet to prevent blood from flowing backward when the heart relaxes.

The right side of the heart receives blood from the body's tissues and pumps it at low pressure to the lungs. The left side receives blood from the lungs and pumps it at higher pressure to the tissues. Thus there are two circulations: the **pulmonary circulation,** which goes to the lungs, and the **systemic circulation,** which goes to all the other tissues. In the lungs, oxygen is absorbed into the blood and carbon dioxide (CO_2) is released from it. The body's tissues remove this oxygen from the blood and add carbon dioxide to it.

Oxygen-rich blood that leaves the left ventricle to enter systemic circulation first passes into the **aorta,** the largest of the thick-walled **arteries** that carry blood to the tissues. Blood then enters a network of arteries. As blood travels farther from the heart, it enters arteries that branch and become smaller. Finally they become so small that only one blood cell at a time can pass through. These microscopic blood vessels, called **capillaries,** penetrate all the tissues of the body. Their thin, highly permeable walls allow nutrients and oxygen to enter the tissues and carbon dioxide and other metabolic wastes to pass into the blood. Then blood carrying waste products flows from the capillaries into small thin-walled **veins.** These join to form into larger and larger veins, which lead back to the heart, where blood is pumped into the pulmonary circulation. In its transit through the systemic circulation, the blood passes through the kidneys, where metabolic wastes are removed, and through networks of capillaries in the walls of the gastrointestinal system, where nutrients are absorbed.

Blood itself is a complex tissue. It's a mixture of erythrocytes, leukocytes, platelets, proteins, dissolved gases, and nutrients (see Sharper Focus: Blood: A Complex Body Tissue, in Chapter 16).

The Lymphatic System. The lymphatic system, also called the immune system, was described in detail in Chapter 17. The location and function of lymph vessels, nodes, organs, and tissues are shown in Figures 17.2 and 17.3 and Table 17.2.

Defenses and Normal Biota: A Brief Review

The circulatory and lymphatic systems are normally sterile, but they occasionally they become contaminated by microorganisms for brief periods. One of the functions of lymphatic circulation is to deliver contaminating microorganisms to a lymph node, where they can be phagocytized. In this way, most microbes are eliminated before serious infection begins. Such transient **bacteremias** (bacteria in blood) are usually brief and asymptomatic. But if they persist, a true infection results.

Clinical Syndromes

Infections of the heart are categorized by anatomical syndrome (**Table 27.1**). **Endocarditis** is infection of the **endocardium,** the heart's inner lining. **Myocarditis** is infection of the **myocardium,** the heart muscle itself. **Pericarditis** is infection of the **pericardium,** the heart's surrounding membrane.

TABLE 27.1 Clinical Syndromes of Cardiac Infection

Syndrome	Region Affected	Signs and Symptoms	Causative Agent
Endocarditis	Endocardium—the smooth inner lining of the heart, including valves	Fever, fatigue, anemia, heart murmurs, or heart failure from valve damage	Acute bacterial infections, usually from highly virulent pathogens; subacute bacterial or fungal infections from less virulent or opportunistic pathogens
Myocarditis	Myocardium—the heart muscle	Irregular heartbeat, chest pain, heart failure	Usually viral, particularly coxsackievirus, but may be fungal, bacterial (e.g., diptheria toxin myocarditis), or protozoal (Chagas' disease)
Pericarditis	Pericardium—the smooth membrane surrounding the heart	Few; in severe cases, scarring or pericardial tamponade may interfere with heart function	Usually viral but may be bacterial or fungal

Blood and lymph contact every organ. So when they become infected, the entire body is affected. These body-wide infections, called **systemic infections,** are difficult to diagnose. A patient who says "I can't breathe" probably has a respiratory infection. One who says "I have diarrhea" probably has a gastrointestinal infection. But a patient with a systemic infection is likely to say "I don't feel well." That's not much for the clinician to go on. However, each systemic infection has specific symptoms. These, along with laboratory tests, make it unique and identifiable.

INFECTIONS OF THE HEART

Infections of the heart are uncommon, but they can be serious. Any infection of this vital organ has the potential to be life threatening.

Endocarditis

Most cases of endocarditis are caused by bacteria, but sometimes fungi are responsible. The bacterial endocarditis that killed C. D. is one of many serious infections that can result from intravenous drug abuse, because nonsterile needles introduce microorganisms directly into the bloodstream.

Bacterial Endocarditis. Microorganism must adhere to the endocardial surface to cause endocarditis. Adherence is unlikely because this surface is so smooth. Therefore, with a normal heart, endocarditis occurs only if large numbers of virulent pathogens are in the blood. This occurred in C. D.'s case. But if the endocardial surface is roughened, from a congenital defect for example, smaller numbers of microorganisms can start an infection.

Once bacteria adhere to the endocardium, they multiply and become surrounded by clotlike material from the blood. These **vegetations** (bulky masses of bacteria and clots), like those shown on C. D.'s echocardiogram, are a hallmark of endocarditis. Bacteria within a vegetation are protected from the body's defenses and from bloodborne antibiotics. As a consequence, endocarditis requires a prolonged course of antimicrobial therapy.

Vegetations cause more problems than just protecting imbedded bacteria. As they grow, pieces break off, forming **emboli.** These can lodge in small blood vessels and interrupt the blood supply to organs. Emboli from vegetations on the right side of the heart enter the pulmonary circulation and can damage the lungs. Such pulmonary emboli probably caused C. D.'s cough and breathing problems. Emboli from vegetations on the left side of the heart enter the systemic circulation. They caused the

spots on C. D.'s skin and nails. Emboli can also cause death if they block the blood supply to vital organs such as the kidneys or the brain.

Vegetations also damage the endocardium itself. Heart valves—which are composed of endocardium—are particularly vulnerable. If they don't function properly, the heart cannot maintain adequate blood pressure to supply vital organs. This condition, called **heart failure,** was the immediate cause of C. D.'s death.

Bacterial endocarditis is difficult to diagnose because clinical signs and symptoms depend upon which heart valves are affected and where the emboli lodge. But regardless of which site is affected, endocarditis patients usually suffer fever and anemia. Definitive diagnosis depends upon isolating the infecting microorganism from the patient's blood.

Acute and Subacute Endocarditis. Bacterial endocarditis can be either acute or subacute.

The onset of **acute bacterial endocarditis** is sudden, and it progresses rapidly. C. D.'s illness was typical. Her heart was normal until virulent staphylococci (probably from her skin) were introduced into her bloodstream during an intravenous injection of drugs. In fact, *Staphylococcus aureus* causes more than half the cases of bacterial endocarditis in intravenous drug users. Non–drug users can also get bacterial endocarditis, but only if significant numbers of a virulent pathogen enter the bloodstream, from a postoperative infection, for example. Acute bacterial endocarditis is usually treated with cephalosporins and penicillinase-resistant penicillins. After the infection has been eliminated, the damaged heart valves must be replaced surgically. Even with treatment, extremely ill patients such as C. D. may die a few weeks after symptoms appear.

Subacute bacterial endocarditis usually occurs in people who have an abnormal heart. The abnormality, even if it's minor, provides a spot for bacteria to adhere. A subacute infection starts gradually and progresses slowly. Patients may complain of fever and fatigue for months. Often several blood cultures are needed to isolate the infecting organism, which usually turns out to be a weakly pathogenic bacterium from the respiratory or gastrointestinal microbiota. Small numbers of these bacteria enter the blood during routine activities of daily living. For example, about one-third of people who have a tooth extracted have streptococci in their bloodstream afterward. Normally these bacteria are rapidly eliminated. But people with heart defects are at risk of developing bacterial endocarditis. They are advised to take penicillin before dental work or other minor surgery. With early diagnosis, subacute bacterial endocarditis is usually curable. Without treatment, it is invariably fatal.

Myocarditis

In the United States, myocarditis is usually viral. In South America it is usually protozoal. Fungi and bacteria can also cause myocarditis. Diphtheria toxin, for example, is highly poisonous to the heart. So untreated diphtheria patients often die of myocarditis.

Viral Myocarditis.

Many viruses cause myocarditis. Usually the specific one is not identified, but coxsackievirus is the most common cause of the disease. Patients with viral myocarditis generally have mild symptoms. They experience only a rapid or slightly irregular heartbeat. Chest pain or signs of heart failure may develop, but they are rare. The only treatment is rest. Most patients do well. Viral myocarditis, however, can be life threatening. It's a likely cause of the rare sudden deaths associated with minor viral infections.

American Trypanosomiasis (Chagas' Disease).

The flagellated protozoan *Trypanosoma cruzi* (**Figure 27.2**) causes an infection called **American trypanosomiasis,** or **Chagas' disease,** named after Carlos Chagas, who first described the disease in Brazil in 1909.

T. cruzi has a complicated life cycle during which it passes through two different hosts—a mammal and a blood-sucking insect. The mammal can be any of more than 100 different species, including humans, dogs, and cats. The most common insect hosts are reduviid bugs (*Triatoma* spp.), commonly called kissing bugs because of their tendency to bite humans on the face. Trypanosomes pass from an infected mammal to an insect during a blood meal. They're passed to humans, when the infected insect bites and at the same time defecates on the victim's skin. Protozoa in the feces enter the bite when the person rubs or scratches it. Infection can also enter through the eye.

Trypanosoma cruzi exists in many parts of the Western Hemisphere, including the southern United States. But almost all cases of Chagas' disease occur in rural parts of Latin America, where people live in houses infested by reduviid bugs. In some areas of South America, nearly half the population is infected. Chagas' heart disease is the most common cause of death among young adults in these regions.

Chagas' disease passes through an acute and a chronic stage. The acute form appears one to two weeks after the infecting insect bites. During this period, protozoa enter many types of cells—brain cells, liver cells, and myocardial cells—multiply there, and kill them. The victim suffers fever, fatigue, a tender swelling at the site of the bite, and a characteristic swelling of the face and eyelids. Children are the most vulnerable. As many as 10 percent of them die. But at this acute stage, the infection can be successfully treated with antiparasitic medications. The drug of choice is nifurtimox.

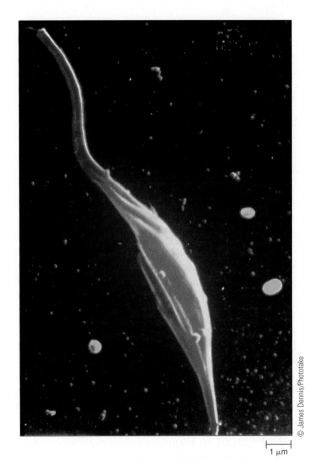

FIGURE 27.2 Scanning electron micrograph of the flagellated, bloodborne form of *Trypanosoma cruzi* that causes American trypanosomiasis.

Those who survive the acute stage without treatment usually suffer the chronic form of the disease. Parasites enter the lymphatic system and from there can spread to many organs, including the esophagus (causing difficulty swallowing), the brain (causing paralysis), and the myocardium (causing death). The only treatment is medication to sustain the failing heart.

In its acute stage, when there are many protozoa in the blood, Chagas' disease can often be diagnosed by examining blood smears. Later, when most parasites are inside cells, diagnosis is more difficult. No serological test is completely satisfactory. Sometimes a suspected patient is allowed to be bitten by an uninfected reduviid bug, and the insect is examined several weeks later.

Pericarditis

Pericarditis is usually viral, but it can be bacterial or protozoal. Most viral pericarditis is a complication of an everyday viral infection. Enteroviruses are the most common cause. Like viral myocarditis, viral pericarditis is usually benign and self-limited.

Bacterial pericarditis is much more serious. It's usually the result of an infection by a *Staphylococcus* or *Streptococcus* spp. that spreads to the pericardium from the lungs or the bloodstream. Pus can fill the space between the pericardium and the myocardium. In so doing it exerts enough pressure on the heart to interfere with its normal beating. This condition, called **pericardial tamponade,** can cause sudden death.

Chronic fungal or mycobacterial infections can also lead to pericarditis. Immunosuppressed patients suffering from opportunistic fungal infections and chronic tuberculosis patients are especially vulnerable. Chronic pericardial inflammation produces scarring, which eventually causes the heart to constrict until it fails. In such cases, **pericardectomy** (surgical removal of the pericardium) may be lifesaving.

SYSTEMIC INFECTIONS

Systemic infections are those in which microorganisms enter the blood or lymphatic circulations or are spread through them. Any bacterial infection—no matter where it begins—can become systemic.

If the bacterial invasion of the bloodstream is brief and harmless, it's called **bacteremia.** If it's persistent and serious, it's called **septicemia.** Usually septicemia is fatal, even with the best medical treatment. Most cases are immunocompromised patients with nosocomial infections.

Bacterial Infections

Bacterial systemic infections include plague, one of the oldest and most dreaded human diseases, and Lyme disease, one of the most recently discovered.

Plague. We tend to associate plague with Europe in the Middle Ages, when it killed between a quarter and a third of the population. But this disease has occurred in many places throughout history, sometimes in devastating epidemics (Chapter 1). For example, 6 million people died in India during an outbreak that began in 1891, and a pandemic was narrowly averted in the early 1900s. Plague is permanently established among rodents, including ground squirrels, prairie dogs, and chipmunks, in North America west of the Rocky Mountains. Fortunately, humans have little contact with these rodents, so transmission is relatively rare. Only about a dozen cases of plague are reported annually in the United States. Worldwide there are 1000 to 2000. About 1 in 7 persons with plague dies.

Plague is caused by *Yersinia pestis*, a Gram-negative coccobacillus (Bergey's Gammaproteobacteria, Chapter 11; **Figure 27.3a**). It's transmitted from one mammalian host to another by infected fleas. Infection blocks the flea's digestive tract, causing it to regurgitate infected material into a bite wound during its blood meal. Because the flea's intestinal blockage keeps it from taking in a complete meal, it bites more frequently and aggressively. Some communities of wild rodents are chronically infected. Because they are highly resistant to plague, few animals die.

Y. pestis is one of the most virulent bacterial pathogens known. The disease is horrible. Human are usually infected when an infected flea bites a person, introducing bacteria into subcutaneous tissues. Then the bacteria enter the lymphatic system and become concentrated in a nearby lymph node. Two to six days later, the person develops a high fever, and the affected lymph node becomes massively enlarged and tender. The diseased node is called a **bubo (Figure 27.3b),** and the fleaborne infection is called **bubonic plague.** Bacteria are phagocytized within the node, but they aren't killed. They continue to multiply within the phagocyte.

Then, in almost all untreated cases, bacteria enter the blood—a stage called **septicemic plague.** Once bloodborne, bacteria spread to almost every organ of the body, usually with fatal consequences. Blood vessels are destroyed, resulting in subcutaneous bleeding that causes black spots on the skin (**Figure 27.3c**). These are the basis of the name **Black Death.**

If the lungs become infected, bacterial cells are aerosolized when the infected person coughs. If another person inhales these organisms, he or she can suffer an infection called **pneumonic plague.** It's even more virulent than bubonic plague. Symptoms begin within 2 to 3 days, and mortality is nearly 100 percent. Pneumonic plague is considered to be a potential biological weapon.

Plague is diagnosed by culturing *Y. pestis* from a bubo, blood, or sputum or by staining these samples with fluorescent antibodies. For treatment, the drug of choice is streptomycin, but a number of other antibiotics are also effective. The best way to prevent plague is to eliminate rats from human housing. A vaccine is also available for high-risk individuals, such as biologists or geologists who work in plague-infested areas.

Tularemia. *Francisella tularensis*, a small Gram-negative coccobacillus (Bergey's Gammaproteobacteria; Chapter 11), was first isolated in 1911 from ground squirrels that appeared to have died of plague in Tulare County, California. Today we know that **tularemia,** the disease it causes, occurs worldwide among wild mammals. Occasionally it affects human beings, but it commonly infects rabbit, squirrels, muskrats, and deer. As many as 1 percent of the wild rabbits

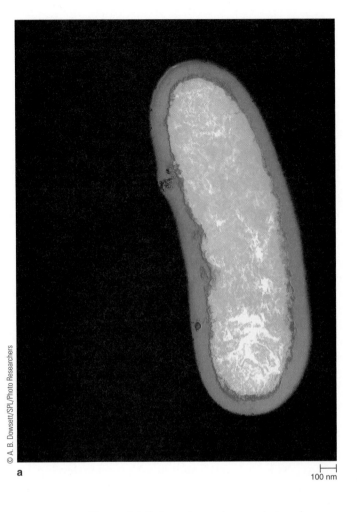

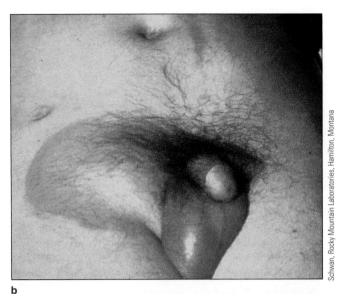

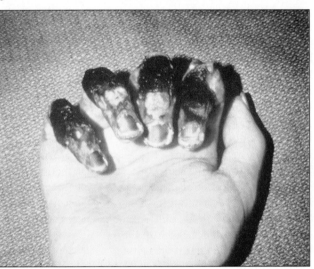

FIGURE 27.3 Plague. (a) Color-enhanced transmission electron micrograph of *Yersinia pestis.* (b) An enlarged lymph node (bubo) in a patient who was probably bitten in the leg by an infected flea. (c) Hand of patient with septicemic plague with characteristic blackened spots.

in North America harbor *F. tularensis.* Ticks are also a significant reservoir. They can transmit the bacterium **transovarially** (through the eggs) from one generation to the next.

Humans acquire tularemia in several ways. Bacteria from a diseased animal enter the body through a break in the skin or across the conjunctiva—when a hunter skins or cleans a diseased animal, for example. Bacteria can also be ingested in infected meat, or they can be inhaled in laboratories where the organism is being cultured. In addition, they can be transmitted by a tick or fly that has recently fed on an infected animal.

The most common clinical syndrome is **ulceroglandular tularemia.** A small sore appears where the bacteria enter the body, and the surrounding lymph nodes become swollen and tender. Bacteria are phagocytized in the lymph node, but they aren't killed. Instead, they multiply within

the phagocyte and enter the bloodstream. Systemic symptoms include high fever, shaking chills, and debilitating headaches. Severely ill persons may develop pneumonia. Only about 10 percent of untreated patients die, but the illness can be serious and persistent. Fever often lasts for more than a month, and relapses usually occur. Less common clinical syndromes of tularemia include **glandular disease** (lymph nodes are swollen but no ulcer is visible), **typhoidal disease** (only systemic symptoms are evident), and **oculoglandular disease** (infection begins in the eye).

Diagnosing tularemia is difficult because the symptoms are so common and nonspecific, but a history of hunting or a tick bite can provide a good hint. Diagnosis by culturing is difficult because most standard media don't support the growth of *F. tularensis.* In addition, culturing is hazardous because of the risk of airborne infection, so diagnosis is

CASE HISTORY

A Near Miss

In 1999 N. E., an 18-month-old boy, was taken to an urgent care pediatric clinic because of fever. He had seemed well the day before, but during that night he woke unexpectedly. His mother noticed he was quite warm. She got him up to give him medication for his fever and noticed he had a shaking chill. When she took his temperature, it was 104°F. Through the rest of the night and the next morning N. E.'s fever remained high. His mother continued to give him medication every 4 hours to relieve the fever, and she gave him lukewarm baths to help cool him. But when the medication wore off, N. E.'s fever continued to be quite high. When his temperature was above 103°F, N. E. was listless and looked ill. When his temperature went down, he perked up a little. The only symptom he had was fever.

When the doctor at the clinic examined N. E., he found a feverish, lethargic toddler. He would look up when his mother spoke to him, but he didn't have the energy to resist being examined. He was offered a bottle of juice, but he wasn't interested. Otherwise he appeared to be fairly normal. He wasn't dehydrated. He did have some nasal congestion, but his ears weren't infected and his throat wasn't red. He was breathing a little more rapidly than normal, due to his fever, but his breathing wasn't labored and he had normal, clear breath sounds wherever the doctor listened. His skin looked normal—no rash—and his abdomen was soft. It didn't seem to hurt when the doctor pressed on it.

In spite of N. E.'s examination being fairly normal, the doctor decided to order some laboratory and x-ray studies. Anytime a child under the age of 2 has a very high fever with no apparent cause, a doctor must consider the possibility of bacteremia (infection of the bloodstream). By far the most common case cause these infections is *Streptococcus pneumoniae*, the pneumococcus.

The doctor ordered several tests: a chest x ray, a urine analysis, and a complete blood cell count with differential. The chest x ray showed no evidence of pneumonia. The urinalysis gave no evidence of a urinary tract infection. The blood count, however, was abnormal. N. E.'s total white blood cell count was extremely elevated at 24,000. Most of these cells were granulocytes, and many were immature. These results indicated that N. E.'s immune system was gearing up to fight some sort of infection. Based on this evidence, N. E. was given an injection of a long-acting, broad-spectrum antibiotic (ceftriaxone) and sent home. A blood culture that had been drawn along with the blood count was also sent to the laboratory. N. E.'s mother was instructed to return to the clinic the next day if he was not better.

N. E.'s fever continued throughout the night. But the next morning he was much better. His mother decided not to return to the clinic because her son was clearly getting over his illness. The next day, however, she received an urgent call from the pediatric clinic. N. E.'s blood culture had grown *Streptococcus pneumoniae*. The doctor wanted him to return to the clinic immediately. When they arrived at the clinic, N. E. looked much better than he had before. He was active, had no fever, and fought with the doctor who was trying to look into his ears and down his throat. His breathing was normal. His neck was not stiff. He wanted to get down and run around the examining room whenever he had the opportunity. A repeat while blood cell count was now 8000—well within the normal range.

The doctors concluded that N. E.'s pneumococcal bacteremia had resolved. The bacteria had been cleared from his bloodstream, and he had not developed complications, such as meningitis or pneumonia, that can occur when bacteria of the bloodstream settle in the meninges and lungs. Maybe the antibiotic helped N. E. clear the bacteria from his bloodstream. Maybe the clearing would have occurred without treatment, as it sometimes does. But pneumococcal bacteremia can be fatal. Children with it often develop meningitis and die, even with the most aggressive medical treatment. Now that *Haemophilus influenzae* has been nearly eliminated by immunization, pneumococcal meningitis is the most common type of fatal meningitis. N. E. was certainly lucky to have come through his pneumococcal infection unscathed.

Case Connections

- *Streptococcus pneumoniae* usually causes specific infections of the lungs, ears, or meninges. What principle of systemic infections does N. E.'s case illustrate?
- Fortunately there won't be many more near misses such as N. E.'s, because in 2000 a vaccine became available that's now given to infants at 2, 4, and 6 months, making such cases quite rare.

usually made by serology. Antibodies against *F. tularensis* rise as the disease progresses. A fluorescent antibody stain is now available.

Treatment is effective if begun early. Streptomycin is considered to be best, but tetracycline is also effective.

Brucellosis. Brucellosis is caused by various species of *Brucella* (Bergey's Alphaproteobacteria; Chapter 11). These Gram-negative coccobacilli cause disease in domestic animals. Different species, which are distinguished by their surface antigens, prefer different animal hosts. The ones that cause most human disease are *Brucella abortus* (which preferentially infects cattle), *Brucella suis* (swine), *Brucella melitensis* (sheep and goats), and *Brucella canis* (dogs). *B. abortus* and *B. canis* cause mild disease in humans, but *B. suis* or *B. melitensis* can be fatal.

Humans get brucellosis by direct contact with infected animals or their secretions. People who handle diseased animals can be infected through a break in the skin, across the conjunctiva, or by inhalation. Drinking infected milk is a particularly likely mode of transmission because *Brucella* cells concentrate in the mammary glands of infected animals. Pasteurization kills *Brucella* spp. Less than 200 cases of brucellosis occur annually in the United States, but more than a half a million cases occur worldwide. In the United States, meat packers, farmers, and veterinarians are the most vulnerable.

Like *Yersinia* and *Francisella*, *Brucella* spp. are able to establish an infection because they survive inside phagocytes. They too pass from lymph to nodes to the blood and then to organs throughout the body, including the liver, spleen, and bone marrow.

As with many other systemic infections, clinical diagnosis is difficult because the typical symptoms—weakness, fatigue, and loss of appetite—are vague and common complaints. At some time during their illness, however, brucellosis patients have a fever, and the fever follows a distinct pattern. It rises late in the day and then declines. This undulating fever (brucellosis is also called **undulant fever**) can be diagnostic. Most patients recover without treatment within weeks, but a few develop a chronic illness that can last for months, causing continuing fever and fatigue. Also, brucellosis can become localized in an organ, often the bones, endocardium, lungs, or brain.

Brucella spp. are a challenge to isolate in the laboratory. They require a complex culture medium, and they must be incubated in an atmosphere with an elevated concentration of CO_2. They are a threat to laboratory technicians, who must take special precautions to avoid becoming infected. Nevertheless, culturing the bacterium is the most reliable way to confirm a diagnosis of brucellosis. But documenting a rising titer of antibodies can also be diagnostic. Treatment is with doxycycline and rifampin over a course of several weeks.

Lyme Disease. Almost all the bacterial pathogens known today were identified in the late 1800s—but not those that cause legionellosis (Chapter 22) and Lyme disease.

Lyme disease itself was discovered in 1975. A woman in Old Lyme, Connecticut, notified the state health

CASE HISTORY

A Fatal Gift

In 1986 a 67-year-old woman died in a New Jersey hospital. The cause was investigated and eventually the following story came to light. A

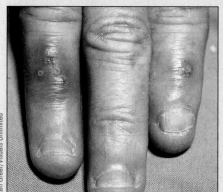

A small sore or ulcer appears where *Francisella tularensis* enter the body.

young friend of the woman had gone rabbit hunting. He killed two rabbits (one of which was losing its fur), cleaned them, and gave them to the victim and her husband. They skinned and froze the rabbits. A few days later the young hunter became ill. An ulcer appeared on his hand. The lymph nodes under his arm became swollen, and he developed a fever. He was examined at a local hospital, but the diagnosis of tularemia was missed. He was treated only with medication to control his fever.

His older neighbor also developed a sore on her hand. Her illness was diagnosed but not until 2 weeks later, when she and her husband were admitted to the hospital. They had

persistent hand ulcers and a severe systemic illness. Within a few days of her admission she was put on streptomycin, but she was already too ill to recover. Then the young hunter was also treated with streptomycin. He recovered rapidly. All three patients showed rises in their *Francisella tularensis* antibody titers. The Centers for Disease Control and Prevention cultured the organism from the bone marrow of the frozen rabbits.

Case Connections

- Why was the young hunter's diagnosis probably missed?
- What question might the attending physician have asked that would have aided the diagnosis?

department that there were an unusual number of cases of childhood arthritis in the town. At about the same time, another woman notified the Yale medical school of an "epidemic" of arthritis in her family. The ensuing investigation led to the clinical description of Lyme disease. In 1982 Willy Burgdorfer, a Montana bacteriologist, identified a spirochete as the causative organism. It was later named *Borrelia burgdorferi* (**Figure 27.4a**). Since then, Lyme disease has become the most commonly diagnosed vector-borne disease in the United States. In 1982 a total of 497 cases were reported in 11 states. In 1997 the total was 12,792 cases in 46 states.

Lyme disease has three clinical stages. Some people experience all of them, others only two, one, or none. The first stage begins 3 to 32 days after having been bitten by an infected tick: A circular red rash resembling a bull's-eye surrounds the bite (**Figure 27.4b**). This distinctive rash is the clinical hallmark of Lyme disease, but it occurs only in 75 percent of patients. The rash disappears spontaneously, and several weeks later the second stage begins. Bacteria spread to nearby lymph nodes, then to the blood, and eventually to other organs, possibly including the brain, joints, heart, liver, spleen, and kidneys. Early symptoms often include headache, stiff neck, muscle aches, and fatigue. Patients often become quite ill. Some may develop menin-

gitis or suffer myocardial damage. Then, about 6 months after the infecting bite, the late stage begins. It's expressed as chronic arthritis that may persist for years. Occasionally joints, especially knees, swell painfully for months.

Lyme disease is transmitted by ticks that belong to the genus *Ixodes*: *Ixodes dammini* in the Northeast and Midwest, *Ixodes pacificus* and *Ixodes neotornae* on the West Coast. *I. dammini* must have three blood meals during its 2-year life cycle. The first two must be on a small animal: almost any small animal will do, although a mouse in preferred. The third must be on a deer. Lyme disease disappears in the absence of deer. Humans become infected when an infected tick bites them instead of a deer. On the West Coast, *I. neotornae* maintains the reservoir in the dusky-footed wood rat and *I. pacificus* transmits the infection to humans. In all regions, ticks reach maturity in the summer, so the high-risk months for Lyme disease are May through July.

Diagnosing Lyme disease is either fairly easy or very difficult. There's no middle ground. No good laboratory test has yet been developed. *Borrelia burgdorferi* is easy to cultivate in the laboratory, but its rarely recovered from the blood of Lyme disease patients. Serological tests are available, but they're not reliable. Both false negative and false positive results occur, even with the most reliable enzyme-linked immunosorbent assay (ELISA). Therefore

SHARPER FOCUS

TICK CHECK

Lyme disease has received a great deal of publicity in recent years, but the relatively simple things we all can do to avoid infection have not. Probably the most important is a daily tick check. The longer a tick is allowed to feed, the more likely it is to transmit infection. If you live in or visit a deer- and *Ixodes*-infested area, it's a good idea to examine yourself and your children every evening and remove the ticks with fine tweezers. If you find a deer tick, save it in a jar to show your doctor. Tick repellents are also helpful. The most effective skin repellent is *N,N*-diethyl-metatoluamide (DEET). It's safe for adults, but repeated use on children can be toxic. Clothing repellents are less toxic, and they also give good protection for about a month. The active ingredient is permethrin. A combination of DEET in low concentrations and permethrin sprayed on clothing once a month probably offers the best protection for adults and children.

Robert Calentine/Visuals Unlimited

The tick that transmits Lyme disease to humans in the Northeast and Midwest, *Ixodes dammini*, is the size of an apple seed and has an orange body with a black circle toward the head.

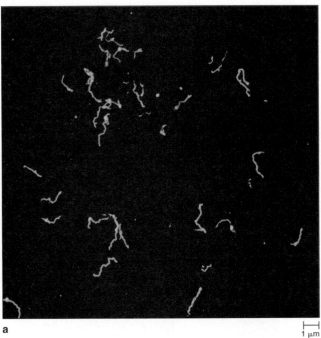

Courtesy Dr. Tom G. Schwan, Rocky Mountain Laboratories, Hamilton, Montana

a

1 µm

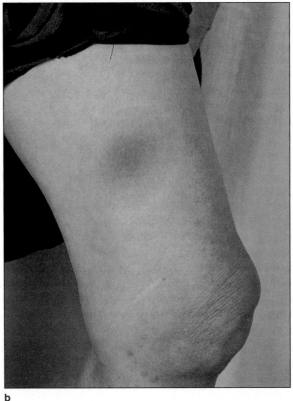

Larry Mulvehill/Photo Researchers

b

FIGURE 27.4   Lyme disease. (a) The causative agent, *Borrelia burgdorferi*. (b) A typical bull's-eye rash that develops where the bite occurred.

clinical criteria are the most common basis for diagnosis. If the patient develops a bull's-eye rash, diagnosis is easy. If he or she doesn't, diagnosis is difficult. Early diagnosis is critical, however, because Lyme disease is cured much more easily during the earliest stages. Later, long-term intravenous therapy may be required. The drugs most commonly used include doxycycline, amoxicillin, and erythromycin.

Why wasn't this relatively common infection recognized earlier? *B. burgdorferi* must have existed long before 1975. Probably the infection has become more common for several reasons. First, the population of the *Ixodes* tick vector is on the rise. Second, the suburban United States has expanded into wooded rural areas. More people now live where the ticks have always lived. And finally, doctors, like everyone else, are more likely to recognize something when they know what to look for.

In 1998 a Lyme disease vaccine composed of a single surface antigen became available. Clinical trials have established that it is safe and effective. However, it is not yet clear how long immunity will last or if *B. burgdorferi* might mutate, rendering the vaccine ineffective.

Relapsing Fever. Besides Lyme disease, *Borrelia* spp. cause relapsing fever. Lyme disease is caused by a single species, *B. burgdorferi*. But relapsing fever is caused by at least three other species. This infection has a sudden onset of fever, chills, headache, and muscle aches. The liver and spleen may become enlarged, and a faint red rash may appear on the trunk. After 3 to 6 days, these symptoms disappear. The patient seems to have recovered. But about a week later, all the symptoms reappear. Cycles of illness and apparent recovery continue as the illnesses gradually becomes less intense. Patients may have as few as 2 relapses or as many as 13.

Illness comes and goes because *Borrelia* spp. are capable of **antigenic variation** (periodically change their surface antigens). During the period of illness, the patient produces antibodies against *Borrelia*'s surface antigens. These clear bacteria from the blood, and the illness seems to resolve. But bacteria in other tissues survive. When they change their surface antigens and enter the bloodstream, illness returns. Eventually, sufficient antibodies are produced to terminate the infection.

There are two forms of relapsing fever: epidemic and endemic. **Epidemic relapsing fever** is caused by *Borrelia recurrentis* and transmitted by the body louse. It tends to occur when crowded, unsanitary conditions exist, such as during a war or a natural disaster. The illness is serious. Mortality in untreated cases can be as high as 40 percent. **Endemic relapsing fever** is caused by *B. hermsii*, *B. turicatae*, and *B. parkeri*, among other species, and is transmitted by ticks. It occurs in many parts of the world, including the western United States. It's milder than epidemic

relapsing fever, and mortality in untreated cases is only about 5 percent.

Relapsing fever can sometimes be diagnosed by observing the bacterium in blood smears of symptomatic patients. Culturing, usually done in immature laboratory mice, is a difficult procedure and is seldom undertaken. Tetracycline, chloramphenicol, penicillin, and erythromycin are all effective.

Anthrax. Anthrax is caused by *Bacillus anthracis*, a Gram-positive, spore-forming rod (Bergey's Firmicutes; Chapter 11). In September 2001, when anthrax was spread intentionally through the mail in letters containing anthrax spores, the status of the disease changed. Before September 2001, anthrax was principally a disease of domestic and wild plant-eating animals such as cattle, goats, sheep, swine, buffalo, and deer. Occasionally, humans became infected with one of three forms of the disease.

The most common form is **cutaneous anthrax (Figure 27.5)**, in which bacteria enter the body through a break in the skin. The bacteria usually come from infected animals or from animal products such as wool or hides. In this disease a blister forms at the site of entry and gradually develops into a blackened crater. In most cases the ulcer heals spontaneously. In 10 to 20 percent of untreated cases, however, bacteria invade the bloodstream, causing a sudden, severe, and fatal septicemia. Cutaneous anthrax can be treated successfully with penicillin during its early stages.

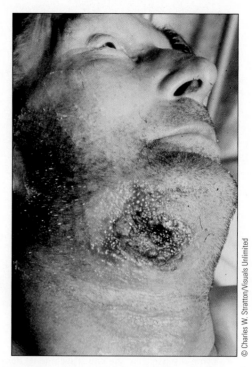

FIGURE 27.5 Cutaneous anthrax. A blackened ulcer at the site of infection.

But when organisms enter the blood and produce substantial quantities of toxin, antibiotics are no longer effective.

A more dangerous form is **respiratory anthrax,** probably because bacteria that land in the lungs are more likely to enter the bloodstream. Naturally occurring respiratory anthrax is quite rare. Before September 2001, there had not been a case of respiratory anthrax in the United States since 1978. Respiratory anthrax was known as **woolsorter's disease** because it was primarily an occupational hazard for people who inhaled bacterial spore while working with anthrax-contaminated wool. It begins with mild respiratory symptoms. Then signs of septicemia develop, and death occurs quickly. Almost none of these patients survive, even with antibiotic treatment. A third, extremely rare form of the disease is **gastrointestinal anthrax.** It occurs when people eat the meat of infected animals.

When anthrax was spread through the mail in late 2001, a number of people contracted cutaneous anthrax and at least 10 people contacted respiratory anthrax. Remarkably, 6 of these 10 patients survived. They had received antibiotics while they were still showing early symptoms of the disease. Early treatment with combinations of antibiotics might have accounted for the unexpected high rate of survival. Most patients received ciprofloxacin (Cipro) or doxycycline along with one or two other antibiotics known to be effective against *B. anthracis* (rifampin, vancomycin, penicillin, ampicillin, chloramphenicol, imipenem, clindamycin, or clarithromycin).

When anthrax spores enter the lungs, they are ingested by phagocytes (macrophages), but they're not killed by them. The spores germinate, producing vegetative cells that reproduce rapidly as they are carried by the macrophages from the lungs to lymph nodes in the chest. There they break out of the macrophages and enter the bloodstream, where they proliferate and produce **anthrax toxin.** The toxin debilitates the host's immune system and eventually kills the patient. Patients die of cardiac arrest or respiratory failure. Early phases of the disease are accompanied by symptoms (fatigue, fever, aches, and cough) typical of many diseases. Later the patient becomes critically ill with high fever, labored breathing, and shock.

The Toxin. Anthrax toxin is complex and deadly (**Figure 27.6**). It alone in the absence of bacterial cells can kill animals. Anthrax toxin consists of three component pro-

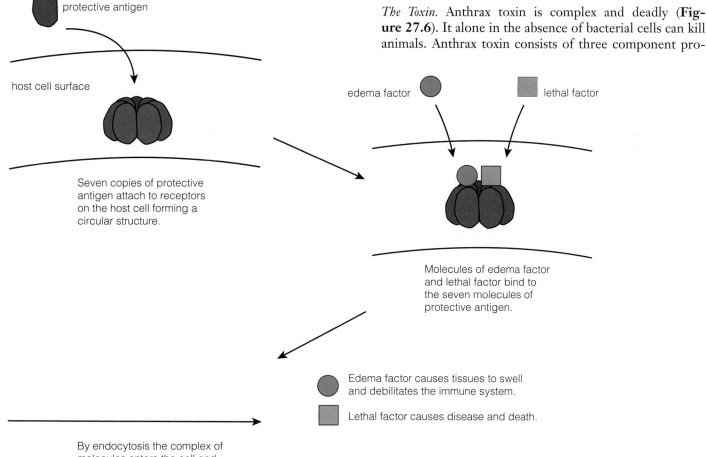

FIGURE 27.6 Anthrax toxin. The roles of its three components: protective antigen, edema factor, and lethal factor.

teins: protective antigen, edema factor, and lethal factor. Protective antigen (so named because antibodies against it render a person immune to the disease) plays the role of bringing the other toxic components into a cell. It does so by binding to specific receptors on the cell surface, forming a complex there to which copies of edema factor and lethal factor bind. Then, by endocytosis, the entire structure is taken into the cell and the components separate. Protective antigen has completed its task; the other two components exert their toxic effects. Edema factor (as its name implies) causes tissue to swell and debilitates the immune system. Lethal factor, by a still unknown mechanism, causes disease and death.

Anthrax as a Weapon. Because *B. anthracis* produces endospores that can remain viable for more than 40 years, large quantities of anthrax can be produced and stored. Because anthrax spores must penetrate deep into the lungs to cause infection, spores must be prepared in the form of an extremely fine powder. It is estimated that between 2500 and 55,000 endospores must enter the lungs to be lethal in half the people infected. Anthrax spores to be used as a weapon would probably have been made resistant to antibiotics.

Because early diagnosis is critical in defending against anthrax, methods for rapid detection of anthrax are being developed. One of the most promising employs **real-time polymerase chain reaction (PCR)**. Like conventional PCR (Chapter 7), primers specific for anthrax DNA are added to the sample to be tested for the presence of anthrax. In real-time PCR, fluorescent labels are also added. If anthrax is in the sample, fluorescence indicating its presence can be detected within about 30 minutes.

Cat Scratch Disease.

The typical patient with cat scratch disease appears in the doctor's office with one extremely swollen lymph node. It's usually under the arm or in the groin. The node is not painful, and usually the patient has no fever or other symptoms. Often there are healing scratch marks on the arm or leg near the affected node and a small sore where the infecting organism entered the body. The patient, usually a child, remembers being scratched by a cat or kitten. Rarely the node remains swollen to the size of a tennis ball for several months. Eventually it returns to its normal size. A few patients are more seriously ill. They complain of fever, fatigue, or even symptoms of meningitis.

Cat scratch disease is caused by *Bartonella henselae*, a pleomorphic Gram-negative rod that's probably part of the normal oral biota of cats and dogs. The infection will resolve on its own, but treatment with rifampin, trimethoprim-sulfamethoxazole, or ciprofloxacin may be useful for patients with systemic symptoms.

Rocky Mountain Spotted Fever.

Rocky Mountain spotted fever (RMSF) and several other systemic diseases discussed here are caused by rickettsiae (Bergey's Chlamydiae; Chapter 11). These small rod-shaped, Gram-negative bacteria multiply only as intracellular parasites. They parasitize arthropods, including ticks, lice, and fleas and infect mammals, including humans, during a blood meal. (Q fever [Chapter 22] is an exception. It's transmitted by inhalation.)

RMSF is caused by *Rickettsia rickettsii*. This severe disease with fever and rash is named for the region where it was first diagnosed, but it occurs in many parts of the Western Hemisphere (**Figure 27.7**). RMSF is not a common disease, but it occurs much more often than do tularemia or brucellosis. In some years as many as 1000 cases are reported to the Centers for Disease Control and Prevention (CDC). The disease is transmitted by a tick. In the Rocky Mountain states it is usually the wood tick, *Dermacentor andersoni*. In the southeastern states it's the dog tick, *Dermacentor variabilis*. In the south-central states it's the Lone Star tick, *Amblyomma americanum*.

Rickettsia rickettsii can be passed transovarially from one generation of ticks to the next. They alone can perpetuate RMSF. In humans, *R. rickettsii* multiplies in the smooth inner lining of blood vessels. In so doing, it damages capillaries in the skin, causing the spotty rash for which the disease is named. In severe cases, internal organs suffer similar capillary damage. The infection becomes overwhelming in about 20 percent of untreated cases. Then abnormal blood clotting leads to shock and death. Only tetracycline and chloramphenicol are effective.

Early diagnosis and treatment of RMSF are difficult because no test exists for rapid identification. The organism isn't cultured in the laboratory because it's so hazardous. Definitive laboratory diagnosis can't wait the several weeks it takes for antibodies to form because prompt antibiotic treatment is essential to avoid serious complications or death. Therefore early effective treatment of RMSF depends upon clinical and epidemiological clues.

A history of exposure to ticks is often the key to diagnosis. The earliest signs and symptoms—fever, headache, and vomiting—are too common to be useful. But within a week a distinctive rash usually appears on the wrists and ankles. Eventually this rash covers the entire body. Children are most likely to be infected, but adults are most likely to die. The disease responds dramatically to antibiotic treatment. Early treatment prevents most fatalities.

Typhus.

The two forms of typhus—epidemic and murine—are also caused by rickettsiae.

Epidemic (Louseborne) Typhus. Epidemic typhus is caused by *Rickettsia prowazekii*. This organism is named after two bacteriologists—Howard Ricketts and Stanislav von Prowazek—who died of typhus while investigating this deadly disease. Epidemic typhus, or **louseborne typhus,** is spread by infected body lice. Lice pass *R. prowazekii*

FIGURE 27.7 Rocky Mountain spotted fever. (a) Incidence of cases within the United States. (b) The characteristic rash that usually appears about a week after infection.

- 1–5 cases
- ▲ 6–10 cases
- ■ >10 cases

a

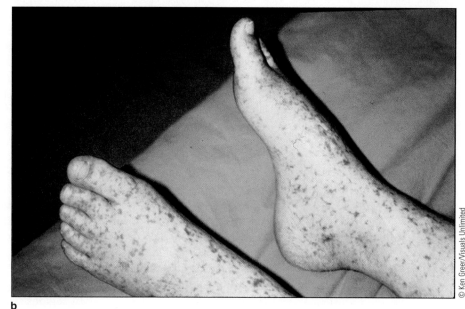

b

© Ken Greer/Visuals Unlimited

from one human to another during a blood meal. When infected lice feed, they defecate on the skin. Then, when the victim scratches, the feces enter the bite wound. *R. prowazekii* multiplies in the wound. Later it enters the bloodstream in large numbers and multiplies in the lining of small blood vessels.

As with other systemic diseases, early diagnosis of typhus is difficult. Only after an 8- to 12-day incubation period do most typhus patients become ill. They suffer an unbearable headache, fever, muscle aches, and shaking chills. Often they become delirious or unconscious. (In Greek, *typhus* means "stuporous.") Several days later, a rash that resembles the rash of Rocky Mountain spotted fever appears in almost all patients, but it does not affect the palms, soles, or face.

Tetracycline and chloramphenicol are remarkably effective. One dose of tetracycline can cure a person who appears to be near death. Untreated, the disease may last as long as 3 weeks. In about 40 percent of patients, shock leads to death. Older adults are especially vulnerable. People who recover from typhus usually acquire long-lasting immunity, but some do not. In them, the infection can become latent, with mild recurrences, called **Brill-Zinsser disease**, years later.

Other than humans and lice, no other animal host is necessary to maintain typhus. Latent human infections sustain it between epidemics. The key to breaking the chain of infection is eliminating human lice. They live in unwashed human clothing and multiply rapidly when people live under crowded, unsanitary conditions. Therefore typhus epidemics often occur during war and other times when people are displaced from their homes. In the nineteenth century, typhus killed many Irish immigrants during their voyage to North America. The disease became known as **ship fever.** Typhus also killed millions of civilians during and after World War I; many died in concentration camps during World War II, but not when it ended. By then the insecticide DDT was available to delouse refugees.

Populations that are free of lice do not suffer from epidemic typhus. The last outbreak in the United States occurred in 1922. But typhus still exists in many parts of the developing world, particularly Africa.

Murine (Fleaborne) Typhus. Rickettsia typhi causes murine, or fleaborne, typhus. This disease is similar to epidemic typhus, but it's less severe and rarely fatal. Murine typhus is mainly a disease of mice and rats (Muridae is the family of rodents that includes rats and mice). It's transmitted from one animal to another by infected fleas. Only occasionally is it transmitted to humans who come in contact with diseased rodents. Most cases begin gradually with headache, muscle ache, and fever. Three to five days later a typical rickettsial rash develops.

Murine typhus is endemic in the United States. People who work in mice- or rat-infested areas (for example, grain storage elevators) are most at risk. Fewer than 100 cases are reported annually, and most of these are in the Southeast and Texas. Murine typhus is sometimes misdiagnosed as Rocky Mountain spotted fever. Because the causative agents are so closely related, serological testing doesn't reveal the error. Like other rickettsial diseases, murine typhus can be effectively treated with tetracycline or chloramphenicol.

Viral Infections

Viruses cause many systemic diseases. Some are mild. Others are deadly.

Yellow Fever. Yellow fever, also known as yellow jack, is caused by the yellow fever virus. It's a small, enveloped RNA virus that belongs to the togavirus family and the flavivirus genus. It's introduced into a new human host by the bite of a mosquito, usually *Aedes aegypti.* Then the virus is transported to nearby lymph nodes, where it multiplies and eventually enters the bloodstream. When it reaches the liver, it multiplies rapidly.

Most people suffer only a few days of mild fever, headache, and weakness. But about 20 percent suffer the classical symptoms of yellow fever—severe fever, chills, headache, and jaundice, which gives the disease its name. In addition to jaundice, liver damage depletes the blood clotting factors that the liver normally produces. This leads to uncontrollable bleeding. About half of severely af-

fected people die. Those who recover suffer no lasting damage, and they are immune thereafter.

Yellow fever has caused devastating epidemics around the world, including cities as far north as Philadelphia. Because *A. aegypti* prefers to feed on human beings, it has adapted to their habits. It lives near people, breeds in the water they store for washing or drinking, and bites them by day.

An endemic form of yellow fever still exists in jungle areas of the Americas, including Panama, where monkeys are the reservoir (Chapter 15). The urban form of the disease has been eliminated from the Western Hemisphere by eradicating *A. aegypti,* but it still occurs in Africa. A live attenuated vaccine developed in the 1930s could prevent most cases, but it's not widely used.

Dengue Fever. Dengue fever resembles yellow fever in many ways. It's caused by several closely related viruses, and it's transmitted principally by the same mosquito, *Aedes aegypti.* Unlike yellow fever, however, dengue fever still occurs widely in most tropical regions, and occasionally it appears in temperate climates, including the United States.

Dengue fever is rarely life threatening, but it can be painful. Two to seven days after being bitten by an infected mosquito, people suffer a high fever and headaches. Soon, a rash and pain in the limbs develop. Limb pain is so intense that the illness is sometimes called **breakbone fever.** The patient appears to recover after a few days, but soon the fever and rash return. Many patients complain of weakness and mental depression long after recovery. No vaccine exists to prevent dengue fever, and treatment is limited to supportive care. The best prevention is control of *A. aegypti.*

A life-threatening syndrome called **dengue hemorrhagic fever** or **dengue shock syndrome** occasionally occurs among children who have had dengue fever. These patients suffer uncontrollable bleeding and shock. The disease is probably a hypersensitivity reaction that occurs when the previously stimulated immune system encounters the pathogen again. These children require intensive supportive medical care. Even then, as many as 40 percent die.

Infectious Mononucleosis. At some time in their lives most humans are infected by the Epstein-Barr virus (EBV). In regions where standards of living and hygiene are poor, infection usually occurs in early childhood, producing a mild illness or none at all. But in the United States and other countries with high standards of living and hygiene, infection typically doesn't occur until 15 to 25. Again the infection can be asymptomatic or it can cause a clinical syndrome called **infectious mononucleosis** (mono). By midlife, more than 90 percent of U.S. residents have been infected by EBV.

EBV is a large DNA-containing herpesvirus surrounded by a lipid envelope (**Figure 27.8**). Like other herpesviruses, EBV produces latent and active infections. Because latent infection occurs in B lymphocytes, EBV's pathogenesis is intimately associated with the immune system.

Pathogenesis. The symptoms of infectious mononucleosis—fever, fatigue, sore throat, and swollen lymph nodes—appear 1 to 2 months after infection. By that time, blood contains an increased number of lymphocytes, many of which are enlarged and atypical in appearance. Half the patients also have an enlarged spleen. Diagnosis is confirmed by serology. Most patients with mononucleosis resume normal activities in 4 to 6 weeks. Major complications involving the central nervous system or fatal rupture of the spleen can occur, but they are rare.

Active infection in epithelial cells near the salivary glands produces large amounts of EBV, which enters oral secretion and is largely responsible for transmission. So infectious mononucleosis is also called the kissing disease. But EBV can also be transmitted by saliva-carrying fomites such as drinking glasses or eating utensils.

EBV establishes a latent infection in B lymphocytes. No new viral particles are produced by them. They are not killed, and they do not change appearance, but they are profoundly altered by EBV. They proliferate without the normal controls on cell division. This abnormal proliferation of EBV-infected B cells initiates a chain of events that results in clinically apparent infectious mononucleosis. When they enter circulation, they cause the early symptoms of fever and fatigue. But they soon stimulate suppresser T cells to bring them back under control. These stimulated T cells enlarge. Their presence in blood is di-

agnostic and gives the disease its name. Proliferation of T cells can also cause lymph nodes, spleen, and liver to swell. Eventually a normal equilibrium between B and T cells is reestablished. Symptoms end, but latent virus persists in the lymphocytes for life.

EBV-associated Cancers. The Epstein-Barr virus is one of the few viruses that have been proved **oncogenic** (cancer-causing). In immunosuppressed patients, EBV can cause a fatal cancer called B-cell lymphoma. If suppresser T cells can't stop the proliferation of infected B cells, runaway multiplication produces tumorlike growths throughout the body and EBV-induced lymphoma. David, the "bubble boy" who survived for many years in a sterile environment despite a severe congenital immunodeficiency, died from EBV-induced lymphoma (see Case History: David—Life in a Germ-Free World, in Chapter 18).

EBV has also been definitively linked to a relatively rare cancer called **Burkitt's lymphoma,** a malignant tumor of the jaw of children of equatorial Africa (**Figure 27.9**). The Epstein-Barr virus was discovered during research on Burkitt's. Moreover, EBV is also associated with **nasopharyngeal carcinoma,** a malignant tumor of the nasopharynx that is common in China. No one is sure why these cancers occur only in certain countries, because EBV infection is worldwide. Development of cancer probably requires an extra stimulus in addition to EBV, maybe early exposure to malaria parasites or genetic factors.

Acquired Immunodeficiency Syndrome. Acquired immunodeficiency syndrome (AIDS) was first identified in 1981. In 1983 its viral cause, human immunodeficiency

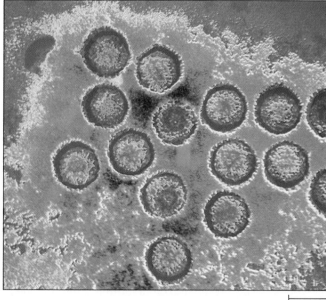

├─────┤ 100 nm

FIGURE 27.8 Epstein-Barr virus.

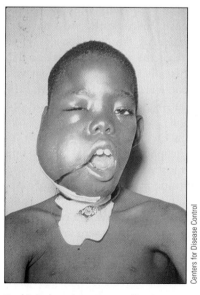

FIGURE 27.9 Burkitt's lymphoma, a malignant tumor of the jaw seen in children of equatorial Africa.

CASE HISTORY

Three Faces of AIDS

Patient 1

A 27-year-old married woman noticed white spots on the inside of her mouth. They were painless, but they didn't go away. Her physician examined scrapings from these lesions and diagnosed thrush, a common infection in babies but unusual in healthy young adults (Chapter 26). Her thrush responded to antifungal medications, but this was only the first of many opportunistic infections. Patient 1, who died 2 years later, reported that her husband had a history of intravenous drug abuse.

Patient 2

An 8-month-old baby suddenly became sick with a high fever and such extreme lethargy that he was hardly arousable. He was rushed to an emergency room, where pneumococcal meningitis was diagnosed. Treatment with antibiotics led to recovery, but he died from another overwhelming bacterial infection only 2 months later. His mother admitted to using illegal intravenous narcotics.

Patient 3

A 32-year-old attorney noticed a cough and fever that became progressively worse until his breathing was rapid and difficult. Because of his respiratory distress, he was admitted to the hospital, where he was diagnosed with *Pneumocystis* pneumonia. In spite of antibiotic treatment and aggressive supportive care, including a mechanical ventilator, his condition continued to worsen. He died 3 weeks later. The man's homosexual companion had died recently of a similar infection.

Case Connections

- All three of these patients had AIDS. How can one disease present three such vastly different sets of symptoms?
- How did Patient 2 probably become infected by HIV?

---

virus (HIV), was isolated. It's now the most thoroughly studied human virus. The biology of HIV is discussed in Chapter 13. Here we'll consider the epidemic disease it causes and how it's transmitted, treated, and prevented.

The Epidemic. Only 20 years after it was first recognized, AIDS threatens to become the most devastating epidemic in human history. A United Nations (UN) report issued in 2002 summarized the present state of epidemic and made predictions. Already, 25 million people have died of AIDS; 40 million people are now infected with HIV. The life expectancy in sub-Saharan Africa has dropped to 47 years, when it would have been 62 years without AIDS. In Botswana, now the most severely affected country, life expectancy has dropped by almost 35 years. Fourteen million children have lost one or both parents.

If we continue the now low level of response in many countries, the future appears bleak: 70 million more people will die from AIDS. In some countries, AIDS will kill half the new mothers. But there are encouraging possibilities. A government program in Brazil to provide medicine to those that need it has turned the epidemic around. Other countries are initiating similar programs. The problem, of course, is money. The UN estimates that $10 billion a year will be needed for poor countries to treat those with HIV and to care for orphans.

The Disease. At first, HIV infection almost inevitably led to AIDS, ending in death within 10 to 11 years. Then in late 1995, with the introduction of a multiple drug regimen called highly active antiretroviral therapy called (**HAART**), prospects changed completely. With such treatment, long-term survival, maybe even a cure, seems possible. Although costs have decreased dramatically, HAART remains extremely costly. It's burdensome (the patient must take three or more drugs twice or more daily), and it probably must be continued indefinitely. So now HIV infection has almost become two separate diseases—one for people in industrialized countries, where HAART is economically feasible, and another for those in countries where it is not. The latter is by far the larger group. Millions of people in sub-Saharan Africa and Southeast Asia are living with HIV or AIDS. In North America and Europe the numbers are hundreds of times less.

First we'll discuss the progress of untreated HIV infection. Then we'll go on to consider the impact of HAART.

Untreated HIV Infections. The most common means of HIV infection are sexual transmission, exposure to contaminated blood, or transmission from mother to fetus or suckling infant (**Table 27.2**). Surface cells, called dendritic cells, usually become infected first. Within 2 days these cells fuse with CD4 lymphocytes, and they carry the infec-

TABLE 27.2 Transmission of HIV

Route	Example
Known Routes of Transmission	
Sexual transmission	Homosexual intercourse between men
	Heterosexual intercourse (men to women and women to men)
Blood	Transfusion of blood or blood products (e.g., clotting factors for hemophiliacs)
	Shared needles among intravenous drug users
	Needle stick, open wound, and mucous-membrane exposure in health-care workers
	Medical injection with an unsterilized needle (common in developing countries)
Mother to infant	Across the placenta before birth
	Passage through birth canal
	Intimate contact around time of birth
Routes Investigated and Shown Not to Mediate Transmission	
Close personal contact	Household contact
	Caring for AIDS patients without exposure to blood or infected secretions
Insect vectors	Mosquitoes, flies
Fomites	Water, soil, humanmade materials

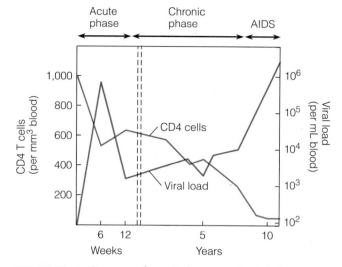

FIGURE 27.10 Progress of a typical untreated HIV infection.
(From Fauci A.S., et al. 1996. Immunopathogenic Mechanisms of HIV Infection, *Annals of Internal Medicine* 124:654–63.)

tion to lymph nodes, where HIV continues to multiply. Within a few days, virus-infected cells enter the bloodstream. Then the virus spreads widely and rapidly to lymph nodes, gut-associated lymphoid tissue, the spleen, and the brain. These first weeks after infection constitute the **acute phase** of HIV infection (**Figure 27.10**). Viral replication and killing of CD4 lymphocytes is massive (**Figure 27.11**). About 3 weeks into this acute phase, most people begin to notice the effects. They have symptoms that resemble those of infectious mononucleosis, including fever, fatigue, rash, and headache. During this acute stage, a lot of things are happening. The virus is multiplying at a high rate, but the disease is difficult to diagnose. The signs and symptoms aren't definitive, and serological tests are still negative in this stage.

After a few more weeks, the **chronic phase** begins. The immune system begins to gain control over the infection. CD8 T lymphocytes start to kill virus-producing cells, and symptoms disappear. The immune system almost never eliminates the infection, but it does pretty well. After about 6 months, **viral load** (the amount of HIV in the bloodstream) declines and becomes relatively constant for the next 8 years or so. During this time, the number of CD4 T cells in the blood begins a gradual and continual decent from its normal of about 800 per mm³. When it reaches 200, people are said to have AIDS. When it falls below 100, the immune system loses its ability to control HIV and almost any other infection. The viral load soars, and other infections, such as those suffered by the patients described in Case History: Three Faces of AIDS, begin. Within a year or two the patient dies. Until recently, it was a great mystery why the chronic phase of HIV infection continued so long in a seemingly quiescent state. Now it's clear that this isn't a quiescent state at all. It's an extremely active phase. Huge amounts of virus are being produced, and huge numbers of CD4 T cells are being killed. The patient survives only because the body is able to make new CD4 T cells at such a prodigious rate.

Now let's consider how HAART changes this pattern.

Treated HIV Infections. Three kinds of drugs have become available to treat HIV infections (**Table 27.3**). They act by inhibiting one or the other of two enzymes—reverse transcriptase and HIV protease—that are essential for HIV's replication (Chapter 11). When the first of these drugs became available, they were used singly and their benefits were quite marginal. Within weeks or months the patient's strain of HIV became resistant to the drug, and it became useless. In retrospect, this probably isn't surprising. It's now known that an infected person makes about 10 billion new HIV virions daily. Chances are high that one of them will undergo mutation to make a drug-resistant enzyme. And that's what happens. But HAART solved the problem. It involves giving several antiviral drugs simultaneously—usually two reverse transcriptase inhibitors and one protease inhibitor. The chances of HIV's becoming simultaneously resistant to all of them are minuscule.

Results with HAART have been quite extraordinary. Often within 8 weeks virus levels in the blood fall tenfold, and within 6 months they become undetectable. The number of CD4 T cells rises. The frequency of secondary infections declines. Between the first half of 1996 and the first half of 1997, death from AIDS in the United States declined by 44 percent.

But if treatment is stopped, viral multiplication begins again. Apparently, HIV in the proviral state (Chapter 11) lies within infected cells waiting to be activated. Possibly, prolonged treatment may prove to cure the infection, but that's not yet clear. It's also possible that drugs can be found to activate the provirus. Then they could be eliminated with antiretroviral drugs.

But, of course, there is a down side. The patient must swallow at lest 8 and often more than 16 pills a day, along with other medicines to control opportunistic infections. And there are unpleasant side effects. They range from rash, nausea, diarrhea, and headaches to anemia, neuropathy, hepatitis, and maybe diabetes. Always there's the frightening cost of medication.

AIDS and Other Diseases. As HIV infection cripples the immune system, a person becomes susceptible to other infections. Thus a secondary infection is usually the immediate cause of death in an AIDS patient. The pathogens that are most threatening to AIDS patients are the ones that are normally controlled by cell-mediated immunity. Among these are *Pneumocystis carinii* (Chapter 22), *Toxoplasma gondii* (discussed later in this chapter), *Cryptococcus neoformans* (Chapter 25), *Histoplasma capsulatum* (Chapter 22), and cytomegalovirus (Chapter 24). AIDS has also sparked a resurgence of tuberculosis, another infection normally controlled by vigorous cell-mediated immunity (Chapter 22).

Children with AIDS are vulnerable to a wider variety of pathogens. They die from infections in which antibody-mediated, as well as cell-mediated, defenses are critical. Thus bacterial infections, such as the pneumococcal meningitis of Patient 2 in Case History: Three Faces of AIDS, are common in pediatric AIDS patients.

HIV itself can become a primary pathogen of the central nervous system when viral particles cross the blood-brain barrier and cause **AIDS dementia** (a deterioration of mental function). The vast majority of HIV-infected babies seem to have retarded neurological de-

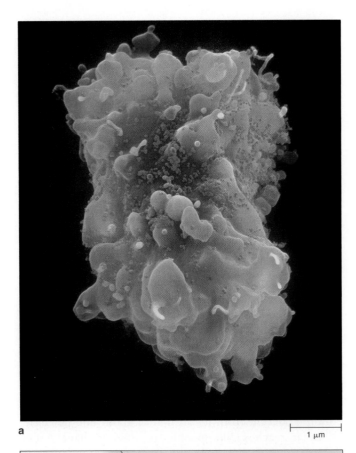

a

1 μm

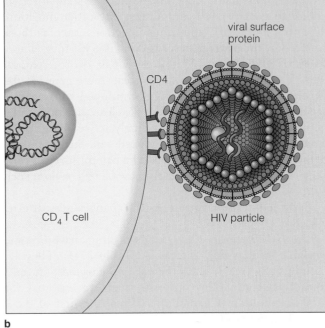

b

FIGURE 27.11   Infection by HIV. (a) Micrograph of a CD4 T cell being attacked by the human immunodeficiency virus (tiny blue particles). (b) Sketch of how viral surface proteins attach to CD4 receptors in the first step of infection.

SHARPER FOCUS

AIDS SCARE

Ordinarily a patient does not expect to be infected by a health-care provider because casual contact does not transmit AIDS. However, after several cases of AIDS were traced to a Florida dentist in 1991, people in the United States worried that there might be a new way to acquire the dreaded infection.

Studies were immediately undertaken to determine the exact mode of transmission from dentist to patient and whether other HIV-infected health-care workers may unknowingly have infected patients. The mode of transmission has never been clearly understood in the Florida case, and six investigations published before November 1991 failed to document additional cases of provider-to-patient transmission. Even a physician working with open lesions on his hands—which is contrary to standard infection-control guidelines—did not transmit his infection to any of the more than 300 patients on whom he performed invasive examinations or vaginal deliveries. The studies unanimously concluded that provider-to-patient transmission of AIDS is extraordinarily rare. The risk is so low it cannot be accurately measured.

velopment. It's a mistake to consider HIV simply as an immune dysfunction, as some texts do.

HIV infection also increases the likelihood of developing certain types of cancer, particularly Kaposi's sarcoma (a rare blood vessel malignancy). Cancers of the lymphatic system, the rectum, and the tongue are also unusually common in AIDS patients.

Transmission. HIV has three main modes of transmission (Table 27.2):

1. By sexual contact with an infected person that transfers infected body fluids—blood, semen, or vaginal secretions

2. By receiving infected blood, usually in transfusions or from sharing intravenous needles with an infected person

3. From mother to infant during pregnancy (bloodborne transmission of HIV across the placenta) or around the time of birth (through infected vaginal secretions or infected breast milk)

Casual contact during normal social interaction in school or the workplace does not spread HIV. Even the more intimate contact that occurs among siblings and family members does not transmit the virus, as long as infected body fluids are not exchanged. Neither insect vectors nor contaminated food or water transmits HIV.

TABLE 27.3 Antiretroviral Drugs

Generic Name	Other Names	Activity
Didanosine	Videx, ddI	Nucleoside analogue inhibitor of reverse transcriptase
Lamivudine	Epivir, 3TC	Nucleoside analogue inhibitor of reverse transcriptase
Stavudine	Zerit, d4T	Nucleoside analogue inhibitor of reverse transcriptase
Zalcitabine	HIVID, ddC	Nucleoside analogue inhibitor of reverse transcriptase
Zidovudine	Retrovir, AZT	Nucleoside analogue inhibitor of reverse transcriptase
Delavirdine	Rescriptor	Nonnucleoside inhibitor of reverse transcriptase
Nevirapine	Viramune	Nonnucleoside inhibitor of reverse transcriptase
Indinavir	Crixivan	Protease inhibitor
Nelfinavir	Viracept	Protease inhibitor
Ritonavir	Norvir	Protease inhibitor

People with ulcer-causing sexually transmitted diseases (STDs) are at increased risk of contracting AIDS because open sores on the genitals provide an easy portal of entry for HIV (Chapter 24).

Epidemiology. When AIDS was first recognized in the United States, it occurred almost exclusively among male homosexuals. It was assumed that homosexuals were particularly susceptible. Now we know that this high incidence occurred because homosexual males had extremely high numbers of sexual contacts within a community in which the virus was already prevalent. Mucous membrane tears during rectal intercourse and a high rate of ulcer-causing STDs in the homosexual community probably contributed. In the early 1980s, AIDS in the United States primarily affected men.

However, this pattern started to change in the late 1980s. AIDS cases became increasing common among drug abusers who shared contaminated needles. This led to spread of the disease in the United States to women and children and increased sexual transmission among heterosexuals. Now most men and women with drug-acquired AIDS are heterosexual. HIV can be transmitted from men to women or women to men during sexual intercourse.

In developing countries the epidemiological pattern of AIDS has been very different. In Africa the infection is spread almost exclusively by heterosexual intercourse. Men and women die from the disease in approximately equal numbers. Many children are infected by HIV at birth.

Prevention. Great progress was made in AIDS research since 1981. The illness was identified. The virus was isolated. Diagnostic tests were developed, and effective treatment became available. In contrast, the pressing problem of prevention remains.

Efforts to contain the pandemic must rely on minimizing HIV transmission through campaigns that promote safer sex and programs to help intravenous drug users stop using dirty needles. Results have been mixed because of difficulties in modifying behavior.

Only a tiny fraction of HIV infections are acquired in the hospital, and efforts to reduce that small number have been encouraging. Most hospital-acquired HIV infections result from a contaminated transfusion or inadvertent exposure of a health-care worker to blood or body secretions of an infected patient. Most infected blood has been eliminated from blood banks since 1985, when tests to identify HIV antibodies became available. Postexposure antiviral drug therapy, although quite new, appears to be effective.

Of course, the most powerful weapon would be a vaccine. This research is being pursued vigorously, but it's difficult work. Killed virus vaccines and attenuated live vaccines are considered dangerous for such a devastating disease. Production of specific HIV antigens by recombinant DNA tech-

nology is being investigated, but HIV's remarkable ability to mutate makes this approach difficult. Most fundamentally, as the Nobel Prize–winning virologist David Baltimore has said, "If the immune system in HIV-infected individuals cannot wipe out the virus, why should a vaccine that activates the same immune response be expected to block infection?" But considering the talent and resources being expended on this effort, we have every reason to be hopeful.

Ebola Hemorrhagic Fever. Ebola hemorrhagic fever burst into the spotlight of public attention in May 1995. An outbreak of hemorrhagic fever had occurred in Kitwit, Democratic Republic of the Congo. Samples of the infectious material were flown to the Centers for Disease Control and Prevention (CDC) in Atlanta, Georgia. They were analyzed in a biosafety level four laboratory (BL-4). (BL-4 is a high-level containment facility for the safe handling of the most dangerous infectious agents.) The BL-4 team rapidly identified Ebola virus in the samples.

The Disease. The high mortality rate (about 92 percent) of this outbreak of Ebola hemorrhagic fever, along with the horror of its symptoms (extensive bleeding and destruction of internal organs), led to great scientific and public concern. Was a new devastating pandemic emerging? The Dustin Hoffman movie *Outbreak* amplified public worry. In spite of its taking 190 lives, the Kitwit outbreak produced some good news: Ebola virus has a restricted reservoir and limited means of transmission. In its present form it will probably be limited to local outbreaks.

The sole reservoir of Ebola virus appears to be an unknown animal that lives in the African savanna or rainforest. Ebola virus is not spread by casual contact such as inhaling virus-laden respiratory secretions. Direct contact with infected blood or body secretions is required, such as reuse of unsterilized needles in hospitals or manual removal of internal organs in preparation for burial. That's how the disease was spread in the Kitwit outbreak.

The Virus. Ebola virus belongs to the Filoviridae. These are enveloped, minus-strand RNA viruses distinguished by their unusual morphology (**Figure 27.12**). They are exceedingly long—sometimes longer than *Escherichia coli*. Ebola virus was discovered in 1976 during two simultaneous outbreaks in Africa—one in Sudan, the other in Congo. Since then, only four known outbreaks have occurred. All were in Africa. The Filoviridae also contains Marburg virus, which causes a hemorrhagic fever. It was discovered in 1967 in Marburg, Germany, when 25 people became infected. They had contact with monkey kidneys imported from Uganda for use in tissue culture. Since then, sporadic cases of Marburg hemorrhagic fever have occurred in Africa.

Protozoal Infections

Worldwide, protozoa are among the most common pathogens of the blood and lymphatic systems.

Malaria. Nobel prize–winner Sir Macfarlane Burnet said, "If we take as our standard of importance the greatest harm to the greatest number, then there is no question that malaria is the most important of all infectious diseases." At any given time, approximately 100 million people are suffering symptoms of malaria. More than 1 million of them— mostly children— die from it every year. Almost all this burden of illness and death is borne by the developing world.

Life Cycle of Plasmodium. Malaria is caused by four species of protozoa that belong to the genus *Plasmodium.* They have a complex life that takes place in mosquitoes and human beings (**Figure 27.13**). Sexual reproduction occurs in female mosquitoes belonging to the genus *Anopheles.* Asexual reproduction occurs in the liver and red blood cells of humans.

When an infected mosquito bites a human, the **sporozoite** form of the protozoan passes from the insect's salivary gland into the person's bloodstream. Within an hour it enters the liver. In liver cells, sporozoites multiply and produce another form, the **merozoite.** Eventually the infected liver cells burst and release huge numbers of merozoites into the bloodstream. Within minutes they invade red blood cells. They multiply there, eventually bursting these cells and releasing even more merozoites. These infect more red blood cells that produce more merozoites. A few merozoites develop into male and female **gametocytes.** If they are consumed by a mosquito during her blood meal, the infectious cycle continues.

All four species of malarial parasites—*Plasmodium falciparum, Plasmodium ovale, Plasmodium malariae,* and *Plasmodium vivax*—have similar life cycles, but there are minor differences. *P. vivax* and *P. ovale* can remain dormant in the liver as **hypnozoites** for months or even years and then cause relapses long after the original illness seems to be over. *P. malariae* and *P. falciparum* do not cause relapsing malaria. *P. falciparum* causes almost all fatal cases.

Clinical Syndrome. Destruction of red cells by merozoites is often synchronized, so many new merozoites are released at the same time. This occurs every 72 hours for *P. malariae* and every 48 hours for the other species. When such mass destruction of red cells occurs, the person experiences a shaking chill followed by a fever as high as 104°F and a drenching sweat. Between attacks, symptoms may be minimal. Huge quantities of malarial antigens are released

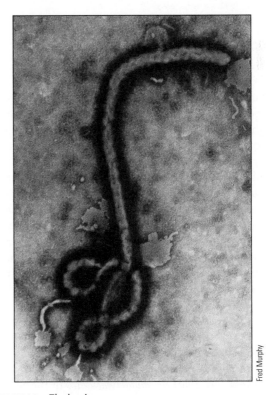

Fred Murphy

FIGURE 27.12 Ebola virus.

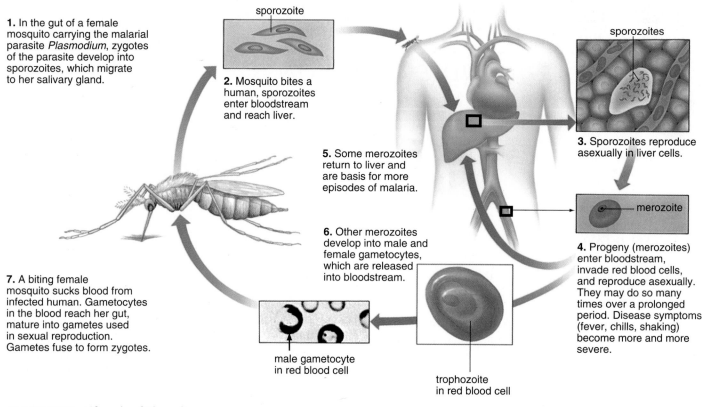

1. In the gut of a female mosquito carrying the malarial parasite *Plasmodium*, zygotes of the parasite develop into sporozoites, which migrate to her salivary gland.

sporozoite

2. Mosquito bites a human, sporozoites enter bloodstream and reach liver.

sporozoites

3. Sporozoites reproduce asexually in liver cells.

merozoite

5. Some merozoites return to liver and are basis for more episodes of malaria.

6. Other merozoites develop into male and female gametocytes, which are released into bloodstream.

4. Progeny (merozoites) enter bloodstream, invade red blood cells, and reproduce asexually. They may do so many times over a prolonged period. Disease symptoms (fever, chills, shaking) become more and more severe.

7. A biting female mosquito sucks blood from infected human. Gametocytes in the blood reach her gut, mature into gametes used in sexual reproduction. Gametes fuse to form zygotes.

male gametocyte in red blood cell

trophozoite in red blood cell

FIGURE 27.13   Life cycle of *Plasmodium*.

(Art by Leonard Morgan from Biology: The Unity and Diversity of Life, 6ᵗʰ ed., by C. Starr and R. Taggart, Brooks/Cole, 1992. All rights reserved.)

when the red blood cell lyse. These stimulate T cells to produce cytokines, including tumor necrotizing factor, which causes the characteristic fever and chills.

Mass destruction of red blood cells has other damaging consequences. It can deplete the number of red blood cells enough to cause anemia. Debris can clog the circulation, cutting off the blood supply to vital organs such as the kidneys or brain. When massive numbers of red cells are destroyed, the hemoglobin released from them turns the urine black. This clinical syndrome, called **blackwater fever,** usually indicates that the infection will be fatal.

The spleen, where macrophages remove damaged red cells from the circulation, often becomes greatly enlarged (**Figure 27.14**). Public health workers can estimate the prevalence of malaria in a community by determining how many have an enlarged spleen.

Epidemiology and Control. Malaria can occur anywhere humans coexist with the *Anopheles* mosquitoes. This includes most tropical and many temperate regions of the world. In the United States there are two species: *Anopheles freeborni* in the West and *Anopheles quadrimaculatus* in the Southeast. Malaria was once common in North America (both George Washington and Abraham Lincoln had it). A handful of do-

mestically acquired malaria cases have been reported in Southern California in the last few years These were caused by parasites brought into the country from Latin American. But malaria is no longer endemic in the United States. Most malaria patients in the United States acquired their infection overseas.

Before 1940, about two-thirds of the world's people lived in malarial areas. During the 1950s and 1960s the disease was eradicated from most temperate regions because DDT made it possible to control mosquitoes inexpensively. Many believed that malaria would soon be conquered. But funding for basic research was cut. Antimalarial programs were relaxed. DDT-resistant mosquitoes emerged, and *Plasmodium falciparum* became drug-resistant. These blows brought a horrifying resurgence during the 1970s. In Sri Lanka, for example, annual cases of malaria were reduced from 3 million to only 18 by the early 1960s. Then it rose again to more than 1 million in the 1970s. Malaria is as prevalent today as it was early in the twentieth century. Only its distribution has changed. Today it is confined almost exclusively to tropical and subtropical countries.

It's now clear that worldwide malaria can't be controlled by insecticides. They've become more expensive, less effective, and harmful to the environment. Moreover, drug therapy today is significantly less effective than it was 30 years ago. Chloroquine, which replaced the centuries-old quinine cure, was extremely successful and inexpensive. It was effective both for treatment and prophylaxis until resistant strains of *P. falciparum* began to emerge (Chapter 21). The newest drug is mefloquine. But drug resistance is a constantly evolving problem, so additional agents are desperately needed. Research is hampered by the fact that the market for these drugs is mainly in poor countries that cannot afford to support expensive research or buy expensive drugs.

A vaccine could be the turning point in the global effort to control malaria. But vaccines are most effective against diseases for which natural immunity is strong and long-lasting. Unfortunately, natural immunity to malaria is weak, slow to develop, and temporary. Still, active research is underway to develop innovative new vaccines.

Until an effective vaccine becomes available, traditional "low-tech" strategies must be used to control malaria. Clinical studies show that one of the best is the use of mosquito netting impregnated with insecticides.

Toxoplasmosis. Toxoplasmosis is a zoonosis that infects many animal species, including reptiles, birds, and mammals. In the United States, about one-third of adults show serological evidence of having had toxoplasmosis.

The protozoan that causes it, *Toxoplasma gondii*, is an obligate intracellular parasite that exists in several forms. **Tachyzoites** are the rapidly multiplying cell type that can invade all mammalian cells except mature red blood cells. Tachyzoites can either cause the host cell to rupture or they can turn into inactive cysts that remain in infected tissue indefinitely. These cysts do not harm the infected

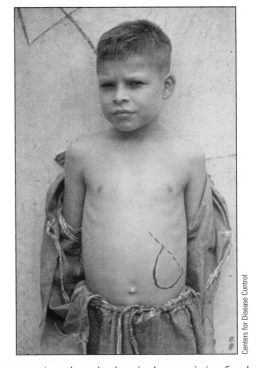

FIGURE 27.14   An enlarged spleen is characteristic of malaria, as shown in this young patient. His massively enlarged spleen is traced on his distended abdomen. Normally the spleen is about the size of a fist.

animal, but they can transmit the infection to other animals that eat them. A different cycle exists in cats. They're the only hosts in which *T. gondii* reproduces sexually. Oocysts are excreted in cat feces. They transmit the infection (**Figure 27.15**).

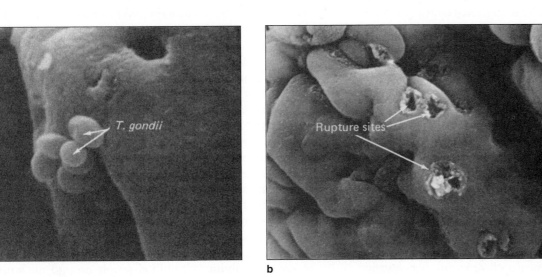

FIGURE 27.15   Electron micrographs showing the intestinal lining of a *Toxoplasma*-infected cat. (a) Protozoal parasites within cells. (b) Rupture site where mature oocytes have passed into the intestine to be excreted.

People become infected in two ways—by eating meat that contains tissue cysts or by eating oocysts that have been excreted by cats. *Toxoplasma* cysts are relatively common in meat. One study reported that 25 percent of pork samples and 10 percent of beef samples were contaminated. Cysts are destroyed by heating above 60°C (140°F), so thoroughly cooked meat is safe. Infection by oocysts from cat feces probably occurs when people inadvertently consume traces of contaminated cat feces from litter boxes, garden soil, or sandboxes.

Most adults with normal immune function suffer little from toxoplasmosis. As the immune response develops, tachyzoites disappear and harmless tissue cysts take their place. Temporary enlargement of lymph nodes may be the only sign of infection. Most people are never aware of their infection.

But people who are immunosuppressed and pregnant women are at risk for serious illness. In immunosuppressed hosts, toxoplasmosis can occur as a newly acquired infection. But more often it is a reactivation of prior infection. Usually the brain, heart, or lungs are affected. Diagnosis is made by identifying the microorganism in samples of infected tissues or fluids or by serology. Antimicrobial treatment with the sulfonamides and pyrimethamine keeps clinical manifestations from worsening. Sometimes it takes as long as a year to bring infection completely under control.

Infection of pregnant women during the early months of gestation can severely damage or kill the embryo, stimulating spontaneous abortion. If infection occurs during later months, the fetus may be born with congenital toxoplasmosis. Some congenitally infected babies are healthy at birth but later develop problems such as seizures, blindness, deafness, or mental retardation. Usually infants also have a rash and enlarged liver and spleen. Congenital toxoplasmosis of a newborn can be diagnosed by the presence of IgM antibodies against *T. gondii*. Because IgM antibodies don't cross the placenta, their presence is proof that the fetus was infected before birth.

No vaccine exists for toxoplasmosis. Prevention depends upon thoroughly cooking potentially infected meat and avoiding cat feces. Because oocysts are not infectious for the first 24 to 48 hours after they are passed, infection can be minimized by emptying cat litter boxes daily. Cats that are always kept indoors have no chance to acquire the infection by eating other infected animals.

Babesiosis. Babesiosis is caused by *Babesia* spp., protozoal parasites that resemble *Plasmodium*. Many species are found around the world—in Africa, Asia, Europe, and North America. Babesiosis is a zoonosis. A tick vector transmits the parasite from one mammalian host to another, usually deer, cattle, or small rodents. Humans become infected when they are bitten by an infected tick. Finding *Babesia* in

a patient's blood smear is diagnostic. Most human infections are relatively mild, but fever, headache, chills, fatigue, and weakness may occur. Infection can be fatal in the elderly or the immunosuppressed. It can be treated with clindamycin and quinine administered together, but most patients recover without treatment. In the United States, babesiosis is seen on the New England coast. The vector is *Ixodes dammini*, the same tick that transmits Lyme disease. Small rodents are the natural reservoir.

Helminthic Infections

Helminthic infections of the blood and lymphatic systems occur primarily in developing countries.

Schistosomiasis. Blood flukes, or schistosomes, are a type of trematode or flatworm that causes schistosomiasis. There are three major species of schistosomes. All have complex life cycles that involve a freshwater snail, so sometimes the infection is called **snail fever.**

The cycle begins when fresh water is contaminated with human feces or urine containing schistosome eggs (**Figure 27.16**). These hatch into larvae called **miracidia,** which parasitize a particular species of snail. The infected snail releases larvae called **cercariae** that can infect people. The cercariae penetrate the skin of people who wade in the contaminated water. Then cercariae enter the blood, where they mature into male and female worms. The female lives in a groove in the male's body. The two live in the bloodstream and produce enormous numbers of eggs. Different schistosomal species live in the blood vessels of different organs: *Schistosoma mansoni* in the colon, *Schistosoma japonicum* in the small intestine, *Schistosoma haematobium* in the bladder. The eggs pass from the blood into the organ and are excreted in the stool or the urine. Finding the eggs in a stool or urine sample is diagnostic.

Once infected, a person can remain infected for decades. That's because the schistosomes disguise themselves. Their surface becomes covered with host proteins, so the immune system does not recognize them as being foreign. As a result, enormous numbers of worms accumulate in the circulation. The patient experiences fever, fatigue, and pain in the affected organ. Eventually the infected tissues become so severely irritated that the body reacts against them as a foreign body, producing granulomas. Thus the extensive tissue damage that occurs with chronic schistosomiasis is a type of hypersensitivity.

Schistosomes are prevalent in Asia, Africa, and the Middle East. They cause a huge burden of human disease. At any time, about 200 million people around the world suffer from schistosomiasis. Ironically, economic development has increased the incidence because dams and irriga-

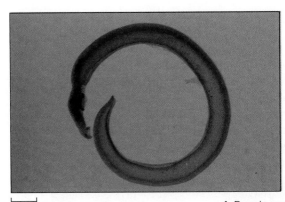

1 mm

FIGURE 27.16 The life cycle of a schistosome, a type of flatworm, requires a certain freshwater snail. An adult *Schistosoma japonicum* worm is shown at left.

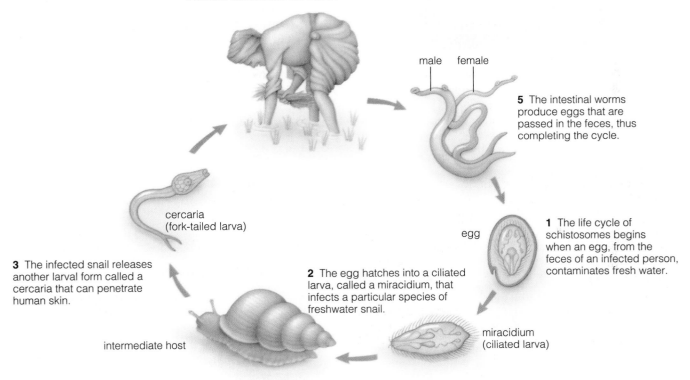

4 People wading in contaminated water who are infected by these cercariae develop the disease schistosomiasis when adult worms develop in veins near the intestine and interfere with normal circulation.

male   female

5 The intestinal worms produce eggs that are passed in the feces, thus completing the cycle.

cercaria (fork-tailed larva)

egg

1 The life cycle of schistosomes begins when an egg, from the feces of an infected person, contaminates fresh water.

3 The infected snail releases another larval form called a cercaria that can penetrate human skin.

2 The egg hatches into a ciliated larva, called a miracidium, that infects a particular species of freshwater snail.

miracidium (ciliated larva)

intermediate host

tion projects create new bodies of fresh water that are ideal habitats for schistosomes and their snail hosts. Although schistosomiasis is seen in the United States among immigrants from endemic countries, the infection can never become endemic in North America because the essential snail hosts are not found here.

Schistosomiasis can be treated with praziquantel or oxamniquine (Chapter 21). Such treatment is relatively expensive and not practical for the millions of victims around the world. Prevention through improved sanitation and controlling the vector snails would be cost effective.

Filariasis. Filariasis is an infection of the blood and lymph by either of two closely related nematodes, *Wuchereria bancrofti* and *Brugia malayi*. It's transmitted by species of *Anopheles, Culex,* and *Aedes* mosquitoes in Africa, Asia, and the Pacific Islands.

During a blood meal, these mosquitoes introduce the larvae of these nematodes beneath the skin. From there the larvae enter the lymphatic vessels, usually in the extremities or the groin. Within the lymph, they mature into adult worms. Nine months to a year later the worms mate and produce microfilariae that reenter the circulation,

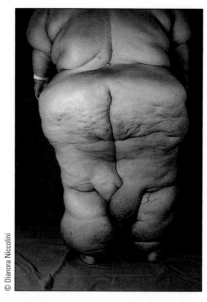

FIGURE 27.17   Elephantiasis.

where they can infect a new mosquito. Presence of micro-filariae in a patient's blood confirms diagnosis.

Many filariasis infections produce only fever and chills when parasites enter the blood. The most obvious manifestation of infection is **elephantiasis,** a swelling of a limb or the scrotum that results when adult worms block lymphatic circulation (**Figure 27.17**). Treatment with diethylcarbamazine is effective if the diagnosis is made before elephantiasis becomes established. Mosquito control is the best prevention.

the cd connection

MICROBIAL DISEASES ➡
Cardiovascular/lymphatic

Summarizes the normal biota, structure/function, clinical syndromes, and infections of the cardiovascular and lymphatic systems.

Table 27.4 summarizes the infections of the cardiovascular and lymphatic systems.

TABLE 27.4 Infections of the Cardiovascular and Lymphatic Systems

Infection	Causative Agent	Mode of Transmission	Symptoms	Prevention and Treatment
Infections of the Heart				
Endocarditis	Bacteria (usually), fungi (rarely)	Acute disease most common in otherwise healthy people who abuse intravenous drugs; subacute disease common in people with previous heart abnormality	Fever, fatigue, anemia, valvular damage, possibly heart failure	Avoid drug abuse; prevent subacute disease with antibiotic prophylaxis; treat with antibiotics, sometimes surgery to replace valves
Myocarditis	Viruses (usually), rarely bacteria, fungi	Usually a complicaiton of an everyday viral illness	Irregular heartbeat, chest pain, heart failure	Supportive care
Trypanosomiasis (Chagas' disease)	*Trypanosoma cruzi*	Bite of reduviid bug	Protozoal myocarditis; fever, fatigue, face swells; can be acute or chronic	Avoid bites; treat acute cases with nifurtimox; treat chronic heart disease with medications to reduce symptoms
Pericarditis	Viruses (usually), baceria or fungi (occasionally)	Usually a complication of an everyday viral disease or a chronic bacterial or fungal infection	Few; may cause scarring or pericardial tamponade and thereby affect heart function	Supportive care; possibly a pericardectomy
Systemic Infections Bacterial				
Plague	*Yersinia pestis*	Infected flea bite, respiratory droplets	Chills, fever, swollen lymph nodes (bubos), black patches of dead skin	Rat abatement programs, quarantine; treat with streptomycin, tetracycline, chloramphenicol

TABLE 27.4 Infections of the Cardiovascular and Lymphatic Systems (continued)

Infection	Causative Agent	Mode of Transmission	Symptoms	Prevention and Treatment
Bacterial				
Tularemia	*Francisella tularensis*	Direct contact with an infected animal (usually rabbit in U.S.); bite of infected tick or fly	Ulcer at site of infection, fever, fatigue, headache	Avoid contact with infected animals; treat with streptomycin or tetracycline
Brucellosis	*Brucella* spp.	Direct contact with infected animals; consuming infected unpasteurized dairy products	Fever (often with characteristic undulating pattern), fatigue, headaches, loss of appetite	Eliminate infected animals from domestic herds; pasteurize milk; treat with tetracycline or streptomycin
Lyme disease	*Borrelia burgdorferi*	Infected tick bite	Bull's-eye rash and fever in acute form; chronic infection can damage joints, heart, nervous system; can cause chronic arthritis	Avoid or remove ticks; treat early with doxycycline, amoxycillin, or erythromycin
Relapsing fever	*Borrelia* spp.	Body louse (epidemic disease) or tick (endemic disease)	Recurrent episodes of fever, headache, muscle aches	Avoid contact with vectors; treat with tetracycline, chloramphenical, penicillin, or erythromycin
Anthrax	*Bacillus anthracis*	Contact with infected wool or hides; rarely, respiratory droplets	Black sore at site of infection, fever, shock, severe pneumonia	Eliminate infected animals from domestic herds; avoid contact with animal products from endemic countries; treat with penicillin
Cat scratch disease	*Bartonella hensalae*	Direct contact with infected animal	Usually one swollen lymph node and healing scratch	Avoid contact with infected animals; treat with trimethoprim-sulfamethoxazole or ciprofloxacin
Rocky Mountain spotted fever	*Rickettsia rickettsii*	Infected tick bite	Characteristic rash over entire body, including palms and soles, fever	Avoid vector; treat with tetracycline or chloramphenicol
Epidemic typhus	*Rickettsia prowazekii*	Infected body louse bite	Unbearable headache, fever, muscle aches, chills; often delirium or unconsciousness; rash over body except palms and soles; shock; can be fatal	Good hygiene; treat with tetracycline or chloramphenicol
Murine typhus	*Rickettsia typhi*	Infected fleabite	Mild fever, headache, muscle aches; rickettsial rash	Avoid rat-infested areas; treat with tetracycline or chloramphenicol
Viral				
Yellow fever	Yellow fever virus	Infected mosquito bite	Fever, headache, jaundice	Avoid mosquito bites; vaccine available; supportive treatment only

continued

TABLE 27.4 Infections of the Cardiovascular and Lymphatic Systems (continued)

Infection	Causative Agent	Mode of Transmission	Symptoms	Prevention and Treatment
Viral				
Dengue fever	Dengue fever virus	Infected mosquito bite	Fever, headache, rash; rarely, dengue hemorrhagic fever	Avoid mosquito bites; supportive treatment only
Infectious mononucleosis	Epstein-Barr virus	Person-to-person	Fever, fatigue, sore throat, enlarged lymph nodes and spleen	No prevention (infection is almost universal); treat symptoms only
Burkitt's lymphoma	Epstein-Barr virus	Person-to-person but only in children in equatorial Africa	Malignant jaw tumor	No prevention; curable with chemotherapy
Acquired immunodeficiency syndrome	Human immunodeficiency virus	Sexual contact, exposure to infected blood or body fluids, mother to infant during pregnancy or near time of birth	Sometimes a mononucleosis-like illness at time of infection; years later, opportunistic infections develop; eventually fatal	Prevent with safer sex practices and avoiding illicit intravenous drugs; HAART very effective, no cure
Protozoal				
Malaria	*Plasmodium* spp.	Infecting mosquito bite	Cyclic chills and fever; tissue damage from red blood cell destruction	Avoid mosquito bites; treat with antimalarials chloroquine or mefloquine; vaccine in field trials
Toxoplasmosis	*Toxoplasma gondii*	Eating meat from an infected animal or ingesting organisms in traces of cat feces	None or few; very serious in immunosuppressed patients and fetus	Cook meat well; avoid infected cat feces; treat with sulfonamides or pyrimethoamine
Babesiosis	*Babesia* spp.	Infected tick bite	Mild fever, headache, chills	Avoid tick bites; treat with clindamycin and quinine together
Helminthic				
Schistosomiasis	*Schistosoma* spp.	Wading in contaminated water; freshwater snails are essential intermediate host	Fever, fatigue, muscle aches, chronic damage to organs; gravely debilitating	Keep water free of human feces and urine; control snail host; treat with praziquantel or oxiaminiquine
Filariasis	*Wuchereria bancrofti* and *Brugia malayi*	Infected mosquito bite	Body parts swell where lymphatic vessels are blocked; debilitating and disfiguring	Avoid mosquito bites; treat with diethylcarbamazine

SUMMARY

Case History: Risky Business (p. 659)

1. Intravenous drug use is a likely cause of endocarditis.

The Cardiovascular and Lymphatic Systems (pp. 659–662)

2. The cardiovascular system consists of the heart, blood vessels, and blood. The lymphatic system consists of lymph vessels and nodes, the spleen, and the thymus. It is also called the immune system.

3. Clinical syndromes of heart infections are endocarditis, myocarditis, and pericarditis. Noncardiac infections of the cardiovascular and lymphatic systems are called systemic infections.

Infections of the Heart (pp. 662–664)

4. Bacterial endocarditis is characterized by vegetations. Emboli are pieces of vegetations that break off, enter the circulation, and interrupt blood flow. They cause death if they enter a vital organ. Vegetations that damage heart valves can cause heart failure. Symptoms of bacterial endocarditis include fever and anemia.

5. Acute bacterial endocarditis is sudden in onset and progresses rapidly. *Staphylococcus aureus* causes most of these infections in intravenous drug users. Treatment is with cephalosporins and penicillinase-resistant penicillins. Heart valves may be surgically replaced.

6. Subacute bacterial endocarditis starts gradually, progresses slowly, and lasts longer than acute disease. Symptoms are fever and fatigue. People with heart defects are at risk. Without treatment, the infection is fatal. Prophylactic treatment with penicillin is routine before oral or other surgery.

7. Viral myocarditis is often caused by coxsackievirus. The only symptom may be an irregular or rapid heartbeat. Treatment consists of rest. The prognosis is good. Rarely the condition can be life threatening.

8. *Trypanosoma cruzi* causes American trypanosomiasis (Chagas' disease). It is transmitted by the reduviid bug. The acute stage is characterized by fever, fatigue, and swelling around the bite. During the chronic stage the heart and other organs become damaged. The acute stage is treated with nifurtimox. The only treatment during the chronic stage is supportive care.

9. Viral pericarditis is usually a complication of a viral infection. It is usually benign and self-limited. The most common agents are enteroviruses.

10. Bacterial pericarditis is usually caused by *Staphylococcus* or *Streptococcus* spp. that spread from a lung or a systemic infection. Pericardial tamponade can cause sudden death.

11. Chronic fungal and mycobacterial infections can also cause pericarditis. The resulting inflammation can cause scarring. It can lead to heart failure. A pericardectomy can be performed.

Systemic Infections (pp. 664–686)

12. Any bacterial infection can spread to the blood. Bacteremias are brief infections. Septicemias are persistent and serious.

Bacterial Infections (pp. 664–674)

13. *Yersinia pestis* causes plague. In the United States, rodents west of the Rocky Mountains are a permanent reservoir. Transmission is by a flea bite. A nearby lymph node becomes massively enlarged and tender (a bubo). Bacteria multiply and enter the blood, spreading to all the organs. Black spots appear on the skin from subcutaneous bleeding. Plague spread by fleas is called bubonic plague. If the lungs become infected, bacterial cells are aerosolized and the infection is spread by the respiratory route (pneumonic plague). Early treatment with streptomycin, tetracycline, or chloramphenicol is usually effective. A vaccine is available for high-risk individuals.

14. *Francisella tularensis* causes tularemia. Rabbits and ticks are the main reservoirs. Bacteria can be ingested or inhaled. They can also enter through a break in the skin. The most common syndrome is ulceroglandular tularemia: A small sore appears where the bacteria enter the body. Diagnosis is difficult. Untreated, the illness can be persistent and serious, with relapses. Streptomycin is the drug of choice.

15. *Brucella* spp. cause brucellosis. Transmission is by direct contact with infected animals or their secretions. Before pasteurization, brucellosis was often milkborne. Brucellosis is also called undulant fever because of characteristic recurrent fever. Recovery is usually spontaneous, but chronic illness can last for months. Diagnosis is difficult. Treatment is with tetracycline and streptomycin.

16. *Borrelia burgdorferi* was discovered as the cause of Lyme disease in 1982. The first sign of infection is often a bull's-eye rash. During the next stage, bacteria spread, causing headache, muscle aches, and fatigue, often followed by serious illness, such as meningitis. The late stage may last for years, with chronic arthritis and painful joint swelling. *Ixodes* ticks are the primary vectors. Without the characteristic bull's-eye rash, diagnosis can be difficult. Treatment is with doxycycline, amoxicillin, and erythromycin.

17. *Borrelia* spp. cause relapsing fever. Onset is sudden, with fever, headache, muscle aches, enlarged liver and spleen, and a faint rash on the trunk. The patient appears to recover in 3 to 6 days but a week later

has a relapse. The cycle continues, with up to 13 relapses. Epidemic relapsing fever is transmitted by the body louse and has a high mortality; endemic relapsing fever is transmitted by ticks and is comparatively mild. Tetracycline, chloramphenicol, and penicillin are effective.

18. Traditionally *Bacillus anthracis* causes anthrax, an animal disease humans can contract through a break in the skin (cutaneous anthrax) or by inhaling bacterial spores (respiratory anthrax). Rare cases of anthrax occur in the United States from anthrax-contaminated hides or wool imported from endemic countries. In 2001 cases of respiratory anthrax were contracted from contaminated letters; 6 of 10 patients died.

19. *Bartonella henselae* causes cat scratch disease, characterized by one extremely swollen lymph node and usually no other symptoms. In rare cases the patient complains of fever, fatigue, or symptoms of meningitis. The infection resolves itself, but trimethoprim-sulfamethoxazole or ciprofloxacin speeds recovery.

20. *Rickettsia rickettsii* causes Rocky Mountain spotted fever. The bacterium is passed transovarially, so no animal reservoir other than ticks is needed. Early signs include fever, headache, vomiting, and a distinctive rash on the wrists and ankles that eventually spreads over the entire body, including the palms and soles. Diagnosis is difficult; there is no test to identify *R. rickettsii*. Untreated, there is risk of overwhelming infection leading to shock and death. Treatment is with tetracycline and chloramphenicol.

21. *Rickettsia prowazekii* causes epidemic (louseborne) typhus. Infection occurs when a person scratches a bite and rubs louse feces, with their rickettsial parasites, into the skin. Onset is sudden, with severe headache, fever, muscle aches, and shaking chills, often leading to delirium and unconsciousness. A rash resembling that of Rocky Mountain spotted

fever appears but does not cover the palms, soles, or face. Tetracycline and chloramphenicol are effective. Mild recurrences of epidemic typhus—called Brill-Zinsser disease—may appear years later.

22. *Rickettsia typhi* causes murine (fleaborne) typhus, which is much milder than epidemic disease. Transmission occurs when humans come in contact with infected rodents. Onset is gradual, followed by a typical rickettsial rash. Treatment is with tetracycline or chloramphenicol.

Viral Infections (pp. 674–681)

23. The yellow fever virus causes yellow fever, transmitted principally by the *Aedes aegypti* mosquito. Symptoms include high fever, severe chills and headache, liver damage causing jaundice, and sometimes uncontrollable bleeding. Mosquito-control programs have eliminated urban yellow fever, but it is endemic in jungles of the Americas. There is a vaccine, but it is seldom used.

24. Several flaviviruses cause dengue fever, also transmitted by the *Aedes aegypti* mosquito. High fever and headache are followed by a rash and intense limb pain; periods of recovery and recurrent illness alternate. Treatment is limited to supportive care. Occasionally children who have recovered from dengue fever become reinfected and develop dengue hemorrhagic fever or dengue shock syndrome, which is fatal in 40 percent of cases.

25. The Epstein-Barr virus (EBV) causes infectious mononucleosis. As a member of the herpesvirus family, EBV establishes an active and latent infection. Transmission is through oral secretion, directly or via fomites. Symptoms include fever, fatigue, sore throat, and swollen lymph nodes. EBV has been proved oncogenic, causing B-cell lymphoma cancer in immunosuppressed cancer patients and organ transplant recipients; Burkitt's lymphoma, a jaw cancer in African children; and na-

sopharyngeal carcinoma, which occurs only in China.

26. The human immunodeficiency virus (HIV) causes acquired immunodeficiency syndrome (AIDS). HIV cripples the immune system—in particular, CD4 T lymphocytes—making a patient vulnerable to other infections. When the CD4 T count falls below 200 per mm^3, opportunistic infections of the skin and mucous membranes appear. When the count falls below 100, life-threatening systemic illnesses occur, sometimes accompanied by mental deterioration. If untreated, death is inevitable. With modern therapy called HAART, life might be prolonged indefinitely, but the therapy is very costly. Transmission occurs through sexual contact that transfers infected body fluids, through infected blood from transfusions or shared needles, and from mother to infant during pregnancy or around the time of birth. During the prolonged asymptomatic period, an individual may unknowingly transmit the virus to others.

27. The epidemiology of HIV in the United States has shifted from homosexual men to drug users sharing contaminated needles; this change has led to increased sexual transmission in the heterosexual community and a rapidly growing number of infected women and children. In Africa the disease is spread almost exclusively by heterosexual intercourse. Diagnosis is by serological testing. Prevention depends upon education on safer sex, campaigns against drug use, and—among health-care providers—following universal precautions. Because there is no cure, treatment is designed to prolong survival.

28. Ebola virus causes a horrible hemorrhagic fever in Africa. It appears to be limited to occasional outbreaks.

Protozoal Infections (pp. 681–684)

29. Four species of *Plasmodium* cause malaria. Transmission is by *Anopheles*

mosquitoes. Sporozoites deposited during a bite travel to the liver, where they produce merozoites. As infected liver cells lyse, merozoites enter red blood cells to multiply. Red blood cells lyse in cycles, every 48 to 72 hours, producing attacks of shaking chills, high fever, and drenching sweat. Patients develop anemia and a swollen spleen; some die from cerebral malaria or blackwater fever. Treatment is with chloroquine and mefloquine, but resistance to both drugs is emerging.

30. *Toxoplasma gondii* causes toxoplasmosis, a zoonosis. Humans acquire the infection by eating meat containing cysts or by eating oocysts excreted by cats. *T. gondii* exists as tachyzoites, as well as inactive cysts. Enlarged lymph nodes may be the only symptom; but in immunosuppressed hosts, the brain, heart, or lungs are often affected. A pregnant woman risks abortion or an infant born with congenital toxoplasmosis, producing seizures, blindness, deafness, or retardation. Prevention depends upon cooking potentially infected meat and avoiding infected cat feces.

31. *Babesia* spp. cause babesiosis, another zoonosis. In the United States, babesiosis is primarily found in New England, where it is transmitted by the same tick that transmits Lyme disease. Symptoms include fever, headache, chills, and fatigue, but recovery is usually spontaneous. Serious cases may be treated with clindamycin and quinine in combination.

Helminthic Infections (pp. 684–686)

32. Schistosomes (blood flukes) are trematodes that cause schistosomiasis. An infected human contaminates water with egg-carrying feces; larvae, called miracidia, hatch from the eggs. Miracidia parasitize a particular species of snail that in turn releases larvae called cercariae. Cercariae penetrate the skin of humans wading in contaminated water and

enter the blood vessels of various organs. Enormous numbers of worms can develop in the circulation over decades. Symptoms include fever, fatigue, and pain in the affected organs. Treatment is with praziquantel or oxamniquine. Prevention through sanitation and eliminating the vector snails is best. Schistosomiasis can never become endemic to the United States because the essential snail host is not found here.

33. *Wuchereria bancrofti* and *Brugia malayi* are nematodes that cause filariasis. Transmission is by species of mosquitoes found in Africa, Asia, and the Pacific Islands. Larvae enter the lymphatic vessels in the groin or extremities and mature; worms mate and produce microfilariae that reenter the circulation. Treatment with diethylcarbamazine is effective only before elephantiasis appears. Mosquito control is the best prevention.

REVIEW QUESTIONS

Case History: Risky Business

1. How can intravenous drug use cause endocarditis?

The Cardiovascular and Lymphatic Systems

2. Describe the organs of the cardiovascular system and their functions. Describe the organs of the lymphatic system and their functions.

3. Define these clinical syndromes: endocarditis, myocarditis, pericarditis. What is a systemic infection, and why is it difficult to diagnose?

Infections of the Heart

4. What's the differences between acute and subacute bacterial endocarditis with respect to causative microorganisms, symptoms, individuals at high risk, and treatment? What are vegetations? What medical problems can they cause?

5. What type of myocarditis is most often found in the United States? What are the most common causative microorganisms, symptoms, and treatment?

6. What type of heart infection is Chagas' disease? What is the causative microorganism? How is it transmitted? Where are the endemic areas? Describe the pathogenesis, diagnosis, and treatment.

7. What are the differences among pericarditis caused by bacteria, viruses, and fungi?

Systemic Infections

8. Define the terms bacteremia and septicemia.

9. How is plague transmitted? What are its symptoms? How is it treated? How can it be prevented? What is pneumonic plague?

10. How is tularemia transmitted? What is its most common clinical syndrome? How is it treated?

11. Who is at highest risk of contracting brucellosis?

12. Is Lyme disease a new disease? Explain. What are the three clinical stages, vectors, and mode of transmission in different parts of the country? What are the diagnosis and treatment?

13. What are the epidemiological patterns of endemic and epidemic relapsing fever? Why must antibiotic treatment be gradual?

14. What's the difference between cutaneous anthrax and respiratory anthrax? Can we expect to eradicate anthrax? Explain.

15. How serious is cat scratch disease? Explain.

16. What is the clinical syndrome of Rocky Mountain spotted fever? How is it treated?

17. What are the differences between the two forms of typhus with respect to causative microorganism, transmission, clinical syndromes, treatment, prevention, and prevalence in the United States?

18. What is the principal vector and mode of transmission of yellow fever? What distinguishes the urban form from the endemic form?

19. Explain this statement: Dengue fever resembles yellow fever in its epidemiology and mode of transmission.

20. What latent and active infections are caused by the Epstein-Barr virus? What is its relation to cancer?

21. What type of virus is HIV and of what clinical significance is this? Which human cells are most damaged by HIV? What is their function? Describe the infective process that begins with infection and ends with death. How is AIDS transmitted? How is AIDS not transmitted?

22. Describe the clinical syndrome of AIDS. Discuss the prevention and treatment of AIDS. How is the epidemiology of AIDS changing in the United States and why?

23. How does the life cycle of *Plasmodium* cause the symptoms of malaria? What is the best current control for malaria?

24. How do humans acquire toxoplasmosis? How serious is this infection? How can it be prevented?

25. What are the similarities and differences between babesiosis and Lyme disease in terms of transmission, symptoms, and virulence?

26. Describe how the schistosome life cycle causes the clinical syndrome of schistosomiasis. What is the treatment? Explain this statement: Schistosomiasis can never become endemic in North America.

27. How do the life cycles of *Wuchereria bancrofti* and *Brugia malayi* cause filariasis and its most obvious clinical manifestation, elephantiasis?

CORRELATION QUESTIONS

1. What are the similarities and differences between the cause of black spots of plague and black urine of malaria?

2. What's the common feature of relapsing fever and epidemic typhus? Of relapsing fever and Lyme disease?

3. What are the features of anthrax that currently make it the most feared agent of biological warfare?

4. Why is it easier to develop a vaccine against yellow fever than AIDS?

5. Is schistosomiasis or malaria a greater threat to become endemic in the United States? Why?

6. What marked similarity is there between cat scratch fever and plague?

ESSAY QUESTIONS

1. Discuss the ethical questions involved in developing and testing an AIDS vaccine.

2. Discuss what you would do or recommend that others do to ensure that an Ebola hemorrhagic fever outbreak doesn't occur in the United States.

SUGGESTED READINGS

AIDS: The unanswered questions. 1993. *Science* May 28, Special Issue.

Aoun, H. 1989. When a house officer gets AIDS. *New England Journal of Medicine* 321:693–96.

Barrett, A. 1996. *Ship Fever and Other Stories.* New York: W.W. Norton.

Defeating AIDS: What will it take? 1998. *Scientific American* July, Special Issue.

Durack, D. T. 1988. Rus in Urbe: Spotted fever comes to town. *New England Journal of Medicine* 318:1388–90.

Kahn, J. O., and B. D. Walker. 1998. Acute human immunodeficiency virus type 1 infection. *New England Journal of Medicine* 339:32–39.

Marshall, E. 1990. Malaria research— what next? *Science* 247:399–402.

McCabe, R., and J. S. Remington. 1988. Toxoplasmosis: The time has come. *New England Journal of Medicine* 318:313–15.

McEvedy, C. 1988. The bubonic plague. *Scientific American* February: 118–23.

Relman, A. S. 1984. Epstein-Barr virus—immortalization and replication. *New England Journal of Medicine* 310:1255–57.

Steere, A. C. 1989. Lyme disease. *New England Journal of Medicine* 321:586–97.

Barcelona: United Nations. 2002. *UN AIDS Report.*

Yung, J. A., and R. J. Collier. 2002. Attacking anthrax. *Scientific American* March: 48–59.

Zinsser, H. 1934. *Rats, lice, and history.* New York: Bantam.

For additional readings, go to InfoTrac College Edition, your online research library at: http://www.infotrac.thomsonlearning.com

TWENTY-EIGHT

Microorganisms and the Environment

© Ken Sakamoto/Black Star

CHAPTER OUTLINE

LEARNING GOALS

To understand:

- *How the evolution of organisms led to our present environment*
- *Soil and water as habitats for the microorganisms that mediate major biogeochemical transformations*

- *The major cycles of matter—the nitrogen cycle, the carbon cycle, the phosphorus cycle, and the sulfur cycle—and the microorganisms involved*
- *The ecological impact of the major cyclic transformations of matter*

- *How knowledge of the cycles of matter is applied to treating wastewater and drinking water*
- *How some humanmade chemicals escape from the carbon cycle and what can be done about it*

When Help Becomes Harm

The patient was a 71-year-old woman with a long history of respiratory problems. After more than 50 years of smoking, her lungs were so damaged by emphysema that she had trouble breathing even during normal activity. When she was admitted to the hospital with a broken hip, breathing became more difficult. Her doctor prescribed an aerosolized medication; she would inhale the drug in a fine mist. These breathing treatments were helpful, but soon the patient became critically ill with what appeared to be pneumonia. Fortunately, legionellosis was quickly diagnosed (Chapter 22). She was immediately treated with tetracycline and given supportive care in the intensive care unit, which allowed her to make a complete recovery. Curiously, water—rather than air or exposure to another person—was the source of this patient's infection. The bacterium that causes legionellosis, *Legionella pneumophila*, grows in water—lakes, streams, air-conditioning towers, and even the water supplies of hospitals. The water used to prepare this patient's breathing treatments was contaminated. As is often the case, other patients in the hospital were also affected.

Case Connection

- Although municipal water supplies are usually free of pathogenic microorganisms, extreme care must be taken, as this case illustrates, to prevent subsequent contamination.

LIFE AND THE EVOLUTION OF OUR ENVIRONMENT

Earth is a rare environment in its suitability for life. The more biologists learn from space probes, the more they think life does not exist elsewhere in our solar system.

Of the many requirements of carbon-based life (life as we know it), liquid water is paramount. For a planet to support life, it must orbit at a suitable distance from its sun to maintain a temperature in the narrow range in which water is liquid. Also, it must have enough mass for its gravity to keep water within its atmosphere. And it must have usable sources of the major biological elements—carbon, hydrogen, nitrogen, oxygen, phosphorus, and sulfur (abbreviated CHNOPS).

The environment determines whether life can evolve. Then life changes the environment. Because microorganisms are so plentiful and their metabolic activities are so diverse, they have a major impact on the environment. Collectively, the biologically mediated changes in Earth's chemical composition are called **biogeochemical transformations** (*bio*, "life"; *geo*, "earth").

Life and Earth's environment evolved together, producing the near-balanced state of coexistence we see today. When Earth was formed about 4.6 billion years ago, there was no oxygen in the atmosphere. Laboratory experiments show that an oxygen-free atmosphere suited the slow formation and accumulation of building blocks of life, including amino acids, nucleic acid bases, carbohydrates, and lipids. In such experiments, electrical discharges (mimicking lightning) provide the energy for the gases in the primitive atmosphere to react. In the first and most famous experiment (**Figure 28.1**) the atmosphere was assumed to be composed of methane, ammonia, hydrogen, and water. Recent findings have cast doubt on this assumption. But other experiments with other chemical mixtures confirm the observation that building blocks of life are made if the atmosphere is reducing. Even if the prelife atmosphere were not sufficiently reducing, building blocks would have come to Earth, because meteors have been shown to contain them. From whatever source, these compounds would have accumulated because no microorganisms were present to metabolize them (see the beginning of Chapter 2).

The first organisms on Earth were probably anaerobic, heterotrophic organisms. They were formed from and lived on the accumulated building blocks. But the oldest known fossils (3.5 billion years old) are of filamentous organisms that look quite like the cyanobacteria found in soil, lakes, and streams today. These ancient cyanobacteria, like modern forms, probably carried out oxygenic photosynthesis. Significant amounts of the oxygen they produced began to accumulate in Earth's atmosphere about 2 billion years ago. Between 800 and 600 million years ago, the concentration of oxygen became high enough to support the metabolism of aerobic organisms. That's when plants and animals appeared. Today oxygen constitutes 21 percent of Earth's atmosphere. This level is

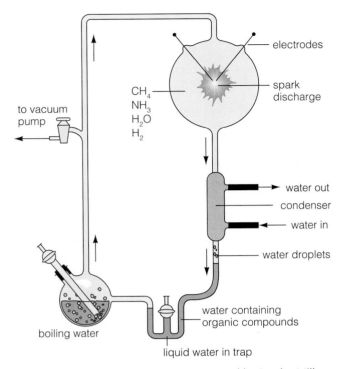

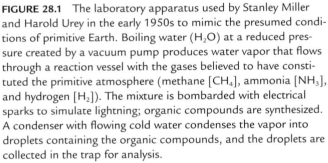

FIGURE 28.1 The laboratory apparatus used by Stanley Miller and Harold Urey in the early 1950s to mimic the presumed conditions of primitive Earth. Boiling water (H_2O) at a reduced pressure created by a vacuum pump produces water vapor that flows through a reaction vessel with the gases believed to have constituted the primitive atmosphere (methane [CH_4], ammonia [NH_3], and hydrogen [H_2]). The mixture is bombarded with electrical sparks to simulate lightning; organic compounds are synthesized. A condenser with flowing cold water condenses the vapor into droplets containing the organic compounds, and the droplets are collected in the trap for analysis.

a balance between oxygen produced by photosynthesis and that used by respiration and combustion.

Over the last 300 to 600 million years, various chemical forms of other life-sustaining elements stabilized in a similar way. Each is formed and used at about the same rate. These biogeochemical cycles of formation and utilization supply organisms with the nutrients they need. They maintain life on Earth.

MICROORGANISMS IN THE BIOSPHERE

All living things contribute to maintaining life. But microorganisms play particularly important roles. Microorganisms transform huge quantities of matter, and only they can carry out certain essential transformations. These transformations occur everywhere in the **biosphere** (the

TABLE 28.1	Microorganisms in a Typical Sample of Soil
Microorganism	**Amount in a Gram of Soil**
Bacteria	10^6 to 10^9 cells
Actinomycetes	10^5 to 10^8 spores
Fungi	10 to 100 meters of hyphae
Yeasts	10^3 cells
Algae and cyanobacteria	10^2 to 10^4 cells
Protozoa	10^4 to 10^6 cells

[a]Not including actinomycetes and cyanobacteria.

Source: Yanagita, T. 1990. *Natural microbial communities.* Tokyo: Japan Scientific Societies Press.

region of Earth that can support life). Many transformations occur in the soil, others in aquatic environments.

Soil

Although it appears inert, soil, particularly its upper layer, teems with microorganisms (**Table 28.1**). They bring about many chemical transformations, and they are vital to soil fertility.

Soil is a complex mixture of chemicals. It consists of inorganic materials, including various minerals. Oxides of iron, aluminum, and silicon usually predominate, but it also contains many anions and cations. Soil also contains organic materials—the residues of dead organisms. Some these organic residues, such as **lignin** (the stable component of woody plants) are long-lasting because they are resistant to microbial decomposition. They accumulate and form the organic fraction of soil called **humus,** which gives soil its brown or black color.

Mineralization. In the transformation known as **mineralization,** microorganisms convert organic material in soil to an inorganic form. The rate and extent of mineralization depends largely upon the availability of oxygen. Compared with anaerobic metabolism, aerobic metabolism is more versatile, so a greater variety of compounds are attacked. Aerobic metabolism is also more complete, producing carbon dioxide and water instead of organic acids and alcohols (Chapter 5). Many organic materials are mineralized only if oxygen is available, but when it is relatively dry and loose, oxygen penetrates soil readily, down to a foot or so. Even then, small regions within soil particles are anaerobic because oxygen-consuming microorganisms use the oxygen

LARGER FIELD

NOT JUST ANOTHER SOIL FUNGUS

Many species of mycorrhizal fungi produce edible mushrooms, including *Amanita* spp. and *Boletus edulis*. Some of these are prized gourmet treats. But the most desirable and certainly the most expensive mushroom is *Tuber melanosporum* and the related species *Tuber* *aestivum, Tuber brumale,* and *Tuber uncinatum*—better known as truffles. Truffles are ascomycetes that form mycorrhizal associations with forest trees, predominantly hazelnut and oak trees. But the edible parts of truffles (their ascocarps; Chapter 12) are produced underground, which makes them difficult to find. Female pigs are extremely sensitive to the smell of truffles because it resembles the odor of a sexually active boar. In France, trained pigs smell the underground truffles and root them up with their snout. Hunting truffles with pigs is so successful and the market so good that over the years there has been a precipitous decline in production. In 1892, 2000 tons of truffles were produced in France. In 1973 only 60 tons were produced. Since then, artificial inoculation of forest soils with *Tuber* spp. has increased production.

FIGURE 28.2 This well-preserved body of a Bronze Age man was buried for more than 4000 years in a bog in Denmark.

faster than it diffuses in. When soils are flooded, they rapidly become anaerobic because water slows diffusion of oxygen to less than the rate that aerobic microorganisms use it. As a result, mineralization proceeds slowly in waterlogged soils such as swamps and bogs. Such slowed mineralization was dramatically demonstrated in the 1960s when the body of a Bronze Age man was found almost intact in a bog in Denmark (**Figure 28.2**). The principle is further illustrated by the fact that waterlogged soils typically contain more than 90 percent organic material, whereas well-aerated agricultural soils usually contain less than 10 percent.

Soil **fertility** (its ability to support plant growth) depends largely upon an adequate supply of inorganic nitrogen, phosphorus, and potassium. The chemical forms of these nutrients that plants can use are produced by microorganisms as they mineralize organic material. Fertilizers are added to enrich a soil's complement of these elements. Potassium is added to fertilizer largely as an inorganic salt. Nitrogen and phosphorus can be added in either organic or inorganic form because the organic forms are readily mineralized by microorganisms. Thus, through mineralization, microorganisms improve soil fertility even when commercial fertilizers are used. Some bacteria add nitrogen to soil. They do so by **nitrogen fixation,** converting gaseous atmospheric nitrogen into solid forms.

Microorganisms in the Soil. Many organisms that live in the soil, including insects, worms, and small vertebrates, are visible to the naked eye. But the vast majority, in terms of weight and metabolic capacity, are microorganisms. Most of these are bacteria (Table 28.1, **Figure 28.3**).

© Ken Wagner/Phototake

FIGURE 28.3 The vast majority of microorganisms living in the soil are bacteria, as shown in this micrograph of sandy soil.

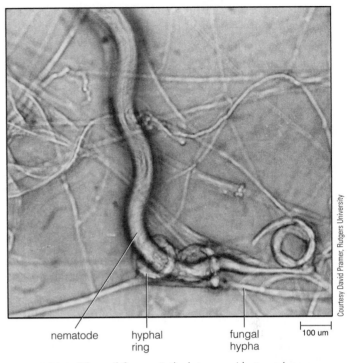

Courtesy David Pramer, Rutgers University

nematode   hyphal   fungal |——| 100 um
              ring   hypha

FIGURE 28.4 The soil fungus *Arthrobotrys conoides* trapping a nematode.

Bacteria. Reflecting the many microenvironments within soil, bacteria found there are extremely diverse. They include representatives with varying response to oxygen (aerobes, anaerobes, and facultative anaerobes), temperature (thermophiles, mesophiles, and psychrophiles), and pH. For example, the presence in soils of thermophilic microorganisms in temperate regions reflects the fact that the soil surface can become quite hot during a sunny day.

Actinomycetes, aerobic Gram-positive bacteria that form branching mycelia (Chapter 11), are important contributors to the ecology of soil. They break down plant and animal remains and keep the soil loose. As many as a hundred million actinomycete colonies, representing more than 20 genera, can be recovered from 1 g of some soils. The most numerous and widely distributed are species of *Streptomyces*. In fact, they give soil its typical odor by producing two volatile substances, geosmin and 2-methyl-isoborneol. Even when grown on laboratory media, most species of *Streptomyces* produce the same earthy odor. Actinomycetes are particularly significant degraders of complex polymers (including chitin) and hydrocarbons, which are relatively resistant to attack by other microorganisms.

A few bacteria that cause human disease are found in soil. They include *Bacillus anthracis*, which causes anthrax (Chapter 27); *Clostridium perfringens*, which causes food poisoning (Chapter 23); *Clostridium tetani*, which causes tetanus (Chapter 25); and *Clostridium botulinum*, which causes botulism (Chapter 25). *Pseudomonas aeruginosa*, which causes opportunistic infections in burn patients and immunologically weakened individuals (Chapter 26), occurs in almost all soils.

The bacterial population of soil changes rapidly as new nutrients become available and existing ones are exhausted.

Fungi. Fungi also degrade organic materials in aerated soils. Most fungi are obligate aerobes. They break down both simple compounds, such as sugars and organic acids, and complex polymers, such as cellulose, starch, pectin, and lignin. Colony counts commonly underestimate the fungal population of soil because a mass of fungal hyphae may produce only a single colony when plated. But total fungal **biomass** (total weight of organisms) is quite impressive. An acre of soil contains between 500 and 5000 pounds of fungi.

Some soil fungi are predators. They trap protozoa or nematodes by producing hyphae that form rings (**Figure 28.4**). Then the fungus puts out hyphae that invade the captured prey and digest it. These predatory fungi significantly limit the populations of soil protozoa and nematodes.

Certain soil fungi, notably species of *Trichoderma* and *Laetisaria*, are mycoparasites. They attack other fungal species, including some that cause plant disease. Treating soil and seeds with mycoparasitic fungi can protect plants from disease. For example, *Trichoderma harzianum* controls **damping-off** (a disease that kills seedlings by blackening and shrinking their stems) of beans, peas, and radishes by killing the fungi—*Rhizoctonia solani* and *Pythium* spp.—that cause these plant diseases. Similarly, *Trichoderma hamatum* improves survival of sugar beet seedlings. *Laetisaria arvalis* protects seedlings of many species from fungal pathogens. Use of mycoparasites for agriculture is not yet commercially feasible.

Soil is also the major reservoir of some fungi that are pathogenic to humans. These include *Blastomyces dermatitidis*, which causes blastomycosis; *Histoplasma capsulatum*, which causes histoplasmosis; and *Coccidioides immitis*, which causes San Joaquin Valley fever (Chapter 22).

Other Microorganisms. Algae are present on the surface of all soils but usually in small numbers. A gram of typical soil contains 100 to 10,000 colony-forming units, amounting to between 7 and 300 pounds of algal biomass per acre. Algae and phototrophic prokaryotes do not contribute significantly to soil fertility except in rice paddies, where cyanobacteria, free-living or in association with plants, fix considerable amounts of nitrogen.

The numbers of protozoa in soil are small, but probably no soil lacks them completely. The numbers of protozoan in soil vary between about 10,000 and 1,000,000 per gram. Their direct effect on biochemical transformations in the soil is minor. Indirectly, however, they play a critical role. They prey on the bacterial population and thus regulate its size and composition.

Many small animals spend part or all of their life cycle in the soil. These include earthworms, slugs, snails, centipedes, millipedes, wood lice, arachnids, insects, flatworms, and roundworms. Together they amount to several hundred pounds of animal tissue in each acre of ordinary soil. Many of the roundworms (nematodes, Chapter 12), attack the roots of plants, sapping their strength and sometimes killing them. They form characteristic knobs on roots. Larvae of the hookworm *Necator americanus*, a nematode that infects humans, are free-living in soil (Chapter 23).

Symbiosis.
Many soil microorganisms live in symbiosis with other organisms. Some of these relationships are **mutualistic** (benefiting both partners). Most of them are with plants. Two important mutualistic symbioses between microorganisms and plants are the mycorrhizae and the rhizosphere.

Mycorrhizae. Some soil fungi form **mycorrhizae** (intimate associations between fungi and roots of plants). The fungi act as additional roots, helping the plant acquire nutrients. Probably, most plants have mycorrhizae, but some mycorrhizal associations between basidiomycetes and beech, birch, and pine trees are particularly abundant in forest soils of temperate regions. Part of the fungal mycelium penetrates the tree root, but most of it remains just outside the root, forming a sheath that can be up to 40 μm thick. Most mycorrhizae-forming fungi cannot be cultivated in the absence of the plant with which they normally associ-

FIGURE 28.5 Effect of mycorrhizae on plant growth. The three juniper seedlings on the left were grown in the presence of fungi that form mycorrhizae. Those on the right were grown in the same soil but without fungi.

ate. Presumably, the plant supplies some essential nutrients to its fungal partner. The plant can survive without the fungus, but not very well. They are yellow and stunted when grown without mycorrhizae. If mycorrhizae are added, they become vigorous and deep green (**Figure 28.5**). The fungus helps the plant acquire mineral nutrients, which are usually in short supply in forest soils.

Some orchids are completely dependent on their mycorrhizal partners. The fungus supplies them organic growth factors, as well as mineral nutrients.

The Rhizosphere. The microbial ecology of the **rhizosphere** (the region of soil immediately surrounding the roots of plants) is significantly different from the rest of the soil. The microorganisms in these regions are not as intimately associated with plants as mycorrhizae are, but they do have a profound effect on plant growth. Specific kinds of microorganisms concentrate in the rhizosphere. This concentration, termed the **rhizosphere effect,** is described quantitatively by the **R:S ratio** (the ratio of the concentration of microorganisms in the rhizosphere to concentration in adjacent soil). For Gram-positive bacteria, actinomycetes, protozoa, and algae, the R:S ratio is relatively small, only about 2 or 3. But for Gram-negative bacteria—particularly species of *Pseudomonas, Flavobacterium,* and *Alcaligenes*—the R:S ratio can be several hundred. Specific microorganisms concentrate in the rhizosphere because plant roots excrete both nutrients and antimicrobial agents. The antimicrobial agents are selective. They inhibit some microorganisms but not others. Thus microorganisms in the rhizosphere benefit from being close to the plant, and the plant also benefits from microorganisms in the rhizosphere. For example, barley associated with a normal rhizosphere microbiota takes up phosphate about twice as effectively as barley grown in sterile soil. Moreover, the rhizosphere microbiota protects against fungal pathogens that attack plant roots.

SHARPER FOCUS

DEEP-SEA HYDROTHERMAL VENTS

Our sun is the ultimate source of energy that maintains life on Earth and drives the cycling of matter. Without sunlight, both phototrophs and heterotrophs would cease to exist (Chapter 5). But chemoautotrophs do not derive their energy from the sun. They get it by oxidizing reduced inorganic compounds.

An ecosystem based on chemoautotrophs is found on the ocean floor near hydrothermal vents—volcano-like places where seawater is heated by lava. Hydrothermal vents occur where Earth's geological crust is constantly being formed. These zones extend through the Atlantic, Pacific, and Indian Oceans. In many places, seawater enters these vents, where it is heated and enriched in inorganic compounds, including hydrogen sulfide.

Almost no light penetrates to these depths, so photosynthesis cannot occur. Yet dense, thriving communities of microorganisms and invertebrate animals live here. Oxidation of H_2S from the vent at the expense of the O_2 dissolved in seawater is the total source of energy for these complex communities. Huge populations of autotrophic sulfur-oxidizing bacteria, up to a billion cells per milliliter of ocean water, make the water turbid in places. Some of these bacteria form dense microbial mats on which fish graze.

The most spectacular vent inhabitants are huge tubeworms, 8 feet long and 15 inches in circumference. They obtain energy in a highly unusual way from bacteria that oxidize hydrogen sulfide. The worms lack an intestinal system. Instead they are filled with spongy tissue, the **trophosome,** that constitutes over half the worm's total weight. The trophosome consists of symbiotic sulfur-oxidizing bacteria that supply the worm with its nutrients.

All members of the hydrothermal vent community are supported directly or indirectly by energy from the oxidation of reduced sulfur compounds. Sunlight plays only an indirect role. Having supported photosynthesis elsewhere on Earth, it forms the oxygen needed by the sulfur-oxidizing bacteria.

© Fred Grassle, Woods Hole Oceanographic Institution

Tubeworms inhabit deep-sea hydrothermal vents.

Water

Seventy percent of Earth is covered by water, and microorganisms live in almost all aqueous environments. They are found in boiling hot springs, in Antarctic waters that rarely rise above the freezing point of saltwater, and in salt-saturated seas. They even grow near hydrothermal vents at the bottoms of oceans (see Sharper Focus: Deep-Sea Hydrothermal Vents).

Streams, rivers, ponds, lakes, and oceans are different aqueous environments, and each of these bodies of water is heterogeneous. The air-water surface, the water column itself, suspended particles in the water column, and the bottom are different environments. Covering so much of Earth's surface, the ocean is a major contributor to the cycles of matter. The biosphere is, on the average, about 38 meters thick over the land masses, but it extends throughout the entire volume of the ocean. In these terms,

99 percent of the earth's biosphere is seawater, although much of it is not a particularly favorable environment for life. More than half the water in the world's oceans is colder than 2°C and under a pressure greater than 100 atmospheres.

Nutrients. In general, aqueous environments support smaller populations of microorganisms than soil because most aqueous environments contain only low concentrations of nutrients.

The concentration of dissolved organic nutrients in oceans is so low that many marine microbiologists believe heterotrophic microorganisms grow only when they are attached to nutrient-containing particles. In contrast, phototrophic microorganisms flourish in the upper regions of the ocean where light penetrates and the inorganic nutrients they require are present in adequate concentrations. These organisms, called **phytoplankton,** constitute a major portion of the world's total photosynthetic capacity. In the lower regions of the ocean, where light does not penetrate, heterotrophs get most of their nutrition from dead phytoplankton drifting down from the top regions. These small particles settle so slowly—at a rate of 0.1 to 1.0 meter per day—that most decompose before they reach the bottom. The 1 percent that does reach the ocean floor supports a teeming microbial activity in the top layer of ocean sediments.

Unless they are polluted, bodies of fresh water are also nutrient-poor, which adds to their clarity and deep blue color. When aqueous environments become nutritionally enriched with nitrogen or phosphorus by runoff water from construction projects and municipal sewage or by industrial or agricultural wastewater, large populations of microorganisms develop (**Figure 28.6**). Such bodies of water become **eutrophic** (enriched in microbial nutrients). Their color changes from blue to green.

Many eutrophic ponds and streams support huge populations of cyanobacteria that grow near the surface. They collect in unsightly masses called **blooms.** Within this rich source of nutrients, aerobic microorganisms use up oxygen more rapidly than it can be supplied from the atmosphere or from the cyanobacteria. Then anaerobic bacteria decompose the masses of microbial cells. The result is unpleasant. The anaerobes make some products of fermentation—including fatty acids, amines, and mercaptans—that have foul odors.

If heavily polluted, even large bodies of water become anaerobic and foul smelling. In England during the early 1800s, the city of London routinely discharged its sewage

© Doug Sokell/Visuals Unlimited

FIGURE 28.6 This eutrophic pond is covered with duckweed. It supports a huge population of cyanobacteria.

into the Thames River. The flow of the river had been sufficient to dilute the sewage and mix in enough oxygen to remain aerobic. Eventually, however, increased discharge overwhelmed the Thames. The river became anaerobic. Londoners referred to this episode as "the big stink." Modern methods of sewage treatment prevent such unpleasant occurrences. We'll discuss sewage treatment later in this chapter.

Pathogens. Fresh water is a significant reservoir for pathogens that cause human disease. Most of these pathogens enter the water in human feces and infect new hosts when the contaminated water is used for drinking. In most cases the diseases are gastrointestinal (**Table 28.2**), but some pathogens that live in water cause respiratory disease (see Case History: When Help Becomes Harm).

Air

Microorganisms do not grow in air, but some types of microorganisms are found in it. Usually they are passengers on **aerosols** (tiny particles of liquid) or dust particles. We make microorganism-bearing aerosols when we cough, sneeze, or talk. Any agitation of water, such as waves breaking, rapids, and sprays, also creates aerosols. Even flaming a wet inoculating loop (Chapter 3) can produce an aerosol bearing live microorganisms. Aerosols and dust are the principal means of transmitting respiratory diseases.

THE CYCLES OF MATTER

Each of the **major bioelements** (nitrogen, carbon, phosphorus, and sulfur) occurs in several different chemical forms. Nitrogen, for example, exists in the atmosphere as nitrogen gas (N_2) and on the earth's surface in organic compounds and as ammonia (NH_3) and nitrate (NO_3^-). These various forms of nitrogen are interconvertible. Nitrogen gas is converted to ammonia, to organic nitrogen, to nitrate, and then back to nitrogen gas. Most of these chemical conversions are carried out by living organisms, principally microorganisms. And only microorganisms can carry out some of these chemical conversions. The chemical conversions that occur in nature are roughly balanced: The rate of production of each form roughly equals its rate of utilization. Most forms of the bioelements are constantly being transformed. For example, virtually all the molecules of nitrogen gas now in our atmosphere were once a part of some living thing, and they will be again. The cyclic chemical interconversion of each of the biological elements is called a **biogeochemical cycle**.

The cycles themselves are interrelated because most natural compounds are made of several bioelements. For simplicity, however, we'll look at each of the major biogeochemical cycles (nitrogen, carbon, phosphorus, and sulfur) separately.

The Nitrogen Cycle

The nitrogen cycle is particularly important to agriculture and ecology because the concentration of nitrogen in most soils is low enough to limit growth of plants and yield of crops. The cycle is complex because nitrogen exists in

TABLE 28.2 Some Human Pathogens and Diseases Transmitted in Water

Pathogen	Disease
Bacteria	
Salmonella typhi	Typhoid fever
Salmonella spp.	Salmonellosis (gastroenteritis)
Escherichia coli	Diarrhea
Legionella pneumophila	Legionellosis
Vibrio cholerae	Cholera
Vibrio parahaemolyticus	Gastroenteritis
Yersinia enterocolitica	Enterocolitis (diarrhea)
Campylobacter jejuni	Diarrhea
Viruses	
Hepatitis A virus	Type A (infectious) hepatitis
Hepatitis E virus	Type E hepatitis
Poliovirus	Poliomyelitis
Protozoa	
Entamoeba histolytica	Amoebic dysentery
Giardia lamblia	Giardiasis (diarrhea)
Balantidium coli	Balantidiasis (diarrhea)
Cryptosporidium	Cryptosporidiosis
Helminths	
Fasciola hepatica (liver fluke)	Hepatitis (liver rot)
Opisthorchis sinensis	Hepatitis (liver rot)

Note: All these pathogens and diseases (except legionellosis and poliovirus) are discussed in Chapter 23. Legionellosis is discussed in Chapter 22 and poliovirus in Chapter 25.

many different forms with different chemical properties. It's convenient to think of the nitrogen cycle beginning with nitrogen gas in the atmosphere. First it's converted to ammonia by nitrogen fixation, then to nitrate by nitrification, and finally back to nitrogen gas by denitrification (**Figure 28.7**). Along the way, intermediates are incorporated into various organisms; for example, nitrate is incorporated into plants, which are consumed by other organisms. But eventually the nitrogen in dead organisms and the waste from organisms are converted by ammonification to ammonia, which reenters the main cycle.

Nitrogen Fixation. Free nitrogen gas (N_2) constitutes about 80 percent of Earth's atmosphere, but it is not avail-

able to most organisms. Only a few prokaryotes are able to fix nitrogen gas, and they pay a high price. They expend more than 16 molecules of adenosine triphosphate (ATP) to fix each molecule of N_2.

There are two classes of nitrogen-fixing bacteria: Those that form symbiotic relationships with plants and free-living (nonsymbiotic) ones that do not (**Table 28.3**). Both classes must maintain an anaerobic environment because **nitrogenase** (the enzyme that catalyzes nitrogen fixation) is rapidly inactivated by oxygen.

The aerobic, free-living nitrogen-fixers have evolved various ways to protect their nitrogenase from oxygen. *Azotobacter* spp. respire oxygen at its surface so rapidly that the interior of the cell, where the nitrogenase is located, is

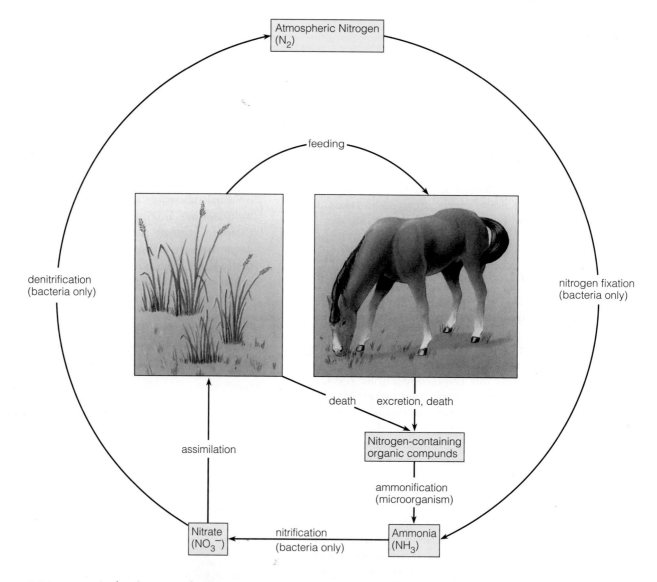

FIGURE 28.7   In the nitrogen cycle, nitrogen gas is converted to ammonia by nitrogen fixation. Ammonia is converted to nitrate by nitrification. Nitrate is converted to nitrogen gas—completing the cycle—by denitrification.

anaerobic. Cyanobacteria must deal with the added complication of producing oxygen themselves by photosynthesis. They fix nitrogen in **heterocysts** (specialized cells with thick walls). Heterocysts do not produce oxygen inside, and they do not let it enter from the outside (Chapter 11). The anaerobic free-living bacteria such as *Clostridium pastorianum* need no special adaptations. They live in oxygen-free environments.

Symbiotic nitrogen-fixers, including *Rhizobium* spp., some cyanobacteria, and *Frankia* spp., depend upon a plant to protect them from oxygen, and the plant supplies nutrients as well. In turn, the bacterium supplies the plant with fixed nitrogen. The complexity of these symbioses between bacteria and plants varies considerably. At one extreme, nitrogen-fixing cyanobacteria simply accumulate in small pockets on the surface of certain lower plants, such as bryophytes, ferns, and cycads.

At the other extreme, *Rhizobium* spp. form complex and intimate relationships with leguminous plants. When the bacterium comes in contact with a **root hair** (tubelike extension of a surface cell), the root undergoes a complex series of morphological changes. Eventually a tube called an infection thread forms, through which the bacterium can penetrate deep into the root tissue. There it forms nodules that become filled with rhizobial cells (**Figure 28.8**). These nodules are small nitrogen-fixing factories. Along with nutrients, the plant supplies the nodule with **leghemoglobin.** This protein binds oxygen and maintains it an optimal concentration: Low enough for nitrogen fixation and high enough for rhizobial metabolism. Within the nodule, rhizobial cells differentiate into **bacteroids,** nongrowing cells that devote their entire metabolic capacity to fixing nitrogen.

Until the twentieth century, almost all nitrogen on Earth was fixed by bacteria. The rest—a relatively minor amount—was fixed by volcanic activity and lightning. However, early in the last century, Fritz Haber, a German chemist, discovered how to convert gaseous nitrogen and hydrogen into ammonia. In 1918 he was awarded a Nobel Prize for his achievement. Now about half the world's supply of nitrogen is fixed by Haber's process. Industrially fixed nitrogen is used mainly to produce fertilizers for agriculture, which has contributed greatly to increased world food production. Certainly the world would be facing mass starvation without it. But fertilizer creates its own problems. The runoff water from fields of fertilized crops

TABLE 28.3 Some Genera of Prokaryotes That Fix Nitrogen	
Free-Living	
Aerobes	Cyanobacteria
Azotobacter	*Anabaena* (also forms symbioses)
Beijerinckia	
Facultative Anaerobes	*Nostoc* (also forms symbioses)
Bacillus	
Enterobacter	*Gleocapsa*
Klebsiella	Archaea
Anaerobes	*Methanococcus*
Clostridium	*Methanobacterium*
Nonoxygenic Phototrophs	**Symbiotic**
	Rhizobium
Rhodobacter	
Rhodospirillum	*Bradyrhizobium*
Chlorobium	*Frankia*

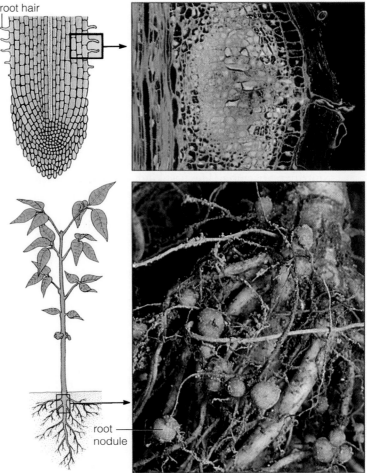

FIGURE 28.8 The nitrogen-fixing bacteria *Rhizobium* and *Bradyrhizobium* form symbiotic relationships with legumes. When bacterial cells contact a root hair, they induce it to form an infection thread. The bacteria travel down the thread into the root tissue, penetrating cells, where they multiply, forming a nodule. In the nodule the bacteria differentiate into bacteroids and devote their metabolic energy to fixing nitrogen.

root hair

root nodule

stimulates algal growth. In Chesapeake Bay, for example, algal growth has diminished the productivity of fisheries.

Increased amounts of fixed nitrogen for agriculture can also be supplied by fostering symbiotic nitrogen fixation. In Western countries, leguminous crops, which harbor symbiotic nitrogen-fixers, are grown as part of crop rotation to replenish the soil's nitrogen content. For centuries rice paddies in Southeast Asia have been enriched with nitrogen by inoculating them with *Azolla*, a small floating water fern. *Azolla* harbors a symbiotic nitrogen-fixing cyanobacterium.

Ammonia, the product of nitrogen fixation, is used to make nitrogen-containing cell components such as amino acids, purines, and pyrimidines for the bacterium itself and for the host plant. With rare exceptions, nitrogen-fixing organisms fix only enough nitrogen for their own and their host's use. Only when they and their hosts die or are consumed does the fixed nitrogen become available to other organisms. Most of the nitrogen compounds they contain are broken down and converted back to ammonia (NH_3) by **ammonification,** which is carried out largely by heterotrophic bacteria.

Nitrification.

Ammonia does not accumulate in the soil. Instead, it's rapidly oxidized by nitrifying bacteria to nitrate ion (NO_3^-) in the process of **nitrification.** Nitrifying bacteria are chemoautotrophs that generate ATP from the energy released in these oxidations. Nitrification occurs in two steps. Each step is brought about by a different kind of nitrifying bacteria. The first kind, typified by *Nitrosomonas* spp., oxidizes ammonia to nitrite ion (NO_2^-).

$$NH_3 + 1\tfrac{1}{2}O_2 \longrightarrow NO_2^- + H_2O + H^+ + \text{Energy}$$
Ammonia   Nitrite Ion

The second kind, typified by *Nitrobacter* spp., oxidizes nitrite ion to nitrate ion (NO_3^-).

$$NO_2^- + \tfrac{1}{2}O_2 \longrightarrow NO_3^- + \text{Energy}$$
Nitrite Ion   Nitrate Ion

Nitrate is the main form of nitrogen that plants use. But nitrate cannot be stored in soil because it's highly soluble and does not adsorb to soil as ammonia does. The nitrate that plants don't use dissolves in groundwater and runs off into streams. This presents two problems. First, the accumulation of nitrate in groundwater can contaminate drinking water. High levels of nitrate are dangerous to humans, particularly infants. Second, the loss of nitrogen in runoff water depletes soil fertility.

Denitrification.

Denitrification (conversion of nitrate to nitrogen gas) is a series of anaerobic respirations (Chapter 5) that only prokaryotes can carry out. Many different prokaryotes can denitrify, including archaea and bacteria, Gram-positives and Gram-negatives. Most oxidize organic compounds and transfer these electrons to nitrate, reducing it stepwise to nitrogen gas. Intermediates in the reductive pathway include nitrite ion and nitrous oxide (N_2O, also known as laughing gas). When high concentrations of nitrate are available to denitrifying bacteria, they release some nitrous oxide along with nitrogen gas. Most denitrifying bacteria are facultative aerobes. That is, when oxygen is available, they use it as the terminal electron acceptor of their electron transport chains. When oxygen is not available, they use nitrate.

The ecological impact of denitrification is complex. On one hand, it is essential to life on the planet. Without denitrification our atmosphere would rapidly (in geological terms) become nitrogen-free. (Half the nitrogen in our atmosphere is fixed every 20 million years.) Earth's supply of nitrogen would accumulate as nitrate in the oceans, while nitrogen starvation would stop plant growth on land. Denitrification also helps purify waste water, impeding eutrophication, and drinking water, making it safe.

On the other hand, denitrification is costly to agriculture. It returns up to 30 percent of nitrogen added as fertilizer to the atmosphere as nitrogen gas. Denitrification also contributes to the destruction of the earth's protective ozone layer because it releases some nitrous oxide. (Internal combustion engines also form nitrous and other nitrogen oxides.) Production of nitrous oxide by denitrification is rising because more fertilizer is being used to feed Earth's growing population.

Denitrification completes the major loop of the nitrogen cycle. The individual steps of this loop—nitrogen fixation, ammonification, nitrification, and denitrification—are all brought about by microorganisms. Three of these steps—nitrogen fixation, nitrification, and denitrification—are brought about exclusively by prokaryotes (although humans now participate in fixing nitrogen by the Haber process).

The Carbon Cycle

The major loop of the carbon cycle is the conversion of carbon dioxide (CO_2) into the organic compounds of living organisms and their conversion back to carbon dioxide.

Like nitrogen, carbon cycles through an atmospheric gas—in this case, carbon dioxide (**Figure 28.9**). But carbon dioxide makes up a much smaller fraction of the atmosphere (0.03 percent) than does nitrogen gas (78 percent). The turnover of carbon dioxide is thus more rapid. Without resupply, the atmosphere's supply of carbon dioxide would last only about 20 years.

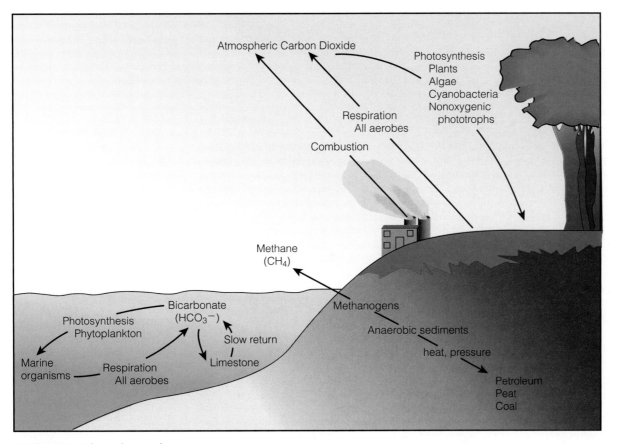

FIGURE 28.9   The carbon cycle.

In the major loop of the carbon cycle, carbon dioxide is removed from the atmosphere by autotrophs, principally phytoplankton and plants. They fix atmospheric CO_2 through photosynthesis, reducing it and incorporating it into organic compounds (Chapter 5). Carbon dioxide is regenerated and returned to the atmosphere by respiration and combustion. Both processes oxidize organic compounds to water and CO_2 (Chapter 5).

Both respiration and combustion require oxygen. Organic materials deposited in anaerobic environments, such as bogs or sediments at the bottoms of bodies of water, are isolated from the major loop of the carbon cycle. Under anaerobic conditions, carbon in organic material returns only slowly to the major loop. Anaerobes convert a vast variety of organic compounds into acetate. Then methanogens convert acetate to methane (CH_4). Methane, being a gas, can escape the anaerobic environment. When it comes in contact with air, it can be oxidized to CO_2 by methane-oxidizing bacteria. But some methane accumulates in the atmosphere.

Over the earth's history, huge quantities of organic materials have been trapped in anaerobic environments. Heat and pressure convert them into the fossil fuels—peat, coal, and petroleum. Oceans and sedimentary rocks also hold vast amounts of carbon in the form of CO_2. The CO_2 in

oceans is in the form of dissolved bicarbonate. Sedimentary rocks such as limestone contain CO_2 in the form of carbonate (CO_3^{2-}). The rate at which the ocean takes up and releases CO_2 is one of the major unknown factors governing the rate of CO_2 accumulation in our atmosphere. The rate of incorporation and release of CO_2 from sedimentary rocks is very slow.

Since the beginning of the industrial revolution in the mid-eighteenth century, humans have burned fossil fuels at an ever-increasing rate. This, along with a decline in photosynthesis because forests are being cut down, has changed the carbon cycle. Instead of being constant, the CO_2 content of the atmosphere has risen sharply (**Figure 28.10**). This is serious because CO_2 is a greenhouse gas. It acts like an atmospheric blanket that prevents some solar heat from radiating into space. The consequence of continued rise in atmospheric carbon dioxide is global warming, which will certainly change climate worldwide. It will cause flooding of coastal regions from the melting of polar ice and expansion of seawater. However, there are many unknown factors in the global warming scenario. What will be the effect on plant growth and increased photosynthesis? What will be the effect on cloud formation? Certainly a warmer world would have more clouds.

SHARPER FOCUS

OFF BALANCE

Nature cleans up its own litter. The materials produced by one organism and eventually the organism itself are broken down and used as nutrients by another. Microorganisms play the major role in this process of cleaning up, and they are good at it because all naturally occurring substances can be broken down by at least one microorganism.

The processes are balanced. Most natural organic compounds are degraded by one microorganism or another at about the same rate they are made. There are exceptions. Some organic compounds—**lignin** (a component of the woody parts of trees), for example—are degraded relatively slowly. Others, because they were preserved in an anaerobic environment, escape degradation for long periods and accumulate in the form of peat, coal, or petroleum. But even these materials are broken down by microorganisms in the right conditions.

The equilibrium between synthesis and degradation of organic compounds, which was maintained for billions of years, has become unbalanced during the past 50 years. We humans now make organic compounds that are extremely resistant to microbial attack. They are called **recalcitrant organic compounds.** Chlorine-containing pesticides are prime examples. Other compounds, including most plastics, are completely resistant to microbial degradation. Recalcitrant and nonbiodegradable compounds are accumulating everywhere in the world and causing significant environmental damage.

There are no simple solutions. But there are some encouraging approaches. The first and most obvious is to restrict the manufacture of compounds that are not susceptible to microbial attack (see Larger Field: Bacteria That Make Plastic). This approach has been successful in the case of certain detergents (alkylbenzene sulfate compounds). In the 1950s this class of detergents contained molecules with branched chains of carbon atoms, which are highly resistant to microbial attack. The detergents passed unaffected through sewage plants and accumulated in fresh water. This caused water in streams and even from household taps to foam. Simply changing the branched chains of carbon atoms to straight chains solved the problem because straight-chain versions are susceptible to microbial attack. And they are still effective detergents.

Another approach is to genetically engineer new microorganisms able to degrade recalcitrant organic compounds. A microorganism can be designed to perform additional or different biochemical reactions. Thus far, microbiologists have altered key enzymes of degradative pathways by mutating encoding genes. They have also used genes from different organisms to construct a new degradative pathway in a single organism. In both cases, more research is needed.

Humans have learned to make organic material that microorganisms can't break down, putting nature out of balance. Undoubtedly we can learn how to restore it. In this chapter, we'll discuss how nature maintains the balance of producing and degrading organic compounds.

But will they act largely as blankets, increasing the rate of warming, or as insulators, decreasing it? No one doubts the world will change if CO_2 in the atmosphere continues to rise.

The Phosphorus Cycle

The phosphorus cycle is comparatively simple. Phosphorus from inorganic phosphate is converted to organic phosphate and back again (**Figure 28.11**). It differs fundamentally from the nitrogen and carbon cycles because there's no significant gaseous intermediate. Thus global redistribution of phosphorus through the atmosphere is not possible. Dissolved phosphate inevitably ends up in the ocean. There are only two significant routes for its returning to land.

First, seabirds that feed on phosphorus-containing sea creatures return the phosphorus to shore in their feces. In some places they form huge guano deposits on land. This route redistributes only a small amount of the ocean's phosphorus. Second, the geological uplift of ocean floors to forms land masses also returns phosphorus from ocean sediments to the land. This route returns enormous quantities of phosphorus, but it is extremely slow.

The phosphorus cycle also differs from the nitrogen and carbon cycles in another important respect. Phosphorus is neither oxidized nor reduced in the cycle. Instead, it remains in the form of phosphate. Because phosphorus is a limiting nutrient in many soils, most fertilizers contain phosphate. Rather than being commercially manufactured like nitrate, however, phosphate is mined. Phosphate-rich deposits are the product of previous geological uplifts.

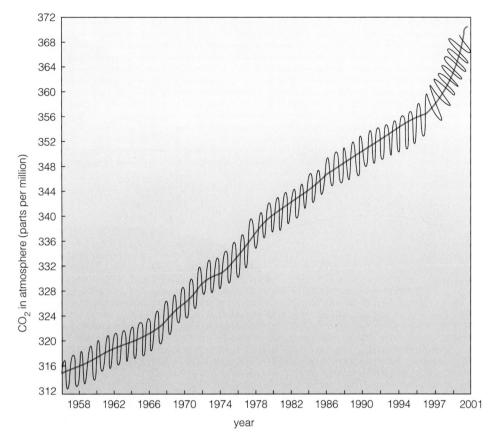

FIGURE 28.10 Accumulation of CO_2 in the atmosphere since 1956. CO_2 levels in the atmosphere continue their upward trend. By 2001 it had reached 370.9 parts per million.

volume is decreased. The pond is allowed to become anaerobic. Then *Acinetobacter* degrades its polyphosphate, releasing phosphate ions into the relatively small pond. The result is a phosphate solution concentrated enough to sell directly to a detergent manufacturer. Thus the process serves two useful purposes: (1) It purifies sewage so it does not cause eutrophication. (2) It produces a valuable product. At the end of the process, the *Acinetobacter* cells in the pond are ready to begin a new cycle of concentrating phosphate.

The Sulfur Cycle

Sulfur, like nitrogen, plays two important roles in biology. (1) Nutrition: Sulfur is a component of certain cell constituents, including amino acids and enzyme cofactors, so all organisms have a nutritional need for sulfur. (2) Energy metabolism: Some autotrophic prokaryotes generate ATP by oxidizing reduced forms of sulfur. Others generate ATP by using oxidized forms of sulfur as terminal electron acceptors of electron transport chains. Together these biological transformations make up the sulfur cycle. The main loop of the cycle consists of reducing sulfate ions (SO_4^{2-}) to hydrogen sulfide gas (H_2S) and reoxidizing it to sulfate. (Sometimes elemental sulfur accumulates during the oxidation of H_2S.) Along the way, sulfur is used by various organisms to meet their nutritional needs (**Figure 28.12**).

Sulfur is the tenth most abundant element on Earth, so few if any natural environments are sulfur deficient. Sulfur compounds are not added to fertilizers because extra sulfur does not stimulate plant growth. As with phosphorus, sulfur does not exist as an atmospheric gas. (Hydrogen sulfide is a gas, but it cannot exist for long in the atmosphere because it reacts spontaneously with oxygen.)

Reduction of Sulfate. It's convenient to begin considering the sulfur cycle at sulfate (SO_4^{2-}). This form of sulfur is abundant in the environment. The legendary White Cliffs of Dover in England, for example, are almost pure gypsum (calcium sulfate). Sulfate is also the form that most

There's no foreseeable prospect of a world phosphorus shortage, but the distribution of commercially feasible phosphorus mines is restricted. North America and Africa are particularly rich in phosphate deposits.

Human intervention has altered the phosphorus cycle by adding phosphate compounds to lakes and streams, causing some ecological damage. As with soils, the productivity of many lakes is limited by availability of phosphorus. When such lakes are enriched, they become eutrophic, with all the negative consequences. Until recently, when many detergents were reformulated to be phosphate-free, sewage was dangerously rich in phosphorus. Still, large amounts remain. Even if sewage is completely mineralized, it can contaminate the receiving lake or stream with phosphate.

Phosphate can be removed from sewage chemically by precipitating it. It can also be removed biologically. Mineralized sewage is diverted into an aerobic pond that is slightly enriched with raw sewage as a carbon source. It is then managed in such a way as to develop a dense culture of the bacterium *Acinetobacter calcoaceticus*. This bacterium stores phosphate in the form of polyphosphate granules (Chapter 4). In the treatment pond, *Acinetobacter,* an obligate aerobe, stores massive quantities of polyphosphate—up to 30 percent of its dry weight. Then conditions are abruptly changed. The carbon source is withdrawn. The

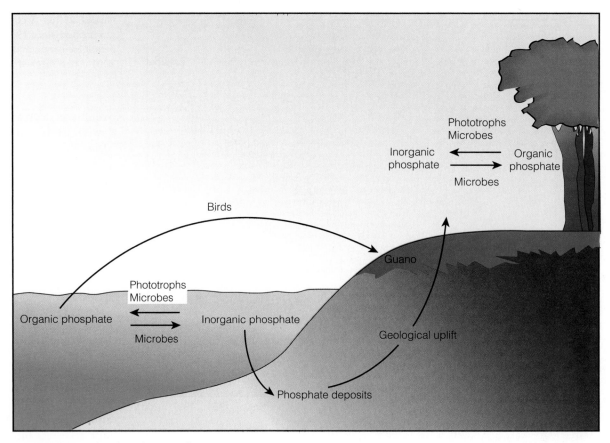

FIGURE 28.11   The phosphorus cycle.

microorganisms and plants use as source of sulfur for nutrition. But sulfate (and sulfite) is present in only a few unusual constituents of cells. Microbes and plants obtain their sulfur by reducing sulfate to sulfide (S^{2-}), which is used to synthesize almost all their sulfur-containing components. Animals derive their sulfide-containing compounds from their diet.

During the process of decomposing dead plants, animals, and their wastes, microorganisms convert organic sulfur compounds to hydrogen sulfide, a process called **desulfurylation.**

Like most microorganisms, sulfate-reducing bacteria reduce sulfate to sulfide when they use it as a nutrient. But they also use it as a terminal electron acceptor of an electron transport chain in a process of anaerobic respiration (Chapter 5). This process uses much more sulfate than does nutritional reduction. Sulfate-reducing bacteria live in sulfate-rich anaerobic environments such as mud flats. The vast amount of H_2S they produce accounts for the rotten-egg smell of these areas. It also accounts for the mud being black because most metal sulfides (for example, iron sulfide) are black. They form as a result of reactions between H_2S and metal ions present in the mud.

Oxidation of H_2S. The H_2S produced by desulfurylation and sulfate-reducing bacteria is reoxidized to sulfate, thereby completing the sulfur cycle.

Two kinds of bacteria oxidize H_2S to sulfate. (1) Chemoautotrophic sulfur-oxidizing bacteria oxidize it in the process of generating ATP. The final oxidation product of these oxidations is sulfate. Elemental sulfur sometimes accumulates as an intermediate. These organisms also oxidize external elemental sulfur when it is available. This microbial oxidation accounts for the common gardening practice of adding sulfur to soil. Sulfur is added to promote the growth of acid-loving plants because the oxidation of sulfur generates sulfuric acid.

$$S + 1\tfrac{1}{2}\, O_2 + H_2O = H_2SO_4$$
Sulfur   Sulfuric Acid

(2) Nonoxygenic phototrophic bacteria also oxidize H_2S to sulfur and sulfate. But they do so in the absence of oxygen. They use H_2S as a source of electrons for photosynthesis (as plants use water).

In most places, populations of sulfur-oxidizing bacteria are limited by the supply of reduced sulfur, which is made by other microorganisms. But in regions of

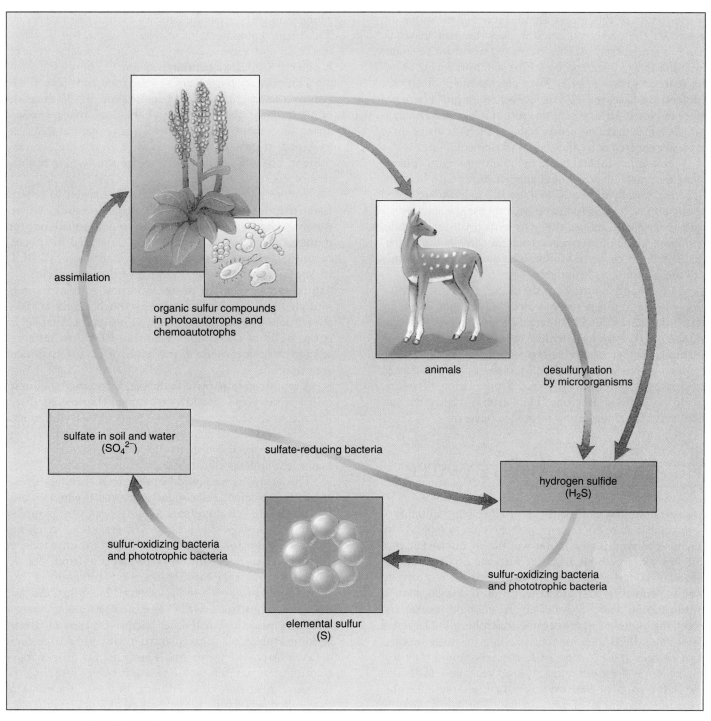

FIGURE 28.12   The sulfur cycle.

(Art by Carlyn Iverson.)

the deep sea near hydrothermal vents, the supply of reduced sulfur is copious. Huge numbers of sulfur-oxidizing bacteria develop and form the basis of a highly unusual ecosystem (see Sharper Focus: Deep-Sea Hydrothermal Vents).

Products and Reactions. One intermediate of the sulfur cycle, elemental sulfur, is a valuable commodity. It's the starting material for producing sulfuric acid, one of the most widely used industrial chemicals. Sulfur in elemental form occurs in huge underground deposits in various parts of the world. Some of the largest are in Texas. These sulfur deposits are the remains of ancient lakes that had anaerobic zones where microorganisms converted sulfate into elemental sulfur in a two-step process. First sulfate-reducing bacteria reduced sulfate to H_2S. Then anoxygenic phototrophic bacteria oxidized the H_2S to elemental sulfur. There are lakes in North Africa where this sort of mixed bacterial metabolism is actively occurring today. The bottoms of these lakes are covered with thick layers of elemental sulfur.

On an experimental basis, the microbiological conversions that make deposits of elemental sulfur have been induced artificially in ponds. Sewage and gypsum are added to the ponds. Sulfate-reducing bacteria use the sewage as a carbon source to reduce the gypsum to hydrogen sulfide, and anoxygenic phototrophic bacteria convert the hydrogen sulfide to elemental sulfur. The process requires only sewage, gypsum, and sunlight.

The oxidative portion of the sulfur cycle (the conversion of sulfide to sulfate) has already been put to use by humans. Many valuable minerals either occur as insoluble sulfides or are trapped within other metal sulfides. For example, copper occurs as copper sulfide. Gold is often trapped within iron pyrite (FeS_2). If these ores are stacked loosely and sprayed with water, a population of sulfur-oxidizing bacteria develops. The bacteria oxidize the sulfide to sulfate. The copper or gold is released.

TREATMENT OF WASTE WATER

Our communities and industries produce huge amounts of sewage that contain human and chemical wastes. Until fairly recently, this wastewater was simply discharged into the nearest large body of water with the expectation that the natural cycling of matter would eventually purify it. But as population and industrialization increased, natural purification did not always occur fast enough. Instead, the receiving bodies of water became anaerobic and blackened by sulfides. They became foul smelling from the production of H_2S and products of fermentation, and they lost their natural flora and fauna. Just 60 years ago, for example, large parts of San Francisco Bay became anaerobic from the discharge of raw sewage. The installation of modern sewage treatment plants has returned the bay to relative health. People can even fish there again.

Sewage Treatment Plants

Sewage is municipal wastewater. It includes all the materials that goes down household and industrial drains. It contains human waste, the residue of soaps and detergents,

ground-up garbage, and chemicals used in manufacturing. The major component of sewage is water, but dissolved materials and some solid organic waste are also present. Large cities produce enormous quantities of sewage.

The primary purpose of sewage treatment plants is to remove the organic material from sewage. That's the component that makes receiving bodies of water become anaerobic. Some sewage treatment plants employ an anaerobic treatment to remove some of organic material present, but most of it is removed by the aerobic respiration of microorganisms.

Sewage plants are designed to add oxygen to sewage faster than it is removed by microbial respiration. Wastewater is classified according to its **biochemical oxygen demand** (BOD, how much oxygen is needed for microorganisms to respire all the organic material in it). BOD is determined by putting a sample of the sewage (with its natural complement of microorganisms) into an oxygen-saturated bottle and measuring how much oxygen remains after 5 days of incubation at $20°C$. The decrease in oxygen is the BOD of the sample. Knowing BOD, operators of sewage plants know how many hours or days of treatment are required.

A typical sewage plant is divided into functional units called primary and secondary treatment (**Figure 28.13**). A few plants also have tertiary treatment. Primary treatment is mechanical rather than biological. Solid material is usually ground up, and primary sludge, the portion that remains insoluble, is allowed to settle out.

The effluent, or liquid coming from primary treatment, is subjected to secondary treatment, which is biological. Some sewage plants aerate sewage by pumping air into a sewage-filled tank. Others use a **trickling filter**—sewage is sprayed in a thin film onto a bed of rocks so that it readily absorbs oxygen from the atmosphere. In either case the sewage is mineralized by a dense population of aerobic bacteria, including *Zoogloea ramigera* (**Figure 28.14**). *Z. ramigera* has a slimy extracellular capsule in which other bacteria and solid material become embedded. This mixture of slime and bacteria is called a **floc**. Floc formation is essential for proper functioning of an aerated tank because it settles rapidly, producing a relatively clear effluent. Some of the mass of floc, called **activated sludge,** is used to inoculate the next batch of sewage entering the aeration tank. The rest is added to the primary sludge.

The combination of primary sludge and excess activated sludge is usually held in an anaerobic sludge digester, a deep unaerated tank in which the mixture is partially degraded by anaerobic bacteria. Digested sludge, the solid portion that remains, is dried and disposed of, often as fertilizer. The anaerobic organisms in an anaerobic sludge digester produce various metabolic end products, including

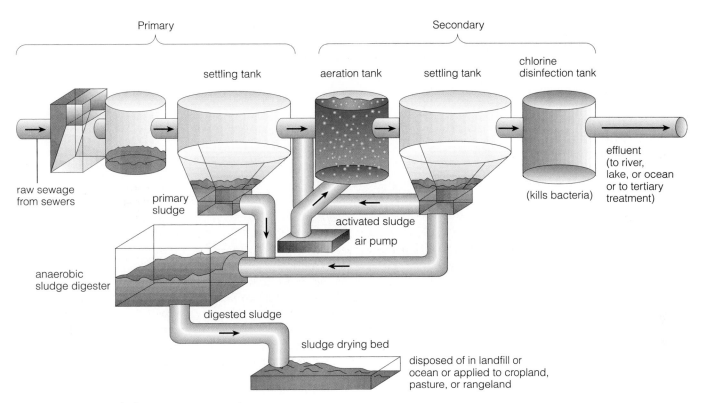

FIGURE 28.13   A typical sewage-treatment plant.

acetate, hydrogen gas, and carbon dioxide. These are converted into methane (CH_4), or natural gas, by **methanogens** (methane-forming archaea). In some sewage plants the methane is used as fuel during the winter to warm the sewage during secondary treatment and thus speed the process. It can also be used to generate electricity. In some developing countries, particularly China and India, many rural households have small anaerobic digesters. They are charged with domestic sewage and manure from farm animals. The methane generated by these digesters is used to cook food, provide light, and heat the house.

The effluent from secondary treatment contains phosphate and nitrate ions, which can cause eutrophication. For this reason, tertiary treatment is sometimes used to remove or reduce the concentration of these ions. Most tertiary treatments are chemical rather than biological. For example, lime, alum (potassium/aluminum sulfate), or ferric chloride is added to remove phosphate. Biological treatments—such as the use of *Acinetobacter* to remove phosphate, described earlier in this chapter—are gaining favor, however. In many places, including the United States, denitrifying bacteria are used to remove nitrate by converting it to nitrogen gas.

The final effluent from a properly functioning sewage plant contains only a small amount of organic material and has a correspondingly low BOD. It can be added to a river or to the ocean without fear that it will make the water anaerobic. For public health reasons, the effluent is treated prior to release with chlorine gas (Cl_2) to kill pathogenic microorganisms that might be present.

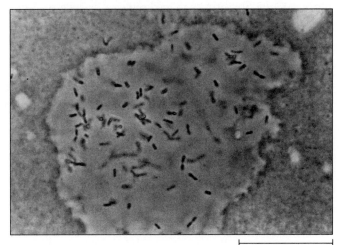

FIGURE 28.14   This micrograph shows a particle of floc formed by *Zoogloea ramigera*.

From *International Journal of Systematic Bacteriology* 21:21–99, 1971. Reprinted by permission of the American Society for Microbiology. Photo courtesy Richard F. Unz, Pennsylvania State University.

Septic Tanks and Oxidation Ponds

Many homes and farms are not linked to a municipal sewage system. Instead they use **septic tanks,** which are essentially small anaerobic digesters (**Figure 28.15**). The sludge settles in the tank, eventually to be pumped out and disposed of. The digested effluent from the tank is distributed to a network of perforated pipes into a **leach field** (the area where the liquid's organic content is mineralized). The effectiveness of the leach field depends upon the capacity of the soil's microbiota to mineralize organic materials. A septic tank system can process small amounts of sewage almost indefinitely, so long as the accumulated sludge is pumped out every few years. It is impractical, however, to chlorinate the effluent, so care must be taken to prevent contamination of drinking water. The septic system must be located away from a well or other source of water.

In arid regions of the world, the soil's vigorous microbial activity can be used to process high volumes of sewage from municipalities or industries. Sewage is distributed into a series of oxidation ponds, earthen ponds stirred with large paddles to promote aeration. Oxygenic phototrophs develop in these ponds, which is helpful because they add more oxygen. Sometimes in arid regions, sewage is simply added to ditches and allowed to seep into the ground, a process called **sewage farming.**

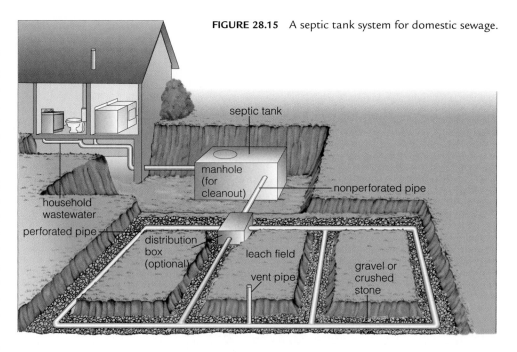

FIGURE 28.15   A septic tank system for domestic sewage.

TREATMENT OF DRINKING WATER

Drinking water from wells usually can be consumed safely, but surface water from most lakes and rivers must be treated. Treatment is designed to clarify the water, to rid it of pathogenic microorganisms, and sometimes to improve its taste (**Figure 28.16**).

Processing Methods

Water to be processed for drinking is pumped into a tank where alum is added. The aluminum ions in alum neutralize charges on the colloidal particles, allowing them to aggregate into sizes that settle out in the flocculation tank. Then the water is clarified by filtering it through beds of sand or diatomaceous earth. Finally, passing water through beds of activated charcoal removes odor-producing compounds and some toxic compounds.

Filtering water through beds of sand 2 to 4 feet deep removes most bacteria that cause waterborne diseases (Chapter 20). Modern sewage treatments remove many viruses as well because they become trapped in alum-created flocs (see Larger Field: How Safe Is Safe? in Chapter 13). But some microorganisms still pass through the filters. Most municipal water treatment plants add chlorine or ozone to kill residual microorganisms. Enough chlorine is added to react with organic material that might be present and have enough left over as free chlorine to kill most microorganisms within 30 minutes.

Testing Methods

Many human diseases are acquired from contaminated drinking water (Table 28.2). To ensure safety, municipal water supplies are routinely tested. It would be impractical to test for all possible pathogens. Instead, indicator organisms are selected. If the water is free of the indicator organism, it's probably free of pathogens.

In many countries, including the United States, **coliform bacteria** (aerobic and facultative anaerobic, Gram-negative, non–spore-forming, rod-shaped bacteria that ferment lactose and form gas) are used as indicators. This practical definition is chosen to include *Escherichia coli*,

LARGER FIELD

AN UNPLANNED EXPERIMENT

More than 100 years ago, in 1892, an unplanned experiment took place in Germany that forever changed world opinion about the importance of treating municipal water supplies. Hamburg, a growing urban center, obtained its drinking water as inexpensively as possible—by pumping it directly from the Elbe River. The adjoining little town of Altona also drew drinking water from the Elbe. But in 1891 Altona had begun filtering it through beds of sand. When a cholera epidemic struck Hamburg but not Altona, the importance of treating drinking water and the effectiveness of filtration were dramatically evident. On the Hamburg side of the street that divided the two municipalities, people died. But people on the Altona side of the street, people did not. Hamburg began to treat its drinking water, and so did other European cities. Cholera has never returned to Europe.

which is almost always present in feces of humans and other animals. In an aqueous environment outside the body, *E. coli* eventually dies, but not faster than other bacterial pathogens. Therefore water that is free of coliform bacteria is most probably safe. Water containing them should not be consumed.

Two methods are used to test for coliform bacteria—the traditional most probable number (**MPN**) test and the now more commonly used membrane filter (**MF**) test (**Figure 28.17**).

The MPN procedure consists of three steps—the presumptive test, the confirmed test, and the completed test. In the presumptive test, tubes of lactose broth are inoculated with water to be tested, incubated, and observed after 24 and 48 hours. If gas is produced, the presumptive test is positive and the presumed number of coliforms can

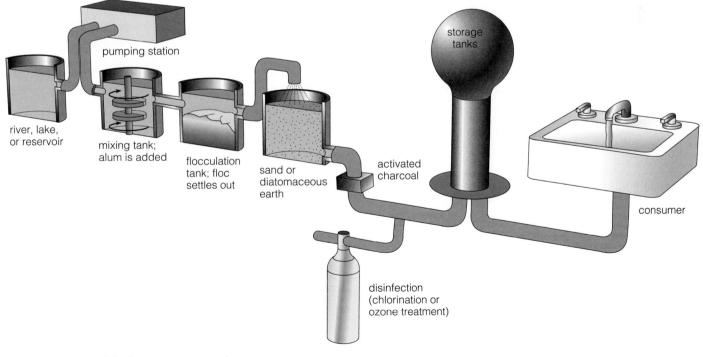

FIGURE 28.16   Municipal water treatment plants.

Most probable number (MPN) test

Membrane filter (MF) test

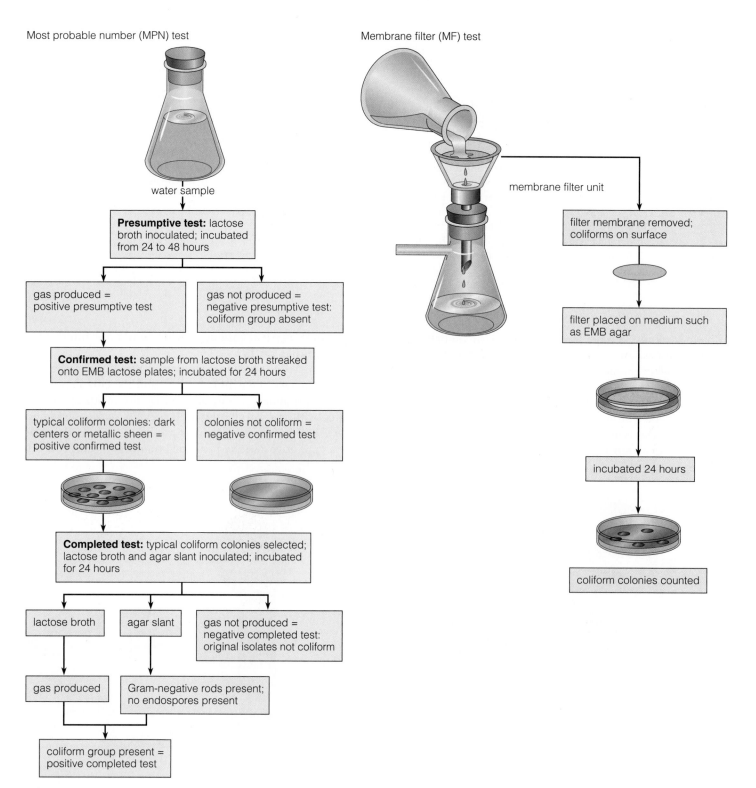

water sample

Presumptive test: lactose broth inoculated; incubated from 24 to 48 hours

gas produced = positive presumptive test

gas not produced = negative presumptive test: coliform group absent

Confirmed test: sample from lactose broth streaked onto EMB lactose plates; incubated for 24 hours

typical coliform colonies: dark centers or metallic sheen = positive confirmed test

colonies not coliform = negative confirmed test

Completed test: typical coliform colonies selected; lactose broth and agar slant inoculated; incubated for 24 hours

lactose broth

agar slant

gas not produced = negative completed test: original isolates not coliform

gas produced

Gram-negative rods present; no endospores present

coliform group present = positive completed test

membrane filter unit

filter membrane removed; coliforms on surface

filter placed on medium such as EMB agar

incubated 24 hours

coliform colonies counted

FIGURE 28.17   Water testing. Two methods are used to test drinking water for the presence of coliform bacteria—the most probable number and membrane filter methods.

SHARPER FOCUS

BACTERIA THAT MAKE PLASTICS

Waste plastic has accumulated almost everywhere on Earth, even on the remote beaches. Plastic litter is becoming a serious ecological problem. Plastics are insoluble in water, and microorganisms have not evolved enzymes to metabolize chemicals that never existed before humans made them. Birds, fish, turtles, and marine mammals become entangled in plastic debris or eat it and die when it blocks their intestines. Plastic in landfills used for garbage disposal remains there permanently.

One solution to the accumulation of plastics in nature is to develop biodegradable plastics. Unfortunately, most plastic products touted as biodegradable are not. They are made of bits of undegradable plastic held together by a matrix of a biodegradable substance such as starch. Microbial action dismantles the matrix but leaves the small bits of plastic intact.

Although humans have so far failed to synthesize completely biodegradable plastics, bacteria have been successful. Some bacteria store their carbon reserves as granules of poly-beta-hydroxyalkanes (Chapter 4), also called polyhydroxyalkanoate (PHA). They accumulate PHAs as we accumulate fat, in huge quantities when a carbon source is available in excess. Then, when starved, they metabolize their PHAs.

PHAs (which, like some synthetic plastics we use, are polyesters) are a biodegradable plastic. The bacteria that make PHAs, and other bacteria as well, readily use PHAs as a carbon source. Some bacteria excrete an extracellular enzyme, a depolymerase, that degrades PHAs into their component beta-hydroxy acids, which they or other microorganisms then use as nutrients.

Bacteria make different forms of PHA, depending upon the source of carbon they are fed. Both the length of the alkane chain and the length of the polymer can be varied by changing bacterial nutrition. PHAs can therefore be fabricated into many different kinds of plastic, suitable for making plastic bags, squeeze bottles, clear glasslike material, and even fabrics.

The idea of using PHAs for commercial plastics has been around for some time. Patents were filed in the United States in 1962, but the plastics were not made industrially until 1982, when Imperial Chemical Industries of Britain marketed them under the trade name of Biopol. The first consumer product, bottles for biodegradable shampoos, was marketed in 1990 by Wella AG in Germany.

The hydrogen-oxidizing bacterium *Alcaligenes eutrophus* is used to make PHAs. The substrates for autotrophic growth—hydrogen gas, carbon dioxide, and air—are inexpensive but potentially explosive, so *A. eutrophus* is grown in a glucose-salts medium. When growth ceases after about 60 hours because the phosphate is exhausted, little PHA has been made. Then more glucose is added, and during the next 48 hours, massive amounts of PHA accumulate, up to 75 percent of the total biomass. PHA is extracted with hot nonpolar solvents (either chloroform or methylene chloride), and cells are filtered out. On cooling, PHA precipitates from the solvent.

On the market, PHAs cost several times more than ordinary plastics, which are inexpensive because they are made from petroleum. But when environmental costs are taken into account, these 100 percent biodegradable plastics might be a bargain.

Plastic debris litters the shore of a beach on one of the Hawaiian Islands.

© Ken Sakamoto/Black Star

be calculated from most probable number tables (Chapter 8). To confirm that the gas-producing organisms are coliforms, samples from the highest dilution showing a positive result are streaked on eosin-methylene blue (EMB) agar. On these plates, coliforms produce distinctive colonies with a metallic sheen and a dark center. If such colonies develop, the confirmed test is positive. To complete the test, these colonies are used to inoculate lactose broth and **slants** (test tubes with solid medium that are allowed to solidify at an angle). If gas is produced in the broth and Gram-negative, non–spore-forming, rod-shaped bacteria develop on the slants, the completed test is positive. A positive result means that coliform bacteria are present and the water is not safe to consume.

In the MF procedure, at least 100 ml of water is filtered and the filter is placed on the surface of a plate, such as an EMB plate, that identifies coliform bacteria. After incubation, typical coliform colonies are counted. Standards of drinking water safety for the United States are set by the Environmental Protection Agency (EPA). They require that the number of coliform bacteria shall not exceed any of the following: (1) an average of 1 per 100 ml in all samples examined in a month, (2) 4 per 100 ml in any one sample if fewer than 20 samples are examined each month, or (3) 4 per 100 ml in 5 percent of the samples if 20 or more samples are examined each month.

SUMMARY

Life and the Evolution of Our Environment (pp. 694–695)

1. Microorganisms play particularly important roles in the cyclic transformations (cycles of matter) that perpetuate life.

Microorganisms in the Biosphere (pp. 695–701)

2. Many microbial transformations take place in the soil. Bacteria improve soil fertility through mineralization and through fixing nitrogen.

3. Soil bacteria are extremely diverse. Actinomycetes are among the most important. A few bacteria that live in the soil cause such human diseases as anthrax, food poisoning, tetanus, and botulism.

4. An acre of soil typically contains between 500 and 5000 pounds of fungi that degrade organic materials. Fungi break down organic compounds in soil. Some are mycoparasites. Others cause such human diseases as blastomycosis, histoplasmosis, and San Joaquin Valley fever.

5. Algae are present on the surface of all soils in small numbers. Protozoa control bacterial soil populations. Larvae of the hookworm *Necator americanus* are free-living in the soil.

6. Some soil fungi form mycorrhizae, intimate associations with the roots of plants. The fungi act as additional roots to help the plant acquire nutrients. The rhizosphere effect, described by the R:S ratio, is another symbiotic association between plants and microorganisms in the region of the rhizosphere.

7. Phytoplankton in the world's oceans constitute a major portion of the world's total photosynthetic capacity. Eutrophic bodies of water support huge populations of cyanobacteria called blooms. Fresh water is a significant reservoir of human pathogens.

8. Microorganisms do not grow in air, but all are found as passengers on aerosols or dust particles.

The Cycles of Matter (pp. 701–710)

9. The major biological elements occur in several different chemical forms that are interconvertible.

10. In the nitrogen cycle, nitrogen gas is converted to ammonia by nitrogen fixation. Ammonia is converted to nitrate by nitrification, and nitrate is converted to nitrogen gas by denitrification.

11. All the steps of the nitrogen cycle are mediated by microorganisms. Two classes of bacteria, symbiotic (such as *Rhizobium* spp.) and free-living (such as *Azotobacter*), fix nitrogen. About half the world's supply of nitrogen is now fixed by the Haber process. Heterotrophic bacteria carry out ammonification, the conversion of nitrogen compounds back to ammonia. Nitrifying bacteria convert ammonia to nitrate.

12. In the carbon cycle, carbon dioxide in the atmosphere is fixed by autotrophs; carbon dioxide is regenerated and returned to the atmosphere through respiration and combustion. The carbon cycle is out of balance today because of burning of fossil fuels. Global warming will probably result.

13. The phosphorus cycle has no significant gaseous intermediate, and phosphorus is neither oxidized nor reduced. The phosphorus cycle converts phosphorus from inorganic phosphate to organic phosphate and back.

14. The bacterium *Acinetobacter calcoaceticus* can be used to remove phosphate from sewage.

15. In the sulfur cycle, sulfate ions are reduced to hydrogen sulfide by sulfate-reducing bacteria and reoxidized to sulfate oxidizing bacteria and certain phototrophs.

16. Sulfate-reducing bacteria live in sulfate-rich anaerobic environments such as mud flats.

Treatment of Wastewater (pp. 710–712)

17. Sewage is municipal wastewater. Sewage plants add oxygen to sewage at a faster rate than it is removed by microbial respiration. Wastewater is classified by its biochemical oxygen demand (BOD).

18. The primary treatment unit in a sewage plant grinds the solid material; the primary sludge settles out, while the effluent proceeds to secondary treatment. The sewage is aerated and then mineralized by aerobic bacteria, including *Zoogloea ramigera*, forming a floc. The rapidly settling floc is called activated sludge. The primary sludge and the activated sludge are sent to an anaerobic sludge digester, where they are partially degraded by anaerobic bacteria. Methanogens convert by-products of the anaerobic bacteria into methane, which can be used as a fuel. The digested sludge is dried and often used as fertilizer. Effluent may be given tertiary treatment. Increasingly, this is biological.

19. Septic tanks are small anaerobic digesters used where municipal sewage systems are not available. Digested effluent is oxidized in leach fields.

Treatment of Drinking Water (pp. 712–716)

20. After processing by flocculation, drinking water is passed through sand beds. This clarifies the water, removing most bacteria and some viruses. Chemicals kill most remaining microorganisms.

21. Municipal drinking water is checked for safety by testing for coliform bacteria, which are indicators of contamination.

22. The two tests for coliforms are the most probable number (MPN) test and the more commonly used membrane filter (MF) test.

REVIEW QUESTIONS

Life and the Evolution of Our Environment

1. Explain this statement: Although microorganisms cause disease, it would be a mistake to think of them only as adversaries.

2. What are biogeochemical transformations? What are the oldest known fossils, and what do they tell us about the evolution of life and Earth's atmosphere?

Microorganisms in the Biosphere

3. By what processes do microorganisms improve soil fertility? Give some examples of microorganisms found in the soil and their functions. What are mycorrhizae and the rhizosphere?

4. What kinds of microorganisms are found in water, and what are their ecological roles?

The Cycles of Matter

5. How do the cycles of matter show the interdependence of living things?

6. Describe the nitrogen cycle and the role of bacteria in nitrogen fixation, nitrification, ammonification, and denitrification. What different ways have *Azotobacter* and *Rhizobium* spp. evolved to protect their nitrogenase from oxygen?

7. What is the Haber process, and why is it significant? How is the water fern *Azolla* used to increase fixed nitrogen for agriculture?

8. What is the major loop of the carbon cycle. Is the carbon cycle balanced? What are some possible ecological consequences?

9. How does the phosphorus cycle differ from the nitrogen and carbon cycles? Describe the phosphorus cycle.

10. Describe the sulfur cycle. What roles do bacteria play? What industrially valuable product is a by-product of the sulfur cycle?

Treatment of Wastewater

11. What is sewage? What is biochemical oxygen demand (BOD)? Why it is important in sewage treatment?

12. What do the primary, secondary, and tertiary treatment units do in a municipal sewage plant? What roles do *Zoogloea ramigera* and the methanogens play?

13. What aspects of sewage treatment occur in a septic tank? In a leach field?

Treatment of Drinking Water

14. How is drinking water processed? How is it tested? What are indicator organisms?

CORRELATION QUESTIONS

1. If you saw gas bubbles rising from a pond with a heavy layer of dead leaves on the bottom, what do you think the gas might be? What simple test could you do to verify your suspicion?

2. To conserve fertilizer, farmers sometimes apply chemicals that inhibit nitrification. How could such a treatment work?

3. What would you expect black mud in an estuary to smell like? Why?

4. Biological removal of phosphate and nitrate from treated sewage prevent eutrophication. Could both of these processes be done at the same time in the same pond? Why?

5. What do N_2 and H_2S production by bacteria have in common?

6. Which do you think evolved first— nitrogen fixation or aerobic respiration? Explain.

ESSAY QUESTIONS

1. Discuss the properties of a world without denitrification.

2. Discuss the properties of a world without nitrification.

SUGGESTED READINGS

Anderson, A. J., and E. A. Dawes. 1990. Occurrence, metabolism, metabolic role, and industrial uses of bacterial polyhydroxyalkanoates. *Microbiological Reviews* 54:450–72.

Atlas, R. M., and R. Bartha. 1998. *Microbial ecology: Fundamentals and applications*, 4th Edition. Menlo Park, Calif.: Benjamin/Cummings.

Goodfellow, M., and S. T. Williams. 1983. Ecology of actinomycetes. *Annual Reviews of Microbiology* 37:189–216.

Racke, K. D., and J. R., Coats, eds. 1990. *Enhanced biodegradation of pesticides in the environment*. Washington, D.C.: American Chemical Society.

Wackett, L. P., and C. D. Hershberger. 2001. *Biocatalysis and biodegradation: Microbial transformation of organic compounds*. Washington, D.C. ASM Press.

For additional readings, go to InfoTrac College Edition, your online research library at: http://www.infotrac.thomsonlearning.com

TWENTY-NINE

Microbial Biotechnology

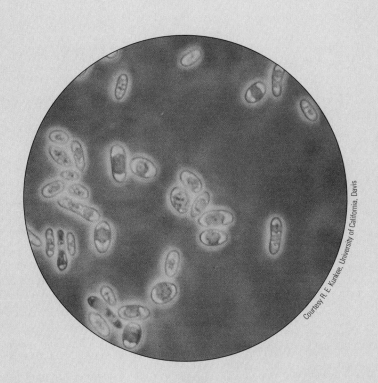

Courtesy R. E. Kunkee, University of California, Davis

CHAPTER OUTLINE

LEARNING GOALS

To understand:

- The traditional uses of lactic acid bacteria to preserve food and make cheese and other dairy products

- How yeasts are used to make wine, beer, and bread, and, with acetic acid bacteria, vinegar

- The principal methods of slowing or preventing growth of microorganisms to preserve food

- How microorganisms are used as insecticides

- The industrial fermentations that produce solvents, amino acids, antibiotics, and enzymes

- How genetically engineered microbes are used to make new products for medicine, industry, and agriculture

The Germ Theory of Swiss Cheese

The germ theory of disease says certain microorganisms cause certain diseases. They also cause certain cheeses. For example, *Propionibacterium* causes Swiss cheese. You couldn't make Swiss cheese without it. *Propionibacterium* grows in unripened cheese by fermenting lactic acid. The lactic acid it needs is present in unripened cheeses because ever-present lactic acid bacteria made from the lactose in milk. The end products of *Propionibacterium*'s fermentation—acetic acids, propionic acid, and CO_2—convert the bland unripened cheese into Swiss cheese. These two acids, along with small amounts of many other products of the fermentation, give Swiss cheese its special flavor. The gaseous product, CO_2, makes the holes that characterize Swiss cheese.

Making Swiss cheese is easy. Making good Swiss cheese is a little more difficult. The strains of *Propionibacterium* must be carefully chosen, and the balance of their fermentation products must be just right. If there is too much or too little propionic or acetic acid, the cheese will not have its identifiable nutty flavor. There must be just enough carbon dioxide to make holes, but not enough to split the cheese. The *Propionibacterium* strain should make enough of the amino acid proline to give the cheese a sweet taste and enough diacetyl for a buttery taste.

The products of *Propionibacterium* do more than just add flavor. Propionic acid is an excellent preservative. It has long been added to bread to prevent the growth of molds. But the mixture of products of a *Propionibacterium* fermentation does better. Such a natural preservative can be made by cultivating *P. shermanii* in skim milk and drying the culture to a powder. The product, called Microgard, is used to preserve cottage cheese. It inhibits Gram-negative bacteria and some yeasts, as well as molds. Other preparations of *Propionibacterium* may be used to preserve grains and bread.

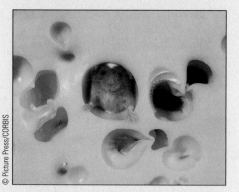

© Picture Press/CORBIS

Propionibacterium is the microorganism used to make Swiss cheese.

Even before humans knew that microorganisms existed, they used them to make food (Swiss cheese is one of many examples), improve crop yields, and dispose of waste. As knowledge of microorganisms grew, humans improved and expanded their uses. Early trial-and-error processes developed into technologies we collectively called **industrial microbiology.** Managing microorganisms became a science. It profoundly changed our lives by giving us the capacity to produce antibiotics, vitamins, food supplements, and industrial chemicals.

Traditional industrial microbiology produced the same products that microbes normally make in their natural environment. It merely produced them efficiently on a large scale. But during the past 30 years, a sea change has occurred. With the emergence of recombinant DNA technology, microorganisms can be engineered to produce new products, products they don't normally make. This capacity has led to the creation of a new industry called **biotechnology** (Chapter 7). Over the years, use of the term has expanded. It now applies to all practical uses of organisms, including traditional ones. Although the term biotechnology encompasses all organisms, microorganisms are and probably will continue to be its primary focus. In this chapter we'll discuss the microbial aspects of biotechnology.

We'll start our discussion of microbiotechnology with the traditional uses of microorganisms to make and preserve human and animal food. Then we'll consider the use of microorganisms as chemical factories to make chemicals and drugs. Finally we'll look at the uses of genetically engineered microbes to make products specified by genes from other organisms.

TRADITIONAL USES OF MICROORGANISMS

The first human use of microorganisms was probably to make and preserve foods and beverages. Although our ancestors didn't know it, they were principally using lactic acid bacteria and yeasts. We still use them. Let's consider why these organisms were and continue to be so important to us.

Lactic Acid Bacteria

Lactic acid bacteria are important because they make lots of acid. They ferment various sugars to form enough lactic acid to inhibit or kill most other microorganisms in their environment (Chapter 11). When their environment is a food, it is protected from attack by other microorganisms and therefore preserved. In addition, the lactic acid and other flavors lactic acid bacteria add to food improve its taste. Moreover, with very few exceptions, including some streptococci, lactic acid bacteria are harmless to humans. These properties make lactic acid bacteria almost ideal agents for preserving food. There are a few limitations. The food must contain enough sugars for the lactic acid bacteria to produce inhibiting amounts of lactic acid, but most plant materials and dairy products do contain adequate amounts. Also, air must be excluded. If it isn't, aerobic microorganisms will metabolize the sugar and some will metabolize the lactic acid. Usually it isn't even necessary to add lactic acid bacteria to the food. Most plant materials and dairy products contain an adequate natural population to start the preservation process.

Plant Foods. Making sauerkraut (meaning acid cabbage in German) illustrates how lactic acid bacteria can be used to preserve food. It's only necessary to shred the cabbage, pack it firmly into a nonmetal container such as a ceramic crock, and add sufficient water to fill pockets of air. The lactic acid bacteria naturally present on the cabbage multiply and convert the sugars in the cabbage into lactic acid. The sauerkraut becomes so acidic that it lasts almost indefinitely if air is excluded. The same principles are used to prepare Spanish-style green olives.

Huge quantities of livestock feed are similarly preserved annually by making it into **silage.** Chopped plants, principally corn, are deposited in large airtight cylindrical silos or plastic lined pits. The lactic acid bacteria present, mainly *Lactobacillus plantarum*, ferment the sugars in the plant material and thereby preserve it (**Figure 29.1**).

Cheese and Other Dairy Products.
Like plant foods, most traditional fermented dairy products were made using the lactic acid bacteria that just happened to be present. Usually the bacteria came from the tools and vessels used. Sometimes a portion of a previous batch was used as an inoculum. Today highly selected cultures are used as inocula. The distinctive properties and tastes of various soured milk products—including buttermilk, acidophilus milk, and yogurt—depend upon the kind of lactic acid bacterium used.

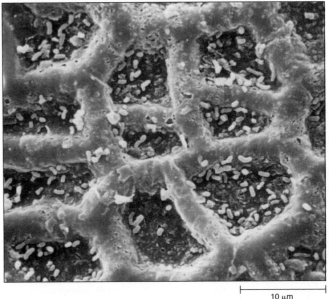

Courtesy Henry P. Fleming, North Carolina State University

10 µm

FIGURE 29.1   The lactic acid bacterium *Lactobacillus plantarum* on the surface of a cucumber fruit.

Different species produce different products (**Table 29.1**). Even different strains of the same species affect the quality and flavor of the product. A great deal of research has been done to produce superior strains for particular products.

Cheese making is also largely a microbiological process (**Figure 29.2**). It consists of two steps—**curdling** and **ripening.** Curdling is the conversion of milk into a solid mass, the curd. The protein fraction—casein—is precipitated, bringing the fat with it. **Whey** (the liquid fraction) is drained off. Curdling is accomplished by allowing lactic bacteria to develop (the resulting lactic acid denatures the casein, causing it to precipitate) and by adding **rennin** (also called chymosin). Rennin is an enzyme previously extracted from a calf stomach but now produced by recombinant DNA technology. It cleaves one form of casein, causing it to precipitate. Whether a lactic acid fermentation proceeds depends upon the type of cheese being made. For example, cottage cheese, which is not acidic, is made with rennin alone.

After curdling, water is pressed out of the curd, salt is added, and the solid mass is allowed to **ripen** (develop it characteristic taste and texture). The species of microorganism involved in ripening depends upon the type of cheese being made (Table 29.1). In traditional cheese manufacture, ripening microorganisms come from utensils or tools used to prepare previous batches. Even so, cheeses produced in the same region by the same procedures differed little from batch to batch. Today selected microorganisms are usually added.

Ripening is a complex process that changes the cheese. Some of the changes come from flavor-giving compounds

TABLE 29.1 Microbiology of Dairy Products and Cheese

Product	Microorganism(s)	Microbial Activity
Acidophilus milk	*Lactobacillus acidophilus*	Lactic acid fermentation
Butter	*Streptococcus diacetilactis, Leuconostoc cremoris*	Lactic acid fermentation
Sour cream	*Streptococcus diacetilactis, Streptococcus Lactis, Leuconostoc cremoris*	Lactic acid fermentation
Buttermilk	*Lactobacillus bulgaricus*	Lactic acid fermentation
Kefir	*Streptococcus lactis, Lactobacillus bulgaricus,* lactose-fermenting yeasts	Combined lactic acid–alcoholic fermentation
Yogurt	*Lactobacillus bulgaricus, Streptococcus thermophilus*	Lactic acid fermentation
Cheese		
Cheddar	Lactic acid bacteria	Lactic acid fermentation
Roquefort	*Penicillium roquefortii*	Production of blue pigment
Swiss	Lactic acid bacteria, *Propionibacterium* spp.	Lactic/propionic acid fermentation

FIGURE 29.2   Curdled milk being stirred in the process of making cheese.

made by the microorganisms. Others come from hydrolyzing milk proteins into a mixture of small soluble peptides and amino acids. In general the more hydrolysis, the softer the cheese. Only about a quarter of the protein in hard cheeses such as cheddar and Swiss is hydrolyzed. Almost all the protein in soft cheeses such as Camembert and Limburger is hydrolyzed.

Yeasts

Yeasts are as important as lactic acid in preparing and preserving food. They are widely spread in nature, occurring naturally on most fruits and flowers and in the exudates of plants. Most yeasts ferment sugars to produce CO_2 and ethanol. Such yeasts are used to make wine, beer, whiskey, and bread. Almost always strains of *Saccharomyces cerevisiae* and closely related species are used. Probably the oldest use of yeast was making wine.

Wine. The origins of **enology** (winemaking) must be as ancient as **viticulture** (grape growing), and viticulture is almost as ancient as agriculture itself. Wine was a well-established article of commerce by the third century B.C. The traditional method of making wine is simple, almost inevitable. Crush the grapes and let them stand in a container with restricted contact with air. Still, modern enology has become a highly sophisticated science (**Figure 29.3**), taught as a distinct field in some universities.

The price of wines depends largely upon the variety and quality of the grapes used. High-quality, flavorful varieties often have low crop yield, and farming practices

FIGURE 29.3   A winery. The large steel tanks are used to ferment and store wine. The wood barrels are being used to age wine. They can also be used as fermentors.

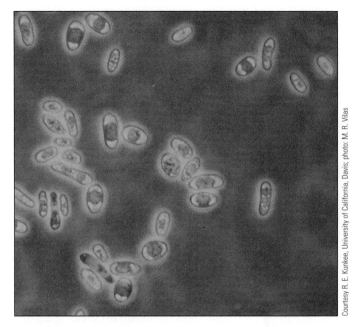

FIGURE 29.4   *Saccharomyces cerevisiae.*

that enhance quality often decrease yield. But the microbiological aspects of winemaking are also critically important. Traditionally, and still in some regions, the **must** (crushed grapes) is not inoculated. Instead, the natural yeast biota of the grape is allowed to carry out the fermentation. But today, in most regions of the world, the natural inoculum is virtually eliminated by treating the must with bisulfite. Then a pure culture of a reliable wine yeast strain is added (**Figure 29.4**).

Sometimes different kinds of yeasts and other microorganisms are used to make different kinds of wine (**Table 29.2**). But the most important distinction—red or white—depends upon the grapes and winemaking practice. Only white wine can be made from white grapes, but

TABLE 29.2 Microbiology of Wine

Organisms	Roles in Winemaking
Saccharomyces cerevisiae	*Primary alcoholic fermentation*—glucose and fructose in must are fermented to ethanol and carbon dioxide.
Saccharomyces bayanus	*Secondary alcoholic fermentation*—in a pressurized tank or bottle, added sugar is fermented to ethanol and carbon dioxide to produce sparkling wines such as champagne.
	Clouding—yeast grows in finished sweet wine, producing undesirable turbidity.
Torulaspora delbrueckii	*Flor layer on sherry*—yeast grows and oxidizes some ethanol to acetaldehyde, producing typical sherry flavor.
Botrytis cinerea	*Production of naturally sweet, flavorful wines*—mold raises sugar content by desiccating grapes, decreases acidity by oxidizing malic acid to carbon dioxide and water, and adds flavor and color.
Oenococcus oeni	*Malolactic fermentation*—malic acid in wine is converted to lactic and carbon dioxide, decreasing acidity and increasing flavor complexity.
Acetobacter spp.	*Aerobic spoilage*—oxidizes some ethanol to acetic acid, giving wine an undesirable vinegar taste.

either red or white wine can be made from red grapes. The red color is in the grape skins. If skins are left in the fermenting must, the red color they contain will be extracted (by alcohol) into the wine.

Spanish sherry (sherry is a British corruption of the name of the Spanish town, Jerez, where much Spanish sherry is made) is made by two microbial steps. First, *Saccharomyces cerevisiae* ferments all the sugar in the must. Then, after **brandy** (distilled wine) has been added to increase the alcohol content to about 15 percent, a layer of the yeast *Saccharomyces fermentati* develops on the surface of the wine, which is stored in half-filled barrels. This layer thickens and becomes crinkly, resembling flower petals, accounting for its Spanish name, *flor* (meaning "flower"). The metabolic products of *S. fermentati*, including large amounts of acetaldehyde, give sherry its distinctive taste.

Certain sweet wines from the Sauternes district of France and parts of Germany are products of an even more complicated microbiological process. In these relatively damp regions, the grapes become infected by the fungus *Botrytis cinerea*, which enters the grape by means of shortened hyphae called haustoria. The haustoria act as small wicks that dry the grape, concentrating its sugars so much that they cannot be fermented completely by the yeasts when the must is made into wine. The fungus produces a distinctive color and flavor and some glycerol, which contributes a thickness and richness to the wine. *Botrytis* does not metabolize the sugar needed by the yeast. Instead it uses malic acid (an organic acid found in all grapes), making the wine less acid. This fungal infection—called "noble rot" by winemakers—produces dessert wines that are highly valued for their distinctive soft, sweet flavor and bright golden color.

Bacteria also contribute to winemaking. In most wine regions the malic acid of premium red wine is removed by a traditionally spontaneous (now inoculated) malolactic fermentation—mediated by the lactic acid bacterium *Oenococcus oeni*. This secondary fermentation converts malic acid, which contains two acid groups, into lactic acid, which contains only a single acid group.

$$HCOOH{-}CH_2{-}CHOH{-}COOH \longrightarrow$$
Malic Acid

$$CH_3{-}CHOH{-}COOH + CO_2$$
Lactic Acid   Carbon Dioxide

The result is a less acidic and more flavorful wine.

Microorganisms can also spoil wine. Some lactic acid bacteria grow in sweet wines, giving it a foul taste. Acetic acid bacteria in wine is exposed to air, oxidizing the alcohol to acetic acid and giving the wine a vinegary taste.

Other Alcoholic Beverages. Most yeasts can ferment only sugars. That's not a complication for fermenting sugar-containing fruit juices, including grape juice (it contains a mixture of glucose and fructose). But grain products contain starch which yeast cannot ferment. First the starch must be **saccharified** (hydrolyzed to its sugar monomer, glucose) and then fermented by yeasts. In the Western world, **amylase** (an enzyme that hydrolyzes starch) from **malt** (heat-dried germinated grains of barley) is used to saccharify cereal grains.

Beer is made by a yeast fermentation of saccharified cereal grain. Traditionally barley was used. Now rice is often added. Cereal grains are cooked and saccharified with malt. The saccharified product, called **wort,** is fermented in large vats (**Figure 29.5**). Finally beer is flavored and to an extent preserved by adding hops, the flowers of the vine *Humulus lupulus*.

Whiskey is distilled, or extracted, from fermented cereal grains. Potatoes, which are used to make vodka, contain largely starch, so they too must be saccharified.

In Asia, amylase from fungi is used to saccharify rice before it is fermented to produce Japanese sake and Chinese rice wine. The moistened rice is heaped on the floor and inoculated with a previous batch of rice on which mixtures of fungi have grown. Within a few days the fungi grow through the pile of rice, producing enough amylase to saccharify all the starch.

Vinegar. Vinegar is a solution of acetic acid made in a two-step process. First, yeasts ferment the sugar in a

FIGURE 29.5 A brewing vat in which cereal grains and malt are being cooked to make wort.

fruit juice, usually apple (to make cider vinegar) or grape (to make wine vinegar), to ethanol. The yeast is usually *Saccharomyces cerevisiae*. Then *Acetobacter* spp., obligately aerobic bacteria, oxidize the ethanol incompletely to acetic acid and water, instead of completely to CO_2 and water as most microorganisms would do.

$$CH_3\!-\!CH_2OH + O_2 \longrightarrow CH_3\!-\!COOH + H_2O$$
Ethanol   Acetic Acid

There are a number of ways to make vinegar. All are designed to provide enough oxygen to oxidize alcohol to acetic acid. The oldest and slowest—but some say best—way is to fill a barrel half full of the alcohol solution. A film of *Acetobacter* develops on the surface, oxidizing the ethanol at a rate set by the diffusion of oxygen into the solution. A more rapid way is to use a vinegar generator (**Figure 29.6**). Vinegar generators are large columns, usually filled with beech-wood chips that become covered with a layer of *Acetobacter* cells. The alcohol solution is fed into the top and trickles down through the chips while air is forced in at the bottom of the column.

Bread. Yeasts are used to raise bread (or **leaven** it, a term that comes from an old word for yeast). Leavening bread before baking is a relatively recent human innovation. Brewer's yeast, a by-product of brewing beer, was first used, but now specially selected strains called baker's yeast are used. Both brewer's yeast and baker's yeast are strains of *Saccharomyces cerevisiae*.

Baker's yeast is produced commercially and dried under conditions that maintain viability and enzyme activity (see Larger Field: Bread in the Desert, in Chapter 2). It can be stored for long periods at room temperature without loss of activity. CO_2 is the fermentation product needed for baking. CO_2 produced in dough creates small bubbles that lighten the texture. Usually the natural sugar in flour produces enough CO_2 to raise the dough. A small amount of sugar is sometimes added.

Mixed Cultures

In some parts of the world, mixed cultures are used to make fermented foods. In Europe a mixture of lactic acid bacteria and yeasts is used to produce mildly alcoholic, soured milk beverages such as kefir. In Asia the production of foods such as soy sauce and miso (made from rice and soybeans) from mixed cultures of microorganisms is a huge industry (**Table 29.3**). In Japan, mixed-culture food industries produce 20 times more revenue than the industrial alcohol industry. Usually, starter cultures containing complex mixtures of bacteria and fungi are used to inoculate food. Because the methods of producing starter cultures have not changed for centuries, these mixtures of microorganisms are highly adapted to coexist and produce a uniform product. Some microorganisms in these starters are found nowhere else and would not survive in nature.

Stable mixed cultures have certain advantages over pure cultures. They have high growth rates and enhanced yields. They are highly resistant to contamination, and they bring about simultaneous multistep transformations of complex mixtures of substrates.

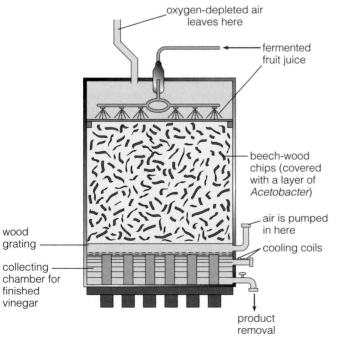

FIGURE 29.6 A vinegar generator.

oxygen-depleted air leaves here

fermented fruit juice

beech-wood chips (covered with a layer of *Acetobacter*)

air is pumped in here

cooling coils

wood grating

collecting chamber for finished vinegar

product removal

TABLE 29.3 Mixtures of Microbial Cultures Used to Make Asian Foods

Food	Microorganisms in Mixture
Soy sauce	*Aspergillus oryzae*
	Aspergillus soyae
	Saccharomyces rouxii
	Candida etchellsii
	Pediococcus halophilus
	Lactobacillus delbrueckii
Miso	*Aspergillus oryzae*
	Aspergillus soyae
	Saccharomyces rouxii
	Candida etchellsii
	Pediococcus halophilus

MICROBES AS INSECTICIDES

Certain species of the endospore-forming bacterial genus *Bacillus*—including *Bacillus larvae*, *Bacillus popilliae*, *Bacillus thuringiensis*—are insect pathogens. These species form crystals of protein, called parasporal bodies, beside their endospores (**Figure 29.7**). When a susceptible insect eats the protein, it is split in the insect's alkaline gut into a highly destructive protein fragment, which ulcerates the intestinal wall and kills the insect. For more than 30 years, *B. thuringiensis* has been used to produce a bioinsecticide called Bt. Bt is sprayed or dusted on plant leaves to control caterpillars and other insects. Within an hour after consuming the endospores, caterpillars stop feeding, and they die several days later.

B. popilliae produces a bioinsecticide called Bp, which is used to control Japanese beetles. Distributed over a lawn, the endospores persist in the soil and control Japanese beetles for 15 to 20 years.

Spores of the protozoan *Nosema locustae* are sold commercially as a bait to combat grasshoppers, locusts, and crickets. They act slowly, requiring 4 to 6 weeks for control. Baculoviruses are being developed as another bioinsecticide (Chapter 13).

FIGURE 29.7   Parasporal bodies of *Bacillus thuringiensis*. Note the bipyramidal shape of these protein crystals.

100 nm

© J. R. Adams/Visuals Unlimited

Unlike some chemical insecticides, Bt, Bp, *Nosema locustae*, and baculoviruses are all ecologically safe. They kill only certain insects and have no effect on plants or animals, including humans.

MICROBES AS CHEMICAL FACTORIES

With their rapid rates of metabolism and growth, microorganisms can synthesize cells and metabolic end products at prodigious rates. Because of this, microorganisms are used industrially to manufacture many commercial products in addition to food. These include organic solvents, vitamins, amino acids, antibiotics, enzymes, proteins with medical uses, and the microorganisms themselves.

The first industrially produced microbial products were manufactured anaerobically by fermentation. Somehow the term stuck, so that today, whether the manufacturing process is aerobic or anaerobic, industrial transformations by microorganisms are called fermentations. The vessels in which these transformations take place are called fermentors.

Anaerobic Fermentations

Because fermentations (in the scientific sense) generate only small amounts of adenosine triphosphate (ATP; Chapter 5), anaerobic microorganisms process large quantities of substrate and produce large quantities of end products when they ferment. Some fermentative end products have commercial value, and producing them is relatively uncomplicated. Of course, the culture does not have to be aerated. But more importantly, rigorous aseptic procedures are unnecessary. That's because contamination by aerobes is impossible and most competing anaerobes grow too slowly to displace the established population.

The economic feasibility of an anaerobically produced industrial product is determined by the price of the product, the price of the substrate, and the cost of recovering the product from the culture. The most common substrates for industrial fermentations are **blackstrap molasses** (the dark residual liquid remaining after sugar has been recovered from cane or beet syrup) and cereal grains.

The price/cost margin is almost always narrow because most anaerobically produced products are relatively inexpensive industrial chemicals. They can also be made by other processes—for example, chemically from petroleum. Price or cost changes often, so once-successful industrial fermentations may be abandoned only to be reinitiated at a later date. The food fermentations discussed earlier con-

SHARPER FOCUS

MICROBES VERSUS PETROLEUM

Making solvents by fermentation first became an industry in England. Initially the English made butanol, a starting material for synthetic rubber, because natural rubber was in short supply. One of the products of the butanol process was acetone, a solvent for nitrocellulose, which is an ingredient of cordite (smokeless powder). With World War I, the market for acetone grew and English solvent manufacturing expanded quickly. After the war, the industry continued to expand because butanol was an ingredient of the rapid-drying nitrocellulose paints used by the expanding automobile industry. A third product of the process was riboflavin (vitamin B$_2$), which met still another market need.

Yet by the early 1960s the solvent industry had virtually disappeared in Britain (and the United States). The petrochemical industry could make butanol-acetone less expensively, and the pharmaceutical industry had developed better microbiological methods for making riboflavin. Today, however, the butanol-acetone process is being intensively reexamined. It may again offer economic advantages because genetic modifications can be made in the producing organism and cheaper waste substrates such as whey are available.

tinue in spite of economic fluctuations. By law, beverage alcohol and acetic acid for vinegar must be microbiological products.

Ethanol. As we have discussed, ethanol (ethyl alcohol) is a constituent of alcoholic beverages. It is also used as a solvent and a fuel. Before World War II, large amounts of ethanol were made by yeast fermentation for industrial purposes. Later, with the rise of the petrochemical industry, the price of ethanol dropped so low that the fermentation product was not economic. Within the past decade, however, the prospects of ethanol have again improved. Ethanol made by fermentation is the major source of automobile fuel in Brazil. In the United States, ethanol is being added to gasoline to make gasohol (90 percent gasoline, 10 percent ethanol). Gasohol will help extend the world's petroleum supply and is less polluting than conventional gasoline. Ethanol is also a renewable resource. It can be made from sugar crops, saccharified cereal grains, and even acid-treated wood, straw, or newsprint.

Acetone and Butanol. Acetone and butanol are commercially useful as organic solvents. They are fermentation products of various species of *Clostridium*. One species, *Clostridium acetobutylicum*, has been used on and off since before World War I to make a mixture of these solvents. The economic ups and downs of this industry, along with some of the technological innovations that it sparked, have captured the attention of industrial microbiology (see Sharper Focus: Microbes versus Petroleum).

The technological advances associated with acetone-butanol fermentation stem from the fact that the process is successful only if other microorganisms can be excluded from the relatively rich medium. As a result, the technology of pure culture on a mass scale had to be developed. These advances were essential to the later development of the antibiotic industry. This industry introduced a new set of complications, because antibiotics were produced by aerobes.

Aerobic Processes

Like antibiotic production, most other industrial microbiological processes now in use are aerobic. Supplying adequate oxygen to aerobic fermentors is an engineering challenge because oxygen is sparingly soluble in water, and it is used rapidly by the dense microbial cultures that industrial processes employ. Moreover, to prevent contamination, the massive inflow of air must be sterilized.

The rate at which oxygen can be added to a culture depends upon the surface area of liquid exposed to the gas. To maximize the surface area, modern industrial fermentors are designed to be filled with small bubbles of air. This is done by forcing air through a **sparger** (which looks somewhat like a showerhead) beneath a rapidly turning **impeller** (a rotating shaft with paddles). The impeller shatters the streams of air into minute bubbles that follow a helical (and therefore prolonged) path to the surface (**Figure 29.8**).

Because a rich broth is used for most aerobic fermentations, contamination is a constant threat. Air sparged into the fermentor is sterilized by filtration. The fermentor itself must be impervious to contamination. Valves through which materials are added to the fermentor or taken out are vulnerable sites. No matter how tightly they

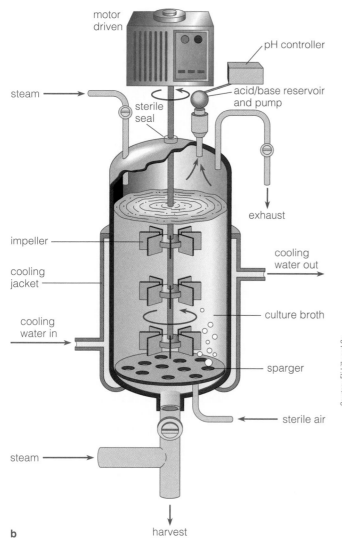

Courtesy Eli Lilly and Co.

a

b

FIGURE 29.8   An industrial fermentor. (a) A fermentor viewed from the top. Note the shaft with attached impellers and cooling coils, which remove heat produced by microbial metabolism. (b) Essential components of a fermentor.

are closed, foreign microorganisms can pass through them and enter the fermentor. Therefore most fermentor valves are bathed in steam.

Today aerobic fermentations are the source of many valuable products, including antibiotics, amino acids, and enzymes. They are also used to bring about particular chemical reactions that are difficult by ordinary chemical means. We'll discuss some of these uses of aerobic fermentations in the following sections.

Antibiotics.   The antibiotic industry began with the Scottish microbiologist Alexander Fleming in 1929 (Chapter 1). Fleming was studying wound infections caused by *Staphylococcus aureus* when he noticed that a contaminating mold colony (later identified as *Penicillium notatum*) seemed to be destroying the surrounding staphylococcal colonies. He reasoned that the mold was producing an antimicrobial substance that might have chemotherapeutic value. He named the active substance penicillin, after the mold, and tried to isolate it. Fleming failed, but 11 years later two British chemists, Howard Florey and Ernst Chain, succeeded. Fleming, Florey, and Chain were awarded the Nobel Prize for their work.

Once penicillin had been isolated, chemists believed mass production would soon follow. They only had to determine the structure of penicillin and then synthesize it chemically. However, even though penicillin is a small molecule, it is extremely difficult to synthesize chemically. That meant that procedures had to be developed to grow large amounts of *P. notatum* aerobically in rich media. The culture vessel had to be kept rigorously free of contamination, because many contaminating bacteria produce **penicillinase** (also called beta-lactamase) that destroys penicillin. With the coming of World War II, penicillin was desperately needed. With intense international cooperation, success came quickly and the antibiotics industry was born.

TABLE 29.4 Some Commercially Produced Antibiotics

Antibiotic	Producing Microorganism	Class
Produced by Fungi		
Cephalosporin	*Cephalosporium acremonium*	Broad-spectrum
Griseofulvin	*Penicillium griseofulvum*	Fungi
Penicillin	*Penicillium chrysogenum*	Gram-positive bacteria
Produced by Gram-positive, Spore-forming Bacteria		
Bacitracin	*Bacillus subtilis*	Gram-positive bacteria
Polymyxin B	*Bacillus polymyxa*	Gram-negative bacteria
Produced by Gram-positive Bacterium, Actinomycete		
Amphotericin B	*Streptomyces nodosus*	Fungi
Chloramphenicol	*Streptomyces venezuelae* (now chemical synthesis)	Broad-spectrum
Cycloheximide	*Streptomyces griseus*	Pathogenic yeasts
Cycloserine	*Streptomyces orchidaceus*	Broad-spectrum
Erythromycin	*Streptomyces erythreus*	Mostly Gram-positive bacteria
Kanamycin	*Streptomyces kanomyceticus*	Gram-positive bacteria
Lincomycin	*Streptomyces lincolnensis*	Gram-positive bacteria
Neomycin	*Streptomyces fradiae*	Broad-spectrum
Nystatin	*Streptomyces noursei*	Fungi
Streptomycin	*Streptomyces griseus*	Gram-negative bacteria (*Mycobacterium tuberculosis*)
Tetracycline	*Streptomyces rimosus*	Broad-spectrum

The spectacular success of penicillin, along with its limitations, stimulated a major search for new antibiotics in the years after World War II. In particular, researchers hoped to find an antibiotic that would cure tuberculosis. Soil samples from all over the world were collected and examined for antibiotic-producing microorganisms. In the 1950s, streptomycin was discovered by Albert Schatz and Selman A. Waksman at Rutgers University. Streptomycin was effective against tuberculosis and some Gram-negative bacteria that penicillin could not control (see Larger Field: The Discovery of Streptomycin, in Chapter 21).

Then several broad-spectrum antibiotics, effective against a wide variety of Gram-positive and Gram-negative bacteria, were discovered through research by drug companies. Although chemically quite different from penicillin, these newly discovered antibiotics fit into a relatively small number of chemically distinct types, with specific targets in the bacterial cell (**Table 29.4**). With the exception of chloramphenicol, all these anti-biotics are too chemically complex for commercial chemical synthesis. Instead they are made by aerobic fermentation.

Over the years the search for antibiotics has continued. New ones have been discovered, but the increasing failure to find new chemical types suggests that most classes of antibiotics have already been discovered. As antibiotics become less useful because pathogens become resistant to them, the need for new ways to control pathogens has become critical.

Amino Acids. Several amino acids are made by aerobic fermentation for use in the food industry (**Table 29.5**). Some are used to improve or modify the flavor of a food. Others are used to enhance nutritional value. For example, lysine is added to bread in Japan because, without enrichment, bread does not contain enough of this essential amino acid to meet human nutritional requirements. Lysine is not added to bread in the United States because U.S. residents consume enough milk and meat to meet their needs for lysine. Japan also manufactures

SHARPER FOCUS

A GOOD HOST

Choosing the right host to express a cloned gene is an important decision. To manufacture medically useful human proteins, a human gene must be cloned in an appropriate cloning vector. It must also be expressed in a living host cell. Which host is best? How do you decide?

When recombinant DNA technology was first developed in the late 1970s, there was no choice. Cloned genes could be inserted only into *Escherichia coli*. The technology rapidly improved, however. By developing new cloning vectors and procedures, it became possible to insert cloned genes into many species of bacteria, yeasts and other fungi, cultured plant and animal cells, and intact plants and animals. Suddenly a wide choice of hosts was available.

What properties should you look for when choosing a host?

1. A good host should produce large amounts of the protein product. Producing large amounts of a protein requires more than efficient transcription and translation. Often, foreign proteins are synthesized rapidly only to be destroyed almost as rapidly by the host cell's proteolytic enzymes. Changing the growth medium or using mutant strains that produce fewer proteolytic enzymes often solves the problem. But sometimes it is necessary to use a different host.

2. A good host should be easy to cultivate. A good host should grow rapidly in an inexpensive medium and

should require a relatively simple fermentor.

3. The biology of a good host should be thoroughly understood. The more that is known about a host's biology, the easier it is to improve its potential as a host and to solve problems that arise during production.

4. A good host should produce the correct form of the protein product. After translation in its native host, the new protein folds in a particular way and often is modified. Human proteins are usually glycosylated, meaning specific sugar molecules are attached to the protein at particular locations. Folding and modification occur differently in other hosts. A good host allows the protein to fold correctly and modifies

it in ways not radically different from the way the native host would.

Escherichia coli satisfies the first three requirements better than any other organism. It is the most thoroughly understood cellular organism. Also, it is easy and inexpensive to cultivate, and its growth medium and genetic constitution can be readily and predictably changed. But the fourth requirement can be a problem. Many human proteins are folded incorrectly in *E. coli*. This is a troublesome but not always insurmountable problem. Often the misfolded form can be purified, unfolded chemically, and refolded in vitro into the correct form. Glycosylation is also a problem because *E. coli* lacks the metabolic machinery to glycosylate. Sometimes

TABLE 29.5 Uses of Commercially Produced Amino Acids

Amino Acid	Use
Alanine	Added to fruit juice to improve taste
Aspartate	Added to fruit juice to improve taste
Cysteine	Added to bread and fruit juice to enhance flavor
Glutamate (MSG)	Added to many foods to enhance flavor
Glycine	Enhances flavor in various foods
Histidine + tryptophan	Prevents rancidity in various foods
Lysine	Used in Japan to make bread a more complete protein
Methionine	Makes soybean products a more complete protein

SHARPER FOCUS

A GOOD HOST (continued)

the unglycosylated form of a medically useful protein is as active and useful as the natural, glycosylated form. Glycosylation rarely changes a protein's activity but often changes its solubility. But proteins can be chemically altered to change their solubility.

Largely because of folding and glycosylation, biotechnology is continuously developing new hosts. Cultured mammalian cells are now used commercially to produce certain human proteins. For example, Chinese hamster ovary (CHO) cells are used to make tissue plasminogen activator (tPA), a human protein that saves the lives of many victims of heart attack by dissolving the blood clots that cause them. CHO cells do not glycosylate exactly as human cells do, but closely enough to be effective.

They also grow quite slowly, require expensive media, and are difficult to manipulate genetically.

The search for new and better hosts seems unending because none used so far is ideal. Promising new candidates include cultured insect cells and intact insects, with certain baculoviruses used as cloning vectors (Chapter 13). In nature, virions of these baculoviruses are released from infected cells in packages called polyhedra. Polyhedra consist of many individual virions embedded in copious amounts of a matrix composed of the protein polyhedrin. Polyhedra protect the virions, allowing them to survive in the wild until consumed by a susceptible insect. In the insect's digestive tract, polyhedrin is broken down, releasing the virions, which then infect the insect.

A cell infected with this kind of baculovirus is converted into a factory for making polyhedrin. To use a baculovirus as a vector, its polyhedrin gene is replaced by a cloned gene that is attached to the polyhedrin promoter—the region of the gene where transcription begins (Chapter 6). In this way the infected cell will become a factory for making the product of the cloned gene. The productivity of cells infected with such a vector can be 10,000 times greater than the productivity of a mammalian cell culture. Moreover, host insect cells perform posttranslational modifications, including glycosylation, almost as human cells do. Cultured cells of the fall armyworm infected with engineered baculoviruses are being used to produce proteins. A vaccine for ac-

quired immunodeficiency syndrome (AIDS) was made this way. Intact insects are also being evaluated as hosts. A very promising insect host for the Baculovirus system is the silkworm, *Bombyx mori*. Silkworms have been cultivated in mass for thousands of years to make silk, so the methods for growing and handling them are highly developed. Some steps have even been automated. Early results suggest that silkworms are excellent hosts for Baculovirus-cloned genes. Larvae are infected by injecting engineered virus. Three to four days later each silkworm has accumulated about 1 mg of the protein product of the cloned gene, a large amount compared with other methods—and much more valuable than silk.

---

large amounts of monosodium glutamate (MSG) by fermentation for use as a flavor enhancer in many foods.

Enzymes. Microorganisms are a rich source of enzymes that have many commercial uses (**Table 29.6**). For example, in food manufacture, commercial enzymes are used to keep candy from crystallizing, to clarify juices, and to curdle milk. Medically, commercial enzymes are used for purposes as diverse as digestive aids and treatment of victims of heart attack by dissolving blood clots. Laundry detergents contain proteolytic (protein-destroying) enzymes to remove certain stains, including blood stains, from fabrics by dissolving the proteins that bind the stain to the fibers of the fabric. Most commercial enzymes are produced aerobically.

Chemical Reactions

Microorganisms offer a degree of chemical specificity unmatched by ordinary chemical reactions. The action of their enzymes is specific enough to change one chemical group in a molecule and leave similar ones untouched. For this reason, microorganisms are used to carry out difficult steps in the industrial synthesis of certain compounds.

One example of this use of microorganisms is the synthesis of the corticosteroids, cortisone and prednisone, compounds that suppress inflammation. One step in their synthesis requires that a hydroxyl group be inserted at a position in the steroid molecule that is protected from ordinary chemical reactions (**Figure 29.9**). An enzyme from the fungus *Rhizopus nigricans* can do this. If the steroid is

TABLE 29.6 Some Commercially Useful Enzymes Produced by Microorganisms

Enzyme	Activity	Producing Microorganism	Use
Cellulase	Hydrolyzes cellulose	*Trichoderma konigi*	Digestive aid
Collagenase	Hydrolyzes collagen	*Clostridium histolyticum*	Promotes wound/burn healing
Diastase	Hydrolyzes starch	*Aspergillus oryzae*	Digestive aid
Glucose isomerase	Converts glucose to fructose	*Streptomyces phaeochromogenes*	Converts glucose from hydrolyzed cornstarch to a sweetener
Invertase	Hydrolyzes sucrose	*Saccharomyces cerevisiae*	Candy manufacture
Lipase	Hydrolyzes lipids	*Rhizopus* spp.	Digestive aid
Pectinase	Hydrolyzes pectin	*Sclerotina libertina*	Clarifies fruit juice
Protease	Hydrolyzes protein	*Bacillus subtilis*	Used in detergents

added to a culture of the microorganism, it carries out the desired reaction. Other microorganisms are used industrially to carry out this sort of difficult chemical reaction.

The relatively recent development of technology for aerobic fermentations made it possible to produce products and carry out chemical transformations that have profoundly affected our lives. Antibiotics alone have largely freed humankind from many infectious diseases that terrorized previous generations. But traditional aerobic fermentations can produce only those materials that are products of normal microbial metabolism. That limitation was swept away in the early 1970s with the advent of recombinant DNA (Chapter 7). Then, at least in principle, it became possible to engineer microorganisms to produce any compound normally produced by any organism.

USING GENETICALLY ENGINEERED MICROBES

The ability to engineer organisms to make totally new compounds has tremendously expanded the fermentation industry, now called biotechnology. Scores of new companies have formed to make valuable products (**Table 29.7**).

The impact of recombinant DNA technology is not limited to making new products. All aspects of microbial biotechnology, including the most traditional, have benefited from this powerful new technology. For example, most strains of lactic acid bacteria used to make cheese have been altered by recombinant DNA technology. Most uses of the many new products to come from biotechnology are medical and agricultural, although certain products, such as renin, are used industrially. We'll consider them in sequence.

Medical Uses

Medically useful compounds manufactured by genetically engineered organisms include hormones and other human proteins.

Hormones. Hormones made by genetically engineered organisms can be used to supplement the needs of patients whose bodies are genetically unable to produce sufficient hormones on their own. Two human hormones, insulin and human growth hormone, are particularly valuable. Both are made by strains of *Escherichia coli* containing human genes cloned into plasmids.

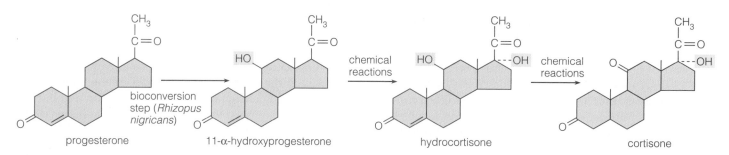

FIGURE 29.9 Steps in the synthesis of cortisone from progesterone. Note that *Rhizopus nigricans* is used to insert a hydroxyl group in the 11-α position of progesterone.

TABLE 29.7 Some Products Made by Genetically Engineered Cells

Product	Use
Medical	
Human growth hormone	To treat pituitary dwarfism
DNase	To treat cystic fibrosis
Insulin	To treat diabetes
Tissue plasminogen activator	To treat heart attacks
Factor VIII	To treat hemophilia
Erythropoietin	Stimulates red blood cell formation
Interferon-alpha	To treat hepatitis C
Industrial	
Rennin	making cheese
Ascorbic acid (vitamin C)	Diet supplement, fruit preservative
Bovine growth hormone	Stimulates milk production
Agricultural	
Bt toxin	Insecticide
Ice-minus bacteria	Protects plants from freezing

Human Growth Hormone. Untreated children born with a pituitary gland that is unable to make sufficient human growth hormone (hGH) become pituitary dwarfs. Before hGH was produced microbially, it was purified from the pituitary glands of human cadavers. Because the supply was limited, many children went untreated. Moreover, hGH obtained from cadavers carried a risk of prion infection.

The methods used to clone and express the hGH gene in bacteria (Chapter 7) are typical of those used to make other human proteins. All human cells carry the gene that encodes hGH, but how do you separate that gene from all others and clone it? One way is to start with messenger RNA (mRNA) instead of DNA. Human pituitary gland tissue makes the body's supply of hGH, so its cells must contain large amounts of the corresponding mRNA. These mRNA molecules can be identified with an appropriate **probe** (a short DNA molecule corresponding to part of the hGH; Chapter 7). The probe, being complementary to its corresponding mRNA, will hybridize (form hydrogen-bonded base pairs). There is a second and more compelling reason for isolating mRNA rather than isolating the gene directly. Genes from eukaryotes, including humans, contain **introns** (stretches of DNA that do not code the protein because they are cut out of mRNA as it matures). Because prokaryotes lack the enzymes to eliminate introns, they would synthesize an incorrect protein from a eukaryotic gene carrying introns.

But how do you make the probe? First you have to know the sequence of amino acids in hGH protein. Then the sequence of bases for the probe can be determined by referring to the genetic code (Chapter 6). But the genetic code is redundant—most amino acids are encoded by several different codons. The genetic code states precisely which amino acid sequence is encoded by a particular sequence of bases in DNA but not the reverse. Because of redundancy of the code, all the DNA sequences that designate the desired amino acid sequence must be synthesized. The mixture is used as a probe.

Once the correct mRNA has been isolated, **reverse transcriptase** (the enzyme from retroviruses that uses RNA as a template to make DNA; Chapter 13) is used to make DNA. This DNA product is called complementary DNA (cDNA) to indicate that it is a copy of mRNA, not the DNA in the gene itself. Then the cDNA is cut with restriction endonucleases and ligated into a bacterial plasmid. The plasmid is inserted into the bacterial host by transformation (Chapter 6). In the host the recombinant DNA molecule is replicated and its genes, including the hGH gene, are expressed. The transformed cell becomes an hGH factory.

The method used to clone hGH is not the only way it could be done. Many variations are possible, and new ones are developed almost daily. The most fundamental changes are in the cloning vector and the host cell. A plasmid can be used to clone hGH, but viral DNA can also be used. Introducing native or recombinant DNA into a host is termed **transformation**. Almost any kind of cell or intact organism can now be used as a host for the recombinant DNA molecule (see Sharper Focus: A Good Host). But no matter which host cell is finally selected to make the protein product, bacteria are almost always used in the cloning process.

After the hGH-producing bacterial strain is obtained, the job of producing hGH commercially has just begun. It is not enough to produce some hGH. The plasmid must be modified and cultural conditions must be developed to produce commercially feasible amounts of hGH. For example, methods have been devised to trigger *Escherichia coli* to make hGH only after a dense culture develops, because large amounts of hGH (and other foreign proteins as well) are detrimental to the cell. That is, the growth phase is separated from the production phase. Methods have also been devised to purify properly folded hGH from the bacterial culture.

Insulin. Insulin is a hormone secreted by the pancreas that lowers glucose levels in the blood. Insufficient insulin production creates imbalances in the body's metabolism, leading to dehydration, excessive urination, and sometimes coma and death. The disorder is called **diabetes mellitus.** The most serious type of diabetes appears in childhood and requires insulin injections throughout the person's life.

In the United States, bovine insulin (from cattle pancreas) was traditionally used for treatment; in Europe, porcine insulin (from pig pancreas) was used. Although lifesaving, animal insulin differs from human insulin, suggesting that it is less effective than the human hormone. Moreover, the increasing use of feedlots (rather than range grazing) to fatten cattle threatened the insulin supply because it decreased the insulin content of cattle pancreases. Now an abundant supply of human insulin is produced by *Escherichia coli* carrying the human gene cloned into a plasmid.

Research is underway to develop microbial production of other human hormones. They include factor VIII (the hormone that most hemophiliacs need) and relaxin (a hormone that causes the cervix to dilate during delivery of a baby; some women do not produce an adequate amount). Along with fermentation, recombinant DNA technology holds the promise that adequate supplies of all needed human hormones will one day be available.

Other Proteins. In addition to hormones, other medically useful human proteins are being produced by fermentation, or procedures for their production are being developed. One of the most important is tissue plasminogen activator (tPA), which binds to blood clots and dissolves them. It is lifesaving for victims of heart attack. By dissolving the clot that cuts off the blood supply to the heart, it prevents heart damage and death. If tPA is administered soon enough after the attack, the patient may recover with no lasting damage.

The procedures used to clone and produce tPA are much like those used for human growth hormone. Although some tPA is produced by bacteria, most is pro-

duced by a mammalian cell line derived from Chinese hamster ovaries (CHO cells). CHO cells are grown in huge fermentors much like microbial cells, using many of the techniques and procedures originally developed for penicillin production. Streptokinase, a bacterial protease (Chapter 15), is also used to treat victims of heart attack.

Human DNase, an enzyme that degrades DNA, is also produced in CHO cells. It is used to treat patients with cystic fibrosis, a genetic disease that causes the lungs to secrete unusually thick mucus. Affected children rarely survive to adulthood, partly because they become infected with heavily encapsulated strains of the opportunistic bacterial pathogen *Pseudomonas aeruginosa*. Macrophages, which accumulate to fight the infection by phagocytizing the bacteria, become packed with bacterial cells. When the bacteria-packed macrophages die, they release their DNA into the lungs and form a mucus that makes breathing difficult. The lungs must be cleared. Traditional treatment involves pounding on the back while the patient tries to cough up mucus. Inhaling human DNase breaks up the mucus painlessly. Nonhuman DNase cannot be used because the immune system would recognize it as foreign protein and produce antibodies. Subsequent treatments might cause anaphylactic shock (Chapter 18).

Agricultural Uses

Microorganisms have long been valuable to agriculture. Genetic engineering has increased their value. Microorganisms have been engineered to protect plants from pests or the environment, to control decay after harvest, and to improve animal husbandry. We'll describe four of these—Bt toxin, ice-minus bacteria, better silage makers, and epidermal growth factor—here.

Bt Toxin. The capacity to produce the insect-killing Bt toxin has been transferred from bacilli into *Pseudomonas fluorescens*, a species of soil bacteria that inhabits the rhizosphere (Chapter 28). Producing Bt toxin right at the root surface protects plant roots from destructive soil insects. Inoculating soil with engineered strains of *Pseudomonas fluorescens* stimulates plant growth.

Ice-Minus Bacteria. In regions of the semiarid western United States, rainfall is heavier over areas with denser vegetation. But which is cause and which is effect? Microbiologists and meteorologists working together in Montana found that certain bacteria, including *Pseudomonas syringae*, growing on the leaves of plants nucleate water and cause rain. *P. syringae* acts much like silver iodide crystals, which are used to seed clouds, but is more effective. It's swept from leaves into the clouds by wind that usually precedes a storm. A protein on the bacterial surface that mimics the shape of ice crystals causes nucleation. Killed *P. syringae* cells are now used for snowmaking at ski resorts. The dead cells are mixed with water and sprayed into the air when the temperature is below freezing.

But *P. syringae* is dangerous on plants because it makes frost damage more likely. Normally, pure water such as that in dew can be supercooled to 4°C below its freezing point and remain liquid. But if nucleation sites such as the surface protein of *P. syringae* are present, frost forms right at the freezing point (**Figure 29.10**). Frost, not low temperature alone, damages plants. Frost-sensitive plants cannot tolerate ice forming within their tissues, so the presence of *P. syringae* can damage frost-sensitive crops on cold nights.

Steven Lindow, a plant pathologist, has developed a method for ridding crops of frost-stimulating *P. syringae*. Using genetic engineering, he deleted the gene encoding the nucleation protein from *P. syringae* to produce a strain he called ice-minus. Spraying young frost-sensitive crops, such as strawberries, with ice-minus bacteria establishes them on the plant, leaving no room for normal frost-inducing strains. Crops are protected up to 95 percent against frost damage. Frost damage in the United States is estimated to cost more than $1 billion annually.

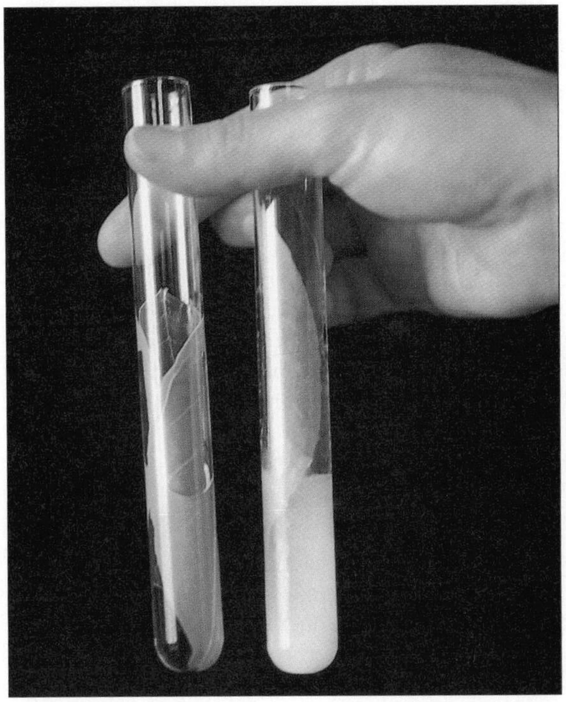
Courtesy Steven Lindow, University of California, Berkeley

FIGURE 29.10 A bean leaf with wild type *Pseudomonas syringae* causes ice-cold water to freeze (right), but one with an ice-minus strain does not (left).

Better Silage Makers. As we discussed earlier in this chapter, silage is protected against decay by lactic acid produced by lactic acid bacteria, principally *Lactobacillus plantarum*. However, it's difficult to preserve whole crop cereals such as barley or wheat this way. They don't have enough sugar to generate protecting levels of lactic acid. Researchers have solved the problem by cloning an a-amylase–encoding gene from *Bacillus amyloliquefaciens* into *Lactobacillus plantarum*. This genetically engineered bacterium can convert the abundant starch in cereal crops into lactic acid. The result is decay-resistant silage.

SUMMARY

Traditional Uses of Microorganisms (pp. 720–725)

1. Biotechnology is all uses of organisms for practical purposes.

2. Lactic acid bacteria ferment sugars to make sauerkraut, silage, green olives, cheese, and other dairy products. Except for some streptococci, lactic acid bacteria are harmless to humans. Their metabolic products have a pleasant taste.

3. Most yeasts ferment sugars to produce carbon dioxide and ethanol. Strains of *Saccharomyces cerevisiae* are usually used for wines. Sauternes and other sweet wines are made from grapes infected by the fungus *Botrytis cinerea*. Lactic acid bacteria growing in sweet wines produce a foul taste, and wine exposed to air allows acetic acid bacteria to develop, giving the wine a vinegary taste.

4. Beer and whiskey are made from fermented cereal grain. Strains of *Saccharomyces cerevisiae*, called brewer's yeast, are used. Because grains contain starch, they must first be saccharified (hydrolyzed to glucose). In Western countries, malt is added because it contains the starch-hydrolyzing enzyme amylase. Potatoes used to make vodka must also be saccharified. In Asia, fungi are used as a source of amylase to saccharify rice to produce sake and rice wine.

5. Vinegar is a solution of acetic acid made in a two-step process. Yeast, usually *S. cerevisiae*, ferments sugar in juice to ethanol. Then *Acetobacter* spp. oxidize the ethanol incompletely to acetic acid and water. Vinegar generators are used to add oxygen at a high rate.

6. Leavened bread was originally made by using brewer's yeast, a by-product of brewing beer. Today we use baker's yeast, specially selected strains of *Saccharomyces cerevisiae*.

7. Mixed cultures are also used to make fermented foods, including kefir, soy sauce, and miso. Starter cultures are complex mixtures of bacteria and fungi used to inoculate food.

Microbes as Insecticides (p. 726)

8. Certain species of *Bacillus* are insect pathogens. They form parasporal bodies (protein crystals) by their endospores that are lethal to insects. *Bacillus thuringiensis* is marketed as a bioinsecticide called Bt to control caterpillars and other insects. *Bacillus popilliae* produces Bp, which kills Japanese beetles.

9. The protozoan *Nosema locustae* is used as bait to kill grasshoppers and locusts. Baculoviruses are also being developed as bioinsecticides.

10. Unlike chemical pesticides, bioinsecticides are ecologically safe.

Microbes as Chemical Factories (pp. 726–732)

11. The transformations by which microorganisms produce antibiotics, vitamins, organic solvents, and so forth are called fermentations, whether they occur aerobically or anaerobically. Similarly, the vessels in which the transformations take place are called fermentors.

12. Anaerobic fermentations often use blackstrap molasses and cereal grains (which are saccharified if necessary) as substrates.

13. Ethanol (also called ethyl alcohol) is a constituent of alcoholic beverages and is used as a solvent and fuel. It is made from saccharified cereal grains.

14. Acetone and butanol, used as solvents, are fermentation products of *Clostridium* spp. To produce large amounts during World War I, the technology of pure cultures on a mass scale had to be developed. This technology was also essential to commercial production of antibiotics.

15. Aerobic processes are used to make antibiotics, amino acids, vitamins, and therapeutically useful proteins. Fermentors with a sparger and impeller keep oxygen in the culture broth. Valves are bathed in steam to prevent foreign microorganisms from entering the fermentor.

16. The antibiotics industry began with penicillin. The search for new antibiotics continues as resistant strains emerge. There is a good chance that most classes of antibiotics have already been discovered.

17. Amino acids are used in the food industry to improve flavor (monosodium glutamate) or enhance nutritional value (lysine).

18. Enzymes are used in food manufacture to keep candy from crystallizing, clarify juices, and curdle milk. They are used medically as digestive aids and to dissolve blood clots. Laundry detergents contain protein-destroying enzymes to remove blood and other stains.

19. Microorganisms are used to carry out difficult steps in the industrial synthesis of certain compounds. For example, the fungus *Rhizopus nigricans* is used in the synthesis of steroids such as cortisone and prednisone.

Using Genetically Engineered Microbes (pp. 732–735)

20. Before recombinant DNA technology, microorganisms could produce only the products of their normal metabolism. Now they can be engineered, at least in principle, to produce any compound produced by any other organism.

21. Medical uses include human growth hormone for treating pituitary dwarfism, insulin for treating diabetes, tissue plasminogen activator (tPA) and streptokinase for dissolving blood clots, and human DNase for treating cystic fibrosis.

22. Ascorbic acid (vitamin C), used as a diet supplement and food preservative, can be produced more cheaply

and in large amounts using genetically engineered *Erwinia*.

23. Through genetic engineering, Bt toxin can be transferred to soil bacteria in the rhizosphere to protect plant roots from destructive soil insects. A genetically engineered strain of *Pseudomonas syringae*, called ice-minus, is used to protect crops from frost damage. *Lactobacillus plantarum* has been genetically engineered to convert starch in crop cereals to sugar to prevent silage from spoiling.

REVIEW QUESTIONS

Traditional Uses of Microorganisms

1. What is biotechnology?
2. Discuss some ways lactic acid bacteria have been used to preserve plant foods and make cheese.
3. What roles, positive and negative, do various microorganisms play in winemaking?
4. Why must cereal grains be saccharified for beer and whiskey making, and how is it done? Which yeasts are most commonly used in beer making?
5. Describe how vinegar is made.
6. How are microorganisms used in making bread?
7. What are starter cultures? What are the benefits of mixed cultures?

Microbes as Insecticides

8. Classify each of these microorganisms and tell how they act as insecticides: *Bacillus thuringiensis*, *Bacillus popilliae*, *Nosema locustae*, baculoviruses.
9. What are the advantages of bioinsecticides over chemical insecticides?

Microbes as Chemical Factories

10. Define fermentation and fermentor.
11. Explain this statement about anaerobic fermentations: Successful industrial fermentations may be abandoned only to be reinitiated at a later date.
12. Name some products made by anaerobic fermentations.
13. Describe the technology that had to be developed before the modern aerobic fermentation industry could be successful. Give some examples of products made by aerobic fermentations.
14. How can microorganisms be used to carry out difficult chemical reactions? Give an example.

Using Genetically Engineered Microbes

15. Explain this statement: Recombinant DNA technology changed the fermentation industry completely.
16. Discuss some medical products of recombinant DNA technology. How is human growth hormone produced?
17. Discuss these examples of genetic engineering in agriculture: Bt toxin, ice-minus bacteria, and better silage makers.

CORRELATION QUESTIONS

1. Why does chopped cabbage undergo a lactic acid fermentation and crushed grapes undergo an alcoholic fermentation?
2. Why does malolactic fermentation make wine less sour?
3. Which fermentation products are necessary for making the following foods: wine, bread, champagne, Swiss cheese?
4. Why are rigorous aseptic procedures more critical for making penicillin than for making ethanol?
5. What is the principle advantage of human growth hormone produced by *Escherichia coli* over that obtained from human cadavers?
6. Is the follow statement technically correct? Cells of *Escherichia coli* that produce human growth hormone contain the human growth hormone gene. Why?

ESSAY QUESTIONS

1. Discuss the advantages and disadvantages of using *Escherichia coli* as a host to make protein products by recombinant DNA technology.
2. Comment on the statement: Products of recombinant DNA technology are not natural.

SUGGESTED READINGS

Demain, A. L., and N. A., Solomon, eds. 1986. *Manual of industrial microbiology and biotechnology.* Washington, D.C.: American Society for Microbiology.

Glazer, A. N. and H. Nikaido. 1995. *Microbial biotechnology.* New York. W. H. Freeman and Company

Halvorson, H. O. 1985. *Engineered organisms in the environment: Scientific issues.* Washington, D.C.: American Society for Microbiology.

Hesseltine, C. W. 1983. Microbiology of oriental fermented foods. *Annual Reviews of Microbiology* 37:575–601.

Jones, D. T., and D. R. Woods. 1986. Acetone-butanol fermentation revisited. *Annual Reviews of Microbiology* 50:484–524.

Knight, P. 1991. Baculovirus vectors for making proteins in insect cells. *ASM News* 57:567–70.

For additional readings, go to InfoTrac College Edition, your online research library at: http://www.infotrac.thomsonlearning.com

APPENDIX I
Metric Measurements and Conversions

Scientific Notation

	Name	Prefix
$10^{-15} = 0.000000000000001$	quadrillionth	femto (f)
$10^{-12} = 0.000000000001$	trillionth	pico (p)
$10^{-9} = 0.000000001$	billionth	nano (n)
$10^{-6} = 0.000001$	millionth	micro (μ)
$10^{-3} = 0.001$	thousandth	milli (m)
$10^{-2} = 0.01$	hundredth	centi (c)
$10^{-1} = 0.1$	tenth	deci (d)
$10^{2} = 100$	hundred	hecto (h)
$10^{3} = 1,000$	thousand	kilo (k)
$10^{6} = 1,000,000$	million	mega (M)
$10^{9} = 1,000,000,000$	billion	giga (G)

Volume

1 microliter (μl)	1 cubic millimeter (mm^3)
	0.000001 liter
1 milliliter (ml)	1 cubic centimeter (cm^3, cc)
	0.001 liter
	0.061 cubic inch
	0.03 fluid ounce
1 liter	1000 milliliters
	1.06 quarts

Length

1 angstrom (Å)	0.1 nanometer
	1.0001 micrometer
1 nanometer (nm)	10 angstroms
	0.001 micrometer
1 micrometer (μm)	1000 nanometers
	0.001 millimeter
1 millimeter (mm)	1,000 micrometers
	0.1 centimeter
	0.0394 inch
1 centimeter (cm)	10 millimeters
	0.394 inch
1 meter (m)	100 centimeters
	39.37 inches
	3.28 feet
1 kilometer (km)	1,000 meters
	3,280 feet
	0.62 mile

Mass

1 picogram (pg)	0.000000000001 g
1 nanogram (ng)	0.000000001 g
1 microgram (μg)	0.000001 gram
1 milligram (mg)	0.001 gram
1 gram (g)	1000 milligrams
	0.353 ounce
1 kilogram (kg)	1000 grams
	2.2 pounds

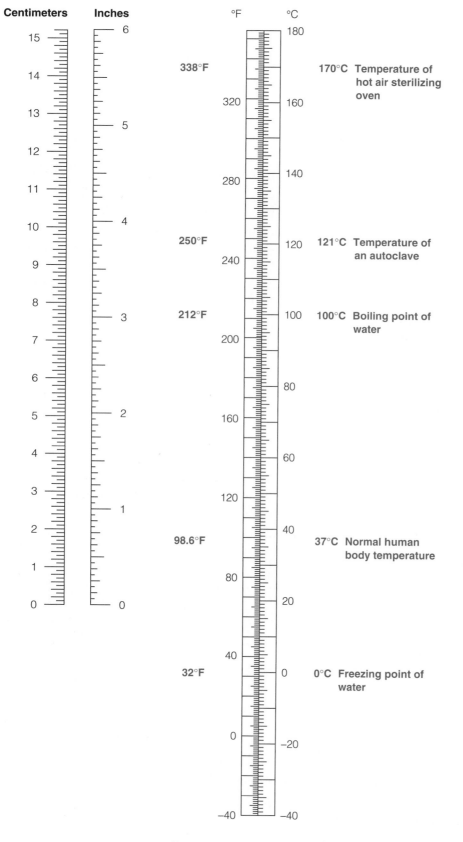

Centimeters   **Inches**

°F   °C

338°F   170°C Temperature of
                               hot air sterilizing
                               oven

250°F   121°C Temperature of
                               an autoclave

212°F   100°C Boiling point of
                               water

98.6°F   37°C   Normal human
                               body temperature

32°F   0°C   Freezing point of
                               water

To convert temperature scales
Fahrenheit to Celsius: $°C = 5/9 \ (°F - 32)$
Celsius to Fahrenheit: $°F = 9/5 \ (°C) + 32$
Kelvin to Celsius: $K = °C + 273.2$

APPENDIX II
Some Word Roots Used in Microbiology

a-, an- not, without, absence. Examples: aseptic, free from infection; anaerobic, in the absence of air.

acet- pertaining to acetic acid. Example: *Acetobacter*, a bacterium that converts ethanol to acetic acid, or vinegar.

actino- having rays. Example: *Actinomyces*, a bacterium that forms star-shaped or rayed colonies.

aer- air. Example: aerobic, in the presence of air.

agglutino- clumped or glued together. Example: agglutination, clumping.

alb- white. Example: *Candida albicans*, a fungus with white colonies.

amphi- around, on both sides. Example: amphitrichous, having flagella at both ends of a cell.

amyl- starch. Example: amylase, an enzyme that degrades starch.

ana- up. Example: anabolic reaction, any synthesis reaction in an organism.

aqua-, aque- water. Examples: aquatic, taking place in water; aqueous, made with water.

arthro- joint. Example: arthropod, an invertebrate having a jointed body and limbs.

asc- sac. Example: ascus, a saclike structure holding spores.

-ase enzyme. Example: polymerase, an enzyme that catalyzes the formation of the polymers DNA or RNA.

aure- gold. Example: *Staphylococcus aureus*, a bacterium with gold-pigmented colonies.

azo-, azoto- nitrogen. Example: *Azospirillum*, a nitrogen-fixing soil bacterium with helical shape.

bacill- small rod. Example: bacillus, a rod-shaped bacterium.

bacteri-, -bacter denoting bacteria. Example: *Agrobacterium*, a plant-infecting bacterium.

bio- life, living organisms. Example: biology, the study of life.

blast- bud. Example: blastomycosis, a disease caused by a yeastlike fungus.

bovi- cow. Example: *Mycobacterium bovis*, a bacterium found in cattle.

butyr- butter. Example: butyric acid, the fatty acid responsible for the odor of rancid butter.

carb- having carbon or carboxyl. Example: carbohydrate, an organic compound formed from carbon, hydrogen, and oxygen.

carcin- cancer. Example: carcinogen, a cancer-causing agent.

cardio- heart. Example: endocarditis, inflammation of the heart's inner lining.

caseo- cheese. Example: casein, a protein produced when milk is curdled by rennet as when making cheese.

caul- stem, stalk. Example: *Caulobacter*, an appendaged or stalked bacteria.

cephalo- of the head or brain. Example: encephalitis, inflammation of the brain.

chlamyd- covered, cloaked. Example: chlamydiospore, spore formed inside a fungal hypha.

chloro- green. Example: chlorophyll, the green pigment in leaves.

-chrome colored. Example: cytochrome, an iron-containing pigment that plays a role in cellular oxidations.

chryso- gold, yellow. Example: *Penicillium chrysogenum*, a mold with yellowish colonies.

-cide killing. Example: bactericide, an agent that kills bacteria.

cili- eyelash. Example: cilia, hairlike organelles.

cocc- berry. Example: coccus, a spherical cell.

coeno- common, shared. Example: coenocytic, having many nuclei not separated by septa.

coli-, colo- colon. Example: coliform bacteria, bacteria found in the large intestine.

con- together. Examples: concentric, having a common center, together in the center.

conidio- dust. Example: conidia, dustlike spores produced by fungi.

coryne- club. Example: *Corynebacterium*, a club-shaped bacterium.

cut- skin. Example: cutaneous, relating to or affecting the skin.

cyan- blue. Example: cyanobacteria, blue-green pigmented bacteria.

cyst- bladder, sac. Example: cystitis, inflammation of the urinary bladder.

cyt-, -cyte cell. Examples: cytoplasm, the fluid within a cell; phagocyte, a cell that consumes foreign material and debris.

dermat- skin. Example: dermatophyte, a fungus parasitic on the skin.

di-, diplo- twice, double. Examples: dimorphic, having two different forms; diplococci, pairs of cocci.

dia- through, apart. Example: diagnosis, the art of distinguishing among diseases by their signs and symptoms.

dys- difficult, abnormal. Example: dysentery, severe abnormal diarrhea.

-emia condition of the blood. Example: septicemia, invasion of the bloodstream by virulent microorganisms.

en-, endo- in, within. Examples: engulf, flow over and enclose; endospore, spore formed inside a cell.

entero- intestine. Example: enterotoxin, a poison affecting the intestine.

epi- upon, over. Example: epidemic, a disease affecting an entire population at once.

erythro- red. Example: erythrocyte, red blood cell.

eu- true, proper, normal. Example: eukaryote, a cell with a true nucleus.

ex-, exo- out of, from, outside. Example: exogenous, from outside the body.

fila- thread. Example: filament, a thin threadlike process or appendage.

flagell- whip. Example: flagellum, a whip-like projection from a cell.

flav- yellow. Example: *Flavobacterium*, a bacterium that produces a yellow pigment.

gamet- to marry. Example: gamete, a reproductive cell.

gastr- stomach. Example: gastritis, inflammation of the stomach.

gen-, -gen, -genesis cause, origin, production. Example: generation, a group of individuals that originated at the same time; pathogen, a microorganism that causes disease; pathogenesis, production of disease by microorganisms.

germin- bud, sprout. Example: germinate, sprout or develop.

gingiv- gum. Example: gingivitis, inflammation of the gums.

-globulin type of protein. Example: immunoglobulin, a protein of the immune system.

glyc- sweet, sugar. Example: glycoprotein, a protein with sugars attached.

gon- reproduction. Example: gonorrhea, pus-producing infection of the reproductive system caused by the bacterium *Neisseria gonorrhoeae*.

hal- salt. Example: halophile, an organism that thrives in high salt concentrations.

haplo- one, single. Example: haploid, half the number of chromosomes or one set.

hemo-, hemat- blood. Examples: hemoglobin, a pigment in red blood cells; hematocrit, an instrument for determining the ratio of red blood cells to whole blood.

hepat- liver. Example: hepatitis, inflammation of the liver.

hetero- different, other. Example: heterotroph, an organism that obtains its nutrients from other organisms.

hist- tissue. Example: histology, the study of tissues.

homo- same, similar. Example: homologous, having the same structure.

hydr-, hydro- water. Example: hydrolysis, breaking down by adding water.

hyper- above, excessive. Example: hypersensitivity, excessively sensitive or susceptible.

hypo- below, deficient. Example: hypotonic, having deficient tension or osmotic pressure.

-iasis disease condition. Example: schistosomiasis, a disease condition caused by trematode worms of the genus *Schistosoma*.

immun- resistance. Example: immunity, the condition of being able to resist a particular infection.

inter- between. Example: interphase, the interval between cycles of mitosis.

intra- within. Example: intracellular, within a cell.

-ism disease or condition. Example: botulism, food poisoning caused by the toxin secreted by *Clostridium botulinum*.

iso- equal, uniform. Example: isotonic, having the same tension or osmotic pressure as another.

-itis inflammation. Example: rhinitis, inflammation of the nose.

-karyo kernel, center. Example: prokaryote, a cell without a true center or nucleus.

kerato- horn. Example: keratin, the horny substance making up skin and nails.

kin- motion. Example: streptokinase, an enzyme produced by streptococci that breaks down or moves fibrin.

lact- milk. Example: *Lactobacillus*, a lactic acid–forming bacterium.

lepto- slender. Example: *Leptospira*, a slender spirochete.

leuko-, leuco- white, colorless. Example: leukocyte, white blood cell.

lip-, lipo- fatty, lipid. Example: lipopolysaccharide, a large molecule composed of fats and sugars.

-logy science, field of study. Example: microbiology, the science that studies microorganisms.

lopho- tufted. Example: lophotrichous, having a tuft of flagella on one side of a cell.

-lysis breaking down, disintegration. Example: hydrolysis, chemical breakdown of a compound into other compounds as a result of taking up water.

macro- large. Example: macromolecule, large molecule.

mening- membrane. Example: meningitis, inflammation of the membranes of the brain.

meso- middle, intermediate. Example: mesophile, an organism that thrives at middle-range temperatures.

meta- beyond, among, change. Example: metabolism, chemical changes occurring within a living organism.

-metry measure. Example: symmetry, measured to be the same or balanced.

micro- very small. Example: microorganism, a very small organism.

mono- single. Example: monotrichous, having a single flagellum.

morph- form. Example: morphology, the study of the form and structure of organisms.

multi- many. Example: multicellular, having many cells.

mur- wall. Example: murein, a polymer characteristic of bacterial cell walls.

muri- mouse. Example: murine typhus, a form of typhus endemic in mice.

mut- change. Example: mutagen, an agent that can cause genetic change.

myco-, -mycete, -myces a fungus. Examples: *Mycobacterium*, a bacterium that often forms filaments like a fungus; *Saccharomyces*, sugar fungus, a genus of yeast.

myxo- mucus, slime. Example: myxobacterium, a slime-producing bacterium.

necro- death. Example: necrosis, cell death or death of an area of tissue.

-nema thread. Example: *Treponema*, a bacterium with long, threadlike cells.

neo- new. Example: neonatal, newborn.

nigr- black. Example: *Rhizopus nigricans*, a mold with black sporangiospores.

nitro- nitrate. Example: *Nitrobacter*, a bacterium that oxidizes nitrite to nitrate.

nitroso- nitrite. Example: *Nitrosomonas*, a bacterium that oxidizes ammonium to nitrite.

noso- disease. Example: nosocomial, a disease acquired in the hospital.

ob- to, toward. Example: obligate, bound or restricted to a particular mode of life.

oculo- eye. Example: ocular lens, the lens closest to the eye in a microscope.

-oid resembling. Example: nucleoid, the region of a procaryote that resembles the eucaryotic nucleus.

-ole small. Example: bronchiole, small, thin-walled branch of a bronchus.

oligo- few, deficient. Example: oligosaccharide, a complex sugar composed of a relatively few simple sugars.

-oma tumor. Example: melanoma, a tumor containing dark pigment.

onco- cancer, tumor. Example: oncogene, a gene that causes tumor formation.

oro- mouth. Example: oropharynx, the region of the throat nearest the mouth.

ortho- straight. Example: orthomyxovirus, a virus with a straight, tubular capsid.

-ose sugar. Example: lactose, milk sugar.

-osis, -sis condition. Example: brucellosis, a condition cased by bacteria of the genus *Brucella*.

-otic relating to a condition. Example: necrotic, relating to local tissue death.

pan- all, completely. Example: pandemic, an epidemic affecting a very large region.

para- beside, abnormal. Example: parasite, an organism that lives at the expense of another.

path- abnormal, diseased. Example: cytopathic, related to abnormal changes in cells.

peri- around. Example: peritrichous, having flagella projecting from all sides.

-phage one that eats. Example: bacteriophage, a virus that digests bacteria.

philo-, -philic liking, having an affinity for. Example: hydrophilic, having an affinity for water.

-phobic fearing, having an aversion to. Example: hydrophobic, having an aversion to water.

-phore bearer of, carrier. Example: conidiophore, a fungal hypha that bears conidia.

photo- light. Example: photosynthesis, the formation of chemical compounds using light as the energy source.

-phyte plant. Example: saprophyte, a plant that obtains nutrients from rotting organic matter.

pil- hair. Example: pilus, a hairlike projection from a cell.

-plast organized granule. Example: chloroplast, an organized granule or organelle containing the green pigment chlorophyll.

pleur- membrane surrounding the lung. Example: pleurisy, inflammation of the membrane surrounding the lung.

pneumo- lung, pulmonary. Example: *Pneumocystis*, a fungus that forms cysts in the lungs.

-pod foot. Example: pseudopod, a footlike structure.

poly- many. Example: polysaccharide, a large molecule composed of many simple sugars.

pre-, pro-, proto- before, in front of. Examples: precursor, substance from which another substance is formed; protozoan, member of Phylum Protozoa, single-celled organisms that existed before animals.

psychro- cold. Example: psychrophile, an organism that thrives at low temperatures.

pyo- pus. Example: *Streptococcus pyogenes*, a pus-producing species of *Streptococcus*.

pyro- fire, heat. Example: pyrogenic, fever-producing.

rhabdo- rod. Example: rhabdovirus, an elongated, bullet-shaped virus.

rhin- nose. Example: rhinitis, inflammation of mucous membranes in the nose.

rhizo- root. Examples: mycorrhiza, the mutualism between a fungus and the roots of a plant.

rhodo- red. Example: *Rhodospirillum*, a red-pigmented, spiral-shaped bacterium.

-rrhage excessive discharge. Example: hemorrhage, excessive bleeding.

-rrhea discharge. Example: diarrhea, abnormal discharge of liquid feces.

sacchar- sugar. Example: disaccharide, a sugar composed of two simple sugars.

sapr- rotten. Example: *Saprolegnia*, a fungus that lives on dead animals.

sarco- fleshy. Example: sarcoma, a tumor of muscle or connective tissues.

schizo- split. Example: schizogony, multiple splitting producing many new cells.

-scope, -scopic see, examine. Example: microscope, an instrument used to examine small things.

semi- half. Example: semipermeable, partially permeable, to small but not large molecules.

-septic rotting. Example: antiseptic, killer of bacteria that can cause rotting.

-some body. Example: ribosome, a small, RNA-rich body in the cytoplasm of a cell.

speci- individual, particular. Example: species, the smallest taxonomic group of organisms with common attributes.

spiro- coil, helix. Example: spirochete, a bacterium with a helical cell.

sporo- spore, seed. Example: sporozoan, an immobile, parasitic protozoan.

-stasis, -static arrest, stop. Example: bacteriostatic, able to stop bacterial growth.

strepto- twined, twisted, knotted. Example: *Streptococcus*, a bacterium that forms chains of connected (knotted) spherical cells.

sub- under, beneath. Example: subcutaneous, just under the skin.

super- above, over. Example: superficial, on or just above the surface.

sym-, syn- together, with. Examples: symbiosis, living together; syndrome, a group of signs and symptoms that occur together.

-taxis, taxon- orderly arrangement, orientation. Examples: chemotaxis, orientation of an organism in relation to chemicals; taxonomy, the arrangement of organisms into natural groups.

therm- heat. Example: thermophile, an organism that thrives at high temperatures.

thio- sulfur. Example: *Thiobacillus*, a bacterium that oxidizes sulfur-containing compounds.

-tome, -tomy to cut. Examples: microtome, an instrument that cuts extremely thin slices; anatomy, cutting up a plant or animal to discover its structure and function.

-tonic tension. Example: hypertonic, having excessive tension or osmotic pressure.

tox- poison. Example: toxic, poisonous.

trans- across, through. Example: transport, movement of substances across a membrane.

trich- hair. Example: *Trichomonas*, a protozoan with several long, hairlike flagella.

-troph food, nutrition. Example: eutrophic, well-nourished or rich in dissolved nutrients.

undul- wave. Example: undulant fever, rising and falling fever caused by brucellosis.

uni- single. Example: universal, affecting the whole.

-uria pertaining to urine. Example: dysuria, difficult or painful urination.

vaccin- from cows. Example: vaccine, a preparation used to produce or increase immunity, the first of which contained matter from cows.

vacu- empty. Example: vacuole, an intracellular structure that appears empty.

vaso- pertaining to vessels. Example: cardiovascular, pertaining to the system of heart and blood vessels.

vesic- bladder, blister. Example: vesicle, a bubble.

-vore devour. Example: detritivore, an animal that eats detritus.

xantho- yellow. Example: *Xanthomonas*, a bacterium with yellow colonies.

xeno- stranger, foreigner. Example: xenograft, a graft from a different species.

zoo- animal. Example: zoonosis, a disease communicable from animals to humans.

zygo- pair, union. Example: zygospore, a spore formed from the fusion of two cells.

-zyme ferment. Example: enzyme, a complex protein that catalyzes biochemical reactions including fermentations.

APPENDIX III
Pronunciation of Scientific Names

Because many scientific names are long and all have the odd look typical of Latinized words, they can be intimidating. But you shouldn't be afraid to use them. Pronunciation of scientific names varies with country, region, and individual scientist. Even the experts on scientific Latin take the issue casually. The authoritative reference (William T. Stearn. 1983. *Botanical Latin*. London: David and Charles) concedes, "How they [scientific names] are pronounced really matters little provided they sound pleasant and are understood by all concerned. . . . "

A few simple guidelines make Stearn's provisions easy to satisfy.

1. Divide the name carefully into syllables (it is safest to assume every vowel belongs to a different syllable) and pronounce each syllable.

   Example: *Thermoactinomyces* is Ther-mo-ac-tin-o-my-ces
   (*not* Ther-moac-tin-o-my-ces)

2. The accent usually falls on the next to the last syllable.

   Example: *Bacillus* is Ba-**cil'**-lus
   (*not* **Ba'**-cil-lus)

In compound names the next to last syllable of both parts is sometimes accented.

   Example: *Acinetobacter* is A-ci-**ne'**-to-**bac'**-ter

But there are exceptions. First, terminal and subterminal double vowels are usually pronounced as two syllables, but the *preceding* syllable is accented.

-eae	is pronounced -e-ae	*Example:* Enterobacteriaceae is En-ter-o-bac-ter-**ac'**-e-ae
-ei	is pronounced -e-i	*Example:* brucei is **bru'**-ce-i
-eus	is pronounced -e-us	*Example:* proteus is **pro'**-te-us
-ia	is pronounced -i-a	*Example:* Nocardia is no-**car'**-di-a
-iae	is pronounced -i-ae	*Example:* malaria is ma-**lar'**-i-ae
-iens	is pronounced -i-ens	*Example:* tumefaciens is tu-me-**fac'**-i-ens
-ii	is pronounced -i-i	*Example:* carinii is car-**in'**-i-i
-io	is pronounced -i-o	*Example:* Desulfovibrio is De-sul-fo-**vib'**-ri-o
-ium	is pronounced -i-um	*Example:* typhimurium is ty-phi-**mur'**-i-um
-ius	is pronounced -i-us	*Example:* acidocaldarius is a-**ci'**-do-cal-**dar'**-i-us

Second, certain genus and species names are based on proper names. In these cases the original sound should be maintained.

   Example: *douglasii* is **doug'**-las-ee-eye
   (not dou-**glass'**-ee-eye)

3. Most consonants are pronounced as in English. The following pronunciation of vowels and certain consonants is preferred by most scientists:

a   It's best to be consistent with your own pronunciation of English words; use the broad **a** only if you use it in normal speech (**a** as in "hat" or **ä** as in "father").

Example: Bacillus is either Bacillus or Bäcillus

ae   Nonterminal ae usually pronounced **ē** as in "see"; terminal -ae pronounced as long **ī** as in "ice," not "ee" or "ay."

Example: *Haemophilus* is Hēmophilus *cholerae* is cholerī

i   Usually short **i** as in "sit."

Example: *utilis* is utilus

-ii   Vowel may be slightly separated, generally pronounced "ee-eye" (and count as one syllable for accent rules).

Example: *leichmannii* is leichman-**ē-ī**

-oea   Usually pronounced "ee-a."

Example: *Zoogloea* is Zooglē-ä

y   Usually pronounced like **i** as in "sit"; sometimes as **ē** as in "see."

Example: *gossypii* is gossipii *Blastomyces* is Blastomēces

ch   Generally pronounced as **k**, not as in "ouch."

Example: *Chlorobium* is **k**lorobium

These guidelines and a little self-confidence are all you need.

APPENDIX IV
Biochemical Pathways

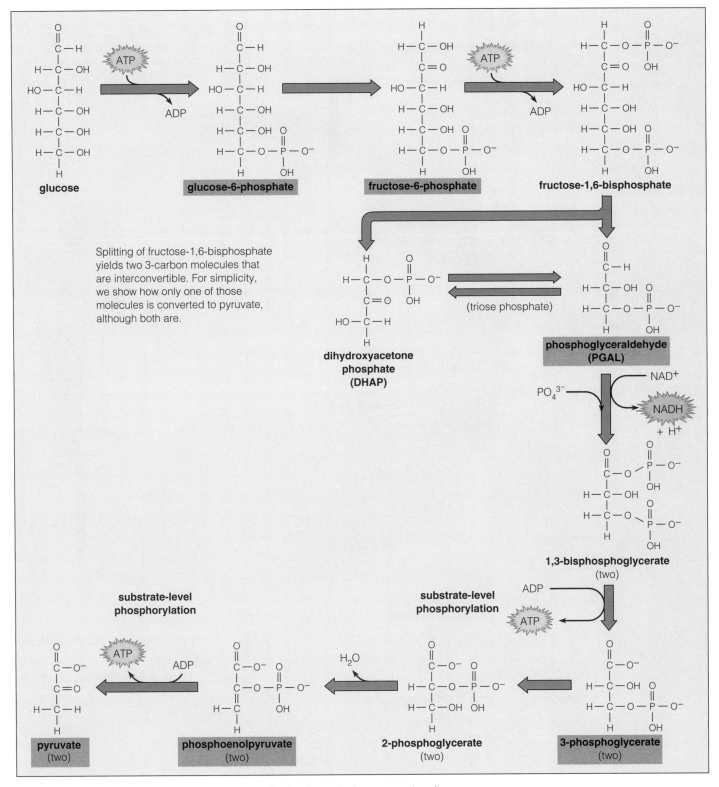

FIGURE A Glycolysis, ending with two pyruvate molecules for each glucose entering the pathway. The pathway produces two molecules of ATP (two are used and four are formed), two molecules of reducing power (as NADH), and six of the twelve precursor metabolites (highlighted in color).

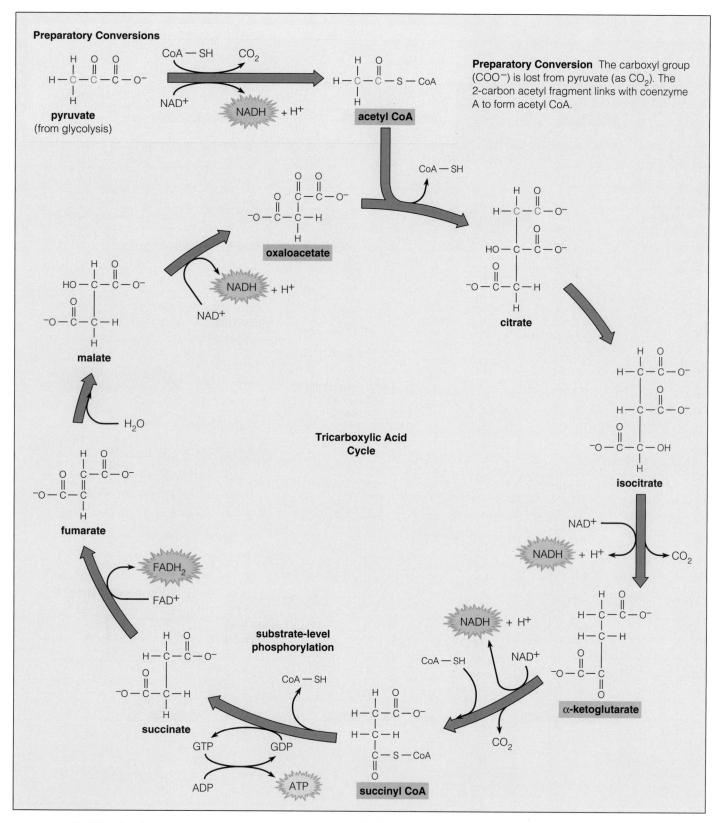

FIGURE B The Tricarboxylic Acid Cycle. With each turn of the cycle, two carbon atoms are added by acetyl-CoA and two are lost as CO_2. Each turn produces one molecule of ATP, four molecules of reducing power (one as $FADH_2$, three as NADH) and three precursor metabolites (shown in color). In addition, one molecule of NADH and one precursor metabolite (in color) are made in the preparatory conversion.

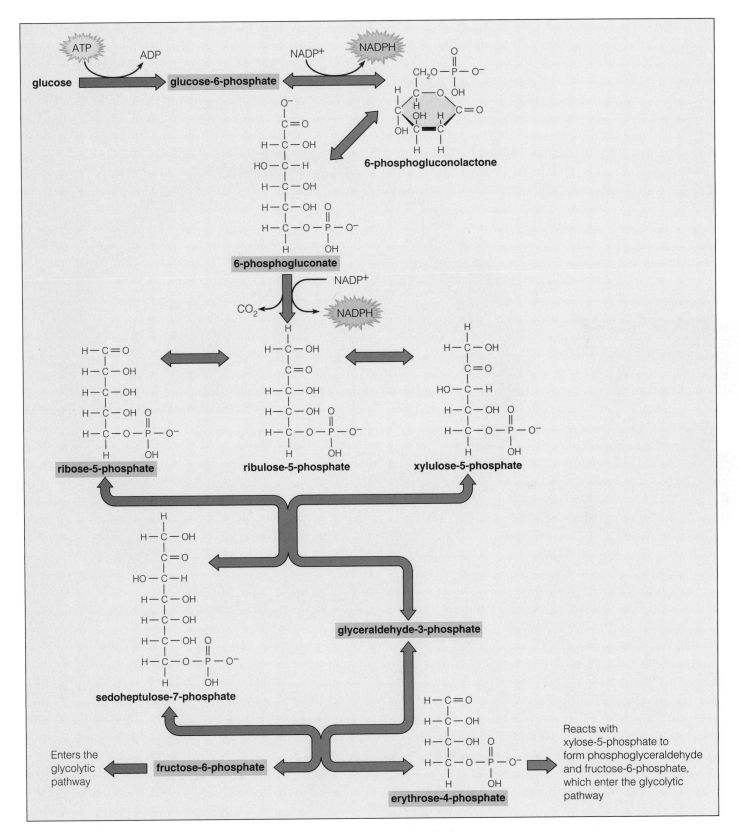

FIGURE C The Pentose Phosphate Pathway. The pathway does not form ATP directly, but it does generate reducing power (in the form of NADPH) in two steps and produces five precursor metabolites. Three of these precursor metabolites (glucose-6-phosphate, phosphoglyceraldehyde, and fructose-6-phosphate) are also made by the glycolytic pathway; two (ribose-5-phosphate and erythrose-4-phosphate) are made only by this pathway.

GLOSSARY

acid fast Property of resisting destaining exhibited by the mycobacterial and nocardioform groups of bacteria.

acquired immunodeficiency syndrome (AIDS) A deadly disease syndrome that develops after infection with human immunodeficiency virus (HIV).

adaptive immunity That portion of the immune system that is activated by exposure to antigens.

adenosine triphosphate (ATP) A compound that conserves metabolic energy in its two high-energy phosphate bonds.

age-adjusted death rate Death rate in a certain age group.

agglutinate To clump together.

agglutination reaction Antigen-antibody reaction that produces a visible clump of particles.

alkaline A greater concentration of OH^- ions than H^+ ions; also called basic.

alkaliphiles Organisms that thrive in alkaline environments.

allergy The common name for type 1 hypersensitivity reactions.

allosteric enzymes Enzymes whose activities change when bound to small molecules, called effectors.

alternate pathway The antibody-independent complement cascade leading to activation of C3.

Ames test Method used to test for mutagens.

amino acids The building blocks of proteins.

ammonification Microbial conversion of nitrogen in organic compounds to ammonia (NH_3).

amoeba (also called Sarcodina) Protozoa that move by pseudopods.

amoeboid Cells that resemble amoebae by forming pseudopods.

anabolic reactions Reactions that build up: biosynthesis and subsequent steps of metabolic assembly line.

anaerobic metabolism Metabolism that occurs in the absence of oxygen: anaerobic respiration and fermentation.

anaerobic respiration A functioning electron transport chain with a terminal electron acceptor other than oxygen.

anaphylactic shock A precipitous fall in blood pressure caused by anaphylaxis.

anaphylaxis Systemic type I hypersensitivity reactions; sensitivity to foreign antigen resulting from previous exposure.

anatomic syndrome Constellation of signs and symptoms implicating specific anatomic structures with an illness.

anions Negatively charged ions.

anneal Term used to describe formation of double-stranded nucleic acid from complementary single strands.

annotation The process of assigning functions to DNA sequences.

anoxygenic photosynthesis The type of photosynthesis carried out by certain prokaryotes that does not produce oxygen as a product.

anti-antibody An antibody produced in response to another antibody acting as an antigen.

antibiotics A metabolic product of an organism that kills or inhibits the growth of microorganisms.

antibody Defensive protein produced by the adaptive immune system in response to exposure to an antigen.

anticodon Three adjacent bases on a tRNA molecule that pair with a complementary codon on mRNA.

antigen-antibody complex The molecular combination between an antigen and an antibody.

antigenic drift Small mutational changes in a virus.

antigenic receptors Antigen-recognizing molecules on lymphocytes.

antigenic shift A sudden change in the properties of a virus resulting from exchange of nucleic acid molecules.

antigenic variation The process by which some pathogens evade recognition by changing their surface antigens.

antigen-presenting cells Lymphocytes that phagocytize foreign materials and display fragments of them as antigens on the cell's surface.

antigens Molecules that provoke an adaptive immune response.

antisepsis A treatment of living tissue that inhibits or destroys microorganisms.

antisera Preparations of sera that inactivate particular microorganisms.

antitoxins Antibodies that bind to and inactivate toxin molecules.

appendages Structures that extend beyond the envelope of microbial cells.

arabinose operon A cluster of genes, including regulatory genes, that encodes the ability to use a sugar, arabinose, as a growth substrate.

archaea A distinct group (domain) of prokaryotes as distantly related to other prokaryotes (bacteria) as they are to eukaryotes.

artificial scheme of classification A scheme not based on phylogenetic relationships.

artificial transformation Introducing DNA into a cell by laboratory manipulations.

artificially acquired active immunity Protection from disease stimulated by vaccines.

artificially acquired passive immunity Protection from disease conferred by administering antibodies formed by an animal or other human(s).

ascospores Sexual spores enclosed within a saclike structure (ascus) produced by one group of fungi (ascomycetes).

atomic force microscope A scanning microscope that exploits the attractive and repulsive forces between atoms to generate images of molecules.

atomic number The number of protons an atom contains.

atomic weight The mass of an atom in atomic weight units (awu); protons and neutrons are each 1 awu.

atoms Smallest component of an element; composed of protons, neutrons, and electrons.

attenuated vaccine Vaccine composed of weakened, nonpathogenic, live microorganisms.

attenuation A mechanism of regulating gene expression in bacteria based on the balance between rates of transcription and translation.

autoclave A pressurized chamber for moist-heat sterilization.

autoimmune disorder A condition in which the immune system launches an attack against the body's own tissues.

autoinfection Continuous reinfection as infective larvae are passed and re-enter the body.

autolysis A process by which destructive enzymes destroy the cell from within.

autotrophs Organisms that obtain all their carbon from carbon dioxide.

axial filament (or **endoflagellum**) The modified flagella of spirochetes that lie within the periplasm; used for movement.

azidothymidine (**AZT**) A thymine analogue used as a drug to control HIV infection.

B lymphocytes (**B cells**) Lymphocytes that produce antibodies.

bacillus (plural, **bacilli**) Lower case: a rod-shaped bacterium; capitalized and italicized: a genus of rod-shaped bacteria.

bacteria One of two domains of prokaryotes; the other is the archaea.

bacterial chromosome The usually single, usually circular, DNA molecule that encodes all a bacterial cell's essential functions.

bactericidal Having a lethal effect on bacteria.

bacteriocins Toxic proteins produced by bacteria that kill other bacteria.

bacteriophages (or **phages**) Viruses that infect bacteria.

bacteriostatic Having an inhibitory effect on bacterial growth.

baker's yeast A specialized strain of *Saccharomyces cerevisiae* used for making bread.

barophiles Organisms that grow only or more rapidly at pressures greater than 1 atmosphere.

Bergey's Manual A book with the scheme that most microbiologists use to identify prokaryotes.

beta sheet (also called a **pleated sheet**) A protein formation in which adjacent peptide chains are held together by hydrogen bonds between amino acids.

binary fission A mode of cell division in which a cross wall forms, producing two cells of approximately equal size.

binomial nomenclature A system of naming organisms in which each organism is identified by a genus designation and a specific epithet.

biochemistry Branch of chemistry that studies molecules made by organisms.

biofilms An irregular layer, 10 to 200 μm thick, of microorganisms imbedded in the extracellular slime they excrete.

biological vector An organism that transports a pathogen from one host to another.

bioremediation Using organisms to detoxify or eliminate toxic materials.

biosynthesis pathways Sequences of metabolic reactions that convert the 12 precursor metabolites into building blocks for synthesis of macromolecules and other essential small molecules.

biotechnology The use of organisms for practical purposes.

black mold A term sometimes applied to *Stachybotrys chartarum*, the fungus responsible for "sick building syndrome."

blue-green algae Former designation of cyanobacteria.

blunt-end ligation Enzymatic joining of two pieces of DNA that lack extending single strands.

bradykinin A substance released during inflammation that causes vasodilation and increased blood vessel permeability.

brightfield illumination Microscopy in which visible light passes through the specimen, causing the background to be brightly lit.

broad-spectrum antibiotics Antibiotics that act against a wide variety of both Gram-positive and Gram-negative bacteria.

broth A liquid complex medium.

budding A form of asexual reproduction common in yeasts in which a bubble forms on the cell surface, grows, and pinches off, forming a new cell.

buffers Mixtures of weak acids or weak bases and their salts that resist changes in pH.

burst period The time interval of the one-step growth curve during which bacterial cells lyse and phage virions are released.

Calvin-Benson cycle The pathway through which most autotrophs incorporate CO_2 into cellular constituents.

capsule The diffuse outermost layer, usually carbohydrate, of many bacterial and some other microbial cells.

carbon cycle The cyclic conversion of carbon-containing compounds that occurs in nature.

carriers Healthy individuals who are reservoirs of infection.

caseation necrosis The dead tissue that looks like cheese, at the center of tubercles caused by tuberculosis.

catabolic reactions The metabolic steps by which certain substrates are broken down into simpler compounds.

cations Positively charged ions.

CD4 T cells (also called **TH** (**helper**) **cells** Lymphocytes that increase immune responsiveness.

CD8 T cells (also called **TC** (**cytotoxic**) **cells** Cytotoxic lymphocytes that kill virus infected cells.

cDNA (complementary DNA) Intron-free DNA synthesized from the mRNA using reverse transcriptase.

cell-mediated immunity Immune responses carried out by T cells.

Centers for Disease Control and Prevention (CDC) The national agency located in Atlanta, Georgia, that does research, collects statistics, and publishes information on infectious diseases.

cercariae The disc-shaped larvae of Trematode flukes; have a tail-like appendage.

chancre A painless weepy ulcer with raised borders that appears at the entry site in primary syphilis.

chemical bonds The forces that hold atoms together in molecules.

chemiosmosis A means of generating ATP by forming an ion gradient across a membrane that drives ATP synthesis from ADP by an enzyme (ATPase) located in the membrane.

chemoautotrophs Microorganisms that generate ATP and reducing power from inorganic chemical reactions.

chemotaxis The process by which cells sense certain chemicals and swim toward regions that contain optimal concentrations of them.

chemotherapy The treatment of disease with chemicals called drugs.

chitin A polymer of *N*-acetylglucosamine.

chlamydiospore A thick-walled, asexual, resting spore.

chloramphenicol A broad-spectrum antibiotic produced by *Streptomyces venezuelae*; now synthesized chemically.

chloroplasts Intracellular organelles in which photosynthesis occurs in phototrophic eukaryotes.

chloroquine Synthetic antimalarial drug.

ciliates A class of protozoa bearing cilia.

cirrhosis Disease of the liver characterized by replacement of functioning liver cells with fibrous scar tissue.

Class I MHC antigens Antigens present on all nucleated cells in the body.

Class II MHC antigens Antigens present only on macrophages, other antigen presenting cells, and certain lymphocytes.

Classical pathway The antibody-activated complement cascade leading to activation of C3.

clonal selection Mechanism by which the presence of an antigen causes proliferation of lymphocytes that recognize that antigen.

clone A population of organisms descended from a single individual by asexual reproduction.

cocci (sing., **coccus**) Spherical-shaped bacterial cells.

codon Three adjacent bases on an mRNA molecule that encode the addition of a particular amino acid to a growing peptide chain or signal it to stop growing.

coenocytic A multinucleate, continuous mass of cytoplasm.

coenzymes Organic molecules that certain enzymes need to be active.

cofactors Inorganic ions that certain enzymes need to be active.

colitis Inflammation of the colon.

colloids Particles ranging in diameter from about 0.001 to 1 μm that remain stably dispersed in water.

colony A clone of cells large enough to be visible on a solid medium.

commensalism A symbiotic relationship in which one partner is neither benefited nor harmed and the other benefits.

communicable disease A disease that can be transmitted from one host to another.

competitive inhibitor A molecule similar enough to an enzyme's normal substrate to bind to its active site, thereby inhibiting the enzymes activity.

complement A family of more than 30 different proteins in serum that function together as a nonspecific defense against infection.

complement fixation assays Tests that detect antigen-antibody reactions by their utilization (fixation) of complement.

complex media Extracts of natural materials used to support growth of microorganisms.

compound Matter composed of molecules that contain more than one type of atom.

confocal microscope A scanning light microscope that uses the computer-processed output of a photodetector to generate an image of one particular slice through a specimen.

conjugative plasmids Circular DNA molecules that encode the ability to transfer a copy of themselves to another bacterial cell by cell-to-cell contact.

conjunctiva (pl., **conjunctivae**) The epithelial covering of the eye.

conjunctivitis Inflammation of the conjunctiva.

contaminate To render impure; e.g., with unwanted microorganisms.

contraindicated Medically inadvisable.

cord factor A mycolic acid found only in virulent strains of *Mycobacterium tuberculosis* that causes these strains to form parallel rows of cells, called cords.

covalent bond The chemical bond formed by sharing pairs of electrons between atoms.

crepitance A crackling sound produced by gas moving through tissue, characteristic of gas gangrene.

culture medium A fluid or gelled solution, containing the nutrients needed for growth of a microorganism.

cyanobacteria The oxygen-producing, phototrophic bacteria.

cyclic AMP (cAMP) $3',5'$-cyclic adenosine monophosphate which acts as a chemical messenger in cells.

cyclic photophosphorylation The process by which phototrophs generate ATP by passing an activated electron from chlorophyll through a membrane-located electron transport chain to chlorophyll in its ground state.

cystitis Inflammation of the bladder.

cytocidal Cell killing.

cytolysins Extracellular enzymes that lyse cells.

cytolysis *See lysis.*

cytomembrane system A complex of membranes that runs through eukaryotic cells.

cytopathic A damaging, nonlethal effect on a cell.

cytoplasm The matrix (ground substance) of a cell, composed primarily of water and protein.

cytoplasmic membrane (also called the **plasma membrane**) The phospholipid membrane that surrounds all cells.

cytosine A pyrimidine base found in DNA and RNA.

cytoskeleton The intracellular structure of eukaryotic cells, composed of microtubules, microfibrils, and intermediate filaments.

Dane particle A component of the hepatitis B virions.

dapsone A sulfone drug; the first antimicrobial agent used to treat leprosy.

darkfield microscopy Method of viewing objects suspended in liquid in which the field of view is illuminated from the side, making the object brilliantly luminous against a dark background.

death phase The phase of microbial growth following the stationary phase in which cells die at an exponential rate.

decontamination A process of rendering a surface that has been heavily exposed to microorganisms safe to handle.

defined medium Culture medium for which the chemical composition is known because it is prepared from pure chemicals.

definitive host The host in which sexual reproduction of a parasite occurs; other hosts are intermediate hosts.

degranulation The process by which lymphocytes release toxic chemicals and inflammatory mediators.

dehydration Medical: depletion of the body's water reserves; chemical: a reaction that removes hydrogen and oxygen atoms from a compound as water.

dehydrogenation reactions Chemical oxidations in which hydrogen atoms are removed from a compound.

delayed hypersensitivity (type IV hypersensitivity; also called cell-mediated hypersensitivity) Reactions initiated by TD cells that produce macrophage-stimulating lymphokines and occur from 12 hours to several weeks after exposure to antigen.

denaturation Destruction of a protein's three-dimensional structure.

denitrification A bacteria-mediated cascade of anaerobic respirations that converts nitrate ion to nitrogen gas.

deoxyribonucleic acid (**DNA**) The macromolecule composed of four deoxynucleotides (A, G, T, and C) that encodes an organism's genetic information.

deoxyribose The five-carbon sugar found in DNA.

dermatitis Inflammation of the skin.

desulfurylation The process of converting sulfur constituents of organic compounds to hydrogen sulfide.

diapedesis The process by which phagocytes escape from blood vessels to enter tissues.

dichotomous key System of answering sequential questions with only two alternatives to identify species.

dideoxynucleoside triphosphate A analogue of naturally occurring nucleoside triphosphates employed in the sequencing of DNA.

differential media Media used to identify microorganisms.

diffraction Bending of light rays which occurs when they pass through a small opening or by the edge of an opaque object.

dikaryon An organism composed of cells, each of which contains two genetically distinct nuclei.

dimorphism The switching between a yeast and a mycelial phase of growth characteristic of some fungi.

diplococci Spherical-shaped bacteria that occur in pairs.

direct microscope count Determining the number of microbial cells in a population by counting under the microscope the number of cells in a chamber of known dimensions filled with diluted sample of the population.

disaccharide A sugar composed of two monosaccharides joined by a glycosidic bond.

disc diffusion method (also called the **Kirby-Bauer method**) Determining the sensitivity of a microorganism to antimicrobial drugs by seeding a plate with the microorganism and placing filter paper discs impregnated with known quantities of different antimicrobial agents on it.

disease A state of functional disequilibrium that is resolved by recovery or death.

disease reservoir Environment where a pathogenic microorganism survives between infections.

disinfection (or **sanitation**) Treatment to reduce the number of pathogens to a level at which they pose no danger of disease.

D-isomer One of the two forms in which a carbon atom attached to four different groups can exist.

DNA ligase An enzyme that seals gaps (missing phosphodiester bonds) in DNA molecules.

DNA melting point The temperature at which double-stranded DNA separates into single strands.

DNA polymerase III The enzyme that catalyzes a reaction between a strand of DNA and a nucleoside triphosphate, producing pyrophosphate and a lengthened DNA strand.

Donovan bodies Intracellular inclusion bodies diagnostic of granuloma inguinale.

double helix Double-stranded DNA.

doubling time (formerly, **generation time**) The period required for a microbial population to produce two new cells for each one that existed before.

driving force The collective term for energy and reducing power.

drug antagonism Interaction between drug action causing the combination to be less effective than either administered alone.

drug resistance A microorganism's being able to grow and reproduce in the presence of a particular drug.

drug synergism An enhanced effect from using drugs in combination.

D-value (**decimal time**) The time required for a particular lethal treatment to kill 90 percent of a microbial population.

dysentery Disease characterized by diarrhea that often contains blood and mucus.

eclipse period The period following phage infection when no intact virions are present.

ectopic pregnancy A life-threatening condition in which a fetus develops inside a fallopian tube.

edema Swelling of tissue by fluid entering spaces between cells.

effector cells Cells that actively fight an infection.

electron A subatomic particle that carries a single negative charge.

electron acceptor A compound or atom that takes up electrons, thereby becoming reduced.

electron donor A compound or atom that loses electrons, thereby becoming oxidized.

electron transport chain A group of compounds embedded in a membrane that undergo sequential oxidation-reduction reactions and in so doing create a proton gradient across the membrane.

electronic count Determining the number of cells (or other particles) in a suspension by electronically scoring decreases in conductivity as the suspension is forced through a small pore.

element Matter composed of only one kind of atom.

elementary bodies (also called **chlamydiospores**) Resistant cell forms of *Chlamydia* that are released when an infected host cell lyses and that transmit infection to a new host.

emulsion A fine suspension of oily droplets in water.

enanthem Redness or blisters on mucous membrane surfaces, such as the inside of the mouth, that are caused by infection.

encephalitis Inflammation of brain tissue.

endemic Diseases that are always present m a population at about the same level.

endocarditis Infection of the heart's inner lining, the endocardium.

endocytosis The process by which a cell engulfs solid material and brings it inside.

endoflagella Spirochete flagella that lie within the periplasm.

endometritis Inflammation of the lining of the uterus.

endoplasmic reticulum (ER) Part of the eukaryotic cells cytomembrane system; a double membrane that folds back upon itself, creating a complex pattern of tubes and layered sacs.

endospores Exceptionally hardy dormant structures that form within the cells of certain species of bacteria.

endotoxin The lipopolysaccharide component of the outer membrane of Gram-negative bacteria that is harmful to humans and other animals; most of the toxicity is mediated by lipid A.

enrichment culture A method of cultivating microorganisms designed to isolate a particular microorganism or type of microorganism from a large, complex natural population.

enteric Intestinal.

enterotoxins Compounds that are harmful to the epithelial cells lining the intestinal tract.

enzyme-linked immunosorbent assays (ELISA) Diagnostic immunological tests that contain an enzyme linked to the indicator antibody.

enzymes Proteins (with the exception of a few RNA molecules) that catalyze specific metabolic reactions.

epidemic A pattern of disease transmission in which many members of a population are affected within a short time.

epidemiology The study of when and where diseases occur and how they are transmitted in human populations.

epitope (also called the **antigenic determinant**) The small region of the antigen molecule that a lymphocyte recognizes.

erythema Abnormal redness of the skin.

erythrocytes Red blood cells.

erythrogenic toxin A substance produced by *Streptococcus pyogenes* that kills cells and causes the rash of scarlet fever.

ester bonds Chemical linkages that form between carboxylic acid and alcohol groups to form esters.

etiologic agent Cause of a disease.

euglenoids A class of protozoa; single celled; motile by two flagella of unequal length.

eukaryotes Organisms composed of cells with internal membranes (all cellular organisms except bacteria and archaea).

evolutionary distance A quantitative estimate of the phylogenetic relatedness of organisms.

exanthem A skin rash, associated with an infectious disease.

exocytosis The process of expelling material from a cell by the reverse process of endocytosis.

exotoxins Highly destructive proteins produced by certain Gram-positive and Gram-negative pathogens; most exotoxins are composed of two subunits, a B, or binding component, and an A, or active component.

exponential growth (also called the **logarithmic growth**) The growth phase during which the number of cells in the population continues to double during the same time interval.

facilitated diffusion Movement of molecules across a membrane from regions of higher to lower concentration mediated by proteins that permit passage only of specific molecules.

facultative anaerobes Organisms that use oxygen to grow when it is available but can also grow without it.

fastidious organisms Organisms that require numerous complex nutrients to grow.

fats Esters that are solid at room temperature formed between a molecule of glycerol and three molecules of fatty acids.

fatty acids Organic acids with a single carboxylate group and a chain of carbon atoms; constituents of fats, oils, phospholipids, and other biochemicals.

fecal-oral route A pattern of disease transmission by which pathogens shed in feces enter a new host through the mouth.

fermentation Scientific: anaerobic generation of ATP from organic compounds totally by substrate-level phosphorylation; industrial: all microbial transformations, either aerobic or anaerobic.

fever Elevated temperature of the body.

flagella Appendages of prokaryotic and eukaryotic cells that confer motility.

flagellates (Mastigophora) A class of protozoa that move by means of flagella.

flavine adenine dinucleotide (FAD) A compound that transfers reducing power by taking up and releasing hydrogen atoms.

floc Material that forms in sewage treatment systems composed largely of particles embedded in extracellular slime produced by bacteria.

fluid mosaic model The proposal that proteins embedded in a phospholipid membrane move freely.

fluorescence microscopy Increases contrast through fluorescence, the property of certain materials to absorb light of one wavelength and give off light of a higher wavelength.

fluorescent-labeled antibodies Antibodies chemically bonded to fluorochromes (fluorescent chemicals).

fomites Inanimate objects such as cups, towels, bedding, and handkerchiefs that transmit disease when contaminated.

freeze-dried Desiccated while frozen.

fruiting bodies Structures that contain or bear spores.

fueling pathways Metabolic pathways that generate precursor metabolites, ATP, and reducing power.

functional groups Parts of organic compounds, composed of specific patterns of atoms; e.g., amino groups.

fungi A large and diverse group of nonphototrophic eukaryotic microorganisms that includes yeasts, molds, and mushrooms.

gamma globulin The antibody-containing protein fraction of serum.

gangrene Tissue death due to impaired blood supply.

gas gangrene An infection caused by *Clostridium perfringens*.

gastritis Inflammation of the stomach.

gastroenteritis Inflammation of the stomach and intestines.

gel electrophoresis Movement of charged molecules through a gel driven by an electric current; molecules of different size and/or charge become separated.

gels Open networks of interconnected colloidal particles.

gene cloning The process of obtaining a set of identical copies of a gene.

gene expression The process of converting the information encoded in DNA into RNA, and then protein.

gene therapy Treating genetic disease by introducing new genes into the affected individual.

genetic code The correspondence between codons in mRNA and amino acids in proteins.

genetic engineering (or **recombinant DNA technology**) A group of techniques for manipulating DNA outside the organism from which it was obtained and reintroducing the recombinant or modified DNA into another cell where it will exert its effect.

genome The totality of genetic information that an organism has.

genomics The study of an organism's DNA.

genotype The form of the genes that an organism has.

German measles Rubella, a rash-producing viral disease that resembles but is unrelated to measles, rubeola.

germicides Chemicals that kill microorganisms.

germinate When a resting structure begins to grow or develop.

germistats Chemicals that inhibit microbial growth.

Ghon complexes Calcified caseous tubercles that indicate a past primary tuberculosis infection.

gingivitis Inflammation of the gums.

gliding A form of nonflagellar motility exhibited by some bacteria.

global regulation Response to a general signal, such as the shortage of any amino acid, that alters expression of many genes.

glycocalyx The slimy or gummy substance that constitutes the outermost layer of the envelope of some bacteria.

glycogen An alpha-linked, branched chain polymer of glucose.

glycolysis The catabolic, metabolic pathway that converts a molecule of glucose into two molecules of pyruvate.

Golgi apparatus An organelle in eukaryotic cells that modifies molecules and sends them to their proper location inside the cell or outside it.

Gram-negative bacteria Bacteria that have a thin cell wall surrounded by an outer membrane.

Gram-positive bacteria Bacteria that have a thick wall and no outer membrane.

Gram-stain A technique that colors Gram-positive bacteria a deep blue and Gram-negative bacteria a light red.

green fluorescent protein A protein from the jellyfish *Aequorea victoria*. used as a probe to determine the cellular location of proteins.

group A Beta-hemolytic streptococcus Also called *Streptococcus pyogenes*, a highly virulent bacterium that causes many different clinical syndromes.

group translocation Entry of a compound into a cell and simultaneous chemical alteration.

growth rate Measurement of how rapidly a microbial population is increasing, usually expressed in doubling times per hour.

gummas Soft granulomas that usually replace skin or bone but may occur in any organ during late syphilis.

halophiles Organisms that grow well in environments with high salt concentrations.

hanging drop preparation A drop of liquid containing microorganisms on a coverslip suspended over a depression slide.

hapten A small molecule, not itself antigenic, that becomes an epitope when attached to a protein.

health A state of relative equilibrium in which the body's organ systems function adequately.

heat labile Easily destroyed by heat.

helminths Worms.

helper T cells (TH) T lymphocytes that activate the functions of other lymphocytes.

hemagglutination Agglutination of red blood cells.

hemagglutinin Proteins that agglutinate red blood cells.

hemolysins Proteins that lyse red blood cells.

hepatitis Inflammation of the liver.

herd immunity The prevention of epidemics due to the scarcity of new susceptible hosts.

heterocysts Specialized, oxygen-impermeable cells of some cyanobacteria in which nitrogen fixation occurs.

heterotrophs Organisms that obtain carbon from organic compounds in their medium or diet.

high-energy bonds Chemical bonds that require considerably less energy to break than is released when new ones form.

higher fungi The Ascomycetes, Basidiomycetes, and Deuteromycetes.

histamine Decarboxylation product of the amino acid histidine that causes vasodilation and increases the permeability of blood vessels.

histidine operon A group of contiguous genes that encodes the enzymes needed to synthesize the amino acid histidine.

horizontal transmission (also called **person-to-person transmission**) The spread of pathogens by direct contact between persons, such as touching, kissing, or sexual intercourse.

host range The spectrum of strains or species that a pathogen attacks.

humoral immunity Protection conferred by antibodies.

hybrid DNA The annealed product of mixing single stands of DNA from different sources.

hydatid cysts Encysted larvae of tape worms of the genus *Echinococcus*.

hydrogen bonds Linkages that form when hydrogen atoms are shared between two molecules or between different parts of the same molecule.

hydrogen ion H^+, a proton.

hydrogenation reactions Addition of hydrogen atoms to a molecule.

hydrolysis Splitting of a molecule by the addition of a molecule of water.

hydrophilic compounds Compounds that dissolve in water.

hydroxide ions OH^-.

hypersensitivity An exaggerated immune response that harms the body.

hypertonic environment A cell's environment with a higher concentration of solutes than the cell's interior.

hyphae The tubelike filaments that constitute a mycelium.

hypotonic environment A cell's environment with a lower concentration of solutes than the cell's interior.

IgA antibodies The second largest class of antibodies; it primarily protects mucous membrane surfaces.

IgD antibodies Constitutes less than 1 percent of the antibody total; it is the main type of antibody found on B cells.

IgE antibodies Constitutes less than 0.01 percent of the antibody total; it causes leukocyte degranulation, which is a primary defense against parasites too large to be eliminated by phagocytosis, such as worms.

IgG antibodies The largest class of antibodies; it activates the complement cascade through the classical pathway.

IgM antibodies Constitute approximately 5 to 10 percent of the antibody total; it is the first antibody class to form during a primary immune response.

imidazoles A major family of antifungal agents that act by inhibiting the synthesis of cytoplasmic membrane sterols.

immune serum globulin An antibody-rich preparation from the pooled serum of many donors.

immune system The network of cells, principally lymphocytes, and organs that extends throughout the body and functions as a defense against infection.

immunization Artificially stimulating the body's immune defenses.

immunocompetence The process by which lymphocytes acquire the capability to function fully in the body's defense.

immunodiffusion test A type of precipitation reaction in which antigens and antibodies are diluted and mixed by diffusion through a gel.

immunoelectrophoresis assay A type of precipitation reaction in which antigens and antibodies are diluted and mixed by electrophoresis through a gel.

immunofluorescence assays Tests in which antigen-antibody reactions are detected by fluorescence because one of the reactants is tagged with a fluorescent dye.

immunogenic Capable of stimulating an immune response.

immunoglobin Synonym for antibody.

immunological memory The ability of memory lymphocytes to recognize an antigen if they encounter it again, greatly accelerating and amplifying the adaptive immune response.

immunological tolerance The immune system's ability not to respond to self antigens.

inactivated vaccines Vaccines containing killed microorganisms.

incidence rate The number of people who develop a disease or condition during a certain period of time divided by the total number of people in the population.

inclusion bodies In prokaryotes, visible structures within the cell other than the nuclear region and ribosomes.

incubate To allow to grow in a warm place.

inducible enzymes Enzymes synthesized in response to an environmental or metabolic signal.

infection The growth of microorganisms in the body.

infectious disease Diseases that can be transmitted from one host to another.

infectious dose (1D) The number of microorganisms that must enter the body to establish an infection in a certain percent of test animals, e.g., $ID_{50} = 50$ percent.

inflammatory mediators Molecular messengers that mediate inflammation.

inflammatory response The body's nonspecific reaction to injury or infection, consisting of redness, pain, heat, swelling, and sometimes loss of function.

interferons Small glycoproteins produced by host cells in response to viral infections.

innate immunity That portion of the immune system that does not depend on exposure to antigens to be active.

innate interior defenses Inflammation, phagocytosis, complement, and interferon—the body's second line of defense against infection.

innate surface defenses Surface features that prevent microbial growth or penetration—the body's first line of defense against infection.

interleukin-l A cytokine produced by white blood cells; one of its many actions is to cause fever.

intoxication A poisoning.

introns Noncoding regions within eukaryotic genes.

invasive Ability of a pathogen to enter host cells or deeper tissues.

ionic bonds Chemical attraction between oppositely charged ions.

ionize To form ions.

ions Charged atoms or groups of atoms.

isomers Molecules with the same kind and number of atoms, but with different arrangement.

isoniazid An antimycobacterial drug.

isotonic environment A cell's environment with the same concentration of solutes as the cell's interior.

isotopes Atoms with the same atomic number but different atomic weight.

jaundice A yellow skin color caused by the buildup of bilirubin when the liver does not function properly.

keratitis Inflammation of the cornea.

killer (K) cells A group of non-B non-T lymphocytes that destroy target cells marked by any of the five antibody classes.

Kirby-Bauer method See *disc diffusion method*.

Koch's postulates Four steps by Robert Koch used to prove that a particular microorganism causes a particular disease.

L isomer See *D isomer*.

lac **operon** Group of contiguous genes that encode enzymes to metabolize lactose.

lactic acid bacteria Aerotolerant bacteria that produce lactic acid as a major product of fermentation.

lactic acid fermentation A fermentation that produces lactic acid as a major product.

lag phase Phase of microbial growth cycle in which metabolism prepares cells to grow.

latent period The period of time following viral infection during which no new virions are produced.

lawn A confluent layer of cells.

lesion An injury, hurt, wound.

lethal dose (LD) The number of microorganisms that must enter the body to kill a certain percent of test animals; e.g., LD_{50} = 50 percent.

leukocidins Enzymes that kill leukocytes.

leukocytes White blood cells.

leukotrienes A class of inflammatory mediators; some increase blood vessel permeability; others attract leukocytes to the inflammation site.

L-forms Strains of bacteria that have lost the ability to form walls.

lichen A symbiotic association between a fungus and an alga or a cyanobacterium.

ligation In recombinant DNA technology, sealing a gap in a DNA molecule with the enzyme DNA ligase.

limiting nutrient The scarcest nutrient in a medium.

line A plastic tube for delivering medications into a vein.

Linnaean scheme Linnaeus's hierarchical scheme of classifying organisms into species, genera, families, classes, order, phyla or divisions, and kingdoms.

lipopolysaccharide (LPS) A compound found only in the outer membrane of Gram-negative bacteria.

local therapy Applying a drug directly to the infected area.

log phase (also called the **logarithmic**, or **exponential phase**) The phase of microbial growth cycle when exponential growth occurs.

lower fungi Coenocytic fungi.

lymph nodes Small bean-shaped organs of the lymphatic system located along lymphatic vessels throughout the body.

lymphatic circulation A system of vessels that collects lymph from tissues and returns it to the bloodstream through the thoracic duct.

lymphocytes A type of leukocyte; part of the body's defense system.

lymphokines Messenger proteins produced by lymphocytes.

lyophilization Freeze-drying.

lysis Rupture of the cytoplasmic membrane and destruction of the cell.

lysogenic cycle One of the two life cycles of temperate phages (along with the lytic cycle) in which phage DNA becomes part of the host cell's genome and is called a prophage.

lysogeny The state in which an infecting phage exists as a prophage.

lysosome Vacuoles that contain enzymes and other chemicals that can destroy most microbial cells.

lytic cycle The life cycle of a phage that lyses the host cell as virions are released.

lytic infections Viral infections that kill the host cell by lysing it.

macrophages Phagocytic lymphocytes that also consume dead microorganisms, dead and dying host cells, and foreign particles; also important as antigen presenting cells.

macroscopic Visible without the aid of a microscope.

magnetotaxis Movement of bacterial cells along magnetic lines of force.

major histocompatibility complex (MHC) The DNA that encodes self antigens of humans.

margination The process of migration of phagocytes to the walls of capillaries.

mast cells Leukocytes that release histamine and heparin, potent inflammatory mediators.

medium See *culture medium*.

meiosis Nuclear division that converts a 2n nucleus into four 1n nuclei.

membrane attack complex A cylinder-like protein complex of the terminal complement pathway that makes a hole through the cytoplasmic membrane and causes cell lysis.

membrane filter A nitrocellulose membrane with holes too small for microbial cells to pass through.

memory B cells Residual B cells formed in response to an infection that allow more rapid response to a subsequent infection.

meningitis Inflammation of the meninges, the membranes that surround the brain and spinal cord.

meningococcus The Gram-negative pathogen *Neisseria meningitidis*.

merozoite The stage of the malarial parasite's life cycle that infects red blood cells.

mesophiles Organisms that grow best at moderate temperatures, around 37°C.

messenger RNA (mRNA) Carries the information from DNA to ribosomes that determines the order of amino acids in a protein.

metabolic intermediates Compounds formed at various steps of metabolic pathways.

metabolism All of the biochemical reactions that take place in a cell.

methanogens The methane-forming archaea.

microaerophiles Microorganisms that need lower concentrations of oxygen than are present in air.

microarray technology A means of determining which of an organism's genes are expressed under various conditions.

microbial antagonism Inhibition of microbial growth by another microorganism.

microbiostatic Inhibition of microbial growth.

minimum bactericidal concentration (MBC) Lowest concentration of a drug that can kill a particular bacterium.

minimum inhibitory concentration (MIC) Lowest concentration of a drug that can inhibit the growth of a particular microorganism.

minus-strand Single-stranded RNA comprising a viral genome that must be transcribed by RNA-dependent RNA polymerase to act as mRNA.

miracidia First larval state in the trematode life cycle that parasitizes a particular species of intermediate host usually a snail.

missense mutation A mutation that changes a codon to one that encodes a different amino acid.

mitosis Cell division in which each daughter cell receives a copy of the same chromosomes that were present in the parent cell.

mixed culture A culture containing more than one kind of microorganism.

molds Filamentous fungi.

mole Avogadro's number (6.02×10^{23}) of molecules.

molecular formula Tells which atoms and how many of each kind constitute a particular molecule.

molecular weight The total of the atomic weights of all of a molecule's atoms.

molecule Two or more atoms joined by chemical bonds.

monera The bacteria.

monoclonal antibodies Antibody molecules produced by a single lymphocyte clone that act against a single epitope.

monocytes A large phagocytic lymphocyte with an oval or horseshoe-shaped nucleus.

monosaccharides A monomer sugar not joined to another by glycosidic bonds.

morbidity Illness and disability.

mordants Compounds or treatments that intensify staining reactions.

mortality Death.

most probable number (MPN) An estimate of numbers of microorganisms in a sample based on a statistical analysis of the probability of cells being present in a diluted sample.

murein The specific kind of peptidoglycan that occurs in bacterial walls.

murine typhus Flea-borne typhus.

mushrooms Fleshy, macroscopic fruiting structures produced by some higher fungi.

mutagens Agents that can induce mutations.

mutation Any chemical change in a cell's genotype.

mutation rate The number of mutations per cell generation.

mutualism A symbiotic relationship in which both partners benefit.

mycelium A mass of hyphae produced by some fungi and actinomycetes.

mycolic acids Long-chain, complex organic acids found in the waxy cell envelope of mycobacteria.

mycoplasmas A group of small, wall-less bacteria.

myelitis Inflammation of the spinal cord.

myocarditis Inflammation of the myocardium (heart muscle).

NAD(P) Designation for NAD or NADP.

naked viruses Viruses not surrounded by a membrane.

narrow-spectrum antimicrobial drugs Drugs that are effective against only a limited number of similar microorganisms; e.g., Gram-negative or Gram-positive bacteria.

natural killer (NK) cells Non-B non-T cells that lyse target human cells by secreting perforins.

naturally acquired active immunity Immunologic protection that follows recovery from an infectious disease.

naturally acquired passive immunity Immunologic protection acquired from antibodies transferred from mother to fetus across the placenta and to the newborn in colostrum.

negative staining Use of a stain that does not penetrate cells or capsules, causing them to appear bright against a dark, stained background.

Negri bodies Inclusion bodies that develop in the brains of rabies victims.

nematodes Roundworms, a phylum of helminths.

neurotoxins Toxic proteins that specifically affect nerve function.

neutron Electrically neutral particle in the nuclei of atoms.

nicotinamide adenine dinucleotide (NAD) A compound that acts as a reservoir of reducing power by accepting and donating pairs of hydrogen atoms.

nicotinamide adenine dinucleotide phosphate (NADP) A phosphorylated form of NAD that serves a similar function.

nitrification The conversion, by bacteria, of ammonia to nitrate.

nitrogen cycle Conversions of nitrogen compounds that occur in nature.

Nomarsky microscopy (also called **differential interference contrast microscopy**) Technique that uses differences in refractive index to produce contrast by interference.

non-B non-T lymphocytes Lymphocytes that function without recognizing antigens.

nonpolar molecules Molecules with no electrically charged regions, water insoluble.

nonsense codons Three codons that do not correspond to the anticodon of any tRNA molecules and therefore stop translation.

nonsense mutation A mutation that changes a codon encoding an amino acid to a nonsense codon.

normal biota The microorganisms that co-exist with humans in a stable relationship on body surfaces.

nosocomial Hospital acquired.

nosocomial infections Infections acquired while in the hospital.

notifiable diseases Diseases that must be reported to government agencies because they affect the public health.

nuclear envelope A double-membrane structure that encloses the eukaryotic nucleus.

nucleoid (or **nuclear region**) The mass of DNA in bacterial cells—not membrane-bound.

nucleoli Dense masses of RNA and protein within the eukaryotic nucleus that manufacture ribosomes.

nucleoplasm The gelatinous matrix of the nucleus.

nucleoside triphosphate A purine or pyrimidine bonded to ribose and three phosphate groups.

nucleotide A purine or pyrimidine bonded to ribose or deoxyribose and one to three phosphate groups.

numerical taxonomy Biological classification based on comparing many characters and using similarities and differences to calculate relatedness among organisms.

nutrient agar Nutrient broth solidified with agar.

nutrient broth A commonly used complex medium.

obligate aerobes Organisms that grow only in the presence of oxygen.

obligate anaerobes Organisms that grow only in the absence of oxygen.

obligate barophiles Organisms that grow only at pressures greater than 1 atmosphere.

obligate intracellular parasites Microorganisms that can reproduce only inside a host cell.

oncogenic Cancer-causing.

one-step growth curve A procedure for simultaneously infecting a bacterial culture with phage virions in order to study their development.

operon A set of contiguous genes that is regulated and transcribed together.

opines Unusual amino acids produced by the crown gall–forming bacterium *Agrobacterium tumefaciens.*

opportunistic infection Infections caused by microorganisms that can infect only debilitated hosts.

opsonin A protein that facilitates phagocytosis.

opsonization The process by which an opsonin facilitates phagocytosis.

ORF Acronym (pronounced as a word) for "open reading frame," applied to a sequence of DNA that is presumed to encode a gene.

organic compounds Carbon-containing compounds.

organic growth factors Organic compounds that certain microorganisms need to grow.

outer membrane The lipopolysaccharide-containing membrane that surrounds Gram-negative bacteria.

oxidation Removal of electrons from an atom or molecule.

oxidation-reduction reaction (**redox reaction**) A reaction in which one atom or molecule loses electrons and another gains them.

oxygenic photosynthesis The type of photosynthesis carried out by plants, algae, and cyanobacteria that produces oxygen as a product.

pandemic A worldwide epidemic.

paper disc method The procedure for determining the effectiveness of an antimicrobial agent: a filter paper disc impregnated with it is placed on a plate seeded with the test microorganism and incubated.

parasitism A symbiotic relationship in which the host is harmed and the parasite benefits.

parasitology The study of protozoan- and helminth-caused disease.

passive immunity Immunity conferred by administering antibodies.

pasteurization Moderate heat treatment to kill pathogens and extend the shelf life of liquid foods.

pathogens Disease-causing microbes.

pellicle Flexible covering that surrounds some protozoa.

pelvic inflammatory disease (PID) Infections of the female upper reproductive tract.

penicillin An antibiotic produced by certain species of *Penicillium.*

pentose phosphate pathway The catabolic metabolic pathway that begins with glucose and produces pentose phosphates.

peptide bonds The bonds that join amino acids in proteins.

peptidoglycan The macromolecule composed of protein and carbohydrate a form of which constitutes the cell walls of bacteria.

pericarditis Inflammation of the pericardium, the membrane surrounding the heart.

perinatal Occurring just before, during, or after birth.

periplasm The organelle between the cytoplasmic membrane and outer membrane of Gram-negative bacteria.

pH scale Quantitative description of acidity or alkalinity.

phage typing Identifying bacterial strains by pattern of their susceptibility to phages.

phages Short for bacteriophages, viruses that infect bacteria.

phagocytosis Engulfment of one cell by another.

pharyngitis Inflammation of the throat.

phase-contrast microscope Microscope that generates contrast by interference between phase-shifted light rays that pass through the specimen and those that do not.

phenotype The expression of a cell's genes.

phosphodiester bonds Chemical linkages that join nucleotides in nucleic acids.

phospholipid bilayer A phospholipid membrane.

phospholipids Constituents of unit membranes; composed of a glycerol, two fatty acids, and a phosphate group to which another group is attached.

phosphorus cycle Chemical conversions of phosphorus that occur in nature.

phosphorylation Chemical addition of a phosphate group to a molecule.

phototaxis Movement of cells toward optimal intensity and quality of light.

phototrophs Organisms that generate ATP and reducing power from light energy.

phycology The study of algae.

phytoplankton Floating, microscopic phototrophic species.

pili Straight hairlike appendages that extend out from surface of a bacterial cell.

pinocytosis Cellular engulfment of liquid.

plaque count A procedure for determining the number of bacteriophages and phage-infected cells in a sample.

plaque-forming units (PFUs) Virions and virus-infected cells.

plaques Circular clear zones on a lawn of cells.

plasma The fluid, cell-free component of blood, including clotting proteins.

plasmids Small usually circular DNA molecules found in some bacteria and other microorganisms that encode nonvital functions.

plasmodium The amorphous slimy mass that constitutes a true slime mold. *Plasmodium:* the genus of protozoa that causes malaria.

plasmolysis The drawing out of water from a cell in a hypertonic environment, decreasing the intracellular volume.

plate count Enumerating microbial cells in a sample by distributing them on an agar plate and counting the colonies that develop after incubation.

platelets Subcellular fragments that participate in blood clotting.

platyhelminths Flatworms, a phylum of helminths.

pleurisy Inflammation of the pleura, the membrane that surrounds the lungs.

plus strand Single-stranded RNA constituting a viral genome that can act directly as mRNA.

pneumococcus *Streptococcus pneumoniae.*

pneumonia Inflammation of the lungs.

polar molecule A molecule that has a positive and a negative region, usually water soluble.

polymerization The process by which monomers are joined together to produce a macromolecule.

polymers Macromolecules built from repeating subunits.

polysaccharides Polymers built from simple sugars.

porins Proteins that form pores in the outer membrane.

portal of entry The anatomic site through which a pathogen enters a host.

portal of exit The anatomic site through which a pathogen leaves its host.

precipitation reaction An antigen-antibody reaction that forms lattices large enough to precipitate.

precursor metabolites The 12 compounds from which all constituents of a cell can be synthesized.

prevalence rate The number of people who have a certain disease at any particular time divided by the number of people in the population.

primary immune response The production of antibody that occurs when a person first encounters a particular antigen.

primary structure of a protein The sequence of amino acids.

prion An infectious agent composed only of protein.

probe In recombinant DNA technology a short, complementary, tagged molecule of nucleic acid used to detect specific pieces of DNA by hybridization.

prokaryotes Bacteria and archaea.

prophage A phage genome integrated into the chromosome of a host cell.

prophylaxis Prevention of disease.

prostaglandins A large family of cytokines that are potent inflammatory mediators.

prosthecae Filamentous or conical extensions present on some prokaryotes.

protein A macromolecule composed of polymerized amino acids.

proteobacteria The largest phylum of the domain Bacteria.

proteomics The study of an organism as revealed by the complete set of proteins it contains.

protists Eukaryotic microorganisms in Haeckel's classification scheme.

proton A subatomic particle that carries a single positive charge; ionized hydrogen atom.

proton acceptors Bases.

proton donors Acids.

protoplast Bacterial cell from which the wall has been removed completely.

protozoa Nonphotosynthetic, unicellular eukaryotes.

provirus A viral genome that has become part of the genome of a host cell.

pseudopods Tubelike structures that amoeboid cells project and withdraw in order to move.

psychrophiles (psychrotrophs) Organisms that grow at low temperatures.

public health A discipline or agency that deals with the development and implementation of plans to prevent and control disease.

puerperal sepsis A bloodborne infection acquired at time of childbirth.

pure culture A culture that contains only a one kind of organism.

purulent Pus-producing.

pus A mixture of dead leukocytes, microorganisms, and host cells.

pyelonephritis Inflammation of the kidneys.

quaternary structure of proteins The way separate polypeptide chains fit together.

radioimmunoassay A test to detect antigen-antibody reactions in which one of the reactants is tagged radioactively.

recombinant DNA technology *See genetic engineering.*

recombination (genetic) Reassortment of genes.

reduction The addition of electrons to an atom or molecule.

refraction Bending that occurs when a ray of light enters an object with a different density at an angle.

refractive index The ratio of the speed of light traveling through a vacuum to the velocity in any particular material.

replica plating Inoculating a fresh plate by pressing it against a piece of velveteen or similar material that has previously been pressed onto a plate with colonies of microorganisms.

replication The biochemical process of making a copy of a DNA molecule.

replication forks The two points within the bubble form in a DNA molecule where replication occurs.

repressible enzymes Enzymes whose synthesis is inhibited in the presence of a signal molecule (repressor).

reservoirs Infectious disease: repositories for pathogens between hosts.

resistance (R) factors Plasmids that carry genes encoding drug resistance.

resolution The capacity to perceive two adjacent parts of an image as distinct.

respiratory burst The event that occurs when a phagocyte's granules produce lethal oxidants.

reticulate bodies Nonvirulent, reproductive cells of *Chlamydia* spp.

retroviruses (retroviridae) Family of viruses that convert their RNA genome into DNA by reverse transcriptase as part of their life cycle.

reverse transcriptase An enzyme that uses RNA as a template to make a complementary strand of DNA.

Rh factor An erythrocyte antigen found in humans and rhesus monkeys.

rhinitis Inflammation of nasal membranes.

ribonucleic acid (RNA) Macromolecular polymer of ribonucleotides.

ribosomes The organelles on which proteins are synthesized.

rifampin A semisynthetic antibiotic that inhibits bacterial RNA polymerases.

ringworm The common name for tinea, a fungal infection of the skin.

RNA polymerase Enzyme that uses DNA as a template to make RNA.

rubella *See German measles.*

rubeola Measles, a rash-producing illness caused by the rubeola virus.

sampling error The inevitable inaccuracy that occurs because no sample is precisely representative of the total population.

sanitation Disinfection: treatment to reduce the number of pathogens to a level at which they pose no danger of disease.

saturated fat Fat composed of fatty acid molecules that do not contain double bonds.

scanning electron microscope (SEM) An electron microscope that generates an image by scanning the surface of the specimen with an electron beam.

scanning tunneling microscope A scanning microscope that views surfaces of metals and semiconducting materials by processing a signal generated by a flow of electrons to a probe held close to the specimen.

secondary immune response Response initiated by memory lymphocytes.

secondary structure of a protein The alpha helix and beta sheet structures that result from hydrogen bonds forming between amino acids.

selective media Media that favor the growth of certain microorganisms over others.

selective toxicity Pharmacology: a drug that harms a pathogen without harming the host.

self-limited illness Illness from which most people recover without medical treatment.

semiconservative replication Refers to DNA replication because each new double helix is composed of one new and one old (conserved) strand.

semipermeable Describes properties of membranes that allow certain molecules to cross freely while blocking the passage of others.

septa Cross walls.

septicemia Persistent and serious infection of the bloodstream.

sequencing The process of determining the succession of monomers in a macromolecule.

serology Diagnostic clinical immunology.

serotype (also called a **serovar**) A taxonomic category identified by serology.

serum killing-power test A procedure in which a patient's drug-containing serum is tested for its ability to kill the infecting microorganism.

serum The cell-free liquid component of blood, not including clotting proteins.

serum resistance Inherent ability of certain bacteria to avoid destruction by serum.

serum sickness A type III hypersensitivity reaction that sometimes occurs when proteins from animal serum are used in medical therapy.

sheaths Long transparent polysaccharide tubes produced by some bacteria.

Shick test An immunological test for immunity to diphtheria.

siderophores Iron-chelating compounds released by microorganisms to obtain iron.

signs Clinical: objective indications of illness.

similarity coefficient (Sj) A number that expresses the relatedness among organisms.

simple stains A single dye used to increase contrast of a specimen.

slide agglutination test Identifying bacteria by mixing a suspension of the unknown bacterium with a known antiserum on a microscope slide and observing if agglutination occurs.

slime layer A thin slimy or gummy layer that surrounds some bacterial cells.

slime molds Two groups of microorganisms: cellular slime molds (Acrasieae) and true slime molds (Myxogastria).

smear Microscopy: a thin film spread on a microscope slide.

special stains Stains that heighten contrast of microbial cells to reveal particular structures, including endospores, flagella, or capsules.

species barrier Factors that restrict pathogens to certain host species.

sporadic diseases Occurring only occasionally in a population.

sporogenesis Development of a spore.

sporozoa Nonmotile protozoa; all sporozoa are parasitic.

sporozoite Form of malarial parasite that passes from the insect's salivary gland to a person's bloodstream.

sporulation The process of forming spores.

stains Microscopy: dyes used to increase contrast.

sterilization Eliminating all microorganisms from an area.

sterols Lipids composed of hydrocarbon rings.

stock cultures Microbial cultures maintained for study and reference.

storage granules Granular inclusions in the cytoplasm that hold reserve supplies of nutrients.

strains Clones that are presumed or known to be generically different.

streak plate method Commonly used method of obtaining a pure culture.

stridor A hoarse sound when a person inhales if the airway near the epiglottis or larynx is narrowed.

substrates Molecules used as nutrients for microorganisms or as reactants for enzyme reactions.

sulfonamides (sulfa drugs) Synthetic antimicrobial agents that act by interfering with the bacterial cell's ability to synthesize folic acid.

sulfur cycle The chemical conversions of sulfur that occur in nature.

surfactants Detergent-like agents that penetrate oily globules in water, producing an emulsion.

symbiosis Two different kinds of organisms living together.

symptoms Subjective reports of illness by a patient.

systemic disease Body-wide infection spread through the bloodstream.

T cells (T lymphocytes) The agents of cell-mediated immunity.

taxis Movement of cells toward favorable environments and away from harmful ones.

taxonomy The science of classifying organisms.

TC (cytotoxic) cells *See CD4 T cells.*

teichoic acid A molecule composed of glycerol or ribitol units linked by phosphate groups that occurs in the walls of Gram-positive bacteria.

temperate phages Phages that can enter the lysogenic state.

terminal electron acceptor Metabolism: compound at the end of an electron transport chain; eg., oxygen for aerobic respiration.

terminal pathway Complement action: the cascade leading from C3 to lethal antimicrobial activity.

tertiary protein structure Structure determined by interactions among the R groups of its various amino acids.

tetracyclines Broad-spectrum antibiotics that interfere with ribosome activity by binding at the A site.

TH (helper) cells *See CD4 T cells.*

thallus The body of a fungus, an alga, or a nonvascular plant.

thermal death point (TDP) The lowest temperature required to kill all microorganisms in a particular liquid suspension in 10 minutes.

thermal death time (TDT) The minimal time required to kill all microorganisms in a particular liquid suspension at a given temperature.

thermoacidophiles A group of archaea that grow at high temperature and low pH.

thermophiles Organisms that grow at high temperature.

thrush A *Candida* infection of the mouth causing white patches on mucous membranes.

tinea Ringworm; infections of the skin, hair, and nails caused by the dermatophyte fungi.

tissue culture Cultivation of eukaryotic cells or tissues in vitro.

titer The highest dilution of a test solution that is active.

toxins Poisonous proteins produced by some microorganisms.

toxoids Treated toxins that have lost their harmful properties but still stimulate the immune system to produce antitoxin.

trace elements Certain inorganic elements that are essential to life but required in only minute amounts.

transcription The first step in gene expression; formation of RNA from a DNA template.

transduction The transfer of chromosomal genes by phage particles containing bacterial DNA.

transformation (1) Entrance of DNA from the environment into a cell. (2) Conversion of a normal cell into a cancer cell.

transfusion reaction A clinical response to a blood transfusion with mismatched blood types.

transgenic organism One that contains genes from another organism.

translation The second step in gene expression, synthesis of protein directed by mRNA.

transmission electron microscopy (TEM) Use of a beam of electrons rather than visible light to form a magnified image of an object.

transovarial transmission Passage of infectious microorganisms from one generation of host to the next through their eggs.

transposable elements Short stretches of DNA that have the capacity to move from one location to another in a genome (jumping genes).

trematodes Flukes, a type of flatworm.

tricarboxylic acid (TCA) cycle A cyclic metabolic pathway that oxidizes acetate, generates ATP, and forms four precursor metabolites.

trophozoite The actively multiplying vegetative stage of a sporozoa.

tubercles Granulomas produced by tuberculosis infection; dense nodules containing mostly activated macrophages and monocytes.

tumor An abnormal tissue growth; may be cancerous or benign.

tumor necrotizing factor A protein secreted by phagocytes that causes loss of fluid from the circulation.

turbidity Cloudiness of a liquid caused by suspended particles.

turgor pressure A cell's internal pressure resulting from a higher intracellular than extracellular osmotic strength.

ulcer An open sore.

ultrastructure Detailed microscopic cell structure.

uncoating Virology: the process by which the capsid and envelope of a virion are removed.

unit membrane A phospholipid bilayer.

universal donor A person with type O blood who can give blood to people with any blood type-O, A, B, or AB.

universal recipient A person with type AB blood who can receive any type of blood—O, A, B, or AB.

unsaturated fat Fat composed of fatty acid molecules that contain double bonds.

urethritis inflammation of the urethra, the outlet of the urinary system.

use-dilution test A specific microorganism is added to dilutions of the germicide in a culture medium to determine which concentrations inhibit growth.

vaccination Using vaccines to produce artificial active immunity.

vaccines Agents that confer immunity without causing disease.

vacuole A membrane-bounded intracellular vesicle.

vaginitis Inflammation of the vagina.

valence electrons The electrons in an atom's outermost shell.

varicella-zoster virus (VZV) A member of the herpesvirus family that causes varicella (chickenpox) and zoster (shingles).

vasodilation Blood vessel enlargement.

vectors Agents that transmit pathogens.

vegetations Abnormal growths on the heart valves that occur in endocarditis.

vegetative cells Cells that grow and reproduce asexually.

vertical transmission Direct transmission of pathogens from mother to fetus or infant.

vesicles Tiny fluid-filled skin lesions.

viable count Measurement of number of living cells in a population.

virions Intact, nonreplicating virus particles.

viroids Small circular molecules of ssRNA without a capsid that cause many plant diseases.

virology The study of viruses.

virulence factors Substances or features of a microorganism that help it infect and cause disease.

virulent phages Phages that always kill their bacterial host.

virus A microscopic packet of nucleic acid usually wrapped in a protein coat.

vital stain A stain for living cells.

vitamins Nutrients required in minute quantities, primarily as precursors of enzyme cofactors.

wet mount A drop of liquid containing microorganisms on a microscope slide covered with a cover slip.

Wirtz-Conklin spore stain A staining technique that selectively colors endospores.

yeast A single-celled fungus.

Ziehl-Neelsen stain A special staining technique for identifying *Mycobacterium tuberculosis* and closely related bacteria.

zoonosis A human disease caused by a pathogen that maintains an animal reservoir.

CREDITS

Chapter 2. 30: From *Biology: The Unity and Diversity of Life*, 6th ed., by C. Starr and R. Taggart, Brooks/Cole, 1992. All rights reserved. **31:** From *Biology: The Unity and Diversity of Life*, 6th ed., by C. Starr and R. Taggart, Brooks/Cole, 1992. All rights reserved. **35:** From *Biology: The Unity and Diversity of Life*, 6th ed., by C. Starr and R. Taggart, Brooks/Cole, 1992. All rights reserved. **36:** Art by Palay/Beaubois from *Biology: The Unity and Diversity of Life*, 6th ed., by C. Starr and R. Taggart, Brooks/Cole, 1992. All rights reserved.

Chapter 3. 65: Part b is from *Biology: The Unity and Diversity of Life*, 6th ed., by C. Starr and R. Taggart, Brooks/Cole, 1992. All rights reserved. **67:** From *Biology: The Unity and Diversity of Life*, 6th ed., by C. Starr and R. Taggart, Brooks/Cole, 1992. All rights reserved.

Chapter 4. 89: Art by Carlyn Iverson. **94:** Art by Carlyn Iverson. **100:** From *Biology: The Unity and Diversity of Life*, 6th ed., by C. Starr and R. Taggart, Brooks/Cole, 1992. All rights reserved. **101:** Art (L) by D. & V. Henninges; art (R) by Leonard Morgan, both from *Biology: The Unity and Diversity of Life*, 6th ed., by C. Starr and R. Taggart, Brooks/Cole, 1992. All rights reserved. **102:** Art by Raychel Ciemma from *Biology: Concepts and Applications*, 2nd ed., by C. Starr, Brooks/Cole, 1994. All rights reserved. **102:** Art by Raychel Ciemma from *Biology: Concepts and Applications*, 2nd ed., by C. Starr, Brooks/Cole, 1994. All rights reserved. **104:** Art (L) by Leonard Morgan; art (R) by Robert Demarest, both from *Biology: The Unity and Diversity of Life*, 6th ed., by C. Starr and R. Taggart, Brooks/Cole, 1992. All rights reserved. **104:** Art (L) by Robert Demarest; art (R) by Leonard Morgan, both from *Biology: The Unity and Diversity of Life*, 6th ed., by C. Starr and R. Taggart, Brooks/Cole, 1992. All rights reserved. **105:** Art (a) by Leonard Morgan; art (b) by Palay/Beaubois, both from *Biology: The Unity and Diversity of Life*, 6th ed., by C. Starr and R. Taggart, Brooks/Cole, 1992. All rights reserved. **107:** Art by Raychel Ciemma from *Biology: Concepts and Applications*, 2nd ed., by C. Starr, Brooks/Cole, 1994. All rights reserved. **108:** From *Biology: The Unity and Diversity of Life*, 6th ed., by C. Starr and R. Taggart, Brooks/Cole, 1992. All rights reserved. **109:** Top art redrawn from *Molecular Biology of the Cell*, 2nd ed., by B. Alberts, et al. Copyright © 1989. Reproduced by permission of Routledge, Inc., part of the Taylor & Francis Group. Below art by Leonard Morgan from *Biology: Concepts and Applications*, 2nd ed., Brooks/Cole, 1994. All rights reserved.

Chapter 5. 117: Information from M. Riley and B. Labedan, Chapter 116 in *Escherichia coli and Salmonella*, F.C. Neidhardt, ed., Washington, D.C., ASM Press.

Chapter 6. 146: Art by Margaret Gerrity. **149:** Art by Margaret Gerrity. **150:** Art by Margaret Gerrity. **151:** Art by Margaret Gerrity. **153:** Art by Margaret Gerrity. **155:** Art by Margaret Gerrity. **159:** Art by Margaret Gerrity. **160:** Art by Margaret Gerrity. **169:** Art by Margaret Gerrity. **173:** Art by Margaret Gerrity. **174:** Art by Margaret Gerrity.

Chapter 7. 182: Art by Margaret Gerrity. **184:** Art by Margaret Gerrity. **188:** Art by Margaret Gerrity. **189:** Art by Margaret Gerrity. **192:** Art by Jeanne Schreiber from *Biology: The Unity and Diversity of Life*, 6th ed., by C. Starr and R. Taggart, Brooks/Cole, 1992. All rights reserved.

Chapter 8. 203: Art from *Biology: The Unity and Diversity of Life*, 6th ed., by C. Starr and R. Taggart, Brooks/Cole, 1992. All rights reserved.

Chapter 9. 229: After M.S. Favero and W.W. Bond, 1991, Sterilization, disinfection, and antisepsis. In Manual of Clinical Microbiology, 5th ed., ed. A. Balows (Washington, D.C.: American Society of Microbiology).

Chapter 10. 252: Courtesy Craig S. Hill, Gen-Probe Transcription-Mediated Amplification: System Principles, Gen-Probe Incorporated, 10210 Genetic Center Drive, San Diego, CA 92121.

Chapter 11. 273: Art by Carlyn Iverson. **277:** Art by Carlyn Iverson. **279:** Art by Carlyn Iverson.

Chapter 12. 287: Art by Carlyn Iverson. **291:** Art by Carlyn Iverson. **292:** Art by Carlyn Iverson. **293:** From *Biology: The Unity and Diversity of Life*, 6th ed., By C Starr and R. Taggart, Brooks/Cole, 1992. All rights reserved. **298:** Art by Palay/Beaubois from *Biology: The Unity and Diversity of Life*, 6th ed., By C Starr and R. Taggart, Brooks/Cole, 1992. All rights reserved. **299:** Art by Carlyn Iverson; adapted from *Biology of Plants*, by Peter H. Raven, Ray F. Evert, and Susan E. Eichhorn 1971, 1975, 1976, 1981, 1986, 1992, 1999 by W.H. Freeman and Company/Worth Publishers. Used with permission. **301:** Art by Carlyn Iverson. **306:** Art by Keith Kasnot from *Biology: The Unity and Diversity of Life*, 6th ed., By C Starr and R. Taggart, Brooks/Cole, 1992. All rights reserved.

Chapter 13. 324: Art by Carlyn Iverson. **325:** Art by Carlyn Iverson. **328:** Art by Carlyn Iverson. **333:** Art by Carlyn Iverson. **335:** Art by Carlyn Iverson.

Chapter 14. 344: Art by Carlyn Iverson. **348:** Art by Carlyn Iverson. **349:** Art by Carlyn Iverson. **352:** Art by Lewis Calver from *Biology: The Unity and Diversity of Life*, 6th ed., By C Starr and R. Taggart, Brooks/Cole, 1992. All rights reserved. **353:** Art by Carlyn Iverson. **354:** Art by Kevin Somerville from *Biology: Concepts and Applications*, 2nd ed., by C. Starr, Brooks/Cole, 1994. All rights reserved. **356:** Art by Kevin Somerville from *Biology: Concepts and Applications*, 2nd ed., by C. Starr, Brooks/Cole, 1994. All rights reserved.

Chapter 16. 390: Art by Carlyn Iverson. **392:** Art by Carlyn Iverson. **402:** Art by Carlyn Iverson.

Chapter 17. 410: Art by Carlyn Iverson. **412:** From *Biology: Concepts and Applications*, 2nd ed., by C. Starr, Brooks/Cole, 1994. All rights reserved.

Chapter 18. 442: Art by Carlyn Iverson.

Chapter 20. 470: *MMWR Weekly Report.*

Chapter 22. 524: Art by Carlyn Iverson.

Chapter 23. 551: Art by Kevin Somerville from *Biology: Concepts and Applications*, 2nd ed., by C. Starr, Brooks/Cole, 1994. All rights reserved. **573:** Art by Carlyn Iverson.

Chapter 24. 584: Art by Kevin Somerville from *Biology: Concepts and Applications*, 2nd ed., by C. Starr, Brooks/Cole, 1994. All rights reserved.

Chapter 25. 614: Art by Kevin Somerville from *Biology: Concepts and Applications*, 2nd ed., by C. Starr, Brooks/Cole, 1994. All rights reserved.

Chapter 26. 633: Art (top) by Robert Demarest from *Biology: The Unity and Diversity of Life*, 6th ed., by C. Starr and R. Taggart, Brooks/Cole, 1992. All rights reserved. Art (bottom) by Carlyn Iverson.

Chapter 27. 660: Art by Carlyn Iverson. **682:** Art by Leonard Morgan from *Biology: The Unity and Diversity of Life*, 6th ed., by C. Starr and R. Taggart, Brooks/Cole, 1992. All rights reserved.

Chapter 28. 695: Source: Yanagita, T. 1990. Natural microbial communities. Tokyo: Japan Scientific Societies Press. **709:** Art by Carlyn Iverson

PHOTO CREDITS

Chapter 1. 1: © Cecil H. Fox/Photo Researchers. **4:** Courtesy the Master and Fellows, Magdalene College, Cambridge. **7:** (left) © Roland Birke/Peter Arnold, Inc. (right) T. Beveridge, S. Schultze, University of Guelph/Biological Photo Service. **8:** (top left) © Jan Hinsch/SPL/Photo Researchers. (top right) Courtesy Catherine Ingraham. (bottom left) © David M. Phillips/Visuals Unlimited. (bottom right) © Hans Reinhard/Bruce Coleman Ltd. **9:** (left) © Fred E. Hossler/Visuals Unlimited. (right) © Heather Davies/SPL/Photo Researchers. **10:** © Cecil H. Fox/Photo Researchers. **11:** The Wellcome Centre Medical Photographic Library, London.**12:** (left) Historical Collections, Armed Forces Institute of Pathology. (right) Reproduced by permission of the President and Council of the Royal Society, London. **13:** © Corbis/Bettmann. **14:** © Corbis/Bettmann. **15:** © The Granger Collection. **17:** © Corbis/Bettmann.

Chapter 2. 22: Thomas A Steitz, Yale University. **23:** © Chesley Bonestell/Space Art International. **31:** Thomas A Steitz, Yale University.

Chapter 3. 49: © Raymond B. Otero/Visuals Unlimited. **57:** (all) © Bruce Iverson. **59:** (both) © Raymond B. Otero/Visuals Unlimited. **60:** (top left) © Jack M. Bostrack/Visuals Unlimited. (top center) © Eric Grave/Phototake. (top right) © John D. Cunningham/Visuals Unlimited. (below) © Peter Lewis, from Bacterial chromosome segregation, *Microbiology* 147:519–526, 2001. **61:** (all) © David M. Phillips/Visuals Unlimited. **62:** Courtesy Mark O. Martin and Tom Pitta, Occidental College. **64:** Courtesy Francis C.Y. Wong. **65:** (left) © George Musil/Visuals Unlimited. (right) Courtesy Gerald C. Johnston, Dalhousie University, Halifax, Nova Scotia. **66:** (right) © J.J. Cardamone & B.A. Phillips/Biological Photo Service. **68:** Courtesy Steven Kowalczykowski. **72:** Courtesy Becton Dickinson Microbiology Systems. **73:** © Lillian Therrien & E.C.S. Chan/Visuals Unlimited. **76:** © G.W. Willis/Biological Photo Service. **77:** Courtesy Forma Scientific, Inc.

Chapter 4. 83: Courtesy D. McLean and M. Kinsey, photo by Paul Baumann, University of California, Davis. **86:** (above) © Ralph A. Slepecky/Visuals Unlimited. **92:** (bottom right) © John McN. Sieburth, University of Rhode Island/Biological Photo Service. **93:** © T.J. Beveridge/Visuals Unlimited. **95:** (far left) © E.C.S. Chan/Visuals Unlimited. (near left) © Carolina Biological Supply/Visuals Unlimited. (near right) © E.C.S. Chan/Visuals Unlimited. **95:** (far right) © Fred Hossler/Visuals Unlimited. **96:** (left) © J.A. Breznak & H.J. Pankratz/Biological Photo Service. (right) © John McN. Sieburth, University of Rhode Island/Biological Photo Service. **97:** D. Balkwill. **98** © T.J. Beveridge,

University of Guelph/Biological Photo Service. **100:** D. McLean & M. Kinsey, photo courtesy Paul Baumann, University of California, Davis. **101:** (left) Courtesy Edward J. Bottone, Mount Sinai Hospital, New York. **104:** (both) © Don W. Fawcett/Visuals Unlimited. **105:** (above) Courtesy Keith A. Porter, University of Pennsylvania. (below) Courtesy L. K. Shumway, College of Eastern Utah. **107:** D. McLean & M. Kinsey, photo courtesy Paul Baumann, University of California, Davis.

Chapter 5. 114: © Carolina Biological Supply/Visuals Unlimited. **116:** © Dr. Linda Stannard, UCT/SPL/Photo Researchers.

Chapter 6. 170: C. C. Brinton, Jr. & J. Carnahan.

Chapter 7. 179: Courtesy Robert Hammer, Howard Hughes Medical Institute, Dallas. **186:** Courtesy Hoefer Scientific Instruments. **188:** Courtesy Robert Hammer, Howard Hughes Medical Institute, Dallas. **190:** Courtesy Norman Lin, Genentech. **197:** Courtesy Prof. Sydney Kustu, University of California, Berkeley.

Chapter 8. 201: © Photodisc. **203:** © George Musil/Visuals Unlimited. **214:** Courtesy Manfred E. Bayer, Fox Chase Cancer Center, Philadelphia. **215:** © K. Talaro/Visuals Unlimited.

Chapter 9. 223: Centers for Disease Control. **234:** (left) Courtesy Stan Lester, Lester Farms, Winters, California.

Chapter 10. 238: © Photodisc. **240:** (far right) © Scott Camazine/Sharon Bilotta Best/Photo Researchers. (far left) © Grant Heilman Photography. (near left) © Runk/Schoenberger/Grant Heilman Photography. (near right) © Dr. Charles Henneghien/Bruce Coleman, Ltd. **242:** (left) Courtesy Princeton University Museum of Natural History. (right) Courtesy Stanley W. Awramik, University of California, Santa Barbara. **243:** © C.A. Henley/Biofotos. **247:** Courtesy Biolog, Inc. **248:** (top left) Christine Case/Visuals Unlimited. (top right) Courtesy BD Diagnostic Systems. (bottom) Courtesy Microbial Diseases Laboratory, Berkeley, California.

Chapter 11. 257: © David M. Phillips/Visuals Unlimited. **260:** (below) © Corale Brierley/Visuals Unlimited. **264:** © Paul W. Johnson/Biological Photo Service.**265:** © Photodisc. **266:** © D.A. Glave/Biological Photo Service. **267:** Courtesy James T. Staley, University of Washington. **268:** © Michael Richard/Visuals Unlimited. **269:** (above) © Carolina Biological Supply/Visuals Unlimited. (bottom left) © Science Photo Library/Custom Medical Stock Photo. (bottom, both) Courtesy William Ormerod,

University of Southern California, photo by E.G. Ruby & M. McFall-Ngai. **270:** (above) Courtesy Dennis Ohman, University of Tennessee. (below) © Dr. K.S. Kim/Peter Arnold, Inc. **271:** Courtesy Patricia Grilione & J. Pangborn, San Jose State University. **272:** (bottom left) © George J. Wilder/Visuals Unlimited. (top right) © David M. Phillips/Visuals Unlimited. (Center) © David M. Phillips/Visuals Unlimited. (bottom right) © R. Kessel & G. Shih/Visuals Unlimited. **274:** (left) © Michael Gabridge/Visuals Unlimited. (below) Courtesy H. Veldkamp, G. Vanden Berg, & LPTM Zevenenhuizen, "Glutamic acid production by *Arthrobacter globiformis*," *Antonie van Leeuwenhoek* 29:35–51, 1963. **277:** (top left) © CDC/Biological Photo Service. (top right) © Koneman/Visuals Unlimited. **278:** © R. Howard Berg/Visuals Unlimited. **279:** © Michael Gabridge/Visuals Unlimited. **280:** (left) © Veronika Burmeister/Visuals Unlimited. (right) © Forsyth Dental Center/Biological Photo Service.

Chapter 12. 284: © Cath Ellis, Dept. of Zoology, University of Hull/SPL/Photo Researchers. **287:** (above) © R.M. Meadows/Peter Arnold, Inc. (below) Penn State University Teaching Collection, Courtesy William Merrill. **288:** Courtesy Alan D.M. Rayner, University of Bath, and the New Phytologist Trust. **290:** © Heather Angel/Biofotos. **291:** (above) © John D. Cunningham/Visuals Unlimited. (below) © David M. Phillips/Visuals Unlimited. **292:** (above) © Larry Jensen/Visuals Unlimited. **294:** (left) Courtesy G. L. Barron, University of Guelph. (right) © Bruce Iverson. **295:** © Eric Crichton/Bruce Coleman, Ltd. **297:** (left) Courtesy Florida Marine Research Institute. (right) © C.C. Lockwood/Cactus Clyde Productions. **298:** Courtesy Greta A. Fryzell, Texas A&M University. **299:** (left) John Ingraham. **301:** © M. Abbey/Visuals Unlimited. **302:** (above) © Arthur M. Siegelman/Visuals Unlimited. (below, both) Courtesy Gary Grimes & Steven L'Hernault, Hofstra University. **304:** (above) © Edward S. Ross. (center) © Carolina Biological Supply Company/Phototake. (below) From "Morphogen Hunting in Dictostelium," Robert R. Kay, Mary Berks & David Traynor, *Development* 1989 Supplement, 81–90, The company of Biologists 1989. **305:** © Cath Ellis, Dept. of Zoology, University of Hull/SPL/Photo Researchers. **307:** (below) Courtesy David Lacomis, University of Pittsburgh School of Medicine.

Chapter 13. 313: © K.G. Murti/Visuals Unlimited. **315:** (above) Kenneth M. Corbett. (below) © Dr. O. Bradfute/Peter Arnold, Inc.**318:** John Ingraham. **322:** © E.C.S. Chan/Visuals Unlimited. **327:** © James Holmes/Cell Tech Ltd/SPL/Photo Researchers. **335:** © K.G. Murti/Visuals Unlimited. **337:** John Ingraham. **338:** (left) USDA

INDEX